Papillomavirus Research

From Natural History to Vaccines and Beyond

Edited by

M. Saveria Campo

Institute of Comparative Medicine
University of Glasgow, UK

Caister Academic Press

Caister Academic Press
32 Hewitts Lane
Wymondham
Norfolk NR18 0JA
England

www.caister.com

British Library Cataloguing-in-Publication Data

A catalogue record for this book is available from the British Library

ISBN: 1-904455-04-2

Description or mention of instrumentation, software, or other products in this book does not imply endorsement by the author or publisher. The author and publisher do not assume responsibility for the validity of any products or procedures mentioned or described in this book or for the consequences of their use.

Printed and bound in Great Britain

Contents

Contributors vi

Introduction 1
M. Saveria Campo

Chapter 1. Linking Human Papillomaviruses to Cervical Cancer: A Long and Winding Road 3
Lutz Gissmann

Chapter 2. Phylogeny and Taxonomy of Papillomaviruses 11
Hans-Ulrich Bernard

Chapter 3. Epidemiology of Human Papillomavirus Infections and Associations with Cervical Cancer: New Opportunities For Prevention 19
F. Xavier Bosch, Silvia de Sanjosé, Xavier Castellsagué, Víctor Moreno and Nubia Muñoz

Chapter 4. Gene Expression of Papillomaviruses 41
Mina Kalantari and Hans-Ulrich Bernard

Chapter 5. Mechanisms and Regulation of Papillomavirus DNA Replication 53
Louise T. Chow and Thomas R. Broker

Chapter 6. The Papillomavirus Transcription/Replication Factor E2: Structure, Function, Cancer and Therapy 73
Iain M. Morgan and Mary M. Donaldson

Chapter 7. The E4 Protein - A Late Starter 83
Sally Roberts

Chapter 8. Biological Activities of Papillomavirus E5 Proteins 97
Frank A. Suprynowicz, M. Saveria Campo and Richard Schlegel

Chapter 9. The Role of the HPV E6 Oncoprotein in Malignant Progression 115
Miranda Thomas, David Pim and Lawrence Banks

Chapter 10. The Biology of the E7 protein of HPV-16 133
Dennis J. McCance

Chapter 11. Organotypic Raft Cultures and the Study of the Natural History of Papillomavirus 145
Jason M. Bodily, Samina Alam, Horng-Shen Chen and Craig Meyers

Chapter 12. Early Events in the Papillomaviral Lifecycle 175
Patricia M. Day and John T. Schiller

Chapter 13. Late Events in the Life Cycle of Human Papillomaviruses 193
Sheila V. Graham

Chapter 14. Transgenic Mouse Models for the *In Vivo* Analysis of Papillomavirus Oncogene Function 213
Paul F. Lambert, Scott J. Balsitis, Anny Shai, Sara J.S. Simonson, and Sybil M.G. Williams

Chapter 15. HPV and Oesophageal Carcinoma 229
Kari Syrjänen

Chapter 16. Recurrent Respiratory Papillomatosis, HPV, and Impact on Host Response 255
Andrea Vambutas and Bettie M. Steinberg

Chapter 17. The Cutaneous Human Papillomavirus Types and Non-melanoma Skin Cancer 269
Veronique Bouvard, Anne-Sophie Gabet, Rosita Accardi, Bakary S. Sylla and Massimo Tommasino

Chapter 18. Papillomavirus-like Particles and Their Applications in Molecular Virology, Human Serology and Vaccines 279
Richard B.S. Roden and Raphael P. Viscidi

Chapter 19. Immunobiology of Papillomaviruses 311
Margaret A. Stanley

Chapter 20. Immune Evasion in Genital Papillomavirus Infection and Cervical Cancer: Role of Cytokines and Chemokines 321
Sigrun Smola-Hess and Herbert Pfister

Chapter 21. HPV Vaccines 341
Michelle Giles and Suzanne M. Garland

Chapter 22. HPV Vaccines in Plants: An Appetising Solution to Control Infection and Associated Cancers 357
Rosella Franconi and Aldo Venuti

Chapter 23. Bovine Papillomavirus: Old System, New Lessons? 373
M. Saveria Campo

Chapter 24. Bovine Papillomaviruses and Equine Sarcoids 389
Lubna Nasir and Stuart W.J. Reid

Chapter 25. Rabbit Viral Papillomas and Carcinomas: Model Diseases for Human Papillomavirus Infections and Associated Carcinogenesis 399
Françoise Breitburd, Mathieu Nonnenmacher, Jérôme Salmon and Gérard Orth

Index 421

Current Books of Interest

Contributors

Rosita Accardi
Infections and Cancer Biology Group
International Agency for Research on Cancer 150 Cours Albert Thomas
69372 Lyon
France

Samina Alam
Department of Microbiology and Immunology
The Pennsylvania State University College of Medicine
Hershey, PA 17033
USA

Scott J. Balsitis
McArdle Laboratory for Cancer Research
1400 University Ave.
Madison, WI 52706
USA

Lawrence Banks
International Centre for Genetic Engineering and Biotechnology
Padriciano 99
I-34012 Trieste
Italy

Hans-Ulrich Bernard
Department of Molecular Biology and Biochemistry
114 Sprague Hall
University of California
Irvine, CA 92697-3900
USA

Jason M. Bodily
Department of Microbiology and Immunology
The Pennsylvania State University College of Medicine
Hershey, PA 17033
USA

F. Xavier Bosch
IDIBELL
Institut Català d'Oncologia (ICO)
Av. Gran Via s/n Km. 2.7
08907 L'Hospitalet de Llobregat
Barcelona
Spain

Veronique Bouvard
Infections and Cancer Biology Group
International Agency for Research on Cancer 150 Cours Albert Thomas
69372 Lyon
France

Françoise Breitburd
Department of Virology
Institut Pasteur
25 rue du Docteur Roux
F-75724 Paris Cedex 15
France

Thomas R. Broker
University of Alabama at Birmingham
Department of Biochemistry and Molecular Genetics
1918 University Boulevard
McCallum Building Room 506
Birmingham, Alabama 35294-0005
USA

M. Saveria Campo
Division of Pathological Sciences
Institute of Comparative Medicine
University of Glasgow
Garscube Estate
Glasgow G61 1QH
Scotland
UK

Xavier Castellsagué
IDIBELL
Institut Català d'Oncologia (ICO)
Av. Gran Via s/n Km. 2.7
08907 L'Hospitalet de Llobregat
Barcelona
Spain

Horng-Shen Chen
Department of Microbiology and Immunology
The Pennsylvania State University College of Medicine
Hershey, PA 17033
USA

Louise T. Chow
University of Alabama at Birmingham
Department of Biochemistry and Molecular Genetics
1918 University Boulevard
McCallum Building Room 506
Birmingham, Alabama 35294-0005
USA

Patricia M. Day
Laboratory of Cellular Oncology
Center for Cancer Research
National Cancer Institute
National Institutes of Health
Bethesda, MD 20892
USA

Mary M. Donaldson
Institute of Comparative Medicine
Division of Pathological Sciences
University of Glasgow
Garscube Estate
Glasgow G61 1QH
Scotland
UK

Rosella Franconi
ENEA
Italian National Agency for New Technologies
Energy and the Environment
BIOTEC
Laboratory of Plant Genetics and Genomics - CR
Casaccia, Rome
Italy

Anne-Sophie Gabet
Infections and Cancer Biology Group
International Agency for Research on Cancer 150 Cours
Albert Thomas
69372 Lyon
France

Suzanne M. Garland
Department of Microbiology and Infectious Diseases
Royal Women's Hospital
132 Grattan St
Carlton VIC 3053
Australia

Michelle Giles
Department of Microbiology and Infectious Diseases
Royal Women's Hospital
132 Grattan St
Carlton VIC 3053
Australia

Lutz Gissmann
Deutsches Krebsforschungszentrum (DKFZ)
FS Infektion und Krebs
Im Neuenheimer Feld 242
D-69120 Heidelberg
Germany

Sheila V. Graham
Institute of Biomedical and Life Sciences
University of Glasgow
Church Street
Glasgow G11 5JR
Scotland
UK

Mina Kalantari
Department of Molecular Biology and Biochemistry
114 Sprague Hall
University of California
Irvine, CA 92697-3900
USA

Paul F. Lambert
McArdle Laboratory for Cancer Research
1400 University Ave.
Madison, WI 52706
USA

Dennis J. McCance
Department of Microbiology and Immunology
University of Rochester
Rochester, NY 14642
USA

Víctor Moreno
IDIBELL
Institut Català d'Oncologia (ICO)
Av. Gran Via s/n Km. 2.7
08907 L'Hospitalet de Llobregat
Barcelona
Spain

Iain M. Morgan
Institute of Comparative Medicine
Division of Pathological Sciences
University of Glasgow
Garscube Estate
Glasgow G61 1QH
Scotland
UK

Nubia Muñoz
Emeritus Scientist
Instituto Nacional de Cancerología
Calle 1ª # 9-85
17158 Bogotá
Colombia

Craig Meyers
Department of Microbiology and Immunology H107
P. O. Box 850
The Pennsylvania State University College of Medicine
Hershey, PA 17033
USA

Lubna Nasir
Institute of Comparative Medicine
University of Glasgow
Bearsden Road
Glasgow G61 1QH
Scotland
UK

Mathieu Nonnenmacher
Department of Virology
Institut Pasteur
25 rue du Docteur Roux
F-75724 Paris Cedex 15
France

Gérard Orth
Department of Virology
Institut Pasteur
25 rue du Docteur Roux
F-75724 Paris Cedex 15
France

Herbert Pfister
Universität zu Köln
Institut für Virologie
Köln
Germany

David Pim
International Centre for Genetic Engineering and Biotechnology
Padriciano 99
I-34012 Trieste
Italy

Stuart W.J. Reid
Institute of Comparative Medicine
University of Glasgow
Bearsden Road
Glasgow G61 1QH
Scotland
UK

Richard B.S. Roden
Department of Pathology
The Johns Hopkins University
Ross Research Building
Room 512B
720 Rutland Avenue
Baltimore, MD 21205
USA

Sally Roberts
Cancer Research UK Institute for Cancer Studies
The University of Birmingham
Vincent Drive
Edgbaston
Birmingham B15 2TT
UK

Jérôme Salmon
Department of Virology
Institut Pasteur
25 rue du Docteur Roux
F-75724 Paris Cedex 15
France

Silvia de Sanjosé
DIBELL
Institut Català d'Oncologia (ICO)
Av. Gran Via s/n Km. 2.7
08907 L'Hospitalet de Llobregat
Barcelona
Spain

John T. Schiller
Laboratory of Cellular Oncology
Center for Cancer Research
National Cancer Institute
National Institutes of Health
Bethesda, MD 20892
USA
Richard Schlegel
Department of Pathology
Georgetown University Medical School
Washington, DC 20057
USA

Anny Shai
McArdle Laboratory for Cancer Research
1400 University Ave.
Madison, WI 52706
USA

Sara J.S. Simonson
McArdle Laboratory for Cancer Research
1400 University Ave.
Madison, WI 52706
USA

Sigrun Smola-Hess
Universität zu Köln
Institut für Virologie
Köln
Germany

Margaret A Stanley
Department of Pathology
Tennis Court Road
Cambridge CB2 1QP
UK

Bettie M. Steinberg
North Shore-Long Island Jewish Research Institute and
Department of Otolaryngology
Long Island Jewish Medical Center
New Hyde Park
New York
USA

Frank A. Suprynowicz
Department of Pathology
Georgetown University Medical School
Washington, DC 20057
USA

Bakary S. Sylla
Infections and Cancer Biology Group
International Agency for Research on Cancer 150 Cours
Albert Thomas
69372 Lyon
France

Kari Syrjänen
Department of Oncology & Radiotherapy
Turku University Central Hospital
Savitehtaankatu 1
FIN-20521 Turku
FINLAND

Miranda Thomas
International Centre for Genetic Engineering and Biotechnology
Padriciano 99
I-34012 Trieste
Italy

Massimo Tommasino
Infections and Cancer Biology Group
International Agency for Research on Cancer 150 Cours Albert Thomas
69372 Lyon
France

Andrea Vambutas
North Shore-Long Island Jewish Research Institute and Department of Otolaryngology
Long Island Jewish Medical Center
New Hyde Park
New York
USA

Aldo Venuti
Laboratory of Virology
Regina Elena Cancer Institute
Via delle Messi d'Oro 156
00158 Rome
Italy

Raphael P. Viscidi
Department of Pediatrics
The Johns Hopkins Hospital
600 North Wolfe Street
Baltimore, MD 21287
USA

Sybil M.G. Williams
McArdle Laboratory for Cancer Research
1400 University Ave.
Madison, WI 52706
USA

From: Papillomavirus Research: From Natural History to Vaccines and Beyond. Edited by: M. Saveria Campo

Introduction

M. Saveria Campo*

Papillomaviruses (PVs) are oncogenic DNA tumour viruses (IARC, 1995). They infect cutaneous and mucous epithelia in a variety of animals, including humans, through cuts or abrasions, and induce papillomas or warts, which generally regress without eliciting any serious clinical problems in the host, but can occasionally persist and progress to malignancy. There are upward of 300 distinct types of PVs, including probably 200 different types of human papillomavirus (HPV), and it is likely that more types will be found, particularly in animals.

Despite their remarkable heterogeneity, all PVs have a very similar genomic organisation. The viral genome is a double stranded circular DNA molecule of approximately 8 kilobases (kb), which can be divided into three regions: the E region coding for early proteins (E1-E7) responsible for the pathogenicity of the virus; the L region coding for late structural proteins (L1, L2), and a non-coding region which contains the cis-elements necessary for replication and transcription of the viral genome. All the known genes are encoded in one DNA stand and transcription is unidirectional (Figure 1).

The life cycle of the virus is tightly coupled with the differentiation process of the epithelial cell: the virus infects the basal keratinocytes, expresses part of its genes in the basal and suprabasal layers, replicates its genome in the differentiating spinous and granular layers, expresses its structural genes and packages its DNA in the squamous layers (there are exception to this general rule, which will be discussed in later chapters), and new infectious virus is finally released with the keratinised squames. During its whole life cycle, the PV genome persists in the infected cell as an episome.

PV does not always go through its full productive life cycle. Occasionally benign tumours persist and may progress to squamous cell carcinoma. From the viral "point of view", neoplastic progression is an accident; cell transformation is a dead-end process for the virus as the transformed cell is no longer permissive for virion production; the viral genome is either incorporated into the cellular genome, as in anogenital squamous cell carcinoma in human (Figure 1); maintained as an extrachromosomal element which replicates in synchrony with the cell cycle, as in urothelial cancers in cattle, or may even be lost by the transformed cells, as in squamous cell carcinomas of the upper gastrointestinal tract in cattle.

One important feature of PVs is their strict species-specificity: even in experimental conditions, PVs do not infect any other host than their natural one. The only known case of naturally occurring cross-species infection is the infection of horses and other equids by bovine papillomavirus (BPV) type 1 and type 2. Given the apparently insurmountable problem of species-

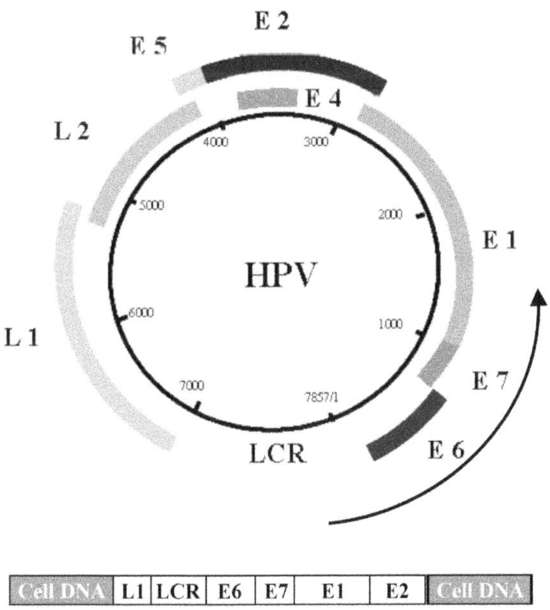

Figure 1. Top. Genomic organization of a typical papillomavirus. The double stranded circular DNA is approximately 8 kilobases long (numbered in the inner circle). The boxes represent early (E) and late (L) open reading frames (ORFs). The E ORFs encode proteins necessary for viral DNA replication and transcription (E1, E2), virion maturation (E4) and cell transformation (E4, E5, E6, E7); the L ORFs encode the major and minor structural proteins (L1 and L2 respectively). The long control region (LCR) contains the cis-elements necessary for replication and transcription of the viral genome. The arrowhead indicates the direction of transcription which is unidirectional. The major transcripts start in the LCR and terminate either at the 3′ end of the E ORFs or at the 3′ end of the L ORFs. Bottom. Diagrammatic representation of a typical integrated HPV genome in anogenital cancer. The integration point in the viral genome is usually within the E2 ORF and often a great proportion of the E2 ORF is lost together with the E4, E5, L2 and L1 ORFs. Loss of E2 leads to unregulated expression of the E6 and E7 oncoproteins. The size of the ORFs is not in scale. For a full description of the viral functions during productive replication or neoplastic transformation, see the individual relevant chapters.

*For correspondence. Email s.campo@vet.gla.ac.uk

specificity, no animal model of HPV infection exists and papillomavirologists have had in the past to rely for direct experimentation on animal papillomavirus systems. The recent advent of organotypic raft cultures, virus-like particles (VLPs) and transgenic mice has revolutionised the PV field revealing new facets of HPV virion structure, life cycle, immunology and oncogenicity.

This book attempts to cover all aspects of PV natural history, from the recognition of PV as a cancer agent in humans and animals, to its complex phylogeny, to the functions of viral proteins during the early and late phases of the productive cycle and during cell transformation, to the role of HPV in cancers other than ano-genital ones, to infection and cancer in animals, to the host immune response, and finally to the development of anti-viral vaccines, both extant and future.

Although the book has been conceived as an organic whole, each chapter contains enough information to stand alone on a particular topic. It is hoped that the readers, whether seasoned papillomavirologists, or new comers to the field, are inspired by the book to conquer ever new summits and explore ever new depths in the pursuit of the common goal to understand the virus and defeat it.

Reference

IARC Monographs on the Evaluation of Carcinogenic Risks to Humans. (1995) Volume 64; Human Papillomaviruses. WHO/IARC, Lyon.

Chapter 1

Linking Human Papillomaviruses to Cervical Cancer: A Long and Winding Road

Lutz Gissmann*

Abstract

Epidemiologic studies suggesting an infectious etiology of cervical cancer have prompted scientist from different areas to search for the responsible agent. The human papillomaviruses (HPV) came into the play as members of a family of animal tumor viruses and because of their known association with genital infections. A long chain of events ranging from the identification of the plurality of HPVs, the regular finding of genomes of certain types, so called high-risk, within tumor biopsies, insights into the transforming mechanism and informative epidemiological surveys finally led to the consensus view within the scientific community of a causative link between these viruses and cervical cancer. This awareness triggered the development of HPV-specific vaccines as means of cancer prevention.

1. Cervical Cancer a Sexually Transmitted Disease

Introducing the link between a sexually transmitted agent and cervical cancer one usually quotes as initial source the report by Domenico Antonio Rigoni-Stern presented in 1842 to Italian scientists (Rigoni-Stern, 1842). It turns out, however, that this article is often misused since, based on the figures on the cancer-related death rates in Verona presented in this report, Dr. Rigoni-Stern was unable - and in fact did not pretend - to conclude that cervical cancer is less frequent in nuns than in married women or widows (Griffiths, 1991). As discussed by Skrabanek and Griffith, also some of the later reports published until the 1960s suggesting an infectious origin of cervical cancer suffered from inaccurate base line data, *e.g.* misclassification of cervical and uterine corpus cancer (Griffiths, 1991; Skrabanek, 1988). However there was a large number of now generally accepted epidemiologic studies that kept the concept of a sexually transmitted agent alive (for review see Brinton, 1992; Doll, 1986; Rotkin, 1973). These studies demonstrated that not only the sexual behavior of a woman but also of her male partner determines her risk for this cancer (Brinton et al., 1989). Further, wives of men who had died from penile cancer were shown to have an increased mortality of cervical cancer (Smith et al., 1980) and circumcision of the partner proved to reduce the risk for this disease (Castellsague et al., 2002). In fact, the importance of the male factor was recognized early on and a putative contribution of sperm or smegma in the carcinogenesis was advocated (for review see Kessler, 1986). Since cervical cancer was reported to often coexist with syphilis and gonorrhea, their causative agents (*Treponema pallidum*, *Neisseria gonorrheae*) were discussed as being responsible for the malignant disease as well (for review see Kessler, 1986). The most intensively studied candidate among the original contenders is the Human Herpes Simplex Virus Type 2 (HSV 2) whose sexual transmission was well documented. Actually a strong association between HSV 2-specific antibodies and cervical cancer had been published (Rawls et al., 1980) but a carefully conducted prospective study failed to show that HSV 2 infection precedes the development of cervical neoplasia (Vonka et al., 1984a; Vonka et al., 1984b) contrary to what was subsequently successfully demonstrated for certain HPV types (Dillner et al., 1997; Kjaer et al., 2002; Koutsky et al., 1992; Sigstad et al., 2002; Wallin et al., 1999). The role of HSV 2 as a player in cervical carcinogenesis was still advocated well after the presence of papillomavirus DNA in tumor cells had been reported (Kessler, 1986; Manjunath et al., 1988). Yet a recently published serological study which included the meta-analysis of literature data came to the conclusion that HSV 2 does not play a role in cervical carcinogenesis and that the previous estimates of its effect found in many cross-sectional studies were biased by confounding factors due to uncontrolled risk factors such as HPV (Lehtinen et al., 2002).

Obviously, epidemiologists and laboratory scientists want to see different criteria fulfilled in order to accept a causal relationship between an infectious agent and a particular cancer (Evans and Mueller, 1990). For instance, for a virologist the regular presence of a candidate virus or traces thereof within the tumor

*For correspondence. Email l.gissmann@dkfz-heidelberg.de

biopsies is mandatory[1]. Unsuccessful in his attempts to detect HSV 2 DNA sequences in tumor cells (zur Hausen et al., 1974b) and not convinced by the hit-and-run concept (Galloway and McDougall, 1983), Harald zur Hausen in the early 1970s started to look out for other viruses. A number of reports on the malignant conversion of genital warts, laryngeal papillomas and epidermodysplasia verruciformis-associated lesions and also the well documented oncogenic properties of the cottontail rabbit papillomavirus raised the hypothesis of a causal relation between human papillomaviruses and cervical cancer (zur Hausen, 1976; 1975). This concept was substantiated when papillomavirus particles were detected in cells ("koilocytes") typically found in preneoplasic cervical lesions (Della Torre et al., 1978; Laverty et al., 1978; Meisels et al., 1977).

2. Papillomavirus Research Has Come a Long Way

Obviously research on papillomaviruses was in its infancy in the early 1970s. Most of the Information available at the time was nicely summarized in a review by Rowson and Mahy (Rowson and Mahy, 1967).

Based on the characteristics of the typical clinical manifestation of papillomavirus infection it was initially referred to as Wart Virus. The infectious nature of warts has been established by numerous reports on individual accidental inoculations from patients (Payne, 1891), with contaminated objects (Watkins, 1961) and by surveys of institutionalized individuals (Massing and Epstein, 1963). Venereal transmission of genital warts was demonstrated in several studies, the best known being the report of transmission of condylomas to the spouses of soldiers returning from service (Barrett et al., 1954).

Proof for the viral etiology of warts dates back to the turn of the 20th century when inoculation with cell-free extracts from various kinds of warts (verrucae vulgares, condylomata acuminata, laryngeal papillomas) demonstrated transmissibility of these tumors (Ciuffo, 1907; Licht, 1894; Serra, 1942; reviewed by Rowson and Mahy, 1967). Since the recipient site rather than the source of the virus inoculum appeared to determine the kind of wart, investigators concluded that there is only one single human papillomavirus. Early studies on transmission of wart extracts also suggested the species-specificity of papillomaviruses (for review see Rowson and Mahy, 1967) which is in accord with the current knowledge, leaving aside the slight promiscuity of the bovine papillomavirus types 1 and 2 that can infect rodent fibroblast *in vitro* and induce tumors in a heterologous species (sarcoids in horses, donkeys and mules) (Chambers et al., 2003; see chapter 24).

The nature of the infectious agent responsible for the induction of warts was first investigated by Strauss and co-workers who described crystalline virus-like structures within intranuclear inclusion bodies of skin warts by electron microscopy (Strauss et al., 1950; Strauss et al., 1949). Similar observations were subsequently also made in genital warts and laryngeal papillomas (Boyle et al., 1973; Dun and Ogilvie, 1968). Introduction of the technique of negative staining for EM analyses permitted a better characterization of the wart virus (Horne and Wildy, 1961) which was later on classified as a (morphological) relative of the polyoma viruses (Klug and Finch, 1965). Over the years it became obvious, however, that the two virus groups are quite different in their biological properties and their molecular structure. Yet it took 35 years before the papillomaviruses were granted the status of an individual virus family now designated as *Papillomaviridae* (van Regenmortel et al., 2000).

Until fairly recently, studying papillomavirus particles and their interactions with host cells depended entirely on the isolation of virus particles from naturally infected epithelia, *i.e.* from the different kinds of warts. The very few reports on replication in simple cell culture 1975 (Eisinger et al., 1975) proved to be not reproducible. This hampered not only the characterization of the virus life cycle and of the natural history of virus infection but also the development of classical (antibody-based) assays for virus classification. Even today most of the papillomaviruses exist only as cloned DNA. In fact, the entire nomenclature of papillomaviruses as it presents itself today (de Villiers et al., 2004; van Regenmortel et al., 2000) as well as much of the knowledge on the structure and the antigenicity of the virus capsids are the result of "reverse virology". An early progress towards the taming of these shrewd viruses was the finding of the ability of the bovine papillomavirus type 1 to transform rodent cells in culture (Dvoretzky et al., 1980). These data laid the ground for a large series of further studies providing today's awareness of the action of the papillomavirus oncoproteins (for review see Munger and Howley, 2002). With the advent of sophisticated three-dimensional cell culture systems that somehow mimic the situation of differentiating epithelia, papillomaviruses can be propagated to a certain degree and genetic analyses of the virus replication functions can be pursued (Chow and Broker, 1997; McBride et al., 2000; Meyers et al., 1992).

[1] In the meantime we have learned that there are exceptions to this rule, i.e. that viruses might contribute to carcinogenesis without being constantly present within the tumour cells by "indirect" mechanisms such as suppression of host cell immune functions (e.g. HIV and Kaposi's Sarcoma or possibly certain cutaneous HPV types and non-melanoma skin cancer) (zur Hausen, 2001).

3. Papillomaviruses and Cervical Cancer: To Make Ends Meet

The initially proclaimed inconsistency with the papillomavirus-cancer hypothesis (zur Hausen, 1976), *i.e.* the failure of regular skin warts to progress into malignancies, was overcome by reports of the heterogeneity of human papillomaviruses (Gissmann and Hausen, 1976; Gissmann et al., 1977; Orth et al., 1977) that had first been suspected when DNA from skin wart virus could not be detected in genital warts (Delap et al., 1976; zur Hausen, et al., 1974a) and when it was reported that sera from skin wart patients failed to react with virus particles from genital warts (Almeida et al., 1969). Subsequent research led to the positive identification of different viruses in lesions other than common warts, *i.e.* warts in Epidermodysplasia verruciformis patients, genital warts, laryngeal papillomas and finally cervical cancer (Dürst et al., 1983; Gissmann et al., 1982; Gissmann and zur Hausen, 1980; Orth et al., 1978). The original strategy for virus hunting was based on the purification of virions from skin warts followed by extraction of their DNA. These molecules were used for restriction enzyme analysis and also as template for *in vitro* synthesis of radioactively labeled complementary RNA as probe for hybridization experiments with total DNA from different warts. It was essential to this strategy that single warts rather than pooled biopsies (even from different patients) were taken as starting material. Otherwise suspected viral clones might have been overlooked and a putative virus heterogeneity been obscured. Consequently, only such warts with a sufficient number of virus particles ($\geq 10^{10}$ theoretically yielding 100ng of viral DNA) were amenable for this kind of analysis. It turned out, however, that the amount of virus particles in genital warts is at least 10 fold lower than in common warts of the skin. Some mileage was gained by taking advantage of the distinct physical properties of the viral DNA, *i.e.* its circular supercoiled form permitting a separation from the bulk of cellular DNA *via* CsCl/ethidium bromide density gradient centrifugation without prior particle purification. However, there appears to be a direct correlation between the amount of virus particle production and of viral DNA replication (Grussendorf-Conen et al., 1983) thus it was not surprising that this DNA (later on classified as HPV 6 genome) was insufficient but for a rough physical restriction enzyme mapping (Gissmann and zur Hausen, 1980). Fortunately, around this time, amplification of DNA in *E. coli* plasmids had become a routine procedure which permitted the preparation of genomic probes to screen for homologous sequences in other biopsies. These studies identified HPV 6 as the prevalent virus in genital warts but failed to detect it in malignant genital tumors (de Villiers et al., 1981; Gissmann et al., 1982).

As first described by Law and colleagues (Law et al., 1979) the genomes of different papillomaviruses cross-hybridize under conditions of low stringency (42°C below the melting temperature of the DNA). These hybridization protocols were used to assay cervical cancer biopsies for novel papillomaviruses finally leading to the identification of HPV 16, 18, 31 and other so called high-risk types (Boshart et al., 1984; Dürst et al., 1983; Lorincz et al., 1986). Initial concerns about the viral nature of the first DNA isolates (Kessler, 1986) were eliminated when their complete sequence became available revealing the typical genomic organization of a papillomavirus (Cole and Danos, 1987; Cole and Streeck, 1986; Seedorf et al., 1985). When the DNA sequence of subsequent HPV 16 isolates cloned from non-malignant cervical lesions was determined it became apparent that the original HPV 16 clone, which had served as starting material for many studies worldwide, contained mutations in genes that are required for virus replication (Dürst, 1987). For example, the initial failure to generate HPV 16 virus-like particles by recombinant expression of the L1 gene turned out be an "artifact" of the original isolate and was narrowed down to a single point mutation (histidine to aspartic acid at amino acid position 202) within the HPV 16 prototype (Kirnbauer et al., 1993).

Soon after the first description of HPV DNAs, their presence in biopsies of cervical cancer or its precursors had been reported by several investigators (Gissmann et al., 1984; Hoepfner and Loning, 1986; Pratili et al., 1986; Tomita et al., 1986; for review see Roman and Fife, 1989). Transforming activity in cell culture (rodent fibroblasts and human keratinocytes) was first demonstrated for the whole HPV genome (Dürst et al., 1987; Kaur and McDougall, 1988; Pirisi et al., 1987; Tsunokawa et al., 1986). Assigning this property to the viral oncogenes E6 and E7 paved the way for an exciting line of research that expanded our awareness of tumor suppressor proteins and cell cycle control in general (for review see Munger and Howley, 2002; see also Chapters 9 and 10 for a detailed description of the properties of the E6 and E7 oncoproteins). Cervical cancer cell lines, some of whom had been in culture for decades, were also found to contain HPV genomes (Boshart et al., 1984; Yee et al., 1985). Analysis of the integrated molecules within such cells led to the hypothesis of a virus-dependent transcriptional regulation ensuring the constitutive expression of the proteins E6 and E7 for maintenance of transformation (Gissmann and Schwarz, 1986; Schwarz et al., 1985). This concept was confirmed by specific inhibition of expression of the viral oncoproteins providing a strong argument against a mere passenger state of HPV within the tumor cells (Butz et al., 2003; DeFilippis et al., 2003; Francis et al., 2000; von Knebel Doeberitz et al., 1988; von Knebel Doeberitz et al., 1992).

Besides this rather convincing evidence from laboratory experiments, there were early epidemiologic studies, which also suggested a link between HPV infection and cervical cancer. For example, in addition to the various studies demonstrating a strong association of viral DNA in biopsies or cervical swabs from cancer patients as compared to controls (for review see Roman and Fife, 1989) the presence of HPV DNA within a cervical lesion was shown to favor its progression (Campion et al., 1986; Fuchs et al., 1988; Schneider et al., 1987; Syrjanen et al., 1985). However these data were not strong enough to warrant the epidemiologists' certificate of a virus-to-tumor link (Munoz et al., 1988). In particular it was argued that HPV is found in a substantial number of healthy women thus cancers appears to be an extremely rare consequence of an virus infection (Meanwell, 1988). At a conference held in 1988 it was argued that the case-control studies conducted thus far "were not planned as full epidemiological investigations and so none of them satisfies the usual criteria of design and analysis..." (Bosch and Munoz, 1989). On this occasion, several types of studies were proposed with particular emphasis on the appropriate selection of the group of controls and on standardized assay that are critical for solid data analysis (Bosch and Munoz, 1989; Franco, 1991). This call was received and as soon as 1992 it was concluded that there is "compelling evidence in favor of a causal interpretation of the association between HPV and cervical cancer" (Munoz and Bosch, 1992). Besides more carefully designed epidemiologic surveys there had also been new lines of evidence arising from serological studies which demonstrated an association of HPV E6- and E7-specific antibodies in tumor patients as compared to controls (Jochmus-Kudielka et al., 1989; for review see Galloway, 1992). In fact, a few years later HPV 16 and 18 were announced as human carcinogens (IARC Working Group, 1995); the other high-risk HPV types were upgraded to this status in 2003 (Munoz et al., 2003). It is now generally accepted that most if not all of the cervical cancer cases develop only as a consequence of persistent HPV infection (Walboomers et al., 1999). Therefore it is fair to assume that other types of squamous cell cancer of the anogenital or upper respiratory tract and of the oral cavity are, at least in some of the cases, also related to HPVs even if the evidence for these tumors may not be equally strong and the mechanisms of transformation may be different (IARC Working Group, 1995).

The general acceptance of the oncogenic functions of papillomaviruses and the availability of virus-like particles, that have been shown to confer protective immunity in animal papillomavirus models (for review see Breitburd and Coursaget, 1999 and Chapter 18), prompted the development of HPV 16- and HPV 18-specific prophylactic vaccines that have yielded very promising results in initial clinical trials (GlaxoSmithKline, 2004; Harro et al., 2001; Koutsky et al., 2002; Chapter 21). Once licensed and publicly available, such vaccines are expected to reduce the incidence of HPV infection, of HPV-induced cervical intraepithelial neoplasia and ultimately of cervical cancer. Successful vaccination campaigns will not only be of significant impact for public health, in particular in high incidence areas of cervical cancer, but will provide the ultimate proof for a the etiologic role of papillomaviruses in this malignancy.

Acknowledgement

The author thanks Dr. Harald zur Hausen for critical reading of the manuscript.

References

Almeida, J. D., Oriel, J. D., and Stannard, L. M. (1969). Characterization of the virus found in genital warts. Microbios *3*, 225-232.

Barrett, T. J., Silbar, J. D., and Mc, G. J. (1954). Genital warts-a venereal disease. J. Amer. Med. Assoc. *154*, 333-334.

Bosch, F. X., and Munoz, N. (1989). Human papillomavirus and cervical neoplasia. In Human papillomavirus and cervical cancer, N. Munoz, F. X. Bosch, and O. M. Jensen, eds. (Lyon, IARC Scientific Publications), pp. 135-151.

Boshart, M., Gissmann, L., Ikenberg, H., Kleinheinz, A., Scheurlen, W., and zur Hausen, H. (1984). A new type of papillomavirus DNA, its presence in genital cancer biopsies and in cell lines derived from cervical cancer. Embo J. *3*, 1151-1157.

Boyle, W. F., Riggs, J. L., Oshiro, L. S., and Lennette, E. H. (1973). Electron microscopic identification of papova virus in laryngeal papilloma. Laryngoscope *83*, 1102-1108.

Breitburd, F., and Coursaget, P. (1999). Human papillomavirus vaccines. Semin Cancer Biol. *9*, 431-444.

Brinton, L. A. (1992). Epidemiology of cervical cancer - overview. In The epidemiolohgy of human papillomavirus and cervical cancer, N. Munoz, F. X. Bosch, K. V. Shah, and A. Meheus, eds. (Lyon, IARC Scientific Publications), pp. 3-23.

Brinton, L. A., Reeves, W. C., Brenes, M. M., Herrero, R., Gaitan, E., Tenorio, F., de Britton, R. C., Garcia, M., and Rawls, W. E. (1989). The male factor in the etiology of cervical cancer among sexually monogamous women. Int. J. Cancer *44*, 199-203.

Butz, K., Ristriani, T., Hengstermann, A., Denk, C., Scheffner, M., and Hoppe-Seyler, F. (2003). siRNA targeting of the viral E6 oncogene efficiently kills human papillomavirus-positive cancer cells. Oncogene *22*, 5938-5945.

Campion, M. J., McCance, D. J., Cuzick, J., and Singer, A. (1986). Progressive potential of mild cervical atypia: prospective cytological, colposcopic, and virological study. Lancet *2*, 237-240.

Castellsague, X., Bosch, F. X., Munoz, N., Meijer, C. J., Shah, K. V., de Sanjose, S., Eluf-Neto, J., Ngelangel, C. A., Chichareon, S., Smith, J. S., et al. (2002). Male circumcision, penile human papillomavirus infection, and cervical cancer in female partners. N. Engl. J. Med. *346*, 1105-1112.

Chambers, G., Ellsmore, V. A., O'Brien, P. M., Reid, S. W., Love, S., Campo, M. S., and Nasir, L. (2003). Association of bovine papillomavirus with the equine sarcoid. J. Gen. Virol. *84*, 1055-1062.

Chow, L. T., and Broker, T. R. (1997). In vitro experimental systems for HPV: epithelial raft cultures for investigations of viral reproduction and pathogenesis and for genetic analyses of viral proteins and regulatory sequences. Clin. Dermatol *15*, 217-227.

Ciuffo, G. (1907). Innesto positivo con filtrato di verucca volgare. Giorn Ital Mal Venereol *48*, 12-17.

Cole, S. T., and Danos, O. (1987). Nucleotide sequence and comparative analysis of the human papillomavirus type 18 genome. Phylogeny of papillomaviruses and repeated structure of the E6 and E7 gene products. J. Mol. Biol. *193*, 599-608.

Cole, S. T., and Streeck, R. E. (1986). Genome organization and nucleotide sequence of human papillomavirus type 33, which is associated with cervical cancer. J. Virol. *58*, 991-995.

de Villiers, E.-M., Fauquet, C., Broker, T. R., Bernard, H.-U., and zur Hausen, H. (2004). Classification of papillomaviruses. Virology *324*, 17-27.

de Villiers, E. M., Gissmann, L., and zur Hausen, H. (1981). Molecular cloning of viral DNA from human genital warts. J. Virol. *40*, 932-935.

DeFilippis, R. A., Goodwin, E. C., Wu, L., and DiMaio, D. (2003). Endogenous human papillomavirus E6 and E7 proteins differentially regulate proliferation, senescence, and apoptosis in HeLa cervical carcinoma cells. J. Virol. *77*, 1551-1563.

Delap, R. D., Friedman-klen, A., and Rush, M. G. (1976). The absence of human papilloma viral DNA sequences in condylomata acuminata. Virology *74*, 268-272.

Della Torre, G., Pilotti, S., de Palo, G., and Rilke, F. (1978). Viral particles in cervical condylomatous lesions. Tumori *64*, 549-553.

Dillner, J., Lehtinen, M., Bjorge, T., Luostarinen, T., Youngman, L., Jellum, E., Koskela, P., Gislefoss, R. E., Hallmans, G., Paavonen, J., et al. (1997). Prospective seroepidemiologic study of human papillomavirus infection as a risk factor for invasive cervical cancer. J. Natl. Cancer Inst. *89*, 1293-1299.

Doll, R. (1986). Implications of epidemiological evidence for further progress. In Viral etiology of cervical cancer, R. Peto, and H. zur Hausen, eds. (Cold Spring Harbor, Cold Spring Harbor Laboratory), pp. 321-326.

Dun, A. E., and Ogilvie, M. M. (1968). Intranuclear virus particles in human genital wart tissue: observations on the ultrastructure of the epidermal layer. J. Ultrastruct. Res. *22*, 282-295.

Dürst, M. (1987). Physical state of papillomavirus DNA in tumors. In Papillomaviruses and human disease, K. Syrjänen, L. Gissmann, and L. G. Koss, eds. (Berlin, Springer Verlag), pp. 393-408.

Dürst, M., Dzarlieva-Petrusevska, R. T., Boukamp, P., Fusenig, N. E., and Gissmann, L. (1987). Molecular and cytogenetic analysis of immortalized human primary keratinocytes obtained after transfection with human papillomavirus type 16 DNA. Oncogene *1*, 251-256.

Dürst, M., Gissmann, L., Ikenberg, H., and zur Hausen, H. (1983). A papillomavirus DNA from a cervical carcinoma and its prevalence in cancer biopsy samples from different geographic regions. Proc. Natl. Acad. Sci. USA. *80*, 3812-3815.

Dvoretzky, I., Shober, R., Chattopadhyay, S. K., and Lowy, D. R. (1980). A quantitative *in vitro* focus assay for bovine papilloma virus. Virology *103*, 369-375.

Eisinger, M., Kucarova, O., Sarkar, N. H., and Good, R. A. (1975). Propagation of human wart virus in tissue culture. Nature *256*, 432-434.

Evans, A. S., and Mueller, N. E. (1990). Viruses and vancer: causal associations. Ann. Epidemiol. *1*, 71-92.

Francis, D. A., Schmid, S. I., and Howley, P. M. (2000). Repression of the integrated papillomavirus E6/E7 promoter is required for growth suppression of cervical cancer cells. J. Virol. *74*, 2679-2686.

Franco, E. L. (1991). The sexually transmitted disease model for cervical cancer: incoherent epidemiologic findings and the role of misclassification of human papillomavirus infection. Epidemiology 2, 98-106.

Fuchs, P. G., Girardi, F., and Pfister, H. (1988). Human papillomavirus DNA in normal, metaplastic, preneoplastic and neoplastic epithelia of the cervix uteri. Int. J. Cancer *41*, 41-45.

Galloway, D. A. (1992). Serological assays for the detection of HPV antibodies. In The epidemiolohgy of human papillomavirus and cervical cancer, N. Munoz, F. X. Bosch, K. V. Shah, and A. Meheus, eds. (Lyon, IARC Scientific Publications), pp. 147-161.

Galloway, D. A., and McDougall, J. K. (1983). The oncogenic potential of herpes simplex viruses: evidence for a 'hit-and-run' mechanism. Nature *302*, 21-24.

Gissmann, L., Boshart, M., Durst, M., Ikenberg, H., Wagner, D., and zur Hausen, H. (1984). Presence of human papillomavirus in genital tumors. J. Invest. Dermatol. *83*, 26s-28s.

Gissmann, L., deVilliers, E. M., and zur Hausen, H. (1982). Analysis of human genital warts (condylomata acuminata) and other genital tumors for human papillomavirus type 6 DNA. Int. J. Cancer *29*, 143-146.

Gissmann, L., and Hausen, H. Z. (1976). Human papilloma virus DNA: physical mapping and genetic heterogeneity. Proc. Natl. Acad. Sci. USA. *73*, 1310-1313.

Gissmann, L., Pfister, H., and Zur Hausen, H. (1977). Human papilloma viruses (HPV): characterization of four different isolates. Virology *76*, 569-580.

Gissmann, L., and Schwarz, E. (1986). Persistence and expression of human papillomavirus DNA in genital cancer. In Ciba Foundation Symposium, P. M. Howley, ed. (Chichester, John Wiley and Sons), pp. 190-207.

Gissmann, L., and zur Hausen, H. (1980). Partial characterization of viral DNA from human genital warts (Condylomata acuminata). Int. J. Cancer *25*, 605-609.

GlaxoSmithKline (2004). http://www.gsk-bio.com/webapp/PressCorner/PressDetail.jsp?PressId=10389.

Griffiths, M. (1991). Nuns, virgins, and spinsters. Rigoni-Stern and cervical cancer revisited. Br. J. Obste. Gynaecol. *98*, 797-802.

Grussendorf-Conen, E. I., Gissmann, L., and Holters, J. (1983). Correlation between content of viral DNA and evidence of mature virus particles in HPV 1, HPV 4 and HPV 6 induced virus acanthoma. L. J. Invest. Dermatol. *81*, 511-513.

Harro, C. D., Pang, Y. Y., Roden, R. B., Hildesheim, A., Wang, Z., Reynolds, M. J., Mast, T. C., Robinson, R., Murphy, B. R., Karron, R. A., et al. (2001). Safety and immunogenicity trial in adult volunteers of a human papillomavirus 16 L1 virus-like particle vaccine. J. Natl. Cancer Inst *93*, 284-292.

Hoepfner, I., and Loning, T. (1986). Human papillomavirus (HPV) infection of cervical lesions detected by immunohistochemistry and *in situ* hybridization. Cancer Detect. Prev. *9*, 293-301.

Horne, R. W., and Wildy, P. (1961). Symmetry in virus architecture. Virology *15*, 348-373.

IARC Working Group (1995). Human papillomaviruses, Vol 64 (Lyon, International Agency for Research on Cancer).

Jochmus-Kudielka, I., Schneider, A., Braun, R., Kimmig, R., Koldovsky, U., Schneweis, K. E., Seedorf, K., and Gissmann, L. (1989). Antibodies against the human papillomavirus type 16 early proteins in human sera: correlation of anti-E7 reactivity with cervical cancer. J. Natl. Cancer Inst. *81*, 1698-1704.

Kaur, P., and McDougall, J. K. (1988). Characterization of primary human keratinocytes transformed by human papillomavirus type 18. J. Virol. *62*, 1917-1924.

Kessler, I. I. (1986). Cervical cancer: Social and sexual correlates. In Viral etiology of cervical cancer, R. Peto, and H. zur Hausen, eds. (Cold Spring Harbor, Cold Spring Harbor Laboratory), pp. 55-64.

Kirnbauer, R., Taub, J., Greenstone, H., Roden, R., Durst, M., Gissmann, L., Lowy, D. R., and Schiller, J. T. (1993). Efficient self-assembly of human papillomavirus type 16 L1 and L1-L2 into virus-like particles. J. Virol. *67*, 6929-6936.

Kjaer, S. K., van den Brule, A. J., Paull, G., Svare, E. I., Sherman, M. E., Thomsen, B. L., Suntum, M., Bock, J. E., Poll, P. A., and Meijer, C. J. (2002). Type specific persistence of high risk human papillomavirus (HPV) as indicator of high grade cervical squamous intraepithelial lesions in young women: population based prospective follow up study. B.M.J. *325*, 572.

Klug, A., and Finch, J. T. (1965). Structure of viruses of the papilloma-polyoma type. 1. Human wart virus. J. Mol. Biol. *11*, 403-423.

Koutsky, L. A., Ault, K. A., Wheeler, C. M., Brown, D. R., Barr, E., Alvarez, F. B., Chiacchierini, L. M., and Jansen, K. U. (2002). A controlled trial of a human papillomavirus type 16 vaccine. N. Engl. J. Med. *347*, 1645-1651.

Koutsky, L. A., Holmes, K. K., Critchlow, C. W., Stevens, C. E., Paavonen, J., Beckmann, A. M., DeRouen, T. A., Galloway, D. A., Vernon, D., and Kiviat, N. B. (1992). A cohort study of the risk of cervical intraepithelial neoplasia grade 2 or 3 in relation to papillomavirus infection. N. Engl. J. Med. *327*, 1272-1278.

Laverty, C. R., Russell, P., Hills, E., and Booth, N. (1978). The significance of noncondylomatous wart virus infection of the cervical transformation zone. A review with discussion of two illustrative cases. Acta Cytol. *22*, 195-201.

Law, M. F., Lancaster, W. D., and Howley, P. M. (1979). Conserved polynucleotide sequences among the genomes of papillomaviruses. J. Virol. *32*, 199-207.

Lehtinen, M., Koskela, P., Jellum, E., Bloigu, A., Anttila, T., Hallmans, G., Luukkaala, T., Thoresen, S., Youngman, L., Dillner, J., and Hakama, M. (2002). Herpes simplex virus and risk of cervical cancer: a longitudinal, nested case-control study in the nordic countries. Amer. J. Epidemiol. *156*, 687-692.

Licht, C. (1894). Om Vorters Smitsomhed. Ugeskrift Laeger *1*, 368-369.

Lorincz, A. T., Lancaster, W. D., and Temple, G. F. (1986). Cloning and characterization of the DNA of a new human papillomavirus from a woman with dysplasia of the uterine cervix. J. Virol. *58*, 225-229.

Manjunath, N., Kaur, H., Bala, S., Kaur, R., Bhargava, V., Rath, G. K., and Seth, P. (1988). Detection of herpes simplex virus type 2 DNA in uterine cervix lesions using cloned Bgl II N. fragment of HSV-2 DNA as a probe. Indian J. Med. Res. *87*, 127-133.

Massing, A. M., and Epstein, W. L. (1963). Natural history of warts. A two year study. Arch. Dermatol. *87*, 306-310.

McBride, A. A., Dlugosz, A., and Baker, C. C. (2000). Production of infectious bovine papillomavirus from cloned viral DNA by using an organotypic raft/xenograft technique. Proc. Natl. Acad. Sci. USA. *97*, 5534-5539.

Meanwell, C. A. (1988). The epidemiology of human papillomavirus infection in relation to cervical cancer. Cancer Surv. *7*, 481-497.

Meisels, A., Fortin, R., and Roy, M. (1977). Condylomatous lesions of the cervix. II. Cytologic, colposcopic and histopathologic study. Acta Cytol. *21*, 379-390.

Meyers, C., Frattini, M. G., Hudson, J. B., and Laimins, L. A. (1992). Biosynthesis of human papillomavirus from a continuous cell line upon epithelial differentiation. Science *257*, 971-973.

Munger, K., and Howley, P. M. (2002). Human papillomavirus immortalization and transformation functions. Virus Res. *89*, 213-228.

Munoz, N., and Bosch, F. X. (1992). Review of case-control and cohort studies. In The Epidemiolohgy of Human Papillomavirus and Cervical Cancer, N. Munoz, F. X. Bosch, K. V. Shah, and A. Meheus, eds. (Lyon, IARC Scientific Publications), pp. 251-261.

Munoz, N., Bosch, F. X., de Sanjose, S., Herrero, R., Castellsague, X., Shah, K. V., Snijders, P. J., and Meijer, C. J. (2003). Epidemiologic classification of human papillomavirus types associated with cervical cancer. N. Engl. J. Med. *348*, 518-527.

Munoz, N., Bosch, X., and Kaldor, J. M. (1988). Does human papillomavirus cause cervical cancer? The state of the epidemiological evidence. Br. J. Cancer *57*, 1-5.

Orth, G., Favre, M., and Croissant, O. (1977). Characterization of a new type of human papillomavirus that causes skin warts. J. Virol. *24*, 108-120.

Orth, G., Jablonska, S., Favre, M., Croissant, O., Jarzabek-Chorzelska, M., and Rzesa, G. (1978). Characterization of two types of human papillomaviruses in lesions of epidermodysplasia verruciformis. Proc. Natl. Acad. Sci. USA. *75*, 1537-1541.

Payne, J. F. (1891). Clinical notes on the contagiousness of common warts. Brit. J. Dermatol *3*, 185-188.

Pirisi, L., Yasumoto, S., Feller, M., Doniger, J., and DiPaolo, J. A. (1987). Transformation of human fibroblasts and keratinocytes with human papillomavirus type 16 DNA. J. Virol. *61*, 1061-1066.

Pratili, M. A., Le Doussal, V., Harvey, P., Laval, C., Bertrand, F., Jibard, N., Croissant, O., and Orth, G. (1986). [Human papillomaviruses in the epithelial cells of the cervix uteri: frequency of types 16 and 18. Preliminary results of a clinical, cytologic and viral study]. J. Gynecol. Obstet. Biol. Reprod. (Paris) *15*, 45-50.

Rawls, W. E., Clarke, A., Smith, K. O., Docherty, J. J., Gilman, S. C., and Graham, S. (1980). Specific antibodies to herpes simplex type 2 virus among women with cervical cancer, Vol 7 (Cold Spring Harbor, NY, Cold Spring Harbor Laboratory).

Rigoni-Stern, D. A. (1842). Statistical facts about cancer on which Doctor Rigoni-Stern based his contribution to the surgeons' subgroup of the IV Congress of the Italian Scientists on 23 September 1842 (Translation). Stat. Med. *6*, 881-884.

Roman, A., and Fife, K. H. (1989). Human papillomaviruses: are we ready to type? Clin. Microbiol. Rev. *2*, 166-190.

Rotkin, I. D. (1973). A comparison review of key epidemiological studies in cervical cancer related to current searches for transmissible agents. Cancer Res. *33*, 1353-1367.

Rowson, K. E. K., and Mahy, B. W. J. (1967). Human Papova (Wart) Virus. Bacteriol. Reviews *31*, 110-131.

Schneider, A., Sawada, E., Gissmann, L., and Shah, K. (1987). Human papillomaviruses in women with a history of abnormal Papanicolaou smears and in their male partners. Obstet. Gynecol. *69*, 554-562.

Schwarz, E., Freese, U. K., Gissmann, L., Mayer, W., Roggenbuck, B., Stremlau, A., and zur Hausen, H. (1985). Structure and transcription of human papillomavirus sequences in cervical carcinoma cells. Nature *314*, 111-114.

Seedorf, K., Krammer, G., Durst, M., Suhai, S., and Rowekamp, W. G. (1985). Human papillomavirus type 16 DNA sequence. Virology *145*, 181-185.

Serra, A. (1942). Studi sul virus della verruca, del papilloma, del condiloma acumonato (etiologia, patogenesi, filtrabilita): nota preventiva. Giorn Ttal Mal Venereol. *65*, 1808-1814.

Sigstad, E., Lie, A. K., Luostarinen, T., Dillner, J., Jellum, E., Lehtinen, M., Thoresen, S., and Abeler, V. (2002). A prospective study of the relationship between prediagnostic human papillomavirus seropositivity and HPV DNA in subsequent cervical carcinomas. Br. J. Cancer *87*, 175-180.

Skrabanek, P. (1988). Cervical cancer in nuns and prostitutes: a plea for scientific continence. J. Clin. Epidemiol. *41*, 577-582.

Smith, P. G., Kinlen, L. J., White, G. C., Adelstein, A. M., and Fox, A. J. (1980). Mortality of wives of men dying with cancer of the penis. Br. J. Cancer *41*, 422-428.

Strauss, M. J., Bunting, H., and Melnick, J. L. (1950). Virus-like particles and inclusion bodies in skin papillomas. J. Invest. Dermatol. *15*, 433-444.

Strauss, M. J., Shaw, E. W., Bunting, H., and Melnick, J. L. (1949). "Crystalline" virus-like particles from skin papillomas characterized by intranuclear inclusion bodies. Proc. Soc. Exptl. Med. *72*, 46-50.

Syrjanen, K., Parkkinen, S., Mantyjarvi, R., Vayrynen, M., Syrjanen, S., Holopainen, H., Saarikoski, S., and Castren, O. (1985). Human papillomavirus (HPV) type as an important determinant of the natural history of HPV infections in uterine cervix. Eur. J. Epidemiol. *1*, 180-187.

Tomita, Y., Kubota, K., Kasai, T., Sekiya, S., Takamizawa, H., and Simizu, B. (1986). Detection of human papillomavirus DNA in genital warts, cervical dysplasias and neoplasias. Intervirology *25*, 151-157.

Tsunokawa, Y., Takebe, N., Kasamatsu, T., Terada, M., and Sugimura, T. (1986). Transforming activity of human papillomavirus type 16 DNA sequence in a cervical cancer. Proc. Natl. Acad. Sci. USA. *83*, 2200-2203.

van Regenmortel, M. H. V., Fauquet, C. M., Bishop, D. H. L., Carstens, E. B., Estes, M. K., Lemon, S. M., Maniloff, J., Mayo, M. A., McGeoch, D. J., Pringle, C. R., and Wickner, R. B. (2000). Virus Taxonomy. Seventh Report of the International Committee on Taxonomy of Viruses (San Diego, Academic Press).

von Knebel Doeberitz, M., Oltersdorf, T., Schwarz, E., and Gissmann, L. (1988). Correlation of modified human papilloma virus early gene expression with altered growth properties in C4-1 cervical carcinoma cells. Cancer Res. *48*, 3780-3786.

von Knebel Doeberitz, M., Rittmuller, C., zur Hausen, H., and Durst, M. (1992). Inhibition of tumorigenicity of cervical cancer cells in nude mice by HPV E6-E7 anti-sense RNA. Int. J. Cancer *51*, 831-834.

Vonka, V., Kanka, J., Hirsch, I., Zavadova, H., Krcmar, M., Suchankova, A., Rezacova, D., Broucek, J., Press, M., Domorazkova, E., and et al. (1984a). Prospective study on the relationship between cervical neoplasia and herpes simplex type-2 virus. II. Herpes simplex type-2 antibody presence in sera taken at enrollment. Int. J. Cancer *33*, 61-66.

Vonka, V., Kanka, J., Jelinek, J., Subrt, I., Suchanek, A., Havrankova, A., Vachal, M., Hirsch, I., Domorazkova, E., Zavadova, H., and et al. (1984b). Prospective study on the relationship between cervical neoplasia and herpes simplex type-2 virus. I. Epidemiological characteristics. Int. J. Cancer *33*, 49-60.

Walboomers, J. M., Jacobs, M. V., Manos, M. M., Bosch, F. X., Kummer, J. A., Shah, K. V., Snijders, P. J., Peto, J., Meijer, C. J., and Munoz, N. (1999). Human papillomavirus is a necessary cause of invasive cervical cancer worldwide. J. Pathol. *189*, 12-19.

Wallin, K. L., Wiklund, F., Angstrom, T., Bergman, F., Stendahl, U., Wadell, G., Hallmans, G., and Dillner, J. (1999). Type-specific persistence of human papillomavirus DNA before the development of invasive cervical cancer. N. Engl. J. Med. *341*, 1633-1638.

Watkins, D. B. (1961). Viral disease in tattoos: verruca vulgaris. Arch. Dermatol *84*, 306-309.

Yee, C., Krishnan-Hewlett, I., Baker, C. C., Schlegel, R., and Howley, P. M. (1985). Presence and expression of human papillomavirus sequences in human cervical carcinoma cell lines. Amer. J. Pathol. *119*, 361-366.

zur Hausen, H. (1976). Condylomata acuminata and human genital cancer. Cancer Res. *36*, 794.

zur Hausen, H. (2001). Oncogenic DNA viruses. Oncogene *20*, 7820-7823.

zur Hausen, H., Gissmann, L., Steiner, L., Dippold, W., and Dregger, J. (1975). Human papillomaviruses and cancer. Bibliotheca Haematologica *43*, 569-571.

zur Hausen, H., Meinhof, W., Scheiber, W., and Bornkamm, G. W. (1974a). Attempts to detect virus-specific DNA sequences in human tumors: I. Nucleic acid hybridizations with complementary RNA of human wart virus. Int. J. Cancer *13*, 650-656.

zur Hausen, J., Schulte-Holthausen, H., Wolf, H., Dorries, K., and Egger, H. (1974b). Attempts to detect virus-specific DNA in human tumors. II. Nucleic acid hybridizations with complementary RNA of human herpes group viruses. Int. J. Cancer *13*, 657-664.

From: Papillomavirus Research: From Natural History to Vaccines and Beyond. Edited by: M. Saveria Campo

Chapter 2

Phylogeny and Taxonomy of Papillomaviruses

Hans-Ulrich Bernard*

Abstract

Papillomaviruses (PVs) have similar genomic organizations, and any pair of two PVs contains at least five homologous genes, although the nucleotide sequence may diverge by more than 50%. Phylogenetic algorithms that permit to compare homologies led to phylogenetic trees that have a similar topology independently of the gene analyzed. Phylogenetic studies strongly suggest that PVs normally evolve together with their mammalian and bird host species, do not change host species, do not recombine, and have maintained their basic genomic organization for a period exceeding 100 million years. These sequence comparisons have laid the foundation for a PV taxonomy, which is now officially recognized by the International Committee on Taxonomy of Viruses. All PVs form the family "*Papillomaviridae*", which is distinct from the "*Polyomaviridae*" thus eliminating the term "*Papovaviridae*". Major branches of the phylogenetic tree of all PVs are considered as "genera", which are identified by Greek letters. Minor branches are considered "species" and unite PV types that are genomically distinct without exhibiting known biological differences. This new taxonomic system does not affect the traditional identification and characterization of PV "types" and their independent isolates with minor genomic differences, referred to as "subtypes" and "variants", all of which are taxa below the level of "species". Understanding the tremendous diversity of PV types, the worldwide distribution of variants of individual human PV (HPV) types, and their potential pathogenic distinctions, continue to be challenges for phylogenetic research and etiological considerations.

1. *Papillomaviridae*, a Homogenous Taxonomic Family in Search of a Phylogenetic Root

Traditionally, PVs were grouped with the polyomaviruses (including the intensely studied simian virus 40 [SV40]) in one taxonomic family, the *Papovaviridae*. This was based on the fact that they have electronmicroscopically similar viral capsids, no envelope and a double-stranded circular DNA genome. The unified taxon "*Papovaviridae*" was difficult to maintain when it became clear in the mid-1980s that polyoma- and papillomaviruses have different genome sizes (about 5 kb vs. 8 kb), different transcriptional strategies (early and late genes confronting one another vs. being oriented in the same direction; see Figure 1 of Introduction), and that their proteins are not homologous. The conclusion that these different taxa of viruses should be considered "families" that are just as distinct from one another as other virus families are was obvious. The negotiations between H. zur Hausen, representing the community of PV scientists, and the International Committee on Taxonomy of Viruses (ICTV) led to the formal recognition of the family "*Papillomaviridae*" (de Villiers et al., 2004).

Presently, there are no known relatives of the *Papillomaviridae*. It has been observed that PVs isolated from hosts that are remotely related to humans are also remotely related to human papillomaviruses (HPVs) and show major genomic idiosynchrasies (Scobie et al., 1997; Terai et al., 2002). This provides one of several arguments that PVs normally evolve in linkage to their hosts, and the observation that bird PVs form a separate remote branch without close linkage to any mammalian PVs suggest that the split between bird PVs and mammalian PVs is as old as that between their hosts, supposedly exceeding 100 million years. One can speculate that a search for PV-like genomes in animals even less related than mammals and birds may detect genomes of even greater diversity. Such considerations led to speculations about the evolutionary origin of PVs. Notably, in spite of the taxonomic separation of the papilloma- and *Polyomaviridae*, these two taxa share an apparently homologous domain of the E1 protein and the T-antigen protein, respectively, responsible for the helicase activity of these proteins. The same homology was detected in a virus-like episome in planaria and in adenoassociated viruses (Rebrikov et al., 2002). It is presently impossible to conclude, but nevertheless attractive to speculate, that such a helicase domain may have been the origin of an ancient episome or virus, that gave rise to all of these elements by becoming linked to genes with different functions. Alternatively, this domain may have been picked up separately by each of these

*For correspondence. Email hbernard@uci.edu.

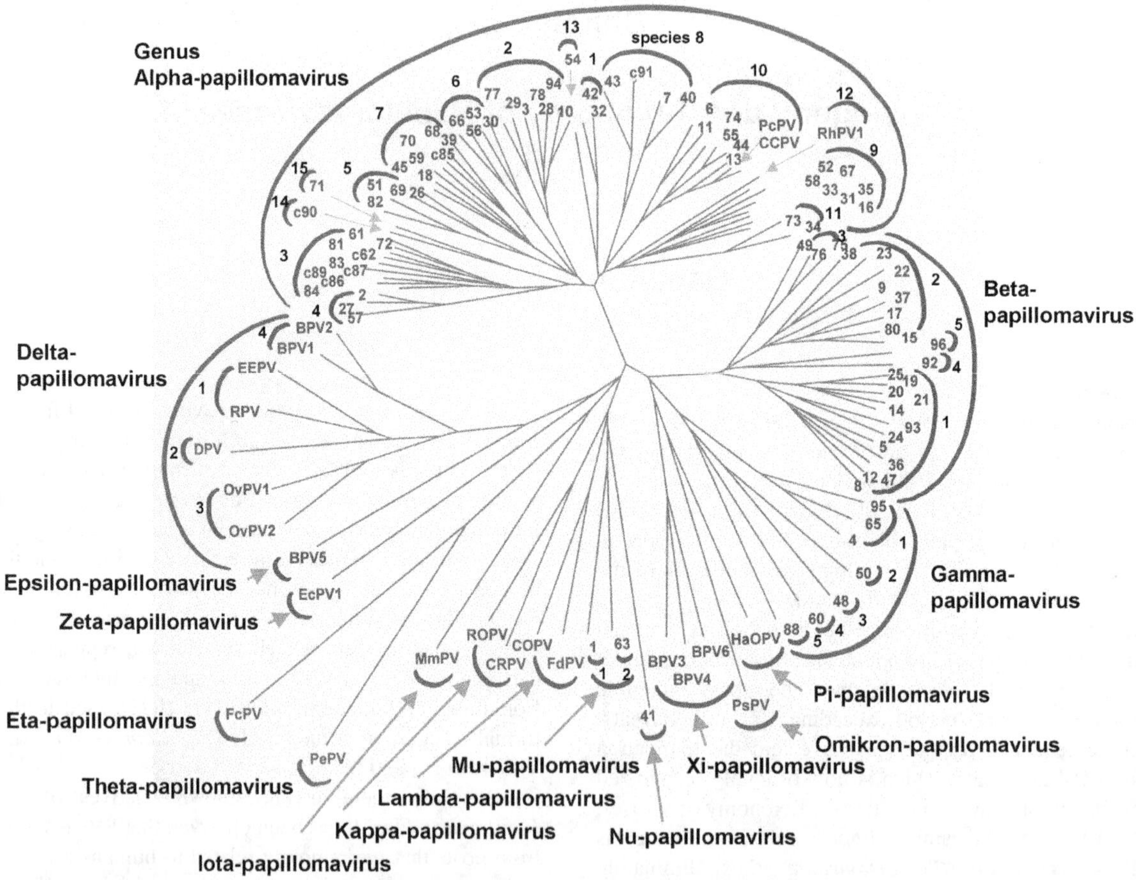

Figure 1. Phylogenetic tree of 118 papillomavirus types. All numbers refer to HPV types, c-numbers to candidate types, *i.e.* HPV genomes isolated as PCR amplicons. All abbreviations including letters refer to animal papillomaviruses. These abbreviations are listed in de Villiers et al., 2004. The outermost semicircular symbols identify papillomavirus genera, for example the genus alpha-papillomavirus. The inner semicircular symbols identify papillomavirus species. For example, the HPV species 8 within the genus alpha-papillomaviruses lumps the HPV types 7, 40, 43, and cand91. Reprinted from "Classification of papillomaviruses" by de Villiers, E.M., Fauquet, C., Broker, T.R., Bernard, H.U., and zur Hausen, H. Virology, 324, 17-27 (2004), with permission from Elsevier.

episomes and viruses by molecular recombination, which would then only support the homology of the domain, but not a common origin of these diverse genomes.

2. "Genus" as a Novel Concept in Papillomavirus Taxonomy

PV genomes were traditionally described as "types" (as further discussed below). The technologies of the 1970s and 1980s, such as DNA hybridization, led to the earliest efforts to determine relationships between different PV types (de Villiers, 1989 and 1997). Today, PV taxonomy is based on phylogenetic algorithms that analyze the similarities between nucleotide sequences. During the research that led to today's taxonomy it became clear that the comparison of whole PV genomes, individual homologous PV genes or even small PV gene segments led to identical or very similar phylogenetic trees (Chan et al., 1992; van Ranst et al., 1992; Chan et al., 1995; Farmer et al., 1995; Matsukura and Sugase, 2001; deVilliers et al., 2004).

When it became possible to compare the sequences of a large number of PV types, most HPV and some animal PV types formed hierarchical assemblages, for which expressions like "major branches" and "minor branches" or "supergroups" and "groups" were coined. In order to give two examples, one major branch was formed by all HPV types that were originally described from genital lesions, and therefore called "genital HPVs", while another one was formed by all PVs from hoofed animals that occurred in epithelial as well as in mesenchymal cells and referred to as "fibropapillomaviruses". Two considerations demanded to reconsider these terms in spite of the stability of the underlying phylogenetic trees. (i) It became difficult to find consistent biological terms for these assemblages. For example, the genital PV group includes among many HPV types that are mostly infecting genital mucosae also some HPV types, *e.g.* HPV-2 and 27, which normally give rise to cutaneous lesions not associated with genital infection. Very similar cutaneous lesions are induced by other unrelated PVs (*e.g.* HPV-4). This shows that a phenotypically based taxonomy, namely one that creates taxa of PV types that give rise to similar pathology, would not reflect the true relationship among these PVs. (ii) The taxonomy of all biological organisms (viruses arguably here considered as being "organisms") recognizes below the taxonomic level of the family the taxa "genus" and "species", and the original terms that were used to describe the topology of phylogenetic trees of PVs would be inconsistent with this.

In order to resolve these problems, the collaboration between the working group on PV taxonomy of the ICTV (consisting of E.M. de Villiers, H. zur Hausen, H. U. Bernard, and T. Broker) and the ICTV led to the result that the term "genus" should be introduced for higher order assemblages, *e.g.* "supergroups" or "major branches" like the genital HPVs. To overcome the problem of the inability to find a coherently descriptive biological name for each genus, it was agreed to introduce Greek letters to name such genera. As a consequence, the "genital HPVs" became the genus "alpha-papillomaviruses". A complete list of genera created by this agreement has recently been published (de Villiers et al., 2004). All genital PVs are referred to as alpha-papillomaviruses, those PV types most often associated with epidermodysplasia verruciformis as beta-PVs. Many other cutaneous HPV types (including HPV-4) belong to the gamma-PVs. HPV-1, 63, and 41, traditionally known to be little related to the other HPVs, form two additional separate genera. Most fibropapillomaviruses belong to the delta-PVs, and the mucosal ungulate PVs to the xi-PVs. For other PV genera and their relationship, see Figure 1. While somewhat unwieldy, this taxonomy took conceptual cues from the taxonomy of herpes viruses, and should be respected as an effort to create consensus and eliminate arbitrariness. It will obviously need a refinement once more PV genera are found than there are letters in the Greek alphabet, which will without doubt happen as PVs are detected in mammals and possibly other vertebrates that are remotely related to those few groups studied so far.

3. Papillomavirus "Species", Biological Facts and Bureaucratic Hurdles

There is a tradition and continuing technical problem that PVs are not identified, maintained and described in cell culture but rather by isolation of their genomes. For PV DNA isolates that were clearly distinct, the term "type" has been introduced in the 1970s, when the multiplicity of these viruses became apparent. With changes of technology, the distinctions between types became redefined. Today, two PV genomes are considered separate types when the nucleotide sequences of their L1 genes, which encode the capsid proteins, have less than 10% similarity. This definition makes HPV types genomically as remotely related as many phenotypically distinct host species, for example humans and the four living ape species. As a consequence, PV scientists became inclined to consider a "PV type" the same as a "PV species", in particular, since there are phylogenetic observations that strongly suggest that fairly closely related PV types evolved into distinct taxa over periods similar to closely related mammals such as humans and their four relatives.

Unfortunately, this reasoning did not become adopted by the ICTV, which is in charge to formally agree on the definition of "virus species" and taxa above the species level. ICTV rules require that each virus species is defined by unique biological and/or pathological characteristics setting it apart from any other virus. This rule can clearly not be fulfilled by many PV types. For example, no biological criteria are known that would distinguish between HPV-2, 27, and 57, which are the cause of common warts, and HPV-6 and 11, which are found in genital warts, laryngeal papillomas, and condylomata acuminata. This bureaucratic hurdle was resolved based on the observation that many of these biologically indistinguishable types formed phylogenetic clusters ("groups", "minor branches"), such as the association between HPV-2, 27 and 57. In order to resolve the impasse over the relationship between "PV types" and "PV species", it was agreed to consider these clusters of related PV types as PV species and identify them by a numbering system. This numbering system has been recently published (de Villiers et al., 2004) and will not be repeated here in detail (for a complete representation of the relationship between all PV species, see Figure 1). Among the PV types grouped into one species, the term "type species" has been coined for the best studied PV type within one species, that is also used to list the biological properties of each species. This new species definition has the advantages (i) to integrate PV taxonomy into the general viral systematics overseen by ICTV, (ii) to be biologically meaningful, (iii) to be based on phylogeny, and (iv) to leave the description of new PV types in the hands of PV researchers (as opposed to larger administrative entities such as the ICTV). It has the disadvantages (i) to introduce a new numbering system that will almost certainly not become widely used in molecular or medical research, and (ii) to place the PV type at a lower phylogenetic level than likely justified by evolutionary time scales and genomic distances.

4. Papillomavirus Types, Natural Taxonomic Entities of Unknown Phylogenetic History and Foci of Clinical Research

The definition of a PV type as a complete PV genome whose L1 nucleotide sequence differs from that of any other one by more than 10% has been mentioned above. No genomes intermediate to two PV types have been found, nor has recombination ever been observed. With the exception of the observation that multiple species of hoofed animals in agricultural settings can be infected by the same PV types, for example certain bovine papillomaviruses (Trenfield et al., 1990; Chambers et al., 2003), all PV types seem to be specific for one host species. The processes that give rise to new PV types are not known. However, genetic drift combined with genetic bottlenecks is a likely source to create diversity and more attractive than alternative explanations. The large numbers of different HPV types that are not biologically distinct does not suggest that each PV type evolved into a novel niche (Chan et al., 1995, Antonsson et al., 2000, 2002, 2003). There is also no indication that immune selection may contribute significantly to the phylogenetic process. While details of the immune response against PV infected cells are still little understood, the absence of inflammation in response to PV infection suggests an absence of ongoing evolutionary processes to evade immune recognition.

Most PV research concentrates on those HPV types that cause malignancies threatening human lives or other important neoplasias, most notably cervical cancer and genital and laryngeal warts. HPV taxonomy may have a major impact on these fields: In order to give an example, about 50% of all cervical carcinoma and its precursor lesions contain HPV-16, a molecularly thoroughly investigated HPV type. HPV-16 is the type species of the HPV species 9, which unifies HPV-16 with HPV-31, 33, 35, 52, 58, and 67 (de Villiers et al., 2004). Each of these HPV types is likely to be a "high-risk" type with an etiological potential similar to HPV-16 (Munoz et al., 2003). While it will be impossible, however, to study each of these types with the same intensity as HPV-16, their close relationship allows to extrapolate that they will have the same or similar molecular properties. Once medical strategies are in place to combat HPV-16 infections, one may hypothesize that the same strategies will be useful to target the six related HPV types.

5. Papillomavirus Subtypes: Why are There So Few?

The term "subtype" is sometimes erroneously used by scientists at the margin of the HPV research instead of the term "type", and this misuse should be avoided. The term "subtype" originally described HPV isolates that were apparently closely related, for example when they cross-hybridized strongly in Southern blots, but had differing restriction patterns. Later, it became clear that many subtypes rather fall into the category of genomic "variants" (see below). Today, subtypes are narrowly defined as PV genomes that differ from another type by 2-10% of their nucleotide sequence. The outcome of this definition was surprising, as it became clear that there are very few PV genomes that fall in this category. There are only five exceptions in the literature: The HPV-2 subtype HPV-2c was found to be identical to HPV-27 (Chan et al., 1997). HPV-67 was found to be a subtype of HPV-34, HPV-46 of HPV-20, and HPV-55 of HPV-44, with the consequence that the type listed first in each pair became eliminated as a type. Lastly, the Chimpanzee-PV and the Pigmy Chimpanzee (Bonobo) PV are related to one another as subtypes (de Villiers et al., 2004).

The reasons for the near-absence of subtypes are not understood, but are likely intrinsic to the question of the evolution of PV types, namely a combination of species specificity, genetic drift and bottlenecks. In the case of human papillomaviruses one may assume that all HPV types were already in existence at the time Homo sapiens came into being, possibly 200,000 years ago, and these last 200,000 years were not enough to allow for an evolution of HPV genomes sufficiently distant from one another to be classified as subtypes. HPV subtypes may be absent for the same little understood reasons that there are no close relatives to Homo sapiens living today, and the viruses may have become extinct together with these ancient relatives.

6. Papillomavirus Variants: Coevolution, Evolutionary Bottlenecks, Spread by Recent Travel, and Medical Relevance?

In contrast to the rarity of subtypes, today's human population is a rich source of variants of each HPV type. Variants are defined as HPV genomes less than 2% different from the original isolate of the respective HPV type, which is called "reference genome" or "prototype". The genomic diversity can exceed this percentage in some parts of the genome, in particular in the long control region, which does not contain genes. Variants have been described from most commonly examined HPV types (Deau et al., 1993; Ho et al., 1993; Ong et al., 1993; Heinzel et al., 1995; Stewart et al., 1996; Chan et al., 1997a; Yamada et al., 1997; Calleja-Macias et al., 2004) and occasionally in animal papillomaviruses (Chambers et al., 2003). It can be generalized that all studies of the variation within a short genomic segment, for example 400 bp of the LCR, identified in all HPV types 20 to 60 variants that formed a small number (two to six) of phylogenetic branches. Among these variants, some seem to prevail, while some apparently rare ones are often added by subsequent studies. There is apparently a high degree of linked evolution between host and variant, as certain variants of many types are highly prevalent in certain ethnic groups (Ho et al., 1993; Ong et al., 1993; Stewart et al., 1996), particularly when one compares European-Caucasian, East Asian, and African people. There are many exceptions to a strict linkage between ethnicity and variants, and it is challenging to ask, but very difficult to resolve, whether these exceptions represent ancient bottlenecks or infections after travel and mixing of populations in historic or recent times. In some cases variants of HPV types spread like chromosomally inherited genes and the viral load of a population reflects their ethnic composition, while other data are difficult to explain. For example, the population of Brazil, which contains about 30% genes of African origin, has only a low prevalence of African HPV-16 and HPV-18 variants (Ho et al., 1993). A mostly European population in Northern Mexico contained 70% HPV-16 variants of native American origin (Calleja-Macias et al., 2004), while a tribe of native Americans in Argentina contained European variants more frequently than native American ones (Picconi et al., 2003). In mixed populations, variants may still be associated with the ethnic groups where they originated, for example African Americans (Ho et al., 1993), but this may only be the result of preferred sexual contacts, and variants can probably easily pass inter-ethnic barriers.

Some investigations suggest that variants of HPV-16 may have different biological (Villa et al., 2000) and pathogenic properties, with non-European variants possibly showing a higher oncogenicity. These questions are still not completely resolved, in particular, since these studies were often done in populations where European HPV-16 variants predominated and epidemiological analyses may not have sufficient statistical power to evaluate the differences between the various non-European variants (Xi et al., 1998; da Costa et al., 2002). In order to gain statistical power, European variants were often compared with non-European variants, which, however, are genetically heterogenous among one another. Some other investigations were performed in European nations lacking any variants except those occurring traditionally in Europeans. The few mutations found in these variants did not suggest differing viral properties due to transcriptional effects from distinct LCR sequences, while some changes in the E6 gene correlated with differing pathological outcome (Nindl et al., 1999; Kammer et al., 2002). In spite of the limitations of this research, intratype molecular diversity should be kept in mind as a potential explanation for the higher incidence of cervical cancer in developing country. While a large fraction of this higher incidence rate is most likely attributed to access to and quality of performance of diagnostic procedures such as Papanecolaou staining and colposcopy, it is also reasonable to assume that populations in Latin America, Africa and parts of Asia are exposed to genomically different variants of HPV types more than populations predominantly made up of people of European ancestry.

7. Evolution of Papillomaviruses: Considerations About the Time Scale

Molecular changes of PV genomes occur clearly at a very low rate and have never been observed *in vitro* nor due to evolution during the last twenty years that PV genome sequences are available. Therefore, the rate of PV genome changes has to be estimated from comparisons of naturally occurring genomes and assumptions about their origin. There are several observations that all lead to estimates that PV genomes change at the same slow rates as the genomes of their vertebrate hosts, namely roughly 1% nucleotide exchange per 100,000 to 1,000,000 years. (1) All ethnic groups of the world, including reproductively highly isolated people such as certain Amazonian Indians (Ho et al., 1993; L. L. Villa, pers. comm.) contain apparently the same complete "set" of high-risk, genital HPV types. This would mean that humankind was never without these viruses, and cervical cancer was always with us, just like all the cancers that are exclusively due to cellular mutations. It is believed that the different ethnic groups of Homo sapiens came into being during the last 200,000 to 1,000,000 years, the time span that would have been available to generate 2% genomic diversity among the different HPV variants (Bernard, 1994). (2) The Chimpanzee and Pigmy chimpanzee PVs are (subtypes of) a PV type within the same species (No. 10), which also contains HPV-6 and 11, suggesting that over a period of a few million years, which led to the divergence of humans and apes, PV types diversified to a level that make ape PVs members of HPV species (de Villiers et al., 1994). (3) All PVs from different species of monkeys (Chan et al., 1997) form branches within the genital HPVs, which are probably the equivalent of species, but not termed this way, as the published phylogeny was based on short amplicons, and whole genomes of most of these monkey viruses are not available. This would mean that it would take 10-25 million years to give rise to the phylogenetic assemblage of PV genomes that qualify as closely related species. Interestingly, an intensive search for genital PVs in neotropical monkeys, which have probably been separated from Old World monkeys for longer than 10-25 million years, was unsuccessful (Bernard and Villa, unpublished observations), possibly suggesting that the origin of the alpha-papillomaviruses postdated this time scale. From all these considerations, one can only conclude that separation of the alpha-, beta-, gamma- and some other HPV genera must have predated the origin of primates. (4) The extremely remote relationship of bird PVs to all mammalian PVs and idiosynchrasies of their genome organization (Terai et al., 2002) suggests that these viruses split from the mammalian PVs at a time when reptilian ancestors gave rise to the separate lineages of mammals and birds. This may suggests that reptiles should have PVs, which however have not been found yet. (5) There are, however, also splits into separate taxa of PVs independent of the speciation of their host as shown by representatives of the beta-papillomaviruses (formerly epidermodysplasia verruciformis PVs) in humans and a remotely related monkey (Chan et al., 1997b).

8. Papillomaviruses: Pathogens or Latent Commensals?

Infections with PVs are well confirmed as the cause of common and genital warts, and infection with HPVs is the most important risk factor in the etiology of cervical cancer. The overwhelming data that support these notions, however, led to the somewhat erroneous characterization of PVs as aggressive neoplastic agents. This is clearly not the case for many little characterized cutaneous HPV types, and even does not strictly apply to the viruses mentioned above. Antonsson and colleagues (Antonsson et al., 2000; 2002; 2003) have recently published several impressive papers that document a vast diversity of not yet fully characterized and isolated PV types in healthy skin, which appear to be very widespread and are rarely found in neoplastic disease (de Villiers, 1998). Similarly, there are data showing that the numerous PV types that give rise in certain genetically predisposed individuals to epidermodysplasia verruciformis, *i.e.* extensive benign and malignant cutaneous neoplasia, are widespread in the general populations, where they normally do not cause disease, with the exception of severely immune suppressed carriers. Beyond these special types, one should be aware of the limitations of disease processes induced by the well studied mucosal and cutaneous HPV types. High-risk HPV types such as HPV-16 most often give rise to neoplasia only in the context of the transformation zone of the cervix, while they lead to subclinical infection elsewhere, for example throughout the vagina and at penile epithelia. And even at the cervix, most infections take a subclinical course (Schiffman and Brinton, 1995). Lastly, even the emergence of a common wart is clearly a unique event, since it is rare that numerous warts emerge from a facial wart, although without doubt the surrounding skin of such a wart frequently becomes wounded and subsequently accessed by HPV particles. All these considerations lead to a picture that PVs apparently evolved to undergo a lifecycle that does not necessarily – and in the case of many HPV types never – lead to clinically detectable neoplasia. As a consequence, one may view the *Papillomaviridae* as a whole as a family of viruses with a commensalic "episome" like lifecycle, including even "high-risk" HPVs in the context of most types of epithelia. This absence of disease, and the inefficiency of PVs to induce inflammatory immune responses, may be a principal reason for the emergence of PV diversity, *i.e.* evolution without immune selection but through genetic drift and bottlenecks of distribution. Relative to the total number of viral genomes that infect epithelial target cells, neoplastic disease is an extremely rare event.

References

Antonsson, A., Forslund, O., Ekberg, H., Sterner, G, and Hansson, B.G. (2000). The ubiquity and impressive genomic diversity of human skin papillomaviruses suggest a commensalic nature of these viruses. J. Virol. *74*, 11636-11641.

Antonsson, A., and Hansson, B.G. (2002). Healthy skin of many species harbours papillomaviruses, which are closely related to their human counterparts. J. Virol. *76*, 12537-12542.

Antonsson, A., Erfurt, C., Hazard, K., Holmgren, V., Simon, M., Kataoka, A., Hossain, S., Hakangard., C, and Hansson, B.G.. (2003). Prevalence and type spectrum of human papillomaviruses in healthy skin samples collected in three continents. J. Gen. Virol. *84*, 1881-1886.

Bernard, H.U. (1994). Coevolution of papillomaviruses and human populations. Trends Microbiol. *2*, 140-143.

Calleja-Macias, I.E., Kalantari, M., Huh, J., Ortiz-Lopez, R., Rojas-Martines, A., Gonzales-Guerrero, J.F., Williamson, A.L., Hagmar, B., Wiley, D.J., Villarreal, L., Bernard, H.U., Barrera-Saldana, H.A. (2004). High prevalence of specific variants of human papillomavirus-16, 18, 31, and 35 in a Mexican population. Virology, *319*, 315-323.

Chambers, G., Ellsmore, V.A., O'Brien, P.M., Reid, S.W., Love, S., Campo, M.S., Nasir, L. (2003). Sequence variants of bovine papillomavirus E5 detected in equine sarcoids. Virus Res. *96,* 141-145.

Chan, S.Y., Bernard, H.U., Ong, C.K., Chan, S.P., Hofmann, B., and Delius H. (1992). Phylogenetic analysis of 48 papillomavirus types and 28 subtypes and variants: a show case for the molecular evolution of DNA viruses. J. Virol. *66,* 5714-5725.

Chan, S.Y., Bernard, H.U., Ratterree, M., Birkebak, T.A., Faras, A. J., and Ostrow, R. S. (1997). Genomic diversity and evolution of papillomaviruses in Rhesus monkeys in Rhesus monkeys. J. Virol. 71, 4938-4943.

Chan, S.Y., Chew, S.H., Egawa, K., Grussendorf-Conen, E.I., Honda, Y., Ruebben, A., Tan, K.C., and Bernard, H.U. (1997)Phylogenetic analysis of the Human papillomavirus type 2 (HPV-2), HPV-27, and HPV-57 group, which is associated with common warts. Virology *239*, 296-302 (1997a).

Chan, S. Y., Delius, H., Halpern, A.L., and Bernard, H.U. (1995). Analysis of genomic sequences of 95 papillomavirus types: Uniting typing, phylogeny, and taxonomy. J. Virol. *69*, 3074-3083.

Chan, S.Y., Ostrow, R. S., Faras, A. J., and Bernard, H.U. (1997b). Genital papillomaviruses (PVs) and epidermodysplasia verruciformis PVs occur in the same monkey species: Implications for PV evolution. Virology *228*, 213-217.

Da Costa, M.M., Hogeboom, C.J., Holly, E.A., Palefsky, J.M. (2002). Increased risk of high-grade anal neoplasia associated with a human papillomavirus type 16 E6 sequence variant. J. Infect. Dis. *185*,1229-1937.

Deau, M.C., Favre, M., Jablonska, S., Rueda, L.A., Orth, G. (1993). Genetic heterogeneity of oncogenic human papillomavirus type 5 (HPV5) and phylogeny of HPV5 variants associated with epidermodysplasia verruciformis. J. Clin. Microbiol. *31*, 2918-2926.

de Villiers, E.M. (1989). Heterogeneity of the human papillomavirus group. J. Virol. 63: 4898-4903.

de Villiers, E-M. (1997). Papillomavirus and HPV typing. Clin. Dermatol. *15*, 199-206.

de Villiers, E-M. (1998). Human papillomavirus infections in skin cancer. Biomed. Pharmacother. *52*, 26-33.

de Villiers, E.M., Fauquet, C., Broker, T.R., Bernard, H.U., and zur Hausen, H. (2004). Classification of papillomaviruses. Virology *324*, 17-27

Farmer, A.D., Calef, C.E., Millman, K., and Myers, G.L.. (1995). The human papillomavirus database. J. Biomed. Sci. *2*, 90-104. (http://hpv-web.lanl.gov/stdgen/virus/hpv/compendium/htdocs/)

Heinzel, P. A., Chan, S. Y., Ho, L., O'Connor, M., Balaram, P., Campo, M. S., Fujinaga, K., Kiviat, N., Kuypers, J., Pfister, H., Steinberg, B.M., Tay, S.K., Villa, L. L., and Bernard, H. U. (1995). Variation of human papillomavirus type 6 (HPV-6) and HPV-11 genomes sampled throughout the world. J. Clin. Microb. *33*, 1746-1754.

Ho, L., Chan, S.Y., Burk, R.D., Das, B.C., Fujinaga, K., Icenogle, J.P., Kahn, T., Kiviat, N., Lancaster, W., Mavromara, P., Labropoulou, V., Mitrani-Rosenbaum, S., Norrild, B., Pillai, M.R., Stoerker, J., Syrjaenen, K., Syrjaenen, S., Tay, S.K., Villa, L.L., Wheeler, C.M., Williamson, A.L., and Bernard, H.U (1993) The genetic drift of human papillomavirus type 16 is a means of reconstructing prehistoric viral spread and movement of ancient human populations. J.Virol. *67*, 6413-6414.

Kammer, C., Tommasino, M., Syrjanen, S., Delius, H., Hebling, U., Warthorst, U., Pfister, H., and Zehbe, I. (2002). Variants of the long control region and the E6 oncogene in European human papillomavirus type 16 isolates: implications for cervical disease. Br. J. Cancer. *86*, 269-273.

Matsukura, T., and Sugase, M. (2001). Relationships between 80 human papillomavirus genotypes and different grades of cervical intraepithelial neoplasia: association and causality. Virology *283*,139-147.

Munoz, N., Bosch, F.X., de Sanjosé, S., Herrero, R., Castellsagué, X., Shah, K.V., Snijders, P.J.F., and Meijer, C.J.L.M. (2003) Epidemiological classification of human papillomavirus types associated with cervical cancer. N. Engl. J. Med. *348*, 518-527.

Nindl, I., Rindfleisch, K., Lotz, B., Schneider, A., and Durst, M. (1999). Uniform distribution of HPV 16 E6 and E7 variants in patients with normal histology, cervical intra-epithelial neoplasia and cervical cancer. Int. J. Cancer *82*, 203-207.

Ong, C.K., Chan, S.Y., Campo, M.S., Fujinaga, K., Mavromara, P., Labropoulou, V., Pfister, H., Tay, S.K., ter Meulen, J., Villa, L.L., and Bernard, H.U. Evolution of human papillomavirus type 18: An ancient phylogenetic root in Africa and intratype diversity reflect coevolution with human ethnic groups. J. Virol. *67*, 6424-6431.

Picconi, M.A., Alonio, L.V., Sichero, L., Mbayed, V., Villa, L.L., Gronda, J., Campos, R., and Teyssie, A. (2003). Human papillomavirus type-16 variants in Quechua aboriginals from Argentina. J. Med, Virol. *69*, 546-552.

Rebrikov, D.V., Bogdanova, E.A., Bulina, M.E., and Lukyanov, S.A. (2002). A new planarian extrachromosomal virus-like element revealed by subtractive hybridization. Mol.Biol. *36*, 813-820.

Schiffman, M.H., and Brinton, L.A. (1995). The epidemiology of cervical carcinogenesis. Cancer *76*, 1888-1901.

Scobie, L., Jackson, M.E., and Campo, M.S. (1997). The role of exogenous p53 and E6 oncoproteins in *in vitro* transformation by bovine papillomavirus type 4 (BPV-4): significance of the absence of an E6 ORF in the BPV-4 genome. J. Gen. Virol. *78*, 3001-3008.

Stewart, A.C., Eriksson, A.M., Manos, M.M., Munoz, N., Bosch, F.X., Peto, J., and Wheeler, C.M. (1996). Intratype variation in 12 human papillomavirus types: a worldwide perspective. J. Virol. *70*, 3127-3136.

Terai, M., DeSalle, R., and Burk, R.D. (2002). Lack of canonical E6 and E7 open reading frames in bird papillomaviruses: Fringilla coelebs papillomavirus and Psittacus erithacus timneh papillomavirus. J. Virol. *76*, 10020-10023.

Trenfield, K., Spradbrow, P.B., and Vanselow, B.A. (1990). Detection of papillomavirus DNA in precancerous lesions of the ears of sheep. Vet. Microbiol. *25*, 103-116.

Van Ranst, M., Kaplan, J.B., and Burk, R.D. (1992). Phylogenetic classification of human papillomaviruses: correlation with clinical manifestations. J. Gen. Virol. *73*, 2653-2660.

Villa, L.L., Sichero, L., Rahal, P., Caballero, O., Ferenczy, A., Rohan, T., and Franco, E.L. (2000). Molecular variants of human papillomavirus types 16 and 18 preferentially associated with cervical neoplasia. J. Gen. Virol. *81*, 2959-2968.

Xi, L.F., Critchlow, C.W., Wheeler, C.M., Koutsky, L.A., Galloway, D.A., Kuypers, J., Hughes, J.P., Hawes, S.E., Surawicz, C., Goldbaum, G., Holmes, K.K., and Kiviat, N.B. (1998). Risk of anal carcinoma in situ in relation to human papillomavirus type 16 variants. Cancer Res. *58*, 3839-3844.

Yamada, T., Manos, M.M., Peto, J., Greer, C.E., Munoz, N., Bosch, F.X., and Wheeler, C.M. (1997). Human papillomavirus type 16 sequence variation in cervical cancers: a worldwide perspective. J. Virol. *71*, 2463-2472.

From: Papillomavirus Research: From Natural History to Vaccines and Beyond. Edited by: M. Saveria Campo

Chapter 3

Epidemiology of Human Papillomavirus Infections and Associations With Cervical Cancer: New Opportunities for Prevention

F. Xavier Bosch*, Silvia de Sanjosé, Xavier Castellsagué, Víctor Moreno and Nubia Muñoz

Abstract

Historical studies on the epidemiology of cervical cancer repeatedly pointed out associations between the sexual behavioral traits of women and men, and the markers of sexually transmitted diseases. When the technology was developed that allowed testing for Human Papillomavirus (HPV) DNA in exfoliated cervical cells, epidemiological studies were able to confirm that most of the observed associations were in fact mediated by the persistent presence of HPV DNA. The finding is of utmost importance because a group of some 15 high-risk HPV types seem to be necessary for the development of all cervical cancer cases worldwide. As a consequence, the strategies for prevention are rapidly evolving towards the incorporation of HPV tests as a screening tool and the testing of novel HPV vaccines. Evidence to date indicates that HPV tests are at least as efficient as conventional cytology as a primary screening test and superior to repeated cytology in the triage of the smears of uncertain significance. Phase II trials have shown that HPV vaccines are highly immunogenic and protect women from persistent type specific HPV infections. The paradigm of cervical cancer prevention is likely to change in the forthcoming years.

1. Introduction

A major discovery in human cancer etiology has been the recognition that cervical cancer is a rare consequence of an infection by some mucosatropic types of Human Papillomavirus (HPV). In Public Health terms, the finding is of importance comparable to the unveiling of the association between cigarette smoking and lung cancer, or between chronic infections with Hepatitis B or Hepatitis C viruses and the risk of liver cancer. Moreover, like in the Hepatitis B disease model, intense efforts are currently being put into the development and testing of vaccines that may prevent and / or treat the relevant HPV infections and, presumably, contribute to the reduction the incidence and the mortality due to cervical cancer.

HPV natural history studies have revealed that HPVs are the commonest of the sexually transmitted infections in most populations. Most HPV exposures result in spontaneous clearance without clinical manifestations and only a small fraction of the infected persons, known as chronic or persistent carriers, will retain the virus and progress to cancer. HPV related-cancers can develop in different sites in the ano-genital tract including the cervix, vagina, vulva penis and anal canal (International Agency for Research on Cancer, 1996). Further, research has indicated that HPV has been significantly associated with cancers at other organ sites such as the skin, the oropharynx and the oesophagus, and these are areas of intensive research (International Agency for Research on Cancer, 1996; Bosch et al., 2003).

2. The Burden of Cervical Cancer and of Human Papillomavirus (HPV) Infections

2.1. Cervical cancer

Population based cancer registries provide reasonable measurements of cervical cancer incidence and mortality statistics are reliable in many parts of the world. Data on invasive cervical cancer extracted from routine sources such *Globocan* and vol. VII of the series *Cancer Incidence in Five Continents* is presented in Table 1 (Ferlay et al., 2001; Parkin et al., 1997). The table includes the estimated number of cases by region, the crude rates and the age adjusted rates. Of notice is the impact of the adjustment procedure, which tends to reduce adjusted rates in developed countries and increase the rates in the developing parts of the world. This phenomena is due to the choice of a world standard population which artifitially averages the age structure of the population. Developed countries have older populations than the standard whereas developing countries have younger populations and, as a consequence, adjusted rates tend to decrease in the former and to increase in the latter. Other difficulties that limit the interpretation of the available literature on cervical cancer incidence include:

*For correspondence. Email x.bosch@ico.scs.es

Table 1. Estimated incidence of cervical cancer in the world

	No Cases Per Year	Incidence Rate		Age Specific Incidence Rate			
		CR	ASRW	15-44	45-54	55-64	65+
World	470,606	15.7	16.1	9.5	44.9	51.8	41.9
More Developed	91,451	15	11.3	11.9	22.4	23.8	26.3
Less Developed	379,153	15.8	18.7	9.0	53.6	65.0	53.8
Africa	67,078	17.1	27.3	11.0	71.5	100.5	95.4
Eastern	30,206	24.4	44.3	16.1	114.8	174.4	153.9
Middle	6,947	14.4	25.1	8.5	54.0	73.3	137.4
Northern	10,479	12.2	16.8	6.2	49.0	68.5	45.9
Southern	5,541	23.2	30.3	15.5	67.8	98.5	118.2
Western	13,903	12.5	20.3	9.5	57.4	70.6	60.3
America	92,136	22	21	15.1	55.2	57.8	55.0
Caribbean	6,670	34.8	35.8	17.7	82.7	102.1	155.6
Central	21,596	31.7	40.3	22.5	111.7	109.9	136.1
South	49,025	28.1	30.9	16.8	85.5	90.2	101.4
Northern	14,845	9.5	7.9	9.0	15.4	16.8	14.2
United States	13,230	9.4	7.8	8.8	15.5	17.2	13.8
Europe	64,928	17.2	13	14.1	26.3	26.5	28.1
Eastern	35,482	21.9	16.8	17.8	34.5	34.9	36.6
Northern	6,049	12.6	9.8	12.0	17.6	16.7	20.2
Southern	10,116	13.7	10.2	10.5	20.8	23.7	20.9
Western	13,282	14.2	10.4	11.3	20.5	19.6	25.1
Asia	245,670	13.6	14.9	7.2	44.0	52.8	39.6
Eastern	51,266	7.1	6.4	2.6	18.4	18.9	25.4
South-Eastern	39,648	15.3	18.3	9.1	59.0	58.2	45.9
South-Central	151,297	20.9	26.5	11.9	79.2	100.8	65.6
Western	3,458	3.8	4.8	2.6	13.1	15.3	14.1
Japan	11,681	18.1	11.1	8.8	21.2	25.6	42.2
ASIA (Excluding Japan)	234,335	13.5	15.2	7.2	45.4	55.1	39.4
Oceania	2,156	14.2	12.6	12.3	27.4	28.2	29.0
Pacific Island*	1,078	29.1	40.3	23.1	91.2	107.1	167.8
Australia & N. Zealand	1,077	9.4	7.7	8.6	15.5	14.8	16.6

* Melanesia, Micronesia and Polynesia. CR: Crude Rate. ASRW: Age standardized rate – world standard. References: (Ferlay et al., 2001; Parkin et al., 1997).

a) the powerful reduction of cervical cancer incidence and mortality achieved by screening, either in organized programs or (to a lesser extend) in spontaneous case-finding settings; b) the dependency of cancer registries on the quality of health services, the availability of diagnostic and registration resources and the accessibility of the population at large to these services; c) the absence of cancer registries in extensive areas of the world, notably in developing countries at high risk of cervical cancer; and d) the lack of registries of pre- invasive cervical lesions in most countries.

Figure 1 shows the estimated age specific incidence rates of invasive cancer in selected populations. Cervical cancer is rare in the young age groups and there is a striking difference in the advanced age groups between developed and developing countries. Figures 1a and 1b are based on individual countries with well developed cancer registries. Figure 1a shows that, for four countries in Europe, the incidence rises steeply between the ages 20 and 35. In the middle age groups, the incidence decreases in relation to the intensity of screening in Sweden or The Netherlands or stays stable in countries with non centralized screening programs in Spain or France. A second mode in incidence occurs in the older age groups, ages 60 and above, in relation to some decrease in screening coverage in these ages and possibly related to the second mode in HPV DNA prevalence observed in some countries. Figure 1b shows an extreme example of

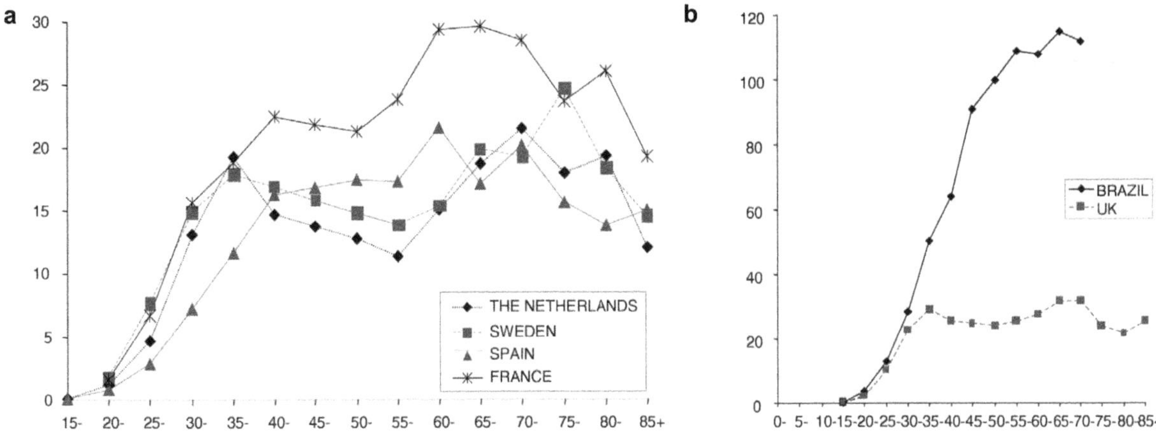

Figure 1 a and b. Age specific incidence of cervical cancer in selected populations (Parkin et al., 1997; Ferlay et al., 2001). Figure adapted from Bosch and de Sanjose (2003).

the age specific incidence of cervical cancer in countries with and without organized screening programs in Brazil and United Kingdom. The incidence before the age of 30 is very similar, suggesting that background exposure to HPV is similar in both countries, however, in the subsequent age groups, the incidence in Brazil continues to rise to levels that are 3 to 4 fold the incidence in United Kingdom.

Some cancer registries provide information on the incidence of pre-neoplasic cervical lesions (cervical intraepithelial neoplasia grade 3 (CIN 3), carcinoma *in-situ*). The development of CIN 3 is a very common event in the young age groups (rates for CIN 3 reached values of 70 and 110 per 100 000 in the age groups 30 to 40 in screening programs) and strongly suggests that an important proportion of these lesions are not true precursors of invasive cancer. Observation in some registries indicate a slight increase of CIN 3 lesions and of cervical cancer in the age group 55 and above (Parkin et al., 1997).

2.2. Human papillomavirus in the general population

In spite of the large number of reports already published, standardized data on genital HPV DNA prevalence in representative samples of populations in different countries are limited. Reports on the HPV DNA prevalence in women participating in screenig programs in Europe or the United States show a remarkable variability from populations within the same countries. Sources of variability include the age of the study group (*i.e.* the ages recommended by the local screening protocol), the variation in the HPV detection systems employed, and the selection bias of the samples investigated in each individual study (*i.e.* if the survey includes patients from clinics or very young age groups). Preliminary results of an on-going literature review indicate that the adjusted point estimates of the prevalence of HPV DNA ranges from 14 to 37% in Africa, close to 20% in Latin America, 8 to 15% in Europe and North America and 7-8% in Asia. Prevalence in the male external genitals is only available for a few countries and the evidence suggests that may be roughly similar to the prevalence in women. On going critical reviews will provide more accurate and relevant data in the future

Further limitations in our estimates of the burden of HPV refer to the lack of registries of cases of genital warts and the limited number of surveys reporting on the prevalence of HPV DNA in representative samples of the male population.

3. Natural History of HPV Infections

3. 1. Descriptive data

Cross sectional studies have repeatedly reported that sub-clinical HPV infections are highly prevalent in young individuals, whereas invasive cervical cancer typically develops in the third decade and above. The proportions of HPV DNA positive women have been placed in the 15-40% range in the young, sexually active, age groups and between 3-10% range in the 35 and above age groups. Figure 2 shows the age-specific cross-sectional prevalence of high-risk HPV DNA in a screening program in The Netherlands and the corresponding age-specific incidence rates of cervical cancer in that country. In countries where intensive screening of young women takes place, part of the HPV prevalence reduction could be attributable to aggressive treatment of HPV related cervical lesions. Women who remain chronic HPV carriers are currently interpreted as the true high-risk group for cervical cancer. In some populations, a second mode of HPV DNA prevalence has been observed for older women (*i.e.* 50 and above) and in some countries (Herrero et al., 2000; Lazcano Ponce et al., 2001; Molano et al., 2002) a second mode in the incidence of CIN 3 /

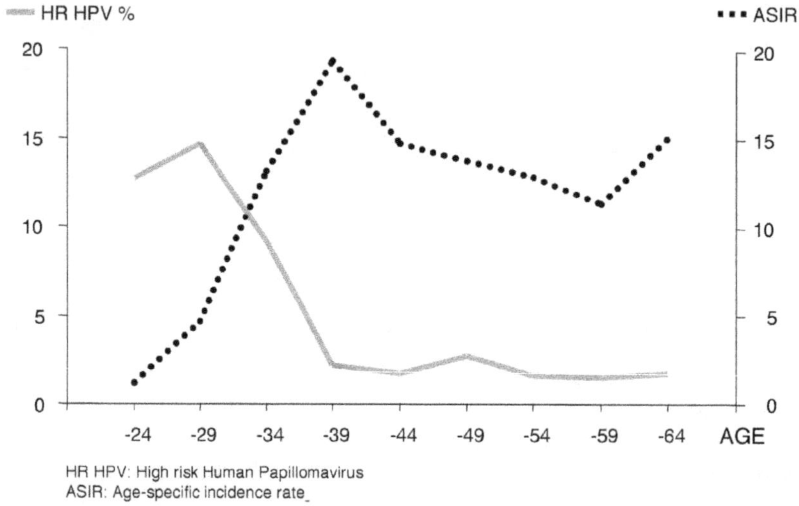

Figure 2. Age specific prevalence of high risk Human Papillomavirus DNA in 3700 women entering a screening program and age-specific incidence rate of cervical cancer in The Netherlands (Jacobs et al., 2000; Parkin et al., 1997).

Carcinoma *in situ* lesions and of invasive cervical cancer has also been observed. Recently, a bimodal pattern has also been described for the incidence of HPV infections in a high-risk population for cervical cancer from Colombia. A first peak of HPV infections by high-risk HPV types is observed in young women and a second in women over 45 years of age (Muñoz et al., 2004). In all of the settings investigated, the prevalence of HPV DNA in the young age groups is strongly related to the sexual behavior patterns that are dominant in each population and particularly on the age at sexual debut of the female population. This is of importance to make local or regional policy recommendations as to age at which HPV vaccines are to be promoted and the age groups in which the addition of HPV testing to screening programs would increase its efficiency.

3.2. Follow-up studies

Numerous cohort studies have monitored women from cytological normalcy to the stage of high-grade CIN (high squamous intraepithelial lesions (HSIL) or CIN 2/3). Monitoring of women to invasive disease is not acceptable on ethical grounds and thus that information is unlikely to ever be available.

Repeated sampling of women being followed for viral persistence and cervical abnormalities has shown that the median duration of the infections tend to be longer for high risk HPV (HR HPV) types as compared to low risk HPV types. In studies in high-risk populations from Brazil and Colombia, the mean duration of HPV detection was 13.5 and 14.8 months for HR HPV types and 8.2 and 11.1 months for the non-oncogenic types (Franco et al., 1999; Muñoz et al., 2004). HPV 16 tended to persist longer than the average for high-risk types and this observation is consistent with a selective advantage that would explain the predominant role of HPV 16 over any other type worldwide (Franco et al., 1999). The results tend to be similar in a student population in the United States and in the United Kingdom (Woodman et al., 2001; Ho et al., 1998a). The self-limiting course of most HPV infections is consistent with the cross-sectional profile displayed in Figure 2. However, the currently observed duration intervals may still suffer from imprecision in the estimates of time at first exposure, from the variability in the end-point definition and from censoring due to treatment of the early lesions.

Follow-up studies of women with and without cervical abnormalities have indicated that the continuous presence of HR-HPV is necessary for the development, maintenance and progression of CIN lesions (Ho et al., 1998a; Koutsky et al., 1992; Ho et al., 1995; Remmink et al., 1995; Nobbenhuis et al., 1999). A substantial fraction (*i.e.* 15-30%) of women with HR-HPV DNA who show normal cytology at recruitment will develop CIN 2 or CIN 3 within the subsequent 4-year interval (Koutsky et al., 1992; Rozendaal et al., 1996; Rozendaal et al., 2000). Conversely, among women found to be HR-HPV DNA negative and in which the cytology is identified as either atypical squamous cells of undetermined significance (ASCUS), borderline or mild dysplasia, CIN 2/3 is unlikely to develop during a follow-up of two years and their cytology is likely to return to normal (Zielinski et al., 2001b; Nobbenhuis et al., 2001a). Women found positive for low risk HPVs rarely become persistent carriers and their probability of progression to CIN 2/3 is extremely low (Zielinski et al., 2001b; Nobbenhuis et al., 2001a; Manos et al., 1999).

As on going cohorts expand their follow up time, more precise estimates are being provided on the predictive value of viral persistence as defined by repeated measurements of viral types. One of such cohorts in Sao Paulo has shown that the incidence of cervical lesions in women who were HPV negative twice was 0.73 per 1000 women-months. The corresponding incidence among women with repeated HPV 16 or HPV 18 positive results was 8.68, a 12-fold increased incidence. The Odds ratios (OR) for HPV persistence among women who were twice HPV positive for the same oncogenic types was OR= 41.2 (95% confidence interval (CI)= 10.7-158.3) (Schlecht et al., 2001). Retrospective assessment of HPV status using archival smears from cases of cervical cancer and controls has provided evidence that HPV DNA preceded the development of invasive disease, and showed its value in signaling false negatives smears (Zielinski et al., 2001b). An interesting observation from the same research group, suggests that clearance of HR HPV in otherwise established cytological lesions, is a marker associated with regression of CIN lesions (Nobbenhuis et al., 2001a; Zielinski et al., 2001a). The latter, however, is not found consistently (Schiffman et al., 2002).

Finally, persistence of HPV DNA detection after treatment for CIN 2/3 is an accurate predictor of relapse, at least as sensitive as repeated vaginal cytology (Nobbenhuis et al., 2001b). A meta analyses of the literature has been published showing that at 6 month following conization the sensitivitity of an HPV test to predict relapse was 91.2% as compared to 64.7% for a Pap smear. At 24 months, the sensitivity of both tests was equivalent at 93% (Paraskevaidis et al., 2004).

3.3. Retrospective cohorts

A particularly interesting approach to conducting follow-up studies of invasive cancer (as opposed to studies of CIN 3) without ethical and time constraints is provided by so called "nested case control studies". These are studies initiated several years in the past that assembled and stored large banks of biological specimens from healthy individuals. Linkage studies can then identify cases of cervical cancer (or any other condition) that have occurred in the interval and the original specimens can then be analyzed for the presence of HPV biomarkers. HPV DNA prevalence can then be compared to the corresponding prevalence in specimens of individuals from the same cohort who did not develop the condition under otherwise equivalent exposures. These studies have documented the existence of HPV exposure years before the development of the disease, thus reproducing the conditions of a longitudinal study. With this approach, a relative risk estimate of 16.4 (95% CI= 4.4-75.1) was observed for invasive cervical cancer in Sweden using DNA extracted from stored Pap smears (Wallin et al., 1999) and of 32 (95% CI= 6.8-153) in the Netherlands (Zielinski et al., 2001b). In a similar study design, an OR of 2.4 (95% CI= 1.6-3.7) was obtained using serologic markers of HPV exposure (Dillner et al., 1997). The lower magnitud of these odds ratios compared to those obtained inretrospective case control studies (see below) are explained, at least in part, by the lower sensitivity of the PCR and serological assays used.

3.4. Determinants of HPV infection: The critical role of the sexual behavior of women and of their male's sexual partners

Epidemiological studies investigating risk factors for HPV infection clearly and consistently have shown that the key determinants among women are the number of her sexual partners, the age at which sexual intercourse was initiated and the likelihood that each of her sexual partners was an HPV carrier.

The role of males as possible vectors of HPV was measured in the early epidemiological studies by questionnaires (often completed by women) that addressed the sexual behavior of the husbands of cervical cancer cases and controls. More recent studies interviewed the husbands or current partners directly and had, in addition, the ability to measure HPV DNA in exfoliated cells from the penile shaft, the coronal sulcus and from the distal urethra (Barrasso et al., 1987; Kjaer, 1991; Bosch et al., 1996; Castellsagué et al., 1997; Kjaer, 1991; Bergman and Nalick, 1992; Muñoz et al., 1996).

These investigations consistently showed that the risk of cervical cancer for a given woman is predictable by the sexual behavior of her husband as much as by her own sexual behavior. In populations where female monogamy is dominant, the population of female sex workers plays an important role in the maintenance and transmission of HPV infections. It is now clear that the probability that a woman is an HPV carrier and her risk of developing cervical cancer is related to the presence of HPV DNA in the penis or the urethra of her husband or sexual partner (Bosch et al., 1994; de Sanjosé et al., 1993; Thomas et al., 2001c; Juarez-Figueroa et al., 2001). More recently, it has been possible to confirm that male circumcision protected males from being HPV carriers and their wives from developing cervical cancer (Castellsague et al., 2002).

These observations confirmed, in terms of HPV infections, century-old observations and scientific hypothesis formulated almost 30 years ago that male sexual behavior is a central determinant of the incidence of cervical cancer. Natural history studies of HPV infections satisfy in biological terms most of the epidemiological and clinical observations that historically linked cervical cancer to a profile of a sexually transmitted infection.

4. The Causal Association Between HPV DNA and Cervical Cancer

The evidence relating HPV infections to cervical cancer includes a large and consistent body of studies indicating a strong and specific role of the viral infection in all settings where investigations have taken place. The association has been recognized as causal in nature by a number of international review parties since the early 90's (Bosch et al., 2002; International Agency for Research on Cancer, 1995; NIH Consensus Development Panel, 1996).

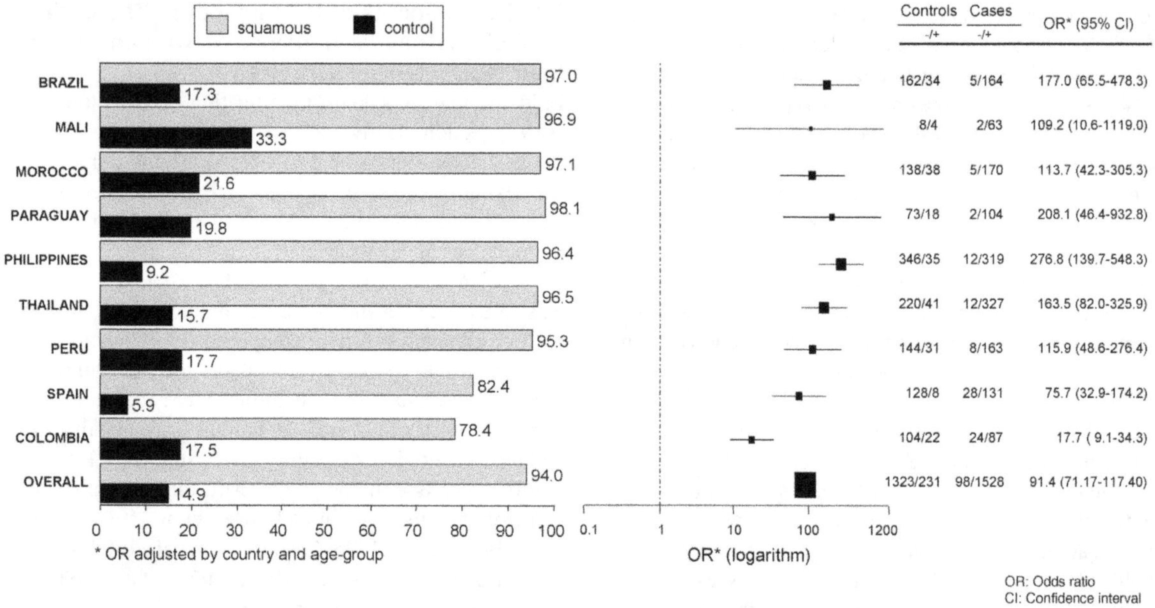

Figure 3. Human Papillomavirus DNA prevalence by case-control status and odds ratios for the association between Human Papillomavirus DNA detection and risk of squamous cell carcinoma of the cervix by country. Figure adapted from Muñoz et al. (2003).

4.1. HPV DNA prevalence in cervical cancer specimens

State-of-the art amplification techniques used in case-control studies, case-series and prevalence surveys have unequivocally shown that, in adequate specimens of cervical cancer, HPV DNA can be detected in 90 to 100% of the cases as compared to a prevalence of some 5-20% from cervical specimens of women identified as suitable epidemiological controls. Detailed investigations of the few cervical cancer specimens that appear as HPV DNA negatives in most series has been occasionally conducted and the results strongly suggest that these are largely false negatives. As a consequence, the claim has been made that this is the first *necessary cause* of a human cancer ever identified, providing a strong rationale for the use of HPV tests in screening programs and for the development of HPV vaccines (Bosch et al., 1995; Walboomers et al., 1999).

4.2. Risk estimates from case control studies

Observations on pre-invasive disease and cohort studies on the HPV natural history have intrinsic limitations for making inferences on cervical cancer causality. This is because in controlled settings, biological progresion is not allowed to continue beyond the stage of high-grade squamous intraepithelial lesions (HSIL/CIN 3) or carcinoma *in situ*. It is thus of importance to notice that the information applicable to cervical cancer would primarily come from case control studies in which the target disease is directly investigated.

In an effort to simplify the vast literature of case control studies, the results of the International Agency for Research on Cancer (IARC) multicenter case control study on invasive cervical cancer will be used as example. In brief, this project included nine case-control studies in different parts of the world, mostly in high-risk countries. A common protocol and questionnary were used, and HPV DNA testing was done in two central research laboratories using the MYO9/11 (Gravitt and Manos, 1992) and the General Primer (GP) GP5+/6+ (de Roda Husman et al., 1995; van Den Brule et al., 1990) polymerase chain reaction (PCR) testing systems. Figure 3 show the summary results on the HPV DNA prevalence in case and control specimens, the risk estimates and their confidence intervals for squamous cell carcinomas (Muñoz et al., 2003).

The figures show very high ORs with estimates in the range of 50 to 150 with several estimates in the several hundred range. These risk estimates lead to calculations of attributable fractions for the entire study greater than 95% (Bosch et al., 2001).

The results are strikingly consistent in the literature for pre-invasive lesions, for squamous cell carcinomas and adenocarcinomas and for studies that tested for HPV DNA as a group or studies that restricted the analyses to HR HPV types. Some of the key contributions have been reported from Costa Rica (Herrero et al., 2000), Thailand (Thomas et al., 2001a), Norway (Olsen et al., 1995), Denmark (Kjaer et al., 1996), and virtually all other countries in which these studies have been conducted.

Studies that have compared risk factors for CIN 3 and invasive cancer have not reported any significant differences in their associations with HPV or with their epidemiological profile (Moreno et al., 1995; Thomas et al., 2001a).

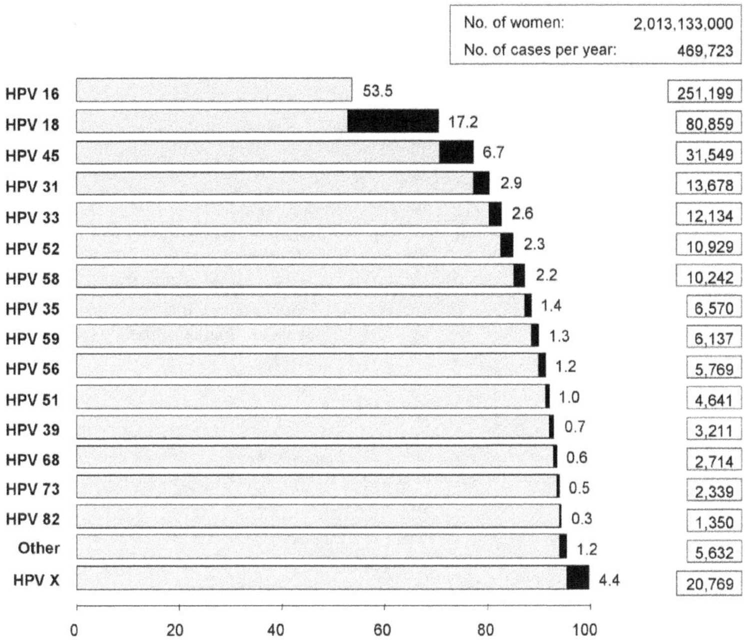

Figure 4. Percentage and number of cervical cancer cases per year attributed to the most frequent Human Papillomavirus genotypes in women 15 years of age and older: All world regions combined. Figure adapted from (Muñoz et al., 2004).

4.3. HPV type distribution in cervical cancer and HPV type-specific risk estimates

Of the more than 35 HPV types found in the genital tract, HPV 16 accounts for some 50% to 60% of the cervical cancer cases in most countries, followed by HPV 18: 10-20%, HPV 45: 4-8% and HPV 31: 1-5 %. Figure 4 shows the cumulative distribution of 15 HPV types in a series of nearly 3000 cervical cancer cases (Muñoz et al., 2004). Of these, the 5 most common (HPVs 16,18, 45, 31 and 33) account for 80% of the distribution in squamous cell cancers and of 94 % in adenocarcinomas. In most studies, HPV 18 predominates in adenocarcinomas in absolute or relative terms. The reasons for such specificity are unknown.

In a series of women without cervical lesions, (corresponding to the controls in most case control studies or ad-hoc HPV prevalence surveys from the general population) the HPV type specific distribution embraces a much larger series of viral types. HPV 16 remains again the most common type (some 20 % of the HPV DNA positive) followed by HPV 18 (some 10% of the HPV DNA positive) HPV 45 (some 8% of the HPV DNA positive) HPV 59 (some 2 % of the HPV DNA positive) and smaller proportions of up to 30 additional HPV types. Many of the rarer types are occasionally found in controls and still convey a high risk for cervical cancer. It is of interest to notice that the geographical variation in type distribution has not been fully documented. Some recent studies from areas where little work has been done in the past suggest that some additional variability could be expected (Castellsague et al., 2001). The selection of HPV 16 in cervical cancer cases translates the biological advantage of this type in fully espressing the oncogenic capacity. An interesting study among women with different degrees of human immunodeficiency virus (HIV) induced immunosupression pointed at the likely increased ability of HPV 16 to escape immunesurveillance as compared to other HPV types as one possible mechanism of such advantage (Strickler et al., 2003).

The pool of IARC studies was large enough to provide type-specific risk estimates for 18 types. Restricting the analyses to the studies that used the GP5+/6+ HPV detection system and to squamous cell carcinomas, the adjusted Odds Ratios for HPV DNA detection (the factor by which the reference risk of cervical cancer is multiplied if HPV DNA is detected) was OR – 158.2 (95% CI– 113.2-220.6). The risk estimates for adenocarcinomas was OR= 76.3 (95% CI= 40.7-143.2) (Muñoz et al., 2003).

Type specific risk estimates for squamous cell cancer were as follows: HPV 16: OR= 435; HPV 18: OR= 248; HPV 45: OR= 198; HPV 31: OR= 124; HPV 52: OR= 200; HPV 33: OR=374; HPV 58: OR= 115; HPV 35: OR=74; HPV 59: OR= 419; HPV 51: OR= 67; HPV 56: OR= 45; HPV 39: infinity; HPV 68: OR= 54. The OR and the 95% confidence limits are displayed in Figure 5 (Muñoz et al., 2003). These studies concluded that the evidence is now sufficient to consider as high-risk carcinogenic types HPV 16, 18, 31, 33, 35, 39, 45, 51, 52, 56, 58, 59, 68, 73 and 82.

A second group of HPV types are rarely found in cases and have been classified as Low-risk types and these include HPVs 6, 11, 40, 42, 43, 44, 54, 61, 70, 72, 81 and CP6108.

From IARC's and other studies, a small group of HPV types remain in the category of uncertain risk. Epidemiologically the group includes the types for

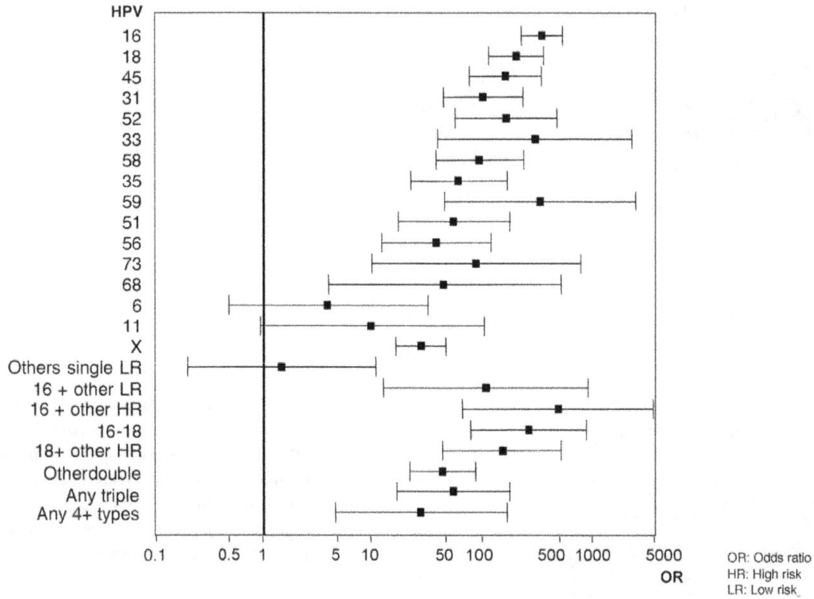

Figure 5. Human Papillomavirus type-specific Odds Ratios and 95% Confidence Intervals for cervical cancer. Source of data: Muñoz et al., 2003.

which the imprecision of the OR estimate is related to very small number of observations (in spite or very large studies). Further, it is now recognized that the type-specific sensitivity of some of the assays used in the different studies vary for some rarer types or in the presence of multiple infections. Moreover, some HPV types (*i.e.* HPV 73 or HPV 70) are classified discordantly according to wheater sequence homology and experimental evidence is used or if epidemiological observations are given priority. Finally, studies that are conducted in different populations reflect the underlying geographical variability and the rarer HPV types may be uninformative in some settings while being clearly associated with disease status in other populations. The types of uncertain nature currently include HPVs 26, 53, 66 and perhaps others.

Figure 5 (Muñoz et al., 2003) shows that the risk for any given high-risk type was not statistically different from the risk reported for HPV 16. Likewise, the risk related to the presence of multiple HPV types in the specimen is no different from the risk linked to a single HPV type.

Under extremely rare circumstances, HPV of the low-risk group (*i.e.* HPV 6 or 11) are found as the only type in specimens of invasive cervical cancer. Although statistically these are largely non significant increases in risk, it should be considered that the ability of the oncoproteins of the low-risk types also show transforming capacity at a very low level. It is plausible that a minute fraction of the population harbors a special susceptibility to HPV and even the presence of a low-risk type is capable of initiating a neoplastic process.

Studies on HPV variants (variation within HPV types affecting down to one nucleotide of the viral genome) are beginning to unveil that the risk of some HPV 16 variants (non-european like) may differ from that of the HPV 16 European prototype (Hildesheim et al., 2001b). The geographical distribution of HPV variants is still being described and its relevance for HPV testing and for vaccine developments is still uncertain.

The proportion of multiple types in a given specimen varies across studies and particularly in relation to the HPV detection method used. Populations at high-risk of cervical cancer and high rates of HIV positivity tend to show higher proportions of multiple types as compared to populations not belonging to these risk groups. Longitudinal studies have suggested that the one-time, cross sectional detection of type specific HPV may underestimate the cumulative lifetime diversity of exposures to HPV (Woodman et al., 2001). However, in all studies of invasive carcinoma the risk linked to multiple HPV types does not vary significantly from the risk linked to single HPV types. Multiple HPV types were detected in the multicentric study on average in 9.4% of the cases and 2.2% of the controls, and did not show a statistically significantly increased risk (OR=114.9 (95% CI= 68.8-191.7)) over women positive for only one HPV type (OR= 172.6 (95% CI= 122.2-243.7)).

4.4. Alternative explanations on the causality of cervical cancer

In the early studies of HPV and cervical cancer, and currently in the majority of real life studies, a fraction of cases are labeled as HPV negative and were compared to HPV positive cases in relation to their epidemiological profile. In broad terms, it was clear that "HPV negative" cases retained the same traits as the rest of the cases (*i.e.* similar age, high number of sexual partners, young age at first sexual intercourse, long-term use of contraceptives, were investigated under the hypothesis that HPV negative cases were a true biological entity (Riou et al., 1990; Viladiu et al., 1997). The proportion of such cases tends to be more important in studies of pre-invasive

Table 2. Environmental risk factors for cervical cancer among human papillomavirus positive women

Risk Factor	Risk exposure	Reference	Key Reference
HPV DNA in cervical exfoliates	Positive for high risk types	Negative	Muñoz et al., 2003
Use of Oral Contraceptives	5 or more years of use	Never	Moreno et al., 2002
Smoking	Ever	Never	Plummer et al., 2003
Parity	5 or more pregnancies	None or 1-2	Muñoz et al., 2002
Clamydia Trachomatis	Ab. positive	Ab. Negative	Smith et al., 2002b
HSV 2	Ab. positive	Ab. Negative	Smith et al., 2002a

HPV: Human Papillomavirus; HSV 2: Herpes simplex virus 2; Ab.: Antibody. References: (Muñoz et al., 2003; 2002; Moreno et al., 2002; Smith et al., 2002b; 2002a; Plummer et al., 2003).

neoplasia (Tabrizi et al., 1999; Burger et al., 1996). As a consequence, for some time there was some uncertainty in the interpretation of the results. In the IARC's studies, HPV negative cases hith parity, etc.). These results suggested, that the apparently HPV negative cases were also related to a sexually transmitted disease (STD) pattern, however, none of the known sexually transmitted agents that had occasionally been associated to cervical cancer, (ie. Herpes simplex virus 2 (HSV 2)) showed a coherent record of results such as the one available for HPV. In the evaluation of the putative HPV negative cases, it has been repeatedly demonstrated that lack of identification of HPV DNA could be attributed primarily to the poor quality of the specimen (tumour necrosis, lack of cancer cells in the specimen, poor preservation) and to the quality of the amplification system used. Walboomers et al. (1999) showed that using histologic verification of the specimen and E7 primers for 14 high-risk HPV types, HPV DNA could be recovered from 99.6% of cases of cervical cancer. Progress made in the preparation of primers for HPV DNA testing have also increased the HPV DNA prevalence in women with normal cervices although at a much lower ceiling.

In the last decade there has not been any hypothesis supported by sound epidemiological or biological data indicating that the etiology of cervical cancer could be independent of HPV. The hypothesis should, however be retained as a scientific and research option. The grounds for such consideration lie in that epithelial cells are capable of developing into cancer cells and cancer growth occurs in all human tissues regardless of a known, viral or non viral, cause. Thus, this may also happen in the human cervix. Secondly, cellular genes that intervene in HPV related carcinogenesis should be able to generate spontaneous or induced mutations leading to cancer in the absence of HPV. Available evidence suggests that this event is rare within the lifespan expectation of the human population and it is thus likely that the non-HPV related cancers currently occur very rarely and probably cluster in very old women. Finally, non-epithelial cancers do occur in the cervix at a low frequency.

4.5. Other risk factors for cervical cancer

Most of the sexual behavior parameters that were linked to cervical cancer in the past are being reevaluated in studies that considered the strong influence of the presence of HPV. Soon after the introduction of HPV testing in research protocols, it became clear that the key risk factors that reflected sexual behavior, such as the number of sexual partners, merely reflected the probability of HPV exposure. Other factors, such as the estimates of age at first exposure (as indicated by age at first sexual intercourse or at first marriage) are still under evaluation. In addition, a number of environmental factors had been historically related to cervical cancer and are currently being re-assessed. These include hormonal factors (use of oral contraceptives and multiparity) other STDs (HSV 2, *Chlamydia trachomatis* and occasionally any other STD), cigarette smoking and dietary factors. A special consideration should be given to exposure to HIV and to other situations of immuno-supression, because of the potential enhancement of the oncogenic progression of HPV in the presence of an immunosupresive virus .

Because of the growing evidence that HPV is a necessary factor in cervical cancer, it soon became a standard procedure in the reports of case control studies to include analyses restricted to HPV positive cases and controls to properly assess the contribution of additional factors to the risk of cervical cancer. In relation to invasive cervical cancer,the IARC's pooled HPV positive restricted analyses included 1768 cases and 262 controls and the key findings concerning environmental risk factors are summarized in Table 2 (Moreno et al., 2002; Muñoz et al., 2002; Smith et al., 2002b; 2002a; Plummer et al., 2003; Muñoz et al., 2003) and discussed briefly.

4.5.1. Long term use of oral contraceptives

Oral contraceptives (OC) use was associated with a significant increase in risk (OR=1.47 (95% CI= 1.02-2.12)). Use of OCs for less than 5 years was not related to cervical cancer (OR=0.77 (95% CI= 0.46-1.29)) but the risk increased significantly for 5-9 years of use (OR= 2.72 (95% CI= 1.36-5.46)) and for 10+ years (OR=4.48 (95%

CI= 2.24-9.36)) (Moreno et al., 2002). The evidence for an association of cervical cancer with the use of oral or other hormonal contraceptives is not entirely consistent. A number of studies that investigated HPV positive women found no associations or only weak associations with HSIL / CIN 3 in subgroup analyses (Deacon et al., 2000; Castle et al., 2002). These apparently conflicting results may reflect the increased cytological surveillance of women that are taking OCs in developed countries and the use of different case definition (from ASCUS up to HSIL / CIN3 as opposed to cervical cancer) in cohort studies. Because of the potential Public Health importance of an interaction between long-term use of OCs and HPV infections in the development of cervical cancer, efforts are now being devoted to verify the results in different populations. A recent meta analyses on the association between hormonal contraceptives and cervical cancer concludes that there is a linear dose response relationship. The duration of the effect after OC cessation remains to be determined (Smith et al., 2003). The relevant units at WHO examined the evidence with an international working group that recognized the significance of the association. However, the group also concluded that, in balance, the benefits currently achieved by the use of OCs in developing countries (*i.e.* unvanted pregnancy avoidance) outbalance the increase in risk and should not force a change in the current family planning strategies (WHO report, in preparation).

4.5.2. High parity

HPV positive women who reported seven or more full term pregnancies had a four-fold increased risk of cervical cancer as compared to similar HPV positive women that were nulliparous (OR= 3.8 (95% CI= 2.7-5.5)). There was still a two-fold increased risk when women reporting 7 or more pregnancies were compared to HPV positive women who reported 1-2 full term pregnancies (Muñoz et al., 2002). Similar results were obtained in Costa Rica (Hildesheim et al., 2001a) and Thailand (Thomas et al., 2001b) as well as among women with pre-invasive disease in the Portland cohort study (Schiffman et al., 1993). In Denmark and in the Manchester cohort study, two populations with low parity, the effects are less visible for preneoplastic lesions (Deacon et al., 2000; Kjaer, 1998).

It has been speculated that the general reduction in the average number of births in developed countries over the last decades may have contributed to the reduction in cervical cancer incidence but formal proof of the hypothesis has not yet been produced.

4.5.3. Cigarette smoking

The pooled results of the IARC's studies found that "ever smoking" was associated with a two-fold, statistically significant, increased risk of cervical cancer with a significant dose response (Plummer et al., 2003).These findings are consistent with those found for "current *vs.* never smoking" among HPV positive women for preneoplastic cervical lesions in the Costa Rica study (OR=2.3) (Hildesheim et al., 2001a), the Portland study (OR=2.7 for CIN 2-3) (Schiffman et al., 1993), the Copenhagen study (OR=1.9), (Kjaer, 1998) and the Manchester study (OR= 2.2) (Deacon et al., 2000). These recent studies are providing growing evidence on the carcinogenic effect of cigarette smoking in women with persistent HPV infection. The Monograph program at IARC reviewed the evidence in 2002 and concluded that smoking was an independent risk factor for cervical cancer (International Agency for Research on Cancer, 2004). However the mechanisms by which cigarette smoking may affect cervical cancer (*i.e.* a direct effect of the tobacco metabolites, an indirect effect related to tobacco-induced immunosupression or to reduced dietary antioxidants) remain elusive.

4.5.4. Co-infection with the Human Immunodeficiency Virus (HIV)

The evidence of a putative interaction between HPV and HIV in the origin of cervical cancer was formally recognized when cervical cancer was recognized as one of the criteria of acquired immune deficiency syndrome (AIDS) among HIV positive women. The subsequent literature largely confirmed the evidence although some major confounders of the epidemiological association tend to obscure the results. In brief these refer to the powerful impact of screening in some populations, the medical surveillance of HIV carriers in developed countries and the short survival time of HIV/AIDS patients in many populations at high risk of cervical cancer in relation to the time intervals between HPV infection and cervical cancer (Bayo et al., 2002; Gichangi et al., 2002; de Sanjose and Palefsky, 2002; International Agency for Research on Cancer, 1996).

4.5.4.1. Cross sectional studies on the prevalence of HPV and CIN lesions in HIV carriers and non carriers

Massad et al. (1999) reported on the baseline cervical cytology from 1713 HIV positive women and 482 HIV negative control women. Cervical cytology was abnormal in 38.3% of HIV positive women compared to 16.2% of the HIV negatives. High-grade lesions, low-grade lesions and ASCUS were all significantly more common among HIV infected women. The risk factors for any abnormal cytology were: a CD4 cell count lower than 200/mm^3 (OR= 2.13 (95% CI= 1.45-3.13)), the presence of HPV DNA and a previous history of abnormal cytology. Ahdieh et al. 2000 (Ahdieh et al., 2000) identified a higher baseline prevalence of cervical abnormalities among HIV infected women (13.4%) compared with HIV negative women (2.4%). In this follow up study, 11 women were identified with CIN in subsequent visits, all of them were HIV positive and had a median CD4 cell counts of 253/mm^3. The risk for CIN was related to HPV persistence in all cases.

4.5.4.2. Risk of CIN among HPV carriers with and without HIV

Thomas et al. (2001c) studied a group of 251 sex workers in Thailand. The HPV DNA prevalence was similar by HIV status. However, the risk of high-grade lesions was two-fold higher in women infected with both HPV and HIV as compared with HIV negative, HPV positive women and 20 times higher than in HIV negative, HPV negative women. Mandelblatt et al. (1999) provided a pooled estimate of 15 studies on the association between HIV and CIN. HIV infected women had an 8-fold increased risk of CIN (OR= 8.8 (95% CI= 6.3-12.5)). Sun et al. (1997) reported that as compared to HIV negative and HPV positive women, women co-infected with HPV and HIV had a decreased regression rate of low-grade lesions and higher rates of the progression from infection to CIN. Ellerbrock et al. (2000) reported that HIV positive women were 4.5-fold more likely than HIV negative women to have or develop CIN within a 54 month interval of follow-up. In the group of HIV carriers, transient HPV infections (rate ratio (RR)=7.4, 95% CI=1.0-57.4), persistent infections with types other than 16 or 18 (RR= 8.9, 95% CI= 1.2-66.2) and persistent infections with HPV type 16 or 18 (RR=11.0, 95% CI= 1.4-88.7) were all significantly associated with CIN.

4.5.4.3. Risk of invasive cervical cancer in HIV carriers

The International Collaboration on HIV and Cancer group (2000) published cancer data from 23 prospective studies that included 47, 936 HIV infected subjects from North America, Europe and Australia for the period 1992-1999. The study concluded that there had not been a significant change in the incidence of invasive cancer (RR= 1.87, 99% CI= 0.77-4.56) during this period.

An overview of early African studies concluded that invasive cancer was not related to exposure to HIV with a summary OR of 0.8 (95% CI= 0.5-1.4) from studies carried out in Rwanda, South Africa and Uganda (Newton et al., 1999). However, recent data from a hospital based case-control study in South-Africa identified an increased risk for cervical cancer (OR=1.6 (95% CI=1.1-2.3)) and for cancer of the vulva (OR= 4.8 (95% CI= 1.9-12.2)) among HIV- infected patients (Sitas et al., 2000).

In the US or in Europe, reports are generally consistent in detecting an increased risk for cervical cancer among HIV infected women. Selik and Rabkin (1998) from the USA reported a relative risk for cervical cancer of 5.5 among HIV positive women. Frisch et al. (2000) used data from the US cancer Match Registry for the period 1978-96 and showed a RR of 5.4 for invasive cervical cancer among HIV positive women compared with the general population in the United States. Similar increases in magnitude were observed for *in situ* cervical cancer (OR=4.6), cancer of the vagina and vulva (OR= 5.8) and anal cancer (OR=6.8). Rates were evaluated at the time prior to AIDS diagnosis, around the period of diagnosis and up to 60 moths after the AIDS diagnosis. The authors identified no major changes in risk before an after AIDS diagnosis. Data is available from the population of New York with similar results (Gallagher et al., 2001).

In southern Europe a strong association between cervical cancer and AIDS has consistently been found. In Italy, the linkage of the National AIDS Registry and the populations cancer registries showed a 15-fold increased risk of ICC for women with AIDS (Franceschi et al., 1998). The joint Italian-French follow-up study of HIV positive women also showed a 13-fold increased rate of cervical cancer for HIV positive women (Serraino et al., 1999). In Spain, the Catalonian AIDS surveillance system detected 58 cases of ICC among 823 HIV positive women, an 18-fold increased risk compared with the general population (Vall Mayans et al., 1999).

In summary, HPV and HIV share some behavioural traits that define a particularly vulnerable high risk group. Progression of HPV infections to CIN lesions and cervical cancer in the context of limited / absent screening seems to be increased among HIV carriers and AIDS patients. Furthermore, evidence is growing suggesting that progression is related to the severity of the immunosuppression as indicated by CD4+ counts.

4.5.5. Coinfection with other sexually transmitted agents (STA)

Markers of exposure to other STA's have been repeatedly found associated to cervical cancer. Results from the IARC's multicenter study found a 2-fold increased risk in cervical cancer for the presence of antibodies to Chlamydia trachomatis (OR= 2.1 (95% CI= 1.1-4.0)) (Smith et al., 2002b) and of antibodies to Herpes simplex type 2 viruses (Smith et al., 2002b; de Sanjosé et al., 1994; Smith et al., 2002b). Non-specific inflammatory changes have also been related to modest increases in risk for preneoplastic cervical lesions among HPV positive women (Castle et al., 2001). The difficulties with the evaluations of such factors lies in the strong colinearity observed among all STD's and the limitations of some of the biomarkers currently used to assess ever exposure or persistent exposure.

4.5.6. Dietary factors and cervical cancer

A systematic review of recent epidemiological evidence about the role of diet and nutrition on the risk of HPV persistence, squamous intraepithelial lesion (SIL) and invasive squamous carcinoma, taking into account HPV, has been completed (García Closas et al., 2005). The review included all published observational studies that controlled for HPV infection and all randomized clinical trials that were published between March 1995 and November 2003. Twenty-nine studies were eligible for this review - 6 clinical trials, 8 observational prospective and 15 case-control studies.

Although there is some epidemiological support for a role of diet and nutritional status on cervical carcinogenesis, the available evidence taking HPV infection into account is insufficient for some nutrients and especially for food items. Conclusions from case-control studies have to be cautious since potential for selection and recall bias needs to be considered. Also, the possible influence of the disease process on nutrient blood concentration has to be evaluated.

Overall, the food group analyses are consistent in showing a protective effect of fruits and vegetables probably by reducing HPV persistence (Sedjo et al., 2002b; Giuliano et al., 2003). In relation to nutrient intake, evidence for a protective effect is consistent for lycopene and vitamin E (Giuliano et al., 1997; Ho et al., 1998b; Nagata et al., 1999; Shannon et al., 2002; Wideroff et al., 1998; Yeo et al., 2000; Sedjo et al., 2002b; Goodman et al., 1998), and moderately consistent for vitamin B12 (Alberg et al., 2000; Goodman et al., 2000; Sedjo et al., 2002a; 2003) and vitamin C (Giuliano et al., 2003; Ho et al., 1998b; Jacobs et al., 2002; Wideroff et al., 1998; Yeo et al., 2000; Goodman et al., 1998). Conclusions for nutrients from studies taking HPV infection into account do not differ substantially from those that did not control for it.

From studies based on prevalent blood levels, the evidence is consistent with an increased risk with increasing levels of homocysteine (Alberg et al., 2000; Goodman et al., 2000; Sedjo et al., 2003). The consistent protective effects found in studies using HPV persistence as endpoint are especially relevant since a possible protective effect of diet and nutrition on the cancer process is more plausible for intermediate endpoints than for more advanced stages of the disease process.

Evidence from strong study designs such as chemoprevention trials and prospective studies is discouraging. In a critical review of all chemoprevention trials available by 2001, it was pointed out that none of them had enough power to detect clinical significant differences in response rates (Follen et al., 2001).

4.5.7. Other environmental risk factors

Early age at exposure has been proven to be a strong determinant of the prognosis of carcinogenic viral infections in relation to cancer development (for example for hepatitis B virus (HBV) induced liver cancer or to Epstein-Barr-Virus (EBV) infection in relation to Burkitt's lymphoma). Likewise, most studies have shown that the risk of cervical cancer is related to age at first sexual intercourse, a proxi for age at first exposure to HPV. Definite evidence that the cervical tissue during menarchy is more vulnerable to HPV infections and perhaps to persistent infections and progression has not been provided. Other poorly understood environmental risk factors under current consideration are the low socio economic level and different measures of general and sexual hygiene.

4.6. Final considerations on the role of cofactors and limitations of the current evidence

Some general considerations in relation to the evaluation of the risk co-factors for cervical cancer considered are worth mentionning. Concerning the statistical evaluation of the associations, the HPV restricted analyses provide slightly higher risk estimates that conventional analyses adjusting for the presence of HPV DNA in cases and controls (Castellsague and Muñoz, 2003). In contrast this approach limits dramatically the number of controls available and introduces an uncertain bias in accepting a fraction of HPV DNA negative cases. The latter is due to the situation in which the reason of the apparent HPV DNA negativity are not determined.

In the analyses restricted to HPV positive women, the magnitude of the relative risk estimates for any of the other risk factors are moderate (range 2 to 6 for extreme categories of exposure) for any of the variables as compared to the risk estimates for HPV DNA (ranges 50 to >500). Moreover, the results in the literature concerning some of these factors remain inconsistent, again as compared to the striking consistency of the associations found for HPV DNA. The interpretation of the risk estimates for co-factors should consider that, as far as we understand them today, these are conditional effects that apply if, and perhaps only if, HPV DNA is present in the specimen. Moreover, none of these factors have been consistently associated to the HPV DNA presence in studies of risk factors for HPV DNA infection. Therefore, it is not convincingly proven that any of the co-factors discussed above can be considered truly independent risk factor for cervical cancer (a factor that could induce cervical cancer in the absence of HPV). Some authors view these results (in conjunction with the risk analyses of HPV exposure) as indicative of a promoter effect of OC use, smoking or multiparity in the transit from HPV infections to HPV related neoplasia.

In spite of considerable progress in study design and HPV testing technology, there are recognized limitations of the available evidence that may explain apparent inconsistencies across studies. One of them is some crudeness in the information that can be obtained trough questionnaires. This affects variables that may indeed modulate the natural course of HPV infections. For example, the inconsistencies in finding associations of cervical cancer with the use of OCs could perhaps be explained by factors related to the OC product composition, the duration of use, the continuity of use, or the ages in which OCs were used. Factors which are largely country-specific, like the variability in time since widespread introduction, availability of combined vs. sequential products, and the estrogen/progesterone doses, could be important determinants yet these are only partially accounted for in the statistical analyses and the associations observed. The same applies to variables like number of partners, or even more difficult to record accurately, the number of partners of the husband.

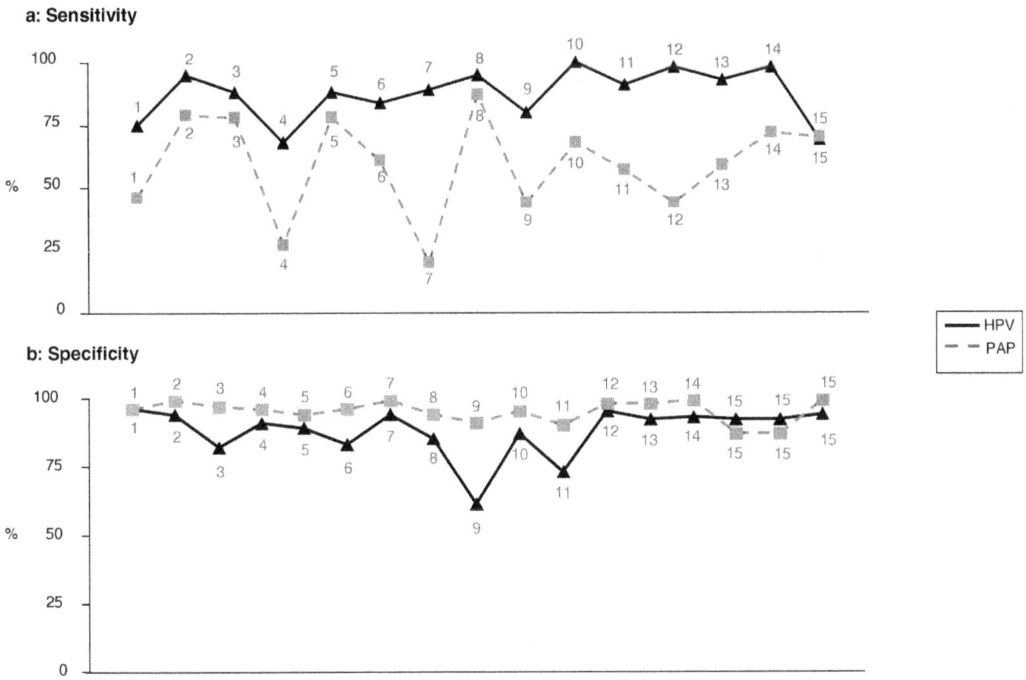

Figure 6. Sensitivity and Specificity of Human Papillomavirus DNA tests and cytology in screening studies. [References: 1: (Cuzick et al., 1995); 2: (Cuzick et al., 1999) ; 3: (Kuhn et al., 2000); 4: (Ratnam et al., 2000); 5: (Schiffman et al., 2000); 6: (Wright, Jr. et al., 2000); 7: (Schneider et al., 2000); 8: (Belinson et al., 2001); 9: (Blumenthal et al., 2001); 10: (Clavel et al., 2001); 11: (Kulasingam et al., 2002); 12: (Petry et al., 2003); 13: (Salmeron et al., 2003); 14: (Cuzick et al., 2003); 15: (Sankaranarayanan et al., 2004)

A second recognized limitation is that many studies did not consider host factors in relation to response to HPV. HLA types and p53 polymorphisms for example, are being actively investigated because of the indication that they may play a role in the natural history of HPV infections. At this stage there is however limited information to substantiate an effect of any of these host factors as independent of HPV or of a strong modulator of the HPV natural history. There is a clear need to pursue research in understanding the factors that determine whether a woman with an HPV infection will clear the infection of become a persistent carrier at high risk of cervical cancer.

5. New Preventive Options: HPV Testing in Screening and Triage Programs

5.1. Screening studies

Cervical cytology has played an important role in screening and clinical management of cervical lesions. A recognized barrier is, however, the limited sensitivity and reproducibility of cervical cytology. It has been suggested that screening based on HPV DNA testing may prove easier to implement and sustain, and considerable efforts are currently being devoted to the testing of this hypothesis. Ideally HPV screening tests should detect all CIN 3 / HSIL and cervical cancer. Either of the current testing systems Hybrid capture 2 (HC 2) and the PCR based GP5+/6+ PCR-enzyme immunoassay (PCR/EIA) have sensitivities for CIN 3 and cervical cancer at least equal and in most studies significantly better than cervical cytology. Specificity of HPV tests is age dependent. In the young age groups the specificity of the HPV tests is lower than cytology and, in the age groups 35 and above (again country-dependent), the specificity of both tests is similar

Since the late 80's, multiple studies have evaluated HPV testing as an adjunct to cytology in screening programs. These have considered HPV testing either as a triage test in cases of mild abnormalities or ASCUS (Cuzick et al., 1994) or as a primary screening test (Cuzick et al., 1995; 2000; Ratnam et al., 2000).

Studies that reflect primary screening conditions (in the absence of fully randomized trials) using in most instances HC 2 have shown that the sensitivity of HPV tests is significantly higher than standard cytology in detecting high-grade lesions whereas the specificity is lower and age dependent. Of importance is the consideration that the predictive value of an HPV test (or any other screening test) should ideally detect prevalent disease and properly classify risk for the next time period. In this respect the Portland cohort showed that the number of HSIL cases-to-be within a 10-year follow up were largely picked up by a single HPV test at entry (some 60% of the cases) as compared to cytology which only detected prevalent disease (some 30% of the cases) without ability to predict risk for the years to come (Lorincz and Richart, 2003).

Figure 6 shows a summary of the literature on HPV screening showing the gains in sensitivity (range: 9 to 69%; average gain: 28%) and the loss of specificity (range: 0 to 30%; average loss: 7%), as compared to cytology. The figure also shows the variability of the sensitivity estimates for cytology across studies in contrast with the consistency achieved by the HPV testing methods (Cuzick et al., 1995; 1999; Kuhn et al., 2000; Ratnam et al., 2000; Schiffman et al., 2000; Wright, Jr. et al., 2000; Schneider et al., 2000; Belinson et al., 2001; Blumenthal et al., 2001; Clavel et al., 2001; Kulasingam et al., 2002; Petry et al., 2003; Salmeron et al., 2003; Cuzick et al., 2003; Sankaranarayanan et al., 2004). In combination, women with both normal cytology and absence of HPV DNA have an extremely low risk of developing cervical cancer in the 10+ subsequent years. By moving to HPV based screening major gains in effectiveness and cost reduction are to be expected from increasing screening intervals and reducing the total number of visits requested per lifetime currently required in most cytology-based screening programs. In contrast, some additional follow up visit would be requiered among HPV positive cytology negative women. In some settings, referral to colposcopy is an important deteminant of the total cost of the program.

In 2004, IARC convened a group of experts to prepare a Monograph on screening for cervical cancer to evaluate the novel evidence generated by advances in screening technology. The working group concluded that there was sufficient evidence of the preventive efficacy of conventional cytology, liquid based cytology and automated cytology. The working group also considered sufficient evidence for HPV testing as a primary screening modality. The implication is that HPV testing, using a validated system, as primary screening modality can be expected to produce preventive results at least as good as the standard Pap smear (International Agency for Research on Cancer, 2005). On going studies are evaluating the cost effectiveness of such proposals and its potential impact in novel protocols, particularly in the age ranges of screening programs and on the length of the screening intervals (Kim et al., 2002; Mandelblatt et al., 2002; Sawaya et al., 2003).

5.2. HPV tests in the triage of minimal cervical abnormalities

One of the first applications of HPV testing in clinical practice was the secondary triage of women referred for colposcopy because of an abnormal Pap smear. In the U.S., the adoption of the Bethesda system (TBS) for cervical cytology reports dramatically increased the proportion of Pap smears with cytological abnormalities that merited clinical attention. It is now accepted that smears with equivocal or minor grade abnormalities represent some 4-7% of all Pap smears in the US, a considerable workload for the purposes of clinical decision-making.

The best evidence on the role of HPV testing as an alternative method to repeated cytology in the presence of an ambiguous abnormal cytology is being provided by the Kaiser Permanente study in 1999 and the ASCUS-LSIL Triage Study (ALTS) in 2000.

The Kaiser Permanente study used concomitant testing focused exclusively on ASCUS as the referral smear's presumptive diagnosis. The sensitivity to detect HSIL or cancer was 89% for HPV and 76% for Pap, with equivalent specificities for both tests (64%) (Manos et al., 1999).

The ALTS trial is a National Cancer Institute (NCI) coordinated randomized clinical trial designed to determine the optimal management plan separately for LSIL and ASCUS cytologic abnormalities. The LSIL component of the ALTS trial had the HPV triage arm terminated prematurely because of the interim observation of a very high rate (83%) of oncogenic HPV positivity. In contrast, the ASCUS component of the trial reinforced the findings of the Kaiser Permanente study. HPV testing yielded 96% sensitivity to detect both CIN 2+ or CIN 3+ histologically-confirmed lesions while a repeat Pap at an ASCUS+ thresholds produced a significantly lower sensitivity of 85% for both definitions of lesion severity. The sensitivity of a repeated cytology at a threshold level of HSIL was as low as 44% for CIN 3+ and 35% for CIN 2+. The proportions of women who would have to be referred for colposcopy due to positivity in these tests were 56% and 59%, respectively. The trial concluded that HPV testing was a viable option in the triage and management of ASCUS smears (Solomon et al., 2001; The ASCUS-LSIL Triage Study (ALTS) Group, 2003a, The ASCUS-LSIL Triage Study (ALTS) Group, 2003b). A meta-analyses of studies of HPV testing *vs.* cytology in cases of ASCUS reached the same conclussions as the ALTS trial (Arbyn et al., 2004). Further trials are evaluating management strategies for women with HPV DNA and normal cytology, a diagnosis that is likely to increase if HPV screening generalizes. The HPV in Addition to Routine Testing (HART) study has reported that annual follow up for these women is a safe strategy avoiding the option of immediate referral to colposcopy (Cuzick et al., 2003).

5.3. HPV vaccines

The identification of HPV as a necessary cause of cervical cancer has prompted the development of HPV prophylactic vaccines so far based upon the properties of one of the most conserved regions in the HPV genome (the L1 region) which is capable of self assemble as a tridimensional structure when expressed *in vitro* (Viral-like particles without the oncogenic parts of the genome) and induce strong immunological responses. Preliminary results are very encouraging. An HPV 16 virus like particle (VLP) based vaccine showed immune responses in virtually all vaccinated women, antibody titers were up to 40-fold the level acquired after natural infection and vaccinated women were fully protected against HPV 16 infections, persistent HPV 16 infections and HPV 16 related CIN (Koutsky et al., 2002). A more developed vaccine based upon similar principles and including HPV 16 and 18 antigens showed equivalent results for both HPV types (Dubin, 2004). The protective efficacy of these HPV vaccines for CIN 2 – CIN 3 lesions is currently under evaluation.

Second generation vaccines are being developed with the objective of including additional HPV types related to cervical cancer, facilitate administration and incorporate other traits that would facilitate introduction into the countries at higher risk.

It is thus likely that the future of cervical cancer prevention will include prophylactic HPV vaccines in the adolescent women followed by a novel type of screening, likely HPV based, with cervical cytology colposcopy and pathology as the triage and diagnostic tools. In developing countries, where regular screening has proven to be largely unrealistic, perhaps a booster vaccination step may be advisable.

Therapeutic vaccines may offer interesting alternatives in populations where a fraction of young adult women are already permanent carriers of HPV DNA. These products incorporate modified fragments of the E6 and/or E7 genes, the viral products consistently expressed in persistent infections and in cervical cancer. Chimeric VLPs have been shown to induce antigen-specific protection in mice from challenge with E7-expressing tumor cells. However, human studies are still in its early phases.

Conclusion

In the last two decades, the etiology of cervical cancer has become a coherent description that includes the identification of any of a limited group of HPV types as the necessary, sexually-transmitted etiologic agents. Exploration of the clinical implications of these findings has resulted in novel screening and vaccination strategies for the prevention of cervical cancer.

Acknowledgements

The statistical contribution of Mireia Díaz in extracting and elaborating cancer statistics and of Cristina Rajo in the preparation of the manuscript are deeply recognized. Partial support has been received from the Fondo de Investigaciones Sanitarias, Spain (FIS PI030240 and FIS 01/1237), and from the Instituto de Salud Carlos III, Spain (Red de Centros RCESP C03/09 and RTICESP C03/10). Dr. Nubia Muñoz was recipient of a grant from the Catalan Government - Agència de Gestió d'Ajuts Universitaris i de Recerca, for short term visiting scientists (2003PIVB 00029).

Reference

Ahdieh, L., Muñoz, A., Vlahov, D., Trimble, C.L., Timpson, L.A., and Shah, K. (2000). Cervical neoplasia and repeated positivity of human papillomavirus infection in human immunodeficiency virus-seropositive and - seronegative women. Amer. J. Epidemiol. *151*, 1148-1157.

Alberg, A.J., Selhub, J., Shah, K.V., Viscidi, R.P., Comstock, G.W., and Helzlsouer, K.J. (2000). The risk of cervical cancer in relation to serum concentrations of folate, vitamin B12, and homocysteine. Cancer Epidemiol. Biomarkers Prev. *9*, 761-764.

Arbyn, M., Buntinx, F., Van Ranst, M., Paraskevaidis, E., Martin-Hirsch, P., and Dillner, J. (2004). Virologic versus cytologic triage of women with equivocal Pap smears: a meta-analysis of the accuracy to detect high-grade intraepithelial neoplasia. J. Natl. Cancer Inst. *96*, 280-293.

Barrasso, R., de Brux, J., Croissant, O., and Orth, G. (1987). High prevalence of papillomavirus-associated penile intraepithelial neoplasia in sexual partners of women with cervical intraepithelial neoplasia. N. Engl. J. Med. *317*, 916-923.

Bayo, S., Bosch, F.X., de Sanjose, S., Muñoz, N., Combita, A.L., Coursaget, P., Diaz, M., Dolo, A., van Den Brule, A.J., and Meijer, C.J.L.M. (2002). Risk factors of invasive cervical cancer in Mali. Int. J. Epidemiol. *31*, 202-209.

Belinson, J., Qiao, Y.L., Pretorius, R., Zhang, W.H., Elson, P., Li, L., Pan, Q.J., Fischer, C., Lorincz, A., and Zahniser, D. (2001). Shanxi Province Cervical Cancer Screening Study: a cross-sectional comparative trial of multiple techniques to detect cervical neoplasia. Gynecol. Oncol. *83*, 439-444.

Bergman, A. and Nalick, R. (1992). Prevalence of human papillomavirus infection in men: comparison of the partners of infected and uninfected women. J. Reprod. Med. *37*, 710-712.

Blumenthal, P.D., Gaffikin, L., Chirenje, Z.M., McGrath, J., Womack, S., and Shah, K. (2001). Adjuntive testing for cervical cancer in low resource settings with visual inspection, HPV, and the Pap smear. Int. J. Gynecol. Obstet. *72*, 47-53.

Bosch, F.X., Castellsagué, X., Muñoz, N., de Sanjosé, S., Ghaffari, A.M., González, L.C., Gili, M., Izarzugaza, I., Viladiu, P., Navarro, C., Vergara, A., Ascunce, N., Guerrero, E., and Shah, K.V. (1996). Male sexual behavior and Human Papillomavirus DNA: key risk factors for cervical cancer in Spain. J. Natl. Cancer Inst. *88(15)*, 1060-1067.

Bosch, F.X. and de Sanjose, S. (2003). Chapter 1: Human papillomavirus and cervical cancer-burden and assessment of causality. J. Natl. Cancer Inst. Monogr. 3-13.

Bosch, F.X., Lorincz, A., Muñoz, N., Meijer, C.J.L.M., and Shah, K.V. (2002). The causal relation between human papillomavirus and cervical cancer. J. Clin. Pathol. *55*, 244-265.

Bosch, F.X., Manos, M.M., Muñoz, N., Sherman, M., Jansen, A., Peto, J., Schiffman, M., Moreno, V., Shah, K.V., and the IBSCC study group (1995). Prevalence of human papillomavirus in cervical cancer: a worldwide perspective. J. Natl. Cancer Inst. *87*, 796-802.

Bosch, F.X., Muñoz, N., de Sanjosé, S., Guerrero, E., Ghaffari, A.M., Kaldor, J., Castellsagué, X., and Shah, K.V. (1994). Importance of human papillomavirus endemicity in the incidence of cervical cancer: an extension of the hypothesis on sexual behavior. Cancer Epidemiol. Biomark Prev. *3*, 375-379.

Bosch, F.X., Rohan, T., Schneider, A., Frazer, I., Pfister, H., Castellsagué, X., de Sanjosé, S., Moreno, V., Puig-Tintoré, Ll.M., Smith, P.G., Muñoz, N., and zur Hausen, H. (2001). Papillomavirus research update: highlights of the Barcelona HPV 2000 international papillomavirus conference. J. Clin. Pathol. *54*, 163-175.

Bosch, F.X., Schiffman, M., Solomon, D., and (eds) (2003). Future Directions in Epidemiologic and Preventive Research on Human Papillomaviruses and Cancer. J. Natl. Cancer Inst. Monographs Nr. 31).

Burger, M.P.M., Hollema, H., Pieters, W.J.L.M., Schröder, F.P., and Quint, W.G.V. (1996). Epidemiological evidence of cervical intraepithelial neoplasia without the presence of human papillomavirus. Brit. J. Cancer *73*, 831-836.

Castellsague, X., Bosch, F.X., Muñoz, N., Meijer, C.J., Shah, K.V., de Sanjose, S., Eluf-Neto, J., Ngelangel, C.A., Chichareon, S., Smith, J.S., Herrero, R., Moreno, V., and Franceschi, S. (2002). Male circumcision, penile human papillomavirus infection, and cervical cancer in female partners. N. Engl. J. Med. *346*, 1105-1112.

Castellsagué, X., Ghaffari, A., Daniel, R.W., Bosch, F.X., Muñoz, N., and Shah, K.V. (1997). Prevalence of penile human papillomavirus DNA in husbands of women with and without cervical neoplasia: A study in Spain and Colombia. J. Infect. Dis. *176*, 353-361.

Castellsague, X., Menendez, C., Loscertales, M.P., Kornegay, J.R., dos Santos, .F., Gomez-Olive, F.X., Lloveras, B., Abarca, N., Vaz, N., Barreto, A., Bosch, F.X., and Alonso, P. (2001). Human papillomavirus genotypes in rural Mozambique. Lancet *358*, 1429-1430.

Castellsague, X. and Muñoz, N. (2003). Chapter 3: Cofactors in human papillomavirus carcinogenesis--role of parity, oral contraceptives, and tobacco smoking. J. Natl. Cancer Inst. Monogr 20-28.

Castle, P.E., Hillier, S.L., Rabe, L.K., Hildesheim, A., Herrero, R., Bratti, M.C., Sherman, M.E., Burk, R.D., Rodriguez, A.C., Alfaro, M., Hutchinson, M.L., Morales, J., and Schiffman, M. (2001). An association of cervical inflammation with high-grade cervical neoplasia in women infected with oncogenic human papillomavirus (HPV). Cancer Epidemiol. Biomark Prev. *10*, 1021-1027.

Castle, P.E., Wacholder, S., Lorincz, A.T., Scott, D.R., Sherman, M.E., Glass, A.G., Rush, B.B., Schussler, J.E., and Schiffman, M. (2002). A prospective study of high-grade cervical neoplasia risk among human papillomavirus-infected women. J. Natl. Cancer Inst. *94*, 1406-1414.

Clavel, C., Masure, M., Bory, J.P., Putaud, I., Mangeonjean, C., Lorenzato, M., Nazeyrollas, P., Gabriel, R., Quereux, C., and Birembaut, P. (2001). Human papillomavirus testing in primary screening for the detection of high-grade cervical lesions: a study of 7932 women. Br. J. Cancer *84*, 1616-1623.

Cuzick, J., Beverley, E., Ho, L., Terry, G., Sapper, H., Mielzynska, I., Lorincz, A., Chan, W.K., Krausz, T., and Soutter, P. (1999). HPV testing in primary screening of older women. Brit. J. Cancer *81(3)*, 554-558.

Cuzick, J., Sasieni, P., Davies, P., Adams, J., Normand, C., Frater, A., Van Ballegooijen, M., and Akker-van Marle, E. (2000). A systematic review of the role of human papilloma virus (HPV) testing within a cervical screening programme: summary and conclusions. Br. J. Cancer *83*, 561-565.

Cuzick, J., Szarewski, A., Cubie, H., Hulman, G., Kitchener, H., Luesley, D., McGoogan, E., Menon, U., Terry, G., Edwards, R., Brooks, C., Desai, M., Gie, C., Ho, L., Jacobs, I., Pickles, C., and Sasieni, P. (2003). Management of women who test positive for high-risk types of human papillomavirus: the HART study. Lancet *362*, 1871-1876.

Cuzick, J., Szarewski, A., Terry, G., Ho, L., Hanby, A., Maddox, P., Anderson, M., Kocjan, G., Steele, S.T., and Guillebaud, J. (1995). Human papilloma virus testing in primary cervical screening. Lancet *345*, 1533-1536.

Cuzick, J., Terry, G., Ho, L., Hollingworth, T., and Anderson, M. (1994). Type-specific human papillomavirus DNA in abnormal smears as a predictor of high-grade cervical intraepithelial neoplasia. Brit. J. Cancer *69*, 167-171.

de Roda Husman, A.M., Walboomers, J.M.M., Van den Brule, A.J.C., Meijer, C.J.L.M., and Snijders, P.J.F. (1995). The use of general primers GP5 and GP6 elongated at their 3' ends with adjacent highly conserved sequences improves human papillomavirus detection by polymerase chain reaction. J. Gen. Virol. *76*, 1057-1062.

de Sanjosé, S., Muñoz, N., Bosch, F.X., Reimann, K., Pedersen, N.S., Orfilia, J., Ascunce, N., González, L.C., Tafur, L., Gili, M., Lette, I., Viladiu, P., Tormo, M.J., Moreo, P., Shah, K., and Wahren, B. (1994). Sexually transmitted agents and cervical neoplasia in Colombia and Spain. Int. J. Cancer *56*, 358-363.

de Sanjosé, S., Palacio, V., Tafur, L., Vazquez, S., Espitia, V., Vazquez, F., Roman, G., Muñoz, N., and Bosch, F.X. (1993). Prostitution, HIV, and Cervical Neoplasia: A Survey in Spain and Colombia. Cancer Epidemiol. Biomark Prev. *2*, 531-535.

de Sanjosé, S. and Palefsky, J. (2002). Cervical and anal HPV infections in HIV positive women and men. Virus Res. *89*, 201-211.

Deacon, J., Peto, J., Yule, R., Desai, M., Binns, W., and Taylor, C. (2000). Sexual behaviour and smoking as determinants of cervical HPV infection and of CIN3 among those infected: a case-control study nested within the Manchester cohort. Brit. J. Cancer *88*, 1565-1572.

Dillner, J., Lehtinen, M., Bjorge, T., Luostarinen, T., Youngman, L., Jellum, E., Koskela, P., Gislefoss, R.E., Hallmans, G., Paavonen, J., Sapp, M., Schiller, J.T., Hakulinen, T., Thoresen, S., and Hakama, M. (1997). Prospective seroepidemiologic study of human papillomavirus infection as a risk factor for invasive cervical cancer. J. Natl. Cancer Inst. *89*, 1293-1299.

Dubin, G. (2004). A double-blind placebo controlled efficacy trial of an adjuvanted human papillomavirus (HPV) type 16/18 L1 virus like particle vaccine (VLP) vaccine. In 21st International Papillomavirus Conference and Clinical Workshop Programe and Abstracts Book, (Mexico City: p. 313.

Ellerbrock, T.V., Chiasson, M.A., Bush, T.J., Sun, X.W., Sawo, D., Brudney, K., and Wright, T.C., Jr. (2000). Incidence of cervical squamous intraepithelial lesions in HIV-infected women. J. Amer. Med. Assoc. *283*, 1031-1037.

Ferlay, J., Bray, F., Pisani, P., and Parkin, D.M. (2001). Globocan 2000: Cancer incidence, Mortality and Prevalence Worldwide, Version 1.0. IARC CancerBase No. 5. (Lyon: IARC Press).

Follen, M., Meyskens, F.L., Jr., Atkinson, E.N., and Schottenfeld, D. (2001). Why most randomized phase II cervical cancer chemoprevention trials are uninformative: lessons for the future. J. Natl. Cancer Inst. *93*, 1293-1296.

Franceschi, S., dal Maso, L., Arniani, S., Crosignani, P., Vercelli, M., Simonato, L., Falcini, F., Zanetti, R., Barchielli, A., Serraino, D., and Rezza, G. (1998). Risk of cancer other than Kaposi's sarcoma and non-Hodgkin's lymphoma in persons with AIDS in Italy. Cancer and AIDS Registry Linkage Study. Br. J. Cancer *78*, 966-970.

Franco, E.L., Villa, L.L., Sobrinho, J.P., Prado, J.M., Rousseau, M., Désy, M., and Rohan, T.E. (1999). Epidemiology of Acquisition and Clearance of Cervical Human Papillomavirus Infection in Women from a High-Risk Area for Cervical Cancer. J. Infect. Dis. *180*, 1415-1423.

Frisch, M., Biggar, R.J., and Goedert, J.J. (2000). Human papillomavirus-associated cancers in patients with human immunodeficiency virus infection and acquired immunodeficiency syndrome. J. Natl. Cancer Inst. *92*, 1500-1510.

Gallagher, B., Wang, Z., Schymura, M.J., Kahn, A., and Fordyce, E.J. (2001). Cancer incidence in New York State acquired immunodeficiency syndrome patients. Amer. J. Epidemiol. *154*, 544-556.

García-Closas, R. Castellsagué, X., Bosch, F.X., and González, C.A. (2005). The role of diet and nutrition in cervical carcinogenesis: A review of recent evidence. Int. J. Cancer. In press.

Gichangi, P., De Vuyst, H., Estambale, B., Rogo, K., Bwayo, J., and Temmerman, M. (2002). HIV and cervical cancer in Kenya. Int. J. Gynaecol. Obstet. *76*, 55-63.

Giuliano, A.R., Papenfuss, M., Nour, M., Canfield, L.M., Schneider, A., and Hatch, K. (1997). Antioxidant nutrients: associations with persistent human papillomavirus infection. Cancer Epidemiol. Biomarkers Prev. *6*, 917-923.

Giuliano, A.R., Siegel, E.M., Roe, D.J., Ferreira, S., Baggio, M.L., Galan, L., Duarte-Franco, E., Villa, L.L., Rohan, T.E., Marshall, J.R., and Franco, E.L. (2003). Dietary intake and risk of persistent human papillomavirus (HPV) infection: the Ludwig-McGill HPV Natural History Study. J. Infect. Dis. *188*, 1508-1516.

Goodman, M.T., Kiviat, N., McDuffie, K., Hankin, J.H., Hernandez, B., Wilkens, L.R., Franke, A., Kuypers, J., Kolonel, L.N., Nakamura, J., Ing, G., Branch, B., Bertram, C.C., Kamemoto, L., Sharma, S., and Killeen, J. (1998). The association of plasma micronutrients with the risk of cervical dysplasia in Hawaii. Cancer Epidemiol. Biomarkers Prev. *7*, 537-544.

Goodman, M.T., McDuffie, K., Hernandez, B., Wilkens, L.R., and Selhub, J. (2000). Case-control study of plasma folate, homocysteine, vitamin B(12), and cysteine as markers of cervical dysplasia. Cancer *89*, 376-382.

Gravitt, P.E. and Manos, M.M. (1992). Polymerase chain reaction-based methods for the detection of human papillomavirus DNA. In The epidemiology of human papillomavirus and cervical cancer. IARC Scientific Publications, N° 119, N.Muñoz, F.X.Bosch, K.V.Shah, and A.Meheus, eds. (Lyon: International Agency for Research on Cancer), pp. 121-133.

Herrero, R., Hildesheim, A., Bratti, C., Sherman, M.E., Hutchinson, M., Morales, J., Balmaceda, I., Greenberg, M.D., Alfaro, M., Burk, R.D.W.S., Plummer, M., and Schiffman, M. (2000). Population-Based Study of Human Papillomavirus Infection and Cervical Neoplasia in Rural Costa Rica. J. Natl. Cancer Inst. *92(6)*, 464-474.

Hildesheim, A., Herrero, R., Castle, P.E., Wacholder, S., Bratti, M.C., Sherman, M.E., Lorincz, A.T., Burk, R.D., Morales, J., Rodriguez, A.C., Helgesen, K., Alfaro, M., Hutchinson, M., Balmaceda, I., Greenberg, M., and Schiffman, M. (2001a). HPV co-factors related to the development of cervical cancer: results from a population-based study in Costa Rica. Br. J. Cancer *84(9)*, 1219-1226.

Hildesheim, A., Schiffman, M., Bromley, C., Wacholder, S., Herrero, R., Rodriguez, A.C., Bratti, M.C., Sherman, M.E., Scarpidis, U., Lin, Q.Q., Terai, M., Bromley, R.L., Buetow, K., Apple, R.J., and Burk, R.D. (2001b). Human Papillomavirus type 16 variants and the risk of cervical cancer. J. Natl. Cancer Inst. *93*, 315-318.

Ho, G.Y., Bierman, R., Beardsley, L., Chang, C.J., and Burk, R.D. (1998a). Natural history of cervicovaginal papillomavirus infection in young women. N. Engl. J. Med. *338*, 423-428.

Ho, G.Y., Palan, P.R., Basu, J., Romney, S.L., Kadish, A.S., Mikhail, M., Wassertheil-Smoller, S., Runowicz, C., and Burk, R.D. (1998b). Viral characteristics of human papillomavirus infection and antioxidant levels as risk factors for cervical dysplasia. Int. J. Cancer *78*, 594-599.

Ho, G.Y.F., Burk, R.D., Klein, S., Kadish, A.S., Chang, C.J., Palan, P., Basu, Y., Tachezy, R., Lewis, R., and Rommey, S. (1995). Perisitent genital human papillomavirus infection as a risk factor for persistent cervical dysplasia. J. Natl. Cancer Inst. *87*, 1365-1371.

International Agency for Research on Cancer (1995). IARC monographs on the evaluation of carcinogenic risks to humans. Vol. 64. Human Papillomaviruses., IARC, ed. (Lyon: International Agency for Research on Cancer).

International Agency for Research on Cancer (1996). IARC Monographs on the Evaluation of Carcinogenic Risks to Humans. Vol. 67 Human Immunodeficiency Viruses and Human T-cell Lymphotropic Viruses., IARC, ed. (Lyon; IARC Press).

International Agency for Research on Cancer (2004). IARC Monographs on the Evaluation of Carcinogenic Risks to Humans. Volume 83. Tobacco Smoke and Involuntary Smoking. IARC, ed. (Lyon: IARC Press).

International Agency for Research on Cancer (2005). IARC Handbooks of Cancer Prevention. Volume 10. Cervical Cancer Screening. (Lyon: IARC Press).

International Collaboration on HIV and Cancer (2000). Highly active antiretroviral therapy and incidence of cancer in human immnodeficiency virus-infected adults. J. Natl. Cancer Inst. *92*, 1823-1830.

Jacobs, E.J., Henion, A.K., Briggs, P.J., Connell, C.J., McCullough, M.L., Jonas, C.R., Rodriguez, C., Calle, E.E., and Thun, M.J. (2002). Vitamin C and vitamin E supplement use and bladder cancer mortality in a large cohort of US men and women. Amer. J. Epidemiol. *156*, 1002-1010.

Jacobs, M.V., Walboomers, J.M.M., Snijders, P.J.F., Voorhorst, F.J., Verheijen, R.H.M., Fransen-Daalmeijer, N., and Meijer, C.J.L.M. (2000). Distribution of 37 mucosotropic HPV types in women with cytologically normal cervical smears: the age-related patterns for high-risk and low types. Int. J. Cancer *87*, 221-227.

Juarez-Figueroa, L.A., Wheeler, C.M., Uribe-Salas, F.J., Conde-Glez, C.J., Zampilpa-Mejia, L.G., Garcia-Cisneros, S., and Hernandez-Avila, M. (2001). Human papillomavirus: a highly prevalent sexually transmitted disease agent among female sex workers from Mexico City. Sex Transm. Dis. *28*, 125-130.

Kim, J.J., Wright, T.C., and Goldie, S.J. (2002). Cost-effectiveness of alternative triage strategies for atypical squamous cells of undetermined significance. J. Amer. Med. Assoc. *287*, 2382-2390.

Kjaer, S.K. (1991). Case-control study of risk factors for cervical neoplasia in Denmark. I: role of the "male factor" on women with one lifetime sexual partner. Int. J. Cancer *48*, 39-41.

Kjaer, S.K. (1998). Risk factors for cervical neoplasia in Denmark. APMIS Suppl *80*, 1-41.

Kjaer, S.K., Van den Brule, A.J.C., Bock, J.E., Poll, P.A., Engholm, G., Sherman, M.E., Walboomers, J.M.M., and Meijer, C.J.L.M. (1996). Human Papilloma virus - The most significant risk determinant of cervical intraepithelial neoplasia. Int. J. Cancer *65*, 601-606.

Koutsky, L.A., Ault, K.A., Wheeler, C.M., Brown, D.R., Barr, E., Alvarez, F.B., Chiacchierini, L.M., and Jansen, K.U. (2002). A controlled trial of a human papillomavirus type 16 vaccine. N. Engl. J. Med. *347*, 1645-1651.

Koutsky, L.A., Holmes, K.K., Critchlow, C.W., Stevens, C.E., Paavonen, J., Beckmann, A.M., de Rouen, T.A., Galloway, D.A., Vernon, D., and Kiviat, N.B. (1992). A cohort Study of the risk of Cervical intraepithelial neoplasia grade 2 or 3 in relation to papillomavirus infection. New Engl. J. Med. *327*, 1272-1278.

Kuhn, L., Denny, L., Pollack, A., Lorincz, A., Richart, R.M., and Wright, T.C. (2000). Human Papillomavirus DNA testing for cervical cancer screening in low-resource settings. J. Natl. Cancer Inst. *92*, 818-825.

Kulasingam, S.L., Hughes, J.P., Kiviat, N.B., Mao, C., Weiss, N.S., Kuypers, J.M., and Koutsky, L.A. (2002). Evaluation of human papillomavirus testing in primary screening for cervical abnormalities: comparison of sensitivity, specificity, and frequency of referral. J. Amer. Med. Assoc. *288*, 1749-1757.

Lazcano Ponce, E., Herrero, R., Muñoz, N., Cruz, A., Shah, K.V., Alonso, P., Hernández, P., Salmerón, J., and Hernández, M. (2001). Epidemiology on HPV infection among Mexican women with normal cervical cytology. Int. J. Cancer *91*, 412-420.

Lorincz, A.T. and Richart, R.M. (2003). Human papillomavirus DNA testing as an adjunct to cytology in cervical screening programs. Arch. Pathol. Lab. Med. *127*, 959-968.

Mandelblatt, J.S., Kanetsky, P., Eggert, L., and Gold, K. (1999). Is HIV infection a cofactor for cervical squamous cell neoplasia? Cancer Epidemiol. Biomarkers Prev. *8*, 97-106.

Mandelblatt, J.S., Lawrence, W.F., Womack, S.M., Jacobson, D., Yi, B., Hwang, Y.T., Gold, K., Barter, J., and Shah, K. (2002). Benefits and costs of using HPV testing to screen for cervical cancer. J. Amer. Med. Assoc. *287*, 2372-2381.

Manos, M.M., Kinney, W.K., Hurley, L.B., Sherman, M.E., Shieh-Ngai, J., Kurman, R.J., Ransley, J.E., Fetterman, B.J., Hartinger, J.S., McIntosh, K.M., Pawlick, G.F., and Hiatt, R.A. (1999). Identifying Women With Cervical Neoplasia Using Human Papillomavirus DNA Testing for Equivocal Papanicolaou Results. J. Amer. Med. Assoc. *281(17)*, 1605-1647.

Massad, L.S., Riester, K.A., Anastos, K.M., Fruchter, R.G., Palefsky, J.M., Burk, R.D., Burns, D., Greenblatt, R.M., Muderspach, L.I., and Miotti, P. (1999). Prevalence and predictors of squamous cell abnormalities in Papanicolaou smears from women infected with HIV-1. Women's Interagency HIV Study Group. J. Acquir. Immune. Defic. Syndr. *21*, 33-41.

Molano, M., Posso, H., Weiderpass, E., Van den Brule, A.J.C., Ronderos, M., Franceschi, S., Meijer, C.J.L.M., Arslan, A., and Muñoz, N. (2002). Prevalence and determinants of HPV infection among Colombian women with normal cytology. Brit. J. Cancer *3*, 324-333.

Moreno, V., Bosch, F.X., Muñoz, N., Meijer, C.J., Shah, K.V., Walboomers, J.M., Herrero, R., and Franceschi, S. (2002). Effect of oral contraceptives on risk of cervical cancer in women with human papillomavirus infection: the IARC multicentric case-control study. Lancet *359*, 1085-1092.

Moreno, V., Muñoz, N., Bosch, F.X., de Sanjosé, S., González, L.C., Tafur, L., Gili, M., Izarzugaza, I., Navarro, C., Vergara, A., Viladiu, P., Ascunce, N., and Shah, K. (1995). Risk factors for progression of cervical intraepithelial neoplasm grade III to invasive cervical cancer. Cancer Epidemiol. Biomark Prev. *4*, 459-467.

Muñoz, N., Bosch, F.X., Castellsague, X., Diaz, M., de Sanjosé, S., Hammouda, D., Shah, K.V., and Meijer, C.J. (2004). Against. which human papillomavirus types shall we vaccinate and screen? The international perspective. Int. J. Cancer *111*, 278-285.

Muñoz, N., Bosch, F.X., de Sanjosé, S., Herrero, R., Castellsague, X., Shah, K.V., Snijders, P.J., and Meijer, C.J. (2003). Epidemiologic classification of human papillomavirus types associated with cervical cancer. N. Engl. J. Med. *348*, 518-527.

Muñoz, N., Castellsagué, X., Bosch, F.X., Tafur, L., de Sanjosé, S., Aristizabal, N., Ghaffari, A.M., and Shah, K.V. (1996). Difficulty in elucidating the male role in cervical cancer in Colombia, a high-risk area for the disease. J. Natl. Cancer Inst. *88(15)*, 1068-1075.

Muñoz, N., Franceschi, S., Bosetti, C., Moreno, V., Herrero, R., Smith, J.S., Shah, K.V., Meijer, C.J., and Bosch, F.X. (2002). Role of parity and human papillomavirus in cervical cancer: the IARC multicentric case-control study. Lancet *359*, 1093-1101.

Muñoz, N., Méndez, F., Posso, H., Molano, M., van del Brule, A.J.C., Ronderos, M., Meijer, C.J.L.M., and Muñoz, A., for the Insituto Nacional de Cancerología HPV study group. (2004). Incidence, duration and determinants of cervical human papillomavirus infection in a cohort of Colombian women with normal cytological results. J. Infect. Dis. *190*, 2077-2087.

Nagata, C., Shimizu, H., Yoshikawa, H., Noda, K., Nozawa, S., Yajima, A., Sekiya, S., Sugimori, H., Hirai, Y., Kanazawa, K., Sugase, M., and Kawana, T. (1999). Serum carotenoids and vitamins and risk of cervical dysplasia from a case-control study in Japan. Br. J. Cancer *81*, 1234-1237.

Newton, R., Beral, V., and Weiss, R.A. (1999). Human Immunodeficiency Virus Infection and Cancer. In Cancer Surveys. Infections and Human Cancer, R.Newton,

V.Beral, and R.A.Weiss, eds. (USA: Imperial Cancer Research Fund. Cold Spring Harbor Laboratory Press), pp. 237-262.

NIH Consensus Development Panel (1996). National Institutes of Health Consensus Development Conference Statement: Cervical Cancer, April 1-3, 1996. J. Natl. Cancer Inst. Monographs *21*, vii-xiii.

Nobbenhuis, M.A., Helmerhorst, T.J., van Den Brule, A.J., Rozendaal, L., Voorhorst, F.J., Bezemer, P.D., Verheijen, R.H., and Meijer, C.J. (2001a). Cytological regression and clearance of high-risk human papillomavirus in women with an abnormal cervical smear. Lancet *358*, 1782-1783.

Nobbenhuis, M.A., Meijer, C.J., van Den Brule, A.J., Rozendaal, L., Voorhorst, F.J., Risse, E.K., Verheijen, R.H., and Helmerhorst, T.J. (2001b). Addition of high-risk HPV testing improves the current guidelines on follow-up after treatment for cervical intraepithelial neoplasia. Br. J. Cancer *84*, 796-801.

Nobbenhuis, M.A., Walboomers, J.M., Helmerhorst, T.J., Rozendaal, L., Remmink, A.J., Risse, E.K., Van der Linden, H.C., Voorhorst, F.J., Kenemans, P., and Meijer, C.J. (1999). Relation of human papillomavirus status to cervical lesions and consequences for cervical-cancer screening: a prospective study. Lancet *354*, 20-25.

Olsen, A.O., Gjoen, K., Sauer, T., Orstavik, I., Naess, O., Kierulf, K., Sponland, G., and Magnus, P. (1995). Human papilloma virus and cervical intraepithelial neoplasia grade II-III: A population-based case-control study. Int. J. Cancer *61*, 312-315.

Paraskevaidis, E., Arbyn, M., Sotiriadis, A., Diakomanolis, E., Martin-Hirsch, P., Koliopoulos, G., Makrydimas, G., Tofoski, J., and Roukos, D.H. (2004). The role of HPV DNA testing in the follow-up period after treatment for CIN: a systematic review of the literature. Cancer Treat. Rev. *30*, 205-211.

Parkin, D.M., Whelan, S.L., Ferlay, J., Raymond, L., Young, J., and (editors) (1997). Cancer Incidence in Five Continents, vol. VII., D.M.Parkin, S.L.Whelan, J.Ferlay, L.Raymond, and J.Young, eds. (Lyon: International Agency for Research on Cancer).

Petry, K.U., Menton, S., Menton, M., Loenen-Frosch, F., de Carvalho, G.H., Holz, B., Schopp, B., Garbrecht-Buettner, S., Davies, P., Boehmer, G., van den, A.E., and Iftner, T. (2003). Inclusion of HPV testing in routine cervical cancer screening for women above 29 years in Germany: results for 8466 patients. Br. J. Cancer *88*, 1570-1577.

Plummer, M., Herrero, R., Franceschi, S., Meijer, C.J.L.M., Snijders, P., Bosch, F.X., de Sanjose, S., Muñoz, N., and for the IARC Multi-centre Cervical Cancer Study Group (2003). Smoking and cervical cancer: pooled analysis of the IARC multi-centric case-control study. Cancer Causes Control *14*, 805-814.

Ratnam, S., Franco, E.L., and Ferenczy, A. (2000). Human Papillomavirus Testing for Primary Screening of Cervical Cancer Precursors. Cancer Epidemiol. Biomark Prev. *9*, 945-951.

Remmink, A.J., Walboomers, J.M., Helmerhorst, T.J., Voorhorst, F.J., Rozendaal, L., Risse, E.K., Meijer, C.J., and Kenemans, P. (1995). The presence of persistent high-risk HPV genotypes in dysplastic cervical lesions is associated with progressive disease: natural history up to 36 months. Int. J. Cancer *61*, 306-311.

Riou, G., Favre, M., Jeannel, D., Bourhis, J., le Doussal, V., and Orth, G. (1990). Association between poor prognosis in early-stage invasive cervical carcinomas and non-detection of HPV DNA. Lancet *335*, 1171-1174.

Rozendaal, L., Walboomers, J.M., van der Linden, J.C., Voorhorst, F.J., Kenemans, P., Helmerhorst, T.J., Van Ballegooijen, M., and Meijer, C.J. (1996). PCR-based high-risk HPV test in cervical cancer screening gives objective risk assessment of women with cytomorphologically normal cervical smears. Int. J. Cancer *68*, 766-769.

Rozendaal, L., Westerga, J., van der Linden, J.C., Walboomers, J.M.M., Voorhorst, F.J., Risse, E.K.J., Boon, M.E., and Meijer, C.J.L.M. (2000). PCR based high risk HPV testing is superior to neutral network based screening for predicting incident CIN III in women with normal cytology and borderline changes. J. Clin. Pathol. *53*, 606-611.

Salmeron, J., Lazcano-Ponce, E., Lorincz, A., Hernandez, M., Hernandez, P., Leyva, A., Uribe, M., Manzanares, H., Antunez, A., Carmona, E., Ronnett, B.M., Sherman, M.E., Bishai, D., Ferris, D., Flores, Y., Yunes, E., and Shah, K.V. (2003). Comparison of HPV-based assays with Papanicolaou smears for cervical cancer screening in Morelos State, Mexico. Cancer Causes Control *14*, 505-512.

Sankaranarayanan, R., Chatterji, R., Shastri, S.S., Wesley, R.S., Basu, P., Mahe, C., Muwonge, R., Seigneurin, D., Somanathan, T., Roy, C., Kelkar, R., Chinoy, R., Dinshaw, K., Mandal, R., Amin, G., Goswami, S., Pal, S., Patil, S., Dhakad, N., Frappart, L., and Fontaniere, B. (2004). Accuracy of human papillomavirus testing in primary screening of cervical neoplasia: results from a multicenter study in India. Int. J. Cancer *112*, 341-347.

Sawaya, G.F., McConnell, K.J., Kulasingam, S.L., Lawson, H.W., Kerlikowske, K., Melnikow, J., Lee, N.C., Gildengorin, G., Myers, E.R., and Washington, A.E. (2003). Risk of cervical cancer associated with extending the interval between cervical-cancer screenings. N. Engl. J. Med. *349*, 1501-1509.

Schiffman, M., Herrero, R., Hidesheim, A., Sherman, M.E., Bratti, M., Wacholder, S., Alfaro, M., Hutchinson, M., Morales, J., Greenberg, M.D., and Lorincz, A.T. (2000). HPV DNA Testing in Cervical Cancer Screening. Results From Women in a High-Risk Province of Costa Rica. J. Amer. Med. Assoc. *283(1)*, 87-93.

Schiffman, M., Wheeler, C.M., and Castle, P.E. (2002). Human papillomavirus DNA remains detectable longer than related cervical cytologic abnormalities. J. Infect. Dis. *186*, 1169-1172.

Schiffman, M.H., Bauer, H.M., Hoover, R.N., Glass, A.G., Cadell, D.M., Rush, B.B., Scott, D.R., Sherman, M.E., Kurman, R.J., Wacholder, S., Stanton, C.K., and Manos, M.M. (1993). Epidemiologic evidence showing that human papillomavirus infection causes most cervical intraepithelial neoplasia. J. Natl. Cancer Inst. *85*, 958-964.

Schlecht, N.F., Kulaga, S., Robitaille, J., Ferreira, S., Santos, M., Miyamura, R.A., Duarte-Franco, E., Rohan, T.E., Ferenczy, A., Villa, L.L., and Franco, E.L. (2001). Persistent human papillomavirus infection as a predictor of cervical intraepithelial neoplasia. J. Amer. Med. Assoc. *286*, 3106-3114.

Schneider, A., Hoyer, H., Lotz, B., Leistritza, S., Kühne-Heid, R., Nindl, I., Müller, B., Haerting, J., and Dürst, M. (2000). Screening for high-grade cervical intra-epithelial neoplasia and cancer by testing for high-risk HPV, routine cytology or colposcopy. Int. J. Cancer *89*, 529-534.

Sedjo, R.L., Fowler, B.M., Schneider, A., Henning, S.M., Hatch, K., and Giuliano, A.R. (2003). Folate, vitamin B12, and homocysteine status. findings of no relation between human papillomavirus persistence and cervical dysplasia. Nutrition *19*, 497-502.

Sedjo, R.L., Inserra, P., Abrahamsen, M., Harris, R.B., Roe, D.J., Baldwin, S., and Giuliano, A.R. (2002a). Human papillomavirus persistence and nutrients involved in the methylation pathway among a cohort of young women. Cancer Epidemiol. Biomarkers Prev. *11*, 353-359.

Sedjo, R.L., Roe, D.J., Abrahamsen, M., Harris, R.B., Craft, N., Baldwin, S., and Giuliano, A.R. (2002b). Vitamin A, carotenoids, and risk of persistent oncogenic human papillomavirus infection. Cancer Epidemiol. Biomarkers Prev. *11*, 876-884.

Selik, R.M. and Rabkin, C.S. (1998). Cancer death rates associated with human immunodeficiency virus infection in the United States. J. Natl. Cancer Inst. *90*, 1300-1302.

Serraino, D., Carrieri, P., Pradier, C., Bidoli, E., Dorrucci, M., Ghetti, E., Schiesari, A., Zucconi, R., Pezzotti, P., Dellamonica, P., Franceschi, S., and Rezza, G. (1999). Risk of invasive cervical cancer among women with, or at risk for, HIV infection. Int. J. Cancer *82*, 334-337.

Shannon, J., Thomas, D.B., Ray, R.M., Kestin, M., Koetsawang, A., Koetsawang, S., Chitnarong, K., Kiviat, N., and Kuypers, J. (2002). Dietary risk factors for invasive and *in-situ* cervical carcinomas in Bangkok, Thailand. Cancer Causes Control *13*, 691-699.

Sitas, F., Pacella-Norman, R., Carrara, H., Patel, M., Ruff, P., Sur, R., Jentsch, U., Hale, M., Rowji, P., Saffer, D., Connor, M., Bull, D., Newton, R., and Beral, V. (2000). The spectrum of HIV-1 related cancers in South Africa. Int. J. Cancer *88*, 489-492.

Smith, J.S., Green, J., Berrington, d.G., Appleby, P., Peto, J., Plummer, M., Franceschi, S., and Beral, V. (2003). Cervical cancer and use of hormonal contraceptives: a systematic review. Lancet *361*, 1159-1167.

Smith, J.S., Herrero, R., Bosetti, C., Muñoz, N., Bosch, F.X., Eluf-Neto, J., Castellsague, X., Meijer, C.J., van Den Brule, A.J., Franceschi, S., and Ashley, R. (2002a). Herpes simplex virus-2 as a human papillomavirus cofactor in the etiology of invasive cervical cancer. J. Natl. Cancer Inst. *94*, 1604-1613.

Smith, J.S., Muñoz, N., Herrero, R., Eluf-Neto, J., Ngelangel, C., Franceschi, S., Bosch, F.X., Walbomers, J.M.M., and Peeling, R.W. (2002b). Evidence for Chlamydia trachomatis as a Human Papillomavirus Cofactor in the Etiology of Invasive Cervical Cancer in Brazil and the Philippines. J. Infect. Dis. *185*, 324-331.

Solomon, D., Schiffman, M., Tarone, R., and the ALTS group (2001). Comparison of three management strategies for patients with atypical squamous cells of undetermined significance: baseline results from a randomized trial. J. Natl. Cancer Inst. *93*, 293-299.

Strickler, H.D., Palefsky, J.M., Shah, K.V., Anastos, K., Klein, R.S., Minkoff, H., Duerr, A., Massad, L.S., Celentano, D.D., Hall, C., Fazzari, M., Cu-Uvin, S., Bacon, M., Schuman, P., Levine, A.M., Durante, A.J., Gange, S., Melnick, S., and Burk, R.D. (2003). Human papillomavirus type 16 and immune status in human immunodeficiency virus-seropositive women. J. Natl. Cancer Inst. *95*, 1062-1071.

Sun, X.W., Kuhn, L., Ellerbrock, T.V., Chiasson, M.A., Bush, T.J., and Wright jr, T.C. (1997). Human papilloma virusinfection in women infected with the human immunodeficiency virus. New Engl. J. Med. *337*, 1343-1349.

Tabrizi, S.N., Fairley, C.K., Chen, S., Borg, A.J., Baghurst, P., Quinn, M.A., and Garland, S.M. (1999). Epidemiological characteristics of women with high grade CIN who do and do not have human papillomavirus. Br. J. Obstet. Gynaecol. *106*, 252-257.

The ASCUS-LSIL Triage Study (ALTS) Group (2003a). A randomized trial on the management of low-grade squamous intraepithelial lesion cytology interpretations. Am. J. Obstet. Gynecol. *188*, 1393-1400.

The ASCUS-LSIL Triage Study (ALTS) Group (2003b). Results of a randomized trial on the management of cytology interpretations of atypical squamous cells of undetermined significance. Am. J. Obstet. Gynecol. *188*, 1383-1392.

Thomas, D.B., Qin, Q., Kuypers, J., Kiviat, N., Ashley, R.L., Koetsawang, A., Ray, R.M., and Koetsawang, S. (2001a). Human papillomaviruses and cervical cancer in Bangkok. II. risk factors for *in situ* and invasive squamous cell cervical carcinomas. Amer. J. Epidemiol. *153(8)*, 732-739.

Thomas, D.B., Ray, R.M., Koetsawang, A., Kiviat, N., Kuypers, J., Qin, Q., Ashley, R.L., and Koetsawang, S. (2001b). Human papillomaviruses and cervical cancer in Bangkok. I. Risk factors for invasive cervical carcinomas with human papillomavirus types 16 and 18 DNA. Am. J. Epidemiol. *153*, 723-731.

Thomas, D.B., Ray, R.M., Kuypers, J., Kiviat, N., Koetsawang, A., Ashley, R.L., Qin, Q., and Koetsawang, S. (2001c). Human papillomaviruses and cervical cancer in Bangkok. III. Role of husbands and commercial sex workers. Amer. J. Epidemiol. *153(8)*, 740-748.

Vall Mayans, M., Maguire, A., Miret, M, and Casabona, J. Disproportionate high incidence of invasive cervical cancer as an AIDS-indicative disease among young women in Catalonia, Spain. Sexually Transmitted Diseases 26, 500-503. 1999.

van Den Brule, A.J., Meijer, C.J., Bakels, V., Kenemans, P., and Walboomers, J.M. (1990). Rapid detection of human papillomavirus in cervical scrapes by combined general primer-mediated and type-specific polymerase chain reaction. J. Clin. Microbiol. *28*, 2739-2743.

Viladiu, P., Bosch, F.X., Castellsagué, X., Muñoz, N., Escribà, J.M., Hamsikova, E., Hofmannova, V., Guerrero, E., Izquierdo, A., Navarro, C., Moreo, P., Izarzugaza, I., Ascunce, N., Gili, M., Muñoz, M.T., Tafur, L., Shah, K.V., and Vonka, V. (1997). Human Papillomavirus DNA and antibodies to Human Papilloma virus 16 E2, L2 and E7 peptides as predictors of survival in patients with squamous cell cervical cancer. J. Clin. Oncol. *15*, 610-619.

Walboomers, J.M., Jacobs, M.V., Manos, M.M., Bosch, F.X., Kummer, J.A., Shah, K.V., Snijders, P.J., Peto, J., Meijer, C.J., and Muñoz, N. (1999). Human papillomavirus is a necessary cause of invasive cervical cancer worldwide. J. Pathol. *189*, 12-19.

Wallin, K.L., Wiklund, F., Angstrom, T., Bergman, F., Stendahl, U., Wadell, G., Hallmans, G., and Dillner, J. (1999). Type-specific persistence of human papillomavirus DNA before the development of invasive cervical cancer. N. Engl. J. Med. *341*, 1633-1638.

Wideroff, L., Potischman, N., Glass, A.G., Greer, C.E., Manos, M.M., Scott, D.R., Burk, R.D., Sherman, M.E., Wacholder, S., and Schiffman, M. (1998). A nested case-control study of dietary factors and the risk of incident cytological abnormalities of the cervix. Nutr. Cancer *30*, 130-136.

Woodman, C.B., Collins, S., Winter, H., Bailey, A., Ellis, J., Prior, P., Yates, M., Rollason, T.P., and Young, L.S. (2001). Natural history of cervical human papillomavirus infection in young women: a longitudinal cohort study. Lancet *357*, 1831-1836.

Wright, T.C., Jr., Denny, L., Kuhn, L., Pollack, A., and Lorincz, A. (2000). HPV DNA testing of self-collected vaginal samples compared with cytologic screening to detect cervical cancer. J. Amer. Med. Assoc. *283*, 81-86.

Yeo, A.S., Schiff, M.A., Montoya, G., Masuk, M., Asselt-King, L., and Becker, T.M. (2000). Serum micronutrients and cervical dysplasia in Southwestern American Indian women. Nutr. Cancer *38*, 141-150.

Zielinski, G.D., Snijders, P.J.F., Rozendaal, L., Voorhorst, F.J., Runsink, A.P., Schipper, F.A., and Meijer, C.J.L.M. (2001a). High-risk HPV testing in women with borderline and mild dyskaryosis: long-term follow-up data and clinical relevance. J. Pathol. *193*, 1-8.

Zielinski, G.D., Snijders, P.J.F., Rozendaal, L., Voorhorst, F.J., Van der Linden, H.C., Runsink, A.P., de Schipper, F.A., and Meijer, C.J.L.M. (2001b). HPV presence precedes abnormal cytology in women developing cervical cancer and signals false negative smears. Br. J. Cancer *85(3)*, 398-404.

From: Papillomavirus Research: From Natural History to Vaccines and Beyond. Edited by: M. Saveria Campo

Chapter 4

Gene Expression of Papillomaviruses

Mina Kalantari and Hans-Ulrich Bernard*

Abstract

All papillomaviruses (PVs) contain a 400-850 bp long noncoding region (referred to as long control region, LCR, or upstream regulatory region, URR), which is positioned between the L1 and E6 genes and contains most of the elements that regulate transcription. All PVs have in common that transcription of early and late genes occurs uni-directionally from one or several promoters located in the LCR or in the E6 and E7 genes. All transcripts undergo differential splicing, which processes the seven to nine genes of PVs to numerous different polycistronic mRNAs. The detailed mechanisms that regulate transcription vary significantly between remotely related PVs such as animal PVs, cutaneous human PVs (HPVs), and alpha (or "genital") PVs. This review concentrates on the latter group, which has been most thoroughly studied. All alpha PVs contain an E6 promoter that is activated by a TATA box and an Sp1 binding site and down-regulated through a feedback loop executed by HPV E2 proteins. An enhancer, centered about 400 bp upstream of the E6 promoter, activates transcription specifically in epithelial cells by the synergism between numerous different transcription factors, with AP-1 playing a major role in the epithelial specificity. The specificity of this enhancer is probably the sole source of the epitheliotropism of HPVs. A silencer, modulated by YY1 and CDP, is positioned between the enhancer and the promoter, and appears to be important to couple HPV transcription with the differentiation of stratified epithelia. The HPV DNA exists in the nucleus of infected cells in the form of chromatin, *i.e.* wrapped around nucleosomes. The nucleosomal structure determines the accessibility of cis-responsive elements and is modulated by histone acetylation and deacetylation that is mediated by some of the factors that bind the enhancer and the silencer, and by DNA methylation. Yet other mechanisms influencing HPV gene expression include mRNA stability and altered transcription after recombination between HPV genomes and host DNA. The interactions between these different regulatory levels are likely a major source that determines the alternative outcomes of HPV infections, namely persistence in a subclinical state or progression to benign and malignant lesions.

1. Introduction

The functions of most genes are executed by the proteins which they encode. The amount of a functional protein is determined by the frequency of transcription of the gene, by splicing, the stability of the resulting transcripts, and the efficiency of translation. For this reason a study of gene expression and specifically of transcriptional regulation is central to understand the overall biology of any form of life including viruses. In PVs, the sites of transcription initiation, the promoters, are located in the LCR, the E6 and the E7 gene, and the intensity of use of these promoters is determined by regulatory elements in and around the LCR. While all PVs share this principle, numerous important details are different in the alpha PVs (the recently coined official term for the "genital" HPV types) (de Villiers et al., 2004) from those in some cutaneous HPVs, *e.g.* HPV-8 (May et al., 1991), and the two unrelated bovine PVs, BPV-1 (Baker et al., 1987; McBride et al., 1991) and BPV-4 (Morgan et al., 1999). This chapter will concentrate on research on the alpha-PVs, where the five types HPV-6, 11, 16, 18, and 31 have been the exclusive model systems, and will only occasionally touch upon some comparisons with unrelated PVs. As introduced in the chapter on taxonomy of PVs, the expression "genital" HPVs will be consistently replaced by the new term alpha-PVs.

A principal problem of all research on PVs, and that includes transcription research, stems from the fact that *in vitro* systems precisely representing any specific part of the viral life cycle in situ are difficult or impossible to obtain. This is due to the diversity of different cellular environments that can house PVs. A PV type such as HPV-16 can exist in simple epithelia of the endocervix, in squamous epithelia of the ectocervix and the vagina, and in cutaneous epithelia of the penis. Within each of these cell types, it can exist in basal stem cells and in the various layers of suprabasal cells, and these cell populations can be in a state of asymptomatic infection, precursor lesions or malignant carcinoma. Data suggest that each of these numerous cell types exposes the virus

*For correspondence. Email hbernard@uci.edu

to different transcriptional mechanisms. The presently available technology does not yet make it possible to get sufficient amounts of cells of the same differentiation state in order to study mechanisms that are specific for each cell type. Cell lines that are derived from cervical carcinomas like CaSki and HeLa, which contain HPV-16 and HPV-18 genomes respectively, are an exception. These cultured cells likely still represent a transcriptional programme characteristic of the conditions in a tumor. There is today agreement about most mechanisms of HPV transcription among the specialists in this field, but when disagreements exist they frequently stem from the use of different cell systems where different transcriptional mechanisms prevail (Sailaja et al., 1999).

The transcription of any eukaryotic protein-encoding gene is a tremendously complex process involving more than 100 different proteins, an apparatus exceeding the size of the ribosome (Hampsey and Reinberg, 1999). It cannot be the scope of this chapter to deal with the principles of the transcriptional process. In short, protein-encoding genes are transcribed by the RNA polymerase II, which has to master four objectives, (i) to identify the start of a gene, namely the promoter, (ii) to initiate the RNA polymerization at this site, (iii) to elongate the RNA at reasonable speed, and (iv) terminate transcription at stop sites. RNA polymerase II does not have by itself the ability to identify promoters, but does so with the help of the general transcription factors, notably TFIID. TFIID recognizes the sequence TATAAA, referred to as TATA box, and initiates the assembly of a complex of factors, including RNA polymerase II, that is competent to initiate transcription. The efficiency of the establishment of this complex, and the ability of RNA polymerase II to start transcription, is catalyzed by many different transcription factors through a variety of mechanisms, including establishment of the complex of general transcription factors, and even effects on the transcript elongation. These transcription factors can bind close to the promoter ("promoter factors") or remotely ("enhancer factors"). The composition of transcription factors that activate any specific promoter determines the idiosynchrasies of the expression of each gene, such as cell type specific expression or expression in response to physiological stimuli.

In mammalian cells DNA is normally not naked but packaged in the form of nucleosomes (chromatin). These nucleosomes are important tools to regulate the accessibility of genes to the transcriptional machinery. Under the influence of histone acetylases (HATs), nucleosomes have a conformation that favors transcription. HAT activity is associated with the general transcription machinery and certain transcription factors. Under the influence of histone deacetylases (HDACs), nucleosomes have a conformation that represses transcription. HDAC activity is associated with certain transcription factors and with certain factors that bind methylated DNA.

In this chapter on the transcription of HPVs, one should keep in mind that many or all of these mechanisms act concomitantly on HPV genomes. This leads unavoidably to problems in the interpretation of most experiments. It is the essence of scientific experiments to measure the mutual influence of a dependent and an independent variable. This is often not possible in transcriptional studies, where the data are the result of the synergism of a large number of variables, and as a consequence, the necessity to dissect these processes from one another unavoidably creates somewhat artificial conditions.

2. Technical Approaches

It is the goal of this chapter to discuss mechanistic models of how transcription of HPVs is regulated. It does not attempt to discuss many experimental details and individual data. Many of the mechanistic models were based on experiments following standard transcription techniques.

1. Transcription initiation sites, indicative of promoter activity, are identified through sequencing the 5′ ends of cDNA clones from the mRNA preparations of HPV containing cells. They are confirmed in situ or after transfection of reporter genes by primer extension and RNAse protection experiments. The coding capacity of transcripts and their alternative spliced mRNAs, which originate from a transcription initiation site, are identified by cloning and sequencing of complete mRNAs.
2. Promoter activities of DNA segments surrounding transcription initiation sites are typically confirmed by fusing these DNA segments with reporter genes such as the luciferase and the chloramphenicol acetyl transferase (CAT) genes, transfection into cell cultures, and quantifying the activity of the reporter protein.
3. Binding sites of transcription factors and repressors that regulate the promoters, enhancers, and silencers are determined by electrophoretic mobility shift assays (EMSAs, or gelshifts) or DNAseI footprints. It is possible to hypothesize about the identity of these factors by mutating the binding sites and comparing them with established binding sites. This gives information about the DNA binding domain of transcription factors. Information about the identity of the protein components that make up a (frequently dimeric or multimeric) transcription factor is obtained in the presence of antibodies that bind to specific transcription factors and change their migration in gelshifts ("supershifts").
4. The functional contribution of particular enhancer and silencer elements is analyzed similarly to the experiments described in point 2, by fusing different short DNA segments and deletions of larger parts of HPV genomes to reporter genes. Very small changes

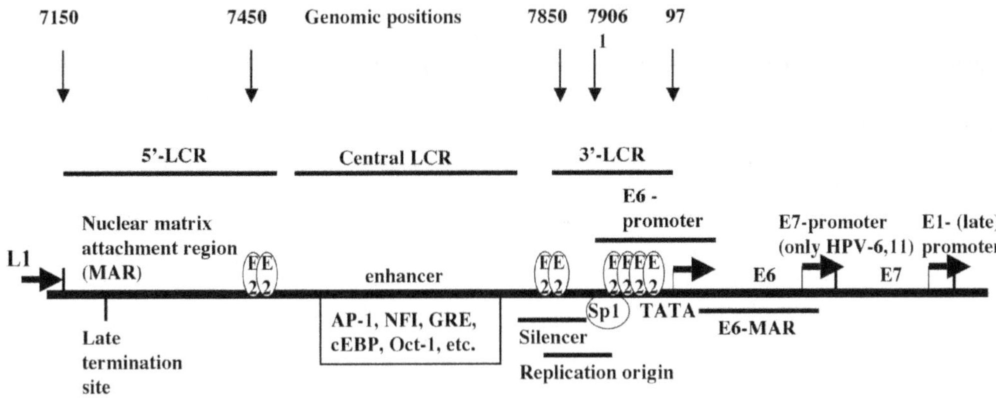

Figure 1. Cis-responsive elements in the long control region, the E6 and E7 genes of alpha ("genital") papillomaviruses. The numbers indicate the genomic positions in HPV-16. Data support that this model applies to HPV-6, 11, 16, 18, and 31, the five paradigms of studying transcription in alpha-PVs, but has to be modified for certain unrelated cutaneous and animal PVs.

of such clones can lead to substantial quantitative differences, a potential source of disagreements between researchers about the origin of the observed effect.

5. Yet another useful approach for qualitative studies of HPV transcription has been in situ hybridization, which has been successfully targeted at clinical material as well as at cell lines. This type of research does not lead to information about factors involved in specific regulatory mechanisms, but is powerful in visualizing transcriptionally active cell populations in complex tissues, and to analyze whether particular chromosomes or domains within the nucleus are the source of HPV transcripts (Iftner et al., 1992; Stoler et al., 1992; van Tine et al., 2003 and 2004).

3. The E6 Promoter

Regulation of the expression of the oncogenes E6 and E7 is arguably the most enticing question in HPV transcriptional research. All alpha-PVs have a highly conserved structure of the E6 promoter that leads to multi-cistronic mRNAs encoding both E6 and E7 (and often yet other genes), and the conservation of these elements makes it likely that they are indispensable for the typical life cycle of these HPV types. The E6 promoter contains a TATA box, which is preceded 31 bp upstream by a G-rich hexamer, a binding site for Sp1, the principal factor that activates the E6 promoter, and contributes to the epithelial specificity of HPV transcription (Apt et al., 1996). Two binding sites for the HPV transcription factor E2, with the sequence ACCGNNNNCGGT, are placed at nearly invariable distances in between the Sp1 site and the TATA box (Gloss and Bernard, 1990; Dong et al., 1993; Tan et al., 1994; Demeret et al., 1997) (Figure 1). This structure has to be understood as prerequisite for a feedback loop that reduces use of the E6 promoter to a minimum through the following mechanism: Many of the E6/E7 transcripts are polycistronic and include sequences transcribed from the E2 gene. Efficient expression of the E6/E7/E2 mRNA leads to high levels of E2 protein. The E2 protein can bind to the E2 sites in between Sp1 and the TATA box and replace Sp1 and TFIID. As E2 is a transcriptional activator, replacement of Sp1 retains E6 promoter function, however at a lower activity level than under the influence of Sp1. Displacement of TFIID abrogates any promoter function. The effect of this negative feedback loop is apparently one of several molecular prerequisites that explain the epidemiological observations that most HPV infections take a subclinical or latent course rather than inducing neoplasia.

4. The E2 Protein

The E2 protein has initially been studied in the bovine PV type 1 (BPV-1), where it binds clusters of sites remote from the promoter and turns these clusters into strong enhancers of the E6 promoter (McBride et al., 1991). No such E2 site clusters exist in the LCRs of alpha-PVs (Figure 1), and it is unclear whether the E2 factor functions in alpha-PVs as an enhancer activating protein. Surprisingly, when tested in appropriate "enhancer-vectors" in cell culture transfections, HPV E2 proteins are just as powerfully activating transcription factors as BPV-1 E2 proteins, and it is difficult to conceive that this factor should have a function that is not used during part of the viral life cycle, although this contradiction has been suggested by some experimental data (Stubenrauch et al., 1998). The principal candidate for activation by E2 in alpha-PVs is the E7 internal, putative "late" promoter (see below), whose regulation is still poorly understood. HPV-6 and 11, but not most of the other alpha-PVs, also have an E6 internal promoter for the expression of E7, and the induction of this and the E7 internal promoter in differentiating cells (diLorenzo and Steinberg, 1995; Rapp et al., 1997) may be based on stimulation by E2.

The transcriptional regulation by E2 is further complicated by the detection of short forms of the E2 protein, which have a DNA binding but no transcription activation domain and solely function as repressors (Stubenrauch et al., 2000 and 2001). In BPV-1, two different short E2 proteins originate from an E8^E2 splice and from an E2 internal promoter (McBride et al., 1991). The first form was also observed in HPVs (Stubenrauch et al., 2000), while the existence of the latter form can be deduced from cDNA sequences (Ozbun and Meyers, 1997). The balance between the full-size and these two short E2 proteins determines whether E2 sites preferentially activate or repress (see Chapter 6, for a fuller description of E2).

The E6 promoters of other PVs than those in the alpha-group have been little studied, but are certainly regulated in quite different ways, as they are normally lacking Sp1 and E2 sites. The beta-PVs are even lacking a proper TATA box, which is replaced by a nearly 50 bp long stretch of Ts and As, a truly exotic promoter structure (May et al., 1991).

The full-size E2 protein has another important function during the initiation of replication (Chiang et al., 1992), which we only want to touch upon briefly in this chapter on transcription. (For a fuller description of E1 and its role in viral DNA replication, see Chapter 5). The E1 protein is the principal initiator of PV replication. It has, however, very poor recognition capabilities for its target sequence, an A and T rich stretch upstream of the E6 promoter (Chen and Stenlund, 2001). The E1 protein can form a complex with the E2 protein, and binding of E2 to either of two E2 sites flanking the E1 sitehelps to target E1 to the proper replication origin.

5. Late Promoters

Polyomaviruses such as SV40 have a characteristic early-late switch of transcription. After infection, the early promoter becomes active under the influence of cellular factors, and leads to transcription and translation of the large T-antigen. When this protein accumulates in the infected cell, it binds to sequences overlapping its own promoter, shuts the early promoter down, and activates the late promoter, From which transcription proceeds in opposite direction to that of the early genes.

Papillomaviruses also have early genes, whose products are involved in transformation, replication, and transcription, and late genes, whose products make up the viral capsid. Transcription of papillomaviruses may happen in three stages, with little transcription detectable in undifferentiated basal cells, increasing transcription of early genes in suprabasal cells, and transcription of late genes increasing in cell layers closer to the epithelial surface (Stoler et al., 1992). While the basal-suprabasal shift from repressed viral genes to an activation of early genes is fairly well understood (see below), the shift from expression of the early to the late genes is still full of puzzles. It is almost certain that no precise switch exists that resembles that of polyomaviruses.

All alpha-PVs express mRNAs from their E6 promoters that also encode the late genes L1 and L2 (Baker and Calef, 1995; Ozbun and Meyers, 1997). Therefore, in principle, late proteins could be translated from the same mRNAs that lead to early proteins. Many of these mRNAs encode early genes upstream of the late genes, but this does not seem to interfere with translation. Most PV proteins are encoded by polycistronic mRNAs, and translation initiation of cistrons downstream of other genes appears not to be an obstacle for PVs as it is the case elsewhere in the world of eukaryotes and their viruses.

In addition to E6 promoter-L2/L1 transcripts, alpha-PVs have a promoter internal to the E7 gene. This promoter is well researched in HPV-31, where it is termed p847 (Ozbun and Meyers, 1997). Limited data are available about its homologue in HPV-11 (where it is called E1 promoter) (Ai et al., 1999) and HPV-16 (p670) (Grassmann et al., 1996). It is not yet understood how this promoter is activated, but as its activity appears to increase during differentiation (Ozbun and Meyers, 1998; Terhune et al., 1999), and since spliced mRNAs derived from its transcripts mostly encode L2 and L1 (and the E4 protein), it is tentatively considered to be the late promoter of alpha-PVs.

Papillomaviruses may have solved the mechanistic problem of late gene expression in a variety of ways, since HPV-8 and BPV-1 have late promoters elsewhere in the genome, namely in the 5′ part of the LCR, upstream of the E6 promoter and all early genes. This is particularly surprising, since it requires the generation of transcripts that include all early genes, which may then be removed by splicing to generate late mRNAs (Baker and Howley, 1987; Barksdale and Baker, 1995; Stubenrauch et al., 1996). HPV-16 and HPV-31 contain promoters in a similar position in the LCR, but their biological role is presently not understood (Ozbun et al., 1999; Tan et al., 2003).

6. The Epithelial Specific Enhancer

The E6 promoters of alpha-PVs have low activities, which can be boosted by an enhancer in the central LCR, about 400 bp 5′ of the promoter elements (Cripe et al., 1987; Gloss et al., 1987; Swift et al., 1987) (Figure 1). The strengths of these enhancers depend on the type of the infected epithelium and modulation by physiological stimuli. Due to variable conditions and divergent reporter constructs stimulation ratios of 10-1000 fold have been reported (Sailaja et al., 1999). These enhancers are specific for epithelial cells, *i.e.* show activity in all cell types that also express keratin genes, but are inactive elsewhere, and determine the epitheliotropism of HPVs. The influence of the epithelial specific enhancer on other promoters has not yet been systematically studied.

The enhancer of each alpha-PV type stretches over a genomic segment of 200-400 bp, which is tightly covered with binding sites that are recognized by a large variety of cellular transcription factors. Many publications from different research groups have dealt with the identification of these factors. It is difficult to do justice to the existing literature on these factors , as some data are agreed by all specialists, while there is disagreement elsewhere, and yet other factors were only studied by single groups. Some of these factors have been thoroughly studied in other organisms and viruses, permitting unequivocal identification, while other factors are not yet isolated and have been described primarily by gelshifts and effects of HPV enhancer mutations. It is certainly clear, however, that the function of HPV enhancers does not stem from a single factor, but from synergy between many different factors, and many discrepancies between the findings of different labs stem from the problems inherent to the dissection of multifactorial systems.

There is broad consensus that the factor AP-1 is quantitatively most important and central to the epithelial specific function of HPV enhancers (Garcia-Carranca et al., 1988; Chan et al., 1990; Sibbet and Campo, 1990; Chong et al., 1991; Dollard et al., 1993; Kyo et al., 1997; Zhao et al., 1999; Sen et al., 2004). AP-1 is intensely researched, and has been described to have a tremendous diversity of functions including some during epithelial differentiation, particularly due to the large variety of its heteromeric subunits of the Jun and Fos family (Angel et al., 2001). Different subunit-composition of AP-1 clearly also plays a role in the cell type and differentiation specific activity of HPV enhancers (Thierry et al., 1992; Kyo et al., 1997; Sen et al., 2004).

HPV enhancers can generally be recognized by sequence inspection, specifically by the search for clusters of five to seven TTGGCT/A sites, which function as half-sites for the normally palindromic targets of the factor NFI (Gloss et al., 1989; Chong et al., 1991; Apt et al., 1993). The overlap of these NFI sites with enhancer function, effects of mutations, and qualitative differences between the composition of NFI factors in epithelial and non-epithelial cells led our group to conclude that NFI is an important contributor to enhancer activity and epithelial specificity. The data from other laboratories confirmed only part of these finding (Butz et al., 1993; Parker et al., 1997; Zhao et al., 1997; Sen et al., 2004), possibly resulting from cell-type specificities and divergent chimeric constructs. Further research may be necessary to refine our knowledge of the exact impact of NFI on HPV biology.

Many HPV types that infect the genital tract contain binding sites for the glucocorticoid and progesterone receptors (Gloss et al., 1987; Pater et al., 1988; Chan et al., 1989; Bromberg-White et al., 2003). As a consequence, these hormones stimulate expression of the HPV oncogenes E6 and E7, and one could speculate, that physiological conditions that lead to an increase of these steroid hormones would increase the transformed state of an HPV infection and possibly the propensity for progression. Interestingly, these molecular findings correlate with the outcome of epidemiological studies indicating that use of anti-ovulants and multiparity, *i.e.* long-term increases of progesterone, are risk factors for the development of cervical neoplasia (Schiffman and Brinton, 1995).

Two transcription factors, PEF-1 (Cuthill and Campo, 1993; Sibbet et al., 1995; Fergusson and Campo, 1998) and KRF-1 (Mack et al., 1991), have so far only been described from binding sites in HPV enhancers, but have not been found to regulate transcription elsewhere. The published data include these factors' cis-responsive effects, gelshifts, and some properties of the proteins, while the proteins themselves or the genes encoding them are not yet available. The well-known factor Oct-1 binds the enhancers of different HPV types (O'Connor et al., 1995; Sibbet et al., 1995; Zhao et al., 1997; Sen et al., 2004), but the reported functional data are in part contradictory. Contributions to cell-type specificity are apparently made by TEF-1 (Ishiji et al., 1991) and the family of C-EBP factors (Sibbet and Campo, 1990; Sen et al., 2004), which are heterodimeric with tissue specific composition similar to AP-1 and NFI.

7. Silencers

A 110 bp segment between the enhancer and the E6 promoter does not contain binding sites for transcriptional activators. It contains the T and A rich sequence which binds the El protein and thereby functions as replication origin. Plasmids constructed to measure the activity of the E6 promoter, which contain this 110 bp segment in between the enhancer and promoter in its natural context, show much less transcriptional activity than chimeric constructs composed only of enhancer and promoter sequences with a deletion of this segment. It had to be concluded that this segment is not only a replication origin but represses transcription and fulfills the definition of a transcriptional silencer (Figure 1).

Dissection of this segment combined with gelshifts, supershifts, and point mutations identified the factors YY1 (Bauknecht et al., 1992; O'Connor et al., 1996; Ai et al., 1999) and CDP (Ai et al., 2000; O'Connor et al., 2000) as the repressors responsible for silencing transcription from the E6 promoter. Both factors have been extensively researched and are known as transcriptional repressors of a variety of cellular genes. Deletion of the silencer may be a tumor progression event, as natural mutants of HPV-16 that had lost the silencer segment, were found in cancer, but not in smears from asymptomatic patients (May et al., 1994).

At first glance, it looks counterintuitive that a viral promoter would have an enhancer that stimulates transcription by one to three orders of magnitude together with a silencer that represses its activity to a similar extent, as these two elements might mutually annihilate one another's functions. This contradiction became resolved by the observation that undifferentiated epithelial cells contain an excess of YY1 and CDP and thereby repress HPV transcription. The concentrations of YY1 and CDP decrease during differentiation, while the concentration of the activator AP-1 increases. Notably, CDP is known from a variety of tissues as a factor that represses certain tissue-specific genes in stem cells and permits their expression by its disappearance during differentiation (for references, see O'Connor et al., 2000). In summary, silencer and enhancer work sequentially and lead to a reduction of HPV gene products in undifferentiated cells and to a stimulation of HPV functions upon differentiation in suprabasal layers of epithelia (Ai et al., 1999 and 2000).

CDP binding sites are also clustered in the 5′ part of the LCR and in the E6 gene of some HPV types (Pattison et al., 1997; Stünkel and Bernard, 2000), and these genomic segments also score as silencers in transient tranfections of chimeric reporter constructs. Additional properties of these genomic segments are discussed below in the section about the binding of HPV genomes to the nuclear matrix.

8. Nucleosomal Organization and Regulation by Chromatin

The chromosomal DNA of eukaryotes and their viruses exists in situ in the form of chromatin, *i.e.* associated with histone octamers in the form of nucleosomes. This fact, which also applies to PV genomes (Rösl et al., 1983), has been known for a long time, but has been traditionally interpreted as a strategy to arrange the unwieldy DNA strings in a physically manageable form. In recent years, the notion emerged that nucleosomes are major tools to regulate gene expression. This is achieved by conformational changes of nucleosomes in response to covalent modification of histones by acetylation, deacetylation and methylation, activities which are associated with specific transcription activators and repressors (Wade et al., 1997).

As expected, HPV genomes are also nucleosomally organized, surprisingly however not in a random fashion but in a sequence specific manner, as found in the HPV-16 genomes of SiHa and CaSki cells. Detailed mapping of nucleosomes assembled *in vitro* on HPV-16 and HPV-18 located two nucleosomes specifically overlapping the E6 promoter elements and the center of the epithelial specific enhancer (Stünkel and Bernard, 1999). The regulatory relevance of this observation was supported by measurements of the transcriptional stimulation of the HPV-16 E6 promoter in response to the histone deacetylase (HDAC) inhibitor trichostatin A. Regulation of HPV transcription by chromatin is probably widespread, since the same compound also stimulated transcription of HPV-11 (Zhao et al., 1999).

The regulation of HPV transcription by chromatin is ultimately under the influence of the silencer and one of the cis-responsive elements of the enhancer. The repressors CDP and YY1 are associated with HDAC activities, and the activator AP-1 with an antagonist histone acetylase (HAT) (for references see Stünkel et al., 1999; Ai et al., 2000; Bernard, 2002). Notably, in HPV-16 and HPV-18, the silencer and one of the AP-1 sites of the enhancer are not included in the nucleosomal structure but positioned between the enhancer- and promoter-nucleosome and thereby freely accessible to these factors. In summary, the regulatory phenomena that we described in the section on the silencer, repression of the early genes of HPVs in undifferentiated cells and activation in differentiated epithelial layers are initiated by the factors CDP, YY1 and AP-1, but ultimately executed at the level of viral chromatin structure. The same mechanism appears to extend to the early-late switch, since the putative late (or E1) promoter is regulated by YY1 (Ai et al., 2000), and chromatin rearrangement occurs during the activation of transcription of the late genes (Klummp and Laimins, 1999; deMar-Pena and Laimins, 2001).

9. Methylation of HPV DNA

The DNA of mammals and their viruses can become methylated; methylation is stably maintained on the cytosine residues of CpG dinucleotides. CpG methylation can lead to transcriptional repression by a variety of mechanisms, including the binding of certain transcriptional repressors like meCP2. meCP2 forms complexes with HDACs with the same transcription inhibitory consequences triggered by CDP and YY1. DNA methylation plays important biological roles, for example in X-chromosome inactivation and genomic imprinting during embryogenesis (Bird, 1992; Burgers et al., 2002; Jaenisch and Bird, 2003), and during the aberrant events of oncogenesis, tumor suppressor genes can not only become inactivated by mutations but also by methylation of their promoters (Herman and Baylin, 2000; Jones, 2003). The recognition processes leading to methylation of specific DNA sequences are unknown, but it was observed that repeated sequences (Yoder et al., 1997) as well as foreign DNA recombining with cellular sequences (Sutter and Doerfler, 1980; Doerfler et al., 2001) trigger methylation events.

The methylation of HPV-1 and cottontail rabbit papillomavirus DNA has been reported long ago (Sugarawa et al., 1983; Wettstein and Stevens, 1983; Burnett and Sleeman, 1984), and this observation was recently extended to alpha HPVs (Badal et al., 2003; Kim et al., 2003; Badal et al., 2004, Kalantari et al. 2004). Heavily methylated HPV-16 DNA is clearly transcriptionally inactive (Rösl et al., 1993; Badal et al., 2003). While these observations are still very recent and need extensive further investigations, the published data support the following hypotheses: Upon infection of undifferentiated epithelial cells, HPV DNA is recognized by the cellular de novo methylation machinery and becomes transcriptionally repressed, a mechanism that may be important for subclinical viral infection. It may induce a state of latent replication, since the replication origin normally remains unmethylated and therefore functional. Once viral genomes escape from methylation through replication during epithelial differentiation, they become transcriptionally active and induce transformation and a full viral life cycle. Upon integration of HPV DNA into cellular DNA during tumorigenesis, many viral genomes become methylated and transcriptionally repressed, but only cells that retain transcriptionally active genomes grow into tumors.

These processes can be interpreted as cellular defenses as well as properties selected during HPV evolution, as they favor the subclinical maintenance and spread of HPVs, providing another likely example of the set of properties that make PVs often commensals rather than pathogens, a hallmark of this virus family.

10. Nuclear Matrix Attachment

The nuclear matrix is a fibrous nuclear structure, whose physical properties are not based on the filamentous nature of DNA and which is structurally separate from the chromatin. Purification of the nuclear matrix has allowed the identification of cellular or viral DNA segments, called nuclear matrix attachment regions (MARs), that have a preferential affinity for the matrix. , . The nuclear matrix appears to organize the nucleus into compartments with specific functions, such as replication, transcription, or splicing (for references, see Tan et al., 1998). HPV genomes adhere to the nuclear matrix through specific MARs, which are in the case of HPV-16 the 5′ part of the LCR, the center of the E6 gene, and the early-late intergenic region (Tan et al., 1998). The MARs contain CDP binding sites and are therefore silencers. Surprisingly, however, when integrated into cellular DNA, they activate viral transcription through a yet unknown mechanism (Stünkel and Bernard, 2000).

Cell lines derived from cervical cancers and *in vitro* modified cell lines, which contain numerous HPV genomes, activate HPV transcription only when the viral genomes reside in a defined nuclear position close to the nucleoli, while viral genomes elsewhere are inactive (van Tine et al., 2003; 2004). In the case of CaSki cells, which contain 500 HPV-16 copies, the active HPV genome(s) is likely unmethylated, while the transcriptionally repressed viral genomes, the vast majority of the total load of viral genomes, are methylated (Badal et al., 2003). It is intriguing to combine all these observations into a hypothesis, that HPV genomes are targeted by methylation activities in many parts of the nucleus, but become or remain demethylated in a selected compartment, and that a sorting process exists that is based on the nuclear matrix and the viral MARs to guide HPV genomes into the respective nuclear location.

11. Recombination With Cellular DNA

In carcinomas, the genomes of high-risk HPVs are normally integrated into the cellular DNA (Daniel et al., 1995), a phenomenon that has never been observed in common warts or genital or laryngeal lesions induced by HPV-6 or 11. The explanations in previous sections of this chapter and additional published data suggest that four different mechanisms target HPV genomes as consequences of recombination with cellular DNA. (1) The MAR associated enhancer stimulates E6 and E7 oncogene expression, and cells with integrated HPV genomes will likely have a growth advantage over cells with only episomal HPV DNA. (2) After loss of episomal HPV genomes, the E6 promoter is not repressed by the HPV E2 protein, if the recombination event occurred in the E1 or E2 genes, placing the E2 gene upstream of the LCR in the linearized HPV genome in a transcriptionally silent or dysregulated environment (Schwarz et al., 1985). (3) E6/E7 transcripts derived from integrated copies show increased stability (Jeon and Lambert, 1995). (4) HPV genomes are targeted by DNA methylation and transcription is repressed. There is no evidence that it is an inherent strategy of HPVs to induce malignant lesions, and one has to conclude that the fortuitous combination of these four events, the first three favoring, the fourth reducing, oncogene expression and therefore neoplastic progression, determines the fate of a high-risk HPV lesion.

12. Transcriptional Termination

All HPV genomes have two transcription termination signals, one between the E2/E5 and the L2 gene (the early-late intergenic region), and one downstream of the L1 gene. cDNA libraries (Baker and Calef, 1995; Ozbun and Meyers, 1997) confirm that both signals are used and are giving rise to the only 3′ termini observed in HPV mRNAs. In the early-late intergenic transcription termination region, a AATAAA polyadenylation signal is linked to three sites recognized weakly by the cleavage stimulatory factor (CstF), and the combination of these sites leads in undifferentiated cells to a series of early mRNAs with heterogenous 3′ termini before the polyA tail. The AATAAA site downstream of L1 is linked to a strong CstF site, leading to a narrow spectrum of late mRNAs in differentiated cells. Read-through from the early to the late genes is permitted by the weak binding of CstF to the three sites in this position, and further facilitated by a decrease of CstF activity upon differentiation. It is even further modulated by yet another element within the L2 gene (Terhune et al., 1999 and 2001). The early-late switch of HPVs is clearly a multifaceted mechanism, and does not only amount to a shift in promoter use, but also to regulation of these termination events and differential splicing and mRNA stability (see below).

13. Translation of Polycistronic mRNAs

Papillomaviruses are unique among eukaryotes and their viruses in that they generate exclusively polycistronic mRNAs that can express multiple proteins from tandemly arranged open reading frames, with highly efficient translation of genes that are encoded downstream of others, such as in the case of the E6-E7 transcripts. In some of these cases, the two linked genes are spaced by some few nucleotides, in others the translation initiation codon of the "downstream-gene" is even within the "upstream-gene". The mechanisms by which this is achieved are unknown and are at odds with the general scanning model of eukaryotic translation, which predicts that the ribosome recognizes the cap structure of the mRNA and then uses the first AUG codon 3′ of the cap for translation initiation (Kozak, 1999). Surprisingly, very few research efforts have addressed this molecularly enigmatic phenomenon. Sequence inspection does not reveal complex cloverleaf structures as they occur in picorna virus RNA that may establish an unconventional translation initiation mechanism. A detailed investigation of the E6-E7 mRNA by Stacey and coll. (2000) came to the conclusion that the translation of the E7 gene requires the cap in front of the E6 gene, and that the AUG of E7 is specifically recognized after negotiating 13 AUGs within the E6 gene. A subsequent study indicated that this "leaky scanning" mechanism does encounter these AUGs as obstacles, as shortening the distance between cap and E7 AUG through the E6* splice led to a significant increase in E7 translation (Zheng et al., 2004).

14. Differential Splicing and Transcript Stability

The diversity of HPV mRNAs has been most thoroughly studied in the case of HPV-31 (Ozbun and Meyers, 1997), where 24 different mRNAs were identified that either started at the E6 promoter, or at the E7 internal ("late") promoter, or at a position within the E2 gene, and terminated either in the early-late intergenic region or downstream of L1. This study is likely representative for all alpha-PVs, as mRNA populations with a similar composition were reported from other HPV types (for a review, see Baker and Calef, 1995).

Aside from differential usage of promoters and termination sites, the diversity of HPV mRNAs originates from differential splicing of the primary transcripts. Most notable are three types of splicing events that (i) occur within the E6 gene (E6* and E6** splices), (ii) fuse the N-terminus of the E1 gene to the E4 open reading frame, and (iii) eliminate the L2 gene and fuse some of the early genes (E6, E7 and E4 or a subset of these) to the L1 gene. The first of these splicing reactions stimulates E7 expression (Zheng et al., 2004), although details of the mechanism and the need for this unconventional reaction are not understood. The objectives of the second and the third splicing event are apparently part of the early-late switch. They seem to be prerequisites for the expression of the E4 protein, which, although not being part of the viral capsid, has become reinterpreted as a late protein (Doorbar et al., 1991), and to assure an excess of the L1 protein, the principal capsid protein, over L2, a minor component of viral particles. Alternative splicing has also been proposed to regulate the relative concentrations of the E1 and E2 proteins, which may alter the viral genomic copy number during differentiation (Deng et al., 2003).

It is likely that the divergent coding capacity of HPV mRNAs requires a tight regulation of the overall composition and the sequential availability of these mRNAs. HPVs have chosen quite diverse and unusual mechanisms to assure an appropriate outcome, which does not only include transcription initiation, termination, and splicing (Zheng et al., 1996), but even the regulation of mRNA stability, which can be affected by elements within the L1 gene (Collier et al., 2002), the L2 gene (Oberg et al., 2003), and in the untranslated region downstream of L1 (Koffa et al., 2000; McPhillips et al., 2004). In the latter case, defined segments downstream of the L1 genes of HPV-16 and HPV-31 have been found to contain splicing donor sites and to bind a variety of factors including a subunit of an auxiliary splicing factor. While these elements do not appear to trigger a splicing reaction, they affect transcript stability in a differentiation dependent manner (see Chapter 13). The temporal effect of splicing on late gene expression has not yet been studied in alpha-PVs, but may be deduced from similar phenomena in BPV-1 (Barksdale and Baker, 1995).

15. Summary

It was the objective of this chapter to summarize the tremendous progress of the understanding of HPV gene expression during the last two decade, beginning in the mid-1980s with the detection of the transcription factor properties of the E2 protein in BPV-1 and an epithelial specific enhancer, activated by cellular proteins, in HPV-16 and HPV-18. The reader, who is not a specialist in transcription research, will be impressed by the complexity of gene expression in a small DNA virus with only eight or nine open reading frames. Specialists should be reminded that in spite of all progress we are still remote from a complete understanding of some of the most important transcriptional aspects of HPVs. In basic research, we have still incomplete insight into the epithelial specificity of HPV transcription, and even only a rudimentary comprehension of the early-late switch. Some of the most important clinical questions, such as the nature of latent infection, the progression of latency to precursor lesions and malignancy, and the preferential occurrence of lesions at specific anatomic sites (transformation zone of the cervix) are likely to be answered – at least in part – by the regulation of the HPV oncogene expression, and provide experimental challenges for the future.

Acknowledgements

This research was supported by NIH grant ROI CA-91964 and by funds from the Chao Family Comprehensive Cancer Center of the University of California Irvine to H.U.B. and M.K.

References

Ai, W., Toussaint, E., and Roman, A. (1999). CCAAT displacement protein binds to and negatively regulates human papillomavirus type 6 E6, E7, and E1 promoters. J. Virol. *73*, 4220-4229.

Ai, W., Narahari, J., and Roman, A. (2000). Yin yang 1 negatively regulates the differentiation-specific E1 promoter of human papillomavirus type 6. J. Virol. *74*, 5198-5205.

Angel, P., Szabowski, A., and Schorpp-Kistner, M. (2001). Function and regulation of AP-1 subunits in skin physiology and pathology. Oncogene *20*, 2413-2423.

Apt, D., Chong, T., Liu, Y., and Bernard, H.U. (1993). Nuclear factor I and epithelial cell specific transcription of human papillomavirus type 16. J. Virol. *67*, 4455-4463.

Apt, D., Watts, R.M., Suske, G, and Bernard, H.U. (1996). High Sp1/Sp3 ratios in epithelial cells during epithelial differentiation and cellular transformation correlate with the activation of the HPV-16 promoter. Virology *224*, 281-291.

Badal, S., Badal, V., Calleja-Macias, I.E., Kalantari, M., Chuang, L.S.H., Li, B.F.L., Bernard, H.U. (2004). The human papillomavirus-18 genome is efficiently targeted by cellular DNA methylation. Virology *324*, 483-492.

Badal, V., Chuang, L.S.H., Badal, S., Tang, E., Villa, L.L., Wheeler, C.M., Li, B.F.L., and H.U. Bernard. (2003). CpG methylation of human papillomavirus-16 DNA in cervical cancer cell lines and in clinical specimens: genomic hypomethylation correlates with carcinogenic progression. J. Virol. *77*, 6227-6234.

Baker, C.C., and Howley, P.M. (1987). Differential promoter utilization by the bovine papillomavirus in transformed cells and productively infected wart tissues. EMBO J. *6*, 1027-1035.

Baker, C.C. and Calef. C. Maps of papillomavirus transcripts. In: Human papillomaviruses 1995 compendium. Edited by Myers, G., Bernard, H.U., Delius, H., Baker, C.C., Icenogle, J., Halpern, A., and Wheeler, C. Los Alamos National Laboratory, Los Alamos, New mexico, part III-A, pp 3-19.

Barksdale, S., and Baker, C.C. (1995). Differentiation-specific alternative splicing of bovine papillomavirus late mRNAs. J. Virol. *69*, 6553-6556.

Bauknecht, T., Angel, P., Royer, H.D., and zur Hausen, H. (1992). Identification of a negative regulatory domain in the human papillomavirus 18 promoter: interaction with the transcriptional repressor YY1. EMBO J. *11*, 4607-4617.

Bernard, H.U. (2002). Gene Expression of Genital Human Papillomaviruses and Potential Antiviral Approaches. Antiviral Therapy *7*, 219-237.

Bird, A. P. (1992). The essentials of DNA methylation. Cell *70*, 5-8.

Bromberg-White, J.L., Sen, E., Alam, S., Bodily, J.M., and Meyers, C. (2003). Induction of the upstream regulatory region of human papillomavirus type 31 by dexamethasone is differentiation dependent. J. Virol. *77*, 10975-10983.

Burgers, W.A., Fuks, F., and Kouzarides, T. (2002). DNA methyltransferases get connected to chromatin. Trends in Genetics 18: 275-277.

Burnett, T.S., and J.P. Sleeman. 1984. Uneven distribution of methylation sites within the human papillomavirus la genome: possible relevance to viral gene expression. Nucleic Acids Res. 12: 8847-8860.

Butz, K., and Hoppe-Seyler, F. (1993). Transcriptional control of human papillomavirus (HPV) oncogene expression: composition of the HPV type 18 upstream regulatory region. J. Virol. *67*, 6476-6486.

Chan, W.K., Klock, G., and H.U. Bernard. (1989). Progesterone and glucocorticoid response elements occur in the long control regions of several human papillomaviruses involved in anogenital neoplasia. J. Virol. *63*, 3261-3269.

Chan, W.K., Chong, T., Bernard, H.U., and Klock, G. (1990). Two AP1 sites in the long control region of human papillomavirus type 16 lead to phorbolester stimulation of the viral E6/E7 promoter. Nucleic Acids Res. *18*, 763-769.

Chen, G., and Stenlund, A. (2001). The E1 initiator recognizes multiple overlapping sites in the papillomavirus origin of DNA replication. J. Virol. *75*, 292-302.

Chiang, C.M., Ustav, M., Stenlund, A., Ho, T.F., Broker, T.R., and Chow, L.T. (1992). Viral E1 and E2 proteins support replication of homologous and heterologous papillomaviral origins. Proc. Natl. Acad. Sci. USA. *89*, 5799-5803.

Chong, T., Apt, D., Gloss, B., Isa, M., Bernard, H.U. (1991). The enhancer of human papillomavirus-16: Binding sites for the ubiquitous transcription factors oct-*1*, NFA, TEF-2, NFI and AP1 participate in the epithelial specific transcription. J.Virol. 65:5933-5943.

Collier, B., Oberg, D., Zhao, X., and Schwartz, S. (2002). Specific inactivation of inhibitory sequences in the 5' end of the human papillomavirus type 16 L1 open reading frame results in production of high levels of L1 protein in human epithelial cells. J. Virol. *76*, 2739-2752.

Cripe, T.P., Haugen, T.H., Turk, J.P., Tabatabai, F., Schmid, P.G., Dürst, M., Gissmann, L., Roman, A., and Turek, L.P. (1987). Transcriptional regulation of the human papillomavirus-16 E6-E7 promoter by a keratinocyte-dependent enhancer, and by viral E2 transactivator and repressor gene products: implications for cervical carcinogenesis. EMBO J. *6*, 1339-1344.

Cuthill, S., Sibbet, G.J., and Campo, M.S. (1993). Characterization of a nuclear factor, papilloma enhancer binding factor-*1*, that binds the long control region of human papillomavirus type 16 and contributes to enhancer activity. Mol. Carcinog. *8*, 96-104.

Daniel, B., Mukherjee, G., Seshadri, L., Vallikad, E., and Krishna, S. (1995). Changes in the physical state and expression of human papillomavirus type 16 in the progression of cervical intraepithelial neoplasia lesions analyzed by PCR. J. Gen. Virol. *76*, 2589-2593.

DelMar-Pena, L.M., and Laimins, L.A. (2001). Differentiation-dependent chromatin rearrangement coincides with activation of human papillomavirus type 31 late gene expression. J. Virol. *75*, 10005-10013.

Demeret, C., Desaintes, C., Yaniv, M., and Thierry, F. (1997). Different mechanisms contribute to the E2 mediated transcriptional repression of human papillomavirus type 18 viral oncogenes. J. Virol. *71*, 9343-9349.

Deng, W., Jin, G., Lin, B.Y., van Tine, B.A., Broker, T.R., and Chow, L.T. (2003). mRNA splicing regulates human papillomavirus type 11 E1 protein production and DNA replication. J. Virol. *77*, 10213-10226.

de Villiers, E.M., Fauquet, C., Broker, T.R., Bernard, H.U., and zur Hausen, H. (2004). Classification of papillomaviruses. Virology, *324*, 17-27.

DiLorenzo, T.P., and Steinberg, B.M. (1995). Differential regulation of human papillomavirus type 6 and 11 early promoters in cultured cells derived from laryngeal papillomas. J. Virol. *69*, 6865-6872.

Doerfler, W., Remus, R., Muller, K., Heller, H., Hohlweg, U., and Schubbert, R. (2001). The fate of foreign DNA in mammalian cells and organisms. Dev. Biol. *106*, 89-97.

Dollard, S.C., Broker, T.R., and Chow, L.T. (1993). Regulation of the human papillomavirus type 11 E6 promoter by viral and host transcription factors in primary human keratinocytes. J. Virol. *67*, 1721-1726.

Dong, G., Broker, T.R., and Chow, L.T. (1994). Human papillomavirus type 11 E2 proteins repress the homologous E6 promoter by interfering with the binding of host transcription factors to adjacent elements. J. Virol. *68*, 6411-6422.

Doorbar, J., Ely, S., Sterling, J., McLean, C., and Crawford, L. (1991). Specific interaction between HPV-16 E1-E4 and cytokeratins results in collapse of the epithelial cell intermediate filament network. Nature *352*, 824-827.

Fergusson, D., and Campo, M.S. (1998). PEF-*1*, an epithelial cell transcription factor which activates the long control region of human papillomavirus type *16*, is glycosylated with N-acetylglucosamine. J. Gen. Virol. *79*, 2753-2760.

Garcia-Carranca, A., Thierry, F., and Yaniv, M. (1988). Interplay of viral and cellular proteins along the long control region of human papillomavirus type 18. J. Virol. *62*, 4321-4330.

Gloss, B., Bernard, H.U., Seedorf, K., and Klock, G. (1987). The upstream regulatory region of the human papillomavirus-16 contains an E2 protein independent enhancer which is specific for cervical carcinoma cells and regulated by glucocorticoid hormones. EMBO J. *6*, 3735-3743.

Gloss, B., Yeo-Gloss, M., Meisterernst, M., Rogge, L., Winnacker, E.L., Bernard, H.U. (1989). Clusters of nuclear factor I binding sites identify enhancers of severa papillomaviruses but alone are not sufficient for enhancer function. Nucleic Acids Res. *17*, 3519-3533.

Gloss, B., and Bernard, H.U. (1990). The E6/E7 promoter of human papillomavirus type 16 is activated in the absence of E2 proteins by a sequence-aberrant Sp1 distal element. J. Virol. *64*, 5577-5584.

Grassmann, K., Rapp, B., Maschek, H., Petry, K.U., and Iftner, T. (1996). Identification of a differentiation-inducible promoter in the E7 open reading frame of human papillomavirus type 16 (HPV-16) in raft cultures of a new cell line containing high copy numbers of episomal HPV-16 DNA. J. Virol. *70*, 2339-2349.

Hampsey, M, and Reinberg, D. (1999). RNA polymerase II as a control panel for multiple coactivator complexes. Curr. Opin. Genes Dev. *9*,132-139.

Herman, J.G., and Baylin, S.B. (2000). Promoter-region hypermethylation and gene silencing in human cancer. In: Jones, P.A. and Vogt, P.K. (eds.) DNA methylation and cancer. Springer, Berlin, pp.35-54.

Iftner, T., Oft, M., Bohm, S., Wilczynski, S.P., and Pfister, H. (1992). Transcription of the E6 and E7 genes of human papillomavirus type 6 in anogenital condylomata is restricted to undifferentiated cell layers of the epithelium. J Virol. *66*, 4639-4646.

Ishiji, T., Lace, M.J., Parkkinen, S., Anderson, R.D. Haugen, T.H., Cripe, T.P., Xiao, J.H., Davidson, I., Chambon, P., and Turek, L. P. (1991). Transcriptional enhancer factor (TEF)-1 and its cell-specific co-activator activate human papillomavirus-16 E6 and E7 oncogene transcription in keratinocytes and cervical carcinoma cells. EMBO J. *11*, 2271-2281.

Jaenisch, R., and Bird, A. (2003). Epigenetic regulation of gene expression: how the genome integrates intrinsic and environmental signals. Nature Gen. *33*, 245-254.

Jeon, S., and Lambert, P.F. (1995). Integration of human papillomavirus type 16 DNA into the human genome leads to increased stability of E6 and E7 mRNAs: implications for cervical carcinogenesis. Proc. Natl. Acad. Sci. USA. *92*, 1654-1658.

Jones, P.A. (2003). Epigenetics in carcinogenesis and cancer prevention. Ann. N. Y. Acad. Sci. *983*, 213-219.

Kalantari, M., Calleja-Macias, I.E., Tewari, D., Hagmar, B., Barrera-Saldana, H.A, Wiley, D.J., and Bernard, H.U. (2004). Conserved methylation patterns of human papillomavirus-16 DNA in asymptomatic infection and cervical neoplasia. J. Virol. *78*, 12762-12772.

Kim, K., Garner-Hamrick, P.A., Fisher, C., Lee, D., and Lambert, P.F. (2003). Methylation patterns of papillomavirus DNA, its influence on E2 function, and implications in viral infection. J. Virol. 77:12450-12459.

Klummp, D.J., and Laimins, L.A. (1999). Differentiation-induced changes in promoter usage for transcripts encoding the human papillomavirus type 31 replication protein E1.

Virology *257*, 239-246.

Koffa, M.D., Graham, S.V, Takagaki, Y, Manley, J.L., and Clements, J.B. (2000) The human papillomavirus type 16 negative regulatory RNA element interacts with three proteins that act at different posttranscriptional levels. Proc. Natl. Acad. Sci. USA *97*, 4677-4682.

Kozak, M. (1999). Initiation of translation in prokaryotes and eukaryotes. Gene *234*, 187-208.

Kyo, S., Klummp, D.J., Inoue, M., Kanaya, T., and Laimins, L.A. (1997). Expression of AP1 during cellular differentiation determines human papillomavirus E6/E7 expression in stratified epithelial cells. J. Gen. Virol. *78*, 401-411.

Mack, D.H., and Laimins, L.A. (1991). A keratinocyte-specific transcription factor, KRF-*1*, interacts with AP-1 to activate expression of human papillomavirus type 18 in squamous epithelial cells. Proc. Natl. Acad. Sci. USA *88*, 9102-9106.

May, M., Helbl, V., Pfister, H., and Fuchs, P.G. (1991). Unique topography of DNA-protein interactions in the non-coding region of epidermodysplasia verruciformis-associated human papillomaviruses. J. Gen. Virol. *72*, 2989-2997.

May, M., Dong, X.P., Beyer-Finkler, E., Stubenrauch, F., Fuchs, P.G., and Pfister, H. (1994). The E6/E7 promoter of extrachromosomal HPV16 DNA in cervical cancers escapes from cellular repression by mutation of Target sequences for YY1. EMBO J. *13*, 1460-1466.

McBride, A.A., Romancsuk, H., and Howley, P.M. (1991). The papillomavirus E2 regulatory proteins. J. Biol. Chem. *266*, 18411-18414.

Mc Phillips, M.G., Veerapradisin, T., Cumming, S.A., Karali, D., Milligan, S.G., Boner, W., Morgan, I.M., and Graham, S.V. (2004) SF2/ASF binds the human papillomavirus type 16 late RNA control element and is regulated during differentiation of virus-infected epithelial cells. J. Virol. *78*, 10598-10605.

Morgan, I.M., Grindlay, G.J., and Campo, M.S. (1999). The bovine papillomavirus type 4 long control region contains an epithelial specific enhancer. J. Gen. Virol. *80*, 23-27.

Oberg, D., Collier, B., Zhao, X., and Schwartz, S. (2003). Mutational inactivation of two distinct negative RNA elements in the human papillomavirus type 16 L2 coding region induces production of high levels of L2 in human cells. J. Virol. *77*,11674-11684.

O' Connor, M.J., and Bernard, H.U. (1995). Oct-1 activates the epithelial specific enhancer of human papillomavirus type 16 via a synergistic interaction with NFI at a conserved composites regulatory element. Virology *207*, 77-88.

O' Connor, M.J., Tan, S.H., Tan, C.H., and Bernard, H.U. (1996). YY1 represses human papillomavirus type 16 transcription by quenching AP-1 activity. J. Virol. *70*, 6529-6539.

O' Connor, M.J., Stünkel, W., Koh, C.H., Zimmermann, H., and Bernard, H.U. (2000).

The differentiation-specific factor CDP/Cut represses transcription and replication of Human papillomaviruses. J. Virol. *74*, 401-410.

Ozbun, M.A., and Meyers, C. (1997). Characterization of late gene transcripts expressed during vegetative replication of human papillomavirus type 31b. J. Virol. *71*, 5161-5172.

Ozbun M.A., and Meyers, C. (1998). Temporal usage of multiple promoters during the life cycle of human papillomavirus 31b. J. Virol. *72*, 2715-2722.

Ozbun, M.A. and Meyers, C. (1999). Two novel promoters in the upstream regulatory region of human papillomavirus type 31b are negatively regulated by epithelial differentiation. J. Virol. *73*, 3505-3510.

Parker, J.N., Zhao, W., Askins, K.J., Broker, T.R., and Chow, L. T. (1997). Mutational analysis of differentiation-dependent human papillomavirus-18 enhancer elements in epithelial raft cultures of neonatal foreskin keratinocytes. Cell Growth Diff. *8*, 751-762.

Pater, M.M., Hughes, G.A., Hyslop, D.E., Nakshatri, H., and A. Pater. (1988). Glucocorticoid-dependent oncogenic transformation by type 16 but not type 11 human papilloma virus DNA. Nature *335*, 832-825.

Pattison, S., Skalnik, D.G., and Roman, A. (1997). CCAAT displacement protein, a regulator of differentiation-specific gene expression, binds a negative regulatory element within the 5′ end of the human papillomavirus type 6 log control region. J. Virol. *71*, 2013-2022.

Rapp, B., Pawellek, A., Kraetzer, F., Schaefer, M., May, C., Purdie, K., Graessmann, K., and Iftner, T. (1997). Cell-type-specific separate regulation of the E6 and E7 promoters of human papillomavirus type 6a by the viral transcription factor E2. J. Virol. *71*, 6956-6966.

Rösl, F., Waldeck, W., and Sauer, G. (1983). Isolation of episomal bovine papillomavirus chromatin and identification of a DNase I-hypersensitive region. J. Virol. *46*, 567-574.

Rösl, F., Arab, A., Klevenz, B., and H. zur Hausen. (1993). The effect of DNA methylation on gene regulation of human papillomaviruses. J. Gen.Virol. *74*, 791-801.

Sailaja, G., Watts, R.M., and Bernard, H.U. (1999). Many different papillomaviruses have low transcriptional activity in spite of strong epithelial specific enhancers. J. Gen. Virol. *80*, 1715-1724.

Schiffman, M.H., and Brinton, L.A. (1995). The epidemiology of cervical carcinogenesis. Cancer *76*, 1888-1901.

Schwarz, E., Freese, U.K., Gissmann, L., Mayer, W., Roggenbuck, B., Stremlau, A., and zur Hausen, H. (1985). Structure and transcription of human papillomavirus sequences in cervical carcinoma cells. Nature *314*, 111-114.

Sen, E., Alam, S., and Meyers, C. (2004). Genetic and biochemical analysis of cis regulatory elements within the keratinocyte enhancer region of the human papillomavirus type 31 upstream regulatory region during different stages of the viral life cycle. J. Virol. *78*, 612-629.

Sibbet, G.J., and Campo, M.S. (1990). Multiple interactions between cellular factors and the non-coding region of human papillomavirus type 16. J. Gen. Virol. *71*, 2699-2707.

Sibbet, G.J., Cuthill, S., and Campo, M.S. (1995). The enhancer in the long control region of human papillomavirus type 16 is up-regulated by PEF-1 and down-regulated by Oct-1. J. Virol. *69*, 4006-4011.

Stacey, S.N., Jordan, D., Williamson, A.J., Brown, M., Coote, J.H., and Arrand, J.R. (2000). Leaky scanning is the predominant mechanism for translation of human papillomavirus type 16 E7 oncoprotein from E6/E7 bicistronic mRNA. J Virol. *74*, 7284-7297.

Stoler, M.H., Rhodes, C.R., Whitbeck, A., Wolinsky, S.M., Chow, L.T, and Broker, T.R. (1992). Human papillomavirus type 16 and 18 gene expression in cervical neoplasias. Hum Pathol. *23*, 117-128.

Stubenrauch, F., Leigh, I.M., and Pfister, H. (1996). E2 represses the late gene promoter of human papillomavirus type 8 at high concentration by interfering with cellular factors. J. Virol. *70*, 119-126.

Stubenrauch, F., Colbert, A.M., and Laimins, L.A. (1998). Transactivation by the E2 protein of oncogenic human papillomavirus type 31 is not essential for early and late viral functions. J. Virol. *72*, 8115-8123.

Stubenrauch, F., Hummel, M., Iftner, T., and Laimins, L.A. (2000). The E8E2C protein, a negative regulator of viral transcription and replication, is required for extrachromosomal maintenance of human papillomavirus type 31 in keratinocytes. J. Virol. *74*, 1178-1186.

Stubenrauch, F., Zobel, T., and Iftner, T. (2001). The E8 domain confers a novel long-distance transcriptional repression activity on the E8E2C protein of high-risk human papillomavirus type 31. J. Virol. *75*, 4139-4149.

Stünkel, W., Huang, Z., Tan, S.H., O□□Connor, M, and Bernard, H.U. (2000). Nuclear matrix attachment regions of human papillomavirus-16 repress or activate the E6 promoter depending on the physical state of the viral DNA. J. Virol. *74*, 2489-2501.

Stünkel, W., and Bernard, H.U. (1999). The chromatin structure of the long control region of human papillomavirus type 16 represses viral oncoprotein expression. J. Virol. *73*, 1918-1930.

Sugawara, K., Fujinaga, K., Yamashita, T., and Ito, Y. (1983). Integration and methylation of shope papilloma virus DNA in the transplantable Vx2 and Vx7 rabbit carcinomas. Virology *131*, 88-99.

Sutter, D., and Doerfler, W. (1980). Methylation of integrated adenovirus type 12 DNA sequences in transformed cells is inversely correlated with viral gene expression. Proc. Natl. Acad. Sci. USA *77*, 253-256.

Swift, F.V., Bhat, K., Younghusband, H.B., and Hamada, H. (1987). Characterization of a cell type-specific enhancer found in the human papilloma virus type 18 genome. EMBO J. *6*, 1339-1344.

Tan, S.H., Leong, L. E.C., Walker, P.A., and Bernard, H.U. (1994). The human papillomavirus type 16 transcription factor E2 binds with low cooperativity to two flanking binding sites and represses the E6 promoter through displacement of Sp1 and TFIID. J. Virol. *68*, 6411-6420.

Tan, S.H., Bartsch, D., Schwarz, E., and Bernard, H.U. (1998). Nuclear Matrix Attachment Regions of Human Papillomavirus Type 16 Point toward Conservation of these Genomic Elements in all Genital Papillomaviruses. J. Virol. *72*, 3610-3622.

Tan, S.H., Baker, C.C., Stünkel, W., and Bernard, H.U. (2003). A transcriptional initiator overlaps with a conserved YY1 site in the long control region of human papillomavirus type 16. Virology, *305*, 486-501.

Terhune, S.S., Milcarek, C., and Laimins, L.A. (1999). Regulation of human papillomavirus type 31 polyadenylation during the differentiation-dependent life cycle. J. Virol. *73*, 7185-7192.

Terhune, S.S., Hubert, W.G., Thomas, J.T., and Laimins, L.A. (2001). Early polyadenylation signals of human papillomavirus type 31 negatively regulate capsid gene expression. J. Virol. *75*, 8147-8157.

Thierry, F., Spyrou, G., Yaniv, M., and Howley, P. (1992). Two AP1 sites binding JunB are essential for human papillomavirus type 18 transcription in keratinocytes. J. Virol. *66*, 3740-3748.

Van Tine, B.A., Knops, J., Broker, T.R., Chow, L.T., and Moen, P.T. (2001). In situ analysis of the transcriptional activity of integrated viral DNA using tyramide-FISH. Dev Biol (Basel)106:381-385.

Van Tine, B.A., Kappes, J.C., Banerjee, N.S., Knops, J., Lai, L., Steenbergen, R.D.M., Meyer, C.L.J.M., Sniejders, P.J.F., Chatis, P., Broker, T.R., Moen, P.T., and Chow, L.T. (2004). Clonal selection for transcriptionally active viral oncogenes during progression to cancer. J. Virol. *78*, 11172-11186.

Wade, P.A., Pruss, D., and Wolffe, A.P. (1997). Histone acetylation: chromatin in action. Trends Biochem Sci. *22*, 128-132.

Wettstein, F.O., and Stevens, J.G. (1983). Shope papilloma virus DNA is extensively methylated in non-virus-producing neoplasms. Virology *126*, 493-504.

Yoder, J.A., Walsh, C. P., and Bestor, T.H. (1997). Cytosine methylation and the ecology of intragenomic parasites. Trends Gen. *13*, 335-340.

Zhao, W., Chow, L.T., and Broker, T.R. (1997). Transcription activities of human papillomavirus type 11 E6 promoter-proximal elements in raft and submerged cultures of foreskin keratinocytes. J. Virol. *71*, 8832-8840.

Zhao, W., Noya, F., Chen, W.Y., Townes, T.M., Chow, L.T., and Broker, T.R. (1999). Trichostatin A up-regulates human papillomavirus type 11 upstream regulatory region-E6 promoter activity in undifferentiated primary human keratinocytes. J. Virol. *73*, 5026-5033.

Zheng, Z.M., He, P., and Baker, C.C. (1996). Selection of the bovine papillomavirus type 1 nucleotide 3225 3′ splice site is regulated through an exonic splicing enhancer and its juxtaposed exonic splicing suppressor. J. Virol. *70*, 4691-4699.

Zheng ZM, Tao M, Yamanegi K, Bodaghi S, and Xiao W. (2004). Splicing of a cap-proximal human papillomavirus 16 E6-E7 intron promotes E7 expression, but can be restrained by distance of the intron from its RNA 5′ cap. J. Mol. Biol. *337*, 1091-108.

From: Papillomavirus Research: From Natural History to Vaccines and Beyond. Edited by: M. Saveria Campo

Chapter 5

Mechanisms and Regulation of Papillomavirus DNA Replication

Louise T. Chow* and Thomas R. Broker

Abstract

Human and animal papillomaviruses each contain a double-stranded circular DNA genome of approximately 7,400 to 8,200 base pairs which replicates as multicopy, extrachromosomal plasmids in the nucleus of infected keratinocytes. To initiate DNA synthesis from the origin of replication (ori), papillomaviruses encode an ori recognition protein E2 and a replicative DNA helicase E1. All other replication proteins as well as anabolic systems that synthesize the deoxynucleotide substrates are supplied by the host. In cycling cells, viral DNA replication maintains a steady copy number of viral genomes per cell. In post-mitotic, differentiated keratinocytes the papillomaviruses induce the host cells back into S-phase and viral DNA undergoes multiple rounds of replication for progeny virion production. In this Chapter, we summarize the findings of numerous laboratories on the mechanisms regulating the initiation of papillomaviral DNA replication as well as the maintenance and amplification of viral DNA, as characterized by *in vitro* assays independent of other viral proteins that are required in the context of the whole genomic sequence in the host tissues.

1. Introduction

The large family of papillomaviruses has stringent host species specificity and tropism for either cutaneous or mucosal squamous epithelia at preferred body sites. Infections may cause hyperproliferative responses and the formation of a wart, condyloma or papilloma. Alternatively, the host might suppress viral activity to a subclinical maintenance state. Such latent infections can reactivate following transient or long term immunosuppression or repeated rounds of wounding and healing. The genome organization, transcriptional patterns, RNA processing, and reproductive programs are generally well conserved among papillomaviruses. Within the basal and parabasal keratinocytes of the squamous epithelia, viral RNAs and DNA are at or below the sensitivity of detection. High levels of viral transcription, DNA amplification, and virion morphogenesis occur in a subset of the keratinocytes undergoing terminal differentiation (reviewed by Chow and Broker, 1997; Chapters 11 and 13). Conversely, in a substantial fraction of the spinous cells, viral DNA amplification is effectively blocked by host defenses (Jian et al., 1999). The E1 and E2 proteins are directly involved in viral DNA replication. The E7 protein of human papillomaviruses (HPVs) targets the retinoblastoma susceptibility protein (pRB)/E2F pathway and promotes S phase reentry by the post-mitotic keratinocytes to support DNA replication (Cheng et al., 1995; Flores et al., 2000; Chapter 10). The viral E6 protein is also required for maintaining extrachromosomal viral plasmids (Thomas et al., 1999; Oh et al., 2004), though its role is not yet understood.

2. Overview of Papillomaviral DNA Replication

There is no simple infectivity test for papillomaviruses. Thus, two complementary assays have been developed to investigate the mechanisms of papillomaviral DNA replication. In the first strategy, origin (ori)-containing recombinant plasmid prepared from *E. coli* is transfected into mammalian cells along with vectors for expression of viral genes. Newly replicated ori plasmid DNA is detected by Southern blots or by polymerase chain reaction amplification as Dpn1 restriction endonuclease-resistant DNA as the newly synthesized DNA is not methylated in mammalian cells at the recognition sites necessary for cleavage (Ustav and Stenlund, 1991; Chiang et al., 1992b; Del Vecchio et al., 1992; Remm et al., 1992; Titolo et al., 1999; Taylor and Morgan, 2003). In the second approach, cell-free replication of ori plasmid is conducted in crude or partially purified mammalian cell extracts, which contain all the cellular replication proteins, supplemented with purified viral replication proteins. Ori-specific DNA replication is revealed as ^{32}P-labeled finished products (supercoiled Form I and relaxed circular Form II) as well as slowly migrating θ-form replication intermediates in neutral agarose gels (Yang et al., 1991; Kuo et al., 1994; Müller et al., 1994; Melendy et al., 1995; Kasukawa et al., 1998; Lee et al., 1998).

Using these two replication assays along with biochemical and biophysical characterization of purified viral proteins and their interactions with the ori sequences, the mechanisms of human (HPV) and bovine

*For correspondence. Email LTChow@uab.edu.

papillomavirus type 1 (BPV-1) DNA replication have been shown to be similar. The viral replication origin sequences consist of an E1 protein binding site (BS) flanked by multiple copies of the E2BS, and replication depends on the virus-encoded E1 and E2 proteins (Ustav and Stenlund, 1991; Yang et al., 1991; Chiang et al., 1992b; Del Vecchio et al., 1992; Remm et al., 1992; Ustav et al., 1993; Kuo et al., 1994; Sverdrup and Khan, 1995; McShan and Wilson, 1997, Lee et al., 1998). E2 is the primary ori recognition protein and plays a critical role in E1 recruitment to the ori (ibid; Seo et al., 1993a; Sedman and Stenlund, 1995; Berg and Stenlund, 1997; Titolo et al., 1999; Chen and Stenlund, 2000; Sanders and Stenlund, 2001). E1 is the replicative DNA helicase (Yang et al., 1993; Seo et al., 1993b; Hughes and Romanos, 1993; Jenkins et al., 1996; Sedman and Stenlund, 1998; Fouts et al., 1999; White et al., 2001; Lin et al., 2002).

Viral DNA replication requires the cellular replication machinery previously identified in the SV40 DNA replication system, including DNA polymerase (pol) α/primase, DNA pol δ, PCNA (the processivity factor or sliding clamp for DNA pol δ), RFC (the PCNA clamp loader), RPA (single-stranded DNA binding protein), and topoisomerases I and II (Yang et al., 1991; Müller et al., 1994; Kuo et al., 1994). In addition, a yet-to-be-identified novel factor is required for HPV ori replication but not for replication from the SV40 ori (Melendy et al., 1995). Moreover, the cyclin E-cyclin dependent kinase (cdk) 2 complex is also necessary for efficient HPV ori replication (Lin et al., 2000). To initiate replication, E1 unwinds the duplex DNA spanning the ori sequences, recruits RPA and interacts directly with the p180 and p70 subunits of DNA polymerase α/primase (Park et al., 1994; Bonne-Andrea et al., 1995b; Masterson et al., 1998; Conger et al., 1999; Amin et al., 2000; Han et al., 1999; Loo and Melendy, 2004). Despite the overall similarities, important differences between BPV-1 and the HPVs are frequently noted. Each event in the replication pathway will be discussed in detail in this Chapter.

In the context of a cell, the HPV-11 E2 protein stably associates with the nuclear matrix (Zou et al., 2000). By microscopic imaging, HPV-11 ori replication foci colocalize with E1 protein, with E2 protein and with RPA. In as much as HPV-11 E2 protein also colocalizes with some of the nuclear foci of promyelocytic leukemia (PML) protein, a major component of ND10 (*i.e.*, PML bodies or PODs) (Swindle et al., 1999), HPV DNA replication very likely takes place in or near ND10 bodies. In contrast, BPV-1 E2 is diffusely distributed in the nucleus, but it can be recruited to PML bodies by the minor capsid protein L2, a late viral protein (Day et al., 1998). Interestingly, upon infection of cultured cells with pseudovirions, L2 and the pseudogenomes are localized to and concentrated in ND10 in PML-positive cells whereas, in PML-negative cells, L2 does not focus into punctate dots. Moreover, when BPV-1 virions are used in these experiments, viral transcription is also higher in PML-positive cells than in PML-negative cells. Collectively, these observations indicate that ND10 are important for establishing virus infection and may indeed be sites of viral DNA replication (Day et al., 2004).

3. Mechanisms of Papillomavirus Replication

The initiation of cellular DNA replication requires the ori recognition complex (ORC) and the presumptive replicative helicase MCM, which is comprised of mcm subunits 2-7. Many additional factors are involved in loading these factors, recruiting DNA polymerases, and initiating DNA replication (reviewed by Bell and Dutta, 2002). Papillomaviral DNA replication parallels closely but is much simpler than cellular DNA replication, making it an excellent model system. Well over 100 human and animal papillomavirus genomes have been cloned, sequenced, aligned, and compared for conserved as well as for divergent protein motifs. Representative virus types have been intensively investigated in replication assays, and portions of E1 and E2 proteins have been crystallized and their structures solved. This body of information and experimental capability make papillomaviral replication ideal for genetic analyses and for mix and match combinatorial assays, shedding additional light on the mechanisms that regulate the initiation of DNA replication.

3.1. The replication origin

By sequence alignment, the origin sequences of papillomaviruses are approximately 80 to 140 base pairs in length and are located within the upstream regulatory region (URR) (alternatively called the long control region or LCR) overlapping the P1 promoter elements immediately preceding the E6 gene (Zhu et al, 1994). They are comprised of several E2 protein binding sites (E2BS) flanking an E1BS where the E1 proteins assemble into a replicative helicase (Figure 1A, 1B). An AT-rich region is located near the E1BS and greatly enhances replication efficiency, but it is not essential. Certain transcription factors that bind to enhancer and promoter sequences within the ori may compete with and modulate the binding of E1 and E2 proteins and the overall replication activity (Lu et al., 1993; Russell and Botchan, 1995; Demeret et al., 1995; Lee et al., 1997).

The E2BS has a consensus palindromic sequence of ACCGN4CGGT to which E2 binds as a dimer. For mucosotropic HPVs, the central 4 base pairs are A: T-rich whereas they are not so in the BPV-1 E2BS. The E1BS of BPV-1 is an imperfect palindrome of 18 bps (Holt et al., 1994; Gillette and Borowiec, 1998) comprised of 2 and possibly 3 overlapping E1-binding hexanucleotide tracts on each half (Sedman et al., 1997; Chen and Stenlund, 2001; reviewed by Stenlund, 2003b) (Figure 1C). The hexanucleotides have a consensus of AACNAT with a relaxed specificity allowing transition

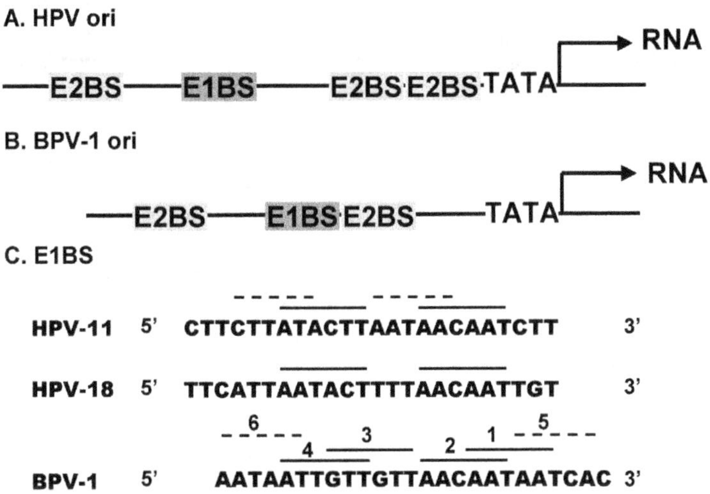

Figure 1. The replication origin of HPVs and BPV-1. The schematics of the origin elements: (A) mucosotropic papillomaviruses. Some cutaneous HPVs have only two E2BSs in the origin region, one on each side of the E1BS. (B) BPV-1. An AT-rich region is usually present at one or both sides of E1BS for HPV and BPV ori sequences. (C) The sequence of the E1BS is comprised of a series of overlapping hexanucleotides (solid lines) that bind the E1 protein, as determined in BPV-1 (Chen and Stenlund, 2001), HPV-11 (Titolo et al., 2003), and HPV-18 (Auster and Joshua-Torr, 2004). In each case, the sequence is the sense strand in the orientation as depicted in (A) and (B). The high affinity binding sites in the HPV E1BSs are aligned with sites 2 and 4 of BPV-1. Dashed lines, low affinity binding sites.

changes in the first 3 positions. Sites 2 and 4 flanking the center of symmetry serve as primary side-by-side binding sites for two E1 monomers on which additional E1 molecules are recruited and assembled. This sequence arrangement is highly conserved in the HPV E1BS, as shown by DNase I footprinting, by binding assays, and by sequence alignments (Frattini and Laimins, 1994a; 1994b; Zhu et al., 1994; Sun et al., 1996; Müller et al., 1997; Titolo et al., 2003; Auster and Joshua-Tor, 2004) (Figure 1C). However, there is a notable difference between the ori organization of HPVs and that of BPV-1. For mucosotropic HPVs, E1BS is well separated from the E2BS on both sides, whereas the BPV-1 ori has one E2BS juxtaposed to the E1BS (Figure 1A, 1B). This may have some functional significance in E1 and E2 protein interactions (discussed below).

The efficiency of ori-specific transient replication increases with the copy number of E2BS. Ori-specific transient or cell-free replication is abolished when each of E2BS's in the URR is mutated (Chiang et al., 1992b; Kuo et al., 1994). A mutated E2 protein which no longer binds to E2BS does not support transient ori replication (Ustav et al., 1993). In contrast, E1BS is dispensable because E1 protein interacts with and is recruited to the ori by the E2 protein (discussed below). Nevertheless, the presence of E1BS increases the efficiency of replication (Lu et al., 1993; Russell and Botchan, 1995; Mendosa et al., 1995; McShan and Wilson, 1997). We suggest that, when the concentrations of the viral proteins are low, such as in basal cells where the native viral transcriptional promoters have low activities, the E1BS may help stabilize E1 protein binding. It is also conceivable that a repressor might recognize this sequence to prevent the loading of the E1 protein, thus controlling the timing and extent of ori replication in the basal cells. Such a repressor might play an important role in maintaining a low viral DNA copy number in these cells in both latent and productive infections.

3.2. E2, the origin recognition and binding protein

3.2.1. The E2 mRNA

The E2 protein of approximately 42-48 kDa is the primary origin binding protein and is critical for recruiting E1 to the ori (for a review, see Stenlund, 2003b). The expression of the E2 gene of all HPVs that have been characterized is regulated at both the transcriptional and post-transcriptional levels. The E2 mRNAs are all derived from a polycistronic primary transcript originating from one or more promoters. The translation of the E2 protein is achieved as a result of an mRNA splice within the E1 gene that deletes almost the entire E1 coding region and positions the E2 ORF close to the 5′-terminus of the message (Rotenberg et al., 1989a; 1989b; Palermo-Dilts et al., 1990; Ozbun and Meyers, 1998; Alloul et al., 1999; Deng et al., 2003). In contrast, the BPV-1 E2 gene has a dedicated promoter and hence its expression is regulated at the transcriptional level (Lambert et al., 1989). The E2 protein consists of three domains, the amino terminal trans-acting (TA) domain (approximately 200 amino acid residues in length), the carboxyl terminal protein dimerization and DNA-binding domain (DBD) (approximately 90 residues long), and a central flexible hinge of 85 or more amino acids (McBride et al., 1989). The DBD and TA domain are relatively well conserved among PVs but the hinge is extremely variable in length and sequence. For both HPV and BPV-1, alternative mRNA splicing or promoters generate several other forms of E2 proteins, including E2C (E2-TR), E8^E2, and E1M^E2C for different viruses (Lambert et al., 1989; Rotenberg et al., 1989a; Chiang et al., 1991; Stubenrauch et al., 2000). Each of these proteins contains the C-terminal dimerization and DNA binding domain and the hinge region, but lacks the amino-terminal trans-acting domain (see also Chapter 6).

3.2.2. The replication function of the full-length E2 protein

All forms of E2 protein bind to the E2BS as a dimer with high specificity and affinity. But only the full-length E2 supports replication, as it alone has the ability to recruit to the ori the E1 protein which otherwise has a low affinity and specificity for the E1BS. The E2 TA domain is critical for this interaction with E1 (to be discussed in a later section), conferring high specificity and efficiency of E1 assembly on the viral ori. The structure of the E2 TA domain of several HPV genotypes has been solved by X-ray crystallography. The full-length TA domain of HPV-16 E2 (residues 1-201) forms a dimer in the crystal and in solution (Antson et al., 2000). This interaction could account for DNA looping observed when BPV-1 E2 binds to widely-separated sites in the URR (Knight et al., 1991). Dimerization presumably would also facilitate interactions with cellular factors that help regulate viral replication and transcription. However, dimerization of the amino terminus might not be a common property. Dimerization was not observed for the TA domain of HPV-18 truncated of residues 1-65, a region encoding two of three prominent amphipathic α-helices at the amino terminus (Harris and Botchan, 1999). Moreover, although the crystal structure of the HPV-11 E2 TA domain is nearly identical to that of HPV-16 E2, there is no evidence for a dimer (Wang et al., 2004).

The crystal structures of the DNA binding domains (DBD) of many papillomaviral E2 proteins alone or in complex with the different E2BS's have also been solved; both similarities and differences are described (reviewed by Hegde, 2002; see also Dell et al., 2003). Briefly, the bound dimer forms an 8-stranded β-barrel adjacent to the two DNA recognition α-helices, one from each monomer, in contact with the palindromic E2BS. Binding of BPV-1 E2 DBD to the E2BS causes the DNA to bend sharply, whereas the mucosotropic HPV E2 DBD favors pre-bent binding sites, such as those with A:A:A:A stacking in the internal 4 base pairs in the consensus E2BS. Binding by HPV E2 is further favored when the E2BS is preceded by A and followed by T, as it is in many copies of the core motif AACCGNNNNCGGTT (Dell et al., 2003). In addition, the flanking sequence amino-terminal to the minimal E2 DBD also contributes to the affinity and stability of the E2:E2BS association (Pepinky et al., 1997).

E2 was initially described as a transcription factor which activates a surrogate promoter by binding to consensus E2BS's (Spalholz et al., 1985; Hirochika et al., 1987; 1988; Phelps and Howley, 1987; Hawley-Nelson et al., 1988). The amino-terminal TA domain of E2 is critical for this transcription regulation. However, when assessed on the native P1 promoter of various HPV genotypes, the homologous viral E2 is either a weak activator or a strong repressor, depending on the levels of E2 protein. In contrast, BPV-1 E2 activates BPV promoters, but it and alternative forms of HPV or BPV-1 E2 always function as transcription repressors of HPV E6 promoters in reporter assays (Lambert et al., 1987; Chin et al., 1988; 1989; Romanczuk et al., 1990; Thierry and Howley., 1991; Dostatni et al., 1991; Chiang et al., 1991; Dong et al., 1994; Tan et al., 1994). Thus, the family of E2 proteins has been hypothesized to impose feedback regulation on viral gene expression and DNA replication (Broker et al., 1989; Chin et al., 1989).

Mutational analyses showed that certain residues in the TA domain function in both activities whereas other residues are only important for transcription or for replication. In particular, residue Ile at position 73 is important for transcription, but not for transient replication, whereas the converse is true for residue Glu at position 39 (Ferguson and Botchan, 1996; Sakai et al., 1996; Brokaw et al., 1996). However, an HPV-31 mutant in residue I73 can support viral DNA maintenance in primary human keratinocytes and viral DNA amplification upon squamous differentiation in raft cultures (Stubenrauch et al., 1998a). Thus, the primary function of the HPV E2 protein appears to be targeting E1 to the origin to initiate viral DNA replication and to maintain viral plasmids in infected cells (to be discussed).

Alternative forms of E2 proteins devoid of the TA domain are inhibitors of replication (Chiang et al., 1992a; Liu et al., 1995; Lim et al., 1998). In particular, the HPV-31 E8^E2 protein is a potent inhibitor of plasmid DNA replication (Zobel et al., 2003), but an HPV-31 mutant which cannot generate the HPV-31 E8^E2 protein cannot be stably maintained in transfected keratinocytes (Stubenrauch et al., 2000). One explanation is that a high copy number of the HPV genome, a condition which normally occurs only in post-mitotic, differentiated keratinocytes, could produce excessive amounts of one or more viral proteins that are incompatible with dividing cells. For instance, high levels of E2 or E1^E4 protein have been reported to cause G2 arrest, senescence, or apoptosis (Frattini et al., 1997; Webster et al., 2000; Davy et al., 2002; Nakahara et al., 2002; Demeret et al., 2003; Raj et al., 2004; Knight et al., 2004). In contrast, BPV-1 mutants in E2 repressors have increased or reduced copy numbers in transfected mouse cell lines (Lambert et al., 1990). Despite the differences between HPVs and BPV, no doubt partially attributable to different assays in different cell types, it is safe to state that the viruses have developed elaborate systems to regulate their genome copy numbers during different phases of the life cycle.

3.2.3. E2-host protein interactions during replication

The full-length E2 protein interacts with a number of transcription factors, including YY1 and the TATA binding protein, and these interactions can affect HPV replication when the cellular proteins are artificially raised to high concentrations (Lee et al., 1998; Hartley and Alexander,

2002). Transcription factors bound to their cognate sites may alter nucleosome formation around the ori (reviewed by Bernard, 2002) and affect the ability of the E2 protein to bind to the ori or to recruit the E1 protein, and thus regulate replication from the ori. On the other hand, BPV-1 E2 protein is able to prevent nucleosome formation on the ori *in vitro*, thus facilitating initiation of replication (Li and Botchan, 1994). It seems plausible that, depending on the relative concentrations of the E2 protein and the relevant cellular proteins in keratinocytes of a particular differentiation state, the ability and extent of the viral DNA to serve as a template for transcription or for replication are regulated.

3.2.4. The regulation of E2 protein localization and function

The hinge domain of all PV types overlaps the E4 gene, and this region is highly divergent in length and in sequence. The evolutionary selection pressure in the region of overlap appears to be primarily on preserving the function of the E4 protein (see Chapter 7). The HPV-11 E2 hinge and the TA domain both associate with nuclear matrix, and the hinge region also contains a nuclear localization sequence (NLS) (RKRAR). Mutation of one or more basic amino acids in the NLS completely abolishes nuclear localization (Zou et al., 2000; Van Tine et al., 2004a). This NLS (K flanked by R residues) appears to be at least partly conserved in many HPV genotypes (Zou et al., 2000). Interestingly, because of the direct interaction between the E1 and E2 proteins, a fraction of the HPV-11 E2 protein mutated in the NLS can hitch-hike on the wild type E1 protein to enter into the nucleus and support transient ori replication. The caveat is that these experiments were conducted in transfected cells where viral proteins were over-expressed. It is probable that, in basal cells where viral proteins are low, the E2 NLS becomes critical for E2 function. In sharp contrast, no functional importance has been ascribed to the hinge domain of the BPV-1 E2 protein, except that large deletions of this domain do not support BPV-1 plasmid replication (Winokur and McBride, 1992; Allikas et al., 2001). Thus, the hinge is thought to provide a flexible linker between the two functional domains for the BPV-1 E2 protein. Also different from the HPV E2, these two BPV-1 E2 domains each contain a nuclear localization sequence (NLS) (Skiadopoulos and McBride, 1996).

The BPV-1 E2 protein is phosphorylated on serine residues by casein kinase II (CKII) in the hinge region. The S301A mutant has elevated plasmid copy number in the transformed cells due to increased protein stability toward degradation by proteasomes (McBride and Howley, 1991; Penrose and McBride, 2000; Penrose et al., 2004). Phosphorylation of BPV-1 E2 also regulates its ability to confer stability on the extrachromosomal viral DNA plasmid when the E1 protein is expressed at high levels (see below). The CKII docking sequence and the substrate sites are not conserved in HPV E2 proteins. Nevertheless, the HPV-18 E2 protein is also degraded by proteasomes, but this degradation is mediated by residues in the amino terminal domain (Bellanger et al., 2001).

3.2.5. The plasmid maintenance function of the E2 protein

The E2 protein is responsible for viral genome persistence in dividing cells. The BPV-1 E2 protein tethers the BPV-1 genome to the mitotic chromosomes (Skiadopoulos and McBride, 1998; Lehman and Botchan, 1998; Ilves et al., 1999) to ensure persistence, similar to the function of the EBNA1 protein of Epstein-Barr herpesvirus, which tethers EBV oriP-containing plasmids to the mitotic chromosomes (Shire et al., 1999; Kanda et al., 2001). The presence of excess BPV-1 E1 protein reduces this association. This E1 effect is more pronounced when the E2 is mutated in all 4 CKII phosphorylation sites in the hinge domain (Voitenleitner and Botchan, 2002). The E2 TA domain mediates the association with mitotic chromosome (Bastien and McBride, 2000). Yeast two hybrid screening and biochemical assays identified the host bromo-domain protein Brd4 as the E2 interacting protein responsible for chromosome tethering (You et al., 2004). Brd4 binds to acetylated histones and thus associates with chromosomes nonspecifically throughout the cell cycle (Dey et al., 2003 and references therein). Interestingly, the presence of BPV-1 E2 protein appears to concentrate Brd4 into small foci on the metaphase chromosomes, based on immunofluorescence assays.

In sharp contrast, the E2 proteins of HPV-11, HPV-16 and HPV-18 associate with mitotic spindles (Van Tine et al., 2004a). This association is observed with either GFP-E2 or with epitope-tagged E2. When the HPV-11 ori plasmid is marked with an array of 40 Gal4 binding sites, Gal4DBD-GFP, which binds to and denotes the location of the ori plasmids, associates with mitotic spindles. This association occurs only when the HPV-11 E2 protein is also present. Interestingly, the region in the E2 C domain responsible for this association was delineated as a stretch of 14 amino acid residues and this sequence is not conserved in the BPV-1 E2 (L.L. Dao, B.A. Van Tine, T.R. Broker and L.T. Chow, unpublished observation). This difference likely accounts for the alternative mechanisms used by HPVs and BPV-1 for maintaining extrachromosomal viral plasmids. Corroborating these differences between HPV and BPV-1 E2 proteins, the HPV-11 E2 protein does not associate with Brd4 in mitotic cells, nor does it affect the distribution of Brd4 (Aaron Duffy, Luan D. Dao, Cheng-Ming Chiang, Shwu-Yuan Wu, T.R. Broker and L.T. Chow, unpublished observations).

For long term maintenance of BPV-1 plasmids in transfected cells, the minimal origin is not sufficient. The upstream sequence in the regulatory region, which contains multiple copies of E2BS's, is also required (Piirsoo et a.,

1996). In the URR of the mucosotropic HPVs, there are only 4 copies of E2BS, three of which reside in the ori proximal to the E6 promoter. Interestingly, deletion of the 5′ region of the URR, which lies further upstream of the distal copy of E2BS, leads to a higher copy number of extrachromosomal DNA in primary keratinocytes (Hubert et al., 1999). It is not known whether this phenotype is attributable to an altered regulation in replication or an indirect effect mediated through transcription. This region of the URR contains alternating purine/thymidine which can form Z-DNA and modulate the superhelical density of the DNA. This region also contains transcription repressor binding sites (reviewed by Bernard, 2002). For HPV-31b, E2BS's 1, 3 and 4 are all required for the long term maintenance of extrachromosomal plasmids in keratinocytes. Unexpectedly, a mutant in E2BS 2, which is normally considered to be part of the ori, did not compromise either short term replication or long term persistence (Stubenrauch et al., 1998b).

3.3. The replicative helicase E1

3.3.1. The E1 mRNA

E1 mRNAs are extremely rare but are thought to originate from a promoter which resides within the E7 ORF. The great majority of the primary transcripts from this promoter are spliced to generate the E1^E4 mRNA or, less frequently, are spliced to generate the E2 mRNA by using the same dominant splice donor but a different splice acceptor within the C-terminal portion of the E1 gene. These splice donor and acceptor sites are highly conserved among all HPVs and the spliced transcripts have been observed both in productively infected lesions (Chow et al., 1987a; 1987b; Nasseri et al., 1987; Rotenberg et al., 1989b; Palermo-Dilts et al., 1990; Sherman et al., 1992a; 1992b) and in cells transiently transfected with an expression vector of the E1 gene or an E1-E2 operon which contains E1 and the overlapping E2/E4 genes (Deng et al., 2003). In the HPV-11 E1-E2 operon, a mutation inactivating the splice donor enhances the expression of the E1 protein by at least 10-fold. The expression of the E1^E4 protein is abolished but the use of alternative or cryptic splice donor sites permits the expression of a small amount of the E2 protein. Interestingly, despite the high abundance of the E1 protein, this mutated operon supports ori replication less efficiently than the wild type plasmid due to a reduced amount of the E2 protein produced from these alternative mRNAs. Detailed replication assays with wild type operon or splice site mutations demonstrate that, not only the absolute amounts of the E1 and E2 proteins, but also their relative ratio, control the efficiency of replication. An elevated ratio of E1/E2 proteins reduces replication whereas increasing the ratio of E2/E1 expression vectors elevates replication (Deng et al., 2003). In contrast, high levels of HPV-1 E2 protein reduce transient replication (Van Horn et al., 2001).

For the high risk HPV genotypes, E1 mRNA can also derive from the E6 promoter. Translation of the E1 protein from the polycistronic mRNA which also contains E6 and E7 ORFs is achieved by ribosome scanning (Remm et al., 1999). The expression of HPV-31b E1 protein is additionally affected by an E6 intragenic splice which occurs only in high-risk HPVs (Hubert and Laimins, 2002). Thus, regulation at the transcriptional level as well as post-transcriptional mRNA splicing are important mechanisms with which the virus could control the extent of viral DNA replication in different strata of the squamous epithelium. For comparison, there is no evidence for an E1 intragenic splice in BPV-1 mRNA. Rather, the various forms of BPV-1 E2 proteins are derived from separate promoters (Lambert et al., 1989), and high levels of E2 reduce transient replication (Männik et al, 2002). These distinctions represent significant differences in the regulation of HPV and BPV DNA replication. As will be discussed, post-translational modifications additionally regulate the activities of the E1 protein.

3.3.2. The E1 protein

The 73 kDa E1 protein forms a dihexameric replicative helicase (for a total MW of approximately 882 kDa) (Liu et al., 1998; Sedman and Stenlund, 1998; Fouts et al., 1999; White et al., 2001; Lin et al., 2002). The functional domains of the E1 protein have been characterized in several HPV genotypes, especially those of HPV-11. The carboxyl terminal half (residues 353-649 in HPV-11 E1) has ATPase/helicase activities and is necessary and sufficient for oligomerization. This domain also interacts with the E2 protein and the DNA polymerase α subunit p70 (Müller and Sapp, 1996; Masterson et al., 1998; Sun et al., 1998; Moscufo et al., 1999; Titolo et al., 1999; Amin et al., 2000; White et al., 2001). However, it is not sufficient to support replication (Amin et al., 2000). A segment of 162 residues (residues 191-352 in HPV-11 E1 protein) immediately upstream of the ATPase/helicase domain is the DNA binding domain, DBD (Titolo et al., 2000; 2003; White et al., 2001). Similar conclusions have also been drawn for the corresponding region of the BPV-1 E1 protein (residues 142-308) (Thorner et al., 1993; Sarafi and McBride, 1995; Leng et al., 1997; Chen and Stenlund, 1998; 2000; Gonzalez et al., 2000; West et al., 2001).

HPV-11 or BPV-1 E1 truncated of the amino terminal segment of approximate 160 residues can support a low level of ori replication, suggestive that this region may have a regulatory function (Ferran and McBride, 1998; Sun et al., 1998; Amin et al., 2000). For BPV-1, a bipartite nuclear localization sequence (NLS) is within the amino terminal domain (Lentz et al., 1993; Leng et al., 1994). We recently defined a stretch of about 50 amino acids within the amino terminal portion of HPV-11 E1 protein as a localization regulatory region (LRR). Briefly, this region contains a dominant nuclear export sequence

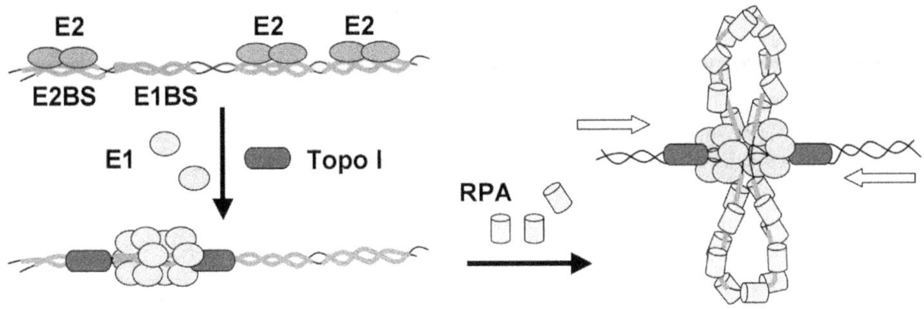

Figure 2. The assembly and helicase activity of the E1 protein. The E2 protein binds to E2BS as a dimer on both sides of the E1BS and recruits E1, which then assembles into a dihexameric complex on the E1BS. The two hexamers stay together and function in coordination as bidirectional helicases. In the presence of topoisomerase I, RPA, and ample supplies of ATP, the DNA on both sides is pumped by the helicases inwardly by axial rotation (unfilled arrows) while being separated into single strands. Topoisomerase I relieves the superhelical strain introduced by DNA strand separation, while the single strands are stabilized by RPA. (Adapted from Lin et al., 2002).

(NES) and a bipartite NLS, both of which are regulated by phosphorylation (Deng et al., 2004b; J.-H. Yu, W. Deng, B.Y. Lin, T.R. Broker and L.T. Chow, unpublished results) (discussed below). The multiple functional motifs of the HPV-11 E1 LRR are highly conserved among HPV E1 proteins, whereas the BPV-1 E1 protein lacks the NES. The reason for the ability of the LRR-truncated E1 proteins to support replication in transfected cells can be attributed to the E1-E2 interaction when the proteins are expressed in high levels in transfected cells, whereby E1 piggy-backs on E2 protein for translocation into the nucleus.

3.3.3. The binding of the E1 protein to the origin

As would be expected for a replicative DNA helicase, which must allow the translocation of the DNA during DNA unwinding, the E1 protein has relatively low affinity and specificity for the E1BS (Frattini and Laimins, 1994b; Sedman and Stenlund, 1995; Liu et al., 1995; Sun et al., 1996; Müller et al., 1997; Dixon et al., 2000). Efficient recruitment of the E1 protein to the ori relies on the interaction with the E2 protein (Figure 2, left). In the presence of E2 protein, the propensity of E1 for ori-specific association is unmasked, and non ori-specific binding to DNA is suppressed (Stenlund, 2003a), leading to a greatly enhanced ori-specific initiation of replication and a corresponding suppression of non-ori replication. Consequently, efficient initiation of ori replication absolutely depends on the presence of E1 and E2 proteins. HPV-1 E1 is something of an exception, as it has a high affinity for E1BS and can promote replication in the absence of E2 protein in transfected cells; not withstanding, replication is greatly stimulated by E2 (Gopalakrishnan and Khan, 1994). The BPV-1 E1 alone has been reported to initiate cell-free replication, and the presence of BPV-1 E2 did not increase ori-specific replication; rather it suppressed replication from non-ori templates (Bonne-Andrea et al., 1995a; 1997). We note that these experiments were conducted with very high concentrations of the BPV-1 E1 protein. This condition could explain the difference with other reports of cell-free replication using either BPV-1 or HPV proteins, in which E2 not only suppressed non-ori replication, but also greatly stimulated the ori-specific replication (Yang et al., 1991; Kuo et al., 1994; Lee et al., 1998; Kasukawa et al., 1998). There are other exceptions where the cellular counterparts might be employed *in vivo* in lieu of the viral proteins (Kim and Lambert, 2002; Angeletti et al., 2002).

The assembly of the BPV-1 E1 on the ori has been investigated in great detail (reviewed by Stenlund, 2003b). In brief, in the presence of E2, the BPV-1 E1 assembles on the ori initially as an (E1)2-(E2)2 complex. As additional E1 molecules are recruited, trimer, hexamer and dihexamer are formed (Sedman and Stenlund, 1996, 1998; Chen and Stenlund, 2002, reviewed by Stenlund, 2003b).

X-ray crystallography of the DBD's of BPV-1 E1 bound to E1BS has revealed that one or two dimers each bind to pairs of overlapping consensus hexanucleotides in different angular phases of the DNA duplex. Binding of each E1 monomer involves a peptide helix and a loop that interact with two halves of the hexanucleotide on opposite strands (Enemark et al., 2000, 2002). Residues V202R and A206R in the BPV-1 DBD are critical for dimer formation, in as much as mutated E1 binds to E1BS as a monomer (Enemark et al., 2000). It was proposed that the E1 dimer nucleates the assembly of the higher E1 oligomers. Insertional mutations into the 18 bp palindromic E1BS also indicate a phasing effect on the efficiency of transient replication (Mendoza et al., 1995). The X-ray crystallographic structure of the HPV-18 E1 DBD suggests a number of differences in the interacting amino acid residues and nucleotides relative to the BPV-1 counterparts (Auster and Joshua-Tor, 2004). However, binding assays of HPV-11 E1 and HPV-18 E1 support the formation of a dimer on the oligonucleotides containing the consensus E1BS through amino acid residues corresponding to those identified in BPV-1 E1 (Titolo et al., 2003; Auster and Joshua-Tor, 2004).

Examination of HPV-11 E1 assembled on the ori by electron microscopy reveals that nearly 90% of the complex contains a single E1 hexamer, with the balance being a dihexamer (Liu et al., 1998). The host heat shock/chaperone proteins Hsp70 and Hdj (Hsp40) greatly increase the E1 affinity for the ori, as revealed by highly elevated formation of complexes in the electrophoretic mobility shift assay. The stimulation by Hsp70 is ATP-dependent and that by Hdj2 requires the J domain, which is conserved from *E. coli* to mammalian cells. However, unlike the synergistic stimulation traditionally observed for chaperone (Hsp70) and its co-chaperone (Hsp40) during protein folding or unfolding, the stimulation of E1 binding to the ori by these two cellular proteins is additive, indicative of a process distinct from refolding denatured proteins. Indeed, nearly 90% of the E1/ori complexes observed in the presence of Hsp40 consists of dihexamer, which is the active helicase, with the balance being a hexamer (Liu et al., 1998; Lin et al., 2002). Hsp70 does not show this effect; its role might be to dissociate preformed hexamer rings to allow them to reassemble around the ori sequence DNA strand as a hexamer. These two modes of E1 assembly on the ori as revealed by different approaches need not be mutually exclusive, as the dimer is the intermediate to hexamer and eventually dihexamer.

3.3.4. The E1-E2 interactions on the ori

The interactions amongst the BPV-1 E1 protein, E2 protein, and the BPV-1 ori have been examined in great detail (reviewed by Stenlund, 2003b). The E2 TA domain of BPV-1 and several HPV genotypes interacts with the helicase domain of the E1 protein (Mohr et al., 1990; Hibma et al., 1995; Winokur and McBride, 1996; Müller and Sapp, 1996; Berg and Stenlund, 1997; Yasugi et al., 1997; Sun et al., 1998; Masterson et al., 1998; Moscufo et al., 1999; Titolo et al., 1999; Chen and Stenlund, 2000). These interactions would explain why only the full-length E2 protein can support ori replication. In addition, the DBD of BPV-1 E2 also interacts with the E1 DBD, creating a sharply bent DNA which facilitates the interaction between the E2 TA domain and the E1 helicase domain (Berg and Stenlund, 1997; Gillitzer et al., 2000). In contrast, no interaction between the DBD's of HPV-11 E1 and E2 protein can be demonstrated (Titolo et al., 1999). It is not known whether this difference is a result of divergent evolution coincidental with the juxtaposition of one of the E2 protein binding sites adjacent to the E1BS in the BPV-1 ori. Some reports suggest that E2 also binds to the amino terminal domain of the E1 protein. However, this association has not been observed by other labs and is not the main association domain.

Because of the general conservation of the ori sequences and the viral replication proteins, E1 and E2 proteins from any virus can initiate replication from ori's of homologous as well as heterologous papillomaviruses. However, mixed pairs of E1 and E2 proteins may not do so efficiently due to undoubtedly less than optimal E1-E2 interactions (Chiang et al., 1992b; Del Vecchio et al., 1992; Berg and Stenlund, 1997; Zou et al., 1998; Gopalakrishnan et al., 1999). For instance, HPV-11 E1 collaborates with the HPV-16 E2 protein, whereas the HPV-16 E1 does not collaborate with the HPV-11 E2 protein (Zou et al, 1998). Employing replication assays with reciprocal chimeric E1 proteins, the E1 domain which is critically involved in productive interactions with E2 protein has been localized to the carboxyl terminus (ibid), in agreement with biochemical assays. We note that, in all the papillomavirus genomes, the carboxyl terminus of the E1 gene invariably overlaps the amino-terminus of the E2 genes in different reading frames. We suggest that these two genes have co-evolved to stay compatible in this essential region of protein-protein interaction.

Although required for the assembly of the pre-replicative complex, E2 protein is not present in the replication complex (Lusky et al., 1994; Liu et al., 1995) (Figure 2, left). The tight association between E2 and the E2BS flanking the E1BS would have blocked translocation of the flanking DNA through the dihexameric, bidirectional helicases. For BPV-1, ATP is required for the displacement of E2 from the E1/ori complex (Sanders and Stenlund, 1998). The presence of HPV-11 E2 protein reduces the E1 helicase activity in the DNA unwinding assay. The inclusion of the heat shock/chaperone proteins Hsp70 and Hdj2 overcomes this inhibition (Lin et al., 2002). The requirement for chaperone proteins in the assembly and activation of the replicative helicase was previously demonstrated for bacterial phages (reviewed by Zylicz, 1993; Polissi et al., 1995). The effects of chaperone proteins on the assembly and activation of the HPV-11 E1 helicase are the first example in mammalian cells. Recently the structure of the HPV-18 helicase domain in complex with the E2 TA domain has been solved by X-ray crystallographic analysis. Molecular modeling suggests that binding of E2 would interfere with the formation of the E1 hexamer (Abbate et al., 2004). Nevertheless, an E1/E2 complex has been demonstrated on the ori for HPV-11, HPV-31b and HPV-33 (Frattini and Laimins, 1994a; 1994b; Sun et al., 1996; Müller et al., 1997; Titolo et al., 1999). Unlike BPV-1, this complex is not affected by ATP (Titolo et al., 1999). The presence of HPV E2 usually enhances the E1-E1BS interaction, but the ratios of the two proteins can affect this association. In none of these latter studies was the state of E1 oligomerization examined.

3.3.5 The E1 helicase activity

When supercoiled DNA is used as substrate *in vitro*, the HPV-11 E1 protein is a potent helicase in a strand unwinding assay (Lin et al., 2002). Unwinding requires topoisomerase I, single-stranded DNA binding protein RPA or SSB (the bacterial equivalent of RPA), ATP and an ATP-regenerating system. As visualized by transmission electron microscopy, unwinding generates two single-strand loops emerging from a central protein complex. Thus, the two E1 hexamer rings stay in contact and likely operate in coordination. This molecular motor pumps the flanking DNA inwardly and the paired parental strands are separated as they pass through both ports of the bidirectional helicase (Figure 2, right). It is not known whether the rings work synchronously by racheting (alternating conformational state model) or via a rotary motion (hand-over-hand model). Had the functional helicase separated into two independent, diverging hexamers during DNA unwinding, the unwinding intermediates would have been a θ-form DNA. The unwinding products (the U form DNA) are completely single-stranded circular DNA with some unreligated full-length linear DNA. The U form DNA is readily detected by ethidium bromide staining of DNA in chloroquine-containing neutral agarose gels. Unwinding is below detection if the substrate has been pre-relaxed by topoisomerase I or linearized with a restriction enzyme. Thus, the torsional energy in the negatively supercoiled DNA helps initiate the unwinding process. This observation is consistent with the AT-rich character of the ori sequences. For comparison, when the helicase substrates are either linear restriction fragments or oligonucleotides annealed to M13 DNA circles, unwinding by the E1 proteins of HPV-11, HPV-6 or BPV-1 is inefficient, and products can only be detected by using radio-labeled substrates or probes (Yang et al., 1993; Seo et al., 1993b; White et al., 2001). DNA unwinding and helicase assays *in vitro* are typically conducted in the presence of high concentrations of the E1 protein and, under these conditions, there is no ori specificity. This would explain why E1 can replicate non-ori plasmids in cell-free replication, when it is present in high concentration (Yang et al., 1991; Kuo et al., 1994; Bonne-Andrea et al., 1997).

3.3.6. The E1-host replication protein interactions

The HPV-11 and HPV-16 E1 proteins interact with the p180 and p70 subunits of the DNA polymerase α/primase (Masterson et al., 1998; Conger et al., 1999; Amin et al., 2000). BPV-1 E1 also binds to the p180 subunit (Park et al., 1994; Bonne-Andrea et al., 1995). These interactions are crucial to recruiting DNA polymerase α/primase to the ori to initiate replication. Interestingly, the HPV-11 and HPV-16 E2 proteins compete with the p70 subunit for interaction with the carboxyl terminus of the HPV E1 proteins. Thus, incoming polymerase α/primase might help displace E2 from the preinitiation complex (Masterson et al., 1998; Conger et al., 1999), along with the action of chaperone proteins. It remains a formal possibility that phosphorylation of one or both viral proteins might also contribute to the dissociation of E2 from the complex. For instance, the BPV E1 protein preferentially binds to underphosphorylated E2 protein (Lusky and Fontaine, 1991). The E1 proteins of BPV-1 and HPVs interact with RPA (Han et al., 1999; Loo and Melendy, 2004) such that the single-stranded DNA generated by the E1 helicase activity is quickly stabilized. In addition, the HPV-11 E1 protein interacts with histone H1 (Swindle and Engler, 1998). This interaction might play a role in unraveling the viral chromatin by stripping away the H1 histones prior to unwinding the DNA.

4. The Regulation of Papillomaviral DNA Replication by Cellular Proteins

A number of additional cellular proteins control PV DNA replication. A couple of selected examples are elaborated here.

4.1. Regulation of papillomaviral DNA replication by the tumor suppressor protein p53

p53 is a major tumor suppressor protein and is targeted by the HPV E6 proteins for inactivation or degradation (Münger et al., 2004; Chapter 9). p53 can inhibit transient amplification of BPV and HPV ori plasmids in transfected cells in a cell type specific manner. Inhibition is not caused by reduced viral protein expression, cell cycle arrest, or apoptosis (Lepik et al., 1998; Lepik and Ustav, 2000). Extensive mutational analyses of p53 suggest that its DBD, the oligomerization domain, and the RPA-interacting domain, but not the transactivating domain, are necessary for this inhibitory activity. Thus, the p53 induction of apoptosis genes or the $p21^{cip1}$ protein which inhibits cdk2 is not responsible for this inhibition. An interaction between HPV-16 E2 and p53 is thought to mediate this inhibition, as a p53 mutant which binds to E2 with a reduced affinity also has a reduced inhibitory activity (Massimi et al., 1999). Although p53 appears not to affect the low copy maintenance replication of BPV-1 plasmid DNA (Lepik et al., 1998), primary keratinocytes transfected with HPV genomic DNA mutated in E6 failed to maintain viral plasmid DNA in submerged cultures (Thomas et al., 1999; Oh et al., 2004). It has been hypothesized that E6 is required to counterbalance detrimental effects of the E7 protein. One effect of HPV E7 expression is the stabilization of p53. It is conceivable that the inability of the E6 mutants to inactivate p53 protein then leads to the inhibition of plasmid replication (Park and Androphy, 2002). However the interpretation of these experiments is complicated due to the fact that, in the absence of functional E6, the primary keratinocytes senesce after 6 or 7 passages. In any case, there has not been data on whether the interaction with E2 is the only means by which p53 inhibits papillomaviral DNA replication. Elucidation of the mechanisms by which p53 suppresses unscheduled viral DNA synthesis is an important objective for future investigation.

4.2. Regulation of the E1 helicase by cyclin E/cdk2

Since HPV DNA replication depends heavily on the cellular DNA replication machinery, it makes sense that there is a cell-cycle based regulation of viral replication proteins. Indeed, evidence suggests that the key to understanding the three phases of HPV replication - establishment, maintenance and vegetative amplification – and the regulation of DNA copy number lies primarily with post-translational modifications of the E2 protein (already discussed in Section 3.2.5) as well as the E1 helicase. The function of E1 proteins of BPV-1, HPV-11 and, most likely, other HPVs are post-translationally controlled by the cdks, in particular, by cyclin E/cdk2 or cyclin A/cdk2. These kinases are known to regulate the cellular MCM helicase complex (reviewed by Bell and Dutta, 2002). The cyclin E/cdk2 complex is critical for the G1 to S phase transition and hence the initiation of DNA replication, whereas cyclin A is required throughout the S phase and for the transition into the G2 phase (reviewed by Sherr and Roberts, 1999).

4.2.1. Cyclin E/cdk2 is essential for efficient initiation of replication

In the *Xenopus* egg extract system, BPV-1 E1 binds to and is stabilized by cyclin E. Following replication, the E1 protein is quick destabilized and degraded. Biochemical evidence suggests that degradation is mediated by the anaphase promoting complex (Cueille et al., 1998; Malcles et al., 2002; Mechali et al., 2004). The E1 proteins of HPVs also bind *in vitro* to multiple cyclins and are substrates of multiple cdk complexes, except for cyclin D/cdk4 (Ma et al., 1999). It is likely that the stability of HPV E1 protein could also be subjected to similar regulation. For instance, in the cell-free replication from the HPV-11 ori, the majority of the replication products is semi-methylated, indicative of one round of semi-conservative replication. Thus, the replication complex dissembles after each round of DNA replication (Liu et al., 1995).

Replication of HPV-11 DNA is regulated by cdks through additional mechanisms. HeLa cell extracts, which support efficient replication from the SV40 ori in the presence of the SV40 T antigen, replicate from the HPV ori poorly. The deficiency can be complemented by adding to the assay cocktail a small amount of 293 cell extract, which provides at least two important components (Lin et al., 2000). One could be the novel factor described by Melendy et al. (1995). The other is cyclin E/cdk2. Recombinant cyclin E/cdk2 complex, but not cyclin A/cdk2, cyclin A/cdc2 or cyclin B/cdc2, restores the replication activity to as much as 50% of that obtained with 293 cell extracts (Lin et al., 2000). HeLa cells abundantly express the amino terminal portion of the HPV-18 E1 protein from integrated viral genomes. The deficiency in the active cyclin E/cdk2 complex in HeLa cell extracts is attributed to its sequestration by this E1 fragment, making it unavailable for the exogenous E1 protein (Ma et al., 1999; Lin et al., 2000). This interpretation is consistent with the inefficient transient replication assays conducted in cervical carcinoma cell lines that harbor integrated HPV DNA but not in a carcinoma cell line which does not (Chiang et al., 1992b).

There are four putative (Ser/Thr-pro) cdk substrates in the HPV-11 E1 protein, three of which (S89, S93 and S107) are located in the amino-terminal domain, while the fourth one (T468) is in the helicase domain. Mutated forms of E1 which cannot be phosphorylated by cyclin/cdk complexes *in vitro* are impaired in their activity in both replication assays (Ma et al., 1999). One of the mechanisms of this regulation by cyclin E/cdk has been elucidated (see Section 4.2.2). HPV-11 E2 is also a substrate of multiple cyclin/cdk complexes *in vitro*, but only in the presence of the HPV-11 E1 protein (Ma et al., 1999). The functional significance of this post-translational modification of E2 is not understood.

4.2.2 CRM-1 mediated nuclear export

Through a series of experiments with a replication-competent HPV-11 E1 protein fused to enhanced green fluorescent protein, a region of 50 amino acids in E1 (residues 80-130) has been determined to be the Localization Regulatory Region (the LRR) because it controls the nucleo-cytoplasmic localization of the protein. This process is regulated by phosphorylation, and the kinase complexes are cyclin E/cdk2 and cyclin A/cdk2, as described below (Deng et al., 2004b).

The HPV-11 LRR contains four interdigitated regulatory elements: a bipartite nuclear import sequence (NLS), the consensus cyclin binding motif RxL, the three serine residues that are substrates for cyclin-dependent kinases, and a nuclear export sequence (NES), which has been delineated to span 20 residues (Deng et al., 2004b; J.-H. Yu, W. Deng, B.Y. Lin, T.R. Broker and L.T. Chow, unpublished observations). Nuclear export of E1 is mediated by a CRM-1 exportin-dependent, leucine-rich NES, the activity of which dominates over the NLS. Phosphorylation of S107 by cdk2 is primarily responsible for inactivating the NES to allow the E1 protein to remain in the nucleus, whereas phosphorylation of S89 and S93 by yet-to-be-identified cytoplasmic kinases enhances nuclear import. Once in the nucleus, presumably cdk2 can keep all three serine residues in the phosphorylated state as long as the cells are in the S phase. The four LRR elements are highly conserved among E1 proteins of mucosotropic HPVs, but the leucine-rich NES cannot be found in the corresponding LRR region of the BPV-1 E1 protein. This may be a critical difference in the regulation of the E1 proteins between HPVs and BPV-1.

In support of the mutational analysis and in agreement with its phosphorylation by multiple cyclin/cdk complexes

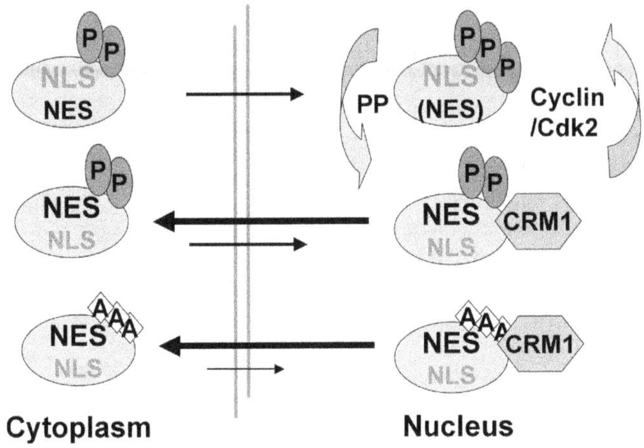

Figure 3. The nucleo-cytoplasmic localization of HPV-11 E1 is controlled by the activities of its nuclear export sequence (NES) and nuclear localization sequence (NLS), both of which are regulated by kinases (a summary of data by Deng et al. 2004b). Top row: Nuclear import mediated by NLS is enhanced (medium arrow) by phosphorylation on the first two of the three serine residues (each followed by a proline residue) in the localization regulatory region (LRR) near the amino terminus. When the cell enters S-phase, cyclin E/cdk2 or cyclin A/cdk2 phosphorylates the third serine residue, which is located within the NES, inactivating it (parenthesis). E1 is then retained in the nucleus to replicate viral DNA. Middle row: Once the cell exits S-phase, cdk2 is no longer active and E1 dephosphorylation by phosphatases activates the NES, causing efficient E1 nuclear export to the cytoplasm (large arrow) through the action of the CRM-1 exportin. Bottow row: Mutations in all three phosphorylation sites in the LRR reduce nuclear import (small arrow) and prohibit NES inactivation by the cdk2 kinases. Thus, E1 nuclear retention is abolished, and effective E1 nuclear export (large arrow) results in a cessation of viral DNA replication. Since the regulatory elements (NES, NLS, several SerPro phosphorylation sites, and consensus cyclin binding motif) in HPV-11 E1 LRR are highly conserved among HPV genotypes, the regulatory mechanisms depicted here could be generally applicable to other HPV E1 proteins.

in vitro, HPV-11 E1 protein is phosphorylated in insect cells and in mammalian cells (Deng et al., 2004b). The phosphorylated E1 protein is reactive with the MPM2 antibody which recognizes phospho-serine and phospho-threonine, followed by proline in the peptide chain. The MPM2 antibody reactivity is abolished *in vitro* by pretreatment of E1 with protein phosphatase 1. Antibody reactivity is also lost *in vivo* following exposure of the cells to the drug roscovitine, which inhibits cdks. HPV-11 E1 mutated in all four cdk phosphorylation sites or in the cyclin binding motif RxL is not detected by MPM2. Inhibition of cdk2 kinases by roscovitine or elevated p21^{cip1} protein leads to cytoplasmic HPV-11 E1. Collectively, these data demonstrate that one or more of these cdk candidate sites are indeed phosphorylated *in vivo* by cdk2. Only when cyclin E or cyclin A/cdk2 are active and the cells enter S phase is the NES function inactivated and the E1 protein retained in the nucleus (Deng et al., 2004b). The regulation of the nucleo-cytoplasmic localization of HPV-11 E1 by kinases is summarized in Figure 3.

It can be reasonably inferred that the loss of cyclin E and cyclin A as cells return to G phases or to quiescence is quickly followed by dephosphorylation and export of the E1 helicase from the nucleus, resulting in the cessation of viral DNA replication. Thus, the low level of viral gene expression coupled with the cyclical E1 phosphorylation and dephosphorylation are key factors in the steady state maintenance of the viral genome in persistent infections. On the other hand, during the productive phase in differentiated cells, the high levels of viral gene expression and nuclear retention of the viral E1 protein during S phase (which is reestablished by the E7 protein) would then allow viral DNA to amplify. In this aspect, the regulation of viral DNA replication by cyclin E/cdk2 is dissimilar to that of cellular DNA replication, where this kinase not only promotes the initiation of replication but also prevents reinitiation by negatively regulating the assembly of the Pre-RC (Bell and Dutta, 2002). HPV has modified this scheme to overcome the tight regulation of once-per-cell cycle licensing of origins in DNA replication and thus gained the crucial ability to achieve multiple rounds of reinitiation without an intervening cell division, enabling vegetative amplification.

5. Concluding Remarks

The relative simplicity of the papillomaviral replication proteins together with their dependence on host replication factors make them an appealing model system with which to explore mammalian replication mechanisms. Investigations into PV DNA replication have revealed many novel findings and highlighted important differences between the mucosotropic HPV genotypes and BPV-1. The similarities as well as the differences also extend to comparisons with host cellular replication requirements. Consideration of the unique and special properties of E2 and E1 replication proteins has defined appealing targets for antiviral drug discovery with minimal or no risk to normal host replication. To this end, peptides, peptide nucleic acids, covalently linked polyamides, and small molecules that specifically inhibit HPV-16 or HPV-11 E1 and E2 interactions or E2 ori interaction have been reported (Kasukawa et al., 1998; Kurg et al., 2000; Schaal et al., 2003; White et al., 2003; Fujii et al., 2003; Deng et al., 2004a; Davidson et al., 2004). Similarly, the shuttling of the HPV E1 protein into and out of the nucleus (Deng et al., 2004b) might also offer a cell-based assay to search for inhibitors of E1 nuclear localization capable of blocking HPV DNA replication and virus persistence.

There are still many unanswered questions concerning the roles of additional host proteins that regulate viral DNA replication. The novel factor which is required for HPV but not SV40 replication remains to be identified (Melendy et al., 1995). The activities of BPV-1 E1 (McShan and Wilson, 2000; Lentz, 2002; Lentz et al., 2002) and HPV-11 E1 are regulated by CKII kinase (J.-H. Yu, B.Y. Lin, T.R. Broker and L.T. Chow, unpublished observations). Transient replication is also regulated by CDP/Cut, a transcription factor (O'Connor et al., 2000). The mechanisms of regulation are not understood in these cases. Of course, the roles of the viral E6 protein and its target tumor suppressor protein p53 on viral DNA replication are most intriguing and remain to be elucidated. In addition, HPV E1 interacts with hSNF5, a chromatin remodeling factor (Lee et al., 1999), and HPV E2 interacts with TopBP1 (Boner et al., 2002), an essential protein involved in DNA repair (Makiniemi et al., 2001; Yamane et al., 2002 and references therein), the *Xenopus* and *Drosophila* homologues of which have been implicated in the initiation of DNA replication (Yamamoto et al., 2000; Van Hatten et al., 2002; Hashimoto and Takisawa, 2003). Moreover, numerous host proteins interact with and affect E2-dependent transcription, including CBP/p300, AMF-1, p/CAF and the nucleosome assembly factor 1 (Lee et al., 2000, 2002; Peng et al., 2002; Müller et al., 2002; Rehtanz et al., 2004). Whether such interactions also have a role in regulating DNA replication directly is not known. Finally, DNA methylation in the E2BS prevents E2 binding (Thain et al., 1996). In this connection, investigations of the regulation of viral activities by DNA methylation and histone modification have only begun, and none have yet dealt with viral DNA replication (Zhao et al., 1999; Kim et al., 2003; Van Tine et al., 2004b; Badal et al., 2004; Kalantari et al., 2004). Future studies should yield exciting additional insights into mechanisms that regulate viral DNA replication, especially during different stages of the viral life cycle.

Acknowledgments

The authors are supported by USPHS grants CA36200, CA83679 and CA107338. We appreciatively recognize the invaluable contributions of past and present lab members as well as our many collaborators who have shared in our investigations into diverse aspects of HPV DNA replication.

References

Abbate, E.A., Berger, J.M., and Botchan, M.R. (2004). The X-ray structure of the papillomavirus helicase in complex with its molecular matchmaker E2. Genes Dev. *18*, 1981-1996.

Allikas, A., Ord, D., Kurg, R., Kivi, S., and Ustav, M. (2001). Roles of the hinge region and the DNA binding domain of the bovine papillomavirus type 1 E2 protein in initiation of DNA replication. Virus Res. *75*, 95-106.

Alloul, N., and Sherman, L. (1999). The E2 protein of human papillomavirus type 16 is translated from a variety of differentially spliced polycistronic mRNAs. J. Gen. Virol. *80*, 29-37.

Amin, A.A., Titolo, S., Pelletier, A., Fink, D., Cordingley, M.G., and Archambault, J. (2000). Identification of domains of the HPV11 E1 protein required for DNA replication *in vitro*. Virology *272,* 137-150.

Angeletti, P.C., Kim, K., Fernandes, F.J., and Lambert, P.F. (2002). Stable replication of papillomavirus genomes in Saccharomyces cerevisiae. J. Virol. *76*, 3350-3358.

Antson, A.A., Burns, J.E., Moroz, O.V., Scott, D.J., Sanders, C.M., Bronstein, I.B., Dodson, G.G., Wilson, K.S., and Maitland, N.J. (2000). Structure of the intact transactivation domain of the human papillomavirus E2 protein. Nature *403*, 805-809.

Auster, A.S., and Joshua-Tor, L. (2004). The DNA-binding domain of human papillomavirus type 18 E1. Crystal structure, dimerization, and DNA binding. J. Biol. Chem. *279*, 3733-3742.

Badal, S., Badal, V., Calleja-Macias, I.E., Kalantari, M., Chuang, L.S., Li, B.F., and Bernard, H.-U. (2004). The human papillomavirus-18 genome is efficiently targeted by cellular DNA methylation. Virology *324*, 483-492.

Bastien, N., and McBride, A.A. (2000). Interaction of the papillomavirus E2 protein with mitotic chromosomes. Virology *270*, 124-134.

Bell, S.P., and Dutta, A. (2002). DNA replication in eukaryotic cells. Annu. Rev. Biochem. *71*, 333-374.

Bellanger, S., Demeret, C., Goyat, S., and Thierry, F. (2001). Stability of the human papillomavirus type 18 E2 protein is regulated by a proteasome degradation pathway through its amino-terminal transactivation domain. J. Virol. *75*, 244-251.

Berg, M., and Stenlund, A. (1997). Functional interactions between papillomavirus E1 and E2 proteins. J. Virol. *71*, 3853-3863.

Bernard, H.-U. (2002). Gene expression of genital human papillomaviruses and considerations on potential antiviral approaches. Antiviral Ther. *7*, 219-237.

Boner, W., Taylor, E.R., Tsirimonaki, R., Yamane, K., Campo, M.S., and Morgan, I.M. (2002). A functional interaction between the human papillomavirus 16 transcription/replication factor E2 and the DNA damage response protein TopBP1. J. Biol. Chem. *277*, 22297-22303.

Bonne-Andrea, C., Santucci, S., and Clertant, P. (1995a). Bovine papillomavirus E1 protein can, by itself, efficiently drive multiple rounds of DNA synthesis *in vitro*. J. Virol. *69*, 3201-3205.

Bonne-Andrea, C., Santucci, S., Clertant, P., and Tillier, F. (1995b). Bovine papillomavirus E1 protein binds specifically DNA polymerase alpha but not replication protein A. J. Virol. *69*, 2341-2350.

Bonne-Andrea, C., Tillier, F., McShan, G.D., Wilson, V.G., and Clertant, P. (1997). Bovine papillomavirus type 1 DNA replication: the transcriptional activator E2 acts *in vitro* as a specificity factor. J. Virol. *71*, 6805-6815.

Brokaw, J.L., Blanco, M., and McBride, A.A. (1996). Amino acids critical for the functions of the bovine papillomavirus type 1 E2 transactivator. J. Virol. *70*, 23-29.

Broker, T.R., Chow, L.T., Chin, M.T., Rhodes, C.R., Wolinsky, S.M., Whitbeck, A., and Stoler, M.H. (1989). A molecular portrait of human papillomavirus carcinogenesis. Cancer Cells *7*, 197-208.

Chen, G., and Stenlund, A. (1998). Characterization of the DNA-binding domain of the bovine papillomavirus replication initiator E1. J. Virol. *72*, 2567-2576.

Chen, G., and Stenlund, A. (2000). Two patches of amino acids on the E2 DNA binding domain define the surface for interaction with E1. J. Virol. *74*, 1506-1512.

Chen, G., and Stenlund, A. (2001). The E1 initiator recognizes multiple overlapping sites in the papillomavirus origin of DNA replication. J. Virol. *75*, 292-302.

Chen, G., and Stenlund, A. (2002). Sequential and ordered assembly of E1 initiator complexes on the papillomavirus origin of DNA replication generates progressive structural changes related to melting. Mol. Cell. Biol. *22*, 7712-7720.

Cheng, S., Schmidt-Grimminger, D.-C., Murant, T., Broker, T.R., and Chow, L.T. (1995). Differentiation-dependent up-regulation of the human papillomavirus E7 gene reactivates cellular DNA replication in suprabasal differentiated keratinocytes. Genes Dev. *9*, 2335-2349.

Chiang, C.-M., Broker, T.R., and Chow, L.T. (1991). An E1M^E2C fusion protein encoded by human papillomavirus type 11 is a sequence specific transcription repressor. J. Virol. *65*, 3317-3329.

Chiang, C.-M., Dong, G., Broker, T.R., and Chow, L.T. (1992a). Control of human papillomavirus type 11 origin replication by the E2 family of transcriptional regulatory proteins. J. Virol. *66*, 5224-5231.

Chiang, C.-M., Ustav, M., Stenlund, A., Ho, T.F., Broker, T.R., and Chow, L.T. (1992b). Viral E1 and E2 proteins support replication of homologous and heterologous papillomaviral origins. Proc. Natl. Acad. Sci. USA *89*, 5799-5803.

Chin, M.T., Broker, T.R., and Chow, L.T. (1989). Identification of a novel constitutive enhancer element and its associated binding protein: Implications for human papillomavirus type 11 enhancer regulation. J. Virol. *63*, 2967-2976.

Chin, M.T., Hirochika, R., Hirochika, H., Broker, T.R., and Chow, L.T. (1988). Regulation of the human papillomavirus type 11 enhancer and E6 promoter by activating and repressing proteins from the E2 open reading frame: functional and biochemical studies. J. Virol. *62*, 2994-3002.

Chow, L.T., and Broker, T.R. (1997). Small DNA tumor viruses. In, Viral Pathogenesis, Ed., N. Nathanson. Lippincott-Raven Publishers, Philadelphia, PA. pp. 267-302.

Chow, L.T., Nasseri, M., Wolinsky, S.M., and Broker, T.R. (1987a). Human papilloma virus type 6 and 11 mRNAs from genital condylomata acuminata. J. Virol. *61*, 2581-2588.

Chow, L.T., Reilly, S., Broker, T.R., and Taichman, L. (1987b). Identification and mapping of human papillomavirus type 1 RNA transcripts recovered from plantar warts and infected epithelial cell cultures. J. Virol. *61*, 1913-1918.

Conger, K.L., Liu, J.-S., Kuo, S.-R., Chow, L.T., and Wong, T. S.-F. (1999). Human papillomavirus DNA replication: Interactions between the viral E1 protein and two subunits of human DNA polymerase α / primase. J. Biol. Chem. *274*, 2696-2705.

Cueille, N., Nougarede, R., Mechali, F., Philippe, M., and Bonne-Andrea, C. (1998). Functional interaction between the bovine papillomavirus virus type 1 replicative helicase E1 and cyclin E-Cdk2. J. Virol. *72*, 7255-7262.

Davidson, W., McGibbon, G.A., White, P.W., Yoakim, C., Hopkins, J.L., Guse, I., Hambly, D.M., Frego, L., Ogilvie, W.W., Lavallee, P., and Archambault, J. (2004). Characterization of the binding site for inhibitors of the HPV11 E1-E2 protein interaction on the E2 transactivation domain by photoaffinity labeling and mass spectrometry. Anal. Chem. *76*, 2095-2102.

Davy, C.E., Jackson, D.J., Wang, Q., Raj, K., Masterson, P.J., Fenner, N.F., Southern, S., Cuthill, SW., Millar, J.B., and Doorbar, J. (2002). Identification of a G(2) arrest domain in the E1^E4 protein of human papillomavirus type 16. J. Virol. *76*, 9806-9818.

Day, P.M., Bake,r C.C., Lowy, D.R., and Schiller, J.T. (2004). Establishment of papillomavirus infection is enhanced by promyelocytic leukemia protein (PML) expression. Proc. Natl. Acad. Sci. USA *101*, 14252-14257.

Day, P.M., Roden, R.B., Lowy, D.R., and Schiller, J.T. (1998). The papillomavirus minor capsid protein, L2, induces localization of the major capsid protein, L1, and the viral transcription/ replication protein, E2, to PML oncogenic domains. J. Virol. *72*, 142-150.

Dell, G, Wilkinson, K.W., Tranter, R., Parish, J., Brady, R.L., and Gaston, K. (2003). Comparison of the structure and DNA-binding properties of the E2 proteins from an oncogenic and a non-oncogenic human papillomavirus. J. Mol. Biol. *334*, 979-991.

Del Vecchio, A.M., Romanczuk, H., Howley, P.M., and Baker, C.C. (1992). Transient replication of human papillomavirus DNAs. J. Virol. *66*, 5949-5958.

Demeret, C., Garcia-Carranca, A, and Thierry. F. (2003). Transcription-independent triggering of the extrinsic pathway of apoptosis by human papillomavirus 18 E2 protein. Oncogene *16*, 168-175.

Demeret, C., Le Moal, M., Yaniv, M., and Thierry, F. (1995). Control of HPV 18 DNA replication by cellular and viral transcription factors. Nucl. Acids Res. *23*, 4777-4784.

Deng, S.J., Pearce, K.H., Dixon, E.P., Hartley, K.A., Stanley, T.B., Lobe, D.C., Garvey, E.P., Kost, T.A., Petty, R.L., Rocque, W.J., Alexander, K.A., and Underwood, M.R. (2004a). Identification of peptides that inhibit the DNA binding, trans-activator, and DNA replication functions of the human papillomavirus type 11 E2 protein. J. Virol. *78*, 2637-2641.

Deng, W., Jin, G., Lin, B.Y., Van Tine, B.A., Broker, T.R., and Chow, L.T. (2003). mRNA splicing regulates human papillomavirus type 11 E1 protein production and DNA replication. J. Virol. *77*, 10213-10226.

Deng, W., Lin, B.Y., Jin, G., Wheeler, C., Ma, T., Harper, J.W., Broker, T.R., and Chow, L.T. (2004b). Cyclin/CDK regulates the nucleo-cytoplasmic localization of the human papillomavirus E1 DNA helicase. J. Virol. *78*, 13954-13965.

Dey, A., Chitsaz, F., Abbasi, A., Misteli, T., and Ozato, K. (2003). The double bromodomain protein Brd4 binds to acetylated chromatin during interphase and mitosis. Proc. Natl. Acad. Sci. USA *100*, 8758-8763.

Dixon, E.P., Pahel, G.L., Rocque, W.J., Barnes, J.A., Lobe, D.C., Hanlon, M.H., Alexander, K.A., Chao, S.F., Lindley, K., and Phelps, W.C. (2000). The E1 helicase of human papillomavirus type 11 binds to the origin of replication with low sequence specificity. Virology *270*, 345-357.

Dong, G., T.R. Broker and L.T. Chow. (1994). Human papillomavirus type 11 E2 proteins repress the homologous E6 promoter by interfering with the binding of host transcription factors to adjacent elements. J. Virol. *68*, 1115-1127.

Dostatni, N., Lambert, P.F., Sousa, R., Ham, J., Howley, P.M., and Yaniv, M. (1991). The functional BPV-1 E2 trans-activating protein can act as a repressor by preventing formation of the initiation complex. Genes Dev. *5*, 1657-1671.

Enemark, E.J., G. Chen, D.E. Vaughn, A. Stenlund, and L. Joshua-Tor. (2000). Crystal structure of the DNA binding domain of the replication initiation protein E1 from papillomavirus. Mol. Cell *6*, 149-158.

Enemark, E.J., Stenlund, A., and Joshua-Tor, L. (2002). Crystal structures of two intermediates in the assembly of the papillomavirus replication initiation complex. EMBO J. *21*, 1487-1496.

Ferguson, M.K., and Botchan, M.R.. (1996). Genetic analysis of the activation domain of bovine papillomavirus protein E2: its role in transcription and replication. J. Virol. *70*, 4193-4199.

Ferran, M.C., and McBride, A.A. (1998). Transient viral DNA replication and repression of viral transcription are supported by the C-terminal domain of the bovine papillomavirus type 1 E1 protein. J. Virol. *72*, 796-801.

Flores, E.R., Allen-Hoffmann, B.L., Lee, D., and Lambert, P.F. (2000). The human papillomavirus type 16 E7 oncogene is required for the productive stage of the viral life cycle. J. Virol. *74*, 6622-6631.

Fouts, E.T., Yu, X., Egelman, E.H., and Botchan, M.R. (1999). Biochemical and electron microscopic image analysis of the hexameric E1 helicase. J. Biol. Chem. *274*, 4447-4458.

Frattini, M.G., Hurst, S.D., Lim, H.B., Swaminathan, S., and Laimins, L.A. (1997). Abrogation of a mitotic checkpoint by E2 proteins from oncogenic human papillomaviruses correlates with increased turnover of the p53 tumor suppressor protein. EMBO J. *16*, 318-331.

Frattini, M.G., and Laimins, L.A. 1994a. The role of the E1 and E2 proteins in the replication of human papillomavirus type 31b. Virology *204*, 799-804.

Frattini, M.G., and Laimins, L.A. 1994b. Binding of the human papillomavirus E1 origin-recognition protein is regulated through complex formation with the E2 enhancer-binding protein. Proc. Natl. Acad. Sci. USA *91*, 12398-12402.

Fujii, T., Austin, D., Guo, D., Srimatkandada, S., Wang, T., Kubushiro, K., Masumoto, N., Tsukazaki, K., Nozawa, S., and Deisseroth, A.B. (2003). Peptides inhibitory for the transcriptional regulatory function of human papillomavirus E2. Clin. Cancer Res. *9*, 5423-5428.

Gillitzer, E., Chen, G., and Stenlund, A. (2000). Separate domains in E1 and E2 proteins serve architectural and productive roles for cooperative DNA binding. EMBO J. *19*, 3069-3079.

Gonzalez, A., Bazaldua-Hernandez, C., West, M., Woytek, K., and Wilson, V.G. (2000). Identification of a short, hydrophilic amino acid sequence critical for origin recognition by the bovine papillomavirus E1 protein. J. Virol. *74*, 245-253.

Gillette, T.G., and Borowiec, J.A. (1998). Distinct roles of two binding sites for the bovine papillomavirus (BPV) E2 transactivator on BPV DNA replication. J. Virol. *72*, 5735-5744.

Gopalakrishnan, V., and Khan, S.A. (1994). E1 protein of human papillomavirus type 1a is sufficient for initiation of viral DNA replication. Proc. Natl. Acad. Sci. USA *91*, 9597-9601.

Gopalakrishnan, V., Sheahan, L., and Khan, S.A. (1999). DNA replication specificity and functional E2 interaction of the E1 proteins of human papillomavirus types 1a and 18 are determined by their carboxyl-terminal halves. Virology *256*, 330-339.

Han, Y., Loo, Y.M., Militello, K.T., and Melendy, T. (1999). Interactions of the papovavirus DNA replication initiator proteins, bovine papillomavirus type 1 E1 and simian virus 40 large T antigen, with human replication protein A. J. Virol. *73*, 4899-4907.

Hartley, K.A., and Alexander, K.A. (2002). Human TATA binding protein inhibits human papillomavirus type 11 DNA replication by antagonizing E1-E2 protein complex formation on the viral origin of replication. J. Virol. *76*, 5014-5023.

Harris, S.F., and Botchan, M.R. (1999). Crystal structure of the human papillomavirus type 18 E2 activation domain. Science *284*, 1673-1677.

Hashimoto, Y., and Takisawa, H. (2003). Xenopus Cut5 is essential for a CDK-dependent process in the initiation of DNA replication. EMBO J. *22*, 2526-2535.

Hawley-Nelson, P., Androphy, E.J., Lowy, D.R., and Schiller, J.T. (1988). The specific DNA recognition sequence of the bovine papillomavirus E2 protein is an E2-dependent enhancer. EMBO J. *7*, 525-531.

Hegde, R.S. (2002). The papillomavirus E2 proteins: structure, function, and biology. Annu. Rev. Biophys. Biomol. Struct. *31*, 343-360.

Hibma, M.H., Raj, K., Ely, S.J., Stanley, M., and Crawford, L. (1995). The interaction between human papillomavirus type 16 E1 and E2 proteins is blocked by an antibody to the N-terminal region of E2. Eur. J. Biochem. *229*, 517-525.

Hirochika, H., Broker, T.R., and Chow, L.T. (1987). Enhancers and trans-acting E2 transcriptional factors of papilloma viruses. J. Virol. *61*, 2599-2606.

Hirochika, H., Hirochika, R., Broker, T.R., and Chow, L.T. (1988). Functional mapping of the human papillomavirus transcriptional enhancer and its complexes with the trans-acting E2 protein. Genes Dev. *2*, 54-67.

Holt, S.E., Schuller, G., and Wilson, V.G. (1994). DNA binding specificity of the bovine papillomavirus E1 protein is determined by sequences contained within an 18-base-pair inverted repeat element at the origin of replication. J. Virol. *68*, 1094-1102.

Hubert, W.G., Kanaya, T., and Laimins, L.A. (1999). DNA replication of human papillomavirus type 31 is modulated by elements of the upstream regulatory region that lie 5′of the minimal origin. J. Virol. *73*, 1835-1845.

Hubert, W.G., and Laimins, L.A. (2002). Human papillomavirus type 31 replication modes during the early phases of the viral life cycle depend on transcriptional and posttranscriptional regulation of E1 and E2 expression. J. Virol. *76*, 2263-2273.

Hughes, F.J., and Romanos, M.A. (1993). E1 protein of human papillomavirus is a DNA helicase/ATPase. Nucl. Acids Res. *21*, 5817-5823.

Ilves, I., Kivi, S., and Ustav, M. (1999). Long-term episomal maintenance of bovine papillomavirus type 1 plasmids is determined by attachment to host chromosomes, which is mediated by the viral E2 protein and its binding sites. J. Virol. *73*, 4404-4412.

Jenkins, O., Earnshaw, D., Sarginson, G., Del Vecchio, A., Tsai, J., Kallender, H., Amegadzie, B., and Browne, M. (1996). Characterization of the helicase and ATPase activity of human papillomavirus type 6b E1 protein. J. Gen. Virol. *77*, 1805-1809.

Jian, Y., Van Tine, B.A., Chien, W.-M., Shaw, G.M., Broker, T.R., and Chow, L.T. (1999). Concordant induction of cyclin E and p21cip1 in differentiated keratinocytes by the HPV E7 protein inhibits cellular and viral DNA synthesis. Cell Growth Diff. *10*, 101-111.

Kalantari, M., Calleja-Macias, I.E., Tewari, D., Hagmar, B., Lie, K., Barrera-Saldana, H.A., Wiley, D.J., and Bernard, H.-U. (2004). Conserved methylation patterns of human papillomavirus type 16 DNA in asymptomatic infection and cervical neoplasia. J. Virol. *78*, 12762-12772.

Kanda, T., Otter, M., and Wahl, G.M. 2001. Coupling of mitotic chromosome tethering and replication competence in Epstein-Barr virus-based plasmids. Mol. Cell. Biol. *21*, 3576-3588.

Kasukawa, H., Howley, P.M., and Benson, J.D. (1998). A fifteen-amino-acid peptide inhibits human papillomavirus E1-E2 interaction and human papillomavirus DNA replication *in vitro*. J. Virol. *72*, 8166-8173.

Kim, K., Garner-Hamrick, P.A., Fisher, C., Lee, D., and Lambert, P.F. (2003). Methylation patterns of papillomavirus DNA, its influence on E2 function, and implications in viral infection. J. Virol. *77*, 12450-12459.

Kim, K., and Lambert, P.F. (2002). E1 protein of bovine papillomavirus 1 is not required for the maintenance of viral plasmid DNA replication. Virology *293*, 10-14.

Knight, G., Grainger, J.R., Gallimore, P.H., and Roberts, S. 2004. Cooperation between different forms of the human papillomavirus type 1 E4 protein to block cell cycle progression and cellular DNA synthesis. J. Virol. *78*, 13920-13933.

Knight, J.D., Li, R., and Botchan, M. (1991). The activation domain of the bovine papillomavirus E2 protein mediates association of DNA-bound dimers to form DNA loops. Proc. Natl. Acad. Sci. USA *88*, 3204-3208.

Kuo, S.-R., Liu, J.-S., Broker, T.R., and Chow, L.T. (1994). Cell-free replication of the human papillomavirus DNA with homologous viral E1 and E2 proteins and human cell extracts. J. Biol. Chem. *269*, 24058-24065.

Kurg, R., Langel, U., and Ustav, M. (2000). Inhibition of the bovine papillomavirus E2 protein activity by peptide nucleic acid. Virus Res. *66*, 39-50.

Lambert, P.F., Hubbert, N.L., Howley, P.M., and Schiller, J.T. (1989). Genetic assignment of multiple E2 gene products in bovine papillomavirus-transformed cells. J. Virol. *63*, 3151-3154.

Lambert, P.F., Monk, B.C., and Howley, P.M. (1990). Phenotypic analysis of bovine papillomavirus type 1 E2 repressor mutants. J. Virol. *64*, 950-956.

Lambert, P.F., Spalholz, B.A., and Howley, P.M. (1987). A transcriptional repressor encoded by BPV-1 shares a common carboxy-terminal domain with the E2 transactivator. Cell *50*, 69-78.

Lee, K.-Y., Broker, T.R., and Chow, L.T. (1998). Transcription factor YY1 inhibits cell-free replication from the HPV-18 origin. J. Virol. *72*, 4911-4917.

Lee, D., Hwang, S.G., Kim, K, and Choe, J. (2002). Functional interaction between p/CAF and human papillomavirus E2 protein. J. Biol. Chem. *277*, 6483-6489.

Lee, D., Kim, H., Lee, Y., and Choe, J. (1997). Identification of sequence requirement for the origin of DNA replication in human papillomavirus type 18. Virus Res. *52*, 97-108.

Lee, D., Lee, B., Kim, J., Kim, D.W., and Choe, J. (2000). cAMP response element-binding protein-binding protein binds to human papillomavirus E2 protein and activates E2-dependent transcription. J. Biol. Chem. *275*, 7045-7051.

Lee, D., Sohn, H., Kalpana, G.V., and Choe, J. (1999). Interaction of E1 and hSNF5 proteins stimulates replication of human papillomavirus DNA. Nature *399*, 487-491.

Lehman, C.W., and Botchan, M.R. (1998). Segregation of viral plasmids depends on tethering to chromosomes and is regulated by phosphorylation. Proc. Natl. Acad. Sci. USA *95*, 4338-4343.

Leng, X., Ludes-Meyers, J.H., and Wilson, V.G. (1997). Isolation of an amino-terminal region of bovine papillomavirus type 1 E1 protein that retains origin binding and E2 interaction capacity. J. Virol. *71*, 848-852.

Leng, X., Wilson, V.G., and Xiao, X.L. (1994). Genetically defined nuclear localization signal sequence of bovine papillomavirus E1 protein is necessary and sufficient for the nuclear localization of E1-beta-galactosidase fusion proteins. J. Gen. Virol. *75*, 2463-2467.

Lentz, M.R., Pak, D., Mohr. I., and Botchan, M.R. (1993). The E1 replication protein of bovine papillomavirus type 1 contains an extended nuclear localization signal that includes a p34cdc2 phosphorylation site. J. Virol. *67*, 1414-1423.

Lentz, M.R. (2002). A carboxyl-terminal serine of the bovine papillomavirus E1 protein is phosphorylated *in vivo* and *in vitro*. Virus Res. *83*, 213-219.

Lentz, M., Zanardi, T., Filzen, R., Carter, J., and Hella, M. (2002). Functional analysis of a carboxyl-terminal phosphorylation mutant of the bovine papillomavirus E1 protein. J. Mol. Biol. *316*, 599-609.

Lepik, D., Ilves, I., Kristjuhan, A., Maimets, T., and Ustav, M. (1998). p53 protein is a suppressor of papillomavirus DNA amplificational replication. J. Virol. *72*, 6822-6831.

Lepik, D., and Ustav, M. (2000). Cell-specific modulation of papovavirus replication by tumor suppressor protein p53. J. Virol. *74*, 4688-4697.

Li, R., and Botchan, M. (1994). Acidic transcription factors alleviate nucleosome-mediated repression of DNA replication of bovine papillomavirus type 1. Proc. Natl. Acad. Sci. USA *91*, 7051-7055.

Lim, D.A., Gossen, M., Lehman, C.W., and Botchan, M.R. (1998). Competition for DNA binding sites between the short and long forms of E2 dimers underlies repression in bovine papillomavirus type 1 DNA replication control. J. Virol. *72*, 1931-1940.

Lin, B. Y., Ma, T., Liu, J.-S., Kuo, S.-R., Jin, G., Broker, T.R., Harper, J.W., and Chow, L.T. (2000). HeLa cells are phenotypically limiting in cyclin E / CDK2 for efficient human papillomavirus DNA replication. J. Biol. Chem. *275*, 6167-6174.

Lin, B. Y., Makhov, A.M., Griffith, J.D., Broker, T.R., and Chow, L.T. (2002). Chaperone proteins abrogate inhibition of the human papillomavirus (HPV) E1 helicase by the HPV E2 protein. Mol. Cell. Biol. *18*, 6592-6604.

Liu, J.-S., Kuo, S.-R., Broker, T.R., and Chow, L.T. (1995). The functions of human papillomavirus type 11 E1, E2 and E2C proteins in cell-free DNA replication. J. Biol. Chem. *270*, 27283-27291.

Liu, J.-S., Kuo, S.-R., Makhov, A.M., Cyr, D.M., Griffith, J.D., Broker, T.R., and Chow, L.T. (1998). Human Hsp70 and Hsp40 chaperone proteins facilitate HPV-11 E1 protein binding to the origin and stimulate cell-free replication. J. Biol. Chem. *273*, 30704-30712.

Loo, Y.M., and Melendy, T. (2004). Recruitment of replication protein A by the papillomavirus E1 protein and modulation by single-stranded DNA. J. Virol. *78,* 1605-1615.

Lu, J., Sun, Y., Rose, R., Bonnez, W., and McCance, D. (1993). Two E2 binding sites (E2BS) alone or one E2BS plus an A/T-rich region are minimal requirements for the replication of the human papillomavirus type 11 origin. J. Virol. *67*, 7131-7139.

Lusky, M., and Fontane, E. (1991). Formation of the complex of bovine papillomavirus E1 and E2 proteins is modulated by E2 phosphorylation and depends upon sequences within the carboxyl terminus of E1. Proc. Natl. Acad. Sci. USA *88*, 6363-6367.

Lusky, M., Hurwitz, J., and Seo, Y.S. (1994). The bovine papillomavirus E2 protein modulates the assembly of but is not stably maintained in a replication-competent multimeric E1-replication origin complex. Proc. Natl. Acad. Sci. USA. *91*, 8895-8899.

Ma, T., Zou, N., Lin, B. Y., Chow, L.T., and Harper, J.W. (1999). Interaction between cyclin-dependent kinases and human papillomavirus replication initiation protein E1 is required for efficient viral replication. Proc. Natl. Acad. Sci. USA *96*, 382-387.

Makiniemi, M., Hillukkala, T., Tuusa, J., Reini, K., Vaara, M., Huang, D., Pospiech, H., Majuri, I., Westerling, T., Makela, T.P., and Syvaoja, J.E. (2001). BRCT domain-containing protein TopBP1 functions in DNA replication and damage response. J. Biol. Chem. *276*, 30399-30406.

Malcles, M.H., Cueille, N., Mechali, F., Coux, O., and Bonne-Andrea, C. (2002). Regulation of bovine papillomavirus replicative helicase E1 by the ubiquitin-proteasome pathway. J. Virol. *76*, 11350-11358.

Männik, A., Runkorg, K., Jaanson, N., Ustav, M., and Ustav, E. (2002). Induction of the bovine papillomavirus origin "onion skin"-type DNA replication at high E1 protein concentrations *in vivo*. J. Virol. *76*, 5835-5845.

Massimi, P., Pim, D., Bertoli, C., Bouvard, V., and Banks, L. (1999). Interaction between the HPV-16 E2 transcriptional activator and p53. Oncogene *18*, 7748-7754.

Masterson, P.J., Stanley, M.A., Lewis, A.P., and Romanos, M.A. (1998). A C-terminal helicase domain of the human papillomavirus E1 protein binds E2 and the DNA polymerase alpha-primase p68 subunit. J. Virol. *72*, 7407-7419.

McBride, A.A., and Howley, P.M. (1991). Bovine papillomavirus with a mutation in the E2 serine 301 phosphorylation site replicates at a high copy number. J. Virol. *65*, 6528-6534.

McBride, A.A., Byrne, J.C., and Howley, P.M. (1989). E2 polypeptides encoded by bovine papillomavirus type 1 form dimers through the common carboxyl-terminal domain: transactivation is mediated by the conserved amino-terminal domain. Proc. Natl. Acad. Sci. USA *86*, 510-514.

McShan, G.D., and Wilson, V.G. (2000). Contribution of bovine papillomavirus type 1 E1 protein residue 48 to replication function. J. Gen. Virol. *81*, 1995-2004.

McShan, G.D., and Wilson, V.G. (1997). Reconstitution of a functional bovine papillomavirus type 1 origin of replication reveals a modular tripartite replicon with an essential AT-rich element. Virology *237*, 198-208.

Mechali, F., Hsu, C.Y., Castro, A., Lorca ,T., and Bonne-Andrea, C. (2004). Bovine papillomavirus replicative helicase E1 is a target of the ubiquitin ligase APC. J. Virol. *78*, 2615-2619.

Melendy, T., Sedman, J., and Stenlund, A. (1995). Cellular factors required for papillomavirus DNA replication. J. Virol. *69*, 7857-7867.

Mendoza, R., Gandhi, L., and Botchan, M.R. (1995). E1 recognition sequences in the bovine papillomavirus type 1 origin of DNA replication: interaction between half sites of the inverted repeats. J. Virol. *69*, 3789-3798.

Mohr, I.J., Clark, R., Sun, S., Androphy, E.J., MacPherson, P., and Botchan, M.R. (1990). Targeting the E1 replication protein to the papillomavirus origin of replication by complex formation with the E2 transactivator. Science *250*, 1694-1699.

Moscufo, N., Sverdrup, F., Breiding, D.E., and Androphy, E.J. (1999). Two distinct regions of the BPV1 E1 replication protein interact with the activation domain of E2. Virus Res. *65*, 141-154.

Müller, F., Giroglou, T., and Sapp, M. (1997). Characterization of the DNA-binding activity of the E1 and E2 proteins and the E1/E2 complex of human papillomavirus type 33. J. Gen. Virol. *78*, 911-915.

Müller, A., Ritzkowsky, A., and Steger, G. (2002). Cooperative activation of human papillomavirus type 8 gene expression by the E2 protein and the cellular coactivator p300. J. Virol. *76*, 11042-11053.

Müller, F., and Sapp, M. (1996). Domains of the E1 protein of human papillomavirus type 33 involved in binding to the E2 protein. Virology *219*, 247-256.

Müller, F., Seo, Y.S., and Hurwitz, J. (1994). Replication of bovine papillomavirus type 1 origin-containing DNA in crude extracts and with purified proteins. J. Biol. Chem. *269*, 17086-17094.

Münger, K., Baldwin, A., Edwards, K.M., Hayakawa, H., Nguyen, C.L., Owens, M., Grace, M., Huh, K. (2004). Mechanisms of human papillomavirus-induced oncogenesis. J Virol. *78*, 11451-11460.

Nakahara, T., Nishimura, A., Tanaka, M., Ueno, T., Ishimoto, A., and Sakai, H. (2002). Modulation of the cell division cycle by human papillomavirus type 18 E4. J. Virol. *76*, 10914-10920.

Nasseri, M., Hirochika, R., Broker, T.R., and Chow, L.T. (1987). A human papillomavirus type 11 transcript encoding an E1^E4 protein. Virology *159*, 433-439.

O'Connor, M.J., Stunkel, W., Koh, C.H., Zimmermann, H., and Bernard, H.-U. (2000). The differentiation-specific factor CDP/Cut represses transcription and replication of human papillomaviruses through a conserved silencing element. J. Virol. *74*, 401-410.

Oh, S.T., Longworth, M.S., and Laimins, L.A. (2004). Roles of the E6 and E7 proteins in the life cycle of low-risk human papillomavirus type 11. J. Virol. *78*, 2620-2626.

Ozbun, M.A. and Meyers, C. (1998). Human papillomavirus type 31b E1 and E2 transcript expression correlates with vegetative viral genome amplification. Virology *248*, 218-230.

Palermo-Dilts, D.A., Broker, T.R., and Chow, L.T. (1990). Human papillomavirus type 1 produces redundant as well as polycistronic messenger RNAs in plantar warts. J. Virol. *64*, 3144-3149.

Park, R.B., and Androphy, E.J. (2002). Genetic analysis of high-risk E6 in episomal maintenance of human papillomavirus genomes in primary human keratinocytes. J. Virol. *76*, 11359-11364.

Park, P., Copeland, W., Yang, L., Wang, T., Botchan, M.R., and Mohr, I.J. (1994). The cellular DNA polymerase alpha-primase is required for papillomavirus DNA replication and associates with the viral E1 helicase. Proc. Natl. Acad. Sci. USA *91*, 8700-8704.

Peng, Y.C., Breiding, D.E., Sverdrup, F., Richard, J., and Androphy, E.J. (2002). AMF-1/Gps2 binds p300 and enhances its interaction with papillomavirus E2 proteins. J. Virol. *74*, 5872-5879.

Penrose, K.J., and McBride, A.A. (2000). Proteasome-mediated degradation of the papillomavirus E2-TA protein is regulated by phosphorylation and can modulate viral genome copy number. J. Virol. *74*, 6031-6038.

Penrose, K.J., Garcia-Alai, M., de Prat-Gay, G., and McBride, A.A. (2004). Casein kinase II phosphorylation-induced conformational switch triggers degradation of the papillomavirus E2 protein. J. Biol. Chem. *279*, 22430-22439.

Pepinsky, R.B., Prakash, S.S., Corina, K., Grossel, M.J., Barsoum, J., and Androphy, E.J. (1997). Sequences flanking the core DNA-binding domain of bovine papillomavirus type 1 E2 contribute to DNA-binding function. J. Virol. *71*, 828-831.

Phelps, W.C., and Howley, P.M. (1987). Transcriptional trans-activation by the human papillomavirus type 16 E2 gene product. J. Virol. *61*, 1630-1638.

Piirsoo, M., Ustav, E., Mandel, T., Stenlund, A., and Ustav, M. (1996). Cis and trans requirements for stable episomal maintenance of the BPV-1 replicator. EMBO J. *15*, 1-11.

Polissi, A., Goffin, L., and Georgopoulos, C. (1995). The *Escherichia coli* heat shock response and bacteriophage lambda development. FEMS Microbiol. Rev. *17*, 159-169.

Raj, K., Berguerand, S., Southern, S., Doorbar, J., and Beard, P. 2004. E1^E4 protein of human papillomavirus type 16 associates with mitochondria. J. Virol. *78*, 7199-7207.

Rehtanz, M., Schmidt, H.M., Warthorst, U., and Steger, G. (2004). Direct interaction between nucleosome assembly protein 1 and the papillomavirus E2 proteins involved in activation of transcription. Mol. Cell. Biol. *24*, 2153-2168.

Remm, M., Brain, R., and Jenkins, J.R. (1992). The E2 binding sites determine the efficiency of replication for the origin of human papillomavirus type 18. Nucl. Acids Res. *20*, 6015-6021.

Remm, M., Remm, A., and Ustav, M. (1999). Human papillomavirus type 18 E1 protein is translated from polycistronic mRNA by a discontinuous scanning mechanism. J. Virol. *73*, 3062-3070.

Romanczuk, H., Thierry, F., and Howley, P.M. (1990). Mutational analysis of cis elements involved in E2 modulation of human papillomavirus type 16 P97 and type 18 P105 promoters. J. Virol. *64*, 2849-2859.

Rotenberg, M.O., Chiang, C.-M., Ho, M.S., Broker, T.R. and Chow, L.T. (1989a). Characterization of cDNAs of spliced HPV-11 E2 mRNA and other HPV mRNAs recovered via retrovirus-mediated gene transfer. Virology *172*, 468-477.

Rotenberg, M.O., Chow, L.T., and Broker, T.R.. (1989b). Characterization of rare human papillomavirus type 11 mRNAs coding for regulatory and structural proteins using the polymerase chain reaction. Virology *172*, 489-497.

Russell, J., and Botchan, M.R. (1995). cis-Acting components of human papillomavirus (HPV) DNA replication: linker substitution analysis of the HPV type 11 origin. J. Virol. *69*, 651-660.

Sakai, H., Yasugi, T., Benson, J.D., Dowhanick, J.J., and Howley, P.M. (1996). Targeted mutagenesis of the human papillomavirus type 16 E2 transactivation domain reveals separable transcriptional activation and DNA replication functions. J. Virol. *70*, 1602-1611.

Sanders, C.M., and Stenlund, A. (1998). Recruitment and loading of the E1 initiator protein: an ATP-dependent process catalyzed by a transcription factor. EMBO J. *17*, 7044-7055.

Sanders, C.M., and Stenlund, A (2001). Mechanism and requirements for bovine papillomavirus, type 1, E1 initiator complex assembly promoted by the E2 transcription factor bound to distal sites. J. Biol. Chem. *276*, 23689-23699.

Sarafi, T.R., and McBride, A.A. (1995). Domains of the BPV-1 E1 replication protein required for origin-specific DNA binding and interaction with the E2 transactivator. Virology *211*, 385-396.

Schaal, T.D., Mallet, W.G., McMinn, D.L., Nguyen, N.V., Sopko, M.M., John, S., and Parekh, B.S. (2003). Inhibition of human papilloma virus E2 DNA binding protein by covalently linked polyamides. Nucl. Acids Res. *31*, 1282-1291.

Sedman, T., Sedman, J. and Stenlund, A. (1997). Binding of the E1 and E2 proteins to the origin of replication of bovine papillomavirus. J. Virol. *71*, 2887-2896.

Sedman, J., and Stenlund, A. (1995). Co-operative interaction between the initiator E1 and the transcriptional activator E2 is required for replicator specific DNA replication of bovine papillomavirus *in vivo* and *in vitro*. EMBO J. *14*, 6218-6228.

Sedman, J., and Stenlund, A. (1996).The initiator protein E1 binds to the bovine papillomavirus origin of replication as a trimeric ring-like structure. EMBO J. 15, 5085-5092.

Sedman, J., and Stenlund, A. (1998). The papillomavirus E1 protein forms a DNA-dependent hexameric complex with ATPase and DNA helicase activities. J. Virol. *72*, 6893-6897.

Seo, Y.S., Müller, F., Lusky, M., Gibbs, E., Kim, H.Y., Phillips, B., and Hurwitz, J. (1993a). Bovine papilloma virus (BPV)-encoded E2 protein enhances binding of E1 protein to the BPV replication origin. Proc. Natl. Acad. Sci. USA *90*, 2865-2869.

Seo, Y.S., Müller, F., Lusky, M., and Hurwitz, J. (1993b). Bovine papilloma virus (BPV)-encoded E1 protein contains multiple activities required for BPV DNA replication. Proc. Natl. Acad. Sci. USA *90*, 702-706.

Sherman, L., Alloul, N., Golan, I., Durst, M., and Baram, A.(1992a). Expression and splicing patterns of human papillomavirus type-16 mRNAs in pre-cancerous lesions and carcinomas of the cervix, in human keratinocytes immortalized by HPV 16, and in cell lines established from cervical cancers. Int. J. Cancer *50*, 356-364.

Sherman, L., Golan, Y., Mitrani-Rosenbaum, S., and Baram, A. (1992b). Differential expression of HPV types 6 and 11 in condylomas and cervical preneoplastic lesions. Virus Res. *25*, 23-36.

Sherr, C.J., and Roberts, J.M. (1999). CDK inhibitors: positive and negative regulators of G1-phase progression. Genes Dev. *13*, 1501-1512.

Shire, K., Ceccarelli,D.F., Avolio-Hunter, T.M., and Frappier, L.(1999). EBP2, a human protein that interacts with sequences of the Epstein-Barr virus nuclear antigen 1 important for plasmid maintenance. J. Virol. *73*, 2587-2595.

Skiadopoulos, M.H., and McBride, A.A. (1996). The bovine papillomavirus type 1 E2 transactivator and repressor proteins use different nuclear localization signals. J. Virol. *70*, 1117-1124.

Skiadopoulos, M.H., and McBride, A.A. (1998). Bovine papillomavirus type 1 genomes and the E2 transactivator protein are closely associated with mitotic chromatin. J. Virol. *72*, 2079-2088.

Spalholz, B.A., Yang, Y.C., and Howley, P.M. (1995). Transactivation of a bovine papilloma virus transcriptional regulatory element by the E2 gene product. Cell *42*, 183-191.

Stenlund, A. (2003a). E1 initiator DNA binding specificity is unmasked by selective inhibition of non-specific DNA binding. EMBO J. *22*, 954-963.

Stenlund, A. (2003b). Initiation of DNA replication: lessons from viral initiator proteins. Nat. Rev. Mol. Cell. Biol. *4*, 777-785.

Stubenrauch, F., Colber, A.M., and Laimins, L.A. (1998a). Transactivation by the E2 protein of oncogenic human papillomavirus type 31 is not essential for early and late viral functions. J. Virol. *72*, 8115-8123.

Stubenrauch, F., Hummel, M., Iftner, T., and Laimins, L.A. (2000). The E8E2C protein, a negative regulator of viral transcription and replication, is required for extrachromosomal maintenance of human papillomavirus type 31 in keratinocytes. J. Virol. *74*, 1178-1186.

Stubenrauch, F., Lim, H.B., and Laimins, L.A. (1998b). Differential requirements for conserved E2 binding sites in the life cycle of oncogenic human papillomavirus type 31. J. Virol. *72*, 1071-1077.

Sun, Y, Han, H., and McCance, D.J. (1998). Active domains of human papillomavirus type 11 E1 protein for origin replication. J. Gen. Virol. *79*, 1651-1658.

Sun, Y.N., Lu, J. Z., and McCance, D.J. (1996). Mapping of HPV-11 E1 binding site and determination of other important cis elements for replication of the origin. Virology *216*, 219-222.

Sverdrup, F., and Khan, S.A. (1995). Two E2 binding sites alone are sufficient to function as the minimal origin of replication of human papillomavirus type 18 DNA. J. Virol. *69*, 1319-1323.

Swindle, C.S., and Engler, J.A. (1998). Association of the human papillomavirus type 11 E1 protein with histone H1. J. Virol. *72*, 1994-2001.

Swindle, C.S., Zou, N., Van Tine, B.A., Shaw, G.W., Engler, J.A., and Chow, L.T. (1999). HPV-11 DNA replication compartments in the nuclei of a transient replication system. J. Virol. *73*, 1001-1009.

Tan, S.H., Leong, L.E., Walker, P.A., and Bernard, H.-U. (1994). The human papillomavirus type 16 E2 transcription factor binds with low cooperativity to two flanking sites and represses the E6 promoter through displacement of Sp1 and TFIID. J. Virol. *68*, 6411-6420.

Taylor, E.R., and Morgan, I.M. (2003). A novel technique with enhanced detection and quantitation of HPV-16 E1- and E2-mediated DNA replication. Virology *315*, 103-109.

Thain, A., Jenkins, O., Clarke, A.R., and Gaston, K. (1996). CpG methylation directly inhibits binding of the human papillomavirus type 16 E2 protein to specific DNA sequences. J. Virol. *70*, 7233-7235.

Thierry. F., and Howley, P.M. (1991). Functional analysis of E2-mediated repression of the HPV18 P105 promoter. New Biol. *3*, 90-100.

Thomas, J.T., Hubert, W.G., Ruesch, M.N., and Laimins, L.A. (1999). Human papillomavirus type 31 oncoproteins E6 and E7 are required for the maintenance of episomes during the viral life cycle in normal human keratinocytes. Proc. Natl. Acad. Sci. USA *96*, 8449-8454.

Thorner, L.K., Lim, D.A., and Botchan, M.R. (1993). DNA-binding domain of bovine papillomavirus type 1 E1 helicase: structural and functional aspects. J. Virol. *67*, 6000-6014.

Titolo, S., Brault, K., Majewski, J., White, P.W., and Archambault, J. (2003). Characterization of the minimal DNA binding domain of the human papillomavirus E1 helicase: fluorescence anisotropy studies and characterization of a dimerization-defective mutant protein. J. Virol. *77*, 5178-5191.

Titolo, S., Pelletier, A., Pulichino, A.M., Brault, K., Wardrop, E., White, P.W., Cordingley, M.G., and Archambault, J. (2000). Identification of domains of the human papillomavirus type 11 E1 helicase involved in oligomerization and binding to the viral origin. J. Virol. *74*, 7349-7361.

Titolo, S., Pelletier, A., Sauve, F., Brault, K., Wardrop, E., White, P.W., Amin, A., Cordingley, M.G., and Archambault, J. (1999). Role of the ATP-binding domain of the human papillomavirus type 11 E1 helicase in E2-dependent binding to the origin. J. Virol. *73*, 5282-5293.

Ustav, M., and Stenlund, A. (1991). Transient replication of BPV-1 requires two viral polypeptides encoded by the E1 and E2 open reading frames. EMBO J. *10*, 449-457.

Ustav, E., Ustav, M., Szymanski, P., and Stenlund, A. (1993). The bovine papillomavirus origin of replication requires a binding site for the E2 transcriptional activator. Proc. Natl. Acad. Sci. USA *90*, 898-902.

Van Hatten, R.A., Tutter, A.V., Holway, A.H., Khederian, A.M., Walter, J.C., and Michael, W.M. (2002). The Xenopus Xmus101 protein is required for the recruitment of Cdc45 to origins of DNA replication. J. Cell Biol. *159*, 541-547.

Van Horn, G., Sheikh, S., and Khan, S.A. (2001). Regulation of human papillomavirus type 1 replication by the viral E2 protein. Virology *287*, 214-224.

Van Tine, B.A., Dao, L.D., Wu, S.-Y., Sonbuchner, T.M., Lin, B.Y., Zou, N., Chiang, C.-M., Broker, T.R., and Chow, L.T. (2004a). HPV origin binding protein associates with mitotic spindles to enable viral DNA partitioning. Proc. Natl. Acad. Sci. USA *101*, 4030-4035.

Van Tine, B.A, Kappes, J.C., Banerjee, N.S., Knops, J., Lai, L., Steenbergen, R.D.M., Meijer, C.L.J.M., Snijders, P.J.F., Chatis, P., Broker, T.R., Moen, Jr., P.T., and Chow, L.T. (2004b). Clonal selection for transcriptionally active viral oncogenes during progression to cancer by DNA methylation-mediated silencing. J. Virol. *78,* 11172-11186.

Voitenleitner, C., and Botchan, M. (2002). E1 protein of bovine papillomavirus type 1 interferes with E2 protein-mediated tethering of the viral DNA to mitotic chromosomes. J. Virol. *76*, 3440-3451.

Wang, Y., Coulombe, R., Cameron, D.R., Thauvette, L., Massariol, M.J., Amon, L.M., Fink, D., Titolo, S., Welchner, E., Yoakim, C., Archambault, J., and White, P.W. (2004). Crystal structure of the E2 transactivation domain of human papillomavirus type 11 bound to a protein interaction inhibitor. J. Biol. Chem. *279*, 6976-6985.

Webster, K., Parish, J., Pandya, M., Stern, P.L., Clarke, A.R., and Gaston, K. (2000). The human papillomavirus (HPV) 16 E2 protein induces apoptosis in the absence of other HPV proteins and via a p53-dependent pathway. J. Biol. Chem. *275*, 87-94.

West, M., Flanery, D., Woytek, K., Rangasamy, D., and Wilson, V.G. (2001). Functional mapping of the DNA binding domain of bovine papillomavirus E1 protein. J. Virol. *75*, 11948-11960.

White, P.W., Pelletier, A., Brault, K., Titolo, S., Welchner, E., Thauvette, L., Fazekas, M., Cordingley, M.G. and Archambault, J. (2001). Characterization of recombinant HPV6 and 11 E1 helicases: effect of ATP on the interaction of E1 with E2 and mapping of a minimal helicase domain. J. Biol. Chem. *276*, 22426-22438.

White, P.W., Titolo, S., Brault, K., Thauvette, L., Pelletier, A., Welchner, E., Bourgon, L., Doyon, L., Ogilvie, W.W., Yoakim, C., Cordingley, M.G., and Archambault, J. (2003). Inhibition of human papillomavirus DNA replication by small molecule antagonists of the E1-E2 protein interaction. J. Biol. Chem. *278*, 26765-26772.

Winokur, P.L., and McBride, A.A. (1992). Separation of the transcriptional activation and replication functions of the bovine papillomavirus-1 E2 protein. EMBO J. *11*, 4111-4118.

Winokur, P.L., and McBride, A.A. (1996). The transactivation and DNA binding domains of the BPV-1 E2 protein have different roles in cooperative origin binding with the E1 protein. Virology *221*, 44-53.

Yamamoto, R.R., Axton, J.M., Yamamoto, Y., Saunders, R.D., Glover, D.M., and Henderson, D.S. (2000). The Drosophila mus101 gene, which links DNA repair, replication and condensation of heterochromatin in mitosis, encodes a protein with seven BRCA1 C-terminus domains. Genetics *156*, 711-721.

Yasugi, T., Benson, J.D., Sakai, H., Vidal, M., and Howley, P.M. (1997). Mapping and characterization of the interaction domains of human papillomavirus type 16 E1 and E2 proteins. J. Virol. *71*, 891-899.

Yamane, K., Wu, X., and Chen, J. (2002). A DNA damage-regulated BRCT-containing protein, TopBP1, is required for cell survival. Mol. Cell. Biol. *22*, 555-566.

Yang, L., Li, R., Mohr, I.J., Clark, R., and Botchan, M.R. (1991). Activation of BPV-1 replication *in vitro* by the transcription factor E2. Nature *353*, 628-632.

Yang, L., Mohr, I., Fouts, E., Lim, D.A., Nohaile, M., and Botchan, M. (1993). The E1 protein of bovine papilloma virus 1 is an ATP-dependent DNA helicase. Proc. Natl. Acad. Sci. USA *90*, 5086-5090.

You, J., Croyle, J.L., Nishimura, A., Ozato, K., and Howley, P.M. (2004). Interaction of the bovine papillomavirus E2 protein with Brd4 tethers the viral DNA to host mitotic chromosomes. Cell *117*, 349-360.

Zhao, W., Noya, F., Chen, W.Y., Townes, T.M., Chow, L.T., and Broker, T.R. (1999). Trichostatin A up-regulates HPV-11 URR-E6 promoter activity in undifferentiated primary human keratinocytes. J. Virol. *73*, 5026-5033.

Zhu, Q.-L., Smith, T.F., Lefkowitz, E., Chow, L.T., and Broker, T.R. (1994). Nucleic Acid and Protein Sequence Alignments of Human and Animal Papillomaviruses Constrained by Functional Sites. University of Alabama at Birmingham. 137 pages.

Zobel, T., Iftner. T., and Stubenrauch, F. (2003). The papillomavirus E8-E2C protein represses DNA replication from extrachromosomal origins. Mol. Cell. Biol. *23*, 8352-8362.

Zou, N., Lin, B.Y., Duan, F., Lee, K.-Y., Jin, G., Guan, R., Yao, G., Lefkowitz, E.J., Broker, T.R, and Chow, L.T. (2000). The hinge region of the human papillomavirus type 11 E2 protein contains major determinants for nuclear localization and nuclear matrix association. J. Virol. *74*, 3761-3770.

Zou, N., Liu, J.-S., Kuo, S.-R., Broker, T.R., and Chow, L.T. (1998). The carboxyl terminal region of the human papillomavirus type 16 E1 protein determines E2 specificity during DNA replication. J. Virol. *72*, 3426-3441.

Zylicz, M. (1993). The Escherichia coli chaperones involved in DNA replication. Philos. Trans. R. Soc. London B Biol. Sci. *339*, 271-277.

From: Papillomavirus Research: From Natural History to Vaccines and Beyond. Edited by: M. Saveria Campo

Chapter 6

The Papillomavirus Transcription/Replication Factor E2: Structure, Function, Cancer and Therapy

Iain M. Morgan* and Mary M. Donaldson

Abstract

As with all papillomavirus (PV) proteins, E2 has been studied in a number of systems with BPV1 providing the original paradigm. Extensive work has also been carried out investigating HPV 16, 18, 11, 6, and 31 E2 proteins amongst others. This chapter will begin with an introduction to the E2 protein and its general structure and function and will then go on to discuss more specific roles. The role of E2 and interacting partners in viral replication and transcription will be discussed, as well as their role in segregating viral genomes and regulating the host cell cycle. Finally, the status of E2 expression in cancer will be discussed and also strategies for therapeutically targeting the E2 protein for the treatment of PV related diseases.

1. Background

The E2 gene of papillomaviruses encodes a product of around 40-45kD, depending upon type. This protein is tripartite in nature (Desaintes and Demeret, 1996). Firstly, in the carboxyl terminus there is a dimerisation domain resulting in the formation of homodimers that then recognise and bind to 12bp palindromic DNA sequences within the transcriptional control region of PV genomes. The crystal structure of this domain has been solved (Liang et al., 1996) and the only other protein with a similar structure is the Epstein Barr Virus (EBV) protein EBNA1 (Epstein Barr Nuclear Antigen 1), although there is no primary sequence similarity between E2 and EBNA1 (Bochkarev et al., 1995). EBNA1 is the viral origin recognition factor for EBV and as E2 carries out the same function for PV (see the E2-E1 section below) this suggests an evolutionary conservation of this structure that is related to function. Secondly, the middle region of E2, called the hinge, has rather indeterminate function, although important for regulating the stability of some E2 proteins (see below) and determining their nuclear localisation in others (Zou et al., 2000). Thirdly, the amino terminal domain is essential for regulation of transcription and also regulation of viral DNA replication via interaction with the E1 protein (Desaintes and Demeret, 1996). The crystal structures for both the HPV16 and HPV18 E2 proteins have been solved and suggest that this domain may also form homodimers (Harris et al., 1999; Antson et al., 2000). There are many conserved amino acids within the amino terminal domain of E2. Mutation studies have demonstrated that many of these conserved residues are essential for E2 transcriptional activation function and also interaction with E1 and therefore DNA replication activity (Sakai et al., 1996). However, there are other conserved residues within this domain that are not involved in the transcription or replication properties of the E2 proteins; the function of these residues remains to be discovered. Figure 1 represents the domains of E2 and their roles in regulating the various functions of the E2 protein that are discussed below.

As well as a full length E2 protein, BPV1 encodes a truncated version of the protein that lacks the amino terminal transactivation and DNA replication domain (Choe et al., 1989). This truncated protein can act as a competitor for the full length E2 protein and prevent its function due to interfering with DNA target sequence recognition (Lim et al., 1998). The exact role for this truncated E2 protein (produced by internal translation of the E2 message) in the BPV1 life cycle is not known and there are as yet no similar truncations described for HPV E2 proteins.

2. Regulation of Viral Replication by the PV E1 and E2 Proteins

Following infection the PV genome must replicate its own genome in order to execute a successful life cycle. This is carried out by the viral factors E1 and E2 interacting with cellular factors and recruiting them for this function (Desaintes and Demeret, 1996). The E1 protein is described in Chapter 5 in this book and therefore only a brief sketch of the E1-E2 interaction is provided here. Following expression, the E2 protein recognises and binds to 12bp palindromic sequences in the viral origin of replication (ori) that is located within the transcriptional control region for the virus. Following binding to the ori, there is a physical interaction between the amino terminal domain of E2 and the E1 protein. E1 is then recruited to the ori

*For correspondence. Email i.morgan@vet.gla.ac.uk

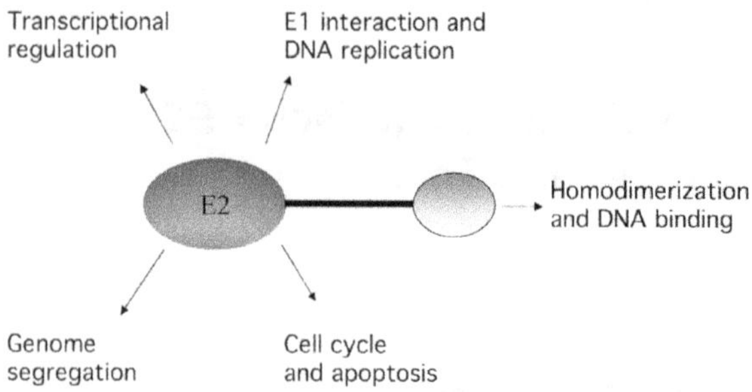

Figure 1. The functions of the E2 protein. The E2 amino terminal domain is involved in regulating several of the biological functions of the E2 protein and is shown in red. This domain is linked to the carboxyl terminus homodimerization and DNA binding domain (grey circle) via a variant hinge region (black line). The role of the E2 protein in regulating these functions is discussed in the text along with the possible involvement of loss of E2 expression in regulating HPV induced carcinogenesis.

whereupon it forms a hexameric complex responsible for initiating and executing viral DNA replication. Therefore the E2 protein is the viral origin recognition complex (ORC) and the E1 protein is the helicase responsible for the initiation of DNA replication. This is functionally analogous to the mammalian ORC proteins that recruit the MCM helicase complex to mammalian origins of DNA replication (Prasanth et al., 2004).

3. E2 and Transcriptional Regulation

Following interaction with target DNA sequences, the E2 protein can either activate or repress transcription from adjacent promoters dependent upon promoter type, cell type and the level of E2 protein (Bouvard et al., 1994; Steger and Corbach, 1997; Vance et al., 1999). The BPV1 long control region (LCR, region responsible for regulating transcription from the viral genome and also containing the viral origin of replication) contains 12 E2 DNA target sequences. E2 interaction with these sequences activates transcription from the adjacent promoter resulting in increased levels of E6 (Sandler et al., 1996). HPV genomes have a different organisation of E2 DNA binding sequences when compared with that of the BPV1 LCR. Only 3-4bp upstream from the TATA box are two E2 target sequences that are in turn separated by 3-4bp. Another 80-100bp upstream is another E2 target sequence; between this site and the two adjacent to the TATA box is an A/T rich region that constitutes the viral origin of DNA replication. Four to five hundred base pairs upstream from the third E2 sequence is another target sequence (see, for example, Steger and Corbach, 1997). Between this target sequence and those surrounding the origin of replication is an epithelial specific enhancer responsible for mediating cell type specific transcriptional activation from the adjacent promoter (see Chapter 4 and for a review Hoppe-Seyler and Butz, 1994). This organisation of E2 target DNA sequences is retained in all HPV, whether cutaneous or mucosal in their specificity, and is also retained in the epitheliotropic BPV4 genome (Vance et al., 1999), demonstrating a close conservation throughout the evolution of PV genomes, and that the organisation of the E2 target sequences plays an important role in regulating the life cycle of the virus.

In human cells, the majority of studies have demonstrated that expression of HPV E2 protein at various levels results in repression of transcription from the adjacent promoter. In one study HPV16 E2 was shown to activate transcription in primary human epithelial cells at low levels of E2 protein, but repression occurred at elevated E2 levels (Bouvard et al., 1994). One of the proposed mechanisms for repression by E2 is an ability of E2 to bind to the TATA-adjacent binding sequence and interfere with TBP binding (Steger and Corbach, 1997). In support of this hypothesis, mutation of the TATA adjacent E2 site partially relieves repression of transcription from the HPV promoters by E2. One of the restrictions in the study of the HPV LCR sequences was the inability to routinely investigate E2 function in the target cell type for viral infection. Primary human epithelial cells are difficult to grow and also difficult to transiently transfect for transcription assays. Certainly in the case of HPV that infect cervical epithelial cells, it is not easy to carry out such studies in the target tissue. To overcome this barrier, studies were carried out with the BPV4 LCR in primary bovine palate keratinocytes. The BPV4 LCR has an identical organisation of E2 binding sequences and epithelial specific enhancer compared with the HPV LCRs (Morgan et al., 1999) and palate keratinocytes are the target cell type for BPV4 infection, resulting in papillomas in the alimentary canal of cattle that can, on occasion, progress to carcinomas (Campo, 2002). The results from these experiments demonstrated that low levels of E2 protein activated transcription from the LCR, while at higher E2 levels transcription transactivation was repressed (Morgan et al., 1998). Mutation of the TATA-adjacent E2 DNA binding site restored the ability of elevated levels of E2 protein to activate transcription from the LCR promoter. These studies also demonstrated that E2 could not activate transcription from the LCR in fibroblast cells; this was not exclusively due to the LCR not operating in fibroblasts but was also partially a property of E2 itself. Therefore as well as the LCR having an epithelial specific enhancer sequence, the E2 protein also activates transcription from the LCR in an epithelial specific manner.

The conclusion from all of these studies is that it seems likely (at least in the early part of the viral life cycle) that following infection, low levels of E2 can activate transcription from the adjacent promoter following interaction with the target sequences in the LCR. At later time points in the infectious cycle, or following transformation of the infected cells, the E2 protein is no longer able to activate transcription from the LCR but operates solely as a transcriptional repressor. This repression may well be beneficial at elevated levels of E2 as it would result in the shut down of transcription, perhaps assisting in the DNA replication process. It also provides a negative feedback loop to control the levels of the E6 and E7 oncoproteins; elevated transcription will result in elevated levels not only of E6 and E7 but also of E2, and this would then act to repress transcription from the LCR. This would result in reduction of the E2 protein levels that would eventually result in elevated levels of transcription once more.

To activate transcription the E2 protein must interact with a host of cellular factors. It is still not clear exactly which cellular proteins are essential for E2 transcriptional function but the known interactors are discussed below. As well as requiring cellular transcriptional co-activators, E2 interacts with cellular proteins that bind to the LCR sequences. There are three main types of factors that interact with the LCR and can also regulate E2 function. Firstly, cellular proteins that interact with the LCR and also with E2; examples of these factors are Sp1 and TBP. An interaction between Sp1 and E2, when bound to the LCR, results in a synergistic activation of transcription (Li et al., 1991), while an interaction between E2 and TBP has been implicated in transcriptional repression from the LCR as discussed above (Steger and Corbach 1997, Morgan et al., 1998). Secondly, E2 can compete for cellular factors that interact with sequences adjacent to or overlapping E2 target sequences. Interestingly, Sp1 and TBP again fall into this category where elevated levels of E2 have been implicated in preventing these factors from interacting with their target sequences and regulating transcription (Tan et al., 1992; Tan et al., 1994). Therefore the interaction between E2 and Sp1/TBP may be extremely complex and be dependent upon the E2 protein levels and the situation of the E2, Sp1 and TBP target sequences on the transcriptional control unit under investigation. As well as Sp1 and TPB, other factors have been identified that interact directly with E2 target sequences and competition between E2 and these factors regulate transcriptional activation from adjacent promoters (Dong et al., 1994; Lewis et al., 1999). Ordinarily, the proteins that recognise the E2 sequences are positive effectors of transactivation and therefore E2 competition for the target sequences of these factors results in transcriptional repression from adjacent promoters. The third type of cellular proteins functionally interacting with E2 comprises those that bind the LCR and regulate E2 function. For example, the BPV4 LCR has several sequences that, by themselves, make no difference to transcriptional activity from adjacent promoters when they are mutated (Vance et al., 2001). However, removal of these sequences results in the promoter being much more responsive to E2 levels and thus in enhanced transcriptional activation. The proteins interacting with these elements are therefore responsible for mediating E2 function even though they probably do not interact directly with E2, nor do they interact with E2 DNA binding sequences.

4. E2 and Regulation of the Host Cell Cycle

Several reports have detailed the ability of BPV1 (Goodwin et al., 1998), HPV16 (Webster et al., 2000) and HPV18 E2 (Desaintes et al., 1999; Demeret et al., 2003) proteins to regulate the cell cycle in a negative manner. Many of these studies have been carried out in cell lines derived from HPV induced cervical cancers and have demonstrated that transient over-expression of the E2 protein results in suppression of E6 and E7 expression promoting re-activation of the p53 and Rb pathways (Goodwin et al., 1998). This reactivation results in cell cycle arrest in G1. Both E6 and E7 are required for the continued proliferation of these cell lines, and disruption of either the E6 or the E7 pathway is potentially a good therapeutic avenue for treating HPV infections (Psyrri et al., 2004). As well as inducing a G1 cell cycle arrest, the E2 protein from both HPV16 and 18 can induce cell death via apoptosis. The precise mechanism of how these two proteins carry out this function may be different; for HPV16 E2 the induction of apoptosis has been reported to be p53-dependent (Webster et al., 2000), while the induction of apoptosis by HPV18 E2 has been described as p53-independent (Desaintes et al., 1999). These differences may well be due to an as yet unclear functional difference between the 16 and 18 E2 proteins. However, the studies on HPV16 and 18 E2 both described induction of apoptosis by an E2 protein that cannot interact with the E2 DNA target sequence, demonstrating that apoptosis is not simply a result of repression of E6 promoting re-activation of the p53 pathway via this route. These studies also revealed that the transcriptional activation and DNA replication functions of E2 are dispensable for the induction of apoptosis, but that the amino terminus of the E2 protein is required (Webster et al., 2000; Desaintes et al., 1999). Therefore the exact mechanism and cellular partner proteins for E2 required for the induction of apoptosis remain to be elucidated, although for HPV18 E2 the activation of apoptosis is mediated via the caspase 8 pathway (Demeret et al., 2003). Interestingly, HPV16 E2 can interact directly with p53 although there has been no functional activity described for this interaction (Massimi et al., 1999).

5. E2 and Viral Genome Segregation

Upon infection it is essential that viral genomes become segregated into daughter cells following cell division.

The E2 protein is an essential mediator of this process and extensive investigations have been carried out with the BPV1 E2 protein, and to a lesser extent those of HPV11, 18 and 16, to determine the mechanism used by E2 to carry out this function. The BPV1 E2 protein associates with mitotic chromatin via the amino terminal domain of the protein and once bound is then able to interact with the viral genomes via the carboxyl terminus dimerization and DNA binding domain (Lehman and Botchan, 1998; Ilves et al., 1999; Abroi et al., 2004; Bastien and McBride, 2000; Skiadopoulos and McBride, 1998). In this way, the BPV1 E2 protein acts as a bridge to link the viral genome to the mitotic chromatin ensuring that following mitosis the viral genomes are equally distributed into the daughter cells. The mechanism of how BPV1 E2 carries out this function has been studied, and while it has been shown that the phosphorylation status of the E2 protein is not essential for association with mitotic chromatin (Bastien and McBride, 2000), phosphorylation mutants were defective in transformation in the context of the whole viral genome (Lehman et al., 1997). This failure was due to the inability of the mutant genomes to efficiently segregate into the daughter cells. This apparent paradox was resolved by the discovery that the phosphorylation mutant E2 proteins had a higher affinity for the E1 protein and that this interaction prevented the mutant E2 protein associating with the mitotic chromatin (Lehman and Botchan, 1998). These results raise the interesting hypothesis that phosphorylation of E2, perhaps at the G2/M transition, regulates the interaction between the E1 and E2 proteins and that this is important for E2 to segregate the viral genomes.

Recently a cellular protein, Brd4, has been proposed as the protein responsible for the interaction of E2 with the mitotic chromatin (You et al., 2004). Brd4 contains Bromo domains that are present in proteins associating with chromatin (Zeng and Zhou, 2002). The domain of Brd4 responsible for interaction with E2 has been identified and this domain, when expressed by itself, acts as a dominant negative to prevent association between E2 and the mitotic chromatin. This mutant Brd4 protein also prevents cellular transformation by the BPV1 genome. These results demonstrate that Brd4 interacts with BPV1 E2 and is a candidate protein for mediating association with mitotic chromatin. Further studies of the E2 and Brd4 proteins will be required to prove that Brd4 is directly responsible for mediating the association between BPV1 E2 and mitotic chromatin. Interaction between Brd4 and HPV16 E2 has also been described (You et al., 2004) and it was postulated that Brd4 therefore might also be responsible for the genome partitioning function of HPV16 E2. However this is debatable as the mechanism by which HPV16 E2 (and other HPV E2 proteins tested to date) segregates the viral genomes is distinct from that of BPV1 E2. In mitosis the HPV E2 proteins are associated with the centrosomes and the mitotic spindles and not with the mitotic chromatin (Van Tine et al., 2004). This is clearly different from the mechanism that BPV1 E2 uses to partition the viral genome. The HPV 11 E2 protein can directly interact with tubulin proteins responsible for forming the centrosomes and the spindle, therefore a direct interaction between E2 and these proteins may be responsible for partitioning the viral genome into daughter cells. Another clear difference between the HPV E2 proteins and BPV1 E2 is that the HPV protein carboxyl terminus domains associate with the spindle and centrosomes, whereas with BPV1 E2 the carboxyl terminus of the protein is not required for association with mitotic chromatin. Therefore the cellular proteins used by HPV E2 proteins to segregate the viral genomes remain to be identified.

It is clear from the studies described above that the E2 protein of all PVs is required for segregating the viral genome into daughter cells and this is therefore an essential part of the viral life cycle. The transactivation and transient DNA replication function of E2 is not required for this process and it is of interest to note that there are several residues on E2 that are conserved in all E2 proteins that have no assigned function; it will be interesting to determine whether some of these residues play an essential role in the genome segregation function of E2.

6. E2 Gene Status in Cervical Cancer

In many human cancers the viral genome is integrated into that of the host (Furumoto and Irahara, 2002). For HPV18 induced tumours this is true in all cases whereas for HPV16 there is a mixture of viral states ranging from episomal, episomal plus integrated, to fully integrated (Badaracco et al., 2002). However, even for HPV16, there is a correlation between viral integration and stage of tumour, with more advanced lesions more likely to have integrated HPV16 (Kalantari et al., 1998). It is not known why there is a difference between HPV16 and HPV18 with regard to viral integration and cancer. In all cases, when the viral DNA integrates, it loses expression of the E2 protein as the integration site is often found to be in the middle of the E2 encoding sequence resulting in loss of expression (Graham and Herrington, 2000), and often fragments of the E2 gene are lost entirely following viral integration into the host genome. It is likely that viral DNA also integrates into the host genome at other viral genome locations therefore maintaining the potential to express the E2 protein, but in the tumours, which are clonal, the selection has been for those cells which lack E2 expression. There have been several suggestions as to why the E2 protein is lost in the HPV induced cancers and these are related to the properties of E2 discussed above. Firstly, loss of the E2 protein will result in loss of repression of the E6 and E7 encoding genes. Increased expression of E6 and E7 will result in an increased interference with the p53 and Rb proteins respectively, and this would result in enhanced proliferation and increased genetic instability (for example, see Goodwin et al., 1998). Indeed, genetic instability of the host genome is related to viral integration (Pett et al., 2004). Secondly, the E2 protein itself has been shown to be anti-proliferative and can result in apoptosis and therefore loss of E2 expression may contribute towards cell proliferation and cell survival (Desaintes et al., 1999; Webster et al., 2000). It is impossible to differentiate between these two possibilities and it may be that a combination of these properties enhances progression to carcinoma following loss of E2 expression.

7. Regulation of E2 Protein Turnover

The stability of the BPV1 E2 protein is regulated by phosphorylation of residues within the hinge region. Phosphorylation of serine 301 by casein kinase II causes a conformational change resulting in degradation of the E2 protein via the proteasome (Penrose et al., 2004). Mutation of serine 301 to alanine results in extension of the protein half life from 50 to 160 minutes as the mutant protein is ubiquitinated to a lesser extent and its turnover is slowed down (Penrose and McBride, 2000). Although the HPV18 E2 protein is degraded via the ubiquitin/proteasome pathway, the mechanism is distinct from that of the BPV1 E2 protein as the amino terminal domain of HPV18 E2 is responsible for mediating degradation (Bellanger et al., 2001). The HPV16 E2 protein is also degraded via the ubiquitin/proteasome pathway and this degradation is enhanced following UVB irradiation; the protein half-life decreases from 80 to around 30 minutes following irradiation (Taylor et al., 2003a). However the domains and residues of HPV16 E2 responsible for this degradation are not known. Therefore more work is required to identify the residues and mechanism responsible for the degradation of the different E2 proteins.

8. E2 Interacting Partners and Their Role in Mediating E2 Function

There are several types of E2 interacting proteins: viral proteins; basal transcription factors; transcriptional co-activators; proteins implicated in cellular transcription/replication processes or other known E2 functions; and those factors that have no obvious functional significance. A list of these interactors is given in Table 1 along with the possible functional significance of the interaction.

8.1. Viral proteins

The E2 interaction with the viral replication factor E1 has already been discussed. Another viral protein that interacts with E2 is the capsid protein L2 that is involved in construction of the viral coat. BPV1 L2 is expressed in the nucleus and can re-locate the BPV1 E2 protein from a general nuclear location into ND10 domains (Day et al., 1998). The L1 protein is also recruited into ND10 by the L2 protein. It is thought that E2 recruitment into ND10 in turn locates the viral DNA into ND10 (via the E2 DNA binding domain) and this may facilitate encapsidation of the viral genome. This is supported by the observation that the E2 protein can enhance encapsidation of DNA into pseudo L1-L2 virions and that the E2 protein itself is also packaged (Zhao et al., 2000). Functional studies demonstrate that the BPV1 L2 protein represses both transcription and DNA replication mediated by BPV1 E2 and BPV1 E1-E2 respectively (Heino et al., 2000). Recent studies with HPV16 L2 and E2 have also demonstrated that L2 can repress the ability of E2 to activate transcription, and that, while L2 induces degradation of E2, degradation is not necessary for repression of E2 transcriptional transactivation. Furthermore, HPV16 L2 does not repress the ability of HPV16 E1-E2 to replicate DNA (Okoye et al., 2004).

In conclusion, the role of the L2-E2 interaction in the papillomavirus life cycle is not clear, but evidence suggests that L2 can modulate E2 function and that E2 may assist encapsidation of the viral genome into infectious particles.

8.2. Basal transcription factors

The E2 protein of BPV1 interacts with two basal transcriptional activation factors, TBP and TFIIB (Rank and Lambert,, 1995; Yao et al., 1998) and HPV16 E2 interacts with TFIIB (Benson et al., 1997). The interaction between BPV1 E2 and TFIIB is conserved from human to yeast where E2 can also act as a transcriptional activator (Benson et al., 1997). Therefore it is likely that the interaction between E2 and TFIIB is functional and may contribute to the ability of E2 to activate transcription (as is the case for other transcriptional activators). However, the interaction between E2 and TBP and its role in transcriptional activation is less clear and, as described above, may be directly related to the ability of E2 to repress transcriptional activation. Although many transcriptional regulators have been shown to interact with basal factors, these interactions by themselves are often insufficient to activate transcription. Instead, the recruitment of chromatin modifying co-activators is also involved in the ability of proteins to activate transcription and E2 is no exception.

8.3. Transcriptional co-activators

The classical histone modification proteins involved in transcriptional co-activation are the histone acetyltransferases (HAT) p300 and pCBP (Chan and LaThangue, 2001) and the E2 protein interacts directly with both these proteins (Lee et al., 2000; Muller et al., 2002), and with another HAT, pCAF (Lee et al., 2002a). Although p300, pCBP and pCAF have been implicated in transcriptional activation by E2, it is almost impossible to determine which of these factors may be responsible for mediating E2 activation from adjacent promoters.

Table 1. E2-interacting proteins as described in the text. When considering the functional relevance of the interaction with E2 a question mark indicates that there remains some doubt over the actual functional relevance of the interaction. These doubts are greater for some of the interactors than for others; ongoing research should remove many of the question marks in the future. The number of the interactors and their breadth of function emphasises the diverse nature of the roles of the E2 protein.

Interacting protein (IP)	Function of IP	Function of E2 interaction	Reference
E1	PV replication factor	PV replication initiation	Desaintes and Demeret, 1996
L2	PV capsid protein	Viral genome encapsidation?	Day et al., 1998; Zhou et al., 2000; Heino et al., 2000; Okoye et al., 2004
TBP	Transcriptional initiation	Inhibition of PV transcription?	Rank and Lambert, 1995; Bouvard et al., 1994; Steger and Corbach, 1997; Vance et al., 1999
TFIIB	Basal transcription factor	E2 transcriptional activation	Rank and Lambert, 1995; Yao et al., 1998; Benson et al., 1997
p300 and CBP	Transcriptional co-activators and histone deacetylases	E2 transcriptional activation?	Lee et al., 2000; Muller et al., 2002
pCAF	Transcriptional co-activator and histone deacetylase	E2 transcriptional activation?	Lee et al., 2002a
AMF1	Transcriptional co-activator interacting with p300	E2 transcriptional activation?	Breiding et al., 1997; Peng et al., 2000
TopBP1	DNA damage response and DNA replication initiation. Transcriptional co-activator.	E2 DNA replication initiation? Protection of viral genome? E2 transcriptional activation?	Boner et al., 2002; Boner and Morgan, 2002
BRCA1	DNA damage response and transcriptional co-activation	Protection of viral genome? E2 transcriptional activation?	Kim et al., 2003
PARP	DNA damage response, protein turnover	Protection of viral genome? E2 turnover?	Lee et al., 2002b
p53	DNA damage response. Control of cellular growth and apoptosis.	Regulation of E2 mediated apoptosis?	Massimi et al., 1999; Webster et al., 2000
SMN	Spinal muscular atrophy determining gene. Unknown function.	E2 transcriptional activation?	Strasswimmer et al., 1999
Brd4	Chromatin binding protein	PV genome segregation?	You et al., 2004
α, β, γ tubulin	Architectural proteins	PV genome segregation?	Van Tine et al., 2004
Several	Several	Several	Boner and Morgan, 2002

This is partially due to the fact that co-expression of E2 and these factors always enhances E2 transcriptional activation, perhaps not surprising as HAT proteins, when over expressed in isolation, result in generic activation of promoters.

Other cellular co-activators have been discovered for E2 and amongst these is the protein AMF1 (Breiding et al., 1997). Mutants of BPV1 E2, which failed to interact with AMF1, failed to activate transcription, strongly suggesting that this was an essential factor for E2 transcriptional activation. However, the structure of E2 suggests that individual mutations can dramatically alter the transcriptional activation domains and therefore mutants that fail to interact with AMF1 may also fail to interact with other essential transcriptional co-activators. Therefore there is still some uncertainty on whether the AMF1 protein is essential for BPV1 E2 function. Interestingly, the AMF1 protein can also interact with p300 (Peng et al., 2000) and could therefore serve as a bridge to recruit this histone modifying protein to the E2 transcriptional activation domain.

Other E2 interacting proteins have also been identified that can act as transcriptional co-activators. Amongst these are three proteins involved in the DNA damage response; TopBP1 (Boner et al., 2002), BRCA1 (Kim et al., 2003) and PARP1 (Lee et al., 2002b). TopBP1 and BRCA1 contain transcriptional activation domains (Makiniemi et al., 2001; Boner et al., 2002; McEachern et al., 2003) and it is possible that chromatin modification functions associated with these proteins are directly involved in altering DNA structure following recruitment to sites of DNA damage. Indeed, the BRCA1 protein can directly interact with CBP/p300 and these act as transcriptional co-activators for BRCA1 (Pao et al., 2000). Therefore it is clearly possible that in some situations these proteins can act as transcriptional co-activators of E2. Another cellular E2 interacting protein, nucleosome assembly protein 1, also interacts with p300 again implicating HAT activity in transcriptional activation by E2 (Rehtanz et al., 2004).

8.4. Other E2 interacting proteins

As well as protein implicated in transcriptional regulation, E2 interacts with other cellular proteins. These include the Survival Motor Neuron (SMN) protein that can act as a transcriptional co-activator for BPV1 E2 (although the essential nature of this interaction is not clear) (Strasswimmer et al., 1999). There are also interacting proteins that serve as candidates for mediating other E2 functions such as the Brd4 and tubulin proteins implicated in mediating genome segregation, as discussed above. A number of other HPV16 E2 interacting proteins were discovered using a yeast two-hybrid screen although the role, if any, of these proteins in regulating E2 functions remains to be determined (Boner and Morgan, 2002).

Several conclusions can be drawn from the E2 interacting proteins that have been identified. It seems that recruitment of HAT activity, either directly or via another interacting bridging protein, is a feature of E2 transcriptional activation. This is not surprising as it is a general feature of many transcriptional activators operating in eukaryotic cells. It is also of interest that the E2 protein interacts with several proteins involved in the DNA damage response. These interactions may be involved in regulating viral DNA replication and may also be involved in protecting the viral genome when the cell is damaged. Recruiting the viral genome to sites of DNA damage via interaction with TopBP1 or BRCA1 may provide a quick way of repairing the viral genome and suppressing replication when the cell is exposed to DNA damaging agents.

Clearly a great deal of work remains to be done to identify the cellular proteins that are essential for mediating E2 function(s). To date there are no interacting proteins that we know are absolutely essential for mediating the function of E2, except the role of the E2-E1 interaction in mediating viral DNA replication. One of the reasons for this is the complicated non-linear nature of the E2 amino terminus where individual point mutations can result in the proposed tertiary structure falling apart. This makes interpretation of point mutant results extremely difficult as they may be affecting other interacting partners, not just the individual one under study. Future studies should concentrate on more in depth mutation-function experiments with the E2 protein and interacting partners, combined with RNAi approaches to knock out expression of the interactors.

9. Therapeutic Targeting of E2

The E2 protein is essential for the viral life cycle. Disruption of E2 protein function would disrupt this process and therefore be an appropriate approach for targeting HPV infections. An argument against disturbing the E2 protein in the high-risk cancer-causing HPV is that loss of E2 function is associated with the progression of HPV lesions towards carcinoma (see above). However, recent studies have demonstrated that substrates for viral integration occur during viral DNA replication where there are double strand DNA breaks in the replicated genomes; these broken genomes can then be incorporated into the host genome by the process of non-homologous end joining (Taylor et al., 2003b). Therefore prevention of DNA replication in the first place would prevent formation of substrates for viral integration. As E2 is essential for viral replication this makes it a good target. The other attractive feature of E2 is that all HPV encode an E2 protein that has many functions and properties in common. Therefore, understanding and targeting the E2 protein could potentially provide a therapeutic approach for the treatment of all HPV related diseases from genital warts to cervical lesions.

One way of inhibiting E2 function is to develop peptides or small molecules that inhibit the interaction between the E1 and E2 proteins. Such molecules have been identified and interfere with DNA replication and require further development to provide therapeutic treatments for HPV infection (Kasukawa et al., 1998; White et al., 2003). Another approach to targeting E2 is to understand the pathways that regulate E2 protein turnover. It is clear that the E2 protein turns over via the proteasome, but the mechanisms targeting it to this pathway are unknown for the HPV E2 proteins. A clue has come from the finding that UVB irradiation increases turnover of the HPV16 E2 protein, suggesting that a UVB induced modification of E2 is responsible for the enhanced turnover (Taylor et al., 2003a). Much work remains to be carried out to identify the cellular pathways controlling turnover of the HPV E2 proteins and how these pathways may be targeted for degradation of the E2 proteins.

A further mechanism for exploiting the E2 protein for therapy is re-administering the protein to HPV transformed cells. Expression of the E2 protein in cell lines established from HPV infected carcinomas results in cell cycle arrest and apoptosis as described above. The barrier to using this approach for the treatment of HPV related cancers is being able to deliver the E2 protein to the target cells. A possible solution to this barrier is the recent development of VP22-E2 proteins (Roeder et al., 2004). The VP22 protein is a herpes simplex virus protein that allows intercellular transport of proteins (Jarver and Langel, 2004). Therefore, when fused to E2, it allows the fusion protein to cross the cell membrane and enter the cell. When this occurs E2 can execute the desired function which in this case is repression of viral oncogene expression and induction of growth arrest and apoptosis. Production of this protein on a mass scale, combined with the correct formulation, could provide an exciting future treatment for HPV related cancers.

10. Conclusions and Future Directions

The E2 protein is involved in a number of processes that are essential for the viral life cycle. These include activation of viral DNA replication in association with E1; control of viral transcription and mediating viral genome segregation into daughter cells. The E2 protein can also interfere with the host cell cycle and this potentially plays a part in a productive infection, although this remains to be determined. There are many residues within the E2 protein that are conserved throughout evolution and studies have demonstrated that some of these contribute towards interaction with E1 (and therefore are essential for the DNA replication properties of the protein) and some are conserved and essential for the transcriptional control properties of E2. However there are several other conserved residues within E2 that have no known function. This is particularly interesting in the amino terminal domain of the E2 protein that is essential for interacting with cellular factors that mediate the E2 functions. Future work on E2 should focus on understanding the cellular proteins interacting with E2 and how these are responsible for mediating E2 function. By this approach, not only will we learn more about the E2 protein and how it functions; insight into cellular processes ranging from transcription and replication to mitosis will also be provided. Identification of functional interacting proteins for E2 will also provide targets and pathways for inhibiting E2 function. Such inhibition represents an opportunity for treating HPV related infections.

Acknowledgements

Apologies to all of our colleagues whose work we have not referenced here due to the breadth of the subject and limited space. MMD is supported by Cancer Research UK.

References

Abroi, A., Ilves, I., Kivi, S., and Ustav, M. (2004). Analysis of chromatin attachment and partitioning functions of bovine papillomavirus type 1 E2 protein. J. Virol. *78*, 2100-2113.

Antson, A.A., Burns, J.E., Moroz, O.V., Scott, D.J., Sanders, C.M., Bronstein, I.B., Dodson, G.G., Wilson, K.S., and Maitland, N.J. (2000). Structure of the intact transactivation domain of the human papillomavirus E2 protein. Nature *403*, 805-809.

Badaracco, G., Venuti, A., Sedati, A., and Marcante, M.L. (2002). HPV16 and HPV18 in genital tumors: Significantly different levels of viral integration and correlation to tumor invasiveness. J. Med. Virol. *67*, 574-582.

Bastien, N, and McBride, A.A. (2000). Interaction of the papillomavirus E2 protein with mitotic chromosomes. Virology *270*, 124-134.

Bellanger, S., Demeret, C., Goyat, S., and Thierry, F. (2001). Stability of the human papillomavirus type 18 E2 protein is regulated by a proteasome degradation pathway through its amino-terminal transactivation domain. J. Virol. *75*, 7244-7251.

Benson, J.D., Lawande, R., and Howley, P.M. (1997). Conserved interaction of the papillomavirus E2 transcriptional activator proteins with human and yeast TFIIB proteins. J. Virol. *71*, 8041-8047.

Bochkarev, A., Barwell, J.A., Pfuetzner, R.A., Furey, W. Jr., Edwards, A.M., and Frappier, L. (1995). Crystal structure of the DNA-binding domain of the Epstein-Barr virus origin-binding protein EBNA 1. Cell *83*, 39-46.

Boner, W., Taylor, E.R., Tsirimonaki, E., Yamane, K., Campo, M.S., and Morgan, I.M. (2002). A Functional interaction between the human papillomavirus 16 transcription/ replication factor E2 and the DNA damage response protein TopBP1. J. Biol. Chem. *277*, 22297-22303.

Boner, W., and Morgan, I.M. (2002). Novel cellular interacting partners of the human papillomavirus 16 transcription/ replication factor E2. Virus Res. *90*, 113-118.

Bouvard, V., Storey, A., Pim, D. and Banks, L. (1994). Characterization of the human papillomavirus E2 protein: evidence of trans-activation and trans-repression in cervical keratinocytes. EMBO J. *13*, 5451-5459.

Breiding, D.E., Sverdrup, F., Grossel, M.J., Moscufo, N., Boonchai, W., and Androphy, E.J. (1997). Functional interaction of a novel cellular protein with the papillomavirus E2 transactivation domain. Mol. Cell. Biol. *17*, 7208-7219.

Campo, M.S. (2002). Animal models of papillomavirus pathogenesis. Virus Res., *89*, 249-261.

Chan, H.M., and La Thangue, N.B. (2001). p300/CBP proteins: HATs for transcriptional bridges and scaffolds. J. Cell Sci. *114*, 2363-2373.

Choe, J., Vaillancourt, P., Stenlund, A., and Botchan, M. (1998). Bovine papillomavirus type 1 encodes two forms of a transcriptional repressor: structural and functional analysis of new viral cDNAs. J. Virol. *63*, 1743-1755.

Day, P.M., Roden, R.B., Lowy, D.R., and Schiller, J.T. (1998). The papillomavirus minor capsid protein, L2, induces localization of the major capsid protein, L1, and the viral transcription/replication protein, E2, to PML oncogenic domains. J. Virol. *72*, 142-150.

Demeret, C., Garcia-Carranca, A., and Thierry, F. (2003). Transcription-independent triggering of the extrinsic pathway of apoptosis by human papillomavirus 18 E2 protein. Oncogene *22*, 168-175.

Desaintes, C., Goyat, S., Garbay, S., Yaniv, M., and Thierry, F. (1999). Papillomavirus E2 induces p53-independent apoptosis in HeLa cells. Oncogene *18*, 4538-4545.

Desaintes, C., and Demeret C. (1996). Control of papillomavirus DNA replication and transcription. Semin. Cancer Biol. *7*, 339-347.

Dong, G., Broker, T.R., and Chow, L.T. (1994). Human papillomavirus type 11 E2 proteins repress the homologous E6 promoter by interfering with the binding of host transcription factors to adjacent elements. J. Virol. *68*, 1115-1127.

Furumoto, H., and Irahara, M. (2002). Human papilloma virus (HPV) and cervical cancer. J. Med. Invest. *49*, 124-133.

Goodwin, E.C., Naeger, L.K., Breiding, D.E., Androphy, E.J., and DiMaio, D. (1998). Transactivation-competent bovine papillomavirus E2 protein is specifically required for efficient repression of human papillomavirus oncogene expression and for acute growth inhibition of cervical carcinoma cell lines. J. Virol. *72*, 3925-3934.

Graham, D.A., and Herrington, C.S. (2000). HPV-16 E2 gene disruption and sequence variation in CIN 3 lesions and invasive squamous cell carcinomas of the cervix: relation to numerical chromosome abnormalities. Mol. Pathol. *53*, 201-206.

Harris, S.F., and Botchan, M.R. (1999). Crystal structure of the human papillomavirus type 18 E2 activation domain. Science *284*, 1673-1677.

Heino, P., Zhou, J., and Lambert, P.F. (2000). Interaction of the papillomavirus transcription/replication factor, E2, and the viral capsid protein, L2. Virology *276*, 304-314.

Hoppe-Seyler, F., and Butz, K. (1994). Cellular control of human papillomavirus oncogene transcription. Mol. Carcinog. *10*, 134-141.

Ilves, I., Kivi, S., and Ustav, M. (1999). Long-term episomal maintenance of bovine papillomavirus type 1 plasmids is determined by attachment to host chromosomes, which Is mediated by the viral E2 protein and its binding sites. J. Virol. *73*, 4404-4412.

Jarver, P., and Langel, U. (2004). The use of cell-penetrating peptides as a tool for gene regulation. Drug Discov. Today *9*, 395-402.

Kalantari, M., Karlsen, F., Kristensen, G. Holm, R., Hagmar, B., and Johansson, B. (1998). Disruption of the E1 and E2 reading frames of HPV 16 in cervical carcinoma is associated with poor prognosis. Int. J. Gynecol. Pathol. *17*, 146-153.

Kasukawa, H., Howley, P.M., and Benson, J.D. (1998). A fifteen-amino-acid peptide inhibits human papillomavirus E1-E2 interaction and human papillomavirus DNA replication *in vitro*. J. Virol. *72*, 8166-8173.

Kim, J., Lee, D., Gwan Hwang, S., Hwang, E.S., and Choe, J. (2003). BRCA1 associates with human papillomavirus type 18 E2 and stimulates E2-dependent transcription. Biochem. Biophys. Res. Commun., *305*, 1008-1016.

Lee, D., Lee, B., Kim, J., Kim, D.W., and Choe, J. (2000). cAMP response element-binding protein-binding protein binds to human papillomavirus E2 protein and activates E2-dependent transcription. J. Biol. Chem. *275*, 7045-7051.

Lee, D., Hwang, S.G., Kim, J., and Choe, J. (2002a). Functional interaction between p/CAF and human papillomavirus E2 protein. J. Biol. Chem. *277*, 6483-6489.

Lee, D., Kim, J.W., Kim, K., Joe, C.O., Schreiber, V., Menissier-De Murcia, J., and Choe, J. (2002b) Functional interaction between human papillomavirus type 18 E2 and poly(ADP-ribose) polymerase 1. Oncogene *21*, 5877-5885.

Lehman, C.W., King, D.S., and Botchan, M.R. (1997). A papillomavirus E2 phosphorylation mutant exhibits normal transient replication and transcription but is defective in transformation and plasmid retention. J. Virol. *71*, 3652-3665.

Lehman, C.W., and Botchan, M.R. (1998). Segregation of viral plasmids depends on tethering to chromosomes and is regulated by phosphorylation. Proc. Natl. Acad. Sci. USA *95*, 4338-4343.

Lewis, H., Webster, K., Sanchez-Perez, A.M., and Gaston, K. (1999). Cellular transcription factors regulate human papillomavirus type 16 gene expression by binding to a subset of the DNA sequences recognized by the viral E2 protein. J. Gen. Virol. *80*, 2087-2096.

Li, R., Knight, J.D., Jackson, S.P., Tjian, R., and Botchan, M.R. (1991). Direct interaction between Sp1 and the BPV enhancer E2 protein mediates synergistic activation of transcription. Cell *65*, 493-505.

Liang, H., Petros, A.M., Meadows, R.P., Yoon, H.S., Egan, D.A., Walter, K., Holzman, T.F., Robins, T., and Fesik, S.W. (1996). Solution structure of the DNA-binding domain of a human papillomavirus E2 protein: evidence for flexible DNA-binding regions. Biochemistry *35*, 2095-2103.

Lim, D.A., Gossen, M., Lehman, C.W., and Botchan, M.R. (1998). Competition for DNA binding sites between the short and long forms of E2 dimers underlies repression in bovine papillomavirus type 1 DNA replication control. J. Virol. *72*, 1931-1940.

Makiniemi, M., Hillukkala, T., Tuusa, J., Reini, K., Vaara, M., Huang, D., Pospiech, H., Majuri, I., Westerling, T., Makela, T.P., and Syvaoja, J.E. (2001). BRCT domain-containing protein TopBP1 functions in DNA replication and damage response. J. Biol. Chem. *276*, 30399-30406.

Massimi, P., Pim, D., Bertoli, C., Bouvard, V., and Banks, L. (1999). Interaction between the HPV-16 E2 transcriptional activator and p53. Oncogene *18*, 7748-5774.

McEachern, K.A., Archey, W.B., Douville, K., and Arrick, B.A. (2003). BRCA1 splice variants exhibit overlapping and distinct transcriptional transactivation activities. J. Cell. Biochem. *89*, 120-132.

Morgan, I.M., Grindlay, G.J., and Campo, M.S. (1998). Epithelial specific transcriptional regulation of the bovine papillomavirus 4 promoter by E2. J. Gen. Virol. *79*, 501-508.

Morgan, I.M., Grindlay, G.J., and Campo, M.S. (1999). The bovine papillomavirus type 4 long control region contains an epithelial specific enhancer. J. Gen. Virol. *80*, 23-27.

Muller, A., Ritzkowsky, A., and Steger, G. (2002). Cooperative activation of human papillomavirus type 8 gene expression by the E2 protein and the cellular coactivator p300. J. Virol. 76, 11042-11053.

Okoye, A., Cordano, P., Taylor, E.R., Morgan, I.M., Everett, R., and Campo, MS. (2004). Human papillomavirus 16 L2 inhibits the transcriptional activation function, but not the DNA replication function, of HPV16 E2. Virus Res., 108, 1-14. Published online 27 August 2004.

Pao, G.M., Janknecht, R., Ruffner, H., Hunter, T., and Verma, I.M. (2000). CBP/p300 interact with and function as transcriptional coactivators of BRCA1. Proc. Natl. Acad. Sci. USA, *97*, 1020-1025.

Peng, Y.C., Breiding, D.E., Sverdrup, F., Richard, J., and Androphy, E.J. (2000). AMF-1/Gps2 binds p300 and enhances its interaction with papillomavirus E2 proteins. J. Virol. *74*, 5872-9587.

Penrose, K.J., Garcia-Alai, M., de Prat-Gay, G., and McBride, A.A. (2004). Casein Kinase II phosphorylation-induced conformational switch triggers degradation of the papillomavirus E2 protein. J. Biol. Chem. *279*, 22430-22439.

Penrose, K.J., and McBride, A.A. (2000). Proteasome-mediated degradation of the papillomavirus E2-TA protein is regulated by phosphorylation and can modulate viral genome copy number. J. Virol. *74*, 6031-6038.

Pett, M.R., Alazawi, W.O., Roberts, I., Dowen, S., Smith, D.I., Stanley, M.A., and Coleman, N. (2004). Acquisition of high-level chromosomal instability is associated with integration of human papillomavirus type 16 in cervical keratinocytes. Cancer Res. *64*, 1359-1368.

Prasanth, S.G., Mendez, J., Prasanth, K.V., and Stillman, B. (2004). Dynamics of pre-replication complex proteins during the cell division cycle. Philos. Trans. R. Soc. Lond. (B Biol. Sci.) *359*, 7-16.

Psyrri, A., DeFilippis, R.A., Edwards, A.P., Yates, K.E., Manuelidis, L., and DiMaio, D. (2004). Role of the retinoblastoma pathway in senescence triggered by repression of the human papillomavirus E7 protein in cervical carcinoma cells. Cancer Res. *64*, 3079-3086.

Rank, N.M., and Lambert, P.F. (1995). Bovine papillomavirus type 1 E2 transcriptional regulators directly bind two cellular transcription factors, TFIID and TFIIB. J. Virol. *69*, 6323-6334.

Rehtanz, M., Schmidt, H.M., Warthorst, U., and Steger, G. (2004). Direct interaction between nucleosome assembly protein 1 and the papillomavirus E2 proteins involved in activation of transcription. Mol. Cell. Biol. *24*, 2153-2168.

Roeder, G.E., Parish, J.L., Stern, P.L., and Gaston, K. (2004). Herpes Simplex Virus VP22-Human Papillomavirus E2 fusion proteins produced in mammalian or bacterial cells enter mammalian cells and induce apoptotic cell death. Biotechnol. Appl. Biochem. *40*, 157-165.

Sakai, H., Yasugi, T., Benson, J.D., Dowhanick, J.J., and Howley, P.M. (1996). Targeted mutagenesis of the human papillomavirus type 16 E2 transactivation domain reveals separable transcriptional activation and DNA replication functions. J Virol. *70*, 1602-1611.

Sandler, A.B., Baker, C.C., and Spalholz, B.A. (1996). Sp1 is critical for basal and E2-transactivated transcription from the bovine papillomavirus type 1 P89 promoter. J. Gen. Virol. *77*, 189-198.

Skiadopoulos, M.H., and McBride, A.A. (1998). Bovine papillomavirus type 1 genomes and the E2 transactivator protein are closely associated with mitotic chromatin. J. Virol. *72*, 2079-2088.

Steger, G., and Corbach, S. (1997). Dose-dependent regulation of the early promoter of human papillomavirus type 18 by the viral E2 protein. J. Virol. *71*, 50-58.

Strasswimmer, J., Lorson, C.L., Breiding, D.E., Chen, J.J., Le, T., Burghes, A.H., and Androphy, E.J. (1999). Identification of survival motor neuron as a transcriptional activator-binding protein. Hum. Mol. Genet. *8*, 1219-1226.

Tan, S.H., Leong, L.E., Walker, P.A., and Bernard, H.U. (1994). The human papillomavirus type 16 E2 transcription factor binds with low cooperativity to two flanking sites and represses the E6 promoter through displacement of Sp1 and TFIID. J. Virol. *68*, 6411-6420.

Tan, S.H., Gloss, B., and Bernard, H.U. (1992). During negative regulation of the human papillomavirus-16 E6 promoter, the viral E2 protein can displace Sp1 from a proximal promoter element. Nucleic Acids Res. *20*, 251-256.

Taylor, E.R., Boner, W., Dornan, E.S., Corr, E.M., and Morgan, I.M. (2003a). UVB irradiation reduces the half-life and transactivation potential of the human papillomavirus 16 E2 protein. Oncogene *22*, 4469-4477.

Taylor, E.R., Dornan, E.S., Boner, W., Connolly, J.A., McNair, S., Kannouche, P., Lehmann, A.R., and Morgan, I.M. (2003b). The fidelity of HPV16 E1/E2-mediated DNA replication. J. Biol. Chem. *278*, 52223-52230.

Van Tine, B.A., Dao, L.D., Wu, S.Y., Sonbuchner, T.M., Lin, B.Y., Zou, N., Chiang, C.M., Broker, T.R., and Chow, L.T. (2004). Human papillomavirus (HPV) origin-binding protein associates with mitotic spindles to enable viral DNA partitioning. Proc. Natl. Acad. Sci. USA *101*, 4030-4035.

Vance, K.W., Campo, M.S., and Morgan, I.M. (2001). A novel silencer element in the bovine papillomavirus type 4 promoter represses the transcriptional response to papillomavirus E2 protein. J. Virol. *75*, 2829-2838.

Vance, K.W., Campo, M.S., and Morgan, I.M. (1999). An enhanced epithelial response of a papillomavirus promoter to transcriptional activators. J. Biol. Chem. *74*, 27839-27844.

Webster, K., Parish, J., Pandya, M., Stern, P.L., Clarke, A.R., and Gaston, K. (2000). The human papillomavirus (HPV) 16 E2 protein induces apoptosis in the absence of other HPV proteins and via a p53-dependent pathway. J. Biol. Chem. *275*, 87-94.

White, P.W., Titolo, S., Brault, K., Thauvette, L., Pelletier, A., Welchner, E., Bourgon, L., Doyon, L., Ogilvie, W.W., Yoakim, C., Cordingley, M.G., and Archambault, J. (2003). Inhibition of human papillomavirus DNA replication by small molecule antagonists of the E1-E2 protein interaction. J. Biol. Chem. *278*, 26765-26772.

Yao, J.M., Breiding, D.E., Androphy, E.J. (1998). Functional interaction of the bovine papillomavirus E2 transactivation domain with TFIIB. J. Virol. *72*, 1013-1019.

You, J., Croyle, J.L., Nishimura, A., Ozato, K., and Howley, P.M. (2004). Interaction of the bovine papillomavirus E2 protein with Brd4 tethers the viral DNA to host mitotic chromosomes. Cell *117*, 349-360.

Zeng, L., and Zhou, M.M. (2002). Bromodomain: an acetyl-lysine binding domain. FEBS Lett. *513*, 124-128.

Zhao, K.N., Hengst, K., Liu, W.J., Liu, Y.H., Liu, X.S., McMillan, N.A., and Frazer, I.H. (2000). BPV1 E2 protein enhances packaging of full-length plasmid DNA in BPV1 pseudovirions. Virology *272*, 382-393.

Zou, N., Lin, B.Y., Duan, F., Lee, K.Y., Jin, G., Guan, R., Yao, G., Lefkowitz, E.J., Broker, T.R., and Chow, L.T. (2000). The hinge of the human papillomavirus type 11 E2 protein contains major determinants for nuclear localization and nuclear matrix association. J. Virol. *74*, 3761-3770.

From: Papillomavirus Research: From Natural History to Vaccines and Beyond. Edited by: M. Saveria Campo

Chapter 7

The E4 Protein – A Late Starter

Sally Roberts*

Abstract

E4, expressed as a fusion protein (E1^E4), is the major protein expressed in papillomavirus (PV) infections. Its role in the PV life cycle is still unknown although emerging evidence suggests that it acts as a major regulator of the productive phase of the life cycle. However, an early role in the infectious cycle cannot be dismissed. E4 interacts with, and in some cases disrupts the organization of, a number of cytoplasmic and nuclear structures and organelles, including keratin intermediate filaments, the cornified cell envelope, mitochondria and ND10 domains. E4 can interfere with the normal progression of the cell cycle, and for some PV types this function seems to be regulated partly by posttranslational modification of the E4 protein. Few cellular partners of E4 have been identified, but one of these interactors is a cellular factor involved in mRNA metabolism. Thus, E4 is likely to have a pleiotropic role in the PV life cycle. To date, E4 functions suggest a role in facilitating and supporting viral genome amplification, the regulation of late gene expression, control of virus maturation and mediation of virus release.

1. Introduction

The papillomavirus (PV) life cycle is divided into three distinct stages. Virus entry and establishment of replication of the viral genome, maintenance of the extrachromosomal genome in the dividing basal epithelial cells, and finally, the switch, in differentiating keratinocytes, to viral genome amplification, expression of late genes and assembly of new virions. The final phase of the PV life cycle is accompanied by expression of the E4 early gene product, actually expressed as a spliced E1^E4 fusion protein. However, despite its abundance in natural infections, unravelling the role of this protein in the PV life cycle has not been straightforward. Only by using systems that enable reintroduction of mutant PV genomes, in which E4 expression has been eliminated, back into keratinocytes has the importance of this protein in PV infections been realised. These are exciting times for E4 research. The protein, which is not well conserved between PVs, displays pleiotropic functions and activities that, in part, are controlled by a complex series of posttranslational modifications. In natural infections, the appearance of E4 protein at the onset of viral genome amplification and its persistence in these cells as they migrate up through the layers of the wart, indicates that E4 is present at all stages of virion morphogenesis. Thus, E4 functions might well be necessary for the commencement, progression and completion of the virus-producing phase of the PV life cycle. Even though the E4 open-reading frame (ORF) lies in the early region of the genome its pattern of expression in warts has meant that it has been referred to as an intermediate or late PV protein. However, early functions for this protein cannot be ruled out. In this chapter, I will therefore concentrate on the biological activities and functions of the enigmatic E4 protein that have been defined so far and where appropriate, discuss what role these functions might have in the PV life cycle. Our understanding of the molecular interactions that orchestrate these activities however, is largely undefined. To date, only a handful of cellular proteins that complex with E4 have been identified. The association between human (H)PV16 E4 and a DEAD-box protein, an interaction identified through yeast two-hybrid analysis, represents the best-characterized molecular interaction, although its significance in the HPV16 life cycle is unknown. First, I will detail what is known of E4 expression in natural infections and describe some of the properties of the protein. Throughout the chapter, the E1^E4 protein will be referred to as E4, unless a distinction needs to be made between the full-length fusion protein and truncated forms of the protein.

2. E4 and the PV Life Cycle

A foremost feature of E4 expression in natural PV infections is the close relationship between E4 and the virus-producing phase of the virus life cycle. Induction of E4 protein expression coincides with the onset of vegetative genome replication (Breitburd et al., 1987; Peh et al., 2002; Figure 1). These two events generally initiate in cells that have moved up from the basal cell layer and begun to differentiate (Doorbar et al., 1997; Peh et al., 2002). However, in lesions caused by canine oral (CO) PV, E4 expression and viral genome amplification begin in the non-differentiating basal cells (Nicholls et al., 2001). The interdependent relationship between E4 and viral genome amplification is also evident from

*For correspondence. Email s.roberts@bham.ac.uk.

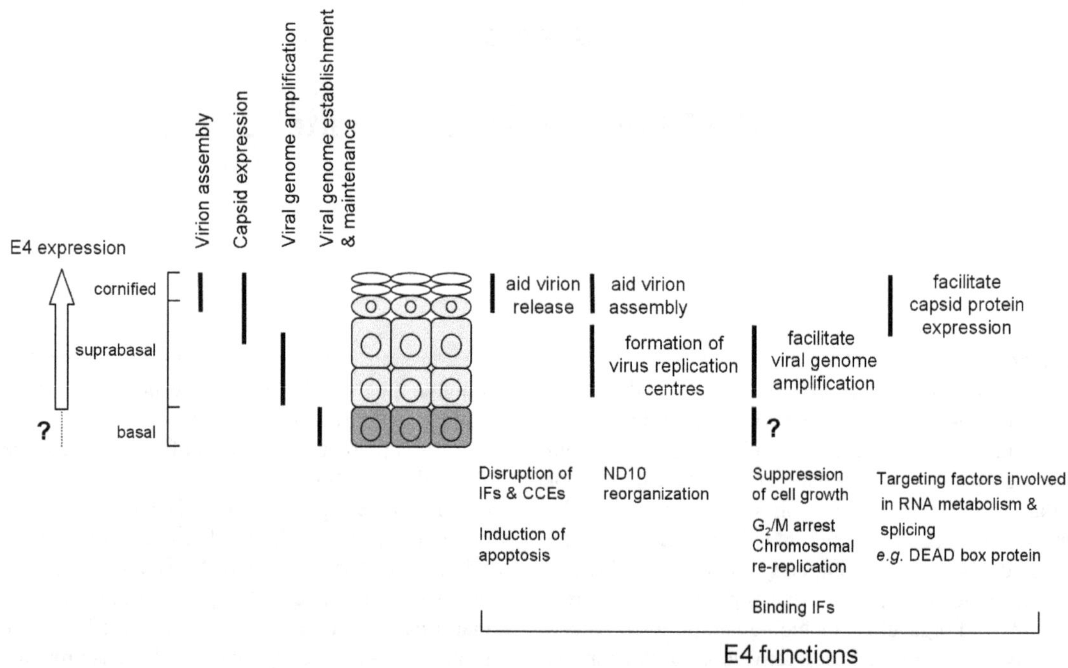

Figure 1. E4 probably has a pleiotropic role in the virus life cycle. The timing of virus events in relation to E4 expression and epithelial differentiation is depicted on the left hand side of the figure. E4 protein is detected in suprabasal keratinocytes at the onset of viral genome amplification. Low levels of E4 protein may be expressed from E1^E4-containing transcripts produced in the early stages of the infectious cycle. Biological activities of E4 that have been described so far, and the roles these might play in the virus life cycle, are described on the right hand side of the figure.

the use of *in vitro* PV genome systems, that are able to switch from early stage genome replication to late stage DNA amplification (Jareborg et al., 1991; Ruesch et al., 1998). Expression of E4 protein persists in cells as they migrate up through the wart layers, and so, as well as being present during amplification of the viral genome, it is also abundant in highly differentiated cells that express the capsid genes and synthesize new progeny virions (Figure 1). Recent studies, intended to evaluate the role of E4 in the virus life cycle, using PV genomes that contain mutations within the E4 ORF that disrupt E4 protein expression (but preserve integrity of the E2 protein), support a strong correlation between the expression of this viral protein and initiation of the productive phase of the virus life cycle. Differentiated keratinocytes harbouring cottontail rabbit (CR) PV or HPV31 "E4 knockout" genomes were significantly impaired in their ability to support viral genome amplification or late gene expression (Peh et al., 2004; Wilson et al., 2005). Loss of E4 expression however, did not affect establishment or maintenance of the viral genomes (Neary et al., 1987; Peh et al., 2004; Wilson et al., 2005). Therefore, E4's role in the virus life cycle seems to be intimately linked to the commencement, and possibly also progression, of the late phase of the PV life cycle in differentiating cells.

Whilst E4 is generally referred to as an intermediate or late PV protein an earlier role in the PV life cycle cannot be ruled out at this stage of our understanding of E4 function. During the early phase of the infectious cycle, E4 may well be expressed at levels significantly lower than those observed during the productive phase, from E1^E4-containing polycistronic mRNAs initiated from the early promoter, or even from the differentiation-dependent promoter (activity of this promoter has been observed in undifferentiated cells [Hummel et al., 1992]). Intriguingly, there is an increase in rabbit papilloma formation at epithelial sites inoculated with "E4 knockout" CRPV genomes, suggesting that E4 might be involved in negatively regulating cell proliferation during the early stages of papilloma formation (Peh et al., 2004). In fact, the ability of E4 to interfere with the host cell's replication machinery (see below) could be linked to such an E4 early function.

The E4 protein plays no role in promoting PV transformation (Neary et al., 1987), and expression of E4 is progressively lost from neoplastic lesions during their progression to cancer (Crum et al., 1990; Middleton et al., 2003).

E4 is the major PV protein expressed during the virus life cycle. In HPV1 warts, 30% of total wart protein is E4 (Doorbar et al., 1986; Breitburd et al., 1987). Thus, it can easily be visualized in the suprabasal compartment of naturally-occurring warts and tumours by immunohistochemical analysis. The intracellular distribution of E4 between the different PV types is quite heterogenous (Table 1). Most studies have found E4 to be

Table 1. Properties of E4 proteins in natural infections and in cell lines

Virus type	Subcellular localization *in vivo*	Subcellular localization in cell lines	Post translational modifications	References
HPV1	Multiple cytoplasmic and nuclear inclusion granules. ND10 domain protein PML associated with nuclear inclusions. Diffuse cytoplasmic distribution	Associated with IFs. *In vivo* like inclusions form along E4-IF structures. PML sequestered to nuclear E4 inclusion. Associated with actin fibres in fibroblasts	Mutliple species. Proteolytic cleavage of full-length protein.Self-multimerization. Serine and threonine phosphorylation. Phosphorylated by PKA *in vitro*. Binds metal ion zinc	Doorbar et al., 1986; 1988 Breitburd et al, 1987 Grand et al., 1989 Roberts et al., 1993; 1994a; 1994b; 2003
HPV2	Multiple cytoplasmic (and nuclear?) inclusion granules	Associated with IFs and induces collapse of the keratin cytoskeleton	Multiple species. Self-multimerization Proteolysis?	Doorbar et al., 1989 Roberts et al., 2003
HPV4	Single large cytoplasmic inclusion body	ND	Multiple species Self-multimerization Proteolysis?	Doorbar et al., 1989
HPV6/11	Concentrates at cell periphery. Co-purifies with CCE (HPV11)	Associated with collapsed keratin cytoskeleton.	Multiple species Self-multimerization Serine and threonine phosphorylation. Phosphorylated by PKA and MAPK *in vitro*	Tomita et al., 1991 Bryan et al., 1998; 2000. Bryan and Brown, 2000
HPV16	Associated with IFs. In superficial cells, occasionally forms a single juxtanuclear IF-containing inclusion. Nuclear structures (25-35 nm).	Associated with IFs and induces collapse of the keratin cytoskeleton. Mitochondria sequestered to collapsed-E4 structures. In differentiated HPV16-containing W12 cells, associated with IF-like structures, but not collapsed	Multiple species Self-multimerization	Doorbar et al., 1991. Roberts et al., 1993; 1997. Raj et al., 2004 Palefsky et al., 1991 Sterling et al., 1993. Wang et al., 2004
HPV31	ND	Punctate cytoplasmic in suprabasal cells of CIN-612 cells in organotypic rafts	ND, but multiple species identified in CIN-612 organotypic rafts	Pray and Laimins, 1993
HPV59	Concentrated near the CCE	Associated with IFs and induces collapse of the keratin cytoskeleton in COS cells. In differentiated (NIKS) cells, primarily localized at the cell periphery	ND	Brown et al., 2004
BPV4	Cytoplasmic.	Cytoplasmic filamentous structures formed in cell lines	ND	Anderson et al., 1997. Roberts, S., unpublished data
COPV	Distributed throughout cell, associated with periphery of cell and nucleus	Associated with nucleoli in cells of lower layers of experimental lesions	ND	Nicholls et al., 2001. Peh et al., 2002
CRPV	Distributed throughout cell	ND	ND	Peh et al., 2002
ROPV	Cytoplasmic inclusions	ND	ND	Peh et al., 2002

IF, intermediate filaments; CCE, cornified cell envelope; HPV, human papillomavirus; BPV, bovine papillomavirus, COPV, canine oral papillomavirus, CRPV, cottontail rabbit papillomavirus, ROPV, rabbit oral papillomavirus, ND, not determined.

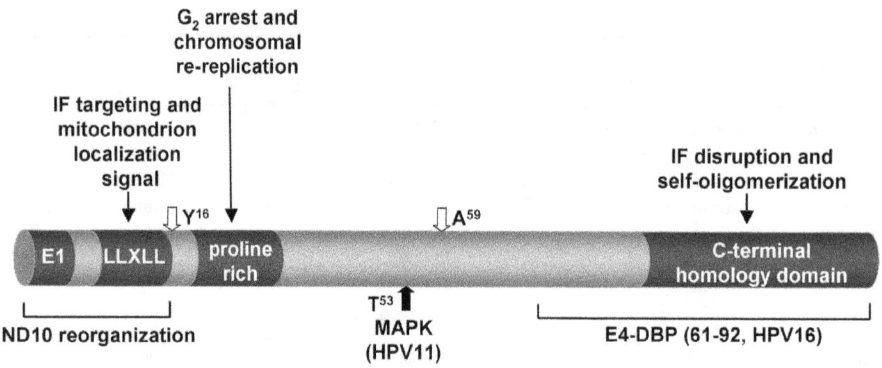

Figure 2. Schematic diagram of the E4 protein is shown with regions involved in known biological functions and interaction with cellular targets. Open arrows indicate the position of proteolytic cleavage sites (HPV1). A shaded arrow identifies the threonine phosphorylated by MAPK (HPV11).

located largely in the cytoplasm, but certainly detectable in the nucleus and at perinuclear sites in some infections. In cutaneous lesions, E4 assembles into prominent inclusion granules, whose size and ultrastructure is characteristic of the infecting virus type. For example, in HPV4 infections, E4 is associated with a single inclusion with a fine fibrillar structure that progressively fills the cytoplasm and distorts the nucleus, forming the typical "signet ring" cell (Croissant et al., 1985; Doorbar et al., 1989). In contrast, multiple small and irregularly shaped, homogenously electron dense bodies form in the lower suprabasal cells of HPV1 warts, but coalesce into larger structures in highly differentiated cells (Croissant et al., 1985; Doorbar et al., 1986; Breitburd et al., 1987). Little is known about the composition of the E4 inclusions, or their role in E4 function. Cytoplasmic inclusions formed *in vivo* and in cell lines are closely associated with keratin tonofilaments, but they do not appear to be aggregates of E4 and keratin proteins (Croissant et al., 1985; Rogel-Gaillard et al., 1993; Doorbar et al., 1996; Roberts et al., 1994a; 2003). It has recently been shown that the promyelocytic leukaemia (PML) protein, the defining component of nuclear domain 10 (ND10) bodies (also known as PML oncogenic domains [PODs]), is found concentrated at intranuclear E4 inclusions in HPV1 warts (Roberts et al., 2003). Because ND10 domains are implicated in PV replication and virion assembly, sequestration of PML to E4 inclusions might facilitate these viral functions (this is discussed later in the chapter).

Association of E4 with inclusion bodies is not a characteristic of PV infections at mucosal surfaces, although rabbit oral (RO) PV is an exception (Peh et al., 2002) (Table 1). In HPV16 lesions, E4 primarily associates with the keratin network of cells in the upper half of cervical lesions (Wang et al., 2004), while HPV11 E4 is located at the cornified cell envelope (CCE) of infected cells (Bryan and Brown, 2000).

3. General Features of E4

The most abundant class of viral mRNA transcript that encodes the spliced E1^E4 gene is a bicistronic message (E1^E4, E5), generated from the late promoter in the E7 ORF. Several other E1^E4-containing polycistronic transcripts that also encode the major and/or the minor capsid proteins are expressed from this late promoter (see Chapter 4 for further information on PV transcription). The E1^E4 splice fuses a short region of the E1 ORF (encoding five N-terminal residues of E1) to the E4 ORF (Chow et al., 1987; Nasseri et al., 1987; Palermo-Dilts et al., 1990). The E4 ORF, which overlaps with the E2 ORF, corresponds to a region of the genome that exhibits considerable sequence divergence between different PVs. The length of the E4 ORF is also variable, ranging from 73 (HPV34) to 304 (HPV47) amino acids. Similarly, the sequence homology among E4 proteins is poor, with proteins from PV types with a similar pathology showing the greatest sequence conservation. Even in these cases, similarity is most often restricted to the amino- and carboxy-terminal domains, with the central region of the protein showing maximum diversity. The reader should refer to Doorbar and Myers (1996) for more information on E4 gene and protein sequence alignments.

Sequences within the amino- and carboxy-termini of E4 are important in mediating biological and biophysical characteristics of the protein (Figure 2). At the N terminus, a leucine rich motif, LLXLL, characteristic of many, but not all, E4 proteins, is the major sequence determinant in the association between E4 and the keratin cytoskeleton (Roberts et al., 1994a), and is also involved in binding mitochondria (Raj et al., 2004). Another common sequence near to the N terminus has a high preponderance of proline residues and this region has been linked to E4's ability to interfere with G_2/M progression of the cell cycle (Davy et al., 2002; Nakahara et al., 2002; Knight et al., 2004). Sequences at the C terminus of E4 mediate molecular interactions between self (Doorbar et al., 1988; Roberts et al., 1997; Ashmole et al., 1998; Bryan et al., 1998; Wang et al., 2004), and this region in the HPV16 protein is also reported to be involved in binding a cellular RNA helicase (Doorbar et al., 2000). A characteristic of these carboxy-terminal sequences is conservation of hydrophobic residues between those viruses that have a similar site of infection *i.e.* skin or anogenital epithelium (Roberts et al., 1994a; 1997). It is these hydrophobic residues that are of particular importance in self-multimerization of the HPV1 and 16 proteins (Roberts et al., 1997; Ashmole et al., 1998). As yet, no function has been assigned to sequences in the more diverse central region of the protein. Perhaps this central region serves as a linker separating the amino- and carboxy-terminal domains (Roberts et al., 1997). The three-dimensional structure of E4 has yet to be resolved, but presumptive helical and β-sheet regions occur primarily at the N and C terminus, while the central domain is largely unstructured (Roberts et al., 1997).

The region of overlap between E4 and E2 ORFs corresponds to the hinge region of the E2 protein that separates the transactivation and DNA binding domains. Sequence analysis of the E2/E4 region in a group of HPV16 intratypic variants showed that far more amino acid variation was apparent in the E2 hinge than in the E4 sequence and that the majority of non-conservative changes in E4 occurred outside those regions suggested to be important in E4 biology (Eriksson et al., 1999). Thus, there may be selective pressure to maintain specific E4 sequences within the amino- and carboxy-terminal domains that are important in E4 structure and/or function (Roberts et al., 1997).

3.1. Posttranslational modifications

E4 protein isolated from human lesions exists as multiple species, formed by a combination of proteolysis, oligomerization and phosphorylation (Table 1). Studies, largely of the HPV1 protein, have shown that the nature of these posttranslational E4 modifications is progressively altered as the infected cell migrates upwards in the epithelial tissue (Breitburd et al., 1987; Doorbar et al., 1988; Grand et al., 1989).

In HPV1 warts, amino-terminal sequences are sequentially cleaved from the full-length E1^E4 polypeptide of 17-kDa to generate peptides of 16-, 11- and 10-kDa (Breitburd et al., 1987; Doorbar et al., 1988; Roberts et al., 1994a) (Figure 2). The full-length E1^E4 molecule appears first, in the parabasal wart layers, coincident with viral genome amplification. The 16-kDa species (that lacks the 15 amino-terminal residues of E1^E4 [Roberts et al., 1994a]) accumulates prior to the 11- and 10-kDa polypeptides (Breitburd et al., 1987). The most processed forms of E4 (11/10-kDa) are the major E4 species present in the most superficial cells, where progeny virions are being assembled. Cellular proteases that cleave E4 have not been identified, although it is unlikely that they are keratinocyte-specific (Roberts et al., 1994a; 1994b). The proteolysed peptides derived from the full-length HPV1 E4 protein do not contain sequences necessary for keratin association or ND10 domain reorganization suggesting that these functions are not required throughout the productive phase of the HPV1 life cycle (Figure 2; Roberts et al., 1994a; 2003). Furthermore, these truncated E4 polypeptides display biological activities that are distinct from those mediated by the full-length E1^E4 protein, supporting the hypothesis that E4 function is regulated by posttranslational events (Knight et al., 2004). These aspects of E4 function are discussed in more detail in a later section.

Although E4 proteins of other PV types are less well characterized than the type 1 protein, there is evidence that some of these proteins might also be subjected to proteolysis during the virus life cycle (Doorbar et al., 1989; Tomita et al., 1991; Pray and Laimins, 1993).

The E4 protein self-associates and exists as multiple oligomers - dimers, trimers/tetramers and higher orders of complexity, possibly up to 24-mers (HPV1) (Doorbar et al., 1988, 1996; Pray and Laimins, 1993; Roberts et al., 1997; Ashmole et al., 1998; Bryan et al., 1998; Wang et al., 2004). Although disulphide linkages do not seem to be necessary for the formation of larger E4 complexes, they are remarkably stable, with HPV1 oligomers persisting in high concentrations of urea (6M) (Roberts et al., 1997; Ashmole et al., 1998; Wang et al., 2004). Phosphorylation of E4, at multiple serine and threonine residues (Breitburd et al., 1987; Grand et al., 1989; Bryan et al., 2000) might promote HPV1 and 11 E4 multimerization (Grand et al., 1987; Bryan et al., 2000) and E4-bound zinc ions may influence the stability of HPV1 E4 oligomers (Roberts et al., 1994b, I. Ashmole and S. Roberts, unpublished data). As already mentioned, the C terminus of E4 is involved in self-oligomerization with the conserved hydrophobic amino acids being of particular importance (Roberts et al., 1997; Ashmole et al., 1998). Computational analysis of this region in the HPV1 E4 protein suggests that self-oligomerization could be mediated through interactions between hydrophobic faces of α-helices (Ashmole et al., 1998). Oligomerization of E4 might be involved in mediating interactions with cellular targets, or even other PV proteins. Interestingly, HPV16 E4 multimerization is suggested to be a key event in the collapse of E4/keratin filaments (Roberts et al., 1997; Roberts, 2002; Wang et al., 2004) (see below).

4. E4 and the Cellular Cytoskeleton

The cellular cytoskeleton is composed of three main networks formed by the intermediate filaments (IFs), microtubules and microfilaments. The keratins form the IF cytoskeleton of epithelial cells. Several groups have found that E4 proteins associate with the keratin IF network in cell lines and in natural infections, and in some cases, the association results in reorganization of the keratin cytoskeleton (Doorbar et al., 1991; Pray and Laimins, 1993; Roberts et al., 1993; 2003). There is also evidence that E4 associates with actin microfilaments in fibroblasts (Roberts et al., 2003). Whilst the E4-IF association is the best characterized, and is discussed in more detail below, the physiological function of either interaction is still one of conjecture.

4.1. E4 and keratin IFs

The function of the keratin IFs has long been thought to be primarily a structural one - to maintain cellular integrity and protect epithelial tissues from mechanical stress. The discovery of a link between keratin mutations and various human epithelial fragility diseases finally proved this hypothesis (reviewed in Lane and McLean, 2004). However, the keratin IFs have been implicated in cellular processes such as cell migration, modulation of apoptosis and regulation of cell growth suggesting that keratins also perform non-structural roles in the cell (for a review see Coulombe and Wong, 2004).

In epithelial cell lines, E4 proteins from PVs that infect cutaneous or mucosal surfaces interact with the keratin networks and, in some cases, induce the reorganization of the keratin cytoskeleton to a perinuclear fibrous clump (Doorbar et al., 1991; Pray and Laimins, 1993; Roberts et al., 1993, 2003 and unpublished observations) (Table 1). The interaction between E4 and the keratin IFs is mediated by a leucine-rich motif (LLXLL) that lies close to the N terminus of E4 (Figure 2) (Roberts et al., 1994a). As already mentioned, not all E4 proteins contain a LLXLL motif (*eg.* HPV types 5 and 8), but whether these proteins associate with the keratin networks has not been determined. Studies to elucidate the nature of the E4-keratin association have shown that the full-length HPV16 E1^E4 protein binds directly to the simple type 1 keratin, K18 (Wang et al., 2004), and although the keratin binding site on the E4 molecule was not identified the leucine-rich cluster LLKLL is a possible candidate binding motif (Roberts et al., 1994a). A hydrophobic sequence at the C terminus of HPV16 E4, conserved between other E4 proteins of the alpha-PVs, mediates reorganization of the cytoskeleton (Roberts et al., 1997). Notably, mutation of the hydrophobic residues in this motif (VX[V/L/I]X[L/V]XL) is sufficient to abrogate HPV16 E4-mediated keratin disruption. Since this motif is also known to mediate HPV11 and HPV16 E4 self-association, perhaps once bound to the keratins *via* its amino terminus, the E4 molecules self-associate thereby cross-linking the keratin fibres and disrupting the cytoskeleton (Roberts et al., 1997; Bryan et al., 1998; Roberts, 2002; Wang et al., 2004).

There are two key issues concerning the association between E4 and the keratin cytoskeleton that have yet to be fully resolved. First of these, is whether reorganization of the keratin cytoskeleton is an anomaly of E4 overexpression in cultured epithelial cells that do not recapitulate the phenotype of the differentiating keratinocyte – the natural site of E4 expression. In HPV16-infected tissue, E4 is colocalized with suprabasal keratins (*e.g.* K13), but perinuclear clumps of E4 and keratin are only observed in a few superficial cells (Wang et al., 2004). Therefore, although HPV16 E4 does appear to be a keratin IF-associated protein *in vivo*, the collapse of this cellular structure is not a common event during the replicative cycle of HPV16. Furthermore, there is no evidence of E4 forming aggregates in suprabasal keratinocytes of organotypic rafts formed from HPV31-containing cells (Pray and Laimins, 1993) or in differentiated cells expressing HPV59 E4 (Brown et al., 2004). The apparent infrequency of keratin collapse in differentiating keratinocytes in comparison to monolayer cell cultures might be a reflection of the different keratin subsets expressed. The simple glandular keratins K8 and K18 are the primary keratin subsets expressed in cultured cells and these appear to be more sensitive to E4-induced collapse than keratins expressed in keratinocytes (*e.g.* K5/K14) (Roberts et al., 1993 and unpublished data).

Not all E4 proteins cause the collapse of the cytoskeleton in epithelial cell lines. HPV1 E4 aligns with the keratin filaments but, as its levels rise, assembles into *in vivo*-like inclusion bodies that form along the E4/keratin filaments (Rogel-Gaillard et al., 1993; Roberts et al., 1993; 2003). Eventually, the cells contain large, multiple inclusions, and although these structures are often surrounded by a cage of keratin protein, there is no longer evidence of E4-keratin co-localization. These studies in cell cultures support findings of HPV1 E4 expression in natural infections, where E4 inclusions are often found in close proximity to keratin bundles, but are not aggregates of E4 and keratin (Rogel-Gaillard et al., 1993; Doorbar et al., 1996). Perhaps, therefore, the interaction between HPV1 E4 and keratins is only transient, forming for a short time in the infectious cycle when E4 protein levels are low, possibly prior to the appearance of E4 inclusions or just at an early stage in their formation (Roberts et al., 2003). A role for the E4-keratin cytoskeleton association only in the early stages of the HPV1 productive cycle is supported by the finding that this activity is only mediated by the full-length form E1^E4 of the protein (Roberts et al., 1994a; 1997), whose expression is restricted to the lower suprabasal cells of warts (Breitburd et al., 1987).

Differences in the behaviour of HPV E4 proteins in epithelial cells might reflect type specific functions, and this might be related to the differences in the organization of the life cycles (Peh et al., 2002). It should be noted however, that the E4 protein of HPV2, a cutaneous virus, when expressed in cultured cells, has a similar effect on the integrity of the keratin networks as the mucosal E4 proteins (Roberts et al., 2003), but after prolonged expression, assembles into multiple *in vivo* like inclusions (S. Roberts, unpublished data). Clearly, the mode and level of E4 expression has a significant impact on its intracellular behaviour.

The second issue relates to the physiological significance of the E4-keratin interaction in the PV life cycle. As already mentioned, keratins IFs encode both structural and non-structural functions and E4 might influence some, or all, of these functions to facilitate the execution of the PV life cycle. Since there is a strong link between the keratin cytoskeleton and maintenance of cell and tissue architecture, E4's ability to disrupt the cytoskeleton, albeit more efficiently in epithelial cell lines than *in vivo*, might compromise the structural integrity of the infected cell in the upper layers of the wart, enabling these cells to rupture readily and release the newly synthesized virus particles into the environment (Doorbar et al., 1991). Indeed, keratinocytes in individuals suffering from the skin blistering disorders readily rupture when subjected to mechanical stress (reviewed in Lane and McClean, 2004). These individuals carry mutations within their keratin genes that inhibit the formation of a cytoskeleton that is able to withstand physical trauma and that consequently breaks down. Epithelial cells lacking the simple keratins (K8/K18) or containing a disrupted cytoskeleton are also more sensitive to specific apoptotic-signals (Caulin et al., 2000; Gilbert et al., 2001; Inada et al., 2001). High levels of HPV16 E4 protein can induce significant level of apoptosis in human cervical cancer cells and this effect correlates with a collapsed K8/K18 keratin network (Raj et al., 2004). It is therefore conceivable that E4's induced disruption of the keratin cytoskeleton has compromised its ability to regulate cell death. Recent work has also found that keratins can influence cell growth (Paramio et al., 1999). Suprabasal keratin K10 sequesters critical signalling molecules leading to a disruption in the pathways that promote cell proliferation (Paramio et al., 2001). E4 too has been found to interfere with cell growth (see below) and maybe the E4-keratin association in some way facilitates these effects. As keratin networks radiate throughout the cytoplasm of the cell they might act as a scaffold, anchoring E4-containing complexes in the cytoplasm (Roberts et al., 1994a). Sequestration of cellular targets to E4-keratin structures might modify or inhibit their function. Indeed, Davy et al. (2005) proposed that as cyclin B/cdk1-E4 complexes are tethered to keratin filaments, this might be the mechanistic basis for HPV16 E4-induced G_2 arrest of epithelial cells (Davy et al., 2002), by preventing nuclear translocation of active cyclinB/cdk1 complexes. As the G_2 arrest function of HPV16 E4 occurs independently of the E4-keratin association (Davy et al., 2002), the importance of keratin tethering of cyclinB/cdk1 complexes to E4 function has yet to be clarified.

Whilst the true function of E4's association with the keratin cytoskeleton has not been determined it does represent a conserved feature of the E4 proteins and is therefore likely to have an important role in E4 function (Figure 1).

4.2. Does E4 aid release of new virions?

During terminal differentiation of keratinocytes, the process of keratinization forms the CCE, an insoluble, highly resistant structure beneath the plasma membrane of the corneocyte. The CCE consists of various crosslinked precursor proteins, including loricrin and involucrin. Filaggrin-crosslinked keratin filaments fill the CCE to finally form the enucleated squame. Newly synthesized papillomavirus virions accumulate in these well-differentiated cells and for effective transmission to other hosts, the cells must break apart to allow the virions to escape into the environment (Bryan and Brown, 2001). The keratinization process does appear to be compromised in HPV-infected keratinocytes (Rogel-Gaillard et al., 1993; Lehr et al., 2004) and this might be linked to E4 functions. The cornified squames isolated from HPV11-infected material are morphologically abnormal and more fragile compared to those from uninfected material (Brown and Bryan, 2000). E4 is present in these aberrant squames, and associated with fragments of the CCE (Brown and Bryan, 2000). By crosslinking to constituents of these structures, E4 could disrupt their normal assembly rendering them fragile and able to release progeny virions easily. Given the mechanical role of keratin IFs in defining cell strength, disturbance of the organization of the keratin cytoskeleton by E4 could also contribute to the fragility of the HPV-infected cell (Doorbar et al., 1991), although, this aspect of E4 biology remains uncertain until the physiological authenticity of E4-mediated disruption of IFs can be fully resolved.

Keratinocyte terminal differentiation is viewed as a specialized form of cell death, although the terminally differentiating cells do not undergo cytolysis. HPV16 E4 can induce apoptosis of cervical tumour keratinocytes after prolonged expression of the viral protein, perhaps as a result of an association between E4 and mitochondria (Raj et al., 2004) (see below). E4-dependent apoptosis, if this occurs *in vivo,* may promote cell lysis in terminally differentiated cells. Thus, E4 could target more than one aspect of corneocyte physiology to compromise structural integrity and, at the same time, induce apoptosis to enable newly-produced virions to be released into the environment to infect other individuals.

Studies that analyse infectivity of virus particles formed in PV-containing cells, plus or minus E4, would be an interesting route of investigation.

5. E4 and Mitochondria

Prolonged expression of HPV16 E4 in epithelial and non-epithelial cell lines is associated with redistribution of mitochondria from the microtubule networks to perinuclear E4 aggregates (Raj et al., 2004). However, the investigators could not verify whether an interaction between E4 and mitochondria occurred in HPV16-induced cervical lesions. E4-mediated reorganization is specific to mitochondria as the intracellular localization of other organelles such as Golgi and lysosomes is unaffected. The mechanism of mitochondrial disruption by E4 is unknown. Raj and colleagues (2004) were able to demonstrate that the LLXLL motif, the sequence also required for E4-keratin association, was involved (Figure 2), although the keratin cytoskeleton itself is not necessary for mitochondrial aggregation. Disruption of mitochondria was followed by a reduction in their membrane potential and eventually apoptosis. Breakdown of the cytokeratin network and redistribution of cytoplasmic organelles are early events in programmed cell death, and, as discussed above, it is possible that E4 triggers these events in order to facilitate breakdown of the infected cells in the more superficial cell layers to aid virus egress.

6. E4 and ND10 Domains

ND10 bodies are discrete nuclear substructures dispersed throughout the nucleus. They have been intensively studied because of their association with aspects of promyelocytic leukemia, virus infection, gene expression, and apoptosis (for a review of these structures see Negorev and Maul, 2001 and Borden, 2002). The structural integrity of ND10 is dependent on the PML protein (Ishov et al., 1999). Modification of PML by the ubiquitin-like protein SUMO-1 is required for proper assembly of these structures (Ishov et al., 1999). Intriguingly, in the lower suprabasal cells of naturally-occurring HPV1 warts, the normal distribution of PML protein is disturbed and it is not found in ND10 bodies (Roberts et al., 2003). Instead, PML is concentrated at intranuclear E4 inclusions, but not at cytoplasmic E4 inclusions (Roberts et al., 2003). Redistribution of PML is E4-mediated, as similar E4-PML structures assembled in the nuclei of cultured keratinocytes expressing E4 alone. Sp100, another integral component of the ND10 domains is similarly reorganized to E4 intranuclear inclusion bodies (S. Roberts, unpublished data). How E4 reorganizes ND10 domains is not yet known, although the full-length E1^E4 protein is required for efficient disruption of the domains. Disruption of ND10 body organization may also be a common function of other E4 proteins (S. Roberts, unpublished data). What is the significance to PV infection of recruitment of ND10 body components to nuclear E4 structures? Studies of other viruses including herpes simplex virus type 1 (HSV-1) and cytomegalovirus (CMV) have shown a strong association between these domains and virus transcription and replication strategies (Doucas et al., 1996; Ishov and Maul, 1996; Ishov et al., 1997; Everett, 2001; Sourvinos and Everett, 2002). Many of these viruses encode early gene products that induce ND10 domain reorganization or disruption *e.g.* HSV-1 ICP0, Epstein-Barr virus (EBV) BLZF-1, CMV IE1 and Adenovirus 5 (Ad5) E4orf3, and HPV1 E4 may be analogous to these otherwise unrelated virus proteins. As PVs also engage with ND10 domains during their synthetic cycle, the ND10 bodies could be sites of early viral transcription and replication following L2 mediated deposition of the incoming viral genome to ND10 domains (Swindle et al., 1999; Guccione et al., 2004; Day et al., 2004). Why viruses need to disrupt ND10 domains is not altogether clear, but it may be a viral strategy to intervene in cell regulatory processes that would otherwise inhibit or limit virus replication – ND10 bodies appear to constitute an interferon–mediated defence mechanism that impedes virus infection (Regad and Chelbi-Alix, 2001). Therefore, E4's reorganization of ND10 bodies might relieve their repressive effects that would otherwise inhibit or limit PV replication. Disruption might also facilitate the formation of viral replication centres, by sequestration of viral and/or cellular proteins that are necessary for transcription and replication (Bell et al., 2000; Wilkinson and Weller, 2004). In fact, HPV1 E2, a protein involved in both PV transcription and replication, is sequestered to HPV1 E4-PML inclusions in cultured keratinocytes (S. Roberts, unpublished data).

Given that E4 redistributes ND10 proteins in the nuclei of cells that have switched from the early, non-productive maintenance stage of replication to the virus-producing stage, reorganization of the domains may be only necessary when the virus changes its replication strategy (Roberts et al., 2003). Indeed, this appears to be the case for in EBV infections where the lytic, but not latent, phase occurs at ND10 remnants (Bell et al., 2000). Disorganization of ND10 bodies by E4 might also be relevant to virion morphogenesis as L2 recruits L1 – the major capsid protein - to ND10 domains (Day et al., 1998; Florin et al., 2002a; 2002b).

Even though we are still at a stage of speculation on the significance of HPV1 E4-mediated disruption of ND10 bodies, the fact that this occurs in natural infections indicates that it is likely to be important, and may well be essential for the progression of the productive phase of the HPV life cycle.

7. Effects of E4 on Cell Growth

It has been known for several years that stable expression of E4 in epithelial cell lines is unachievable (Roberts et al., 1993), but the biological basis for the refractory nature of E4 on cell growth is only just beginning to be understood. Several studies have now found that E4 expression in cultured cells can influence the progression of the cell cycle. Recently, it has been shown that expression of several different E4 proteins (HPV types 1, 11, 16 and 18) in human keratinocytes induced a G_2 arrest in the cell cycle (Davy et al., 2002; Nakahara et al., 2002; Knight et al., 2004). While this occurred independently of other HPV protein functions, the cell cycle arrest function of E4 was active in HPV transformed cells (Davy et al., 2002; Nakahara et al., 2002; Knight et al., 2004). Cyclin/cdk complexes drive G_2/M progression and commitment to cell division is mediated by activation of the cyclin B1-cdk complex, the core component of the mitosis-promoting factor. Reconstitution of cyclin B1 expression in HPV1 E4-expressing cells reversed the G_2 arrest, suggesting that the cell cycle block was likely due to an aberration in the pathways that regulate this cyclin complex (Knight et al., 2004). Activated cyclin B1 complexes are sequestered to "collapsed" HPV16 E4/keratin bundles in epithelial cells suggesting that failure of cells to complete the G_2 to M transition might be due to an E4-induced cytoplasmic retention of this cyclin complex (Davy et al., 2005). However, cytoplasmic sequestration of the cyclin B1 complex to cytoplasmic E4-keratin filaments did not occur in cells arrested in G_2 by the HPV1 E4 protein (Knight et al., 2004). On the contrary, the G_2/M arrest by HPV1 E4 expression appears to be mediated by inhibitory phosphorylation of cdk1, the kinase partner of cyclin B1 (G. Knight and S. Roberts, unpublished data).

The G_2 arrest function of HPV1 and 16 E4 proteins is dependent on a threonine amino acid residue contained is within the proline–rich domain found near the N-termini of the protein (Davy et al., 2002; Knight et al., 2004) (Figure 2). In the anogenital PV E4 proteins the threonine residue lies in a potential cdk phosphorylation site (S/TPXK/R) with a cyclin binding motif (Cy or RXL motif) downstream (Davy et al., 2002). As yet, there is no direct evidence to show that this site is indeed a target of cdk phosphorylation, and the Cy motif of HPV16 E4 is not necessary for G_2 arrest (Davy et al., 2002). However, cyclin E/cdk1 and cyclin A/cdk2 complexes have been shown to interact, *via* the Cy motif, with the HPV18 E4 protein (Peter Whyte, personal communication).

An E4-induced block in G_2/M progression was accompanied by a significant level of chromosomal re-replication and this was dependent on E4's G_2 arrest function (Nakahara et al., 2002; Knight et al., 2004). It is therefore possible that the G_2/M arrested cells eventually re-renter the cell cycle and undergo additional rounds of cellular DNA replication, but without intervening cytokinesis.

A surprising finding was that the G_2 arrest function of HPV1 E4 was mediated not by the full-length E1^E4 protein, but a truncated E4 polypeptide lacking the extreme amino terminus of the full-length protein (Roberts et al., 1994a; Knight et al., 2004). The truncated protein is equivalent to a 16-kDa species that is derived from the full-length E1^E4 protein by proteolysis (Figure 2, Y16), and appears after E1^E4 synthesis has been initiated in HPV1 warts. Why this cutaneous E4 should adopt a different strategy to the anogenital E4s might be a reflection of differences in the timing of life cycle events between the cutaneous and anogenital viruses (Peh et al., 2002). Initiation of viral genome amplification and E4 expression are significantly delayed in the HPV16 life cycle, occurring in the more differentiated cells where early gene activity (E6, E7 and E5) is soon down regulated. In contrast, in HPV1 infections both events occur as soon as cells have moved up from the basal cell layers, and continue to proceed together with expression of early genes such as E7 through several layers of suprabasal keratinocytes.

Intriguingly, Knight and colleagues (2004) also found that co-expression of the HPV1 full-length E1^E4 protein and the truncated 16-kDa polypeptide had a far greater deleterious effect on keratinocyte growth than expression of the individual polypeptides. Further analysis revealed that although co-expression of the polypeptides maintained a strong G_2/M arrest, the cells did not exhibit chromosomal re-replication. Interestingly, a reduction in cellular DNA synthesis was also apparent in the diploid population of cells co-expressing the two E4 polypeptides. The findings of this study suggest that these two forms of the HPV1 protein are able to functionally cooperate to negatively influence keratinocyte proliferation (Knight et al., 2004).

The underlying molecular basis for the impact of HPV1 E4 on cellular proliferation and rereplication of chromosomal DNA is not known, but it has been suggested that HPV1 E4 can directly interfere with the process of cellular DNA synthesis. In fact, recombinant HPV1 E4 protein is a very effective inhibitor of initiation of chromosomal replication when introduced into an *in vitro* mammalian cellular DNA replication system (Stoeber et al., 1998; Roberts et al., 2000). The studies described above have shown that E4 can have a dramatic effect on the growth of cultured cells. However, because of the lack of information on E4's role in the virus life cycle, it is not possible to be sure at what stage in the PV infectious cycle these activities might be relevant. One possibility is that the inhibitory actions of E4 on the mitotic cycle might stimulate suprabasal cells, already driven by E7 functions into a pseudo S phase (Cheng et al., 1995), to remain in this state to maximize PV genome amplification. A high level of viral genome amplification would be dependent on nucleotides and essential replication enzymes acquired from the cellular pools. Continuous unscheduled replication of the host chromosome would however limit the availability of these factors to the virus. Inhibition of host cell DNA synthesis, by E4, might therefore allow the host cell factors to be sequestered to viral DNA replication centres. Thus, the overall effect of E4 would be to keep the infected keratinocyte in a metabolically active state without competing host DNA synthesis in order to maximize virus genome replication. Intriguingly, herpesviruses may have evolved to execute a similar mechanism - to maintain high S phase kinase activity but without competing replication of the host DNA - to support high-level virus production (Wiebusch and Hagemeir, 2001; Biswas et al., 2003; Kudoh et al., 2003; Wiebusch et al., 2003).,

Collectively, these studies have shown that E4 can interfere with the normal progression of the cell cycle. Clearly, further experimentation is necessary to define the mode of action of the E4 proteins and, just as importantly, the consequence of these actions on PV replication *in vivo*.

8. E4 and Late Gene Expression

DEAD-box proteins participate in a number of cellular processes that involve RNA interactions such as transcription, ribosome biogenesis, pre-mRNA splicing and editing, RNA export to the cytoplasm, translation and RNA degradation (for a review of DEAD-box functions, see Rocak and Linder [2004]). DEAD-box proteins use the energy from ATP hydrolysis to dissociate RNA-protein interactions or rearrange intra- or intermolecular RNA structures. Full-length HPV16 E1^E4 interacts with a novel DEAD-box protein E4-DBP *via* C terminal sequences (Figure 1) (Doorbar et al., 2000). The interaction partially reduces both ATP binding and ATPase activity of E4-DBP *in vitro*. Although no function for E4-DBP has been defined, E4-DBP has homology with bacterial and yeast proteins (Rrp3, SrmB, RhlB and DeaD) that are involved in the regulation of mRNA stability. A negative regulatory element (NRE) present in the 3' region of late PV transcripts plays a major role in the regulation of late gene expression. A number of cellular factors involved in RNA splicing, mRNA transport and metabolism bind to the NRE of HPV16 late mRNAs (Dietrich-Goetz et al., 1997; Koffa et al., 2000; Cumming et al., 2003; McPhillips et al., 2004; see Chapter 13)) and DEAD box proteins are known to bind to some of these proteins. It is conceivable therefore that E4-DBP function might be linked to the activity of the HPV16 NRE, although an association between E4-DBP and the major late HPV16 RNA transcript is not dependent on the presence of the NRE (Doorbar et al., 2000). DEAD box proteins also function in protein translation initiation (Rocak and Linder, 2004). The hepatitis virus core protein interacts with the DEAD box protein DBX and inhibits translation of RNA transcripts (Mamiya and Worman, 1999). In a similar fashion, an HPV16 E4:E4-DBP interaction might inhibit translation of specific cellular and/or viral mRNAs. However, we can only speculate on the biological relevance of this interaction to the HPV16 life cycle. The finding that an E4:E4-DBP association was not common to different E4 proteins, even those closely related to HPV16 (Doorbar et al., 2000) suggests that this interaction might only be relevant to the HPV16 life cycle. Perhaps of greater significance is the finding that E4 proteins of several unrelated types (including HPV1 and 16) interact with the splice factor kinase SRPK1 (I. Bell and S. Roberts, unpublished data). SRPK1 phosphorylates serine and arginine rich splice factors, some of which have been implicated in HPV late gene expression (McPhillips et al., 2004). Thus E4, via its association with SRPK1, may facilitate late gene expression by modifying the activity of host cell RNA splice factors.

9. Conclusions

Since the E4 gene product was identified in the late 1980s, the role of this highly divergent protein in the virus life cycle has intrigued workers in the PV field. Even though the precise role of this protein in the virus life cycle is still elusive, many of the biological functions and activities that have now been assigned to E4 suggest that it might have a role at more than one stage of the PV life cycle (Figure 1). ND10 domains may play a central role in the infectious cycle of PVs, and the link between E4 and these nuclear structures might indicate a role for E4 in virus transcription, replication and virion assembly. E4 actions impinge on cell growth by interacting with the cell cycle. This E4 function may facilitate efficient genome amplification, or might be important in a role for E4 in the early stages of the life cycle. Disturbance of cellular integrity by targeting those cell structures that impart mechanical strength to the cell (IFs and CCE), as well as promoting cell death, may facilitate escape of newly synthesized virions from the superficial keratinocyte. As found with other PV proteins, there may be differences in the biological actions of E4 proteins between virus types, and this may reflect the heterogeneity within their life cycles. E4 function will probably be regulated through posttranslational modifications, occurring as the infected cells migrate up through the wart layer. Biological activities of E4 map primarily to sequences that lie at the amino and carboxy termini of the proteins, regions that exhibit the highest levels of similarity between PV types. Identification of molecular partners of E4, coupled with the elucidation of the three-dimensional structure of the viral protein should enable a fuller picture of the molecular actions of E4 to be established. It is important to note that studies of E4 function have largely been confined to monolayer cultures of epithelial cells. On the basis that *in vivo* E4 is overexpressed in differentiating keratinocyte, it could be argued that the monolayer systems may not reveal the true functions of E4. Despite these reservations, however, much progress has been made, and some of E4s activities, identified initially in monolayer cell cultures, have been found to occur in natural infections *e.g.* ND10 reorganization, association with the CCE. Studies using intact PV genomes in differentiating keratinocytes in which the E4 ORF is mutated are beginning to identify E4 as a major regulator of the productive phase of the virus life. Whether E4 has a pleiotropic role in the virus life cycle, with functions that support virus functions prior and post the switch from the non-productive phase to the virus producing phase should be revealed by the use of such tissue culture systems.

Acknowledgements

I would like to thank past and present members of the laboratory for their contributions to some of the research, supported by Cancer Research UK, described here.

References

Anderson, R.A., Scobie, L., O' Neil, B.W., Grindlay, G.J., and Campo, M.S. (1997). Viral proteins of bovine papillomavirus type 4 during the development of alimentary canal tumours. Vet. J. *154*, 69-78.

Ashmole, I., Gallimore, P.H., and Roberts, S. (1998). Identification of conserved hydrophobic C-terminal residues of the human papillomavirus type 1 E1^E4 protein necessary for oligomerization *in vivo*. Virology *240*, 221-231.

Bell, P., Lieberman, P.M., and Maul, G.G. (2000). Lytic but not latent replication of Epstein-Barr virus is associated with PML and induces sequential release of nuclear domain 10 proteins. J. Virol. *74*, 11800-11810.

Biswas, N., Sanchez, V., and Spector, D.H. (2003). Human cytomegalovirus infections leads to accumulation of geminin and inhibition of the licensing of cellular DNA replication. J. Virol. *77*, 2369-2376.

Borden, K.L.B. (2002). Pondering the promyelocytic leukaemia protein (PML) puzzle: possible functions for PML nuclear bodies. Mol. Cell Biol. *22*, 5259-5269.

Breitburd, F., Croissant, O., and Orth, G. (1987). Expression of human papillomavirus type-1 E4 gene products in warts. Cancer Cells (New York: Cold Spring Harbor Laboratory Press) *5*, 115-122.

Brown, D.R., Brown, C.R., and Lehr, E.E. (2004). Intracellular expression patterns of the human papillomavirus type 59 E1^E4 protein in COS cells, keratinocytes and genital epithelium. Intervirology *47*, 321-327.

Brown, D.R., and Bryan, J.T. (2000). Abnormalities of cornified cell envelopes isolated from human papillomavirus type 11-infected genital epithelium. Virology *271*, 65-70.

Bryan, J.T., and Brown, D.R. (2000). Association of the human papillomavirus type 11 E1^E4 protein with cornified cell envelopes derived from infected genital epithelium. Virology *277*, 262-269.

Bryan, J.T., and Brown, D.R. (2001). Transmission of human papillomavirus type 11 infection by desquamated cornified cells. Virology *281*, 35-42.

Bryan, J.T., Fife, K.H., and Brown, D.R. (1998). The intracellular expression pattern of the human papillomavirus type 11 E1^E4 protein correlates with its ability to self associate. Virology *241*, 49-60.

Bryan, J.T., Han, A., Fife, K.H., and Brown, D.R. (2000). The human papillomavirus type 11 E1^E4 protein is phosphorylated in genital epithelium. Virology *268*, 430-439.

Caulin, C., Ware, C.F., Magin, T.M., and Oshima, R.G. (2000). Keratin-dependent, epithelial resistance to tumor necrosis factor-induced apoptosis. J. Cell Biol. *149*, 17-22.

Cheng, S., Schmidt-Grimminger, D.C., Murant, T., Broker, T.R., and Chow, L.T. (1995). Differentiation-dependent up-regulation of the human papillomavirus E7 gene reactivates cellular DNA replication in suprabasal differentiated keratinocytes. Genes Dev. *9*, 2335-2349.

Chow, L.T., Reilly, S.S., Broker, T.R., and Taichman, L.B. (1987). Identification and mapping of human papillomavirus type 1 RNA transcripts recovered from plantar warts and infected epithelial cell cultures. J. Virol. 61, 1913-1918.

Coulombe, P.A., and Wong, P. (2004). Cytoplasmic intermediate filaments revealed as dynamic and multipurpose scaffolds. Nature Cell Biol. *6*, 699-706.

Croissant O., Breitburd, F., and Orth, G. (1985). Specificity of cytopathic effect of cutaneous human papillomaviruses. Clinics in Dermatol. *3*, 43-55.

Crum, C.P., Barber, S., Symbula, M., Snyder, K., Saleh, A.M., and Roche, J.K. (1990). Coexpression of the human papillomavirus type 16 E4 and L1 open reading frames in early cervical neoplasia. Virology *178*, 238-246.

Cumming, S.A., McPhillips, M.G., Veerapraditsin, T., Milligan, S.G., and Graham, S.V. (2003). Activity of the human papillomavirus type16 late negative regulatory element is partly due to four weak consensus 5' splice sites that bind a U1 snRNP-like complex. J. Virol. *77*, 5167-5177.

Davy, C.E., Jackson, D.J., Raj, K., Peh, W.L., Southern, S.A., Das, P., Sorathia, R., Laskey, P., Middleton, K., Nakahara, T., Wang, Q., Masterson, P.J., Lambert, P.F., Cuthill, S., Millar, J.B., and Doorbar, J. (2005). Human papillomavirus type 16 E1 E4-induced G2 arrest is associated with cytoplasmic retention of active Cdk1/cyclin B1 complexes. J. Virol. *79*, 3998-4011.

Davy, C.E., Jackson, D.J., Wang, Q., Raj, K., Masterson, P.J., Fenner, N.F., Southern, S., Cuthill, S., Millar, J.B.A., and Doorbar, J. (2002). Identification of a G2 arrest domain in the E1^E4 protein of human papillomavirus type 16. J. Virol. *76*, 9806 9818.

Day, P.M., Baker, C.C., Lowy, D.R., and Schiller, J.T. (2004). Establishment of papillomavirus infection is enhanced by promyelocytic leukaemia protein (PML) expression. Proc. Natl. Acad. Sci. USA. *101*, 14252-14257.

Day, P.M., Roden, R.B.S., Lowy, D.R., and Schiller, J.T. (1998). The papillomavirus minor capsid protein, L2, induces localization of the major capsid protein L1, and the viral transcription/replication protein, E2, to PML oncogenic domains. J. Virol. *72*, 142-150.

Dietrich-Goetz, W., Kennedy, I.M., Levins, B., Stanley, M.A., and Clements, J.B. (1997). A cellular 65-kDa protein recognizes the negative regulatory element of human papillomavirus late mRNA. Proc. Natl. Acad. Sci. USA *94*, 163-168.

Doorbar, J., Campbell, D., Grand, R.J.A., and Gallimore, P.H. (1986). Identification of the human papillomavirus-1a E4 gene products. EMBO. J. *5*, 355-365.

Doorbar, J., Coneron, I., and Gallimore, P.H. (1989). Sequence divergence yet conserved physical characteristics among the E4 proteins of cutaneous human papillomaviruses. Virology *172*, 51-62.

Doorbar, J., Elston, R.C., Napthine, S., Raj, K., Medcalf, E., Jackson, D., Coleman, N., Griffin, H.M., Masterson, P., Stacey, S., Mengistu, Y., and Dunlop, J. (2000). The E1^E4 protein of human papillomavirus type 16 associates with a putative RNA helicase through sequences in its C terminus. J. Virol. *74*, 10081-10095.

Doorbar, J., Ely, S., Sterling, J., McLean, C., and Crawford, L. (1991). Specific interaction between HPV-16 E1-E4 and cytokeratins results in collapse of the epithelial cell intermediate filament network. Nature *352*, 824-827.

Doorbar, J., Evans, H.S., Coneron, I., Crawford, L.V., and Gallimore, P.H. (1988). Analysis of HPV-1 E4 gene expression using epitope-defined antibodies. EMBO. J. *7*, 825-833.

Doorbar, J., Foo, C., Coleman, N., Medcalf, L., Hartley, O., Prospero, T., Napthine, S., Sterling, J., Winter, G., and Griffin, H. (1997). Characterization of events during the late stages of HPV16 infection *in vivo* using high affinity synthetic Fabs to E4. Virology *238*, 40-52.

Doorbar, J., Medcalf, E., and Napthine, S. (1996). Analysis of HPV1 E4 complexes and their association with keratins *in vivo*. Virology *218*, 114-126.

Doorbar, J., and Myers, G. (1996). The E4 protein. In Human papillomaviruses Vol III, G. Myers, H. Delius, J. Icenogel, H-U. Bernard, C. Baker, A. Halpern, C. Wheeler, ed. (Los Alamos National Laboratory, New Mexico), pp, 58-80.

Doucas, V., Ishov, A.M., Romo, A., Juguilon, H., Weitzman, M.D., Evans, R.M., and Maul, G.G. (1996). Adenovirus replication is coupled with the dynamic properties of PML nuclear structure. Genes Dev. *10*, 196-207.

Eriksson, A., Herron, J.R., Yamada, T., and Wheeler, C.M. (1999). Human papillomavirus type 16 variant lineages characterized by nucleotide sequence analysis of the E5 coding segment and the E2 hinge region. J. Gen. Virol. *80*, 595-600.

Everett, R. D. 2001. DNA viruses and viral proteins that interact with PML nuclear bodies. Oncogene 20, 7266-7273.

Florin, L., Sapp, C., Streeck, R.E., and Sapp, M. (2002a). Assembly and translocation of papillomavirus capsid proteins. J. Virol. *76*, 10009-10014.

Florin, L., Schafer, F., Sotlar, K., Streeck, R.E., and Sapp, M. (2002b). Reorganization of nuclear domain 10 induced by papillomavirus capsid protein L2. Virology *295*, 97-107.

Gilbert, S., Loranger, A., Daigle, N., and Marceau, N. (2001). Simple epithelium keratins 8 and 18 provide resistance to Fas-mediated apoptosis. The protection occurs through a receptor-targeting modulation. J. Cell Biol. *154*, 763-773.

Grand, R.J.A., Doorbar, J., Smith, K.J. Coneron, I., and Gallimore, P.H. (1989). Phosphorylation of the human papillomavirus type 1 E4 proteins *in vivo* and *in vitro*. Virology *170*, 201-213.

Guccione E., Lethbridge K.J., Killick, N., Leppard, K.N., and Banks, L. (2004). HPV E6 proteins interact with specific PML isoforms and allow distinctions to be made between different POD structures. Oncogene *23*, 4662-4672.

Hummel, M., Hudson, J.B., and Laimins, L.A. (1992). Differentiation-induced and constitutive transcription of human papillomavirus type 31b in cell lines containing viral episomes. J. Virol. *66*, 6070-6080.

Inada, H., Izawa, I., Nishizawa, M., Fujita, E., Kiyono, T., Takahashi, T., Momoi, T., and Inagaki, M., (2001). Keratin attenuates tumor necrosis factor-induced cytotoxicity through association with TRADD. J Cell Biol. 155, 415-426.

Ishov, A.M., and Maul, G.G. (1996). The periphery of nuclear domain 10 (ND10) as site of DNA virus deposition. J. Cell Biol. *134*, 815-826.

Ishov, A.M., Sotnikov, A.G., Negorev, D., Vladimirova, O.V., Neff, N., Kamitani, T., Yeh, E.T., Strauss III, J.F., and Maul, G.G. (1999). PML is critical for ND10 formation and recruits the PML-interacting protein Daax to this nuclear structure when modified by SUMO-1. J. Cell Biol. *147*, 221-234.

Ishov, A.M., Stenberg, R.M., and Maul, G.G. (1997). Human cytomegalovirus immediate early interaction with host nuclear structures: definition of an immediate transcript environment. J. Cell Biol. *138*, 5-16.

Jareborg, N., and Burnett, S. (1991). Immunofluorescent detection of bovine papillomavirus E4 antigen in the cytoplasm of cells permissive in vitro for viral DNA amplification. J. Gen. Virol. *72*, 2269-2274.

Koffa, M.D., Graham, S.V., Takagaki, Y., and Clements, J.B. (2000). The human papillomavirus type 16 late negative regulatory element interacts with three proteins that act at different post-transcriptional levels. Proc. Natl. Acad. Sci. USA *96*, 4677-4682.

Knight, G.L, Grainger, J.R. Gallimore, P.H., & Roberts, S. (2004). Cooperation between different forms of the human papillomavirus type 1 E4 protein to block cell cycle progression and inhibit cellular DNA synthesis. J. Virol. *78,* 13920-13933.

Kudoh, A., Fujita, M., Kiyono, T., Kuzushima, K., Sugaya, Y., Izuta, S., Nishiyama, Y., and Tsurumi, T. (2003). Reactivation of lytic replication from B cells latently infected with Epstein-Barr virus occurs with high S phase cyclin-dependent kinase activity while inhibiting cellular DNA replication. J. Virol. *77*, 851-861.

Lane, E.B., and McLean, W.H. (2004). Keratins and skin disorders. J. Pathol. *204*, 355-366.

Lehr, E., Hohl, D., Huber, M., and Brown, D. (2004). Infection with human papillomavirus alters expression of the small proline rich proteins 2 and 3. J. Med. Vir. *72*, 478-483.

Mamiya, N., and Worman, H.J. (1999). Hepatitis C virus core protein binds to a DEAD box RNA helicase. J. Biol. Chem. *274*, 15751-15756.

McPhillips M.G., Veerapraditsin, T., Cumming, S.A., Karali, D., Milligan, S., Boner, W., Morgan, I.M., and Graham, S.V. (2004). SF2/ASF binds the human papillomavirus type 16 late RNA control element and is regulated during differentiation of virus-infected epithelial cells. J. Virol. *78*, 10598-10605.

Middleton, K., Peh, W., Southern, S., Griffin, H., Sotlar, K., Nakahara, T., El-Sherif, A., Morris, L., Seth, R., Hibma, M., Jenkins, D., lambert, P., Coleman, N., and Doorbar, J. (2003). Organization of human papillomavirus productive cycle during neoplastic progression provides a basis for selection of diagnostic markers. J. Virol. *77*, 10186-10201.

Nakahara, T., Nishimura, A., Tanaka, M., Ueno, T., Ishimoto, A., and Sakai, H. (2002). Modulation of the cell division cycle by human papillomavirus type 18 E4. J. Virol. *76*, 10914-10920.

Nasseri, M., Hirochika, T., Broker, T.R., and Chow, L.T. (1987). A human papillomavirus type 11 transcript encoding an E1^E4 protein, Virology *159*, 433-439.

Neary, K., Horwitz, B.H., and DiMaio, D. (1987). Mutational analysis of open-reading frame E4 of bovine papillomavirus type 1. J. Virol. *61*, 1248-1252.

Negorev, D., and Maul, G.G. (2001). Cellular proteins localized at and interacting within ND10/PML nuclear bodies/PODs suggest functions of a nuclear depot. Oncogene *20*, 7234-7242.

Nicholls, P.K., Doorbar,J., Moore, R.A., Peh, W., Anderson, D.M., and Stanley, M.A. (2001). Detection of viral DNA and E4 proteins in basal keratinocytes of experimental canine oral papillomavirus lesions. Virology *284*, 82-98.

Palefsky, J.M., Winkler, B., Rabanus, J-P., Clark, C., Chan, S., Nizet, V., and Schoolnik, G.K.. (1991). Characterization of *in vivo* expression of the human papillomavrius type 16 E4 protein in cervical biopsy tissues. J. Clin. Invest. *87*, 2132-2141.

Palermo-Dilts, D.A., Broker, T.R., Chow, L.T. (1990). Human papillomavirus type 1 produces redundant as well as polycistronic mRNAs in plantar warts. J. Virol. *64*, 3144-3149.

Paramio, J.M., Casanova, M.L., Segrelles, C., Mittnacht, S., Lane, E.B. and Jorcano, J.L. (1999). Modulation of cell proliferation by cytokeratins K10 and K16. Mol. Cell Biol. *19*, 3086-3094.

Paramio, J.M., Segrelles, C., Ruiz, S., and Jorcano, J.L. (2001). Inhibition of protein kinase B (PKB) and PKCζ mediates keratin K10-induced cell cycle arrest. Mol. Cell Biol. *21*, 7449-7459.

Peh, W.L., Brandsma, J.L., Christensen, N.D., Cladel, N.M., Wu, X., and Doorbar, J. (2004). The viral E4 protein is required for the completion of the cottontail rabbit papillomavirus productive cycle *in vivo*. J. Virol. *78*, 2142-2151.

Peh, W.L., Middleton, K., Christensen, N., Nicholls, P., Egawa, K., Sotlar, K., Brandsma, J., Percival, A., Lewis, J., Liu, W.J., and Doorbar, J. (2002). Life cycle heterogeneity in animal models of human papillomavirus-associated disease. J. Virol. *76*, 10401-10416.

Pray, T.R., and Laimins, L. A. (1995). Differentiation-dependent expression of E1^E4 proteins in cell lines maintaining episomes of human papillomavirus type 31b. Virology 206, 679-685.

Raj, K., Berguerand, S., Southern, S., Doorbar, J., and Beard, P. (2004). E1^E4 protein of human papillomavirus type 16 associates with mitochondria. J. Virol. *78*, 7199-7207.

Regad, T., and Chelbi-Alix, M.K. (2001). Role and fate of PML nuclear bodies in response to interferon and viral infections. Oncogene 20, 7274-7286.

Roberts, S. (2002). Biology of the E4 protein. In Human Papillomaviruses, D. J. McCance, ed. (Elsevier Science B. V. Amsterdam), pp. 117-140.

Roberts, S., Ashmole, I., Gibson, L.J., Rookes, S.M., Barton, G.J., and Gallimore, P.H. (1994a). Mutational analysis of human papillomavirus E4 proteins: Identification of structural features important in the formation of cytoplasmic E4/cytokeratin networks in epithelial cells. J. Virol. *68*, 6432-6445.

Roberts, S., Ashmole, I., Johnson, G.D., Kreider, J.W., and Gallimore. P.H. (1993). Cutaneous and mucosal papillomavirus E4 proteins form intermediate filament-like structures in epithelial cells. Virology 197, 176-187.

Roberts, S., Ashmole, I., Rookes, S.M., and Gallimore, P.H. (1997). Mutational analysis of the human papillomavirus type 16 E1^E4 protein shows that the C terminus is dispensable for keratin cytoskeleton association but is involved in inducing disruption of the keratin filaments. J. Virol. *71*, 3554-3562.

Roberts, S., Ashmole, I., Sheehan, T.M.T., Davies, A.H., and Gallimore, P.H. (1994b). Human papillomavirus type 1 E4 protein is a zinc-binding protein. Virology *202*, 865-874.

Roberts, S., Hillman, M.L., Knight, G.L., and Gallimore, P.H. (2003). The ND10 component promyelocytic leukemia protein relocates to human papillomavirus type 1 E4 intranuclear inclusion bodies in cultured keratinocytes and in warts. *77*, 673-684.

Roberts, S., Stoeber, K., Gallimore, P.H., and Williams, G.H. (2000). The human papillomavirus E1^E4 protein inhibits initiation of cellular DNA replication. International Papillomavirus Conference, Barcelona, Spain.

Rocak, S., and Linder, P. (2004). DEAD-box proteins: the driving forces behind RNA metabolism. Nature Rev. *5*, 232-241.

Rogel-Gaillard, C., Pehau-Arnaudet,G., Breitburd, F., and Orth, G. (1993). Cytopathic effect in human papillomavirus type 1-induced inclusion warts: in vitro analysis of the contribution of two forms of the viral E4 protein. J. Invest. Dermatol. *101*, 843-851.

Ruesch, M.N., Stubenrauch, F., and Laimins. L.A. (1998). Activation of papillomavirus late gene transcription and genome amplification upon differentiation in semisolid medium is coincident with expression of involucrin and transglutaminase but not keratin-10. J. Virol. *72,* 5016-5024.

Sterling, J.C., Skepper, J.N., and Stanley, M.A. (1993). Immunoelectron microscopical localization of human papillomavirus type 16 L1 and E4 proteins in cervical keratinocytes cultured *in vivo*. J. Invest. Dermatol. *100*, 154-158.

Sourvinos, G., and Everett, R.D. (2002). Visualization of parental HSV-1 genomes and replication compartments in association with ND10 in live infected cells. EMBO. J. *21*, 4989-4997.

Stoeber, K., Mills, A.D., Kubota, Y., Krude, T., Romanowski, P., Marheineke, K., Laskey, R.A., and Williams, G.H. (1998). Cdc6 protein causes premature entry into S phase in a mammalian cell-free system. EMBO J. *17*, 7219-7229.

Swindle, C.S., Zou, N., Van Tine, B.A., Shaw, G.M., Engler, J.A., and Chow, L.T. (1999). Human papillomavirus DNA replication compartments in a transient DNA replication system. J. Virol. *73*, 1001-1009.

Tomita, Y., Fuse, A., Sekine, H., Shirasawa, H., Simizu, B., Sugimoto, M., and Funahashi. S. (1991). Human papillomavirus type 6 and 11 E4 gene products in condyloma acuminata. J. Gen Virol. *72*, 731-744.

Wang, Q., Griffin, H., Southern, S., Jackson, D., Martin, A., McIntosh, P., Davy, C., Masterson, P.J., Walker, P.A., Laskey, P., Omary, M.B. and Doorbar, J. (2004). Functional analysis of the human papillomavrus type 16 E1^E4 protein provides a mechanism for *in vivo* and in vitro keratin filament reorganization. J. Virol. *78*, 821-833.

Wiesbusch, L., and Hagemeier, C. (2001). The human cytomegalovirus immediate early 2 protein dissociates cellular DNA synthesis from cyclin-dependent kinase activation. EMBO J. *20*, 1086-1098.

Wiesbusch, L., Uecker, R., and Hagemeier, C. (2003). Human cytomegalovirus prevents replication licensing by inhibiting MCM laoding onto chromatin. EMBO Rep. *4*, 42-46.

Wilkinson, D.E., and Weller, S.K. (2004). Recruitment of cellular recombination and repair proteins to sites of herpes simplex virus type 1 DNA replication is dependent on the composition of viral proteins within prereplicative sites and correlates with the induction of the DNA damage response. J. Virol. *78*, 4783-4796.

Wilson R., Fehrman, F., and Laimins, L. A. (2005). Role of the E1--E4 protein in the differentiation-dependent life cycle of human papillomavirus type 31. J. Virol. *79*, 6732-6740.

Chapter 8

Biological Activities of Papillomavirus E5 Proteins

Frank A. Suprynowicz, M. Saveria Campo and Richard Schlegel*

Abstract

This chapter discusses biological activities of the bovine papillomavirus type 1 E5 protein (BPV-1 E5), the closely-related E5 protein of bovine papillomavirus type 4 (BPV-4 E5) and the E5 protein of high-risk human papillomavirus type 16 (HPV-16 E5). BPV-1 E5 transforms fibroblasts by inducing constitutive, ligand-independent activation of the β-type platelet-derived growth factor receptor (PDGF-R_β), and by constitutively activating c-Src and interfering with Golgi acidification independently of PDGF-R_β signaling. BPV-4 E5 similarly transforms fibroblasts and disrupts gap junction-mediated intercellular communication. Both BPV proteins down-regulate surface expression of major histocompatibility complex (MHC) class I molecules, which should enhance the ability of infected cells to evade detection by the host immune system. HPV-16 E5 is only weakly oncogenic, but acts in a number of ways to increase the duration and scope of HPV-16 infections, which in turn may increase the probability of malignant progression. These actions include a reduction of MHC class I and class II surface expression, the disruption of gap junctional intercellular communication and the enhancement of ligand-dependent epidermal growth factor receptor activation.

1. Introduction

Interest in papillomavirus E5 proteins has been fueled by the finding that the 44-amino acid product of the bovine papillomavirus type 1 E5 gene (BPV-1 E5) can induce oncogenic transformation in immortalized tissue culture cells, and by the causative role of high-risk human papillomaviruses (HPVs) in cervical cancer. This review provides a detailed analysis of the known biological activities of the E5 protein from BPV-1 (a Delta-BPV), the closely-related E5 protein of bovine papillomavirus type 4 (BPV-4; a Xi-BPV), and the E5 protein of high-risk HPV type 16 (HPV-16; an Alpha-HPV), since these three E5 proteins have been studied most extensively during the past 20 years (For papillomavirus classification, see Chapter 2). In particular, the review will assess how the various biological activities of E5 proteins may promote malignant transformation.

2. BPV-1 E5

E5, the major transforming gene of BPV-1 (Yang et al., 1985; DiMaio et al., 1986; Groff and Lancaster, 1986; Rabson et al., 1986; Schiller et al., 1986; Schlegel et al., 1986), encodes a 44-amino acid polypeptide (Figure 1) that is the smallest known viral oncoprotein (Schlegel et al., 1986; Burkhardt et al., 1987). BPV-1 E5 can transform immortalized murine cells *in vitro*, enabling the cells to form foci, grow in an anchorage-independent manner in soft agar, and form tumors in nude mice (Schiller et al., 1984; 1986; Yang et al., 1985; Bergman et al., 1988; Schlegel et al., 1986; Goldstein et al., 1994). BPV-1 E5 also can transform mortal human diploid fibroblasts to form foci, although the cells do not exhibit anchorage-independence (Petti and Ray, 2000). These fibroblasts are not immortalized by BPV-1 E5, but rather are induced to undergo apoptotic cell death during prolonged periods of confluence (Petti and Ray, 2000; Zhang et al., 2002a). In animals, BPV E5 expression is correlated with the transformation of fibroblasts and epithelial cells to form benign and malignant tumors. BPV types 1 and 2 are widely believed to participate in the pathogenesis of equine sarcoids, which are locally invasive, benign fibroblastic skin tumors (Chambers et al., 2003a). These lesions contain BPV E5 transcripts (Nasir and Reid, 1999; Carr et al., 2001a) and express E5 protein (Carr et al., 2001b; Chambers et al., 2003b). BPV E5 also may

10

NH_2-**M-P-N-L-W-F-L-L-F-L-G-L-V-A-A-M-Q-L-L-**

20 30

L-L-L-F-L-L-L-F-F-L-V-Y-W-D-H-F-E-C-S-C-

40

T-G-L-P-F-COOH

Figure 1. Primary amino acid sequence of BPV-1 E5.

*For correspondence. Email schleger@georgetown.edu

be a causative agent in the formation of bovine urinary bladder tumors. BPV E5 DNA has been detected in both neoplastic and normal bovine bladder epithelium, however only biopsies taken from urinary bladder tumors show evidence of E5 protein expression (Borzacchiello et al., 2003).

2.1. Structure

BPV-1 E5 consists of a highly-hydrophobic, alpha-helical transmembrane domain (amino acids 4-31) and hydrophilic C-terminal domain (amino acids 32-44) (Schlegel et al., 1986; Burkhardt et al., 1987; Horwitz et al., 1988). It is a type II transmembrane protein that is localized in the Golgi apparatus with its C terminus oriented inside the Golgi lumen (Burkhardt et al., 1989). The targeting of BPV-1 E5 to late (*medial*- and *trans*) Golgi compartments appears to be critical for its transforming activity, since appending the endoplasmic reticulum (ER) retrieval sequence, KDEL, to its C-terminus causes BPV-1 E5 to be sequestered in the ER and early (*cis*-) Golgi, and renders it transformation-incompetent (Sparkowski et al., 1995).

BPV-1 E5 that is isolated from cells under non-reducing conditions exists as a mixture of monomers, dimers and tetramers, however, dimers are the most abundant species (Burkhardt et al., 1987; Horwitz et al., 1988; Sparkowski et al., 1996; Adduci and Schlegel, 1999). The fact that reducing agents can convert E5 dimers and tetramers to monomers indicates that the oligomers are stabilized by intermolecular disulfide bonding between cysteine residues (Burkhardt et al., 1987; Horwitz et al., 1988). Cysteines are present at positions 37 and 39 of the hydrophilic BPV-1 E5 C-terminus, and both residues participate in the stabilization of E5 oligomers. Single substitutions of arginine, serine or methionine for either cysteine 37 or cysteine 39 reduce the proportion of BPV-1 E5 that is oligomeric, and the substitution of serine or methionine for both cysteines completely abrogates the ability of E5 to form stable oligomers (Horwitz et al., 1988 and R. Schlegel, unpublished). E5 dimers are additionally stabilized by specific hydrophobic interactions in the transmembrane domain, since transmembrane alanine point mutants have been identified that can form tetramers but are defective for dimerization (Adduci and Schlegel, 1999).

Oligomerization appears to be essential for the transforming activity of BPV-1 E5, since C37S/C39S and C37M/C39M double cysteine mutations that render E5 monomeric are non-transforming (Horwitz et al., 1988 and R. Schlegel, unpublished). The existence of transformation-competent BPV-1 E5 mutants that dimerize but cannot form stable tetramers, or that form tetramers but not stable dimers (Adduci and Schlegel, 1999), suggests that both types of E5 oligomers contribute to the transforming activity of wild-type (wt) BPV-1 E5.

2.2. Transformation via PDGF receptor activation

BPV-1 E5 has no intrinsic enzymatic activity, but forms stable complexes *in vivo* with the β-type platelet-derived growth factor receptor (PDGF-R_{β}) (Petti and DiMaio, 1992) and 16 kDa pore-forming subunit of the vacuolar H^+-ATPase (V-ATPase) (Goldstein and Schlegel, 1990; Goldstein et al., 1991). Structural and functional aspects of the interaction between BPV-1 E5 and the PDGF-R_{β} have been studied in detail, and PDGF-R_{β} activation has been shown to play an important role in BPV-1 E5-mediated cellular transformation (DiMaio and Mattoon, 2001).

A useful approach for studying the interaction of BPV-1 E5 with the PDGF-R_{β} has involved co-expressing the two proteins in murine hematopoietic cells that lack endogenous PDGF-Rs (Goldstein et al., 1994; Drummond-Barbosa et al., 1995; Staebler et al., 1995; Petti et al., 1997; Klein et al., 1998; Lai et al., 1988). Normally, the proliferation of these cells is completely dependent on the presence of interleukin-3 (IL-3) in the culture medium, but this requirement can be circumvented by ectopically expressing growth factor receptors and supplementing cultures with the corresponding growth factor (Pierce et al., 1988). Hematopoietic cells that co-express BPV-1 E5 and the PDGF-R_{β} require neither IL-3 nor PDGF for growth; thereby demonstrating that E5 can constitutively activate the PDGF-R_{β} to a level that generates proliferative signals (Goldstein et al., 1994; Drummond-Barbosa et al., 1995). Both mature and precursor forms of the PDGF-R_{β} are activated by BPV-1 E5, since both are associated with E5 and both undergo constitutive tyrosine phosphorylation indicative of activation (Goldstein et al., 1994; Drummond-Barbosa et al., 1995). Importantly, experiments in hematopoietic cells, and in mouse mammary gland epithelial cells that also lack endogenous PDGF-Rs, indicate that the proliferative signals resulting from BPV-1 E5-mediated PDGF-R_{β} activation induce cellular transformation. In both types of cells, BPV-1 E5 expression alone is not sufficient to induce transformation; however cells that co-express E5 and the PDGF-R_{β} are capable of forming tumors in nude mice (Nilson and DiMaio, 1993; Goldstein et al., 1994).

Investigations of exogenous PDGF-R_{β} activation by BPV-1 E5 in hematopoietic cells frequently have been carried out in parallel with similar studies of endogenous PDGF-R_{β} activation by E5 in murine fibroblasts. Sustained PDGF-R_{β} activation has been observed both in fibroblast cell lines that stably express BPV-1 E5, and in fibroblasts that acutely express E5 as a result of retroviral infection (Petti et al., 1991). The activation of endogenous PDGF-R_{β}s is characterized by the formation of stable BPV-1 E5-receptor complexes (Petti and DiMaio, 1992), persistent tyrosine phosphorylation of the receptor (Petti et al., 1991), receptor dimerization (Lai et al., 1998), serum-independent proliferation (Suprynowicz et al., 2000) and cellular transformation (Klein et al., 1998; 1999).

The ability of BPV-1 E5 to activate growth factor receptors appears to be restricted to the PDGF-R_β. Co-expressing BPV-1 E5 with the closely-related α-type PDGF-R, epidermal growth factor receptor (EGF-R), or colony-stimulating factor 1 receptor in hematopoietic cells does not induce receptor-E5 complex formation, receptor activation, IL-3-independent proliferation or tumor formation in nude mice (Goldstein et al., 1994). Similarly, BPV-1 E5 does not activate endogenous EGF-Rs, insulin receptors, basic fibroblast growth factor receptors or insulin-like growth factor receptors in NIH 3T3 fibroblasts (Suprynowicz et al., 2002). Nevertheless, it should be noted that over-expressing 40-fold higher levels of exogenous human EGFRs or colony-stimulating factor 1 receptors in NIH 3T3 cells does enhance BPV-1 E5-mediated transformation (Martin et al., 1989).

A detailed mutational analysis of the mechanism of PDGF-R_β activation by BPV-1 E5 has been carried out in hematopoietic cells by co-expressing PDGF-R_β and BPV-1 E5 mutants in the absence of endogenous wt PDGF-Rs. These studies have demonstrated that BPV-1 E5-mediated activation of the PDGF-R_β occurs independently of PDGF, since E5 constitutively activates PDGF-R_β deletion mutants that lack the extracellular domain responsible for ligand binding (Drummond-Barbosa et al., 1995; Staebler et al., 1995). Moreover, the co-expression of BPV-1 E5 with chimeric α and β-type PDGF-Rs has shown that a short region of the PDGF-R_β consisting of only the transmembrane domain and juxtamembrane amino acids is responsible for binding E5. Replacing this region in the α-type PDGF-R with the corresponding region from the PDGF-R_β enables the chimeric α-type PDGF-R to stably associate with BPV-1 E5, undergo constitutive tyrosine phosphorylation and support IL-3-independent proliferation (Staebler et al., 1995). Further analysis of BPV-1 E5 and PDGF-R_β point mutants in hematopoietic cells and fibroblasts indicates that hydrogen bonding between glutamine 17 and threonine 513 in the transmembrane domains of E5 and the PDGF-R_β, respectively, and electrostatic attraction between aspartate 33 and lysine 499 in the juxtamembrane regions of the two proteins, are critical for complex formation (Petti et al., 1997; Klein et al., 1998; Klein et al., 1999). These studies also show that PDGF-R_β activation requires a PDGF-R_β molecule stably binding to each subunit of an E5 homodimer. BPV-1 E5 mutants that are unable to bind to the PDGF-R_β as a result of amino acid substitutions at positions 17 and 33 are unable to induce receptor activation (Petti et al., 1997; Klein et al., 1998). The substitution of alanine at positions 7, 13, 18, 25 or 26 in the BPV-1 E5 transmembrane domain generates mutants that stably bind to the PDGF-R_β, but do not cause receptor activation because they cannot form dimers (Adduci and Schlegel, 1999).

Lastly, mutational studies have shown that the constitutive tyrosine phosphorylation and activation of PDGF-R_βs that associate with BPV-1 E5 dimers directly result from receptor autophosphorylation. The K634R PDGF-R_β point mutant, which lacks protein tyrosine kinase (PTK) activity, forms complexes with BPV-1 E5 in hematopoietic cells, but does not undergo tyrosine phosphorylation or activation (Drummond-Barbosa et al., 1995). In contrast, this mutant is tyrosine phosphorylated when co-expressed with BPV-1 E5 and an enzymatically active PDGF-R_β deletion mutant that lacks most of the extracellular domain (Lai et al., 1998). Since the truncated PDGF-R_β co-immunoprecipitates with E5 and the K634R receptor mutant in these experiments, it is apparent that BPV-1 E5 brings pairs of PDGF-R_βs into sufficiently close proximity so that intermolecular trans-phosphorylation of the receptor cytoplasmic PTK domains occurs (Lai et al., 1988). Therefore, constitutive activation of the PDGF-R_β by BPV-1 E5 occurs in a manner that is mechanistically similar to ligand-induced PDGF-R_β activation: both PDGF and E5 induce receptor dimerization and trans-phosphorylation.

2.3. Transformation via c-Src activation

An extensive characterization of BPV-1 E5 transmembrane mutants has identified a number of point mutants (Table 1) that transform fibroblasts even though they minimally activate the PDGF R_β due to defects in receptor binding (Sparkowski et al., 1996; Adduci and Schlegel, 1999) or dimerization (Adduci and Schlegel, 1999). These mutants induce neither the PDGF-R_β tyrosine phosphorylation nor enhanced growth in reduced-serum medium that are mediated by wt BPV-1 E5 (Adduci and Schlegel, 1999; Suprynowicz et al., 2000). The existence of two different classes of BPV-1 E5 mutants that dissociate transformation from PDGF-R_β activation implies that additional, PDGF-R_β-independent signal transduction pathway(s) are activated by the wt BPV 1 E5 protein and that these signals contribute to cellular transformation.

Recent work has shown that BPV-1 E5 can transform fibroblast cell lines independently of PDGF-R_β activation by inducing sustained activation of the non-receptor PTK c-Src, an important regulator of cell proliferation and survival (Martin, 2001). Both wt BPV-1 E5 and transforming E5 mutants that are defective for PDGF-R_β activation constitutively elevate endogenous c-Src PTK activity in NIH 3T3 cells to a level normally associated with acute serum or growth factor treatment (Suprynowicz et al., 2002). In contrast, c-Src is not activated by the non-transforming Q17G E5 mutant in these cells. Several lines of evidence indicate that constitutive c-Src activation is responsible for the transforming activity of L26A BPV-1 E5, which is highly defective for PDGF-R_β activation. L26A E5 cannot transform c-Src-deficient ($Src^{-/-}$) SYF cells unless c-Src expression is reconstituted, and cannot transform NIH 3T3 cells in which endogenous

Table 1. Characteristics of E5 proteins and their mutants

E5	FF*	SI*	AI*	PDGF-Rβ activation	16K binding	Gap junction inhibition	EGF-R activation	PI3-K activation	c-Src activation	Cyclin A promoter/ Cyclin A-cdk2 activation	MHC I down-regulation
BPV-1	+	+	++	+	+	+	-	+	+	+++	+
Q17G	-	-	-	-	-	n.d.	-	-	-	n.d.	n.d.
L24A	++	-	++	-	n.d.	n.d.	-	++	++	n.d.	n.d.
L26A	++	-	+	-	n.d.	n.d.	-	+	+	n.d.	n.d.
C37S/C39S C37M/C39M	-	n.d.	n.d.	-	n.d.	n.d.	n.d.	n.d.	n.d.	n.d.	n.d.
E5-KDEL	-	n.d.	n.d.	+	+	n.d.	n.d.	n.d.	n.d.	n.d.	n.d.
BPV-4	+	+	+	n.d	+	+	n.d.	n.d.	+^	+++	+
N17S	-	-	+	n.d	+	-	n.d	n.d	n.d	-	n.d
N17Y	-	-	-	n.d	+	-	n.d	n.d	n.d	-	-
N17A	++	+	++	n.d	+	+	n.d	n.d	n.d	++	+
E5ΔC	-	-	-	n.d	+	-	n.d	n.d	n.d	-	-
4E5N-1E5C	-	-	-	n.d	+	-	n.d	n.d	n.d	-	n.d
1E5N-4E5C	-	-	+	n.d	n.e.	-	n.d	n.d	n.d	+	n.d
HPV-16	-	-	+	-	+	+	+	-	-	n.d.	+

*FF, focus formation; SI, serum independence; AI, anchorage independence. The plus sign represents the extent of the characteristic, the minus sign its absence. ^, unpublished observations; n.d., not done; n.e., not established as the chimera was not efficiently translated in vitro (Ashrafi et al., 2000).

c-Src PTK activity is maintained at a basal level due to over-expression of the kinase-defective K295R c-Src dominant-negative mutant (Suprynowicz et al., 2002).

Interestingly, wt BPV-1 E5 and transforming E5 mutants that persistently activate c-Src do not activate c-Fyn, a closely-related Src PTK family member that is also present in NIH 3T3 cells (Suprynowicz et al., 2002). Since transforming E5 proteins are localized in membranes of the *trans*-Golgi compartment, the differential activation of c-Src by E5 may be related to differences in the subcellular targeting of c-Fyn and c-Src. While c-Fyn is localized exclusively in the plasma membrane of NIH 3T3 cells (van't Hof and Resh, 1997), a significant subset of c-Src is present in late endosomal and *trans*-Golgi membranes (Kaplan et al., 1992), where it could more readily be influenced by E5.

Wt BPV-1 E5 and transforming E5 mutants that constitutively activate c-Src also induce persistent activation of heterodimeric phosphoinositide 3-kinase (PI 3-K), through a mechanism that involves increased tyrosine phosphorylation of the PI 3-K 85 kDa regulatory subunit (Suprynowicz et al., 2000). The activation of PI 3-K by these E5 proteins does not require constitutive c-Src or PDGF-R_β activation, but is not by itself sufficient to induce transformation (Suprynowicz et al., 2000, 2002). However, persistent PI 3-K activation may be required in combination with constitutive c-Src activation to enable BPV-1 E5-mediated transformation independently of PDGF-R_β signaling.

2.4 Transformation via V-ATPase inactivation

V-ATPases are membrane-spanning proton pumps that acidify the lumen of intracellular compartments, including endosomes, lysosomes and the Golgi apparatus. The ATP-driven acidification of these organelles is important for cellular processes such as receptor-mediated endocytosis, protein glycosylation, protein trafficking and protein degradation (Stevens and Forgac, 1997; Forgac, 1999). In cells, BPV-1 E5 stably associates with the 16 kDa pore-forming c subunit of the V-ATPase (16K) (Goldstein and Schlegel, 1990; Goldstein et al., 1991) in complexes that also can contain PDGF-R_β subunits (Goldstein et al., 1992). An analysis of various BPV-1 E5 and 16K deletion mutants has mapped the binding site for 16K to the transmembrane domain of E5 (Goldstein et al., 1992), and the binding site for BPV-1 E5 to the fourth transmembrane domain of 16K (Andresson et al., 1995). The interaction between positively-charged glutamine 17 in the transmembrane domain of E5 and negatively-charged glutamic acid 143 in the fourth transmembrane domain of 16K is particularly important for the stable association of these two proteins. BPV-1 E5-16K complex formation is dramatically reduced by the substitution of arginine at position 143 of 16K, but can be restored by complementary substitutions of glutamic acid or aspartic acid at position 17 of E5 (Andresson et al., 1995).

The association of BPV-1 E5 with 16K induces persistent alkalinization of the *trans*-Golgi lumen by inhibiting V-ATPase H^+ pumping. Alkalinization is not induced by BPV-1 E5 mutants that cannot bind 16K, or by E5 mutants that do not localize to the *trans*-Golgi (Schapiro et al., 2000). Since BPV-1 E5 does not reduce levels of V-ATPase subunits in the Golgi, it likely impairs H^+ pumping by interfering with the assembly of V-ATPase subunits into a functional holoenzyme (Schapiro et al., 2000). The strict correlation that exists between the transforming potential of BPV-1 E5 mutants and their ability to induce *trans*-Golgi alkalinization (Schapiro et al., 2000), suggests that Golgi alkalinization may play a significant role in E5-mediated transformation. This hypothesis is supported by the finding that the E143R 16K mutant partially alkalinizes the *trans*-Golgi lumen and transforms fibroblasts to grow in soft agar independently of BPV-1 E5 (Andresson et al., 1995; Schapiro et al., 2000). A 16K deletion mutant that completely lacks the fourth transmembrane domain is even more oncogenic in these cells (Andresson et al., 1995). Further evidence linking BPV-1 E5-mediated transformation to V-ATPase dysfunction is provided by the observation that the over-expression of wt 16K inhibits E5-mediated transformation in NIH 3T3 and C127 fibroblasts by 75% in focus formation assays (Andresson et al., 1995).

Since BPV-1 E5 mutants that induce *trans*-Golgi alkalinization and transform cells also constitutively activate c-Src, it is possible that Golgi pH and c-Src are involved in a common mechanism of E5-dependent transformation (Schapiro et al., 2000; Suprynowicz et al., 2002). Currently, it is not known how BPV-1 E5 activates c-Src, and further work is needed to determine whether Golgi alkalinization plays a role in this process. However, it is known that BPV-1 E5-dependent Golgi alkalinization and PDGFR_β activation are distinct mechanisms of transformation, since the Q17S and L26A E5 mutants induce trans-Golgi alkalinization and transform NIH 3T3 cells without activating the PDGFR_β (Schapiro et al., 2000).

2.5. Down-regulation of MHC class I expression

Papillomavirus infections typically are eliminated after several months by a cytotoxic T-lymphocyte (CTL)-mediated host immune response to viral antigens. However, more persistent viral infections can occur, and these lesions are more likely to undergo malignant progression (Frazer, 1996; O'Brien and Campo, 2002). The ability of papillomaviruses to avoid immune detection for prolonged periods has been attributed to the occurrence of viral infections in basal epithelia, where viral antigens may not be accessible to CTLs, to the low levels of viral antigens that are produced during infection, and to the absence of inflammation and virus-induced cell lysis (Frazer et al., 1999).

Recent studies suggest that papillomavirus proteins may enable infected cells to evade immune recognition by CTLs by interfering with the presentation of viral antigens on the cell surface (O'Brien and Campo, 2002). The presentation of antigenic viral peptides on the surface of infected cells is mediated by major histocompatibility complex class I (MHC I) molecules, which bind proteolytic fragments of viral proteins in the ER and transport them through the Golgi apparatus to the plasma membrane (Cresswell et al., 1999). In primary embryonic bovine palate (PalF) cells, BPV-1 and BPV-4 E5 proteins abrogate MHC I expression at the plasma membrane by causing MHC I to remain sequestered inside the cell, as well as by decreasing total cellular levels of MHC I heavy chain mRNA and protein (Ashrafi et al., 2002). The transport of MHC I is selectively inhibited by BPV E5, since surface expression of another glycosylated plasma membrane protein, the transferrin receptor, is not affected in these cells (Ashrafi et al., 2002). Sequestered MHC I molecules co-localize with BPV E5 in the Golgi apparatus, and their retention appears to be a consequence of E5-mediated alkalinization of the Golgi lumen (Marchetti et al., 2002). Treatment of non-E5-expressing cells with the H^+ ionophore monensin induces Golgi alkalinization by dissipating the H^+ gradient that normally exists between the cytoplasm and Golgi lumen, and induces a fourfold decrease in surface MHC I expression without significantly reducing the total amount of cellular MHC I. In contrast, surface expression of the transferrin receptor is not affected by monensin treatment in these experiments (Marchetti et al., 2002). Additionally, Golgi alkalinization may destabilize MHC I-antigen complexes, since the association of peptides with MHC I is known to be pH dependent (Reich et al., 1997; Gromme et al., 1999).

The functions of E5 in BPV-induced lesions are not well characterized. However, it has become clear that the down-regulation of MHC I by BPV E5 takes place not only in cultured cells but crucially also *in vivo*. In BPV-4-induced papillomas from the palate, rumen and esophagus of cattle, MHC I is absent in cells expressing E5, whereas MHC I is detected on the surface of adjacent cells that do not express E5 (Araibi et al., 2004). This is the first example of a defined action of E5 during natural infection, and it may be an important biological activity of BPV E5 proteins as it would prolong infection by interfering with the ability of the host immune system to detect infected cells.

3. BPV-4 E5

In the epitheliotropic Xi-BPVs (BPV-3, -4 and -6), an E5-like ORF, originally designated E8, has replaced the missing E6 ORF in the 5′ portion of the early region (Jackson et al., 1991). The protein encoded by this ORF has been most extensively studied in BPV-4. Given that the functional similarities between BPV-1 E5 and BPV-4 E8 are greater that their differences (see below), and not withstanding the different genome location, both the E8 ORF and its product have been renamed BPV-4 E5 ORF and BPV-4 E5 protein (Morgan and Campo, 2000; O'Brien et al., 2001) and this nomenclature has been used since.

10
NH_2-**M-S-L-W-L-I-Y-V-L-L-L-F-W-C-A-F-N-F**

20 30
L-A-L-L-F-A-I-I-V-Y-L-L-L-I-S-T-I-T-R-L-D

40
G-W-D-COOH

Figure 2. Primary amino acid sequence of BPV-4 E5.

BPV-4 E5 consists of 42 amino acids and has a similar putative structure to that of BPV-1 E5, with a major hydrophobic transmembrane domain and hydrophilic C-terminal domain (Figure 2). It is expressed early in infection (Anderson et al., 1997) in the basal and suprabasal layers of stratified epithelia and, like BPV-1 E5 (Burnett et al., 1992), also can be expressed in the upper keratinized layers of the papilloma (Anderson et al., 1997; Araibi et al., 2004). Like BPV-1 E5, it is localized in the endomembrane compartments of cultured cells, primarily the Golgi apparatus (Burkhardt et al., 1989; Pennie et al., 1993; Marchetti et al., 2002; Zago et al., 2004). BPV-4 E5 transforms established mouse NIH 3T3 fibroblasts on its own (O'Brien and Campo, 1998; Zago et al., 2004) and confers the ability of growing in low serum and independently of anchorage to primary embryonic bovine palate (PalF) cells that have been partially transformed by BPV-4 E7 and activated ras (Pennie et al., 1993; O'Brien et al., 1999).

In both NIH 3T3 and PalF cells, BPV-4 E5 activates the transcriptional promoter of the cyclin A gene, and cyclin A-cdk2 activity is increased (O'Brien and Campo, 1998; O'Brien et al, 1999). These effects have been deemed responsible for the anchorage-independent growth of the transformed cells: E5 mutants incapable of transformation do not activate the cyclin A promoter or increase cyclin A-cdk2 protein kinase activity (Table 1). In NIH 3T3 cells activation of the cyclin A promoter requires CAAT box integrity, functional NF-Y transcriptional complexes and a CAAT box binding factor (p110 CBF) that associates with NF-YA only in E5-expressing cells (Grindlay et al, 2005). As BPV-4 E5 does not activate E2F (Grindlay et al, 2005), and the activation of cyclin A-cdk2 is independent of ERK (Zago et al., 2004), it appears that, contrary to BPV-1 E5, BPV-4 E5 can achieve cell transformation independently from the constitutive activation of tyrosine kinase growth factor receptors in mouse fibroblasts. In addition to the up-regulation of cyclin A, BPV-4 E5 up-regulates the cyclin-dependent kinase inhibitor $p27^{kip1}$ through increased protein synthesis (O'Brien and Campo, 1998; O'Brien at al., 2001). However, despite the increased levels of $p27^{kip1}$ the cells continue cycling as E5 prevents $p27^{kip1}$ from acting as a cell cycle inhibitor by promoting its association with an elevated pool of cyclin D1-cdk4 complexes (O'Brien et al., 2001). The up-regulation of $p27^{kip1}$ is not required for E5-mediated anchorage-independent growth, as a version of E5, which allows cells to grow independently of attachment, does not elevate $p27^{kip1}$ (Zago et al., 2004).

Both BPV-4 and BPV-1 E5 proteins have dramatic effects on the morphology of PalF cells. The cells are enlarged and vacuolated (Ashrafi et al., 2000) and have a distended and fragmented Golgi apparatus (Ashrafi et al., 2002). At least some of these characteristics are due to alkalinization of the Golgi lumen (Marchetti et al., 2002), an effect mediated by the binding of BPV-1 and BPV-4 E5 proteins to the 16K subunit of the V-ATPase H^+ pump (Goldstein and Schlegel, 1990; Goldstein et al., 1991; Faccini et al., 1996). Another outcome of the interaction between the E5 proteins and 16K is the strong inhibition of gap junction-mediated intercellular communication, as assayed by the diffusion of microinjected fluorescent dye molecules in monolayer cultures of E5-transformed cells (Faccini et al., 1996; Ashrafi et al., 2000). Gap junctions are responsible for tissue homeostasis, and impairment of their function leads to the withdrawal of a transformed/infected cell from the control of its healthy neighboring cells. By inhibiting gap junctional communication, E5 would contribute to the development of a transformed clone.

3.1. Mutational analysis

As with BPV-1 E5, the sequestration of BPV-4 E5 in the ER/cis Golgi via a C-terminal KDEL motif inhibits protein function: KDEL-appended BPV-4 E5 is no longer capable of trans-activating the cyclin A promoter in non-adherent cells (O'Brien and Campo, 1998). Furthermore, the 17th amino acid residue of BPV-4 E5 (N rather than the Q in BPV-1 E5) is important for cell transformation (Table 1). Mutation of N17 to Y completely abrogates the ability of BPV-4 E5 to transform cells, activate the cyclin A promoter, increase cyclin A-cdk2 kinase activity (O'Brien et al., 1999), inhibit gap junctional intercellular communication (Ashrafi et al., 2000) and down-regulate MHC I expression (Ashrafi et al., 2002). Substitution of S for N17 partially abolishes E5-mediated cell transformation, maintaining the ability to confer anchorage independent growth but not growth in low serum, and mutation of N17 to A renders E5 hypertransforming (O'Brien et al., 1999). The C-terminal domain of BPV-4 E5 is equally important. Deletion mutants of both BPV-4 E5 and BPV-1 E5 that lack the C-terminal domain are incapable of inducing cell transformation, and the truncated BPV-4 E5 mutant (E5ΔC) exhibits none of the known functions of the wt protein (Table 1).

Despite the functional similarities between BPV-4 E5 and BPV-1 E5, domain-swap studies indicate that the N-terminal hydrophobic and C-terminal hydrophilic domains of the two proteins are not fully equivalent (O'Brien et al., 1999). A chimera consisting of the N-terminal domain of BPV-4 E5 and C-terminal domain of BPV-1 E5 (4E5N-1E5C) can neither transform bovine PalF cells nor activate the cyclin A promoter, whereas the reciprocal chimera (1E5N-4E5C) is still capable of inducing anchorage independent growth and activating the cyclin A promoter (O'Brien et al., 1999). However, PalF cells that express the 1E5N-4E5C protein do not show the typical vacuolated cytoplasm and prominent pseudopodes characteristic of PalF cells which have been transformed by wt BPV-4 E5 or BPV-1 E5 (Ashrafi et al., 2000). Neither E5 chimera is able to down-regulate gap junction-mediated intercellular coupling, showing that the overall integrity of E5 is required for this activity (Ashrafi et al., 2000). Surprisingly, all the mutants and chimeras of BPV-4 E5, including those that do not interfere with gap junctions, bind to 16K *in vitro*, indicating that 16K binding may be necessary but not sufficient for BPV-4 E5-mediated cell transformation (Ashrafi et al., 2000).

4. HPV-16 E5

Epidemiological and experimental studies have identified high-risk HPVs as the primary causative agent of cervical cancer worldwide (zur Hausen and de Villiers, 1994; zur Hausen, 2002; see Chapters 1 and 3). High-risk HPV DNA is present in approximately 99% of malignant cervical biopsies (Walboomers et al., 1999); most frequently this DNA is HPV type 16 (HPV-16), followed by HPV types 18, 31 and 33 (Laimins, 1993; zur Hausen and de Villiers, 1994). The major transforming protein of HPV-16 is E7, which in conjunction with the E6 gene product promotes cell proliferation, immortalization and malignant progression by interfering with the function of numerous regulatory host proteins (McMurray et al., 2001; Fehrmann and Laimins, 2003; see Chapters 9 and 10 for a detailed description of the functions of E6 and E7). The biological activities of HPV-16 E5, however, are not as well characterized and their roles in HPV-mediated pathogenesis are less understood.

In contrast to BPV-1 E5, HPV-16 E5 exhibits only weak transforming activity *in vitro* (DiMaio and Mattoon, 2001; McMurray et al., 2001; Fehrmann and Laimins, 2003). HPV-16 E5 enhances the transformation of primary murine epithelial cells by HPV-16 E7 (Valle and Banks, 1995), and can induce anchorage-independent growth in immortalized murine fibroblasts, but cannot induce focus formation (Leechanachai et al., 1992; Pim et al., 1992; Straight et al., 1993). The primary biological activity of HPV-16 E5 appears to be stimulating EGF-dependent proliferation in keratinocytes and other cells that express large numbers of EGF-Rs (Leechanachai et al., 1992; Straight et al., 1993; Bouvard et al., 1994; Valle

10
NH_2-**M-T-N-L-D-T-A-S-T-T-L-L-A-C-F-L-L-C-F-**
20 30 40
C-V-L-L-C-V-C-L-L-I-R-P-L-L-L-S-V-S-T-Y-T-
50 60
S-L-I-I-L-V-L-L-L-W-I-T-A-A-S-A-F-R-C-F-I-
70 80
V-Y-I-I-F-V-Y-I-P-L-F-L-I-H-T-H-A-R-F-L-I-T-COOH

Figure 3. Primary amino acid sequence of HPV-16 E5.

and Banks, 1995). Additional activities that have been described for HPV-16 E5 include increasing cellular resistance to apoptosis (Kabsch and Alonso, 2002a; Zhang et al., 2002b), inhibiting gap junction-mediated cell-cell communication (Oelze et al., 1995; Tomakidi et al., 2000), down-regulating surface expression of MHC (HLA) class I and II molecules (Zhang et al., 2003; Ashrafi et al., 2005) and enhancing keratinocyte immortalization by HPV-16 E6 and E7 (Valle and Banks, 1995; Stoppler et al., 1996).

4.1. Structure

HPV-16 E5 is a small (83 amino acids), hydrophobic protein (Figure 3) that localizes to intracellular membranes and shares little sequence homology with BPV-1 E5 (Bubb et al., 1988; Halbert and Galloway, 1988; Conrad et al., 1993). When over-expressed in COS cells, epitope-tagged HPV-16 E5 is present in the ER and nuclear envelope and, to a lesser extent, in the Golgi apparatus (Conrad et al., 1993). However, HPV-16 E5 that is expressed at a lower, more physiological level in primary human foreskin keratinocytes (HFKs) localizes almost exclusively to the ER; very little E5 is detected in the Golgi apparatus and in early endosomes (Disbrow et al., 2003). The orientation of HPV-16 E5 in membranes is unknown, but amino acid sequence analysis indicates that the protein is composed of three distinct hydrophobic domains that are capable of spanning a membrane three times (Bubb et al., 1988). Alternatively, HPV-16 E5 may exist in a β-hairpin conformation, in which the N- and C-termini are embedded in the membrane and a subset of amino acids 30-55 are exposed on the cytoplasmic or luminal surface (Alonso and Reed, 2002).

Despite initial reports to the contrary (Conrad et al., 1993; Hwang et al., 1995), HPV-16 E5 can self-associate to form dimers and higher-order oligomers *in vitro* (Yang et al., 2003) and *in vivo* (Disbrow et al., 2003; Gieswein et al., 2003). Oligomerization is mediated by multiple hydrophobic interactions (Gieswein et al., 2003) and by disulfide bonding (Disbrow et al., 2003) between E5 monomers. Hydrophobic interactions also are responsible for the stable association of HPV-16 E5 with the 16K subunit of the V-ATPase H^+ pump (Conrad et al., 1993; Adam et al., 2000; Rodriguez et al., 2000; Ashby et al., 2001; Gieswein et al., 2003). An analysis of the binding competency of various HPV-16 E5 and 16K mutants in COS cells indicates that the association of these proteins is mediated primarily by the fourth transmembrane domain of 16K and the region of E5 between amino acids 41 and 54, although binding is enhanced by E5 amino acids 54-71 (Adam et al., 2000). In contrast, amino acids 54-78 of HPV-16 E5 are required for binding to 16K when the two proteins are co-translated *in vitro* in the presence of canine pancreatic microsomes (Rodriguez et al., 2000). HPV-16 E5 binds to the PDGF-R_β, EGF-R and colony stimulating factor-1 receptor when E5 and the receptors are over-expressed at high levels in COS cells (Hwang et al., 1995). However, endogenous growth factor receptors, which are present at lower levels, have not been detected in complexes with HPV-16 E5.

4.2. EGF receptor activation

A number of studies have shown that EGF can act synergistically with HPV-16 E5 to stimulate adherent and anchorage-independent proliferation in keratinocytes, and in fibroblasts that over-express EGF-Rs (Leechanachai et al., 1992; Pim et al., 1992; Straight et al., 1993; Bouvard et al., 1994; Valle and Banks, 1995). These observations imply that HPV-16 E5 stimulates mitogenic signaling through the EGF-R; however, ligand-independent EGF-R activation has not been reported. Rather, HPV-16 E5 enhances ligand-dependent tyrosine phosphorylation and activation of the EGF-R, especially when ligand-starved cells are acutely stimulated with EGF, and at low concentrations of EGF (Pim et al., 1992; Straight et al., 1993; Crusius et al., 1998; Rodriguez et al., 2000; Zhang et al., 2002b). In keratinocytes, HPV-16 E5 does not increase the affinity of the EGF-R for EGF or alter rates of [^{125}I]EGF internalization and degradation. However, E5 increases the total number of EGF-Rs per cell 2.5-4-fold and significantly delays the proteolytic degradation of EGF-Rs that bind EGF and are internalized in endosomes (Straight et al., 1993). If the internalized receptors remain active for an extended period of time, the ability of HPV-16 E5 to delay EGF-R degradation could account for enhanced EGF-R activation and mitogenic signaling. Although it is unclear how HPV-16 E5 can differentially inhibit the degradation of internalized EGF-Rs and not the degradation of internalized EGF, anti-EGF-R immunofluorescence microscopy of E5-expressing HFKs that have been treated with fluorescent EGF shows that internalized EGF moves to perinuclear lysosomes, where it is degraded, while a significant number of EGF-Rs remain in endosomes for hours and eventually reappear on the cell surface. These receptors evidently are recycled to the plasma membrane, since their appearance does not require protein synthesis (Straight et al., 1993).

The finding that HPV-16 E5 binds to the 16K subunit of the V-ATPase H^+ pump (Conrad et al., 1993; Adam et al., 2000; Rodriguez et al., 2000; Ashby et al., 2001; Gieswein et al., 2003) that is responsible for endosome acidification (Stevens and Forgac, 1997; Forgac, 1999), suggests that E5 may extend the lifespan of internalized EGF-Rs by interfering with the acidification of endosomes that is necessary for ligand dissociation and receptor proteolysis (Straight et al., 1993). This theory is supported by ratio-imaging measurements of endosomal pH in normal and HPV-16 E5-expressing HFKs using fluorescein-dextran conjugates (Straight et al., 1995). The measurements show that endosome acidification and the degradation of internalized EGF-Rs are delayed at least 4-fold in HFKs that express HPV-16 E5, and that EGF-R degradation is similarly delayed when chloroquine or bafilomycin A1 is used to inhibit endosome acidification in control HFKs that do not express E5 (Straight et al., 1995). In yeast, HPV-16 E5 interferes with the assembly and overall stability of V-ATPase complexes, and induces pH-related growth and protein trafficking defects that are characteristic of impaired V-ATPase function (Briggs et al., 2001). This phenotype can partially be reversed by over-expressing the wt yeast 16K protein (Briggs et al., 2001). Apparently, the ability of HPV-16 E5 to disrupt yeast V-ATPase complexes requires binding interactions between multiple regions of E5 and 16K, since HPV-16 E5 mutants have been identified that bind to 16K but do not impair V-ATPase activity (Adam et al., 2000; Briggs et al., 2001).

Since the vast majority of HPV-16 E5 in HFKs is localized in the ER (Disbrow et al., 2003), it is important to consider how E5 binding to 16K in the ER could reduce V-ATPase activity in endosomes. In mammalian cells and in yeast, 16K is a component of the V_0 integral domain of the V-ATPase that is responsible for H^+ translocation (Stevens and Forgac, 1997; Forgac, 1999; Graham et al., 2000). The V_0 domain is assembled in the ER and subsequently is transported to acidic organelles and to the plasma membrane, where it can become internalized in endosomes (Myers and Forgac, 1993; Sato and Toyama, 1994; Stevens and Forgac, 1997; Graham et al., 2000). The V_1 peripheral domain of the V-ATPase that is responsible for ATP hydrolysis (Stevens and Forgac, 1997; Forgac, 1999; Graham et al., 2000) appears to be recruited from the cytoplasm to V_0-containing endosomes in conjunction with a 50 kDa subunit of the AP-2 clathrin assembly complex. The association of V_1 and V_0 domains and the 50 kDa protein on endosomal membranes reconstitutes the V-ATPase holoenzyme that mediates endosome acidification (Liu et al., 1994). An analysis of V-ATPase assembly in yeast indicates that the V_0 domain cannot be assembled if one of the V_0 subunits is missing. In addition, the remaining V_0 subunits are destabilized and no V-ATPase subunits are transported or recruited to the vacuolar membrane (Stevens and Forgac, 1997; Graham et al., 2000). Therefore, HPV-16 E5 may disrupt V-ATPase complexes and inhibit vacuolar acidification (yeast) and endosomal acidification (mammalian cells) by binding to 16K in the ER and preventing it from participating in the assembly of the V-ATPase V_0 domain, or by interfering with the function of ER membrane proteins that are required for V_0 assembly but are not V-ATPase subunits (Stevens and Forgac, 1997; Forgac, 1999; Graham et al., 2000).

Alternatively, it has been proposed that HPV-16 E5 may delay endosome acidification and the degradation of internalized EGF-Rs without inhibiting V-ATPase-mediated H^+ transport (Thomsen et al., 2000). In fibroblasts, HPV-16 E5 expression actually increases the number of acidic intracellular compartments that accumulate the membrane-permeable, acidotropic probe, LysoTracker Red. These compartments include late endosomes, since they exhibit the same pattern of localization as the late endosomal proteins Rab7 and cation-independent mannose 6-phosphate receptor. In contrast, membrane-impermeable pH probes that are internalized via endocytosis do not become acidified until they fuse with perinuclear late endosomes. Normally, the transport and fusion of early endosomes with acidic late endosomes occurs within 30 minutes; however, this process is severely delayed (>3 hours) in cells that express HPV-16 E5 (Thomsen et al., 2000). The implication of these observations is that HPV-16 E5 delays EGF-R degradation by interfering with membrane trafficking that is responsible for the fusion of early and late endosomes. Support for such a model is provided by the finding that HPV-16 E5 disrupts actin filaments in fibroblasts (Thomsen et al., 2000), and by the inability of HPV-16 E5 to inhibit the yeast V-ATPase in one study (Ashby at al., 2001). In addition, an HPV-16 E5 mutant that lacks the last C-terminal 5 amino acids fails to enhance ligand-dependent EGF-R activation in keratinocytes even though it binds to the 16K V-ATPase subunit (Rodriguez et al., 2000).

Interestingly, the scenarios in which HPV-16 E5 delays endosomal acidification by inhibiting V-ATPase function or by disrupting endocytic trafficking may not be mutually exclusive. It has been reported that the V-ATPase-dependent acidification of endosomes isolated from kidney proximal tubule epithelial cells is required for recruitment of the ADP-ribosylation factor, ARF6, and its guanine nucleotide exchange factor, ARNO, from the cytoplasm (Maranda et al., 2001). Since ARF6 has been identified as a regulator of the receptor-mediated endocytic pathway (D'Souza-Schorey et al., 1995, 1998), it is possible that HPV-16 E5 disrupts endocytic trafficking as a result of inhibiting endosome acidification.

A third mechanism that may augment ligand-dependent EGF-R activation in HPV-16 E5-expressing keratinocytes is suggested by the observation that increased receptor activation occurs immediately after EGF addition; too soon to be explained by delayed receptor degradation (Crusius et al., 1998). Furthermore, acute EGF stimulation also activates the ErbB2 receptor in HPV-16 E5-expressing cells (but not in control cells) even though EGF is not a ligand for this receptor (Crusius et al., 1998). These results argue that HPV-16 E5 may promote weak binding interactions between active EGF-R dimers and inactive EGF-R and ErbB2 receptor monomers, so that the monomeric receptors are activated as a result of trans-phosphorylation of the receptor cytoplasmic domains. This type of E5-dependent receptor association would be more likely to occur in cells that contain greater numbers of EGF-Rs, such as fibroblasts that over-express exogenous EGF-Rs and keratinocytes (Collins et al., 1988; Gottlieb et al., 1992). However, since HPV-16 E5 does not induce EGF-independent activation of the EGF-R or ErbB2 receptor, it apparently cannot mediate productive associations of inactive receptor monomers. Indeed, cross-linking experiments find similar numbers of EGF-R dimers in normal and HPV-16 E5-expressing keratinocytes in the absence of EGF (Crusius et al., 1998).

4.3. Other biological activities

The enhanced EGF-R activation and EGF-dependent cell proliferation that are induced by HPV-16 E5 occur concomitantly with the increased activation of downstream intermediates in the EGF-R signaling pathway. Mitogenic signaling through the EGF-R involves sequential activation of the Ras GTPase and the Raf, MEK and mitogen-activated protein (MAP) kinases, ERK1 and ERK2. Activated ERK1/2 form homodimers and translocate into the nucleus, where they phosphorylate transcription factors that control the expression of genes involved in regulating cell proliferation, including c-fos and c-jun (Robinson and Cobb, 1997; Lewis et al., 1998). HPV-16 E5 potentiates both the amplitude and duration of EGF-induced MAP kinase activation in human fibrosarcoma cells (Gu and Matlashewski, 1995) and murine fibroblasts that over-express the EGF-R (Crusius et al., 1997). HPV-16 E5 also enhances EGF-dependent transactivation of the c-fos and c-jun promoters (Chen et al., 1996a; 1996b) and increases levels of c-fos and c-jun mRNA in fibroblasts in response to serum or EGF (Leechanachai et al., 1991; Bouvard et al., 1994; Crusius et al., 1997). A study by Tsao et al. (1996) provides evidence that elevated c-jun expression in HPV-16 E5-expressing cells can additionally promote cell proliferation by repressing transcription of the $p21^{WafI/CipI}$ gene, which encodes an inhibitor of cyclin-dependent kinases (CDK) that regulate cell cycle progression.

The ability of HPV-16 E5 to increase c-fos and c-jun expression in host cells during viral infection may stimulate transcription of the HPV-16 E6 and E7 genes. The upstream regulatory region (URR/LCR) of the HPV genome contains an AP-1 binding site which is responsive to c-fos/c-jun dimers, and EGF-mediated activation of the URR promoter is two-fold greater in keratinocytes that express HPV-16 E5 (Bouvard et al., 1994).

Although most of the known biological activities of HPV-16 E5 are related to enhanced ligand-dependent EGF-R activation, several activities have been described that occur independently of EGF-R signaling. An analysis of gap junction-mediated intercellular communication in monolayer cultures of immortalized human keratinocytes shows that HPV-16 E5 strongly inhibits cell coupling (Oelze et al., 1995). This impairment of cell coupling does not involve gross alterations of the structure or localization of gap junctions, or the differentiation status of the cells, but is correlated with decreased phosphorylation of the major gap junction protein, connexin 43 (Oelze et al., 1995). Interestingly, HPV-16 E5 expression leads to an almost complete disappearance of connexin 43 in immortalized human keratinocytes that are maintained in raft cultures (Tomakidi et al., 2000). In fibroblasts, HPV-16 E5 constitutively activates phospholipase C-γ in an EGF-independent manner and increases intracellular levels of the phospholipase C-γ products, diacylglycerol and inositol phosphates (Crusius et al., 1999).

Several reports have ascribed both pro-apoptotic and anti-apoptotic activities to HPV-16 E5 under various conditions that do not involve EGF-R activation. HPV-16 E5 sensitizes immortalized human keratinocytes to apoptosis induced by osmotic stress (Kabsch and Alonso, 2002b), but inhibits apoptosis induced by the Fas ligand (FasL) and the tumor necrosis factor-related apoptosis-inducing ligand (TRAIL) (Kabsch and Alonso, 2002a). Increased resistance to FasL-induced apoptosis appears to result from an E5-dependent reduction in expression of the Fas receptor, while increased resistance to TRAIL-induced apoptosis is correlated with impaired assembly of TRAIL-containing signaling complexes (Kabsch and Alonso, 2002a). In contrast, the ability of HPV-16 E5 to significantly reduce apoptosis induced by UV B irradiation in primary HFKs has been attributed to enhanced ligand-dependent EGF-R activation and the increased downstream activation of ERK1/2, PI 3-K and protein kinase B/Akt (Zhang et al., 2002b).

Like BPV E5, HPV-16 E5 reduces the expression of MHC I on the cell surface by promoting its retention in the Golgi apparatus, which renders infected cells less susceptible to CTL-mediated immune detection. In immortalized human keratinocytes, HPV-16 E5 selectively reduces surface expression of the MHC I molecules HLA-A and HLA-B without affecting levels of the non-classical MHC molecules HLA-C and HLA-E (Ashrafi et al., 2004). This selectivity is significant, since cells with little or no MHC I on their surface are recognized and killed by natural killer (NK) cells, important components of the innate immune response.

NK cells are prevented from killing cells with low surface MHC I by the presence on the cell surface of non-classical MHC molecules, which engage the inhibitory receptors of NK cells (Yokoyama, 1998). Therefore, HPV-16 E5-expressing cells are potentially capable of escaping both CTL killing, by down-regulating classical MHC I expression, and NK killing, by not interfering with non-classical MHC expression. Since HPV-16 E5 binds to the 16K subunit of the V-ATPase H^+ pump, it is worth pointing out that the HIV-1 Nef protein, which interacts with the Vma13p V-ATPase subunit (Lu et al., 1998), down-regulates classical MHC I but not non-classical MHC surface expression, and that Nef-expressing cells evade both CTL and NK killing (Cohen et al., 1999). A similar situation is observed in HCMV-infected cells (Furman et al., 2002; Tomasec et al., 2000). It would therefore appear that HPV-16 E5, like other viral proteins, has evolved similar mechanisms for escaping both the innate (NK) and adaptive (CTL) immune response (O'Brien and Campo, 2002).

MHC I is not the only antigen-presenting complex affected by HPV-16 E5. Primary HFKs express MHC class II (MHC II) molecules on the cell surface following treatment with interferon-γ, which elicits an immune response by presenting antigenic peptides to $CD4^+$ T lymphocytes. MHC II surface expression is significantly reduced in interferon-γ-treated HFKs that express HPV-16 E5 due to the impaired assembly of peptide-loaded, mature MHC II dimers. MHC II assembly requires the sequential proteolytic processing of an associated chaperone protein, called the invariant chain, in acidic endosomal compartments and this processing is inhibited by E5-mediated endosome alkalinization (Zhang et al., 2003).

4.4 Role of E5 in HPV-mediated pathogenesis

HPVs enter stratified epithelia through microlesions and infect basal keratinocytes that are capable of proliferation. During this non-productive stage of the HPV life cycle, the E1, E2, E5, E6 and E7 early viral proteins are expressed, and the viral genome is maintained at approximately 50-100 episomal copies per cell by replication in synchrony with the host cell DNA. In addition, the infected population of basal keratinocytes undergoes lateral expansion as a result of enhanced proliferation (McMurray et al., 2001; zur Hausen, 2002; Fehrmann and Laimins, 2003). The demonstrated ability of HPV-16 E5 to enhance ligand-dependent activation of the EGF-R (Straight et al., 1993; Crusius et al., 1998; Rodriguez et al., 2000; Zhang et al., 2002b) and stimulate EGF-dependent proliferation in cultured human keratinocytes (Straight et al., 1993; Bouvard et al., 1994; Valle and Banks, 1995) suggests that HPV-16 E5 may play a major role in expanding populations of HPV-16-infected basal keratinocytes *in vivo* by enhancing ligand-dependent EGF-R activation.

The productive stage of the HPV life cycle is triggered by the onset of differentiation in infected basal keratinocytes that migrate into suprabasal layers of the stratified epithelium. The stage is characterized by expression of the L1 and L2 late viral genes that encode capsid proteins, amplification of the viral genome to thousands of copies per cell, and the assembly of infectious virus particles (Chapters 11 and 13). Normally, keratinocytes that begin to differentiate lose the capacity to replicate their DNA. However, active host DNA replication machinery is required for viral DNA amplification and HPV E7 proteins can reprogram differentiating suprabasal cells to support DNA synthesis (McMurray et al., 2001; zur Hausen, 2002; Fehrmann and Laimins, 2003; Chapter 10). Recent work suggests that HPV-16 E5 and the high-risk HPV type 31 (HPV-31) E5 protein also promote DNA synthesis in differentiating suprabasal keratinocytes to facilitate amplification of the viral genome. Keratinocytes in organotypic raft cultures can be forced to exit the cell cycle and undergo terminal differentiation, but will continue to replicate DNA under these conditions if transfected with the HPV-16 or HPV-31 circular genome. Importantly, a 50% reduction in the number of suprabasal keratinocytes undergoing DNA synthesis in raft cultures is observed if the keratinocytes are transfected with mutant HPV-16 or HPV-31 genomes that harbor a frameshift mutation or altered initiation or stop codon in the E5 gene that abrogates E5 protein expression (Fehrmann et al., 2003; Genther et al., 2003; see Chapter 13 for a discussion of the role of E5 in the late stages of the virus life cycle). Moreover, a two-fold increase in the number of suprabasal keratinocytes engaged in DNA synthesis has been observed in transgenic mice that express HPV-16 E5 in the basal layer of stratified squamous epithelia (Genther-Williams et al., 2005).

In high-risk HPV infections that progress to cancer, the viral DNA typically integrates into the host genome, which often results in the disruption and loss of the E2 and E5 genes (zur Hausen and de Villiers, 1994; Fehrmann and Laimins, 2003). For this reason, HPV E5 proteins are not thought to play a role in malignant progression. In contrast to other high-risk HPVs, HPV-16 DNA can exist in integrated, episomal, or integrated and episomal forms in malignant lesions of the cervix. Since the E5 gene is present in lesions that harbor episomal HPV-16 DNA, one study found that 60% of HPV-16-positive cervical cancers express the HPV-16 E5 protein (Chang et al., 2001). Nevertheless, the fact that 40% of the HPV-16-positive cancers in this study do not express E5 indicates that HPV-16 E5 is not required for the progression stage of HPV-16-mediated carcinogenesis.

Although HPV-16 E5 may not contribute to the malignant progression of low-grade epithelial lesions, there is evidence suggesting that E5 functions during the early stages of HPV-16 infections to increase the likelihood of subsequent oncogenic transformation. (1) Sequence analysis of the wt HPV-16 E5 gene reveals a high incidence (40%) of infrequently-used codons that cause the HPV-16 E5 protein to be expressed at a low level in mammalian cells. Replacement of these codons in the HPV-16 E5 gene with common mammalian codons greatly increases E5 protein expression in cultured cells (Disbrow et al., 2003). An analysis of HPV-16 E5 sequence variants in 100 HPV-16-positive women with or without cervical neoplasia indicates that E5 variants with common mammalian codons, likely to be expressed at a higher level, are more pathogenic (Bible et al., 2000). (2) Estrogen-treated transgenic mice that express HPV-16 E6 and E7 in basal epithelial cells develop fewer cancers than similarly-treated transgenic mice that express HPV-16 E5, E6 and E7 (Riley et al., 2003). (3) The targeted expression of HPV-16 E5 (without HPV-16 E6 and E7) in basal epithelial cells of transgenic mice causes skin tumors to arise at a much higher frequently than is associated with the targeted expression of HPV-16 E6 or E7 (Genther-Williams et al., 2005).

One or more of the demonstrated biological activities of HPV-16 E5 may be responsible for early post-infection events that enhance the probability of subsequent carcinogenesis. The ability of HPV-16 E5 to down-regulate surface expression of MHC I (Ashrafi et al., 2005) and MHC II (Zhang et al., 2003) may allow HPV-16-infected cells to avoid detection by the host immune system. Resulting increases in the duration and size of the HPV infection are likely to increase the probability that some infected cells become transformed. Indeed, viral persistence is considered to be an important factor in the neoplastic progression of a premalignant lesion (zur Hausen, 2002). The ability of HPV-16 E5 to enhance ligand-dependent activation of the EGF-R (Straight et al., 1993; Crusius et al., 1998; Rodriguez et al., 2000; Zhang et al., 2002b), and to promote DNA synthesis in differentiating keratinocytes (Genther et al., 2003), may further expand the number of infected cells. Evidence for the importance of EGF-R activation in the expansion of HPV-16 infections is provided by transgenic mice that express HPV-16 E5 and are heterozygous for a dominant negative mutant of the EGF-R. These mice do not exhibit the epidermal hyperplasia that is seen in HPV-16 E5-expressing transgenic mice that are homozygous for the wt EGF-R (Genther-Williams et al., 2005). Finally, the ability of HPV-16 E5 to inhibit gap junction-mediated communication between epithelial cells in monolayer (Oelze et al., 1995) and in raft cultures (Tomakidi et al., 2000) by interfering with connexin 43 suggests that E5 may promote oncogenic transformation *in vivo* by rendering infected cells more insensitive to homeostatic growth control signals from adjacent, non-infected cells (Tomakidi et al., 2000). Decreased connexin 43 expression has been correlated with malignancy (Fitzgerald et al., 1994; Budunova et al., 1995; Cesen-Cummings et al., 1998), and the transformed phenotype of human mammary carcinoma cells can be suppressed by connexin 43 transfection (Hirschi et al., 1996).

5. Conclusions

While E5 is the best understood protein of BPV and its functions have been elucidated in some detail at least in cultured cells, HPV E5 has remained the "Cinderella" among the HPV transforming proteins and its mode of action is still far from understood, although great progress has been made recently. The analysis of the functions of BPV and HPV E5 proteins *in vivo* is in its infancy but it is already revealing fascinating aspects of its biology. E5 expression in pre-malignant and some malignant cervical lesions (Chang et al., 2001), the presence of E5 variants with higher codon usage in high grade cervical lesions (Bible et al., 2000) and in equine sarcoids (Chambers et al., 2003b; Chapter 24), the E5-associated down-regulation of MHC I in papillomas (Araibi et al., 2004), and the high frequency of skin tumors in E5-transgenic mice (Genther-Williams et al., 2005) all point to an important role of E5 during PV-induced pathogenesis. This conclusion is confirmed by the recent successful eradication of HPV-induced tumors in mice by vaccination with E5 peptides (Chen et al., 2004). Further investigations into E5 functions will undoubtedly help elucidate important early events in HPV infection that determine persistence and disease development.

References

Adam, J.L., Briggs, M.W., and McCance, D.J. (2000). A mutagenic analysis of the E5 protein of human papillomavirus type 16 reveals that E5 binding to the vacuolar H^+-ATPase is not sufficient for biological activity, using mammalian and yeast expression systems. Virology *272,* 315-325.

Adduci, A.J., and Schlegel, R. (1999). The transmembrane domain of the E5 oncoprotein contains functionally discrete helical faces. J. Biol. Chem. *274,* 10249-10258.

Alonso, A., and Reed, J. (2002). Modeling of the human papillomavirus type 16 E5 protein. Biochim. Biophys. Acta *1601,* 9-18.

Anderson, R.A., Scobie, L., O'Neil, B.W., Grindlay, G.J., and Campo, M.S. (1997). Viral proteins of bovine papillomavirus type 4 during the development of alimentary canal tumours. Vet. J. *154,* 69-78.

Andresson, T., Sparkowski, J., Goldstein, D.J., and Schlegel, R. (1995). Vacuolar H^+-ATPase mutants transform cells and define a binding site for the papillomavirus E5 oncoprotein. J. Biol. Chem. *270,* 6830-6837.

Araibi, E.H., Marchetti, B., Ashrafi, G.H., and Campo, M.S. (2004). Down-regulation of MHC class I in bovine papillomas. J. Gen. Virol. *85,* 2809-2814.

Ashby, A.D., Meagher, L., Campo, M.S., and Finbow, M.E. (2001). E5 transforming proteins of papillomaviruses do not disturb the activity of the vacuolar H^+-ATPase. J. Gen. Virol. *82,* 2353-2362.

Ashrafi, G.H., Haghshenas, M.R., Marchetti, B., O'Brien, P.M., and Campo, M.S. (2005). The E5 protein of human papillomavirus type 16 selectively down-regulates surface HLA class I. Int. J. Cancer, *113*, 276-283. Published online: 25 Aug 2004.

Ashrafi, G.H., Pitts, J.D., Faccini, A.M., McLean, P., O'Brien, V., Finbow, M.E., and Campo, M.S. (2000). Binding of BPV-4 E8 to ductin, down-regulation of gap junction intercellular communication and cell transformation are independent events. J. Gen. Virol. *81*, 689-694.

Ashrafi, G.H., Tsirimonaki, E., Marchetti, B., O'Brien, P.M., Sibbet, G.J., Andrew, L., and Campo, M.S. (2002). Down-regulation of MHC class I by bovine papillomavirus E5 oncoproteins. Oncogene *21*, 248-259.

Bergman, P., Ustav, M., Sedman, J., Moreno-Lopez, J., Vennstrom, B., and Pettersson, U. (1988). The E5 gene of bovine papillomavirus type 1 is sufficient for complete oncogenic transformation of mouse fibroblasts. Oncogene 2, 453-459.

Bible, J.M., Mant, C., Best, J.M., Kell, B., Starkey, W.G., Raju, K.S., Seed, P., Biswas, C., Muir, P., Banatvala, J.E., and Cason, J. (2000). Cervical lesions are associated with human papillomavirus type 16 intratypic variants that have high transcriptional activity and increased usage of common mammalian codons. J. Gen. Virol. *8*, 1517-1527.

Borzacchiello, G., Iovane, G., Narcante, M.L., Poggiali, F., Roperto, S., and Venuti, A. (2003). Presence of bovine papillomavirus type 2 DNA and expression of the viral oncoprotein E5 in naturally occurring urinary bladder tumours in cows. J. Gen. Virol. *84,* 2921-2926.

Bouvard, V., Matlashewski, G., Gu, Z.M., Storey, A., and Banks, L. (1994). The human papillomavirus type 16 gene cooperates with the E7 gene to stimulate proliferation of primary cells and increases viral gene expression. Virology *203,* 73-80.

Briggs, M.W., Adam, J.L., and McCance, D.J. (2001). The human papillomavirus type 16 E5 protein alters vacuolar H+-ATPase function and stability in *Saccharomyces cerevisiae*. Virology *280,* 169-175.

Bubb, V., McCance, D.J., and Schlegel, R. (1988). DNA sequence of the HPV-16 E5 ORF and the structural conservation of its encoded protein. Virology *163*, 243-246.

Budunova, I.V., Carbajal, S., and Slaga, T. (1995). The expression of gap junctional proteins during different stages of mouse skin carcinogenesis. Carcinogenesis *16*, 2717-2724.

Burkhardt, A., DiMaio, D., and Schlegel, R. (1987). Genetic and biochemical definition of the bovine papillomavirus E5 transforming protein. EMBO J. *6,* 2381-2385.

Burkhardt, A., Willingham, M., Gay, C., Jeang, K.T., and Schlegel, R. (1989). The E5 oncoprotein of bovine papillomavirus is oriented asymmetrically in Golgi and plasma membranes. Virology *170*, 334-339.

Burnett, S., Jareborg, N., and DiMaio, D. (1992). Localization of bovine papillomavirus type 1 E5 protein to transformed basal keratinocytes and permissive differentiated cells in fibropapilloma tissue. Proc Natl Acad Sci U S A. *89*, 5665-5669.

Campo, M.S. (2002). Animal models of papillomavirus pathogenesis. Virus Res. *89*, 249-261.

Carr, E.A., Theon, A.P., Madewell, B.R., Griffey, S.M., and Hitchcock, M.E. (2001a). Bovine papillomavirus DNA in neoplastic and nonneoplastic tissues obtained from horses with and without sarcoids in the western United States. Am. J. Vet. Res. *62,* 741-744.

Carr, E.A., Theon, A.P., Madewell, B.R., Hitchcock, M.E., Schlegel, R., and Schiller, J.T. (2001b). Expression of a transforming gene (E5) of bovine papillomavirus in sarcoids obtained from horses. Am. J. Vet. Res. *62,* 1212-1217.

Cesen-Cummings, K., Fernstrom, M.J., Malkinson, A.M., and Ruch, R.J. (1998). Frequent reduction of gap junctional intercellular communication and connexin 43 expression in human and mouse lung carcinoma cells. Carcinogenesis *19*, 61-67.

Chambers, G., Ellsmore, V.A., O'Brien, P.M., Reid, S.W.J., Love, S., Campo, M.S., and Nasir, L. (2003a). Association of bovine papillomavirus with the equine sarcoid. J. Gen. Virol., *84,* 1055-1062.

Chambers, G., Ellsmore, V.A., O'Brien, P.M., Reid, S.W.J., Love, S., Campo, M.S., and Nasir, L. (2003b). Sequence variants of bovine papillomavirus E5 detected in equine sarcoids. Virus Res. *96,* 141-145.

Chang, J.L., Tsao, Y.P., Liu, D.W., Huang, S.J., Lee, W.H., and Chen, S.L. (2001). The expression of HPV-16 E5 protein in squamous neoplastic changes in the uterine cervix. J. Biomed. Sci. *8*, 206-213.

Chen, S.L., Huang, C.H., Tsai, T.C., Lu, K.Y., and Tsao, Y.P. (1996a). The regulation mechanism of c-jun and junB by human papillomavirus type 16 E5 oncoprotein. Arch. Virol. *141*, 791-800.

Chen, S.L., Lin, Y.K., Li, L.Y., Tsao, Y.P., Lo, H.Y., Wang, W.B., and Tsai, T.C. (1996b). E5 proteins of human papillomavirus types 11 and 16 transactivate the c-*fos* promoter through the NF1 binding element. J. Virol. *70*, 8558-8563.

Chen Y.F., Lin, C.W., Tsao, Y.P. and Chen, S.L. (2004) Cytotoxic-T-lymphocyte human papillomavirus type 16 E5 peptide with CpG-oligodeoxynucleotide can eliminate tumor growth in C57BL/6 mice. J. Virol. 78, 1333-1343.

Cohen, G.B., Gandhi, R.T., Davis, D.M., Mandelboim, O., Chen, B.K., Strominger, J.L., and Baltimore, D. (1999). The selective downregulation of class I major histocompatibility complex proteins by HIV-1 protects HIV-infected cells from NK cells. Immunity *10*, 661-671.

Collins, M.K.L., Downward, J., Miyajima, A., Maruyama, K., Arai, K.I., and Mulligan, R.C. (1988). Transfer of functional EGF receptors to an IL3-dependent cell line. J. Cell. Physiol. *137*, 293-298.

Conrad, M., Bubb, V.J., and Schlegel, R. (1993). The human papillomavirus type 6 and 16 E5 proteins are membrane-associated proteins which associate with the 16-kilodalton pore-forming protein. J. Virol. *67,* 6170-6178.

Cresswell, P., Bangia, N., Dick, T.,and Dietrich, G. (1999). The nature of the MHC class I peptide loading complex. Immunol. Rev. *172*, 21-28.

Crusius, K., Auvinen, E., and Alonso, A. (1997). Enhancement of EGF- and PMA-mediated MAP kinase activation in cells expressing the human papillomavirus type 16 E5 protein. Oncogene *15*, 1437-1444.

Crusius, K., Auvinen, E., Steuer, B., Gaissert, H., and Alonso, A. (1998). The human papillomavirus type 16 E5-protein modulates ligand-dependent activation of the EGF receptor family in the human epithelial cell line HaCaT. Exp. Cell Res. *241*, 76-83.

Crusius, K., Kaszkin, M., Kinzel, V., and Alonso, A. (1999). The human papillomavirus type 16 E5 protein modulates phospholipase C-γ-1 activity and phosphatidyl inositol turnover in mouse fibroblasts. Oncogene *18,* 6714-6718.

D'Souza-Schorey, C., Li, G., Colombo, M.I., and Stahl, P.D. (1995). A regulatory role for ARF6 in receptor-mediated endocytosis. Science *267*, 1175-1178.

D'Souza-Schorey, C., van Donselaar, E., Hsu, V.W., Yang, C., Stahl, P.D., and Peters, P.J. (1998). ARF6 targets recycling vesicles to the plasma membrane: insights from an ultrastructural investigation. J. Cell Biol. *140*, 603-616.

DiMaio, D., and Mattoon, D. (2001). Mechanisms of cell transformation by papillomavirus E5 proteins. Oncogene *20,* 7866-7873.

DiMaio, D., Guralski, D., and Schiller, J.T. (1986). Translation of open reading frame E5 of bovine papillomavirus is required for its transforming activity. Proc. Natl. Acad. Sci. USA *83,* 1797-1801.

Disbrow, G.L., Sunitha, I., Baker, C.C., Hanover, J., and Schlegel, R. (2003). Codon optimization of the HPV-16 E5 gene enhances protein expression. Virology *311,* 105-114.

Drummond-Barbosa, D., Vaillancourt, R.R., Kazlauskas, A., and DiMaio, D. (1995). Ligand-independent activation of the platelet-derived growth factor □ receptor: requirements for bovine papillomavirus E5-induced mitogenic signaling. Mol. Cell. Biol. *15,* 2570-2581.

Faccini, A.M., Cairney, M., Ashrafi, G.H., Finbow, M., Campo, M.S., and Pitts, J.D. (1996). The bovine papillomavirus type 4 E8 protein binds to ductin (16kDa) and causes loss of gap junctional intercellular communication in primary fibroblasts. J. Virol. *70*, 9041-9045.

Fehrmann, F., and Laimins, L.A. (2003). Human papillomaviruses: targeting differentiating cells for malignant transformation. Oncogene 22, 5201-5207.

Fehrmann, F., Klumpp, D.J., and Laimins, L.A. (2003). Human papillomavirus type 31 E5 protein supports cell cycle progression and activates late viral functions upon epithelial differentiation. J. Virol. *77*, 2819-2831.

Fitzgerald, D.J., Fusenig, N.E., Boukamp, P., Piccoli, C., Mesnil, M., and Yamasaki, H. (1994). Expression and function of connexin in normal and transformed human keratinocytes in culture. Carcinogenesis *15*, 1859-1865.

Forgac, M. (1999). Structure and properties of the vacuolar (H^+)-ATPases. J. Biol. Chem. *274*, 12951-12954.

Frazer, I.H. (1996). Immunology of papillomavirus infection. Curr. Opin. Immunol. *8*, 484-491.

Frazer, I.H., Thomas, R., Zhou, J., Leggatt, G., Dunn, L., McMillan, N., Tindle, R.W., Filgueira, L., Manders, P., Barnard, P., and Sharkey, M. (1999). Potential strategies utilised by papillomavirus to evade host immunity. Immunol. Rev. *168*, 131-142.

Furman, M.H., Dey, N., Tortorella, D., and Ploegh, H.L. (2002). The human cytomegalovirus US10 gene product delays trafficking of major histocompatibility complex class I molecules. J. Virol. *76*, 53-56.

Genther-Williams, S.M, Disbrow, G.L., Schlegel, R., Lee, D., Threadgill, D.W., and Lambert, P.F. (2005). Requirement of epidermal growth factor receptor for hyperplasia induced by E5, a high risk HPV oncogene. Cancer Res. In press.

Genther, S.M., Sterling, S., Duensing, S., Munger, K., Sattler, C., and Lambert, P.F. (2003). Quantitative role of the human papillomavirus type 16 E5 gene during the productive stage of the viral life cycle. J. Virol. *77*, 2832-2842.

Gieswein, C.E., Sharom, F.J., and Wildeman, A.G. (2003). Oligomerization of the E5 protein of human papillomavirus type 16 occurs through multiple hydrophobic regions. Virology *313,* 415-426.

Goldstein, D.J., and Schlegel, R. (1990). The E5 oncoprotein of bovine papillomavirus binds to a 16 kD cellular protein. EMBO J. *9*, 137-145.

Goldstein, D.J., Andresson, T., Sparkowski, J.J., and Schlegel, R. (1992). The BPV-1 E5 protein, the 16 kDa membrane pore-forming protein and the PDGF receptor exist in a complex that is dependent on hydrophobic transmembrane interactions. EMBO J. *11*, 4851-4859.

Goldstein, D.J., Finbow, M.E., Andresson, T., McLean, P., Smith, K., Bubb, V., and Schlegel, R. (1991). Bovine papillomavirus E5 oncoprotein binds to the 16K component of vacuolar H+-ATPases. Nature *352,* 347-349.

Goldstein, D.J., Li, W., Wang, L.M., Heidaran, M.A., Aaronson, S., Shinn, R., Schlegel, R., and Pierce, J.H. (1994). The bovine papillomavirus type 1 E5 transforming protein specifically binds and activates the β-type receptor for the platelet-derived growth factor but not other related tyrosine kinase-containing receptors to induce cellular transformation. J. Virol. *68,* 4432-4441.

Gottlieb, A.B., Khandke, L., Krane, J.F., Staiano-Coico, L., Ashinoff, R., and Krueger, J.G. (1992). Anthralin decreases keratinocyte TGF-alpha expression and EGF-receptor binding *in vitro*. J. Invest. Dermatol. *98*, 680-685.

Graham, L.A., Powell, B., and Stevens, T.H. (2000). Composition and assembly of the yeast vacuolar H^+-ATPase complex. J. Exp. Biol. *203*, 61-70.

Grindlay GJ, Campo MS and O'Brien V. (2005). Transactivation of the cyclin A promoter by bovine papillomavirus type 4 E5 protein. Virus Research, Virus Research, 108, 29-38. Published online 17 September 2004.

Groff, D.E., and Lancaster, W.D. (1986). Genetic analysis of the 3′ early region transformation and replication functions of bovine papillomavirus type 1. Virology *150,* 221-230.

Gromme, M., Uytdehaag, F.G., Janssen, H., Calafat, J., van Binnendijk, R.S., Kenter, M.J., Tulp, A., Verwoerd, D., and Neefjes, J. (1999). Recycling MHC class I molecules and endosomal peptide loading. Proc. Natl. Acad. Sci. USA *96*, 10326-10331.

Gu, Z., and Matlashewski, G. (1995). Effect of human papillomavirus type 16 oncogenes on MAP kinase activity. J. Virol. *69*, 8051-8056.

Halbert, C.L., and Galloway, D.A. (1988). Identification of the E5 open reading frame of human papillomavirus type 16. J. Virol. *62*, 1071-1075.

Hirschi, K.K., Xu, C.E., Tsukamoto, T., and Sager, R. (1996). Gap junction genes Cx26 and Cx43 individually suppress the cancer phenotype of human mammary carcinoma cells and restore differentiation potential. Cell Growth Differ. *7*, 861-870.

Horwitz, B.H., Burkhardt, A.L., Schlegel, R., and DiMaio, D. (1988). 44-amino-acid E5 transforming protein of bovine papillomavirus requires a hydrophobic core and specific carboxyl-terminal amino acids. Mol. Cell. Biol. *8*, 4071-4078.

Hwang, E.S., Nottoli, T., and DiMaio, D. (1995). The HPV16 E5 protein: expression, detection, and stable complex formation with transmembrane proteins in COS cells. Virology *211,* 227-233.

Jackson, M.E., Pennie, W.D., McCaffery, R.E., Smith, K.T., Grindlay, G.J., and Campo, M.S. (1991). The B subgroup bovine papillomaviruses lack an identifiable E6 open reading frame. Molecular Carcinogenesis *4*, 382-387.

Kabsch, K., and Alonso, A. (2002a). The human papillomavirus type 16 E5 protein impairs TRAIL- and FasL-mediated apoptosis in HaCaT cells by different mechanisms. J. Virol. *76,* 12162-12172.

Kabsch, K., and Alonso, A. (2002b). The human papillomavirus type 16 (HPV-16) E5 protein sensitizes human keratinocytes to apoptosis induced by osmotic stress. Oncogene *21*, 947-953.

Kaplan, K.B., Swedlow, J.R., Varmus, H.E., and Morgan, D.O. (1992). Association of p60$^{c\text{-}src}$ with endosomal membranes in mammalian fibroblasts. J. Cell Biol. *118*, 321-333.

Klein, O., Kegler-Ebo, D., Su, J., Smith, S., and DiMaio, D. (1999). The bovine papillomavirus E5 protein requires a juxtamembrane negative charge for activation of the platelet-derived growth factor β receptor and transformation of C127 cells. J. Virol. *73*, 3264-3272.

Klein, O., Polack, G.W., Surti, T., Kegler-Ebo, D., Smith, S.O., and DiMaio, D. (1998). Role of glutamine 17 of the bovine papillomavirus E5 protein in platelet-derived growth factor β receptor activation and cell transformation. J. Virol. *72,* 8921-8932.

Lai, C.C., Henningson, C., and DiMaio, D. (1998). Bovine papillomavirus E5 protein induces oligomerization and trans-phosphorylation of the platelet-derived growth factor β receptor. Proc. Natl. Acad. Sci. USA *95,* 15241-15246.

Laimins, L.A. (1993). The biology of human papillomaviruses: from warts to cancer. Infect. Agents Dis. *2*, 74-86.

Leechanachai, P., Banks, L., Moreau, F., and Matlashewski, G. (1992). The E5 gene from human papillomavirus type 16 is an oncogene which enhances growth factor-mediated signal transduction to the nucleus. Oncogene *7,* 19-25.

Lewis, T.S., Shapiro, P.S., and Ahn, N.G. (1998). Signal transduction through MAP kinase cascades. Adv. Cancer Res. *74*, 49-139.

Liu, Q., Feng, Y., and Forgac, M. (1994). Activity and *in vitro* reassembly of the coated vesicle (H^+)-ATPase requires the 50-kDa subunit of the clathrin assembly complex AP-2. J. Biol. Chem. *269*, 31592-31597.

Lu, X., Yu, H., Liu, S.H., Brodsky, F.M., and Peterlin, B.M. (1998). Interactions between HIV1 Nef and vacuolar ATPase facilitate the internalization of CD4. Immunity *8*, 647-656.

Maranda, B., Brown, D., Bourgoin, S., Casanova, J.E., Vinay, P., Ausiello, D.A., and Marshansky, V. (2001). Intra-endosomal pH-sensitive recruitment of the Arf-nucleotide exchange factor ARNO and Arf6 from cytoplasm to proximal tubule endosomes. J. Biol. Chem. *276*, 18540-18550.

Marchetti, B., Ashrafi, G.H., Tsirimonaki, E., O'Brien, P.M., and Campo, M.S. (2002). The bovine papillomavirus oncoprotein E5 retains MHC class I molecules in the GA and prevents their transport to the cell surface. Oncogene *21*, 7808-7816.

Martin, G.S. (2001). The hunting of the Src. Nature Rev. *2*, 467-475.

Martin, P., Vass, W.C., Schiller, J.T., Lowy, D.R., and Velu, T.J. (1989). The bovine papillomavirus E5 transforming protein can stimulate the transforming activity of EGF and CSF-1 receptors. Cell *59*, 21-32.

McMurray, H.R., Nguyen, D., Westbrook, T.F., and McCance, D.J. (2001). Biology of human papillomaviruses. Int. J. Exp. Path. *82*, 15-33.

Morgan, I.M., and Campo, M.S. (2000). Recent Developments in Bovine Papillomaviruses. Papillomavirus Report *11*, 127-132.

Myers, M., and Forgac, M. (1993). Assembly of the peripheral domain of the vacuolar H(+)-adenosine triphosphatase. J. Cell. Physiol. *156*, 35-42.

Nasir, L., and Reid, S.W.J. (1999). Bovine papillomaviral gene expression in equine sarcoid tumours. Virus Res. *61,* 171-175.

Nilson, L.A., and DiMaio, D. (1993). Platelet-derived growth factor receptor can mediate tumorigenic transformation by the bovine papillomavirus E5 protein. Mol. Cell. Biol. *13*, 4137-4145.

O'Brien, P.M., and Campo, M.S. (2002). Evasion of host immunity directed by papillomavirus-encoded proteins. Virus Research *88*, 103-117.

O'Brien, V., and Campo, M.S. (1998). BPV-4 E8 transforms NIH 3T3 cells, up-regulates cyclin A and cyclin A-associated kinase activity and de-regulates expression of the cdk inhibitor p27^{KIP1}. Oncogene *17*, 293-301.

O'Brien, V., Ashrafi, G.H., Grindlay, G.J., Anderson, R.A., and Campo, M.S. (1999). A mutational analysis of the transforming functions of the E8 protein of bovine papillomavirus type 4. Virology *254*, 385-394.

O'Brien, V., Grindlay, J.G., and Campo, M.S. (2001). Cell transformation by the E5/E8 protein of bovine papillomavirus type 4: p27^{kip1}, elevated through increased protein synthesis, is sequestered by cyclin D1 CDK4 complexes. J. Biol. Chem. *276*, 33861-33868.

Oelze, I., Kartenbeck, J., Crusius, K., and Alonso, A. (1995). Human papillomavirus type 16 E5 protein affects cell-cell communication in an epithelial cell line. J. Virol. *69*, 4489-4494.

Pennie, W.D., Grindlay, G.J., Cairney, M., and Campo, M.S. (1993). Analysis of the transforming functions of BPV-4. Virology *193*, 614-628.

Petti, L., and DiMaio, D. (1992). Stable association between the bovine papillomavirus E5 transforming protein and activated platelet-derived growth factor receptor in transformed mouse cells. Proc. Natl. Acad. Sci. USA *89*, 6736-6740.

Petti, L., Nilson, L.A., and DiMaio, D. (1991). Activation of the platelet-derived growth factor receptor by the bovine papillomavirus E5 transforming protein. EMBO J. *10,* 845-855.

Petti, L.M., and Ray, F.A. (2000). Transformation of mortal human fibroblasts and activation of a growth inhibitory pathway by the bovine papillomavirus E5 oncoprotein. Cell Growth Differ. *11,* 395-408.

Petti, L.M., Reddy, V., Smith, S.O., and DiMaio, D. (1997). Identification of amino acids in the transmembrane and juxtamembrane domains of the platelet-derived growth factor receptor required for productive interaction with the bovine papillomavirus E5 protein. J. Virol. *71*, 7318-7327.

Pierce, J.H., Ruggiero, M., Fleming, T.P., Di Fiore, P.P., Greenberger, J. S., Varticovski, L., Schlessinger, J., Rovera, G., and Aaronson, S.A. (1988). Signal transduction through the EGF receptor transfected in IL-3-dependent hematopoietic cells. Science *239*, 628-631.

Pim, D., Collins, M., and Banks, L. (1992). Human papillomavirus type 16 E5 gene stimulates the transforming activity of the epidermal growth factor receptor. Oncogene *7,* 27-32.

Rabson, M.S., Yee, C., Yang, Y.C., and Howley, P.M. (1986). Bovine papillomavirus type 1 3′ early region transformation and plasmid maintenance functions. J. Virol. *60*, 626-634.

Reich, Z., Altman, J.D., Boniface, J.J., Lyons, D.S., Kozono, H., Ogg, G., Morgan, C., and Davis, M.M. (1997). Stability of empty and peptide-loaded class II major histocompatibility complex molecules at neutral and endosomal pH: comparison to class I proteins. Proc. Natl. Acad. Sci. USA *94*, 2495-2500.

Riley, R.R., Duensing, S., Brake, T., Munger, K., Lambert, P.F., and Arbeit, J.M. (2003). Dissection of human papillomavirus E6 and E7 function in transgenic mouse models of cervical carcinogenesis. Cancer Res. *63*, 4862-4871.

Robinson, M.J., and Cobb, M.H. (1997). Mitogen-activated protein kinase pathways. Curr. Opin. Cell Biol. *9*, 180-186.

Rodriguez, M.I., Finbow, M.E., and Alonso, A. (2000). Binding of human papillomavirus 16 E5 to the 16 kDa subunit c (proteolipid) of the vacuolar H^+-ATPase can be dissociated from the E5-mediated epidermal growth factor receptor overactivation. Oncogene *19,* 3727-3732.

Sato, S.B., and Toyama, S. (1994). Interference with the endosomal acidification by a monoclonal antibody directed toward the 116(100)-kD subunit of the vacuolar type proton pump. J. Cell Biol. *127*, 39-53.

Schapiro, F., Sparkowski, J., Adduci, A., Suprynowicz, F., Schlegel, R., and Grinstein, S. (2000). Golgi alkalinization by the papillomavirus E5 oncoprotein. J. Cell Biol. *148,* 305-315.

Schiller, J.T., Vass, W.C., and Lowy, D.R. (1984). Identification of a second transforming region in bovine papillomavirus DNA. Proc. Natl. Acad. Sci. USA *81*, 7880-7884.

Schiller, J.T., Vass, W.C., Vousden, K.H., and Lowy, D.R. (1986). The E5 open reading frame of bovine papillomavirus type 1 encodes a transforming gene. J. Virol. *57,* 1-6.

Schlegel, R., Wade-Glass, M., Rabson, M.S., and Yang, Y.-C. (1986). The E5 transforming gene of bovine papillomavirus encodes a small hydrophobic polypeptide. Science *233*, 464-467.

Sparkowski, J., Anders, J., and Schlegel, R. (1995). E5 oncoprotein retained in the endoplasmic reticulum/cis Golgi still induces PDGF receptor autophosphorylation but does not transform cells. EMBO J. *14,* 3055-3063.

Sparkowski, J., Mense, M., Anders, J., and Schlegel, R. (1996). E5 oncoprotein transmembrane mutants dissociate fibroblast transforming activity from 16-kilodalton protein binding and platelet-derived growth factor receptor binding and phosphorylation. J. Virol. *70,* 2420-2430.

Staebler, A., Pierce, J.H., Brazinski, S., Heidaran, M.A., Li, W., Schlegel, R., and Goldstein, D.J. (1995). Mutational analysis of the β-type platelet-derived growth factor receptor defines the site of interaction with the bovine papillomavirus type 1 E5 transforming protein. J. Virol. 69, 6507-6517.

Stevens, T.H., and Forgac, M. (1997). Structure, function and regulation of the vacuolar (H^+)-ATPase. Ann. Rev. Cell Dev. Biol. *13,* 779-808.

Stoppler, M.C., Straight, S.W., Tsao, G., Schlegel, R., and McCance, D.J. (1996). The E5 gene of HPV-16 enhances keratinocyte immortalization by full-length DNA. Virology *223*, 251-254.

Straight, S.W., Herman, B., and McCance, D.J. (1995). The E5 oncoprotein of human papillomavirus type 16 inhibits the acidification of endosomes in human keratinocytes. J. Virol. *69,* 3185-3192.

Straight, S.W., Hinkle, P.M., Jewers, R.J., and McCance, D.J. (1993). The E5 oncoprotein of human papillomavirus type 16 transforms fibroblasts and effects the downregulation of the epidermal growth factor receptor in keratinocytes. J. Virol. *67,* 4521-4532.

Suprynowicz, F.A., Baege, A., Sunitha, I., and Schlegel, R. (2002). c-Src activation by the E5 oncoprotein enables transformation independently of PDGF receptor activation. Oncogene *21,* 1695-1706.

Suprynowicz, F.A., Sparkowski, J., Baege, A., and Schlegel, R. (2000). E5 oncoprotein mutants activate phosphoinositide 3-kinase independently of platlet-derived growth factor receptor activation. J. Biol. Chem. *275,* 5111-5119.

Thomsen, P., van Deurs, B., Norrild, B., and Kayser, L. (2000). The HPV16 E5 oncogene inhibits endocytic trafficking. Oncogene *19,* 6023-6032.

Tomakidi, P., Cheng, H., Kohl, A., Komposch, G., and Alonso, A. (2000). Connexin 43 expression is downregulated in raft cultures of human keratinocytes expressing the human papillomavirus type 16 E5 protein. Cell Tissue Res. *301*, 323-327.

Tomasec, P., Braud, V.M., Rickards, C., Powell, M.B., McSharry, B.P., Gadola, S., Cerundolo, V., Borysiewicz, L.K., McMichael, A.J., and Wilkinson, G.W. (2000). Surface expression of HLA-E, an inhibitor of natural killer cells, enhanced by human cytomegalovirus gpUL40. Science *287*, 1031-1033.

Tsao, Y.P., Li, L.Y., Tsai, T.C., and Chen, S.L. (1996). Human papillomavirus type 11 and 16 represses $p21^{WafI/SdiI/CipI}$ gene expression in fibroblasts and keratinocytes. J. Virol. *70*, 7535-7539.

Valle, G.F., and Banks, L. (1995). The human papillomavirus (HPV)-6 and HPV-16 proteins co-operate with HPV-16 E7 in the transformation of primary rodent cells. J. Gen. Virol. *76,* 1239-1245.

van't Hof, W., and Resh, M.D. (1997). Rapid plasma membrane anchoring of newly synthesized $p59^{fyn}$: selective requirement for NH_2-terminal myristoylation and palmitoylation at cysteine-3. J. Cell Biol. *136*, 1023-1035.

Walboomers, J.M., Jacobs, M.V., Manos, M.M., Bosch, F.X., Kummer, J.A., Shah, K.V., Snijders, P.J., Peto, J., Meijer, C.J., and Munoz, N. (1999). Human papillomavirus is a necessary cause of invasive cervical cancer worldwide. J. Pathol. *189*, 12-19.

Yang, D.H., Wildeman, A.G., and Sharom, F.J. (2003). Overexpression, purification, and structural analysis of the hydrophobic E5 protein from human papillomavirus type 16. Protein Expr. Purif. *30*, 1-10.

Yang, Y.C., Spalholz, B.A., Rabson, M.S., and Howley, P.M. (1985). Dissociation of transforming and trans-activation functions for bovine papillomavirus type 1. Nature *318*, 575-577.

Yokoyama, W.M. (1998). Natural killer cell receptors. Curr. Opin. Immunol. *10*, 298-305.

Zago, M., Campo, M.S., and O'Brien, V. (2004). Cyclin A expression and growth in suspension can be uncoupled from p27 de-regulation and ERK activity in cells transformed by BPV-4 E5. J. Gen. Virol. *85*, 3585-3595.

Zhang, B., Li, P., Wang. E., Brahmi, Z., Dunn, K.W., Blum, J.S., and Roman, A. (2003). The E5 protein of human papillomavirus type 16 perturbs MHC class II antigen maturation in human foreskin keratinocytes treated with interferon-γ. Virology *310*, 100-108.

Zhang, B., Spandau, D.F., and Roman, A. (2002b). E5 protein of human papillomavirus type 16 protects human foreskin keratinocytes from UV B-irradiation-induced apoptosis. J. Virol. *76*, 220-231.

Zhang, Y., Lehman, J.M., and Petti, L.M. (2002a). Apoptosis of mortal human fibroblasts transformed by the bovine papillomavirus E5 oncoprotein. Mol. Cancer Res. *1*, 122-136.

zur Hausen, H. (2002). Papillomaviruses and cancer: from basic studies to clinical application. Nature Rev. Cancer *2*, 342-350.

zur Hausen, H. and de Villiers, E.M. (1994). Human papillomaviruses. Ann Rev. Microbiol. *48*, 427-447.

From: Papillomavirus Research: From Natural History to Vaccines and Beyond. Edited by: M. Saveria Campo

Chapter 9

The Role of the HPV E6 Oncoprotein in Malignant Progression

Miranda Thomas, David Pim and Lawrence Banks*

Abstract

It is now well established that the continued expression of the two major viral oncoproteins, E6 and E7, is essential for the maintenance of the transformed phenotype in cells derived from cervical tumours. Numerous *in vitro* and *in vivo* systems have been used to address the respective contributions of each protein to the development of malignancy. Recent studies have begun to highlight a role for E6 in the later, more malignant, stages of disease progression. Therefore, understanding the mechanisms by which E6 brings this about has important implications both for virus-induced cancers and for cancer progression in general. Interestingly, a recently identified class of cellular E6 targets has been implicated in this activity. These proteins control processes regulating cell proliferation, cell polarity, and cell-cell contact, all of which are perturbed during metastatic progression. In this chapter we shall review the activities of E6 with respect to those cellular targets that may be responsible for driving the later stages of malignant progression.

1. Introduction

There is now overwhelming evidence of the direct contribution of the HPV E6 and E7 oncoproteins in the development and maintenance of the transformed phenotype in cervical cancer (Chapter 3 of this book). The E6 and E7 genes are invariably retained and expressed, both in cervical cancers and in the cell lines isolated from them, many years after the initial immortalising events. This indicates the importance of these proteins for the maintenance of the transformed phenotype (Schwarz et al., 1985; Smotkin and Wettstein, 1986; Androphy et al., 1987; Banks et al., 1987). Indeed, E6 and E7 expression in tumour-derived cell lines has been blocked using a number of approaches: blocking peptides (Butz et al., 2000; Nauenburg et al., 2001), antisense RNA (von Knebel Döberitz et al., 1992; Steele et al., 1992; Hu et al., 1995; Venturini et al., 1999), ribozymes (Alvarez-Salas et al., 1998) and RNA interference (Jiang and Milner, 2002; 2004; Butz et al., 2003; Yoshinouchi et al., 2003). In each case this resulted in the senescence and/or apoptosis of these tumour-derived cell lines. This suggests that blocking E6 or E7 might prove the most effective strategy for chemotherapeutic intervention, and thus provides a fundamentally important reason for determining the modes of action of the viral oncoproteins.

Although more than 100 different HPV types have been identified, only a small number of them have been associated with the development of human cancers. Understanding the differences between the "high-risk" virus types and the vast majority of "low-risk" types has been one of the main challenges in HPV research for many years. During the following discussion, we will highlight those aspects of E6 function that we believe to be responsible for the differences in pathology observed between different HPV types, and which most probably contribute to the malignant progression of HPV-induced tumours.

2. HPV E6 and Cell Transformation

Although the continued presence of E6 and E7 within cervical tumour-derived cell lines suggests that they play a role in the transformation process, the first direct evidence that they possess intrinsic transforming activity came from studies in NIH3T3 cells (Tsunokawa et al., 1986; Yasumoto et al., 1986). In these studies it was found that the E7 gene encodes the major transforming activity (Kanda et al., 1988; Vousden et al., 1988; see Chapter 10 for the role of E7 in cell transformation and carcinogenesis), and that E6 has only weak transforming activity in established rodent cells (Bedell et al., 1989; Sedman et al., 1991). However, HPV-16 and HPV-18 E6s were shown to cooperate with an activated ras oncogene to transform primary rodent cells (Storey and Banks, 1993; Pim et al., 1994; Liu et al., 1994). These studies provided the earliest evidence of clear biological differences between the oncoproteins derived from high- and low-risk virus types. In these assays, high-risk HPV E6 and E7 proteins showed strong transforming activity, whereas the oncoproteins from low-risk HPVs had, at best, only weak transforming activity (Storey et al., 1988; Phelps et al., 1988; Storey and Banks, 1993; Pim et al., 1994; Liu et al., 1994).

Papillomaviruses are species-specific and therefore assays using human cells are the gold standard for

*For correspondence. Email banks@icgeb.org

assessing HPV oncoprotein function. Late-passage mammary epithelial cells are susceptible to immortalisation by HPV E6 (Band et al., 1991; Wazer et al., 1995); however, the low-risk HPV E6 proteins are also active in this assay (Band et al., 1993) so it is not clear what transformation-relevant activities of E6 are assayed by this system. The most meaningful *in vitro* assays of HPV transformation are obviously those using primary human keratinocytes, the target cells of the virus *in vivo*. Several studies have shown that E6 and E7 from high-risk HPV types can cooperate to immortalise human foreskin, oral or cervical keratinocytes (Dürst et al., 1987; Kaur and McDougall, 1988; Schlegel et al., 1988; Sexton et al., 1993; Garner-Hamrick et al., 2004). It is also clear that, as in primary rodent cells, the low-risk HPV E6 and E7 genes have no activity in such assays (Woodworth et al., 1989; Pecoraro et al., 1989). It is important to note, moreover, that the immortalised keratinocytes are not fully transformed by E6 and E7, and cannot form tumours in nude mice. The cells become tumourigenic only after extended passage in culture, or upon transfection of activated oncogenes (DiPaolo et al., 1989; Dürst et al., 1989; Hurlin et al., 1991), thereby reflecting the multistep nature of cancer development *in vivo*.

More recently, further insights into the activities of the two major HPV oncogenes have been provided by an elegant series of studies in transgenic mice, in which E6 and E7 were expressed from the keratin 14 promoter. This promoter is activated in basal epithelium, and thus results in the E6 and E7 proteins being expressed in physiologically relevant tissue (Arbeit et al., 1994; Herber et al., 1996; Song et al., 1999; see Chapter 14 for a full description of HPV-transgenic mice). Mice expressing E6 and E7 developed epithelial neoplasias at a high frequency, as well as cervical cancer in mice treated with œstrogen. More detailed analysis of the respective roles of E6 and E7 in these transgenic mice proved to be most revealing. When E6 and E7 were expressed individually, each protein gave rise to skin tumours: when E7 was expressed alone, mainly benign, differentiated tumours were induced with high frequency. In contrast, mice expressing only E6 developed significantly fewer tumours, which were however much more aggressive and prone to progression into metastatic cancer (Song et al., 2000). Studies in these mice using chemical carcinogens indicate that E7 promotes immortalisation, whereas E6 promotes progression into full malignancy (Song et al., 2000). More recent work with transgenic mice has shown that the PDZ ligand of HPV-16 E6 (see below), which is specific for the high-risk mucosal virus types, is absolutely required for E6's induction of epithelial hyperplasia (Nguyen et al., 2003a). Since the malignancy of cervical cancer renders it fatal, it would seem that E6 might, therefore, be the better target for chemotherapeutic intervention. In the following discussion we propose to review the activities of E6 that might be relevant to its induction of transformation, and to focus on those that appear to contribute to the metastatic tendency of HPV-induced tumours.

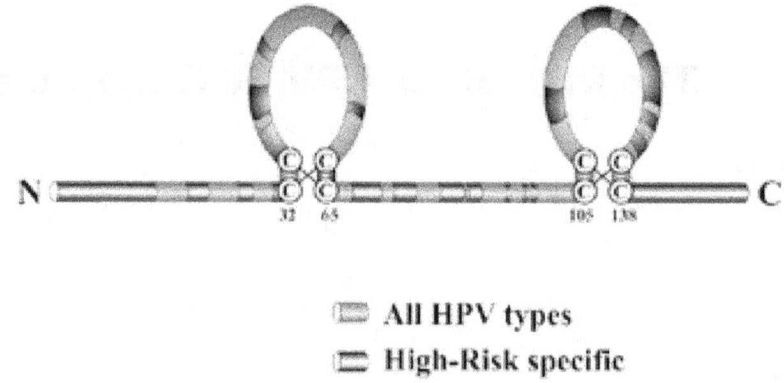

Figure 1. Comparison of the HPV E6 amino acid sequences of high- and low-risk virus types. Regions in dark grey show high-risk specific sequences. The remainder of the molecule shows a high degree of conservation across many different HPV types.

3. Interaction of E6 with its Cellular Targets

The HPV E6 proteins are approximately 150 amino acids in length, having an apparent molecular weight of around 18kDa. They are characterised by the strict conservation of four metal binding Cys-x-x-Cys motifs, which permit the formation of two zinc fingers (Cole and Danos, 1987; Barbosa et al., 1989; Grossman and Laimins, 1989). Mutants of E6 whose zinc fingers are disrupted are defective in almost all assays (Kanda et al., 1991; Sherman and Schlegel, 1996), indicating that this structural feature of E6 is vital for its effective activity. The high- and low-risk E6 proteins are highly homologous, particularly in the central portion of the molecule, reflecting their many common activities. Figure 1 shows a representation of the E6 sequence, highlighting regions that are high-risk E6-specific, and those regions that are conserved amongst many different HPV E6 types. However, the extreme amino- and carboxy- terminal portions of the protein are more divergent, and of particular interest is the PDZ-binding motif found in the extended C-terminus of the E6 proteins from high-risk, mucosal-specific HPV types: the implications of that will be discussed below.

Many cellular targets of the high risk E6 proteins have now been described and it is impossible to catalogue them all, since many are themselves the subjects of investigation and their roles in the transformation process remain unknown. However, many and various viruses have been implicated in the promotion of cellular transformation (see zur Hausen, 2001, for review) and there appear to be a number of common pathways of virus-induced transformation. Rigorous analysis of these pathways and comparison with the cellular targets of other known viral oncoproteins reveals a more limited subset of E6-interacting partners. These are shown schematically in

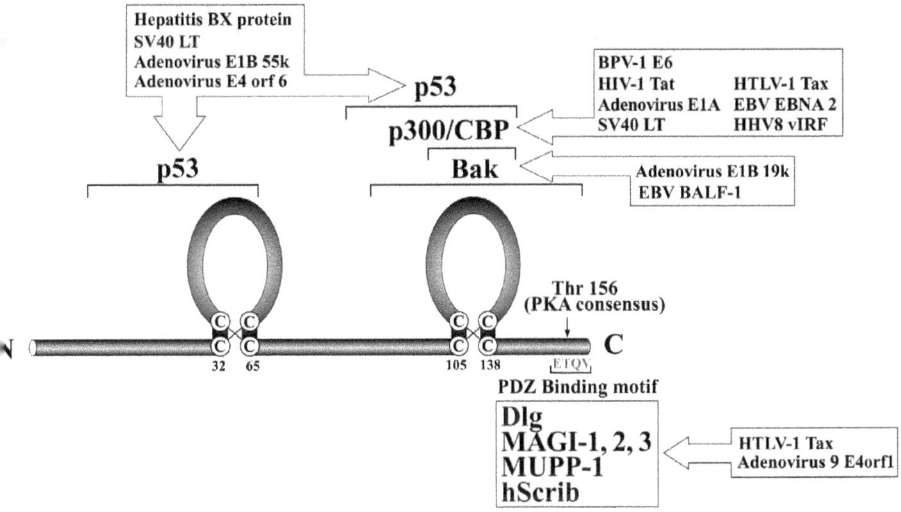

Figure 2. A schematic diagram of the HPV-18 E6 protein and regions involved in interactions with some of its known cellular targets. Highlighted are those proteins also known to be targets of other viral transforming proteins, suggesting common pathways of viral transformation.

Figure 2, together with their apparent sites of interaction on the HPV-18 E6 protein. We propose to concentrate our discussion upon the cellular targets of E6 that are known to have roles in regulating the processes required for the cell transformation induced by different viral families.

4. p53 as a Target for Viral Oncoproteins

The tumour suppressor p53 is the best known target of high-risk HPV E6 (See Thomas et al., 1999, for review). It is also targeted by a number of other viral oncoproteins: the SV40 large tumour antigen (TAg) (Lane and Crawford, 1979; Linzer and Levine, 1979) binds to p53's DNA binding domain, thus blocking its transactivation of target genes (Ruppert and Stillman, 1993). The Adenovirus (Ad) E1B-55k protein (Sarnow et al., 1982), binds the transactivation domain of p53 (Lin et al., 1994), while the Hepatitis B virus X protein prevents p53 transactivation by sequestering it in the cytoplasm (Elmore et al., 1997). Each of these interactions results in the accumulation of p53 in stable and inactive complexes, however no such complexes and little or no p53 protein can be detected in HPV-containing cells, despite there being abundant p53 mRNA present (Matlashewski et al., 1986; Scheffner et al., 1991). *In vitro* studies revealed that HPV-16 and HPV-18 E6s can bind to p53 and induce its degradation via the ubiquitin pathway (Werness et al., 1990; Scheffner et al., 1990). This activity of E6 explains why the levels of p53 in HPV-transformed cell lines are often so low (Band et al., 1991). It also explains the observations that p53 is wild type in such cell lines, and in HPV-positive cervical tumours, while mutant p53 is detected in over 50% of all other tumours (Crook et al., 1991; Scheffner et al., 1991).

A further conservation of function between Adenoviruses and HPVs is demonstrated by the finding that, in addition to its ability to block p53 transactivation, Ad E1B-55k can associate with the Ad E4orf6 gene product, resulting in p53 degradation through the proteasome pathway (Steegenga et al., 1998; Querido et al., 2001). Additionally, both high and low-risk HPV E6 proteins can complex efficiently with p53 without targeting it for degradation (Li and Coffino, 1996) and can thus inhibit p53 transactivation by blocking its DNA binding activity, (Lechner and Laimins, 1994; Pim et al., 1994; Thomas et al., 1995). Thus, since not all p53 appears to be degraded in either system, these data suggest the possibility that HPVs and Adenoviruses might actually require the presence of p53 at certain stages of the viral life cycle. In support of this, p53 has also been found in association with the replication complexes of a number of other viruses, including herpes simplex virus (Wilcock and Lane, 1991; Zhong and Hayward, 1997) and cytomegalovirus (Muganda et al., 1994; Fortunato and Spector, 1998), and p53 has also been shown to enhance the replicative fidelity of the HIV-1 reverse transcriptase (Bakhanashvili, 2001) all of which data suggests that p53 might be required for viral DNA replication. In addition, p53 has a 3′-5′ exonuclease function, which could be used in a proofreading capacity (Huang, 1998) since HPV replication uses the cellular DNA Polymerase α, which lacks this (Kuo et al., 1994; Masterson et al., 1998; Conger et al., 1999). It should also be noted that HPV-16 E2, the origin binding protein, has also been shown to interact with p53 (Massimi et al., 1999; Chapter 6), and this might serve to recruit p53 to centres of HPV replication in a manner analogous to the activity of the HSV major DNA binding protein (Wilcock and Lane, 1991; Zhong and Hayward, 1997).

The tumour suppressor activities of p53 are well documented (see Oncogene Reviews issue no 18(53), 1999), and it is clear that the loss of p53 would tend to result in an increased risk of transformation and malignant progression (Donehower et al., 1992; Kemp et al., 1993). Moreover, E6 has been shown to induce numerous chromosomal abnormalities (Reznikoff et al., 1994; White et al., 1994) similar to those seen in cells lacking p53, suggesting that E6-induced inactivation of p53 might contribute to this genomic instability. One of the major functions of p53 is the induction of growth arrest and/or apoptosis (El Deiry et al., 1993; Lowe et al., 1994; Wu and Levine, 1994), conditions that are non-permissive for viral replication, thus the logic of E6-induced p53 inactivation is inescapable. Papillomaviruses have very small genomes and do not express most of the proteins required for their replication; they are therefore dependent upon the replication machinery of the cell. However, their life cycle requires that they replicate only in terminally differentiating epithelial cells, which have exited the cell cycle and ceased DNA replication. The viral E7 protein acts to re-start the cell cycle through release of E2F by its interaction with pRb; through blocking the inhibitory activities of the cyclin-dependent kinase inhibitors p21 and p27; and by activating the cyclin A and cyclin E promoters (see Pim et al., 2001; Münger et al., 2001 for reviews). The unintended consequence of E7's reactivation of DNA replication in a differentiating cell is the activation of p53 to initiate apoptosis, which, in turn, is circumvented by E6's degradation of p53 (see Chapter 10).

The low-risk HPV E6 proteins either do not induce p53 degradation, or do so at much lower efficiency (Storey et al., 1998). The reason for this is probably related to the finding that low-risk HPV types initiate their replicative cycle lower in the stratified epithelium than the high-risk viruses (Doorbar et al., 1997). At this position the keratinocytes are probably still expressing at least some of the proteins required for viral DNA replication. The high-risk viruses do not initiate replication until later in the differentiation of the epithelium, when the cells have already switched off the expression of DNA replication proteins. Thus the high-risk virus E7 proteins would need to exert stronger effects upon the cells to induce apparently inappropriate DNA replication (Sato et al., 1989; Banks et al., 1990), thus activating p53, which E6 is then required to counteract.

For a long time it was widely assumed that the E6-p53 interaction was central to the ability of E6 to bring about cell transformation. However there is growing evidence that many other targets of E6 play a vital role in E6-induced transformation and malignancy (Pim et al., 1994; Pan and Griep, 1995; Nakagawa et al., 1995; Liu et al., 1999). Although blocking the ability of E6 to target p53 for degradation should have a negative impact on the growth of HPV transformed cells (Beer-Romero et al., 1997), it is not clear how relevant this would be to cancer. In many cells derived from cervical tumours, the blocking of E6-induced degradation of p53 is not, alone, sufficient to induce high levels of p53 expression, indicating that the signalling pathways upstream of p53 may be compromised in some cervical tumour-derived cell lines (Mantovani and Banks 1999). High levels of wild type p53 in cervical lesions at different stages of disease progression have also been reported (Cooper et al., 1993; Lie et al., 1999), suggesting that E6 does not degrade all p53 all of the time, again highlighting the possibility that the presence of some p53 might be beneficial to viral replication.

5. Targeting Apoptotic Effectors

Apoptosis, or programmed cell death, is a process that is hard-wired in the cell. It is essential for the correct development and homeostasis of multicellular organisms (Jacobson et al., 1997; Strasser et al., 1994), and acts as a defence against infection and neoplasia (Thompson, 1995). The proteins of the Bcl-2 family are critical death regulators operating immediately upstream of the mitochondria and endoplasmic reticulum (Adams and Cory, 1998; Wei et al., 2001; Scorrano et al., 2003). Bcl-2 family members, of which there are many, are thought to be present in all multicellular eukaryotes (Zamzami et al., 1998; Reed, 1998). They have, variously, pro- or anti-apoptotic effects, and they can modulate these by forming hetero- or homo-dimers (Gross et al., 1999), which then form larger complexes. Monomeric inactive Bak resides in the mitochondrial membrane (Wolter et al., 1997); upon activation Bak forms active homo-dimers and higher order complexes (Hsu et al., 1997; Goping et al., 1998; Wei et al., 2000) which then accelerate the opening of the voltage dependent anion channel (VDAC). This allows the rapid release of cytochrome C into the cytosol, which in turn leads to activation of the caspase cascade (Shimizu et al., 1999). Analysis of the conservation of function between various DNA tumour virus proteins shows, perhaps not surprisingly, that both Adenoviruses and HPVs impinge on this pathway. The Ad E1B-19k protein can act as an anti-apoptotic member of the Bcl2 family (Huang et al., 1997; White 2001, for review), while the HPV E6 proteins target the pro-apoptotic Bak for degradation (Thomas and Banks, 1998; 1999). Mutants of Bak that cannot bind E6 are resistant to degradation and have a considerably extended half-life even in the absence of HPV E6, it therefore seems likely that E6 enhances the normal turnover of Bak in the cell. Interestingly, the equally pro-apoptotic Bax is not a target for degradation. The Bax gene is induced by p53 (Miyashita and Reed, 1995) and is thus probably found at very low levels in E6-expressing cells. In addition, Bak, but not Bax, is expressed at high levels at the site of HPV replication in the upper epithelial layers (Krajewski et al., 1994; 1996). It has been shown that the E6 proteins of cutaneous HPV types can also target Bak for degradation (Jackson et al., 2000), and that Bak levels do not rise in HPV-containing epithelium upon UVB irradiation. Thus Bak is a target of high- and low-risk, mucosal and cutaneous HPV types, indicating that this is a highly conserved function of the HPV E6 proteins.

Since many of the Bcl2 family proteins' functions are controlled by alterations in hetero-dimerisation, it is clear that E6 activity, by slightly altering the balance of key equilibria might have quite fundamental effects upon the fate of the host cell. Recent data shows that the HPV-16 E6 can also inhibit apoptosis by inducing the degradation of the FADD (Fas associated death domain) death effector, thus inhibiting Fas signalling (Filippova et al., 2004), suggesting multiple ways in which E6 may inhibit apoptosis.

At present, it is not clear how relevant the inhibition of Bak activity by E6 might be for the development of malignancy, since Bak is highly expressed in the upper layers of the epithelium, at a stage of differentiation that is not reached during HPV-induced immortalisation and transformation. This suggests that Bak may be a valuable target of E6 during the normal viral life-cycle, but that it may not be relevant during malignant progression. However, further studies in both mucosal and cutaneous HPV-induced cancers are required to investigate this further

6. Transcriptional Regulation Subverted

Transcription from eukaryotic genes requires the assembly of multi-protein complexes, in addition to RNA polymerase II. The basal transcriptional machinery is targeted to individual promoters through interaction between promoter-specific factors and transcriptional co-activators. Prominent among the transcriptional co-activators are the proteins of the CBP/p300 family, and these also possess intrinsic histone acetylase activity, which is required for chromatin re-modelling (Berger, 1999; Goodman and Smolik, 2000). Given their importance in the control and correct functioning of the transcriptional machinery, it is perhaps not surprising that they are targeted by a number of DNA tumour virus proteins (see Figure 2). Both SV40 TAg (Eckner et al., 1996) and Ad E1A protein (Smith and Ziff, 1988; Stein et al., 1990) target CBP/p300 and this interaction has been shown to contribute to their transforming activities. Thus reports that HPV-16 E6 and BPV-1 E6 proteins also interact with CBP/p300, and reduce its transcriptional activity (Patel et al., 1999; Zimmermann et al., 1999; 2000) were extremely interesting, although there was little biological evidence to connect the observations with an effect upon transformation. However, it has been shown more recently that both HPV-16 E6 and HPV-11 E6 can complement a mutant of Ad E1A that is defective in p300 binding in the transformation of primary epithelial cells (Bernat et al., 2002). This interaction also appears to occur in a p53-independent manner. Since the E6 proteins from HPV-11, HPV-16 and BPV-1 all interact with p300/CBP, it seems unlikely that this interaction is uniquely responsible for the abilities of HPV E6 to drive malignant progression. However, the demonstration that E6 can complement a p300 binding-defective Ad E1A in transformation assays suggests that it might, nonetheless, be a contributing factor.

7. E6-associated Protein

The E6 associated protein (E6-AP) is a ubiquitin protein ligase and is the prototype HECT (homologous to E6-AP carboxy terminus) domain protein. It is encoded by the UBE3A gene, mutations in which are responsible for Angelman syndrome, a serious developmental disease (Kishino et al., 1997; Matsuura et al., 1997). E6-AP was originally identified by its association with high-risk E6 proteins in the degradation of p53 (Huibregtse et al., 1991; 1993a). In uninfected cells p53 levels are normally controlled by proteasomal degradation using the MDM2 ubiquitin-protein ligase (Honda et al., 1997), but this pathway appears to be non-functional in HPV positive cells (Hengstermann et al., 2001) and, indeed, the pattern of p53 ubiquitination induced by MDM2 and by E6/E6AP appears to be completely different (Camus et al., 2003). E6-AP does not appear to be involved in the degradation of p53 in the absence of E6: a number of studies using dominant-negative mutants of E6-AP (Talis et al., 1998), or antisense oligonucleotides to block E6-AP function (Beer-Romero et al., 1997) only detected increased p53 levels in HPV positive cells. In contrast, E6-AP has been shown to recognise both Bak and c-Myc, and binding of E6 to E6-AP appears to enhance the normal pathways by which the levels of these proteins are controlled (Thomas and Banks, 1998; Gross-Mesilaty et al., 1998). Interestingly E6's binding to E6-AP appears to enhance the self-ubiquitination of E6-AP itself (Kao et al., 2000), perhaps as a feedback control.

HPV E6 binds to E6-AP through a linear helical motif on E6-AP, with a hydrophobic patch along one side that is essential for E6 binding (Elston et al., 1998; Chen et al., 1998; Be et al., 2001). Recent work using synthetic peptides has shown, however, that the core motif binds only weakly to E6, and that the binding is probably stabilised by contacts made in other parts of the protein (Sterlinko Grm et al., 2004). The same study further showed that these peptides blocked the E6-induced degradation of p53, but had no effect upon the binding of E6 to p53, suggesting that E6-AP binding by E6 is not required for interaction with p53. This was indeed confirmed using E6 and p53 expressed in wheatgerm extract, in which no E6-AP is present (Sterlinko Grm et al., 2004). In addition, these peptides also blocked the E6-induced degradation of Dlg (see below), which does not require E6-AP (Pim et al., 2000). This is particularly interesting as the ligase involved in Dlg degradation is currently unknown, but presumably its binding may involve the same region of E6.

The helical binding motif has been found on a number of other E6 binding partners, including E6BP/ERC55, MCM7 and paxillin (Chen et al., 1995; Kuhne and Banks

Consensus PDZ domain-binding sequence . . . . X(T/S)XV

High-risk mucosal HPVs	HPV-16 E6	CC.....RSSRTRRETQL
	HPV-18 E6	CCNRARQERLQRRRETQV
	HPV-45 E6	CCDQARQERLRRRRETQV
	HPV-31 E6	CW......R.RPRTETQV
	HPV-33 E6	CW......R.SRRRETAL
	HPV-35 E6	CW......K.PTRRETEV
Low-risk mucosal HPVs	HPV-6 E6	CWTTCMEDMLP.......
	HPV-11E6	CWTTCMEDLLP.......
High-risk cutaneous HPV	HPV-8 E6	CKHLYHDW..........
Low-risk cutaneous HPV	HPV-2a E6	CGSSCTATDPASRTLH..

Figure 3. Alignment of the carboxy terminal amino acids of HPV E6 proteins. Note the complete absence of PDZ-binding motif in cutaneous and in low-risk mucosal E6s. In addition, note the differences in PDZ ligand sequence between HPV types 18, 45, 31 and 35 which have a Valine residue at the carboxy terminus, whereas HPV types 16 and 33 have a Leucine residue.

1998; Tong and Howley, 1997), indicating that it is a common method by which E6 can interact with a number of cellular partners. None of these proteins appears to play a part in the oncogenic activity of high risk HPV. However, it is possible that targeting MCM7 (which is part of the replication licensing complex) to increase the rate of viral DNA replication might contribute to the chromosomal instability seen in E6-expressing cells (White et al. 1994; Filatov et al., 1998) which, in the context of high-risk E6-induced p53 degradation, could result in a secondary oncogenic effect.

8. Targets of E6 with PDZ Domains

As can seen from Figures 2 and 3, the extreme carboxy terminal four amino acids of E6 (XTQV/L) are absolutely specific to high-risk, mucosal HPV E6s. These four amino acids form a consensus PDZ-binding domain, and of all the protein interactions of E6 this is the only one that is strictly high-risk specific. All the other activities of E6 can be detected, to some extent, in both high- and low-risk HPV E6 proteins.

PDZ domains are stretches of 80-90 amino acids (Ponting and Philips, 1995; Fanning and Anderson, 1999) that are bound with high affinity by the sequence T/SXV (Songyang et al., 1997), which is usually found at the extreme carboxy terminal of their interacting proteins. MAGUK proteins (membrane-associated guanylate kinase homologues) are large polypeptides containing multiple protein-protein interaction domains, including PDZ domains, and they are thought to act as molecular scaffolds to mediate the formation of multiple-protein complexes. MAGUKs are detected on the cell membrane, often at regions of cell-cell contact, including subapical tight junctions (Ide et al., 1999; Mino et al., 2000; Kawabe et al., 2001), lateral adherens junctions (Nishimura et al., 2000; Firestein and Rongo, 2001) and the apical junctions of *C. elegans* (McMahon et al., 2001; Köppen et al., 2001). Indeed, cells expressing E6 were shown to form weaker cell contacts and grow in a more disorganised fashion, whereas cells expressing E6 mutants defective for PDZ protein interactions retained normal aspects of cell polarity (Nakagawa and Huibregtse, 2000). Because of their location, it is thought that they may be involved in both inter-cell and intra-cellular signalling (see Figure 4), particularly since some also have nuclear localisation signals (Dobrosotskaya et al., 1997). As well as high-risk E6, PDZ-binding motifs have been found in a number of other viral oncoproteins, including the HTLV-1 Tax protein and the Ad9 E4-ORF1 gene product (Lee et al., 1997; Kiyono et al., 1997).

9. The MAGI Family

The MAGI family of proteins is one of the more recently identified group of PDZ domain-containing targets for Ad9 E4ORF1 and HPV E6 (Glaunsinger et al., 2000; Pim et al., 2000; Thomas et al., 2001). These are MAGUKs with Inverted domain structure, having their guanylate kinase homology domains at the amino terminal end of the polypeptide. In addition MAGI-1, MAGI-2 and MAGI-3 each have WW domains and at least 5 PDZ domains (Dobrosotskaya et al., 1997; Wu et al., 2000a; Wu et al., 2000b) giving them the potential to interact with a large number of other proteins, including β-catenin (Dobrosotskaya and James, 2000), mNET-1 (Dobrosotskaya 2001) and PTEN (Wu et al., 2000a; Wu et al., 2000b). As with other PDZ-containing targets, MAGI-1 is aberrantly sequestered in the cytoplasm by Ad9 E4-ORF1, and is targeted by high-risk HPV E6 proteins for ubiquitin-mediated degradation (Glaunsinger et al., 2000). MAGI-2 and MAGI-3 are also targets for E6 induced degradation (Thomas et al., 2002) and it is possible that they too are targets for Ad9 E4ORF1. The functions of the MAGI proteins are not yet entirely clear, although MAGI-2, and MAGI-3 have been reported to interact with the PTEN tumour suppressor (Wu et al., 2000a; Wu et al., 2000b), and MAGI-1 can suppress immortalisation in an oncogene cooperation assay in primary rodent cells (Massimi et al., 2004).

10. MUPP1

MUPP1 is another PDZ containing target of Ad9 E4ORF1 and HPV E6 (Lee et al., 2000). MUPP1 is a multiple

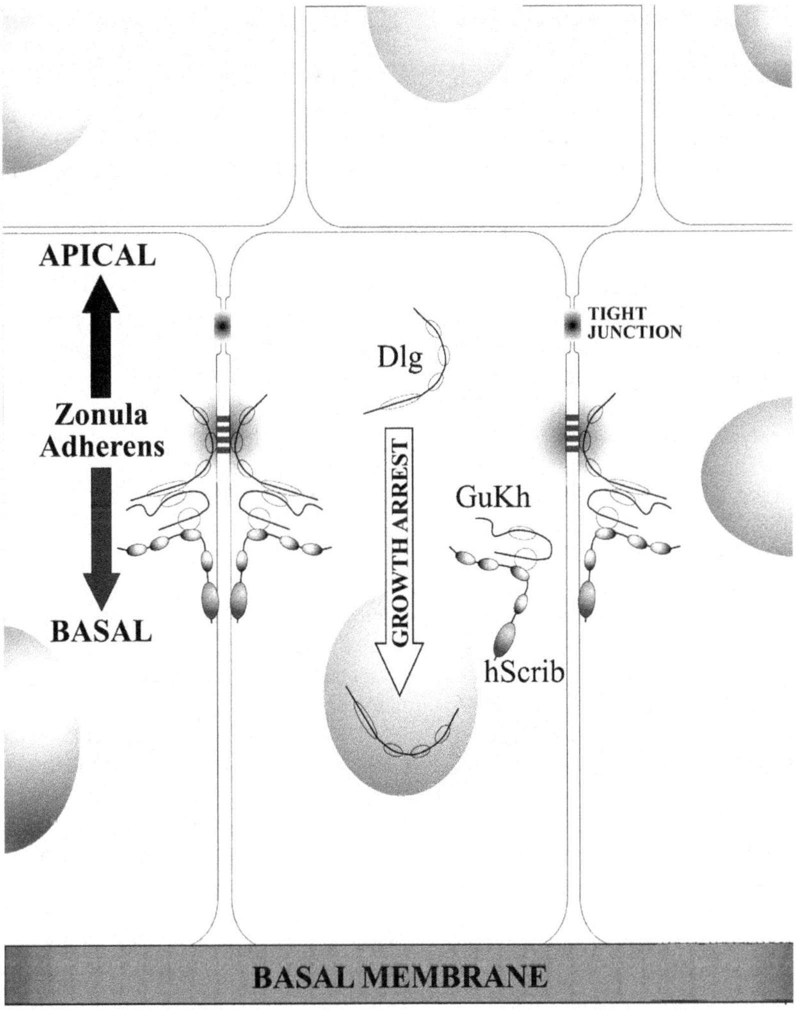

Figure 4. Dlg and hScrib play important roles in control of apico-basal polarity and in growth arrest signalling within the cell. Through their mutual interaction with GuKholder (GuKh), both localise to the same region within the cell, just basal to the zonula adherens, where they cooperate to regulate apico-basal polarity. A subset of Dlg can be found in the nucleus and may function in control of cell proliferation.

PDZ-containing protein, having 13 PDZ domains (Ullmer et al., 1998), which is thought to be involved in the selective targeting and assembly of signalling complexes (Becamel et al., 2000). In polarised epithelial cells it is localised exclusively at tight junctions, interacting with the Claudin-1 and JAM-1 (junctional adhesion molecule) proteins at adjacent PDZ domains – PDZ10 and PDZ9, respectively- and is also thought to be involved in cell polarity signalling (Hamazaki et al., 2001).

11. hDlg and hScrib

Perhaps the most important of E6's known PDZ-containing targets are Dlg and hScrib, both of which are classified as potential tumour suppressor proteins, based on their growth suppressor activities in Drosophila, as well as in mammalian cells (Goode and Perrimon, 1997; Bilder, 2004; Massimi et al., 2004; Thomas et al., unpublished observations). Dlg is classified as a tumour suppressor since its deletion in Drosophila results in aberrant cytoskeletal organisation and in disruption of polarity in columnar epithelium (Goode and Perrimon, 1997; Bilder et al., 2000). Its role in maintaining cell polarity has been separated, by mutational analysis, from its role in controlling cellular proliferation (Woods et al., 1996). In complex with the adenomatous polyposis coli (APC) protein, human Dlg represses the G1 to S transition (Ishidate et al., 2000), and its overexpression results in downregulation of β-catenin, and hence cell growth, via an APC-independent pathway (Hanada et al., 2000). Dlg also colocalises with APC at the tips of cellular protrusions in epithelial cells in culture (Iizuka-Kogo et al., 2004). HTLV-1 Tax protein has also been shown to bind Dlg and to counteract these activities (Suzuki et al., 1999). Unlike the Tax and E4-ORF1 proteins, E6 is expressed at very low levels and cannot sequester Dlg, however it can target Dlg for ubiquitin-mediated degradation (Gardiol et al., 1999). Dlg possesses three PDZ domains and, interestingly, the ability of E6 to target Dlg for degradation requires PDZ domain 2, which is the PDZ domain recognised by APC (Gardiol et al., 2002). This raises the possibility that E6, by degrading Dlg, might interfere with the normal functioning of the APC pathway which, together with p53, is one of the commonest pathways to be perturbed during the development of colon cancer (Morin et al., 1996; Kinzler and Vogelstein, 1996). Work on transgenic mice shows that expression of HPV-16 E6 in the ocular lens results in multilayering and disorganisation of the epithelium and an increased number of undifferentiated, nucleate cells; a similar effect to that seen upon the expression of mutant Dlg in the same system (Nguyen et al., 2003b). Interestingly, E6 would appear to preferentially target nuclear phosphorylated forms of Dlg (Mantovani et al., 2001; Massimi et al., 2004) suggesting that this signalling subset of Dlg is the true target of E6, while the disruption of the polarity control complex is a, fortunately rare, side-effect (see Figure 4). Indeed it would appear that the role of the PDZ ligand of E6 is directed towards a function regulating cell proliferation since both transgenic mice and keratinocytes expressing E6 mutated within the PDZ ligand exhibit greatly reduced proliferation compared with those expressing wild type E6 (Nguyen et al., 2003a; Lee and Laimins 2004).

Dlg plays a part in a highly complex series of protein interactions that regulate the development and control of epithelial cell polarity (Bilder et al., 2000; 2003; Bilder 2004; Tanentzapf and Tepass; 2003). Proteins of this group are highly conserved in higher eukaryotes and the mammalian homologues of Dlg, Scribble and Lgl have been shown to complement their absence in mutant Drosophila (Thomas et al., 1997; Dow et al., 2003; Grifoni et al., 2004).

The high-risk HPV E6s also induce the degradation of hScrib (Nakagawa and Huibregtse, 2000), the human homologue of Drosophila Scribble. hScrib is a member of the LAP (leucine-rich and PDZ domain) family of proteins, having four PDZ domains and a Leucine-rich domain that is required for its basolateral localisation in epithelial cells (Kim, 1997; Bilder and Perrimon, 2000; Bryant and Huwe, 2000; Legouis et al., 2003). Studies in Drosophila have shown that Scribble and Dlg can at least partly complement each other (Bilder et al., 2000) and it has also been shown that Dlg and hScrib form a complex that is required for the maintenance of epithelial polarity (reviewed in Humbert et al., 2003 and see Figure 4) and expression of both proteins is altered in HPV-induced tumours (Watson et al., 2002; 2003; Cavatorta et al., 2003; Nakagawa et al., 2004). The correct functioning of the complex absolutely requires both Dlg and hScrib and it seemed somewhat redundant that both should be targeted by the normally very efficient HPV E6 proteins (see Figure 4). However, we have recently shown that hDlg is predominantly targeted by HPV-18 E6, while HPV-16 E6 preferentially targets hScrib (Thomas et al., 2005). This is interesting for two reasons: it is the first description of a PDZ-containing target that is preferentially targeted by HPV-16 E6. Secondly, it implies that the target cells of the two virus types may not be the same or that the outcome of transformation might be different. Data supporting this latter suggestion comes from studies showing that cervical cancers induced by different polymorphic variants in HPV-16 and HPV-18 exhibit different proportions of adenocarcinoma and squamous cell carcinomas (De Boer et al., 2004). This may in turn be related to differences in the abilities of diverse E6 proteins to activate MAP kinase and Notch signalling, both of which have been reported to promote papillomavirus mediated neoplastic progression (Talora et al., 2002; Chakrabarti et al., 2004;Veeraraghavalu et al., 2004; Six et al., 2004).

12. What Ligase?

Comparison of the E6 induced degradation of target proteins has shown that the MAGI proteins are even more susceptible to degradation than p53, and certainly more so than Dlg, hScrib or MUPP1 (Lee et al., 2000; Pim et al., 200; Thomas et al., 2001; 2005). Apart from obvious differences in kinetics of degradation, other evidence suggests that E6 does not degrade these proteins through identical pathways, invoking the possible involvement of other, as yet unidentified, ubiquitin ligases. We had previously shown (Pim et al., 2000) that HPV-18 E6 can induce the degradation of Dlg in wheatgerm extract, which is known to lack E6AP (Huibregtse et al., 1993a). Furthermore, construction of chimaeric E6 molecules, in which the carboxy terminal region of HPV-18 E6 was fused to the amino terminus of HPV-11 E6 conferred PDZ protein-binding activity on the low-risk E6 protein, and allowed it to target Dlg for degradation. In contrast, in the same assay, MAGI-1 was completely resistant to the chimaeric E6 protein (Pim et al., 2000). However, since HPV-11 E6 only binds weakly, if at all to E6AP (Huibregtse et al., 1993a), this supports the suggestion that the ability of the chimaeric protein to degrade Dlg is indeed E6AP independent. A recent study of *in vitro* degradation assays performed in reticulocyte lysate from which the E6AP had been immunodepleted showed interesting differences in the target proteins' susceptibilities to degradation in the presence of E6 protein (Sterlinko Grm et al., 2004; Sterlinko Grm and Banks, 2004). Thus, immunodepletion of E6AP blocked p53 degradation, but had no effect upon the degradation of Dlg, MAGI-1, MAGI-2 or MAGI-3 induced by either HPV-16 or HPV-18 E6s. Taken together with previous studies that have shown marked differences in the kinetics of degradation of Dlg and the MAGI proteins (Thomas et al., 2001; 2002), these data suggest that at least one, and possibly more, ubiquitin-protein ligases are

PDZ CLASS	CONSENSUS TARGET SEQUENCE	EXAMPLES
I	[S/T] - X - V - COOH	Dlg, Scribble, PSD-95, MUPP1, MAGI 1, 2, 3
II	[F/Y] - X - [F/V/A] - COOH	hCASK
III	D - X - V - COOH	nNOS
IV	Ψ- [D/E] - COOH	hINADL
V	[L/I] - X - Φ- [I/L/V] - COOH	omi/HtrA2

Ψ -non-charged   Φ – hydrophobic/non-polar   X-any

Table 1. Differences between the consensus target sequences for different PDZ domain classes.. Adapted from Junqueira et al., 2003.

employed by the HPV E6 proteins, in addition to E6AP. This proposition is supported by work in transgenic mice expressing HPV-16 E6 in the ocular lens. The number of BrdU-positive cells seen in the lens of mice expressing mutant E6, lacking the PDZ-binding motif, was indistinguishable from non-transgenic mice. In contrast, mice expressing E6 defective in its α-helical binding domain (and thus defective in binding E6AP) showed an increased level of BrdU-positive cells, comparable with that found in mice expressing the wild type virus (Nguyen et al., 2003b). Thus a PDZ-containing target of E6 is required to promote cell cycle progression, and its degradation does not involve E6AP.

13. PDZ Ligand Specificity

It has been shown that HPV-18 E6 binds to Dlg and to MAGI-1 more efficiently than HPV-16 E6, and induces their degradation with concomitantly greater efficiency (Pim et al., 2000; Thomas et al., 2001). This is caused by a single amino acid difference in the PDZ-binding motif at the carboxy terminus of E6, where HPV-18 E6 has a perfect motif (ETQV), and that of HPV-16 E6 was thought to be sub-optimal (ETQL) (see Figure 3 for a comparison of high-risk E6 PDZ binding domains). However, we have recently found that the HPV-16 E6 binds to hScrib more efficiently than does HPV-18 E6 and, equally, is more effective in inducing the degradation of hScrib (Thomas et al., 2005); this difference is directly related to the single amino acid difference in the PDZ-binding domain. This finding is very interesting in that it provides an explanation for the apparent redundancy of the HPV E6 proteins targeting mutually dependent components of the same complex. It also defines a new class of PDZ-containing proteins – those that are bound more efficiently by ETQL, and are thus more susceptible to HPV-16 E6. It will obviously be of interest to determine whether this has broader implications for the natural cellular targets of these motifs, since it would also imply that the number of PDZ ligands is greater than was originally thought.

Although Dlg and hScrib are both required for the fully effective functioning of the polarity-determining complex that they form (Bilder et al., 2003; Humbert et al., 2003), there appear to be differences in their recruitment. In synaptic development, which is very similar to epithelia in many respects, the Dlg can localise correctly to the region just basal of the zonula adherens where the Dgl/Scribble/Lgl complex forms (Tanentzapf and Tepass, 2003; Johnson and Wodarz, 2003) (see Figure 4). However, Scribble, although it localises via its Leucine-rich domain to the basolateral membrane, is not able to colocalise with Dlg in the absence of a protein called GuKholder, which binds to the PDZ2 of Scribble and the GuK domain of Dlg and brings them together in the complex (Mathew et al., 2002). Thus in the absence of Scribble (equivalent to the presence of HPV-16 E6), Dlg can form a partial complex; whereas in the absence of Dlg (equivalent to the presence of HPV-18 E6), no complex can form. If the complex forms similarly in epithelial cells it throws new light on studies indicating that HPV-18-containing cervical tumours are more prone than HPV-16-containing tumours to recurrence and to metastasis (reviewed in Franco 1992). This is the first finding that correlates different clinical phenotypes with the defined biochemical activities of proteins from different, but closely related, HPV types, and indicates that the E6-induced control of PDZ protein levels within the cell could be fundamental in the progression to malignancy.

An important feature of the E6-PDZ domain interaction is that it is highly defined. Many of E6's other protein interactions involve large regions of the E6 protein (see Figure 2) and make molecular modelling and the rational design of inhibitors extremely difficult. In contrast the E6-PDZ interaction is mediated by a four amino acid motif, which can be disabled for binding either by point mutations (Gardiol et al., 1999; Glaunsinger et al., 2000), or by simple phosphorylation of the threonine residue within the motif by protein kinase A (Kuhne et al., 2000). There are at least five classes of PDZ binding motif, and these are shown in Table 1. Of these, the known targets of the high-risk mucosal HPV types belong only to Class 1 and we appear to have begun to define an additional level of specificity within that class. Given these observations, and the fact that the structures of a number of PDZ domains have been solved (Doyle et al., 1996; Morais Cabral, 1996; Kozlov et al., 2000), the design of molecules that could interfere with these interactions seems highly feasible.

14. A Question of Specificity

Since the PDZ domain-containing proteins impinge upon such a large number of important pathways, there is a risk that blocking the interactions between PDZ domains and PDZ ligands might have extremely toxic effects. However, it is clear from the discussion above that there are additional controls on the specificity of the E6/PDZ interaction. Detailed analysis of the PDZ domains from a number of E6 target proteins has shown that only one domain per protein is recognised by HPV E6 (Lee et al., 2000; Gardiol et al., 2002; Thomas et al., 2001; 2002; 2005). This indicates the importance of peripheral amino acid sequences, both in the PDZ domain and its ligand. Specific binding to PDZ domains has also been shown with β-catenin, which binds to PDZ5 of MAGI-1 (Dobrosotskaya and James, 2000), with the tumour suppressor PTEN, which binds to PDZ2 of MAGI-2 and MAGI-3 (Wu et al., 2000a; 2000b), and with mNET1 which binds to PDZ1 of MAGI-1 (Dobrosotskaya, 2001). The GuKholder also binds specifically only to the PDZ2 domain of the Scribble protein (Mathew et al., 2002). This indicates that very specific blocking of E6's PDZ binding might be possible, without affecting other PDZ domain interactions. Indeed, it has been shown that exogenously expressed MAGI-3 PDZ1 domain can block the *in vitro* and *in vivo* E6-induced degradation of Dlg, MAGI-2 and MAGI-3, thereby confirming proof-of-principle for this approach (Thomas et al., 2002).

It is also becoming clearer that there are other factors that can determine the specificity of E6 interactions. We discussed earlier the possibility that a certain amount of p53 might be required during the replication of viral DNA; obviously if that is the case there must be a mechanism for preventing certain E6 interactions. The E6* proteins provide a potential means of doing this; they are abundant, alternatively spliced transcripts of the high-risk E6 gene (Smotkin et al., 1989). They can bind to E6-AP or hetero-dimerise with E6, blocking their interaction (Pim et al., 1997; Pim and Banks 1999) and thus preventing their inducing, or enhancing, the degradation of E6/E6-AP target proteins, of which p53 is the only one known to be resistant to E6-AP in the absence of E6.

Another important method of altering protein activity is by post-translational modifications such as phosphorylation, acetylation, glycosylation, sumoylation and ubiquitination. The PKA phosphorylation of E6 within its PDZ-binding domain has a profoundly inhibitory effect upon its interaction with Dlg (Kuhne et al., 2000) and probably with the other PDZ-containing targets. This implies that when phosphorylated by PKA, E6 will not recognise PDZ domain-containing substrates, but will instead be available to carry out other activities, including p53 degradation. Since the E6 protein is expressed at very low levels it is clear that it cannot interact with all its targets simultaneously. Therefore this type of post-translational modification again produces an additional layer of specificity to the regulation of E6 function.

15. Conclusions

Over the last few decades, some of the most powerful tools for dissecting certain pathways have been DNA tumour viruses. Serendipitous findings from the study of viral transformation have thrown new light on the pathways required for cellular growth control. As their usefulness in this field becomes more fully acknowledged, we can expect that the fields of molecular virology and molecular cell biology will become more integrated in the elucidation of interconnections between apparently unconnected aspects of cellular homeostasis. Work with DNA tumour virus oncoproteins over the years has provided a wealth of information on the central pathways regulating cell division. The data discussed here suggest that targeting HPV E6 might allow this information to be used in the treatment of a major human cancer.

Acknowledgements

The authors gratefully acknowledge research support provided by the Associazione Italiana per la Ricerca sul Cancro.

References

Adams, J.M. and Cory, S. (1998). The Bcl-2 protein family: arbiters of cell survival. Science *281*, 1322-1326.

Alvarez-Salas, L.M., Cullinan, A.E., Siwkowski, A., Hampel, A. and DiPaolo. J.A. (1998). Inhibition of HPV-16 E6/E7 immortalisation of normal keratinocytes by hairpin ribozymes. Proc. Natl. Acad. Sci. USA *95*, 1189-1194.

Androphy, E.J., Hubbert, N.L., Schiller, J.T. and Lowy, D.R. (1987). Identification of the HPV-16 E6 protein from transformed mouse cells and human cervical carcinoma cell lines. EMBO J. *6*, 989-992.

Arbeit, J.M., Münger, K., Howley, P. and Hanahan, D. (1994). Progressive squamous epithelial neoplasia in K-14 human papillomavirus type -16 transgenic mice. J. Virol. *68*, 4358-4368.

Bakhanashvili, M. (2001). p53 enhances the replicative fidelity of DNA synthesis by human immunodeficiency virus type 1 reverse transcriptase. Oncogene *20*, 7635-7644.

Band, V., DeCaprio, J.A., Delmolino, L., Kulesa, V. and Sager, R. (1991). Loss of p53 protein in human papillomavirus type 16-immortalised human mammary epithelial cells. J. Virol. *65*, 6671-6676.

Band, V., Dalal, S., Delmolino, L. and Androphy, E. (1993). Enhanced degradation of p53 protein in HPV-6 and BPV-1 E6-immortalised human mammary epithelial cells. EMBO J. *12*, 1847-1852.

Banks, L., Spence, P., Androphy, E., Hubbert, N., Matlashewski, G., Murray, A. and Crawford, L. (1987). Identification of human papillomavirus type 18 E6 polypeptide in cells derived from human cervical carcinomas. J. Gen. Virol. *68*, 1351-1359.

Banks, L., Edmonds, C. and Vousden, K.H. (1990). Ability of the HPV-16 E7 protein to bind RB and induce DNA synthesis is not sufficient for efficient transforming activity in NIH3T3 cells. Oncogene *5*, 1383-1389.

Barbosa, M., Lowy, D. and Schiller, J. (1989). Papillomavirus polypeptides E6 and E7 are zinc binding proteins. J. Virol. *63*, 1404-1407

Be, X., Hong, Y., Wei, J., Androphy, E.J., Chen, J.J. and Baleja, J.D. (2001). Solution structure determination and mutational analysis of the papillomavirus E6 interacting peptide of E6AP. Biochemistry *40*, 1293-1299.

Becamel, C., Figge, A., Poliak, S., Dumuis, A., Peles, E. Bockaert, J., Lubbert, H. and Ullmer, C. (2000). Interaction of serotonin 5-hydroxytryptamine type 2C receptors with PDZ10 of the multi-PDZ domain protein, MUPP1. J. Biol. Chem. *276*, 12974-12982.

Bedell, M.A., Jones, K.H., Grossman, S.R. and Laimins, L.A. (1989). Identification of human papillomavirus type 18 transforming genes in immortalised and primary cells. J.Virol. *63*, 1247-1255.

Beer-Romero, P., Glass, S. and Rolfe, M. (1997). Antisense targeting of E6AP elevates p53 in HPV-infected cells but not in normal cells. Oncogene *14*, 595-602.

Berger, S.L. (1999). Gene activation by histone and factor acetyltransferases. Curr. Opin. Cell Biol. *11*, 336-341.

Bernat, A., Massimi, P. and Banks, L. (2002). Complementation of a p300/CBP defective binding mutant of Adenovirus E1a by the HPV E6 proteins. J. Gen. Virol. *83*, 829-33.

Bilder, D. (2004). Epithelial polarity and proliferation control: links from the Drosophila neoplastic tumor suppressors. Genes Dev. *18*, 1909-25.

Bilder, D. and Perrimon, N. (2000). Localisation of apical epithelial determinants by the basolateral PDZ protein Scribble. Nature *403*, 676-680.

Bilder, D., Li, M. and Perrimon, N. (2000). Cooperative regulation of cell polarity and growth by *Drosophila* tumor suppressors. Science *289*, 113-116.

Bilder, D., Schober, M. and Perrimon, N. (2003). Integrated activity of PDZ protein complexes regulates epithelial polarity. Nat Cell Biol. *5*, 53-58.

Bryant PJ and Huwe A. (2000). LAP proteins: what's up with epithelia Nat Cell Biol. 2, E141-3.

Butz, K., Denk, C., Ullmann, A., Scheffner, M. and Hoppe-Seyler, F. (2000). Induction of apoptosis in human papillomavirus-positive cancer cells by peptide aptamers targeting the viral E6 oncoprotein. Proc. Natl. Acad. Sci. USA *97*, 6693-6697.

Butz K, Ristriani T, Hengstermann A, Denk C, Scheffner M, Hoppe-Seyler F. (2003). siRNA targeting of the viral E6 oncogene efficiently kills human papillomavirus-positive cancer cells. Oncogene *22*, 5938-45.

Camus, S., Higgins, M., Lane, D. and Lain, S. (2003). Differences in the ubiquitination of p53 by Mdm2 and the HPV protein E6. FEBS Lett. *536*, 220-224.

Cavatorta AL, Fumero G, Chouhy D, Aguirre R, Nocito AL, Giri AA, Banks L, Gardiol D. (2004). Differential expression of the human homologue of drosophila discs large oncosuppressor in histologic samples from human papillomavirus-associated lesions as a marker for progression to malignancy. Int J Cancer *111*, 373-80.

Chakrabarti, O., Veeraraghavalu, K., Tergaonkar, V., Liu, Y., Androphy, E., Stanley, M. and Krishna, S. (2004). Human papillomavirus type 16 E6 amino acid 83 variants enhance E6-mediated MAPK signalling and differentially regulate tumorigenesis by Notch signalling and oncogenic ras. J. Virol. *78*, 5934-5945.

Chen, J.J., Reid, C., Band, V. and Androphy, E. (1995). Interaction of papillomavirus E6 oncoproteins with a putative calcium binding protein. Science *269*, 529-531.

Chen, J.J., Hong, Y., Rustumzadeh, E., Baleja, J.D. and Androphy, E.J. (1998). Identification of an alpha helical motif sufficient for association with papillomavirus E6. J. Biol. Chem. *273*, 13537-13544.

Cole, S.T. and Danos, O. (1987). Nucleotide sequence and comparative analysis of the human papillomavirus type 18 genome. Phylogeny of papillomaviruses and repeated structure of the E6 and E7 gene products. J. Mol. Biol. *193*, 599-608.

Conger, K.L., Lin, J.S., Kuo, S.R., Chow, L.T. and Wang, T.S. (1999). Human papillomavirus DNA replication. Interactions between the viral E1 protein and two subunits of human DNA polymerase alpha/primase. J. Biol. Chem. *274*, 2696-2705.

Cooper, K., Herrington, C.S., Evans, M.F., Gatter, K.C. and McGee, J.O. (1993). p53 antigen in cervical condylomata, intraepithelial neoplasia, and carcinoma: relationship to HPV infection and integration. J. Pathol. *171*, 27-34.

Crook, T., Wrede, D. and Vousden K. (1991). p53 point mutation in HPV negative human cervical carcinoma cell lines. Oncogene *6*, 873-875.

De Boer MA, Peters LA, Aziz MF, Siregar B, Cornain S, Vrede MA, Jordanova ES, Fleuren GJ. 2005. Human papillomavirus type 18 variants: Histopathology and E6/E7 polymorphisms in three countries. Int J Cancer. *114*, 422-425.

DiPaolo, J., Woodworth, C., Popescu, M.C., Notario, V. and Doniger, J. (1989). Induction of human cervical squamous cell carcinoma by sequential transfection with human papillomavirus 16 DNA and viral Harvey ras. Oncogene *4*, 395-399.

Dobrosotskaya, I.Y., (2001). Identification of mNET1 as a candidate ligand for the first PDZ domain of MAGI-1. Biochem. Biophys. Res. Commun. *283*, 969-975.

Dobrosotskaya, I.Y., and James, G.L. (2000). MAGI-1 interacts with beta-catenin and is associated with cell-cell adhesion structures. Biochem. Biophys. Res. Commun. *270*, 903-909.

Dobrosotskaya, I., Guy, R.K. and James, G.L. (1997). MAGI-1, a membrane-associated guanylate kinase with a unique arrangement of protein-protein interaction domains. J. Biol. Chem. *272*, 31589-31597.

Donehower, L.A., Harvey, M., Slagle, B.L., McArthur, M.J., Montgomery, C.A. Jr., Butel, J.S. and Bradley, A. (1992). Mice deficient for p53 are developmentally normal but susceptible to spontaneous tumours. Nature *356*, 215-221.

Doorbar, J., Foo, C., Coleman, N., Medcalf, L., Hartley, O., Prospero, T., Napthine, S., Sterling, J., Winter, G. and Griffin, H. (1997). Characterisation of events during the late stages of HPV16 infection *in vivo* using high affinity synthetic Fabs to E4. Virology *238*, 40-52.

Dow LE, Brumby AM, Muratore R, Coombe ML, Sedelies KA, Trapani JA, Russell SM, Richardson HE, Humbert PO. (2003). hScrib is a functional homologue of the Drosophila tumour suppressor Scribble. Oncogene *22*, 9225-9230.

Doyle, D., Lee, A., Lewis, J., Kim, E., Sheng, M. and MacKinnon, R. (1996). Crystal structures of a complexed and peptide-free membrane protein-binding domain: molecular basis of peptide recognition by PDZ. Cell *85*, 1067-1076.

Dürst, M., Dzarlieva-Petrusevka, R., Boukamp, P., Fusenig, N. and Gissmann, L. (1987). Molecular and cytogenetic analysis of immortalised human primary keratinocytes obtained after transfection with human papillomavirus type 16 DNA. Oncogene *1*, 251-256.

Dürst, M., Gallahan, D., Gilbert, J. and Rhim, J.S. (1989). Glucocorticoid enhanced neoplastic transformation of human keratinocytes by human papillomavirus type 16 and an activated ras oncogene. Virology *173*, 767-771.

Eckner, R., Ludlow, J.W., Lill, N.L. Oldread, E. Arany, Z, Modjtahedi, N., DeCaprio, J.A., Livingston, D.M. and Morgan J.A. (1996). Association of p300 and CBP with simian virus 40 large T antigen. Mol. Cell. Biol. *16*, 3454-3464.

El Deiry, W., Tokino, T., Velculescu, V., Levy, D., Parsons, R., Trent, J., Lin, D., Mercer, W., Kinzler, K and Vogelstein, B. (1993). WAF-1, a potential mediator of p53 tumor suppression. Cell *75*, 817-825.

Elmore, L.W., Hancock, A.R., Chang, S.F., Wang, X.W., Chang, S. Callahan, C.P., Geller, D.A., Will, H. and Harris, C.C. (1997). Hepatitis B virus X protein and p53 tumor suppressor interactions in the modulation of apoptosis. Proc. Natl. Acad. Sci. USA. *94*, 14707-14712.

Elston, R., Napthine, S. and Doorbar, J. (1998). The identification of a conserved binding motif within human papillomavirus binding peptides. J. Gen. Virol. *79*, 371-374.

Fanning, A.S., and Anderson, J.M. (1999). PDZ domains: fundamental building blocks in the organisation of protein complexes at the plasma membrane. J. Clin. Invest. *103*, 767-772.

Filatov, L., Golubovskaya, V., Hurt, J., Byrd, L., Phillips, J. and Kaufman, W. (1998). Chromosomal instability, is correlated with telomerase erosion and inactivation of G2 checkpoint function in human fibroblasts expressing human papillomavirus type 16 E6 oncoprotein. Oncogene *16*, 1825-1838.

Filippova M, Parkhurst L, Duerksen-Hughes PJ. (2004). The human papillomavirus 16 E6 protein binds to Fas-associated death domain and protects cells from Fas-triggered apoptosis. J Biol Chem. *279*, 25729-44.

Firestein, B.L. and Rongo, C. (2001). DLG-1 is a MAGUK similar to SAP97 and is required for adherens junction formation. Mol. Biol. Cell *12*, 3465-3475.

Fortunato, E.A. and Spector, D.H. (1998). p53 and RPA are sequestered in viral replication centers in the nuclei of cells infected with human cytomegalovirus. J.Virol. *72*, 2033-2039.

Franco, E.L. (1992). Prognostic value of human papillomavirus in the survival of cervical cancer patients: an overview of the evidence. Cancer Epidemiol. Biomarkers Prev. *1*, 499-504.

Gardiol, D., Kühne, C., Glaunsinger, B., Lee, S.S., Javier, R. and Banks, L. (1999). Oncogenic human papillomavirus E6 proteins target the discs large tumour suppressor for proteasome-mediated degradation. Oncogene *18*, 5487-5496.

Gardiol, D., Galizzi, S. and Banks, L. (2002). Mutational analysis of the Discs Large tumour suppressor identifies domains responsible for HPV-18 E6 mediated degradation. J. Gen. Virol. *83*., 283-289.

Garner-Hamrick PA, Fostel JM, Chien WM, Banerjee NS, Chow LT, Broker TR, Fisher C. (2004). Global effects of human papillomavirus type 18 E6/E7 in an organotypic keratinocyte culture system. J Virol. *78*, 9041-9050.

Glaunsinger, B., Lee, S. S., Thomas, M., Banks, L. and Javier, R. (2000). Interactions of the PDZ-protein MAGI-1 with adenovirus E4-ORF1 and high-risk papillomavirus E6 oncoproteins. Oncogene *19*, 1093-1098.

Goode, S. and Perrimon, N. (1997). Inhibition of patterned cell shape change and cell invasion by discs large during Drosophila oogenesis. Genes Dev. *11*, 2532-2544.

Goodman, R.H. and Smolik, S. (2000). CBP/p300 in cell growth, transformation and development. Genes Dev. *14*, 1553-1577.

Goping, I., Gross, A., Lavoie, J., Nguyen, M., Jemmerson, R., Roth, K., Korsmeyer, S. and Shore, G. (1998). Regulated targeting of Bax to mitochondria. J. Cell Biol. *143*, 207-215.

Grifoni, D., Garoia, F., Schimanski, C.C.,Schmitz, G., Laurenti, E., Galle, P.R., Pession, A., Cavicchi, S and Strand, D. (2004). The human protein Hugl-1 substitutes for Drosophila lethal giant larvae tumour suppressor function *in vivo*. Oncogene *23*, 8688-8694.

Gross, A., McDonnell, J.M. and Korsmeyer, S.J. (1999). BCL-2 family members and the mitochondria in apoptosis. Genes Dev. *13*, 1899-1911.

Grossman, S.R. and Laimins, L.A. (1989). E6 protein of human papillomavirus type 18 binds zinc. J.Virol. *65*, 1247-1255.

Gross-Mesilaty, S., Reinstein, E., Bercovich, B., Tobias, K., Schwartz, A., Kahana, C. and Ciechanover, A. (1998). Basal and human papillomavirus E6 oncoprotein-induced degradation of Myc proteins by the ubiquitin pathway. Proc. Natl. Acad. Sci. USA *95*, 8058-8063.

Hamazaki, Y., Itoh, M., Sasaki, H., Furuse, M. and Tsukita, S. (2001). Multi PDZ-containing protein 1 (MUPP1) is concentrated at tight junctions through its possible interaction with claudin-1 and junctional adhesion molecule (JAM). J. Biol. Chem. *277*,455-461.

Hanada, N., Makino, K., Koga, H., Morisaki, T., Kuwahara, H., Masuko, N., Tabira, Y., Hiraoka, T., Kitamura, N., Kikuchi, A. and Saya, H. (2000). NE-dlg, a mammalian homolog of Drosophila dlg tumor suppressor, induces growth suppression and impairment of cell adhesion: possible involvement of down-regulation of beta-catenin by NE-dlg expression. Int. J. Cancer *86*, 480-488.

Hengstermann A Linares LK, Ciechanover A, Whitaker NJ, Scheffner M. (2001) Complete switch from MDM2 to human papillomavirus E6-mediated degradation of p53 in cervical cancer cells. Proc. Natl. Acad. Sci. USA *98*, 1218-1223.

Herber, R., Liem, A., Pitot, H. and Lambert, P.F. (1996). Squamous epithelial hyperplasia and carcinoma in mice transgenic for the human papillomavirus type 16 E7 oncogene. J.Virol. *70*, 1873-1881.

Honda, R., Tanaka, H. and Yasuda, H. (1997). Oncoprotein MDM2 is a ubiquitin ligase E3 for tumor suppressor p53. FEBS Lett. *420*, 25-27.

Hsu, Y., Wolter, K. and Youle, R. (1997). Cytosol-to-membrane redistribution of Bax and Bcl-XL during apoptosis. Proc. Natl. Acad. Sci. USA *94*, 3668-3672.

Hu, G., Liu, W., Hanania, E.G., Fu, S., Wang, T. and Deisseroth, A.B. (1995). Suppression of tumorigenesis by transcription units expressing the antisense E6 and E7 messenger RNA (mRNA) for the transforming proteins of the human papillomavirus and the sense mRNA for the retinoblastoma gene in cervical carcinoma cells. Cancer Gene Ther. *2*, 19-32.

Huang, D.C., Cory, S. and Strasser, A. (1997). Bcl-2, Bcl-XL and adenovirus protein E1B19kD are functionally equivalent in their ability to inhibit cell death. Oncogene *14*, 405-414.

Huang, P. (1998). Excision of mismatched nucleotides from DNA: a potential mechanism for enhancing DNA replication fidelity by the wild type p53 protein. Oncogene *17*, 261-270.

Huibregtse, J., Scheffner, M. and Howley, P. (1991). A cellular protein mediates association of p53 with the E6 oncoprotein of human papillomavirus types 16 or 18. EMBO J. *10*, 4129-4135.

Huibregtse, J., Scheffner, M. and Howley, P. (1993a). Cloning and expression of the cDNA for E6-AP, a protein that mediates the interaction of the human papillomavirus E6 oncoprotein with p53. Mol. Cell. Biol. *13*, 775-784

Humbert P, Russell S, Richardson H. (2003) Dlg, Scribble and Lgl in cell polarity, cell proliferation and cancer. Bioessays *6*, 542-53.

Hurlin, P.J., Kaur, P., Smith, P.P., Peres-Reyes, N., Blanton, R. and McDougall., J.K. (1991). Progression of human papillomavirus type 18-immortalised human keratinocytes to a malignant phenotype. Proc. Natl. Acad. Sci. USA *88*, 570-574.

Ide, N., Y. Hata, H. Nishioka, K. Hirao, I. Yao, M. Deguchi, A. Mizoguchi, H. Nishimori, T. Tokino, Y. Nakamura, and Y. Takai. (1999). Localization of membrane–associated guanylate kinase (MAGI)-1/BAI-associated protein (BAP)1 at tight junctions of epithelial cells. Oncogene *18*, 7810-7815.

Iizuka-Kogo A, Shimomura A and Senda T. 2005. Colocalization of APC and DLG at the tips of cellular protrusions in cultured epithelial cells and its dependency on cytoskeletons. Histochem Cell Biol. *123*, 67-73.

Ishidate, T., Matsumine, A., Toyoshima, K. and Akiyama, T. (2000). The APC/-hDLG complex negatively regulates cell cycle progression from the G0/G1 to S phase. Oncogene *19*, 365-372.

Jackson, S., Harwood, C., Thomas, M., Banks, L. and Storey, A. (2000). Role of Bak in UV-induced apoptosis in skin cancer and abrogation by HPV E6 proteins. Genes Dev. *14*, 3065-3073.

Jacobson, M.D., Weil, M. and Raff, M.C. (1997). Programmed cell death in animal development. Cell. *88*, 347-354.

Jiang M and Milner J. (2002). Selective silencing of viral gene expression in HPV-positive human cervical carcinoma cells treated with siRNA, a primer of RNA interference. Oncogene *21*, 6041-6048.

Jiang, M and Milner, J. (2004). Selective Silencing of Viral Gene E6 and E7 Expression in HPV-Positive Human Cervical Carcinoma Cells Using Small Interfering RNAs. Methods Mol Biol. *292*, 401-20.

Johnson, K and Wodarz, A. (2003). A genetic hierarchy controlling cell polarity. Nature Cell Biol. *5*, 12-14.

Junqueira, D., Cilenti, L., Musumeci, L., Sedivy, J.M., and Zervos, A.S. (1993). Random mutagenesis of PDZ(Omi) domain and selection of mutants that specifically bind the Myc proto-oncogene and induce apoptosis. Oncogene *22*, 2772-2781.

Kanda, T., Furuno, A. and Yoshiike, K. (1988). Human papillomavirus type 16 open reading frame E7 encodes a transforming gene for rat 3Y1 cells. J.Virol. *62*, 610-613.

Kanda, T., Watanabe, S., Zanma, S., Sato, H., Furuno, A. and Yoshiike, K. (1991). Human papillomavirus type 16 E6 proteins with glycine substituted for cysteine in metal binding motifs. Virology *185*, 536-543.

Kao, W.H., Beaudenon, S.L., Talis, A.L., Huibregtse, J.M. and Howley, P.M. (2000). Human papillomavirus type 16 E6 induces self-ubiquitination of the E6-AP ubiquitin-protein ligase. J.Virol. *74*, 6408-6417.

Kaur, P. and McDougall, J.K. (1988). Characterisation of primary human keratinocytes transformed by human papillomavirus type 18. J.Virol. *62*, 1917-1924.

Kawabe, H., Nakanishi, H., Asada, M., Fukuhara, A., Morimoto, K., Takeuchi, M. and Takai, Y. (2001). Pilt: a novel peripheral membrane protein at tight junctions in epithelial cells. J. Biol. Chem. *276*, 48350-48355.

Kemp, C.J., Donehower, L.A., Bradley, A. and Balmain, A. (1993). Reduction of p53 gene dosage does not increase initiation or promotion, but enhances malignant progression of chemically induced skin tumours. Cell *74*, 813-822.

Kim, S.K. (1997). Polarized signaling: basolateral receptor localization in epithelial cells by the PDZ-containing proteins. Curr. Opin. Cell Biol. *9*, 853-859.

Kinzler, K.W. and Vogelstein, B. (1996). Lessons from hereditary colorectal cancer. Cell *87*, 159-170.

Kishino, T., Lalande, M. and Wagstaff, J. (1997). UBE3A/E6-AP mutations cause Angelman syndrome. Nat. Genet. *15*, 70-73.

Kiyono, T., Hiraiwa, A., Ishii, S., Takahashi, T. and Ishibashi, M. (1997). Binding of high-risk human papillomavirus E6 oncoproteins to a human homologue of the Drosophila discs large tumour suppressor protein. Proc. Natl. Acad. Sci. USA *94*, 11612-11616.

Köppen, M., Simske, J.S., Sims, P.A., Firestein, B.L., Hall, D.H., Radice, A.D., Rongo, C. and Hardin, J.D. (2001). Cooperative regulation of AJM-1 controls junctional integrity in *Caenorhabditis elegans* epithelia. Nature Cell Biol. *3*, 983-991.

Kozlov, G., Gehring, K. and Ekiel, I. (2000). Solution structure of the PDZ2 domain from human phosphatase hPTP1E and its interactions with C-terminal peptides from the Fas receptor. Biochemistry *39*, 2572-2580.

Krajewski, S., Krajewska, M., Shabaik, A., Miyashita, T., Wang, H.G. and Reed, J.C. (1994). Immunohistochemical determination of *in vivo* distribution of Bax, a dominant inhibitor of Bcl-2. Am. J. Path. *145*, 1323-1336.

Krajewski, S., Krajewska, M. and Reed, J.C. (1996). Immunohistochemical analysis of *in vivo* patterns of Bak expression, a pro-apoptotic member of the Bcl-2 protein family. Cancer Res. *56*, 2849-2855.

Kühne, C. and Banks, L. (1998). E3-ubiquitin ligase/E6-AP links multicopy maintenance protein 7 to the ubiquitination pathway by a novel motif, the L2G box. J. Biol. Chem. *273*, 34302-34309.

Kühne, C., D. Gardiol, C. Guarnaccia, H. Amenitsch, and L. Banks. (2000). Differential regulation of human papillomavirus E6 by protein kinase A: conditional degradation of human discs large protein by oncogenic E6. Oncogene *19*, 5884-5891.

Kuo, S.R., Liu, J.S., Broker, T.R. and Chow, L.T. (1994). Cell-free replication of the human papillomavirus DNA with homologous viral E1 and E2 proteins and human cell extracts. J. Biol. Chem. *269*, 24058-24065.

Lane, D.P. and Crawford, L.V. 1979. T antigen is bound to a host protein in SV40 transformed cells. Nature *278*, 261-263.

Lechner, M. and Laimins, L. (1994). Inhibition of p53 DNA binding by human papillomavirus E6 proteins. J. Virol. *68*, 4262-4275.

Lee, S., Weiss, R. and Javier, R. (1997). Binding of human virus oncoproteins to hDlg/SAP97, a mammalian homologue of the Drosophila discs large tumour suppressor protein. Proc. Natl. Acad. Sci. USA *94*, 6670-6675.

Lee, S., Glaunsinger, B., Mantovani, F., Banks, L. and Javier, R. (2000). The multi-PDZ domain protein MUPP1 is a cellular target for both human adenovirus E4-ORF1 and high-risk papillomavirus type 18 E6 oncoproteins. J. Virol. *74*, 9680-9693.

Legouis, R., Jaulin-Bastard, F., Schott, S., Navarro, C., Borg, J-P. and Labouesse, M. (2003). Basolateral targeting by leucine-rich repeat domains in epithelial cells. EMBO Rep. *4*, 1096-1102.

Li, X. and Coffino, P. (1996). High risk human papillomavirus E6 protein has two distinct binding sites within p53, of which only one determines degradation. J. Virol. *70*, 4509-4516.

Lie, A.K., Skarsvag, S., Skomedal, H., Haugen, O.A. and Holm, R. (1999). Expression of p53, MDM2, and p21 proteins in high grade cervical intraepithelial neoplasia and relationship to human papillomavirus infection. Intl J. Gynecol. Pathol. *18*, 5-11.

Lin, J., Chen, J., Elenbaas, B. and Levine, A.J. (1994). Several hydrophobic amino acids in the p53 amino terminal domain are required for transcriptional activation, binding to mdm-2 and the adenovirus 5 E1B 55-kD protein. Genes Dev. *8*, 1235-1246.

Linzer, D.I. and Levine, A.J. 1979. Characterisation of a 54k dalton cellular SV40 tumour antigen present in SV40-transformed cells and uninfected embryonal carcinoma cells. Cell *17*, 43-52.

Liu, Y., Chen, J.J., Gao, Q., Dalal, S., Hong, Y., Mansur, C.P., Band, V. and Androphy, E.J. (1999). Multiple functions of human papillomavirus type 16 E6 contribute to the immortalisation of human mammary epithelial cells. J.Virol. *73*, 7297-7307.

Liu, Z., Ghai, J., Ostrow, R.S., McGlennen, R.C. and Faras, A.J. (1994). The E6 gene of human papillomavirus type 16 is sufficient for transformation of baby rat kidney cells in cotransfection with activated Ha-ras. Virology *201*, 388-396.

Lowe, S.W., Jacks, T., Houseman, D.E. and Ruley, H.E. (1994). Abrogation of oncogene associated apoptosis allows transformation of p53-deficient cells. Proc. Natl. Acad. Sci. USA *91*, 2026-2030.

Mantovani, F. and Banks, L. (1999). Inhibition of E6 induced degradation of p53 is not sufficient for stabilisation of p53 protein in cervical tumour derived cell lines. Oncogene *18*, 3309-3315.

Mantovani F, Massimi P and Banks L. (2001). Proteasome-mediated regulation of the hDlg tumour suppressor protein. J Cell Sci. *114*, 4285-4292.

Massimi, P., Pim, D., Bertoli, C., Bouvard, V. and Banks, L. (1999). Interaction between the HPV-16 E2 transcriptional activator and p53. Oncogene *18*, 7748-7754.

Massimi P, Gammoh N, Thomas M, Banks L. (2004). HPV E6 specifically targets different cellular pools of its PDZ domain-containing tumour suppressor substrates for proteasome-mediated degradation. Oncogene *23*, 8033-8039.

Masterson, P.J., Stanley, M.A., Lewis, A.P. and Romanos, M.A. (1998). A C-terminal helicase domain of the human papillomavirus E1 protein binds E2 and the DNA polymerase alpha-primase p68 subunit. J.Virol. *72*, 7407-7419.

Mathew, D., Gramates, L.S., Packard, M., Thomas U., Bilder, D., Perrimon, N., Gorczyca, M and Budnik, V. (2002). Recruitment of scribble to the synaptic scaffolding complex requires GUK-holder, a novel DLG binding protein. Curr. Biol. *12*, 531-539.

Matlashewski, G., Banks, L., Pim, D. and Crawford, L. (1986). Analysis of human p53 proteins and mRNA levels in normal and transformed cells. Eur. J. Biochem. *154*, 665-672.

Matsuura, T., Sutcliffe, J., Fang, P., Galjaard, R., Jiang, Y., Benton, C., Rommens, J. and Beaudet, A. (1997). *De novo* truncating mutations in E6-AP ubiquitin-protein ligase (UBE3A) in Angelman syndrome. Nat. Genet. *15*, 74-77.

McMahon, L., Legonis, R., Vonesch, J.L. and Labouesse, M. (2001). Assembly of C elegans apical junctions involves positioning and compaction by LEET-413 and protein aggregation by the MAGUK protein DLG-1. J. Cell Sci. *114*, 2265-2277.

Mino, A., Ohtsuka, T., Inoue, E. and Takai, Y. (2000). Membrane-associated guanylate kinase with inverted orientation (MAGI)-1/brain angiogenesis inhibitor-1 associated protein (BAP1) as a scaffolding molecule for Rap small G protein GDP/GTP exchange protein at tight junctions. Genes Cells *5*, 1009-1016.

Miyashita, T. and Reed, J.C. (1995). Tumor suppressor p53 is a direct transcriptional activator of the human bax gene. Cell *80*, 293-299.

Morais Cabral, JH., Petosa, C., Sutcliffe, MJ., Raza, S., Byron, O., Poy, F., Marfatia, SM., Chishti, AH. and Liddington, RC. (1996). Crystal structure of a PDZ domain. Nature *382*, 649-652.

Morin, P.J., Vogelstein, B. and Kinzler, K.W. (1996). Apoptosis and APC in colorectal tumorigenesis. Proc. Natl. Acad. Sci. USA *93*, 7950-7954.

Muganda, P., Mendoza, O., Hernandez, J. and Qian, Q. (1994). Human cytomegalovirus elevates levels of the cellular protein p53 in infected cells. J.Virol. *68*, 8028-8034.

Münger, K., Basile, J.R., Duensing, S., Eichten, A., Gonzalez, S.L., Grace, M. and Zacny, V.L. (2001). Biological activities and molecular targets of the human papillomavirus E7 oncoprotein. Oncogene Reviews *20*, 7888-7898.

Nakagawa, S. and Huibregtse, J. (2000). Human scribble (Vartul) is targeted for ubiquitin-mediated degradation by the high-risk papillomavirus E6 proteins and the E6-AP ubiquitin-protein ligase. Mol., Cell., Biol. *20*, 8244-8253.

Nakagawa, S., Watanabe, S., Yoshikawa, H., Taketani, Y., Yoshiike, K. and Kanda, T. (1995). Mutational analysis of human papillomavirus type 16 E6 protein: transforming function for human cells and degradation of p53 *in vitro*. Virology *212*, 535-542.

Nakagawa S, Yano T, Nakagawa K, Takizawa S, Suzuki Y, Yasugi T, Huibregtse JM, Taketani Y. (2004). Analysis of the expression and localisation of a LAP protein, human scribble, in the normal and neoplastic epithelium of uterine cervix Br. J. Cancer *90*, 194-9.

Nauenburg, S., Zwerschke, W. and Jansen-Dürr, P. (2001). Induction of apoptosis in cervical carcinoma cells by peptide aptamers that bind to the HPV-16 E7 protein. FASEB J. *15*, 592-594.

Nguyen, ML, Nguyen, MM Lee, D, Griep, AE and Lambert, PF. (2003a). The PDZ ligand domain of the human papillomavirus type 16 E6 protein is required for E6's induction of epithelial hyperplasia *in vivo*. J.Virol. *77*, 6957-6964.

Nguyen, MM, Nguyen, ML, Caruana, G, Bernstein, A, Lambert, PF and Griep, AE. (2003b). Requirement of PDZ-containing proteins for cell-cycle regulation and differentiation in the mouse lens epithelium. Mol. Cell. Biol. *23*, 8970-8981.

Nishimura, W., Iizuxa, T., Hirabayashi, S., Tanaka, N. and Hata, Y. (2000). Localisation of BAI-associated protein1/ membrane associated guanylate kinase-1 at adherens junctions in normal rat kidney cells: polarised targeting mediated by the carboxy-terminal PDZ domains. J. Cell Physiol. *185*, 358-365.

Pan, H. and Griep, A.E. (1995). Temporally distinct patterns of p53-dependent and p-53-independent apoptosis during mouse lens development. Genes and Dev. *9*, 2157-2169.

Patel, D., Huang, S.M., Baglia, LA and McCance, D.J. (1999). The E6 protein of human papillomavirus type 16 binds to and inhibits co-activation by CBP and p300. EMBO J. *18*, 5061-5072.

Pecoraro, G., Morgan, D. and Defendi, V. (1989). Differential effects of human papillomavirus type 6, 16 and 18 DNAs on immortalisation and transformation of human cervical epithelial cells. Proc. Natl. Acad. Sci. USA *86*, 563-567.

Phelps, W.C., Yee, C.L., Münger, K. and Howley, P.M. (1988). The human papillomavirus type 16 E7 gene encodes transactivation and transformation functions similar to those of adenovirus E1a. Cell *53*, 539-547.

Pim, D. and Banks, L. (1999). HPV-18 E6*I protein modulates the E6-directed degradation of p53 by binding to full length HPV-18 E6. Oncogene *18*, 7403-7408.

Pim, D., Storey, A., Thomas, M., Massimi, P. and Banks, L. (1994). Mutational analysis of HPV-18 E6 identifies domains required for p53 degradation *in vitro*, abolition of p53 transactivation *in vivo* and immortalisation of primary BMK cells. Oncogene *9*, 1869-1876.

Pim, D., Massimi, P. and Banks, L. (1997). Alternatively spliced HPV-18 E6* protein inhibits E6-mediated degradation of p53 and suppresses transformed cell growth. Oncogene *15*, 257-264.

Pim, D., Thomas, M., Javier, R., Gardiol, D. and Banks, L. (2000). HPV E6 targeted degradation of the discs large protein: evidence for the involvement of a novel ubiquitin-ligase. Oncogene *19*, 719-725.

Pim, D., Thomas, M. and Banks, L. (2001). The function of the human papillomavirus oncogenes, p145-192. *In* R.J.A. Grand (ed.),Viruses, Cell Transformation and Cancer. Elsevier, Amsterdam.

Ponting, C., and Phillips, C. (1995). DHR domains in syntrophins, neural NO synthases and other intracellular proteins. Trends Biochem. Sci. *20*, 102-103.

Querido, E., Morrison, M.R., Chu-Phan-Dang, H., Thirlwell, S.W., Boivin, D., Branton, P.E. and Morrison, M.R. (2001). Identification of three functions of Ad E4ORF6 protein that mediate p53 degradation by the E4ORF6-E1B55k complex. J. Virol. *75*, 699-709.

Reed, J. (1998). Bcl-2 family proteins. Oncogene *17*, 3225-3236.

Reznikoff, CA, Belair, C, Savelieva, E, Zhai, Y, Pfeifer, K, Yeager, T, Thompson, KJ, DeVries, S, Bindley, C and Newton, MA. (1994). Long term stability and minimal genotypic and phenotypic alterations in HPV16 E7-, but not E6-, immortalised human uroepithelial cells. Genes Dev. *8*, 2227-2240.

Ruppert, J.M. and Stillman, B. (1993). Analysis of a protein-binding domain of p53. Mol. Cell. Biol. *13*, 3811-3820.

Sarnow, P., Ho, Y.S., Williams, J and Levine, A.J. (1982). Adenovirus E1b-58kd tumor antigen and SV40 large tumor antigen are physically associated with the same 54kd cellular protein in transformed cells. Cell *28*, 387-394.

Sato, H., Furuno, A. and Yoshiike, K. (1989). Expression of human papillomavirus type 16 E7 gene induces DNA synthesis of rat 3Y1 cells. Virology *168*, 195-199.

Scheffner M., Werness, B., Huibregtse, J., Levine, A. and Howley, P.. (1990). The E6 oncoprotein encoded by human papillomavirus types 16 and 18 promotes the degradation of p53. Cell *63*, 1129-1136.

Scheffner, M., Münger, K., Byrne, J.C. and Howley, P.M. (1991). The state of the p53 and retinoblastoma genes in human cervical carcinoma cell lines. Proc. Natl. Acad. Sci. USA *88*, 5523-5527.

Schlegel, R., Phelps, W.C., Zhang, Y.L. and Barbosa, M. (1988). Quantitative keratinocyte assay detects two biological activities of human papillomavirus DNA and identifies viral types associated with cervical carcinoma. EMBO J. *7*, 3181-3187.

Schwarz, E., Freese, U., Gissmann, L., Mayer, W., Roggenbuck, B., Stremlau, A. and zur Hausen, H. (1985). Structure and transcription of human papillomavirus sequences in cervical carcinoma cells. Nature *314*, 111-114.

Scorrano L, Oakes SA, Opferman JT, Cheng EH, Sorcinelli MD, Pozzan T, Korsmeyer SJ. (2003). BAX and BAK regulation of endoplasmic reticulum Ca2+: a control point for apoptosis. Science *300*, 135-9.

Sedman, S.A., Barbosa, M.S., Vass, W.C., Hubbert, N.L., Haas, J.A., Lowy, D.R. and Schiller, J.T. (1991). The full length, E6 protein of human papillomavirus type 16 has transforming and transactivating activities and cooperates with E7 to immortalise keratinocytes in culture. J.Virol. *65*, 4860-4866.

Sexton, C.J., Proby, C.M., Banks, L., Stables, J.N., Powell, K., Navsaria, H. and Leigh, I.M. (1993). Characterisation of factors involved in human papillomavirus type 16-mediated immortalisation of oral keratinocytes. J. Gen. Virol. *74*, 755-761.

Sherman, L. and Schlegel, R. (1996). Serum- and calcium-induced differentiation of human keratinocytes is inhibited by the E6 oncoprotein of human papillomavirus type 16. J. Virol. *70*, 3269-3279.

Shimizu, S., Narita, M. and Tsujimoto, Y. (1999). Bcl-2 family proteins regulate the release of apoptogenic cytochrome C by the mitochondrial channel VDAC. Nature *399*, 483-487.

Six EM, Ndiaye D, Sauer G, Laabi Y, Athman R, Cumano A, Brou C, Israel A, Logeat F. (2004). The Notch Ligand Delta1 Recruits Dlg1 at Cell-Cell Contacts and Regulates Cell Migration. J. Biol. Chem. *279*, 55818-55826.

Smith, D.H and Ziff, E.B. (1988). The amino terminal region of adenovirus serotype 5 E1a protein performs two separate functions when expressed in primary baby rat kidney cells. Mol. Cell. Biol. *8*, 3882-3890.

Smotkin, D. and Wettstein, F. (1986). Transcription of human papillomavirus type 16 early genes in a cervical cancer and a cancer-derived cell line, and identification of the E7 protein. Proc. Natl. Acad. Sci., USA *83*, 4680-4684.

Smotkin, D., Prokoph, H. and Wettstein, F.O. (1989). Oncogenic and non-oncogenic human genital papillomaviruses generate E7 mRNA by different mechanisms. J.Virol. *63*, 1441-1447.

Song, S., Pitot, H.C. and Lambert, P.F. (1999). The human papillomavirus type 16 E6 gene alone is sufficient to induce carcinomas in transgenic animals. J.Virol. *73*, 5887-5893.

Song, S., Liem, A., Miller, J.A. and Lambert, P.F. (2000). Human papillomavirus type 16 E6 and E7 contribute differently to carcinogenesis. Virology *267*, 141-150.

Songyang, Z., A.S. Fanning, C. Fu, J. Xu, S.M. Marfatia, A.H. Chisti, A. Crompton, A.C. Chan, J.M. Anderson, and L.C. Cantley. (1997). Recognition of unique carboxyl-terminal motifs by distinct PDZ domains. Science *275*, 73-77.

Steegenga, W.T., Riteco, N., Jochemsen, A.G., Fallaux, F.J. and Bos, J.L. (1998). The large E1B protein together with the E4orf6 protein target p53 for active degradation in adenovirus infected cells. Oncogene *16*, 349-357.

Steele, C., Sacks, P.G., Adler-Storthz, K. and Shillitoe, E.J. (1992). Effect on cancer cells of plasmids that express antisense RNA of human papillomavirus 18. Cancer Res. *52*, 4706-4711.

Stein, R.W., Corrigan, M., Yaciuk, P., Whelan, J. and Moran, E. (1990). Analysis of E1A-mediated growth regulation functions: binding of the 300-kilodalton cellular product correlates with E1A enhancer repression function and DNA synthesis inducing activity. J.Virol. *64*, 4421-4427.

Sterlinko Grm, H. and Banks, L. (2004). Degradation of hDlg and MAGIs by human papillomavirus E6 is E6-AP-independent. J. Gen. Virol. *85*, 2815-2819.

Sterlinko Grm H, Weber M, Elston R, McIntosh P, Griffin H, Banks L, Doorbar J. (2004). Inhibition of E6-induced degradation of its cellular substrates by novel blocking peptides. J. Mol. Biol. *335*, 971-985.

Storey, A. and Banks, L. (1993). Human papillomavirus type 16 E6 gene cooperates with EJ-ras to immortalise primary mouse cells. Oncogene *8*, 919-924.

Storey, A., Pim, D., Murray, A., Osborn, K., Banks, L. and Crawford, L. (1988). Comparison of the *in vitro* transforming activities of human papillomavirus types. EMBO J. *7*, 1815-1820.

Storey, A. Thomas, M., Kalita, A., Harwood, C., Gardiol, D., Mantovani, F., Breuer, J., Leigh, I., Matlashewski, G. and Banks, L. (1998). Role of a p53 polymorphism in the development of a human papillomavirus-associated cancer. Nature *393*, 229-234.

Strasser, A., Harris, A.W., Corcoran, L.M. and Cory, S. (1994). Bcl-2 expression promotes B- but not T-lymphoid development in SCID mice. Nature *368*, 457-460.

Suzuki, T., Ohsugi, Y., Uchida-Toita, M., Akiyama, T. and Yoshida, M. (1999). Tax oncoprotein of HTLV-1 binds to the human homologue of Drosophila discs large tumor suppressor protein, hDLG, and perturbs its function in cell growth control. Oncogene *18*, 5967-5972.

Talis, A.L., Huibregtse, J.M. and Howley, P.M. (1998). The role of E6AP in the regulation of p53 protein levels in human papillomavirus (HPV)-positive and HPV-negative cells. J. Biol. Chem. *273*, 6439-6445.

Talora C, Sgroi DC, Crum CP, Dotto GP. (2002). Specific down-modulation of Notch1 signaling in cervical cancer cells is required for sustained HPV-E6/E7 expression and late steps of malignant transformation. Genes Dev. *16*, 2252-2263.

Tanentzapf G. and Tepass U. (2003). Interactions between the crumbs, lethal giant larvae and bazooka pathways in epithelial polarization. Nat. Cell Biol. *5*, 46-52.

Thomas, M. and Banks, L. (1998). Inhibition of Bak-induced apoptosis by HPV-18 E6. Oncogene 17, 2943-2954.

Thomas, M. and Banks, L. (1999). Human papillomavirus (HPV) E6 interactions with Bak are conserved amongst E6 proteins from high and low risk HPV types. J. Gen. Virol. *80*, 1513-1517.

Thomas, M., Massimi, P., Jenkins, J. and Banks, L. (1995). HPV-18 E6 mediated inhibition of p53 DNA binding activity is independent of E6 induced degradation. Oncogene *10*, 261-268.

Thomas, M., Pim, D. and Banks, L. (1999). Role of the E6-p53 interaction in the molecular pathogenesis of HPV. Oncogene *18*, 7690-7700.

Thomas, M., Glaunsinger, B., Pim, D., Javier, R. and Banks, L. (2001). HPV E6 and MAGUK protein interactions: determination of the molecular basis for specific protein recognition and degradation. Oncogene *20*, 5431-5439.

Thomas, M., Laura, R., Hepner, K., Guccione, E., Sawyers, C., Lasky, L and Banks, L. (2002). Oncogenic human papillomavirus E6 proteins target the MAGI-2 and MAGI-3 proteins for degradation. Oncogene, *21*, 5088-96.

Thomas, M., Massimi, P., Borg, J.-P., and Banks, L. (2005). The hScrib/Dlg apico-basal control complex is differentially targeted by HPV-16 and HPV-18 E6 proteins. Oncogene. In press.

Thomas U, Phannavong B, Muller B, Garner CC, Gundelfinger ED. (1997). Functional expression of rat synapse-associated proteins SAP97 and SAP102 in Drosophila dlg-1 mutants: effects on tumor suppression and synaptic bouton structure. Mech. Dev. *62*, 161-74.

Thompson, C.B. (1995). Apoptosis in the pathogenesis and treatment of disease. Science *267*, 1456-1462.

Tong, X. and Howley, P.M. (1997). The bovine papillomavirus E6 oncoprotein interacts with paxillin and disrupts the actin cytoskeleton. Proc. Natl. Acad. Sci. USA *94*, 4412-4417.

Tsunokawa, Y., Takebe, N., Kasamatsu, T., Terada, M. and Sugimura, T. (1986). Transforming activity of human papillomavirus type 16 DNA sequence in a cervical cancer. Proc. Natl. Acad. Sci. USA *83*, 2200-2203.

Ullmer, C., Schmuck, K., Figge, A. and Lubbert, H. (1998). Cloning and characterisation of MUPP1, a novel PDZ domain protein. FEBS Lett. *424*, 63-68.

Veeraraghavalu K, Pett M, Kumar RV, Nair P, Rangarajan A, Stanley MA, Krishna S. (2004). Papillomavirus-mediated neoplastic progression is associated with reciprocal changes in JAGGED1 and manic fringe expression linked to notch activation. J. Virol. *78*, 8687-8700.

Venturini, F., Braspenning, J., Homann, M., Gissmann, L. and Sczakiel, G. (1999). Kinetic selection of HPV-16 E6 /E7-directed antisense nucleic acids: antiproliferative effects on HPV-16 transformed cells. Nucleic Acids Res. *27*, 1585-1592.

von Knebel Döberitz, M., Rittmuller, C., zur Hausen, H. and Dürst, M. (1992). Inhibition of tumorigenicity of cervical cancer cells in nude mice by HPV E6-E7 antisense RNA. Int. J. Cancer *51*, 831-834.

Vousden, K.H., Doniger, J., DiPaolo, J.A. and Lowy, D.R. (1988). The E7 open reading frame of human papillomavirus type 16 encodes a transforming gene. Oncogene Res. *3*, 167-175.

Watson RA, Rollason TP, Reynolds GM, Murray PG, Banks L, Roberts S. (2002). Changes in expression of the human homologue of the Drosophila discs large tumour suppressor protein in high-grade premalignant cervical neoplasias. Carcinogenesis *11*, 1791-1796.

Watson RA, Thomas M, Banks L, Roberts S. (2003). Activity of the human papillomavirus E6 PDZ-binding motif correlates with an enhanced morphological transformation of immortalized humankeratinocytes. J. Cell. Sci. *116*, 4925-4934.

Wazer, D.E., Liu, X.L., Chu, Q., Gao, Q. and Band, V. (1995). Immortalisation of distinct human mammary epithelial cell types by human papillomavirus 16 E6 or E7. Proc. Natl. Acad. Sci. USA *92*, 3687-3691.

Wei, M.C., Lindsten, T., Moothka, V.K., Weiler, S., Gross, A., Ashiya, M., Thompson, C.B. and Korsmeyer, S.J. (2000). tBID, a membrane-targeted death ligand, oligomerises BAK to release cytochrome c. Genes Dev. *14*, 2060-2071.

Wei MC, Zong WX, Cheng EH, Lindsten T, Panoutsakopoulou V, Ross AJ, Roth KA, MacGregor GR, Thompson CB, Korsmeyer SJ. (2001). Proapoptotic BAX and BAK: a requisite gateway to mitochondrial dysfunction and death. Science *292*, 727-730.

Werness, B., Levine, A. and Howley, P. (1990). Association of human papillomavirus types 16 and 18 E6 proteins with p53. Science *248*, 76-79.

White, A.E., Livanos, E.M. and Tlsty, T.D. (1994). Differential disruption of genomic integrity and cell cycle regulation in normal human fibroblasts by the HPV oncoproteins. Genes Dev. *8*, 666-677.

White, E. (2001). Regulation of the cell cycle and apoptosis by the oncogenes of adenovirus. Oncogene Reviews *20*, 7836-7846.

Wilcock, D. and Lane, D.P. (1991). Localisation of p53, retinoblastoma and host replication protein at sites of viral replication in herpes-infected cells. Nature *349*, 429-431.

Wolter, K.G., Hsu, Y.T., Smith, C.L., Nechushtan, A., Xi, X.G. and Youle, R.J. (1997). Movement of Bax from the cytosol to mitochondria during apoptosis. J. Cell Biol. *139*, 1281-1292.

Woods, D.F., Wu, J.W. and Bryant, P.J. (1996). Localisation of proteins to the apico-lateral junctions of Drosophila epithelia. Dev. Genet. *20*, 111-118.

Woodworth, C.D., Doniger, J. and DiPaolo, J.A. (1989). Immortalisation of human foreskin keratinocytes by various human papilloma virus DNAs corresponds to their association with cervical carcinoma. J. Virol. *63*, 159-164

Wu, X and Levine A.J (1994) p53 and E2F-1 cooperate to mediate apoptosis. Proc. Natl. Acad. Sci. USA *91*, 3602-3606.

Wu, X., Hepner, K., Castelino-Prabhu, S., Do, D., Kaye, M.B., Yuan, X-J., Wood, J., Ross, C., Sawyers, C.L. and Whang, Y.E. (2000). Evidence for regulation of the PTEN tumor suppressor by a membrane-localized multi-PDZ domain containing protein, MAGI-2. Proc. Natl. Acad. Sci. USA *97*, 4233-4238.

Wu Y., Dowbenko, D., Spencer, S., Laura, R., Lee, J., Gu, Q. and Laskey, L. (2000). Interaction of the tumor suppressor PTEN/MMAC with a PDZ domain of MAGI-3, a novel membrane associated guanylate kinase. J. Biol. Chem. *275*, 21477-21485.

Yasumoto, S., Burkhardt A.L., Doniger, J. and DiPaolo J.A. (1986). Human papilloma virus type 16 DNA-induced malignant transformation of NIH3T3 cells. J. Virol. *57*, 572-577.

Yoshinouchi M, Yamada T, Kizaki M, Fen J, Koseki T, Ikeda Y, Nishihara T, Yamato K. (2003). In vitro and *in vivo* growth suppression of human papillomavirus 16-positive cervical cancer cells by E6 siRNA. Mol Ther. *8*, 762-768.

Zamzami, N., Brenner, C., Marzo, I., Susin, S.A. and Kroemer, G. (1998). Subcellular and submitochondrial mode of action of Bcl2-like oncoproteins. Oncogene *16*, 2265-2282.

Zhong L and Hayward, G.S. (1997). Assembly of complete, functionally active herpes simplex virus DNA replication compartments and recruitment of associated viral and cellular proteins in transient cotransfection assays. J. Virol. *71*, 3146-3160.

Zimmermann, H., Degenkolbe, R., Bernard, H.U. and O'Connor, M.J. (1999). The human papillomavirus type 16 E6 oncoprotein can downregulate p53 activity by targeting the transcriptional coactivator CBP/p300. J. Virol. *73*, 6209-6219.

Zimmermann, H., Koh, C.H., Degenkolbe, R., O'Connor, M.J., Muller, A., Steger, G., Chen, J.J., Lui, Y., Androphy, E.J. and Bernard, H.U. (2000). Interaction with CBP/p300 enables the bovine papillomavirus type 1 E6 oncoprotein to downregulate CBP/p300-mediated transactivation by p53. J. Gen. Virol. *81*, 2617-2623.

zur Hausen, H. (2001). Oncogenic DNA viruses. Oncogene Reviews *20*, 7820-7823.

From: Papillomavirus Research: From Natural History to Vaccines and Beyond. Edited by: M. Saveria Campo

Chapter 10

The Biology of the E7 Protein of HPV-16

Dennis J. McCance*

Abstract

The E7 protein of oncogenic human papillomaviruses (HPV), such as HPV-16, is essential for the replication of the virus in the stratified epithelium. Replication of the viral genome requires that the terminally differentiating keratinocytes are induced back into the cell cycle to provide the necessary environment for DNA propagation. An unwanted outcome of this cell cycle stimulation is the immortalization of the infected cells, which along with unknown mutations in the genome of the infected cells, may lead to a malignant phenotype. In this Chapter, I will describe properties of E7 that are consistent with the ability of the protein to inhibit keratinocyte differentiation, stimulate cell cycle progression and cause immortalization of keratinocytes.

1. Introduction

Papillomaviruses replicate in stratified epithelium and the major amplification step takes place in the upper part of the epithelium. This is the region of the epithelium where there is terminal differentiation, and as such the cellular environment would not be conducive to viral DNA replication since components of the cellular replicative machinery would be at low or undetectable levels. In concert with other early region proteins, such as E6 and E5, the role of E7 is to stimulate cells, which are programmed to differentiate, to re-enter the cell cycle and progress through to S-phase resulting in an adequate supply of the cell's replicative machinery for viral DNA replication. One of the consequences of prolonged stimulation is that certain HPV types can immortalize primary keratinocytes in culture and bypass senescence, a normal end-point for primary cells. High-risk HPVs can immortalize cultures of primary cells including the natural target cell of HPV, the keratinocyte. Inhibiting the activity of E6 or E7 by anti-sense, or interfering RNA molecules in the form of oligonucleotides (RNAi), or short hairpin loops (shRNA), will abrogate the immortalized phenotype of primary cultures leading to senescence and indicating that these proteins are necessary for the continuance of the immortalized state.

The E6 and E7 proteins of the high risk human papillomaviruses are important for many of the biological properties of the whole genome, especially in relation to the bypass of cellular arrest signals, such as differentiation, and immortalization of primary human keratinocytes. While it is possible to determine the individual properties of the viral proteins, these two proteins act in concert during infection and complement one another. Therefore, this complementary nature needs be appreciated when describing the biological properties of the individual proteins. For instance, E7 is the major immortalizing protein of the high-risk types such as HPV-16 and –18, and can by itself immortalize primary human keratinocytes, although the process is much more efficient in the presence of HPV-16 or -18 E6. A virus, HPV-6, which produces benign lesions, must also stimulate cells into S-phase to replicate successfully, yet in culture the E7 protein from HPV-6 or even the full length genome is unable to immortalize keratinocytes and fails to display other characteristics of oncogenic HPVs. One "black hole" of papilloma virology is how does HPV-6 achieve the same goals as HPV-16 of regulating cells so that an environment for viral genome replication is achieved, yet shows none of the tissue culture characteristics of the high-risk types.

HPV-16 E7 message is transcribed early in infection, producing low levels of protein. E7 has been shown to interact with a number of cellular proteins, although the biological consequences of such interactions are not always clear. In this first section, the interaction of E7 with various cellular proteins and the known biological consequences will be discussed. In addition, the work described will concentrate on the activities of HPV-16 E7, which is typical of the high risk viruses, but where appropriate will compare the functions of E7 from low risk HPV types, such as HPV-6.

2. E7 Interactions and the Resulting Biology

HPV -16 E7 has been shown to bind to a number of cellular proteins, although the biological outcome of these interactions is not always clear. The first major interaction described was with the retinoblastoma (Rb) family of proteins, which includes the prototypical

*For correspondence. Email dennis_mccance@urmc.rochester.edu

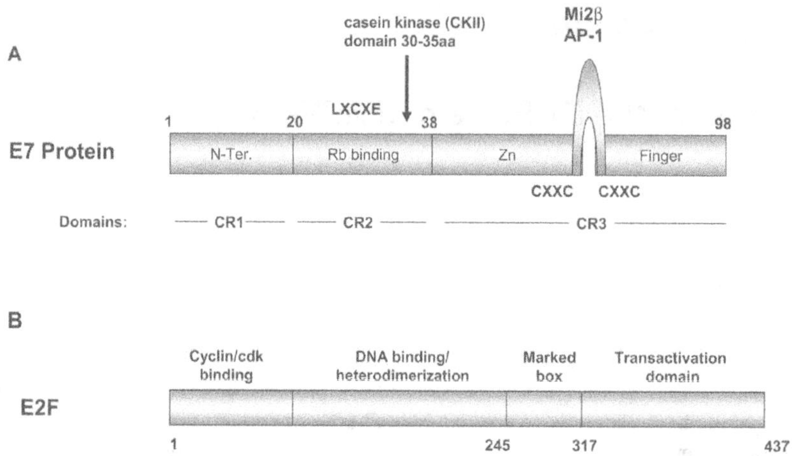

Figure 1. Diagrammatic representation of the domains of (A) HPV-16 E7 and (B) a typical E2F (E2F1-5).

retinoblastoma protein, p105Rb and the other members, p107 and p130 (Davies et al., 1993; Dyson et al., 1989).

2.1. Interaction with the retinoblastoma family

The E7 protein can be divided into three domains (Figure 1A), with the Rb binding domain contained within the second region between amino acids 20-40 (Barbosa et al., 1990; Dyson et al., 1992). This region contains the typical LXCXE motif found in many proteins which bind Rb, and a casein kinase consensus sequence (Barbosa et al., 1990), which can be phosphorylated *in vitro*, and although the biological activity of this phosphorylation has not been clearly defined, there are changes in the extent of phosphorylation of E7 during the cell cycle (Massimi and Banks, 2000). The N-terminal domain from amino acids 1–20 has not been extensively investigated and, as of writing, only one cellular protein has been shown to bind to this region, namely the histone acetyl transferase, protein CBP associated factor (P/CAF, (Huang et al., 1993)). Deletion and point mutations within this region abrogate most of the biological properties of the protein, indicating an important function for this region, although all these mutants still bind the Rb family (Brokaw et al., 1994; Phelps et al., 1992). Whether function(s) of the N-terminal domain depends on binding cellular proteins, or it is important for the structural integrity of E7, is at present unclear. Finally there is the zinc finger domain in the C-terminal region and the structural integrity of this region is required for most, if not all, of the biological activity of E7 (Antinore et al., 1996; Chesters et al., 1990; Edmonds and Vousden, 1989; Helt and Galloway, 2001; Jewers et al., 1992; McIntyre et al., 1993; Phelps et al., 1992). This is also the region responsible for the dimerization of E7. Although the structure of E7 has not been solved, due to the insolubility of the protein, the dimerization of E7 appears to stabilize the protein. However, the conformation changes substantially when the pH shifts from 7.0 to 5.0, suggesting that this may help in the ability of E7 to bind a number of different cellular proteins (Alonso et al., 2002).

The Rb family inhibits the activity of a number of transcription factors, which are important for the expression of proteins involved in G1 to S-phase progression, DNA synthesis, and in the G2/M transition (Dyson, 1998; Ren et al., 2002) during the cell cycle. Most information on the repressive activity of the family concerns the regulation of the E2F family by the prototypical Rb member, p105Rb. The E2F family, which consists of seven members that bind to DNA as a dimer with DP-1 and -2, has been shown to bind to different Rb family members. For instance, E2F-1, -2 and -3 preferentially bind p105Rb, E2F4 binds p107, while

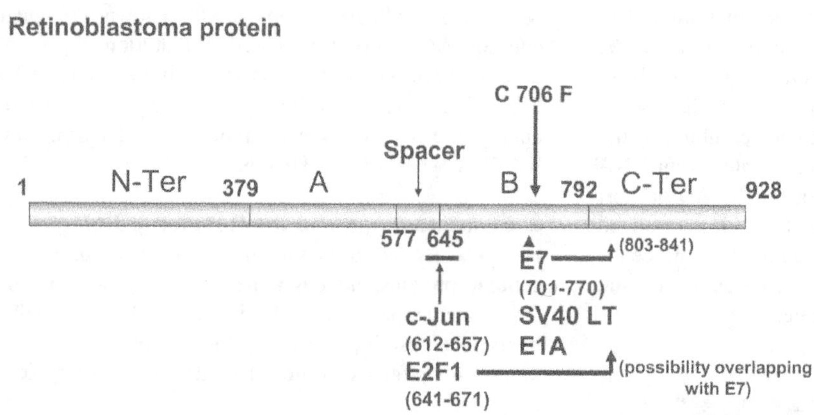

Figure 2. Domains of the prototypical Rb family member, p105Rb. E2F1 has been shown to contact several parts of the interface between pockets A and B of p105Rb as well as a region in the C-terminus. E7 binds to the B-pocket and also to a region in the C-terminus which may overlap with E2F1.

E2F-4 and 5 bind p130 (Dyson, 1998). E2F6 does not have a transactivation domain and therefore does not bind Rb members and E2F7 has not been well characterized, although it appears to be a repressor of a subset of genes regulated by the E2F family (de Bruin et al., 2003; Di Stefano et al., 2003). Normally, during early and mid G1, the hypophosporylated p105Rb binds to and represses E2F1-3 family members, while p107 binds E2F4. Then in late in G1, the p105Rb and p107 are hyperphosphorylated by the cyclin D/cdk4/6 and cyclin E/cdk2 complexes, resulting in the abrogation of the repressive function and the activation of transcription by the E2F family (Dyson, 1998; Ludlow et al., 1989; Ludlow et al., 1990). E2F1 binds to p105Rb through a LXCXE containing peptide (between 409 and 426 amino acids) in the C-terminal transactivation domain and to a different region through the marked box region of E2F (245 to 317 amino acids; Figure 1b)). The LXCXE motif of E2F binds to a series of alpha-helices contained in the A- and B-pockets in regions that oppose each other and the marked box binds an ill defined region in the C-terminus, probably overlapping with the second E7 binding domain (see below). The marked box of E2F1 increases the binding to p105Rb by 12-fold over the transactivation domain alone (Xiao et al., 2003), so it appears to be a very significant interaction.

The binding domains on p105Rb for E2F and E7 are both separate and overlapping (Figure 2). For instance, a mutation at amino acid 706 (C706P) in the B-pocket of p105Rb, inhibits binding of E7 and other viral proteins to p105Rb (Lee et al., 1998), but not E2F binding. The major binding domain of E7 on p105Rb is in the B-pocket, between amino acids 700 and 772 (Huang et al., 1993), more than 30 angstroms apart from the E2F binding site (Lee et al., 1998; Xiao et al., 2003). E7 binds through the LXCXE motif contained in the CR2 domain (Figure 1A; (Huang et al., 1993)) and also through a region in the CR3 domain that binds to p105Rb in the C-terminus between amino acids 803 and 841. The CR3 domain of E7 stabilizes the interaction of E7 with p105Rb, as binding studies have demonstrated that the interaction of the CR3 with the C-terminal domain of p105Rb causes a 15-fold increase in binding affinity (Patrick et al., 1994; Xiao et al., 2003). In fact the disruption of the binding of E2F for p105Rb by E7 appears to be mediated by the CR3 domain which may inhibit binding of the marked box or create a structure which no longer allows the marked box to bind. This disruption of the marked box binding by E7, may then cause dissociation of E2F and p105Rb, even in the presence of LXCXE binding.

While it would appear that the binding of Rb family members by E7 from high-risk types and the reduced binding of low risk types may account for the ability of HPV-16 to produce malignant disease, it should be pointed out that E7 from HPV-1, a virus infecting cutaneous surfaces, binds to Rb as well as HPV-16 E7, surprisingly, this virus only produces benign disease and can not immortalize primary cells (Schmitt et al., 1994). These results indicate that other functions of E7, in combination with the p105Rb interaction, may be important for the phenotypes produced by high-risk viruses. The LXCXE motif is found within other viral oncoproteins such as E1A and SV40 LTAg, as well as many cellular Rb binding proteins including cyclin D1, cyclin D2, cyclin D3, BRG1 (a human homolog of SWI/SNF proteins in yeast), and histone deacetylases-1 and –2 (Brehm et al., 1998; Brehm et al., 1999; Dowdy et al., 1993; Dunaief et al., 1994; Magnaghi-Jaulin et al., 1998). The high level of conservation of this motif among viral and cellular proteins suggests that viral oncoproteins such as E7 may compete directly with several cellular proteins for binding to Rb family members.

E7 from low risk types such as HPV-6 can also bind to the Rb family, although the level of binding is approximately 10% to 20% of that seen with the high-risk types (Gage et al., 1990; Munger et al., 1991). It is unclear whether this level of binding has any biological relevance; however, experiments in tissue culture cells have shown that HPV-6 E7, unlike HPV-16 E7, does not activate E2F. The variance in binding efficiency can be accounted for by a single difference in the sequences next to their respective LXCXE motifs: HPV-6 E7 contains a glycine at amino acid 20, while HPV-16 E7 possesses an aspartic acid at the equivalent position (Heck et al., 1992; Sang and Barbosa, 1992).

So in summary, the LXCXE is required for binding to the Rb family, but it is not sufficient to abrogate Rb function. There is an additional requirement for the CR3 domain for derepression of the Rb family, leading to transcriptional activation. These results suggest that both the CR2 and CR3 domains are required for high affinity binding to Rb and the subsequent disruption of Rb-E2F complexes and may help to explain some of the biological functions of E7 described below.

2.2. p105Rb, cellular transcription and E7 function

The mechanism by which Rb represses transcription has, until recently, not been defined. Since Rb binds to the transcriptionally active domain of E2F, it was thought that Rb might repress transcription by inhibiting the ability of E2F to interact with the basal transcription machinery of the cell. However, recently Rb has been shown to interact with histone deacetylases of the class 1 type, such as HDAC1 and 2 (Brehm et al., 1998; Magnaghi-Jaulin et al., 1998) and the methylase SUV39H1 (Nielsen et al., 2001), enzymes that remodel chromatin and so affect transcription. These interactions are mediated through the B-pocket of the Rb family. Actively transcribed regions of DNA show a high level of histone acetylation and low methylation while repressed genes exhibit the opposite profile. There is evidence of a histone code, which determines the structure of chromatin and therefore whether a gene is transcribed or not (Jenuwein and Allis, 2001). Recent work has shown an interesting Rb-dependent sequence of events in this histone code hypothesis. This work has shown that for repression of transcription, lysine 9 on the tail of histone 3 is deacetylated and then methylation takes place on the same amino acid (Nielsen et al., 2001). There is evidence that Rb/methylase complex cannot methylate histone 3 on lysine 9 when the lysine is already acetylated. Therefore, to repress an active promoter Rb/HDAC may deacetylate lysine 9 on the histone 3 tail and once deacetylated, Rb may recruit the methylase to methylate lysine 9. Methylation of lysine 9 then specifically binds the Heterochromatin Protein 1 (HP-1), which is known to be involved in the condensation of sections of DNA. Therefore, there is a sequence of events, mediated by Rb, which leads to chromatin condensation and repression of promoters.

As well as binding to the Rb family, E7 has been shown to interact with a complex containing HDAC activity independent of binding to Rb (Brehm et al., 1999). E7 does not bind directly to the HDAC enzyme, but through a protein contained in the complex called Mi2-beta. Mi2-beta was first described as part of a large transcriptionally repressive complex called NURD (*Nu*cleosome *R*emodeling histone *D*eacetylase complex, (Zhang et al., 1998)) and has not, so far, been shown to be part of any Rb complex. The CR3 domain of E7 is required for interaction with Mi2-beta and mutations in this domain (zinc finger), which do not affect Rb binding, or the integrity of the zinc finger structure, abrogate binding to Mi2-beta and thus to the HDAC complex (Brehm et al., 1999). It is not clear if E7 derepresses Rb repression by competing away an HDAC activity from the B-pocket and also by disrupting a separate complex containing Mi2-beta and HDAC, or whether the proteins are part of one large repressive complex.

While the exact mechanism of E7 derepression remains unclear, the targeting of histone deacetylases provides yet another method by which E7 can de-repress gene transcription, and may explain the essential nature of the CR3 domain in activating E2F-regulated genes. The fact that E7 binds the Rb family and that this property is shared with other viral oncoproteins, such as the adenovirus E1A and simian virus 40 large T antigen, suggests that tumor viruses possess evolutionary conserved attributes, and underscores the importance of Rb binding in the natural history of virus infection. The retinoblastoma protein family members play a central role in the regulation of the eukaryotic cell cycle and so disruption of this function will have consequences for the controlled growth of cells.

2.3. pRb acetylation and the involvement in keratinocyte differentiation

As well as stimulating cell cycle progression by interacting with the Rb family, E7 can also inhibit normal differentiation of keratinocytes, a function that may be required for subsequent stimulation of cell cycle progression. There is increasing evidence that p105Rb, unlike the other Rb family members, is involved in the terminal differentiation of a number of tissues. Therefore, does E7 inhibit differentiation through the interaction with p105Rb? Previous work has shown that pRb-/- cells do not differentiate into muscle cells, although they are able to arrest under the stimuli that would differentiate normal cells (Schneider et al., 1994; Zacksenhaus et al., 1996). This arrest is due to other Rb family members and is temporary, as cells can be stimulated to enter the cell cycle when given growth factors. Therefore, p105Rb is required for permanent cell cycle arrest and subsequent induction of differentiation-specific transcription. More recent work has shown that p105Rb is acetylated (Chan et al., 2001), at least in part by the histone acetyl transferase, P/CAF, (pCBP/p300 associated factor) binding directly to p105Rb (Nguyen et al., 2004). Acetylated p105Rb was observed during muscle and keratinocyte cell differentiation and was absent in cycling cells (Nguyen et al., 2004). The acetylation of p105Rb occurs on amino acids 873 and 874 and has been shown to be required for the permanent arrest of muscle cells as a preliminary step in terminal differentiation (Nguyen et al., 2004). The acetylation was also required for the ability of p105Rb to promote the subsequent steps in differentiation as determined by the production of differentiation specific markers, such as the myosin heavy chain. The acetylase, P/CAF, was shown to bind to and acetylate p105Rb both *in vitro* and *in vivo*. p105Rb is also acetylated during keratinocyte differentiation and using interfering RNA (RNAi), p105Rb has been shown to be required for permanent cell cycle arrest prior to differentiation (D. Nguyen, L. Baglia, D.J. McCance, unpublished data). In addition, in the absence of p105Rb, there is a reduction in differentiation-specific markers and morphological changes, in organotypic cultures, consistent with inhibition of differentiation. This was also observed in mouse skin (Balsitis et al., 2003), when p105Rb was specifically knocked out in skin of the animals, using the cre-lox technology. In the p105Rb knockout mouse skin, there was a delay in differentiation, with extensive parakeratosis (the retention of nuclei in the horny layer of the skin). Additionally, as observed with human organotypical culture *in vitro*, in the mouse skin there is a significant reduction of differentiation-specific markers such as keratin K10 (suprabasal) and an increase of basal markers, such as keratin K14, in the suprabasal cells. If acetylated p105Rb is required for the differentiation of keratinocytes, what effect does E7 have on acetylation? Initial results suggest that E7 can inhibit the activity of

P/CAF and reduce the acetylation of p105Rb. Therefore, this is the first indication that E7 may have an effect on p105Rb, not only by binding to the protein, but by inhibiting an important modification required for keratinocyte differentiation.

What is the function of the acetylated p105Rb? Two papers (MacLellan et al., 2000; Miyake et al., 2000) described the interaction of p105Rb with the ubiquitin ligase MDM2 and a protein of unknown function E1A-like inhibitor of differentiation 1 (EID-1). This latter protein, when expressed in cells inhibits differentiation and maintains the cells in cycle. It has also been shown to bind to p300, a transcriptional co-activator and inhibit its ability to activate MyoD, which is a major regulator of muscle differentiation. When muscle cells are induced to differentiate, MDM2 binds to the C-terminus of pRbp105 and EID-1 to the pocket region, resulting in the ubiquitination and degradation of EID-1 by MDM2 and successful passage through to terminal differentiation. It was later found that the acetylation of p105Rb, at amino acids 873 and 874 in the C-terminus, during early differentiation was required for the binding of hdm2/MDM2 to the C-terminus and the subsequent degradation of EID-1 (Nguyen et al., 2004). There are three members of the EID-1 family, but their function in the cells is still unclear, and no specific functions for any one of them has been delineated. This work has shown that as well as an effect of transcription, p105Rb can affect the stability of proteins at a post-translational level. Since it appears that E7 can inhibit acetylation of p105Rb, it will be important to determine the affect of E7 on this pathway. The prediction would be that E7 would stabilize EID-1, leading to inhibition of keratinocyte differentiation and cell cycle progression.

2.4. Degradation of the Rb family mediated by E7

Another mechanism to abrogate the function of the Rb family would be to alter the stability of the proteins. Several groups have reported that the half-life of the Rb family is significantly reduced in the presence of E7 (Berezutskaya and Bagachi, 1997; Giarre et al., 2001; Helt and Galloway, 2001; Jones et al., 1997b). This would be another obvious way in which E7 could reduce the repressive effects of the Rb family. There is no doubt that the half-life of Rb is reduced in the presence of E7, but in the situation where E6 is also expressed, the ability of E7 to destabilize Rb is reduced. Therefore, since E6 is usually expressed with E7, the significance of the stability of the Rb family is not so clear. In addition, some mutants of E7 that can still cause degradation of Rb family members, do not bypass arrest signals and do not extend the life span of keratinocytes (Helt and Galloway, 2001). HPV-16 E7 was shown in yeast to bind to the subunit S4-ATPase component of the 26S proteasome and the C-terminal zinc finger was required for binding. However, all the experiments were either in yeast or *in vitro* and no interaction was shown in mammalian cells. Therefore, it still remains unclear how E7 facilitates the degradation of Rb.

2.5. Interaction with AP-1 factors

Further attempts to identify Rb independent functions of E7 have led to the discovery that E7 can interact with members of the AP-1 family of transcription factors, including c-Jun, JunB, JunD and c-Fos (Antinore et al., 1996). AP-1 transcription factors appear to mediate early mitogenic effects and are also implicated in keratinocyte and myeloid cell differentiation (Angel and Karin, 1991; Eckert and Welter, 1996). Specific mutational analysis and binding data indicate that the E7 zinc finger motif, but not the Rb binding domain, is involved in this interaction. Moreover, E7 binds to a region between amino acids 224 and 249 of c-Jun and can trans-activate transcription from a Jun responsive promoter. The E7/c-Jun interaction was further demonstrated to be important in the ability of E7 to transform rat embryo fibroblasts (REFs) in the presence of an activated *ras*.

It has also been observed that c-Jun binds to the Rb family, although most of the work has been carried out on the interaction with p105Rb. The interaction of c-Jun with p105Rb results in a Rb-dependent regulation of promoters containing AP-1 sites (Nead et al., 1998; Nishitani et al., 1999). The interaction between p105Rb and c-Jun is mediated via the leucine zipper of c-Jun and the B pocket and C-terminus of p105Rb. c-Jun does not bind in the shallow groove of the p105Rb B-pocket, which binds E7, since the mutation C706F, which abrogates binding of E7 and other viral proteins, does not inhibit the binding of c-Jun (Lee et al., 1998). c-Jun binds in a highly conserved domain which acts as an interface between the A and B-pockets of p105Rb ((Lee et al., 1998), Figure 2). This area is also the binding site of the E2F family of transcription factors. p105Rb causes an increase in c-Jun binding to an AP-1 consensus site resulting in activated transcription from c-Jun responsive promoters. This transcriptional activation is probably due to the ability of p105Rb to compete with a c-Jun specific inhibitory function that represses binding and subsequent transcription. Interestingly, it was demonstrated that the presence of E7 inhibited p105Rb activation of c-Jun transcription, and that this effect was mediated through the LXCXE motif of E7, suggesting that transcriptional down-regulation by E7 in this case is p105Rb dependent. However, since E7 and c-Jun bind to different regions of the B-pocket this inhibition cannot simply be due to steric hindrance. A similar situation is true for E7 and E2F in that the abrogation of E2F repression by Rb is more complex than just competition for the binding site on Rb. It was also shown that hypophosphorylated Rb is complexed to c-Jun in terminally differentiating keratinocytes but not in cycling cells. Finally, in terminally

Table 1. Immortalization of human primary keratinocytes by HPV-16 E7 and mutants[a]

E7	Binding Rb	Binding HDAC activity	Degradation of Rb	Immortalization by E7 alone		Immortalization in co-operation with E6	
				Mass	Clones	Mass	Clones
Wild type	++++	++++	++++	3/3	7/14	4/4	17/17
H2P	++++	++++	-[b]	-[c]	NT[d]	NT	NT
C24G	-	++++	-[b]	0/3	0/16	1/4	2/12
L67R	++++	-	NT	0/3	0/16	3/3	11/11
S71C	++++	++++	NT	2/3	3/16	3/3	11/11
R77G	++++	++++	NT	3/3	9/16	3/3	10/11
LL82/83RR	++++	-	NT	0/3	0/16	3/3	11/11

[a] The number indicates the amino acid mutated.
[b] Taken from reference Helt and Galloway (2001)
[c] The E7 mutant H2P has been shown in mass cultures to be unable to immortalize keratinocytes (Helt and Galloway, 2001). No information in co-operation with E6 is available.
[d] Not Tested

differentiating keratinocytes, the presence of E7 seems to cause a significant reduction in c-Jun levels.

The data discussed above suggest an intricate mode of action whereby HPV E7 can modulate both the process of cell cycle progression as well as cell differentiation, through interactions with AP-1 factors and p105Rb. On the one hand, by binding to c-Jun independently of p105Rb, E7 may potentiate the activation of genes involved in early cell cycle progression and thus promote S phase entry. On the other hand, by associating with both p105Rb and c-Jun, E7 may de-regulate keratinocyte differentiation via disruption of p105Rb/c-Jun complexes. It should be noted that p105Rb, unlike the other family members, has been implicated in cell differentiation. As such, the differential targeting of AP-1 factors and p105Rb provides a potential mechanism used by E7 to uncouple differentiation from cell cycle progression.

2.6. Interaction with TBP and TAFs

The interactions described so far implicate the association of E7 with proteins that regulate the transcription of specific genes; namely those involved in cell cycling and differentiation. E7 however has been found to associate with members of the basal transcriptional machinery. E7 can bind to the TATA-binding protein (TBP; (Phillips and Vousden, 1997)) and to TBP-associated factor-110 (TAF-110; (Mazzarelli et al., 1995)). Binding to TBP seems to require three regions of E7: the Rb binding domain, the CKII domain and the CR3 region (Phillips and Vousden, 1997). The importance of these interactions is unclear, but they suggest that E7 may mediate its effects on gene transcription by targeting the more general transcription machinery.

2.7. Other E7 interactions

The E7 from the high risk HPV-16 was shown in different cell types to bind directly to E2F1 and activate transcription from different promoters known to be activated by E2F-1 such as the dihydrofolate reductase (DHFR) and the cyclin E promoters (Hwang et al., 2002). The activation was independent of the ability of either protein to bind to the Rb family. Therefore, it would appear that E7 can relieve repression of E2F-1 by binding to Rb and also activate E2F-1 transcription by binding directly to the transcription factor .

Cells expressing E7 have an increased potential for growth and to accommodate this there presumably has to be an increase in metabolism of the cells. The glycolytic phosphometabolites, which are necessary for the biosynthesis of nucleic acid, phospholipids and complex carbohydrates are up-regulated in G1 phase of the cell cycle. The key enzymes are hexokinases, 6-phosphofructo-1-kinase and especially pyruvate kinase (PK). HPV-16 E7 has been shown to interact with M2 isoenzyme of PK and to shift the conformation of the enzyme to the dimeric form from the more active tetrameric form (Zwerschke et al., 1999). Why shifting the conformation to a less active form is beneficial for the virus is unclear. E7 appears to bind to the enzyme, although it is difficult to imagine that the amount of E7 in a cell could possibly bind all the M2-PK enzyme, unless a subset of the enzyme complex is targeted. In addition, the enzyme is cytoplasmic yet most of E7 is nuclear, which again casts doubt on the biological significance of the findings.

Over expression of the mouse ARF protein (p19ARF) was shown to relocalize E7 from the nucleoplasm to the nucleolus in human cells (Pan et al., 2003). One consequence of this relocation was the inhibition of the ability of E7 to stimulate cellular DNA synthesis. There is no detectable interaction between E7 and ARF and so the mechanism of this relocation is unclear and whether it is the consequence of over-expressing ARF has not been clarified.

3. Effects of E7 on Cell Growth and Immortalization:

E7 has been demonstrated to alter growth phenotypes in a variety of cell types, the most biologically relevant of which is primary human foreskin keratinocytes (HFK), or keratinocytes from other sites, such as the cervix or oral cavity. Immortalization of HFK correlates with oncogenicity of HPVs, with high-risk HPVs (*e.g.* HPV-16, -18, -31) able to efficiently induce immortalization and low-risk HPVs (*e.g.* HPV-6, -11) incapable of significantly extending HFK lifespan (Barbosa and Schlegel, 1989; Halbert et al., 1992; Hudson et al., 1990; McCance et al., 1988; Pirisi et al., 1987). Expression of E7 alone from oncogenic HPV types is sufficient to immortalize HFK, although in our hands immortalization only occurs in medium containing serum and feeder cells, and not in defined media such as Keratinocyte Growth Medium (KGM, Clonetics) or Epilife (Cascade Biologicals, H.R. McMurray, D. Nguyen, and D.J.McCance, unpublished data). When E7 is present alone, it appears that the CR1, CR2 and CR3 domains are important for immortalization (Helt and Galloway, 2001). Both p105Rb binding and the ability to associate with histone deacetylases (HDAC) are therefore important properties of E7 for immortalization. However, the immortalizing activity of E7 is more efficient in the context of E6 expression or the entire HPV genome and not all the domains are required, compared to when E7 is expressed alone (Table 1).

Most of the studies on immortalization have been carried out using mass cultures rather than clonal lines and these former studies have shown that the CR1 and CR3 regions of E7 appear critical for its immortalizing activity, in co-operation with E6 (Jewers et al., 1992; Takami et al., 1992). For instance, a mutation at the second amino acid, or one which disrupts the zinc finger in the CR3 domain and abrogates the association of E7 with HDAC, abolish the ability of E7 to immortalize HFK in the context of the whole genome (Jewers et al., 1992). This latter mutation reduces the stability of E7 and probably disrupts the structure of the protein and so both scenarios may account for the dramatic reduction in immortalization. This conclusion is strengthened by the results showing that other mutations in the zinc finger which inhibit the ability of E7 to associate with HDAC, but do not reduce stability are still able to efficiently immortalize HFK (Table 1), in co-operation with E6.

When using the ability of clones to grow and become immortalized as a measure of the immortalization process, it appears that the p105Rb binding domain (CR2) is also important for efficient immortalization, even in the presence of E6. The immortalization ability of a selection of E7 mutants is shown in Table 1. There is also reduced immortalization frequency when the same E7 mutant is in the context of the whole genome (Jewers et al., 1992) and the results as a whole suggest that the interaction with the Rb family may be important for efficient immortalization in cell culture. It is possible that in cells where the E7 mutant C24G which does not bind Rb family members, is present and the cells become immortal, that there are other spontaneous mutations in other parts of the Rb pathway, such as p16^{INK4a} (Dickson et al., 2000). Such mutations have been observed in spontaneously immortalized keratinocytes in culture. What is not clear, is if Rb binding is required in the epithelium for viral DNA replication and production of premalignant lesions. Work with the cottontail rabbit papillomavirus (CRPV) showed that the full length genome, with an E7 mutation that cannot bind Rb family members, was able to produce wart lesions when injected into the skin of domestic rabbits (Defeo-Jones et al., 1993) at the same frequency as the wild type genome. Warts produced by CRPV DNA become carcinomas and frequently metastasize, but in this experiment no information was presented on whether these lesion progressed to malignancy. These results suggest that the binding of Rb may be dispensable for the production of premalignant disease, but leave open the question whether binding is necessary for malignant conversion.

E7 can also induce cellular transformation in other cell types as measured by different assays. HPV-16 E7 can induce anchorage-independent growth in NIH3T3 and, in cooperation with an activated *ras*, induce focus formation in REFs, baby rat kidney cells, and various rodent fibroblast cell lines (Munger and Phelps, 1993). Similar to HFK immortalization, E7 transformation of rodent cells correlates with oncogenic potential of HPVs, and chimeric molecules (utilizing domains of E7 from high risk HPV-16 and low risk HPV-6 in fusion constructs) indicate that the CR1 and CR2 regions of HPV-16 E7 are essential for this phenotype (Munger et al., 1991; Takami et al., 1992). Mutational analysis has corroborated the importance of the CR1 and CR2 domains and shown that Rb binding is required for transformation (Banks et al., 1990; Brokaw et al., 1994; Chesters et al., 1990; Heck et al., 1992; Phelps et al., 1992). In addition, mutation of the zinc-finger (CR3) reduces the transformation potential of E7; however, since the CR3 mutations used in these studies disrupted the zinc finger structure, it is not clear whether these mutations disrupt CR3-specific activities necessary for transformation or simply alter the structure of E7 (Chesters et al., 1990; Edmonds and Vousden, 1989; McIntyre et al., 1993). It appears that overlapping domains of E7 are essential for rodent cell transformation and HFK immortalization.

In addition to its immortalizing and transforming activities, E7 from HPV-16 but not from HPV-6, allows cells to bypass several growth arrest signals including exposure to transforming growth factor-β, DNA damage, serum deprivation, anchorage-independent suspension, and suprabasal differentiation (Banks et al., 1990; Demers et al., 1996; Schulze et al., 1998). Although the precise mechanisms by which these signals lead to arrest are not completely defined, E7 has been shown to interact with several cellular factors which control transcription and cell cycle regulation.

4. Effect of E7 on Cyclin/kinase Activity

As previously discussed, viral genome amplification occurs in the terminally differentiated part of the epithelium, which would normally not contain components of the DNA replicative machinery. Therefore, for HPV to amplify its genome, it must stimulate the cell to provide replication components and activities in the absence of growth signals. The G1/S cyclins and cyclin-dependent kinases (cyclin D-cdk4, cyclin E-cdk2, and cyclin A-cdk2) are essential regulators of progression from the first gap phase G1 through DNA synthesis (Sherr, 1993; Sherr and Roberts, 1999). In a proliferating cell, mitogenic stimulation signals the synthesis and assembly of cyclin D-cdk4 complexes which contribute to inactivation of Rb via phosphorylation, leads to expression of cyclin E, and sequesters inhibitors of cdk2 (Cip/Kip inhibitors). Subsequently, cyclin E-cdk2 continues to inactivate Rb and phosphorylates other substrates essential for initiation of DNA replication and S phase entry. During S phase, cyclin A-cdk2 is assembled and remains active through the second gap G2 phase. As integral components of the G1/S and S/G2 transitions, cyclin-cdks are logical targets of HPV. The effect of E7 on each of these cyclin-cdks is discussed below.

Cyclin E-cdk2 is essential for initiation of DNA synthesis and activates cell cycle progression even in the absence of cyclin D-cdk4 (Geng et al., 1999; Krude et al., 1997, Ohtsubo et al., 1995), underlining this complex as a potential target by which E7 propels cell cycle re-entry. Interestingly, several groups have demonstrated E7-mediated increases in cyclin E-cdk2 activity in asynchronous HFK (Funk et al., 1997; Jones et al., 1997a; Ruesch and Laimins, 1998; Ruesch and Laimins, 1997; Westbrook et al., 2002). In addition, E7-expressing cells maintain cyclin E-cdk2 activity in the presence of growth arrest signals such as epithelial differentiation, serum deprivation, and anchorage-independent growth (Ruesch and Laimins, 1998; Schulze et al., 1998). This phenotype is abrogated by mutation of the CR1 and CR2 region of E7, which are important for cellular immortalization and transformation (see above). The mechanism by which E7 affects cyclin E-cdk2 activity *in vivo* has not been clearly defined, though several models have been proposed. The presence of E7 dysregulates cyclin E expression at the transcriptional (via relief of E2F-dependent repression) and post-transcriptional levels, suggesting that E7 may, in part, affect cyclin E-cdk2 activity via increased synthesis of cyclin E (Botz et al., 1996; Martin et al., 1998; Zerfass et al., 1995). However, the contribution of elevated cyclin E expression is, alone, unlikely to be sufficient to increase cyclin E-cdk2 activity, because E7-expressing cells also exhibit dramatically elevated total and cdk2-associated levels of p21^{CIP-1}, a potent inhibitor of cdk2 (Funk et al., 1997; Jones et al., 1997a; Westbrook et al., 2002). High exogenous or endogenous cyclin E expression has been shown to be unable to maintain cyclin E-cdk2 activity in the presence of high levels of the Cip/Kip family of cdk inhibitors, p21^{CIP-1} and p27^{KIP-1} (Alevizopoulos et al., 1998; Sewing et al., 1997).

E7 has been detected in complex with p107-cyclin E-cdk2, suggesting that E7 may affect cyclin E-cdk2 through physical interaction (McIntyre et al., 1993). Indeed, recent evidence also suggests that E7 interacts with p21^{CIP-1} and p27^{KIP-1} *in vitro* and *in vivo*; *in vitro* experiments have shown that E7 can bind to and derepress the kinase activity of p21^{CIP-1}-cyclin E-cdk2 complexes (Funk et al., 1997; Jones et al., 1997a). However, it is not clear whether active E7-p21^{CIP-1}-cyclin E-cdk2 complexes exist *in vivo*. Other reports refute the interaction between E7 and p21^{CIP-1} (Hickman et al., 1997; Ruesch and Laimins, 1997). Our own research would concur with the latter findings and shows that E7 upregulates the level of p21^{CIP-1}, but that the cellular localization is altered in the presence of E7 with p21^{CIP-1} found in the cytoplasm rather than the nucleus (Westbrook et al., 2002). This is achieved by an increase in Akt (protein kinase B) activity in the presence of E7 where Akt has been shown to phosphorylate p21^{CIP-1} in the nuclear localization domain of the protein, causing accumulation in the cytoplasm (Zhou et al., 2001). An E7 mutant unable to bind Rb family members is defective in regulating Akt kinase activity, although the significance of this is unclear. Therefore, E7 indirectly down-regulates the ability of p21^{CIP-1} to abrogate the cyclinE/cdk2 activity by altering its cellular localization such that it is confined to the cytoplasm and thus unable to inhibit cyclinE/cdk2 kinase activity in the nucleus.

Another mechanism by which E7 may activate cyclinE/cdk2 is through cdc25A. This protein is a phosphatase, which removes an inhibitory phosphate group from amino acid 15 on cdk2 (Jinno et al., 1994), allowing it to be activated by a cdk2-associated kinase (CAK) on a threonine residue at amino acid 160 (Nagahara et al., 1999). E7 has been shown to up-regulate cdc25A at the transcriptional level by relieving Rb and HDAC repression (Katich et al., 2001; Nguyen et al., 2002). This regulation is independent of cell cycle progression (Nguyen et al., 2002), indicating that E7 is having a direct effect on the cdc25A promoter and the up-regulation is not due to an indirect effect of cells progressing through the cycle. Therefore E7 may regulate cyclinE/cdk2 activity through transcriptional up-regulation of (1) cdc25A which will in turn activate cyclin E/cdk2 complexes and (2) Akt activity which will cause the cytoplasmic localization of p21^{CIP} and so prevent it from abrogating the activity of the cyclinE/cdk2 kinase complexes.

The effects of E7 on cyclin A-cdk2 (which regulates the progression of S phase) are similar to those observed

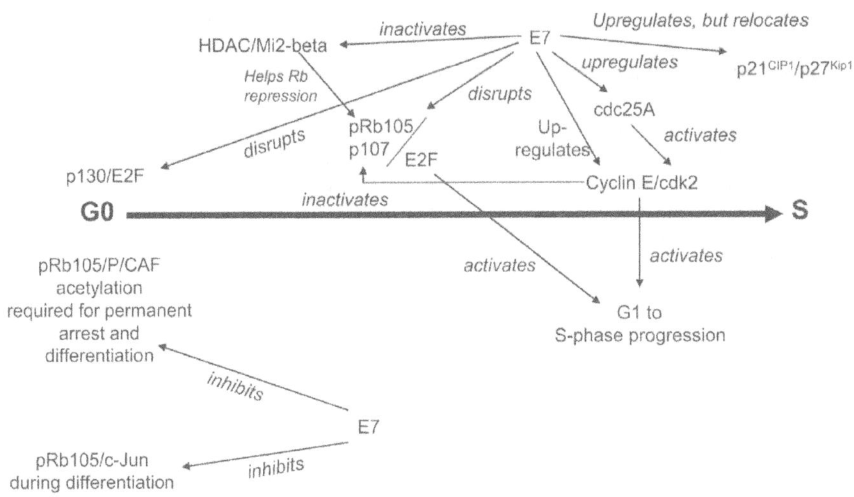

Figure 3. Summary of the functions of E7 during differentiation and G1 to S-phase of the cell cycle.

for cyclin E-cdk2. Cyclin A-cdk2 activity is elevated in asynchronous HFK expressing E7 (Martin et al., 1998). In supporting data, E7-expressing fibroblasts synthesize cyclin A in the presence of growth arrest signals such as anchorage-independent suspension and serum starvation (Schulze et al., 1998; Zerfass et al., 1995). E7 has also been demonstrated to bind cyclin A-cdk2 in a complex involving E2F and p107 (Arroyo et al., 1993; Pagano et al., 1992). The effect of E7 on the activities of the cyclin A-cdk2-E2F-p107 complex has not been elucidated, although E7 appears unable to disrupt the complex. However, it is noteworthy that cyclin A-cdk2 inhibits the DNA-binding activity of E2F complexes as cells undergo DNA replication, and the expression of mutant E2F proteins lacking the cdk-binding site generates stable DNA-binding E2F complexes and arrests cells in S phase (Krek et al., 1995). Therefore any alteration in the activity of the cyclinA-cdk2-p107 by E7 may have an effect on S phase length and because HPV requires the cellular DNA replication machinery for its own DNA synthesis, prolonging the time during which the terminally differentiating host cell is competent to synthesize DNA would benefit the replicative requirements of the virus. In concert, the effects of E7 on dysregulation of cyclin E-cdk2 (which regulates initiation of S phase) and cyclin A-cdk2 (which regulates the progression of S phase) may contribute to extending the duration of "replicative competence" in the host cell.

5. Conclusions

HPV papillomaviruses replicate and complete their life cycle in terminally differentiating cells. Since this cellular environment will not be conducive to viral genome replication, the early viral proteins need to stimulate cells into the S-phase of the cycle to create an environment for efficient DNA replication. The biological functions of E7 are important for this transition and the protein acts on pathways that are involved in cell arrest, differentiation and those that are essential for G1 to S-phase progression. These pathways include those that are regulated by the Rb family which have effects on cell arrest, differentiation and transcription of proteins regulating the G1 to S-phase transition, the pathways that control the chromatin structure which in turn will determine the transcriptional pattern in the cells and pathways stimulated by the G1 cyclin kinases. All of these pathways interact at different stages and so E7 has multiple input signals to bypass arrest and stimulate G1 to S-phase progression. An attempt to summarize the affects of E7 on the cell cycle is shown in Figure 3.

Acknowledgements

The author would like to acknowledge the funding from National Institutes of Health, USA.

References

Alevizopoulos, K., Catarin, B., Vlach, J., and Amati, B. (1998). A novel function of adenovirus E1A is required to overcome growth arrest by the CDK2 inhibitor p27(Kip1). EMBO J. *17*, 5987-5997.

Alonso, L. G., Garcia-Alai, M. M., Nadra, A. D., Lapena, A. N., Almeida, F. L., Gualfetti, P., and Prat-Gay, G. D. (2002). High-risk (HPV16) human papillomavirus E7 oncoprotein is highly stable and extended, with conformational transitions that could explain its multiple cellular binding partners. Biochem. *41*, 10510-10518.

Angel, P., and Karin, M. (1991). The role of Jun, Fos and the AP-1 complex in cell-proliferation and transformation. Biochim. Biophys. Acta *1072*, 129-157.

Antinore, M. J., Birrer, M. J., Patel, D., Nader, L., and McCance, D. J. (1996). The human papillomavirus type 16 E7 gene product interacts with and trans-activates the AP1 family of transcription factors. EMBO J. *15*, 1950-1960.

Arroyo, M., Bagchi, S., and Raychaudhuri, P. (1993). Association of the human papillomavirus type 16 E7 protein with the S-phase-specific E2F-cyclin A complex. Mol. Cell. Biol. *13*, 6537-6546.

Balsitis, S. J., Sage, J., Duensing, S., Munger, K., Jacks, T., and Lambert, P. F. (2003). Recapitulation of the effects of the human papillomavirus type 16 E7 oncogene on mouse epithelium by somatic Rb deletion and detection of pRb-independent effects of E7 *in vivo*. Mol. Cell Biol. *23*, 9094-9103.

Banks, L., Edmonds, C., and Vousden, K. (1990). Ability of the HPV16 E7 protein to bind RB and induce DNA synthesis is not sufficient for efficient transforming activity in NIH3T3 cells. Oncogene *5*, 1383-1389.

Barbosa, M. S., Edmonds, C., Fisher, C., Schiller, J. T., Lowy, D. R., and Vousden, K. H. (1990). The region of the HPV E7 oncoprotein homologous to adenovirus E1a and Sv40 large T antigen contains separate domains for Rb binding and casein kinase II phosphorylation. EMBO J. *9*, 153-160.

Barbosa, M. S., and Schlegel, R. (1989). The E6 and E7 genes of HPV-18 are sufficient for inducing two-stage *in vitro* transformation of human keratinocytes. Oncogene *4*, 1529-1532.

Berezutskaya, E., and Bagachi, S. (1997). The human papillomavirus E7 oncoprotein functionally interacts with the S4 subunit of the 26S proteasome. J. Biol. Chem. *272*, 30135-30140.

Botz, J., Zerfass-Thome, K., Spitkovsky, D., Delius, H., Vogt, B., Eilers, M., Hatzigeorgiou, A., and Jansen-Durr, P. (1996). Cell cycle regulation of the murine cyclin E gene depends on an E2F binding site in the promoter. Mol. Cell Biol. *16*, 3401-3409.

Brehm, A., Miska, E. A., McCance, D. J., Reid, J. L., Bannister, A. J., and Kouzarides, T. (1998). Retinoblastoma protein recruits histone deacetylase to repress transcription [see comments]. Nature *391*, 597-601.

Brehm, A., Nielsen, S. J., Miska, E. A., McCance, D. J., Reid, J. L., Bannister, A. J., and Kouzarides, T. (1999). The E7 protein associates with Mi2 and histone deacetylase activity to promote cell growth. EMBO J. *18*, 2449-2458.

Brokaw, J. L., Yee, C. L., and Munger, K. (1994). A mutational analysis of the amino terminal domain of the human papillomavirus type 16 E7 oncoprotein. Virology *205*, 603-607.

Chan, H. M., Krstic-Demonacos, M., Smith, L., Demonacos, C., and La Thangue, N. B. (2001). Acetylation control of the retinoblastoma tumour-suppressor protein. Nat. Cell Biol. *3*, 667-674.

Chesters, P. M., Vousden, K. H., Edmonds, C., and McCance, D. J. (1990). Analysis of human papillomavirus type 16 open reading frame E7 immortalizing function in rat embryo fibroblast cells. J. Gen. Virol. *71*, 449-453.

Davies, R., Hicks, R., Crook, T., Morris, J., and Vousden, K. (1993). Human papillomavirus type 16 E7 associates with a histone H1 kinase and with p107 through sequences necessary for transformation. J. Virol. *67*, 2521-2528.

de Bruin, A., Maiti, B., Jakoi, L., Timmers, C., Buerki, R., and Leone, G. (2003). Identification and characterization of E2F7, a novel mammalian E2F family member capable of blocking cellular proliferation. J. Biol. Chem. *278*, 42041-42049.

Defeo-Jones, D., Vuocolo, G. A., Haskell, K. M., Hanobik, M. G., Kiefer, D. M., McAvoy, E. M., Ivey-Hoyle, M., Brandsma, J. L., Oliff, A., and Jones, R. E. (1993). Papillomavirus E7 protein binding to the retinoblastoma protein is not required for viral induction of warts. J. Virol. *67*, 716-725.

Demers, G. W., Espling, E., Harry, J. B., Etscheid, B. G., and Galloway, D. A. (1996). Abrogation of growth arrest signals by human papillomavirus type 16 E7 is mediated by sequences required for transformation. J. Virol. *70*, 6862-6869.

Di Stefano, L., Jensen, M. R., and Helin, K. (2003). E2F7, a novel E2F featuring DP-independent repression of a subset of E2F-regulated genes. Embo J. *22*, 6289-6298.

Dickson, M. A., Hahn, W. C., Ino, Y., Ronfard, V., Wu, J. Y., Weinberg, R. A., Louis, D. N., Li, F. P., and Rheinwald, J. G. (2000). Human keratinocytes that express hTERT and also bypass a p16(INK4a)-enforced mechanism that limits life span become immortal yet retain normal growth and differentiation characteristics. Mol. Cell. Biol. *20*, 1436-1447.

Dowdy, S. F., Hinds, P. W., Louie, K., Reed, S. I., Arnold, A., and Weinberg, R. A. (1993). Physical interaction of the retinoblastoma protein with human D cyclins. Cell *73*, 499-511.

Dunaief, J. L., Strober, B. E., Guha, S., Khavari, P. A., Alin, K., Luban, J., Begemann, M., Crabtree, G. R., and Goff, S. P. (1994). The retinoblastoma protein and BRG1 form a complex and cooperate to induce cell cycle arrest. Cell *79*.

Dyson, N. (1998). The regulation of E2F by Rb-family proteins. Genes Dev. *12*, 2245-2262.

Dyson, N., Guida, P., Munger, K., and Harlow, E. (1992). Homologous sequences in adenovirus E1A and human papillomavirus E7 proteins mediate interaction with the same set of cellular proteins. J. Virol. *66*, 6893-6902.

Dyson, N., Howley, P. M., Munger, K., and Harlow, E. (1989). The human papilloma virus-16 E7 oncoprotein is able to bind to the retinoblastoma gene product. Science *243*, 934-937.

Eckert, R. L., and Welter, J. F. (1996). Transcription factor regulation of epidermal keratinocyte gene expression. Mol. Biol. Reports *23*, 59-70.

Edmonds, C., and Vousden, K. H. (1989). A point mutational analysis of human papillomavirus type 16 E7 protein. J. Virol. *63*, 2650-2656.

Funk, J. O., Waga, S., Harry, J. B., Espling, E., Stillman, B., and Galloway, D. A. (1997). Inhibition of CDK activity and PCNA-dependent DNA replication by p21 is blocked by interaction with the HPV-16 E7 oncoprotein. Genes Dev. *11*, 2090-2100.

Gage, J. R., Meyers, C., and Wettstein, F. O. (1990). The E7 proteins of the nononcogenic human papillomavirus type 6b (HPV-6b) and of the oncogenic HPV-16 differ in retinoblastoma protein binding and other properties. J. Virol. *64*, 723-730.

Geng, Y., Whoriskey, W., Park, M. Y., Bronson, R. T., Medema, R. H., Li, T., Weinberg, R. A., and Sicinski, P. (1999). Rescue of cyclin D1 deficiency by knock in cyclin E. Cell *97*, 767-777.

Giarre, M., Caldeira, S., Malanchi, I., Ciccolini, F., Joao Leao, M., and Tommasino, M. (2001). Induction of pRb degradation by human papillomavirus type 16 E7 protein is essential to efficiently overcome $p16^{INK4a}$- imposed G1 cell cycle arrest. J. Virol. *75*, 4705-4712.

Halbert, C. L., Demers, G. W., and Galloway, D. A. (1992). The E6 and E7 genes of human papillomavirus type 6 have weak immortalizing activity in human epithelial cells. J. Virol. *66*, 2125-2134.

Heck, D. V., Yee, C. L., Howley, P. M., and Munger, K. (1992). Efficiency of binding the retinoblastoma protein correlates with the transforming capacity of the E7 oncoproteins of the human papillomaviruses. Proceedings of the National Academy of Sciences of the United States of America *89*, 4442-4446.

Helt, A. M., and Galloway, D. A. (2001). Destabilization of the retinoblastoma tumor suppressor by human papillomavirus type 16 E7 is not sufficient to overcome cell cycle arrest in human keratinocytes. J. Virol. *75*, 6737-6747.

Hickman, E. S., Bates, S., and Vousden, K. H. (1997). Pertubation of the p53 response by human papillomavirus type 16 E7. J. Virol. *71*, 3710-3718.

Huang, P. S., Patrick, D. R., Edwards, G., Goodhart, P. J., Huber, H. E., Miles, L., Garsky, V. M., Oliff, A., and Heimbrook, D. C. (1993). Protein domains governing interactions between E2F, the retinoblastoma gene product, and human papillomavirus type 16 E7 protein. Mol. Cell. Biol. *13*, 953-960.

Hudson, J. B., Bedell, M. A., McCance, D. J., and Laiminis, L. A. (1990). Immortalization and altered differentiation of human keratinocytes *in vitro* by the E6 and E7 open reading frames of human papillomavirus type 18. J. Virol. *64*, 519-526.

Hwang, S. G., Lee, D., Kim, J., Seo, T., and Choe, J. (2002). Human papillomavirus type 16 E7 binds to E2F1 and activates E2F1-driven transcription in a retinoblastoma protein-independent manner. J. Biol. Chem. *277*, 2923-2930.

Jenuwein, T., and Allis, C. D. (2001). Translating the histone code. Science *293*, 1074-1080.

Jewers, R. J., Hildebrandt, P., Ludlow, J. W., Kell, B., and McCance, D. J. (1992). Regions of human papillomavirus type 16 E7 oncoprotein required for immortalization of human keratinocytes. J. Virol. *66*, 1329-1335.

Jinno, S., Suto, K., Nagata, A., Igarashi, M., Kanaoka, Y., Nojima, H., and Okayama, H. (1994). cdc25A is a novel phosphatase functioning early in the cell cycle. EMBO J. *13*, 1549-1556.

Jones, D. L., Alani, R. M., and Munger, K. (1997a). The human papillomavirus E7 oncoprotein can uncouple cellular differentiation and proliferation in human keratinocytes by abrogating p21Cip1-mediated inhibition of cdk2. Genes Dev *11*, 2101-2111.

Jones, D. L., Thompson, D. A., and Munger, K. (1997b). Destabilization of the RB tumor suppressor protein and stabilization of p53 contribute to HPV type 16 E7-induced apoptosis. Virology *239*, 97-107.

Katich, S. C., Zerfass-Thome, K., and Hoffmann, I. (2001). Regulation of the Cdc25A gene by the human papillomavirus Type 16 E7 oncogene. Oncogene *20*, 543-550.

Krek, W., Xu, G., and Livingston, D. M. (1995). Cyclin A-kinase regulation of E2F-1 DNA binding function underlies suppression of an S phase checkpoint. Cell *83*, 1149-1158.

Krude, T., Jackman, M., Pines, J., and Laskey, R. A. (1997). Cyclin/cdk-dependent initiation of DNA replication in a human cell-free system. Cell *88*, 109-119.

Lee, J.-O., Russo, A. A., and Pavletich, N. P. (1998). Structure of the retinoblastoma tumor-suppressor pocket domain bound to a peptide from HPV E7. Nature *391*, 859-865.

Ludlow, J. W., DeCaprio, J. A., Huang, C. M., Lee, W. H., Paucha, E., and Livingston, D. M. (1989). SV40 large T antigen binds preferentially to an underphosphorylated member of the retinoblastoma susceptibility gene product family. Cell *56*, 57-65.

Ludlow, J. W., Shon, J., Pipas, J. M., Livingston, D. M., and DeCaprio, J. A. (1990). The retinoblastoma susceptibility gene product undergoes cell cycle-dependent dephosphorylation and binding to and release from SV40 large T. Cell *60*, 387-396.

MacLellan, W. R., Xiao, G., Abdellatif, M., and Schneider, M. D. (2000). A novel Rb- and p300-binding protein inhibits transactivation by MyoD. Mol. Cell Biol. *20*, 8903-8915.

Magnaghi-Jaulin, L., Groisman, R., Naguibneva, I., Robin, P., Lorain, S., Le Villain, J. P., Troalen, F., Trouche, D., and Harel-Bellan, A. (1998). Retinoblastoma protein represses transcription by recruiting a histone deacetylase [see comments]. Nature *391*, 601-605.

Martin, L. G., Demers, G. W., and Galloway, D. A. (1998). Disruption of the G1/S transition in human papillomavirus type 16 E7-expressing human cells is associated with altered regulation of cyclin E. J. Virol. *72*, 975-985.

Massimi, P., and Banks, L. (2000). Differential phosphorylation of the HPV-16 E7 oncoprotein during the cell cycle. Virology *276*, 388-394.

Mazzarelli, J. M., Atkins, G. B., Geisberg, J. V., and Ricciardi, R. P. (1995). The viral oncoproteins Ad5 E1A, HPV16 E7 and SV40 TAg bind a common region of the TBP-associated factor-110. Oncogene *11*, 1859-1864.

McCance, D. J., Kopan, R., Fuchs, E., and Laimins, L. A. (1988). Human papillomavirus type 16 alters human epithelial cell differentiation *in vitro*. Proceedings of the National Academy of Sciences of the United States of America *85*, 7169-7173.

McIntyre, M. C., Frattini, M. G., Grossman, S. R., and Laimins, L. A. (1993). Human papillomavirus type 18 E7 protein requires intact Cys-X-X-Cys motifs for zinc binding, dimerization, and transformation but not for Rb binding. J. Virol. *67*, 3142-3150.

Miyake, S., Sellers, W. R., Safran, M., Li, X., Zhao, W., Grossman, S. R., Gan, J., DeCaprio, J. A., Adams, P. D., and Kaelin, W. G., Jr. (2000). Cells degrade a novel inhibitor of differentiation with E1A-like properties upon exiting the cell cycle. Mol. Cell Biol. *20*, 8889-8902.

Munger, K., and Phelps, W. C. (1993). The human papillomavirus E7 protein as a transforming and transactivating factor. Biochim. Biophys. Acta *1155*, 111-123.

Munger, K., Yee, C. L., Phelps, W. C., Pietenpol, J. A., Moses, H. L., and Howley, P. M. (1991). Biochemical and biological differences between E7 oncoproteins of the high- and low-risk human papillomavirus types are determined by amino-terminal sequences. J. Virol. *65*, 3943-3948.

Nagahara, H., Ezhevsky, S. A., Vocero-Akbani, A. M., Kaldis, P., Solomon, M. J., and Dowdy, S. F. (1999). Transforming growth factor beta targeted inactivation of cyclin E:cyclin-dependent kinase 2 (Cdk2) complexes by inhibition of Cdk2 activating kinase activity. Proc. Natl. Acad. Sci. USA *96*, 14961-14966.

Nead, M. A., Baglia, L. A., Antinore, M. J., Ludlow, J. W., and McCance, D. J. (1998). Rb binds c-Jun and activates transcription. EMBO J. *17*, 2342-2352.

Nguyen, D. X., Baglia, L. A., Huang, S. M., Baker, C. M., and McCance, D. J. (2004). Acetylation regulates the differentiation-specific functions of the retinoblastoma protein. Embo J. *23*, 1609-1618.

Nguyen, D. X., Westbrook, T. F., and McCance, D. J. (2002). HPV-16 E7 maintains elevated levels of the cdc25A tyrosine phosphatase during deregulation of cell cycle arrest. J. Virol. *76*, 619-632.

Nielsen, S. J., Schneider, R., Bauer, U. M., Bannister, A. J., Morrison, A., O'Carroll, D., Firestein, R., Cleary, M., Jenuwein, T., Herrera, R. E., and Kouzarides, T. (2001). Rb targets histone H3 methylation and HP1 to promoters. Nature *412*, 561-565.

Nishitani, J., Nishinaka, T., Cheng, C. H., Rong, W., Yokoyama, K. K., and Chiu, R. (1999). Recruitment of the retinoblastoma protein to c-Jun enhances transcription activity mediated through the AP-1 binding site. J. Biol. Chem. *274*, 5454-5461.

Ohtsubo, M., Theodoras, A. M., Schumacher, J., Roberts, J. M., and Pagano, M. (1995). Human cyclin E, a nuclear protein essential for the G1-to-S phase transition. Mol. Cell. Biol. *15*, 2612-2624.

Pagano, M., Pepperkok, R., Verde, F., Ansorge, W., and Draetta, G. (1992). Cyclin A is required at two points in the human cell cycle. EMBO J. *11*, 961-971.

Pan, W., Datta, A., Adami, G. R., Raychaudhuri, P., and Bagchi, S. (2003). P19ARF inhibits the functions of the HPV16 E7 oncoprotein. Oncogene *22*, 5496-5503.

Patrick, D. R., Oliff, A., and Heimbrook, D. C. (1994). Identification of a novel retinoblastoma gene product binding site on human papillomavirus type 16 E7 protein. J. Biol. Chem. *269*, 6842-6850.

Phelps, W. C., Munger, K., Yee, C. L., Barnes, J. A., and Howley, P. M. (1992). Structure-function analysis of the human papillomavirus type 16 E7 oncoprotein. J. Virol. *66*, 2418-2427.

Phillips, A. C., and Vousden, K. H. (1997). Analysis of the interaction between human papillomavirus type 16 E7 and the TATA-binding protein, TBP. J. Gen. Virol. *78*, 905-909.

Pirisi, L., Yasumoto, S., Feller, M., Doniger, J., and DiPaolo, J. A. (1987). Transformation of human fibroblasts and keratinocytes with human papillomavirus type 16 DNA. J. Virol. *61*, 1061-1066.

Ren, B., Cam, H., Takahashi, Y., Volkert, T., Terragni, J., Young, R. A., and Dynlacht, B. D. (2002). E2F integrates cell cycle progression with DNA repair, replication, and G(2)/M checkpoints. Genes Dev. *16*, 245-256.

Ruesch, M., and Laimins, L. A. (1998). Human papillomavirus oncoproteins alter differentiation-dependent cell cycle exit on suspension in semisolid medium. Virology *250*, 19-29.

Ruesch, M. N., and Laimins, L. A. (1997). Initiation of DNA synthesis by human papillomavirus E7 oncoproteins is resistant to p21-mediated inhibition of cyclin E-cdk2 activity. J. Virol. *71*, 5570-5578.

Sang, B. C., and Barbosa, M. S. (1992). Single amino acid substitutions in "low-risk" human papillomavirus (HPV) type 6 E7 protein enhance features characteristic of the "high-risk" HPV E7 oncoproteins. Proc. Natl. Acad. Sci. USA *89*, 8063-8067.

Schmitt, A., Harry, J. B., Rapp, B., Wettstein, F. O., and Iftner, T. (1994). Comparison of the properties of the E6 and E7 genes of low- and high-risk cutaneous papillomaviruses reveals strongly transforming and high Rb-binding activity for the E7 protein of the low-risk human papillomavirus type 1. J. Virol. *68*, 7051-7059.

Schneider, J. W., Gu, W., Zhu, L., Mahdavi, V., and Nadal-Ginard, B. (1994). Reversal of terminal differentiation mediated by p107 in Rb-/- muscle cells. Science *264*, 1467-1471.

Schulze, A., Mannhardt, B., Zerfass-Thome, K., Zwerschke, W., and Jansen-Durr, P. (1998). Anchorage-independent transcription of the cyclin A gene induced by the E7 oncoprotein of human papillomavirus type 16. J. Virol. *72*, 2323-2334.

Sewing, A., Wiseman, B., Lloyd, A. C., and Land, H. (1997). High-intensity Raf signal causes cell cycle arrest mediated by p21Cip1. Mol. Cell. Biol. *17*, 5588-5597.

Sherr, C. J. (1993). Mammalian G1 cyclins. Cell *73*, 1059-1065.

Sherr, C. J., and Roberts, J. M. (1999). CDK inhibitors: positive and negative regulators of G1-phase progression. Genes Dev. *13*, 1501-1512.

Takami, Y., Sasagawa, T., Sudiro, T. M., Yutsudo, M., and Hakura, A. (1992). Determination of the functional difference between human papillomavirus type 6 and 16 E7 proteins by their 30 N-terminal amino acid residues. Virology *186*, 489-495.

Westbrook, T. F., Nguyen, D., Thrash, B., and McCance, D. J. (2002). E7 abolishes Raf-induced arrest via mislocalization of $p21^{CIP1}$. Mol. Cell. Biol. *22*, 7041-7052.

Xiao, B., Spencer, J., Clements, A., Ali-Khan, N., Mittnacht, S., Broceno, C., Burghammer, M., Perrakis, A., Marmorstein, R., and Gamblin, S. J. (2003). Crystal structure of the retinoblastoma tumor suppressor protein bound to E2F and the molecular basis of its regulation. Proc. Natl. Acad. Sci. USA. *100*, 2363-2368.

Zacksenhaus, E., Jiang, Z., Chung, D., Marth, J. D., Phillips, R. A., and Gallie, B. L. (1996). pRb controls proliferation, differentiation, and death of skeletal muscle cells and other lineages during embryogenesis. Genes Dev. *10*, 3051-3064.

Zerfass, K., Schulze, A., Spitkovsky, D., Friedman, V., Henglein, B., and Jansen-Durr, P. (1995). Sequential activation of cyclin E and cyclin A gene expression by human papillomavirus type 16 E7 through sequences necessary for transformation. J. Virol. *69*, 6389-6399.

Zhang, Y., LeRoy, G., Seelig, H. P., Lane, W. S., and Reinberg, D. (1998). The dermatomyositis-specific autoantigen Mi2 is a component of a complex containing histone deacetylase and nucleosome remodeling activities. Cell *95*, 279-289.

Zhou, B. P., Liao, Y., Xia, W., Spohn, B., Lee, M.-H., and Hung, M.-C. (2001). Cytoplasmic localization of $p21^{CIP1/WAF1}$ by Akt-induced phosphorylation in *HER-2*/neu-overexpressing cells. Nature Cell Biol. *3*, 245-252.

Zwerschke, W., Mazurek, S., Massimi, P., Banks, L., Eigenbrodt, E., and Jansen-Durr, P. (1999). Modulation of type M2 pyruvate kinase activity by the human papillomavirus type 16 E7 oncoprotein. Proc. Natl. Acad. Sci. USA. *96*, 1291-1296.

From: Papillomavirus Research: From Natural History to Vaccines and Beyond. Edited by: M. Saveria Campo

Chapter 11

Organotypic Raft Cultures and the Study of the Natural History of Papillomavirus

Jason M. Bodily, Samina Alam, Horng-Shen Chen and Craig Meyers*

Abstract

Organotypic (raft) cultures have proven to be a valuable tool for understanding the natural history of papillomaviruses (PVs). Through generation of fully differentiated and permissive epithelium *in vitro*, insights have been gained into the response of PVs to differentiation throughout their normal life cycles, as well as how the viruses contribute to loss of differentiation and progression toward malignancy. In this chapter, we will focus entirely on studies of HPV biology that have used raft culture to arrive at important conclusions. We first review the history of raft culture and the use of rafts in the study of PVs and genital cancer. Studies of normal and altered differentiation, production of virions *in vitro*, gene expression and function during the normal life cycle, interactions with the cell cycle machinery, and malignant progression will be discussed. We will also examine studies that use raft culture to understand the immunology of PVs, interaction with other virus types and cofactors, and therapeutic testing. Finally, we will discuss areas of papillomavirus biology that remain largely unexplored but that are accessible for study through the use of this important technology.

1. Introduction

In 1933 Shope and Hurst showed that papillomaviruses (PVs) were pathogenic agents and that they were the causative agent of infectious papillomatosis of cottontail rabbits (Shope and Hurst, 1933). Two years later Rous and Beard reported that benign virus-associated papillomatosis of the rabbit could progress to carcinomas (Rous and Beard, 1935). Interest in PV research waned when numerous attempts to develop systems for the propagation of the virus *in vitro* were unsuccessful. In the late 1970's the discipline of molecular biology was developed with the ability to clone PV genomes. Once the genome was cloned sufficient quantities of materials were available to begin a systematic study of PVs. A third major event that stimulated interest in studying PVs occurred in the 1980's, when collection of human neoplastic tissue specimens allowed for a correlation to be made between PVs and human neoplastic lesions of the anogenital region (Chapter 1).

Today PVs have been established as important and widespread pathogens involved in the etiology of benign and malignant lesions of cutaneous and mucosal epithelium. Until 1992 the study of PVs was hindered by the lack of reproducible cell culture system permissive for the viral life cycle. The development of the organotypic raft culture system has allowed researchers to begin a detailed study of papillomavirus-induced lesions *in vitro*. This chapter is solely dedicated to studies on the biology and natural history of PVs that have used the raft culture system as a foundation. We have previously written several "how to" articles on the techniques to make raft cultures, therefore we will not report that here. Interested readers are recommended to check these articles for practical information on making raft cultures (Meyers, 1996; Meyers, 2002; Meyers et al., 1994; Meyers et al., 1998). We first review the history of raft culture and the use of rafts in the study of PVs and genital cancer. Studies of normal and altered differentiation, production of virions *in vitro*, gene expression and function during the normal life cycle, interactions with the cell cycle machinery, and malignant progression will be discussed. We will also examine studies that use raft culture to understand the immunology of PVs, interaction with other virus types and cofactors, and therapeutic testing. Finally, we will discuss areas of papillomavirus biology that remain largely unexplored but that are accessible for study through the use of this important technology.

2. Organotypic Raft Cultures: A Brief History

Numerous techniques have been used by investigators to accurately culture three-dimensional epithelium *in vitro*. These efforts have led to the development of several systems for the investigation of problems relating to the epidermis (Figure 1). Organotypic culture has proven to accurately mimic the *in vivo* physiology of the epidermis. Organotypic cultures are constructed by placing epithelial cells on top of a dermal equivalent component and then raising the cells to the air-liquid interface. The dermal

*For correspondence. Email cmm10@psu.edu

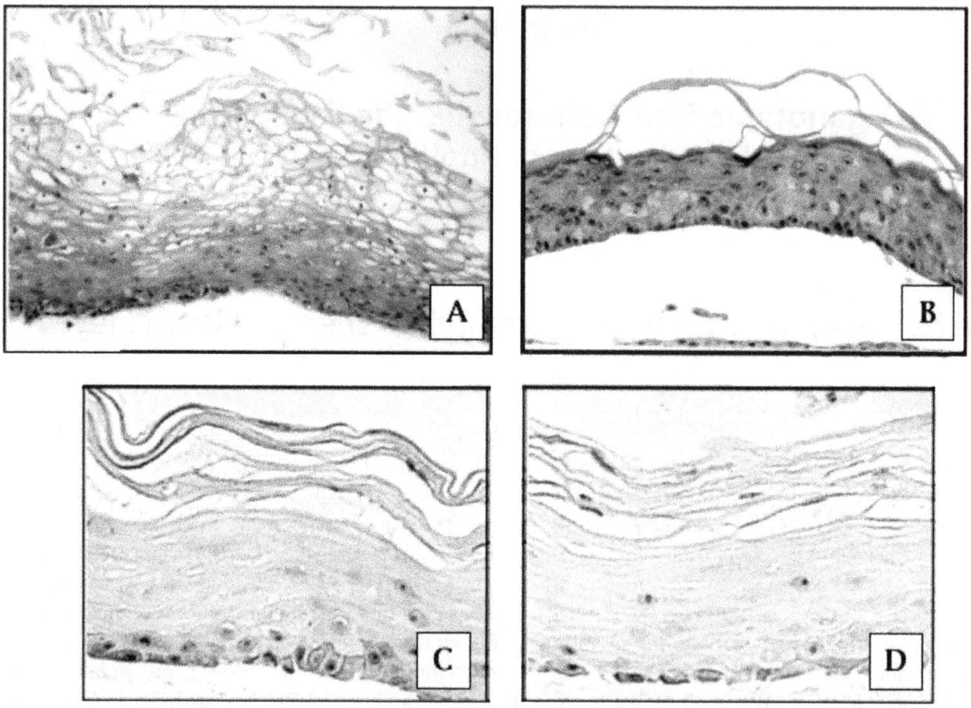

Figure 1. Hematoxylin and eosin staining pattern of three-dimensional raft culture tissues. Primary keratinocytes from different anatomical sites were allowed to fully stratify and differentiate in raft cultures. A) Ectocervix; B) Foreskin; C-D) Buccal mucosa.

equivalent has consisted of a natural dermal elements or a synthetic collagen matrix maintained on a rigid support. Use of either of these techniques has resulted in a more complete differentiation program than is observed in simple monolayer culture. One of the first successful methods used the reticular aspect of split-thickness sections of pig skin as the dermal equivalent and placing human skin explants on top (Freeman et al., 1976). By day 14, the epithelial cells had stratified and differentiated into characteristic basal, spinous, granular, and squamous layers. Successful grafting onto patients with third-degree burns proved that these constructs were functional in place of authentic skin. When deepidermized human skin flaps were used instead of the pig skin, similar results were obtained (Didierjean et al., 1981; Regnier et al., 1984; Woodley et al., 1982). Introduction of collagen matrices as the dermal equivalent simplified the use of organotypic cultures.

The term "raft" culture probably originated when differentiation of liver epithelium was first observed on a collagen matrix that was raised to the air-liquid interface by allowing them to float on the media surface (Michalopoulos and Pitot, 1975). The liver epithelium grown on these raft cultures showed morphological features, extended viability, and functional features comparable to the *in vivo* state. Raft culture systems were then developed for mammary epithelium (Emerman and Pitelka, 1977), rat lingual epithelial cells (Fusenig et al., 1978), and mouse epidermal cells (Lillie et al., 1980), all faithfully reproducing differentiation *in vitro*. Further ease in the use of raft cultures came from the use of silicon chambers (Fusenig et al., 1978) and stainless steel grids (Lillie et al., 1980) to hold the rafts. An improvement in the accuracy of the differentiated state came from the introduction of fibroblasts into the collagen matrix (Bell et al., 1979; 1983). In general, laboratories today use an adaptation of a technique described by Asselineau and Prunieras (Asselineau and Prunieras, 1984). This technique includes type I rat tail collagen mixed with fibroblasts, media, and buffer at 0°C and then placed in tissue culture dishes and incubated at 37°C until the collagen mixture has solidified. The collagen/fibroblast solution forms a lattice or raft that will act as the dermal equivalent. Epithelial cells of choice are seeded on top of the collagen/fibroblast matrix, and allowed to settle, attach, and then grow to confluency while remaining submerged. The raft is then lifted to the air-liquid interface and feeding occurs from below so that the culture media only comes in contact with the bottom of the raft. Incubation time is determined empirically and is dependent on the nature of the epithelial cells used. During the incubation the epithelial cells stratify and differentiate into full-thickness tissue mimicking the *in vivo* situation. Following stratification the rafts can be harvested and examined by standard histological, molecular, and biochemical techniques.

The ability of raft cultures to faithfully mimic the *in vivo* situation is not restricted to non-transformed

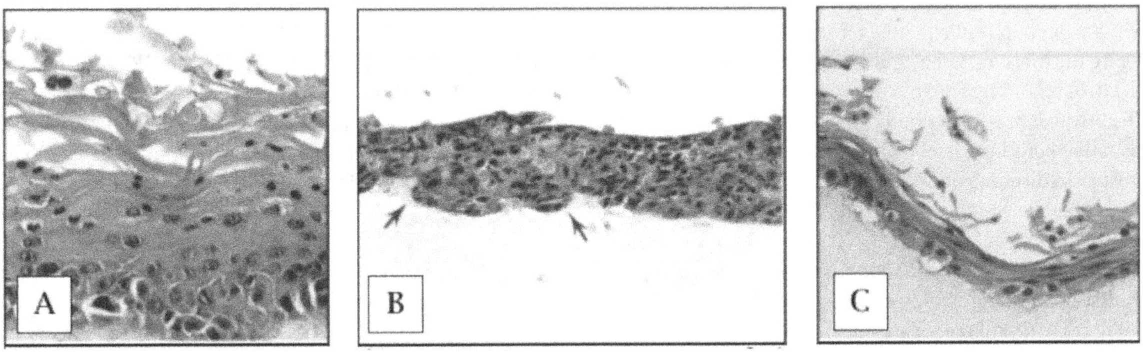

Figure 2. Hematoxylin and eosin staining pattern of three-dimensional raft culture tissues from A) Cervical intraepithelial neoplasia type 1 (CIN-1); B) Invasive carcinoma of the cervix; C) Laryngeal papillomatosis.

epithelium, but rafts can also mimic tumoral architecture *in vitro* using cells derived from neoplasias of all grades (Regnier et al., 1988). Therefore raft cultures are a powerful tool not only for the study of normal epithelia but also for epithelial neoplasia and tumorigenesis (Figure 2). Many important human pathogens, including many viruses, target the epithelium during at least some part of their natural history. Pathogenic viruses use human epithelium as the initial site of infection, the site for replication, a staging area for transportation to other tissues, and a site of latency or persistence. The three-dimensional characteristics of raft cultures provide the unique advantage that they can be studied temporally and spatially. Perhaps most importantly, rafts allow the cells to undergo the authentic program of differentiation, and it is this feature that makes them unique and essential tools for the study of differentiation-dependent biology of PVs.

The majority of raft cultures used to study human papillomaviruses (HPVs) and their associated diseases have used keratinocytes originating from the anogenital region (Figure 2A-B). Additionally, raft culture model systems have been developed using oral cavity epithelial cells for the study of HPV-associated head and neck neoplasias (Figure 1C and 2C) (Aaltonen et al., 1998; Chen et al., 2003; DiLorenzo and Steinberg, 1995; Oda et al., 1996a; Vambutas et al., 1993; Yoo et al., 2000).

3. Differentiation-dependent Molecular Biology of the Life Cycle

In simplest terms, keratinocyte differentiation consists of two parallel and interconnected processes – cell cycle withdrawal and differentiation-specific gene expression – which together result in histological differentiation. HPV responds to differentiation at multiple levels including productive genome replication (Bedell et al., 1991; Flores and Lambert, 1997; Ozbun and Meyers, 1998a), transcription (Ai et al., 2000; Barksdale and Baker, 1993; Grassmann et al., 1996; Hummel et al., 1992; Ozbun and Meyers, 1998b), splicing (Barksdale and Baker, 1995; Ozbun and Meyers, 1997), polyadenylation (Baker and Noe, 1989; Hummel et al., 1995; Terhune et al., 2001; Terhune et al., 1999), translation (Cumming et al., 2003; 2002; Koffa et al., 2000; McPhillips et al., 2004), and virion morphogenesis (Meyers et al., 2002; 1992; 1997). However, the virus is not a passive passenger along the differentiation pathway, particularly in higher grade lesions (Figure 2A-B). Rather, infection alters the behavior of infected cells to increase the number of basal-like layers, induce mitoses in the suprabasal layers by induction (or retention) of DNA synthesis in the suprabasal tissue strata, and alter normal appearance and distribution of differentiation markers (Bedell et al., 1991; McCance et al., 1988). Koilocytes and pyknotic nuclei are also induced by HPV, and nuclei are often retained in the uppermost strata, in contrast to normal differentiation patterns (McCance et al., 1988). However, even while supporting HPV infection, cell lines from low grade lesions or lines derived experimentally that maintain episomal genomes, in which the virus life cycle is proceeding according to the viral genetic program, still retain the capacity to sustain differentiation-specific gene expression (Cheng et al., 1995; Flores et al., 1999; Grassmann et al., 1996; Meyers et al., 1992; Ozbun and Meyers, 1996; 1998a; Rader et al., 1990; Zhao et al., 1997; 1999). The presence of histological differentiation, even in the face of alteration by the virus, illustrates that differentiation is not a single, defined, linear pathway but rather consists of a network of interrelated events that can be separated to some extent. Thus it is important when discussing "differentiation," especially in the context of HPV infection, to specify which aspect of differentiation is referred to and that the experimenters choose carefully to assay what they actually want to determine.

The molecular biology of PVs centers on manipulating, preserving, and/or responding to these differentiation processes. In particular, the virus preserves (or reactivates) the cell cycle in suprabasal strata while allowing the program of differentiation-induced gene expression to proceed (Cheng et al., 1995; Meyers et al.,

1992; Ozbun and Meyers, 1996; 1998a). In this way, the virus is able to replicate its DNA while at the same time using the transcriptional and translational environment of the differentiating cell to control its own gene expression. If this balance is properly maintained, the virus can usually reproduce itself with little more damage to the host than discomfort or cosmetic embarrassment (except the cases in which the benign lesion itself becomes a physical obstruction such as laryngeal papillomatosis). It is only when the replication/gene expression balance is disrupted that pathology develops that is dangerous to both the host and the virus. Thus understanding the ways the virus responds to and manipulates differentiation in the context of normal infection is largely what is meant by understanding the virus life cycle, and this understanding is needed in turn to understand the virus-induced pathology in a biologically meaningful way. Raft cultures are the only system currently available to generate fully differentiated squamous epithelia *in vitro*, and therefore hold an important place in studies of HPV molecular biology.

Much of our understanding of the molecular biology of HPV in the late or productive stage of the life cycle has been gained through studies of raft cultures of the CIN-612 cell line. The CIN-612 cell line was established in collaboration with Dr. George Wilbanks at the Rush Presbyterian-St. Lukes Medical Center in Chicago from a CIN I lesion from a 22 year old woman and contains episomal HPV31b genomes (Bedell et al., 1991; Rader et al., 1990). When grown in raft culture, the morphology of the CIN-612 cells resembled that seen in biopsies from low grade CIN I lesions *in vivo*. Growth properties were first examined by De Geest et al. (De Geest et al., 1993), and the line was first used in molecular studies of HPV31b by Bedell et al. (Bedell et al., 1991). The 9E subclone of the CIN-612 line was used to examine the effects of differentiation on viral gene expression (Hummel et al., 1992). In this section, studies of this line will be discussed prominently along with those of other lines.

3.1. HPV protein functions

Raft cultures have proven valuable in understanding the functions of PV proteins in the context of the viral life cycle. (For a more detailed description of PV protein functions, refer to the chapters dealing with the individual proteins). Using retroviruses containing the HPV18 and 11 URR-driven E6 and E7 genes, it was shown that although these proteins did not radically alter the expression of differentiation markers in the appropriate strata in raft tissue, the E7 protein was necessary and sufficient for suprabasal DNA synthesis and expression of the proliferation marker PCNA (Cheng et al., 1995). What function this serves in the life cycle has not been proven, but it is likely that expression of the DNA synthesis machinery in differentiated cells is needed for the amplification of the viral genome. The expression of E6 and E6* did not affect suprabasal DNA synthesis (Cheng et al., 1995). HPV31 genomes containing null mutations in E6 or E7 were incapable of episomal maintenance in human foreskin keratinocytes (HFKs) (Thomas et al., 1999). Using immortalized cell lines harboring HPV16 E7 mutant genomes, it was found that markers of suprabasal DNA synthesis were also markedly reduced in E7 mutant rafts, as were cell cycle control proteins such as p53, p21, and mdm2 (Flores et al., 2000). Not surprisingly, amplification as assayed by DNA *in situ* hybridization, which is expected to depend on or result from suprabasal DNA synthesis, was abolished in these rafts, and L1 expression was reduced (Flores et al., 2000). E7 null rafts also had a less dysplastic differentiation pattern than wild type rafts and a lower level of apoptosis in suprabasal cells (Flores et al., 2000). Further studies on the action of E6 and E7 will be discussed in detail in later sections of this chapter.

Genomes containing transactivation mutants of E2 were able to undergo episomal replication, amplification, and transcription from the late promoter p742, although with somewhat reduced efficiently, depending on the particular mutation (Stubenrauch et al., 1998a), suggesting that E2 transactivation may not be necessary for many aspects of the life cycle. However, the designation of "transactivation null" for the E2 mutations examined in this study was determined using an artificial E2-dependent luciferase construct in monolayer culture, and not any of the native viral promoters. Thus it remains a question whether these mutations were actually defective for transactivation of viral promoters in the context of the whole viral genome. Whether transactivation by E2 affects cellular promoters in a way meaningful to the virus is not known but may be amenable to study using organotypic cultures.

The small ORF E8, which is imbedded in the E1 ORF, can be spliced to the C terminus of E2 to form the E8^E2C protein, which was shown to be necessary for episomal maintenance of HPV31 genomes despite this protein's inhibitory effect on transient replication (Stubenrauch et al., 2000). Levels of transcripts containing this ORF remain constant upon culture in rafts (Stubenrauch et al., 2000).

The expression of E1^E4 was increased dramatically in raft cultures with a negligible effect of the addition of the synthetic diacylglycerol, 1,2-dioctanoly-sn-diacylglycerol (C8) (Pray and Laimins, 1995). E1^E4 was highly expressed in rare cells in monolayer cultures of 9Es and frequently in suprabasal layers of 9E rafts, regardless of C8 treatment with a diffuse cytoplasmic staining and slight granularity (Pray and Laimins, 1995). Occasional E1^E4 positive cells were observed in the basal layer, as well. Some cells that were not positive for L1 expressed E1^E4, but most cells positive for L1 expressed E1^E4 (Pray and Laimins, 1995) in concordance with studies *in vivo* (Doorbar et al., 1997; Nicholls et al., 2001; Peh et al., 2002). Collapse of the E1^E4 signal into large inclusions was not observed (Pray and Laimins, 1995). This is in contrast to the punctate patterns seen upon transient transfection of E1^E4 into immortalized cell lines (Pray and Laimins, 1995).

In raft cultures of 9E cells, although E5a was present in most early transcripts (Ozbun and Meyers, 1997), its actual protein expression began at around 2 days of culture and peaked around 10-12 days regardless of C8 treatment and diminishing thereafter (Mayer and Meyers, 1998). E5a was concentrated in the basal and granular layers and colocalized with filaggrin at the level of individual cells (Mayer and Meyers, 1998). Filaggrin was occasionally observed in the basal layer of rafts (Mayer and Meyers, 1998). Rafts of HaCaT cells stably expressing high levels of E5 in all cells by means of a heterologous promoter were still able to differentiate, but E5 increased the expression and activation of EGFR and reduced the expression of the gap junctional protein Cx43 (Tomakidi et al., 2000a; 2000b). The physiological relevance of such high levels of E5 is unknown, given that the pattern of expression in virally infected rafts was lower and more sporadic (Mayer and Meyers, 1998). Using a translation terminator approach in the context of the whole viral genomes, Fehrmann et al. (2003) and Genther et al. (2003) showed that the E5 proteins of HPV31 and HPV16 were not necessary for episomal maintenance and did not have an effect on tissue growth or differentiation, but amplification, late promoter activation, and protein expression were reduced relative to wild type in E5 mutants. The production of infectious virus progeny was not assayed.

3.2. Promoters and transcripts

Transcriptional regulation of PV genes has been extensively studied in monolayer cells (see Chapter 4). Here the analysis of transcription regulation is confined to raft cultures.

CIN-612 9E cells have been subjected to extensive transcriptional analyses using raft-derived RNAs to determine the pattern of transcription throughout the viral life cycle (Hummel et al., 1992; Ozbun and Meyers, 1997; 1998b). Generally similar transcripts containing early and late genes in HPV16 and 18 have also been observed (Baker and Calef, 1997; Frattini et al., 1997; Meyers et al., 1997), but a systematic analysis of transcription throughout the viral life cycle of these viruses has not been performed using raft cultures.

In studies of transcription from the stably maintained episomal HPV31b in 9E cells, several promoters have been described, including p99, and p3320 which were constitutive, and p742 which was induced by differentiation, each named for the nucleotide of the major start site (Ozbun and Meyers, 1998b). There is a possible transcriptional start site at nucleotide 834 as assayed by primer extension, but it was not confirmed by S1 or ExoVII analysis (Ozbun and Meyers, 1998b). There also may be a promoter in HPV16 around nucleotide 480 that responds to differentiation (Grassmann et al., 1996). Several minor promoters were found in the URR of HPV31b upstream of p99, including p77, p49, p7855, p7783, and p7375 (Ozbun and Meyers, 1999b). Start sites a few nucleotides upstream of p97 of HPV16 have also been observed (Grassmann et al., 1996). p49 and p7375, like p99 and p77, were constitutive, with activity present in both monolayers and rafts (Ozbun and Meyers, 1999a). p7855 and p7783 were repressed by differentiation in raft culture (Ozbun and Meyers, 1999b). p7783 has the potential to drive expression from two small ORFs in the URR, potentially encoding proteins of 29 and 27 amino acids but which lack apparent signals for efficient translation (Ozbun and Meyers, 1999b). The transcripts initiated from p77 appear to be generally the same as those initiated from p99 except for the additional ~20 nucleotides at the 5′ end (Ozbun and Meyers, 1998b). p3320 was used at relatively high levels throughout the life cycle with E5a and E5b being the first ORFs downstream of the promoter (Ozbun and Meyers, 1998b) That E5a was expressed at relatively constant levels throughout the life cycle (Mayer and Meyers, 1998) may be due to the constitutive activity of this promoter.

Rafts have provided much information about the differentiation-dependent late promoter of HPV31, p742. p742 has been shown to initiate transcription from a family of start sites located in the E7 ORF (Hummel et al., 1992; Ozbun and Meyers, 1998b). Several studies, some of which used raft cultures, have described numerous start sites by RNAse protection assays (RPA) or primer extension, and 6-8 of these were the major sites, with one cluster around nucleotides 737-767 and another cluster around nucleotides 695-710 (del Mar Pena et al., 2001; Hummel et al., 1992; Ozbun and Meyers, 1997). Although activity from p742 was detectable in monolayer cultures, suggesting a basal transcriptional activity not dependent on differentiation, activity increased dramatically upon differentiation of infected cells in rafts (Hummel et al., 1992; Ozbun and Meyers, 1997; 1998b), or upon suspension of the cells in semisolid medium (Ruesch et al., 1998). Given the heterogeneity of monolayer cultures, it is not clear whether the low level of p742 activity seen in monolayer was due to a basal transcriptional activity in all cells or to a high level of activity in a few spontaneously differentiating cells. The late promoter for HPV16 is called p670 and was also increased upon differentiation in rafts (Grassmann et al., 1996; Rohlfs et al., 1991). This promoter also had multiple start sites but unlike p742 had a major start site at nucleotide 667 with the other sites being relatively minor species, at least in the cell line investigated (Grassmann et al., 1996).

From raft cultures of 9E cells, transcripts containing both early and late ORFs have been cloned and analyzed.

For a review of transcripts expressed throughout the viral life cycle, see Ozbun and Meyers (Ozbun and Meyers, 1999a). Transcripts containing E1 and E2 originated from p99 or p77, with a small population originating from p742 (Ozbun and Meyers, 1998a). Transcripts for late genes were initiated at several promoters including p99, p77, p742, and p3320 (Ozbun and Meyers, 1998b). No late gene transcripts initiated upstream of p77 (Ozbun and Meyers, 1999b). The E1 splice donor at nucleotide 877 was used in all of the late transcripts detected (Ozbun and Meyers, 1997). L1 and L2 ORFs could exist in the same transcript, although many transcripts containing L1 but not L2 (using the 5552 splice acceptor) were described (Ozbun and Meyers, 1997). Transcripts encoding L1 and L2 were expressed at extremely low levels in monolayer cultures, but by both RPA and northern blot analysis, L1 and L2 transcript expression was easily detectable in 9E rafts, peaking at around 12 days of growth with a decrease in expression by day 16 (Bedell et al., 1991; Ozbun and Meyers, 1997).

Interestingly, monolayer cultures expressed more late transcripts than rafts at 4 days, when there was a single, basal-like layer of cells present (Ozbun and Meyers, 1997), indicating that monolayer culture is not strictly analogous to basal cells. Similar discordance between monolayers and the basal layer has been observed in studies of E6/E7 promoter activity (Zhao et al., 1997; 1999). HFK rafts may have several cell layers by this time, although both infected and uninfected rafts were fully stratified and mature by day 10 (Ozbun and Meyers, 1997; Steele et al., 2002).

Transcripts containing the E6/E7 ORFs did not increase significantly upon growth in 9E raft cultures, but those encoding all of the other viral ORFs did increase, and this increase could be largely accounted for by the activation of the late promoter (Hummel et al., 1992; Ozbun and Meyers, 1998b). Whether the relative quantity of proteins produced in an infection is regulated significantly at transcription level has yet to be studied in detail, but the diversity of transcripts containing these ORFs suggests that such regulation may be complex in time, space, and function. Differential splicing and the presence of multiple ORFs resulted in cDNAs with the potential to encode a variety of fusion proteins, truncated forms of known proteins, and alternately initiated proteins (Ozbun and Meyers, 1997). Whether the diverse coding potential discovered using raft-derived cDNAs will lead to the discovery of additional HPV protein species, and whether these putative proteins have functions in the viral life cycle, remains to be seen. It also has not been determined which of the many polycistronic RNAs found in differentiating cultures are actually responsible for the expression of any given protein. It cannot automatically be assumed that differentiating keratinocytes use the normal rules of translation, so raft cultures may turn out to be a critical tool in answering these questions.

Major splice sites used throughout the life cycle were those from nucleotides 210 to 413 (E6*) and nucleotides 377 to 3295 (E1^E4), with other splices used more or less often in many combinations (Hummel et al., 1992; Ozbun and Meyers, 1997). Transcripts using the E6* and E1^E4 splice sites outnumbered those that did not, so that full-length E6 and E1 encoding transcripts were relatively less common (Hummel et al., 1992; Ozbun and Meyers, 1997). A splice site was observed internal to the E1 ORF which removed the bulk of the gene (Ozbun and Meyers, 1998a), perhaps facilitating the expression of downstream ORFs such as E2.

From an experimental point of view, it is important to remember that the biology of differentiated cells may differ significantly from that seen in monolayer culture. For example, a variety of transcripts, such as GAPDH or β-actin, are used in RNA analyses as internal controls. Using raft cultures of both infected and uninfected cells, it was found that that β-actin levels vary considerably through the process of differentiation and between infected and uninfected cells, with expression dropping to undetectable levels in 9E rafts (Steele et al., 2002). GAPDH also fluctuated to a moderate degree but was comparable between infected and uninfected cells (Steele et al., 2002). 28S rRNA and cyclophilin levels remained relatively unchanged through the differentiation process and between cell types (Steele et al., 2002).

3.3. Cis elements and mechanisms of regulation

Using retroviral transduction to deliver a LacZ reporter under the control of wild type or mutant HPV18 or 11 URR, several authors showed that the E6/E7 promoters were activated in suprabasal cells and that binding sites for AP-1, Oct-1, Sp1, and AP-2 were needed for full activity of the URR in this context (Parker et al., 1997; Zhao et al., 1997; 1999). However, in the context of an integrated retrovirus including both the URR and the E6/E7 ORFs, the E6/E7 promoters of HPV18 and 11 were not activated by differentiation in suprabasal strata (Cheng et al., 1995). These data suggest that the E6 and E7 ORFs may contain elements that influence E6/E7 promoter activity, so that when these sequences are present, the inducibility of the URR disappears. Lack of induction of the E6/E7 promoter when downstream elements are present is consistent with the observation that the activity of the analogous p99 promoter of HPV31 was not increased by differentiation in the context of the whole episomally maintained genome (Ozbun and Meyers, 1998b). Using raft cultures of CIN-612 cells, in which the viral genome is complete and episomal, it was found that E6/E7 expression was highest in the lower strata with a declining gradient in upper layers correlating with c-Jun, JunB, and JunD expression (Kyo et al., 1997). This is in spite of the known amplification of viral genomes in the differentiated layers (Bedell et al.,

1991; Ozbun and Meyers, 1998b), suggesting that early promoter activity is not directly controlled by genome amplification or copy number.

C/EBP binding sites in the 5′ URR of HPV11 repressed URR-driven LacZ expression from retroviral vectors in basal cells (Zhao et al., 1999), although the binding of C/EBP was not demonstrated. Because of similarity in binding sequence, these sites may also mediate binding of CDP to the 5′ URR to repress the E6 promoter (Pattison et al., 1997) or may correspond to the nuclear matrix association region described for HPV16 in a similar position (Stunkel et al., 2000; Tan et al., 1998). It is also interesting that the 5′ URR of HPV31 had a negative effect on viral replication (Hubert et al., 1999b). Trichostatin A (TSA), a histone deacetylase inhibitor, was found to upregulate the E6/E7 promoter of HPV11 in basal cells in the context of an integrated retrovirus vector, even above the level seen by mutation of the C/EBP binding sites (Zhao et al., 1999), suggesting that the status of histone acetylation may be an important regulatory mechanism of this promoter. The involucrin and keratin 14 (K14) promoters were not upregulated by TSA, suggesting that this effect is specific to the HPV11 URR (Zhao et al., 1999). The activity of TSA on promoter activity from episomal templates was not examined in this study, but it has been shown to be sufficient to activate p99 in 9E cells in monolayer culture (del Mar Pena et al., 2001).

The URR of HPV31 contains four E2 binding sites (E2BS1-4) that are conserved in the mucosal HPVs. Of these, three were required for episomal maintenance in a study by Stubenrauch et al. (1998b), but using a linker scanning strategy, our laboratory has found that a mutant genome containing an 18 base substitution including the E2BS1 could maintain itself episomally but was defective for late functions (Meyers, unpublished data). Perhaps other elements in the same area are differentially affected by the different mutations used in the two studies. One site (E2BS2) had apparently no effect on any of the viral processes studied when mutated in the context of the whole viral genome, including episomal maintenance, amplification, late promoter activation, and L1 transcript expression (Stubenrauch et al., 1998b). It was not determined, however, whether infectious virions were produced from cells infected with E2BS2 mutant genomes. It is possible that this site is involved in viral packaging, for instance by bringing the genome into proximity with the late proteins (Day et al., 1998; Swindle et al., 1999). Rafts are perhaps the only existing system to test this possibility.

HPV31 genomes forced to integrate by a mutation in E1 failed to produce E1^E4 and L1 protein (Frattini et al., 1996), from which it was concluded that episomal genomes are needed for late gene expression. Only one line with integrated HPV DNA was reported in that study, however, and whether the late ORFs were still present in an expressible form was not determined. We have found that episomality was not necessary for p742 activity (Bodily and Meyers, unpublished data), but p742 has less stringent requirements for activity than expression of L1 and L2, so the role of episomal maintenance, amplification, and epigenetic factors such as chromatin structure, methylation, and subnuclear positioning that may be affected by integration remain to be fully investigated. Using raft cultures of HPV6 and 11 positive cells, as well as low and high calcium cultures, it was found that the amounts of HPV RNA and DNA were not correlated (DiLorenzo and Steinberg, 1995), suggesting that gene expression is not quantitatively tied to genome amplification.

3.4. Replication

Along with studies of gene expression, raft cultures have proven valuable in studying differentiation-dependent replication. Viral DNA amplification in CIN-612 rafts occurred in the nuclei of koilocytes in the upper strata of the epithelium, although not all koilocytes were positive for amplification (Bedell et al., 1991). Only a few nuclei showed signal for viral DNA in 9E rafts by *in situ* hybridization at 4 days but many nuclei showed a strong signal by day 8 throughout the tissue (Ozbun and Meyers, 1998a). Viral genome copy number in 9E cells increased in C8-treated raft cultures and peakd at around day 10, with a subsequent decline (Ozbun and Meyers, 1998a). After 12-16 days, although many nuclei showed a strong signal, areas of the rafts without signal increased (Ozbun and Meyers, 1998a), suggesting that the decline of copy number as measured by Southern blot analysis was probably related to a decrease in the percentage of amplifying cells in the culture, rather than a decrease in the level of amplification per cell. Cell lines containing HPV31 mutants null for E1 or for the replication function of E2 were unable to maintain the genomes episomally, but contained high levels of apparently integrated viral DNA (Stubenrauch et al., 1998a). E1- and E2-containing transcripts in rafts peaked at around day 12 with a decline thereafter (Ozbun and Meyers, 1998a). The relative increase in the level of E1 transcripts was greater than that of E2 (Ozbun and Meyers, 1998a), suggesting that the ratio of these two proteins changes through the life cycle and may have regulatory significance. Even though they must be expressed early in the life cycle to support the maintenance phase of infection, because of the late increase in the transcripts for these proteins it may be appropriate to also consider E1 and E2 as late genes.

In the context of the whole HPV31 genome, the keratinocyte enhancer in the URR, and to a lesser extent the auxiliary enhancer, were necessary in *cis* for stable episomal replication (Hubert et al., 1999b). Interestingly, mutation of the Oct-1 element in the keratinocyte enhancer or deletion of the 5′ URR domain both increased transient and stable replication efficiency and neither mutation interfered with amplification in raft cultures (Hubert et al., 1999b).

3.5. Late protein expression

Differentiation dependence of genome amplification and late transcription was first observed *in vitro* by Bedell et al. (Bedell et al., 1991), but expression of late proteins, which is required for virion production was not observed. Differentiation to a degree sufficient to activate p742 was not sufficient, by itself, to result in the appearance of late transcripts in 9E cells (Hummel et al., 1995). Expression of L1 protein in raft culture was first accomplished by the addition of the phorbol ester, 12-O-tetradecanoylphorbol 13-acetate (TPA) to differentiating rafts, and L1 expression correlated closely with the induction of filaggrin and K10 by TPA (Meyers et al., 1992). The expression of L1 opened the door to virion propagation *in vitro*. Synthetic C8 was shown to work similarly to TPA. Protein kinase C (PKC) activation was not necessary for L1/L2 transcript expression in 9E rafts, but the transcript level increased substantially when C8 was added (Ozbun and Meyers, 1997). The addition of PKC activators to HPV-infected rafts did not result in the further activation of p742 (an approximately spinous layer function) but did result in increased read-through to the late genes (a granular layer function) (Hummel et al., 1995). This observation suggests that there may be an as-yet undefined PKC responsive translational control mechanism in infected cells, since L1 protein expression requires PKC activation (Meyers et al., 1992), although translation of L1 protein may simply require a threshold level of late transcripts provided by the activation of p742. PKC activation was not necessary to induce L1 protein expression in cell lines derived by electroporation of HFKs with HPV31a (Frattini et al., 1997) or BC-1-Ep/SL cells with HPV16 (Flores et al., 1999), perhaps because these cells were able to more completely differentiate in the absence of PKC activation relative to patient-derived 9E or W12E cells (Flores et al., 1999). An explanation for the requirement for PKC activators appears to be read-through past the early poly-A signal to the late transcripts (Hummel et al., 1995).

That L1 (as well as K10 and filaggrin) was not expressed in rafts without the addition of TPA, even though the cultures were stratified and morphologically differentiated to some extent, indicates that the differentiation dependency on L1 is greater than that on stratification per se (Meyers et al., 1992), and that stratification alone is not indicative that the full program of differentiation has been completed. Thus if expression of L1 or production of infectious virions is a desired experimental endpoint, they should be directly measured and confirmed.

4. Propagation of Infectious Virus

The raft culture system mimicking the *in vivo* three-dimensional tissue morphology was a breakthrough for papillomavirologists allowing them to propagate infectious viruses *in vitro*. PV replication is intimately connected to the differentiation program of its host tissue, squamous epithelium. This requirement posed a major obstacle for the propagation of HPV *in vitro*. In 1992 using raft culture systems, two laboratories reported the production of infectious HPV *in vitro*. By careful manipulations of the raft culture system combined with the addition to the culture media of TPA, the propagation of virions from a latent cell line (CIN-612 9E) was accomplished (Meyers et al., 1992). Examination of raft culture cross-sections by immunohistochemistry and electron microscopy demonstrated the presence of the major capsid protein and viral particles in the suprabasal layers. Viral particles were subsequently purified by isopycnic sedimentation centrifugation and shown by Southern blot analysis and electron microscopy to be complete virions (Meyers et al., 1992). Subsequently, viral stocks were prepared and shown to contain infectious HPV. In a second important study, Dollard et al. showed that a different raft culture model could sustain HPV productive program (Dollard et al., 1992). Recovered HPV11 condylomatous cysts grown in the mouse renal capsule xenograft system (Kreider et al., 1986; 1987; 1985) were planted onto collagen matrices. Epithelial sheets grew out from the cysts, stratified and differentiated, and virions were identified (Dollard et al., 1992).

The organotypic raft culture system has permitted the study of a number of differentiation-specific aspects of the PV life cycle. Most of these investigations were initially limited to a single virus type, HPV31b. Additionally, the HPV31b CIN-612 9E raft system was the only system that could reproducibly recreate the complete viral life cycle that included the propagation of infectious viral stocks. In order to expand the repertoire of HPV types capable of producing infectious viral stocks, linearized HPV18 genomic DNA was electroporated into primary human keratinocytes, followed by clonal expansion and induction of epithelial stratification and differentiation in raft culture. Virus particles approximately 50 nm in diameter were observed in the raft tissue by electron microscopy. HPV18 virions purified by gradient centrifugation were capable of infecting keratinocytes *in vitro*, as shown by the expression of multiple HPV18-specific, spliced transcripts (Meyers et al., 1997). This study showed that infectious HPV stocks could be produced for a second round of infection. Another study electroporated recircularized HPV18 genomes into primary keratinocytes (Frattini et al., 1997). The resulting cell lines maintained episomal copies and when grown in raft culture induced late viral functions such as viral genome amplification and late gene expression, but propagation of infectious virus was not investigated. The repertoire of PV types for which infectious stocks could be made was further expanded by electroporating linearized HPV16, 31a, 39, and 45 genomic DNAs into primary keratinocytes, where, like HPV18, they recircularized and were maintained episomally at a range of 50-100 copies of viral genomic DNA per cell. Following epithelial stratification and differentiation in raft culture, late viral functions, including viral DNA amplification, capsid gene expression, and virion morphogenesis, occurred (McLaughlin-Drubin et al., 2004; McLaughlin-Drubin and Meyers, 2004; McLaughlin-Drubin et al., 2003). By testing a panel of PV VLP-derived MAbs, it was shown that HPV45 infection was neutralized not only with HPV45 VLP-specific MAbs, but also with HPV18 VLP-specific antiserum, demonstrating serological cross-reactivity between HPV18 and HPV45 (McLaughlin-Drubin et al., 2003). In contrast, HPV16 infection was only neutralized with HPV16 VLP-derived MAbs (McLaughlin-Drubin et al., 2004). Lee et al., using a cre/lox adenovirus transduction system established episomal copies of HPV16 in primary keratinocytes (Lee et al., 2004). Derived cell lines were shown to propagate infectious HPV16 following growth and differentiation in raft culture.

The ability to introduce viral DNA into keratinocytes and derive HPV-positive cell lines that when grown in raft culture transit the complete viral life cycle, including virion morphogenesis, allows questions particular to the natural history and clinical presentation of PVs to be asked. Studies on the presence of HPV DNA in cervical samples show that 10% or more of all clinical lesions contain at least two different HPV types. Using techniques of electroporation and raft cultures the question of whether multiple HPV types can exist in the same cell and interact with one another could be tested *in vitro*. Using electroporation, combinations of genital HPV genomes were introduced into primary keratinocytes and a subset of viral genomes were shown to be stably maintained and to replicate episomally in the same cell. Interactions between types occurred, resulting in repression of replication, integration, and recombination (McLaughlin-Drubin and Meyers, 2004). These studies provided insight into the interactions that may occur between HPV types in naturally occurring lesions.

Since the techniques to derive HPV positive cell lines begin with viral DNA, mutational studies can be performed on all aspects of the viral life cycle. Many of these studies are discussed elsewhere in this chapter. One area where genetic analyses have great potential when coupled to raft culture technology is the study of virion morphogenesis. Raft cultures allow for the virion morphogenesis to proceed *in vitro* along similar natural pathways that occur *in vivo*. The generation of chimeric viruses has been a useful tool in other virus systems to analyze infection and replication. Chimeric virus techniques have been applied to PV virion morphogenesis asking the question whether the nonstructural genes of HPV type can functionally interact with the structural genes of another type. The L2 and L1 of HPV18 were replaced with the L2 and L1 of HPV16. The chimeric genomic viral DNA was capable of transiting the complete life cycle producing infectious titers equivalent to wild-type HPV18 (Meyers et al., 2002). Infection was specifically neutralized with human HPV16 VLP-specific antiserum and not with human HPV18 VLP-specific antiserum (Meyers et al., 2002). More studies are needed to determine the extent of the evolutionary conservancy of the interaction between nonstructural and structural genes among evolutionary diverse PV types.

The animal PVs have not been forgotten in using raft cultures to develop *in vitro* systems to study propagation of infectious progeny. Bovine papillomavirus type 1 (BPV1) has long been a prototype for understanding PV DNA replication, gene expression, and transformation. Comparatively little is known concerning BPV1's life cycle in host bovine epithelial tissue. In one provocative study, raft cultures were made using BPV1 transformed fibroblasts as the mesenchymal cells in the dermal equivalents (McBride et al., 2000). Bovine keratinocytes were placed on top of the dermal equivalents and raised to the air-liquid interface where they stratified and differentiated. The keratinocytes had been infected with virus particles or transfected with BPV1 DNA (McBride et al., 2000). A few days after growing at the air-liquid interface the rafts were grafted on to the backs of nude mice. In 6-8 weeks large xenografts grew exhibiting a hyperplastic and hyperkeratotic epithelium. Further analyses showed that the complete BPV1 life cycle had occurred with the isolation of virus particles and demonstration of infection (McBride et al., 2000).

The power of rafts to study the very late steps in the life cycle that depend on completion of the entire differentiation program is well established (Hummel et al., 1995; Meyers et al., 2002; 1992; Ozbun and Meyers, 1997; 1998a). Although not using raft cultures directly, the studies of Ozbun (2002a; 2002b) employed raft-generated virus stocks to examine the cascade of transcripts expressed initially upon infection of a variety of cells. Because raft culture remains the only way to generate wild type or mutant virus stocks *in vitro*, rafts have great indirect potential in elucidating the very early steps in the infection pathway. Raft-derived virus stocks have been shown to infect HaCaT cells and to produce late transcripts upon cell growth in raft culture (Ozbun, 2002a).

In summary, raft cultures have been well established as an *in vitro* system for the study of PV virus propagation. Raft culture technologies provide the potential for many important investigations concerning PV morphogenesis. These investigations can provide important insight into the natural history of PV infections and provide targets for new therapeutic interventions.

5. Human Papillomavirus Cell Cycle Control

5.1. Reactivation of host DNA synthesis in differentiating cells

Raft cultures derived from both primary and HPV positive keratinocytes have been indispensable as a model system in which to study HPV-host interactions. The ability of both high risk HPV18 as well as low risk HPV11 E7 oncoprotein to reactivate host DNA replication machinery in post-mitotic, non-cycling, terminally differentiated cells was demonstrated using the HPV E7 protein expressed from its native URR promoter (Cheng et al., 1995). In raft cultures derived from primary human keratinocytes DNA polymerase α and PCNA are detected only in the parabasal cells (Cheng et al., 1995). PCNA is the processivity factor for DNA polymerase δ and is indicative of active DNA synthesis (Mushika et al., 1988). Retrovirus mediated gene transfer of the intact E7 gene of either HPV18 or HPV11 into raft cultures derived from primary human keratinocytes was shown to induce PCNA protein expression in the suprabasal cells (Cheng et al., 1995). However, in raft cultures expressing either the HPV18 or the HPV11 E7 protein, PCNA expression was shown to be induced in the differentiated compartment (Cheng et al., 1995). Induction of PCNA in differentiated cells was accompanied by extensive incorporation of [^{3}H]thymidine in the nuclei of a fraction of suprabasal cells, indicative of reactivation of host DNA synthesis (Cheng et al., 1995). Both these features have consistently served as hallmarks of HPV activity. Thus, expression of E7 was shown to be necessary and sufficient to reactivate unscheduled host DNA synthesis in a subset of post-mitotic suprabasal cells, equivalent to DNA synthesis in the basal or parabasal cells.

5.2. Deregulation of cell cycle proteins

The virus-host interactions leading to early stages of immortalization were shown to involve proteins that control cell cycle checkpoints. It was hypothesized that expression of the E6 and E7 transcripts in the basal cells was an essential step in HPV mediated immortalization of keratinocytes (Steenbergen et al., 1998b). This idea was based on the fact that critical steps in cellular immortalization require inactivation of multiple proteins which regulate key cell cycle check points. Expression of the E6 and E7 transcripts correlated with the degree of dysplastic morphology of raft cultures as well as the expression of the universal cyclin-dependent kinase (CDK) inhibitor p21^{cip1} (Steenbergen et al., 1998b). Raft cultures derived from keratinocytes containing HPV16 and HPV18 whole genomes, were negative for p21^{cip1} protein expression in the basal layer, and this was coincidental with high levels of E6 and E7 expression (Steenbergen et al., 1998b). The absence of p21^{cip1} in the basal layers was hypothesized to result from E6 dependent degradation of the tumor suppressor p53, which is a transcriptional regulator of p21^{cip1} under certain cellular conditions (Sherr and Roberts, 1995). In addition, E7 interferes with p21^{cip1}-mediated inhibition of cyclin/CDK complexes and mediates unscheduled entry into S phase of the cell cycle (Sherr and Roberts, 1995). Expression of E6 and E7 in the basal cells was correlated with increased proliferation as assessed by Ki-67 (MIB-1) expression, dysplastic morphology, and decreased expression of differentiation markers K10 and profilaggrin (Steenbergen et al., 1998b). These studies demonstrated that basal expression of HPV E6 and E7 resulting in the loss of p21^{cip1} protein may be an important step in HPV-mediated transformation, but is not sufficient for immortalization or loss of terminal differentiation (Steenbergen et al., 1998b).

In contrast to HPV-induced loss of p21^{cip1} in the basal layer, HPV was shown to induce p21cip1 protein expression in differentiated keratinocytes via convergence of multiple cellular mechanisms. The E7 oncoprotein of HPV18 expressed from its native promoter using retroviral vectors in primary human keratinocytes induced p21^{cip1} protein in high levels in a subset of suprabasal cells while it was still absent from the basal layer (Jian et al., 1998). The differential expression of p21^{cip1} was intriguing given that expression of HPV18 E7 alone gave rise to two opposite results in the same raft epithelium. Although the abundance of p21^{cip1} RNA increased in primary human keratinocyte raft cultures irrespective of the presence of HPV18 E7, p21^{cip1} protein expression was strongly induced only in HPV18 E7 expressing raft cultures (Jian et al., 1998). These results could not be correlated with p53 expression, suggesting that the accumulation of p21cip1 was post-transcriptional in nature. These results further suggested that upregulation of p21cip1 may act to block unscheduled DNA synthesis reactivated initially upon expression of E7.

In order to understand the complex nature of p21^{cip1} expression in response to E7 one needs to consider the role of p21^{cip1} in cell cycle control. Under normal conditions, p105Rb is inactivated in G1 phase via phosphorylation by a number of CDKs, which are themselves targets for inactivation via CDK inhibitors such as p21^{cip1} (Sherr and Roberts, 1995). In addition, p21^{cip1} interferes with S phase entry and progression by inhibiting Cyclin E/CDK2 complexes. Induction of p21^{cip1} can occur via both p53-mediated as well as p53-independent pathways. Induction of p21^{cip1} is p53 independent during cellular differentiation and is associated with exit from the cell cycle (Huang et al., 2000). p21^{cip1} also binds to PCNA and interferes with DNA synthesis (Sherr and Roberts, 1995). These observations were used to predict that the p21^{cip1} protein may be a target for HPV mediated degradation during all phases of productive viral growth and life cycle (Jian et al., 1998). Thus upregulated p21^{cip1} protein in a fraction of the post-mitotic differentiating suprabasal cells was intriguing (Cheng et al., 1995; Jian et al., 1998) since this is the compartment where productive

HPV replication and virion synthesis is also known to occur (Meyers et al., 1992). It was expected that $p21^{cip1}$ expression in these compartments would be low. The surprising induction of $p21^{cip1}$ protein expression in the HPV18 E7 expressing differentiated raft cultures was thus thought to be attributable to increased formation of cyclin/CDK/PCNA/$p21^{cip1}$ quaternary complexes (Jian et al., 1998; Sherr and Roberts, 1995). If so, S phase entry would be prevented in that fraction of suprabasal cells expressing upregulated $p21^{cip1}$ in E7 expressing raft cultures (Jian et al., 1998; 1999). In agreement with this prediction, induction of $p21^{cip1}$ was shown to correlate with inhibition of unscheduled host DNA synthesis in differentiated keratinocytes also resulting in low levels of viral DNA amplification and RNA expression (Jian et al., 1998). Thus, it was suggested that the fraction of cells undergoing $p21^{cip1}$ induction may constitute protective host response which effectively inhibits unscheduled DNA synthesis reactivated by HPV E7 expression upon infection (Jian et al., 1998).

The $p27^{kip1}$ CDK inhibitor was shown to lead to the heterogeneous accumulation of $p21^{cip1}$ and inactivation of DNA synthesis in a fraction of post mitotic cells (Noya et al., 2001). In general, both $p21^{cip1}$ and $p27^{kip1}$ play similar roles in the cell cycle as they both inhibit the kinase activity of Cyclin E/CDK2 by sequestering the protein in inactive complexes (Sherr and Roberts, 1995). The $p27^{kip1}$ protein was not observed in basal or parabasal cells but was shown to be expressed in differentiated cells regardless of the presence of E7 (Noya et al., 2001). In the same study, and as shown earlier (Jian et al., 1998), in the presence of E7, $p21^{cip1}$ protein was shown to accumulate in a posttranscriptional manner and was further stabilized by binding with E7-induced Cyclin E protein (Noya et al., 2001). In contrast to $p21^{cip1}$, $p27^{kip1}$ protein expression was not further affected by E7. However, both $p21^{cip1}$ and $p27^{kip1}$ were found to sequester cyclin E/CDK2 complex into kinase inactive complexes (Noya et al., 2001). This was a significant observation implicating host specific responses to HPV in significantly delaying S-phase re-entry in HPV infected cells. Interestingly, in this study, E7 was also shown to induce Cyclin E expression. The authors proposed that the presence of stable $p27^{kip1}$ protein would be important in preventing unscheduled DNA synthesis under conditions where the $p21^{cip1}$ protein may be expressed in low levels. Inadequate $p21^{cip1}$ levels would be ineffective in inhibiting the resultant increase in kinase activity of Cyclin E/CDK2 complexes (Noya et al., 2001). E7 induced Cyclin E expression was also attributed to p21cip1 protein stabilization due to the formation of kinase-inactive complexes and resistance to degradation by proteosome specific pathways (Noya et al., 2001). In contrast, the other subset of cells which successfully underwent reactivation of host DNA synthesis was shown not to divide but rather undergo endoreduplication by executing another round of S phase (Chien et al., 2000).

E7-mediated binding to p105Rb and activation of E2F specific S phase genes is necessary but not sufficient to induce S-phase entry (Chien et al., 2000). E7 mutants blocked in their ability to be phosphorylated by casein kinase II, bound to p105Rb *in vitro* but failed to induce DNA polymerase α and Cyclin E in raft cultures and failed to induce PCNA expression. Thus, casein kinase II mediated phosphorylation of E7 was shown to be critical for reactivating S phase (Chien et al., 2000).

5.3. Cell cycle regulation by high-risk vs. low-risk papillomaviruses

A comprehensive study of cellular pathways altered by E6 and E7 oncogenes in primary human keratinocyte raft cultures has been recently published (Garner-Hamrick et al., 2004). In this study, microarray analysis of raft cultures infected with retroviruses expressing the high-risk wild-type HPV18 E6 and E7 genes revealed significant alterations in the expression of a multitude of genes including transcripts associated with cell cycle regulation and differentiation. One prominent observation of this study was that E6/E7 expression dramatically upregulated transcripts involved in DNA synthesis with a concomitant downregulation of transcripts involved in protein synthesis. These results strongly confirm numerous earlier studies which showed the importance of tipping the balance in favor of host and viral DNA replication in differentiated, post-mitotic keratinocytes, with concurrent diversion from cellular protein synthetic activities required in cornification of suprabasal cells (Garner-Hamrick et al., 2004). In this study, the clarity of the microarray analyses was further confirmed with the observation that PCNA transcript expression was upregulated, coupled with observable bromodeoxyuridine incorporation in E6/E7 expressing raft cultures.

Although microarray analysis of changes induced by high-risk HPVs has been successfully performed in organotypic cultures, similar studies using low-risk HPVs may be problematic. Microarray analysis has been reported looking at the alterations in cellular genes affected when low risk HPV11 whole genomes were transfected into keratinocytes and raised to the air-liquid interface (Thomas et al., 2001). Raft cultures derived in this manner demonstrated their ability to maintain episomal HPV11 genomes and an extended life span, accompanied by alteration in tissue morphology and stratification, which appeared less severe than those generated by HPV31-positive raft cultures (Thomas et al., 2001). However, HPV11 transfected cells lines suspended in methylcellulose were used for the microarray analysis, suggesting the inability of low risk viruses to sufficiently immortalize raft cultures for these types of analyses. In the meantime, difficulties in studying low-risk HPVs in raft cultures have been partially circumvented by using patient derived HPV positive lesions and propagating them in organotypic cultures. Laryngeal papilloma cells infected with HPV11 have been used to generate raft cultures (Aaltonen et al., 1998; Vambutas et al., 1993). Tonsillar epithelial cell infected with HPV16 have also been used to study tonsillar carcinomas (Chen et al., 2003). Raft cultures derived from patient lesions have been used as models to study head and neck cancers (Oda et al., 1997; Yoo et al., 2000).

Weak immortalizing activities of low risk HPVs was unmasked by expressing them in combination with high risk HPV encoded E6 and E7 proteins, culminating in a range of effects. The E7 gene of high-risk HPVs such as HPV16 or HPV18 alone was found to be sufficient to immortalize HFKs and efficiency was increased with in presence of the E6 protein of either HPV type (Halbert et al., 1992; Hudson et al., 1990). Although low risk HPV6 E6 gene, alone or in concert with HPV6 E7 gene was not able to immortalize primary keratinocytes, HPV6 E6 together with HPV16 E7 increased the frequency of immortalization, thus revealing weak immortalizing action of HPV6 E6 gene (Halbert et al., 1992). Similarly, weak immortalizing activities of the HPV6 E7 gene was revealed in combination with expression of HPV16 E6 in raft cultures. The HPV6 E7/HPV16 E6 combination was able to reactivate host cellular DNA synthesis in suprabasal layers of raft cultures, as well as increase PCNA expression and formation of disorganized epithelium (Halbert et al., 1992), whereas expression of HPV6 E7 on its own did not result in a dysplastic morphology or aberrant PCNA staining, presumably because HPV6 E7 binds p105Rb with less affinity than HPV16 E7 (Gage et al., 1990; Halbert et al., 1992).

6. Progression

In contrast to the relatively minor changes that accompany the viral life cycle in a low-grade lesions, neoplastic progression of HPV associated lesions can occur over a period of time and results from accumulated changes which occur in the host genome which translate to changes in host protein expression as well as activity. Organotypic cultures have been essential in determining the mechanisms of HPV-associated progression in lesions arising from molecular changes in host genetic material as well as host cellular proteins. Malignant progression of HPV associated carcinoma has been recapitulated in raft cultures based on changes in morphological and histological markers of stratified epithelium. Cumulatively, these changes result from genetic instability and changes in protein expression, as related to changes in programmed differentiation of keratinocytes in HPV infected tissue. Although the E6 and E7 oncogenes are sufficient for immortalization, it is widely accepted that other genetic events must accumulate for high-risk HPV specific malignant progression. One of the biological characteristics of HPV infected cells is their tendency to develop genetic instability, which appears as forms of chromosome imbalances, or aneuploidy, in early steps of carcinogenic progression (Hashida and Yasumoto, 1991; White et al., 1994).

6.1. Chromosomal abnormalities

One avenue of studies has shown that HPV induced chromosome imbalances are attributable to abnormal centrosome duplication, and are an important contributor to malignant progression (Duensing et al., 2001). Overexpression of HPV16 E6 and E7 has been shown to be directly responsible for abnormal multipolar mitoses *in vitro* (Duensing et al., 2000) similar to those observed in pre-invasive lesions from patient biopsies (Crum et al., 1984). Organotypic raft cultures derived from spontaneously immortalized keratinocyte cell line (NIKS) transfected with wild-type HPV16 DNA have been used to demonstrate that such centrosome abnormalities and chromosome imbalances can be recapitulated *in vitro* (Duensing et al., 2001). Expression of episomal HPV16 genome was found to be sufficient to induce centrosome abnormalities and chromosomal aberrations in basal cells of raft cultures, strengthening the hypothesis that progressive genomic destabilization leads to loss of cell cycle check-points, eventual integration of episomal HPV DNA, and manifestation of carcinogenic progression (Kessis et al., 1996).

In addition to centrosome abnormalities, high-risk HPVs have also been shown to induce basal cell tetrasomy in low grade cervical lesions (Southern et al., 1997). Whereas tetraploidy is the reduplication of the cellular DNA to achieve 4N content, tetrasomy refers to sister chromatid separation after DNA re-duplication (Southern et al., 2001). Both events have the potential to disrupt normal cell cycle regulation (Southern et al., 2001). Tetrasomy was identified to occur in the suprabasal strata of raft cultures expressing HPV18 E7 and E6/E7, but not in raft cultures expressing E6 alone, suggesting that suprabasal tetrasomy occurs in association with expression of the E7 gene (Southern et al., 2001). On the other hand, tetrasomy was shown to occur in the basal cells of raft cultures expressing the entire genome of HPV18 similar to the *in vivo* situation. These results suggest that conditions leading to basal cell tetrasomy are different from conditions leading to suprabasal tetrasomy and may be determined by such differences as cell growth conditions, differentiation and viral gene expression during the complete viral life cycle (Southern et al., 2001).

Because the E7 oncoprotein appeares to be a prominent player in the induction of tetrasomy (Southern et al., 2001) and its ability to bind p105Rb has been well characterized, another study asked the question whether p105Rb binding was necessary for E7 to induce tetrasomy (Southern et al., 2004). This study utilized a series of mutations in the p105Rb binding site and/or the CKII phosphorylation sites of HPV16 E7 and tested its ability to induce tetrasomy in raft cultures (Southern et al., 2004). The mutants induced tetrasomy to the same extent as the wild-type E7, and there was no correlation between the ability of E7 to bind p105Rb and its ability to induce tetrasomy.

Specific losses in chromosomal loci have been correlated with the loss of cellular tumor suppressors. A comprehensive study of chromosomal alterations which accompany HPV mediated immortalization has been published (Steenbergen et al., 1998a). In this study, organotypic raft cultures derived from primary human keratinocytes transfected with HPV16 and HP18 genomes were utilized. Chromosomal analysis was performed using comparative genomic hybridization. These studies determined that chromosomal imbalances were shown to increase through the process of repeated passaging and subsequent cellular immortalization. Clonal loss of heterozygosity was detected in immortal descendents and occurred at chromosomal loci 3p, 11p and 13q (Steenbergen et al., 1998a). All immortalized cell lines displayed clonal allelic losses at these loci and displayed severely dysplastic phenotype, accompanied by loss of terminal differentiation. These *in vitro* studies were correlated with *in vivo* data showing that changes at these loci were also identified in HPV immortalized cells (Oda et al., 1996b). Another study showed that immortalized HPV16 and HPV18 clones in raft cultures displayed allelic losses at chromosomal loci 3p combined with 11q or 18q, or at 10p with allele losses at 3p or 10p which were correlated with the onset of activity of telomerase, the telomere-lengthening enzyme (Steenbergen et al., 1996), suggesting that 3p and 10p may encode genes that are involved in telomerase regulation. In an earlier study, HPV16 and HPV18 transfected immortal cell lines expressed strong telomerase activity as compared with mortal precursor cells (Steenbergen et al., 1996). Immortalization of human cells has been linked to reactivation of telomerase (Harley and Kim, 1996). Telomerase activity has been suggested to compensate for telomerase shortening during progressive cell division and may act to overcome signals that trigger senescence (Colgin and Reddel, 1999). Thus, telomerase activity has been considered a marker for immortalization (Krupp et al., 2000).

It was postulated that repeated passaging of epithelial cells infected with the oncogenic PVs in culture would give rise to tumorigenic cells. Raft cultures derived from early passage of HPV18 immortalized keratinocytes resembled premalignant intraepithelial neoplasias *in vivo*, whereas later passages had features consistent with squamous cell carcinoma (Hurlin et al., 1991). Populations which arose as a result of repeated passaging were then transferred subcutaneously into BALB/c athymic mice and proved to be tumorigenic. Karyotyping of cells derived from these tumors revealed aberrations involving deletion in the q arm of chromosome 5, together with loss of chromosome 21 and chromosome 4, suggesting that secondary events must occur for tumorigenic progression.

Episomal HPV16 genome of the W12 cell line was shown to undergo integration with repeated passaging of monolayer cultures, which could be correlated with the generation of a higher grade neoplasia when such cells were grown in raft cultures (Pett et al., 2004). Raft cultures generated from cells maintaining integrated HPV16 genome also displayed invasion of the underlying collagen matrix. In contrast, raft cultures derived from monolayer cultures maintaining episomal copies of HPV16 morphologically resembled low grade lesions (Pett et al., 2004).

6.2. Histopathological changes accompanying progression

Chromosomal aberrations translate to host specific changes at the protein level which subsequently translate to dysplastic progression of lesions. When HPV positive keratinocytes are passaged over a period of time, organotypic raft cultures derived from such keratinocytes mimic characteristics of *in vivo* lesions by undergoing gradual loss of stratification, expression of differentiation markers, and terminal differentiation, and increased dysplastic progression. In contrast, these changes occur over a period of decades in HPV infected patients. It has been shown that HPV16 and HPV18 play an important role in the early stages of cervical cancer development (Zheng et al., 1994). DNA from either type when transfected into human ectocervical cells, caused immortalization, and when grown in raft cultures, the cells underwent altered differentiation as in cervical preneoplastic lesions which were reported to be highly unstratified (Zheng et al., 1994). Likewise, HPV33 DNA transfected keratinocytes grown in organotypic raft cultures resembled CIN III lesions and showed altered expression of keratin markers (Gilles et al., 1994). Transfected cell lines at low passage levels maintained epithelial morphology of low grade lesions in raft cultures but at high passage resembled CIN III lesions, supporting the hypothesis that secondary cellular events may be involved in the acquisition of this morphology (Blanton et al., 1991; Hudson et al., 1990; McCance et al., 1988).

Premalignant changes of the cervical epithelium resulting from HPV oncogene expression and host-cell interaction, precede the appearance of invasive cervical cancer (Rader et al., 1990). Organotypic cultures derived from patient tissue biopsies were shown to display histological features of neoplastic cervical epithelium. When allowed to differentiate, the raft cultures of cells isolated from condyloma acuminatum as well as invasive cervical carcinoma recapitulated the same morphological features found in *in vivo* lesions (Rader et al., 1990). To demonstrate that this property was not unique to raft cultures derived from patient samples, another study demonstrated that HFKs immortalized with HPV16 and HPV18 also stratified in raft cultures which were also morphologically similar to premalignant lesions (Blanton et al., 1991).

Most carcinomas derive from the transition zone between the endocervix and ectocervix (zur Hausen, 1989; zur Hausen, 1991), although the biological or molecular basis for this is unknown. One study examined the effect of E6/E7 expression in rafts of cells from endocervix, ectocervix, and foreskin and found differences in expression and the apparent histological grade of the rafts, with endocervix showing the most dysplasia upon E6/E7 expression (Sun et al., 1997).

6.3. Deregulation of specific host proteins

As mentioned above, changes in host chromosome composition further affect cellular proteins, which in combination determine malignant progression. Gap junction intercellular communication (GJIC) and gap junction proteins may act as tumor suppressors (Trosko and Ruch, 1998). Disruption of intercellular communication through gap junctions has been reported in malignant cell lines and tumors (Aasen et al., 2003; Saitoh et al., 1997), and although normal cervical tissue have abundant gap junctions, these structures are deficient in cervical carcinoma (McNutt and Weinstein, 1969) and severe dysplasia (McNutt et al., 1971). Specifically, the gap junction protein connexin (Cx43) expression was found to be reduced in dysplastic cervical tissue biopsies (King et al., 2000). To determine whether changes in gap junction proteins from patient biopsy material correlate with expression of gap junction in raft cultures, the W12 HPV16 positive keratinocyte line was used to examine changes in the expression of connexins (Aasen et al., 2003). W12-derived rafts containing episomal or integrated forms of the HPV16 genome were found to have no abnormalities in GJIC. On the other hand, a W12-derived cell line designated GPX4, which was selected for growth independence of feeder cells, presented dysplastic morphology when grown in organotypic raft cultures. Dysplastic appearance was accompanied by a complete loss of differentiation and showed decreased cell-cell coupling but this could not be clearly correlated with Cx43 expression (Aasen et al., 2003). Thus, downregulation of GJIC frequently observed in tumor and malignant cell lines is not directly regulated by HPV expression but is established gradually over a period time and may be a consequence (or cause) of dysplastic progression (Aasen et al., 2003).

In addition to the ordered expression of keratins in differentiating epidermis, epidermal differentiation is also characterized by the expression of a family of adhesion receptors termed integrins (Albelda et al., 1990). Basal epidermal cells adhere to the basement membrane via specific cell integrins. Organotypic raft cultures generated using FEPEIL-8 cells (HPV16 positive) exhibiting disorganized stratification and were limited in their ability to differentiate (Kaur and Carter, 1992). Additionally, the keratinocytes underwent invasion into the collagen layer together with upregulated expression of $\alpha3\beta1$ and $\alpha2\beta1$ integrins and underexpression of $\alpha6\beta4$ integrins compared to raft cultures derived from primary keratinocytes (Albelda et al., 1990). On the other hand, deregulated integrin expression may itself influence the presentation of HPV specific dysplasia. In another study, ordered stratification of epithelium and regulated integrin expression could be induced by substituting human foreskin fibroblasts for Swiss 3T3 fibroblasts (Kaur and Carter, 1992). Human embryonic skin fibroblasts also developed polarity towards HPV16 DNA containing keratinocytes, *i.e.* immortalized cells appeared to be chemotactic to fibroblasts and fibroblasts presumably stimulated growth of epithelial cells (Zheng and Vaheri, 1995). The immortalized cells were thought to produce and release factors which influenced the movement of fibroblasts and invasion into the collagen layer (Kaur and Carter, 1992). These results suggest that fibroblasts may also contribute to HPV induced carcinogenesis and dysplasia.

One secondary event involved in HPV associated carcinogenesis is altered regulation of the c-*myc* protooncogene. This protein normally functions as a regulator of cell proliferation but is expressed in a wide range of neoplasms (Amati and Land, 1994). Deregulation of c-*myc* is thought to be linked to its ability to interfere with terminal differentiation. Transcriptional switching between Myc:Max to Mad:Max complexes resulting from downregulation of c-*myc* expression is necessary for the transition from proliferation to differentiation, and is thought to regulate epithelial terminal differentiation. Myc:Max complexes activate transcription and cell proliferation, whereas Mad:Max complexes repress transcription and downregulate c-*myc*-mediated cell transformation. Since the complete life cycle of HPV is dependent on keratinocyte differentiation, one report examined the possibility that c-*myc* regulation may be disrupted in HPV associated tumorigenesis (Hurlin et al., 1995). In primary keratinocyte derived raft cultures, Myc expression was found to be restricted to the basal layer where PCNA expression was also observed, and Mad was expressed in terminally differentiated suprabasal layers (Hurlin et al., 1995). In contrast, both Myc and Mad were expressed in both basal and differentiating layers of raft cultures derived from HPV18 containing keratinocytes, and the cultures resembled *in vivo* dysplastic squamous lesions (Hurlin et al., 1995). The coordinated interaction between Myc, Mad and Max subunits is yet to be examined in raft cultures.

7. Differentiation

7.1. Host proteins as markers of differentiation

Raft cultures derived from primary foreskin keratinocytes develop histological strata consisting of basal, spinous, granular and cornified layers accompanied by the production of differentiation-specific keratins (Asselineau et al., 1986; Kopan et al., 1987; Wilson et al., 1992). The basic-acidic partner keratins K5 and K14 proteins are expressed in basal layer whereas K1 and K10 proteins are expressed in cells which have left the basal layers and have undergone the commitment to differentiate (Fuchs et al., 1987). Morphological stratification of organotypic raft cultures has been correlated with the expression of differentiation markers, prominent ones of which are the keratins and cell cycle markers such as PCNA (Merrick et al., 1992a). Raft cultures derived from normal keratinocytes were shown to lack K8 expression and PCNA expression was found to be restricted to the basal layers as expected, whereas K1 and filaggrin were expressed in the suprabasal layers. Expression of keratins in a differentiation dependent manner has been used as a marker to assess the histological grade of raft cultures derived from HPV infected cells. For example keratin K8 has been found to be abundantly expressed in squamous cell carcinoma of the cervix (Ivanyi et al., 1990). Similar K8 expression was recapitulated in raft cultures derived from HPV18 containing cell lines (Merrick et al., 1992a). In normal epithelium, filaggrin is expressed only in the granular layer (Dale et al., 1985; Steven et al., 1990). K1, involucrin and profilaggrin are considered markers of terminal differentiation (Banks-Schlegel and Green, 1981; Dale et al., 1985; Fuchs et al., 1987). In contrast, filaggrin expression was downregulated in raft cultures derived from HPV infected cell lines similar to CIN I lesions (Meyers et al., 1992). Another study found that HPV18 derived raft cultures expressed decreasing amounts of K10 and profilaggrin with increasing degree of dysplasia (Steenbergen et al., 1998b).

Human epithelial cells transfected with HP16 DNA retained the ability to stratify and express differentiation-specific keratin markers, but the histopathological changes in epithelial morphology were similar to intraepithelial neoplasia (McCance et al., 1988). These cells exhibited altered differentiation patterns such as lack of ordered differentiated layers, koilocytosis, abnormal chromosomal staining and reduced cytoplasmic/nuclear ratio in the suprabasal cells of rafts (McCance et al., 1988). In raft cultures derived from HPV infected cells, the degree of epithelial differentiation was found to be dependent on the nature of the HPV infected cell lines and ranged from ability to cornify and produce more differentiated epithelial layers, to highly disorganized epithelial layers that resembled carcinoma *in situ* (Blanton et al., 1991). In addition, after repeated passaging of few hundred generations, raft cultures of HPV16-infected keratinocytes displayed histological changes similar to those observed in transition from grade I to grade III neoplasia in patients. These changes are presumably due to the integration of the HPV16 genome and are indicative of malignant lesions (Boshart et al., 1984; Durst et al., 1983; Pett et al., 2004) whereas in premalignant lesions *in vivo* the viral DNA is present as an episome. The E7 gene of HPV16 alone has been shown to stimulate keratinocyte proliferation in organotypic cultures which results in a disorganized epithelial layer and loss of differentiation (Halbert et al., 1992). Cumulatively, these studies pointed to the importance of correlating phenotypic alterations with the degree of differentiation, and hence progressive dysplasia (Merrick et al., 1992a).

Cell lines derived from vaginal intraepithelial neoplasias (VIN) expressing HPV33 or HPV16 have been used in generating raft cultures (Hietanen et al., 1998). Organotypic cultures derived using early passages of these cell lines resembled VIN. Using cells from later passages, the raft cultures resembled full-thickness intraepithelial neoplasia and microinvasive cancer, indicating progression towards a more transformed phenotype. Abnormal differentiation patterns in raft cultures were accompanied by aberrant expression of K18 (which is normally absent from vaginal keratinocytes) and increased K17 expression, both of which were associated with loss of orderly differentiation and correlated with the progressive nature of the lesion (Syrjanen et al., 1988). Upregulation of K8 and K18 was also observed in keratinocytes maintaining oncogenic PVs (Merrick et al., 1992b). K19 was upregulated in HPV16 containing VIN, but interestingly, aberrant keratin expression and abnormal morphology of raft cultures could not be correlated with changes in cell-cycle specific proteins such as p53, p21 and PCNA, which are also indicators of proliferation (Hietanen et al., 1998).

Loss of differentiation has also been correlated with specific cell cycle regulatory proteins. The HPV16 E6 oncoprotein binds p53 and promotes its degradation via the ubiquitin-dependent pathway and is an important step in HPV mediated immortalization (Scheffner et al., 1990). In rafts of normal keratinocytes, overexpression of p53 delayed the expression of the differentiation markers K10 and profilaggrin, in agreement with the inhibition of the expression of several keratins by p53 (Woodworth et al., 1993). In addition, normal raft cultures underwent a decrease in their growth rate and showed premature cell flattening and involucrin expression; p53 was detected in the basal cells but not in the differentiating compartments suggesting that downregulation of p53 during differentiation may be required for expression of specific squamous layer genes. In contrast, HPV raft cultures expressed low levels of endogenous p53 due to the presence of E6, and inactivation of p53 correlated with alterations of the differentiation of the HPV infected keratinocytes. However, overexpression of p53 did not reverse aberrant differentiation of these cultures (Woodworth et al., 1993), suggesting that loss of p53 via E6 mediated degradation is not necessarily the primary cause of dysplastic progression, and that severe changes in differentiation patterns in HPV infected cells develop progressively over time due to secondary cellular changes.

7.2. Loss of response to growth factors signalling pathways

The loss of orderly stratification and differentiation markers accompanying changes during progression from CINI to CINIII is also accompanied by loss of cellular responsiveness to growth associated signaling pathways. TGF-β1 is a cellular protein with growth-inhibitory properties which modulates cellular growth and differentiation. It is expressed in both normal human epithelial cells as well as normal cervical epithelium (Dagnino et al., 1993). Loss of responsiveness to TGF-β1 plays a role in HPV associated carcinogenesis (Woodworth et al., 1990). TGF-β1 treatment inhibits growth of HPV infected monolayers together with downregulation of E6 and E7 oncogenic transcription (Braun et al., 1992; Woodworth et al., 1990). TGF-β1 treatment of neoplastic lines containing integrated HPV16, 18 and 31 DNA in organotypic cultures resulted in the expression of normal markers of differentiation (Ozbun and Meyers, 1996) and the cultures acquired a more normal morphology. TGF-β1 treatment also inhibited deregulation of K8 expression in the HPV18 positive invasive carcinoma-derived cell line AWCA (Rader et al., 1990). These results point to the importance of studying the growth regulatory pathways affecting the life cycle of HPV in the differentiation dependent environment. Clearly, the actions of TGF-β1 are greatly determined by the ability of the tissue to differentiate.

Treatment of HPV DNA positive raft cultures with all-*trans*-retinoic acid resulted in growth inhibition of ectocervical lines transfected with HPV16 or 18 (Behbakht et al., 1996). Growth inhibition was accompanied by loss of both morphological stratification and failure to express differentiation markers suggesting that retinoic acid induces growth inhibition in cervical neoplastic cell lines (Behbakht et al., 1996). It was hypothesized that retinoic acid stimulated increases in TGF-β secretion and this would lead to the observed cell growth inhibition. A related study found that terminal differentiation of HPV-immortalized keratinocytes was resistant to the effect of retinoic acid (Merrick et al., 1993). This study found that between 10 to 30-fold higher concentrations of retinoic acid were required to block terminal differentiation of HPV positive organotypic cultures compared with control cultures (Merrick et al., 1993). In contrast, retinoic acid treatment of raft cultures derived from non-malignant HPV16 containing lines were prevented from progressing to severe dysplasia and resembled the epithelium of normal cervix (Sarma et al., 1996). Interestingly, raft cultures derived from tumorigenic cell lines which morphologically resembled severe dysplasia/carcinoma *in situ* were resistant to retinoic acid treatment (Sarma et al., 1996). Thus, retinoic acid-activated pathways have the ability to regulate differentiation pattern depending on the grade of the lesion.

8. Cofactors

Infection with oncogenic HPVs is necessary for cellular immortalization, but is not sufficient for progression to carcinoma. Numerous cofactors have been identified as contributing to the carcinogenic progression of HPV infections to malignancy. Cofactors for the progression to cervical cancer include copathogens such as herpes simplex viruses (HSV) and Chlamydia (Anttila et al., 2001; Clarke and Clements, 1991; DiPaolo et al., 1996; Hildesheim et al., 1991; Luostarinen et al., 2004; Matsumoto et al., 2003; Smith et al., 2004; 2002a; 2002b; zur Hausen, 1982), cigarette smoke (Ali et al., 1994; Clarke et al., 1982; Hildesheim et al., 1990; Kjellberg et al., 2000), immunosuppression (Halpert et al., 1986; Shokri-Tabibzadeh et al., 1981), and oral contraceptive use (Anonymous 1991; 1992; Beral et al., 1988; 1999; Brinton, 1991; Brinton et al., 1990; Hildesheim et al., 1990; Ursin et al., 1994; Vessey et al., 1983). Of these cofactors, as far as we can tell, there has been only a few publications where the role of HSV, cigarette smoke, and contraceptive hormones have been investigated as to their effect on the natural history of HPV using raft culture model systems (Bromberg-White et al., 2003; Khare et al., 1997; Meyers et al., 2003). Studies on the interactions of HSV with HPV in raft cultures are discussed elsewhere in this chapter, therefore we will restrict our discussion here to cigarette and hormonal effects on HPV.

It is widely accepted that carcinogens in cigarette smoke act as in initiators of numerous cancer types by inducing multiple genetic alterations, mainly through the formation of DNA adducts (Hecht, 1999; Kato et al., 1993; Shields, 2002). Mutagenic insults conferred by tobacco related carcinogens alter gene expression, cell proliferation and differentiation of cells (Osada and Takahashi, 2002). During the establishment of HPV infection and multistage carcinogenic progression, many tumor suppressor genes are inactivated and functions of many cell cycle checkpoints are deregulated (Funk and Galloway, 1998). Epidemiological studies strongly suggest that cigarette smoking is a cofactor positively linked with the development of HPV-associated cervical cancer (Nischan et al., 1988; Stockwell and Lyman, 1987). HPV18 immortalized ectocervical (Nakao et al., 1996) and HPV16 immortalized endocervical (Yang et al., 1996) cell lines were treated with cigarette smoke condensate to produce cell lines that were tumorigenic in nude mice. When grown in raft culture the HPV18-immortalized, cigarette smoke condensate-treated ectocervical cells displayed severe dysplasia reminiscent of cervical carcinoma whereas the untreated cells displayed only a mild dysplasia (Nakao et al., 1996). The HPV16-immortalized, cigarette smoke condensate-treated, endocervical cells exhibited a thicker highly disorganized epithelia resembling a CIN III lesion's morphology when grown in organotypic raft culture whereas untreated lines did not (Yang et al., 1996).

Epidemiological studies have provided evidence that oral contraceptives and steroid hormones play a role in the development of cervical carcinogenesis. These studies have indicated an increased risk of cervical cancer with increased duration of use (1991; 1992; Beral et al., 1988; Brinton, 1991; 1990; Hildesheim et al., 1990; Thomas, 1991; Vessey et al., 1983). There is also evidence that steroid hormones affect HPV infection and the development of HPV-associated cancer. The incidence of squamous cell carcinoma of the genital tract is much higher in women than men (Ciocca et al., 1992). HPV prevalence is at its highest during the reproductive years (Ciocca et al., 1992), and pregnancy is a permissive environment for a persistent HPV infection (Arbeit et al., 1996). Additionally, benign and malignant lesions of the genital tract occur predominately in the estrogen-sensitive cells (Auborn et al., 1991). The general hypothesis concerning the relationship between steroid hormones and HPV infection is that steroid hormones contribute to the persistence of HPV, and to neoplastic progression by increasing viral gene expression (Arbeit et al., 1996). While there have been numerous studies on the affects of steroid hormones on the biology and oncogenic potential of HPV, only one published study utilized raft culture technology in their study. Human ectocervical cells immortalized with a genetically modified HPV16 to contain a triplet of its glucocorticoid response element demonstrated in the presence of hormone a more abnormal epithelial differentiation pattern when grown in raft culture (Khare et al., 1997). This suggests that the hormones may impact directly on HPV gene regulation to induce a more transformed state.

Although raft cultures provide a more realistic model and are very amenable to cofactor studies, relatively few studies have been published. The role of cofactors in HPV-associated carcinogenic progression is very important and has a major impact on human disease. We feel that in the quest for greater insight into the interplay of cofactors and HPV raft cultures have been underutilized.

9. Interaction of HPV with Other Epitheliotrophic Viruses

The tropism of HPV for stratified squamous epithelium is clinically significant, as it presents opportunity for coinfection with other epitheliotropic viruses, which also infect the cervical epithelium. Both pathogenic and non-pathogenic viral interactions with HPV have been studied in the differentiation dependent organotypic cultures. Epidemiological studies strongly correlate the protective effect of the adeno-associated virus infections with the development of cervical cancer (Georg-Fries et al., 1984; Mayor et al., 1976). Adeno-associated virus type 2 (AAV2) is classified as a helper-dependent parvovirus, which requires helper functions from such viruses as adenovirus (Casto et al., 1967), herpes simplex virus (Buller et al., 1981), and pox virus (Schlehofer et al., 1986). In turn, AAV inhibits the replication and progeny formation of its helper virus. The negative correlation between the prevalence of anti-AAV2 antibodies and the development of cervical cancer would suggest that AAV2 may inhibit HPV functions. Although the tumor suppressive properties of AAV2 have been well documented against bovine papillomavirus in monolayer cultures, evidence for a direct molecular interaction between HPV and AAV2 was lacking until recently. Raft cultures derived from CIN-612 9E keratinocytes with episomal HPV31b were infected with AAV2 at increasing multiplicities of infection (MOI) (Meyers et al., 2001). AAV2 was shown to upregulate HPV31b replication at low MOI, but significantly inhibited HPV31b replication at high MOI. Inhibition of HPV replication could be correlated with high levels of active AAV2 replication, suggesting that HPV is a typical helper virus for AAV2 (Meyers et al., 2001). Also, the ability of AAV2 to inhibit HPV replication was shown to be differentiation dependent as this inhibition was not observed in monolayer cultures. The MOI dependence of AAV2 on its effect on HPV replication was also demonstrated in another study (Agrawal et al., 2002).

The herpes simplex viruses (HSV) are known to be sexually transmitted and are associated with recurring, painful ulcerations (Sherman et al., 1991). Since both HPV and HSV are sexually transmitted and infect the same cell type, the two viruses have the potential to interact in the cervical epithelium. HSV infection was shown to induce significant cytopathic effects in HPV positive raft cultures, and could be correlated with expression of HSV encoded ICP4 and VP5 proteins. Interestingly, HPV genome copy numbers were maintained in steady amounts through periods of crisis generated during HSV induced cytopathogenicity. Cycles of tissue regeneration were documented, suggesting that HPV could induce cellular mechanisms for tissue repair to protect its own replication and gene expression (Meyers et al., 2003).

Human adenoviruses have been increasingly used as vectors for gene transfer and gene therapy of various diseases. Recently, there has been interest in generating replication-competent adenoviruses as oncolytic agents for neoplasias of epithelial origin such as head and neck cancers (Nemunaitis et al., 2000). Using the raft culture system it was possible to study differentiation dependency of some of the early promoters of adenoviruses, which may prove useful in terms of understanding adenovirus gene expression and host interaction in a clinical setting (Noya et al., 2003). Deletion of specific domins of adenovirus E1A protein in replication competent adenoviruses affect their ability to selectively target and kill HPV18 E6E7 expressing cells (Balague et al., 2001).

10. HPV Immunology Using Raft Culture

A small number of studies have used organotypic cultures to examine the interaction of infected keratinocytes with cells and soluble factors of the immune system.

Allogeneic peripheral blood mononuclear cells (PBMCs) added to organotypic cultures of cervical carcinoma cell lines could infiltrate the keratinocyte layers and cause apoptosis of the keratinocytes (Jacobs et al., 1998). Most of the infiltrating cells were T cells, but a small number of NK cells were also present (Jacobs et al., 1998). *In vitro* stimulation of the PBMCs before coculture increased the rates of both infiltration and killing, and the cytotoxic activity was more effective against SiHa cells than normal cervical keratinocytes (Delvenne et al., 2001; Jacobs et al., 1998). ICAM-1, which facilitates cell-mediated killing, was also upregulated on SiHa cells in the presence of activated PBMCs (Jacobs et al., 1998).

Like T cells and NK cells, dendritic cells (DCs), of which Langerhans cells are a subset, were able to infiltrate raft cultures and differentiate (Hubert et al., 1999b). This infiltration was facilitated by the secretion of GM-CSF by the keratinocytes (Hubert et al., 1999b). The expression of GM-CSF by HPV containing carcinoma cell lines was very low compared to human cervical keratinocytes (HCKs), so the infiltration and differentiation of DCs into carcinoma rafts was poor unless GM-CSF was added (Delvenne et al., 2001; Hubert et al., 1999b). Upon infiltration into rafts of HPV-containing carcinoma cell lines, DCs were also able to induce apoptosis of the keratinocytes without death of the DCs themselves. On the contrary, infiltration into normal HCK rafts was not altered by GM-CSF nor was keratinocyte death a result (Hubert et al., 1999a). Expression of E-cadherin, which facilitates retention of DCs in the epithelium, was reduced in rafts of HPV-containing carcinomas (Delvenne et al., 2001). These data suggest that reduced levels of lymphocyte infiltration into lesions *in vivo*, and the resulting sluggish immune response, may partly result from viral alteration of the chemotactic or survival signals upon which the lymphocytes normally depend. IFN-γ and TNF-α reduced the growth of both normal and carcinoma rafts (Delvenne et al., 2001).

11. Therapeutic Testing

Raft cultures provide a tool for creating a three-dimensional model of the differentiating epithelium, in which the authentic architecture of a differentiated tissue is accessible to a wide variety of experimental manipulations. The spatial and temporal effects of a treatment can be examined using essentially any histological, biochemical, and molecular assay to which an *in vivo* tissue specimen could be subjected. Therefore, raft cultures are strong candidates for testing therapeutic intervention against HPV-associated human neoplasias. Only a few published studies have exploited the use of raft cultures to study the efficacy of therapeutic interventions.

The effect of retinoic acid on the development of severe dysplasia or carcinoma *in situ* has been investigated using raft culture models. Retinoic acid treatment of HPV16 transformed endocervical raft cultures resulted into columnar-like cells two to three layers thick resembling simple epithelium of the endocervix (Shindoh et al., 1995). Additionally, retinoic acid could reverse the dysplastic phenotype of HPV18-immortalized endo- and ectocervical cell in raft culture tissue (Yokoyama et al., 2001). However, raft cultures of the CIN-612 cell line lost the top cornified layer when were treated with retinoic acid (Behbakht et al., 1996). This would suggest that retinoic acid increased neoplastic progression with apparent decrease in differentiation potential, but, given that the top cornified or keratin layers are often lost during histological processing of raft cultures, this observation needs to be repeated. A growth inhibitory effect of interferon-γ (IFNγ) and tumor necrosis factor α (TNFα) treatment of HPV-transformed cervical keratinocyte rafts was demonstrated by an overall thinning of the tissue and a reduction in the density of proliferating cells (Delvenne et al., 2001). This result was less pronounced when the tissue was treated with IFNγ alone. In general, treatment of dysplastic raft tissue with biologicals such as retinoic acid, IFNγ, and TNFα appears to induce a less dysplastic, more normal differentiation program. Similar results were obtained when CIN I, CIN III, and invasive carcinoma raft cultures were treated with TGF-β (Ozbun and Meyers, 1996). The phenotype of raft cultures of HPV18-infected invasive carcinoma cells resembled the phenotype of CIN-I when treated with TGF-β (Ozbun and Meyers, 1996). Studies using monolayer culture models to determine the efficacy of IFN therapy for the treatment of cervical cancer have been highly inconsistent with controversial conclusions. Sen et al., tested the effects of two commercially available IFNs, interferon-α-n3 and interferon-α-2b, on vegetative HPV replication in raft cultures capable of sustaining the complete viral life cycles of HPV16, 18, and 31b (Sen et al., 2004b). These studies showed that the two IFNs varied in their ability to affect viral load of the different HPV types.

Expression of EGF receptor (EGFR) was abundant throughout the thickness of rafts of HPV-containing cervical carcinoma lines, while it was expressed at a lower level and only in the basal layer of rafts of normal cervical keratinocytes (Renard et al., 2002). Because of this difference, bispecific antibodies, which bind both EGFR and CD3, were found to facilitate the infiltration of T cells into carcinoma rafts and increase killing of the keratinocytes by the lymphocytes, while the antibodies had no effect on T cell infiltration into rafts of normal cervical keratinocytes (Renard et al., 2002).

Several experimental and epidemiological studies suggest that antioxidant vitamins can inhibit carcinogenesis. Studies were initiated to test whether or not the antioxidant vitamins A, C, and E could have this effect on raft cultures (Mason et al., 1999). In undifferentiated monolayer cultures all three vitamins induced a reduction of proliferation of both normal oral epithelial cells and HPV16-immortalized oral epithelial cells, and flow cytometry showed that the cells were arresting in greater numbers in S phase following exposure to the vitamins (Mason et al., 1999). However, when the cells were grown in raft culture, there was little difference between vitamin-treated and control cultures, except for a reduction in the number of layers in the vitamin E-treated cultures (Mason et al., 1999). This study is a good example of how the effects of treating undifferentiated monolayer cultures do not always correlate to the effects seen when fully stratified and differentiated raft cultures are treated.

The major tea polyphenol, (-)-epigallocatechin gallate (EGCG), a component in green tea shown to have efficacy in cancer prevention, was tested for its effects on primary endocervical, HPV18-immortalized and HPV18-transformed endocervical and ectocervical raft culture growth. Raft cultures treated with EGCG showed a decrease in the thickness of the tissues and an inhibition of telomerase activity regardless of which cell type was used (Yokoyama et al., 2004). However, EGCG treatment of rafts grown from HPV18-immortalized cells resulted in apoptosis while rafts grown from transformed lines did not (Yokoyama et al., 2004).

Raft culture models have also been used test for adverse affects of vaginal microbicides (Aebischer and McDougall, 1997). Although this is not testing a therapeutic effect, it is testing potential adverse affects of a prophylactic agent on human tissue. Repeated treatment of HFKs and HPV-immortalized cervical keratinocytes with the microbicides resulted in harmful effects (histological alterations) seen in raft tissues but undetectable in undifferentiated monolayer culture models (Aebischer and McDougall, 1997).

12. Conclusions and Some Unanswered Questions

As can be seen from the variety and scope of the studies reviewed here, raft cultures have yielded enormous insights into the biology of PVs both in their normal life cycles and in the development of neoplasia. Many questions have been answered but many still remain.

The E6 and E7 mediated abrogation of major cell cycle regulatory proteins have been extensively studied in monolayer cultures. However, only a limited number of studies have addressed the role of cell cycle proteins in HPV life cycle under differentiating conditions. From the available studies, it is apparent that HPV specific changes in the cell cycle machinery generally result in an environment conducive to high levels of host DNA synthesis and thereby viral DNA synthesis. Studies using monolayer cultures have revealed complexities of HPV-regulated cell cycle control mechanisms that may or may not turn out to be true in differentiating cells. Since the HPV life cycle is intricately connected to the host cell cycle, it will be necessary to perform more systematic studies to elucidate how different protein complexes and activities of the cell cycle machinery affect each stage of the HPV life cycle. Much remains to be understood about the molecular mechanisms of suprabasal DNA synthesis, and raft cultures, perhaps employing genetic techniques on viral or cellular genes, could be used to work out these mechanisms.

As compared to induction of DNA synthesis and cell cycle dysregulation, less appears to be understood about how the virus alters other aspects of cellular differentiation. How the relatively mild disruption seen in low grade lesions becomes the severe disruption seen in cancers, and how gene expression and DNA synthesis components of the differentiation program are related and affected by infection, are not well understood but rafts will continue to be helpful.

Much attention has been paid to understanding the viral oncogenes and their effects on the host tissue differentiation and morphology, as well as on the features of tissues that are undergoing malignant progression. Many of these studies have expressed the viral oncogenes at non-physiological levels or out of context of the other viral genes and control elements. While such expression may mimic the situation in late stages of cancer progression, it is reasonable to suppose that at earlier stages, in which other components of the virus or the cell may still be active, there may be important interactions along the progression pathway that have been overlooked. Effects of E6 and E7 on cellular gene expression are becoming better appreciated, but the possibility that the transactivator E2 may function in a normal infection by altering the activity of cellular promoters is a largely unexplored idea, and rafts provide means to test this potential regulation through all the stages of the differentiation pathway.

Some molecular correlates of high and low cancer risk have been described (Munger and Howley, 2002), but the actual biology underlying this difference in malignant potential between virus types in the context of the infected host tissue is unknown. Further, to what extent different virus types, even closely related ones, are similar or different in the molecular control of their life cycles is unknown. The genetic constitution of the individual host also appears to be an important influence on the outcome of infection (Pfister, 1987a; 1987b; 1996; zur Hausen, 1989; 1991; 1996; 1999), but details of how host background affects virus biology are fuzzy at best. Rafts may provide a tool to dissect these relevant host and viral genetic factors.

Although much effort has been expended to understand mechanisms of progression, the observation that integration almost always accompanies progression remains unexplained mechanistically. What causes integration, and whether integration causes or is an effect of cancer progression is a major unanswered question.

It is clear that HPV is not able to induce cancer alone, and although some cofactors have been described, raft cultures have been sorely underutilized to understand their effects, both in relation to the virus and to the cells themselves. Because the molecular environment of a differentiating tissue is so different from monolayer culture, in order to understand how these cofactors contribute to disease, raft cultures will probably be indispensable.

HFKs are often used as host cells in experimental studies of viral life cycles, but this is less to ensure authenticity than for convenience in obtaining the material for use. It certainly cannot be assumed that foreskin cells, which are male and cutaneous, are in every way analogous to cervical cells, which are mucosal and female. Furthermore, even within the mucosal histotype, biological differences between endocervical, transition zone, ectocervical, vaginal, vulval, oral, and laryngeal keratinocytes have not been extensively characterized, and whether any possible differences are reflected in normal or pathological virus biology is not known. Why, for instance, do transition zone keratinocytes tend to be cancer prone? What cells harbor HPV in males? To what extent do hormones affect HPV biology and how? This last question in particular will require careful choices of model cells in terms of gender and tissue of origin.

In contrast to studies of progression towards cancer, our understanding of the normal life cycle of HPV in productive infection is fairly limited. Since rafts are at present the only system for studying the HPV life cycle *in vitro*, such studies have been possible only recently, and much of what we know, such as transcript mapping, analysis of protein expression, replication, and virion morphogenesis represents description of the events in the life cycle rather than experimental manipulation. Although these observations have been important, systematic, and elegant, the molecular basis for the observed events, as well as the significance of what has been observed, remains, for the most part, unknown. Mutational analyses of the virus proteins in the context of the whole genome have only scratched the surface in terms of what remains to be understood.

None of the life cycle studies reviewed here using protein mutagenesis in the context of the whole genome have attempted to rescue observed defects by providing the protein in trans. This would be important, not only to establish that the observations are actually due to loss of the protein, but also because by providing mutant proteins in trans one could dissect at the molecular level what can and cannot complement the mutation. Of those studies that have been performed, very few have actually measured the production of infectious virus progeny, so even among the poorly understood late processes, virion morphogenesis remains even less understood still. Except for the involvement of PKC in the early to late polyadenylation switch, the cellular signals to which the virus responds in each stratum along the differentiation pathway, and the events in the virus life cycle that these signals control, remain to be characterized. Also, since HPV capsid synthesis in the suprabasal layers requires translation in differentiating conditions, alteration of protein synthesis activities may have regulatory significance. Thus, control of protein synthesis in a differentiating environment will be an important aspect of future studies.

Studies of HPV immunology using raft cultures are, so far, rather primitive, although they show great promise of further development in the future. Difficulties matching the HLA types of keratinocyte and immune cells for coculture are probably not insurmountable with ingenuity and careful preparation of material. It is possible that a combined strategy using rafts grafted onto mice (such as that of McBride et al., 2000) may open the world of mouse genetics and immunology to studies of PVs.

It has been observed in a variety of PV types, both *in vivo* and in raft culture, that not all cells at the same stage of maturity (in terms of tissue strata) support HPV replication, transcription, or protein expression to the same degree, even (in the case of cultured cells) when the raft cultures are derived from a clonal cell population (Barksdale and Baker, 1993; Bedell et al., 1991; Cheng et al., 1995; Doorbar et al., 1997; Frattini et al., 1997; Jian et al., 1998; Jian et al., 1999; Mayer and Meyers, 1998; McBride et al., 2000; Middleton et al., 2003; Nicholls et al., 2001; Ozbun and Meyers, 1998a; Peh et al., 2002). The majority of cells appear to be relatively quiescent, while an occasional cell supports viral activity to a high level. The basis of this heterogeneity is unknown but would undoubtedly lead to better understanding of the factors necessary for supporting late HPV functions.

The very early stages of the life cycle are also mysterious. It is often suggested that the virus gains entry to the basal cells of the epithelium via "microtraumas in the epithelium," but real evidence for this common assertion is scant. Mimicking epithelia susceptible to experimental trauma, rafts may provide a strategy for actually testing hypotheses about pathways of tissue entry and establishment of infection.

An enormous diversity of transcripts has been observed using rafts, originating from several promoters with a variety of splices and having the potential to encode numerous putative proteins (E6* (I, II, III), E6^E4, E1* (I, II), E1^E4*, E5a*, L1s, X, URR Y, Z, etc.). Why there are so many different transcripts, how they are generated, what their functions are, and whether any of these potential alternative proteins exist and with what functions, are intriguing but totally unanswered questions. Even in the case of known proteins, which transcripts among the many possibilities are actually responsible for their translation is generally a matter of speculation.

Finally, we may have things to learn about raft growing technology itself. For example, peak followed by decline has been observed in raft cultures in the expression of L1 and L2 transcripts (Ozbun and Meyers, 1997), E1 and E2 transcripts, viral DNA amplification (Ozbun and Meyers, 1998a), and E5a protein expression (Mayer and Meyers, 1998). It was also seen in HaCaT cell rafts in the expression of K10 (Tomakidi et al., 2000b). The basis of this late decline is unknown. It has been hypothesized that this decline may be due to an inability of the cells of the basal layer to replenish the suprabasal strata (McBride et al., 2000). It would be interesting to understand whether the composition of the seeding population with respect to stem cells vs. transit amplifying cells could alter the behavior or lifespan of a raft, or if progress through the viral life cycle would be different. McBride et al. showed that rafts that would ordinarily degenerate could live for months when grafted onto mice (McBride et al., 2000). The type of fibroblasts in the dermal equivalent may also influence the vigor of raft growth (Dollard et al., 1992). Clearly growth factor or other signaling interactions between the keratinocytes and fibroblasts or other cell types remain to be better understood and may have an impact on HPV biology.

In conclusion, the power of raft cultures to increase our understanding of HPV biology has been employed extensively but remains immense. The future promises many additional insights, not only to enrich our understanding, but also to contribute to possible treatments or prevention strategies for these important pathogens.

References

Anonymous (1991). Depot-medroxyprogesterone acetate (DMPA) and risk of epithelial ovarian cancer. The WHO Collaborative Study of Neoplasia and Steroid Contraceptives. Int. J. Cancer *49*, 191-195.

Anonymous (1992). Depot-medroxyprogesterone acetate (DMPA) and risk of invasive squamous cell cervical cancer. The WHO Collaborative Study of Neoplasia and Steroid Contraceptives. Contraception *45*, 299-312.

Aaltonen, L.M., Wahlstrom, T., Rihkanen, H., and Vaheri, A. (1998). A novel method to culture laryngeal human papillomavirus-positive epithelial cells produces papilloma-type cytology on collagen rafts. Eur. J. Cancer *34*, 1111-1116.

Aasen, T., Hodgins, M.B., Edward, M., and Graham, S.V. (2003). The relationship between connexins, gap junctions, tissue architecture and tumour invasion, as studied in a novel *in vitro* model of HPV-16-associated cervical cancer progression. Oncogene *22*, 7969-7980.

Aebischer, F., and McDougall, J.K. (1997). Organotypic raft cultures for the *in vitro* evaluation of vaginal microbicidal agents. Microbicides have epithelium-specific effects when repeatedly added onto organotypic human keratinocyte cultures. Sex. Transm. Dis. *24*, 69-76.

Agrawal, N., Mane, M., Chiriva-Internati, M., Roman, J.J., and Hermonat, P.L. (2002). Temporal acceleration of the human papillomavirus life cycle by adeno-associated virus (AAV) type 2 superinfection in natural host tissue. Virology *297*, 203-210.

Ai, W., Narahari, J., and Roman, A. (2000). Yin yang 1 negatively regulates the differentiation-specific E1 promoter of human papillomavirus type 6. J. Virol. *74*, 5198-5205.

Albelda, S.M., Mette, S.A., Elder, D.E., Stewart, R., Damjanovich, L., Herlyn, M., and Buck, C.A. (1990). Integrin distribution in malignant melanoma: association of the beta 3 subunit with tumor progression. Cancer Res. *50*, 6757-6764.

Ali, S., Astley, S.B., Sheldon, T.A., Peel, K.R., and Wells, M. (1994). Detection and measurement of DNA adducts in the cervix of smokers and non-smokers. Int. J. Gynecol. Cancer *4*, 188-193.

Amati, B., and Land, H. (1994). Myc-Max-Mad: a transcription factor network controlling cell cycle progression, differentiation and death. Curr. Opin. Genet. Dev. *4*, 102-108.

Anttila, T., Saikku, P., Koskela, P., Bloigu, A., Dillner, J., Ikaheimo, I., Jellum, E., Lehtinen, M., Lenner, P., Hakulinen, T., et al. (2001). Serotypes of Chlamydia trachomatis and risk for development of cervical squamous cell carcinoma. Jama *285*, 47-51.

Arbeit, J.M., Howley, P.M., and Hanahan, D. (1996). Chronic estrogen-induced cervical and vaginal squamous carcinogenesis in human papillomavirus type 16 transgenic mice. Proc. Natl. Acad. Sci. USA *93*, 2930-2935.

Asselineau, D., Bernard, B.A., Bailly, C., Darmon, M., and Prunieras, M. (1986). Human epidermis reconstructed by culture: is it "normal"? J. Invest. Dermatol. *86*, 181-186.

Asselineau, D., and Prunieras, M. (1984). Reconstruction of 'simplified' skin: control of fabrication. Br. J. Dermatol. *111 Suppl. 27*, 219-222.

Auborn, K.J., Woodworth, C., DiPaolo, J.A., and Bradlow, H.L. (1991). The interaction between HPV infection and estrogen metabolism in cervical carcinogenesis. Int. J. Cancer *49*, 867-869.

Baker, C., and Calef, C. (1997). Maps of Papillomavirus mRNA Transcripts (Los Alamos National Laboratory).

Baker, C.C., and Noe, J.S. (1989). Transcriptional termination between bovine papillomavirus type 1 (BPV- 1) early and late polyadenylation sites blocks late transcription in BPV-1-transformed cells. J. Virol. *63*, 3529-3534.

Balague, C., Noya, F., Alemany, R., Chow, L.T., and Curiel, D.T. (2001). Human papillomavirus E6E7-mediated adenovirus cell killing: selectivity of mutant adenovirus replication in organotypic cultures of human keratinocytes. J. Virol. *75*, 7602-7611.

Banks-Schlegel, S., and Green, H. (1981). Involucrin synthesis and tissue assembly by keratinocytes in natural and cultured human epithelia. J. Cell Biol. *90*, 732-737.

Barksdale, S., and Baker, C.C. (1995). Differentiation-specific alternative splicing of bovine papillomavirus late mRNAs. J. Virol. *69*, 6553-6556.

Barksdale, S.K., and Baker, C.C. (1993). Differentiation-specific expression from the bovine papillomavirus type 1 P2443 and late promoters. J. Virol. *67*, 5605-5616.

Bedell, M.A., Hudson, J.B., Golub, T.R., Turyk, M.E., Hosken, M., Wilbanks, G.D., and Laimins, L.A. (1991). Amplification of human papillomavirus genomes *in vitro* is dependent on epithelial differentiation. J. Virol. *65*, 2254-2260.

Behbakht, K., DeGeest, K., Turyk, M.E., and Wilbanks, G.D. (1996). All-trans-retinoic acid inhibits the proliferation of cell lines derived from human cervical neoplasia. Gynecol. Oncol. *61*, 31-39.

Bell, E., Ivarsson, B., and Merrill, C. (1979). Production of a tissue-like structure by contraction of collagen lattices by human fibroblasts of different proliferative potential *in vitro*. Proc. Natl. Acad. Sci. USA *76*, 1274-1278.

Bell, E., Sher, S., Hull, B., Merrill, C., Rosen, S., Chamson, A., Asselineau, D., Dubertret, L., Coulomb, B., Lapiere, C., et al. (1983). The reconstitution of living skin. J. Invest. Dermatol. *81*, 2s-10s.

Beral, V., Hannaford, P., and Kay, C. (1988). Oral contraceptive use and malignancies of the genital tract. Results from the Royal College of General Practitioners' Oral Contraception Study. Lancet *2*, 1331-1335.

Beral, V., Hermon, C., Kay, C., Hannaford, P., Darby, S., and Reeves, G. (1999). Mortality associated with oral contraceptive use: 25 year follow up of cohort of 46 000 women from Royal College of General Practitioners' oral contraception study. Bmj *318*, 96-100.

Blanton, R.A., Perez-Reyes, N., Merrick, D.T., and McDougall, J.K. (1991). Epithelial cells immortalized by human papillomaviruses have premalignant characteristics in organotypic culture. Am. J. Pathol. *138*, 673-685.

Boshart, M., Gissmann, L., Ikenberg, H., Kleinheinz, A., Scheurlen, W., and zur Hausen, H. (1984). A new type of papillomavirus DNA, its presence in genital cancer biopsies and in cell lines derived from cervical cancer. EMBO J *3*, 1151-1157.

Braun, L., Durst, M., Mikumo, R., Crowley, A., and Robinson, M. (1992). Regulation of growth and gene expression in human papillomavirus- transformed keratinocytes by transforming growth factor-beta: implications for the control of papillomavirus infection. Mol. Carcinog. *6*, 100-111.

Brinton, L. (1991). Oral Contraceptives and Cervical Neoplasia. Contraception *43*, 581-595.

Brinton, L., Reeves, W., Brenes, M., Herrero, R., de Britton, R., Gaitan, E., Tenorio, F., Garcia, M., and Rawls, W. (1990). Oral contraceptive use and risk of invasive cerivcal cancer. Int. J. Epidemiol. *19*, 4-11.

Bromberg-White, J.L., Alam, S., Sen, E., and Meyers, C. (2003). The induction of the upstream regulatory region of human papillomavirus type 31 by dexamethasone is differentiation dependent. J. Virol. *77*, 10975-10983.

Buller, R.M., Janik, J.E., Sebring, E.D., and Rose, J.A. (1981). Herpes simplex virus types 1 and 2 completely help adenovirus- associated virus replication. J. Virol. *40*, 241-247.

Casto, B.C., Atchison, R.W., and Hammon, W.M. (1967). Studies on the relationship between adeno-associated virus type I (AAV- 1) and adenoviruses. I. Replication of AAV-1 in certain cell cultures and its effect on helper adenovirus. Virology *32*, 52-59.

Chen, R.W., Aalto, Y., Teesalu, T., Durst, M., Knuutila, S., Aaltonen, L.M., and Vaheri, A. (2003). Establishment and characterisation of human papillomavirus type 16 DNA immortalised human tonsillar epithelial cell lines. Eur. J. Cancer *39*, 698-707.

Cheng, S., Schmidt-Grimminger, D.C., Murant, T., Broker, T.R., and Chow, L.T. (1995). Differentiation-dependent up-regulation of the human papillomavirus E7 gene reactivates cellular DNA replication in suprabasal differentiated keratinocytes. Genes Dev. *9*, 2335-2349.

Chien, W.M., Parker, J.N., Schmidt-Grimminger, D.C., Broker, T.R., and Chow, L.T. (2000). Casein kinase II phosphorylation of the human papillomavirus-18 E7 protein is critical for promoting S-phase entry. Cell Growth Differ. *11*, 425-435.

Ciocca, D.R., Lo Castro, G., Alonio, L.V., Cobo, M.F., Lotfi, H., and Teyssie, A. (1992). Effect of human papillomavirus infection on estrogen receptor and heat shock protein hsp27 phenotype in human cervix and vagina. Int. J. Gynecol. Pathol. *11*, 113-121.

Clarke, E.A., Morgan, R.W., and Newman, A.M. (1982). Smoking as a risk factor in cancer of the cervix: additional evidence from a case-control study. Am. J. Epidemiol. *115*, 59-66.

Clarke, P., and Clements, J.B. (1991). Mutagenesis occurring following infection with herpes simplex virus does not require virus replication. Virology *182*, 597-606.

Colgin, L.M., and Reddel, R.R. (1999). Telomere maintenance mechanisms and cellular immortalization. Curr. Opin. Genet. Dev. *9*, 97-103.

Crum, C.P., Ikenberg, H., Richart, R.M., and Gissman, L. (1984). Human papillomavirus type 16 and early cervical neoplasia. N. Engl. J. Med. *310*, 880-883.

Cumming, S.A., McPhillips, M.G., Veerapraditsin, T., Milligan, S.G., and Graham, S.V. (2003). Activity of the human papillomavirus type 16 late negative regulatory element is partly due to four weak consensus 5′ splice sites that bind a U1 snRNP-like complex. J. Virol. *77*, 5167-5177.

Cumming, S.A., Repellin, C.E., McPhillips, M., Radford, J.C., Clements, J.B., and Graham, S.V. (2002). The human papillomavirus type 31 late 3′ untranslated region contains a complex bipartite negative regulatory element. J. Virol. *76*, 5993-6003.

Dagnino, L., Pietenpol, J.A., Yang, E.Y., and Moses, H.L. (1993). Transforming growth factor regulation of keratinocyte growth. Recent Results Cancer Res. *128*, 15-29.

Dale, B.A., Resing, K.A., and Lonsdale-Eccles, J.D. (1985). Filaggrin: a keratin filament associated protein. Ann. N.Y. Acad. Sci. *455*, 330-342.

Day, P.M., Roden, R.B., Lowy, D.R., and Schiller, J.T. (1998). The papillomavirus minor capsid protein, L2, induces localization of the major capsid protein, L1, and the viral transcription/replication protein, E2, to PML oncogenic domains. J. Virol. *72*, 142-150.

De Geest, K., Turyk, M.E., Hosken, M.I., Hudson, J.B., Laimins, L.A., and Wilbanks, G.D. (1993). Growth and differentiation of human papillomavirus type 31b positive human cervical cell lines. Gynecol. Oncol. *49*, 303-310.

del Mar Pena, L.M., Laimins, L.A., Doorbar, J., Medcalf, E., and Napthine, S. (2001). Differentiation-dependent chromatin rearrangement coincides with activation of human papillomavirus type 31 late gene expression. J. Virol. *75*, 10005-10013.

Delvenne, P., Hubert, P., Jacobs, N., Giannini, S.L., Havard, L., Renard, I., Saboulard, D., and Boniver, J. (2001). The organotypic culture of HPV-transformed keratinocytes: an effective *in vitro* model for the development of new immunotherapeutic approaches for mucosal (pre)neoplastic lesions. Vaccine *19*, 2557-2564.

Didierjean, L., Woodley, D., Regnier, M., Prunieras, M., and Saurat, J.H. (1981). Skin explant cultures: expression of cytoplasmic differentiation antigens in outgrowth cells. J. Invest. Dermatol. *76*, 38-41.

DiLorenzo, T.P., and Steinberg, B.M. (1995). Differential regulation of human papillomavirus type 6 and 11 early promoters in cultured cells derived from laryngeal papillomas. J. Virol. *69*, 6865-6872.

DiPaolo, J.A., Popescu, N.C., Woodworth, C.D., and Zimonjic, D.B. (1996). Papillomaviruses and potential copathogens. Toxicol. Lett. *88*, 1-7.

Dollard, S.C., Wilson, J.L., Demeter, L.M., Bonnez, W., Reichman, R.C., Broker, T.R., and Chow, L.T. (1992). Production of human papillomavirus and modulation of the infectious program in epithelial raft cultures. Genes Dev. *6*, 1131-1142.

Doorbar, J., Foo, C., Coleman, N., Medcalf, L., Hartley, O., Prospero, T., Napthine, S., Sterling, J., Winter, G., and Griffin, H. (1997). Characterization of events during the late stages of HPV16 infection *in vivo* using high-affinity synthetic Fabs to E4. Virology *238*, 40-52.

Duensing, S., Duensing, A., Flores, E.R., Do, A., Lambert, P.F., and Munger, K. (2001). Centrosome abnormalities and genomic instability by episomal expression of human papillomavirus type 16 in raft cultures of human keratinocytes. J. Virol. *75*, 7712-7716.

Duensing, S., Lee, L.Y., Duensing, A., Basile, J., Piboonniyom, S., Gonzalez, S., Crum, C.P., and Munger, K. (2000). The human papillomavirus type 16 E6 and E7 oncoproteins cooperate to induce mitotic defects and genomic instability by uncoupling centrosome duplication from the cell division cycle. Proc. Natl. Acad. Sci. USA *97*, 10002-10007.

Durst, M., Gissmann, L., Ikenberg, H., and zur Hausen, H. (1983). A papillomavirus DNA from a cervical carcinoma and its prevalence in cancer biopsy samples from different geographic regions. Proc. Natl. Acad. Sci. USA *80*, 3812-3815.

Emerman, J.T., and Pitelka, D.R. (1977). Maintenance and induction of morphological differentiation in dissociated mammary epithelium on floating collagen membranes. In Vitro *13*, 316-328.

Fehrmann, F., Klumpp, D.J., and Laimins, L.A. (2003). Human papillomavirus type 31 E5 protein supports cell cycle progression and activates late viral functions upon epithelial differentiation. J. Virol. *77*, 2819-2831.

Flores, E.R., Allen-Hoffmann, B.L., Lee, D., and Lambert, P.F. (2000). The human papillomavirus type 16 E7 oncogene is required for the productive stage of the viral life cycle. J. Virol. *74*, 6622-6631.

Flores, E.R., Allen-Hoffmann, B.L., Lee, D., Sattler, C.A., and Lambert, P.F. (1999). Establishment of the human papillomavirus type 16 (HPV-16) life cycle in an immortalized human foreskin keratinocyte cell line. Virology *262*, 344-354.

Flores, E.R., and Lambert, P.F. (1997). Evidence for a switch in the mode of human papillomavirus type 16 DNA replication during the viral life cycle. J. Virol. *71*, 7167-7179.

Frattini, M.G., Lim, H.B., Doorbar, J., and Laimins, L.A. (1997). Induction of human papillomavirus type 18 late gene expression and genomic amplification in organotypic cultures from transfected DNA templates. J. Virol. *71*, 7068-7072.

Frattini, M.G., Lim, H.B., and Laimins, L.A. (1996). In vitro synthesis of oncogenic human papillomaviruses requires episomal genomes for differentiation-dependent late expression. Proc. Natl. Acad. Sci. USA *93*, 3062-3067.

Freeman, A.E., Igel, H.J., Herrman, B.J., and Kleinfeld, K.L. (1976). Growth and characterization of human skin epithelial cell cultures. In Vitro *12*, 352-362.

Fuchs, E., Tyner, A.L., Giudice, G.J., Marchuk, D., RayChaudhury, A., and Rosenberg, M. (1987). The human keratin genes and their differential expression. Curr. Top. Dev. Biol. *22*, 5-34.

Funk, J.O., and Galloway, D.A. (1998). Inhibiting CDK inhibitors: new lessons from DNA tumor viruses. Trends in Biochemical Sciences *23*, 337-341.

Fusenig, N.E., Amer, S.M., Boukamp, P., and Worst, P.K. (1978). Characteristics of chemically transformed mouse epidermal cells *in vitro* and *in vivo*. Bull. Cancer *65*, 271-279.

Gage, J.R., Meyers, C., and Wettstein, F.O. (1990). The E7 proteins of the nononcogenic human papillomavirus type 6b (HPV- 6b) and of the oncogenic HPV-16 differ in retinoblastoma protein binding and other properties. J. Virol. *64*, 723-730.

Garner-Hamrick, P.A., Fostel, J.M., Chien, W.M., Banerjee, N.S., Chow, L.T., Broker, T.R., and Fisher, C. (2004). Global effects of human papillomavirus type 18 E6/E7 in an organotypic keratinocyte culture system. J. Virol. *78*, 9041-9050.

Genther, S.M., Sterling, S., Duensing, S., Munger, K., Sattler, C., and Lambert, P.F. (2003). Quantitative role of the human papillomavirus type 16 E5 gene during the productive stage of the viral life cycle. J. Virol. *77*, 2832-2842.

Georg-Fries, B., Biederlack, S., Wolf, J., and zur Hausen, H. (1984). Analysis of proteins, helper dependence, and seroepidemiology of a new human parvovirus. Virology *134*, 64-71.

Gilles, C., Piette, J., Peter, W., Fusenig, N.E., and Foidart, J.M. (1994). Differentiation ability and oncogenic potential of HPV-33- and HPV-33 + ras-transfected keratinocytes. Int. J. Cancer *58*, 847-854.

Grassmann, K., Rapp, B., Maschek, H., Petry, K.U., and Iftner, T. (1996). Identification of a differentiation-inducible promoter in the E7 open reading frame of human papillomavirus type 16 (HPV-16) in raft cultures of a new cell line containing high copy numbers of episomal HPV-16 DNA. J. Virol. *70*, 2339-2349.

Halbert, C.L., Demers, G.W., and Galloway, D.A. (1992). The E6 and E7 genes of human papillomavirus type 6 have weak immortalizing activity in human epithelial cells. J. Virol. *66*, 2125-2134.

Halpert, R., Fruchter, R.G., Sedlis, A., Butt, K., Boyce, J.G., and Sillman, F.H. (1986). Human papillomavirus and lower genital neoplasia in renal transplant patients. Obstet. Gynecol. *68*, 251-258.

Harley, C.B., and Kim, N.W. (1996). Telomerase and cancer. Important Adv. Oncol. 57-67.

Hashida, T., and Yasumoto, S. (1991). Induction of chromosome abnormalities in mouse and human epidermal keratinocytes by the human papillomavirus type 16 E7 oncogene. J. Gen. Virol. *72 (Pt 7)*, 1569-1577.

Hecht, S.S. (1999). DNA adduct formation from tobacco-specific N-nitrosamines. Mutat. Res. *424*, 127-142.

Hietanen, S., Syrjanen, K., and Syrjanen, S. (1998). Characterization of keratin and cell cycle protein expression in cell lines from squamous intraepithelial lesions progressing towards a malignant phenotype. Br. J. Cancer *77*, 766-775.

Hildesheim, A., Mann, V., Brinton, L.A., Szklo, M., Reeves, W.C., and Rawls, W.E. (1991). Herpes simplex virus type 2: a possible interaction with human papillomavirus types 16/18 in the development of invasive cervical cancer. Int. J. Cancer *49*, 335-340.

Hildesheim, A., Reeves, W.C., Brinton, L.A., Lavery, C., Brenes, M., De La Guardia, M.E., Godoy, J., and Rawls, W.E. (1990). Association of oral contraceptive use and human papillomaviruses in invasive cervical cancers. Int. J. Cancer *45*, 860-864.

Huang, L., Sowa, Y., Sakai, T., and Pardee, A.B. (2000). Activation of the p21WAF1/CIP1 promoter independent of p53 by the histone deacetylase inhibitor suberoylanilide hydroxamic acid (SAHA) through the Sp1 sites. Oncogene *19*, 5712-5719.

Hubert, P., van den Brule, F., Giannini, S.L., Franzen-Detrooz, E., Boniver, J., and Delvenne, P. (1999). Colonization of *in vitro*-formed cervical human papillomavirus- associated (pre)neoplastic lesions with dendritic cells: role of granulocyte/macrophage colony-stimulating factor. Am. J. Pathol. *154*, 775-784.

Hubert, W.G., Kanaya, T., and Laimins, L.A. (1999). DNA replication of human papillomavirus type 31 is modulated by elements of the upstream regulatory region that lie 5′ of the minimal origin. J. Virol. *73*, 1835-1845.

Hudson, J.B., Bedell, M.A., McCance, D.J., and Laiminis, L.A. (1990). Immortalization and altered differentiation of human keratinocytes *in vitro* by the E6 and E7 open reading frames of human papillomavirus type 18. J. Virol. *64*, 519-526.

Hummel, M., Hudson, J.B., and Laimins, L.A. (1992). Differentiation-induced and constitutive transcription of human papillomavirus type 31b in cell lines containing viral episomes. J. Virol. *66*, 6070-6080.

Hummel, M., Lim, H.B., and Laimins, L.A. (1995). Human papillomavirus type 31b late gene expression is regulated through protein kinase C-mediated changes in RNA processing. J. Virol. *69*, 3381-3388.

Hurlin, P.J., Foley, K.P., Ayer, D.E., Eisenman, R.N., Hanahan, D., and Arbeit, J.M. (1995). Regulation of Myc and Mad during epidermal differentiation and HPV- associated tumorigenesis. Oncogene *11*, 2487-2501.

Hurlin, P.J., Kaur, P., Smith, P.P., Perez-Reyes, N., Blanton, R.A., and McDougall, J.K. (1991). Progression of human papillomavirus type 18-immortalized human keratinocytes to a malignant phenotype. Proc. Natl. Acad. Sci. USA *88*, 570-574.

Ivanyi, D., Groeneveld, E., Van Doornewaard, G., Mooi, W.J., and Hageman, P.C. (1990). Keratin subtypes in carcinomas of the uterine cervix: implications for histogenesis and differential diagnosis. Cancer Res. *50*, 5143-5152.

Jacobs, N., Moutschen, M.P., Franzen-Detrooz, E., Boniver, V., Boniver, J., and Delvenne, P. (1998). Organotypic culture of HPV-transformed keratinocytes: a model for testing lymphocyte infiltration of (pre)neoplastic lesions of the uterine cervix. Virchows Arch. *432*, 323-330.

Jian, Y., Schmidt-Grimminger, D.C., Chien, W.M., Wu, X., Broker, T.R., and Chow, L.T. (1998). Post-transcriptional induction of p21cip1 protein by human papillomavirus E7 inhibits unscheduled DNA synthesis reactivated in differentiated keratinocytes. Oncogene *17*, 2027-2038.

Jian, Y., Van Tine, B.A., Chien, W.M., Shaw, G.M., Broker, T.R., and Chow, L.T. (1999). Concordant induction of cyclin E and p21cip1 in differentiated keratinocytes by the human papillomavirus E7 protein inhibits cellular and viral DNA synthesis. Cell Growth Differ. *10*, 101-111.

Kato, S., Bowman, E.D., Harrington, A.M., Blomeke, B., and Shields, P.G. (1995). Human lung carcinogen-DNA adduct levels mediated by genetic polymorphisms *in vivo*. J. Natl. Cancer Inst. *87*, 902-907.

Kaur, P., and Carter, W.G. (1992). Integrin expression and differentiation in transformed human epidermal cells is regulated by fibroblasts. J. Cell Sci. *103 (Pt 3)*, 755-763.

Kessis, T.D., Connolly, D.C., Hedrick, L., and Cho, K.R. (1996). Expression of HPV16 E6 or E7 increases integration of foreign DNA. Oncogene *13*, 427-431.

Khare, S., Pater, M.M., Tang, S.C., and Pater, A. (1997). Effect of glucocorticoid hormones on viral gene expression, growth, and dysplastic differentiation in HPV16-immortalized ectocervical cells. Exp. Cell Res. *232*, 353-360.

King, T.J., Fukushima, L.H., Hieber, A.D., Shimabukuro, K.A., Sakr, W.A., and Bertram, J.S. (2000). Reduced levels of connexin43 in cervical dysplasia: inducible expression in a cervical carcinoma cell line decreases neoplastic potential with implications for tumor progression. Carcinogenesis *21*, 1097-1109.

Kjellberg, L., Hallmans, G., Ahren, A.M., Johansson, R., Bergman, F., Wadell, G., Angstrom, T., and Dillner, J. (2000). Smoking, diet, pregnancy and oral contraceptive use as risk factors for cervical intra-epithelial neoplasia in relation to human papillomavirus infection. Br. J. Cancer *82*, 1332-1338.

Koffa, M.D., Graham, S.V., Takagaki, Y., Manley, J.L., and Clements, J.B. (2000). The human papillomavirus type 16 negative regulatory RNA element interacts with three proteins that act at different posttranscriptional levels. Proc. Natl. Acad. Sci. USA *97*, 4677-4682.

Kopan, R., Traska, G., and Fuchs, E. (1987). Retinoids as important regulators of terminal differentiation: examining keratin expression in individual epidermal cells at various stages of keratinization. J. Cell Biol. *105*, 427-440.

Kreider, J.W., Howett, M.K., Lill, N.L., Bartlett, G.L., Zaino, R.J., Sedlacek, T.V., and Mortel, R. (1986). In vivo transformation of human skin with human papillomavirus type 11 from condylomata acuminata. J. Virol. *59*, 369-376.

Kreider, J.W., Howett, M.K., Stoler, M.H., Zaino, R.J., and Welsh, P. (1987). Susceptibility of various human tissues to transformation *in vivo* with human papillomavirus type 11. Int. J. Cancer *39*, 459-465.

Kreider, J.W., Howett, M.K., Wolfe, S.A., Bartlett, G.L., Zaino, R.J., Sedlacek, T., and Mortel, R. (1985). Morphological transformation *in vivo* of human uterine cervix with papillomavirus from condylomata acuminata. Nature *317*, 639-641.

Krupp, G., Bonatz, G., and Parwaresch, R. (2000). Telomerase, immortality and cancer. Biotechnol. Annu. Rev. *6*, 103-140.

Kyo, S., Klumpp, D.J., Inoue, M., Kanaya, T., and Laimins, L.A. (1997). Expression of AP1 during cellular differentiation determines human papillomavirus E6/E7 expression in stratified epithelial cells. J. Gen. Virol. *78*, 401-411.

Lee, J.H., Yi, S.M., Anderson, M.E., Berger, K.L., Welsh, M.J., Klingelhutz, A.J., and Ozbun, M.A. (2004). Propagation of infectious human papillomavirus type 16 by using an adenovirus and Cre/LoxP mechanism. Proc. Natl. Acad. Sci. USA *101*, 2094-2099.

Lillie, J.H., MacCallum, D.K., and Jepsen, A. (1980). Fine structure of subcultivated stratified squamous epithelium grown on collagen rafts. Exp. Cell Res. *125*, 153-165.

Luostarinen, T., Lehtinen, M., Bjorge, T., Abeler, V., Hakama, M., Hallmans, G., Jellum, E., Koskela, P., Lenner, P., Lie, A.K., et al. (2004). Joint effects of different human papillomaviruses and Chlamydia trachomatis infections on risk of squamous cell carcinoma of the cervix uteri. Eur. J. Cancer *40*, 1058-1065.

Mason, B., Ghanee, N., Haigh, W.G., Lee, S.P., and Oda, D. (1999). Effect of vitamins A, C and E on normal and HPV-immortalized human oral epithelial cells in culture. Anticancer Res. *19*, 5469-5474.

Matsumoto, K., Yasugi, T., Oki, A., Hoshiai, H., Taketani, Y., Kawana, T., and Yoshikawa, H. (2003). Are smoking and chlamydial infection risk factors for CIN? Different results after adjustment for HPV DNA and antibodies. Br. J. Cancer *89*, 831-833.

Mayer, T.J., and Meyers, C. (1998). Temporal and spatial expression of the E5a protein during the differentiation-dependent life cycle of human papillomavirus type 31b. Virology *248*, 208-217.

Mayor, H.D., Drake, S., Stahmann, J., and Mumford, D.M. (1976). Antibodies to adeno-associated satellite virus and herpes simplex in sera from cancer patients and normal adults. Am. J. Obstet. Gynecol. *126*, 100-104.

McBride, A.A., Dlugosz, A., and Baker, C.C. (2000). Production of infectious bovine papillomavirus from cloned viral DNA by using an organotypic raft/xenograft technique. Proc. Natl. Acad. Sci. USA *97*, 5534-5539.

McCance, D.J., Kopan, R., Fuchs, E., and Laimins, L.A. (1988). Human papillomavirus type 16 alters human epithelial cell differentiation *in vitro*. Proc. Natl. Acad. Sci. USA *85*, 7169-7173.

McLaughlin-Drubin, M.E., Christensen, N.D., and Meyers, C. (2004). Propagation, infection, and neutralization of authentic HPV16 virus. Virology *322*, 213-219.

McLaughlin-Drubin, M.E., and Meyers, C. (2004). Evidence for the coexistence of two genital HPV types within the same host cell *in vitro*. Virology *321*, 173-180.

McLaughlin-Drubin, M.E., Wilson, S., Mullikin, B., Suzich, J., and Meyers, C. (2003). Human papillomavirus type 45 propagation, infection, and neutralization. Virology *312*, 1-7.

McNutt, N.S., Hershberg, R.A., and Weinstein, R.S. (1971). Further observations on the occurrence of nexuses in benign and malignant human cervical epithelium. J. Cell. Biol. *51*, 805-825.

McNutt, N.S., and Weinstein, R.S. (1969). Carcinoma of the cervix: deficiency of nexus intercellular junctions. Science *165*, 597-599.

McPhillips, M.G., Veerapraditsin, T., Cumming, S.A., Karali, D., Milligan, S.G., Boner, W., Morgan, I.M., and Graham, S.V. (2004). SF2/ASF binds the human papillomavirus type 16 late RNA control element and is regulated during differentiation of virus-infected epithelial cells. J. Virol. *78*, 10598-10605.

Merrick, D.T., Blanton, R.A., Gown, A.M., and McDougall, J.K. (1992a). Altered expression of proliferation and differentiation markers in human papillomavirus 16 and 18 immortalized epithelial cells grown in organotypic culture. Am. J. Pathol. *140*, 167-177.

Merrick, D.T., Blanton, R.A., Gown, A.M., and McDougall, J.K. (1992b). Altered expression of proliferation and differentiation markers in human papillomavirus 16 and 18 immortalized epithelial cells grown in organotypic culture [published erratum appears in Am. J. Pathol. 1992 Mar;140(3):following Table of Contents]. Am. J. Pathol. *140*, 167-177.

Merrick, D.T., Gown, A.M., Halbert, C.L., Blanton, R.A., and McDougall, J.K. (1993). Human papillomavirus-immortalized keratinocytes are resistant to the effects of retinoic acid on terminal differentiation. Cell Growth Differ. *4*, 831-840.

Meyers, C. (1996). Organotypic (raft) epithelial tissue culture system for the differentiation-dependent replication of papillomavirus. Methods in Cell Science *18*, 1-10.

Meyers, C. (2002). Epithelial cell culture: three-dimensional cervical system. In Methods of Tissue Engineering, A. Atala, and R. P. Lanza, eds. (San Diego, Academic Press), pp. 263-271.

Meyers, C., Alam, S., Mane, M., and Hermonat, P.L. (2001). Altered biology of adeno-associated virus type 2 and human papillomavirus during dual infection of natural host tissue. Virology *287*, 30-39.

Meyers, C., Andreansky, S.S., and Courtney, R.J. (2003). Replication and interaction of herpes simplex virus and human papillomavirus in differentiating host epithelial tissue. Virology *315*, 43-55.

Meyers, C., Bromberg-White, J.L., Zhang, J., Kaupas, M.E., Bryan, J.T., Lowe, R.S., and Jansen, K.U. (2002). Infectious virions produced from a human papillomavirus type 18/16 genomic DNA chimera. J. Virol. *76*, 4723-4733.

Meyers, C., Frattini, M.G., Hudson, J.B., and Laimins, L.A. (1992). Biosynthesis of human papillomavirus from a continuous cell line upon epithelial differentiation. Science *257*, 971-973.

Meyers, C., Frattini, M.G., and Laimins, L.A. (1994). Tissue culture techniques for the study of human papillomaviruses in stratified epithelia., Vol. 1, 1st edn. (San Diego, Academic Press).

Meyers, C., Frattini, M.G., and Laimins, L.A. (1998). Tissue culture techniques for the study of human papilloma viruses in stratified epithelia., Vol. 1, 2nd edn. (San Diego, Academic Press).

Meyers, C., Mayer, T.J., and Ozbun, M.A. (1997). Synthesis of infectious human papillomavirus type 18 in differentiating epithelium transfected with viral DNA. J. Virol. *71*, 7381-7386.

Michalopoulos, G., and Pitot, H.C. (1975). Primary culture of parenchymal liver cells on collagen membranes. Morphological and biochemical observations. Exp. Cell Res. *94*, 70-78.

Middleton, K., Peh, W., Southern, S., Griffin, H., Sotlar, K., Nakahara, T., El-Sherif, A., Morris, L., Seth, R., Hibma, M., et al. (2003). Organization of human papillomavirus productive cycle during neoplastic progression provides a basis for selection of diagnostic markers. J. Virol. *77*, 10186-10201.

Munger, K., and Howley, P.M. (2002). Human papillomavirus immortalization and transformation functions. Virus Res. *89*, 213-228.

Mushika, M., Miwa, T., Suzuoki, Y., Hayashi, K., Masaki, S., and Kaneda, T. (1988). Detection of proliferative cells in dysplasia, carcinoma in situ, and invasive carcinoma of the uterine cervix by monoclonal antibody against DNA polymerase alpha. Cancer *61*, 1182-1186.

Nakao, Y., Yang, X., Yokoyama, M., Pater, M.M., and Pater, A. (1996). Malignant transformation of human ectocervical cells immortalized by HPV 18: *in vitro* model of carcinogenesis by cigarette smoke. Carcinogenesis *17*, 577-583.

Nemunaitis, J., Ganly, I., Khuri, F., Arseneau, J., Kuhn, J., McCarty, T., Landers, S., Maples, P., Romel, L., Randlev, B., et al. (2000). Selective replication and oncolysis in p53 mutant tumors with ONYX-015, an E1B-55kD gene-deleted adenovirus, in patients with advanced head and neck cancer: a phase II trial. Cancer Res. *60*, 6359-6366.

Nicholls, P.K., Doorbar, J., Moore, R.A., Peh, W., Anderson, D.M., and Stanley, M.A. (2001). Detection of viral DNA and E4 protein in basal keratinocytes of experimental canine oral papillomavirus lesions. Virology *284*, 82-98.

Nischan, P., Ebeling, K., and Schindler, C. (1988). Smoking and invasive cervical cancer risk. Results from a case-control study. Am. J. Epidemiol. *128*, 74-77.

Noya, F., Balague, C., Banerjee, N.S., Curiel, D.T., Broker, T.R., and Chow, L.T. (2003). Activation of adenovirus early promoters and lytic phase in differentiated strata of organotypic cultures of human keratinocytes. J. Virol. *77*, 6533-6540.

Noya, F., Chien, W.M., Broker, T.R., and Chow, L.T. (2001). p21cip1 Degradation in differentiated keratinocytes is abrogated by costabilization with cyclin E induced by human papillomavirus E7. J. Virol. *75*, 6121-6134.

Oda, D., Bavisotto, L.M., Schmidt, R.A., McNutt, M., Bruckner, J.D., Conrad, E.U., 3rd, and Weymuller, E.A., Jr. (1997). Head and neck osteosarcoma at the University of Washington. Head Neck *19*, 513-523.

Oda, D., Bigler, L., Lee, P., and Blanton, R. (1996a). HPV immortalization of human oral epithelial cells: a model for carcinogenesis. Exp. Cell Res. *226*, 164-169.

Oda, D., Bigler, L., Mao, E.J., and Disteche, C.M. (1996b). Chromosomal abnormalities in HPV-16-immortalized oral epithelial cells. Carcinogenesis *17*, 2003-2008.

Osada, H., and Takahashi, T. (2002). Genetic alterations of multiple tumor suppressors and oncogenes in the carcinogenesis and progression of lung cancer. Oncogene *21*, 7421-7434.

Ozbun, M., and Meyers, C. (1999). Human papillomavirus type 31b transcription during the differentiation-dependent viral life cycle. Current Topics in Virology *1*, 203-217.

Ozbun, M.A. (2002a). Human papillomavirus type 31b infection of human keratinocytes and the onset of early transcription. J. Virol. *76*, 11291-11300.

Ozbun, M.A. (2002b). Infectious human papillomavirus type 31b: purification and infection of an immortalized human keratinocyte cell line. J. Gen. Virol. *83*, 2753-2763.

Ozbun, M.A., and Meyers, C. (1996). Transforming growth factor beta1 induces differentiation in human papillomavirus-positive keratinocytes. J. Virol. *70*, 5437-5446.

Ozbun, M.A., and Meyers, C. (1997). Characterization of late gene transcripts expressed during vegetative replication of human papillomavirus type 31b. J. Virol. *71*, 5161-5172.

Ozbun, M.A., and Meyers, C. (1998a). Human papillomavirus type 31b E1 and E2 transcript expression correlates with vegetative viral genome amplification. Virology *248*, 218-230.

Ozbun, M.A., and Meyers, C. (1998b). Temporal usage of multiple promoters during the life cycle of human papillomavirus type 31b. J. Virol. *72*, 2715-2722.

Ozbun, M.A., and Meyers, C. (1999). Two novel promoters in the upstream regulatory region of human papillomavirus type 31b are negatively regulated by epithelial differentiation. J. Virol. *73*, 3505-3510.

Parker, J.N., Zhao, W., Askins, K.J., Broker, T.R., and Chow, L.T. (1997). Mutational analyses of differentiation-dependent human papillomavirus type 18 enhancer elements in epithelial raft cultures of neonatal foreskin keratinocytes. Cell Growth Differ. *8*, 751-762.

Pattison, S., Skalnik, D.G., and Roman, A. (1997). CCAAT displacement protein, a regulator of differentiation-specific gene expression, binds a negative regulatory element within the 5′ end of the human papillomavirus type 6 long control region. J. Virol. *71*, 2013-2022.

Peh, W.L., Middleton, K., Christensen, N., Nicholls, P., Egawa, K., Sotlar, K., Brandsma, J., Percival, A., Lewis, J., Liu, W.J., and Doorbar, J. (2002). Life cycle heterogeneity in animal models of human papillomavirus-associated disease. J. Virol. *76*, 10401-10416.

Pett, M.R., Alazawi, W.O., Roberts, I., Dowen, S., Smith, D.I., Stanley, M.A., and Coleman, N. (2004). Acquisition of high-level chromosomal instability is associated with integration of human papillomavirus type 16 in cervical keratinocytes. Cancer Res. *64*, 1359-1368.

Pfister, H. (1987a). Human papillomaviruses and genital cancer. Adv. Cancer Res. *48*, 113-147.

Pfister, H. (1987b). Relationship of papillomaviruses to anogenital cancer. Obstet. Gynecol. Clin. North Am. *14*, 349-361.

Pfister, H. (1996). The role of human papillomavirus in anogenital cancer. Obstet. Gynecol. Clin. North Am. *23*, 579-595.

Pray, T.R., and Laimins, L.A. (1995). Differentiation-dependent expression of E1--E4 proteins in cell lines maintaining episomes of human papillomavirus type 31b. Virology *206*, 679-685.

Rader, J.S., Golub, T.R., Hudson, J.B., Patel, D., Bedell, M.A., and Laimins, L.A. (1990). In vitro differentiation of epithelial cells from cervical neoplasias resembles *in vivo* lesions. Oncogene *5*, 571-576.

Regnier, M., Desbas, C., Bailly, C., and Darmon, M. (1988). Differentiation of normal and tumoral human keratinocytes cultured on dermis: reconstruction of either normal or tumoral architecture. In Vitro Cell Dev. Biol. *24*, 625-632.

Regnier, M., Pautrat, G., Pauly, G., and Prunieras, M. (1984). Natural substrates for the reconstruction of skin *in vitro*. Br. J. Dermatol. *111 Suppl. 27*, 223-224.

Renard, I., Mezzanzanica, D., Canevari, S., Ferrini, S., Boniver, J., Delvenne, P., and Jacobs, N. (2002). Anti-CD3/anti-epidermal growth factor receptor-bispecific antibody retargeting of lymphocytes against human neoplastic keratinocytes in an autologous organotypic culture model. Am. J. Pathol. *160*, 113-122.

Rohlfs, M., Winkenbach, S., Meyer, S., Rupp, T., and Durst, M. (1991). Viral transcription in human keratinocyte cell lines immortalized by human papillomavirus type-16. Virology *183*, 331-342.

Rous, P., and Beard, J.W. (1935). The progression to carcinoma of virus-induced rabbit papillomas (Shope). J. Exp. Med. *62*, 523-548.

Ruesch, M., Stubenrauch, F., and Laimins, L. (1998). Activation of papillomavirus late gene transcription and genome amplification upon differentiation in semisolid medium is coincident with expression of involucrin and transglutaminase but not keratin 10. J. Virol. *72*, 5016-5024.

Saitoh, M., Oyamada, M., Oyamada, Y., Kaku, T., and Mori, M. (1997). Changes in the expression of gap junction proteins (connexins) in hamster tongue epithelium during wound healing and carcinogenesis. Carcinogenesis *18*, 1319-1328.

Sarma, D., Yang, X., Jin, G., Shindoh, M., Pater, M.M., and Pater, A. (1996). Resistance to retinoic acid and altered cytokeratin expression of human papillomavirus type 16-immortalized endocervical cells after tumorigenesis. Int. J. Cancer *65*, 345-350.

Scheffner, M., Werness, B.A., Huibregtse, J.M., Levine, A.J., and Howley, P.M. (1990). The E6 oncoprotein encoded by human papillomavirus types 16 and 18 promotes the degradation of p53. Cell *63*, 1129-1136.

Schlehofer, J.R., Ehrbar, M., and zur Hausen, H. (1986). Vaccinia, herpes simpex, and carcinogen iduced DNA amplification in a human cell line and support replication of a helpervirus dependent parvovirus. Virology *152*, 110-117.

Sen, E., McLaughlin-Drubin, M.A., and Meyer, C. (2004b). Efficacy of two commercial preparations if interferon-α on human papillomavirus replication. Anticancer Res. *25*, 765-778.

Sherman, K.J., Daling, J.R., Chu, J., Weiss, N.S., Ashley, R.L., and Corey, L. (1991). Genital warts, other sexually transmitted diseases, and vulvar cancer. Epidemiology *2*, 257-262.

Sherr, C.J., and Roberts, J.M. (1995). Inhibitors of mammalian G1 cyclin-dependent kinases. Genes Dev. *9*, 1149-1163.

Shields, P.G. (2002). Molecular epidemiology of smoking and lung cancer. Oncogene *21*, 6870-6876.

Shindoh, M., Sun, Q., Pater, A., and Pater, M.M. (1995). Prevention of carcinoma in situ of human papillomavirus type 16-immortalized human endocervical cells by retinoic acid in organotypic raft culture. Obstet. Gynecol. *85*, 721-728.

Shokri-Tabibzadeh, S., Koss, L.G., Molnar, J., and Romney, S. (1981). Association of human papillomavirus with neoplastic processes in the genital tract of four women with impaired immunity. Gynecol. Oncol. *12*, S129-140.

Shope, R.E., and Hurst, E.W. (1933). Infectious papillomatosis of rabbits. J. Exp. Med. *58*, 607-623.

Smith, J.S., Bosetti, C., Munoz, N., Herrero, R., Bosch, F.X., Eluf-Neto, J., Meijer, C.J., Van Den Brule, A.J., Franceschi, S., and Peeling, R.W. (2004). Chlamydia trachomatis and invasive cervical cancer: a pooled analysis of the IARC multicentric case-control study. Int. J. Cancer *111*, 431-439.

Smith, J.S., Herrero, R., Bosetti, C., Munoz, N., Bosch, F.X., Eluf-Neto, J., Castellsague, X., Meijer, C.J., Van den Brule, A.J., Franceschi, S., and Ashley, R. (2002a). Herpes simplex virus-2 as a human papillomavirus cofactor in the etiology of invasive cervical cancer. J. Natl. Cancer Inst. *94*, 1604-1613.

Smith, J.S., Munoz, N., Herrero, R., Eluf-Neto, J., Ngelangel, C., Franceschi, S., Bosch, F.X., Walboomers, J.M., and Peeling, R.W. (2002b). Evidence for Chlamydia trachomatis as a human papillomavirus cofactor in the etiology of invasive cervical cancer in Brazil and the Philippines. J. Infect. Dis. *185*, 324-331.

Southern, S.A., Evans, M.F., and Herrington, C.S. (1997). Basal cell tetrasomy in low-grade cervical squamous intraepithelial lesions infected with high-risk human papillomaviruses. Cancer Res. *57*, 4210-4213.

Southern, S.A., Lewis, M.H., and Herrington, C.S. (2004). Induction of tetrasomy by human papillomavirus type 16 E7 protein is independent of pRb binding and disruption of differentiation. Br. J. Cancer *90*, 1949-1954.

Southern, S.A., Noya, F., Meyers, C., Broker, T.R., Chow, L.T., and Herrington, C.S. (2001). Tetrasomy is induced by human papillomavirus type 18 E7 gene expression in keratinocyte raft cultures. Cancer Res. *61*, 4858-4863.

Steele, B.K., Meyers, C., and Ozbun, M.A. (2002). Variable expression of some "housekeeping" genes during human keratinocyte differentiation. Anal. Biochem. *307*, 341-347.

Steenbergen, R.D., Hermsen, M.A., Walboomers, J.M., Meijer, G.A., Baak, J.P., Meijer, C.J., and Snijders, P.J. (1998a). Non-random allelic losses at 3p, 11p and 13q during HPV-mediated immortalization and concomitant loss of terminal differentiation of human keratinocytes. Int. J. Cancer *76*, 412-417.

Steenbergen, R.D., Parker, J.N., Isern, S., Snijders, P.J., Walboomers, J.M., Meijer, C.J., Broker, T.R., and Chow, L.T. (1998b). Viral E6-E7 transcription in the basal layer of organotypic cultures without apparent p21cip1 protein precedes immortalization of human papillomavirus type 16- and 18-transfected human keratinocytes. J. Virol. *72*, 749-757.

Steenbergen, R.D., Walboomers, J.M., Meijer, C.J., van der Raaij-Helmer, E.M., Parker, J.N., Chow, L.T., Broker, T.R., and Snijders, P.J. (1996). Transition of human papillomavirus type 16 and 18 transfected human foreskin keratinocytes towards immortality: activation of telomerase and allele losses at 3p, 10p, 11q and/or 18q. Oncogene *13*, 1249-1257.

Steven, A.C., Bisher, M.E., Roop, D.R., and Steinert, P.M. (1990). Biosynthetic pathways of filaggrin and loricrin--two major proteins expressed by terminally differentiated epidermal keratinocytes. J. Struct. Biol. *104*, 150-162.

Stockwell, H.G., and Lyman, G.H. (1987). Cigarette smoking and the risk of female reproductive cancer. Am. J. Obstet. Gynecol. *157*, 35-40.

Stubenrauch, F., Colbert, A.M., and Laimins, L.A. (1998a). Transactivation by the E2 protein of oncogenic human papillomavirus type 31 is not essential for early and late viral functions. J. Virol. *72*, 8115-8123.

Stubenrauch, F., Hummel, M., Iftner, T., and Laimins, L.A. (2000). The E8E2C protein, a negative regulator of viral transcription and replication, is required for extrachromosomal maintenance of human papillomavirus type 31 in keratinocytes. J. Virol. *74*, 1178-1186.

Stubenrauch, F., Lim, H.B., and Laimins, L.A. (1998b). Differential requirements for conserved E2 binding sites in the life cycle of oncogenic human papillomavirus type 31. J. Virol. *72*, 1071-1077.

Stunkel, W., Huang, Z., Tan, S.H., O'Connor, M.J., and Bernard, H.U. (2000). Nuclear matrix attachment regions of human papillomavirus type 16 repress or activate the E6 promoter, depending on the physical state of the viral DNA. J. Virol. *74*, 2489-2501.

Sun, Q., Tang, S.C., Pater, M.M., and Pater, A. (1997). Different HPV16 E6/E7 oncogene expression patterns in epithelia reconstructed from HPV16-immortalized human endocervical cells and genital keratinocytes. Oncogene *15*, 2399-2408.

Swindle, C.S., Zou, N., Van Tine, B.A., Shaw, G.M., Engler, J.A., and Chow, L.T. (1999). Human papillomavirus DNA replication compartments in a transient DNA replication system. J. Virol. *73*, 1001-1009.

Syrjanen, S., Cintorino, M., Armellini, D., Del Vecchio, M.T., Leoncini, P., Bugnoli, M., Pallini, V., Silvestri, S., Tosi, P., and Mantyjarvi, R. (1988). Expression of cytokeratin polypeptides in human papillomavirus (HPV) lesions of the uterine cervix: 1. Relationship to grade of CIN and HPV type. Int. J. Gynecol. Pathol. *7*, 23-38.

Tan, S.H., Bartsch, D., Schwarz, E., and Bernard, H.U. (1998). Nuclear matrix attachment regions of human papillomavirus type 16 point toward conservation of these genomic elements in all genital papillomaviruses. J. Virol. *72*, 3610-3622.

Terhune, S.S., Hubert, W.G., Thomas, J.T., and Laimins, L.A. (2001). Early polyadenylation signals of human papillomavirus type 31 negatively regulate capsid gene expression. J. Virol. *75*, 8147-8157.

Terhune, S.S., Milcarek, C., and Laimins, L.A. (1999). Regulation of human papillomavirus type 31 polyadenylation during the differentiation-dependent life cycle. J. Virol. *73*, 7185-7192.

Thomas, D.B. (1991). The WHO Collaborative Study of Neoplasia and Steroid Contraceptives: The influence of combined oral contraceptives on risk of neoplasms in developing and developed countries. Contraception *43*, 695-712.

Thomas, J.T., Hubert, W.G., Ruesch, M.N., and Laimins, L.A. (1999). Human papillomavirus type 31 oncoproteins E6 and E7 are required for the maintenance of episomes during the viral life cycle in normal human keratinocytes. Proc. Natl. Acad. Sci. USA *96*, 8449-8454.

Thomas, J.T., Oh, S.T., Terhune, S.S., and Laimins, L.A. (2001). Cellular changes induced by low-risk human papillomavirus type 11 in keratinocytes that stably maintain viral episomes. J. Virol. *75*, 7564-7571.

Tomakidi, P., Cheng, H., Kohl, A., Komposch, G., and Alonso, A. (2000a). Connexin 43 expression is downregulated in raft cultures of human keratinocytes expressing the human papillomavirus type 16 E5 protein. Cell Tissue Res. *301*, 323-327.

Tomakidi, P., Cheng, H., Kohl, A., Komposch, G., and Alonso, A. (2000b). Modulation of the epidermal growth factor receptor by the human papillomavirus type 16 E5 protein in raft cultures of human keratinocytes. Eur. J. Cell Biol. *79*, 407-412.

Trosko, J.E., and Ruch, R.J. (1998). Cell-cell communication in carcinogenesis. Front. Biosci. *3*, d208-236.

Ursin, G., Peters, R.K., Henderson, B.E., d'Ablaing, G., 3rd, Monroe, K.R., and Pike, M.C. (1994). Oral contraceptive use and adenocarcinoma of cervix. Lancet *344*, 1390-1394.

Vambutas, A., Di Lorenzo, T.P., and Steinberg, B.M. (1993). Laryngeal papilloma cells have high levels of epidermal growth factor receptor and respond to epidermal growth factor by a decrease in epithelial differentiation. Cancer Res. *53*, 910-914.

Vessey, M.P., Lawless, M., McPherson, K., and Yeates, D. (1983). Oral contraceptives and cervical cancer. Lancet *2*, 1358-1359.

White, A.E., Livanos, E.M., and Tlsty, T.D. (1994). Differential disruption of genomic integrity and cell cycle regulation in normal human fibroblasts by the HPV oncoproteins. Genes Dev. *8*, 666-677.

Wilson, J.L., Dollard, S.C., Chow, L.T., and Broker, T.R. (1992). Epithelial-specific gene expression during differentiation of stratified primary human keratinocyte cultures. Cell Growth Differ. *3*, 471-483.

Woodley, D., Saurat, J.H., Prunieras, M., and Regnier, M. (1982). Pemphigoid, pemphigus and Pr antigens in adult human keratinocytes grown on nonviable substrates. J. Invest. Dermatol. *79*, 23-29.

Woodworth, C.D., Notario, V., and DiPaolo, J.A. (1990). Transforming growth factors beta 1 and 2 transcriptionally regulate human papillomavirus (HPV) type 16 early gene expression in HPV- immortalized human genital epithelial cells. J. Virol. *64*, 4767-4775.

Woodworth, C.D., Wang, H., Simpson, S., Alvarez-Salas, L.M., and Notario, V. (1993). Overexpression of wild-type p53 alters growth and differentiation of normal human keratinocytes but not human papillomavirus-expressing cell lines. Cell Growth Differ. *4*, 367-376.

Yang, X., Jin, G., Nakao, Y., Rahimtula, M., Pater, M.M., and Pater, A. (1996). Malignant transformation of HPV 16-immortalized human endocervical cells by cigarette smoke condensate and characterization of multistage carcinogenesis. Int. J. Cancer *65*, 338-344.

Yokoyama, M., Nakao, Y., Iwasaka, T., Pater, A., and Sugimori, H. (2001). Retinoic acid and interferon-alpha effects on cell growth and differentiation in cervical carcinoma cell lines. Obstet. Gynecol. *98*, 332-340.

Yokoyama, M., Noguchi, M., Nakao, Y., Pater, A., and Iwasaka, T. (2004). The tea polyphenol, (-)-epigallocatechin gallate effects on growth, apoptosis, and telomerase activity in cervical cell lines. Gynecol. Oncol. *92*, 197-204.

Yoo, G.H., Washington, J., Piechocki, M., Ensley, J., Shibuya, T., Oda, D., and Wei, W.Z. (2000). Progression of head and neck cancer in an *in vitro* model. Arch. Otolaryngol. Head Neck Surg. *126*, 1313-1318.

Zhao, W., Chow, L.T., and Broker, T.R. (1997). Transcription activities of human papillomavirus type 11 E6 promoter-proximal elements in raft and submerged cultures of foreskin keratinocytes. J. Virol. *71*, 8832-8840.

Zhao, W., Chow, L.T., and Broker, T.R. (1999). A distal element in the HPV-11 upstream regulatory region contributes to promoter repression in basal keratinocytes in squamous epithelium. Virology *253*, 219-229.

Zheng, J., and Vaheri, A. (1995). Human skin fibroblasts induce anchorage-independent growth of HPV-16-DNA-immortalized cervical epithelial cells. Int. J. Cancer *61*, 658-665.

Zheng, J., Wahlstrom, T., Paavonen, J., and Vaheri, A. (1994). Altered growth behavior of human cervical epithelial cells transfected by HPV type 16 and 18 DNA. Int. J. Cancer *58*, 713-720.

zur Hausen, H. (1982). Human genital cancer: synergism between two virus infections or synergism between a virus infection and initiating events? The Lancet *2*, 1370-1372.

zur Hausen, H. (1989). Papillomaviruses in anogenital cancer as a model to understand the role of viruses in human cancers. Cancer Res. *49*, 4677-4681.

zur Hausen, H. (1991). Human papillomaviruses in the pathogenesis of anogenital cancer. Virology *184*, 9-13.

zur Hausen, H. (1996). Papillomavirus infections--a major cause of human cancers. Biochim. Biophys. Acta *1288*, F55-78.

zur Hausen, H. (1999). Papillomaviruses in human cancers. Proc. Assoc. Am. Physicians *111*, 581-587.

From: Papillomavirus Research: From Natural History to Vaccines and Beyond. Edited by: M. Saveria Campo

Chapter 12

Early Events in the Papillomaviral Lifecycle

Patricia M. Day* and John T. Schiller

Abstract

The chapter follows the chronological transit of the PV virion from the initial interaction with the cell surface until delivery into the nucleus. The interaction of the particle with the cell surface occurs largely via interaction of the major capsid protein, L1, with cell surface heparan sulfate proteoglycans. It is unclear if a secondary receptor is involved. It is also unclear if the minor capsid protein, L2, plays a role in cell surface interactions. Most PV types that have been examined appear to enter the cell via a clathrin-dependent endocytic pathway. Infection can be abrogated by inhibitors of this pathway and colocalization of capsid proteins with markers of the early and late endosomal compartments has been observed. It has recently been proposed that L2 can mediate the escape of the virus from the endosome by lysis of the limiting membrane. L2 and the genome traffic into the nucleus and colocalize at ND10. Localization to this domain is critical for efficient transcription of the PV genome.

1. Introduction

The elucidation of the earliest events in the PV (PV) lifecycle has, until recently, lagged in comparison to the knowledge obtained for many other groups of viruses. In large measure this can be attributed to the inability of PV to replicate under normal cell culture conditions. However, this situation has changed rather dramatically over the past decade, with the advent of technologies for the *in vitro* production of authentic virions, pseudovirions and non-infectious virus-like particles (VLPs) for animal and human PV types. This chapter focuses on the multiple early steps in PV infection, from cell recognition through transit of the genome into the nucleus in preparation for the initial transcription events. We also provide comparisons with the infection strategies used by other viruses, to highlight the common and distinct properties of the mechanisms employed to accomplish similar goals.

The virus capsid has evolved to fulfill numerous roles that are critical to the establishment of the viral infection. In the case of nonenveloped icosahedral virions, including PVs, the proteinaceous coat surrounds the viral nucleic acid and protects it from degradation by external factors. The particle interacts with the cell surface, mediates adsorption and is internalized. Following penetration into the host cell, the virus must undergo a controlled process of uncoating to render the genome accessible to the cellular transcription and replication machinery. For DNA viruses, excluding poxviruses, this necessitates the delivery of the genetic material into the nucleus.

2. The Tools: VLPs and Pseudoviruses Versus Authenic Virions

The observation that the major capsid protein, L1, can efficiently self-assemble into particles that resemble authentic virions in a number of tractable cell culture systems allowed the field access to a useful tool for analysis of binding and entry. Much of the work examining these parameters has been performed with such particles, termed virus-like particles or VLPs. VLPs resemble authentic virions both by antigenic and immunologic measures and by EM analysis (Christensen et al., 1994; Kirnbauer et al., 1993; Rose et al., 1993; Rose et al., 1994) (see Chapter 18 for a fuller description of VLPs). VLPs also competitively inhibit cell association of authentic virions. By these criteria, VLPs were viewed as an acceptable surrogate for authentic virions for studying virion/cell interactions. However, subsequent work found that the degree of disulfide bond formation can differ between VLPs and virions (Fligge et al., 2001). Authentic virions were more trypsin resistant, and this increase in stability seemed to be dependent upon DNA encapsidation. It is not clear that these differences affect the events of cell surface interaction, internalization and trafficking, although it should be a concern especially as discrepant internalization kinetics have been reported for VLPs and pseudovirions (Selinka et al., 2003). A more recent study demonstrated that VLPs produced in mammalian cells can also be "matured" *in vitro* to assume a highly disulfide cross-linked and protease resistant conformation similar to authentic virions (Buck, 2005).

Test tube assembled DNA-VLP complexes have been used to extend the VLP entry analyses (Bousarghin et al., 2002; 2003). In this case, plasmids encoding quantifiable marker proteins are coupled, either covalently or non-

*For correspondence. Email pmd@nih.gov

covalently, to the outside of VLPs and expression of the marker gene is monitored to indicate the productive delivery of the DNA. However, this approach, with its unnatural configuration of the DNA and its high and variable particle to infectivity ratio, is being replaced by the advent of efficient pseudovirus technology (described below).

Laboratory-produced viral capsids that have encapsidated heterologous DNAs are termed pseudoviruses. They have been obtained by a number of methodologies. Initially, the papillomaviral capsid proteins were expressed via infection with recombinant viral vectors, such as Semliki Forest virus (Roden et al., 1996) or Vaccinia virus (Unckell et al., 1997), in cell lines that harbored the BPV1 episomal genome, or a marker plasmid. Encapsidation has also been accomplished by the *in vitro* dissociation and re-association of VLPs in the presence of a marker plasmid (Touze and Coursaget, 1998; Yeager et al., 2000). In most systems, co-expression of the L2 minor capsid protein greatly increased the production of infectious capsids. Although the particle to infectivity ratio was seldom specifically measured for these pseudovirus preparations, the relatively low number of infectious events observed implies that the ratio was quite high, even when L2 was incorporated into the capsids.

Viral expression systems were utilized in the early studies due to very poor expression of the L1 and L2 proteins after transfection of plasmid-based eukaryotic expression vectors. Increased levels of L1 and L2 protein expression after simple DNA transfection was not achieved until the codon composition of the genes was altered to remove rarely used codons that might limit translation and/or to remove negative regulatory elements in the RNA (Leder et al., 2001; Mossadegh et al., 2004; Zhou et al., 1999). Recently the use of codon-modified capsid genes has allowed the use of conventional mammalian cell transfection methods to generate high-titer PV pseudovirus (Buck et al., 2004). The replication of L1 and L2 expression plasmids containing the SV40 origin of replication after transfection of cells over-expressing SV40 T antigen along with an SV40-origin containing marker plasmid results in reproducible and efficient pseudovirus production. Pseudovirions produced from this system have been validated by multiple assays and, at present, appear to be indistinguishable from authentic virions.

Organotypic (raft) culture systems are capable of producing infectious virions after keratinocytes harboring extrachromosomal PV genomes are induced to undergo terminal differentiation at a liquid-air interface (Meyers et al., 1992) (see Chapter 11 for a full description of raft cultures and their use). This mode of virion production clearly mirrors *in vivo* virion production most closely. However, the production capacity of the current raft culture methods is limited. In addition, with the exception of BPV1, there are not easy quantitative assays for infection by authentic virions. Therefore, investigators that study the establishment of infection by authentic HPVs generally utilize RT-PCR assays to monitor the expression of spliced viral mRNA (Meyers et al., 1997).

Table 1. Summary of the proposed roles of the two PV capsid proteins in the process of cellular binding and entry.

POSSIBLE ROLES OF PV CAPSID PROTEINS DURING ESTABLISHMENT OF INFECTION

L1	EXTRACELLULAR PROTECTION OF GENOME ATTACHMENT TO CELL SURFACE HSPG
L2	SECONDARY RECEPTOR ACTIN INTERACTION ENDOSOME ESCAPE TUBULIN INTERACTION CHAPERONING GENOME TO ND10

L1 and L1/L2 VLPs will continue to be useful reagents for PV infection studies investigating the individual roles of L1 and L2 in binding and internalization (Table 1). However, it is our opinion that pseudovirions will become the standard reagents for most studies of PV infection, because they can now be reproducibly generated in virtually limitless amounts in a simple cell culture system. Nevertheless, the possibility remains that capsids assembled in terminally differentiating keratinocytes may have subtle differences from those produced in replicating cells and that these differences may influence some aspects of virus infection. It, therefore, would be reassuring if major findings involving PV infection that are discovered using pseudovirions were confirmed with authentic virions derived from papillomas or rafts.

3. Cell Surface Binding

Considerable effort has been expended to identify the moiety (ies) that mediate cell surface attachment of PV. The initial work in this area was performed with VLPs. VLPs were found to interact with a wide variety of cell lines of divergent species origins (Roden et al., 1994); (Muller et al., 1995; Qi et al., 1996), including many epithelial and fibroblast-derived lines. VLPs derived from a number of different PV types also inhibited the association of VLPs from other types or authentic BPV virions. These studies, in total, suggested that PVes bind to a common widely expressed and evolutionarily conserved cell surface receptor. Additionally, the attachment of VLPs to the cell surface could be abrogated by pretreatment of the cells with trypsin, suggesting a receptor with a protein component (Muller et al., 1995; Volpers et al., 1995).

Binding to the cell surface depends primarily on L1, since BPV VLPs composed of L1 alone or composed of

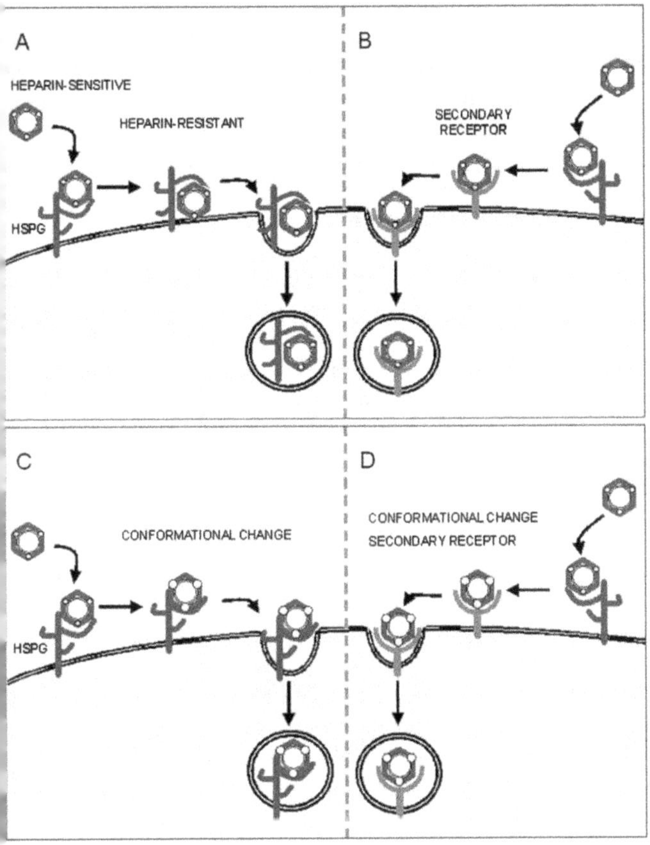

Figure 1. This figure illustrates the potential interactions of PV capsids with cell surface proteins that occur prior to entry into the cell. It is possible that cell post-binding surface events are critical for endocytosis to occur. The resistance to neutralization of virus by exogenous heparin can be accounted for by either conformational changes in the HSPG after capsid binding (panel A), transfer to a secondary receptor (panel B), conformational changes in the capsid, (panel C) or a combination of these events (panel D). Subtle conformational changes in the capsid (panels C and D) may trigger endocytosis or allow for interaction with a secondary receptor. L2 may be involved in this process.

both L1 and L2 were equally effective in inhibiting the binding of BPV virions (Roden et al., 1994). However, later evidence suggested that L2 interactions at the cell surface are also likely to play a role in infection. In one study, an epitope located within amino acid residues 108 to 120 of HPV-16 L2 was found to be recognized by a neutralizing monoclonal antibody (Kawana et al., 1999). It was later shown that exogenous addition of this peptide could interfere with infectivity of the virus. The peptide itself could bind to the cell surface and be internalized (Kawana et al., 2001). In a separate study, a second region of HPV-16 L2, residues 13 to 31, was shown to interact with a widely expressed cell surface molecule after the initial binding of capsid has occurred (Yang et al., 2003a). These observations could indicate PV cell surface binding is mediated by multiple receptors. Based on analogy with other viral/receptor interactions, a reasonable hypothesis is that productive infection would require an initial low specificity interaction mediated by L1 binding, followed by the interaction of a more specific protein component with L2.

A few years after the initial characterization of the cell surface interactions of VLPs, it was demonstrated that PV VLPs have a high binding affinity for glycosaminoglycans (GAGs) (Joyce et al., 1999). GAGs are unbranched polysaccharide chains composed of repeating disaccharide units (reviewed in Sawitzky, 1996). GAGs are divided into four major groups: 1) heparin and heparan sulfate 2) chondroitin sulfate and dermatan sulfate, 3) keratan sulfate, and 4) hyaluronic acid. The core structure of GAGs contains limited information due to its repetitive character. Nevertheless, because both units of the repeated disaccharide can be variably N- and O- sulfated, the resultant macromolecules are complex and heterogeneous. Most GAG species are covalently attached to a protein core to form a proteoglycan. The initial study by Joyce et al. demonstrated that VLPs can interact with immobilized heparin (Joyce et al., 1999). This interaction could be inhibited by soluble heparin, dextran sulfate, and to a lesser extent, chondroitin sulfate C. Likewise, sulfated polysaccharides inhibited the binding of VLPs to the cell surface. Prior treatment of the cells with heparinase or heparitinase also resulted in reduced VLP association. Undersulfation of the GAG chains, produced by cholate treatment of the cells, showed comparable effects. Finally, the cell line pgsA-745, a CHO-derived mutant deficient in xylosyltransferase (the enzyme necessary for the first sugar transfer in GAG synthesis), showed reduced VLP binding. This work established the ability of VLPs to interact with heparin and cellular GAGs. It introduced the idea that proteoglycans are the broadly expressed, trypsin-sensitive cellular surface receptor that the earlier studies had alluded to. Heparan sulfate proteoglycan, HSPGs, are well described receptors for growth factors, lipoproteins and numerous bacteria and viruses (reviewed in Bernfield et al., 1999).

Giroglou et al. (2001) used a pseudovirion assay to demonstrate that the interaction with cell surface HSPGs is critical for infection. This is a critical point, as attachment of a virus to the cell surface does not necessarily imply internalization and entry into a productive pathway. This study also suggested that, after binding to the HSPG, the capsids shift from a form that is sensitive to competition by exogenous heparin to a form that is resistant. This could be due to a stabilization of the virion's interaction with the HSPG matrix. Alternatively, it may reflect the interaction of the virion with a second receptor molecule. These alternatives are illustrated in Figure 1.

The two major families of cell surface HSPGs are the syndecans and glypicans (reviewed in: Bernfield et al., 1992; Fransson, 2003). They differ in core protein size and the mode of attachment to cell surface. Syndecans have an extended ectodomain that holds the heparan sulfate side chains distal from the cell membrane. Syndecans also contain transmembrane and cytoplasmic domains. Glypicans have a more globular protein core that is attached to the cell membrane via a glycosyl-phosphatidyl-inositol anchor. Due to the condensed protein core, the glypicans hold the heparan sulfate close to the cell membrane.

Generally the interaction between HSPGs and their ligands occurs via the heparan sulfate portions of the molecule with little contribution from the core protein (reviewed in Tumova et al., 2000). Enzymatic stripping of the HS moieties usually prevents ligand interaction. However, in some cases the HS chains might not be required for HSPG function. For instance, mutation of the GAG attachment sites within betaglycan does not interfere with its ability to bind and present TGF-beta to the type II receptor (Lopez-Casillas et al., 1993). Many viruses use HSPGs as primary attachment molecules (reviewed in Spillmann, 2001). The binding capabilities of the GAG side chains for these interactions are well established, but the role of the core proteins is not well understood.

Shafti-Keramat et al. (2003) examined the roles of the various syndecan and glypican core proteins in PV binding and infection. Authentic HPV11 infection of an HSPG deficient cell line increased 8-fold after transfection of the syndecan-1 gene and 4-fold after transfection of the syndecan-4 or glypican-1 gene. Cell surface binding of VLPs and PV infection correlated with the HS expression levels. Therefore, there does not appear to be a strong requirement for a specific HSPG protein core for infection. Syndecan-1 is the predominant HSPG in epithelial tissue and may serve as the primary attachment receptor *in vivo*, solely due to its high level of expression in the appropriate target cell (Bernfield, 1992). Exposure of the basal epidermal cells to the virus by wounding is generally thought to facilitate PV infection, and syndecan-1 expression is normally upregulated in both the dermis and the epithelium after skin wounding (Elenius et al., 1991; Gallo and Trowbridge, 2003). Multiple studies have demonstrated the importance of syndecan-1 and its GAG chains in wound healing and delayed wound healing due to the impaired functioning of keratinocytes was observed in syndecan-1 knock-out mice (Stepp et al., 2002).

Mammalian herpesviruses adsorb strongly to proteoglycans, especially HSPGs (reviewed in: Spear et al, 2000; Campadelli-Fiume et al., 2000). For HSV, this high affinity attachment step enhances infectivity, but does not appear to be an absolute requirement for the virus to infect a cell. Cells that are deficient in expression of both heparan sulfate and chondroitin sulfate proteoglycans demonstrate poor association of HSV virions, but can still be infected (Gruenheid et al., 1993). Conversely, binding of the virus to heparan sulfate on certain cell types did not lead to virus entry (Geraghty et al., 1998; Shieh et al., 1992). It is apparent that for cell penetration HSV usually interacts with co-receptors that are distinct from the proteoglycan attachment receptor (reviewed in Spear et al., 2000). However, under some circumstances, specific sites in heparan sulfate generated by O-sulfation can serve as entry receptors for HSV-1 in the absence of any additional protein receptor (Shukla et al., 1999).

How closely the HSV story parallels that of PV is unclear. To date the requirement for the expression of HSPGs for PV binding and infection is absolute, although the research in this area is nowhere near as extensive for PV as that for HSV. The existence of a secondary receptor for PV is appealing, but is still largely inferential. Studies to identify a secondary receptor are hampered by the inability to identify epithelial cell lines that bind PV capsids strongly but are resistant to infection.

Alpha 6 integrin was initially proposed to act as a PV receptor in 1997 (Evander et al., 1997). HPV6 VLPs were shown to interact with α6 integrin by co-immunoprecipitation. α6-negative cell lines bound VLPs poorly and anti-α6 integrin antibodies diminished the binding of VLPs to the cell surface. However, as the HSPG story was developing, it was also shown that α6 expression is not essential for PV attachment and infection. For instance, α6 integrin-deficient human keratinocytes are permissive for BPV 4 infection (Sibbet et al., 2000) and HPV11 infection (Shafti-Keramat et al., 2003). Given the close association of proteoglycans and integrins as matrix components, it is possible that the experimental association of α6 integrin with PV binding and entry is a secondary effect due to its interaction with HSPGs.

Interestingly, a two-step entry mechanism with integrins providing the secondary receptor function has been described for other viruses. The initial interaction of murine polyomavirus with host cells occurs through capsid binding of sialic acid-containing cell membrane molecules (Fried et al., 1981). However, it was shown recently that expression of alpha4 integrin enhances permissivity to polyomavirus infection by acting as a post-attachment receptor (Caruso et al., 2003). Also, rotavirus strain RRV uses sialic acid as a primary cell receptor and subsequently interacts with integrins at a post-attachment step to infect host cells (Arias et al.); (Guerrero et al., 2000).

Unexpectedly, PV VLPs have been shown to bind strongly to a variety of immunocytes, including myeloid-derived dendritic cells, B cells, monocytes, and macrophages, but interestingly not T cells (Da Silva et al., 2001). Importantly, this interaction can also occur *in vivo*, as BPV-1/2 DNA was found to be present in circulating lymphocytes in cattle (Campo et al., 1994). There is no published data on whether any of the immune cells that bind VLPs can actually be infected by PV. We have found that these cell types are poorly transducible with PV pseudovirions (unpublished data). Therefore, it is critical to realize that the trafficking pathway in cells of these lineages may be distinct and not intrinsically informative as to the infective entry mechanism. It remains to be resolved whether the strong interactions with immunocytes primarily involve HSPGs. The kinetics of uptake of PV capsids by dendritic cells is much faster than it is for epithelial cells (discussed below), which may indicate the involvement of a distinct receptor (Fausch et al., 2003; Lenz et al., 2001; Yan et al., 2004). The interaction of VLPs with these cells can lead to phenotypic maturation and release of various mediators of innate immunity (Lenz et al., 2001; Yang et al., 2004). It is unlikely that α6 integrin is the primary receptor on immunocytes, as antibodies to α6 integrin did not specifically block VLP binding, and T cells express high levels of cell surface α6 integrin and do not bind significant amounts of VLPs (Da Silva et al., 2001). The finding that addition of IgG antibodies, regardless of specificity, to the cells resulted in partial inhibition of VLP binding suggested that a cell surface receptor for IgG might serve as an accessory molecule in VLP binding to immunocytes. Consistent with this conjecture, monoclonal antibodies to CD16 (FcγRII) specifically inhibited VLP binding to immunocytes, and VLP binding to immunocytes from FcγRIII knockout mice was reduced (Da Silva et al., 2001). However, transfection of the FcγRIII gene into a nonbinding B cell line did not increase VLP binding, and PV capsids efficiently bind and infect FcγRIII negative lines, such as HeLa. Therefore FcγRIII is neither a sufficient, nor necessary, PV receptor. Interaction with FcγRIII is likely to be more important in the immune response to PV capsids than in their infectivity.

It is important to note that viral entry into cells is not the only determinant of cell tropism. This is especially true for PVs, which exhibit promiscuous cell association while only completing their life cycle in differentiating squamous epithelium (Lowy and Howley, 2001; Meyers et al., 1992). Therefore, while the early events of infection may be similar in permissive and nonpermissive cell types, there is, nevertheless, a restriction of viral replicative functions and virion production that is determined by factors tied to the keratinocyte differentiation program.

4. Surface Events

For some viruses, virion interactions at the cell surface can elicit a substantial conformational change in the viral particle. For PVs, a conformational shift could be reflected in the described shift from a heparin sensitive to heparin resistant form (Giroglou et al., 2001) (Figure 1). Supporting this are additional data from the same group which describe the improved recognition of a neutralizing epitope after virion attachment to the cell surface (Selinka et al., 2003). However, for HPV11, no change in conformation was observed after cell surface attachment when a panel of L1 monoclonal antibodies was examined (Culp and Christensen, 2004). In this type of experiment, positive results may be more conclusive, as the immmunodominant epitopes recognized by the majority of monoclonal antibodies may not reveal a subtle alteration in capsid conformation. Our unpublished data indicate a change in the antibody accessibility of a subdominant HPV16 L1 epitope after cell surface attachment of the virus. However, this could be due to either a conformational change in the virion particle after binding or to steric effects of receptor protein(s) binding on the accessibility of the epitope.

The entry of enveloped viruses into cells requires the fusion of the viral membrane with the target cell membrane. This fusion may occur at the cell surface or after internalization of the virus. The most characterized cell surface viral fusion is that mediating the entry of HIV (reviewed in Colman and Lawrence, 2003). The HIV envelope glycoprotein gp160 has both receptor-binding and fusion activities, with the latter function dependent upon the cleavage of gp160 to the gp120 and gp41 subunits. These two subunits promote viral entry through a coordinated series of conformational transitions that lead to the fusion of the HIV-1 and target cell membranes.

Non-enveloped viruses cannot employ membrane fusion to enter the host cell, and there is considerable variety in the mechanisms that these viruses have evolved as successful alternatives. However, relatively little is known about what cell surface events prelude the entry of non-enveloped viruses. In the best understood example, dramatic structural changes of picornavirus capsids are induced by adsorption to its receptor at the cell surface (De Sena and Mandel, 1977; Joklik and Darnell, 1961). These receptor-mediated conformational changes are necessary for subsequent uncoating events and result in the formation of a transmembrane pore (Kolatkar et al., 1999; Tosteson et al., 2004). The viral RNA genome is probably released directly into the cytoplasm through this channel (Danthi et al., 2003). However, it is possible that this conformation change occurs at a later step and the genome is released through a transmembrane pore after endocytosis. It is thought that nodaviruses enter via an analogous mechanism (Bong et al., 1999). There is considerably less information on receptor-induced conformational changes in other non-enveloped viruses.

Both adenovirus and reovirus use extended, icosahedrally arranged fibers as attachment proteins (Chroboczek et al., 1995; Stehle and Dermody, 2004; Xia et al., 1995). Adenoviral capsids do not undergo conformational changes at the cell surface, excepting the dissociation of the fiber protein following engagement of the integrin receptor by the penton base (Greber et al., 1993). The subsequent interaction of the icosahedral capsid promotes the rapid internalization of virus particles into clathrin-coated vesicles (Wang et al., 1998). Reoviruses are interesting, in this respect, as the viral protein that mediates cell surface attachment, σ1, undergoes a reversible structural change upon cell binding or after *in vitro* interaction with sialic acid, the primary cellular receptor (Fernandes et al., 1994).

Based on the observations to date, it is unlikely that PVs undergo as extensive a surface rearrangement as the picornaviruses. Also, the cytoplasmic delivery of the papillomaviral genome would undoubtedly be more proximal to the nuclear pore as observed with other DNA viruses. PVs do not possess a specialized attachment protein analogous to the fiber proteins of adenoviruses and reoviruses. The predominant capsid structural protein, L1, is also the viral protein responsible for the initial cell interaction, although, as discussed earlier, the L2 protein may contribute to surface interactions. One of the regions of L2 that has been described to mediate this appears not to be displayed on the virion surface in solution. Because this region, residues 13 to 31 of HPV16 L2, is not necessary for virion binding, it was proposed to interact with a cell surface molecule after the virion has attached to a primary receptor (Yang et al., 2003a). This interpretation suggests that there is a conformational change in the viral particle at the cell surface that unmasks this domain of L2. Consistent with this idea, L2 can induce cross-type neutralizing antibodies as a free protein immunogen, but not when it is found in its typical context, assembled in a PV capsid (Roden et al., 2000). The various possible surface events are illustrated in Figure 1.

It is critical to note at this juncture that a wide range of protocols have been used to obtain the PV capsids for binding and entry studies. Some of the variable results in these studies could, at least in part, be due to the "maturity" state of the VLPs and pseudovirus used in the studies. PV capsids extracted from replicating cultured cells can exist in two forms (Buck, 2005). "Immature" capsids are larger, less regular, and less protease resistant than "mature" capsids", this indicates that a substantial change in conformation occurs during the maturation process. Maturation is associated with an ordered process of intramolecular L1 disulfide bond formation. It can be induced by the inclusion of a prolonged 37°C incubation under oxidizing conditions during the purification procedure. Therefore, it is conceivable that the omission of a maturation step could result in assay variability due to particle heterogeneity. Additionally, *in situ* (at the cell surface) maturation of immature particles could result in an apparent conformational change that is not related to receptor engagement or the commencement of uncoating.

This phenomenon might help to explain the disparate results reported for internalization kinetics. PV particles enter the cell relatively slowly after receptor engagement, although discrepant half-times have been reported. The kinetics were reported to be different for VLPs versus pseudovirions (Selinka et al., 2003), with half-times of 3 hours and 7 hours respectively. By contrast, a microscopic study comparing the entry of VLPs and authentic virions could not discern any gross differences in the time course of internalization and intracellular trafficking (Day et al., 2003). It may also be relevant that these studies cannot distinguish infective virions from defective particles, and capsid preparations can have particle to infectivity ratios that range from slightly less than 10:1 to more than 1000:1. Thus, microscopic studies do not rule out the possibility that the trafficking observed for the majority of the particles does not reflect the trafficking of the subset of particles that lead to infection.

Neutralization experiments with cultured cells have demonstrated an unusually long period of susceptibility of infectious PV capsids to the addition of virion-specific monoclonal antibodies. Christensen and colleagues showed that substantial neutralization of BPV-1 could occur by application of neutralizing MAbs even 24 hours after addition of the virus (Christensen et al., 1995). His laboratory expanded on these observations by demonstrating a similar susceptibility for HPV-11, HPV-40 and CRPV (Culp and Christensen, 2004). Heparin neutralization of HPV-11 pseudovirus infection several hours post-infection has also been reported (Selinka et al., 2002). It is not clear that all of the observed neutralization occurs at the cell surface. As both antibodies (Gollins and Porterfield, 1986a; Rojas and Apodaca, 2002) and heparin (Barzu et al., 1996; Letourneur et al., 1995) are known to be endocytosed, it is conceivable that the ostensible surface neutralization actually occurs in an intracellular compartment. Nevertheless, the cumulative evidence is consistent with an unusually extended residence on the cell surface for PVs. Most ligands, including the majority of viruses, are internalized rapidly (within minutes) after the initial receptor encounter and engagement. The reason for the delayed kinetics observed for PVs is unknown, but it is noteworthy that syndecans have been reported also to have a slow rate of internalization after ligand binding (Williams and Fuki, 1997; Yanagishita, 1992). Alternatively, it might be due to the existence of a secondary receptor, if the purported receptor is sparsely arrayed or exhibits particular requirements for endocytosis.

5. Signaling

The interaction of a viral particle with the cell surface can trigger intracellular signaling pathways (reviewed in Greber, 2002). The possible responses by the cell include stimulation of innate and adaptive immunity, growth, proliferation, survival and apoptosis. For adenoviruses and AAV, capsid-activated cell signaling increases the efficiency of viral entry and gene delivery (Meier and Greber, 2003).

The Erk pathway, which regulates cell growth and differentiation, is activated by many DNA tumor viruses, including HCMV, hepatitis B and C viruses, SV40 and adenoviruses (reviewed in Greber, 2002). In the case of adenovirus, Erk activation has post-entry effects on infection leading to the activation of viral transcription (Whalen et al., 1997).

Endocytic sorting is regulated by signaling from surface molecules (reviewed in Benmerah, 2004). The ability of the virus to manipulate these signals may help determine the early course of infection. For instance, adenovirus-2 induces two separate signaling pathways at the cell surface. The first pathway requires αv integrins and activates protein kinase A (Wickham et al., 1993). The second pathway is independent of integrins and activates the p38/MAP kinase pathway (Suomalainen et al., 2001). If the latter pathway is disrupted, adenovirus particles are unable to traffic towards the nucleus and are retained at the cell periphery.

There is very little known about PV-induced cell signaling events. In the only study addressing the issue, it was found that high concentrations of L1-only VLPs induced cell proliferation through activation of the Ras-MAP kinase pathway (Payne et al., 2001). VLP binding caused phosphorylation of the β4 integrin subunit, recruitment of Shc and the activation of Ras, Raf and Erk2. The effect of this signaling on later viral events has not been explored.

Given the affinity of PV particles for HSPGs, it is reasonable to suspect that this interaction could mediate signaling events. HPSGs have been demonstrated to promote cell-cell signaling indirectly by acting as a co-receptor (reviewed in: Couchman, 2003; Tumova et al., 2000). These studies found that the HSPG itself lacked an intrinsic kinase activity. However, recent results suggest that HSPGs can have a more direct role in mediating signal transduction by using cytoplasmic kinases, adapter proteins, or even the HSPG cytoplasmic domain itself to mediate signal transduction events (Chua et al., 2004; Lim et al., 2003; Oh et al., 1997).

Recent studies have demonstrated a direct syndecan-4 involvement in FGF-2 signaling and internalization (Horowitz et al., 2002; Volk et al., 1999). The internalization is initiated by a clustering of syndecan-4 that causes Rac1 activation and proceeds in a caveolae-, clathrin- and dynamin-independent manner (Tkachenko et al., 2004). The capacity of syndecan-4 clustering to activate Rac1 is consistent with the known syndecan-4 oligomerization-dependent activation of PKCα (Horowitz and Simons, 1998; Oh et al., 1997). This activation in turn induces macropinocytosis (Grimmer et al., 2002).

Considering the dearth of information about signaling events during PV infection, many basic questions remain unanswered. Is the transmission of a signal a necessary surface event to mediate endocytosis of the viral particle? Is the successful completion of downstream steps in a productive infection contingent upon signaling events during the endocytic and trafficking process? In examining these questions, it will be important to consider whether effects seen after addition of relatively large numbers of capsids per cell *in vitro* are biologically relevant to PV infections *in vivo*, where relatively few virions might normally be expected to interact with a given target cell.

6. Endocytosis and Vesicular Trafficking

Highly sensitive nested RT-PCR techniques have detected early viral transcripts by 4 hours post-infection for HPV-31b (Ozbun, 2002), 2 hours post-infection for BPV-4 (Sibbet et al., 2000), 6 hours post-infection for HPV-11 (Culp and Christensen, 2004) and 12 hours post-infection for BPV-1 (Day et al., 2003). However, these times represent the leading edge of infection, as less sensitive, but more quantitative techniques show a gradual increase in the number of viral transcripts in the 2-4 days following infection in most susceptible cell lines (Culp and Christensen, 2004). As detailed above, the average residence on the cell surface for a PV particle is remarkably long, although slow internalization alone cannot entirely account for the slow infection rate of the virus.

Following binding to the cell surface, viruses are internalized into the intracellular milieu. Several pathways have been well described, and the majority of viruses appear to enter via one of these routes. However, it is likely that alternative pathways, that are yet to be characterized, are utilized by some viruses. Nonenveloped viruses generally enter the cell through either clathrin-coated vesicles or the clathrin-independent caveolae system. The former pathway is used by the cell for uptake of many nutrients and growth factors (reviewed in Gruenberg, 1991). It is frequently hijacked by intracellular pathogens as a portal to the cytosol. The vesicular microenvironment may facilitate viral infection through protease-dependent degradation of the viral capsid, as for reoviruses (Ebert et al., 2002). Some viruses, *e.g.* adenoviruses, are dependent upon the progressive acidification encountered during retrograde vesicular transport (Svensson, 1985). Various molecules and macromolecular complexes enter the cells via caveolar receptors, including folate, cholesterol, serum lipoproteins, and liposomes (reviewed in Pelkmans and Helenius, 2002). Under normal conditions, caveolae do not separate from the plasma membrane, but their association with certain viruses can induce their internalization. SV40, mouse polyomavirus, echovirus 1 and some filoviruses enter the cell by this mechanism (reviewed in Pelkmans and Helenius, 2003). The internalized caveolae fuse with caveosomes, and the caveosomes bypass the acidic endocytic compartment to deliver their contents

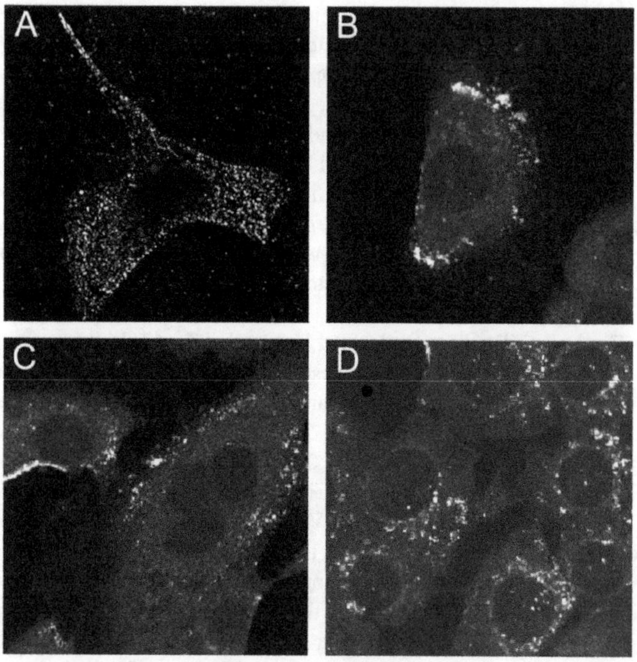

Figure 2. This figure shows the internalization of BPV-1 authentic virions as detected with an antibody directed against the L1 capsid protein. The cells shown in panel A were incubated with virions for 60 minutes at 4°C to prevent internalization and lateral membrane movement. The cells were subsequently incubated at 37°C to allow progress of entry events. Panel B shows the fixation after 90 minutes of internalization. At this time many cells show an obvious coalescence of viral particles at the cell surface. This is similar to the capping phenomenon observed prior to endocytosis of many polyvalent ligands. The cells in panel C were fixed 4 hours post-internalization. The viral particles are largely located in peripheral vesicles which are known to be early endosomes, although some residual surface particles are still evident. Panel D shows the pattern of internalization virions at 8 hours. The staining shows a more perinuclear distribution of vesicles, although the heterogeneity of the entry process is obvious.

into the smooth ER. The distinction between utilization of the clathrin-dependent and caveolar pathways can be achieved through analysis of biochemical inhibition of ligand uptake.

We used this approach to examine the pathway by which PV infect cells (Day et al.). The assay was based upon the detection of spliced mRNA at 12 hours post-infection with authentic BPV-1 virions. It was found that chlorpromazine, an agent that has been shown to block clathrin-mediated endocytosis, could prevent PV infection. Infection was also affected by addition of the actin-disrupting drug, latrunculin B. By addition of the drugs at set times after infection had commenced, it was determined that both of these agents acted at an early step in entry. Accumulation of the viral transcript was prevented if they were added up to 2 hours after infection, but not when added 4 hours after infection or later.

The addition of bafilomycin A also prevented infection when added early in infection. Bafilomycin A can affect early endosomal trafficking in addition to its well established inhibition of vacuolar ATPases (Bayer et al., 1998; Clague et al., 1994). Therefore, this effect indicates that PV infection is dependent upon exposure to the low pH of the untreated endosomal compartment and/or the virus must enter the late endocytic compartment to establish a productive infection. In accord with the former conjecture, other lysosomotropic agents, chloroquine and NH_4Cl, also prevented infection. Traffic into the late endosomes is prevented with the microtubule disrupting agent nocodazole, which also abrogated infection. It is an intriguing possibility that there may also be a dependence upon microtubule function at the step of transferring the genome from the vesicular compartment to the nucleus (discussed below). In addition to the inhibitor analysis, the capsids were sequentially colocalized with established markers of endocytic compartments, including: the clathrin adaptor protein, AP-2; transferrin receptor, a marker of early endosomes; and Lamp-2, a marker of the late endosomes and lysosomes. The overall conclusion from this study was that BPV-1 utilizes the clathrin-dependent endocytic pathway for infectious uptake. A typical microscopic analysis is shown in Figure 2 and the proposed entry pathway is illustrated in Figure 3.

Selinka et al. also examined a set of biochemical inhibitors for effects on HPV-33 pseudovirus infection (Selinka et al., 2002). They also found a dependence upon an intact actin cytoskeleton and microtubules. Neither of these studies could demonstrate an effect of caveolae disruption, via nystatin or filipin treatment, on either BPV-1 or HPV-33 infection, reinforcing the conclusion that uptake of these two PV types is via a clathrin-dependent rather than caveolae-dependent pathway.

However, a study utilizing infectious VLP-DNA complexes, generated by mixing VLPs with naked DNA, found unexpectedly that HPV-31 infection was inhibited by nystatin, a caveolae inhibitor (Bousarghin et al., 2003). In this assay, linearized reporter plasmid DNA is bound to the outside of VLP particles. It is unclear how the exterior DNA might affect cell surface intramolecular interactions and the subsequent trafficking of the particles. However, in the same study, the entry of HPV-16, which is closely related phylogenetically to HPV-31, and HPV-58 was found to be insensitive to caveolar inhibition and was blocked by chlorpromazine,

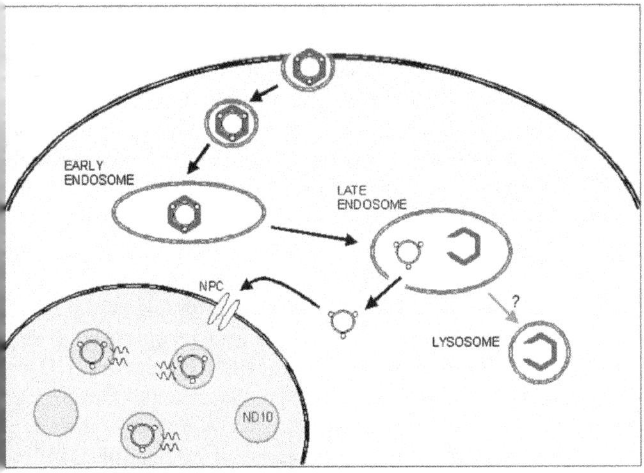

Figure 3. PV capsids enter the cell via an endocytic pathway that is likely to be clathrin-dependent. Viral particles traverse the early endosomes and uncoat in a late endosomal compartment. It is postulated that L2 mediates escape from the limiting membrane and travels with the genome into the nucleus. L2 and the genome colocalize at the nuclear subdomain, ND10. Intact ND10 have been found to be critical for efficient viral transcription. Residual L1 shows continued association with Lamp + compartments. No evidence of L1-containing subviral particles exiting the endosomal compartment has been found.

an inhibitor of clathrin-mediated uptake. To date, with the exception of HPV-31, all the data on the uptake of other PV types is consistent with internalization via clathrin-mediated endocytosis. It will be interesting to further examine the HPV-31 uptake pathway, in comparison to that of other PV types. The different PV capsids are indistinguishable at the ultrastructural level. Possibly HPV-31 interacts differently with HSPGs or has a unique co-receptor. There is precedence for closely related naked icosohedral viruses using distinct uptake pathways. As noted above, SV40 enters via caveolae, while the related JC polyomavirus enters via clathrin coated pits (Pho et al., 2000).

As detailed above, PVs are internalized slowly. Typically ligand-receptor complexes transit via clathrin-mediated entry to peripheral endosomes in fewer than 10 minutes (Hanover et al., 1985; Hanover et al., 1984; Schmid, 1988). It is also apparent that the subsequent trafficking steps of PV capsids are also unusual in their delayed kinetics. We estimate that the virions do not exit the early endosomes within the first 2 hours after entry. The transit into the late endosomes begins by approximately 4 hours, and the commencement of their egress from the late endosomes occurs in the 6 to 8 hour window. Although the biochemical and microscopic data support clathrin-dependent uptake, the kinetics are unusual for internalization by this mechanism. Therefore, the possibility that PV utilize a novel endocytic pathway for entry must still be considered.

In general the complexities of vesicular trafficking are still not well understood, and the various mechanisms by which viruses exploit this system is even more unclear. There are a number of instances in which viral trafficking does not conform to the established kinetics of known endocytic pathways. Canine parvovirus and mouse minute virus are both internalized quickly by clathrin-mediated endocytosis, but show an unexplained lag between cell entry and escape of the virus from the endosome (Parker and Parrish, 2000). SV40 and hepatitis B viruses also show a prolonged residence at the cell surface. The subsequent intracellular trafficking of hepatitis viruses is not well defined. The post-internalization steps are also quite slow, although hepatitis viruses are unlikely to be utilizing a similar pathway to PVes, as an intact actin cytoskeleton is not required for their entry (Funk et al., 2004).

7. Endosome Escape

After endocytic uptake, access of most ligands to the cytosol is limited and tightly regulated. Some enveloped viruses escape from the endocytic compartment by fusion of the viral envelope with the endosomal membrane. For instance, the flavivirus envelope protein is activated by the low pH of the endosome (Gollins and Porterfield, 1986b). The change in pH promotes conformational changes in the viral protein that catalyzes merging of the viral membrane with the endosomal membrane, thereby allowing the viral RNA to escape into the cell (Gollins and Porterfield, 1986b; Heinz and Allison, 2001; Monath et al., 2002). Likewise, influenza hemagglutinin undergoes a low pH-induced conformational change that leads to the extrusion of apolar regions of the protein (Doms et al., 1985; White et al., 1982). This causes insertion of the hemagglutinin protein into the endosomal membrane forming a fusion pore (Gething et al., 1986).

To allow escape of the viral genome into the cytosol, nonenveloped viruses either lyse the limiting vesicular membrane or generate a pore through it. There are several descriptions of viral proteins that have evolved mechanisms for endosome escape, with adenovirus being the best understood example (reviewed in: Medina-Kauwe, 2003; Meier and Greber, 2003). At low pH, the virus penetrates the endosomal membrane and escapes into the cytosol. This lytic activity of the virus is attributed to the penton, as studies show that the penton undergoes a low pH-induced conformational change that exposes hydrophobic regions (Seth et al., 1984). The cytoplasmic viral particles appear morphologically intact, although biochemical investigations have indicated that stepwise dismantling of the capsid begins in the vesicular compartment (Greber et al., 1993). It has been found that the viral capsid itself promotes the cytosolic entry processes, as inactive virus can augment transferrin-mediated gene transfer. This is the basis for the AVET (adenovirus-enhanced transferrin-mediated) transfection system in which inactivated adenovirus is complexed with plasmid DNA and polylysine-transferrin (Zatloukal et al., 1992).

Pore formation by a nonenveloped virus has been definitively demonstrated only for reoviruses. In this case, acid-dependent proteolytic disassembly within endosomes results in the production of infectious subvirion particles. These activated particles can induce membrane channels that resemble the fusion pores created by some of the enveloped viruses (Tosteson et al., 1993; Chandran et al., 2002). This membrane penetration delivers a partially uncoated particle directly into the cytoplasm.

There is good evidence that a similar mechanism is at play in the uncoating of the minor-group human rhinovirus (HRV2). This virus enters the cell via clathrin-dependent endocytosis (Hofer et al., 1994). Low pH mediates a conformational change in capsid morphology leading to uncoating in a late endosomal compartment (Bayer et al., 1998). The viral RNA genome appears to be translocated through a pore across the endosomal membrane (Prchla et al., 1994), while the empty rhinovirus capsids are trafficked into the lysosomal compartment where they are rapidly degraded.

Parvoviruses appear to use a different strategy for endosome escape. A well conserved domain within the capsid protein, VP1up, has phospholipase A_2 activity (Suikkanen et al., 2003; Zadori et al., 2001). This domain is sequestered inside the capsid prior to endocytosis, but is externalized during a conformational transition. Viruses that are deficient in phospholipase A_2 activity are impaired in the transfer of the viral genome from perinuclear vesicles into the nucleus. Intriguingly, phospholipase A_2 has been shown to adversely affect the integrity of lysosomal membranes (Marone et al., 1983).

The mechanism by which PVes gain access to the cytosol is only now being examined in detail. Recent data suggest that the minor capsid protein, L2, has a critical role in mediating the endosome escape of PVs. A short, hydrophobic peptide derived from the carboxyl terminus of HPV-33 L2 has been shown to have membrane disrupting capacity (Martin Sapp, personal communication). Pseudovirions whose L2 proteins lack this region are unable to establish infection, even though DNA encapsidation and cell surface interactions are unaffected. Microscopic analysis of non-infectious capsids containing a similar deletion in BPV-1 L2 shows that the mutant L2 protein and the viral genome are retained together within the endosomal compartment at late timepoints post-infection (Day, unpublished data). Initial analysis also indicates the integrity of endosomal membranes is disrupted during PV infection, to the extent that large FITC-conjugates of dextran (100 kD) are no longer retained (Day, unpublished data). These data are suggestive of a global disruption of the endosomal membrane, more similar to that seen with adenovirus, than that observed during reovirus or rhinovirus infection.

Our lab has recently developed two assays to examine the uncoating of papillomaviral pseudoviruses: one that detected exposure to an interior region of L2 and another that detected the viral pseudogenome (Day et al., 2004). An L2 gene encoding a carboxyl-terminally HA-tagged protein was constructed. Placement of the epitope at this site allows accessibility to HA antibody only after uncoating has begun, as the carboxyl terminus of L2 is not displayed on the surface of intact PV capsids (Liu et al., 1997). The determination of when this epitope became accessible to antibody staining allowed us to assess the timing and localization of disassembly events. At least partial disassembly of the viral particle was expected to occur before nuclear import, as it has been reported that intact HPV capsids are unable to enter the nucleus (Merle et al., 1999; Nelson et al., 2000; 2002). Consistent with this expectation, the initial uncoating was detected in cytoplasmic vesicles, by 6 hours of infection, with an increase in vesicular staining developing over the next few hours. The detection of uncoated virus colocalized with Lamp-1, a marker of the late endosomal and lysosomal compartments (unpublished data).

To examine the uncoating of the incoming viral pseudogenome, pseudovirus stocks were generated in the presence of BrdU. The purified, labeled pseudovirus particles were then added to target cells, and the input pseudogenome was detected with an anti-BrdU antibody. Similarly to the tagged L2 protein, the genome was not accessible to antibody detection while enclosed in the intact viral particle and was first detected in endocytic vesicles at 6 hours post-infection (Day et al., 2004). Therefore, these results also support the conclusion that at least partial uncoating occurs in a late endosomal compartment (Figure 3).

8. Traversing the Cytoplasm

The issue of how the papillomaviral genome transits from the endosome to the nucleus has not been systematically addressed. The cytoplasm imposes a substantial barrier. The presence of organelles, the cytoskeleton, and high protein concentrations inhibits the free diffusion of macromolecules. The diffusion of DNA in the cytoplasm is strongly size-dependent, with little or no diffusion of DNA molecules larger than 2000 bp (Lukacs et al., 2000). Therefore, the 8000 bp papillomaviral genome would be unable to traverse the cytosol from the endosome into the nucleus without a chaperoning molecule to direct its transit. The translational diffusion of other macromolecules in the cytoplasm has also been examined. Free diffusion is decreased relative to that in water, which suggests a cytoplasmic "sieving" mechanism that was proposed to involve the skeletal mesh (Luby-Phelps et al., 1987). Steric hindrance largely prevents the free movement of solutes over 500 kD (Seksek et al., 1997). Therefore, the movement of virions and viral genome-protein complexes would have to be an active process. We favor a model in which L2 mediates the delivery of the viral genome from the endosome to the nucleus perhaps in conjunction with a cell-encoded chaperone. A role for L2 is supported by the colocalization of L2 and the pseudogenome in the nucleus, as described in the next section. However, it is still unclear how the L2-DNA complex negotiates the intervening cytosol.

Cytoplasmic transport normally depends on a network of three principal types of protein filaments: microtubules, intermediate filaments, and actin microfilaments (reviewed in Sodeik, 2000). Intermediate filaments are not polarized and would not provide the directionality necessary for transport (Fuchs and Yang, 1999). Transport along actin or microtubules is mediated by motor protein complexes that use cellular energy to perform the work of moving cargo along the filaments.

There is growing evidence that F-actin directly participates in most membrane internalization events (reviewed in Engqvist-Goldstein and Drubin, 2003). It is essential for the formation of membrane protrusions for macropinocytosis and phagocytosis, as well as for caveolae-mediated uptake. The role of actin in clathrin-dependent endocytosis is more controversial. However, there is increasing evidence that actin polymerization is involved in clathrin-coated vesicle formation (Fujimoto et al., 2000; Merrifield et al., 2002). The actin binding motor protein, myosin VI, is the only myosin that directs minus-ended traffic, the movement in a retrograde direction, away from the plasma membrane (Cramer, 2000; Wells et al., 1999). This myosin is enriched in clathrin vesicles and interference with myosin VI function prevents transferrin internalization (Buss et al., 2001).

Therefore, although many viruses have been found to be dependent upon actin for establishment of infection, in the majority of cases, this dependence probably reflects the effect of actin destabilization on the endocytic machinery. The sole demonstration of direct utilization of actin for virus entry is for the baculovirus, *Autographa californica* M nucleopolyhedrovirus (AcMNPV). AcMNPV has been shown to utilize actin to propel its free cytoplasmic capsids towards the nucleus (Lanier and Volkman, 1998), after they have been released from the endosome by a fusion event. A direct interaction of the capsid proteins with actin mediates the nucleation of thick actin cables (Lanier et al., 1996). This polymerization of actin, in conjunction with a myosin-like motor, drives the transport of the capsids to the nucleus.

As noted above, an intact actin network has been shown to be required for PV infection (Day et al., 2003; Selinka et al., 2002; Zhou et al., 1995). However, it has been demonstrated to be critical at an early stage of infection (Day et al., 2003), unlike what was shown for AcMNPV. Therefore, this dependence is likely to be a manifestation of the requirement of polymerized actin filaments for endocytosis. The only report that does not support this interpretation suggested a role for actin in the translocation of a viral complex across the cytoplasm (Yang et al., 2003b). The authors reported that VLPs composed of only L1 trafficked differently from those that contained both capsid proteins. The L1+L2 VLPs accumulated in the perinuclear region within 2 hours, whereas the L1 only VLPs were more peripheral. It was proposed that the direct interaction of L2 with actin filaments directed these particles towards the nucleus. However, it is somewhat difficult to reconcile these results with the bulk of the observations on PV uptake and trafficking. The kinetics of capsid internalization have reproducibly been determined to be substantially slower than that observed in this study. Further investigation of the role of L2-actin interaction in infectivity is warranted.

In many viral systems, it has been shown that interaction between viral capsid proteins and the microtubular network allows for the efficient nuclear targeting of viral particles. Depolymerization of microtubules prevents the transition of cargo from early to late endosomes. Therefore, the inhibition of viral infection by microtubule inhibitors may reflect a requirement for transit into late endosomes or lysosomes. The observation that nocodazole seemed to inhibit PV infection at a later step than this transition suggests that microtubules may be required for the post-endosomal step of delivering the viral genome into the nucleus (Day et al., 2003). Interestingly, an interaction of L1 and tubulin was found by Liu et al. (2001).

We have also observed colocalization of L2 and γ-tubulin after viral uncoating, indicating L2 localization at the microtubule-organizing center (MTOC) (unpublished observations). The MTOC determines the intracellular organization of microtubules, and serves as the initiation site for the assembly of microtubules, which grow outward from the MTOC to the periphery of the cell. The kinesin motor proteins move cargo towards the plus end of the microtubule at the cell periphery and the dynein motor proteins move cargo towards the minus end of the microtubule, the MTOC.

There are numerous examples of viruses usurping the minus-ended transport along microtubules to mediate delivery of the capsid proximal to the nucleus. After fusion of the viral envelope with the plasma membrane, herpes simplex virus type 1 (HSV1) capsids are transported along microtubules from the cell periphery to the nucleus (Dohner et al., 2002; Sodeik et al., 1997). This retrograde transport is mediated by interaction of the capsid protein VP26 with cytoplasmic dynein (Douglas et al., 2004).

The interaction of adenovirus-5 with an intact microtubule cytoskeleton and functional cytoplasmic dynein occurs after escape from endosomes and is a necessary prerequisite to viral nuclear localization (Leopold et al., 2000). The transport of adenovirus-2 to the minus end of microtubules proximal to the nucleus is mediated by the dynein/dynactin motor complex (Suomalainen et al., 1999). Microtubule depolymerization prevents polyomavirus migration toward the nucleus, while disruption of the actin microfilaments appears to have no detrimental effect on the ability of the virus to reach the nucleus (Sanjuan et al., 2003). Intact microtubules and dynein are also required for the cytoplasmic transport of canine parvovirus capsids and contribute to the accumulation of the capsid in the nucleus (Suikkanen et al., 2003). Incoming foamy retroviral proteins are trafficked to the MTOC prior to nuclear translocation of the viral genome and this is mediated by a direct interaction between Gag and dynein light chain 8 (Petit et al., 2003).

9. Nuclear Entry

DNA viruses, excluding poxviruses, must deliver their genetic material into the nucleus, to allow access to the host cell transcriptional machinery. The nuclear homing of karyophilic viral particles during natural infection can occur by various mechanisms (reviewed by Whittaker et al., 2000). Diffusion of metabolites and small non-nuclear molecules as well as active, mediated import of protein through the nuclear envelope occurs through nuclear pore complexes (NPCs). Small virions such as those from hepatitis B virus (Rabe et al., 2003) can pass intact directly through the NPC. However, the interior diameter of the central NPC channel prevents the capsids of most nuclear viruses from entering without at least partial disassembly of the virion. Adenoviral capsids complete their dissociation at the NPC, with nuclear delivery of protein VII and the viral DNA (Greber et al., 1997). The capsids of HSV-1 also dock at the nuclear pore, and extrude their DNA into the nucleus. SV40 may transit the NPC in intact form despite its 50 nm size (Clever et al., 1991), though an alternate route into the nucleus cannot be excluded.

Intact PV capsids exceed the size capacity for transit across the NPC. It has also been formally demonstrated that intact particles are not able to cross this barrier (Merle et al., 1999; Nelson et al., 2000; Nelson et al., 2002). Until recently, it has been unclear whether, after disassembly, either of the PV capsid proteins accompanies the viral genome into the nucleus. We found that HeLa cells infected with BPV-1 virus or pseudovirus showed no evidence of nuclear L1 with any of several tested L1 antibodies or fixation conditions. However, the accumulation of nuclear L2 derived from input virions was detected with a number of polyclonal antisera after acetone fixation (Day et al., 2004). In addition, we were able to demonstrate colocalization in the nucleus of HA-tagged L2 and BrdU-labeled viral pseudogenome, using the techniques described above to investigate uncoating. Interestingly, the two viral components colocalized in a punctate pattern reminiscent of the L2 localization at ND10 observed after expression of de novo synthesized protein (Day et al., 1998), although in this instance the L2 protein is derived from input virions. The accumulation of both L2 and pseudogenome at ND10 was confirmed by colocalization with PML, the defining ND10 protein. It was first detectable approximately 12 hours post-infection and was abundant by 20 hours after initiation of infection.

The colocalization of both L2 and the pseudogenome adjacent to ND10 led us to speculate whether the presence of ND10 might play a role in the establishment of infection. As PML is required for ND10 formation (Ishov et al., 1999), we compared PML -/- cells with those expressing PML (PML -/- cells transfected with the PML3 gene) for their ability to support early events in infection. Infection by BPV pseudovirus with an encapsidated GFP expression plasmid was an order of magnitude more efficient in PML-expressing cells compared to the PML-null cells. Using a quantitative RT-PCR assay, we also found that early mRNA expression via authentic BPV virus was 10-fold lower in PML -/- cells than in the PML3 reconstituted cells. In the cells lacking PML, the distribution of L2, although nuclear, was more diffuse. These data demonstrate the dependence of efficient early PV transcription on either intact ND10 or expression of the PML protein.

Several other DNA viruses also initiate their synthetic processes in the vicinity of ND10 (reviewed in Everett, 2001; Maul, 1998), but a distinct role for ND10 in their transcription has not been established. For example, although parental HCMV and SV40 genomes are found in association with ND10 (Ahn et al., 1999; Ishov et al., 2002; Tang et al., 2000), the levels of viral RNA were unaltered in cells lacking PML and ND10 (Marshall et al., 2002; Tang et al., 2000). This is in contrast to our findings for PV.

The positive function we have identified for PML or ND10 in PV infection also contrasts with the postulated negative role for ND10 in many other viral infections. Most DNA viruses have evolved a means to disrupt ND10 soon after the immediate early viral transcription has begun. PVs have not been demonstrated to disrupt ND10, but L2 has been shown to more modestly affect the structure by increasing the localization of Daxx to this domain and decreasing the levels of Sp100 found at ND10 (Florin et al., 2002). The functional consequences of these changes in ND10 composition for PV infection have not been determined.

In the nucleus, DNA fragments are immobile (Lukacs et al., 2000), and L2 is not known to possess any transcriptional activity, suggesting that the main function of L2 in this process is to localize the DNA to a site that favors transcription. Therefore, the findings outlined above suggest a model involving L2-dependent chaperoning of the viral genome through the cytoplasm, with the subsequent deposition of both the L2 protein and the DNA at ND10, which promotes efficient viral transcription (Figure 3, Table 1).

10. Future Directions

Although the last decade has witnessed substantial progress in the understanding of the steps in PV infection, many aspects of the process remain unresolved. Is there a critical secondary L1-specific cell surface receptor? What is the identity of the putative L2-specific receptor(s) and does it have an obligatory role in infection? Answers to these questions may require the identification of epithelial cell lines that bind capsids normally but are resistant to infection. They may also require the generation of subtle L1 and L2 mutants that are defective for interaction with the specific receptors identified. Additionally the application of a whole genome siRNA library may yield important information as to what cellular proteins are essential for PV infection.

What types of signals are transmitted by cell-surface binding of the virion? Do they occur under biologically relevant virion concentrations and are they important for infection? Are the innate immune responses induced by capsid binding of immunocytes mediated by the same or different cell-surface receptors as those on epithelial cells? Are any of the immunocytes that avidly bind PV capsids readily infected by them? If so, does this infection play a role in the immune response to the virus?

What accounts for the remarkably slow infection kinetics of PVs, especially the slow rate of internalization and vesicular trafficking, despite apparently being mediated by clathrin-dependent endocytosis? The finding that several steps in the infectious process are notably delayed for PV raises the possibility that the viruses have specifically evolved to delay the onset of viral gene expression. Is this a mechanism to inhibit the initial immune surveillance of virus infection? This is an interesting possibility, since the sites of wounding or trauma that are thought to be the primary sites of most PV infections are also sites to which immunocytes would be attracted. To address these, as well as other, questions, it will important to extend infectivity studies to normal keratinocytes. It remains a formal possibility that the slow kinetics of infection observed in studies to date is due to the use of cells that are not biologically relevant to PV infection. Alternatively, infection of keratinocytes *in vivo* may be even slower, if the initial escape from immune surveillance hypothesis is correct. In this regard, it would be interesting to examine the kinetics of keratinocyte infection during wound healing in animal models. This analysis should be feasible using pseudovirions that have encapsidated quantifiable marker genes such as GFP or beta-galactosidase.

What conformational changes occur in the capsid during uncoating and escape from the endosome? What triggers these changes? Finally, how and in what form does the viral genome complex traverse the cytoplasm and enter the nucleus? Does the complex contain cellular factors, in addition to L2? What happens to L1 during and after uncoating? Is it simply degraded or does it accompany the L2-genome complex in a form that is not recognized by the existing battery of L1-specific antibodies?

The studies to answer many of these questions will be technically challenging. However, the impressive technological advances in PV virion analysis achieved over the last decade, in addition to improvements in the general methodologies for studying viral infections, provide reasons to be optimistic about the further advancement of this field.

References

Ahn, J. H., Jang, W. J., and Hayward, G. S. (1999). The human cytomegalovirus IE2 and UL112-113 proteins accumulate in viral DNA replication compartments that initiate from the periphery of promyelocytic leukemia protein-associated nuclear bodies (PODs or ND10). J. Virol. *73*, 10458-10471.

Arias, C. F., Isa, P., Guerrero, C. A., Mendez, E., Zarate, S., Lopez, T., Espinosa, R., Romero, P., and Lopez, S. (2002). Molecular biology of rotavirus cell entry. Arch. Med. Res. *33*, 356-361.

Barzu, T., Pascal, M., Maman, M., Roque, C., Lafont, F., and Rousselet, A. (1996). Entry and distribution of fluorescent antiproliferative heparin derivatives into rat vascular smooth muscle cells: comparison between heparin-sensitive and heparin-resistant cultures. J. Cell. Physiol. *167*, 8-21.

Bayer, N., Schober, D., Prchla, E., Murphy, R. F., Blaas, D., and Fuchs, R. (1998). Effect of bafilomycin A1 and nocodazole on endocytic transport in HeLa cells: implications for viral uncoating and infection. J. Virol. *72*, 9645-9655.

Benmerah, A. (2004). Endocytosis: signaling from endocytic membranes to the nucleus. Curr. Biol. *14*, R314-316.

Bernfield, M., Gotte, M., Park, P. W., Reizes, O., Fitzgerald, M. L., Lincecum, J., and Zako, M. (1999). Functions of cell surface heparan sulfate proteoglycans. Annu. Rev. Biochem. *68*, 729-777.

Bernfield, M., Kokenyesi, R., Kato, M., Hinkes, M. T., Spring, J., Gallo, R. L., and Lose, E. J. (1992). Biology of the syndecans: a family of transmembrane heparan sulfate proteoglycans. Annu. Rev. Cell Biol. *8*, 365-393.

Bong, D. T., Steinem, C., Janshoff, A., Johnson, J. E., and Reza Ghadiri, M. (1999). A highly membrane-active peptide in Flock House virus: implications for the mechanism of nodavirus infection. Chem. Biol. *6*, 473-481.

Bousarghin, L., Combita-Rojas, A. L., Touze, A., El Mehdaoui, S., Sizaret, P. Y., Bravo, M. M., and Coursaget, P. (2002). Detection of neutralizing antibodies against human papillomaviruses (HPV) by inhibition of gene transfer mediated by HPV pseudovirions. J. Clin. Microbiol. *40*, 926-932.

Bousarghin, L., Touze, A., Sizaret, P. Y., and Coursaget, P. (2003). Human papillomavirus types 16, 31, and 58 use different endocytosis pathways to enter cells. J. Virol. *77*, 3846-3850.

Buck, C. B., Pastrana, D. V., Lowy, D. R., and Schiller, J. T. (2004). Efficient intracellular assembly of papillomaviral vectors. J. Virol. *78*, 751-757.

Buck, C. B., Thompson, C.D., Pang, Y-Y. S., Lowy, D.R., Schiller, J.T. (2005). Maturation of papillomavirus capsids. J. Virol. *79*, 2839-2846.

Buss, F., Arden, S. D., Lindsay, M., Luzio, J. P., and Kendrick-Jones, J. (2001). Myosin VI isoform localized to clathrin-coated vesicles with a role in clathrin-mediated endocytosis. EMBO J. *20*, 3676-3684.

Campadelli-Fiume, G., Cocchi, F., Menotti, L., and Lopez, M. (2000). The novel receptors that mediate the entry of herpes simplex viruses and animal alphaherpesviruses into cells. Rev. Med. Virol. *10*, 305-319.

Campo, M. S., Jarrett, W. F., O'Neil, W., and Barron, R. J. (1994). Latent papillomavirus infection in cattle. Res. Vet. Sci. *56*, 151-157.

Caruso, M., Cavaldesi, M., Gentile, M., Sthandier, O., Amati, P., and Garcia, M. I. (2003). Role of sialic acid-containing molecules and the alpha4beta1 integrin receptor in the early steps of polyomavirus infection. J. Gen. Virol. *84*, 2927-2936.

Chandran, K., Farsetta, D. L., and Nibert, M. L. (2002). Strategy for nonenveloped virus entry: a hydrophobic conformer of the reovirus membrane penetration protein micro 1 mediates membrane disruption. J. Virol. *76*, 9920-9933.

Christensen, N. D., Cladel, N. M., and Reed, C. A. (1995). Postattachment neutralization of papillomaviruses by monoclonal and polyclonal antibodies. Virology *207*, 136-142.

Christensen, N. D., Hopfl, R., DiAngelo, S. L., Cladel, N. M., Patrick, S. D., Welsh, P. A., Budgeon, L. R., Reed, C. A., and Kreider, J. W. (1994). Assembled baculovirus-expressed human papillomavirus type 11 L1 capsid protein virus-like particles are recognized by neutralizing monoclonal antibodies and induce high titres of neutralizing antibodies. J. Gen. Virol. *75 (Pt 9)*, 2271-2276.

Chroboczek, J., Ruigrok, R. W., and CUSA ck, S. (1995). Adenovirus fiber. Curr. Top. Microbiol. Immunol. *199 (Pt 1)*, 163-200.

Chua, C. C., Rahimi, N., Forsten-Williams, K., and Nugent, M. A. (2004). Heparan sulfate proteoglycans function as receptors for fibroblast growth factor-2 activation of extracellular signal-regulated kinases 1 and 2. Circ. Res. *94*, 316-323.

Clague, M. J., Urbe, S., Aniento, F., and Gruenberg, J. (1994). Vacuolar ATPase activity is required for endosomal carrier vesicle formation. J. Biol. Chem. *269*, 21-24.

Clever, J., Yamada, M., and Kasamatsu, H. (1991). Import of simian virus 40 virions through nuclear pore complexes. Proc. Natl. Acad. Sci. USA *88*, 7333-7337.

Colman, P. M., and Lawrence, M. C. (2003). The structural biology of type I viral membrane fusion. Nat. Rev. Mol. Cell. Biol. *4*, 309-319.

Couchman, J. R. (2003). Syndecans: proteoglycan regulators of cell-surface microdomains? Nat. Rev. Mol. Cell. Biol. *4*, 926-937.

Cramer, L. P. (2000). Myosin VI: roles for a minus end-directed actin motor in cells. J. Cell Biol. *150*, F121-126.

Culp, T. D., and Christensen, N. D. (2004). Kinetics of *in vitro* adsorption and entry of papillomavirus virions. Virology *319*, 152-161.

Da Silva, D. M., Velders, M. P., Nieland, J. D., Schiller, J. T., Nickoloff, B. J., and Kast, W. M. (2001). Physical interaction of human papillomavirus virus-like particles with immune cells. Int. Immunol. *13*, 633-641.

Danthi, P., Tosteson, M., Li, Q. H., and Chow, M. (2003). Genome delivery and ion channel properties are altered in VP4 mutants of poliovirus. J. Virol. *77*, 5266-5274.

Day, P. M., Baker, C. C., Lowy, D. R., and Schiller, J. T. (2004). Establishment of papillomavirus infection is enhanced by promyelocytic leukemia protein (PML) expression. Proc. Natl. Acad. Sci. USA *101*, 14252-14257.

Day, P. M., Lowy, D. R., and Schiller, J. T. (2003). Papillomaviruses infect cells via a clathrin-dependent pathway. Virology *307*, 1-11.

Day, P. M., Roden, R. B., Lowy, D. R., and Schiller, J. T. (1998). The papillomavirus minor capsid protein, L2, induces localization of the major capsid protein, L1, and the viral transcription/replication protein, E2, to PML oncogenic domains. J. Virol. *72*, 142-150.

De Sena, J., and Mandel, B. (1977). Studies on the *in vitro* uncoating of poliovirus. II. Characteristics of the membrane-modified particle. Virology *78*, 554-566.

Dohner, K., Wolfstein, A., Prank, U., Echeverri, C., Dujardin, D., Vallee, R., and Sodeik, B. (2002). Function of dynein and dynactin in herpes simplex virus capsid transport. Mol. Biol. Cell *13*, 2795-2809.

Doms, R. W., Helenius, A., and White, J. (1985). Membrane fusion activity of the influenza virus hemagglutinin. The low pH-induced conformational change. J. Biol. Chem. *260*, 2973-2981.

Douglas, M. W., Diefenbach, R. J., Homa, F. L., Miranda-Saksena, M., Rixon, F. J., Vittone, V., Byth, K., and Cunningham, A. L. (2004). Herpes simplex virus type 1 capsid protein VP26 interacts with dynein light chains RP3 and Tctex1 and plays a role in retrograde cellular transport. J. Biol. Chem. *279*, 28522-28530.

Ebert, D. H., Deussing, J., Peters, C., and Dermody, T. S. (2002). Cathepsin L and cathepsin B mediate reovirus disassembly in murine fibroblast cells. J. Biol. Chem. *277*, 24609-24617.

Elenius, K., Vainio, S., Laato, M., Salmivirta, M., Thesleff, I., and Jalkanen, M. (1991). Induced expression of syndecan in healing wounds. J. Cell Biol. *114*, 585-595.

Engqvist-Goldstein, A. E., and Drubin, D. G. (2003). Actin assembly and endocytosis: from yeast to mammals. Annu. Rev. Cell Dev. Biol. *19*, 287-332.

Evander, M., Frazer, I. H., Payne, E., Qi, Y. M., Hengst, K., and McMillan, N. A. (1997). Identification of the alpha6 integrin as a candidate receptor for papillomaviruses. J. Virol. *71*, 2449-2456.

Everett, R. (2001). DNA viruses and viral proteins that interact with PML nuclear bodies. Oncogene *20*, 7266-7273.

Fausch, S. C., Da Silva, D. M., and Kast, W. M. (2003). Differential uptake and cross-presentation of human papillomavirus virus-like particles by dendritic cells and Langerhans cells. Cancer Res. *63*, 3478-3482.

Fernandes, J., Tang, D., Leone, G., and Lee, P. W. (1994). Binding of reovirus to receptor leads to conformational changes in viral capsid proteins that are reversible upon virus detachment. J. Biol. Chem. *269*, 17043-17047.

Fligge, C., Schafer, F., Selinka, H. C., Sapp, C., and Sapp, M. (2001). DNA-induced structural changes in the papillomavirus capsid. J. Virol. *75*, 7727-7731.

Florin, L., Schafer, F., Sotlar, K., Streeck, R. E., and Sapp, M. (2002). Reorganization of nuclear domain 10 induced by papillomavirus capsid protein l2. Virology *295*, 97-107.

Fransson, L. A. (2003). Glypicans. Int. J. Biochem. Cell Biol. *35*, 125-129.

Fried, H., Cahan, L. D., and Paulson, J. C. (1981). Polyoma virus recognizes specific sialyligosaccharide receptors on host cells. Virology *109*, 188-192.

Fuchs, E., and Yang, Y. (1999). Crossroads on cytoskeletal highways. Cell *98*, 547-550.

Fujimoto, L. M., Roth, R., Heuser, J. E., and Schmid, S. L. (2000). Actin assembly plays a variable, but not obligatory role in receptor-mediated endocytosis in mammalian cells. Traffic *1*, 161-171.

Funk, A., Mhamdi, M., Lin, L., Will, H., and Sirma, H. (2004). Itinerary of hepatitis B viruses: delineation of restriction points critical for infectious entry. J. Virol. *78*, 8289-8300.

Gallo, R., and Trowbridge, J. M. (2003). Proteoglycans and Glycosaminoglycans of Skin. In Dermatology in General Medicine, E. Freedberg I.M., A.Z., Wolff, K., Austen, K.F., Goldsmith, L.A., Katz, S.I., ed. (New York, McGraw-Hill), pp. 210-216.

Geraghty, R. J., Krummenacher, C., Cohen, G. H., Eisenberg, R. J., and Spear, P. G. (1998). Entry of alphaherpesviruses mediated by poliovirus receptor-related protein 1 and poliovirus receptor. Science *280*, 1618-1620.

Gething, M. J., Doms, R. W., York, D., and White, J. (1986). Studies on the mechanism of membrane fusion: site-specific mutagenesis of the hemagglutinin of influenza virus. J. Cell Biol. *102*, 11-23.

Giroglou, T., Florin, L., Schafer, F., Streeck, R. E., and Sapp, M. (2001). Human papillomavirus infection requires cell surface heparan sulfate. J. Virol. *75*, 1565-1570.

Gollins, S. W., and Porterfield, J. S. (1986a). A new mechanism for the neutralization of enveloped viruses by antiviral antibody. Nature *321*, 244-246.

Gollins, S. W., and Porterfield, J. S. (1986b). The uncoating and infectivity of the flavivirus West Nile on interaction with cells: effects of pH and ammonium chloride. J. Gen. Virol. *67 (Pt 9)*, 1941-1950.

Greber, U., Suomalainen, M., Stidwell, R. B., K, Ebersold, M., and Helenius, A. (1997). The role of the nuclear pore complex in adenovirus DNA entry. EMBO J. *16*, 5998-6007.

Greber, U. F. (2002). Signalling in viral entry. Cell Mol. Life Sci. *59*, 608-626.

Greber, U. F., Willetts, M., Webster, P., and Helenius, A. (1993). Stepwise dismantling of adenovirus 2 during entry into cells. Cell *75*, 477-486.

Grimmer, S., van Deurs, B., and Sandvig, K. (2002). Membrane ruffling and macropinocytosis in A431 cells require cholesterol. J. Cell Sci. *115*, 2953-2962.

Gruenberg, J. (1991). *In vitro* studies of endocytic membrane traffic. Infection *19 Suppl 4*, S210-215.

Gruenheid, S., Gatzke, L., Meadows, H., and Tufaro, F. (1993). Herpes simplex virus infection and propagation in a mouse L cell mutant lacking heparan sulfate proteoglycans. J. Virol. *67*, 93-100.

Guerrero, C. A., Mendez, E., Zarate, S., Isa, P., Lopez, S., and Arias, C. F. (2000). Integrin alpha(v)beta(3) mediates rotavirus cell entry. Proc. Natl. Acad. Sci. USA *97*, 14644-14649.

Hanover, J. A., Beguinot, L., Willingham, M. C., and Pastan, I. H. (1985). Transit of receptors for epidermal growth factor and transferrin through clathrin-coated pits. Analysis of the kinetics of receptor entry. J. Biol. Chem. *260*, 15938-15945.

Hanover, J. A., Willingham, M. C., and Pastan, I. (1984). Kinetics of transit of transferrin and epidermal growth factor through clathrin-coated membranes. Cell *39*, 283-293.

Heinz, F. X., and Allison, S. L. (2001). The machinery for flavivirus fusion with host cell membranes. Curr. Opin. Microbiol. *4*, 450-455.

Hofer, F., Gruenberger, M., Kowalski, H., Machat, H., Huettinger, M., Kuechler, E., and Blass, D. (1994). Members of the low density lipoprotein receptor family mediate cell entry of a minor-group common cold virus. Proc. Natl. Acad. Sci. USA *91*, 1839-1842.

Horowitz, A., and Simons, M. (1998). Phosphorylation of the cytoplasmic tail of syndecan-4 regulates activation of protein kinase Calpha. J. Biol. Chem. *273*, 25548-25551.

Horowitz, A., Tkachenko, E., and Simons, M. (2002). Fibroblast growth factor-specific modulation of cellular response by syndecan-4. J. Cell Biol. *157*, 715-725.

Ishov, A. M., Sotnikov, A. G., Negorev, D., Vladimirova, O. V., Neff, N., Kamitani, T., Yeh, E. T., Strauss, J. F., 3rd, and Maul, G. G. (1999). PML is critical for ND10 formation and recruits the PML-interacting protein daxx to this nuclear structure when modified by SUMO-1. J. Cell Biol. *147*, 221-234.

Ishov, A. M., Vladimirova, O. V., and Maul, G. G. (2002). Daxx-mediated accumulation of human cytomegalovirus tegument protein pp71 at ND10 facilitates initiation of viral infection at these nuclear domains. J. Virol. *76*, 7705-7712.

Joklik, W. K., and Darnell, J. E., Jr. (1961). The adsorption and early fate of purified poliovirus in HeLa cells. Virology *13*, 439-447.

Joyce, J. G., Tung, J. S., Przysiecki, C. T., Cook, J. C., Lehman, E. D., Sands, J. A., Jansen, K. U., and Keller, P. M. (1999). The L1 major capsid protein of human papillomavirus type 11 recombinant virus-like particles interacts with heparin and cell-surface glycosaminoglycans on human keratinocytes. J. Biol. Chem. *274*, 5810-5822.

Kawana, K., Yoshikawa, H., Taketani, Y., Yoshiike, K., and Kanda, T. (1999). Common neutralization epitope in minor capsid protein L2 of human papillomavirus types 16 and 6. J. Virol. *73*, 6188-6190.

Kawana, Y., Kawana, K., Yoshikawa, H., Taketani, Y., Yoshiike, K., and Kanda, T. (2001). Human papillomavirus type 16 minor capsid protein l2 N-terminal region containing a common neutralization epitope binds to the cell surface and enters the cytoplasm. J. Virol. *75*, 2331-2336.

Kirnbauer, R., Taub, J., Greenstone, H., Roden, R., Durst, M., Gissmann, L., Lowy, D. R., and Schiller, J. T. (1993). Efficient self-assembly of human papillomavirus type 16 L1 and L1-L2 into virus-like particles. J. Virol. *67*, 6929-6936.

Kolatkar, P. R., Bella, J., Olson, N. H., Bator, C. M., Baker, T. S., and Rossmann, M. G. (1999). Structural studies of two rhinovirus serotypes complexed with fragments of their cellular receptor. EMBO J. *18*, 6249-6259.

Lanier, L. M., Slack, J. M., and Volkman, L. E. (1996). Actin binding and proteolysis by the baculovirus AcMNPV: the role of virion-associated V-CATH. Virology *216*, 380-388.

Lanier, L. M., and Volkman, L. E. (1998). Actin binding and nucleation by Autographa californica M nucleopolyhedrovirus. Virology *243*, 167-177.

Leder, C., Kleinschmidt, J. A., Wiethe, C., and Muller, M. (2001). Enhancement of capsid gene expression: preparing the human papillomavirus type 16 major structural gene L1 for DNA vaccination purposes. J. Virol. *75*, 9201-9209.

Lenz, P., Day, P. M., Pang, Y. Y., Frye, S. A., Jensen, P. N., Lowy, D. R., and Schiller, J. T. (2001). Papillomavirus-like particles induce acute activation of dendritic cells. J. Immunol. *166*, 5346-5355.

Leopold, P. L., Kreitzer, G., Miyazawa, N., Rempel, S., Pfister, K. K., Rodriguez-Boulan, E., and Crystal, R. G. (2000). Dynein- and microtubule-mediated translocation of adenovirus serotype 5 occurs after endosomal lysis. Hum. Gene Ther. *11*, 151-165.

Letourneur, D., Caleb, B. L., and Castellot, J. J., Jr. (1995). Heparin binding, internalization, and metabolism in vascular smooth muscle cells: II. Degradation and secretion in sensitive and resistant cells. J. Cell. Physiol. *165*, 687-695.

Lim, S. T., Longley, R. L., Couchman, J. R., and Woods, A. (2003). Direct binding of syndecan-4 cytoplasmic domain to the catalytic domain of protein kinase C alpha (PKC alpha) increases focal adhesion localization of PKC alpha. J. Biol. Chem. *278*, 13795-13802.

Liu, W. J., Gissmann, L., Sun, X. Y., Kanjanahaluethai, A., Muller, M., Doorbar, J., and Zhou, J. (1997). Sequence close to the N-terminus of L2 protein is displayed on the surface of bovine papillomavirus type 1 virions. Virology *227*, 474-483.

Liu, W. J., Qi, Y. M., Zhao, K. N., Liu, Y. H., Liu, X. S., and Frazer, I. H. (2001). Association of bovine papillomavirus type 1 with microtubules. Virology *282*, 237-244.

Lopez-Casillas, F., Wrana, J. L., and Massague, J. (1993). Betaglycan presents ligand to the TGF beta signaling receptor. Cell *73*, 1435-1444.

Lowy, D. R., and Howley, P. M. (2001). Papillomaviruses. In Fields Virology, D. M. Knipe, and P. M. Howley, eds. (Lippincott Williams and Wilkins), pp. 2231-2264.

Luby-Phelps, K., Castle, P. E., Taylor, D. L., and Lanni, F. (1987). Hindered diffusion of inert tracer particles in the cytoplasm of mouse 3T3 cells. Proc. Natl. Acad. Sci. USA *84*, 4910-4913.

Lukacs, G. L., Haggie, P., Seksek, O., Lechardeur, D., Freedman, N., and Verkman, A. S. (2000). Size-dependent DNA mobility in cytoplasm and nucleus. J. Biol. Chem. *275*, 1625-1629.

Marone, G., Fimiani, B., Torella, G., Poto, S., Bianco, P., and Condorelli, M. (1983). Possible role of arachidonic acid and of phospholipase A2 in the control of lysosomal enzyme release from human polymorphonuclear leukocytes. J. Clin. Lab. Immunol. *12*, 111-116.

Marshall, K. R., Rowley, K. V., Rinaldi, A., Nicholson, I. P., Ishov, A. M., Maul, G. G., and Preston, C. M. (2002). Activity and intracellular localization of the human cytomegalovirus protein pp71. J. Gen. Virol. *83*, 1601-1612.

Maul, G. (1998). Nuclear domain 10, the site of DNA virus transcription and replication. Bioessays *20*, 660-667.

Medina-Kauwe, L. K. (2003). Endocytosis of adenovirus and adenovirus capsid proteins. Adv. Drug Deliv. Rev. *55*, 1485-1496.

Meier, O., and Greber, U. F. (2003). Adenovirus endocytosis. J. Gene Med. *5*, 451-462.

Merle, E., Rose, R., LeRoux, L., and Moroianu, J. (1999). Nuclear import of HPV11 L1 capsid protein is mediated by karyopherin alpha2beta1 heterodimers. J. Cell. Biochem. *74*, 628-637.

Merrifield, C. J., Feldman, M. E., Wan, L., and Almers, W. (2002). Imaging actin and dynamin recruitment during invagination of single clathrin-coated pits. Nat. Cell Biol. *4*, 691-698.

Meyers, C., Frattini, M. G., Hudson, J. B., and Laimins, L. A. (1992). Biosynthesis of human papillomavirus from a continuous cell line upon epithelial differentiation. Science *257*, 971-973.

Meyers, C., Mayer, T. J., and Ozbun, M. A. (1997). Synthesis of infectious human papillomavirus type 18 in differentiating epithelium transfected with viral DNA. J. Virol. *71*, 7381-7386.

Monath, T. P., Arroyo, J., Levenbook, I., Zhang, Z. X., Catalan, J., Draper, K., and Guirakhoo, F. (2002). Single mutation in the flavivirus envelope protein hinge region increases neurovirulence for mice and monkeys but decreases viscerotropism for monkeys: relevance to development and safety testing of live, attenuated vaccines. J. Virol. *76*, 1932-1943.

Mossadegh, N., Gissmann, L., Muller, M., Zentgraf, H., Alonso, A., and Tomakidi, P. (2004). Codon optimization of the human papillomavirus 11 (HPV 11) L1 gene leads to increased gene expression and formation of virus-like particles in mammalian epithelial cells. Virology *326*, 57-66.

Muller, M., Gissmann, L., Cristiano, R. J., Sun, X. Y., Frazer, I. H., Jenson, A. B., Alonso, A., Zentgraf, H., and Zhou, J. (1995). Papillomavirus capsid binding and uptake by cells from different tissues and species. J. Virol. *69*, 948-954.

Nelson, L., Rose, R., LeRoux, L., Lane, C., Bruya, K., and Moroianu, J. (2000). Nuclear import and DNA binding of human papillomavirus type 45 L1 capsid protein. J. Cell. Biochem. *79*, 225-238.

Nelson, L., Rose, R., and Moroianu, J. (2002). Nuclear import strategies of high risk HPV16 L1 major capsid protein. J. Biol. Chem. *277*, 23958-23964.

Oh, E. S., Woods, A., and Couchman, J. R. (1997). Syndecan-4 proteoglycan regulates the distribution and activity of protein kinase C. J. Biol. Chem. *272*, 8133-8136.

Ozbun, M. A. (2002). Human papillomavirus type 31b infection of human keratinocytes and the onset of early transcription. J. Virol. *76*, 11291-11300.

Parker, J. S., and Parrish, C. R. (2000). Cellular uptake and infection by canine parvovirus involves rapid dynamin-regulated clathrin-mediated endocytosis, followed by slower intracellular trafficking. J. Virol. *74*, 1919-1930.

Payne, E., Bowles, M. R., Don, A., Hancock, J. F., and McMillan, N. A. (2001). Human papillomavirus type 6b virus-like particles are able to activate the Ras-MAP kinase pathway and induce cell proliferation. J. Virol. *75*, 4150-4157.

Pelkmans, L., and Helenius, A. (2002). Endocytosis via caveolae. Traffic *3*, 311-320.

Pelkmans, L., and Helenius, A. (2003). Insider information: what viruses tell us about endocytosis. Curr. Opin. Cell Biol. *15*, 414-422.

Petit, C., Giron, M. L., Tobaly-Tapiero, J., Bittoun, P., Real, E., Jacob, Y., Tordo, N., De The, H., and Saib, A. (2003). Targeting of incoming retroviral Gag to the centrosome involves a direct interaction with the dynein light chain 8. J. Cell Sci. *116*, 3433-3442.

Pho, M. T., Ashok, A., and Atwood, W. J. (2000). JC virus enters human glial cells by clathrin-dependent receptor-mediated endocytosis. J. Virol. *74*, 2288-2292.

Prchla, E., Kuechler, E., Blaas, D., and Fuchs, R. (1994). Uncoating of human rhinovirus serotype 2 from late endosomes. J. Virol. *68*, 3713-3723.

Qi, Y. M., Peng, S. W., Hengst, K., Evander, M., Park, D. S., Zhou, J., and Frazer, I. H. (1996). Epithelial cells display separate receptors for papillomavirus VLPs and for soluble L1 capsid protein. Virology *216*, 35-45.

Rabe, B., Vlachou, A., Pante, N., Helenius, A., and M, K. (2003). Nuclear import of hepatitis B virus capsids and release of the viral genome. Proc. Natl. Acad. Sci. USA *100*, 9849-9854.

Roden, R. B., Greenstone, H. L., Kirnbauer, R., Booy, F. P., Jessie, J., Lowy, D. R., and Schiller, J. T. (1996). *In vitro* generation and type-specific neutralization of a human papillomavirus type 16 virion pseudotype. J. Virol. *70*, 5875-5883.

Roden, R. B., Kirnbauer, R., Jenson, A. B., Lowy, D. R., and Schiller, J. T. (1994). Interaction of papillomaviruses with the cell surface. J. Virol. *68*, 7260-7266.

Roden, R. B., Yutzy, W. H. t., Fallon, R., Inglis, S., Lowy, D. R., and Schiller, J. T. (2000). Minor capsid protein of human genital papillomaviruses contains subdominant, cross-neutralizing epitopes. Virology *270*, 254-257.

Rojas, R., and Apodaca, G. (2002). Immunoglobulin transport across polarized epithelial cells. Nat. Rev. Mol. Cell Biol. *3*, 944-955.

Rose, R. C., Bonnez, W., Reichman, R. C., and Garcea, R. L. (1993). Expression of human papillomavirus type 11 L1 protein in insect cells: *in vivo* and *in vitro* assembly of viruslike particles. J. Virol. *67*, 1936-1944.

Rose, R. C., Reichman, R. C., and Bonnez, W. (1994). Human papillomavirus (HPV) type 11 recombinant virus-like particles induce the formation of neutralizing antibodies and detect HPV-specific antibodies in human sera. J. Gen. Virol. *75 (Pt 8)*, 2075-2079.

Sanjuan, N., Porras, A., and Otero, J. (2003). Microtubule-dependent intracellular transport of murine polyomavirus. Virology *313*, 105-116.

Sawitzky, D. (1996). Protein-glycosaminoglycan interactions: infectiological aspects. Med. Microbiol. Immunol. (Berl) *184*, 155-161.

Schmid, S. L., R. Fuchs, P. Male, I. Mellman (1988). Two distinct subpopulations of endosomes involved in membrane recycling and transport to lysosomes. Cell *52*, 73-83.

Seksek, O., Biwersi, J., and Verkman, A. S. (1997). Translational diffusion of macromolecule-sized solutes in cytoplasm and nucleus. J. Cell Biol. *138*, 131-142.

Selinka, H. C., Giroglou, T., Nowak, T., Christensen, N. D., and Sapp, M. (2003). Further evidence that papillomavirus capsids exist in two distinct conformations. J. Virol. *77*, 12961-12967.

Selinka, H. C., Giroglou, T., and Sapp, M. (2002). Analysis of the infectious entry pathway of human papillomavirus type 33 pseudovirions. Virology *299*, 279-287.

Seth, P., Fitzgerald, D., Ginsberg, H., Willingham, M., and Pastan, I. (1984). Evidence that the penton base of adenovirus is involved in potentiation of toxicity of Pseudomonas exotoxin conjugated to epidermal growth factor. Mol. Cell. Biol. *4*, 1528-1533.

Shafti-Keramat, S., Handisurya, A., Kriehuber, E., Meneguzzi, G., Slupetzky, K., and Kirnbauer, R. (2003). Different heparan sulfate proteoglycans serve as cellular receptors for human papillomaviruses. J. Virol. *77*, 13125-13135.

Shieh, M. T., WuDunn, D., Montgomery, R. I., Esko, J. D., and Spear, P. G. (1992). Cell surface receptors for herpes simplex virus are heparan sulfate proteoglycans. J. Cell Biol. *116*, 1273-1281.

Shukla, D., Liu, J., Blaiklock, P., Shworak, N. W., Bai, X., Esko, J. D., Cohen, G. H., Eisenberg, R. J., Rosenberg, R. D., and Spear, P. G. (1999). A novel role for 3-O-sulfated heparan sulfate in herpes simplex virus 1 entry. Cell *99*, 13-22.

Sibbet, G., Romero-Graillet, C., Meneguzzi, G., and Campo, M. S. (2000). alpha6 integrin is not the obligatory cell receptor for bovine papillomavirus type 4. J. Gen. Virol. *81 Pt 2*, 327-334.

Sodeik, B. (2000). Mechanisms of viral transport in the cytoplasm. Trends Microbiol. *8*, 465-472.

Sodeik, B., Ebersold, M. W., and Helenius, A. (1997). Microtubule-mediated transport of incoming herpes simplex virus 1 capsids to the nucleus. J. Cell Biol. *136*, 1007-1021.

Spear, P. G., Eisenberg, R. J., and Cohen, G. H. (2000). Three classes of cell surface receptors for alphaherpesvirus entry. Virology *275*, 1-8.

Spillmann, D. (2001). Heparan sulfate: anchor for viral intruders? Biochimie *83*, 811-817.

Stehle, T., and Dermody, T. S. (2004). Structural similarities in the cellular receptors used by adenovirus and reovirus. Viral Immunol. *17*, 129-143.

Stepp, M. A., Gibson, H. E., Gala, P. H., Iglesia, D. D., Pajoohesh-Ganji, A., Pal-Ghosh, S., Brown, M., Aquino, C., Schwartz, A. M., Goldberger, O., et al. (2002). Defects in keratinocyte activation during wound healing in the syndecan-1-deficient mouse. J. Cell Sci. *115*, 4517-4531.

Suikkanen, S., Antila, M., Jaatinen, A., Vihinen-Ranta, M., and Vuento, M. (2003). Release of canine parvovirus from endocytic vesicles. Virology *316*, 267-280.

Suomalainen, M., Nakano, M. Y., Boucke, K., Keller, S., and Greber, U. F. (2001). Adenovirus-activated PKA and p38/MAPK pathways boost microtubule-mediated nuclear targeting of virus. EMBO J. *20*, 1310-1319.

Suomalainen, M., Nakano, M. Y., Keller, S., Boucke, K., Stidwill, R. P., and Greber, U. F. (1999). Microtubule-dependent plus- and minus end-directed motilities are competing processes for nuclear targeting of adenovirus. J. Cell Biol. *144*, 657-672.

Svensson, U. (1985). Role of vesicles during adenovirus 2 internalization into HeLa cells. J. Virol. *55*, 442-449.

Tang, Q., Bell, P., Tegtmeyer, P., and Maul, G. G. (2000). Replication but not transcription of simian virus 40 DNA is dependent on nuclear domain 10. J. Virol. *74*, 9694-9700.

Tkachenko, E., Lutgens, E., Stan, R. V., and Simons, M. (2004). Fibroblast growth factor 2 endocytosis in endothelial cells proceed via syndecan-4-dependent activation of Rac1 and a Cdc42-dependent macropinocytic pathway. J. Cell Sci. *117*, 3189-3199.

Tosteson, M. T., Nibert, M. L., and Fields, B. N. (1993). Ion channels induced in lipid bilayers by subvirion particles of the nonenveloped mammalian reoviruses. Proc. Natl. Acad. Sci. USA *90*, 10549-10552.

Tosteson, M. T., Wang, H., Naumov, A., and Chow, M. (2004). Poliovirus binding to its receptor in lipid bilayers results in particle-specific, temperature-sensitive channels. J. Gen. Virol. *85*, 1581-1589.

Touze, A., and Coursaget, P. (1998). *In vitro* gene transfer using human papillomavirus-like particles. Nucleic Acids Res. *26*, 1317-1323.

Tumova, S., Woods, A., and Couchman, J. R. (2000). Heparan sulfate proteoglycans on the cell surface: versatile coordinators of cellular functions. Int. J. Biochem. Cell Biol. *32*, 269-288.

Unckell, F., Streeck, R. E., and Sapp, M. (1997). Generation and neutralization of pseudovirions of human papillomavirus type 33. J. Virol. *71*, 2934-2939.

Volk, R., Schwartz, J. J., Li, J., Rosenberg, R. D., and Simons, M. (1999). The role of syndecan cytoplasmic domain in basic fibroblast growth factor-dependent signal transduction. J. Biol. Chem. *274*, 24417-24424.

Volpers, C., Unckell, F., Schirmacher, P., Streeck, R. E., and Sapp, M. (1995). Binding and internalization of human papillomavirus type 33 virus-like particles by eukaryotic cells. J. Virol. *69*, 3258-3264.

Wang, K., Huang, S., Kapoor-Munshi, A., and Nemerow, G. (1998). Adenovirus internalization and infection require dynamin. J. Virol. *72*, 3455-3458.

Wells, A. L., Lin, A. W., Chen, L. Q., Safer, D., Cain, S. M., Hasson, T., Carragher, B. O., Milligan, R. A., and Sweeney, H. L. (1999). Myosin VI is an actin-based motor that moves backwards. Nature *401*, 505-508.

Whalen, S. G., Marcellus, R. C., Whalen, A., Ahn, N. G., Ricciardi, R. P., and Branton, P. E. (1997). Phosphorylation within the transactivation domain of adenovirus E1A protein by mitogen-activated protein kinase regulates expression of early region 4. J. Virol. *71*, 3545-3553.

White, J., Kartenbeck, J., and Helenius, A. (1982). Membrane fusion activity of influenza virus. EMBO J. *1*, 217-222.

Whittaker, G., Kann, M., and Helenius, A. (2000). Viral entry into the nucleus. Annu. Rev. Cell Dev. Biol. *16*, 627-651.

Wickham, T. J., Mathias, P., Cheresh, D. A., and Nemerow, G. R. (1993). Integrins alpha v beta 3 and alpha v beta 5 promote adenovirus internalization but not virus attachment. Cell *73*, 309-319.

Williams, K. J., and Fuki, I. V. (1997). Cell-surface heparan sulfate proteoglycans: dynamic molecules mediating ligand catabolism. Curr. Opin. Lipidol. *8*, 253-262.

Xia, D., Henry, L., Gerard, R. D., and Deisenhofer, J. (1995). Structure of the receptor binding domain of adenovirus type 5 fiber protein. Curr. Top. Microbiol. Immunol. *199 (Pt 1)*, 39-46.

Yan, M., Peng, J., Jabbar, I. A., Liu, X., Filgueira, L., Frazer, I. H., and Thomas, R. (2004). Despite differences between dendritic cells and Langerhans cells in the mechanism of papillomavirus-like particle antigen uptake, both cells cross-prime T cells. Virology *324*, 297-310.

Yanagishita, M. (1992). Metabolism of plasma membrane-associated heparan sulfate proteoglycans. Adv. Exp. Med. Biol. *313*, 113-120.

Yang, R., Day, P. M., Yutzy, W. H. t., Lin, K. Y., Hung, C. F., and Roden, R. B. (2003a). Cell surface-binding motifs of L2 that facilitate papillomavirus infection. J. Virol. *77*, 3531-3541.

Yang, R., Murillo, F. M., Lin, K. Y., Yutzy, W. H. t., Uematsu, S., Takeda, K., Akira, S., Viscidi, R. P., and Roden, R. B. (2004). Human papillomavirus type-16 virus-like particles activate complementary defense responses in key dendritic cell subpopulations. J. Immunol. *173*, 2624-2631.

Yang, R., Yutzy, W. H. t., Viscidi, R. P., and Roden, R. B. (2003b). Interaction of L2 with beta-actin directs intracellular transport of papillomavirus and infection. J. Biol. Chem. *278*, 12546-12553.

Yeager, M. D., Aste-Amezaga, M., Brown, D. R., Martin, M. M., Shah, M. J., Cook, J. C., Christensen, N. D., Ackerson, C., Lowe, R. S., Smith, J. F., et al. (2000). Neutralization of human papillomavirus (HPV) pseudovirions: a novel and efficient approach to detect and characterize HPV neutralizing antibodies. Virology *278*, 570-577.

Zadori, Z., Szelei, J., Lacoste, M. C., Li, Y., Gariepy, S., Raymond, P., Allaire, M., Nabi, I. R., and Tijssen, P. (2001). A viral phospholipase A2 is required for parvovirus infectivity. Dev. Cell. *1*, 291-302.

Zatloukal, K., Wagner, E., Cotten, M., Phillips, S., Plank, C., Steinlein, P., Curiel, D. T., and Birnstiel, M. L. (1992). Transferrinfection: a highly efficient way to express gene constructs in eukaryotic cells. Ann. N Y Acad. Sci. *660*, 136-153.

Zhou, J., Gissmann, L., Zentgraf, H., Muller, H., Picken, M., and Muller, M. (1995). Early phase in the infection of cultured cells with papillomavirus virions. Virology *214*, 167-176.

Zhou, J., Liu, W. J., Peng, S. W., Sun, X. Y., and Frazer, I. (1999). Papillomavirus capsid protein expression level depends on the match between codon USA ge and tRNA availability. J. Virol. *73*, 4972-4982.

From: Papillomavirus Research: From Natural History to Vaccines and Beyond. Edited by: M. Saveria Campo

Chapter 13

Late Events in the Life Cycle of Human Papillomaviruses

Sheila V. Graham*

Abstract

Late events in the life cycle of human papillomaviruses (HPVs) include vegetative viral DNA replication, capsid protein production, virion assembly and virus transmission.

Such late events are tightly linked to differentiation of the epithelium that the virus infects. The molecular mechanisms underpinning these events are not completely understood as yet but clearly involve contributions from both the host cell and the virus genome itself. This chapter summarises the main late events in the infectious life cycle. Current knowledge of virus late gene expression and its regulation is detailed. In particular, the interplay between viral early and late promoters and the important role of negative regulatory elements that reside in viral late RNAs are discussed. Finally, the putative trafficking and assembly pathways involving host cell and viral gene products that result in production of infectious virus particles are outlined.

1. Introduction

The genome of papillomavirus C5 encodes early proteins, which function in viral transformation, episomal maintenance of the genome and regulation of viral gene expression, and late proteins, which form the viral capsid (Figure 1) (Howley, 1996). The E1^E4 protein transcribed from the early coding region functions also as a late protein (Chapter 7), and the same may be true for E5 (Chapter 8).

The productive life cycle of papillomaviruses is tightly linked to the differentiation programme of the epithelium that the virus infects (zur Hausen, 1996; Stubenrauch and Laimins, 1999) (Figure 2). In particular, late events in the virus life cycle that are completed in the suprabasal layers of the epithelium are co-ordinately controlled with epithelial differentiation. These events include, sequentially, vegetative viral DNA replication, synthesis of virus capsid proteins, assembly of virus capsids and virus release and transmission to other cells. Although these actual processes are reasonably well understood, the various cellular pathways required to accomplish them, and the regulatory contributions from the virus, are only just beginning to be elucidated.

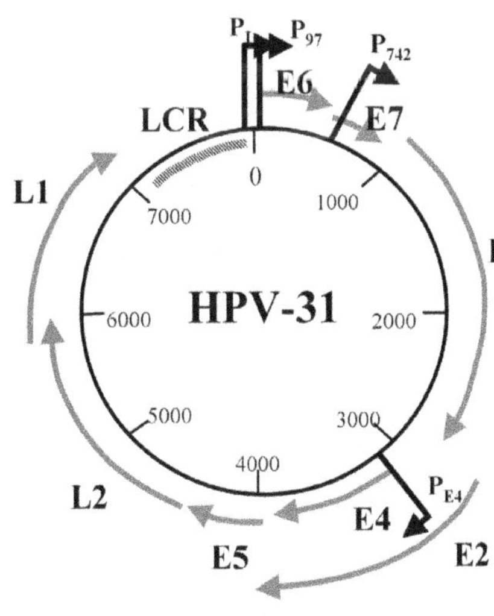

Figure 1. Schematic diagram of the HPV-31 genome. Arrows indicate open reading frames. Arrow heads indicate the four promoters. LCR, long control region, indicated by a curved bar.

Abbreviations

HPV, human papillomavirus; BPV, bovine papillomavirus; W12E, W12 cervical epithelial cells with episomal copies of the HPV-16 genome; LCR, long control region; URR, upstream regulatory region; ATP, adenosine triposphate; E6AP, E6-associated proteins; PML, promyelocytic leukaemia protein; PODs, PML-oncogenic domains; ND-10, nuclear domain (s) 10; DEAD/H, amino acids DEAD/H-motif-containing protein; E4-DBP, E4 DEAD box protein; EGF, epidermal growth factor; TRAIL, tumour necrosis factor-related apoptosis-inducing ligand; Cx43, connexin 43; GJ, gap junction; GJIC, gap junction intercellular communication; hDlg, human homologue of *Drosophila* Disc Large; IFA, immunofluorescence; PDZ, PSD95/Dlg/ZO-1; NLS, nuclear localisation signal; mRNA, messenger ribonucleic acid; hnRNP, heterogeneous ribonucleoprotein particle; snRNP, small nuclear ribonucleoprotein particle; UTR, untranslated region; PKC, protein kinase C; EIAV, equine infectious anaemia virus; CTE, constitutive transport element; CMV, cytomegalovirus ; HIV; human immunodeficiency virus; RRE, rev-response element; ARE, AU-rich element; VLP, virus-like particle.

*For correspondence. Email s.v.graham@bio.gla.ac.uk

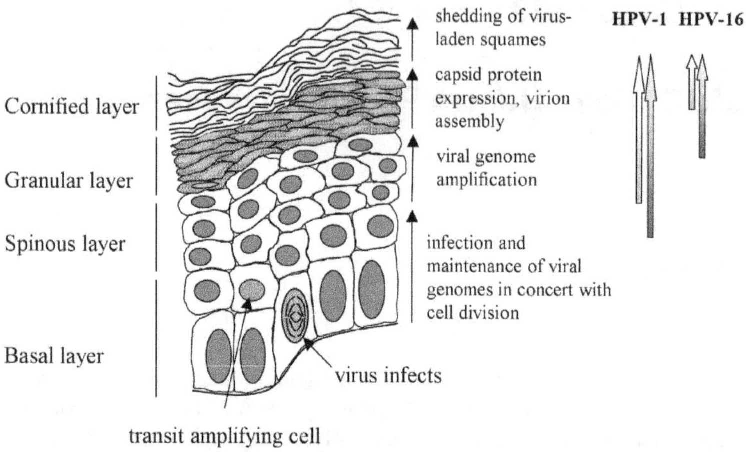

Figure 2. Schematic diagram of cutaneous epithelia. Dark shaded arrows indicate the extent of late events, and light shades arrows the areas of L1 protein expression, during the life cycles of HPV-1 and HPV-16.

Much of the research on papillomaviruses focuses on the roles of the "high risk" types in cell transformation and tumour progression rather than on the virus life cycle. Studies of transformed cells transfected with plasmids expressing virus proteins do not always elucidate the roles of these proteins in a normal infection, rather they reveal regulatory aspects of these proteins in the transformed cell (Table 1). Extensive study of the roles of the virus proteins in a normal infection has been precluded because it is extremely difficult to grow papillomaviruses in the laboratory; they absolutely require a differentiating epithelium for completion of their life cycle. Recent technical advances that will be discussed below have made such studies more accessible.

This chapter will focus mainly on discussion of late events in the life cycles of the high-risk mucosal types human papillomavirus type 16 (HPV-16) and 31 (HPV-31) that infect cervical epithelia. The former virus is extremely important medically as it has been found associated with around 60% of cervical lesions (Walboomers et al., 1999). However, there is evidence that this virus is not very productive *in vivo* perhaps due to restriction of the late events of its life cycle to the very uppermost layers of the infected epithelium (Peh et al., 2002) and so it may be imprudent to extrapolate results from studies using this virus to other papillomaviruses. An animal model that is beginning to provide extremely useful insights into late life cycle events is cottontail rabbit papillomavirus. This is because of ease of access to lesions and the ability to perform experiments on whole animals as well as the finding that late events commence lower in the epithelium than those for HPV-16 (Peh et al., 2002). HPV-11, a "low risk" mucosal type, will also be discussed where data are available. Finally, some studies on HPV-1, as a model cutaneous type, will be described.

2. The Life Cycle

Human papillomaviruses infect stem cells in the basal epithelial layer, or the transit amplifying cells that can be products of division of the stem cells (Watt, 2001) (Figure 2). Transit amplifying cells can go on to divide several times in the lowest epithelial layers and so a reservoir of virus genomes can be maintained. At this stage virus early gene expression is activated and leads to limited amplification of the virus genome, to 20-100 copies, in the infected cell (Ruesch et al., 1998a). As the infected cell divides, the virus E2 protein ensures equal distribution of the virus genomes between cells by tethering the genomic episomes to host cell chromosomes (Skiadopoulos and McBride, 1998; You et al., 2004). An uninfected transit amplifying cell will, upon division, produce one daughter cell that remains in the basal epithelial layers, and another cell that at some point undergoes commitment to differentiation. This daughter cell withdraws from the cell cycle, and begins to move to the upper epithelial layers (Watt, 2001; Alonso

Table 1. Protein functions late in the virus life cycle.

Protein	Role in late events in the life cycle
E1	Virus genome replication: ATP-dependent DNA helicase
E2	Virus genome replication and transcription; virus genome segregation; genome encapsidation; regulation of cellular gene expression; cell cycle regulation.
E1^E4	Remodels cytokeratin network; alters ND 10s; involved in virus assembly, mediates G_2 cell cycle arrest.
E5	Abrogates differentiation; facilitates replication and late gene expression
E6	Inhibits apoptosis and differentiation; maintains episomal genomes; may regulate cell architecture and signalling
E7	Allows cell cycle progression in differentiating, virus-infected cells
L1	Structural protein, major capsid protein
L2	Minor capsid protein; recruitment of L1; virus assembly

and Fuchs, 2002). However, if the cell is infected, the viral oncoproteins, E6 and E7 cooperate to inhibit exit from the cell cycle, and the cell remains poised in G_1 until vegetative viral genome amplification is activated in the differentiated cells of the upper epithelial layers (Longworth and Laimins, 2004). E6 and E7 are the well-known oncoproteins of the high risk papillomavirus types. However, in a normal infection, even the high risk type oncoproteins are expressed at a level low enough not to cause transformation of the infected cell. Transformation and tumour progression are rare events caused by some papillomavirus infections and occur most often when the virus genome is no longer episomal in the nucleus but has integrated into the host genome. Under these conditions the virus oncoproteins can become very highly overexpressed both by up-regulation of the major early promoter of the virus and by increased stability of the mRNAs encoding these proteins (zur Hausen, 1996).

Once an infected epithelial cell reaches the suprabasal layers, viral late proteins begin to be produced, starting with the E1^E4 protein. Coincident with this, virus genomes undergo one to two logs of amplification. Then the virus genome-containing cells differentiate further and move to the granular layer where the virus capsid proteins are produced and the genome is packaged. Finally, the terminally differentiated epithelial cells are shed from the upper layer of the epithelium, and the virus particles they contain are released as the dead squames disintegrate. These late events will be described in detail in this chapter. Restriction of production of the highly immunogenic capsid proteins to the upper layers of epithelia may allow the virus to persist in the undifferentiated basal epithelial cells undetected by the host immune response. In addition, expression of the capsid proteins in terminally differentiated cells is coordinated with completion of the virus life cycle. Assembly of the capsids in cells that are about to undergo shedding may increase efficiency of transmission to surrounding tissue.

3. Temporal Control of Late Events in the Virus Life Cycle

Relatively little is known about how late events in the virus life cycle are accomplished in space and time in a stratified epithelium. One recent comprehensive study examined the timing of three key stages in the life cycle of various animal papillomaviruses by using molecular markers in immunocytochemistry staining of infected tissue (Peh et al., 2002). The three stages analysed were early life cycle, onset of virus genome amplification and capsid protein expression. These studies revealed that although each stage is accomplished sequentially in different epithelial layers, as expected, there is surprising heterogeneity in the life cycles of different papillomaviruses that may relate to the types of infections the different virus sub-types cause (Figure 2). Mucosal papillomavirus types seem to restrict late gene expression to the superficial layers of the epithelium while cutaneous papillomaviruses generally begin these events lower down the epithelium. Within the mucosal sub-type the anogenital-infective human papillomaviruses 11 and 16, in particular, restrict the late stages of their life cycles to the upper epithelial layers. In contrast, cutaneous HPVs, causing warts (*e.g.* HPV-2) and verrucae (HPV-1), begin late events lower down in the spinous layer of the epithelium. The infection target site and the need to delay late events in the life cycle, especially production of the highly immunogenic capsid proteins may have important implications for infectivity and avoidance of the immune response for some papillomaviruses. However, the molecular mechanisms underlying these changes are still under investigation.

4. Models of Study of the Human Papillomavirus Life Cycle

Immunocytochemistry studies give considerable insight into life cycle events for different papillomaviruses. However, they rely on analysis of tissue isolated from lesions and such samples are often difficult to obtain due to their rarity or the need for ethical permission for the study. Moreover, use of lesion tissue precludes experiments designed to determine effects of altering the cellular environment or mutating the virus genome. Previously, study of the papillomavirus life cycle had been precluded due to the inability to grow these viruses in culture and a lack of a cell culture system that adequately supported the virus life cycle. However, derivation of two cell lines has revolutionised our ability to study, in particular, late events in the life cycle of high risk anogenital human papillomaviruses. The first line produced was the W12 cell line, harbouring HPV-16 genomes, established by Margaret Stanley in Cambridge, UK (Stanley et al., 1989). The second was the CIN612-9E cell line harbouring HPV-31 genomes, derived in the Laimins laboratory in Chicago, USA (Meyers et al., 1992). Both cell lines were established from biopsies of low grade cervical lesions. The cells are immortalised but not transformed and require J2-3T3 fibroblast feeder cells, and various mitogens, including epidermal growth factor, hydrocortisone and cholera toxin, to grow in culture. If low passage numbers of these non-continuous cell lines are used, the HPV-16 and HPV-31 genomes are maintained as 50-100 episomal copies. However, integration of the episomal genomes is observed after a number of passages in culture (Jeon et al., 1995) and this event results in loss of the ability of the virus to complete its life cycle (Frattini et al., 1996). These cell lines can be grown in monolayer culture and can be induced to differentiate by addition of phorbol esters (Meyers et al., 1992), by culturing in methylcellulose (Ruesch and Laimins, 1998b) and by culturing in 1.2 mM Ca^{++} at high density (Jeon et al., 1995). Expression of the major L1 virus capsid protein in monolayer differentiated W12 cells and observation of virus-like particles has indicated that the virus life cycle may be completed in these cells (Flores et al., 1999). More importantly, these cell lines can be grown efficiently in organotypic raft culture, an *in vitro* tissue culture system consisting of keratinocytes seeded into a collagen matrix maintained on a solid support. This culture system mimics the normal three-dimensional epithelium that the virus infects and sustains important morphological, physiological and biochemical characteristics of epithelial cell differentiation *in vitro* through lifting of the tissue to the air-liquid interface (see Chapter 11). Excellent progress has been achieved recently using the episomal HPV31b-containing biopsy-derived cell line CIN 612 9E, and the W12E subclone of the HPV-16-containing line, to establish raft cultures wherein full expression of virus capsid proteins and production of virus particles occurred when the tissue was raised to the air-liquid interface (Meyers et al., 1992; Flores et al., 1999). Virion synthesis from the cell lines in raft culture required the addition of protein kinase C (PKC) activators to induce terminal keratinocyte differentiation (Meyers et al., 1992; Ozbun and Meyers, 1997; McLaughlin-Drubin et al., 2004). Another approach has been to establish cell lines from HPV-containing biopsy-derived material and explanting the cells onto the flanks of immuno-compromised mice where they can expand into tumours. One study demonstrated the production of HPV-16 virions in such a system (Sterling et al., 1990) and a raft culture has been established from an explant of one tumour infected with HPV 11 (Stoler et al., 1989). Finally, freshly established monolayer cultures of normal human primary keratinocytes in serum-free medium, can be co-transfected with HPV genomes (Meyers et al., 1992). Following clonal selection of transformed cells by culturing in the presence of the drug, the viral genomes seemed to be stably maintained episomally in all clones examined and these cloned cell lines have been differentiated in organotypic raft cultures with synthesis of both HPV 31 and HPV 18 virions in the suprabasal layers of the raft epithelia (Meyers et al., 1992; Frattini et al., 1997). Recently, significant progress has been made by the Meyers laboratory in propagating a number of high risk anogenital human papillomaviruses in organotypic raft culture demonstrating that these culture systems are excellent models in which to study the full life cycle of the high risk anogenital human papillomaviruses (Chapter 11; McLaughlin-Drubin et al., 2004).

5. E1 and E2 and Virus Genome Amplification

The early proteins E1 and E2 are both required for viral DNA replication and the latter also for regulation of the early virus promoter (P_{97} in HPV-16; P_{99} in HPV-31). Virus genome amplification upon initial infection of a basal epithelial cell is limited to production of 20-100 copies. Following this the genomes are maintained in concert with the DNA replication cycle of the undifferentiated host cells. The virus E1 and E2 proteins accomplish this by recruiting the cellular DNA replication machinery of these mitotically-active cells. E1 is an ATP-dependent DNA helicase that can unwind the papillomavirus origin of replication to allow replication to commence (Wilson et al., 2002), while E2 is a transcription factor (Ham et al., 1991) that recruits E1 to the origin by tethering to an E2 DNA binding site adjacent to the origin (You et al., 2004) (see Chapters 5 and 6). E1 and E2 are among the first proteins to be expressed during a productive infection and continue to be expressed throughout the virus life cycle (Steele et al., 2002). For a productive infection, virus genomes require to be amplified further to yield thousands of genomes in the suprabasal layers of the infected epithelium. However, differentiated epithelial cells no longer progress through the cell cycle and have terminated expression of DNA replication factors. The virus circumvents this problem using the oncoproteins, E6 and E7 that act to allow cell cycle progression and to block apoptosis of these inappropriately cycling cells (Longworth and Laimins, 2004). Expression of E6 and E7 is largely from the "early" virus promoter and this promoter is regulated by E2. There is evidence that E2 expression increases during the life cycle of the genital papillomaviruses (Maitland et al., 1998) and evidence of an increase in levels of E1 and E2 mRNAs upon methylcellulose-induced differentiation of HPV-31 containing cells in monolayer culture (Klumpp and Laimins, 1999). In these latter experiments, a strong induction of E1 transcripts initiating from the differentiation-specific promoter was observed and this induction was much greater than for E2 transcripts implying that these two proteins are not co-ordinately regulated despite their tightly linked roles in virus genome replication. High levels of E2 have been proposed to reduce expression of E1 RNAs through repression of the "early" virus promoter. However, transcription of E1 encoding RNAs from the differentiation-specific promoter active in the suprabasal layers would circumvent E2 repression. These transcripts could direct more efficient translation of E1 protein as the E1 open reading frame would be translated first in these mRNAs.

Using W12 cells as a model system it was shown that, in undifferentiated epithelial cells, HPV-16 genome replication is by the conventional "theta" mode of replication of circular molecules. However, virus genome amplification was found to switch to a "rolling circle" mode in differentiated W12 cells (Flores and Lambert, 1997). It is possible that the ratio of E1 to E2 proteins during differentiation of virus-infected cells influences the mode of replication. A theta mode of replication requires assembly of replication complexes at every round of initiation including binding of E2 upstream of the origin and tethering of E1 to the origin. In a rolling circle mode one initiation event generates many daughter DNA molecules and thus is more economical on replication proteins, which includes E1 as the ATP-dependent DNA helicase. It may be possible that E1 is brought to the origin of replication during the rolling circle mode by a mechanism that is not so dependent upon E2 recruitment.

One extensive study of spatial organisation of replication using HPV-11 E1, E2 and an HPV-11 replication origin plasmid in keratinocytes showed localisation of E1, E2 and host replication factor A (RPA) to specific nuclear foci. Labelling with bromodeoxyuridine proved that these sites were active in DNA replication (Swindle et al., 1999). In fact, these sites seemed to be the predominant sites in the transfected cell nuclei for replication, although this may be due to the nature of the transfection approach rather than a reflection of what normally occurs in an infected cell. The co-localisation of the viral replication proteins and RPA was observed even in the absence of the replication origin plasmid suggesting that the virus reorganises the host replication machinery into special replication compartments in virus infected cells or that cellular replication sites present during S-phase are the targets for E1 and E2 proteins.

E2 is a typical transcription factor with a DNA binding domain, an activation domain, and a hinge region that separates these two domains (Hegde, 2002). Apart from its key role in virus genome replication, E2 regulates RNA polymerase II-mediated transcription of papillomavirus genomes by binding to E2 response elements in the virus long control region (Ham et al., 1991). E2 binds a number of cellular transcription factors, including p300/CBP, (Lee et al., 2000), p/CAF (Lee et al., 2002a), Sp1 (Li et al., 1991), AMF-1 (Breiding et al., 1997), p53 (Massimi et al., 1999), TFIID and TFIIB (Rank and Lambert, 1995), and C/EBP factors (Hadaschik et al., 2003) implying that it has the potential to regulate cellular gene expression (for a more detailed description of E2 functions, see Chapter 6). In addition, our recent data indicate that E2 can also regulate expression of a key RNA processing factor, SF2/ASF (McPhillips et al., 2004) revealing that E2 may have a number of general effects on cellular gene expression. Terminally differentiated keratinocytes are programmed to shut down nuclear gene expression (Fuchs and Byrne, 1994). In papillomavirus-infected cells the cellular gene expression machinery must remain active in order to transcribe viral genomes and process and translate the late transcripts produced. It is not known whether, following vegetative viral genome amplification, all viral genomes begin to be transcribed to yield late gene products or whether only the parental genomes are transcriptionally competent. If the former situation were true, the considerable increase in transcription and RNA processing events would require increased abundance of nuclear proteins. Our survey of levels of a wide range of RNA processing factors indicated that only factors involved in regulating alternative splicing were up-regulated significantly, so any effect of E2 on regulation of host gene expression is selective (McPhillips et al., 2004). Another recent study demonstrated a cooperative role for E2 with the keratinocyte differentiation-responsive C/EBP transcription factor family in the activation of the involucrin promoter. Involucrin is a major differentiation-specific cellular marker as are keratins 1 and 10, which are also expressed in the suprabasal epithelial layers and under control of C/EBPβ (Agarwal et al., 1999). Thus E2 could synergise with C/EBP factors to promote cellular differentiation perhaps antagonising the anti-differentiation effects of the viral oncoproteins and thus facilitating late events in the virus life cycle.

Finally, E2 can induce senescence and repress cell growth when expressed in cervical cancer cell lines, indicating that E2 can feed into the various cellular pathways that cause these (Francis et al., 2000; Goodwin et al., 1998; Webster et al., 2000). Indeed, the cyclin/cdk inhibitor $p21^{CIP1}$ was found to be downstream in the E2-mediated signalling pathway (Wells et al., 2000). These studies were carried out in tumour cell lines, but in a normal infection, expression of the virus oncoproteins is effectively repressed by E2. In this case, E2 may provide a check to expression of the virus oncoproteins to ensure that they do not over-activate the cell cycle. In accordance with this hypothesis, E2 has been shown to directly down-regulate the human telomerase reverse transcriptase promoter thus interfering with host cell growth regulation, perhaps leading to senescence (Lee et al., 2002b).

6. Roles of the E6 and E7 Oncoproteins in Late Events in the Virus Life Cycle

The virus E6 and E7 proteins are expressed early in infection from the virus "early" promoter (see Chapters 9 and 10 for the functions of E6 and E7 in cell transformation). However, these proteins continue to be expressed throughout the life cycle implying that they have important functions late in the virus life cycle. Mutation analysis, in the context of the whole HPV-31 genome, of the genes encoding both proteins, followed by transfection into keratinocytes, has revealed that they are required for stable maintenance of episomal genomes although nothing is known about the mechanism controlling this (Thomas et al., 1999; Flores et al., 2000; Oh et al., 2004). Both high risk E6 and E7 proteins are also required for disruption of the normal balance between cell division and differentiation that is required to support a productive papillomavirus infection. E6 binds the tumour suppressor p53 while E7 binds the cell cycle controller, p105Rb. Both high and low risk types E6 proteins have similar structures and bind p53. However, the low risk types do not induce degradation probably because of low levels of binding to E6-associated protein (E6AP), the ubiquitin ligase that targets p53 to the proteosome (Huibregtse et al., 1993; Storey et al., 1998). Low and high risk E7 proteins both bind p105Rb but often with low affinity (Ciccolini et al., 1994). HPV-1 unexpectedly binds p105Rb with high affinity, but there is no resultant transcriptional activation of E2F-regulated, growth-related genes (Schmitt et al., 1994). Other roles of these proteins in low risk papillomavirus infections has not been fully elucidated and these may be just as important during the virus life cycle as the p53 and p105Rb interactions.

One result of E7 binding to members of the Rb family, including p105Rb, p107 and p130 (Dyson et al., 1989; Edmonds and Vousden, 1989) is constitutive activation of the E2F transcription factors, which regulate expression of various cell cycle regulators (Weintraub et al., 1995). E7 also interacts with the cyclin/cdk inhibitors $p21^{Cip1}$ and $p27^{Kip1}$, the cyclins A and E and abrogates $p21^{Cip1}$ and $p27^{Kip1}$ inhibition of cyclin-dependent kinase activity (Davies et al., 1993; Tommasino et al., 1993; Zerfass-Thome et al., 1996; Funk et al., 1997; Jones et al., 1997). High risk E7 proteins can also up-regulate cyclins A and E (Martin et al., 1998). Thus E7 can create conditions that are favourable for cells to bypass the G_1 restriction point of the cell cycle, enter S-phase and re-activate the cellular DNA replication machinery to allow viral DNA synthesis. This situation may also inhibit differentiation as normal keratinocytes require p105Rb-mediated cell cycle exit to differentiate. Simultaneous regulation of cyclin/cdk inhibitors as well as p105Rb may be a two-pronged approach to ensure that the differentiating cell remains in the cell cycle.

The E6 protein forms a ternary complex with the tumour suppressor p53 together with E6A6 that ubiquitinates E6 and targets it for proteasome-mediated degradation. E6 can also indirectly regulate p53 transcription. These actions of E6 block p53-mediated cell cycle arrest and any subsequent apoptosis of the infected cell (zur Hausen, 1996). E6 also has an inhibitory association with the transcriptional co-activator, p300/CBP, which regulates expression of p53, also resulting in down-regulation of levels of this cell cycle regulator (Patel et al., 1999). With reduced p53 levels the infected cell has a reduced ability to terminally differentiate. This is important, because the cell must retain biosynthetic capability in order to facilitate completion of the virus life cycle functions. Experiments with transgenic mice that express HPV-16 E6 showed reduced differentiation of the ocular lens (Pan and Griep, 1994), while an expansion of the basal compartment of the epithelia was observed in other transgenic mice wherein E6 is expressed from a keratin 14 promoter (Song et al., 2000). However, E6 has other p53-independent functions. For example, the protein is required for stable maintenance of episomal genomes upon genome amplification (Thomas et al., 1999; Oh et al., 2004). In particular, E6 can interact with a series of PDZ domain-containing proteins. PDZ domains are defined as the sites of association with PSD-95, Dlg and ZO-1 proteins giving rise to the acronym. The proteins that interact with E6 include hDlg (Kiyono et al., 1997; Lee et al., 1997), MUPP-1 (Ullmer et al., 1998), and hScrib (Nakagawa and Huibregtse, 2000), that are associated with cell-cell junctions. Binding of E6 to motifs in the C-terminal domain of these proteins results in their degradation, at least for hDlg and hScrib by targeting them for proteosome-mediated degradation (Gardiol et al., 1999; Nakagawa and Huibregtse, 2000). PDZ domain-containing proteins are often capable of forming molecular scaffolds in the cell and are known to regulate cell signalling, differentiation and motility (Hung and Sheng, 2002; Roh and Margolis, 2003). Although it has been suggested that E6 binding of such proteins may be one way in which overexpressed E6 protein transforms cells, it is also possible that expression of low levels of E6 in HPV-infected epithelial cells could modulate temporal and spatial aspects of cell differentiation to accommodate vegetative virus genome amplification, virion formation and release from terminally differentiated cells.

7. E1^E4, the First Virus Late Protein

E1^E4 is not an "early" virus protein as it is first expressed in the suprabasal layers of the infected epithelium and is the first late protein expressed in lesions (for a more detailed description of E4 functions, see Chapter 7). It is the most abundant viral protein in the upper layers of the infected epithelium and its expression coincides with the onset of vegetative viral genome amplification (Peh et al., 2002). Despite the fact that the protein possesses a good nuclear localisation signal, it is found mainly in the cytoplasmic compartment, distributing both diffusely and with cytoplasmic intermediate filaments, with some perinuclear staining. In differentiated infected keratinocytes the protein is seen in perinuclear bundles (Doorbar et al., 1991). The most abundant E1^E4 RNAs also encode the capsid protein L1 (Hummel et al., 1995; Ozbun and Meyers, 1997). However, there is a distinct temporal and spatial gap between E1^E4 expression in the spinous layer and L1 expression in the granular layer of the epithelium (Doorbar et al., 1997). The reason for the linked expression of these proteins is not known. E1^E4 is extensively post-translationally modified by phosphorylation and also by proteolytic cleavage processing at its N-terminus (Doorbar et al., 1988; Grand et al., 1989; Rogel-Gaillard et al., 1992) and these modifications are thought to control protein functions. HPV-16 E1^E4, has been shown to associate with keratin filaments in the cytoplasm and cause them to collapse both *in vitro* and *in vivo* forming perinuclear bundles (Doorbar et al., 1991; Roberts et al., 1993). Intriguingly, HPV-1 E1^E4 does not bind cytoplasmic keratin filaments but forms nuclear and cytoplasmic inclusion bodies in suprabasal epithelial cells (Roberts et al., 1993). More recently it has been shown that, following cytokeratin network collapse, E1^E4 binds to mitochondria (Raj et al., 2004) causing them to detach from microtubules and relocate to a large single perinuclear cluster. Following this, mitochondrial membrane potential is severely reduced and apoptosis ensues, a situation that may predispose the infected cell to be more susceptible to breakage, leading to more efficient release of newly synthesised virions (Raj et al., 2004).

Release of infectious virions from the epithelial surface is the final phase in the virus life cycle and one that ultimately determines the success of transmission and further infection. HPV-11 E1^E4 is cross-linked with the cornified cell envelopes (CCE) of terminally differentiated HPV-11-infected epithelial cells (Bryan and Brown, 2000) and this seems to lead to modifications of CCE which are subtly morphologically altered and more fragile. For example, loricrin, the major component of CCE, is transcriptionally repressed and lost from the envelopes in HPV-11 infected cells and the presence of E1^E4 and loricrin in CCE is mutually exclusive (Lehr et al., 2002). A similar situation exists for keratin 10, a minor component of the CCE, that is thought to have a role in linking the proteins in the envelope. Such changes may be necessary to facilitate virus release from the infected cell. However, it has been estimated that E1^E4 only contributes to around 5% of the mass of the cornified envelope and its association may be non-functional and due simply to high levels of expression in the terminal stages of the virus life cycle.

E1^E4 can relocate PML (promyelocytic leukaemia protein) away from PML-oncogenic domains (PODs), or nuclear domain(s) 10 (ND10 domains), of which this protein is a major component (Ascoli and Maul, 1991). ND-10 domains are though to regulate transcription, growth and apoptosis as well as being holding sites for nuclear factors (Doucas, 2000). The reorganisation of ND-10 domains was observed both in keratinocytes transfected with E1^E4 expression constructs and in cells of the suprabasal layers of wart tissue (Roberts et al., 2003). This indicates that this nuclear reorganisation may be an initial event of the late phase of the virus life cycle. ND-10 domains have an association with the replication of a number of DNA and RNA viruses. Indeed, there is good evidence that this is also true for papillomaviruses, as will be discussed later, and this study provides further evidence that E1^E4 is an important viral protein for the replicative cycle, at least of HPV-1.

Another function of E1^E4 in the infected cell seems to be linked to an ability of the protein to cause G2 arrest. Remarkably E1^E4 causes cell cycle arrest not only in mammalian cells but also in fission yeast (Davy et al., 2002; Nakahara et al., 2002). It has been suggested that E1^E4-mediated arrest in G_2 could mimic S-phase to allow replication of virus genomes. Cellular DNA replication would not occur due to differentiation induced phosphorylation and degradation of MCM and Cdc6 that are required to initiate cellular DNA replication, but the presence of the viral E1 and E2 proteins would allow initiation of viral genome replication. One interesting possible link that has yet to be explored is that E2 can also inhibit mitosis by arresting cells in G_2 (Fournier et al., 1999). The E1^E4 open reading frame is wholly contained within the E2 coding region, but expressed from a different open reading frame and it may be that expression of the two proteins is linked, for example by expression from the same promoter or translation from the same mRNA, to allow synergistic action on cell cycle arrest.

Finally, E1^E4 also associates with a protein with DEAD/H box helicase activity (Doorbar et al., 2000). DEAD/H box helicases are thought to unwind RNA in an ATP-dependent manner, although this has only been proved experimentally for a small number of proteins, including the human p68 and p72 RNA helicases (Tanner and Linder, 2001). In addition, p68 has been implicated in the replication of hepatitis C virus (Tanner and Linder, 2001) and appears to have a regulatory involvement in keratinocyte proliferation and gene expression (Kahlina et al., 2004). The E4-DBP is a nucleolar protein capable of nucleocytoplasmic shuttling. This is particularly interesting given that E1^E4 possesses a nuclear localisation signal and has been shown to locate, in part, to the nucleolus, suggesting that it may also be a shuttling protein. Any RNA(s) with which E4-DBP associates have not been identified. It was proposed that this putative RNA helicase may regulate HPV-16 late gene expression at a post-transcriptional level (Doorbar et al., 2000), but the E4-DBP does not bind the HPV-16 late NRE (see below) (Graham and Doorbar, unpublished results). Any role of the E4-DBP in the life cycle of HPV-16 clearly warrants further investigation.

8. E5, a Late Expressed Oncoprotein

The E5 protein is a small, (83 amino acids) hydrophobic cytoplasmic protein that localises to the endomembrane compartments (see Chapter 8). BPV E5 has a major role in bovine fibroblast transformation but, for HPVs, its role in transformation is minor (Leptak et al., 1991). As is the case with most viral proteins, E5 is expressed only at very low levels in infected cells. However, immunostaining studies with an anti-E5 antibody demonstrated that E5 is expressed at early and late times during the HPV life cycle. The HPV-31b E5 protein has two forms, E5a and E5b. Studies on HPV-31 E5a have demonstrated that it is most abundant in the basal and particularly the granular layers of the epithelium (Mayer and Meyers, 1998). This is likely to be true also for E5b as the proteins are co-expressed. The dual expression pattern results from the expression of transcripts that encode E5 early in infection that appear to originate from the early promoter and others that appear late in infection that are transcribed from the late promoter of HPV-31 (Fehrmann et al., 2003).

The first analysis of late functions of the E5 protein involved injecting E5 mutant CRPV genomes into the epithelia of rabbits. A reduction in the numbers of papillomas with time was observed indicating that E5 is required for a productive virus infection (Brandsma et al., 1992). Two subsequent studies in human papillomaviruses have demonstrated that E5 probably does not have a significant role in the non-productive phase of the HPV life cycle but does have some role in the productive stages. In both studies, one on HPV-31 (Fehrmann et al., 2003) and the other on HPV-16 (Genther et al., 2003) the approach taken was to stably transfect wild type and E5 mutant HPV genomes into keratinocytes, differentiate these in monolayer and organotypic raft culture, and examine the phenotype of the E5 mutant HPV transfected cells. In the HPV-31 studies a reduction in viral genome amplification, and late gene expression was observed, less late RNA and a lower level of E1^E4 protein was detected in the cells transfected with the mutant E5 genomes. There was little morphological difference in organotypic raft cultures established from keratinocytes transfected with wild type and E5 mutant HPV-31 genomes showing E5 could not be regulating differentiation of the infected cell. E5 could be affecting cell cycle progression because a reduction in levels of cyclins A and B was observed in the E5 mutant HPV-31 genome transfected cells. E5 may have a cooperative role with E7 in abrogating differentiation-induced cell cycle in virus-infected cells. The effects noted in the experiments with HPV-16 were much more subtle, though displaying similar trends, except that no reduction in E1^E4 expression was noted in cells transfected with the E5 mutant virus genome.

Other studies have demonstrated that E5 binds the 16 kDa subunit of the vacuolar proton-ATPase (Conrad et al., 1993; Valle and Banks, 1995), and, by disrupting this complex, inhibits endosomal acidification (Straight et al., 1993; 1995; Valle and Banks, 1995). Loss of a pH gradient in the endosomes could inhibit degradation of internalised growth factor receptors leading to their recycling to the plasma membrane. However, there is a block to endocytic transport in W12 cells and in fibroblasts transfected with an E5 expression construct, perhaps in response to an altered actin cytoskeleton, and retention of an acidified late endocytic compartment (Thomsen et al., 2000). The membrane-bound proton ATPase is a component of gap junction plaques. When E5 was overexpressed in keratinocytes, gap junction communication was inhibited (Oelze et al., 1995) by depleting the membrane of connexin 43, the most abundant epithelial gap junction protein (Tomakidi et al., 2000). Gap junction communication is known to regulate cell growth and differentiation (Kelsell et al., 2001), and so regulation of connexin 43 by virus proteins could influence the delicate balance between cell cycle progression and arrest in virus-infected cells. Of course, in a situation of E5 overexpression, the cell may begin to progress towards transformation and so the effects on connexin 43 may be related to tumour progression as opposed to cell differentiation. Finally, the observation of expression of E5 in concert with expression of filaggrin, a marker of terminally differentiated epithelial cells, raises the possibility that expression of the protein has some role in capsid formation or genome encapsidation (Mayer and Meyers, 1998).

9. Assembly and Synthesis of Virus Capsids

The icosahedral virus capsid consists of 72 capsomeres, which are pentamers of the 57 kDa L1 major capsid protein, and probably 12 molecules of the larger, 78 kDa, L2 minor capsid protein, per virion (Trus et al., 1997). *In vitro*, virus like particles (VLPs) that are indistinguishable from true virus capsids can self assemble from L1 protein (Zhou et al., 1991; Kirnbauer et al., 1992; 1993; Heino et al., 1995). However, when co-expressed with L1, L2 is incorporated into the VLPs and a hydrophobic domain near the carboxyl terminus of L2 mediates its interaction with L1 (Finnen et al., 2003). Disulphide bonds between L1 monomers hold the capsid together but no such linkage has been found for L2 proteins in the capsid (Doorbar and Gallimore, 1987). Although L2 appears dispensable for formation of capsids it is essential for virus infectivity and capsids lacking L2 do not incorporate DNA (Kawana et al., 2001; Stauffer et al., 1998; Unckell et al., 1997; Zhou et al., 1994). Interestingly, L1 and L2 proteins do not seem to be able to bind DNA in a sequence-specific manner so how the replicated virus genome is recognised for packaging is not known (Mallon et al., 1987; Zhou et al., 1994).

Virus capsid proteins are synthesised in the cytoplasm so require to be translocated to the nucleus for assembly of virus particles. L1 and L2 proteins both possess good nuclear localisation signals (NLSs). L1 contains two classic NLSs in its C-terminus (Zhou et al., 1991) while L2 contains a non-classical arginine-rich NLS (Becker et al., 2003). NLSs are recognised by nucleo-cytoplasmic shuttling proteins of two classes, adapters of the karyopherin α/importin α family and by import receptors of the karyopherin β/importin β family. HPV-16 L2 enters the nucleus via several different pathways, including those mediated by karyopherin (Kap)$\alpha_2\beta_1$, Kapβ_2 and Kapβ_3 (Darshan et al., 2004). HPV-16 L1 capsomers enter the nucleus using the Kap$\alpha_2\beta_1$-mediated nuclear import pathway (Nelson et al., 2002). Indeed, L1 physically interacts with Kapα2 and can inhibit nuclear import of Kapβ_2 and its cargo. This cargo is predominantly the RNA processing-related factors hnRNP A1 and A2. Both are shuttling proteins that have nuclear and cytoplasmic functions. hnRNP A1 has been implicated in regulation of alternative splicing, nuclear export of RNAs to which it binds, as well as regulation of mRNA stability and translation (Pinol-Roma, 1997). If retained in the cytoplasm in differentiated epithelial cells this protein could enhance the translation of late mRNAs and loss of the protein in the nucleus could inhibit export of cellular

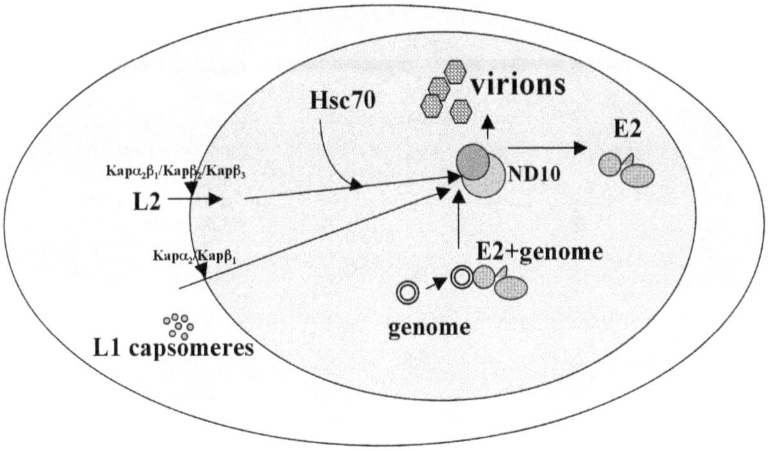

Figure 3. Possible pathway of virion assembly. The diagram shows possible interactions between cellular and viral proteins involved in recruitment of capsid proteins and virion assembly, the terminal events in the virus life cycle.

mRNAs. However, it is likely that there is much reduced cellular mRNA expression in differentiated epithelial cells as these cells are shutting down nuclear function (Fuchs and Byrne, 1994) so there may be limited significance of a block to nuclear export of mRNAs. In addition, the experiments demonstrating inhibition of nuclear import of Kapβ_2 were carried out in HeLa cells, which are undifferentiated epithelial cells. Recent data demonstrate a high nuclear abundance of hnRNP A1 in differentiated HPV-16 genome-containing epithelial cells where capsid protein production is occurring (Veerapraditsin, Zhang and Graham, unpublished results).

There is, as yet, no data suggesting a role, other than structural, for L1. It is possible that it has no non-structural function during the virus life cycle as this protein constitutes the major virus capsid protein. It has a propensity for self-association so it may act as a scaffold for virion formation. Future studies to elucidate any interaction with virus genomes, or any other non-structural virus protein should provide interesting results.

There is emerging evidence that L2 has a number of non-structural roles late in the virus life cycle (Figure 3). Virus capsids do not assemble in the cytoplasm. Instead L2 protein is expressed first and translocated to the nucleus where it has been reported to enter ND-10s (Ascoli and Maul, 1991). Chaperones are often essential for efficient protein folding and assembly of multicomponent protein structures. A search for a possible chaperone for L2 in virus assembly uncovered a role for Hsc70 (Florin et al., 2004). Hsc70 is required for incorporation of L2 into virus capsids. The protein forms a complex with L2 in the cytoplasm and aids its translocation into the nucleus. However, the seemingly complex nature of the interaction of the two proteins may hint at a series of transient associations that could perform different functions in the cytoplasm and the nucleus depending on what other chaperones or L2 interacting proteins are present. L1 protein forms into capsomeres in the cytoplasm. These are translocated to the nucleus after L2. It has been proposed that L2 recruits L1 capsomeres to ND-10 domains for virion assembly (Day et al., 1998). L2 has been shown to reorganise ND-10 resulting in recruitment of the transcriptional repressor, Daxx and loss of the activator, Sp100 (Florin et al., 2002). The major ND-10 protein, PML, is not necessarily involved in this process because in PML homozygous knock-out mouse epithelial cells L2 still forms foci to which it recruits L1 and Daxx and L2-containing VLPs are produced (Becker et al., 2004). Studies on the nuclear sites of HPV-11 replication compartments revealed that PML was recruited to E1/E2-containing replication foci but this may indicate that virus replication is carried out adjacent to ND-10 domains as has been demonstrated for other small DNA viruses such as SV40 and adenovirus (Ishov and Maul, 1996). However, in a hamster fibroblast cell line harbouring multiple copies of the BPV-1 genome, infected with a Semliki Forest virus expressing BPV-1 E2, analysis of the nuclear distribution of BPV-1 E2 showed only a diffuse nuclear localisation (Day et al., 1998). Expression of L2, but not L1, was found to be required for moving E2 into discrete nuclear foci identified as ND-10 domains. Compartmentalisation of viral DNA replication and virion assembly may make the process more efficient and preclude the incorporation of partially replicated viral DNA into virions. A full understanding of the pathway of virion assembly is likely to lead to exciting new avenues for design of anti-viral therapies

10. Regulation of Virus Gene Expression

Transcripts expressed during the life cycle of HPV-31 in monolayer and organotypic raft culture have been mapped (Hummel et al., 1995; Ozbun and Meyers, 1997). For HPV-16, only a limited analysis of transcripts produced in monolayer culture of differentiated W12 cells or lesion tissue has been published (Doorbar et al., 1990; Rohlfs et al., 1991; Higgins et al., 1992; Grassman et al., 1996). Three HPV-31 promoters active through out the life cycle have been identified; P_{97}, the so called "early"

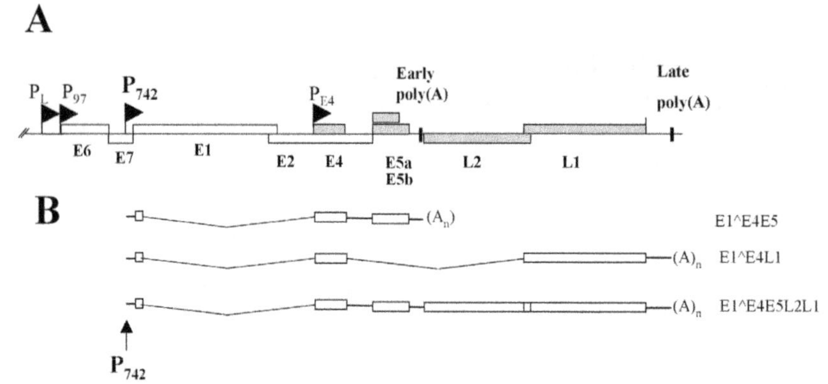

Figure 4. Late transcripts expressed from the HPV-31 genome. A, linearised representation of the HPV-31 genome showing the four promoters with the late promoter in bold type. Stippled bars indicate late-expressed genes. Vertical bars indicate polyadenylation sites. B. The three main transcripts expressed at late times during the HPV-31 life cycle (del Mar Pena and Laimins, 2001).

promoter; P_L or P_{77} situated 20 nucleotides upstream of P_{97}, and P_{3320}, a promoter found within the E4 gene and similarly situated promoters appear also to be present in the HPV-16 genome (Doorbar et al., 1990; Rohlfs et al., 1991; Grassman et al., 1996). Promoters in the LCR, upstream of P_{77} and P_{97} have been identified in a number of papillomaviruses (Baker and Howley, 1987; Wettstein et al., 1987; Palermo-Dilts et al., 1990; Stubenrauch et al., 1992). Most recently, two such promoters in HPV-31 have been shown to be active mainly in the basal layers of CIN 612 9E raft cultures (Ozbun and Meyers, 1999). A single differentiation-specific promoter is located within the E7 ORF (P_{742} for HPV-31 and P_{670} for HPV-16) but transcription start sites are scattered over a 200 base pair region (Ozbun and Meyers, 1998; Grassman et al., 1996). The P_{77}, P_{97} and P_{3320} promoters are typical TATA box and initiator element core promoters (Ozbun and Meyers, 1998). The first is a weak promoter but the latter two are predicted to be strong. Differentiation-dependent *cis*-acting elements in the upstream regulatory region of the virus genome that could regulate P_{97} have been delineated (Sen et al., 2002) but similar, detailed studies have not been completed for the other promoters.

The HPV-31 P_{742} late promoter is rather different. Transcription from these clustered P_{742} start sites was strongly induced in the uppers layers of organotypic rafts cultures derived from HPV-31-containing CIN612 9E cells (Ozbun and Meyers, 1998). Some of the start sites are used for both early and late transcripts and some are late transcript-specific. P_{742} displays a C/EBP factor binding site located 34 base pairs upstream of nucleotide 742 but no initiator element. Several weak consensus TATA boxes are located upstream of the various initiation sites around nucleotide 742 and these may direct low level transcription initiation (Ozbun and Meyers, 1998; Del Mar Pena and Laimins, 2001). Studies of DNase I hypersensitive sites have demonstrated that, upon differentiation, the nucleosome profile covering the E7 open reading frame alters to produce sensitivity to DNase I. Presumably, an alteration in expression of certain cellular chromatin remodelling factors during epithelial differentiation results in this change, or changes in chromatin density over the "late" promoter are induced by the mechanics of vegetative viral DNA replication (Del Mar Pena and Laimins, 2001). Epithelial differentiation stage-specific activation of all these promoters is not fully understood and there may be life cycle stage switching between all four promoters as there is evidence that the HPV-31 P_{742} promoter may also be used very early in infection (Steele et al., 2002).

Upon epithelial differentiation and vegetative viral genome amplification, a very significant increase in the abundance of late gene transcripts is detected (Hummel et al., 1995). This may be due to transcription from amplified genomes but also from up-regulation of the virus "late" promoter. The latter could be controlled by changes in the chromatin environment brought about by the process of DNA replication, epithelial differentiation-specific expression of transcription factors (Ruesch and Laimins, 1998a) or increased efficiency of RNA processing of the polycistronic late transcripts. Up-regulation of late gene expression could be induced by protein kinase C (PKC) in the differentiating epithelial cell (Hummel et al., 1995), as PKC has been demonstrated to coordinate gene expression changes during epithelial differentiation (Dlugosz and Yuspa, 1993). PKC could induce the expression of cellular transcription factors that induce the late promoter, or RNA processing factors that regulate late transcript stability and/or translatability. Alternatively, PKC could alter levels of phosphorylation of such factors. One key RNA processing factor whose abundance and levels of phosphorylation in differentiating epithelial cells is altered by the presence of HPV-16 genomes, is SF2/ASF (McPhillips et al., 2004). Virus late transcripts undergo extensive alternative splicing and a splicing decision between synthesis of transcripts encoding only L1 and transcripts encoding both L1 and L2 must be a regulated process (Ozbun and Meyers, 1997). SF2/ASF is a constitutive splicing factor but is also pivotal in regulating alternative splicing (Caceres and Krainer, 1997). Therefore, increased levels of this protein in the nucleus may be required for efficient late pre-mRNA processing.

In PKC inducer-treated organotypic raft culture of CIN612 9E cells, 19 different RNAs containing the L1 coding region, and 4 RNAs containing the L2 coding region have been identified (Ozbun and Meyers, 1997). HPV-31 late transcripts encode not just L1 and L2 but also E6, E6*, E7, E1^E4, E5a and E5b from an array of alternatively spliced transcripts whose abundance is dramatically increased in differentiated epithelial tissue (Figure 4). Notably, many of these transcript encode both E1^E4 and L1 or E1^E4, E5, L2 and L1 (Grassman et al., 1996; Higgins et al., 1992; Hummel et al., 1995). In the HPV-31 study, L1-encoding transcripts were twice as abundant as L2-encoding transcripts in line with the evidence that, at least in an HPV-1a productive infection, L1 protein is expressed at a higher level that L2 protein (Doorbar and Gallimore, 1987). However, the polycistronic nature of the L2 encoding transcripts poses a problem for translation as the L2 open reading frame is always downstream of the E1^E4 and E5 open reading frame and would thus be translated very inefficiently. No transcription initiation sites just upstream of the L2 open reading frame have ever been found. Some translation enhancer element, that may be differentiation-specific, must be present to allow expression of this protein that appears to be required, not only as part of the virus capsid but also in a regulatory role for virion assembly. A common feature of the late transcripts is that they do not encode E1 and E2. E1 and E2 proteins are also required at late times of infection because they are pivotal in directing replication of the virus genome and genome amplification occurring in the suprabasal layers of the infected epithelium. Almost certainly these proteins are transcribed from a different population of RNAs that terminate upstream of the late region at the early polyadenylation signal.

All late transcripts mapped to date are polyadenylated at the single late polyadenylation site in HPV-31 (Ozbun and Meyers, 1997) and at one of two HPV-16 late polyadenylation sites (Kennedy et al., 1990). In addition, a weak late polyadenylation site has been identified in the 5′ end of the L1 coding region. This site can be utilised in HeLa cells transfected with an expression plasmid for L1 but its use in HPV-16-infected cells *in vivo* has not been detected (Zhao et al., 2004). For HPV-31 and –16, differential use of the early and late polyadenylation signals during the life cycle has been demonstrated (Hummel et al., 1992; Ozbun and Meyers, 1998; 1999; Rohlfs et al., 1991). The early polyadenylation signal in each genome is a poor fit to the consensus sequence and is probably a weak signal. Although HPV-31 and HPV-16 transcripts have been shown to use this site for 3′ end formation of early region transcripts, read-through into the late region clearly occurs upon epithelial differentiation and may also occur early in the life cycle as late transcripts have been detected even in undifferentiated epithelial cells (Stoler et al., 1989; Klumpp and Laimins, 1999; Koffa et al., 2000). HPV-31 transcripts that read-through the early poly(A) site mostly terminate in the L2 open reading frame and are rapidly degraded (Terhune et al., 1999). Read-through of the HPV-31 early polyadenylation signal has been demonstrated directly by assaying transcription read-through of a *Renilla luciferase* reporter gene, with the early polyadenylation site downstream, into a firefly luciferase gene. Differentiation only induced a 50% increase in read-through possibly due to the weak nature of the early polyadenylation site demonstrating the possibility of read-through even in undifferentiated cells. Read-through could be increased by a decrease in expression of cellular polyadenylation enhancer factors that modulate the activity of poor consensus poly(A) signals (Terhune et al., 1999). Mutation analysis revealed that not only the early polyadenylation site itself but also sequences in the 5′ end of the L2 open reading frame regulated read-through, although the mechanism of action remains unclear (Terhune et al., 2001).

11. Late rRgulatory RNA Elements

Elements that may regulate late gene expression have been mapped in the late region of a number of papillomaviruses. These elements fall into two types, those that are located within the L1 and L2 open reading frames and those that are situated in the late 3′ untranslated region (3′ UTR) (Figure 5). The paradigm for these elements is the late regulatory element of bovine papillomavirus type 1. This 53 nucleotide element in the late 3′ UTR can repress expression of the bovine papillomavirus type 1 (BPV-1) L1 mRNA in undifferentiated fibroblasts, the cell type that this virus infects (Furth and Baker, 1991; Furth et al., 1994). Analysis of cellular proteins that bind this element has demonstrated that the cellular splicing factor U1 snRNP (U1 small ribonucleoprotein particle) recognises a good consensus binding site within the element. Binding of this protein complex to the element is essential for the repression of gene expression (Furth et al., 1994). Subsequent *in vitro* polyadenylation studies demonstrated an inhibitory interaction of the U1 70 kDa subunit of U1 snRNP with poly(A) polymerase (Gunderson et al., 1998). However, no follow-up studies have revealed how the repression is alleviated in differentiated epithelial cells to allow virus late gene expression. Similar studies in HPV-16 identified a similar, though more complex element, also in the late 3′ UTR that was also able to repress gene expression in undifferentiated epithelial cells (Kennedy et al., 1990). This element is 79 nucleotides in length. Its 5′ portion contains not one, but four binding sites for U1 snRNP (Dietrich-Goetz et al., 1997). However, all four are weak binding sites and mutation analysis has demonstrated that all four are required for full repression of gene expression (Cumming et al., 2003). This 5′ portion of the HPV-16 element does indeed bind U1 snRNP. However, the 3′ portion of the element also makes an important contribution to repression of gene expression

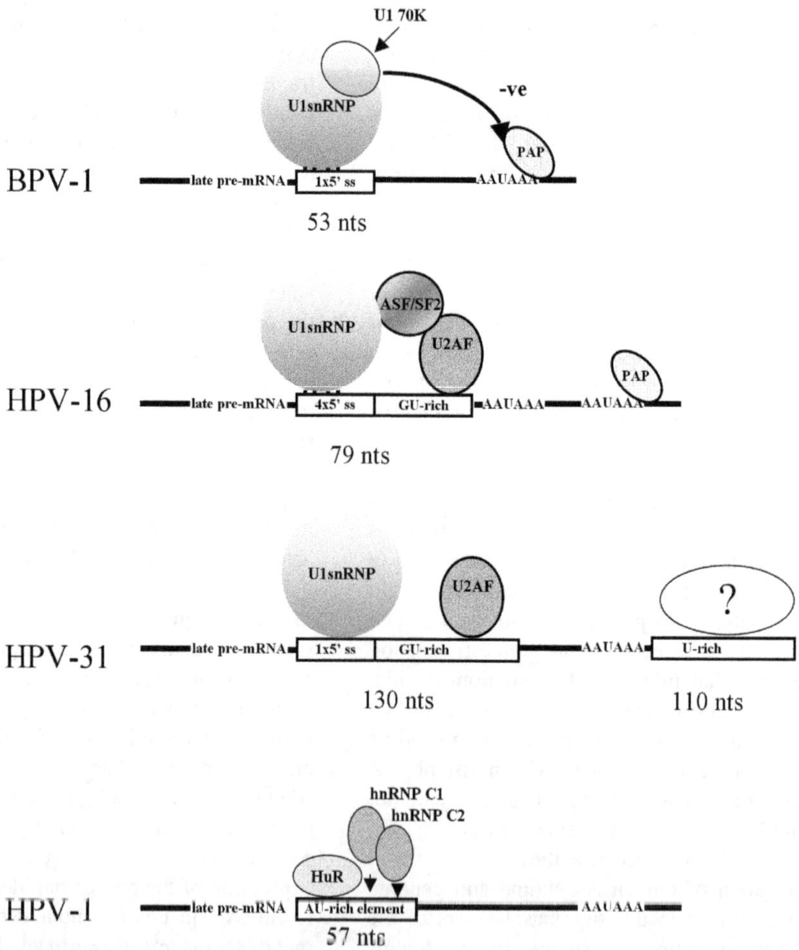

Figure 5. Diagrams of the late 3′ untranslated region inhibitory elements of BPV-1, HPV-1, -16, and –31 and the proteins they bind. 5′ss, 5′ splice site; PAP, poly(A) polymerase; GU-rich, guanosine and uridine-rich RNA element; U-rich, uridine-rich RNA element.

(Cumming et al., 2003). The GU-rich nature of the region makes it an efficient binding site for a number of other cellular RNA processing factors, including the auxiliary splicing factor $U2AF^{65}$ (Dietrich-Goetz et al., 1997), hnRNP A1 (Veerapraditisn and Graham, unpublished), the polyadenylation factor CstF-64 and the RNA stability regulator HuR (Koffa et al., 2000). It is clear that all of these proteins cannot bind simultaneously to the element. However, in the nucleus of cells transfected with a reporter gene construct containing the element $U2AF^{65}$ is strongly associated with it and can co-precipitate another key splicing factor, SF2/ASF (McPhillips et al., 2004) which is known to bridge U1 snRNP and $U2AF^{65}$ in various splicing complexes (Caceres and Krainer, 1997). The HPV-16 element has not been shown thus far to inhibit polyadenylation, although the element is able to direct use of a weak over a strong polyadenylation site (McGuire and Graham, unpublished). In addition, *in vitro* RNA decay assays revealed that RNAs containing the element displayed enhanced degradation (Kennedy et al., 1991). Finally, the element may cause nuclear retention of transcripts (Koffa et al., 2000).

Inhibitory RNA elements have also been mapped in the L1 and L2 coding regions of HPV-16 and these can act independently of the NRE in the 3′ UTR. The first 514 nucleotides of HPV-16 L1 contains a number of inhibitory elements that act independently of each other (Collier et al., 2002). The major element is located within the first 129 nucleotides of the L1 coding region. The activity of these elements was assayed by placing the L1 coding region downstream of a strong promoter, the CMV immediate early promoter, and determining the levels of L1 RNA produced in transient transfection by northern blotting. Furthermore, the region responsible for inhibition was mapped by fusing truncated L1 sequences to the equine infectious anaemia virus (EIAV) p55gag coding sequence, in order to generate high mRNA levels for northern blot analysis, then by mutating the inhibitory elements in the context of the L1 gene driven by the cytomegalovirus (CMV) promoter and assaying mRNA levels. The inhibitory sequences down-regulated the levels of L1 RNA in the nucleus but did not inhibit translation in a coupled rabbit reticulocyte lysate system.

Earlier studies showed that the human immunodeficiency virus type 1 (HIV-1) nuclear export factor, Rev, and its corresponding rev-responsive element (RRE) or a simian retrovirus type 1 constitutive transport element (CTE) could abrogate the inhibitory effect (Tan and Schwartz, 1995). These factors inhibit splicing and facilitate nuclear export of unspliced transcripts. This implies that the inhibition was exerted on a nuclear component of the mRNA biogenesis pathway and that mRNA splicing may be targeted. Alternatively, as Rev and the RRE stabilise HIV capsid protein transcripts, these data could indicate that the HPV inhibitory elements act at the level of mRNA stability.

Most recently, the L1 element has been shown to bind the splicing regulator, hnRNP A1 and its role as a splicing silencer has been clearly demonstrated (Zhao et al., 2004). The silencing effect suppressed the use of the splice acceptor site located directly upstream of the L1 open reading frame. Introduction of mutations that altered the sequence of the RNA but did not change the coding potential of the L1 transcript resulted in use of this 3′ splice site. Splice site suppression would favour transcripts that include both L2 and L1 open reading frames over a spliced transcript encoding only L1. Because excision of the long intron would effectively by-pass the early polyadenylation site, the latter transcript could potentially be expressed at early times of infection. A splicing silencer in the 5′ end of the L1 coding region that was active specifically at early times of infection, perhaps by binding abundant hnRNP A1, would ensure that this event was inefficient and would abrogate inappropriate production of L1 protein.

The situation for the L2 inhibitory elements is rather more complicated. Two large regions displaying inhibitory activity have been mapped, one to the 5′ end of the L2 gene, termed inhibitory region 1, and the other, termed inhibitory region 2, to the 3′ end (Öberg et al., 2003). These elements have so far resisted accurate localisation and definition but have been demonstrated to bind heterogeneous ribonucleoprotein K and poly(rC)-binding proteins 1 and 2 (PCBP-1, -2) (Collier et al., 1998). These proteins seem to be able to inhibit translation of L2 mRNA *in vitro*, and, for PCBP-1, also *in vivo*. However, like the L1 inhibitory element, these L2 elements also regulate RNA stability because wild type L2 mRNA could not be detected in transient transfection experiments in HeLa cells but mutations in the L2 coding region resulted in high levels of transcripts. The Rev and RRE or CTE elements could not increase L2 protein levels from retrovirus-derived eukaryotic expression plasmids so the L1 and L2 inhibitory elements must operate by different mechanisms (Öberg et al., 2003). The question remains why the coding sequence inhibitory elements appear so large and are composed of multiple sub-elements when the 3′ UTR RNA regulatory elements are generally quite short and simple.

An inhibitory element in the 3′ UTR of HPV-31 was studied as a comparison with the HPV-16 element (Cumming et al., 2002). As these two virus types are very closely related, it was expected that a very similar element would be present. However, although there is considerable sequence similarity (68%) over the 79 nucleotide region of homology with the HPV-16 NRE, and this region does exert repressive activity on reporter gene expression in transient transfection studies in HeLa cells, a 130 nucleotide region encompassing this region of homology is required for full repressive effect. Surprisingly, an additional element of 110 nucleotides, located downstream of the late polyadenylation site was also found to possess inhibitory activity. As expected, due to its sequence similarity, the upstream element bound similar proteins to the HPV-16 NRE *in vitro*. However, the downstream element also bound a subset of the proteins that bound the upstream element but also some novel proteins whose identity is not yet known (Cumming et al., 2002). Such an inhibitory element that surrounds the HPV-31 late polyadenylation site might be expected to regulate late transcript polyadenylation.

HPV-1, a cutaneous type virus, also contains a late regulatory element in its late 3′ UTR. This element is a little different to those analysed in other papillomaviruses. It is also short, at 57 nucleotides, but is very AU-rich (93%) (Tan and Schwartz, 1995). There are two motifs within the element, AUUUA and UUUUU. AU-rich RNA elements (AREs) are well known to regulate decay of the mRNAs that contain them (Wilson and Brewer, 1999). For example, the *c-fos* mRNA is extremely unstable having a half life of less than 30 minutes due to the presence of an ARE in its 3′UTR. When the *c-fos* ARE is removed or mutated the mRNA is stabilised. The *c-fos* ARE can cause destabilisation of a heterologous stable transcript such as β-globin mRNA. The element seems to act in a biphasic manner, first enhancing poly(A) tail shortening then decay of the body of the message (Mitchell and Tollervey, 2000). These elements can bind a number of proteins *in vitro* and the human homologue of the *Drosophila* ELAV gene, HuR, when overexpressed in the cytoplasm inhibits ARE-mediated decay of *c-fos* transcripts (Peng et al., 1998; Fan and Steitz, 1998). There is significant sequence homology between the HPV-1 inhibitory element and the *c-fos* ARE, they bind a similar range of proteins and the HPV-1 ARE reduces mRNA stability (Sokolowski et al., 1997). The UUUUU-motifs binds hnRNP C1/C2 while the AUUUA-motifs binds HuR, as expected. Nuclear HuR is necessary for the degradative effect (Sokolowski et al., 1999). This may be through inhibition of nuclear export, leading to degradation of nucleus-confined transcripts. Recently, another layer of complexity has been added to the mechanism of action of this element due to the finding that the element inhibits translation perhaps by interfering with poly(A) binding protein, which can also bind the element (Wilkund et al., 2002).

The diversity and complexity of the RNA elements that appear to regulate papillomavirus late gene expression is extraordinary. It is clearly important that the virus takes care not to allow expression of its capsid proteins before the infected epithelial cells differentiates to the upper layers of the epithelium where the host immune response cannot reach. All the studies thus far have been carried out in HeLa cells as a model for undifferentiated epithelial cells and it is known that the concentration of gene expression regulatory factors, such as RNA processing factors, is elevated in these highly transformed cells. The next stage in these analyses will be to decipher what these regulatory elements are doing in undifferentiated and differentiated virus genome-containing epithelial cells.

Acknowledgements

I am grateful to Louise Yule and Sarah Mole for critical reading of the manuscript and to John Doorbar for many interesting discussions on the life cycles of papillomaviruses.

References

Agarwal, C., Efimova, T., Welter, J.F., Crish, J.F., and Eckhert, R.L. (1999). CCAAT/Enhancer-binding proteins: A role in regulation of human promoter response to phorbol ester. J. Biol. Chem. *274*, 6190-6194.

Alonso, L., and Fuchs, E. (2003). Stem cells of the skin epithelium. Proc Natl. Acad. Sci. USA. *100*, 11830-11835.

Ascoli, C.A. and Maul, G.G. (1991). Identification of a novel nuclear domain. J. Cell Biol. *112*, 785-795.

Baker, C.C. and Howley, P.M. (1987). Differential promoter utilization by the bovine papillomavirus in transformed cells and productively infected wart tissues. EMBO J. *6*, 1027-1035.

Becker, K. A., Florin, L., Sapp, C., and Sapp, M. Dissection of human papillomavirus type 33 L2 domains involved in nuclear domains (ND) 10 homing and reorganization. Virol. 314, 161-167. 2003.

Becker, K.A., Florin, L., Sapp, C., Maul, G.G., and Sapp, M. (2004). Nuclear localisation but not PML protein is required for incorporation of papillomavirus minor capsid protein L2 into virus-like particles. J. Virol. *78*, 1121-1128.

Brandsma, J.L., Yang, Z.H., DiMaio, D., Barthold, S.W., Johnson, E., and Xiao, W. (1992). The putative E5 open reading frame of cottontail rabbit papillomavirus is dispensible for papillomavirus formation in domestic rabbits. J. Virol. *66*, 6204-6207.

Breiding, D.E., Sverdrup, F., Grossel, M.J., Moscufo, N., Boonchai, W., and Androphy, E.J. (1997). Functional interaction of a novel cellular protein with the papillomavirus E2 transactivation domain. Mol. Cell. Biol. *17*, 7208-7219.

Bryan, J. and Brown, D.R. (2000). Association of the human papillomavirus type 11 E1^E4 protein with cornified cell envelopes derived from infected genital epithelium. Virol. *277*, 262-269.

Caceres, J.F. and Krainer, A.R. (1997). Mammalian pre-mRNA splicing factors. In Eukaryotic mRNA processing, A.R.Krainer, ed. (Cold Spring Harbor. NY: IRL Press), pp. 174-241.

Ciccolini, F., Di Pasquale, G., Carlotti, F., Crawford, L., and Tommasino, M. (1994). Functional studies of E7 proteins from different HPV types. Oncogene *9*, 2633-2638.

Collier, B., Goobar, L., Sokolowski, M., and Schwartz, S. (1998). Translational inhibition *in vitro* of human papillomavirus type 16 L2 mRNA mediated through interaction with heterogenous ribnucleoprotein K and poly (rC)-binding proteins 1 and 2. J. Biol. Chem. *273*, 22648-22656.

Collier, B., Őberg, D., Zhao, X., and Schwartz, S. (2002). Specific inactivation of inhibitory sequences in the 5‘ end of the human papillomavirus type 16 L1 open reading frame results in production of high levels of L1 protein in human epithelial cells. J. Virol. *76*, 2739-2752.

Conrad, M., Bubb, V.J., and Schlegel, R. (1993). The human papillomavirus type 6 and 16 E5 proteins are membrane-associated proteins which associate with the 16-kilodalton pore-forming protein. J. Virol. *67*, 6170-6178.

Cumming, S.A., Repellin, C.E., McPhillips, M., Radford, J.C., Clements, J.B., and Graham, S.V. (2002). The human papillomavirus type 31 late 3′ untranslated region contains a complex bipartite negative regulatory element. J. Virol. *76*, 5993-6003.

Cumming, S.A., McPhillips, M.G., Veerapraditsin, T., Milligan, S.G., and Graham, S.V. (2003). Activity of the human papillomavirus type 16 late negative regulatory element is partly due to four weak consensus 5′ splice sites that bind a U1 snRNP-like complex. J. Virol. *77*, 5167-5177.

Darshan, M.S., Lucchi, J., Harding, E., and Moroianu, J. (2004). The l2 minor capsid protein of human papillomavirus type 16 interacts with a network of nuclear import receptors. J. Virol. *78*, 12179-12188.

Davies, R., Hicks, R., Crook, T., Morris, J., and Vousden, K. (1993). Human papillomavirus type 16 E7 associates with a hostone H1 kinase and with p107 through sequences necessary for transformation. J. Virol. *67*, 2521-2528.

Davy, C.E., Jackson, D.J., Wang, Q., Raj, K., Masterson, P.J., Fenner, N.F., Southern, S., Cuthill, S., Millar, J.A.B., and Doorbar, J. (2002). Identification of a G_2 arrest domain in the E1^E4 protein of human papillomavirus type 16. J. Virol. *76*, 9806-9818.

Day, P.M., Roden, R.B.S., Lowy, D.R., and Schiller, J.T. (1998). The papillomavirus minor capsid protein, L2, induces localisation of the major capsid protein, L1, and the viral transcription/replication protein, E2, to PML oncogenic domains. J. Virol. *72*, 142-150.

Del Mar Pena, L. and Laimins, L.A. (2001). Differentiation-dependent chromatin rearrangement coincides with activation of human papillomavirus type 31 late gene expression. J. Virol. *75*, 10005-10013.

Dietrich-Goetz, W., Kennedy, I.M., Levins, B., Stanley, M.A., and Clements, J.B. (1997). A cellular 65kDa protein recognizes the negative regulatory element of human papillomavirus late mRNA. Proc. Natl. Acad. Sci. USA *94*, 163-168.

Dlugosz, A.A. and Yuspa, S.H. (1993). Coordinate changes in gene expression which mark the spinous to granular cell transition in epidermis are regulated by protein kinase C. J. Cell Biol. *120*, 217-225.

Doorbar, J. and Gallimore, P.H. (1987). Identification of proteins encoded by the L1 and L2 open reading frames of human papillomavirus 1a. J. Virol. *61*, 2793-2799.

Doorbar, J., Evans, H.S., Coneron, I., Crawford, L., and Gallimore, P.H. (1988). Analysis of HPV1 E4 gene expression using epitope-defined antibodies. EMBO J. *7*, 825-833.

Doorbar, J., Parton, A., Hartley, K., Banks, L., Crook, T., Stanley, M.A., and Crawford, L. (1990). Detection of novel splicing patterns in a HPV-16-containing keratinocyte cell line. Virol. *178*, 254-262.

Doorbar, J., Ely, S., Sterling, J., McLean, C., and Crawford, L. (1991). Specific interaction between HPV16 E1^E4 and cytokeratins results in collapse of the epithelial cell intermediate filament network. Nature *352*, 824-827.

Doorbar, J., Foo, C., Coleman, N., Medcalf, E., Hartley, O., Prospero, T., Napthine, S., Sterling, J., Winter, G., and Griffin, H. (1997). Characterisation of events during the late stages of HPV16 infection *in vivo* using high-affinity synthetic Fabs to E4. Virol. *238*, 40-52.

Doorbar, J., Elston, R.C., Napthine, S., Raj, K., Medcalf, E., Jackson, D., Coleman, N., Griffin, H.M., Masterson, P., Stacey, S., Mengistu, Y., and Dunlop, J. (2000). The E1^E4 protein of human papillomavirus type 16 associates with a putative RNA helicase through sequences in its C terminus. J. Virol. *74*, 10081-10095.

Doucas, V. (2000). The promyelocytic (PML) nuclear compartment and transcriptional control. Biochem. Pharmacol. *60*, 1197-1201.

Dyson, N., Howley, P.M., Munger, K., and Harlow, E. (1989). The human papilloma virus-16 E7 oncoprotein is able to bind to the retinoblastoma gene product. Science *243*, 934-937.

Edmonds, C., and Vousden, K.H. (1989). A point mutational analysis of human papillomavirus type 16 E7 protein. J. Virol. *63*, 2650-2656.

Fan, X.C. and Steitz, J.A. (1998). Overexpression of HuR, a nuclear-cytoplasmic shuttling protein, increases the *in vivo* stability of ARE-containing mRNAs. EMBO J. *17*, 3448-3460.

Fehrmann, F., Klumpp, D.J., and Laimins, L.A. (2003). Human papillomavirus type 31 E5 protein supports cell cycle progression and activates late viral functions upon epithelial differnetiation. J. Virol. *77*, 2819-2831.

Finnen, R.L., Erickson, K.D., Chen, X.S., and Garcea, R.L. (2003). Interactions between papillomavirus L1 and L2 capsid proteins. J. Virol. *77*, 4818-4926.

Flores, E.R. and Lambert, P.F. (1997). Evidence for a switch in the mode of human papillomavirus type 16 DNA replication during the viral life cycle. J. Virol. *71*, 7167-7179.

Flores, E.R., Allen-Hoffman, B.L., Lee, D., Sattler, C., and Lambert, P.F. (1999). Establishment of the human papillomavirus type 16 (HPV-16) life cycle in an immortalized human foreskin keratinocyte cell line. Virol. *262*, 344-354.

Flores, E.R., Allen-Hoffman, B.L., Lee, D., and Lambert, P.F. (2000). The human papillomavirus type 16 E7 oncogene is required for the productive stage of the viral life cycle. J. Virol. *74*, 6622-6631.

Florin, L., Schäfer, F., Sotlar, K., Streeck, R.E., and Sapp, M. (2002). Reorganization of nuclear domain 10 induced by papillomavirus capsid protein L2. Virol. *295*, 97-107.

Florin, L., Becker, K.A., Sapp, C., Lambert, C., Sirma, H., Muller, M., Streeck, R.E., and Sapp, M. (2004). Nuclear translocation of papillomavirus minor capsid protein L2 requires Hsc70. J. Virol. *78*, 5546-5553.

Fournier, N., Raj, K., Saudan, P., Utzig, S., Sahli, R., Simanis, V., and Beard, P. (1999). Expression of human papillomavirus 16 E2 protein in *Schizosaccharomyces pombe* delays the initiation of mitosis. Oncogene *18*, 4015-4021.

Francis, D.A., Schmid, S.I., and Howley, P.M. (2000). Repression of the integrated papillomavirus E6/E7 promoter is required for growth suppression of cervical cancer cells. J. Virol. *74*, 2679-2686.

Frattini, M.G., Lim, H.B., and Laimins, L.A. (1996). *In vitro* synthesis of oncogenic human papillomaviruses requires episomal genomes for differentiation-dependent late gene expression. Proc. Natl. Acad. Sci. USA *93*, 3062-3067.

Frattini, M.G., Lim, H.B., Doorbar, J., and Laimins, L.A. (1997). Induction of human papillomvirus type 18 late gene expression and genomic amplification in organotypic cultures from transfected DNA templates. J. Virol. *71*, 7068-7072.

Fuchs, E. and Byrne, C. (1994). The epidermis: rising to the surface. Curr. Opin. Gen. Dev. *4*, 725-736.

Funk, J.O., Waga, S., Harry, J.B., Espling, E., Stillman, B., and Galloway, D.A. (1997). Inhibition of CDK activity and PCNA-dependent DNA replication by p21 is blocked by interaction with the HPV-16 E7 oncoprotein. Genes Dev. *11*, 2090-2100.

Furth, P.A. and Baker, C.C. (1991). An element in the bovine papillomavirus late 3′ untranslated region reduces polyadenylated cytoplasmic RNA levels. J. Virol. *65*, 5806-5812.

Furth, P.A., Choe, W.-T., Rex, J.H., Byrne, J.C., and Baker, C.C. (1994). Sequences homologous to 5′ splice sites are required for the inhibitory activity of papillomavirus late 3′ untranslated regions. Mol. Cell. Biol. *14*, 5278-5289.

Gardiol, D., Kühne, C., Glausinger, B., Lee, S., Javier, R., and Banks, L. (1999). Oncogenic human papillomavirus E6 proteins target the discs large tumour suppressor for proteasome-mediated degradation. Oncogene *18*, 5487-5496.

Genther, S.M., Sterling, S., Duensing, S., Munger, K., Sattler, C., and Lambert, P.F. (2003). Quantitative role of the human papillomavirus type 16 E5 gene during the productive stage of the viral life cycle. J. Virol. *77*, 2832-2842.

Goodwin, E.C., Naeger, L.K., Breiding, D.E., Androphy, E.J., and DiMaio, D. (1998). Transactivation-competent bovine papillomavirus E2 protein is specifically required for efficient repression of human papillomavirus oncogene expression and for acute growth inhibition of cervical carcinoma cell lines. J. Virol. *72*, 3925-3934.

Grand, R.J.A., Doorbar, J., Smith, K.J., Coneron, I., and Gallimore, P.H. (1989). Phosphorylation of the human papillomavirus type 1 E4 proteins *in vivo* and *in vitro*. Virol. *170*, 201-213.

Grassman, K., Rapp, B., Maschek, H., Petry, K.U., and Iftner, T. (1996). Identification of a differentiation-inducible promoter in the E7 open reading frame of human papillomavirus type 16 (HPV-16) in raft cultures of a new cell line containing high copy numbers of episomal HPV-16 DNA. J. Virol. *70*, 2339-2349.

Gunderson, S.I., Polycarpou-Schwarz, M., and Mattaj, I.W. (1998). U1 snRNP inhibits pre-mRNA polyadenylation through a direct interaction between U1 70K and poly(A) polymerase. Mol. Cell *1*, 255-264.

Hadaschik, D., Hinterkeuser, K., Oldak, M., Pfister, H.J., and Smola-Hess, S. (2003). The papillomavirus E2 protein binds to and synergizes with C/EBP factors involved in keratinocyte differentiation. J. Virol. *77*, 5253-5265.

Ham, J., Dostatni, N., Gauthier, J.M., and Yaniv, M. (1991). The papillomavirus E2 protein: a factor with many talents. Trends Biochem. Sci. *16*, 440-444.

Hegde, R.S. (2002). The papillomavirus E2 proteins. Annu. Rev. Biophys. Biomol. Struct. *31*, 343-360.

Heino, P., Dillner, J., and Schwartz, S. (1995). Human papillomavirus type 16 capsid proteins produced from recombinant Semliki Forest virus assemble into virus-like particles. Virol. *214*, 349-359.

Higgins, G.D., Uzelin, D.M., Phillips, G.E., McEvoy, P., and Burrel, C.J. (1992). Transcription patterns of human papillomavirus type 16 in genital intraepithelia neoplasia: evidence for promoter usage within the E7 open reading frame during epithelial differentiation. J. Gen. Virol. *73*, 2047-2057.

Howley, P.M. (1996). *Papillomavirinae*: the viruses and their replication. In Fields Virology, B.N.Fields, D.M.Knipe, and P.M.Howley, eds. (Philadelphia, Pa.: Lippincott-Raven Publishers), pp. 2045-2076.

Huibregtse, J.M., Scheffner, M., and Howley, P.M. (1993). Cloning and expression of the cDNA for E6-AP, a protein that mediates the interaction of the human papillomavirus E6 oncoprotein with p53. Mol. Cell. Biol. *13*, 775-784.

Hummel, M., Hudson, J.B., and Laimins, L.A. (1992). Differentiation-induced and constitutive transcription of human papillomavirus type 31b in cell lines containing viral episomes. J. Virol. *66*, 6070-6080.

Hummel, M., Lim, H.B., and Laimins, L.A. (1995). Human papillomavirus type 31b late gene expression is regulated through protein kinase C-mediated changes in RNA processing. J. Virol. *69*, 3381-3388.

Hung, A.Y. and Sheng, M. (2002). PDZ domains: structural modeules for protein complex assembly. J. Biol. Chem. *277*, 5699-5702.

Ishov, A.M. and Maul, G.G. (1996). The periphery of nuclear domain 10 (ND10) as sites of DNA virus deposition. J. Cell Biol. *134*, 815-826.

Jeon, S., Allen-Hoffman, B.L., and Lambert, P.F. (1995). Integration of human papillomavirus type 16 into the human genome correlates with a selective growth advantage of cells. J. Virol. *69*, 2989-2997.

Jones, D.L., Alani, R.M., and Munger, K. (1997). The human papillomavirus E7 oncoprotein can uncouple cellular differentiation and proliferation in human keratinocytes by abrogating p21Cip1-mediated inhibition of cdk2. Genes Dev. *11*, 2101-2111.

Kahlina, K., Goren, I., Pfeilschifter, J., and Frank, S. (2004). p68 DEAD box RNA helicase expression in keratinocytes. J. Biol. Chem. *279*, 44872-44882.

Kawana, Y., Kawana, K., Yoshikawa, Y., Taketani, Y., Yosjiike, K., and Kanda, T. (2001). Human papillomavirus type 16 minor capsid protein L2 N-terminal region containing a common neutralization epitope binds to the cell surface and enters the cytoplasm. J. Virol. *75*, 2331-2336.

Kelsell, D.P., Dunlop, J., and Hodgins, M.B. (2001). Human diseases: clues to cracking the connexin code? Trends Cell Biol. *11*, 2-6.

Kennedy, I.M., Haddow, J.K., and Clements, J.B. (1990). Analysis of human papillomavirus type 16 late mRNA 3′ processing signals *in vitro* and *in vivo*. J. Virol. *64*, 1825-1829.

Kennedy, I.M., Haddow, J.K., and Clements, J.B. (1991). A negative regulatory element in the human papillomavirus type 16 genome acts at the level of late mRNA stability. J. Virol. *65*, 2093-2097.

Kirnbauer, R., Booy, F.P., Cheng, N., Lowy, D.R., and Schiller, J.T. (1992). Papillomavirus L1 major capsid protein self-assembles into virus-like particles that are highly immunogenic. Proc. Natl. Acad. Sci. USA *89*, 12180-12184.

Kirnbauer, R., Taub, J., Greenstone, H.L., Roden, R.B.S., Dürst, M., Gissman, L., Lowy, D.R., and Schiller, J.T. (1993). Efficient self-assembly of human papillomavirus type 16 L1 and L1-L2 into virus-like particles. J. Virol. *67*, 6929-6936.

Kiyono, T., Hiraiwa, A., Fujita, M., Hayashi, Y., Akiyama, T., and Ishibashi, M. (1997). Binding of high-risk human papillomavirus E6 oncoproteins to the human homologue of the *Drosophila* discs large tumor suppressor protein. Proc. Natl. Acad. Sci. USA *94*, 11612-11616.

Klumpp, D.J. and Laimins, L.A. (1999). Differentiation-induced changes in promoter usage for transcripts encoding the human papillomavirus type 31 replication protein E1. Virol. *257*, 239-246.

Koffa, M.D., Graham, S.V., Takagaki, Y., Manley, J.L., and Clements, J.B. (2000). The human papillomavirus type 16 negative regulatory element interacts with three proteins that act at different posttranscriptional levels. Proc. Natl. Acad. Sci. USA *97*, 4677-4682.

Lee, D., Lee, B., Kim, J., Kim, D.W., and Choe, J. (2000). cAMP response element-binding protein binds to human papillomavirus E2 protein and activates E2-dependent transcription. J. Biol. Chem. *275*, 7045-7051.

Lee, D., Hwang, S.G., Kim, J., and Choe, J. (2002a). Functional interaction between p/CAF and human papillomavirus E2 protein. J. Biol. Chem. *277*, 6483-6489.

Lee, D., Kim, H.-Z., Jeong, K.W., Shim, Y.S., Horikawa, I., Barrett, J.C., and Choe, J. (2002b). Human papillomavirus E2 down-regulates the human telomerase reverse transcriptase promoter. J. Biol. Chem. *277*, 27748-27756.

Lee, S.S., Weiss, R.S., and Javier, R.T. (1997). Binding of human virus oncoproteins to hDlg/SAP97, a mammalian homolog of the *Drosophila* discs large tumor suppressor protein. Proc. Natl. Acad. Sci. USA *94*, 6670-6675.

Lehr, E., Jarnik, M., and Brown, D.R. (2002). Human papillomavirus type 11 alters the transcription and expression of loricrin, the major cell envelope protein. Virol. *298*, 240-247.

Leptak, C., Ramon y Cajal, S., Kulke, R., Horwitz, B.H., Riese, I., Dotto, G.P., and Doorbar, J. (1991). Tumorigenic transformation of murine keratinocytes by the E5 genes of bovine papillomavirus type 1 and human papillomavirus type 16. J. Virol. *65*, 7078-7083.

Li, R., Knight, .JD., Jackson, S.P., Tjian, R., and Botchan, M.R. (1991). Direct interaction between Sp1 and the BPV enhancer E2 protein mediates synergistic activation of transcription. Cell *65*, 493-505.

Longworth, M.S. and Laimins, L.A. (2004). Pathogenesis of human papillomaviruses in differentiating epithalia. Microbiol. Mol. Biol. Rev. *68*, 362-372.

Maitland, N.J., Conway, S., Wilkinson, N.S., Ramsdale, J., Morris, J.R., Sanders, C.M., Burns, J.E., Stern, P.L., and Wells, M. (1998). Expression patterns of the human papillomavirus type 16 transcription factor E2 in low- and high-grade cervical intraepithelial neoplasia. J. Pathol. *186*, 275-280.

Mallon, R.G., Wojciechowicz, D., and Defendi, V. (1987). DNA-binding activity of papillomavirus proteins. J. Virol. *61*, 1655-1660.

Martin, L.G., Demers, G.W., and Galloway, D.A. (1998). Disruption of the G1/S transition in human papillomavirus type 16 E7-expressing human cells is associated with altered regulation of cyclin E. J. Virol. *72*, 975-985.

Massimi, P., Pim, D., Bertoli, C., Bouvard, V., and Banks, L. (1999). Interaction between the HPV-16 E2 transcriptional activator and p53. Oncogene *18*, 7748-7754.

Mayer, T.J. and Meyers, C. (1998). Temporal and spatial expression of the E5a protein during the differentiation-dependent life cycle of human papillomavirus type 31b. Virol. *248*, 208-217.

McLaughlin-Drubin, M.E., Christensen, N., and Meyers, C. (2004). Propagation, infection, and neutralization of authentic HPV16 virus. Virol. *322*, 213-219.

McPhillips, M.G., Veerapraditsin, T., Cumming, S.A., Karali, D., Milligan, S.G., Boner, W., Morgan, I.M., and Graham, S.V. (2004). SF2/ASF binds the human papillomavirus type 16 late RNA control element and is regulated during epithelial differentiation. J. Virol. *78*, 10598-10605.

Meyers, C., Frattini, M.G., Hudson, J.B., and Laimins, L.A. (1992). Biosynthesis of human papillomavirus from a continuous cell line upon epithelial differentiation. Science *257*, 971-973.

Mitchell, P. and Tollervey, D. (2000). mRNA stability in eukaryotes. Curr. Opin. Gen. Dev. *10*, 193-198.

Nakagawa, S. and Huibregtse, J.M. (2000). Human scribble (Vartul) is targeted for ubiquitin-mediated degradation by the high-risk papillomavirus E6 proteins and the E6AP ubiquitin-protein ligase. Mol. Cell. Biol. *20*, 8244-8253.

Nakahara, T., Nishimura, A., Tanaka, M., Ueno, T., Ishimoto, A., and Sakai, H. (2002). Modulation of the cell division cycle by human papillomavirus type 18 E4. J. Virol. *76*, 10914-10920.

Nelson, L.M., Rose, R.C., and Moroianu, J. (2002). Nuclear import strategies of high risk HPV16 L1 major capsid protein. J. Biol. Chem. *277*, 23958-23964.

Oelze, I., Kartenbeck, J., Crusius, K., and Alonso, A. (1995). Human papillomavirus type 16 E5 protein affects cell-cell communication in an epithelial cell line. J. Virol. *69*, 4489-4494.

Oh, S.T., Longworth, M.S., and Laimins, L.A. (2004). Roles of the E6 and E7 proteins in the life cycle of low-risk human papillomavirus type 11. J. Virol. *78*, 2620-2626.

Ozbun, M.A. and Meyers, C. (1997). Characterisation of late gene transcripts expressed during vegetative replication of human papillomavirus type 31b. J. Virol. *71*, 5161-5172.

Ozbun, M.A. and Meyers, C. (1998). Temporal usage of multiple promoters during the life cycle of human papillomavirus type 31b. J. Virol. *72*, 2715-2722.

Ozbun, M.A. and Meyers, C. (1999). Two novel promoters in the upstream regulatory region of human papillomavirus type 31b are negatively regulated by epithelial differentiation. J. Virol. *73*, 3505-3510.

Öberg, D., Collier, B., Zhao, X., and Schwartz, S. (2003). Mutational inactivation of two distinct negative RNA elements in the human papillomavirus type 16 L2 coding region induces production of high levels of L2 in human cells. J. Virol. *77*, 11674-11684.

Palermo-Dilts, D.A., Broker, T.R., and Chow, L.T. (1990). Human papillomavirus type 1 produces redundant as well as polycistronic mRNAs in plantar warts. J. Virol. *64*, 3144-3149.

Pan, H. and Griep, A.E. (1994). Altered cell cycle regulation in the lens of HPV-16 E6 or E7 transgenic mice: implications for tumour suppressor gene function in development. Genes Dev. *8*, 1285-1299.

Patel, D., Huang, S., Baglia, L., and McCance D.J. (1999). The E6 protein of human papillomavirus type 16 binds to and inhibits co-activation by CBP and p300. EMBO J. *18*, 5061-5072.

Peh, W.L., Middleton, K., Christensen, N., Nicholls, P., Egawa, K., Sotlar, K., Brandsma, J., Percival, A., Lewis, J., Liu, W.J., and Doorbar, J. (2002). Life cycle heterogeneity in animal models of human papillomavirus-associated disease. J. Virol. *76*, 10411-10416.

Peng, S.S.-Y., Cehn, C.-Y.A., Xu, N., and Shyu, A.-B. (1998). RNA stabilization by the AU-rich element binding protein, HuR, an ELAV protein. EMBO J. *17*, 3461-3470.

Pinol-Roma, S. (1997). HnRNP proteins and the nuclear export of mRNA. Seminars Cell Dev. Biol. *8*, 57-63.

Raj, K., Berguerand, S., Southern, S., Doorbar, J., and Beard, P. (2004). E1^E4 protein of human papillomavirus type 16 associates with mitochondria. J. Virol. *78*, 7199-7207.

Rank, N.M. and Lambert, P.F. (1995). Bovine papillomavirus type 1 E2 transcriptional regulators directly bind two cellular transcription factors, TFIID and TFIIB. J. Virol. *69*, 6323-6334.

Roberts, S., Ashmole, I., Johnson, G.D., Kreider, J.W., and Gallimore, P.H. (1993). Cutaneous and mucosal human papillomavirus E4 proteins form intermediate filament-like structures in epithelial cells. Virol. *197*, 176-187.

Roberts, S., Hillman, M.L., Knight, G.L., and Gallimore, P.H. (2003). The ND10 component promyelocytic leukemia protein relocates to human papillomavirus type 1 E4 intranuclear inclusion bodies in cultures keratinocytes and in warts. J. Virol. *77*, 673-684.

Rogel-Gaillard, C., Breitburd, F., and Orth, G. (1992). Human papillomavirus type 1 E4 proteins differing by their N-terminal ends have distinct cellular locations when transiently expressed *in vitro*. J. Virol. *66*, 816-823.

Roh, M.H. and Margolis, B. (2003). Composition and function of PDZ protein complexes during cell polarization. Am. J. Physiol. Renal Physiol. *285*, F377-F387.

Rohlfs, M., Winkenbach, S., Meyer, S., Rupp, T., and Dürst, M. (1991). Viral transcription in human keratinocyte cell lines immortalized by human papillomavirus type-16. Virol. *183*, 331-342.

Ruesch, M.N., Stubenrauch, F., and Laimins, L.A. (1998a). Activation of papillomavirus late gene transcription and genome amplification upon differentiation in semisolid medium is coincident with expression of involucrin and transglutaminase but not keratin. J. Virol. *72*, 5016-5024.

Ruesch, M.N. and Laimins, L.A. (1998b). Human papillomavirus oncoproteins alter differentiation-dependent cell cycle exit on suspension in semisolid medium. Virol. *250*, 19-29.

Schmitt, A., Harry, J.B., Rapp, B., Wettstein, F.O., and Iftner, T. (1994). Comparison of the properties of the E6 and E7 genes of low- and high-risk cutaneous papillomaviruses reveals strongly transforming and high Rb-binding activity for the E7 protein of the low-risk human papillomavirus type 1. J. Virol. *68*, 7051-7059.

Sen, E., Bromberg-White, J.L., and Meyers, C. (2002). Genetic analysis of *cis* regulatory elements within the 5′ region of the human papillomavirus type 31 upstream regulatory region during different stages of the viral life cycle. J. Virol. *76*, 4798-4809.

Skiadopoulos, M.H. and McBride, A.A. (1998). Bovine papillomavirus type 1 genomes and the E2 transactivator protein are closely associated with mitotic chromatin. J. Virol. *72*, 2079-2088.

Sokolowski, M., Zhao, C., Tan, W., and Schwartz, S. (1997). AU-rich mRNA instability elements on human papillomavirus type 1 late mRNAs and c-*fos* mRNAs interact with the same cellular factors. Oncogene *15*, 2303-2319.

Sokolowski, M., Furneaux, H., and Schwartz, S. (1999). The inhibitory activity of the AU-rich RNA element in the human papillomavirus type 1 late 3′ untranslated region correlates with its affinity for the elav-like HuR protein. J. Virol. *73*, 1080-1091.

Song, S., Liem, A., Miller, J.A., and Lambert, P.F. (2000). Human papillomavirus types 16 E6 and E7 contribute differently to carcinogenesis. Virol. *267*, 141-150.

Stanley, M.A., Browne, H.M., Appelby, M., and Minson, A.C. (1989). Properties of a non-tumourigenic human cervical keratinocyte cell line. Int. J. Cancer *43*, 672-676.

Stauffer, Y., Raj, K., Masternak, K., and Beard, P. (1998). Infectious human papillomvirus type 18 pseudovirions. J. Mol. Biol. *283*, 529-536.

Steele, B.K., Meyers, C., and Ozbun, M.A. (2002). Variable expression of some "housekeeping" genes during human keratinocyte differentiation. Anal. Biochem. *307*, 341-347.

Sterling, J., Stanley, M.A., Gatward, G., and Minson, T. (1990). Production of human papillomavirus type 16 virions in a keratinocyte cell line. J. Virol. *64*, 6305-6307.

Stoler, M.H., Wolinsky, S.M., Whitbeck, A., Broker, T.R., and Chow, L.T. (1989). Differentiation-linked human papillomavirus types 6 and 11 transcription in genital condylomata revealed by *in situ* hybridisation with message-specific RNA probes. Virol. *172*, 331-340.

Storey, A., Thomas, M., Kalita, A., Harwood, C., Gardiol, D., Mantovani, F., Breuer, J., Leigh, I.M., Matlashweski, G., and Banks, L. (1998). Role of a p53 polymorphism in the development of human papillomavirus-associated cancer. Nature *393*, 229-234.

Straight, S.W., Hinkle, P.M,, Jewers, R.J., and McCance, D.J. (1993). The E5 oncoprotein of human papillomavirus type 16 transforms fibroblasts and effects the downregulation of the epidermal growth factor receptor in keratinocytes. J. Virol. *67*, 4521-4532.

Straight, S.W., Herman, B., and McCance D.J. (1995). The E5 oncoprotein of human papillomavirus type 16 inhibits the acidification of endosomes in human keratinocytes. J. Virol. *69*, 3185-3192.

Stubenrauch, F., Malejczyk, J., Fuchs, P.G., and Pfister, H.J. (1992). Late promoter of human papillomavirus type 8 and its regulation. J. Virol. *64*, 3144-3149.

Stubenrauch, F. and Laimins, L.A. (1999). Human papillomavirus life cycle: active and latent phases. Seminars Cancer Biol. *9*, 379-386.

Swindle, C.S., Zou, N., Van Tine, B.A., Shaw, G.M., Engler, J.A., and Chow, L.T. (1999). Human papillomavirus DNA replication compartments in a transient DNA replication system. J. Virol. *73*, 1001-1009.

Tan, W. and Schwartz, S. (1995). The rev protein of human immunodeficiency virus type 1 counteracts the effect of an AU-rich negative element in the human papillomavirus type 1 late 3′ untranslated region. J. Virol. *69*, 2932-2945.

Tanner, N.K. and Linder, P. (2001). DExD/H box RNA helicases: from generic motors to specific dissociation functions. Mol. Cell *8*, 251-262.

Terhune, S.S., Milcarek, C., and Laimins, L.A. (1999). Regulation of human papillomavirus type 31 polyadenylation during the differentiation-dependent life cycle. J. Virol. *73*, 7185-7192.

Terhune, S.S., Hubert, W.G., Thomas, J.T., and Laimins, L.A. (2001). Early polyadenylation signals of human papillomavirus type 31 negatively regulates capsid gene expression. J. Virol. *75*, 8147-8157.

Thomas, J.T., Hubert, W.G., Ruesch, M.N., and Laimins, L.A. (1999). Human papillomavirus type 31 oncoproteins E6 and E7 are required for the maintenance of episomes during the viral life cycle in normal human keratinocytes. Proc. Natl. Acad. Sci. USA *96*, 8449-8454.

Thomsen, P., van Deurs, B., Norrild, B., and Kayser, L. (2000). The HPV16 E5 oncogene inhibits endocytic trafficking. Oncogene *19*, 6023-6032.

Tomakidi, P., Cheng, H., Kohl, A., Komposch, G., and Alonso, A. (2000). Connexin 43 expression is downregulated in raft cultures of human keratinocytes expressing the human papillomavirus type 16 E5 protein. Cell Tissue Res. *301*, 323-327.

Tommasino, M., Adamczewski, J.P., Carlotti, F., Barth, C.F., Manetti.R., Contorni, M., Cavalieri, F., Hunt, T., and Crawford, L. (1993). HPV16 E7 protein associates with the protein kinase p33CDK2 and cyclin A. Oncogene *8*, 195-202.

Trus, B.L., Roden.R.B.S., Greenstone, H.L., Vrhel, M., Schiller, J.T., and Booy, F.P. (1997). Novel structural features of bovine papillomavirus capsid revealed by three-dimensional reconstruction to 9Å resolution. Nat. Struct. Biol. *4*, 413-420.

Ullmer, C., Schmuck, K., Figge, A., and Lubbert, H. (1998). Cloning and characterization of MUPP1, a novel PDZ domain protein. FEBS Lett. *424*, 63-68.

Unckell, F., Streeck, R.E., and Sapp, M. (1997). Generation and neutralization of pseudovirions of human papillomavirus type 33. J. Virol. *71*, 2934-2939.

Valle, G.F. and Banks, L. (1995). The human papillomavirus (HPV)-6 and HPV-16 E5 proteins co-operate with HPV-16 E7 in the transformation of primary rodent fibroblasts. J. Gen. Virol. *76*, 1239-1245.

Walboomers, J.M.M., Jacobs, M.V., Manos, M.M., Bosch, F.X., Kummer, F.X., Shah, K.V., Snijders, P.J.F., Peto, J., Meijer, C.J.L.M., and Munoz, N. (1999). Human papillomavirus is a necessary cause of invasive cervical cancer worldwide. J. Pathol. *189*, 12-19.

Watt, F. (2001). Stem cell fate and patterning in mammalian epidermis. Curr. Opin. Gen. Dev. *11*, 410-417.

Webster, K., Parish, J., Pandya, M., Stern, P.L., Clarke, A.R., and Gaston, K. (2000). The human papillomavirus (HPV) 16 E2 protein induces apoptosis in the absence of other HPV proteins and via a p53-dependent pathway. J. Biol. Chem. *275*, 87-94.

Weintraub, S.J., Chow, K.N., Luo, R.X., Zhang, S.H., He, S., and Dean, D.C. (1995). Mechanism of active transcriptional repression by the retinoblastoma protein. Nature *375*, 812-815.

Wells, S.I., Francis, D.A., Karpova, A.Y., Dowhanick, J.J., Benson, J.D., and Howley, P.M. (2000). Papillomavirus E2 induces senescence in HPV-positive cells via p105Rb- and p21(CIP)-dependent pathways. EMBO J. *19*, 5762-5771.

Wettstein, F.O., Barbosa, M.S., and Nasseri, M. (1987). Identification of the major cottontail rabbit papillomavirus late RNA cap site and mapping and quantitation of an E2 and minor E6 coding mRNA in papillomas and carcinomas. Virol. *159*, 321-328.

Wilkund, L., Sokolowski, M., Carlsson, A., and Schwartz, S. (2002). Inhibition of translation by UAUUUAU and UAUUUUUAU motifs of the AU-rich RNA instability element in the HPV-1 late 3′ untranslated region. J. Biol. Chem. *277*, 40462-40471.

Wilson, V.G., West, M., Woztek, K., and Rangasamy, D. (2002). Papillomavirus E1 proteins: form, function, and features. Virus Genes *24*, 275-290.

Wilson.G.M. and Brewer, G. (1999). The search for trans-acting factors controlling messenger RNA decay. Prog. Nucleic Acid Res. Mol. Biol. *62*, 257-291.

You, J., Croyle, J.L., Nishimura, A., Ozato, K., and Howley, P.M. (2004). Interaction of the bovine papillomavirus E2 protein with Brd4 tethers the viral DNA to host mitotic chromosomes. Cell *117*, 349-360.

Zerfass-Thome, K., Zwerschke, W., Mannhardt, B., Tindle, R., Botz, J.W., and Jansen-Durr, P. (1996). Inactivation of the cdk inhibitor p27KIP1 by the human papillomavirus type 16 E7 oncoprotein. Oncogene *13*, 2330.

Zhao, X., Rush, M., and Schwartz, S. (2004). Identification of an hnRNP A1-dependent splicing silencer in the human papillomavirus type 16 L1 coding region that prevents premature expression of the late L1 gene. J. Virol. *78*, 10888-10905.

Zhou, J., Sun, X.Y., Stenzel, D.J., and Frazer, I.H. (1991). Expression of vaccinia recombinant HPV 16 L1 and L2 ORF proteins in epithelial cells is sufficient for assembly of HPV virion-like particles. Virol. *185*, 251-257.

Zhou, J., Sun, X.Y., Louis, K., and Frazer, I.H. (1994). Interaction of human papillomavirus (HPV) type 16 capsid proteins with HPV DNA requires an intact L2 N-terminal sequence. J. Virol. *68*, 619-625.

zur Hausen, H. (1996). Papillomavirus infections - a major cause of human cancers. Biochem. Biophys. Acta *1288*, F55-F78.

From: Papillomavirus Research: From Natural History to Vaccines and Beyond. Edited by: M. Saveria Campo

Chapter 14

Transgenic Mouse Models for the *In Vivo* Analysis of Papillomavirus Oncogene Function

Paul F. Lambert*, Scott J. Balsitis, Anny Shai, Sara J.S. Simonson and Sybil M.G. Williams

Abstract

The E6 and E7 oncoproteins of the high risk HPVs are recognized multifunctional proteins that can affect multiple cellular processes implicated in tumorigenesis. *In vivo* models are critical to appreciate fully the oncogenic properties of HPV genes in the context of the appropriate organ sites. As with other cancers, the preferred animal for studying human cancers in the laboratory is the mouse owing to the advent of genetic engineering, the wealth of prior knowledge gained in areas like chemical carcinogenesis, genetic modifiers of cancers, development and pathology, as well as their practical value given their small size, short gestational period and the availability of genetically inbred strains. This chapter reviews the knowledge gained from the study of HPV transgenic mouse models. HPV transgenic mouse models have provided the means for dissecting the role of viral genes in cancer, and biological systems in which to develop and test new modalities for treating HPV-associated disease. Their study has provided compelling proof in support of the original hypothesis that HPVs cause cancer.

1. Introduction

Herein, we summarize the findings of studies assessing the function of papillomaviral genes and their regulatory elements in the context of transgenic mouse models. The advent of transgenic mouse technologies, which allow the introduction of foreign genes into the mouse germ line, have provided scientists with unprecedented insights into the biological properties of genes *in vivo*. This technology, along with the capacity to alter the expression or function of endogenous mouse genes through homologous recombination, has provided papillomavirus biologists with an unique opportunity not only to study the properties of papillomavirus genes *in vivo*, but also to gain insight into their mechanisms of action. In this chapter, we describe transgenic mouse models in which the three papillomaviral oncogenes, E5, E6 and E7, have been characterized (Table 1). For each oncogene, we describe its biological properties *in vivo*, and its mechanism of action in those cases where the mouse models have provided such insight. Separately, we describe the development and use of HPV transgenic mice as models for cervical cancer and for studies on host immunological responses to HPV oncogenes.

2. Biological Properties of the E5 Oncoprotein

Bovine Papillomavirus Type 1 (BPV1) was the prototype papillomavirus for early studies in tissue culture owing to the discovery in the late 1970s that it could transform mouse C127 fibroblasts (Lowy et al., 1980). The E5 gene largely accounts for the transforming activity of BPV1 in these cells (DiMaio et al., 1986; Dvoretzky et al., 1980; Schiller et al., 1986). It encodes one of the smallest known oncogenic proteins, a 44 amino acid, dimeric transmembrane protein that can activate the PDGF receptor in a ligand independent manner (DiMaio and Mattoon, 2001). E5's activation of the PDGF receptor is argued to occur either through a direct interaction with the receptor (Goldstein et al., 1994; Petti and DiMaio, 1992; 1994; Petti et al., 1997) or through its association with the16-kDa subunit of the vacuolar proton ATPase, a complex that is involved in the acidification of intracellular compartments (Goldstein et al., 1992; 1991). Regardless of the mechanism, E5's activation of the PDGFR accounts for E5's potent transforming activity in fibroblasts (Drummond et al., 1995; Goldstein et al., 1994; Nilson and DiMaio, 1993; Riese and DiMaio, 1995) (see Chapter 8 for more details on BPV E5 functions).

Given the focus given to BPV1 in many of the early laboratory studies of papillomavirus biology, it is not surprising that the first transgenic mouse model for papillomaviruses, likewise, was a BPV1 transgenic mouse (Lacey et al., 1986). These BPV1 transgenic mice developed fibrosarcomas at about 8 months of age in areas prone to wounding such as the face, tail and head. The tumors were histologically quite similar to those observed in cattle, the natural host of BPV1, and arose in three distinct stages: mild fibromatosis, aggressive

*For correspondence. Email lambert@oncology.wisc.edu.

Table 1. Transgenic mouse models for papillomaviruses and their roles in cancer

Transgene	Transcriptional Promoter	Papillomavirus Genes Present	Tissue Expression Pattern	Phenotypes	Citations
BPV-1	Homologous	All BPV-1 genes	Fibroblast	Fibrosarcoma	Lacey et al., 1986; Sippola et al., 1989; Lindgren et al., 198 Kandel et al., 1991; Bossy-We et al., 1992
MMTVE6/E7	MMTV	HPV16 E6,E7	Leydig cell, salivary gland, skin, cervix	Tumors of the testicles, salivary gland, skin, and female reproductive tract	Kondoh et al., 1991; 1994; 19 Sasagawa et al., 1994
βactinE6/E7	β actin	HPV16 E6, E7	CNS	Neuroepithelial carcinomas	Arbeit et al., 1993
αAcryE6/E7	αA crystallin	HPV16 E6, E7	Lens, retina, skin	Epithelial hyperplasia. Tumors of the lens, retina and skin	Griep et al., 1993; Lambert et al., 199; Frazer et al., 1995; H et al., 1997; Griep et al., 1998 Hilditch-Maguire et al., 1999
BK10E6/E7	Bovine keratin 10	HPV16 E6, E7	Stratified epithelia	Epithelial hyperplasia	Auewarakul et al., 1994
K1E6/E7	Keratin 1	HPV18 E6, E7	Stratified epithelia	papillomas	Greenhalgh et al., 1994
K14HPV16	Keratin 14	All HPV16 genes	Stratified epithelia	Epithelial hyperplasia. Tumors of the skin and cervix	Arbeit et al., 1994; 1996; Coussens et al., 1996; Jin et al 1999; Elson et al., 2000; Riley al., 2003
K14E6/E7	Keratin 14	HPV16 E6, E7	Stratified epithelia	Epithelial hyperplasia. Tumors of the skin	Melero et al., 1997; Schreiber al., 2004
mK5E6/E7	Mouse keratin 5	HPV16 E6, E7	Lung, thymus	Thymic hyperplasia. Lung tumors	Carraresi et al., 2001
TyrE6E7	Tyrosinase	HPV16 E6, E7	skin	Skin tumors	Kang et al., 2000
bK6E6E7	Bovine keratin 6	HPV16 E6, E7	Hair follicles	Alteration of hair follicle cycling	Escalante-Alcalde et al., 2000
HPV18E6/E7	Homologous	HPV18 E6, E7	Urogenital tract	Enlarged seminal vesicles. Mesenchymal tumors of the female reproductive tract	Comerford et al., 1995
HPV16	homologous	All HPV16 genes	Epithelia, Lymphatic	Carcinoma, lymphoma	Yang et al., 1995; 1998
αAcryE6	αA crystallin	HPV16 E6	lens	Altered differentiation. Tumors when crossed to αAcryE7 mice	Pan and Griep, 1994; 1995
K14E6	Keratin 14	HPV16 E6	Stratified epithelia, lens	Epithelial hyperplasia. Tumors of the skin and cervix	Song et al., 1998; 1999; 2000; Brake et al., 2003; Riley et al., 2003; Schaeffer et al., 2004

K14E6^{I128T}	Keratin 14	HPV16 E6^{I128T}	Stratified epithelia, lens	Epithelial hyperplasia,	Nguyen et al., 2002a; 2003b
K14E6$^{\Delta146\text{-}151}$	Keratin 14	HPV16 E6$^{\Delta146\text{-}151}$	Stratified epithelia, lens	Abrogated DNA damage response. Skin tumors with chemical carciongens	Nguyen et al., 2003a; 2003b Simonson and Lambert, unpublished observations
K14MnE6	Keratin 14	MnPV E6	skin	Skin tumors with chemical carcinogens	Helfrich et al., 2004
αAcryE7	αA crystallin	HPV16 E7	lens	Altered differentiation. Tumors when crossed to αAcryE6 mice	Pan and Griep, 1994; 1995 McCaffrey et al., 1999
K14E7	Keratin 14	HPV16 E7	Stratified epithelia, lens	Epithelial hyperplasia. Tumors of the skin and cervix	Herber et al., 1996; Gulliver et al., 1997; Song et al., 1998; 2000; Frazer et al., 1998; 2001; Doan et al., 1998; 2000; Tindle et al., 2001; Leggatt et al., 2002; Nguyen et al., 2002b; Balsitis et al., 2003; Brake et al., 2003; Malcolm et al., 2003; Riley et al., 2003; Matsumoto et al., 2004; Schaeffer et al., 2004
IRBPE7	Interstitial retinol binding protein	HPV16 E7	retina	Retinal tumors	Howes et al., 1994
K14E5	Keratin 14	HPV16 E5	Stratified epithelia	Epithelial hyperplasia, skin tumors	S. Williams et al., submitted

fibromatosis and fibrosarcoma. Interestingly, these tumors and the preneoplastic lesions that gave rise to these tumors harbored the BPV1 genome extrachromosomally, whereas normal skin and other tissues in the mice contained only the integrated copies found in the germ lineage. Thus the viral DNA had excised itself and replicated as a nuclear plasmid in some somatic cells. This presumably arose because the BPV1 transgene had been engineered to contain 1.69 copies of the BPV1 genome, thereby facilitating homologous recombination-driven excision of the viral genome (Lacey et al., 1986). That the virus could replicate extrachromosomally in the mouse tissue, once excised, was predictable because, in transformed mouse C127 cells, the BPV1 genome persists as a nuclear plasmid (Lowy et al., 1980).

Analysis of cell lines derived from these distinct stages of tumor development revealed that characteristics attributed to each stage were heritable, and that detectable expression of BPV1 genes occurs only in proliferating lesions (Sippola et al., 1989). Among the viral genes that could be detected was E5. Given its potent transforming activities in mouse fibroblasts in tissue culture, E5 was immediately suspected to be responsible for the induction of the fibrosarcomas in these BPV1 transgenic mice. Expression of the viral genes alone, however, was not sufficient to account for progression of the premalignant lesions to malignancy, as the aggressive fibromatoses contained the same levels of viral DNA, viral RNA and E5 oncoprotein as did the fibrosarcomas (Sippola et al., 1989). Therefore, it was postulated that additional cellular events are necessary for progression to malignancy. Tumors and low passage cell lines from normal tissue, fibromatoses, and fibrosarcomas were analyzed in a number of different studies to determine the molecular changes that lead to tumorigenesis in this model. Karyotypic analysis of cell populations derived from fibrosarcomas uncovered a common set of abnormalities in chromosome 8 and/or chromosome 14, whereas even the aggressive fibromatoses did not contain these abnormalities (Lindgren et al., 1989). A significant increase in the level of vascularization in the aggressive, premalignant fibromatoses and the fibrosarcomas as compared to the less-aggressive, premalignant lesions and normal tissue corresponded to a difference in the localization of basic fibroblast growth factor (bFGF), a known angiogenic factor (Kandel et al., 1991). Another study found *junB* and c-*jun* protein levels to be upregulated in aggressive fibromatoses and fibrosarcomas, but not in normal skin or the less aggressive fibromatoses (Bossy-Wetzel et al., 1992). Investigators in this study transfected *junB* and c-*jun* into cell lines derived from the less aggressive fibromatoses and found this to be sufficient to alter the morphology and growth rate of these cells, approximating that observed in the cell lines derived from the aggressive fibromatoses and tumors. However, *junB* and c-*jun* transfected cells were not tumorigenic in immunodeficient mice, in contrast to cell lines derived from fibrosarcomas, indicating that *junB* and c-*jun* can only induce a subset of characteristics of fibrosarcomas. The study of the BPV1 transgenic mice provided not only evidence for the tumorigenic potential of papillomaviruses but also for the concept that multiple genetic or epigenetic alterations contribute to cancer.

3. *In Vivo* Properties of HPV-16 E5

Because of the structural similarities between BPV1 E5 and high-risk HPV E5 proteins, and because BPV1 E5 has strong transforming potential, work soon began to determine if also HPV E5 was oncogenic (see Chapter 8 for a more detailed description of the properties of HPV E5). The high risk HPV found most abundantly in human cervical cancer is HPV type 16 (HPV16). The E5 protein of HPV16 was classified as an oncogene given its ability to transform murine fibroblasts and human keratinocytes *in vitro* (Leechanachai et al., 1992; Leptak et al., 1991; Pim et al., 1992; Straight et al., 1993). Two biochemical properties have been attributed to the HPV16 E5 protein *in vitro*: an ability to enhance epidermal growth factor (EGF) receptor (EGFR) signaling in the presence of ligand (*e.g.* EGF) (Leechanachai et al., 1992; Pim et al., 1992; Straight et al., 1993), and the ability to bind to the 16-kDa subunit of the vacuolar proton ATPase, like BPV1 E5 (Adam et al., 2000; Briggs et al., 2001; Conrad et al., 1993).

To examine the *in vivo* properties of HPV16 E5, K14E5 mice were generated expressing a codon-optimized version of the E5 protein referred to previously as E5* (Williams et al., submitted). This codon-optimized HPV16 E5 gene (Disbrow et al., 2003) was used because codons rare in mammalian genes are found commonly in papillomavirus genes (Leder et al., 2001; Zhou et al., 1999), and comprise approximately 40% of the codons in the HPV16 E5 translational open reading frame. E5* was cloned into an expression plasmid containing the human keratin (K) 14 promoter and 3′ polyadenylation sequences. The K14 promoter was chosen because it directs expression of the transgene to the basal layer of the epithelium where HPV begins its viral life cycle, and from which it is thought that HPV-associated cancers arise.

It was demonstrated that HPV16 E5 exerted potent effects on the epidermis of the K14E5 transgenic mice, inducing gross phenotypes that included neonatal death, runting, wrinkled, scaly skin and tail, and alopecia. Histological phenotypes included hyperplasia, hyperkeratosis, enhanced DNA synthesis, and an aberrant pattern of differentiation (Williams et al., 2005). These phenotypes are very similar to phenotypes that are observed in mice over-expressing ligands of the EGFR in the epidermis, such as TGF-α and AR (Cook et al., 1997; Dominey et al., 1993; Vassar et al., 1992). This similarity in phenotypes led to the hypothesis that E5's activities are dependent upon EGFR signaling. When the K14E5 transgenic mice were crossed to mice that were heterozygous for a dominant negative mutation in the tyrosine kinase domain of the EGFR, $EGFR^{Wa5/+}$ (Lee et al., 2004), there was an attenuation of E5-induced phenotypes, including a complete loss of the epidermal hyperplasia induced by E5. These K14E5; $EGFR^{Wa5/+}$ mice provided the first demonstration that E5's activities require the functional EGFR.

To determine whether HPV16 E5 contributed to tumorigenesis in the mouse epithelia, K14E5 mice were allowed to age and were monitored for spontaneous tumor development. In two lines of K14E5 mice, spontaneous skin tumors developed in 78% and 100% of mice by 40 weeks of age, with tumors developing as early as 7 weeks of age. This data indicates that HPV16 E5 is tumorigenic in murine squamous epithelium *in vivo*. Further studies will be necessary to learn whether the oncogenic potential of E5 is also dependent on EGFR activity, or whether other cellular targets might contribute.

4. *In Vivo* Properties of Co-expressed High Risk HPV E6 and E7 Genes

In cervical cancers, high risk anogenital HPV genomes are commonly found integrated into the host genome and two viral genes, E6 and E7, selectively up-regulated in their expression (Schwarz et al., 1985; Yee et al., 1985). The E6 and E7 genes of the high risk HPVs, most notably HPV16, were subsequently found to be transforming in tissue culture based experiments (Bedell et al., 1987; Matlashewski et al., 1987; Munger et al., 1989; Phelps et al., 1988; Schlegel et al., 1988) and to immortalize epithelial cells in culture (Hawley-Nelson et al., 1989; Kaur and McDougall, 1989; Pirisi et al., 1988; Woodworth et al., 1988). For these reasons E6 and E7 were labeled oncogenes. In many of these tissue culture assays, E6 and E7 acted synergistically. Therefore it was not surprising that many of the first efforts to study the activity of these oncogenes *in vivo* involved the generation of transgenic mice expressing the two viral genes together. In this section we summarize the phenotypes of such mice. More detailed information regarding the individual properties of E6 and E7 is provided in subsequent sections in which are described studies using transgenic mice that express these genes individually.

The earliest HPV transgenic mouse models provided strong support for the oncogenic potential of high risk HPV E6 and E7 genes in multiple tissues. Furthermore they provided evidence for the role of cellular factors in contributing to HPV-associated cancers, including somatic changes in the activity of host factors within the cancer cell as well as the influence of genetic background in determining risk for HPV-associated cancers. The first HPV transgenic mice described in the literature expressed both the E6 and E7 oncogenes of high risk HPV16 from the Mouse Mammary Tumor Virus (MMTV) enhancer/ promoter region (Kondoh et al., 1991). These mice spontaneously developed testicular tumors in males, which were subsequently characterized to be Leydig cell in origin (Kondoh et al., 1994) and to have the c-kit receptor highly active (Kondoh et al., 1995). Female mice from this same line of mice developed salivary gland carcinomas, lymphomas and skin histiocytomas (Sasagawa et al., 1994). Also, some of these female mice developed dysplastic and/or hyperplastic changes in the cervix and vagina (Sasagawa et al., 1994). Expression of HPV16 E6 and E7 from the human β-actin promoter led to neuroepithelial carcinomas (Arbeit et al., 1993). Expression of HPV16 E6 and E7 from the αA crystallin promoter, which normally directs expression of genes to the lens, was found not only to cause epithelial tumors in the lens (Griep et al., 1993), but also, due to ectopic transgene expression in one line, squamous cell carcinomas in the skin (Lambert et al., 1993) and retinoblastomas (Griep et al., 1998). Interestingly, the retinoblastomas only arose when the transgenic mice were crossed to another inbred strain of mice, indicating a role of host genes in determining risk for cancers induced by HPV oncogenes.

By the mid 1990s, transcriptional promoters, such as ones derived from keratin genes, were being used to direct expression of the HPV oncogenes to stratified squamous epithelia. These mouse models provided insight into the effects of high risk HPV E6 and E7 genes in more relevant cell types, in particular the ability of these genes together to induce the proliferation of cells within stratified epithelia. Transgenic expression of the HPV16 E6 and E7 genes in the suprabasal compartment of the epidermis, achieved using the bovine K10 promoter, led to hyperplasia, but not tumorigenesis (Auewarakul et al., 1994). In contrast, expression of the high risk HPV18 E6 and E7 genes in the same compartment, achieved using the human K1 promoter, led to wart-like lesions in the skin of adult mice, and these papillomas carried activating mutations in H-ras (Greenhalgh et al., 1994). Expression of the entire early region of HPV16 genome from the human K14 promoter, which directs expression to the basal compartment of stratified squamous epithelia, led to epithelial hyperplasia (Arbeit et al., 1994), and squamous cell carcinomas were found to arise once this transgene was crossed onto the FVB/N genetic background, again arguing a role of host genes in determining risk for HPV associated cancer (Coussens et al., 1996). In another study transgenic mice expressing the HPV16 E6 and E7 genes from the K14 promoter on the C57bl/6 genetic background again failed to develop skin tumors (Melero et al., 1997) but did so when crossed to FVB/N mice carrying an activated ras oncogene (Schreiber et al., 2004). Here the F1 mice carrying only the HPV16 transgenes failed to develop tumors. No studies have been reported with this line of mice on a pure FVB/N background. When HPV16 E6 and E7 were placed under the control of the mouse K5 promoter in order to direct expression to the basal compartment, the resulting mice developed lung carcinomas and thymic hyperplasia (Carraresi et al., 2001). Expression of HPV16 E6 and E7 genes from the tyrosinase promoter induced epidermal hyperplasia and susceptibility to chemically induced skin tumors (Kang et al., 2000); whereas their expression from the bovine K6 promoter altered the growth phase of hair follicles (Escalante-Alcalde et al., 2000).

Whereas earlier studies had indicated that the transcriptional signals within long control region (LCR) of the HPV16 or HPV18 genome were not effective in directing transgene expression (Choo et al., 1992; Choo et al., 1993), other investigators found that the HPV18 LCR directed expression of its cognate E6 and E7 genes to the urogenital tract, leading to enlarged seminal vesicles in males and tumors of mesenchymal origin in the reproductive tract of females (Comerford et al., 1995). Highly metastatic, poorly differentiated carcinomas and malignant lymphomas arose in mice carrying a dimeric HPV16 genome (Yang et al., 1995). Here, transcripts with the potential to encode E6 and E7 were sometimes but not always present, and in the latter cases, distinct patterns of cellular proto-oncogene expression were found (Yang et al., 1998).

These studies provide an abundance of evidence demonstrating the oncogenic potential of the high risk HPV oncogenes. The individual properties of E6 and E7 when expressed *in vivo* are described in the following sections.

5. Individual Properties of the E6 Oncogene *In Vivo*

The first transgenic mouse model to evaluate the individual properties of E6 were the αAcryE6 mice (Pan and Griep, 1994). In these mice, the E6 genes was targeted for expression to the highly differentiated, lens fiber cell compartment, by placing it under the transcriptional control of the αA crystallin (αAcry) promoter. The lens has proven a valuable tissue in which to characterize papillomaviral gene function because, like tissues that papillomaviruses normally infect, it is an epithelial tissue composed of poorly differentiated, proliferating epithelial cells and terminally differentiated, post-mitotic fiber cells. Three important discoveries made with these mice were that E6 can: 1) inhibit epithelial cell differentiation; 2) inhibit apoptosis, and 3) induce tumors. The αAcryE6 mice presented with lens cataracts once they opened their eyes.

5.1. Inhibition of cell differentiation

Histopathologically, the lens fiber cell differentiation was disrupted. This disruption was first evident during embryogenesis, when the lens tissue begins to develop and primary fiber cell differentiation first arises. At this time, fiber cells elongate inwardly towards the anterior of the lens vesicle from the posterior surface and undergo a maturation process in which all internal organelles including the nucleus are lost. Of these features of fiber cell differentiation, E6 perturbed the latter process specifically. That is, in the fiber cells of αAcryE6 lenses, intracellular organelles including nuclei failed to disappear (Pan and Griep, 1994). This process of organelle loss involves the activation of caspases, specifically caspase 6 or yet-to-be-identified caspases with similar substrate specificity (Foley et al., 2004). How E6 inhibits organelle loss remains unknown; however, it does not relate to E6's inhibition of p53, as p53-null mice failed to display any defect in normal lens fiber cell differentiation (Pan and Griep, 1995).

5.2. Inhibition of apoptosis

The second property discovered with the αAcryE6 mice is E6's ability to inhibit apoptosis. This property was observed when αAcryE6 mice were crossed to αAcryE7 mice. In the latter mouse strain, HPV16 E7 induces apoptosis in the lens fiber cell compartment (Pan and Griep, 1994), a process that involves activation of caspase 3 (Foley et al., 2004). In mice transgenic for both HPV16 E6 and E7, generated by breeding αAcryE6 mice with αAcryE7 mice, E7-induced apoptosis was greatly inhibited. This ability of E6 to inhibit apoptosis was greater than that seen when the αAcryE7 transgene was placed on a p53-deficient background. This led to the conclusion that E6 can inhibit apoptosis through the modulation of p53-dependent as well as p53-independent apoptotic pathways. The nature of the latter pathway remains to be elucidated.

5.3. Induction of tumours

The third major discovery made with the αAcryE6 mice was that E6 itself contributes to cancer formation. This was demonstrated by the fact that, whereas tumors did not arise in either the αAcryE6 or the αAcryE7 mice, tumors did arise in the doubly transgenic offspring expressing both HPV16 E6 and HPV16 E7 genes that were generated by mating these two lines of mice (Pan and Griep, 1994).

5.4. E6 expression in stratified epithelia

The biological properties of HPV16 E6 in stratified squamous epithelia such as the epidermis and the cervical epithelium were elucidated from the study of K14E6 transgenic mice (Song et al., 1999). In these mice, the HPV16 E6 gene was placed under the transcriptional control of the K14 promoter, targeting expression of E6 to the basal compartment of stratified squamous epithelium. These mice also expressed the E6 gene in the poorly differentiated epithelia within the lens, providing the unique opportunity to contrast the effects of E6 expression when expressed in poorly differentiated (lens epithelia: K14E6 mice) versus highly differentiated (lens fibers: αAcryE6 mice) compartments of an epithelial tissue. The study of K14E6 mice provided several important insights into the biological properties of E6 in addition to those made with the αAcryE6 mice. First it was observed that E6 could induce epithelial hyperplasia in the mouse epidermis and the lens epithelia. This hyperplasia was associated with a delay in the normal differentiation as evidenced by a delay in the onset of keratin 10 expression in the suprabasal compartment of the epidermis, and a delay in the onset of expression of markers of lens fiber cell differentiation in the lens. It was also associated with the presence of cells within the suprabasal compartment supporting DNA synthesis in the epidermis. Normally only cells within the basal compartment, where cell proliferation is normally restricted, undergo new rounds of DNA synthesis. The induction of epithelial hyperplasia in the epidermis and the lens by E6 is not a consequence of E6's inactivation of p53, as p53-null mice do not display this phenotype (Pan and Griep, 1995; Song et al., 1999). Furthermore, $K14E6^{I128T}$ mice, expressing a mutant of HPV16E6 that fails to inactivate p53, retain the ability to induce epithelial hyperplasia in the epidermis and the lens; and this phenotype is not further accentuated when the transgene is placed on a p53 deficient background (Nguyen et al., 2002a; Nguyen et al., 2003b). E6's ability to induce epithelial hyperplasia, instead, correlates with its binding to a class of cellular proteins called PDZ domain proteins (Nguyen et al., 2003a; Nguyen et al., 2003b). This conclusion was based upon the characterization of $K14E6^{\Delta 146\text{-}151}$ mice which express a mutant version of HPV16 E6 missing amino acids at the C terminus that mediate wild type E6's interaction with the PDZ domain proteins (Gardiol et al., 1999; Massimi et al., 2004; see Chapter 9 for further details on E6-PDZ proteins interaction). These PDZ domain partners of E6 include the mammalian homologs of two Drosophila proteins, DLG and Scribble, considered to be tumor suppressors because their disruption in Drosophila leads to a hyperplastic and disorganized epithelial compartment during development. Disruption of Dlg function in the mouse, due to a gene trap mutation, causes hyperplastic defects in the lens epithelium in embryonic mice (Nguyen et al., 2003b) that are less obvious in the embryonic epidermis (Simonson and Lambert, unpublished observations). Unfortunately, the epithelial phenotypes resulting from this mutation could only be scored in embryos, because homozygosity for this gene trap mutation leads to severe cleft palate and consequently, rapid onset of death in neonates. While these defects are less severe than that seen in K14E6 mice, they suggest a possible contribution of Dlg inactivation in mediating E6's hyperplastic effects.

Comparison of the phenotypes of the lens in K14E6 and αAcryE6 mice, when expression of E6 was targeted to the lens epithelium versus the lens fiber cells, demonstrates the different consequences that E6 can have depending upon the differentiated state of the cell. In the poorly differentiated lens epithelium, the major consequence is an induction of hyperplasia with an associated delay in the onset of differentiation, whereas, in the highly differentiated compartment, hyperplasia is absent, however specific events in late steps in terminal differentiation, *i.e.* denucleation, are inhibited. These observations may reflect different roles of E6 in modulating cellular properties throughout the life cycle of HPVs, which is tied to the differentiation of the host epithelium.

Another important insight gained from the study of the K14E6 mice came from an analysis of DNA damage responses. It had earlier been demonstrated that HPV16 E6 can inhibit DNA damage responses in tissue culture (Foster et al., 1994; Kessis et al., 1993). As seen in tissue culture, expression of E6 *in vivo* led to a complete abrogation of DNA damage responses in the epidermis of K14E6 transgenic mice in response to ionizing radiation (Song et al., 1998). The normal response to ionizing radiation in the epidermis is a transient inhibition of DNA synthesis, which is mediated through the cdk inhibitor p21 (Song and Lambert, 1999). Surprisingly, E6 could more effectively inhibit this response compared to that observed in p53-null mice, arguing that E6 must abrogate DNA damage responses through p53-dependent as well as p53-independent pathways (Song et al., 1998). This p53-independent pathway has not been identified. Whether it relates to the p53-independent pathway by which E6 inhibits apoptosis in the lens fiber cells is also unknown.

Perhaps the most significant insight gained to date from the analysis of the K14E6 mice is that E6 can contribute to multiple stages of carcinogenesis. K14E6 mice spontaneously develop squamous cell carcinomas (Song et al., 1999). Treatment of the mice with the tumour initiator DMBA or the tumour promoter TPA revealed that E6 contributes weakly to the 'promotion' stage of tumorigenesis involved in the formation of benign tumors or papillomas in the skin, and more strongly to the progression stage, that defines the process by which a benign tumor progresses to a malignant carcinoma (Song et al., 2000). The mechanisms by which HPV16 E6 contributes to these distinct stages of carcinogenesis are distinct. This conclusion was drawn from the observation that, in K14E6$^{\Delta 146\text{-}151}$ mice, E6 is defective for contributing to the promotion stage of carcinogenesis but retains the ability to contribute to progression (Simonson and Lambert, 2005). These same mice are defective for E6-induced epidermal hyperplasia, indicating a correlation between the ability of E6 to contribute to promotion and to induce hyperplasia. Furthermore, this observation indicates that the interaction of E6 with PDZ domain partners likely contributes to promotion. How E6 contributes to tumor progression has yet to be defined unambiguously; however, the reduced ability of K14E6^{I128T} mice, which encode a form of HPV16 E6 protein that is unable to inactivate p53 *in vivo*, to contribute to progression compared to K14E6 mice encoding wild type E6 protein supports the hypothesis that E6's inactivation of p53 contributes to its role in tumor progression (Nguyen et al., 2002a).

It is clear that HPV16 E6 is not sufficient for tumor formation, as only a small percentage of K14E6 mice develop spontaneous skin tumors late in life, even though the gene is expressed throughout the epidermis of the mouse. Consistent with a requirement for additional factors contributing to E6-asssociated cancers, specific karyotypic changes were found in tumors arising in K14E6 mice (Schaeffer et al., 2004), most notably an increased copy number of mouse chromosome 6, which is partially orthologous to human chromosome 3q, a region often found amplified in human cervical cancer (Heselmeyer et al., 1996).

Recently, investigators have generated and characterized transgenic mice expressing the E6 gene of Mastomys natalensis papillomavirus (MnPV) from the K14 promoter. MnPV causes keratoacanthomas of the skin in its natural host (Helfrich et al., 2004). The MnPV E6 transgenic mice spontaneously developed skin tumors in their adult life and when treated topically on their skin with chemical carcinogens had a higher rate of progression than did nontransgenic mice. But unlike the K14E6 transgenic mice expressing the HPV16 E6 gene, the MnPV E6 transgenic mice failed to display either epidermal hyperplasia or an increased rate of promotion in the context of the skin co-carcinogenesis studies. Perhaps this reflects the fact that only the high risk HPVs have the capacity to bind PDZ domain proteins, an activity of HPV16 E6 that was shown to correlate with epidermal hyperplasia and tumor promotion (Nguyen et al., 2003a; Nguyen et al., 2003b).

6. Individual Properties of the E7 Oncogene *In Vivo*

As is the case for E6, much of our understanding of the biological properties of E7 *in vivo* come from studies of the HPV16 E7 gene when placed under the transcriptional control of the αAcry promoter, directing its expression to the lens fiber cell compartment, or the K14 promoter, directing its expression primarily to the basal compartment of stratified squamous epithelia. Below is described the properties that have been identified through the study of these αAcryE7 and K14E7 animals.

6.1. Induction of DNA Synthesis

One of the most striking consequences of E7 expression is induction of DNA synthesis both at a higher rate than in control mice and in differentiating, post-mitotic cells. Expression of E7 in the developing mouse lens using the αAcry promoter resulted in the appearance of many mitotic cells in the posterior and posterior-equatorial regions of the lens, which normally contain differentiated, post-mitotic cells in nontransgenic mice (Pan and Griep, 1994). Similarly, targeting E7 to the basal layer of stratified squamous epithelia using the K14 promoter resulted in hyperplasia of the ear and trunk skin, mouth palate, esophagus, and forestomach (Herber et al., 1996). Assays for the presence of the DNA synthesis marker PCNA or incorporation of the nucleotide analog BrdU showed that hyperplasia coincided with an increased rate of DNA synthesis in the epithelium. Furthermore, DNA synthesis occurred not only in the normally proliferative basal epithelial layer, but also in the normally quiescent suprabasal epithelial layers of ear and trunk skin (Gulliver et al., 1997; Herber et al., 1996).

6.2. Inhibition of differentiation

In many epithelial tissues, the process of differentiation is linked to cell cycle withdrawal. Thus, it is perhaps not surprising that E7-induced cell cycle progression is accompanied by alteration of the affected tissue's normal differentiation program. This is evidenced not only by the induction of DNA synthesis in normally post-mitotic cells, but also by alterations in the morphology of differentiating cells in both E7 transgenic lens and skin (Gulliver et al., 1997, Herber et al., 1996, Pan and Griep, 1994). Examination of differentiation markers indicates that the extent of E7's effects on differentiation vary with tissue site. K14E7 transgenic skin and ear exhibit altered expression patterns of differentiation-specific keratins (Balsitis et al., 2003; Griep et al., 1993; Gulliver et al., 1997), while αAcryE7 transgenic lenses express the lens specific differentiation markers β– and γ–crystallin at reduced levels in the lens fiber cells (Griep, personal communication).

6.3. Induction of apoptosis

E7 can also induce cell death, however the extent and mechanism of this effect appears to vary depending on the targeted tissue. When E7 was expressed in murine retina using the interstitial retinol-binding protein (IRBP) promoter, apoptosis was induced on such a massive scale that the entire photoreceptor layer was lost by 4 weeks of age (Howes et al., 1994). Similarly, αAcryE7 mice exhibit degenerated and misshapen lenses associated with extensive apoptosis in the lens (Pan and Griep, 1994). By contrast, while K14E7 transgenic mice show some increased apoptosis in the skin, hair follicle, sebaceous glands, and the lens epithelium, the net effect of E7 expression is hyperplasia rather than tissue degeneration (Herber et al., 1996; Nguyen et al., 2002b). One interpretation of these findings is that the degree to which E7 induces apoptosis is context dependent and relate to the access to survival factors (Stolen et al., 1997) and/or the differentiation state of the cells in which E7 is expressed.

6.4. Abrogation of DNA damage responses

HPV16 E7 can abrogate DNA damage induced responses. When nontransgenic mice are exposed to 5 Gy ionizing radiation from a ^{137}Cs source, DNA damage responses halt the cell cycle to avoid replication of damaged DNA and allow time for DNA repair to occur. Consequently, little to no DNA synthesis is detectable in the epidermis 24 hours later (Song et al., 1998). By contrast, K14E7 mice show no decrease in DNA synthesis rates in the epidermis after exposure to ionizing radiation. Rather, E7-expressing epidermis continues to replicate DNA exposed to ionizing radiation (Song et al., 1998). This effect may accelerate the accumulation of additional mutations necessary for a high-risk HPV infected cell to become malignant.

6.5. Mutational activity

E7 may accelerate the accumulation of mutations. In cultured cells *in vitro*, E7 is capable of inducing supernumerary centrosomes (more than 2 centrosomes per cell), and these centrosome abnormalities result in multipolar mitoses (Duensing et al., 2001a; Duensing and Munger, 2003). These centrosome abnormalities occur prior to the development of aneuploidy, and it has been hypothesized that centrosome abnormalities may contribute to tumor development by promoting chromosome mis-segregation or breakage during multipolar cell divisions (Duensing et al., 2001a). Examination of K14E7 transgenic mice confirmed that E7 induces centrosome abnormalities *in vivo* (Balsitis et al., 2003). However, examination of multiple tumor cell lines derived from K14E7 mice showed no correlation between frequency of centrosome abnormalities and severity of karyotypic instability (Schaeffer et al., 2004). Thus, while E7 clearly induced supernumerary centrosomes both *in vitro* and *in vivo*, the significance of this effect in tumor development requires further investigation.

6.6. Tumour induction

Multiple mouse models have demonstrated that E7 has potent tumorigenic activity. However, E7 is not sufficient to induce tumorigenesis alone in any tissue tested. Due to the extensive apoptosis and tissue degeneration observed in the retina of IRBP-E7 mice and the lens of αAcryE7 mice, tumors were not observed in these mice except for a single tumor in one αAcryE7 mouse (Howes et al., 1994; Pan and Griep, 1994). However, apoptosis was reduced and tumors occurred when E7 was expressed in the lens or retina of p53-null mice (Howes et al., 1994; Pan and Griep, 1995).

Consistent with the less severe apoptotic effect observed in the skin of K14E7 mice, spontaneous tumors occur in the skin of these mice without a requirement for p53 inactivation. However, tumors occur in only about 10% of these mice and only after a long latency period, indicating additional genetic changes are necessary for tumors to develop (Herber et al., 1996). The role of E7 in skin tumorigenesis was further elucidated by treating mice with DMBA or TPA. In these studies, E7 was found to increase the formation of benign tumors in DMBA/TPA treated mice or mice treated with DMBA alone. E7, however, was not able to initiate tumors in mice treated with TPA alone, or to induce progression to malignancy at a rate higher than seen in control mice. These observations show that E7 acts primarily at the tumor promotion stage in mouse skin, rather than at the tumor initiation or progression stages (Song et al., 2000). The co-carcinogenesis studies lead to the conclusion that E7 promotes the formation of benign tumors, and E6 acts primarily to accelerate progression of benign tumors to the malignant stage.

A subsequent study found that expression of dominant-negative c-jun in the skin of K14E7 mice inhibits E7's promotion activity in DMBA/TPA carcinogensis studies, but does not affect E7's ability to induce DNA synthesis or hyperplasia or disrupt DNA damage responses. These observations indicate no role for c-jun in the acute effects of E7 expression, but c-jun activation may be an important component of E7's tumor promotion activity (Young et al., 2002).

Although all of these studies suggest additional genetic changes are required for E7-expressing cells to form tumors, little additional information is available to identify the nature of specific mutations involved in the development of E7-expressing tumors. Karyotypic analysis of tumor cell lines derived from DMBA/TPA-treated K14E7 mice revealed no consistent chromosomal changes in the tumor cell lines (Schaeffer et al., 2004). Thus, many questions remain about the transition from E7-expressing cell to E7-expressing tumor.

6.7. Functional interactions with tumour suppressors

p53 inactivation attenuates E7-induced apoptosis in the retina and lens, indicating that E7 induces apoptosis in part through p53-dependent pathways in these tissues (Howes et al., 1994; Pan and Griep, 1995). Clearly p53-independent pathways also contribute to E7's induction of apoptosis (Pan and Griep, 1995). In several studies, mice transgenic for E7 mutants which cannot bind or cannot degrade the pRb tumor suppressor protein failed to induce any of the tested phenotypes of E7, including DNA synthesis, hyperplasia, altered differentiation, impaired DNA damage responses, or tumor formation (Gulliver et al., 1997; Pan and Griep, 1995; Song et al., 1998). These results are consistent with pRb inactivation playing a critical role in all of these activities of E7. However, these studies are not definitive because these E7 mutants are also deficient for a number of other activities and their expression could not be verified due to the lack of good antibodies against E7.

Additional evidence comes from comparisons of the effects of pRb inactivation to the effects of E7 expression. Germline inactivation of pRb recapitulates the apoptotic effect of E7 in the mouse lens, and somatic pRb inactivation in the epidermis reproduces E7's effects on DNA synthesis, differentiation, centrosome abnormalities, and DNA damage responses (Balsitis et al., 2003; Pan and Griep, 1995). Also consistent with a role of Rb in mediating E7's effects, the biological activities of E7 were found to be greatly attenuated when αAcryE7 transgenic mice were placed on a genetic background nulligenic for E2F1, a transcription factor normally regulated by pRb, and deregulated in the presence of E7 (McCaffrey et al., 1999). While these correlations support a major role for pRb inactivation in the phenotypes of E7, E7 was also shown to produce increased hyperplasia and dysplasia when expressed in pRb-deficient skin (Balsitis et al., 2003). Combined, these observations suggest that pRb inactivation is very important in the functioning of E7, but other molecular targets of E7 are likely to play additional roles.

7. Mouse Models for Assessing the Role of HPV in Cervical Cancer

High-risk HPVs, such as types 16 and 18, account for over 90% of human cervical cancers. This section highlights the mouse models that target the female reproductive tract as well as discussing the role of estrogen in cervical cancer progression.

Prior to the development of HPV transgenic mice, a study employed the use of a retroviral delivery system to transfer the HPV16 E6 and E7 oncogenes to the murine uterine cervix (Sasagawa et al., 1992). The recombinant retrovirus containing the E6 and E7 genes was delivered to the vaginal cavity of CD-1 nude or C57BL/6J mice. Reproductive tract analysis of CD-1 nude mice, at 9 months, and C57BL/6J, at 15 months, revealed a percentage of the mice had developed high grade dysplasias. None of these mice developed squamous cell carcinomas. Cancer developed only when the mice were treated with the E6/E7 recombinant retrovirus in combination with either the tumor promoter TPA or the mutagen MNNG. The study also revealed that immunosuppression promoted HPV-induced cervical carcinogenesis as CD-1 nude mice had a shorter latency in neoplastic progression.

7.1. HPV and estrogen

Models for studying cervical cancer improved with the generation of a HPV16 transgenic mouse (Arbeit et al., 1994). In this model, the human K14 promoter targets HPV gene expression to the basal layer of the squamous epithelium including the uterine cervix. These mice express the entire early region of the HPV genome minus the LCR region. Virgin female mice were treated with 17-β-estradiol chronically in the form of a 60 day continual release pellet starting at 4 weeks of age for a period of up to 6 months (Arbeit et al., 1996). Chronic estrogen treatment places the mouse in constant estrus and prevents the mouse from estrogen cycling. Estrogen treatment of nontransgenic mice increased the proliferative capacity in the epithelia of the reproductive tract, resulting in a thickened or hyperplastic epithelium that retained the ability to differentiate normally. When treated with estrogen, K14HPV16 transgenic mice progressed from low grade dysplasia to high grade dysplasia, or carcinoma *in situ* (CIS) of the reproductive tract by 6 months of treatment. At the dose of 0.25 mg of estrogen, 60% of transgenic mice developed CIS or frank cancers. Untreated transgenic mice never developed spontaneous cervical cancer when monitored up to 1.5 years of age. The levels of transgene expression were monitored throughout the course of estrogen treatment and neoplastic progression. E6 and E7 mRNA levels remained constant in both cases and did not increase in frank cancer. Estrogen treatment, which induces persistent estrus, did not alter the levels of transgene expression when the treated mice were compared to untreated transgenic mice in estrus; however, levels of transgene expression were not detectable in diestrus in untreated mice.

After testing various doses of estrogen, 0.05 mg was determined to be the lowest dose which still produced physiological levels of estrogen in the mouse epithelium that mimicked estrus and prevented cycling. Six month treatment of transgenic mice at this dose produced a range of neoplastic progression, with a slight majority of cancers occurring in the transformation zone (Elson et al., 2000), the same region in which cervical cancers commonly arise in women. An early event in neoplastic progression is the appearance of squamous metaplasia in the transformation zone, which demarks the boundary between columnar and stratified epithelium. This metaplasia could arise from the alteration of the differentiation state of columnar epithelial cells. Alternatively, metaplastic events could arise directly from the multipotent glandular reserve stem cells, which have the potential to differentiate into either squamous or columnar cells depending on the cue. These glandular reserve cells express low levels of K14 and K18 (Tsutsumi et al., 1993), markers for squamous and columnar cells respectively. In K14HPV transgenic mice, these glandular reserve cells can also express HPV oncogenes, E6 and E7, because of the promoter driving their expression. Possible theories on glandular reserve cell fate favoring squamous metaplasia are E7's interaction with M2 pyruvate kinase resulting in a drop in pH, a persistence in cell cycle dysregulation by E6 and E7 oncogenes, and estrogen receptor α (ER α) signaling.

The individual roles of HPV16 E6 and E7 in cervical carcinogenesis were elucidated by the further study of K14E6 (Song et al., 1999) and K14E7 (Herber et al., 1996) mice. These transgenic mice were treated with 0.05mg of estrogen for 6 months, similar to K14HPV16 transgenic mice, in order to identify the individual roles of E6 and E7 (Riley et al., 2003). A similar percentage of K14E7 and K14HPV16 mice developed cervical cancers. All of the K14E7 mice developed high-grade dysplasias or microinvasive cancers (MIC). K14E6 mice, however, did not develop any cancer and only displayed CIN I (cervical intraepithelial neoplasia) to CIN II lesions. Higher E6 mRNA levels relative to K14HPV16 mice ruled out the possibility that E6 levels were insufficient for cervical carcinogenesis. E6 function was confirmed by scoring for centrosome abnormalities and p53 protein expression. Tumors were larger in estrogen treated K14E6/K14E7 doubly transgenic mice, leading to the conclusion that E6 contributes late in the cervical carcinogenesis development. The results from the cervical carcinogenesis studies with K14E6 and K14E7 mice differ from the findings made in the skin of these same mice. In the cervix, E7 clearly is the more potent oncogene. In contrast, E6 is the more potent oncogene in the skin, as K14E6 mice spontaneously develop primarily skin carcinomas, whereas K14E7 mice develop primarily benign skin papillomas (Song et al., 2000).

The effect of prolonged treatment with estrogen has been investigated (Brake and Lambert, 2005). K14E7 and K14E6/K14E7 mice were treated with the same dose of estrogen for nine months, instead of six. In addition, cohorts of mice were treated with estrogen for an initial six-month period and then estrogen was withdrawn for 3 months to study the dependence on estrogen on tumor growth. Results from this long-term study confirmed that the presence of E6 greatly contributed to tumor size and severity. Interestingly, when exogenous estrogen was withdrawn for 3 months, there was a reduction in the percentage of mice developing cervical cancer from 100% in both E7 and E6E7 mice to 38% and 82% respectively. In addition, the area of tumor invasion severely decreased by approximately 85% in the highest case. These results argue for the necessity of estrogen in the induction, maintenance and progression of cervical carcinogenesis in combination with HPV oncogenes.

7.2. New biomarkers

Studies using K14E6 and K14E7 mice have helped the identification of a new biomarker for human cervical cancer. Current biomarkers for cervical cancer include Ki67 and cyclin E (Keating et al., 2001). Immunohistological analysis of the estrogen treated HPV transgenic mice with antibodies to these known biomarkers for cervical cancer showed similar staining patterns as in human cervical cancer, thus further confirming the validity of this mouse model for human cervical cancer. MCM7, an E2F responsive gene, is up-regulated following E7's inactivation of Rb, and staining of mouse cervical lesions for MCM7 showed that MCM7 expression correlated with neoplastic progression. This was found to be the case also for human cervical cancer, offering the possibility of using MCM7 as a new biomarker for human cervical cancer (Brake et al., 2003).

7.3. Potential treatments

The ultimate goal in cervical cancer research is to identify potential targets for either chemoprevention or develop therapeutics. One potential therapeutic stemming from these animal studies is the anti-estrogen phytochemical, indole-3-carbinol (I3C) found in cruciferous vegetables or in the form of supplements. Tumor incidence was drastically decreased from 76% to 8% when transgenic mice were fed a diet supplemented with I3C (Jin et al., 1999). I3C caused a reduction of the estrogen-induced hyperproliferation of the reproductive tract epithelium. Although estrogen did not have an effect on ear epithelium, transgenic mice on the I3C diet showed a reduction in the severity of skin dysplasia and skin cancer lesions. Thus I3C was effective in opposing the effects of estrogen.

I3C is metabolized into diindolylmethane (DIM) as a result from acid-catalyzed condensation in the stomach (Grose and Bjeldanes, 1992) after a meal of cruciferous vegetables. DIM was also tested to see its effectiveness in reducing the severity of lesions. DIM was more potent and effective at inducing apoptosis in cervical cancer cell lines than I3C (Chen et al., 2001). These chemicals induced apoptosis via DNA strand breaks specifically in transformed cells in the cervical epithelium. This specificity, however, did not necessitate the presence of HPV, as both were effective in inducing apoptosis in HPV negative cervical cancer cell lines and MCF breast carcinoma cell lines. The mechanism by which I3C/DIM work is still somewhat unclear, but several possibilities exist. I3C has been shown to inhibit transcription of cdk6 (Cram et al., 2001) resulting in cell cycle inhibition (Cover et al., 1998). Furthermore, the condensation products of I3C, including DIM, are ligands for the aryl hydrocarbon receptor (Chen et al., 1996) and alter the metabolism of estrogen in both breast and cervical cancer cell lines. In cervical cancer cell lines, I3C has been shown to influence estrogen metabolism to produce an anti-estrogen derivative, 2-hydroxyestrone (Yuan et al., 1999) competing with estrogen for receptor binding and reducing the production of 16α-hydroxyestrone, a metabolite that has been linked to carcinogenic activity (Auborn et al., 1991; Yuan et al., 1999). Based on these *in vivo* studies, I3C could potentially serve as a potential therapeutic for cervical cancer.

8. Use of HPV Transgenic Mice for Immunological Studies

Multiple investigators have made use of HPV transgenic mice to investigate host immune responses to viral antigens, and develop protocols for inducing antigen specific therapeutic immune responses. We summarize those studies in this section.

Even though viral genes expressed as transgenes could be considered self-antigens, multiple studies have demonstrated that HPV transgenic mice can mount new immune responses to E7 when appropriately stimulated. When immunized with E7 protein, the HPV transgenic mouse line, αAcryE6E7 line 19, which develop squamous cell carcinomas (Griep et al., 1993), mounted antibody and T helper (Th) responses to E7 that were indistinguishable from those seen in immunized syngenic, nontransgenic mice (Frazer et al., 1995; Herd et al., 1997). Cytotoxic T Lymphocyte (CTL) responses could not be evaluated in line 19 αAcryE6E7 mice, as there are no identifiable E7 CTL epitopes on the FVB (H-2^q) genetic background (Khammanivong et al., 2003). Interestingly, non immunized line 19 mice, when allowed to age and develop skin lesions, developed spontaneous E7 specific immune responses (Frazer et al., 1995). While these antigen specific responses neither contributed nor modulated the skin disease that arises in these mice, nonspecific, local inflammatory responses do contribute positively to skin disease (Hilditch-Maguire et al., 1999). K14E7 mice expressing E7 from the keratin 14 promoter display a split tolerance to E7 characterized by an ability to mount normal Th and B cell responses to E7, but an inability to mount E7-specific CTL responses on genetic backgrounds (*i.e.* H-2b) in which E7 epitopes for CTL do exist (Doan et al., 1998; 2000; Frazer et al., 1998).

The absence of CTL responses in K14E7 mice was associated with a non-specific down-regulation of CD8-positive T cells (Tindle et al., 2001) probably attributable to transgene associated disturbance of the thymic architecture (Malcolm et al., 2003). Similar split tolerance was observed in HPV transgenic mice in which E7 expression was directed by the K10 promoter (Borchers et al., 1999). The underlying reason for this split tolerance is not understood, but could reflect the fact that keratin promoters are active in thymic epithelial cells (Frazer et al., 1998).

The expression of HPV antigens in the skin of transgenic mice has allowed the investigation of immune recognition of viral antigens in immunological naïve mice. Surprisingly, E7 was found not to function as a classical minor transplantation antigen when expressed in mouse epidermis. Skin from K14E7 mice, when grafted onto nontransgenic mice, was not rejected even when these mice were immunized against E7 and developed E7-specific CTL responses (Dunn et al., 1997; Frazer et al., 2001). Similarly, grafts in which E6 antigen was expressed from the K14 promoter were also accepted

(Matsumoto et al., 2004). However, recipient animals could be induced to reject the skin grafts in an E7-specific manner by stimulating systemic proinflammatory responses (Frazer et al., 2001), or through passive transfer of E7 specific CTLs in combination with E7-specific immunization (Matsumoto et al., 2004).

Why it is so difficult for the immune system to see E7 antigen in the context of the skin remains unclear. Perhaps E7 can inhibit host immune responses. In one study, a comparison of levels of expression of various IFN responsive genes in E7 transgenic versus non-transgenic mice suggested that E7 can suppress innate immune responses through its interaction with IRF1 (Um et al., 2000). In another study, however, no effect by E7 on interferon's induction of MHC class I (a major responder to IFN) was noted when keratinocytes from K14E7 transgenic mice were compared to those from non-transgenic mice (Leggatt et al., 2002). An alternative explanation for why E7 is so poorly seen by the host immune response is that E7, a non-secreted antigen, is poorly cross-presented by professional antigen presenting cells and/or that the local immune environment of the skin is not conducive to the host immune system recognizing keratinocyte-expressed antigens. Regardless of the mechanism, the insights gained from these grafting studies have clear implications for the design of effective therapeutic vaccines for treating patients infected with HPV.

9. Summary

The use of transgenic mice has provided further understanding of the biological properties of papillomavirus oncogenes, unequivocal demonstration of their cancer causing potential, and useful animal models with which to investigate novel approaches to preventing, detecting and treating HPV-associated cancers. This area of research is likely to continue to provide important new insights for the papillomavirus field.

Acknowledgements

We thank Drs. Anne Griep and Ian H. Frazer for editorial comments.

References

Adam, J. L., Briggs, M. W., and McCance, D. J. (2000). A mutagenic analysis of the E5 protein of human papillomavirus type 16 reveals that E5 binding to the vacuolar H+-ATPase is not sufficient for biological activity, using mammalian and yeast expression systems. Virol. *272*, 315-325.

Arbeit, J., Howley, P. M., and Hanahan, D. (1996). Chronic estrogen-induced cervical and vaginal squamous carcinogenesis in human papillomavirus type 16 transgenic mice. Proc. Natl. Acad. Sci. *USA 93*, 2930-2935.

Arbeit, J., Munger, K., Howley, P. M., and Hanahan, D. (1993). Neurorpithelial carcinomas in mice transgenic with human papillomavirus type 16 E6/E7 ORFs. Am. J. Pathol. *142*, 1187-1197.

Arbeit, J., Munger, K., Howley, P. M., and Hanahan, D. (1994). Progressive squamous epithelial neoplasia in K14-Human papillomavirus type 16 transgenic mice. J. Virol. *68*, 4358-4368.

Auborn, K. J., Woodworth, C., DiPaolo, J. A., and Bradlow, H. L. (1991). The interaction between HPV infection and estrogen metabolism in cervical carcinogenesis. Int. J. Cancer *49*, 867-869.

Auewarakul, P., Gissmann, L., and Cid, A. A. (1994). Targeted expression of the E6 and E7 oncogenes of human papillomavirus type 16 in the epidermis of transgenic mice elicits generalized epidermal hyperplasia involving autocrine factors. Mol. Cell. Biol. *14*, 8250-8258.

Balsitis, S. J., Sage, J., Duensing, S., Munger, K., Jacks, T., and Lambert, P. F. (2003). Recapitulation of the effects of the HPV-16 E7 oncogene on mouse epithelium by somatic Rb deletion and detection of pRb-independent effects of E7 *in vivo*. Mol. Cell. Biol. *23*, 9094-9103.

Bedell, M. A., Jones, K. H., and Laimins, L. A. (1987). The E6-E7 region of human papillomavirus type 18 is sufficient for transformation of NIH 3T3 and rat-1 cells. J. Virol. *61*, 3635-3640.

Borchers, A., Braspenning, J., Meijer, J., Osen, W., Gissmann, L., and Jochmus, I. (1999). E7-specific cytotoxic T cell tolerance in HPV-transgenic mice. Arch. Virol. *144*, 1539-1556.

Bossy-Wetzel, E., Bravo, R., and Hanahan, D. (1992). Transcription factors junB and c-jun are selectively up-regulated and functionally implicated in fibrosarcoma development. Genes Dev. *6*, 2340-2351.

Brake, T., Connor, J. P., Petereit, D. G., and Lambert, P. F. (2003). Comparative analysis of cervical cancer in women and in a human papillomavirus-transgenic mouse model: identification of minichromosome maintenance protein 7 as an informative biomarker for human cervical cancer. Cancer Res. *63*, 8173-8180.

Brake, T., and Lambert, P.F. (2005). Estrogen contributes to the onset, persistence, and malignant progression of cervical cancer in a human papillomavirus-transgenic mouse model.Proc. Natl. Acad. Sci. USA *102*, 2490-2495.

Briggs, M. W., Adam, J. L., and McCance, D. J. (2001). The human papillomavirus type 16 E5 protein alters vacuolar H(+)-ATPase function and stability in *Saccharomyces cerevisiae*. Virol. *280*, 169-175.

Carraresi, L., Tripodi, S. A., Mulder, L. C., Bertini, S., Nuti, S., Schuerfeld, K., Cintorino, M., Bensi, G., Rossini, M., and Mora, M. (2001). Thymic hyperplasia and lung carcinomas in a line of mice transgenic for keratin 5-driven HPV16 E6/E7 oncogenes. Oncogene *20*, 8148-8153.

Chen, D. Z., Qi, M., Auborn, K. J., and Carter, T. H. (2001). Indole-3-carbinol and diindolylmethane induce apoptosis of human cervical cancer cells and in murine HPV16-transgenic preneoplastic cervical epithelium. J. Nutr. *131*, 3294-3302.

Chen, I., Safe, S., and Bjeldanes, L. (1996). Indole-3-carbinol and diindolylmethane as aryl hydrocarbon (Ah) receptor agonists and antagonists in T47D human breast cancer cells. Biochem. Pharmacol. *51*, 1069-1076.

Choo, K. B., Chong, K. Y., Liew, L. N., Hsu, H. C., and Cheng, W. T. (1992). Unregulated and basal transcriptional activities of the regulatory sequence of the type 18 human papillomavirus genome in transgenic mice. Virol. *188*, 378-383.

Choo, K. B., Liew, L. N., Liew, S. J., and Cheng, W. T. (1993). Hypermethylation of the human papillomavirus transgenome in transgenic mice. Proc. Natl. Sci. Council, Republic of China Part B, Life Sciences *17*, 159-163.

Comerford, S., Maika, S., Laimins, L. A., Messing, A., Elsasser, H., and Hammer, R. (1995). E6 and E7 expression from the HPV 18 LCR: development of genital hyperplasia and neoplasia in transgenic mice. Oncogene *10*, 587-597.

Conrad, M., Bubb, V. J., and Schlegel, R. (1993). The human papillomavirus type 6 and 16 E5 proteins are membrane-associated proteins which associate with the 16-kilodalton pore-forming protein. J. Virol. *67*, 6170-6178.

Cook, P. W., Piepkorn, M., Clegg, C. H., Plowman, G. D., DeMay, J. M., Brown, J. R., and Pittelkow, M. R. (1997). Transgenic expression of the human amphiregulin gene induces a psoriasis-like phenotype. J. Clin. Invest. *100*, 2286-2294.

Coussens, L., Hanahan, D., and Arbeit, J. (1996). Genetic predisposition and parameters of malignant progression in K14-HPV16 transgenic mice. Am. J. Pathol. *149*, 1899-1917.

Cover, C. M., Hsieh, S. J., Tran, S. H., Hallden, G., Kim, G. S., Bjeldanes, L. F., and Firestone, G. L. (1998). Indole-3-carbinol inhibits the expression of cyclin-dependent kinase-6 and induces a G1 cell cycle arrest of human breast cancer cells independent of estrogen receptor signaling. J. Biol. Chem. *273*, 3838-3847.

Cram, E. J., Liu, B. D., Bjeldanes, L. F., and Firestone, G. L. (2001). Indole-3-carbinol inhibits CDK6 expression in human MCF-7 breast cancer cells by disrupting Sp1 transcription factor interactions with a composite element in the CDK6 gene promoter. J. Biol. Chem. *276*, 22332-22340.

DiMaio, D., Guralski, D., and Schiller, J. T. (1986). Translation of open reading frame E5 of bovine papillomavirus is required for its transforming activity. Proc. Natl. Acad. Sci. USA *83*, 1797-1801.

DiMaio, D., and Mattoon, D. (2001). Mechanisms of cell transformation by papillomavirus E5 proteins. Oncogene *20*, 7866-7873.

Disbrow, G. L., Sunitha, I., Baker, C. C., Hanover, J., and Schlegel, R. (2003). Codon optimization of the HPV-16 E5 gene enhances protein expression. Virol. *311*, 105-114.

Doan, T., Chambers, M., Street, M., Fernando, G. J., Herd, K., Lambert, P., and Tindle, R. (1998). Mice expressing the E7 oncogene of HPV16 in epithelium show central tolerance, and evidence of peripheral anergising tolerance, to E7-encoded cytotoxic T-lymphocyte epitopes. Virol. *244*, 352-364.

Doan, T., Herd, K. A., Lambert, P. F., Fernando, G. J., Street, M. D., and Tindle, R. W. (2000). Peripheral tolerance to human papillomavirus E7 oncoprotein occurs by cross-tolerization, is largely Th-2-independent, and is broken by dendritic cell immunization. Cancer Res. *60*, 2810-2815.

Dominey, A. M., Wang, X. J., King, L. E., Jr., Nanney, L. B., Gagne, T. A., Sellheyer, K., Bundman, D. S., Longley, M. A., Rothnagel, J. A., Greenhalgh, D. A., and et al. (1993). Targeted overexpression of transforming growth factor alpha in the epidermis of transgenic mice elicits hyperplasia, hyperkeratosis, and spontaneous, squamous papillomas. Cell Growth Differ. *4*, 1071-1082.

Drummond, B. D., Vaillancourt, R. R., Kazlauskas, A., and DiMaio, D. (1995). Ligand-independent activation of the platelet-derived growth factor beta receptor: requirements for bovine papillomavirus E5-induced mitogenic signaling. Mol. Cell. Biol. *15*, 2570-2581.

Duensing, S., Duensing, A., Crum, C. P., and Munger, K. (2001a). Human papillomavirus type 16 E7 oncoprotein-induced abnormal centrosome synthesis is an early event in the evolving malignant phenotype. Cancer Res. *61*, 2356-2360.

Duensing, S., and Munger, K. (2003). Human Papillomavirus Type 16 E7 oncoprotein can induce abnormal centrosome duplication through a mechanism independent of inactivation of retinoblastoma protein family members. J. Virol. *77*, 12331-12335.

Dunn, L. A., Evander, M., Tindle, R. W., Bulloch, A. L., de Kluyver, R. L., Fernando, G. J., Lambert, P. F., and Frazer, I. H. (1997). Presentation of the HPV16E7 protein by skin grafts is insufficient to allow graft rejection in an E7-primed animal. Virol. *235*, 94-103.

Dvoretzky, I., Shober, R., Chattopadhyay, S. K., and Lowy, D. R. (1980). A quantitative *in vitro* focus assay for bovine papilloma virus. Virol. *103*, 369-375.

Elson, D., Riley, R., Lacey, A., Thordarson, G., Talamantes, F., and Arbeit, J. (2000). Sensitivity of the Cervical Transformation Zone to Estrogen-Induced Squamous Carcinogenesis. Cancer Res. *60*, 1267-1275.

Escalante-Alcalde, D., Recillas-Targa, F., Valencia, C., Santa-Olalla, J., Chavez, P., Marroquin, A., Gutierrez, X., Gariglio, P., and Covarrubias, L. (2000). Expression of E6 and E7 papillomavirus oncogenes in the outer root sheath of hair follicles extends the growth phase and bypasses resting at telogen. Cell Growth Differ. *11*, 527 539.

Foley, J. D., Rosenbaum, H., and Griep, A. E. (2004). Temporal regulation of VEID-7-amino-4-trifluoromethylcoumarin cleavage activity and caspase-6 correlates with organelle loss during lens development. J. Biol. Chem. *279*, 32142-32150.

Foster, S. A., Demers, G. W., Etscheid, B. G., and Galloway, D. A. (1994). The ability of human papillomavirus E6 proteins to target p53 for degradation *in vivo* correlates with their ability to abrogate actinomycin D-induced growth arrest. J. Virol. *68*, 5698-5705.

Frazer, I. H., Fernando, G. J., Fowler, N., Leggatt, G. R., Lambert, P. F., Liem, A., Malcolm, K., and Tindle, R. W. (1998). Split tolerance to a viral antigen expressed in thymic epithelium and keratinocytes. Eur. J. Immunol. *28*, 2791-2800.

Frazer, I. H., Kluyver, R. D., Leggatt, G. R., Yang Guo, H., Dunn, L., White, O., Harris, C., Liem, A., and Lambert, P. (2001). Tolerance or immunity to a tumor antigen expressed in somatic cells can be determined by systemic proinflammatory signals at the time of first antigen exposure. J. Immunol. *167*, 6180-6187.

Frazer, I. H., Leippe, D. M., Dunn, L. A., Liem, A., Tindle, R. W., Fernando, G. J., Phelps, W. C., and Lambert, P. F. (1995). Immunological responses in human papillomavirus 16 E6/E7-transgenic mice to E7 protein correlate with the presence of skin disease. Cancer Res. *55*, 2635-2639.

Gardiol D, Kuhne C, Glaunsinger B, Lee SS, Javier R, Banks L. (1999) Oncogenic human papillomavirus E6 proteins target the discs large tumour suppressor for proteasome-mediated degradation. Oncogene *18*, 5487-5496.

Goldstein, D. J., Andresson, T., Sparkowski, J. J., and Schlegel, R. (1992). The BPV-1 E5 protein, the 16 kDa membrane pore-forming protein and the PDGF receptor exist in a complex that is dependent on hydrophobic transmembrane interactions. Embo J. *11*, 4851-4859.

Goldstein, D. J., Finbow, M. E., Andresson, T., McLean, P., Smith, K., Bubb, V., and Schlegel, R. (1991). Bovine papillomavirus E5 oncoprotein binds to the 16K component of vacuolar H(+)-ATPases. Nature *352*, 347-349.

Goldstein, D. J., Li, W., Wang, L. M., Heidaran, M. A., Aaronson, S., Shinn, R., Schlegel, R., and Pierce, J. H. (1994). The bovine papillomavirus type 1 E5 transforming protein specifically binds and activates the beta-type receptor for the platelet-derived growth factor but not other related tyrosine kinase-containing receptors to induce cellular transformation. J. Virol. *68*, 4432-4441.

Greenhalgh, D. A., Wang, X. J., Rothnagel, J. A., Eckhardt, J. N., Quintanilla, M. I., Barber, J. L., Bundman, D. S., Longley, M. A., Schlegel, R., and Roop, D. R. (1994). Transgenic mice expressing targeted HPV-18 E6 and E7 oncogenes in the epidermis develop verrucous lesions and spontaneous, rasHa-activated papillomas. Cell Growth and Differentiation *5*, 667-675.

Griep, A., Herber, R., Jeon, S., Lohse, J., Dubielzig, R., and Lambert, P. F. (1993). Tumorigenicity by human papillomavirus type 16 E6 and E7 in transgenic mice correlates with alterations in epithelial cell growth and differentiation. J. Virol. *67*, 1373-1384.

Griep, A., Krawcek, J., Lee, D., Liem, A., Albert, D., Carabeo, R., Drinkwater, N., McCall, M., Sattler, C., Lasudry, J., and Lambert, P. F. (1998). Multiple genetic loci modify risk for retinoblastoma in transgenic mice. Invest. Ophth Vis. Sci. *39*, 2723-2732.

Grose, K. R., and Bjeldanes, L. F. (1992). Oligomerization of indole-3-carbinol in aqueous acid. Chem. Res. Toxicol. *5*, 188-193.

Gulliver, G. A., Herber, R. L., Liem, A., and Lambert, P. F. (1997). Both conserved region 1 (CR1) and CR2 of the human papillomavirus type 16 E7 oncogene are required for induction of epidermal hyperplasia and tumor formation in transgenic mice. J. Virol. *71*, 5905-5914.

Hawley-Nelson, P., Vousden, K. H., Hubbert, N. L., Lowy, D. R., and Schiller, J. T. (1989). HPV16 E6 and E7 proteins cooperate to immortalize human foreskin keratinocytes. Embo J. *8*, 3905-3910.

Helfrich, I., Chen, M., Schmidt, R., Furstenberger, G., Kopp-Schneider, A., Trick, D., Grone, H. J., Zur Hausen, H., and Rosl, F. (2004). Increased incidence of squamous cell carcinomas in Mastomys natalensis papillomavirus E6 transgenic mice during two-stage skin carcinogenesis. J. Virol. *78*, 4797-4805.

Herber, R., Liem, A., Pitot, H., and Lambert, P. F. (1996). Squamous epithelial hyperplasia and carcinoma in mice transgenic for the human papillomavirus type 16 E7 oncogene. J. Virol. *70*, 1873-1881.

Herd, K., Fernando, G. J., Dunn, L. A., Frazer, I. H., Lambert, P., and Tindle, R. W. (1997). E7 oncoprotein of human papillomavirus type 16 expressed constitutively in the epidermis has no effect on E7-specific B- or Th-repertoires or on the immune response induced or sustained after immunization with E7 protein. Virol. *231*, 155-165.

Heselmeyer, K., Schrock, E., du, M. S., Blegen, H., Shah, K., Steinbeck, R., Auer, G., and Ried, T. (1996). Gain of chromosome 3q defines the transition from severe dysplasia to invasive carcinoma of the uterine cervix. Proc. Natl. Acad. Sci. USA *93*, 479-484.

Hilditch-Maguire, P. A., Lieppe, D. M., West, D., Lambert, P. F., and Frazer, I. H. (1999). T cell-mediated and non-specific inflammatory mechanisms contribute to the skin pathology of HPV 16 E6E7 transgenic mice. Intervirol. *42*, 43-50.

Howes, K., Ransom, N., Papermaster, D., Lasudry, J., Albert, D., and Windle, J. (1994). Apoptosis or retinoblastoma: alternative fates of photoreceptors expressing HPV-16 E7 gene in the presence or absence of p53. Genes Dev. *8*, 1300-1310.

Jin, L., Qi, M., Chen, D. Z., Anderson, A., Yang, G. Y., Arbeit, J. M., and Auborn, K. J. (1999). Indole-3-carbinol prevents cervical cancer in human papilloma virus type 16 (HPV16) transgenic mice. Cancer Res. *59*, 3991-3997.

Kandel, J., Bossy-Wetzel, E., Radvanyi, F., Klagsbrun, M., Folkman, J., and Hanahan, D. (1991). Neovascularization is associated with a switch to the export of bFGF in the multistep development of fibrosarcoma. Cell *66*, 1095-1104.

Kang, J. K., Kim, J. H., Lee, S. H., Kim, D. H., Kim, H. S., Lee, J. E., and Seo, J. S. (2000). Development of spontaneous hyperplastic skin lesions and chemically induced skin papillomas in transgenic mice expressing human papillomavirus type 16 E6/E7 genes. Cancer Lett. *160*, 177-183.

Kaur, P., and McDougall, J. K. (1989). HPV-18 immortalization of human keratinocytes. Virol. *173*, 302-310.

Keating, J. T., Ince, T., and Crum, C. P. (2001). Surrogate biomarkers of HPV infection in cervical neoplasia screening and diagnosis. Adv. Anat. Pathol. *8*, 83-92.

Kessis, T. D., Slebos, R. J., Nelson, W. G., Kastan, M. B., Plunkett, B. S., Han, S. M., Lorincz, A. T., Hedrick, L., and Cho, K. R. (1993). Human papillomavirus 16 E6 expression disrupts the p53 mediated cellular response to DNA damage. Proc. Natl. Acad. Sci. USA *90*, 3988-3992.

Khammanivong, V., Liu, X. S., Liu, W. J., Rodda, S. J., Leggatt, G. R., Tindle, R. W., Frazer, I. H., and Fernando, G. J. (2003). Paucity of functional CTL epitopes in the E7 oncoprotein of cervical cancer associated human papillomavirus type 16. Immunol Cell Biol *81*, 1-7.

Kondoh, G., Hayasaka, N., Li, Q., Nishimune, Y., and Hakura, A. (1995). An *in vivo* model for receptor tyrosine kinase autocrine/paracrine activation: auto-stimulated KIT receptor acts as a tumor promoting factor in papillomavirus-induced tumorigenesis. Oncogene *10*, 341-347.

Kondoh, G., Murata, Y., Aozasa, K., Yutsudo, M., and Hakura, A. (1991). Very high incidence of germ cell tumorigenesis (seminomagenesis) in human papillomavirus type 16 transgenic mice. J. Virol. *65*, 3335-3339.

Kondoh, G., Nishimune, Y., Nishizawa, Y., Hayasaka, N., Matsumoto, K., and Hakura, A. (1994). Establishment and further characterization of a line of transgenic mice showing testicular tumorigenesis at 100% incidence. J. Urol. *152*, 2151-2154.

Lacey, M., Alpert, S., and Hanahan, D. (1986). Bovine papillomavirus genome elicits skin tumours in transgenic mice. Nature *322*, 609-612.

Lambert, P. F., Pan, H., Pitot, H., Liem, A., Jackson, M., and Griep, A. (1993). Epidermal cancer associated with expression of human papillomavirus type 16 E6 and E7 oncogenes in the skin of transgenic mice. Proc. Natl. Acad. Sci. *USA 90*, 5583-5587.

Leder, C., Kleinschmidt, J. A., Wiethe, C., and Muller, M. (2001). Enhancement of capsid gene expression: preparing the human papillomavirus type 16 major structural gene L1 for DNA vaccination purposes. J. Virol. *75*, 9201-9209.

Lee, D., Cross, S. H., Strunk, K. E., Morgan, J. E., Bailey, C. L., Jackson, I. J., and Threadgill, D. W. (2004). Wa5 is a novel ENU-induced antimorphic allele of the epidermal growth factor receptor. Mamm. Genome *15*, 525-536.

Leechanachai, P., Banks, L., Moreau, F., and Matlashewski, G. (1992). The E5 gene from human papillomavirus type 16 is an oncogene which enhances growth factor-mediated signal transduction to the nucleus. Oncogene *7*, 19-25.

Leggatt, G. R., Dunn, L. A., De Kluyver, R. L., Stewart, T., and Frazer, I. H. (2002). Interferon-gamma enhances cytotoxic T lymphocyte recognition of endogenous peptide in keratinocytes without lowering the requirement for surface peptide. Immunol. Cell. Biol. *80*, 415-424.

Leptak, C., Ramon, y, Cajal, S, Kulke, R., Horwitz, B. H., Riese, D. d., Dotto, G. P., and DiMaio, D. (1991). Tumorigenic transformation of murine keratinocytes by the E5 genes of bovine papillomavirus type 1 and human papillomavirus type 16 [published erratum appears in J. Virol. 1992 Mar;66(3):1833]. J. Virol. *65*, 7078-7083.

Lindgren, V., Sippola, T. M., Skowronski, J., Wetzel, E., Howley, P. M., and Hanahan, D. (1989). Specific chromosomal abnormalities characterize fibrosarcomas of bovine papillomavirus type 1 transgenic mice. Proc. Natl. Acad. Sci. USA *86*, 5025-5029.

Lowy, D. R., Dvoretzky, I., Shober, R., Law, M. F., Engel, L., and Howley, P. M. (1980). In vitro tumorigenic transformation by a defined sub-genomic fragment of bovine papilloma virus DNA. Nature *287*, 72-74.

Malcolm, K. M., Gill, J., Leggatt, G. R., Boyd, R., Lambert, P., and Frazer, I. H. (2003). Expression of the HPV16E7 oncoprotein by thymic epithelium is accompanied by disrupted T cell maturation and a failure of the thymus to involute with age. Clin. Dev. Immunol. *10*, 91-103.

Massimi P, Gammoh N, Thomas M, Banks L. (2004) HPV E6 specifically targets different cellular pools of its PDZ domain-containing tumour suppressor substrates for proteasome-mediated degradation. Oncogene *23*, 8033-8039.

Matlashewski, G., Schneider, J., Banks, L., Jones, N., Murray, A., and Crawford, L. (1987). Human papillomavirus type 16 DNA cooperates with activated ras in transforming primary cells. Embo J. *6*, 1741-1746.

Matsumoto, K., Leggatt, G. R., Zhong, J., Liu, X., de Kluyver, R. L., Peters, T., Fernando, G. J., Liem, A., Lambert, P. F., and Frazer, I. H. (2004). Impaired antigen presentation and effectiveness of combined active/passive immunotherapy for epithelial tumors. J. Natl. Cancer Inst. *96*, 1611-1619.

McCaffrey, J., Yamasaki, L., Dyson, N. J., Harlow, E., and Griep, A. E. (1999). Disruption of retinoblastoma protein family function by human papillomavirus type 16 E7 oncoprotein inhibits lens development in part through E2F-1. Mol. Cell. Biol. *19*, 6458-6468.

Melero, I., Singhal, M. C., McGowan, P., Haugen, H. S., Blake, J., Hellstrom, K. E., Yang, G., Clegg, C. H., and Chen, L. (1997). Immunological ignorance of an E7-encoded cytolytic T-lymphocyte epitope in transgenic mice expressing the E7 and E6 oncogenes of human papillomavirus type 16. J. Virol. *71*, 3998-4004.

Munger, K., Phelps, W. C., Bubb, V., Howley, P. M., and Schlegel, R. (1989). The E6 and E7 genes of the human papillomavirus type 16 together are necessary and sufficient for transformation of primary human keratinocytes. J. Virol. *63*, 4417-4421.

Nguyen, M., Song, S., Liem, A., Androphy, E., Liu, Y., and Lambert, P. F. (2002a). A mutant of human papillomavirus type 16 E6 deficient in binding alpha-helix partners displays reduced oncogenic potential *in vivo*. J. Virol. *76*, 13039-13048.

Nguyen, M. L., Nguyen, M. M., Lee, D., Griep, A. E., and Lambert, P. F. (2003a). The PDZ ligand domain of the human papillomavirus type 16 E6 protein is required for E6's induction of epithelial hyperplasia *in vivo*. J. Virol. *77*, 6957-6964.

Nguyen, M. M., Nguyen, M. L., Caruana, G., Bernstein, A., Lambert, P. F., and Griep, A. E. (2003b). Requirement of PDZ-containing proteins for cell cycle regulation and differentiation in the mouse lens epithelium. Mol. Cell. Biol. *23*, 8970-8981.

Nguyen, M. M., Potter, S. J., and Griep, A. E. (2002b). Deregulated cell cycle control in lens epithelial cells by expression of inhibitors of tumor suppressor function. Mech. Dev. *112*, 101-113.

Nilson, L. A., and DiMaio, D. (1993). Platelet-derived growth factor receptor can mediate tumorigenic transformation by the bovine papillomavirus E5 protein. Mol. Cell. Biol. *13*, 4137-4145.

Pan, H., and Griep, A. (1994). Altered cell cycle regultion in the lens of HPV-16 E6 or E7 transgenic mice: implications for tumor suppressor gene function in development. Genes Dev. *8*, 1285-1299.

Pan, H., and Griep, A. (1995). Temporally distinct patterns of p53-dependent and p53-independent apoptosis during mouse lens development. Genes Dev *9*, 2157-2169.

Petti, L., and DiMaio, D. (1992). Stable association between the bovine papillomavirus E5 transforming protein and activated platelet-derived growth factor receptor in transformed mouse cells. Proc. Natl. Acad. Sci. USA *89*, 6736-6740.

Petti, L., and DiMaio, D. (1994). Specific interaction between the bovine papillomavirus E5 transforming protein and the beta receptor for platelet-derived growth factor in stably transformed and acutely transfected cells. J. Virol. *68*, 3582-3592.

Petti, L. M., Reddy, V., Smith, S. O., and DiMaio, D. (1997). Identification of amino acids in the transmembrane and juxtamembrane domains of the platelet-derived growth factor receptor required for productive interaction with the bovine papillomavirus E5 protein. J. Virol. *71*, 7318-7327.

Phelps, W. C., Yee, C. L., Munger, K., and Howley, P. M. (1988). The human papillomavirus type 16 E7 gene encodes transactivation and transformation functions similar to those of adenovirus E1A. Cell *53*, 539-547.

Pim, D., Collins, M., and Banks, L. (1992). Human papillomavirus type 16 E5 gene stimulates the transforming activity of the epidermal growth factor receptor. Oncogene *7*, 27-32.

Pirisi, L., Creek, K. E., Doniger, J., and DiPaolo, J. A. (1988). Continuous cell lines with altered growth and differentiation properties originate after transfection of human keratinocytes with human papillomavirus type 16 DNA. Carcinogenesis 9, 1573-1579.

Riese, D. J., 2nd, and DiMaio, D. (1995). An intact PDGF signaling pathway is required for efficient growth transformation of mouse C127 cells by the bovine papillomavirus E5 protein. Oncogene *10*, 1431-1439.

Riley, R., Duensing, S., Brake, T., Munger, K., Lambert, P. F., and Arbeit, J. (2003). Dissection of human papillomavirus E6 and E7 function in transgenic mouse models of cervical carcinogenesis. Cancer Res. *63*, 4862-4871.

Sasagawa, T., Inoue, M., Inoue, H., Yutsudo, M., Tanizawa, O., and Hakura, A. (1992). Induction of uterine cervical neoplasias in mice by human papillomavirus type 16 E6/E7 genes. Cancer Res. *52*, 4420-4426.

Sasagawa, T., Kondoh, G., Inoue, M., Yutsudo, M., and Hakura, A. (1994). Cervical/vaginal dysplasias of transgenic mice harbouring human papillomavirus type 16 E6-E7 genes. J. Gen. Virol..

Schaeffer, A., Nguyen, M., Liem, A., Lee, D., Montagna, C., Lambert, P. F., Ried, T., and Difilippantonio, M. (2004). E7 and E7 oncoproteins induce distinct patterns of chromosomal aneuploidy in skin tumors from transgenic mice. Cancer Res. *64*, 538-546.

Schiller, J. T., Vass, W. C., Vousden, K. H., and Lowy, D. R. (1986). E5 open reading frame of bovine papillomavirus type 1 encodes a transforming gene. J. Virol. *57*, 1-6.

Schlegel, R., Phelps, W. C., Zhang, Y. L., and Barbosa, M. (1988). Quantitative keratinocyte assay detects two biological activities of human papillomavirus DNA and identifies viral types associated with cervical carcinoma. Embo J. *7*, 3181-3187.

Schreiber, K., Cannon, R. E., Karrison, T., Beck-Engeser, G., Huo, D., Tennant, R. W., Jensen, H., Kast, W. M., Krausz, T., Meredith, S. C., et al. (2004). Strong synergy between mutant ras and HPV16 E6/E7 in the development of primary tumors. Oncogene *23*, 3972-3979.

Schwarz, E., Freese, U. K., Gissmann, L., Mayer, W., Roggenbuck, B., Stremlau, A., and zur, H. H. (1985). Structure and transcription of human papillomavirus sequences in cervical carcinoma cells. Nature *314*, 111-114.

Simonson, S.J.S., Difilippantonio, M. and Lambert, P.F. (2005). Two distinct activities contribute to HPV16 E6's oncogenic potential. cancer Res. In press

Sippola, T. M., Hanahan, D., and Howley, P. M. (1989). Cell-heritable stages of tumor progression in transgenic mice harboring the bovine papillomavirus type 1 genome. Mol. Cell. Biol. *9*, 925-934.

Song, S., Gulliver, G. A., and Lambert, P. F. (1998). Human papillomavirus type 16 E6 and E7 oncogenes abrogate radiation-induced DNA damage responses *in vivo* through p53-dependent and p53-independent pathways. Proc. Natl. Acad. Sci. *USA 95*, 2290-2295.

Song, S., and Lambert, P. F. (1999). Different responses of epidermal and hair follicular cells to radiation correlate with distinct patterns of p53 and p21 induction. Am J Pathol *155*, 1121-1127.

Song, S., Liem, A., Miller, J. A., and Lambert, P. F. (2000). Human Papillomavirus Types 16 E6 and E7 Contribute Differently to Carcinogenesis. Virol. *267*, 141-150.

Song, S., Pitot, H. C., and Lambert, P. F. (1999). The human papillomavirus type 16 E6 gene alone is sufficient to induce carcinomas in transgenic animals. J. Virol. *73*, 5887-5893.

Stolen, C. M., Jackson, M. W., and Griep, A. E. (1997). Overexpression of FGF-2 modulates fiber cell differentiation and survival in the mouse lens. Dev. *124*, 4009-4017.

Straight, S. W., Hinkle, P. M., Jewers, R. J., and McCance, D. J. (1993). The E5 oncoprotein of human papillomavirus type 16 transforms fibroblasts and effects the downregulation of the epidermal growth factor receptor in keratinocytes. J. Virol. *67*, 4521-4532.

Tindle, R. W., Herd, K., Doan, T., Bryson, G., Leggatt, G. R., Lambert, P., Frazer, I. H., and Street, M. (2001). Nonspecific down-regulation of CD8+ T-cell responses in mice expressing human papillomavirus type 16 E7 oncoprotein from the keratin-14 promoter. J. Virol. *75*, 5985-5997.

Tsutsumi, K., Sun, Q., Yasumoto, S., Kikuchi, K., Ohta, Y., Pater, A., and Pater, M. M. (1993). In vitro and *in vivo* analysis of cellular origin of cervical squamous metaplasia. Am. J. Pathol. *143*, 1150-1158.

Um, S. J., Kim, E. J., Hwang, E. S., Kim, S. J., Namkoong, S. E., and Park, J. S. (2000). Antiproliferative effects of retinoic acid/interferon in cervical carcinoma cell lines: cooperative growth suppression of IRF-1 and p53. Int. J. Cancer *85*, 416-423.

Vassar, R., Hutton, M. E., and Fuchs, E. (1992). Transgenic overexpression of transforming growth factor alpha bypasses the need for c-Ha-ras mutations in mouse skin tumorigenesis. Mol. Cell. Biol. *12*, 4643-4653.

Williams, S.M.G., Disbrow, G.L., Schlegel, R., Lee, D., Threadgill, D.W. and Lambert, P.F. (2005). Epidermal growth factor receptor mediates hyperplasia induced by E5, a high risk HPV oncogene. Cancer Res. In press

Woodworth, C. D., Bowden, P. E., Doniger, J., Pirisi, L., Barnes, W., Lancaster, W. D., and DiPaolo, J. A. (1988). Characterization of normal human exocervical epithelial cells immortalized *in vitro* by papillomavirus types 16 and 18 DNA. Cancer Res. *48*, 4620-4628.

Yang, J. T., Liu, C. Z., Domer, P., and Iannaccone, P. (1998). Expression of lymphomagenic oncogenes in T-cell lymphomas of HPV 16 transgenic mice. Cancer Det. Prevent. *22*, 405-415.

Yang, J. T., Liu, C. Z., and Iannaccone, P. (1995). The HPV 16 genome induces carcinomas and T-cell lymphomas in transgenic mice. Am. J. Pathol. *147*, 68-78.

Yee, C., Krishnan, H. I., Baker, C. C., Schlegel, R., and Howley, P. M. (1985). Presence and expression of human papillomavirus sequences in human cervical carcinoma cell lines. Am. J. Pathol. *119*, 361-366.

Young, M., Farrell, L., Lambert, P. F., Awasthi, P., and Colburn, N. (2002). Protection against human papillomavirus type 16-E7 oncogene-induced tumorigenesis by *in vivo* expression of dominant negative c-jun. Mol. Carcinogen *34*, 72-77.

Yuan, F., Chen, D. Z., Liu, K., Sepkovic, D. W., Bradlow, H. L., and Auborn, K. (1999). Anti-estrogenic activities of indole-3-carbinol in cervical cells: implication for prevention of cervical cancer. Anticancer Res *19*, 1673-1680.

Zhou, J., Liu, W. J., Peng, S. W., Sun, X. Y., and Frazer, I. (1999). Papillomavirus capsid protein expression level depends on the match between codon usage and tRNA availability. J. Virol. *73*, 4972-4982.

From: Papillomavirus Research: From Natural History to Vaccines and Beyond. Edited by: M. Saveria Campo

Chapter 15

HPV and Oesophageal Carcinoma

Kari Syrjänen*

Abstract

Oesophageal squamous cell carcinoma (SCC) has a peculiar geographic distribution, with up to 500-fold variations in incidence between the low- and high-risk regions. The first reports suggesting HPV involvement in both benign and malignant squamous cell tumours of the esophagus date back to 1982. The abundant literature accumulated on this subject is summarised in this review. To date, 322 Oesophageal squamous cell papillomas (SCP) have been analysed using different HPV detection methods in 32 separate studies. HPV has been detected in 26.1% (84) of these cases. Until now, 1,485 SCCs have been examined by *in situ* hybridisation, of which 22.9% (341/1.485) were HPV DNA positive. This is almost identical with the 22.3% prevalence rate calculated among 2,645 SCCs analysed by PCR (589/2,645). HPV prevalence bears a close correlate to the incidence of SCC, being low (0-3%) and high (up to 80%) in the respective geographic regions. In large-scale seroepidemiological studies, the increased risk for SCC among HPV16-seropositives has reached OR (odds ratios) up to 15, even in low-incidence countries. Screening studies in high-incidence areas of China using balloon cytology sampling report high prevalence of HPV DNA among asymptomatic subjects. There are close human parallels to both SCP and SCC in the cattle, linked with BPV4 and alimentary carcinogens, present in bracken fern. *In vitro* experiments on HPV-positive cancer cell lines implicate similar molecular mechanisms to those involved in HPV-associated genital carcinogenesis. The current data suggest that the (multifactorial) aetiology of oesophageal cancer differs between the low- and high-incidence geographic areas. Oncogenic HPV types probably play an important causal role only in the high-incidence regions.

1. Introduction

The association of human papillomavirus (HPV) infections and squamous cell precancer lesions of the uterine cervix has been firmly established (Syrjänen et al., 1987; IARC, 1995; Aubin et al., 2003; Chapters 1 and 3 of this book). Today, there is no doubt whatsoever that oncogenic HPV types are the single most important etiological factors of cervical squamous cell carcinoma (IARC, 1995; Syrjänen et al., 2000; Bosch et al., 2002; zur Hausen, 2003; Aubin et al., 2003). The early 1980's witnessed a rapid expansion of HPV research from the genital tract to cover other squamous cell epithelia, thus significantly widening the scope of HPV-associated human tumours (Syrjänen et al., 2000; Aubin et al., 2003). This includes the squamous cell lesions in the upper respiratory tract (sino-nasal papilloma and carcinoma, laryngeal papilloma and carcinoma), eye (conjunctival papilloma and carcinoma), oral mucosa (condyloma, papilloma, wart, carcinoma, focal epithelial hyperplasia [FEH]), bronchus (papilloma, carcinoma) as well as the upper digestive tract (oesophagus) (Syrjänen et al., 1987; zur Hausen, 1999a; Syrjänen et al., 2000a; Syrjänen, 2002a; 2003a; Aubin et al., 2003).

The squamous cell lining of the esophagus is in direct continuity with the oral mucosa, and the first descriptions of oral HPV lesions (reviewed by S. Syrjänen, 1987) and other preliminary preliminary reports suggested that this virus might be involved in the development of both benign (Syrjänen et al., 1982) and malignant (Syrjänen, 1982) squamous cell lesions of the esophagus as well. These early observations were based on the discovery of morphological similarities between HPV induced lesions in the genital tract (condyloma) and squamous cell tumours (papillomas and carcinomas) of the oesophageal mucosa. These purely morphological findings were soon substantiated by the demonstration of HPV structural proteins in these lesions using immunohistochemistry (IHC) (Syrjänen et al., 1982; Goldsmith, 1984; Morris et al., 1986; 1987; Brooks et al., 1987; Kashima et al., 1987; Syrjänen, 1987).

Following these pioneering observations, HPV research has resulted in a rapidly expanding literature on both benign and malignant oesophageal lesions in different geographic regions. The evidence accumulated during the past 20 years strongly implicates a causal role for HPV at least in a subset of oesophageal carcinomas (Togawa *at al.*, 1995; Snijders et al., 1997; Poljak et al., 1998; Sur et al., 1998; zur Hausen, 1999b; Matsha et al., 2002; Syrjänen, 2002b; 2003b; Gillison et al., 2003; Cervantes, 2004), and the data derived from different lines of research are reviewed here. This discussion is strictly limited to HPV and only briefly addresses other potential aetiological agents and the intriguing global

*For correspondence. Email kari.syrjanen@smwconsultants.fi.

epidemiology of this disease, which are reviewed elsewhere (Syrjänen et al. 1996; 2000a; 2000b; Gillison et al., 2003; Cervantes, 2004). Because HPV has been linked only with squamous cell carcinoma (SCC), this discussion will be restricted to this histological type of oesophageal cancer and its precursors (dysplasia) as well as to benign squamous cell papilloma.

2. Squamous Cell Papilloma

Oesophageal squamous cell papilloma (SCP) is a benign epithelial proliferation, regarded as a rare tumour until the early 1980's. Since then, however, increasing attention has been paid to these lesions, following the first evidence of their possible association with HPV (Syrjänen et al., 1982). Thus, a couple of recent reviews counted a total of 200 cases reported in the world literature until the year 2000 (Gomez et al., 1997; Syrjänen, 2000; Mosca et al., 2001).

Morphologically, SCPs are exophytic growths with papillary extensions covered by the squamous epithelium and supported by connective tissue stalks (Syrjänen et al., 1982; Toet et al., 1985; Winkler et al., 1985; Polit, 1990). Some authors prefer to classify oesophageal papillomas histologically into three types (exophytic, 50%, endophytic, 37%, and spiked, 13%), based on the predominant shape of the squamous papillae (Odze et al., 1993). However, we have not found such a morphological distinction to serve any useful practical purposes (Chang et al., 1991).

Bearing resemblance to SCPs at other mucosal sites, the epithelial cells in oesophageal SCPs also frequently contain koilocytotic changes. Indeed, it was in these cells that HPV structural proteins were localized by IHC staining in the original case report implicating HPV involvement (Syrjänen et al., 1982). However, according to our experience, koilocytotic changes are not invariably as prominent as in the exophytic genital HPV lesions (Chang et al., 1991). This morphological difference might reflect either a) a lower viral load in oesophageal SCPs, or b) a different aetiology (HPV-related and non-related) of these benign epithelial proliferations.

2.1. Epidemiology

Oesophageal SCPs are rare tumours, first reported in 1928 (Patterson, 1928), and histologically verified as a separate entity in 1959 (Adler et al., 1959). Since then, a continuos flow of reports on single cases and small case series has steadily increased the number of documented cases. From the late 1970's, the literature on these lesions has been repeatedly reviewed (Miller et al., 1978; Parnell et al., 1978; Waterfall et al., 1978; Lombardi et al., 1980; Javdan et al., 1984; Quitadamo et al., 1988; Politoske, 1992; Orlowska et al., 1994; Gomez et al., 1997; Mosca et al., 2001). Not unexpectedly, discrepant numbers of total cases found in the literature are given in these reviews, depending on the coverage of the literature search. By 1994, however, a credible total of 141 reported cases was achieved (Orlowska et al., 1994). Subsequently, a number of additional cases were published (Carr et al., 1994a; 1994b; Ferrari et al., 1995; Poljak et al., 1995; van Cutsem et al., 1995; Yamada et al., 1995; Sandvik et al., 1996), and this author calculated a total of 218 reported cases until February 1998 (Syrjänen, 2000).

The incidence figures of oesophageal SCP have not been systematically established (Syrjänen, 2000; 2003b). Some rough estimates are available, however, substantiating the distinct rarity of these lesions. Thus, only 15 SCPs were found among 20,000 consecutive gastro-intestinal endoscopies (Franzin et al., 1983), giving an incidence of 0.075% ($75/10^5$). Figures of the same order were recorded among 14,900 upper gastrointestinal endoscopies reviewed in 8 years, with 6 SCPs being detected (0.04%; $40/10^5$) (Fernandez-Rodriguez et al., 1986). In another series of 4,100 consecutive double-contrast studies, 14 SCPs were found, resulting in a substantially higher incidence rate (0.3%; $341/10^5$) (Montesi et al., 1983). Even higher incidence figures (0.43%; $432/10^5$) were obtained when 35 cases were diagnosed among 8,095 consecutive endoscopies (Sablich et al., 1988). Our own experience is more consistent with the lower figures. We recently reviewed our 18,000 endoscopies and found 12 patients with histologically documented SCPs (0.07%; $66/10^5$) (Chang et al., 1991). Strikingly identical figures (0.07%; $66/10^5$) were reported from Saudi Arabia, where 10 SCPs were found among 14,232 patients examined in Riyadh (Al-Sohaibani et al., 1995). More recently, even lower incidence rates have been calculated, with only 9 cases were detected among 7,618 endoscopies (0.01%) (Mosca et al., 2001), and another 9 cases among 124,000 patients in Taiwan (0.0073%) (Lio et al., 1996).

2.2. Clinical characteristics

The clinical characteristics of oesophageal SCPs have been exhaustively described in a series of reviews, and a detailed discussion of these data falls outside the scope of this review (Quitadamo et al., 1988; Politoske, 1992; Orlowska et al., 1994; Gomez et al., 1997; Syrjänen, 2000; Mosca et al., 2001). Most papillomas do occur as single lesions (85%) and seem to be located in the distal oesophagus (70%) (Odze et al., 1993). However, a series of well documented multiple SCPs has been reported as well (Chatelain et al., 1968; Benisch et al., 1974; Nuwayhid et al., 1977; Parnell et al., 1978; Nuss et al., 1978; Brinson et al., 1987; Fekete et al., 1988; Bretagne et al., 1989; Hörding et al., 1989; Sandvik et al., 1996).

Interestingly, many of these cases have occurred in children, not infrequently, associated with HPV-induced papillomatosis of the upper respiratory tract (Nuwayhid et al., 1977; Frootko et al., 1978; Hörding et al., 1989; Sandvik et al., 1996; Chistiakova, et al., 1998; Batra et al., 2001). The first case report on synchronous and metachronous multicentric squamous cell carcinomas in the upper aero-digestive tract was published recently, describing a patient with 14 foci of primary SCC and one severe dysplasia, developed in succession in the thoracic oesophagus, oral floor, soft palate, uvula, lingual radix, piriform recess, hypopharynx, cervical oesophagus, trachea and lingual body (Shan et al., 1997). Pillai et al. (2001) suggested the development of a "condemned mucosa" syndrome in pathogenesis of HPV-associated upper aero-digestive tract tumors. This possibility has received some epidemiological support from a recent study on the Swedish family-cancer database, analyzed for second cancers after oral, oesophageal, rectal, cervical, genital and skin cancers. A strong and consistent association of second cancers was observed at all these sites, in men and women (Hemminki et al., 2001).

2.2.1. Natural history

Sometimes, benign SCP may be confused with squamous cell carcinoma, verrucous cancer (VC) in particular, emphasising the important role of trans-endoscopic resection of this tumour (Ravry, 1979; Biemont *at al.*, 1991; Yuki et al., 1993). So far, however, only two cases have been documented, the probable malignant transformation of a benign SCP to a low-grade VC (Starr et al., 1979), and the malignant degeneration of an oesophageal SCP with HPV involvement (van Cutsem et al., 1991; 1992). Waluga et al. (2002) reported oesophageal papillomatosis being complicated by carcinoma, and a recurrent proximal oesophageal stricture was shown to be associated with dysplasia in another case of papillomatosis (Narayani et al., 2002). All these data suggest that malignant transformation of characteristic benign SCPs is extremely rare (Gomez et al., 1997; Mosca et al., 2001; Syrjänen, 2000; 2003b).

Despite their benign histology, oesophageal papillomas may prove to be fatal, as demonstrated by the case reported by Hörding et al. (1989). They described an otherwise healthy 27-year-old man with extensive papillomatosis originating in the distal part of the oesophagus and spreading into the main and intermediate bronchus. Dot blot hybridisation disclosed HPV11 DNA. Despite intense treatment with CO_2-laser vaporization, systemic and topical interferon and bleomycin, papillomatosis progressed unremittingly during the next 2 years, and finally resulted in the death of the patient.

On the other hand, there is now some anecdotal evidence that oesophageal papillomatosis can regress spontaneously. Kato et al. (2003) described a 83-year-old man with diffuse polyposis of the entire length of the oesophagus and stenosis in the antrum of the stomach. Histological examination confirmed the diagnosis of SCP and poorly differentiated adenocarcinoma of the stomach. HPV type-specific PCR disclosed HPV16 and HPV 33 in the biopsies. Surprisingly, one month after total gastrectomy, follow-up endoscopy revealed a complete regression of oesophageal papillomatosis. According to the authors, the natural history of this rare case provides another indirect evidence suggesting that oesophageal SCP is caused by HPV.

2.3.Evidence for HPV involvement

Until the early 1980's, few cases of oesophageal SCPs had been published (Miller et al., 1978; Parnell et al., 1978; Lombardi et al., 1980). At that time, gastro-oesophageal reflux and chronic irritation were the two mechanisms implicated in the pathogenesis of these lesions. The appearance in 1982 of the first evidence suggesting that HPV might be another potential etiological agent of these lesions has been followed by an increasing number of reports addressing this issue using different techniques.

During the early 1980's, morphology and IHC were used to document HPV involvement; these were replaced by different hybridisation assays and PCR in the studies run since the early 1990's. The data accumulated until 1998 have been summarised (Syrjänen, 2000) and are updated here to September 2004. All the studies reporting HPV detection in oesophageal papillomas are summarized in Table 1.

2.3.1. Morphology, cytology, immunohistochemistry

The preliminary evidence on HPV involvement in oesophageal squamous cell tumours has its roots in a single case of SCP, submitted to this author for histopathological examination in 1981. Based on remarkable morphological similarities to exophytic HPV lesions in the genital tract, the idea was raised that HPV might be involved in the pathogenesis of these oesophageal proliferations as well (Syrjänen et al., 1982). This lesion was studied by light microscopy and using an indirect immunoperoxidase-PAP technique for demonstration of HPV antigens. HPV antigens were detected in the nuclei of koilocytes, cells characteristic to HPV lesions in the genital tract. In the preliminary report, we raised the possibility that these benign SCPs might possess malignant potential, and also hypothesized the possible role of HPV in human oesophageal carcinogenesis. As evident from the following discussion, this prompted an entirely new branch of HPV research, subsequently extended to cover practically all extra-genital squamous epithelia (Syrjänen et al., 2000a; Aubin et al., 2003).

In another early report, 75 oesophageal lesions were analysed (including 2 SCPs and 73 FEH) for histological evidence of HPV (Winkler et al., 1985). Thirteen of the cases (2 SCPs and 11 FEH) showed morphological evidence of HPV infection, and HPV antigens were demonstrated by IHC in 4/13 cases (31%). At the same time, another report from France showed by IHC that a SCP, with a morphology suggestive of HPV infection, contained HPV antigens (Lesec et al., 1985). Subsequently, a condylomatous leukoplakia lesion was reported, which was also HPV antigen-positive (Tomasino et al., 1988).

Table 1. Detection of HPV in oesophageal squamous cell papillomas

Detection Method	Lesion	HPV Types Detected	HPV Positive		Authors	Year
			Number	%		
IHC, HB	SCP	HPV Ag	1/1	100	Syrjänen et al.	1982
IHC, HB	SCP	HPV Ag	2/2	100	Winkler et al.	1985
IHC, HB	SCP	HPV Ag	1/1	100	Lesec et al.	1985
PAP	SCP, CIS*	Koilocytes	1/1	100	De Borges et al.	1986
IHC, DB	SCP	HPV Ag; DNA-	1/1	100	Fekete et al.	1988
DB	SCP	6, 11	1/1	100	Hörding et al.	1989
HB	SCP	Koilocytosis	1/5	20	Valerdiz et al.	1989
IHC, EM	SCP, SCC**	HPV Ag, virions	1/1	100	Van Cutsem et al.	1991
ISH, PCR	SCP	6,11,16,18	0/12	0.0	Chang et al.	1991
ISH	SCP	31,33,35	6/33	18.1	Fontolliet et al.	1991
ISH	SCP	6,11	1/1	100	Janson et al.	1991
ISH	SCP	6,11	1/1	100	Politoske	1992
PCR	SCP	6/11, 16/18	1/26;12/26***	46.1	Odze et al.	1993
ISH, PCR	SCP	6,11	1/17	5.8	Carr et al.	1994a
ISH, SB	SCP	6,11,16,18	0/1	0.0	Nakano et al.	1994
PCR	SCP	6	1/7	14.2	Poljak et al.	1994
IHC	SCP	HPV Ag	0/10	0.0	Al-Sohaibani et al.	1995
PCR	SCP	6,11,16,18	0/3	0.0	Ferrari et al.	1995
ISH, PCR	SCP	6	1/29	3.4	Poljak et al.	1995
PCR	SCP	16,18	1/1	100	Van Cutsem et al.	1995
PCR	SCP	16	1/1	100	Yamada et al.	1995
ISH	SCP	6,11,16,18	0/1	0.0	Mastour et al.	1996
PCR	SCP	6,11,16,18	0/1	0.0	Sandvik et al.	1996
HB	SCP	Koilocytosis	3/3	100	Van den Borre et al.	1996
ISH, HB	SCP	Wide spectrum	1/10	10.0	Woo et al.	1996
PCR	SCP	45	1/1	100	Ratoosh et al.	1997
PCR	SCP	NS	1/1	100	Ravakhah et al.	1998
PCR	SCP	6,20,DL284,DL436	7/11	63.6	Lavergne et al.	1999
PCR	SCP	16,18	2/42	4.8	Talamini et al.	2000
ISH	SCP	6,11,16,18	0/9	0.0	Mosca et al.	2001
PCR	SCP	Wide spectrum	32/82	30.0	Szentirmay et al.	2002
PCR	SCP	16,18	1/1	100	Kato et al.	2003

Ag, HPV antigens; DB, dot blot hybridization; EM, electron microscopy; HB, histological biopsy; IHC, Immunohistochemistry; ISH, *in situ* hybridization; SB, Southern blot hybridization; PAP, Papanicolaou smear; PCR, polymerase chain reaction; SCP, squamous cell papilloma; SCC, squamous cell carcinoma; *CIS contiguous to SCP; **Malignant transformation of SCP to SCC; ***(n=38 cases) double infections by HPV16 & 18 in 3 cases; NS, not specified

The latest report relying only on IHC for HPV detection in these lesions comes from Saudi-Arabia, where 10 SCPs were analysed, but IHC failed to demonstrate HPV antigen expression in any of these (Al-Sohaibani et al., 1995).

There are some additional studies, where HPV involvement in these lesions has been suggested by cytology or morphology alone (Table 1). Cytological smears contained characteristic koilocytotic cells, similar to those described in genital condylomas, and the diagnosis of oesophageal SCP was confirmed by biopsy. An additional biopsy from the lower oesophagus revealed a squamous carcinoma *in situ* adjacent to the papilloma (de Borges et al., 1986). In a series of 5 SCPs, one presented with typical koilocytosis in the biopsy (Valerdiz et al., 1989). Another recent report of 3 SCPs used only histology to assess the evidence for HPV (van den Borre et al., 1996).

2.3.2. Dot Blot and in situ hybridization

Hybridization technology gradually replaced IHC in diagnosis of HPV by the end of the 1980's (IARC, 1995; Syrjänen et al., 2000a). Hybridization assays were soon applied also to HPV detection in oesophageal SCPs, starting with dot blot (DB) and *in situ* hybridisation (ISH) techniques. In the first of these studies, a 60 year-old woman was described with multiple warty tumours found at endoscopy (Fekete et al., 1988). On light microscopy, the morphology was consistent with SCP; IHC suggested HPV involvement, but search for HPV DNA was negative. The authors discussed the possible association of SCP with verrucous carcinoma (Fekete et al., 1988).

During the 1990's, we conducted a systematic survey of oesophageal squamous cell lesions including both malignant and benign lesions from one of the high-incidence areas (Henan province) in China. For comparison, we also analysed HPV in SCPs collected from Finland, which is a low-risk country for this disease. Among 18,000 endoscopies performed in Finland between 1985-1990, a total of 14 patients clinically diagnosed as having a SCP could be identified. By light microscopy, 12 of these fulfilled the histological criteria of SCP. ISH and PCR were used to analyse the samples for HPV types 6, 11, 16, and 18. Despite their characteristic HPV-suggestive morphology, HPV DNA could not be found in any of these lesions. These findings were compared with the frequent detection of HPV DNA in oesophageal samples from China, and the suggestion was made that these proliferations might have a different etiology in the low- and high-risk areas (Chang et al., 1991). This view is clearly substantiated by the highly variable HPV detection rates in malignant squamous cell lesions as well (See below).

The second largest series of SCPs analysed by ISH involved 33 cases diagnosed between 1973 and 1988 using the histological criteria of Winkler et al. (1985). HPV typing was made for HPV 6/11, 16/18, 31/33/35. Although none of the lesions fulfilled all the histological criteria of HPV infection, HPV 31/33/35 DNA was found in 6/33 (18.1%) of these papillomas. The authors could not establish any clinical impact of SCPs in oesophageal carcinogenesis however (Fontolliet et al., 1991). These data are similar to another report, where only 1/10 SCPs analysed by ISH was HPV positive (Woo et al., 1996).

These series were followed by three case reports, all using ISH for HPV analysis (Janson et al., 1991; Politoske 1992; Mastour et al., 1996). HPV 6/11 DNA signals were detected in the first two cases, but the lesion was negative for HPV 6/11 and HPV16/18 in the third report. The difficulty in making a comprehensive review of the literature at that time is well illustrated by the fact that Politoske (1992) considered this case to be only the second one with documented HPV involvement (Table 1).

After completion of the recent review of these data (Syrjänen 2000), only one additional study used ISH to analyse 9 cases of SCPs (Mosca et al., 2001) (Table 1). Consistent with our experience on SCPs in a low-risk country (Chang et al., 1991), no HPV was detected in any of the 9 lesions.

2.3.3. Polymerase chain reaction

The early 1990's witnessed the rapid development of PCR technology, extensively applied to HPV detection in different lesions (Syrjänen et al., 2000a; Aubin et al., 2003). To evaluate between the two suggested etiologies of papillomas, a) chronic mucosal irritation, and b) HPV infection, the clinical data and histological features of 38 SCPs were analysed, including PCR in 26 cases (Odze et al., 1993). This is the second largest series of SCPs studied for HPV until today (Table 1). Fifty percent of the papillomas (13/26) tested, from 57% of the patients (12/21), were positive for HPV, most frequently for type 16 (9/13), type 16/18 together (3/13), and only rarely for type 6/11 (1/13). These authors proposed a multi-factorial etiology for SCPs, implicating synergistic actions between mucosal irritation and HPV in their pathogenesis (Odze et al., 1993).

Considerably lower HPV prevalence was reported in another sizable series of papillomas, analysed by ISH, PCR and SB (Carr et al., 1994a). In 23 lesions from 17 patients analysed for HPV types 6/11, 16/18, 18, and 31/33/51 by ISH (followed by PCR and SB), HPV DNA was found in only one lesion, which was positive for HPV 6/11. Failing to find substantial evidence for HPV, the authors concluded that other pathogenetic mechanisms, such as mucosal injury and repair, are more important than HPV in the aetiology of these lesions (Carr et al., 1994a). Similar conclusions were reached in another study, where 29 SCPs were analysed for HPV DNA by ISH and PCR (Poljak et al., 1995). No evidence of HPV DNA was found using ISH, but HPV6 DNA was amplified by PCR in one of the lesions.

These controversial observations on HPV DNA detection in SCPs persist in the most recent PCR studies as well. In a papilloma at the site of endoscopic injection sclerotherapy for varices, integrated HPV16 was demonstrated, whereas the surrounding normal mucosa was negative (Yamada et al., 1995). PCR analysis disclosed HPV16 and 18 in another SCP, which completely disappeared after intra-tumoural injection of (S)-1-(3-hydroxy-2-phosphonyl-methoxypropyl) cytosine (HPMPC), in a 69-year-old woman with concomitant hypo-pharyngeal papillomatosis (van Cutsem et al., 1995). In another study, HPV could not be confirmed by PCR in 3 SCPs, leading the authors to suggest that there is no association between oesophageal SCP and HPV (Ferrari et al., 1995). In a 28-year-old woman with massive oesophageal papillomatosis, the lesions regressed after a 3.5-year observation period, and repeated PCR analyses were negative, despite the typical histological features and concomitant pharyngeal papillomas (Sandvik et al., 1996). The most exotic mechanism of HPV spread was described in a child, in whom HPV45 was detected in both a palmar wart and in oesophageal SCP, suggesting mastication of the former as a feasible route of oesophageal infection (Ratoosh et al.,

1997). This is the first report of this HPV type in either of these lesions, and a more systematic search for the newer HPV types in oesophageal squamous cell lesions would be clearly warranted.

2.3.4. Conclusions on HPV and papillomas

Reviewing the available literature (Table 1), it is apparent that the data on HPV detection in benign oesophageal SCPs are contradictory at the moment. When previously reviewed, there were 24 published studies, with the total number of 166 papillomas analysed by any of the detection techniques (histology, cytology, IHC, ISH, DB, SB, and PCR). In those studies, HPV has been detected in 40 cases, giving an overall prevalence of 24.1% (Syrjänen, 2000). Between 1998-2004, six additional studies were published, now updated in Table 1. Together with one additional report from 1996 (Woo et al., 1996), these comprise a total of 156 new cases, with HPV detection rate of 44/156 (28.2%). Taken together, 322 papillomas have been analysed for HPV, of which 84/322 (26.1%) are reported positive. Due to difficulties in defining the precise numbers of SCP in some of these studies, however, as well as the suspected double-reporting of some cases, these figures may not be totally accurate.

Interestingly, however, the overall prevalence of HPV (26.1%) in oesophageal papillomas is very similar to that calculated from the published literature of almost all other extra-genital squamous cell lesions analysed for HPV until now, as recently discussed (Syrjänen et al., 2000a; Aubin et al., 2003). In oesophageal SCPs, there is a tendency towards higher detection rates when using PCR, although even then, a wide variation (from 0% to 50%) in the figures is evident. This is not unusual, however, when PCR is used in HPV detection at different mucosal sites.

Because of the rarity of these lesions, the number of cases analysed is still too small to draw definite conclusions on whether HPV has a role in their aetiology. Beyond any doubt, there are well documented series where HPV definitely seems to be involved. On the other hand, there are series and individual cases where all papillomas were completely negative even by PCR. This was true with 12 cases of our own (Chang et al., 1991), and with the series of Carr et al. (1994a) and that of Poljak et al. (1995). On the other hand, 50% of the 26 SCPs reported by Odze et al. (1993) contained HPV DNA.

Also the distribution of the HPV types in the positive cases is different from benign papillomas at other mucosal sites, in that HPV16/18 seems to be much more frequent in the oesophagus, as compared with HPV 6/11. Other high-risk types (HPV31/33/35) were found in 6 HPV-positive cases (Fontolliet et al. 1991), and there is an anecdotal case of HPV 45 (Ratoosh et al., 1997). According to our traditional thinking, HPV6 and 11 are the predominant types in benign papillomas, *e.g.* in the respiratory tract, oral and genital mucosa (zur Hausen, 1999a; 1999b). Thus, the detection of high-risk types 16 or 18 in such lesions is puzzling and may reflect our incomplete knowledge of HPV biology (Galloway et al., 1996).

There is no doubt that oesophageal SCPs are truly benign lesions, and their malignant degeneration is an extreme rarity (de Borges et al., 1986; van Cutsem et al., 1991; 1992; Waluga et al., 2002). This is in contrast to the early concepts of SCP as a precursor of SCC (Ottenjann et al., 1984). According to our own experience, all papillomas seen by us were histologically benign with no signs of dysplasia (Chang et al., 1991). This makes it highly unlikely that the sometimes exophytic oesophageal carcinomas would be preceded by benign exophytic proliferations left unnoticed at that stage.

To be cautious, the pathogenesis of oesophageal SCPs must be considered an unresolved issue at the moment. Although HPV seems to be involved in some, the majority of these lesions do not seem to present any evidence of HPV involvement. It may well be that lesions with different etiology and pathogenesis exist in the oesophagus, like in some other squamous epithelia (Syrjänen, 2002b; 2003b). An interesting concept was proposed some years ago, suggesting that multiple papillomas in the proximal oesophagus seem to favour HPV involvement, whereas the etiology of isolated lesions in the distal oesophagus could be better explained by chronic gastro-oesophageal reflux and mucosal irritation (Politoske, 1992). In the light of the well established HPV involvement in papillomatosis of the upper respiratory tract and in different lesions of the oral mucosa, this concept is certainly worth exploring further. As discussed above, there is already some anecdotal evidence on the spread of HPV down the oesophagus from respiratory papillomatosis lesions (Nuwayhid et al., 1977; Frootko et al., 1978; Hörding et al., 1989; Sandvik et al., 1996; Chistiakova, et al., 1998; Batra et al., 2001).

Because of the rarity of oesophageal SCPs, a joint effort would be necessary to collect a representative series of these lesions, derived from both the proximal and distal oesophagus, and with extensive clinical records on any existing gastro-oesophageal disorders. When meticulously sampled and analysed by the most sensitive PCR techniques, some more light could be hopefully shed on the role of HPV in the pathogenesis of these benign but highly interesting squamous cell tumours.

3. Squamous Cell Carcinoma

The vast majority of oesophageal carcinomas are of the squamous cell (SCC) type. Among these SCCs, verrucous carcinoma (VC) represents a special entity. This is an extreme rarity, however, only 13 well documented cases having been reported by 1994 (Garrard et al., 1994). Adenocarcinomas (AC) usually develop in the pre-existing Barrett's oesophagus, which is an established risk condition for this malignancy (Baehr et al., 1994).

Because HPV has only been implicated in aetiology of SCC, the discussion will be restricted only to this histological type of oesophageal carcinoma.

Oesophageal SCC is a major health problem in certain high-risk areas of the world, and this disease continues to be the subject of intense research, including clinical and epidemiological studies as well as extensive basic research and animal experimentation (Chang et al., 1992a; Togawa *at al.*, 1995; Benamouzig et al., 1996; Franceschi et al., 1996; Syrjänen et al., 1996; Snijders et al., 1997; Poljak et al., 1998; Sur et al., 1998; zur Hausen, 1999b; Matsha et al., 2002; Gillison et al., 2003; Shen et al., 2003a; Cervantes, 2004). This has generated a massive flow of reports, and over 23,000 papers on oesophageal cancer are currently available in Medline. Thus, an attempt to cover all these divergent areas of oesophageal cancer research would not be feasible in this context, and the discussion is strictly limited to the subjects related to the implicated aetiological role of HPV. In particular, the extensive experimental carcinogenesis in animal models (Asamoto et al., 2002; Shen et al., 2002b) falls outside the scope of this text.

3.1. Epidemiology

Oesophageal cancer is the eighth most common cancer in the world with an uneven ethnic and geographical distribution (Ferlay et al., 2001). Indeed, one of the most intriguing features of this disease is the considerable regional variation in its incidence (Parkin et al., 1988). There are several well-defined high-risk areas (China, Iran and South Africa) and low-risk areas (Europe and North America). In all these areas, however, the disease is exceedingly rare in individuals under the age of 30 and the median age is around 65 for both sexes (Parkin et al., 1998; Ferlay et al., 2001).

In most countries, the incidence rates are around 2.5 to 5.0 per 100,000 for males and 1.5 to 2.5 for females (Ferlay et al., 2001). In distinct areas, however, the incidence rates are remarkably higher, varying up to 500-fold from one area to another, from one country to another, and even between different ethnic groups within the same country (Parkin et al., 1988; Chang et al., 1992a). Numerous epidemiological studies have identified the high-risk countries for oesophageal cancer: the People's Republic of China (Li et al., 1980; Zheng et al., 1993; He et al., 2003), Singapore (Sons, 1987), Iran (Dowlatshahi et al., 1985; Ghavamzadeh et al., 2001), former USSR, Puerto Rico, Chile, Brazil, Switzerland, France (Sons, 1987), and South Africa (Jaskiewicz et al., 1987; Somdyala et al., 2003). Among these high-risk countries, the highest incidence rates have been reported in the northern parts of China, the Caspian littoral of Iran, and the Transkei area of South Africa (Jaskiewicz et al., 1987; Ghavamzadeh et al., 2001; He et al., 2003; Somdyala et al., 2003).

In the People's Republic of China, the annual deaths due to oesophageal cancer account for 27% of all cancer deaths among males and 20% among females, ranking among the two leading causes of cancer deaths, second only to cancer of the stomach (Li et al., 1980; Lu et al. 1985; Chang et al., 1992a; Dong et al., 2002). In Linxian (a county with a population of 800,000 in Henan Province of North China), the age-adjusted average incidence rates were $161.33/10^5$ in males and $102.88/10^5$ in females during 1971-1974 (Chang et al., 1992a; Dong et al., 2002). The deaths due to oesophageal cancer in this area account for 16% of all deaths and 65% of all cancer deaths (Chang et al., 1992a). Similar figures have been reported from Iran and also from South Africa, where up to $180/10^5$ and $246/10^5$ new cases of SCC have been recorded per year, respectively (Ghavamzadeh et al., 2001; Sitas et al., 1998).

In general, the incidence of oesophageal cancer is higher in males than females. This sex difference disappears in the high-incidence areas, but curiously enough, the incidence in females is generally much lower in the adjacent low-risk areas. The majority of cancer cases are encountered in the older age group, showing a peak between 50 and 70 years. However, in the high-incidence areas, the disease may occur 5-15 years earlier than in low-risk countries (Chang et al., 1992a). Although the incidence and mortality rates slightly fluctuate from year to year, they do not appear to decline to any significant extent (Parkin et al., 1998; Ferlay et al., 2001).

3.2. Risk factors

The reasons for these major regional variations in the incidence of this disease are poorly understood. It seems clear, however, that no single etiological factor could account for such a dramatic variation in the frequency of oesophageal cancer in the distinct geographic areas (Baehr et al., 1994; Franceschi et al., 1996; Ribeiro et al., 1996; Syrjänen et al., 1996; Lin et al., 2002; Xibib et al., 2003). The disease preferentially occurs in groups with a low socio-economic status or those afflicted with poverty (Tollefson et al., 1985; Franceschi et al., 1996), which seems to be a common denominator in all high-incidence areas.

These highly variable incidence rates between the high- and low-risk areas as well as the dramatic differences between sexes and ethnic groups even within the distinct geographic areas, suggest a predominant aetiological role for environmental factors (Chang et al., 1994; 1995). Furthermore, the accumulated recent data strongly suggest that the aetiology of this disease can be different in low- and high-risk areas. In industrialized countries, consumption of alcohol and tobacco, especially in combination, are the major risk factors (Syrjänen, 2000; 2003b; Znaor et al., 2003; Xibib et al., 2003). In high-incidence areas, nutritional deficiencies, ingestion of hard foods and/or hot liquids and infectious agents are also risk factors (Chang et al., 1992a; Syrjänen et al., 1996; Sharp et al., 2001; Chitra et al., 2004). HPV has been listed among these infectious agents relatively recently, supported by the increasingly growing literature on this subject (Togawa *at al.*, 1995; Snijders et al., 1997; Poljak et al., 1998; Sur et al., 1998; zur Hausen, 1999a; 1999b; Matsha et al., 2002; Syrjänen, 2000; 2002b; 2003b; Gillison et al., 2003; Cervantes, 2004).

The oesophagus is one of the mucosal sites most frequently in contact with environmental factors, and as such it is also a significant route of entry for foreign agents into the human body. Such potentially harmful agents include pathogenic micro-organisms, chemical irritants, environmental pollutants or food additives (Baehr et al., 1994; Ribeiro et al., 1996; Sur et al., 1998; Cervantes, 2004). Therefore, it seems likely that the pathogenesis of oesophageal cancer is, in some way or another, associated with these factors. This is also strongly suggested by the fact that in high-incidence areas of China (*e.g.* Linxian), oesophageal (and pharyngeal) cancer incidence in domestic chickens (which usually share the food sources and the natural environment of their hosts) is 10-fold higher than that of chickens in the low-incidence areas (Li et al., 1980; Lu et al., 1985; Ghadirian et al., 1988; Rubio et al., 1989; Chang et al., 1992a).

3.2.1. Infectious agents

The implicated and suspected risk factors of oesophageal cancer have been extensively studied (Chang et al., 1992a; Ribeiro et al., 1996; Syrjänen, 2000; 2002b; 2003b; Gillison et al., 2003; Xibib et al., 2003; Znaor et al., 2003, Cervantes, 2004). However, little new insight in risk factors has been provided since the classical works of Wynder et al. (1961), and Silber (1985), suggesting that some chemicals (*e.g.*, nitrosamines, mycotoxins, cigarette smoke, excessive alcohol intake and opium abuse), nutritional deficiencies (particularly that of vitamins A, B, C and certain trace elements, *e.g.*, molybdenum and zinc), as well as physical factors (*e.g.*, coarse and hot food), are associated with the development of this malignancy. The detailed discussion of these vast data is not possible in this context.

The evidence suggesting an etiological role of certain micro-organisms in oesophageal carcinogenesis has received increasing attention recently (Chang et al., 1992a; Baehr et al., 1994; Ribeiro et al., 1996; Syrjänen et al., 1996; Li et al., 2001; Hendricks et al., 2002; Crew et al., 2004; Ye et al., 2004). The carcinogenic effects of the implicated micro-organisms can be mediated through two principal mechanisms: a) by producing direct carcinogens or promoters, or b) by acting directly on the host cells (Chang et al., 1992a).

Accordingly, fungal contamination and infection may be involved in oesophageal carcinogenesis by producing nitrosamines and/or their precursors as well as mycotoxins. Among the latter, fumonisins B1 and B2 are a family of mycotoxins that have been identified and classified as possible human carcinogens (Gelderblom et al., 1988). Most maize and other foodstuffs are likely to be contaminated with detectable levels of fumonisins, produced by *Fusarium monoliforme* and *Fusarium proliferatum*, the two most prevalent moulds associated with maize worldwide (Marasas, 1996). The mutagenic and carcinogenic effects of the extracts from several fungi isolated from the grains and foodstuffs in the high-risk areas have been demonstrated by *in vitro* and *in vivo* studies. Thus, *Fusarium, Geotrichum, Aspergillus* and other genera not only could reduce nitrates to nitrites, but could also decompose proteins and increase the amount of amines in food, consequently promoting the formation of nitrosamines (Chang et al., 1992a; Crew et al., 2004; Ye et al., 2004). Extracts of fumonisin cultures have also been shown to induce the formation of novel DNA adducts. Not unexpectedly, studies in the high-risk areas have suggested that the ingestion of mouldy foodstuffs and pickled vegetables is closely related to oesophageal cancer both in China (Zhen et al., 1984; Chang et al., 1992a) and other high-incidence areas (Marasas, 1996).

In addition to fungi, also bacteria may be associated with oesophageal cancer, by producing carcinogenic chemicals and increasing cell proliferation while stimulating the inflammatory process. Recently, Helicobacter pylori has received increasing attention as a possible causative agent of upper gastrointestinal cancer. The available evidence suggests that H. pylori may reduce the risk of oesophageal adenocarcinoma (AC), whereas infection with CagA-positive strains of H. pylori may increase the risk for oesophageal SCC (Crew et al., 2004; Ye et al., 2004). These data are controversial, however, because some other studies have reported a protective effect for H. pylori both in AC and SCC (Henrik-Siman et al., 2001).

Of the viruses, Herpes simplex virus (HSV), Cytomegalovirus (CMV) and Epstein-Barr virus (EBV) have been shown to infect the oesophageal mucosa, but so far, no firm evidence on their role in oesophageal carcinogenesis has been provided (Chang et al., 1992a; Syrjänen et al., 1996; Awerkiew et al., 2003; Ohnuma et al., 2003). On the other hand, the evidence linking HPV to oesophageal SCC is increasing in strength, and the number of studies on this subject is increasing constantly.

3.3. Evidence of HPV involvement

Following the preliminary report suggesting HPV involvement in benign oesophageal SCPs (Syrjänen et al., 1982), the author of this review made the first systematic survey for similar evidence in oesophageal carcinomas (Syrjänen, 1982). Indeed, this led to the first report in a long list of papers published since 1982 to explore this issue. The updated (September 2004) list of these papers reporting HPV detection in oesophageal carcinomas is given in Table 2. Before discussing these data in more detail, other lines of evidence are briefly summarised.

3.3.1. Evidence from animal studies

Substantial evidence for the involvement of papillomavirus in oesophageal carcinogenesis has been obtained from studies on cattle, particularly in the Scottish Highlands, which is a high-incidence area for upper GI tract papillomas and carcinomas (Jarrett et al., 1978; Campo et al., 1980; 1990; Campo, 1987; Jarrett, 1987). Persistent and widespread papillomatosis and carcinomas can be experimentally induced with bovine papillomavirus 4 (BPV 4) infection in these animals (Campo, 1987; Jarrett, 1987; Campo et al., 1994). Bracken fern is the environmental cofactor of BPV4. Field studies in this region have revealed that up to 96% of the cancer-bearing animals have concomitant papillomas, and 40% had more than 15 papillomas in the alimentary tract. On electron microscopy, viral particles were readily identified in these lesions, morphologically indistinguishable from papillomaviruses (Hamada et al., 1989). These cells also stained positively by IHC using anti-BPV serum. Recently, similar evidence was reported for the first time in Europe outside Britain (Borzacchiello et al., 2003). In the south of Italy, where bracken fern is common, examination of 1,133 slaughterhouse cattle revealed oesophageal lesions (exophytic and inverted papillomas) in 147 (13%). The presence of BPV4 was established by PCR in over 60% of histologically verified papillomas, confirmed by nucleotide sequencing (Borzacchiello et al., 2003).

In many instances, the progression from benign papillomas to carcinomas has been clearly documented (Jarrett et al., 1978; Jarrett, 1987). It was soon proved that the ingestion of bracken fern is the crucial factor in the malignant conversion of benign papillomas (Campo et al., 1994). Bracken fern contains carcinogenic agents (radiomimetics) and immunosuppressants (Campo, 1987). While high copy numbers of BPV4 DNA sequences are regularly detected in both naturally occurring and experimentally induced papillomas, no viral DNA or viral antigens are present in their malignant counterparts, indicating that the viral genomes are not required for the maintenance of the malignant state (Campo et al., 1985; Gaukroger et al., 1989; 1991; Anderson et al., 1997).

Taken together, these studies on cattle imply that: 1) BPV4 may execute one of the early events in cell transformation, and its genetic information may not be required for malignant progression; 2) immunosuppression caused by the ingestion of bracken fern allows the spread and persistence of BPV-induced papillomas; and 3) the bracken fern supplies co-carcinogens and carcinogens, leading to cell transformation and progression to malignant squamous cell neoplasia. This animal model has proved extremely useful in the development of vaccines against HPV (Campo et al., 1994).

3.3.2. Morphological features suggesting HPV involvement

In 1982, the author examined a series of 60 oesophageal SCCs looking for morphological evidence of HPV, *i.e.*, cells with koilocytotic morphology (Syrjänen, 1982). Epithelial changes consistent with an exophytic lesion were present in one case, those of inverted lesions in three cases, and features of flat condylomas were observed in 20 cases. In this report the possibility was raised that HPV could be an agent contributing to the evolution of human oesophageal SCC. This prompted a series of reports from other authors substantiating the proposal (Goldsmith, 1984; Hille et al., 1986; Hale et al., 1989; Chang et al., 1990a).

Hille et al. (1986) found morphological evidence for HPV infection in 23 cases (33%) in a series of 70 patients collected from South Africa (a high-risk area). These authors were also the first to demonstrate HPV antigens by IHC in 7/23 of these cases. In another series of 20 cases from the same country, morphological evidence of HPV was described in 13 (65%) specimens (Hale et al., 1989). The material was also examined for viral antigens, but IHC was entirely negative. In a series of 51 cancer biopsies derived from Linxian (a high-incidence area in China), epithelial changes suggesting HPV infection were present in 25 cases (49.0%), either within or adjacent to the malignant lesions (Chang et al., 1990a). Of 22 oesophageal lesions suggesting HPV, two were SCCs, and in both cases, the morphology of biopsies taken from the adjacent epithelium was also consistent with HPV (Matos et al., 1990).

These early morphological observations prompted an intense search for the evidence of HPV in oesophageal carcinomas by more specific diagnostic tools (Table 2). Altogether, these merely morphological reports contain data on 203 SCCs, among which HPV-suggestive morphology was described in 87 cases (42.8%) (Syrjänen, 20002b; 2003b).

3.3.3. HPV antigen expression detected by immunohistochemistry

Hille et al. (1986) were the first to use IHC for demonstration of HPV antigens in these lesions, and found antigen expression in 7/70 (10%) of their cases. In a series of oesophageal carcinomas from Japanese and Chinese patients, the detection rate of HPV antigen was considerably higher in the Chinese patients (23%) than in the Japanese patients (13%) (Mori et al., 1989). This is consonant with a more recent data from 61 carcinomas in Japan, where IHC was positive in 8% of the cases (Nakamura et al., 1995). HPV antigen expression confirmed by IHC was coincident with the detection of HPV particles in a SCC reported from Belgium (van Cutsem et al., 1991).

As updated in a recent review (Syrjänen, 2002b), the total number of oesophageal carcinomas analysed

Table 2. Detection of HPV in oesophageal squamous cell carcinomas

Detection method	Area or country	HPV types detected	HPV positive		Authors	Year
			Number	%		
HB	Finland	–	24/60	40	Syrjänen	1982
HB	S Africa	–	23/70	33	Hille et al.	1986
HB	S Africa	–	13/20	65	Hale et al.	1989
HB	Venezuela	–	2/2	100	Matos et al.	1990
HB	China	–	25/51	49	Chang et al.	1990a
HB, ISH	Finland	-	11/61	18	Hippeläinen et al.	1993
IHC	S Africa	Ag	7/70	10	Hille et al.	1986
IHC	Japan	Ag	2/15	13	Mori et al.	1989
IHC	China	Ag	7/31	23	Mori et al.	1989
IHC	Japan	Ag	5/61	8	Nakamura et al.	1995
IHC, EM	Belgium	Ag, virions	1/1	100	van Cutsem et al.	1991
IHC	Japan	Ag	0/4	0	Kuwano et al.	2001
IHC	China	E6, E7[1]	15/18;16/18	83/89[1]	Xu et al.	2004
FISH	Australia	11, 13, 16, 18	5/10	50	Kulski et al.	1986
FISH	China	11, 16, 18	3/80	66	Chang et al.	1990a
HISTOFISH	Australia	6, 11, 16, 18	9/39	23	Kulski et al.	1990
Slot blot	Hong Kong	6, 11, 16, 18	0/37	0	Loke et al.	1990
Dot blot	France	6/11, 16/18	5/12	42	Benamouzig et al.	1992
SB	China	16	12/24	50	Li et al.	1991
SB	China	11, 16, 18, 30	8/20	40	Chang et al.	1992b
SB, PCR	China	16, 18	0/35	0	Lu et al.	1995
SB, PCR	China	16, 18	37/103	36	He et al.	1996
ISH	China	6, 11, 16, 18	22/51	43	Chang et al.	1990a
ISH	France	6, 11, 16, 18, 31, 33	1/12	8	Benamouzig et al.	1992
ISH	UK	6, 11, 16, 18, 31, 33	0/4	0	Ashworth et al.	1993
ISH	China	6,11,16,18, 30	85/363	23	Chang et al.	1993
ISH	S Africa	6, 7, 16, 18, 30	3/10	30	van Rensburg et al.	1993
ISH	Japan	6, 11, 16, 18, 31, 33	24/71	34	Furihata et al.	1993
ISH	Japan	16, 18	13/42	31	Ono et al.	1994
ISH	S Africa	6, 11, 18, 31, 33	25/48	52	Cooper	1995
ISH	China	6, 11, 16, 18	ND/40		Li et al.	1996
ISH	Korea	Wide spectrum	11/25	44	Woo et al.	1996
ISH	China	Wide spectrum	3/36	8	Chang et al.	1997
ISH, IHC	India	16,18	ND/ND	63	Agarwal et al.	1998
ISH	Japan	6, 11, 16, 18	37/123	30	Takahashi et al.	1998
ISH	China	6, 11, 16, 18, 30, 53	117/700	17	Chang et al.	2000a
PCR	USA	16/18	0/13	0	Kiyabu et al.	1989
PCR	S Africa	Various	6/14	43	Williamson et al.	1991
PCR	Korea	16, 18	16/24	67	Kim et al.	1991
PCR	Japan	16, 18, CP	3/45	7	Toh et al.	1992
PCR	China	6, 11, 16, 18	25/51	49	Chang et al.	1992b
PCR	China	CP	24/40	60	Chen	1993
PCR	Slovenia	CP	2/20	10	Poljak et al.	1993
PCR	China	GP	24/40	60	Chen et al.	1994
PCR	Sweden	GP	0/10	0	Lewensohn-Fuchs et al.	1994
PCR	Different	CP, 16, 18	10/72	14	Togawa et al.	1994
PCR	S Africa	E6, GP	6/9	67	Cooper et al.	1995
PCR	France	6, 11, 16, 18, 31, 33	0/75	0	Benamouzig et al.	1995
PCR	Japan	CP	0/31	0	Akutsu et al.	1995
PCR, ISH	Japan	CP	ND/16	ND	Hosone et al.	1995
PCR	Japan	CP	3/45	7	Sugimachi et al.	1995
PCR	Japan	CP	15/72	21	Shibagaki et al.	1995

PCR	Holland	CP	0/61	0	Smits et al.	1995
PCR	Portugal	16, 18	9/16	56	Fidalgo et al.	1995
PCR	China	6, 16, 18	3/70	4	Suzuk et al.	1996
PCR	USA	6, 16, 18	1/23	4	Suzuk et al.	1996
PCR	USA	73	1/1	100	West et al.	1996
PCR	France	–	0/75	0	Benamouzig et al.	1996
PCR	China	16, 18	32/152	21	He et al.	1997
PCR	Holland	CP	0/63	0	Kok et al.	1997
PCR	Hong Kong	CP, SSCP	6/75	9	Lam et al.	1997
PCR	Alaska	CP	10/22	45	Miller et al.	1997
PCR	Japan	18	3/41	7	Mizobuchi et al.	1997
PCR	UK	–	0/22	0	Morgan et al.	1997
PCR	USA	–	0/11	0	Paz et al.	1997
PCR	Italy	–	0/18	0	Rugge et al.	1997
PCR	USA	CP, RFLP, 16	1/51	2	Turner et al.	1997
PCR	Japan	CP, 16, 18	0/103	0	Saegusa et al.	1997
PCR	Slovenia	CP, 6, 16, 18	0/121	0	Poljak et al.	1998
PCR	Japan	CP, 16, 18	17/27	63	Khurshid et al.	1998
PCR	Japan	CP, 16, 18	3/24	12	Takahashi et al.	1998
PCR	China, S Africa	CP	19/63	30	Lavergne et al.	1999
PCR	China	CP	20/117	17	de Villiers et al.	1999
PCR, ISH, IHC	Taiwan	CP	1/31	3.2	Wang et al.	1999
PCR	Italy	CP, 16, 18	0/45	0	Talamini et al.	2000
PCR	Japan	CP	12/75	16	Kawaguchi et al.	2000
Reverse Line PCR	USA	CP	1/22	4.5	Kamath et al.	2000
PCR	Belgium	CP	1/21	2	Lambot et al.	2000
IS PCR, ISH	China	L1, E6, E7[1]	18/30	60[1]	Lu et al.	2001
PCR	China	CP	2/32	6	Peixoto-Guimaraes et al.	2001
PCR	China	CP	17/101	17	Chang et al.	2000a
PCR	India	CP	25/40	63	Sobti et al.	2001
PCR	Italy	CP, RFLP	8/17	47	Astori et al.	2001
PCR	Japan	CP	20/48	42	Hasegawa et al.	2002
PCR	South Africa	CP, SEQ, 11,39,52,	23/50	46	Matsha et al.	2002
PCR	China	CP	115/176	65.5	Shen et al.	2002a
PCR	Germany	CP	0/23	0	Awerkiew et al.	2003
HCII	Brazil	HR& LR	LR: 1/40	2.5	Weston et al.	2003
PCR	China	CP	28/152	18.4[2]	Liu et al.	2003
PCR	China	CP	28/40	70	Xu et al.	2003
PCR	China	CP	31/48	63	Zhou et al.	2003
PCR	China	CP	20/30	66.7	De Villiers et al.	2004
PCR	Mexico	CP	15/17	88	Acevedo-Nuno et al.	2004
PCR	France	CP	0/1	0	Regragui et al.	2004

Ag, HPV antigens; CP, consensus primers; EM, electron microscopy; FISH, filter in situ hybridisation; GP, general primers; HB, histological biopsy; HCII, Hybrid Capture 2; HR, high-risk types; IHC, immunohistochemistry; ISH, in situ hybridization; LR, Low-risk types; ND, not defined; PCR, polymerase chain reaction; RFLP, restriction fragment length polymorphism; SB, Southern blot hybridisation; SEQ, sequencing, [1]E6 and E7 detection rates; [2]20.3% in high- and 8.3% in low-incidence area.

by IHC is 182, of which HPV antigen expression has been demonstrated in 23 cases (12.6%). Recently, in an interesting paper from China, the expression of HPV16 E6 and E7 oncoproteins was analysed using IHC in normal tissues (n=70), dysplastic lesions (n=43) and in 18 carcinomas. HPV16 E6 expression was detected in 59.3%, 88.4%, and 83.3% of the normal mucosa, dysplasia, and carcinoma, respectively, while the corresponding prevalence of HPV16 E7 was 62.1%, 90.7%, and 88.9%, respectively. Double expression of HPV16 E6 and E7 was rare in normal mucosa (25.7%), but highly prevalent in dysplastic lesions (88.3%) and in carcinomas (83.3%) (Xu et al., 2004).

3.3.4. Detection of HPV DNA by filter in situ hybridization

During the late 1980's, filter *in situ* hybridization (FISH) was widely used in detecting HPV DNA at different mucosal sites, the genital tract in particular. This technique was shown to have a sub-optimal sensitivity and specificity making it obsolete by the early 1990's, when only few studies on oesophageal carcinomas were completed using FISH. The HPV detection rates in these studies varied from 23% (Kulski et al., 1990) to 50% and 66% (Kulski et al., 1986; Chang et al., 1990b). Altogether, FISH has been used to analyse 129 carcinomas, showing HPV DNA in 67 cases (51.9%), which is considerably higher than that shown using any other detection techniques (because of the high false positive rate inherent to this detection technique).

3.3.5. Detection of HPV DNA by dot blot and Southern blot hybridization

Following the demise of FISH, dot blot (DB) hybridization gained increasing popularity in HPV detection, before the emergence of more sensitive techniques, like ISH, SB and finally PCR. The two studies using this technique show divergent results. While HPV DNA was not found in any of the 37 cases from Hong Kong examined by Loke et al. (1990), 42% of cases proved to be positive in a series collected from France (Benamouzig et al., 1992). The latter figure is more consistent with the results from Chinese patients examined by SB, where HPV DNA was found in 50% and 40% of cases (Li et al., 1991; Chang et al., 1992b). Indeed, in a series of 20 fresh biopsy specimens derived from the high-incidence area of China, 9 were shown to contain HPV DNA sequences by SB under low-stringency conditions, and 8 remained positive when hybridized with a cocktail probe of HPV 11, 16, 18, and 30 DNA under high-stringency conditions (Chang et al., 1992b). HPV DNA sequences in these carcinomas appeared to be present mainly in an integrated form. The detection rate was even higher in another Chinese series, where HPV16 DNA alone was found in 12/24 cases by SB (Li et al., 1991).

3.3.6. Detection of HPV DNA by in situ hybridization (ISH)

ISH was used for HPV detection in 8 separate studies published during the 1990's (Table 2). The series studied were relatively small, except the one reported by us in 1993 (Chang et al., 1993), comprising 363 cases, which was the largest series analysed for HPV until then. In these ISH studies, HPV detection rates vary considerably, and they tend to be somewhat lower than those reported before by other hybridization techniques. In our series from the high-incidence area of China, 85/363 (23.4%) tumours contained HPV DNA under low-stringency conditions (Chang et al., 1993). Positive signals were found in cancer cells in 71 cases (19.6%), in the surrounding epithelial cells with hyperplastic or dysplastic changes in 13 cases (3.6%), in cancer cells and surrounding epithelial cells in 10 cases (2.8%), and in the resection margins in 1 case (0.3%). Thirty-four (40%) of the 85 HPV-positive tumours were shown to contain at least one of the HPV types 6, 11, 16, 18, or 30 DNA sequences. HPV16 was the predominant type, occurring in 18.8% of the 85 HPV-positive specimens. In addition to the primary tumours, HPV DNA sequences were found in 12.3% (7/57) of the lymph node metastases as well (Chang et al., 1993).

Until 1998, ISH has been applied to study a total of 601 oesophageal SCCs, of which HPV DNA has been found in 28.8% (173/601) of the cases (Syrjänen, 2000). HPV16 was by far the most frequent viral type detected in these studies (Chang et al., 1990a; 1993; Benamouzig et al., 1992; van Rensburg et al., 1993; Furihata et al., 1993; Ono et al., 1994; Cooper et al., 1995). The detection rates in lesions from the high-risk areas are usually significantly higher as compared with those from Europe, which mostly show a low prevalence of HPV (Benamouzig et al., 1992; Ashworth et al., 1993). By the mid 1990's, this led several authors to suggest that oesophageal carcinoma might have a different etiology in the low- and high-risk areas (Chang et al., 1992a; Benamouzig et al., 1996; Syrjänen et al., 1996; Miller et al., 1997; Mizobuchi et al., 1997; Rugge et al., 1997).

Between 1998-2002, a number of important studies using ISH were published (Syrjänen, 2002b; 2003b) (Table 2). These include the largest series (n=700) of oesophageal carcinomas analysed with ISH , following the completion of our systematic survey of these lesions in 2000 (Chang et al., 2000a). In this study, 117/700 (16.9%) SCCs were HPV DNA positive, with HPV types 6, 11, 16, 18, and 30 accounting for 39.8% of the positivity. The involvement of other (possibly new) HPV types in a considerable proportion of the remaining positive lesions was clearly suggested. Indeed, several novel HPV types have been demonstrated in oesophageal carcinomas by subsequent studies using PCR and sequencing (de Villiers et al., 1999; 2004; Lavergne et al., 1999).

Altogether, ISH has been used to study 1,485 oesophageal SCCs, and HPV DNA has been found in

22.9% (341/1.485) of the cases (Table 2). HPV16 is the most frequent single HPV type detected. Furthermore the most recent studies based on the largest series confirm that HPV prevalence is markedly higher in oesophageal carcinomas from high-incidence areas than in low-incidence areas, *i.e.* from Europe. This sustains the concept that oesophageal carcinoma might have a different aetiology in the low- and high risk areas.

3.3.7. Detection of HPV DNA by polymerase chain reaction (PCR)

The vast majority of the studies conducted during the 1990's have utilized PCR as the HPV detection method of choice. Until 1998, 28 such studies had been reported, comprising a total number of 1,183 carcinomas (Syrjänen, 2000). HPV DNA was found in 15.6% of these cases (185/1,183). Since then, an increasing number of PCR-data have been published, increasing the number of studies by 14 (to 42) and the total number of analysed cases by 837 until March 2002, with 14.9% HPV prevalence (123/837) (Syrjänen, 2002b; 2003b). From 2002 onward, 11 additional studies have been published, adding a total of 625 carcinomas, and reporting a HPV prevalence as high as 44.9% (281/625) (Table 2). This is explained by the fact that almost all of these most recent studies are from high-incidence areas (Hasegawa et al., 2002; Matsha et al., 2002; Shen et al., 2002a; Liu et al., 2003; Xu et al., 2003; Zhou et al., 2003; de Villiers et al., 2004; Acevedo-Nuno et al., 2004), while those from the low-risk regions continue to report low HPV prevalence rates (Awerkiew et al., 2003; Weston et al., 2003).

There are now a total of 2,645 oesophageal carcinomas analysed by PCR, of which 589/2,645 (22.3%) are HPV-positive (Table 2). Similar to other lesions at various mucosal sites analysed for HPV by different techniques, the detection rates of HPV DNA in oesophageal cancer samples analysed by PCR seem to be subjected to wide variations. Accordingly, there are series where no HPV-positive cancer was found (Kiyabu et al., 1989; Lewensohn-Fuchs et al., 1994; Akutsu et al., 1995; Benamouzig et al., 1995; 1996; Smits et al., 1995; Kok et al., 1997; Morgan et al., 1997; Paz et al., 1997; Rugge et al., 1997; Saegusa et al., 1997; Poljak et al., 1998; Talamini et al., 2000; Awerkiew et al., 2003; Regragui et al., 2004). On the other hand, HPV DNA has been amplified in up to 60-80% of oesophageal carcinomas in several other studies (case reports excluded) (Kim et al., 1991; Chen et al., 1994; Cooper et al., 1995; Fidalgo et al., 1995; Miller et al., 1997; Khurshid et al., 1998; Lu et al., 2001; Sobti et al., 2001; Shen et al., 2002a; Xu et al., 2003; Zhou et al., 2003; de Villiers et al., 2004; Acevedo-Nuno et al., 2004).

There seems to be a common denominator for these low and high detection rates; namely, the geographic distribution of the material. Accordingly, almost all of the studies where HPV DNA could not be amplified from the tumours were carried out in low risk countries in Europe or in the USA. This also applies to our series from Finland, where evidence for HPV involvement was found in 18% of the 61 patients (Hippeläinen et al., 1993). In contrast, the detection rates of HPV DNA in carcinomas derived from the high-incidence areas (China, South Africa, Japan, Alaska) are significantly higher, and only two studies that include Japanese patients failed to report any evidence of HPV (Akutsu et al., 1995; Saegusa et al., 1997).

Surprisingly, the overall detection rate of HPV DNA (22.3%) by PCR is almost identical to that obtained in all the ISH studies (22.9%), and certainly significantly lower than the figures reported with SB (40–50%) and FISH (51.9%) (Table 2). The reasons for these divergent results may be purely technical (as a result of errors in the interpretation of the data) and, perhaps most importantly, may reflect the diversity of these lesions in different geographical regions. However, PCR is regarded as the most sensitive HPV detection method, and the higher detection rate with two hybridisation methods might indicate a cross hybridisation of HPV probes with human DNA or DNA from other micro-organisms, as discussed in the technology chapters of two recent HPV textbooks (Syrjänen et al., 2000a; Aubin et al., 2003).

3.3.8. Screening of cancer precursors by DNA techniques

There is little doubt that oesophageal SCC develops through distinct precursor lesions such as dysplasia and carcinoma *in situ* (Ushigome et al., 1967; Nagamatsu et al., 1992). The existence of these lesions has been well recognized in the high-incidence regions of China, where screening programmes have been organised to detect these cancer precursors by balloon cytology (Chang et al., 1992a; 1992b; Roth et al., 1997). However, these precancer lesions have not been systematically studied for HPV involvement until relatively recently (Jacob et al., 1993; Yoshikane, et al., 1995; Trottier et al., 1997; Cintorino et al., 2001; Li et al., 2001; 2002; Si et al., 2003). We recently completed a systematic analysis of over 2,000 histological specimens from >700 oesophageal cancer patients from Henan Province (China), one area with the highest risk for this disease (Chang et al., 1990a; 1990b; 1992a; 1992b; 1994; 1995). On microscopic examination, dysplastic lesions frequently accompany invasive SCCs, and these lesions frequently present with koilocytotic changes, as illustrated in a recent review (Syrjänen, 2000).

In a series of 51 biopsies, epithelial changes suggesting HPV within or adjacent to the carcinoma lesions were found in 25 cases (49.0%) (Chang et al., 1990a). HPV DNA sequences were detected in 22/51 (43.1%) of the specimens, being most frequently localized in the areas showing either epithelial hyperplasia (36.1%) or dysplasia (22.2%), adjacent to invasive carcinomas. HPV16 and

18 were the two principal types, found in 16/22 HPV DNA-positive cases (72.7%) (Chang et al., 1990a). In another series of 80 balloon cytology specimens derived from patients with previously diagnosed oesophageal dysplasia, HPV DNA was detected in 22.2% (2/9) of the cases without cytological atypia, in 50% (3/6) with mild dysplasia, in 80.6% (25/31) with moderate dysplasia, and in 67.9% (19/28) of the cases with severe dysplasia (Chang et al., 1990b).

This first report of HPV testing in oesophageal cytology was soon followed by other reports, testing for HPV in balloon cytology samples. HPV DNA was recently detected in oesophageal cytological brushings in a substantial proportion (17%) of HIV-infected patients, without any clinically detectable lesions (Trottier et al., 1997). However, HPV DNA was not detected in any of the oesophageal achalasia lesions by PCR (Birgisson et al., 1997). In a recent study in two areas with different cancer risk in Anyang (China), specimens collected by balloon cytology from volunteers were analysed by PCR and ISH (Li et al., 2001). The prevalence of HPV16 in the high-incidence area was 1.9-fold higher than that of the low incidence area (72% and 37%, respectively, $p<0.01$). Moreover, HPV prevalence closely correlated with the grade of cytological atypia. These data were confirmed in another study from the same group (Li et al., 2002), and similar differences between the regions at different cancer risk were published by others as well (Si et al., 2003).

In our series of 700 SCCs, HPV signals were detected in the surrounding hyperplastic and dysplastic epithelia in 3.6% of the patients, and in 0.2% of the tumour resection margins (Chang et al., 2000a). The presence of HPV in the adjacent normal and dysplastic epithelium is in agreement with the "condemned mucosa" concept, recently proposed to explain the pathogenesis of multifocal HPV-induced carcinogenesis in the genital tract as well as in the upper aero-digestive tract (Pillai et al., 2000), including oesophageal SCC (Cintorino et al., 2001).

3.3.9. Evidence from serological studies

Since their description in 1992, virus-like particles (VLPs) have been extensively used to detect serum HPV antibodies in patients with different malignant tumours, including those of the oesophagus in both low-risk and high-risk areas of this disease (Dillner et al., 1996; Han et al., 1996; Björge et al., 1997; Lagergren et al., 1999). Among the former, IgG antibodies against HPV16 VLPs were measured in a serum bank of 39,268 samples collected in Finland (Dillner et al., 1996). Eight of 39 sera from cases were HPV16 seropositive as compared with 2/78 sera from matched controls ($p<0.001$; OR14.6, 95%CI, 1.8-117). Being a low risk country for this disease, this high OR clearly suggests a significantly increased risk for this malignancy among HPV16 seropositive subjects. Data are surprisingly similar from Norway, where an increased risk of oesophageal cancer among HPV16 seropositive subjects was of the same magnitude (OR, 6.6; 95%CI, 1.1-71) (Björge et al., 1997). Highly unexpectedly, however, no such increased risk could be confirmed in a study from neighbouring Sweden, where the age- and sex-adjusted ORs for SCC were 1.0 (95% CI, 0.5-2.0) for HPV16 seropositive patients and 0.5 (95% CI, 0.2 to 1.1) for HPV18 seropositive cases (Lagergren et al., 1999). The reasons for these evidently discrepant results in three Nordic (low-risk) countries remain unclear, but may be related to differences in some of the technological aspects of the serological methods used for the measurement of HPV antibodies.

In a high-incidence area of Shaanxi Province (China), 90 patients with tumours and 121 matched controls were analysed for the presence of HPV16 antibodies using VLP as the antigen (Han et al., 1996). Using HPV16 seropositivity cut-off values similar to those used in cervical cancer studies, 24% of the patients and only 7% of the controls were seropositive (OR, 4.5; 95%CI, 1.8-11.9), OR increasing with higher levels of HPV16 seroreactivity. Thus, these serological results correlate with the HPV DNA detection data in the balloon cytology samples from the screening populations in these high-incidence areas (Chang et al., 1990b; Li et al., 2001), suggesting that HPV16 infection is an important risk factor for oesophageal carcinoma in this region.

This is in contrast to the serological data from the low-incidence areas in Europe, where HPV antibody titres do not seem to correlate with the risk for oesophageal carcinoma (van Doornum et al., 2003). In a retrospective seroepidemiological study in the Netherlands, prevalence of HPV16 antibodies in the negative control group (18%) was equal or higher than that (14%) of the oesophageal SCC patients ($p=0.7$). These results agree with the HPV DNA data reported by others in that country (Smits et al., 1995; Kok et al., 1997), suggesting a non-significant association between HPV and oesophageal carcinoma in this low-incidence country.

Conflicting as these seroepidemiological results from different countries seem to be, the reported data are concordant with the results of HPV DNA detection. Accordingly, in countries where oesophageal carcinoma incidence is low, HPV DNA detection rates are also very low, the contrary being true in high-incidence countries. Being a rare disease in low-risk countries, serological response to HPV16 usually gives higher OR for oesophageal cancer risk (Dillner et al., 1996; Björge et al., 1997), as compared with high-incidence countries, where HPV prevalence is higher in controls as well (Han et al., 1996). To fully elucidate the value of these complex seroepidemiological data, a multi-center international trial would be necessary, where HPV serology is related to oesophageal cancer incidence and DNA prevalence in a population screening.

3.3.10. In vitro studies on molecular mechanisms

During the 1990's, the loss of function of the p53 tumour suppressor gene by mutation or interaction with the E6 protein of the oncogenic HPV types was established as a frequent event in oesophageal SCC (Chang et al., 1994; 1995). The presence of p53 mutations in HPV positive SCCs suggests that HPV and p53 mutation are not mutually exclusive events (Cooper, 1995; Chang et al., 1995). Similar results have been reported in HPV positive oesophageal cancer cell lines (Reiss et al., 1992), where different types of p53 mutations were a frequent occurrence. Interestingly, p53 seems to be mutated even in early precancer lesions, at least in the high-incidence areas of China (Gao et al., 1994), but p53 mutations are exceptional in benign SCPs (Poljak et al., 1996). The presence of frequent mutations of the p53 gene in both HPV positive and HPV negative carcinomas suggests an important role for environmental carcinogens in oesophageal carcinogenesis (Chang et al., 1994; 1995; Franceschi et al., 1996; Galloway et al., 1996; Montesano et al., 1996a; 1996b; Meltzer, 1996; Lam et al., 1997; Shen et al., 2001). The interactions between HPV16 E7 and the retinoblastoma protein (pRB) in clinical samples have not yet been systematically studied.

Apart from several HPV-negative oesophageal cancer cell lines (Wong et al., 1992; Zheng et al., 1999), HPV16 and HPV18 positive cell lines have been established from oesophageal SCCs (Togawa et al., 1994), and used to study HPV replication (Togawa et al., 1994). HPV16 and HPV18 genomes were independently transiently transfected into HCE-4 and HCE-7 cell lines with and without E1 and E2 genes under the control of heterologous promoters. These cell lines supported viral replication, and heterologous E1 and E2 were not required for HPV replication, suggesting that specific host nuclear factors in oesophageal squamous epithelial cells may support HPV replication (Togawa et al. 1994).

A long series of *in vitro* studies with HPV-positive cell lines was initiated in 1999 by Shen et al. (1999a; 1999b). Malignant transformation of human embryonic epithelial cells was induced *in vitro* by HPV18 E6/E7 in synergy with TPA, and these cells were subsequently grown as an established SHEE (synergistically infected human embryonic oesophageal) cell line (Shen et al., 2000). In their biological behavior, the SHEE cells are close to the basal cells of their original fetal oesophageal mucosa, maintaining their proliferative potential and capability of differentiation. Three different passages of this cell line were used to study cytogenesis, telomere and telomerase, and the c-myc, ras, bcl-2, and p53 genes. The results revealed that changes of chromosomes, telomere length, telomerase activity, and the expression of certain genes occurred as a dynamic progressive process, suggesting that they are important events in the immortalisation of HPV infected Oesophageal epithelial cells (Shen et al., 2001).

The authors subsequently analysed telomere length, telomerase activity and its subunits in different passages of the same (SHEE) cell line (Shen et al., 2002c). At early passages, telomeres were shortened without expression of telomerase and hTERT, leading to senescence and cell death. From the 20th to the 30th passages, however, activation of telomerase and maintenance of telomere length paralleled the progressive process of immortalization of these cells, suggesting the usefulness of telomerase as a potential biomarker of immortalization (Shen et al., 2002b; 2002c). This is consistent with the recent data from Oesophageal cancer biopsies as well (Tominaga et al., 2002).

The same cell line was used to study the promoter effects of sodium butyrate in high or low dosages in the carcinogenesis process (Shen et al., 2002d). The malignant change of immortalized cells induced by sodium butyrate is a two-stage process; in high dosage, sodium butyrate induces cell death and in low dosage it induces cell proliferation, suggesting its importance as a promotor in oesophageal carcinogenesis (Shen et al., 2002d). To follow the process of malignant transformation, the malignant phenotype, tumorigenesis and invasive potential of SHEE cells were further studied at different passages (Shen et al., 2002e; 2002f; 2003c). In continuous culture, cells from the 10th to the 85th passage changed gradually from pre-immortal, immortal, to precancerous, and finally to fully malignantly transformed stages. All these changes comprised a dynamic progressive process. This cell line provides an ideal model to study the molecular mechanisms of oesophageal carcinogenesis.

Fully malignant transformation with a strong invasive potential occurs at the 85th passage (Shen et al., 2002e; 2002f). The immortal phenotype requires both activation of telomerase and genetic alterations that abrogate normal differentiation and promote cellular proliferation (Shen et al., 2003c). At this stage (SHEE85), the cell karyotype is hypotriploid and the cells expressed hTERT, c-myc, k-ras, bcl-2 and p53, together with HPV18 E6/E7 (Shen et al., 2003b). Up-regulated ezrin expression is one of the important factors possibly associated with the invasive phenotype in transformed SHEE cells (Shen et al., 2003a). The most recent paper in this impressive series of *in vitro* studies with the SHEE cell line analyses the events in the multistage process of carcinogenesis, with progressive alterations of cell morphology, telomerase, cytogenesis, oncogenes and tumorigenicity, at the 11th, 35th, 65th and 100th passages (Shen et al., 2004). Cells transplanted at the 65th and 100th passage into SCID mice resulted in tumour formation, but only the 100th passage cells could grow in nude mice. The up-regulated expression of fascin 1 gene seems to be closely correlated with the malignant transformation of SHEE cells (Rong et al., 2004).

Recently, also HPV16 has been used to induce immortal cell lines (CHEK-1), by transduction of HPV16 E6/E7 into the primary culture of human oesophageal keratinocytes, using a recombinant adenovirus (Sashiyama et al., 2001). These cells have an extended lifespan, up-regulated telomerase activity and serum resistant growth, and displayed the M1 and M2 stages of keratinocyte immortalisation. Further studies on cellular changes related to their malignant progression showed that spindle-shaped, rapidly proliferating clones with prominent actin stress fibres had reduced alpha-6, alpha-4 and alpha-2 integrin expression and showed anchorage-independent growth. These data imply that down-regulation of alpha-2 and alpha-6/beta-4 integrins are related to immortalization and malignant progression of this cell line (Sashiyama et al., 2002).

Taken together, these highly interesting *in vitro* data demonstrate that carcinogenesis of the oesophageal epithelial cells induced by HPV16 and HPV18 E6 and E7 oncoproteins is a multistage process. This seems to progress through the initial, immortal, pre-malignant and malignant transformation stages. There is little doubt that an invasive oesophageal squamous cell carcinoma results from the accumulation of multiple cellular, genetic and molecular changes. Results from these recent *in vitro* experiments give additional support to the clinical and molecular epidemiological data, suggesting with increasing strength that oesophageal squamous cell carcinoma is another HPV-related disease, at least in the high-incidence geographic regions.

References

Acevedo-Nuno, E., Gonzalez-Ojeda, A., Vazquez-Camacho, G., Balderas-Pena Luz Ma, A., Moreno-Villa, H., and Montoya-Fuentes, H. (2004). Human papillomavirus DNA and protein in tissue samples of oesophageal cancer, Barrett's oesophagus and oesophagitis. Anticancer Res. *24*, 1319-1323.

Adler, R.H., Carberry, D.M., Ross, C.A. (1959). Papilloma of the esophagus: association with hiatal hernia. J. Thorac. Surg. *37*, 625-635.

Agarwal, S.K., Chatterji, A., Bhambhani, S., and Sharma, B.K. (1998) Immunohistochemical co-expression of human papillomavirus type 16/18 transforming (E6) oncoprotein and p53 tumour suppressor gene proteins in oesophageal cancer. Indian J. Exp. Biol. *36*, 559-563.

Akutsu, N., Shirasawa, H., Nakano, K., Tanzawa, H., Asano, T., and Kobayashi, S. (1995). Rare association of human papillomavirus DNA with esophageal cancer in Japan. J. Infect. Dis. *171*, 425-428.

Al-Sohaibani, M.O., and Al-Rashed, R.S. (1995). Squamous papilloma of the esophagus: A clinicopathologic study of 10 cases and review of the literature. Ann. Saudi Med. *15*, 140-142.

Anderson, R.A., Scobie, L., O'Neil, B.W., Grindlay, G.J., and Campo, M.S. (1997). Viral proteins of bovine papillomavirus type 4 during the development of alimentary canal tumours. Vet. J. *154*, 69-78.

Asamoto, M., Toriyama-Baba, H., Ohnishi, T., Naito, A., Ota, T., Ando, A., Ochiya, T., and Tsuda, H. (2002). Transgenic rats carrying human c-Ha-ras proto-oncogene are highly susceptible to N-nitrosomethylbenzylamine induction of esophageal tumorigenesis. Jpn. J. Cancer Res. *93*, 744-751.

Ashworth, M.T., McDicken, I.W., Southern, S.A., and Nash, J.R. (1993). Human papillomavirus in squamous cell carcinoma of the oesophagus associated with tylosis. J. Clin Pathol *46*, 573-575.

Astori, G., Merluzzi, S., Arzese, A., Brosolo, P., de Pretis, G., Maieron, R., Pipan, C., and Botta, G.A. (2001). Detection of human papillomavirus DNA and p53 gene mutations in esophageal cancer samples and adjacent normal mucosa. Digestion *64* , 9–14

Aubin, F., Prétet, J-L., and Mougin, C. (eds)(2003). Papillomavirus Humains. Biologie et Pathologie Tumorale (London: Editions DEC and DOC, and EM Inter).

Awerkiew, S., Bollschweiler, E., Metzger, R., Schneider, P.M., Holscher, A.H., and Pfister, H. (2003). Esophageal cancer in Germany is associated with Epstein-Barr-virus but not with papillomaviruses. Med. Microbiol. Immunol. (Berl) *192*, 137-140.

Baehr, P.H., and McDonald, G.B. (1994). Esophageal infections: Risk factors, presentation, diagnosis, and treatment. Gastroenterol. *106*, 509-532.

Batra, P.S., Hebert, R.L., Haines, G.K. and Holinger, L.D. (2001). Recurrent respiratory papillomatosis with esophageal involvement. Int. J. Pediatr. Otorhinolaryngol. *58*, 233-238.

Benamouzig, R., Pigot, F., Quiroga, G., Dalidire, P., Chaussade, S., and Catalan, F. (1992). Human papillomavirus infection in esophageal squamous cell carcinoma in Western countries. Int. J. Cancer *50*, 549-552.

Benamouzig, R., Jullian, E., Chang, F.J., Robaskiewicz, M., Flejou, J.F., and Raoul, J.L. (1995). Absence of human papillomavirus DNA detected by polymerase chain reaction in French patients with esophageal carcinoma. Gastroenterol. *109*, 1876-1881.

Benamouzig, R., Jullian, E., Pompidou, A., and Rautureau, J. (1996). Are human papillomaviruses involved in esophageal carcinogenesis? Head and Neck Cancer – Adv. Basic Res. *114*, 391-395.

Benisch, B.M., and Mantell, C. (1974). Multiple squamous papillomas of the esophagus. Arch. Otolaryngol. *100*, 379-382.

Biemond, P., ten Kate, F.J., and van Blankenstein, M. (1991). Esophageal verrucous carcinoma: histologically a low-grade malignancy but clinically a fatal disease. J. Clin. Gastroenterol. *13*, 102-107.

Birgisson, S., Galinski, M.S., Goldblum, J.R., Rice, T.W., and Richter, J.E. (1997). Achalasia is not associated with measles or known herpes and human papilloma viruses. Dig. Dis. Sci. *42*, 300-306.

Björge, T., Hakulinen, T., Engeland, A., Jellum, E., Koskela, P., Lehtinen, M., Luostarinen, T., Paavonen, J., Sapp, M., Schiller, J., Thoresen, S., Wang, Z., Youngman, L., and Dillner, J. (1997). A prospective, seroepidemiological study of the role of human papillomavirus in esophageal cancer in Norway. Cancer Res. *57*,3989–3992.

Borzacchiello, G. Ambrosio, V., Roperto, S., Poggiali, F., Tsirimonakis, E., Venuti, A. Campo, M.S., and Roperto, F. (2003). Bovine papillomavirus type 4 in oesophageal papillomas of cattle from the South of Italy. J. Comp. Pathol. *128*, 203-206.

Bosch, X.F., Lörincz, A., Muñoz, N., Meijer, C.J. and Shah, K.V. (2002). The causal relation between human papillomavirus and cervical cancer. J. Clin. Pathol. *55*, 244–265.

Bretagne, M.C., Derelle, J., Bernard, C., Perrin, C., and Hoeffel, J.C. (1989). Aerodigestive papillomatosis in an infant. J. Radiol. *70*, 487-90.

Brinson, R.R., Schuman, B.M., Mills, L.R., Thigpen, S., and Freedman, S. (1987). Multiple squamous papillomas of the esophagus associated with Goltz syndrome. Am. J. Gastroenterol. *82*, 1177-1179.

Brooks, L.A., Steyn, L.M., and Renan, M.J. (1987). Oesophageal carcinoma. Search for a possible viral aetiology using molecular hybridisation techniques. S. Afr. Med. J. Suppl. 6-8.

Campo, M.S. (1987). Papillomas and cancer in cattle. Cancer Surv *6*, 39–54.

Campo, M.S., Moar, M.H., Jarrett, W.F.H., and Laird, H.M. (1980). A new papillomavirus associated with alimentary cancer in cattle. Nature *280*, 180-182.

Campo, M.S., McCaffery, R.E., Doherty, I., Kennedy, I.M., and Jarrett, W.F.H. (1990). The Harvey ras 1 gene is activated in papillomavirus-associated carcinomas of the upper alimentary canal in cattle. Oncogene *5*, 303-308.

Campo, M.S., and Jarrett, W.F. (1994). Vaccination against cutaneous and mucosal papillomavirus in cattle. Ciba Found. Symp. *187*, 61-73.

Carr, N.J., Bratthauer, G.L., Lichy, J.H., Taubenberger, J.K., Monihan, J.M., and Sobin, L.H. (1994a) Squamous cell papillomas of the esophagus: a study of 23 lesions for human papillomavirus by *in situ* hybridization and the polymerase chain reaction. Hum Pathol *25*, 536-540.

Carr, N.J., Monihan, J.M., and Sobin, L.H. (1994b). Squamous cell papilloma of the esophagus: a clinicopathologic and follow-up study of 25 cases. Am. J. Gastroenterol. *89*, 245-248.

Cervantes, J. (2004). Update on the pathogenesis and immunotherapy of esophageal squamous cell carcinoma. Rev. Gastroenterol. Peru *24*, 165-170.

Cintorino, M., Tripodi, S.A., Santopietro, R., Antonio, P., Lutfi, A., Chang, F., Syrjänen, S., Shen, Q., Tosi, P., and Syrjänen, K. (2001). Cytokeratin expression in oesophageal squamous cell carcinoma: a study of 354 cases from the high-incidence area of Linxian, China. Anticancer Res. *21*, 4195–4201.

Chang, F., Syrjänen, S., Shen, Q., Ji, H.X., and Syrjänen, K. (1990a) Human papillomavirus (HPV) DNA in esophageal precancer lesions and squamous cell carcinomas from China. Int. J. Cancer *45*, 21-25.

Chang, F., Shen, Q., Zhou, J., Wang, C., Wang, D., Syrjänen, S., and Syrjänen, K. (1990b). Detection of human papillomavirus DNA in cytologic specimens derived from esophageal precancer lesions and cancer. Scand. J. Gastroenterol. *25*, 383-388.

Chang, F., Janatuinen, E., Pikkarainen, P., Syrjänen, S., and Syrjänen, K. (1991). Esophageal squamous cell papillomas. Failure to detect human papillomavirus DNA by *in situ* hybridization and polymerase chain reaction. Scand. J. Gastroenterol. *26*, 535-545.

Chang, F., Syrjänen, S.M., Wang, L., and Syrjänen, K. (1992a). Infectious agents in the etiology of esophageal cancer. Gastroenterol. *103*, 1336-1348.

Chang, F., Syrjänen, S., Shen, Q., Wang, L., Wang, D., and Syrjänen, K. (1992b). Human papillomavirus (HPV) involvement in esophageal precancer lesions and squamous cell carcinomas as evidenced by microscopy and different DNA techniques. Scand. J. Gastroenterol. *27*, 553-563.

Chang, F., Syrjänen, S., Shen, Q., Wang, L., and Syrjänen, K. (1993). Screening for human papillomavirus infections in esophageal squamous cell carcinomas by *in situ* hybridization. Cancer 72, 2525-2530.

Chang, F., Syrjänen, S., Tervahauta, A., Kurvinen, K., Wang, L., and Syrjänen, K. (1994). Frequent mutations of p53 gene in oesophageal squamous cell carcinomas with and without human papillomavirus (HPV) involvement suggest the dominant role of environmental carcinogens in oesophageal carcinogenesis. Br. J. Cancer *70*, 346-351.

Chang, F., Syrjänen, S., and Syrjänen, K. (1995). Implications of the p53 tumor-suppressor gene in clinical oncology. J. Clin. Oncol. *13*, 1009-1022.

Chang, F., Syrjänen, S., Wang, L., Shen, Q., and Syrjänen, K. (1997). p53 overexpression and human papillomavirus (HPV) infection in oesophageal squamous cell carcinomas derived from a high-incidence area in China. Anticancer Res. *17*, 709–715.

Chang, F., Syrjänen, S., Shen, Q., Cintorino, M., Santopietro, R., Tosi, P., and Syrjänen, K. (2000a). Detection of HPV DNA in oesophageal squamous cell carcinoma from the high-incidence area of Linxian, China. Scand. J. Gastroenterol. *35*, 123-130.

Chang, F., Syrjänen, S., Shen, Q., Cintorino, M., Santopietro, R., Tosi, P., and Syrjänen, K. (2000b). Presence of HPV, CMV, HSV and EBV in esophageal squamous cell carcinomas from a high-incidence area of China. Anticancer Res. *20*, 3935-3940.

Chatelain, P., Coudry, H., and Vittouris, P. (1968). Apropos of a case of papillomatosis of the esophagus. J. Fr. Otorhinolaryngol. Audiophonol. Chir. Maxillofac. *17*, 509-513.

Chen, B.F. (1993). Polymerase chain reaction in detection of human papillomavirus DNA in esophageal carcinoma. Zhonghua Yi. Xue. Za. Zhi. *73*, 667-669.

Chen, B., Yin, H., and Dhurandhar, N. (1994). Detection of human papillomavirus DNA in esophageal squamous cell carcinomas by the polymerase chain reaction using general consensus primers. Hum. Pathol. *25*, 920-923.

Chistiakova, V.R., Iablonskii, S.V., Kovshenkova, I.D. (1998). Two cases of generalized papillomatosis of the upper respiratory tract and esophagus in children. Vestn. Otorinolaringol. *5*, 56-57.

Chitra, S., Ashok, L., Anand, L., Srinivasan, V., and Jayanthi, V. (2004). Risk factors for esophageal cancer in Coimbatore, southern India: a hospital-based case-control study. Indian J. Gastroenterol. *23*, 19-21.

Cooper, K. (1995). p53 mutations in human papillomavirus-associated oesophageal squamous cell carcinoma. Br. J. Cancer *72*, 1337.

Cooper, K., Taylor, L., and Govind, S. (1995). Human papillomavirus DNA in oesophageal carcinomas in South Africa. J. Pathol. *175*, 273–277.

Crew, K.D., and Neugut, A.I. (2004). Epidemiology of upper gastrointestinal malignancies. Semin. Oncol. *31*, 450-464.

De Borges, R.J., Acevedo, F., Miralles, E., and Mijares, P. (1986). Squamous papilloma of the esophagus diagnosed by cytology. Report of a case with concurrent occult epidermoid carcinoma. Acta Cytol. *30*, 487-490.

de Villiers, E.M., Lavergne, D., Chang, F.,Syrjänen, K., Tosi, P., Cintorino, M., Santopietro, R., and Syrjänen, S. (1999). An interlaboratory study to determine the presence of human papillomavirus DNA in esophageal carcinoma from China. Int. J. Cancer *81*, 225–228.

de Villiers, E.M., Gunst, K., Stein, H., and Scherubl, H. (2004). Esophageal squamous cell cancer in patients with head and neck cancer: Prevalence of human papillomavirus DNA sequences. Int. J. Cancer *109*, 253-258.

Dillner, J., Knekt, P., Schiller, J.T., and Hakulinen, T. (1996). Prospective seroepidemiological evidence that human papillomavirus type 16 infection is a risk factor for oesophageal squamous cell carcinoma. B.M.J. *311*, 346.

Dong, Z., Tang, P., Li, L., and Wang, G. (2002). The strategy for esophageal cancer control in high-risk areas of China. Jpn. J. Clin. Oncol. *32*, 10-20.

Dowlatshahi, K., and Miller, R.J. (1985). Role of opium in esophageal cancer: a hypothesis. Cancer Res. *45*, 1906-1907.

Fekete, F., Chazouilleres, O., Ganthier, V., Molas, G., and Potet, F. (1988). A case of esophageal papillomatosis in adults. Gastroenterol. Clin. Biol. *12*, 66-70.

Ferlay J, Bray F, Pisani P, et al. GLOBOCAN 2000: cancer incidence, mortality and prevalence worldwide, Version 1.0. IARC Cancer Base No. *5*. Lyon: IARC Press, 2001.

Fernandez-Rodriguez, C.M., Badia-Figuerola, N., Ruiz del Arbol, L., Fernandez-Seara, J., Dominguez, F., and Aviles-Ruiz, J.F. (1986). Squamous papilloma of the esophagus: report of six cases with long-term follow-up in four patients. Am. J. Gastroenterol. *81*, 1059-1062.

Ferrari, A.P., Lanzoni, V.P., Kondo, M., Lichtenstein, D.R., and Carr-Locke, D.L. (1995). Esophageal squamous cell papilloma. Report of 3 cases. Diagnostic aspects. Rev. Assoc. Med. Bras. *41*, 266-270.

Fidalgo, P.O., Cravo, M.L., Chaves, P.P., Leitao, C.N., and Mira, F.C. (1995). High prevalence of human papillomavirus in squamous cell carcinoma and matched normal esophageal mucosa: assessment by polmerase chain reaction. Cancer *76*, 1522-1528.

Fontolliet, C., Hurlimann, J., Monnier, P., Ollyo, J.B., Levi, F., and Savary, M. (1991). Is papilloma of the esophagus a preneoplastic lesion? Study of 33 cases. Schweiz. Med. Wochenschr. *121*, 754-757.

Franceschi, S., Munoz, N., Bosch, X.F., Snijders, P.J., and Walboomers, J.M. (1996). Human papillomavirus and cancers of the upper aerodigestive tract: a review of epidemiological and experimental evidence. Cancer Epidemiol. Biomarkers Prev. *5*, 567-575.

Franzin, G., Musola, R., Zamboni, G., Nicolis, A., Manfrini, C., and Fratton, A. (1983). Squamous papillomas of the esophagus. Gastrointest. Endosc. *29*, 104-106.

Frootko, N.J., and Rogers, J.H. (1978). Oesophageal papillomata in the child. J. Laryngol. Otol. *92*, 823-827.

Furihata, M., Ohtsuki, Y., Ogoshi, S., Takahashi, A., Tamiya, T., and Ogata, T. (1993). Prognostic significance of human papillomavirus genomes (type 16, 18) and aberrant expression of p53 protein in human esophageal cancer. Int. J. Cancer *54*, 226-230.

Galloway, D.A., and Daling, J.R. (1996). Is the evidence implicating Human papillomavirus type 16 in esophageal cancer hard to swallow? J. Natl. Cancer Inst. *88*, 1421-1423.

Gao, H., Wang, L.-D., Zhou, Q., Hong,J.Y., Huang, T.Y., and Yang, C.S. (1994). p53 tumor suppressor gene mutation is early esophageal precancerous lesions and carcinoma among high-risk populations in Henan, China. Cancer Res. *54*, 4242–4246.

Garrard, C.L., Sheih, W.J., Cohn, R.A., and Sawyers, J.L. (1994). Verrucous carcinoma of the esophagus: Surgical treatment for an often fatal disease. Am. Surg. *60*, 613-616.

Gaukroger, J., Bradley, A., O'Neil, B., Smith, K., Campo, S., and Jarrett, W. (1989). Induction of virus-producing tumors in athymic nude mice by bovine papillomavirus type 4. Veterin. Record. *125*, 391-392.

Gaukroger, J., Chandrachud, L., Jarrett, W.F.H., McGarvie, G.E., Yeudall, W.A., McCaffery, R.E., Smith, K.T., and Campo, M.S. (1991). Malignant transformation of a papilloma induced by bovine papillomavirus type 4 in the nude mouse renal capsule. J. Gen. Virol. *72*, 1165-1168.

Gelderblom, W.C., Jaskiewicz, K., and Marasas, W.F. (1988). Fumonisins- novel mycotoxins with cancer promoting activity produced by *Fusarium monoliforme*. Appl. Environ. Microbiol. *54*, 1806-1811.

Ghadirian, P., Vobecky, J., Vobecky, J.S. (1988). Factors associated with cancer of the oesophagus: an overview. Cancer Detect. Prev. *11*, 225-234.

Ghavamzadeh, A., Moussavi, A., Jahani, M., Rastegarpanah, M., and Iravani, M. (2001). Esophageal cancer in Iran. Semin. Oncol. *28*, 153-157.

Gillison, M.L., and Shah, K.V. (2003). Role of mucosal human papillomavirus in nongenital cancers. J. Natl. Cancer Inst. Monogr. *31*, 57-65.

Goldsmith, M.F. (1984). Papillomavirus invades esophagus, incidence seems to be increasing. JAMA. *251*, 2185-2187.

Gomez, N.A., Arevalo, C.A., Rojas, J.E., and Camacho, J.F. (1997). Esophageal squamous papilloma: a case report and review of the literature. Acta Gastroenterol. Latinoam. *27*, 131-133.

Hale, M.J., Liptz, T.R., and Paterson, A.C. (1989). Association between human papillomavirus and carcinoma of the oesophagus in South African blacks. A histochemical and immunohistochemical study. S. Afr. Med. J. *76*, 329-330.

Hamada, M., Oyamada, T., Yoshikawa, H., and Yoshikawa, T. (1989). Morphological studies of esophageal papilloma naturally occurring in cattle. Nippon Juigaku. Zasshi. *51*, 345-351.

Han, C., Qiao, G., Hubbert, N.L., Li, L., Sun, C., Wang, Y., Yan, M., Xu, D., Li, Y., Lowy, D.R., and Schiller, J.T. (1996). Serologic association between human papillomavirus type 16 infection and esophageal cancer in Shaanxi Province, China. J. Natl. Cancer Inst. *88*, 1467-1471.

Hasegawa, M., Ohoka, I., Yamazaki, K., Hanami, K., Sugano, I., Nagao, T., Asoh, A., Wada, N., Nagao, K., and Ishida, Y. (2002). Expression of p21/WAF-1, status of apoptosis and p53 mutation in esophageal squamous cell carcinoma with HPV infection. Pathol Int. *52*, 442-450.

He, D., Tsao, S.W., and Bu, H. (1996). Human papillomavirus infection and esophageal squamous cell carcinoma. Zhonghua Bing. Li. Xue. Za. Zhi. *25*, 351–354.

He, D., Zhang, D-K., lam, K-Y., Ma, L., Ngan, H.Y-S., Liu, S.S., and Tsao, S-W. (1997). Prevalence of HPV infection in esophageal squamous cell carcinoma in Chinese patients and its relationship to the p53 gene mutation. Int. J. Cancer *72*, 959-964.

He, Y.T., Hou, J., Qiao, C.Y., Chen, Z.F., Song, G.H., Li, S.S., Meng, F.S., Jin, H.X., and Chen, C. (2003). An analysis of esophageal cancer incidence in Cixian county from 1974 to 1996. World J. Gastroenterol. *9*, 209-213.

Hemminki, K., Jiang, Y., and Dong, C. (2001). Second primary cancers after anogenital, skin, oral, esophageal and rectal cancers: etiological links? Int. J. Cancer *93*, 294-298.

Hendricks, D., and Parker, M.I. (2002). Oesophageal cancer in Africa. IUBMB Life *53*, 263-268.

Henrik-Siman, J., Forsgren, A., Berglund, G., and Floren, C.H. (2001). Helicobacter pylori infection is associated with a decreased risk of developing oesophageal neoplasms. Helicobacter *6*, 310-316.

Hille, J.J., Margolius, K.A., Markowitz, S., and Isaacson, C. (1986). Human papillomavirus infection related to oesophageal carcinoma in black South Africans. A preliminary study. S. Afr. Med. J. *69*, 417-420.

Hippeläinen, M., Eskelinen, M., Lipponen, P., Chang, F., and Syrjänen, K. (1993) Mitotic activity index, volume corrected mitotic index and human papilloma-virus suggestive morphology are not prognostic factors in carcinoma of the oesophagus. Anticancer Res. *13*, 677-681.

Hosone, M., Ishiwata, T., Nomura, N., Kawanami, O. (1995). A molecular pathological study on the localization of the human papillomavirus using nested PCR and *in situ* hybridization methods in paraffin embedded tissues. Nippon Ika. Daigaku. Zasshi. *62*, 131-141.

Hörding, M,, Hörding, U., Daugaard, S., Norrild, B., and Faber, V. (1989). Human papilloma virus type 11 in a fatal case of esophageal and bronchial papillomatosis. Scand. J. Infect. Dis. *21*, 229-231.

IARC (1995). Monographs on the evaluation of carcinogenic risks to humans. Papillomaviruses, Vol. *64*. (Lyon: IARC).

Jacob, J.H., Riviere, A., Mandard, A.M., Munoz, N., Crespi, M., Etienne, Y., Castellsague, X., Marnay, J., Lebigot, G., and Qiu, S.L. (1993). Prevalence survey of precancerous lesions of the oesophagus in a high-risk population for oesophageal cancer in France. Eur. J. Cancer. Prev. *2*, 53-59.

Janson, J.A., Baillie, J., and Pollock, M. (1991). Endoscopic removal of esophageal condylomata acuminatum containing human papilloma virus. Gastrointest. Endosc. *37*, 367-370.

Jarrett, W.F.H. (1987). Environmental carcinogens and papillomaviruses in the pathogenesis of cancer. Proc. R. Soc. Lond. B. Biol. Sci. *231*, 1–11.

Jarrett, W.F.H., NcNeil, P.E., Grimshaw, T.R., Selman, I.E., and NcIntyre, W.I.M. (1978). High incidence area of cattle cancer with a possible interaction between an environmental carcinogen and a papillomavirus. Nature *274*, 215-217.

Jaskiewicz, K., Marasas, W.F.O., and van der Walt, F.E. (1987). Oesophageal and other main cancer patterns in four districts of Transkei, 1981-1984. S. Afr. Med. J. *72*, 27-30.

Javdan, P., and Pitman, E.R. (1984). Squamous papilloma of esophagus. Dig. Dis. Sci. *29*, 317-320.

Kamath, A.M., Wu, T.T., Heitmiller, R., Daniel, R., and Shah K.V. (2000). Investigation of the association of esophageal carcinoma with human papillomaviruses. Dis. Esophagus *13*, 122-124.

Kashima, H., and Mounts, P. (1987). Tumors of the head and neck, larynx, lung and esophagus and their possible relation to HPV. In Papillomaviruses and Human Disease, K. Syrjänen, L. Gissmann, L.G. Koss, eds. (Berlin: Springer Verlag), pp. 138–157.

Kato, H., Orito, E., Yoshinouchi, T., Ueda, R., Koizumi, T., and Yoshinouchi, M. (2003). Regression of esophageal papillomatous polyposis caused by high-risk type human papilloma virus. J. Gastroenterol. *38*, 579-583.

Kawaguchi, H., Ohno, S., Araki, K., Miyazaki, M., Saeki, H., Watanabe, M., Tanaka, S., and Sugimachi, K. (2000). p53 polymorphism in human papillomavirus-associated esophageal cancer. Cancer Res. *60*, 2753–2755.

Khurshid, A., Kazuya, N., Hanae, I., and Manabu, I. (1998). Infection of human papillomavirus (HPV) and Epstein-Barr virus (EBV) and p53 expression in human esophageal carcinoma. J. Pak. Med. Assoc. *48*, 138–142.

Kim, W.H., Song, S.Y., Kang, J.K., Park, I.S., and Choi, H.J. (1991). Detection of human papillomavirus DNA from esophageal cancer tissue using polymerase chain reaction. Gastroenterol. *100*, A1272.

Kiyabu, M.T., Shibata, D., Arnheim, N., Martin, W.J., and Fitzgibbons, P.L. (1989). Detection of human papillomavirus in formalin-fixed, invasive squamous cell carcinoma using polymerase chain reaction. Am. J. Surg. Pathol. *13*, 221-224.

Kok, T.C., Nooter, K., Tjong-A-Hung, S.P., Smits, H.L., and Ter Schegget, J. (1997). No evidence of known types of human papillomavirus in squamous cell cancer of the oesophagus in a low-risk area. Eur. J. Cancer *33*, 1865-1868.

Kulski, J., Demeter, T., Sterret, G.F., and Shilkin, K.B. (1986). Human papillomavirus DNA in esophageal carcinoma. Lancet *ii*, 683-284.

Kulski, J.K., Demeter, T., Mutavdzic, S., Sterrett, G.F., Mitchell, K.M., and Pixley, E.C. (1990). Survey of histologic specimens of human cancer for human papillomavirus types 6/11/16/18 by filter *in situ* hybridization. Am. J. Clin. Pathol. *94*, 566-570.

Kuwano, H., Sumiyoshi, K., Sonoda, K., Kitamura,K., Toh, Y., Nakashima, H., and Sugimachi, K. (2001). Pathogenesis of esophageal squamous cell carcinoma with lymphoid stroma. Hepatogastroenterol. *48*, 458–461.

Lagergren, J., Wang, Z., Bergström, R., Dillner, J., and Nyren, O. (1999). Human papillomavirus infection and esophageal cancer: a nationwide seroepidemiologic case-control study in Sweden. J. Natl. Cancer Inst. *91*, 156–162.

Lam, K.Y., He, D., Ma, L., Zhang, D., Ngan, H.Y., Wan, T.S., and Tsao, S.W. (1997). Presence of human papillomavirus in esophageal squamous cell carcinomas in Hong Kong Chinese and its relationship with p53 gene mutation. Hum. Pathol. *28*, 657-663.

Lambot, M.A., Haot, J., Peny, M.O., Fayt, I., and Noel, J.C. (2000). Evaluation of the role of human papillomavirus in oesophageal squamous cell carcinoma in Belgium. Acta Gastroenterol. Belg. *63*, 154–156.

Lavergne, D., and de Villiers, E.M. (1999). Papillomavirus in esophageal papillomas and carcinomas. Int. J. Cancer *80*, 681-684.

Lesec, G., Gogusev, J., Fermaud, H., Gorce, D., Lemaitre, J.P., and Verdier, A. (1985). Presence d'un antigene viral de groupe papilloma dans un condylome aesophagien chez i'homme. Gastroenterol. Clin. Biol. *9*, 165-168.

Lewensohn-Fuchs, I., Munck-Wikland, E., Berke, Z., Magnusson, K.P., Pallesen, G., Auer, G., Lindholm, J., Linde, A., Aberg, B., and Rubio, C. (1994). Involvement of aberrant p53 expression and human papillomavirus in carcinoma of the head, neck and esophagus. Anticancer Res. *14*, 1281-1285.

Li, M., Li, P., and Li, B. (1980). Recent progress in research on esophageal cancer in China. Adv. Cancer Res. *33*, 173-249.

Li, J., Zhang, Y., and Gao, D. (1996). Study on the interrelationship between human papilloma virus infection and Langerhans cell in carcinogenesis of esophagus. Zhonghua Bing. Li. Xue. Za. Zhi. *25*, 83-85.

Li, T., Lu, Z.M., Chen, K.N., Guo, M., Xing, H.P., Mei, Q., Yang, H.H., Lechner, J.F., and Ke, Y. (2001). Human papillomavirus type 16 is an important infectious factor in the high incidence of esophageal cancer in Anyang area of China. Carcinogenesis *22*, 929-934.

Li, T., Lu, Z.M., Guo, M., Wu, Q.J., Chen, K.N., Xing, H.P., Mei, Q., and Ke, Y. (2002). p53 codon 72 polymorphism (C/G) and the risk of human papillomavirus-associated carcinomas in China. Cancer *95*, 2571-2576.

Li, Y., Huang, G., Xiao, H., Huang, Y., Mao, T., and Deng, W. (1991). The status of human papillomavirus 16 DNA in the tissues of human esophagus carcinoma. Hua. His. I. Ko. Ta. Hsueh. Hsueh. Pao. *22*, 157-160.

Lin, K., Shen, W., Shen, Z., Wu, Y., and Lu, S. (2002). Dietary exposure and urinary excretion of total N-nitroso compounds, nitrosamino acids and volatile nitrosamine in inhabitants of high- and low-risk areas for esophageal cancer in southern China. Int. J. Cancer *102*, 207-211.

Lio, W.J., Huang, C.K., and Chen, G,H. (1996). Squamous papilloma of the esophagus. Zhonghua Yi. Xue. Za. Zhi. (Taipei) *57*, 413-417.

Liu, Y.L., Li, X.M., Jin, G.L., Yan, X., Yang, J.Z., Wang, J.L., Li, Y.H., Wang, F.R., and Zhang, X.H. (2003). HPV detection and FHIT expression in esophageal squamous carcinoma from high incidence area in Cixian County. Ai. Zheng. *22*, 492-495.

Loke, S.L., Ma, L., Wong, M., Srivastava, G., Lo, I., and Bird, C.C. (1990). Human papillomavirus in oesophageal squamous cell carcinoma. J. Clin. Pathol. *43*, 909-912.

Lombardi, J.P., Tang, D., and Myhre, O.A. (1980). Squamous cell papilloma of the esophagus: a case report and review of the literature. Int. Surg. *65*, 459-461.

Lu, J.B., Yang, W.X., Liu, J.M., Li, Y.S., and Qin, Y.M. (1985). Trends in morbidity and mortality for oesophageal cancer in Linxian country. Int. J. Cancer *36*, 643-645.

Lu, S., Luo, F., and Li, H. (1995). Detection of human papilloma virus in esophageal squamous cell carcinoma and adjacent tissue specimens in Linxian. Zhonghua Zhong. Liu. Za. Zhi. *17*, 321–324.

Lu, Z., Chen, K., and Guo, M. (2001). Detection of HPV in human esophageal cancer in high-incidence area and its correlation with p53 expression. Zhonghua Zhong. Liu. Za. Zhi. *23*, 220-223.

Marasas, W.F. (1996). Fumonisins: History, World-Wide Occurrence and Impact in Fumonisins in Food. In Advance in Experimental Medicine and Biology, vol *392*, L.S. Jackson, J.W. DeVries, and L.B Bullerman, eds. (New York: Plenum Press), pp. 1-18.

Mastour, M., Lefebre, A., Dambron, P., Ramdani, M., Duhamel, O., and Gislon, J. (1996). Papillome oesophagien hemorragique associe a la prise d'anti-inflammatoires non steroidiens. Gastroenterol. Clin. Biol. *20*, 406-407.

Matos, M., Golindano, C., Guirola, E., Wright, H., Capozzolo, N., and Nishimura, M. (1990). Esophageal lesions caused by human papillomaviruses (HPV). G E N *44*, 377-384.

Matsha, T., Erasmus, R., Kafuko, A.B., Mugwanya, D., Stepien, A., and Parker, M.I. (2002). Human papillomavirus associated with oesophageal cancer. J. Clin. Pathol. *55*, 587–590.

Meltzer, S.J. (1996). The molecular biology of esophageal carcinoma. Recent Results. Cancer Res. *142*, 1–8.

Miller, B.J., Murphy, F., and Lukie, B.E. (1978). Squamous cell papilloma of esophagus. Can. J. Surg. *21*, 538-540.

Miller, B.A., Davidson, M., Myerson, D., Icenogle, J., Lanier, A.P., Tan, J., and Beckmann, A.M. (1997). Human papillomavirus type 16 DNA in esophageal carcinomas from Alaska natives. Int. J. Cancer *71*, 218-222.

Mizobuchi, S., Sakamoto, H., Tachimori, Y., Kato, H., Watanabe, H., and Terada, M. (1997). Absence of human papillomavirus-16 and -18 DNA and Epstein- Barr virus DNA in esophageal squamous cell carcinoma. Jpn. J. Clin. Oncol. *27*, 1-5.

Montesano, R., Hollstein, M., and Hainaut, P. (1996a). Molecular etiopathogenesis of esophageal cancers. Ann. Ist. Super. Sanita *32*, 73–84.

Montesano, R., Hollstein, M., and Hainaut, P. (1996b). Genetic alterations in esophageal cancer and their relevance to etiology and pathogenesis: a review. Int. J. Cancer *69*, 225–235.

Montesi, A., Pesaresi, A., Graziani, L., Salmistraro, D., Dini, L., and Bearzi, I. (1983). Small benign tumors of the esophagus: radiological diagnosis with double-contrast examination. Gastrointest. Radiol. *8*, 207-212.

Morgan, R.J., Perry, A.C.F., Newcomb, P.V., Hardwick, R.H., and Alderson, D. (1997). Human papillomavirus and oesophageal squamous cell carcinoma in the UK. Eur. J. Surg. Oncol. *23*, 513-517.

Mori, M., Shimono, R., Inoue, T., Kuwano, H., Sugimachi, K., and Zhang, R.G. (1989). Papillomavirus and esophageal cancer in the Japanese and Chinese. Gastroenterol. *84*,1126-1127.

Morris, H., and Price, S. (1986). Langerhans' cells, papillomaviruses and oesophageal carcinoma. A hypothesis. S. Afr. Med. J. *69*, 413-417.

Morris, H.H., and Price, S.K. (1987). Langerhans' cells, papillomaviruses and oesophageal carcinoma--a hypothesis. S. Afr. Med. J. *Suppl*: 15-8.

Mosca, S., Manes, G., Monaco, R., Bellomo, P.F., Bottino, V., and Balzano, A. (2001). Squamous papilloma of the esophagus: long-term follow up. J. Gastroenterol. Hepatol. *16*, 857-861.

Nagamatsu, M., Mori, M., Kuwano, H., Sugimachi, D., and Akiyoshi, T. (1992). Serial histologic investigation of squamous epithelial dysplasia asociated with carcinoma of the esophagus. Cancer *69*, 1094-1098.

Nakamura, T., Ide, H., EguchiI, R., Hayashi, K., Hanyu, F., and Nagasako, K. (1995). Expression of p53 protein related to human papillomavirus and DNA ploidy in superficial esophageal carcinoma. Surg. Today *25*, 591-597.

Nakano, K. (1994). Characteristics of human papillomavirus (HPV) infection in papilloma of the head and neck-detection of HPV according to clinical features and type specificity in the head and neck. Nippon Jibiinkoka. Gakkai. Kaiho. *97*, 1381-1392.

Narayani, R.I., and Young, G.S. (2002). Recurrent proximal esophageal stricture associated with dysplasia in squamous cell papillomatosis. Gastrointest. Endosc. *56*, 591-594.

Nuss, D.D., Aeling, J.L., Clemons, D.E., and Weber, W.N. (1978). Multiple hamartoma syndrome (Cowden's disease). Arch. Dermatol. *114*, 743-746.

Nuwayhid, N.S., Ballard, E.T., and Cotton, R. (1977). Esophageal papillomatosis: case report. Ann. Otol. Rhinol. Laryngol. *86*, 623-626.

Odze, R., Antonioli, D., Shocket, D., Noble-Topham, S., Goldman, H., and Upton, M. (1993). Esophageal squamous papillomas. A clinicopathologic study of 38 lesions and analysis for human papillomavirus by the polymerase chain reaction. Am. J. Surg. Pathol. *17*, 803-812.

Ohnuma, H., Sato, Y., Takayama, T., Takimoto, R., Abe, T., Hagiwara, S., Kukitsu, T., Nobuoka, A., Sato, T., Kogawa, K., Kato, J., and Niitsu, Y. (2003). Esophageal cancer complicated by cytomegalovirus esophagitis during chemoradiotherapy: case report. Gastrointest. Endosc. *57*, 622-626.

Ono, H., Takahashi, A., Ogoshi, S., Furihata, M., and Ohtsuki, Y. (1994). Relationship between h-ras p21 product and p53 protein or high-risk human papillomaviruses in esophageal cancer from Kochi, Japan. Am. J. Gastroenterol. *89*, 646-647.

Orlowska, J., Jarosz, D., Gugulski, A., Pachlewski, J., and Butruk, E. (1994). Squamous cell papillomas of the esophagus: report of 20 cases and literature review. Am. J. Gastroenterol. *89*, 434-437.

Ottenjann, R., Kuhner, W., and Weingart, J. (1984). Esophageal papilloma--potential precursor of squamous cell carcinoma?. Dtsch. Med. Wochenschr. *109*, 613-615.

Parkin, D.M., Läärä, E., and Muir, C.S. (1988). Estimates of the worldwide frequency of sixteen major cancers in 1980. Int. J. Cancer *41*, 184-197.

Parnell, S.A., Peppercorn, M.A., Antonioli, D.A., Cohen, M.A., and Joffe, N. (1978). Squamous cell papilloma of the esophagus. Report of a case after peptic esophagitis and repeated bougienage with review of the literature. Gastroenterol. *74*, 910-913.

Patterson, E.J. (1928). Multiple papillomata of the esophagus report of a case. Transact. Am. Bronchosc. Soc. *5*, 1169-1172.

Paz, I.B., Cook, N., Odom-Maryon, T., Xie, Y., and Wilczynski, S.P. (1997). Human papillomavirus (HPV) in head and neck cancer. An association of HPV 16 with squamous cell carcinoma of Waldeyer's tonsillar ring. Cancer *79*, 595-604.

Peixoto-Guimaraes, D., Hsin Lu, S., Snijders, P., Wilmotte, R., Herrero, R., Lenoir, G., Montesano, R., Meijer, C.J., Walboomers, J., and Hainaut, P. (2001). Absence of association between HPV DNA, TP53 codon 72 polymorphism, and risk of oesophageal cancer in a high-risk area of China. Cancer Lett. *162*, 231–235.

Pillai, M.R., and Nair, M.K. (2000). Development of a condemned mucosa syndrome and pathogenesis of human papillomavirus-associated upper aerodigestive tract and uterine cervical tumors. Exp. Mol. Pathol. *69*, 233-241.

Polit, S.A. (1990). Squamous cell papillomas of the esophagus. J. Pediatr. Gastroenterol. Nutr. *11*, 285-287.

Politoske, E.J. (1992). Squamous papilloma of the esophagus associated with the human papillomavirus. Gastroenterol. *102*, 668-673.

Poljak, M., and Cerar, A. (1993). Human papillomavirus type 16 DNA in oesophageal squamous cell carcinoma. Anticancer Res. *13*, 2113-2116.

Poljak, M., and Cerar, A. (1994). Detection of human papilloma virus type-6 DNA in an esophageal squamous cell papilloma. Eur. J. Clin. Microbiol. Infect. Dis. *13*, 188-189.

Poljak, M., Orlowska, J., and Cerar, A. (1995). Human papillomavirus infection in esophageal squamous cell papillomas: a study of 29 lesions. Anticancer Res. *15*, 965-969.

Poljak, M., Cerar, A., and Orlowska, J. (1996). p53 protein expression in esophageal squamous cell papillomas: a study of 36 lesions. Scand. J. Gastroenterol. *31*, 10–13.

Poljak, M., Cerar, A., Seme, K. (1998). Human papillomavirus infection in esophageal carcinomas: a study of 121 lesions using multiple broad-spectrum polymerase chain reactions and literature review. Hum. Pathol. *29*, 266–271.

Quitadamo, M., and Benson, J. (1988). Squamous papilloma of the esophagus: a case report and review of the literature. Am. J. Gastroenterol. *83*, 194-201.

Ratoosh, S.L., Glombicki, A.P., Lockhart, S.G., Rady, P.L., Chin, R., Arany, I., Hughes, T.K., and Tyring, S.K. (1997). Mastication of verruca vulgaris associated with esophageal papilloma: HPV-45 sequences detected in oral and cutaneous tissues. J. Am. Acad. Dermatol. *36*, 853-857.

Ravakhah, K., Midamba, F., and West, B.C. (1998). Esophageal papillomatosis from human papilloma virus proven by polymerase chain reaction. Am. J. Med. Sci. *316*, 285-288.

Ravry, M.J. (1979). Endoscopic resection of squamous papilloma of the esophagus. Am. J. Gastroenterol. *71*, 398-400.

Regragui, A., Lakhdar, H., Abderrahman Alaoui Belabbas, M., Amrani, M., Gamra, L., Alaoui Belabbas, M. (2004). Esophageal sarcomatoid carcinoma: report of a case with morphological, immunohistochemical and molecular study. Gastroenterol. Clin. Biol. *28*, 487-489.

Reiss, M., Brash, D.E., Munoz-Antonia, T., Vellucci, V.F., and Meltzer, S.J. (1992). Status of the p53 tumor suppressor gene in human squamous carcinoma cell lines. Oncol. Res. *4*, 349–357.

Ribeiro, U., Posner, M.C., Safatle-Ribeiro, A.V., and Reynolds, J.C. (1996). Risk factors for squamous cell carcinoma of the oesophagus. Br. J. Surg. *83*, 1174-1185.

Rong, J., Xu, L.Y., Cai, W.J., Xiong, X.D., Li, J.T., Fang, W.K., Shen, Z.Y., and Li, E.M. (2004). Expression of fascin 1 gene in the process of the immortalized esophageal carcinoma carcinogenesis. Ai Zheng. *23*, 243-248.

Roth, M.J., Liu, S-F., Dawsey, S.M., Zhou, B., Copeland, C., Wang, G-Q., Solomon, D., Baker, S.G., Giffen, C.A., and Taylor, P.R. (1997). Cytologic detection of esophageal squamous cell carcinoma and precursor lesions using balloon and sponge samplers in asymptomatic adults in Linxian, China. Cancer *80*, 2047-2059.

Rubio, C.A., and Liu, F-S. (1989). Spontaneous squamous carcinoma of the esophagus in chickens. Cancer *64*, 2511-2514.

Rugge, M., Bovo, D., Busatto, G., Parenti, A.R., Fawzy, S., Guido, M., Ancona, E., Ninfo, V., Ruol, A., and Shiao, Y.H. (1997). p53 alterations but no human papillomavirus infection in preinvasive and advanced squamous esophageal cancer in Italy. Cancer Epidem. Biomarker Prev. *6*, 171-176.

Sablich, R., Benedetti, G., Bignucolo, S., and Serraino, D. (1988). Squamous cell papilloma of the esophagus. Report on 35 endoscopic cases. Endoscopy *20*, 5-7.

Saegusa, M., Hashimura, M., Takano, Y., Ohbu, M., and Okayasu, I. (1997). Absence of human papillomavirus genomic sequences detected by the polymerase chain reaction in oesophageal and gastric carcinomas in Japan. Mol. Pathol. *50*, 101–104.

Sandvik, A.K., Aase, S., Kveberg, K.H., Dalen, A., Folvik, M., and Naess, O. (1996). Papillomatosis of the esophagus. J. Clin. Gastroenterol. *22*, 35-37.

Sashiyama, H., Shino, Y., Kawamata, Y., Tomita, Y., Ogawa, N., Shimada, H., Kobayashi, S., Asano, T., Ochiai, T., and Shirasawa, H. (2001). Immortalization of human esophageal keratinocytes by E6 and E7 of human papillomavirus type 16. Int. J. Oncol. *19*, 97–103.

Sashiyama, H., Shino, Y., Sakao, S., Shimada, H., Kobayashi, S., Ochiai, T., and Shirasawa, H. (2002). Alteration of integrin expression relates to malignant progression of human papillomavirus-immortalized esophageal keratinocytes. Cancer Lett. *177*, 21-28.

Shan, L., Nakamura, Y., Nakamura, M., Zhang, Z., Jing, X., and Hara, T. (1997). Synchronous and metachronous multicentric squamous cell carcinomas in the upper aerodigestive tract. Pathol. Int. *47*, 68-72.

Sharp, L., Chilvers, C.E., Cheng, K.K., McKinney, P.A., Logan, R.F., Cook-Mozaffari, P., Ahmed, A., and Day, N.E. (2001). Risk factors for squamous cell carcinoma of the oesophagus in women; a case-control study. Br. J. Cancer *85*, 1667-1670.

Shen, Z., Shen, J., and Zeng, Y. (1999a). Biological characteristics of human fetal esophageal epithelial cell line immortalized by the E6 and E7 gene of HPV type 18. Zhonghua Shi. Yan. He. Lin. Chuang. Bing. Du. Xue. Za. Zhi. *13*, 209-212.

Shen, Z., Cen, S., and Zeng, Y. (1999b). Immortalization of human fetal esophageal epithelial cells induced by E6 and E7 genes of human papilloma virus 18. Zhonghua Shi. Yan. He. Lin. Chuang. Bing. Du. Xue. Za. Zhi. *13*, 121-123.

Shen, Z., Cen, S., Shen, J., Cai, W., Xu, J., Teng, Z., Hu, Z., and Zeng, Y. (2000). Study of immortalization and malignant transformation of human embryonic esophageal epithelial cells induced by HPV18 E6E7. J. Cancer Res. Clin. Oncol. *126*, 589–594.

Shen, Z.Y., Xu, L.Y., Chen, X.H., Cai, W.J., Shen, J., Chen, J.Y. Zheng, R.M., Cai, W.J., and Zeng, Y. (2001). The genetic events of HPV-immortalized esophageal epithelium cells. Int. J. Mol. Med. *8*, 537-542.

Shen, Z.Y., Hu, S.P., Lu, L.C., Tang, C.Z., Kuang, Z.S., Zhong, S.P, and Zeng, Y. (2002a). Detection of human papillomavirus in esophageal carcinoma. J. Med. Virol. *68*, 412-416.

Shen, Z.Y., Xu, L.Y., Li, E.M., Shen, J., Zheng, R.M., Cai, W.J., and Zeng, Y. (2002b). Immortal phenotype of the esophageal epithelial cells in the process of immortalization. Int. J. Mol. Med. *10*, 641-646.

Shen, Z.Y., Xu, L.Y., Li, E.M., Cai, W.J., Chen, M.H., Shen, J., and Zeng, Y. (2002c). Telomere and telomerase in the initial stage of immortalization of esophageal epithelial cell. World J. Gastroenterol. *8*, 357-362.

Shen, Z., Shen, J., Cai, W., Chen, M., Wu, X., Zheng, R., and Zeng, Y. (2002d). The promoter effects of sodium butyrate on the malignant transformation of the immortalized esophageal epithelium induced by human papillomavirus. Zhonghua Bing. Li. Xue. Za. Zhi. *31*, 327-330.

Shen, Z., Shen, J., Cai, W., Chen, J., and Zeng, Y. (2002e). Malignant transformation of the immortalized esophageal epithelial cells. Zhonghua Zhong. Liu. Za. Zhi. *24*, 107-109.

Shen, Z.Y., Xu, L.Y., Chen, M.H., Shen, J., Cai, W.J., and Zeng, Y. (2002f). Progressive transformation of immortalized esophageal epithelial cells. World J. Gastroenterol. *8*, 976-981.

Shen, Z.Y., Xu, L.Y., Chen, M.H., Li, E.M., Li, J.T., and Wu, X.Y. (2003a). Upregulated expression of Ezrin and invasive phenotype in malignantly transformed esophageal epithelial cells. World J. Gastroenterol. *9*, 1182-1186.

Shen, Z.Y., Xu, L.Y., Chen, M.H., Cai, W.J., Shen, J., Chen, J.Y., and Zeng, Y. (2003b) Cytogenetic and molecular genetic changes in malignant transformation of immortalized esophageal epithelial cells. Int. J. Mol. Med. *12*, 219-224.

Shen, Z.Y., Cen, S., Xu, L.Y., Cai, W.J., Chen, M.H., Shen, J., and Zeng, Y. (2003c). E6/E7 genes of human papilloma virus type 18 induced immortalization of human fetal esophageal epithelium. Oncol. Rep. *10*, 1431-1436.

Shen, Z.Y., Xu, L.Y., Li, E.M., Cai, W.J., Shen, J., Chen, M.H., Cen, S., Tsao, S.W. and, Zeng, Y. (2004). The multistage process of carcinogenesis in human esophageal epithelial cells induced by human papillomavirus. Oncol Rep. *11*, 647-654.

Shibagaki, I., Tanaka, H., Shimada, Y., Wagata, T., Ikenaga, M., and Imamura, M. (1995). p53 mutation, murine double minute 2 amplification, and human papillomavirus infection are frequently involved but not associated with each other in esophageal squamous cell carcinoma. Clin. Cancer Res. *1*, 769-773.

Si, H.X., Tsao, S.W., Poon, C.S., Wang, L.D., Wong, Y.C., and Cheung, A.L. (2003). Viral load of HPV in esophageal squamous cell carcinoma. Int. J. Cancer *103*, 496-500.

Silber, W. (1985). Carcinoma of the oesophagus: aspects of epidemiology and aetiology. Proc. Nutr. Soc. *44*, 101-110.

Sitas, F., Madhoo, J., and Wessie, J. (1998). Incidence of histologically diagnosed cancer in South Africa. 1993-1995. National Cancer Registry of South Africa. (Johannesburg: South African Institute for Medical Research).

Smits, H.L., Tjong-A-Hung, S.P., ter-Schegget, J., Nooter, K., and Kok, T. (1995). Absence of human papillomavirus DNA from esophageal carcinoma as determined by multiple broad spectrum polymerase chain reactions. J. Med. Virol. *46*, 213-215.

Snijders, P.J., Steenbergen, R.D., Meijer, C.J., and Walboomers, J.M. (1997). Role of human papillomaviruses in cancer of the respiratory and upper digestive tract. Clin Dermatol. *15*, 415-425.

Sobti, R.C., Kochar, J., Singh, K., Bhasin, D., and Capalash, N. (2001). Telomerase activation and incidence of HPV in human gastrointestinal tumors in North Indian population. Mol. Cell. Biochem. *217*, 51–56.

Somdyala, N.I., Marasas, W.F., Venter, F.S., Vismer, H.F., Gelderblom, W.C., and Swanevelder, S.A. (2003). Cancer patterns in four districts of the Transkei region--1991-1995. S. Afr. Med. J. *93*, 144-148.

Sons, H.U. (1987). Etiologic and epidemiologic factors of carcinoma of the esophagus. Surg.. Gynecol. Obstet. *165*, 183-190.

Starr, A.J., Shenoy, B.V., and Ruffolo, E.H. (1979). Squamous papilloma of the esophagus. South Med. J. *72*, 1203-1205.

Sugimachi, K., Sumiyoshi, K., Nozoe, T., Yasuda, M., Watanabe, M., Kitamura, K., Tsutsui, S., Mori, M., and Kuwano, H. (1995). Carcinogenesis and histogenesis of esophageal carcinoma. Cancer *75*, 1440-1445.

Sur, M., and Cooper, K. (1998). The role of the human papilloma virus in esophageal cancer. Pathology *30*, 348–354.

Suzuk, L., Noffsinger, A.E., Hui, Y.Z., and Fenoglio-Preiser, C.M. (1996). Detection of human papillomavirus in esophageal squamous cell carcinoma. Cancer *78*, 704-710.

Syrjänen, K.J. (1982). Histological changes identical to those of condylomatous lesions found in esophageal squamous cell carcinomas. Arch. Geschwulstforsch. *52*, 283–292.

Syrjänen, K.J. (1987). Human papillomavirus (HPV) infections and their associations with squamous cell neoplasia. Arch. Geschwulstforsch. *57*, 417-444.

Syrjänen, K. (2000). HPV infections of the oesophagus. In Papillomavirus infections in human pathology, K. Syrjänen, S. Syrjänen. (New York: J. Wiley and Sons), pp. 413–428.

Syrjanen, K.J. (2002a). HPV infections and lung cancer. J. Clin. Pathol. *55*, 885-891.

Syrjänen, K. (2002b). HPV infections and oesophageal cancer. J. Clin. Pathol. *55*, 721-728.

Syrjänen, K.J. (2003a). HPV infections in benign and malignant sinonasal lesions. J. Clin. Pathol. *56*, 174-181.

Syrjänen, K.J. (2003b). HPV et tumeurs épidermoïdes bénignes et malignes de l'æsophage. Chapter 37. In Papillomavirus Humains. Biologie et pathologie tumorale, F. Aubin, J-L. Pretet, and C. Mougin, Eds. (Paris: Editions TEC and DOC), pp. 523-548.

Syrjänen, K.J., Pyrhönen, S., Aukee, S., and Koskela, E. (1982). Squamous cell papilloma of the oesophagus: a tumour probably caused by Human papilloma virus (HPV). Diagn. Histopathol. *5*, 291-296.

Syrjänen, K., Gissmann, L., and Koss, L.G., (eds.)(1987). Papillomaviruses and Human Disease. (Berlin: Springer Verlag).

Syrjänen, K., Chang, F., and Syrjänen, S. (1996). Infectious agents as etiological factors in esophageal carcinogenesis. In Recent advances in gastroenterological carcinogenesis I: E. Tahara, K. Sugimachi, T. Oohara, eds. (Bologna: Monduzzi Editore), pp. 29–43.

Syrjänen, K., and Syrjänen, S. (2000a). Papillomavirus Infections in Human Pathology. (New York: Wiley and Sons).

Syrjänen, K.J., Chang, F., and Syrjänen, S.M. (2000b). HPV infections in etiology of benign and malignant sinonasal, bronchial and oesophageal squamous cell lesions. In 4th International Multidisciplinary Congress EUROGIN 2000, J. Monsonego, ed. (Bologna: Monduzzi Editore), pp. 169-179.

Syrjänen, S.M. (1987). Papillomavirus infections in the oral cavity. In Papillomaviruses and human disease, K. Syrjänen, L. Gissmann, L.G. Koss, eds. (Berlin: Springer Verlag), pp. 104–137.

Szentirmay, Z., Szanto, I., Balint, I., Polus, K., Remenar, E., Tamas, L. (2002). Causal association between human papilloma virus infection and head and neck and esophageal squamous cell carcinoma. Magy Onkol *46*, 35-41.

Takahashi, A., Ogoshi, S., Ono, H., Ishikawa,T., Toki,T., Ohmori, N., Iwasa, M., Iwasa, Y., Furihata, M., and Ohtsuki, Y. (1998). High-risk human papillomavirus infection and overexpression of p53 protein in squamous cell carcinoma of the esophagus from Japan. Dis. Esophagus *11*, 162–167.

Talamini, G., Capelli, P., Zamboni, G., Mastromauro, M., Pasetto, M., and Castagnini, A. (2000). Alcohol, smoking and papillomavirus infection as risk factors for esophageal squamous-cell papilloma and esophageal squamous-cell carcinoma in Italy. Int. J. Cancer. *86*, 874-878.

Toet, A.E., Dekker, W., Odo Op den Orth, J., and Blok, P. (1985). Squamous cell papilloma of the esophagus: report of four cases. Gastrointest. Endosc. *31*, 77-79.

Togawa, K., Jaskiewicz, K., Takahashi, H., Meltzer, S.J., and Rustgi, A.K. (1994). Human papillomavirus DNA sequences in esophagus squamous cell carcinoma. Gastroenterol. *107*, 128-136.

Togawa, Z., and Rustgi, A.K. (1995). Human papillomavirus and esophageal squamous cell carcinoma. Gastroenterol. *108*, 1605-1606.

Toh, Y., Kuwano, H., Tanaka, S., Baba, K., Matsuda, H., Sugimachi, K., and Mori, R. (1992). Detection of human papillomavirus DNA in esophageal carcinoma in Japan by polymerase chain reaction. Cancer *70*, 2234-2238.

Tollefson, L. (1985). The use of epidemiology, scientific data, and regulatory authority to determine risk factors in cancer of some organs of the digestive system. 2. Esophageal cancer. Regul. Toxicol. Pharmacol. *5*, 255-275.

Tomasino, R., Nuara, R., Napou, G., and Morello, V. (1988). Condylomatous leukoplakic lesion of the esophagus - a case study with demonstration of papilloma virus antigen by the immunoperoxidase technique. Ital. J. Gastroenterol. *20*, 265-268.

Tominaga, T., Kashimura, H., Suzuki, K., Nakahara, A., Tanaka, N., Noguchi, M., Itabashi, M., and Ohkawa, J. (2002). Telomerase activity and expression of human telomerase catalytic subunit gene in esophageal tissues. J. Gastroenterol. *37*, 418-427.

Tripodi, S., Chang, F., Syrjänen, S., Shen, Q., Cintorino, M., Alia, L., Santopietro, R., Tosi, P., and Syrjänen, K. (2000). Quantitative image analysis of oesophageal squamous cell carcinoma from the high-incidence area of China, with special reference to tumour progression and Papillomavirus (HPV) involvement. Anticancer Res. *20*, 3855-3862.

Trottier, A.M., Coutlee, F., Leduc, R., Ghattas, G., Toma, E., and Allaire, G. (1997). Human immunodeficiency virus infection is a major risk factor for detection of human papillomavirus DNA in esophageal brushings. Clin. Infect. Dis. *24*, 565-569.

Turner, J.R., Shen, L.H., Crum, C.P., Dean, P.J. and Odze, R.D. (1997). Low prevalence of human papillomavirus infection in esophageal squamous cell carcinomas from North America: analysis by a highly sensitive and specific polymerase chain reaction-based approach. Hum. Pathol. *28*, 174–178.

Ushigome, S., Spjut, H.J., and Noon, G.P. (1967). Extensive dysplasia and carcinoma *in situ* of esophageal epithelium. Cancer *20*, 1023-1032.

Valerdiz, S., Vazquez, J.J., and Munoz, M. (1989). Papillomas of the esophagus: presentation of 5 cases and review of the literature. Rev. Esp. Enferm. Apar. Dig. *75*, 259-261.

van Cutsem, E., Geboes, K., Visser, L., Devos, R., Jenssens, J., and Lerut, T. (1991). Squamous papillomatosis of the oesophagus with malignant degeneration and demonstration of the human papilloma virus. Eur. J. Gastroenterol. Hepatol. *3*, 561-566.

van Cutsem, E., Geboes, K., and Vantrappen, G. (1992). Malignant degeneration of esophageal squamous papilloma associated with the human papillomavirus. Gastroenterol. *103*, 1119-1120.

van Cutsem, E., Snoeck, R., Van Ranst, M., Fiten, P., Opdenakker, G., Geboes, K., Janssens, J., Rutgeerts, P., Vantrappen, G., and de-Clercq, E. (1995). Successful treatment of a squamous papilloma of the hypopharynx-esophagus by local injections of (S)-1-(3-hydroxy-2-p hosphonylmethoxypropyl) cytosine. J. Med. Virol. *45*, 230-235.

van den Borre, K., Rutgeerts, L., Ghillebert, G., and Tanghe, W. (1996). Squameuze slokdarmpapillomen. Ned. Tijdschr. Geneeskd. *140*, 987-990.

van Doornum, G.J., Korse, C.M., Buning-Kager, J.C., Bonfrer, J.M., Horenblas, S., Taal, B.G., and Dillner, J. (2003). Reactivity to human papillomavirus type 16 L1 virus-like particles in sera from patients with genital cancer and patients with carcinomas at five different extragenital sites. Br. J. Cancer *88*, 1095-1100.

van Rensburg, E.J.V., Venter, E.H., and Simson, I.W. (1993). Human papillomavirus DNA in aerodigestive squamous carcinomas demonstrated by means of *in situ* hybridization. S. Afr. Med. J. *83*, 516-518.

Waluga, M., Hartleb, M., Sliwinski, Z.K., Romanczyk, T., and Wodolazski, A. (2000). Esophageal squamous-cell papillomatosis complicated by carcinoma. Am. J. Gastroenterol. *95*, 1592-1593.

Wang, L.S., Chow, K.C., Wu, Y.C., Li, W.Y., and Huang, M.H. (1999). Detection of Epstein-Barr virus in esophageal squamous cell carcinoma in Taiwan. Am. J. Gastroenterol. *94*, 2834-2839.

Waterfall, W.E., Somers, S., and Desa, D.J. (1978). Benign oesophageal papillomatosis. A case report with a review of the literature. J. Clin. Pathol. *31*, 111-115.

West, A.B., Soloway, G.N., Lizarraga, G., Tyrrell, L., and Longley, J.B. (1996). Type 73 human papillomavirus in esophageal squamous cell carcinoma: A novel association. Cancer *77*, 2440-2444.

Weston, A.C., and Prolla, J.C. (2003). Association between esophageal squamous cell carcinoma and human papillomavirus detected by Hybrid Capture II assay. Dis, Esophagus *16*, 224-228.

Williamson, A.L., Jaskiesicz, K., and Gunning, A. (1991). The detection of human papillomavirus in oesophageal lesions. Anticancer Res. *11*, 263-265.

Winkler, B., Capo, V., Reumann, W., Ma, A., La-Porta, R., Reilly, S., Green, P.M., Richart, R.M., and Crum, C.P. (1985). Human papillomavirus infection of the esophagus. A clinicopathologic study with demonstration of papillomavirus antigen by the immunoperoxidase technique. Cancer *55*, 149-155.

Wong, F.H., Hu, C.P., Chen, S.C., Yu, Y.T., and Chang, C. (1992). Absence of genomes of DNA tumor viruses and expression of oncogenes and growth factors in two esophageal carcinoma cell lines of Chinese origin. Chung Hua. Min. Kuo. Wei. Sheng. Wu. Chi. Mien. I. Hsueh. Tsa. Chih. *25*, 59-68.

Woo, Y.J., Yoon, H.K. (1996). *In situ* hybridization study on human papillomavirus DNA expression in benign and malignant squamous lesions of the esophagus. J. Korean Med. Sci *11*, 467–473.

Wynder, E.L., and Bross, I.J. (1961). A study of etiological factors in cancer of the esophagus. Cancer *14*, 389-413.

Xibib, S., Meilan, H., Moller, H., Evans, H.S., Dixin, D., Wenjie, D., and Jianbang, L. (2003). Risk factors for oesophageal cancer in Linzhou, China: a case-control study. Asian Pac. J. Cancer Prev. *4*, 119-124.

Xu, C.L., Qian, X.L., Zhou, X.S., Zhao, Q.Z., and Li, Y.C. (2004). Expression of HPV16-E6 and E7 Oncoproteins in Squamous Cell Carcinoma Tissues of Esophageal Cancer and Non-cancer Tissues. Ai. Zheng. *23*, 165-168.

Xu, W.G., Zhang, L.J., Lu, Z.M., Li, J.Y., Ke, Y., and Xu, G.W. (2003). Detection of human papillomavirus type 16 E6 mRNA in carcinomas of upper digestive tract. Zhonghua Yi. Xue. Za. Zhi. *83*, 1910-1914.

Yamada, Y., Ninomiya, M., Kato, T., Nagaki, M., Kato, M., Hatakeyama, H., Moriwaki, H., and Muto, Y. (1995). Human papillomavirus type 16-positive esophageal papilloma at an endoscopic injection sclerotherapy site. Gastroenterol. *108*, 550-553.

Yanai, H., Hirano, A., Matsusaki, K., Kawano, T., Miura, O., Yoshida, T., Okita, K., and Shimizu, N. (2003). Epstein-Barr virus association is rare in esophageal squamous cell carcinoma. Int. J. Gastrointest. Cancer *33*, 165-170.

Ye, W., Held, M., Lagergren, J., Engstrand, L., Blot, W.J., McLaughlin, J.K., and Nyren. O. (2004). Helicobacter pylori infection and gastric atrophy: risk of adenocarcinoma and squamous-cell carcinoma of the esophagus and adenocarcinoma of the gastric cardia. J. Natl. Cancer Inst. *96*, 388-396.

Yoshikane, H., Yokoi, T., Suzuki, T., Yoshioka, N., Ogawa, Y., and Hamajima, E. (1995). A case of esophageal dysplasia associated with Human papilloma virus. Am. J. Gastroenterol. *90*, 1369-1370.

Yuki, T., Kawano, S., Ogihara, T., Sato, N., Okamoto, S., and Sato, K. (1993). Esophageal squamous cell papilloma, suspected of squamous cell carcinoma. Internal Med. *32*, 773-776.

Zhen, Y.Z. (1984). Isolation and culture of fungi from the cereals in counties of Henan Province--5 with high and 3 with low incidences of esophageal cancer. Zhonghua Zhong. Liu. Za. Zhi. *6*, 27-29.

Zheng, J., Fu, M., Wang, X., Lin, C., and Wu, M. (1999). Lack of human papillomavirus types 16 and 18 DNA in esophageal squamous carcinoma cell lines. Chin. Med. J. (Engl) *112*, 534-537.

Zheng, W., Jin, F., Devesa, S.S., Blot, W.J., Fraumeni, J.F., Jr., and Gao, Y.T. (1993). Declining incidence is greater for esophageal than gastric cancer in Shanghai, People's Republic of China. Br. J. Cancer *68*, 978-982.

Zhou, X.B., Guo, M., Quan, L.P., Zhang, W., Lu, Z.M., Wang, Q.H., Ke, Y., and Xu, N.Z. (2003). Detection of human papillomavirus in Chinese esophageal squamous cell carcinoma and its adjacent normal epithelium. World J. Gastroenterol. *9*, 1170-1173.

Znaor, A., Brennan, P., Gajalakshmi, V., Mathew, A., Shanta, V., Varghese, C., and Boffetta, P. (2003). Independent and combined effects of tobacco smoking, chewing and alcohol drinking on the risk of oral, pharyngeal and esophageal cancers in Indian men. Int. J. Cancer *105*, 681-686.

zur Hausen, H. (1999a). Yohei Ito memorial lecture: papillomaviruses in human cancers. Leukemia *13*, 1–5.

zur Hausen, H. (1999b). Papillomaviruses in human cancers. Proc. Assoc. Am. Physicians. *111*, 581-587.

zur Hausen, H. (2003). Papillomaviruses and cancer: from basic studies to clinical application. Nat. Rev. Cancer. *2*, 342-350.

From: Papillomavirus Research: From Natural History to Vaccines and Beyond. Edited by: M. Saveria Campo

Chapter 16

Recurrent Respiratory Papillomatosis, HPV, and Impact on Host Response

Andrea Vambutas* and Bettie M. Steinberg

Abstract

Recurrent Respiratory Papillomatosis (RRP) is a devastating disease characterized by multiple recurrent, benign growths that cause significant morbidity by airway obstruction. Adjunct therapy to traditional surgical removal has been sought due to the aggressive nature of the disease. The lesions are caused by human papillomavirus (HPV) type 6 and 11. Although many of the same papillomavirus proteins are expressed in high risk HPV types, the function of these proteins are different in the low risk HPVs 6 and 11. In laryngeal papillomas cells, there is a defect in terminal differentiation and significant alterations in signal transduction. Latent viral infection is commonplace, and the triggers for activation of latency are unknown. Host immune responses demonstrate local immunocompromise, where many components for an appropriate immune response are reduced in papillomas. Ubiquitously expressed viral proteins are obvious targets for vaccine development. Unfortunately, the two most common targets, E6 and E7, either fail to induce any response or initiate inappropriate Th2-like responses. This has led investigators to increase immunogenicity of these proteins. To date, a completely successful therapeutic vaccine has not been identified. An ideal therapeutic vaccine in RRP would require the control of active and latent disease, with a minimal inflammatory response.

1. Clinical Disease

1.1. Clinical course

Recurrent respiratory papillomatosis (RRP) is a devastating disease caused primarily by HPV 6/11 and characterized by multiple recurrences and spontaneous activation of latently infected epithelium by unknown triggers. Growth is often exuberant and obstructive (Figure 1). Patients with severe disease frequently require surgery in excess of three times per year and as often as every several weeks to maintain an airway. Juvenile onset tends to be more aggressive. In one study of 603 children, there was a mean number of surgeries per child of 5.1 per year (Reeves et al., 2003). Although most patients have laryngeal disease, progression beyond the larynx is common. In one study, 69% had subglottic disease and 29% had tracheal disease (Weiss et al., 1983). Although standard therapy is surgical excision, multiple adjunct therapies have been sought out due to the aggressive nature of the disease, surgical risks and morbidity associated with frequent surgeries on the airway.

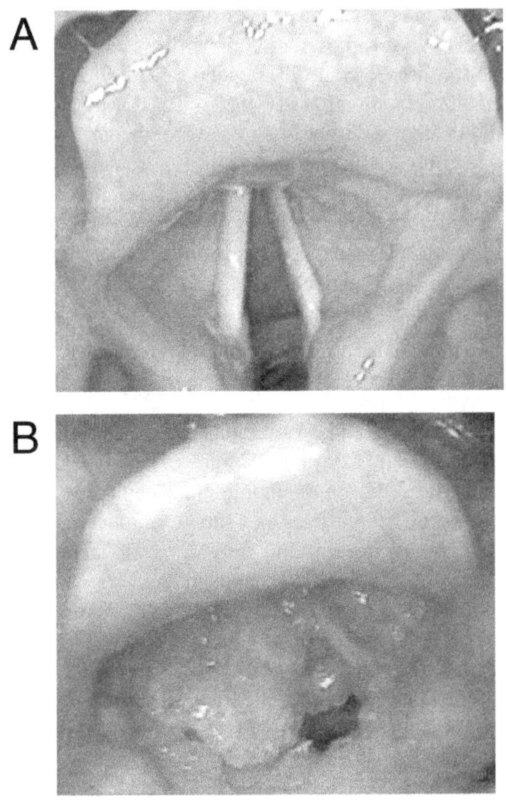

Figure 1. A. Endoscopic view of a normal larynx. B. Endoscopic view of a larynx with RRP. Exuberant papillomas involve the true vocal cord, the false vocal cord and the subglottic region.

*For correspondence. Email vambutas@lij.edu

1.2. Prevalence

RRP affects both children and adults, with 50% of patients presenting as adult onset disease (Abramson et al., 1987). Although the disease is rare, with a prevalence of 1.69-2.59 per 100,000 for juvenile onset papillomatosis (Armstrong et al., 2000), patient costs are in excess of $201,724 per patient and annual cost in the United States is estimated at $40-123 million per year depending on prevalence, making this a rare but costly disease (Bishai et al., 2000). Juvenile onset disease is acquired from an HPV-infected mother, primarily at birth. Approximately 7 per 1000 children born to mothers with vaginal condyloma will develop RRP (Wiatrak, 2003).

1.3. Treatment strategies

Conventional therapy for RRP remains surgical excision. Methods of surgical excision include cold dissection techniques, CO_2 laser, and use of powered shavers (Dedo et al., 2001; Pasquale et al., 2003; Patel et al., 2003). Scheduled clearance of disease has been shown to control disease better than waiting for symptomatic regrowth of the papillomaotus lesions (Dedo et al., 2003). Adjunct therapies to surgical excision have met with variable results (Schraff et al., 2004). Immunomodulators, such as alpha interferon, have been used therapeutically. In the 1980's, α-interferon therapy provided therapeutic hope for disease control (Haglund et al., 1981; Healy et al., 1988; Leventhal et al., 1991), but latent HPV DNA could be detected after treatment (Steinberg et al., 1988) and recurrences were observed during treatment. Indole-3-carbinol, a naturally occurring phytochemical found in cruciferous vegetables, or its metabolite diindolyl methane (DIM) has induced significant regression of lesions in some patients (Newfield et al., 1993; Rosen et al., 2004). This effect was possibly mediated by altering estrogen metabolism, preferentially inducing 2-hydroxylation of estrogen with a net reduction in 16a-hydroxylation (Newfield et al., 1993). The shift in metabolites prevented estradiol induction of anchorage independent growth of HPV-transformed cells *in vitro* (Newfield et al., 1998). Indole-3-carbinol/DIM therapy has very few side effects and shows significant promise as an adjunct therapy, but only approximately half of patients respond. Photodynamic therapy is capable of significantly reducing disease in some patients, but also failed to eradicate latent infection (Abramson et al., 1992; Shikowitz et al., 1998). Uncontrolled studies suggested a beneficial role for anti-reflux treatments (Borkowski et al., 1999) although this may only serve to reduce associated soft tissue injury (Holland et al., 2002). Recent studies have demonstrated that intralesional injection of cidofovir, an antiviral drug that can induce apoptosis in papillomavirus-infected cells, resulted in a significant control of lesions in some patients (Snoeck et al., 1988; Bielamowicz et al., 2002; Pransky et al., 2003). However, in one prospective trial, cidofovir did not decrease severity or frequency of disease (Peyton et al., 2004). Long term results of cidofovir injections showed return to pretreatrment severity once therapy was discontinued (Milczuk et al., 2003), and a combination of surgery and cidofovir was necessary for persistent papillomas (Naiman et al., 2003).

1.4. Progression to malignancy

Although RRP is a benign disease, malignant conversion of lesions is not infrequent, especially as the disease progresses into the lower airways. Published conversion rates are variable, but papillomas spontaneously convert to carcinoma in approximately 3% of patients. As such, many have argued that this is not a benign disease, but rather a pre-malignant state. In this light, many of the early cellular changes seen in RRP may provide insight into the early cellular changes in carcinomas in general. However, in patients with conversion to malignancy, histologic progression with increasing dysplasia was not observed in one study (Go et al., 2003). Interestingly, if the disease progresses down the trachea and into the lungs, the likelihood of malignant conversion escalates rapidly. In the seven malignant conversions described by Go et al. (2003), five were tracheostomy dependent. Similarly, malignant conversion was seen in 3 of 3 RRP patients with broncheopulmonary involvement (A. Abramson, personal communication). Some investigators have suggested that HPV 11 infection results in more aggressive disease than infection with HPV 6 (Rabah et al., 2001). However, in two studies of patients with bronchial RRP that converted to lung carcinoma, one contained HPV 6a and the other HPV 11 (Byrne et al., 1987; DiLorenzo et al., 1992). Both carcinomas contained HPVs with a duplication of the URR Point mutations in the URR do not appear to correlate with clinical course (Gabbott et al., 1997), and deletions of the L1 open reading frame in 2 other RRP patients had no effect on clinical disease status (Suzuki et al., 1995), suggesting that only specific HPV mutations may be associated with malignant progression. In one case of malignant progression, integrated HPV 11 was observed together with mutation in the *p53* tumor surpressor gene (Rady et al., 1998), whereas papillomas usually express or overexpress wild type p53 (Poljak et al., 1997). However, p53 expression levels did not correlate with malignant transformation (Go et al., 2003). Dysplasia was inversely related to the overexpression of p21 (Erdamar et al., 2002), with most papillomas having high levels of p21 in the parabasal layers. In one case of severe dysplasia, co-infection with HPV 11 and HPV 16 was observed, with HPV 16 demonstrating an integrated pattern by in situ hybridization (Lin et al., 1997). The more oncogenic "high risk" HPVs (types16 and 18) may also play a role in a subset of head and neck cancers in patients without a history of RRP, especially in tonsillar cancers (Mineta et al., 1998; Gillison et al., 2001; Herrero et al., 2003).

2. Virology

The regulation of HPV 6/11 in respiratory papillomas appears similar in overall pattern to that described for the more well-studied HPVs 16, 18 and 31 in genital tract infections, but it must be appreciated that there are significant differences between the so-called low risk mucosal HPVs and those more commonly associated with cervical cancer. These differences can result in altered effects of viral proteins on cellular targets, and therefore extrapolation from studies with HPVs 16, 18 and 31 must be made with caution.

2.1. Expression

Although HPVs are capable of infecting the respiratory epithelium, productive viral replication and expression of viral genes required for papilloma formation are restricted to squamous epithelium. To this end, irritation of tracheal epithelium, resulting in conversion from ciliated columnar to stratified squamous epithelium, probably facilitates progression of disease into the trachea. Replication of HPV 6 and 11 is episomal in respiratory papillomas (Auborn et al., 1996), with initiation of replication occurring in the URR and replication forks proceeding bidirectionally (Auborn et al., 1994). Increase in transcript abundance, which is required for viral replication and virion production, is contingent on squamous differentiation (Steinberg et al., 1988; DiLorenzo et al., 1995). HPV 6 and 11 produce both early and late proteins in respiratory papillomas, but only a subset of cells in a papilloma productively replicate viral DNA, and not all papillomas express late proteins.

The roles of the early proteins in the viral life cycle appear to be somewhat different in the low and high-risk viruses, and many of their functions in HPV 6/11 are unclear (Thomas et al., 2001; Oh et al., 2004). The viral replication/transcription proteins E1 and E2 have similar functions in both HPV 6/11 and high risk HPV 16 (Chiang et al., 1992). In contrast to HPV 16, where E7 binds the tumor suppressor PRB with high affinity (Munger et al., 1989), thereby thwarting cell cycle control and allowing for viral DNA replication, HPV 11 E7 binds with a relatively low affinity. However, it is still required to maintain viral DNA as an episome (Oh et al., 2004) and is able to reactivate cellular DNA replication in differentiating keratinocytes (Chiang et al., 1992). Unlike the high risk HPV E6, HPV 11 E6 does not bind to or induce the degradation of p53 (Slebos et al., 1995). It also lacks the PDZ-binding domain required to bind to a number of tumor suppressor proteins (Gardiol et al., 1999), and binding to one of these proteins is required for induction of hyperplasia by HPV 16 (Nguyen et al., 2003). It is not yet clear how HPV 6/11 infection is protected from apoptosis induced by unscheduled DNA synthesis, or how it induces papilloma formation. However, Gupta et al. (2003) recently reported that E6 proteins of both high risk and low risk HPVs can induce the cell cycle by altering cellular transcription, and perhaps that is a critical function of HPV 6/11 E6. Both HPV 16 and HPV 11 E5 proteins are associated with vacuolar ATPases (Conrad et al., 1993). Many functions have been proposed for this protein, including alterations in growth factor receptor signaling. However, most of those studies were based on overexpression of the E5 proteins. When HPV 31 E5 translation was prevented in the context of expression of the whole viral genome, there was no effect on EGF receptor levels or phosphorylation (Fehrmann et al., 2003). The functions of E4 are not well understood, especially for HPV 6/11, but this abundant protein is expressed concomitantly with the late proteins.

2.2. Latency

Most patients with RRP have detectable latent HPV infection in their laryngeal and tracheal epithelium. Although the larynx is the primary site of lesions during active disease, with laryngeal involvement in 96.1% of patients (Reeves et al., 2003), latent infection rates are comparable between the larynx and trachea (Abramson et al., 2004). During latent infection, HPV 11 El and E2 transcripts could be readily identified by reverse-transcriptase PCR, but transcripts for E6 and E7 were below detectable levels (Maran et al., 1995). Thus, latency most likely represents the active suppression of E6 and E7 transcription rather than a passive process. Given that prevalence of latent HPVs in the larynx and trachea are similar, yet the larynx has a significantly higher proportion of active disease than the trachea, it is inferred that the host cell in the trachea is more resistant to spontaneous activation than laryngeal cells are. The cellular factor(s) that mediate this activation are yet to be determined. However, the host cell type may play a critical role in activation of latent viral infection. Additionally, the hallmark of an active viral infection is the presence of E6 and E7 transcripts. The fact that these proteins are consistently expressed in active infections and seem critical for the viral life cycle have led investigators to target these antigens in vaccination strategies (see below).

2.3. Effects on papilloma cell phenotype

Altered signal transduction is evident in laryngeal papillomas. Overexpression of the epidermal growth factor receptor (EGFR) (Vambutas et al., 1993) results in the activation of the MAP kinase pathway (Johnston et al., 1999). This receptor overexpression is a result of receptor recycling without degradation (Johnston et al., 1999). Receptor recycling is a common mechanism in tumors and is felt to correlate with mitogenic potency. The mechanism of this recycling may be a failure of the proteasome. In 3T3 cells transfected with an EGFR expression construct, recycling correlated with decreased EGFR ubiquitination and EGFR degradation, showing that de-ubiquitination must preceed degradation (Alwan et al., 2003). Differences in proteasome inhibition between high and low risk HPV types have been reported, further suggesting differential function of these very homologous viral proteins (Kehmeier et al., 2002).

Respiratory papillomas overexpress the tumor suppressor PTEN, resulting in reduced activation of Akt (Zhang et al., 2000). PTEN also reduces STAT3 activation (Sun et al., 2002). Increased activation of p50 NF-κB may, at least in part, serve to counteract the overexpression of PTEN (Vancurova et al., 2002), while concurrent overexpression of p21Cip1/waf1 would inhibit the cell cycle but may reduce apoptosis. Post-transcriptional increase of p21/Cip1 is seen in a subset of cells in the spinous layer of both respiratory papillomas containing HPV 6/11 and genital lesions induced by high risk HPVs. It can be induced by expression of HPV 18 E7 (Jian et al., 1999) in cultured keratinocytes.

Cyclooxgenase 2 (COX-2) is overexpressed in laryngeal papillomas (Robinson et al., 1999). It can be induced by multiple signal transduction pathways including activation of the EGFR, and levels are regulated transcriptionally, by mRNA stability and by protein turnover (reviewed by: Subbaramaiah et al., 2003; Dannenberg et al., 2003). In murine neuronal cells, inhibition of the ubiquitin/proteasome pathway using cadmium resulted in the specific increase in COX-2 and its product prostaglandin E2 (PGE_2) (Figueiredo-Pereira et al., 2002). In a rat model of autoimmune diabetes, use of the 26S proteasome inhibitor MG132 inhibited IL1β-induced COX-2 expression (Kwon et al., 1998). Proteasome inhibition resulted in an increase and stabilization of ubiquitin conjugated COX-2, thereby suggesting the ubiquitin proteasome pathway is capable of regulating COX-2 levels. The interplay between COX-2 and the ubiquitin-proteasome pathway in HPV-induced lesions is yet to be determined.

2.4. Proliferation/differentiation

Respiratory papilloma tissues are characterized by hyperplastic epithelium with a relatively low proliferative rate and delayed and altered differentiation compared to normal epithelium (Steinberg et al., 1990). In fact, we postulate that the altered differentiation causes the hyperplasia of the spinous layer in the papillomas. Induction of Proliferating Cell Nuclear Antigen (PCNA) in the upper spinous layers of lesions induced by both low risk and high risk HPVs is a marker of reactivation of the host replication machinery to replicate viral DNA, rather than a marker of cellular proliferation (Demeter et al., 1994). The reduction in STAT 3 activation in papillomas may play a role in this failure to differentiate as activated STAT3 is required for differentiation of laryngeal stratified squamous epithelium (Wu et al., 2003). Other signaling pathways that contribute to the defect in differentiation are still being characterized.

3. Host Immune Response

3.1. Immune evasion

Effective antigen presentation to the immune system requires the infected cell to process the foreign peptide and transport a fragment to the cell surface for recognition by the immune system. Both the Major Histocompatibility Complex Class I (MHC-I) and Class II (MHC-II) proteins are required for functional immune recognition. Components required for class I MHC antigen presentation include TAP1/TAP2, the 26S proteasome, MHC-I itself, calreticulin, β2- microglobulin and tapasin (Sadasivan et al., 1996).

Patients with RRP are systemically immunocompetent (Bonagura, 1994), however, these patients are locally immunocompromised, expressing reduced levels of MHC-I on the cell surface (Bonagura et al., 1994). TAP1, in a heterodimeric complex with TAP2, is an ATP dependent peptide transporter that imports antigenic peptides from the cytosol to the endoplasmic reticulum for assembly with MHC-I (Spies et al., 1992). In RRP, apparent loss of TAP1 by immunohistochemical staining correlated significantly with clinical disease severity (Vambutas et al., 2000). Patients with apparent low TAP1 levels returned to the operating room on average within 5-6 weeks after surgery in order to maintain an airway, whereas good TAP1 expressers returned to the operating room in 32 weeks (Figure 2; Vambutas et al., 2000). We subsequently determined that the transporter is not absent in RRP, but rather is unavailable for antibody detection in the presence of the HPV11 E7 protein (Vambutas et al., 2001). HPV 11 E7 was also specifically able to drastically reduce ATP dependent peptide transport through the TAP transporter in an *in vitro* assay (Vambutas et al.,

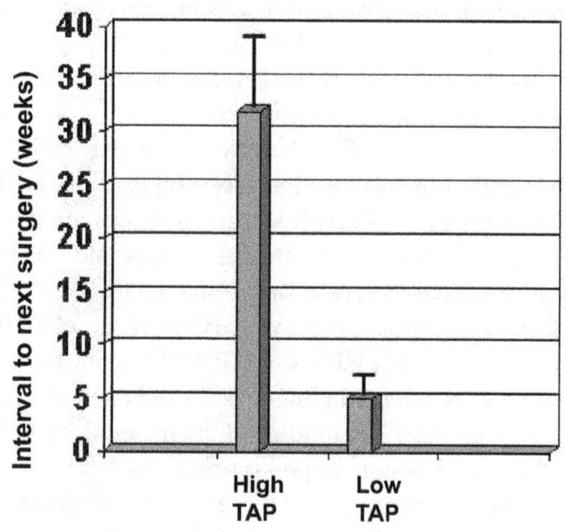

Figure 2. Comparison of TAP1 expression in lesions with clinical course. Low TAP1 expression correlated with an aggressive clinical course. The difference was significant at p=0.0095. Reprinted with permission from Vambutas et al., Clin. Lab. Diag. Immunol, 2000; 7 (1): 79-85.

2001). Interestingly, langerhans cells were also reduced in number in papillomas when compared to uninfected epithelium (Chardonnet et al., 1986; Viac et al., 1987). This loss correlated with the loss of β-2-microglobulin, another critical component required for MHC-I assembly (Viac et al., 1993).

Our studies parallel studies with other HPV-induced lesions. Cromme et al. (1994) reported that loss of TAP1 correlated with loss of surface MHC-I. Defects in multiple components of MHC-I processing and expression have been reported in cervical carcinomas, including proteasome subunits and their activators PA28 α/β (Ritz et al., 2001), TAP1/2, MHC-I itself (Ritz et al., 2001; Evans et al., 2001) and proteasome subunits LMP2 and 7 (Evans et al., 2001). Deficiencies in the TAP transporter have also been associated with many other carcinomas, including prostate, renal cell, melanoma, breast, and small cell lung cancer (Seliger et al., 1997). In some of these cancers, expression of TAP1 following gene transfer was able to restore immunogenicity to the carcinoma, inducing high levels of surface MHC-I and stimulating allogenic T cells (Kallfelz et al., 1999). These results highlight the importance of this key transporter.

By analogy with other tumors, COX-2 and its product PGE_2 may also be an important signaling molecule that regulates the immune system in RRP, since COX-2 is overexpressed in papillomas (Robinson et al., 1999) and papilloma cells express elevated levels of PGE_2 (R. Wu, personal communication). When dendritic cells were cultured in supernatant from lung carcinoma cells, the supernatant induced a failure to generate an antitumor immune response, whereas supernatant from cells treated with a selective COX-2 inhibitor did not inhibit antitumor dendritic cell activity (Sharma et al., 2003). The dendritic cells cultured in the tumor cell supernatant showed losses of multiple components required for antigen presentation, including MHC-I, MHC-II, TAP2 and TAP1, compared to cells cultured in supernatant from COX-2 inhibited cells. Thus, treatment with the selective COX-2 inhibitor celecoxib could potentially improve immune function in RRP patients as well.

3.2. Variability in immune response genes

RRP susceptibility or severity may be determined, in part, by genetic variability of human leukocyte antigen (HLA) genes, expecially those for Class-II alleles, and genotyping might eventually be able to predict clinical outcome. DR antigen expression on epithelial cells has been previously reported in RRP (Viac et al., 1993). DRB*0301 was recently correlated with disease severity (Gelder et al., 2003). Interestingly, in this study the authors tried to correlate DRB1*0301 status with response to an epitope on L1 virus-like particles (VLP), but there was no difference between patients who were DRB1*0301 and those who were DRB1*0301 negative (Gelder et al., 2003). We previously reported enrichment of DQ3 in a small study (Bonagura et al., 1994). This was confirmed in a larger patient population in which we also reported that DRB1*0301- DQB1*0201 is enriched in severe RRP in our patients (Bonagura et al., 2004). This more recent study also began to address cellular function associated with HLA type. DRB1*0301 was associated with reduced IFN-γ expression in patients with RRP, thereby skewing responses to a Th2 type (Bonagura et al., 2004). In one study of adult onset RRP, there was no difference between DQA1 and DQB1 alleles and the reference population (Aaltonen et al., 1999). Surprisingly, HPV DNA could only be detected in 64% of analyzed specimens in that study, perhaps reflecting very low levels of viral DNA in the limited samples, and those patients had a history of less severe disease. Prevalence of DQB1*0501 was enriched in the patients with absence of detectable HPV (Aaltonen et al., 1999). Of interest DQB1*0501 was previously found to be protective against cervical cancer (Gregoire et al., 1994). The TAP gene is located in close proximity to the Class-II locus on chromosome 6. TAP1 is highly conserved, with little genetic variability. Original reports defined only two polymorphic sites in the TAP1 protein (Colonna et al., 1992). One is at position 333, converting isoleucine to valine. The other is at position 637, which converts aspartic acid to glycine (Carrington et al., 1993). The polymorphism at amino acid 637 is located in the ATPase domain (Colonna et al., 1992, Gaudet et al., 2001). Glycine is a strong inhibitor of alpha-helix

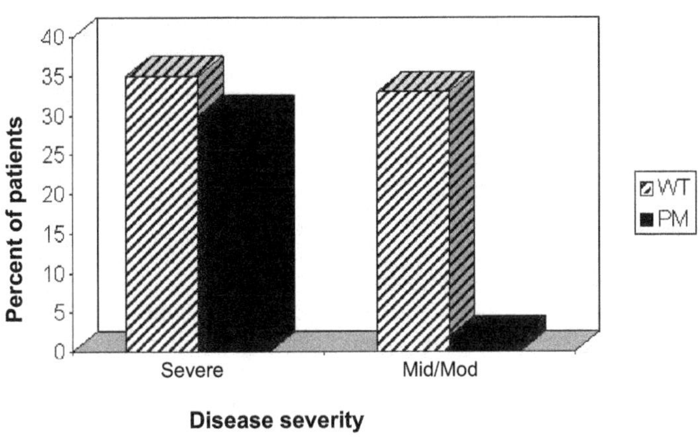

Figure 3. Comparison of disease severity with presence of the TAP1 polymorphism. Severe disease correlated with the presence of a TAP1 polymorphism at amino acid 637 (p=0.015). Reprinted with permission from Vambutas el al., J. Infect. Dis. 2004; 189 (5): 871-879.

formation (Chou et al., 1974), and its presence could potentially induce a large conformational change in the ATPase domain of TAP1 with the polymorphism, thereby changing the transporter's interaction with other proteins or with ATP itself. We have found that the polymorphism in the ATPase domain correlated with severe disease (Figure 3; Vambutas et al., 2004), and that it appeared to be in linkage disequilbrium with DRB*0102. When in linkage, the DRB*0102 allele appeared to mitigate the deleterious effect of the TAP1 polymorphism (Vambutas et al., 2004). However, this allele may independently predispose individuals to RRP, as it was enriched in both the mild-moderate as well as the severe disease group compared to the population as a whole (Bonagura et al., 2004).

3.3. Cytokine responses to HPV proteins

CD4 T cells provide an important role in providing cytokines to help drive cytotoxic T cell responses. Unfortunately, cultured respiratory papilloma-derived CD4+ T cells respond to HPV proteins from autologous papilloma tissue by producing Th2 type cytokines in the relative absence of Th1 cytokines (Bonagura et al., 1999). Presence of these Th2 cytokines correlated with disease severity, and this lack of a Th1 response may, in part, contribute to the inability to prime CD8+ T cells to eradicate disease. Moreover, CD8+, CD28- T cells were enriched in the lesions (Bonagura et al., 1999). Although the function of this T cell subset is unknown, the authors speculated that these T cells could be suppressive in nature. The specific viral proteins eliciting the Th2 response were not discovered until recently. Responses to the E6 protein were the driving force for this Th1/Th2 imbalance by preferentially inducing IL-10 production over gamma-interferon (IFN-γ) (Figure 4), whereas responses to E7 were generally minimal and similar to controls (DeVoti et al., 2004). Additionally, E6 inhibited IL-2 and IL-18 expression, thereby supporting its role as a potent inducer of Th2 responses (De Voti et al., 2004). Interestingly, in studies on cervical cancer, cervical levels of IL-12 and IL-10 are higher than those found in blood, suggesting that local cytokine production may be the predominant driving force for T cell responses (Castle et al., 2002). This correlates with our observations in RRP patients that they appear generally immunocompetent while locally immunocompromised to HPV infection.

Chronic viral infection has been proposed to result in T cell exhaustion and progressive loss of IL-2, TNF-α and finally IFN-γ production (Wherry et al., 2003). This pattern has been observed in RRP. IL-2 is an initiating interleukin capable of driving a Th1 response. Snowden et al. (Snowden et al., 2001) reported that serum IL-2 and IL-2 receptor (IL-2R) levels were significantly lower in a group of RRP patients when compared to age matched controls. Interestingly, and counter-intuitively, they found that IL-2 and IL-2R levels were higher in those

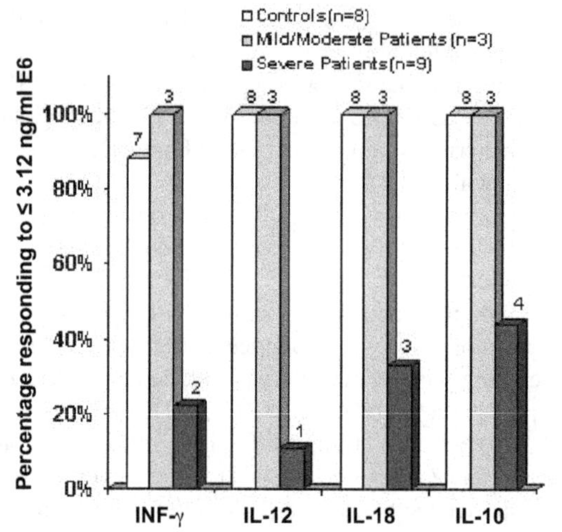

Figure 4. Ability of peripheral blood mononuclear cells to respond to low concentrations of E6 protein. Patients with severe disease required higher concentrations of E6 protein to induce Th1-like cytokine responses compared to controls and patients with mild-moderate disease. Reprinted with permission from DeVoti et al., Clin. Diag. Lab. Immunol. 2004; 11(3): 538-47.

RRP patients with more severe disease. Bonagura et al. (1999) and DeVoti et al. (2004) reported that reduced IFN-γ expression correlated with disease severity. This disparity between studies may reflect either methodology differences or the possibility that increased IL-2 release may represent an attempt to drive a Th1 response in face of continuous antigen exposure. The exhausted T cells would be more likely to be found in non-lymphoid tissues, where they would encounter viral antigen and yet be incapable of mounting appropriate responses (Zhou et al., 2004). However, we have observed a general paucity of infiltrating T cells in papillomas (our unpublished observations). Maintenance of healthy CD8+ T cells requires appropriate help from CD4+ T cells (Melief et al., 2000) and effective vaccine strategies require polarization of naïve CD4 + T cells to Th1 cells. Sustaining a prolonged Th1 response has been a major obstacle in vaccine development (Seder et al., 2003). In fact, clearance of Hepatitis C virus infections has been most strongly correlated to CD4+ Th1 responses (Missale et al., 1996; Rollier et al., 2004). Other significant factors required to generate an effective response to an immunogen include providing sufficient antigen for a sufficient period of time (LeFrancois et al., 2003) and appropriate activation of professional antigen presenting cells (Zwaveling et al., 2002).

3.4. Antibody expression

Antibody responses to papillomavirus proteins during chronic laryngeal infection are generally poor. In one study, neutralizing antibodies against HPV-11 were detected in approximately half of patients with laryngeal papillomas (Christensen et al., 1992). Aaltonen et al. (2001) reported no difference between adult-onset RRP patients and controls for presence of antibodies to HPV 6 and 16, but some increase in likelihood of detecting antibodies to HPV 11 in female patients. However, antibodies did not correlate with HPV DNA type in the larynx, and many of the female patients had concurrent clinical or cytologic evidence of cervical HPV infection. They concluded that antibodies were primarily generated as a result of concurrent cervical infection. In patients with condylomata accuminata induced by HPV 6, presence of antibodies did not predict clinical course but rather served as a marker of viral exposure (Wikstrom et al., 1995).

Juvenile-onset RRP may be acquired from mothers with genital tract infection during passage through the birth canal. In a review of studies examining this transmission, HPV DNA could be found in oral swabs of babies from 4% to 87% of the time, HPV type concordance between mother and child was only 57-69% and HPV antibodies were present 10-57% of the time (Syrjanen et al., 2000). Usually, there was no correlation between seropositivity and HPV DNA detection (Syrjanen et al., 2000).

4. Vaccine Development

4.1. Therapeutic vs preventive vaccines

Great strides have been made in the development of a preventative vaccine for genital HPV infection, with an HPV 16 VLP vaccine shown to be protective in clinical trials (see chapter X). Development of preventative vaccines specifically targeted to RRP are not practical as (1) babies acquire the infection at birth although very few children born to women with vaginal condyloma develop RRP and (2) the mechanism of transmission of HPV for development of adult-onset RRP is unknown. The possibility remains that immunization of women to prevent cervical infection with HPVs 6 and 11 might ultimately reduce airway infection and RRP.

Unfortunately, similar strides have not been achieved in the development of therapeutic vaccines. This is, in part, due to the immune evasion strategies mounted by the lesion, highlighting the challenge of therapeutic vaccines. Healthy uninfected patients vaccinated with a VLP vaccine against HPV 11 were capable of mounting specific antibody responses as well as a lymphoproliferative response characterized by production of IL-2 and IFN-γ (Emeny et al., 2002). Unfortunately, similar responses are unlikely in RRP patients already infected with HPV, as they have shown atretic responses in making either IL-2 (Snowden et al., 2001) or IFN-γ (DeVoti et al., 2004) in response to PV proteins.

There are multiple challenges to development of a therapeutic vaccine for RRP. Activation of latent viral infection is commonplace, and an effective vaccine would have to prevent such activations. Inflammatory responses are poorly tolerated as the cross sectional area of the upper respiratory tract is narrow, especially in children. Initial attempts at crude vaccination strategies have met with variable results. Injection of autogenous tumor lysates was able to significantly control disease in a small group of patients (Stephens et al., 1979), suggesting that enhanced immune response could have some efficacy. General immune stimulation by vaccination with BCG has been tried, but was largely ineffectual (Strauss et al., 1982). Intralesional mumps vaccine in combination with surgical excision appeared to induce remission of disease in 76% of patients (Pashley et al., 2002), but controlled trials to confirm these results have not been performed. Needed is a therapeutic vaccine that specifically enhances host immune response to specific HPV proteins. Presence of the E6 and E7 proteins as specific antigens in infected tissues has lead many groups to target these proteins in vaccine development.

4.2. Animal models for vaccine development

Prior to human clinical trials, vaccine strategies are developed and tested in animal models. While xenograft models in mice may be appropriate for cervical cancer studies, additional factors should be considered in the design and testing of vaccines for RRP. Airways of effected individuals are significantly compromised due to disease. Therefore, only minimal inflammatory responses can be tolerated. A hallmark of this disease is reactivation of latent disease by unknown triggers. Optimal vaccine strategies should evaluate control of latent viral infection as well as active disease.

Two animal models are available to test vaccines against natural papillomavirus infections. The cottontail rabbit papillomavirus model (CRPV) is an ideal model to study papillomavirus latency and therapeutic vaccines. Latent viral infection could be established and increased transcription of E6 and E7 marked the transition to active clinical disease and wart formation (Zhang et al., 1999). Two promising vaccines that have been tested in the rabbit model are the E6-ubiquitin fused vaccine (Leachman et al., 2002) and the DNA vaccine encoding E1, E2, E6 and E7 when used in combination with intralesional cidofovir (Christensen et al., 2001). Alternatively, the canine oral papillomavirus (COPV) is also a suitable model, as natural infections occur in the dog (Nicholls et al., 1999; Jahan-Parwar et al., 2003). Interestingly, oral papillomas in beagles spontaneously regress, making them better models for prophylactic vaccines. In a case of severe non regressing COPV infection where vaccination with L1 VLP produced high levels of antibodies but no clinical effect, the infecting COPV DNA was sequenced and compared to COPV from animals where spontaneous regression occurred. There was no significant variation in the viral DNA from the dog with multiple persistent lesions, highlighting a defect in host immunity rather than viral pathogenicity as the cause for the difference in the infection (Nicholls et al., 1999). In fact, regression of papillomas that is usually observed in this model is associated with abundant infiltrating CD4+ and CD8+ lymphocytes, highlighting the host immune response (Nicholls et al., 2001).

4.3. Antigenic targets and lessons from cervical cancer

The requisite expression of E6 and E7 proteins during productive infections and persisting lesions have made them obvious targets for vaccine development. The N-terminal region of E7 is thought to be the most antigenic (Bauer et al., 2000), as the majority of naturally processed E7 peptides are derived from this region. Short E7 peptide vaccines have not resulted in good cytotoxic T cell responses *in vitro* (Steele et al., 2002) or in phase I clinical trials in cervical cancer (Ressing et al., 2000), suggesting that MHC-I antigen presentation is impaired. Although mapping and use of immunodominant MHC-I epitopes of HPV 16 and 18 E7 for peptide vaccines has been performed, this has not improved efficacy of vaccination strategies (as reviewed in Galloway, 2003). This may be explained by the fact that although CD8+ T cells can be strong mediators of viral clearance, chronic viral infections disrupt the expected immunodominant CD8+ T cell responses and tissue distribution of these T cells (Wherry et al., 2003). The incomplete protection seen with E7 may be explained by the observations of Doan et al. that E7 expressed in the epithelium may induce tolerance (Doan et al., 1999). Furthermore, mice expressing HPV 16 E7 protein non-specifically down-regulated their CD8+ T cell responses, suggesting the response is class I restricted and cellular immune response is impaired (Tindle et al., 2001). Cervical carcinomas (HPV 16+) are known to overexpress E7, and have been found to show losses of both TAP1 and MHC-I, the ligand for the CD8 receptor (Cromme et al., 1994).

4.4. Increasing immunogenicity

Poor immunologic responses to short antigenic peptides of viral proteins have led investigators to test strategies to increase immunogenicity. One approach was to fuse the E7 protein to the Hsp65 heat shock protein (Suzue et al., 1997) making E7 more immunogenic. Stressgen has received orphan drug status for the treatment of RRP with their Hsp-E7 vaccine, but clinical results in a large controlled study the treatment of RRP are yet to be determined. The increased immunogenicity of a fusion with heat shock protein may be attributable to many causes. Induction of a Th1 type response and tumor regression was observed in an animal model following HspE7 vaccination (Chu et al., 2000a), and subsequent studies showed that the response was CD8 dependent/CD4 independent (Chu et al., 2000b). Antigen for MHC-I presentation to CD8 T cells is predominantly processed by the 26S proteasome. If PA28 (the proteasome activator protein) is lost, Hsp90 can compensate. Introduction of IFN-γ into this system induced the up-regulation of PA28 and abrogated the need for Hsp90 (Yamano et al., 2002).

Another method to increase immunogenicity is to harness the professional antigen presenting cell, the dendritic cell, to present antigen to the immune system. By altering the length of the peptides in a vaccine, antigen presenting cells would optimally present MHC-II bound peptides to CD4+ T-cells and MHC-I bound peptides to CD8+ T-cells (Zwaveling et al., 2002). Pharmacodynamics of the long peptides appear to be superior to minimal T-cell epitopes for the presentation of MHC class I peptides, probably because the long peptides require processing, which can only be achieved efficiently by these professional antigen presenting cells. Previous studies in a murine model demonstrated that the long peptide vaccine was able to eradicate HPV 16 positive tumors. When this long peptide strategy was used in conjunction with CpG, a dendritic cell activator, survival of HPV16-induced tumor bearing mice was significantly prolonged compared to untreated controls (Zwaveling et al., 2002). The long peptide vaccine strategy may be able to recruit dendritic cells to the lesions. In a study using this long peptide strategy in a CRPV model, long peptides from E6 and E7 significantly reduced papillomas growth and latent viral infection, although we did not see improvement with the addition of CpG (our unpublished results). In the mouse, CpG activates both myeloid and plasmacytoid dendritic cells. In humans, CpG activates plasmacytoid dendritic cells, but not myeloid dendritic cells, via toll-like receptor 9 which results in the synthesis of type I interferon (Rotherfusser et al., 2002). In a mouse model, CpG used in combination with E7 to stimulate dendritic cells was able to elicit IL-12 production (a Th1 cytokine) and better improve protection against tumor than when either E7 or CpG alone was used (Kim et al., 2004). Thus several different approaches might be possible to enhance dendritic cell presentation of HPV proteins in a therapeutic vaccine for RRP.

An effective vaccine strategy for RRP requires control of latent disease and minimal local inflammatory response. Most likely, any vaccine strategy or modulation of the host immune response may require surgical clearance of disease, as elimination of latent infection has not been achieved to date. The key to disease prevention of RRP is in the elimination of latent disease. As such, treatment efficacy should be measured not only by the control of active disease, but control or eradication of latent disease as well.

References

Aaltonen, L.M., Auvinen, E., Dillner, J., Lehtinen, M., Paavonen, J., Rihkanen-Vaheri, A. (2001). Poor antibody response against human papillomavirus in adult-laryngeal papillomatosis. J. Med. Microbiol. *50*, 468-471.

Aaltonen, L.M., Rihkanen, H., and Vaheri, A. (2002). Human papillomavirus in larynx. Laryngoscope *112*, 700-7.

Aaltonen, L.M., Partnanen, J., Auvinen, E., Rihkanen, H., and Vaheri, A. (1999). HLA-DQ alleles and human papillomavirus DNA in adult-onset laryngeal papillomatosis. J. Infect. Dis. *179*, 682-685.

Abramson, A.L., Steinberg, B.M., and Winkler, B. (1987). Laryngeal papillomatosis: clinical, histopathologic and molecular studies. Laryngoscope *97*, 678-685.

Abramson, A.L., Shikowitz, M.J., Mullooly, V.M., Steinberg, B.M., Amella, C.A., and Rothstein, H.R. (1992). Clinical effects of photodynamic therapy on recurrent laryngeal papillomas. Arch. Otolaryngol. Head Neck Surg. *118*, 25-29.

Abramson, A.L., Nouri, M., Mullooly, V., Fisch, G., and Steinberg, B.M. (2004). Latent human papillomavirus infection is comparable in the larynx and trachea. J. Med. Virol. *72*, 473-477.

Alwan, H.A., VanZoelen, E.J., ad VanLeeuwen, J.E. (2003). Ligand-induced lysosomal epidermal growth factor receptor (EGFR) degradation is preceded by proteasome-dependent EGFR de-ubiquitination. J. Biol. Chem. *278*, 35781-35790.

Armstrong, L.R, Preston, E.J., Reichert, M., Phillips, D.L., Nisenbaum, R., Todd, N.W., Jacobs, I.N., Inglis, A.F., Manning, S.C., and Reeves, W.C. (2000). Incidence and prevalence of recurrent respiratory papillomatosis among children in Atlanta and Seattle. Clin. Infect. Dis. *31*, 107-109.

Auborn, K.J., Little, R.D., Platt, T.H.K., Vaccariello, M.A., and Schildkraut, C.L. (1994). Replicative intermediates of human papillomavirus type 11 in laryngeal papillomas: site of replication initiation and direction of replication. Proc. Natl. Acad. Sci. *91*, 7340-7344.

Auborn, K.J., Wang, H., Vaccariello, M.A., and Taichman, L.B. (1996). Kinetics of HPV 11 DNA replication after infection of keratinocytes with virions. Virus Res. *43*, 85-90.

Bauer M, Wagner H, and Lipford GB. (2000). HPV type 16 protein E7 HLA-A2 binding peptides are immunogenic but not processed and presented. Immunol. Lett. *71*, 55-59.

Berry, J.M., and Palefsky, J.M. (2003). A review of human papillomavirus vaccines: from basic science to clinical trials. Front. Biosci *8*, s333-345.

Bielamowicz, S., Villagomez, V., Stager, S.V., and Wilson, W.R. (2002). Intralesional cidofovir for laryngeal papilloma in an adult cohort. Laryngoscope. *112*, 696-699.

Bishai, D., Kashima, H., and Shah, K. (2000). The cost of juvenile-onset recurrent respiratory papillomatosis. Arch. Otolaryngol. Head Neck Surg. *126*, 935-939.

Bonagura, V.R., Siegal, F.P., Abramson, A.L., Santiago-Schwartz, F., O-Reilly, M.E., Shah, K., Drake, D., and Steinberg, B.M. (1994). Enriched HLA-DQ3 phenotype and decreased class I major histocompatibility complex antigen expression in recurrent respiratory papillomatosis. Clin. Diag. Lab. Immunol. *1*, 357-360.

Bonagura, V.R., Hatam, L., DeVoti, J.D., Zeng, F., and Steinberg, B.M. (1999). Recurrent respiratory papillomatosis: altered CD8+ T-cell subsets and Th1/Th2 cytokine imbalance. Clin. Immunol. *93*, 302-311.

Bonagura, V.R., Vambutas, A., DeVoti, J.A., Rosenthal, D.W., Steinberg, B.M., Abramson, A.L., Shikowitz, M.J., Gjertson, D.W., and Reed, E.F. (2004). HLA alleles, IFN-gamma responses to HPV-11 E6, and disease severity in patients with recurrent respiratory papillomatosis. Human Immunol. *65*, 773-782.

Borkowski, G., Sommer, P., Stark, T. Sudhoff, H., and Luckhaupt, H. (1999). Recurrent respiratory papillomatosis associated with gastroesophageal reflux disease in children. Eur. Arch. Otorhinolaryngol. *256*, 370-372.

Byrne, J.C., Tsao, M.S., Fraser, R.S., and Howley, P.M. (1987). Human papillomavirus-11 DNA in a patient with chronic laryngotracheobronchial papillomatosis and metastatic squamous-cell carcinoma of the lung. N. Engl. J. Med. *317*, 873-878.

Carrington, M., Colonna, M., Spies, T., Stephens, J.C., and Mann, D.L. (1993). Haplotypic variation of the transporter associated with antigen processing (TAP) genes and their extension of HLA class II region haplotypes. Immunogenetics. *37*, 266-273.

Castle, P.E., Hildesheim, A., Bowman, F.P., Strickler, H.D., Walker, J.L., Pustilink, T., Edwards, R.P., and Crowley-Nowick, P.A. (2002). Cervical concentrations of interleukin-10 and interleukin-12 do not correlate with plasma levels. J. Clin. Immunol. *22*, 23-27.

Chardonnet, Y., Viac, J., Leval, J., Begoud, G., Morgon, A., and Thivolt, J. (1986). Laryngeal papillomas: local cellular immune response, keratinization and viral antigen. Virchows Arch. B. Cell. Pathol. Incl. Mol. Patho*l*. *51*, 421-428.

Cheng S., Schmidt-Grimminger D.C., Murant T., Broker T.R., and Chow L.T.. (1995). Differentiation-dependent up-regulation of the human papillomavirus E7 gene reactivates cellular DNA replication in suprabasal differentiated keratinocytes. Genes Dev. *9*, 2335-2349.

Chiang C.M., Dong G., Broker T.R., and Chow L.T. (1992). Control of human papillomavirus type 11 origin of replication by the E2 family of transcription regulatory proteins. J Virol. *66*, 5224-5231.

Christensen, N.D., Kreider, J.W., Shah, K.V., and Rando, R.F. (1992). Detection of human serum antibodies that neutralize infectious human papillomavirus type 11 virions. J. Gen. Virol. *73*, 1261-1267.

Christensen, N.D., Han, R., Cladel, N.M., and Pickel, M.D. (2001). Combination treatment with intralesional cidofovir and viral-DNA vaccination cures cottontail rabbit papillomavirus-induced papillomas and reduces recurrences. Antimicrob. Agents Chemother. *45*, 1201-1209.

Chou, P.Y., and Fasman, G.D. (1974). Conformational parameters for amino acids in helical, β-sheet, and random coil regions calculated from proteins. Biochemistry. *13*, 211-222.

Chu, N.R., Wu, H.B., Wu, T., Boux, L.J., Siegel, M.I., and Mizzen, L.A. (2000a). Immunotherapy of a human papillomavirus (HPV) type 16 E7-expressing tumor by administration of fusion protein comprising Mycobacterium bovis bacilli Calmette-Guerin (BCG) hsp65 and HPV16E7. Clin. Exp. Immunol. 121, 216-225.

Chu, N.R., Wu, H.B., Wu, T.C., Boux, L.J., Mizzen, L.A., and Siegel, M.I. (2000b). Immunotherapy of a human papillomavirus (HPV) type 16 E7-expressing tumor by administration of fusion protein comprising Mycobacterium bovis BCG hsp65 and HPV16E7. Cell Stress Chap. 5, 401-405.

Colonna, M., Bresnahan, M., Bahram, S., Strominger, J.L., and Spies, T. (1992). Allelic variants of the human putative peptide transporter involved in antigen processing. Proc. Natl. Acad. Sci.*89*, 3932-3936.

Conrad, M., Bubb, V.J., and Schlegel, R. (1993). The human papillomavirus type 6 and 16 E5 proteins are membrane-associated proteins which associate with the 16-kilodalton pore-forming protein. J. Virol. *67,* 6170-6178.

Cromme, F.V., Airey, J., Heemels, M.T., Ploegh, H.L., Keating, P.J., Stern, P.L., Meijers, C.J.L.M., and Walboomers, J.M.M.. (1994). Loss of transporter protein, encoded by the TAP-1 gene, is highly correlated with loss of HLA expression in cervical carcinomas. J. Exp. Med. *179,* 335-340.

Dedo, H.H., and Yu, K.C. (2001). CO2 laser treatment in 244 patients with respiratory papillomas. Laryngoscope. *111,*1639-1644.

Demeter, L.M., Stoler, M.H., Broker, T.R., and Chow, L.T. (1994). Induction of proliferating cell nuclear antigen in differentiated keratinocytes of human papillomavirus-infected lesions. Human Pathol. *25,* 1375-1376.

DeVoti, J.A., Steinberg, B.M., Rosenthal, D.W., Hatam, L., Vambutas, A., Abramson, A.L., Shikowitz, M.J., and Bonagura, V.B. (2004). Failure of gamma interferon but not interleukin-10 expression in response to human papillomavirus type 11 E6 protein in respiratory papillomatosis. Clin. Diag. Lab. Immunol. *11,* 538-547.

DiLorenzo, T.P. Tamsen, A., Abramson, A.L., and Steinberg, B.M. (1992). Human papillomavirus type 6a DNA in the lung carcinoma of a patient with recurrent laryngeal papillomatosis is characterized by a partial duplication. J. Gen. Virol. *73,* 423-428.

DiLorenzo, T.P., and Steinberg, B.M. (1995). Differential regulation of human papillomavirus type 6 and 11 early promoters in cultured cells derived from laryngeal papillomas. J. Virol. *69,* 6865-6872.

DiLorenzo, T.P., Chen, D., Zhang, P, and Steinberg, B.M. (1998). Evidence for the separate regulation of the human papillomavirus type 11 E7 and E6 promoters by viral cis sequences near the E6 promoter. Virology *243,* 130-139.

Doan,T., Herd, K., Street, M., Bryson, G., Fernando, G., Lambert, P., and Tindle, R. (1999). Human papillomavirus type 16 E7 oncoprotein expressed in peripheral epithelium tolerizes E7-directed cytotoxic T-lymphocyte precursors restricted through human and mouse major histocompatibility complex class I alleles. J. Virology. *73,* 6166-6170.

Emeny, R.T., Wheeler, C.M., Jansen, K.U., Hunt, W.C., Fu, T.M., Smith, J.F., MacMullen S., Esser, M.T., and Paliard, X. (2002). Priming of human papillomavirus type 11-specific humoral and cellular immune responses in college-aged women with a virus-like particle vaccine. J. Virol. *76,* 7832-7842.

Erdamar, B., Keles, N., Kaur, A., Suoglu, Y., and Kiyak, E. (2002). Expression of the cyclin-kinase inhibitors p21 (WAF1) and p21 (Kip1) and the p53 tumor suppressor genes in adult-onset laryngeal papillomatosis. Eur. Arch. Otorhinolaryngol. *259,* 516-520.

Evans, M., Borysiewicz, L.K., Evans, A.S., Rowe, M., Jones, M., Gileadi, U., Cerundolo, V., and Man, S. (2001). Antigen processing defects in cervical carcinomas limit the presentation of a CTL epitope from human papillomavirus 16 E6. J. Immunol. *167,* 5420-5428.

Fehrmann, F., Klumpp, D.J., and Laimins, L.A. (2003). Human papillomavirus type 31 E5 protein supports cell cycle progression and activates late viral functions upon epithelial differentiation. J Virol. *77,* 2819-2831.

Figueiredo-Pereira, M.E., Li, J., Jansen, M., and Rockwell, P. (2002). N-acetylcysteine and celecoxib lessen cadmium toxicity which is associated with cyclooxygenase-2 up-regulation in mouse neuronal cells. J. Biol. Chem.. *277,* 25283-25289.

Gabbott, M., Cossart, Y.E., Kan, A., Konopka, M., Chan, R., and Rose, B.R. (1997). Human papillomavirus and host variables as predictors of clinical course in patients with juvenile-onset recurrent respiratory papillomatosis. J. Clin. Microbiol. *35,* 3098-3103.

Galloway, D.A. (2003). Papillomavirus vaccines in clinical trials. Lancet Infect. Dis. *3,* 469-475.

Gardiol, D., Kuhne, C., Glaunsinger, B., Lee, S.S., Javier, R., and Banks, L. (1999). Oncogenic human papillomavirus E6 proteins target the discs large tumour suppressor for proteasome-mediated degradation. Oncogene.*18,* 5487-5496.

Garland, S.M. (2002). Human papillomavirus update with a particular focus on cervical disease. Pathology. *34,* 213-224.

Gaudet, R., and Wiley, D.C. (2001). Structure of the ABC ATPase domain of human TAP1, the transporter associated with antigen processing. EMBO. *20,* 4964-4972.

Gelder, C.M., Williams, O.M., Hart, K.W., Wall, S., Williams, G., Ingrams, D., Bunce, M., Welsh, K., Marshall, S.E., and Borysiewicz, L. (2003). HLA class II polymorphisms and susceptibility to recurrent respiratory papillomatosis. J. Virol. *77,* 1927-1939.

Gillison M.L., and Shah K.V. (2001). Human papillomavirus-associated head and neck squamous cell carcinoma: mounting evidence for an etiologic role for human papillomavirus in a subset of head and neck cancers. Curr. Opin. Oncol. *13,* 183-188.

Go, C., Schwartz, M.R., and Donovan, D.T. (2003). Molecular transformation of recurrent respiratory papillomatosis, viral typing and p53 overexpression. Ann. Otol. Rhinol. Laryngol. *112,* 298-302.

Gregoire, L., Lawrence W.D., Kukuruga, K., Einsenbrey, A.B., and Lancaster, W.D. (1994). Association between HLA-DQB! Alleles and risk for cervical cancer in African-American women. Int. J. Cancer *57,* 504-507.

Gupta, S., Takhar, P.P., Degenkolbe, R., Koh, C.H., Zimmermann, H., Yang, C.M., Guan Sim, K., Hsu, S.I., and Bernard H.U. (2003). The human papillomavirus type 11 and 16 E6 proteins modulate the cell-cycle regulator and transcription cofactor TRIP-Br1.Virology. *317,* 155-164.

Healy, G.B., Gelber, R.D., Trowbridge, A.L., Grundfast, K.M., Ruben, R.J., and Price, K.N. (1988). Treatment of recurrent respiratory papillomatosis with human leukocyte interferon. Results of a multicenter randomized clinical trial. N. Engl. J. Med. *319,*401-407.

Herrero, R., Castellsague, X., Pawlita, M., Lissowska, J., Kee, F., Balaram, P., Rajkumar, T., Sridhar, H., Rose, B., Pintos, J., Fernandez, L., Idris, A., Sanchez, M.J., Nieto, A., Talamini, R., Tavani, A., Bosch, F.X., Reidel, U., Snijders, P.J., Meijer, C.J., Viscidi, R., Munoz, N., and Franceschi, S.; IARC Multicenter Oral Cancer Study Group. (2003). Human papillomavirus and oral cancer: the International Agency for Research on Cancer multicenter study. J. Natl. Cancer Inst. *95,* 1772-1783.

Holland, B.W., Koufman, J.A., Postma, G.N., and McGuirt, W.F. (2002). Laryngopharyngeal reflux and laryngeal web formation in patients with pediatric recurrent respiratory papillomas. Laryngoscope. *112,* 1926-1929.

Jahan-Parwar, B., Chhetri, D.K., Hart, S., Bhuta, S., and Berke, G.S. (2003). Development of a canine model for recurrent respiratory papillomatosis. Ann. Otol. Rhinol. Laryngol. *112,* 1011-1013.

Jian, Y., Van Tine, B.A., Chien, W.M., Shaw, G.M., Broker, T.R., and Chow, L.T. (1999). Concordant induction of cyclin E and p21cip1 in differentiated keratinocytes by the human papillomavirus E7 protein inhibits cellular and viral DNA synthesis. Cell Growth Differentialtion. *10,* 101-111.

Johnston, D., Hall, H., DiLorenzo, T.P., and Steinberg, B.M. (1999). Elevation of the epidermal growth factor receptor and dependent signaling in human papillomavirus-infected laryngeal papillomas. Cancer Res. *59,* 968-974.

Kashima, H.K., Mounts, P., and Shah, K. (1996). Recurrent respiratory papillomatosis. Obstet. Gynecol. Clin. North Am. *23,* 699-706.

Kallfelz, M., Jung, D., Hilmes, C., Knuth, A., Jaeger, E., Huber, C., and Seliger, B. (1999). Induction of immunogenicity of a human renal-cell carcinoma cell line by TAP-1 gene transfer. Int. J. Cancer *81,* 125-133.

Kehmeier, E., Ruhl, H., Voland, B., Stoppler, M.C., Androphy, E.,. and Stoppler, H. (2002). Cellular steady-state levels of "high risk" but not "low risk" human papillomavirus (HPV) E6 proteins are increased by inhibition of proteasome-dependent degradation independent of their p53- and E6AP-binding capabilities. Virology *299,*72-87.

Kim, T.G., Kim, C.H., Won, E.H., Bae, S.M., Ahn, W.S., Park, J.B., and Sin, J.I. (2004). CpG-ODN-stimunlated dendritic cells act as a potent adjuvant for protein delivery to induce antigen-specific antitumor immunity HPV 16 E7-associated animal tumour model. Immunol. *112,* 117-125.

Kwon, G., Corbett, J.A., Hauser, S., Hill, J.R., Turk, J., and McDaniel, M.L. (1998). Evidence for involvement of the proteasome complex (26S) and NfkappaB in IL-1beta-induced nitric oxide and prostaglandin production by rat islets and RINm5F cells. Diabetes. *47,* 583-591.

Leachman, S.A., Shylankevich, M., Slade, M.D., Levine, D., Sundaram, R.K., Xiao, W., Bryan, M., Zelterman, D., Tiegelaar, R.E., and Brandsma, J.L. (2002). Ubiquitin-fused and/or multiple early genes from cottontail rabbit papillomavirus as DNA vaccines. J. Virol. *76,* 7616-7624.

Lefrancois, L., Marzo, A., and Williams, K. (2003). Sustained response initiation is required for T cell clonal expansion but not for effector or memory development *in vivo.* J. Immunol. *171,* 2832-2839.

Leventhal, B.G., Kashima, H.K., Mounts, P., Thurmond, L., Chapman, S., Buckley, S., and Wold, D. (1991). Long-term response of recurrent respiratory papillomatosis to treatment with lymphoblastoid interferon alfa-N1. Papilloma Study Group. N. Engl. J. Med. *325,* 613-617.

Lin, K.Y., Westra, W.H., Kashima, H.K., Mounts, P., and Wu, T.C. (1997). Coinfection of HPV-11 and HPV-16 in a case of laryngeal squamous papillomas with severe dysplasia. Laryngoscope *107,* 942-947.

Maran, A., Amella, C.A., DiLorenzo, T.P., Auborn, K.J., Taichman, L.B., and Steinberg, B.M. (1995). Human papillomavirus transcripts are present at low abundance in latently infected respiratory tissues. Virology *212,* 285-294.

Melief, C.J., Toes, R.E., Medema, J.P., van der Burg, S.H., Ossendorp, F., and Offringa, R. (2000). Strategies for immunotherapy of cancer. Adv. Immunol. *75,* 235-282.

Milczuk, H.A. (2003). Intralesional cidofovir for the treatment of severe juvenile recurrent respiratory papillomatosis: long-term results in 4 children. Otolaryngol. Head Neck Surg. *128,* 788-794.

Mineta, H., Ogino, T., Amano, H.M., Ohkawa, Y., Araki, K., Takebayashi, S., and Miura K. (1998). Human papillomavirus (HPV) type 16 and 18 detected in head and neck squamous cell carcinoma. Anticancer Res. *18,* 4765-4768.

Missale, G., Bertoni, R., Lamonaca, V., Valli, A., Massari, M., Mori, C., Rumi, M.G., Houghton, M., Fiaccadori, F., and Ferrari, C. (1996). Different clinical behaviors of acute hepatitis C virus infection are associated with different vigor of the antiviral cell-mediated immune response. J. Clin Invest. *98,* 706-714.

Munger, K., Werness, B.A., Dyson, N., Phelps, W.C., Harlow, E., and Howley, P.M.(1989). Complex formation of human papillomavirus E7 proteins with the retinoblastoma tumor suppressor gene product. EMBO J. *20,* 4099-4105.

Naiman, A.N., Ceruse, P., Coulombeau, B., and Froehlich, P. (2003). Intralesional cidofovir and surgical excision for laryngeal papillomatosis. Laryngoscope. *113,* 2174-2181.

Newfield, L., Goldsmith, A., Bradlow, H.L., and Auborn, K. (1993). Estrogen metabolism and human papillomavirus-induced tumors of the larynx: chemo-prophylaxis with indole-3-carbinol. Anticancer Res. *13,* 337-341.

Newfield, L., Bradlow, H.L., Sepkovic, D.W., and Auborn, K.A. (1998). Estrogen metabolism and the malignant potential of human papillomavirus immortalized keratinocytes. P.S.E.B.M. *217,* 322-326.

Nguyen, M.L., Nguyen, M.M., Lee, D., Griep, A.E., and Lambert, P.F. (2003). The PDZ ligand domain of the human papillomavirus type 16 E6 protein is required for E6's induction of epithelial hyperplasia *in vivo.* J. Virol. *77,*6957-6964.

Nicholls, P.K., Klaunberg, B.A., Moore, R.A., Santos, E.B., Parry, N.R., Gough, G., and Stanley, M.A. (1999). Naturally occurring, nonregressing canine oral papillomavirus infection: host immunity, virus characterization, and experimental infection. Virology *265,* 365-374.

Nicholls, P.K., Moore, P.F., Andersen, D.M., Parry, N.R., Gough, G., and Stanley, M.A. (2001). Regression of canine oral papillomas is associated with infiltrating CD4+ and CD8+ lymphocytes. Virology. *283,* 31-39.

Oh, S.T., Longworth, M.S., and Laimins, L.A. (2004). Roles of the E6 and E7 proteins in the life cycle of low-risk human papillomavirus type 11. J. Virol. *78,* 2620-2626.

Pashley, N.R. (2002). Can mumps vaccine induce remission in recurrent respiratory papilloma? Arch. Otolaryngol. Head Neck Surg. *128,* 783-786.

Pasquale, K., Wiatrak, B., Woolley, A., and Lewis, L. (2003). Microdebrider versus CO2 laser removal of recurrent respiratory papillomas: a prospective analysis. Laryngoscope. *113,* 139-143.

Patel, N., Rowe, M., and Tunkel, D. (2003). Treatment of recurrent respiratory papillomatosis in children with a microdebrider. Ann. Otol. Rhinol. Laryngol. *112,* 7-10.

Peyton Shirley, W., and Wiatrak, B. (2004). Is cidofovir a useful adjunctive therapy for recurrent respiratory papillomatosis in children? Int. J. Pediatr. Otorhinolaryngol. *68,* 413-418.

Poljak, M., Gale, N., Kambic, V., and Luzar, B. (1997) p53 protein overexpression in laryngeal squamous cell papillomas. Anticancer Res. *17,* 2201-2205.

Pransky, S.M., Magit, A.E., Kearns, D.B., Kang, D.R., and Duncan, N.O. (1999). Intralesional cidofovir for recurrent respiratory papillomatosis in children. Arch. Otolaryngol. Head Neck Surg. *125,* 1143-1148.

Pransky, S.M., Albright, J.T., and Magit, A.E. (2003) Long-term follow-up of pediatric recurrent respiratory papillomas managed with intralesional cidofovir. Laryngoscope *113,* 1583-1587.

Rabah, R., Sakr, W., Thomas, R., Lancaster, W.D., and Gregoire, L. (2000). Human papillomavirus type, proliferative activity, and p53: potential markers of aggressive papillomatosis. Arch. Pathol. Lab. Med. *124,* 721-724.

Rabah, R., Lancaster, W.D., Thomas, R., and Gregoire, L. (2001). Human papillomavirus-11-associated recurrent respiratory papillomatosis is more aggressive than human papillomavirus-6-associated disease. Pediatr. Dev. Pathol. *4,* 68-72.

Rady, P.L., Schnadig, V.J., Weiss, R.L., Hughes, T.K., and Tyring, S.K. (1998). Malignant transformation of recurrent respiratory papillomatosis associated with integrated human papillomavirus type 11 DNA and mutation of p53. Laryngoscope *108,* 735-740.

Reeves, W.C., Ruparelia, S.S., Swanson, K.I., Derkay, C.S., Marcus, A., and Unger, E. (2003). National registry for juvenile-onset recurrent respiratory papillomatosis. Arch. Otolaryngol. Head Neck Surg. *129,* 976-982.

Ressing, M.E., van Driel, W.J., Brandt, R.M., Kenter, G.G., de Jong, J.H., Bauknecht, T., Fleuren, G.J., Hoogerhout, P., Offringa, R., Sette, A., Celis, E., Grey, H., Trimbos, B.J., Kast, W.M., and Melief, C.J. (2000). Detection of T helper responses, but not of human papillomavirus-specific cytotoxic T lymphocyte responses, after peptide vaccination of patients with cervical carcinoma. J. Immunotherapy. *23,* 255-266.

Ritz,U., Momberg, F., Pilch, H., Huber, C., Maeurer, M.J., and Seliger B. (2001). Deficient expression of components of the MHC class I antigen processing machinery in human cervical carcinoma. Int. J. Oncol. *19,*1211-1220.

Robinson, A.B., Das, S.K., Bruegger, D.E., Hoover, L.A., and Sanford, T.R. (1999) Characterization of cyclooxygenase in laryngeal papilloma by molecular techniques. Laryngoscope *109,* 1137-1141.

Rollier, C., Depla, E., Drexhage, J.A.R., Verschoor, E.J., Verstrepen, B.E., Fatmi, A., Brinster, C., Fournillier, A., Whelan, J.A., Jacobs, D., Maertens, G., Inchaupse, G., and Heeney, J.L. (2004). Control of heterologous Hepatitis C virus infection in chimpanzees is associated with the quality of vaccine-induced peripheral T helper immune response. J. Virol. *78,* 187-196.

Rosen, C.A., and Bryson, P.C. (2004). Indole-3-carbinol for recurrent respiratory papillomatosis: long-term results. J. Voice. *18,* 248-253.

Rothenfusser, S., Tuma, E., Endres, S., Hartmann. G. (2002). Plasmacytoid dendritic cells: the key to CpG. Human Immunol. *63,* 1111-1119.

Sadasivan, B., Lehner, P.J., Ortmann, B., Spies, T., and Cresswell, P. (1996). Roles of calreticulin and a novel glycoprotein, tapasin, in the interaction of MHC class I molecules with TAP. Immunity. *5,* 103-114.

Schraff, S., Derkay, C.S., Burke, B., and Lawson, L. (2004). American Society of Pediatric Otolaryngology members' experience with recurrent respiratory papillomatosis and the use of adjuvant therapy. Arch. Otolaryngol. Head Neck Surg. *130,* 1039-1042.

Seder, R.A., and Ahmed, R. (2003). Similarities and differences in CD4+ and CD8+ effector and memory T cell generation. Nature Immunol. *4,* 835-842.

Seliger, B., Maeurer, M.J., and Ferrone, S. (1997). TAP off-tumors on. Immunol. Today *18,* 292-299.

Sharma, S., Stolina, M., Yang, S.C., Baratelli, F., Lin, J.F., Atianzar, K., Luo, J., Zhu, L., Lin, Y., Huang, M., Dohadwala, M., Batra, R.K., and Dubinett, S.M. (2003). Tumor cyclooxygenase-2-dependent suppression of dendritic cell function. Clin. Cancer Res. *9,* 961-968.

Shikowitz, M.J., Abramson, A.L., Freeman, K., Steinberg, B.M., and Nouri, M. (1998). Efficacy of DHE photodynamic therapy for respiratory papillomatosis: immediate and long term results. Laryngoscope *108,* 962-967.

Slebos, R.J., Kessis, T.D., Chen, A.W., Han, S.M., Hedrick, L., and Cho K.R. (1995). Functional consequences of directed mutations in human papillomavirus E6 proteins: abrogation of p53-mediated cell cycle arrest correlates with p53 binding and degradation *in vitro*. Virology *208,* 111-120.

Snoeck, R., Wellens, W., Desloovere, C., Van Ranst, M., Naesens, L., DeClercq E., and Feenstra, L. (1998). Treatment of severe laryngeal papillomatosis with intralesional injections of cidofovir [(S)-1-(3-hydroxy-2-phosphonylmethoxypropyl)cytosine]. J. Med. Virol. *54,* 219-225.

Snowden, R.T., Thompson, J., Horwitz, E., and Stocks, R.M. (2001). The predictive value of serum interleukins in recurrent respiratory papillomatosis: a preliminary study. Laryngoscope *111,* 404-408.

Spies, T., Cerundolo, V., Colonna, M., Cresswell, P., Townsend, A., and DeMars, R. (1992). Presentation of viral antigen by MHC class I molecules is dependent on a putative peptide transporter heterodimer. Nature *355,* 644-646.

Steele, J.C., Roberts, S., Rookes, S.M., and Gallimore, P.H.(2002). Detection of CD4+ and CD8+-T-cell responses to Human Papillomavirus type 1 antigens expressed at various stages of the virus life cycle by using an enzyme-linked immunospot assay of gamma interferon release. J. Virol. *76,* 6027-6036.

Steinberg, B.M., Gallagher, T., Stoler, M., and Abramson, A.L. (1988). Persistence and expression of human papillomavirus during interferon therapy. Arch. Otolaryngol. Head Neck Surg. *114,* 27-32.

Steinberg, B.M., Meade, R. Kalinowski, S., and Abramson A.L. (1990). Abnormal differentiation of human papillomavirus-induced laryngeal papillomas. Arch. Otolaryngol. Head Neck Surg. *116,*1167-1171.

Stephens, C.B., Arnold, G.E., Butchko, G.M., and Hardy, C.L. (1979). Autogenous vaccine treatment of juvenile laryngeal papillomatosis. Laryngoscope *89,* 1689-1696.

Strauss, M., Conner, G.H., Harvey, H.A., and White, D.S. (1982). Aerodigestive tract papillomatosis: bacilli Calmette-Guerin (BCG) immunotherapy. Laryngoscope *92,* 971-975.

Subbaramaiah K., and Dannenberg A.J. (2003). Cyclooxygenase 2: a molecular target for cancer prevention and treatment. Trends in Pharmacol. Sci. *24,*96-102.

Sun, S., and Steinberb, B.M. (2002). PTEN is a negative regulator of STAT3 activation in human papillomavirus-infected cells. J. Gen. Virol. *83*, 1651-1658.

Suzue K., Zhou X., Eisen H.N. and Young, R.A. (1997). Heat shock fusion proteins as vehicles for antigen delivery into the major histocompatibility complex class I presentation pathway. P.N.A.S. *94*, 13146-13151.

Suzuki, T., Tomita, Y., Nakano, K., Shirasawa, H., and Simizu, B. (1995). Deletion in the L1 open reading frame of human papillomavirus 6a genomes associated with recurrent laryngeal papilloma. J. Med. Virol. *43*, 191-197.

Swygart C. (1997). Human papillomaviruses: disease and laboratory diagnosis. Br. J. Biomed Sci. *54*, 299-303.

Syrjanen, S., and Puranen, M. (2000). Human papillomavirus infections in children: the potential role of maternal transmission. Crit. Rev. Oral. Biol. Med. *11*, 259-274.

Thomas, J.T., Oh, S.T., Terhune, S.S., and Laimins, L.A.. (2001). Cellular changes induced by low-risk human papillomavirus type 11 in keratinocytes that stably maintain viral episomes. J. Virol. *75*, 7564-7571.

Tindle R.W., Herd K., Doan T., Bryson, G., Leggatt, G.R., Lambert, P., Frazer, I.H., and Street, M. (2001). Non-specific down-regulation of CD8+ T-cell responses in mice expressing Human Papillomavirus type 16 E7 oncoprotein from the keratin 14 promoter. J.Virology. *75*, 5985-5997.

Vambutas, A., DiLorenzo, T.P., and Steinberg, B.M. (1993). Laryngeal papilloma cells have high levels of epidermal growth factor receptor and respond to epidermal growth factor by a decrease in epithelial differentiation. Cancer Res. *53*, 910-914.

Vambutas, A., Bonagura, V.R., and Steinberg, B.M.. (2000). Altered expression of TAP-1 and major histocompatibility complex class I in laryngeal papillomatosis: correlation of TAP-1 with disease. Clin. Diag. Lab. Immunol. *7*,79-85.

Vambutas ,A., DeVoti, J., Pinn, W., Steinberg, B.M., and Bonagura, V.R. (2001). Interaction of Human Papillomavirus Type 11 E7 protein with TAP-1 results in a reduction of ATP dependent peptide transport. Clin. Immunol. *101*, 94-99.

Vambutas, A., Bonagura, V.R., Reed, E.F., Abramson, A.L., Mullooly, V., DeVoti, J., Gjertson, D.W., and Steinberg, B.M.. (2004). A TAP1 polymorphism is a potential determinant for disease severity in HPV-induced recurrent respiratory papillomatosis. J. Infect. Dis.*189*, 871-879.

Vancurova, I., Wu, R., Miskolci, V., and Sun, S. (2002). Increased p50/p50 NF-kB activation in human papillomavirus type 6- or type 11-induced laryngeal papilloma tissue. J. Virol. *76*, 1533-1536.

Viac, J., Chardonnet, Y., Thivolet, J., Leval, J., and Morgon, A. (1987). Epithelial expression of HLA-DR antigens in laryngeal papillomas with a dense inflammatory reaction. Pathol. Res. Pract. *182*, 157-60.

Viac, J., Soler, C., Chardonnet, Y., Euvrard, S., and Schmitt, D. (1993). Expression of immune associated surface antigens of keratinocyte human papillomavirus-derived lesions. Immunobiology *188*, 392-402.

Villa L.L. (1997). Human papillomaviruses and cervical cancer. Adv. Cancer Res. *71*, 321-341.

Weiss, M.D., and Kashima, H.K. (1983). Tracheal Involvement in laryngeal papillomatosis. Laryngoscope. *93*, 45-8.

Wherry, E.J., Blattman, J.N., Murali-Krishna, K. van der Most, R., and Ahmed, R. (2003). Viral persistence alters CD8 T-cell immunodominance and tissue distribution and results in distinct stages of functional impairment. J. Virol. *77*, 4911-4927.

Wiatrak, B.J. (2003). Overview of recurrent respiratory papillomatosis. Curr. Opin. Otolaryngol. Head Neck Surg. *11*, 433-441.

Wikstrom, A., van Doornum, G.J., Kirnbauer, R., Quint, W.G., and Dillner, J. (1995). Prospective study on the development of antibodies against human papillomavirus type 6 among patients with condyloma accuminata new asymptomatic infection. J. Med. Virol. *46*, 368-374.

Wu, R., Sun, S., and Steinberg, B.M. (2003). Requirement of STAT3 activation for differentiation of mucosal stratified squamous epithelium. Mol. Med. *9*, 77-84.

Yamano, T., Murata, S., Shimbara, N., Tanaka, N., Chiba, T., Tanaka, K., Yui and Udono, H. (2002). Two distinct pathways mediated by PA28 and hsp90 in major Histocompatibility complex class I antigen processing. J. Exp. Med. *196*, 185-196.

Zhang, P., Nouri, M., Brandsma, J.L., Iftner, T., and Steinberg, B.M. (1999). Induction of E6/E7 expression in cottontail rabbit papillomavirus latency following UV activation. Virology, *263*, 388-394.

Zhang, P., and Steinberg, B.M. (2000). Overexpression of PTEN/MMAC1 and decreased activation of Akt in human papillomavirus-infected laryngeal papillomas. Cancer Res. *60*, 1457-1462.

Zhou, S., Ou, R., Huang, L., Price, G.E., and Moskophidis, D. (2004). Differential tissue- specific regulation of antiviral CD8+ T cell immune responses during chronic viral infection. J. Virol. *78*, 3578 3600.

Zwaveling, S., Ferreira Mota, S.C., Nouta, J., Johnson, M., Lipford, G.B., Offringa, R., van der Burg, S.H., and Melief, C.J.M.. (2002). Established human papillomavirus type 16-expressing tumors are effectively eradicated following vaccination with long peptides. J. Immunol. *169*, 350-358.

From: Papillomavirus Research: From Natural History to Vaccines and Beyond. Edited by: M. Saveria Campo

Chapter 17

The Cutaneous Human Papillomavirus Types and Non-melanoma Skin Cancer

Veronique Bouvard, Anne-Sophie Gabet, Rosita Accardi, Bakary S. Sylla and Massimo Tommasino*

Abstract

Non-melanoma skin cancer (NMSC) is the most common form of malignancy in Caucasian adult populations. UV light is a key environmental risk factor for NMSC. The fact that impairment of the immune system increases the risk for development of skin cancers strongly suggests that infectious agents are also involved in their aetiology. The epitheliotropic human papillomaviruses (HPVs) are likely candidates, especially those classified in the genus beta of the HPV phylogenetic tree, also known as *Epidermodysplasia verruciformis* (EV) HPV types. Independent investigations have shown that the DNA of several EV HPV types is highly prevalent in NMSC of both immunocompromised and immunocompetent individuals. Despite this association, however, a direct role of the EV HPV types in the development of NMSC remains to be proven. Due to the lack of a universal protocol for detection of EV HPV types, many studies using various methods have found different spectra of HPV types in skin lesions. Thus, it is still not known whether, within the genus beta, certain HPV types are more prevalent than others in malignant skin lesions. In addition, EV HPVs appear to be highly ubiquitous, being commonly detected in healthy skin of both immunocompromised and immunocompetent individuals.

Functional studies on mucosal high-risk HPV types have clearly demonstrated that the products of two early genes, E6 and E7, play a key role in the transformation of infected cells by disrupting the regulation of cell cycle and apoptosis. Very little is known about the biological properties of E6 and E7 of the majority of EV HPV types. Partial characterization of a limited number of skin HPV types (EV and non-EV) has revealed that also their E6 and E7 proteins have the ability to interfere with the regulation of apoptosis and cell cycle. These findings need confirmation in studies in which additional HPV types are included to establish the biological significance of these events.

In conclusion, some cutaneous HPV types may be involved in skin cancer development. However, the initial functional data indicate that the molecular mechanisms leading to cellular transformation are different from those of mucosal HPV types. EV viruses may act only at early stages of carcinogenesis, by potentiating the deleterious effects of other carcinogens such as UV radiation. Further functional and epidemiological studies are required to evaluate this model.

1. Non-melanoma Skin Cancer

Non-melanoma skin cancer (NMSC) is the most common cancer among adult fair-skinned populations (Pisani et al., 2002). Risk factors for NMSC development include UV exposure, fair skin, fair eyes, advancing age and impairment of the immune system (Pfister, 2003). The majority of NMSC occurs mainly on sun-exposed body sites. This points to UV radiation, and especially UV-B, as being a major environmental factor in the pathogenesis of these tumours (Ananthaswamy et al., 1997; Armstrong and Kricker, 2001; Preston and Stern, 1992). Exposure of the skin to UV radiation results in specific DNA lesions such as cyclopyrimidine dimers, which if not properly repaired can lead to the formation of deleterious mutations (Sinha and Hader, 2002).

The incidence of NMSC is rising continuously across the globe due to the ageing of the population and the migration of people with sun-sensitive skin to subtropical regions. Growing prosperity has also had an impact on this disease, by permitting wider travel of fair-skinned individuals to countries with high sun exposure. In the United States, approximately one million new NMSC cases are reported every year and increased NMSC incidence has been observed in other regions such as Europe and Australia (Greenlee et al., 2000; Kiviat, 1999; Marks, 1995).

NMSC comprises two main histological types: basal-cell carcinoma (BCC) and squamous-cell carcinoma (SCC). BCC is about four to five times more frequent than SCC. BCC is a low-grade malignant epithelial

*For correspondence. Email tommasino@iarc.fr

neoplasm with several histological variants that rarely metastasise. In contrast, SCC is a malignant neoplasm of the epidermis showing evidence of squamous cell differentiation and can be highly invasive. Both tumours have a good prognosis and are not normally associated with mortality. However, increased incidence of other invasive cancers and cancer mortality in individuals with a history of NMSC has been reported (Frisch and Melbye, 1995; Kahn et al., 1998; Levi et al., 1998; Levi et al., 1997). In addition, the economic burden is increasing, as progressively more patients require treatment and costly operations.

2. The *Epidermodysplasia verruciformis* HPV Types

It is well accepted that impairment of the immune system favours the development of NMSC. Immunosuppressed organ transplant recipients have a 50–100-fold increased risk of developing NMSC compared to the general population. NMSCs occur 10–20 years earlier in immunosuppressed individuals than in immunocompetent ones (Boyle et al., 1984; Kiviat, 1999; Walder et al., 1976; 1971). The cumulative incidence of skin cancer in patients under immunosuppressive treatment for 10–25 years is approximately 30–40%. The link with immune status strongly supports the role in NMSC of the cutaneous human papillomaviruses (HPVs), in particular those belonging to genus beta of the HPV phylogenetic tree (de Villiers et al., 2004; Chapter 2). These HPV types were first isolated in cancer-prone patients suffering from a rare autosomal recessive genetic disorder called *Epidermodysplasia verruciformis* (EV), characterized by a defect in cell-mediated immunity leading to abnormal susceptibility to widespread and persistent HPV infection of the skin (Orth et al., 1978; 1979). Twenty-five EV HPV types have been fully characterized so far, namely 5, 8, 9, 12, 14, 15, 17, 19–25, 36–38, 47, 49, 75, 76, 80, 92, 93, 96, but it is certain that many more EV types exist and new, previously uncharacterized EV HPV types continue to be reported. Due to their abnormal susceptibility to HPV infection, EV patients develop disseminated pityriasis versicolor-like lesions and flat warts (Jablonska et al., 1972; Majewski et al., 1997; Orth et al., 1979). These skin lesions differ from normal warts and appear as red, brown or achromic plaques. They arise early in life and represent a risk for SCC development. Indeed, in approximately 30–60% of cases, HPV lesions develop into multifocal SCC at sun-exposed regions. HPV types 5 or 8 are the most frequently detected in skin lesions of EV patients, but types 14, 17, 20 and 47 are also occasionally found. EV was recently linked to mutations in either of two related genes EVER1 and EVER2 (also termed TMC6 and TMC8, respectively) (Ramoz et al., 2002). The function of the proteins encoded by these genes is still unknown, but analysis of their primary structure has revealed that they are likely to be integral membrane proteins (Ramoz et al., 2002). It is likely that they are involved in controlling the immune response to viral infections and their lack may favour HPV persistence. The hypothesis that impairment of the immune system favours the development of EV HPV-associated diseases is also supported by other recent findings. After haematopoietic stem-cell transplantation, individuals with a rare primary deficiency of T-cell- and B-cell-mediated immunity are prone to chronic HPV infection and to develop skin lesions similar to those found in EV patients (Laffort et al., 2004).

3. Prevalence of HPV Types in NMSC of non-EV Individuals

As mentioned above, immunosuppressed patients, such as organ transplant recipients, have increased risk for NMSC development. The use of very sensitive detection methods, such as nested polymerase chain reactions (PCRs), has revealed the presence of DNA from several EV HPV types in 70–80% of NMSC in these patients (Arends et al., 1997; Berkhout et al., 2000; de Villiers et al., 1997; Ferrandiz et al., 1998; Harwood et al., 2000; Hopfl et al., 1997; Meyer et al., 2003; Shamanin et al., 1996; Stark et al., 1994; Stockfleth et al., 2004; Tieben et al., 1994). Often, more than one EV HPV type was found in the same skin lesion. Although HPV5 and HPV8 are the major types associated with SCC in EV patients, they appear to be less frequent in skin lesions of immunosuppressed individuals, who are infected with a broad spectrum of different types of the HPV genus beta. Several studies have convincingly proved that the DNA of EV HPV types is also present in NMSCs of immunocompetent individuals, although at a lower rate than in immunosuppressed patients (up to approximately 50%) (Arends et al., 1997; Astori et al., 1998; Biliris et al., 2000; Boxman et al., 2000; Caldeira et al., 2003; Ferrandiz et al., 1998; Forslund et al., 1999; Harwood et al., 2000; Iftner et al., 2003; Meyer et al., 2003; Pfister and Ter Schegget, 1997; Shamanin et al., 1996; Stockfleth et al., 2004). Detection of common cutaneous types (*e.g.*, 2, 3, 4, 7 and 77) and low- and high-risk mucosal types in NMSC of immunosuppressed or immunocompetent patients has also been reported (Biliris et al., 2000; Iftner et al., 2003; Shamanin et al., 1996), but their involvement in skin carcinogenesis is less likely.

EV HPV DNA can be found in healthy skin of immunocompromised or immunocompetent individuals (Antonsson et al., 2000; Astori et al., 1998; Berkhout et al., 2000; de Jong-Tieben et al., 2000; Forslund et al., 1999; Harwood et al., 2004; Stark et al., 1994). A wide variety of skin HPV types was found in skin swabs from five different sites of renal transplant recipients, dialysis patients and healthy individuals (Antonsson et al., 2000). No difference in prevalence or in spectrum of HPV types was observed in healthy skin from sun-exposed and non-exposed sites (Harwood et al., 2004). These findings show that the EV HPV types are highly ubiquitous in healthy skin. However, comparative studies of EV HPV DNA prevalence in healthy skin of individuals with or without NMSC have shown that history of NMSC is significantly associated with EV HPV types, but not with other cutaneous HPV types that are normally responsible for benign skin lesions (Harwood et al., 2004). Boxman and colleagues have shown that EV HPV DNA can be detected in plucked eyebrow hair, suggesting that hair follicles are a natural reservoir of the virus. In agreement with the study in healthy skin, prevalence of EV HPV DNA in plucked eyebrow hairs is higher in patients with solar keratoses (Boxman et al., 1997; 1999; 2000; 2001) or SCC (Struijk et al., 2003) than in healthy individuals. The link between EV HPV types and skin cancer is further supported by the few epidemiological studies based on serological tests. Antibodies against the major capsid protein L1 of EV HPV8 are more prevalent in immunocompromised patients, who have high risk of skin cancer, than in the general population (Stark et al., 1998). In addition, seroreactivity to L1 virus-like particles of EV-HPV types 8 and/or 38 is significantly increased in immunocompetent individuals with SCC (Feltkamp et al., 2003; Masini et al., 2003; Stark et al., 1998).

Although many findings support the association of EV HPV types with NMSC, their direct role in the aetiology of this disease remains to be proven. It is not yet clear whether specific EV HPV types are more prevalent than others in skin cancers. Due to the high heterogeneity of EV HPV types and the lack of a universal protocol for their detection, independent investigations have used different HPV typing methods and obtained inconsistent results. Thus, no high-risk HPV types have been identified so far within the EV group. In addition, all the studies have determined only the presence of viral DNA in skin specimens, which is not necessarily a sign of an active viral infection. Detection of viral mRNA would indicate active infection, but no such data are available for skin lesions. Forslund et al. reported that the prevalence of HPV DNA is much higher in swab skin samples than in biopsies taken after removal of the superficial layers by tape stripping (Forslund et al., 2004). This suggests that the majority of viral DNA is present on the external skin surface, with only a small number of keratinocytes infected by the virus. The same authors showed the presence of viral particles on the skin surface, providing a sign of infection with completion of the viral life cycle (Forslund et al., 2004). On the basis of these findings, it can be hypothesised that EV HPV infections induce inconspicuous or clinically unapparent lesions.

The use of quantitative, type-specific PCR has revealed that far less than one HPV DNA copy per cell is present in NMSC biopsies (Pfister, 2003). Therefore, the presence of the viral genome may not be required in the later stages of carcinogenesis. This may imply a need to consider a new scenario for the role of HPV in NMSC pathogenesis, such as a "hit-and-run" mechanism. In this case, HPV viral DNA is required at the initiation phase, but not for the maintenance of the transforming state. This mode of action is clearly different from that established for mucosal high-risk HPV in cervical cancer, where the viral genome is constantly required for maintenance of the neoplastic phenotype of the infected cell (zur Hausen, 2002). The hypothesis that the EV HPV types are no longer required after development of cancer is consistent with the fact that additional carcinogens are involved in skin carcinogenesis. It is well accepted that sun exposure, one of the main risk factors for skin cancer (Ananthaswamy et al., 1997; Armstrong and Kricker, 2001; Preston and Stern, 1992), induces irreversible DNA damage, possibly leading to inactivation or activation of cellular tumour suppressor or oncoproteins, respectively. In such a scenario, the virus could become dispensable for the maintenance of the cancer.

Recent data indicate that EV HPV types may also be involved in the development of other epithelial cancers. For example, EV HPV types are highly prevalent in squamous-cell carcinoma of the conjunctiva (Ateenyi-Agaba et al., 2004). Interestingly, as shown for NMSC, impairment of the immune system and exposure to sunlight are risk factors for conjunctival cancer (Newton, 1996), underlining the fact that similar events are involved in the development of these two types of cancer.

4. Psoriasis and EV HPV Types

A few recent studies have shown that EV HPV types are highly prevalent in the skin of patients with psoriasis, a T-cell-mediated immunological disorder characterized by epidermal proliferation (Favre et al., 1998; Mahe et al., 2003; Weissenborn et al., 1999). The use of PCR-based methods has revealed the presence of HPV DNA in up to 90% of skin samples from these patients (reviewed in Pfister, 2003). The most commonly detected types are HPV5 and HPV36, but other members of the genus beta have also been found, *e.g.*, HPV38. DNA of genital HPV types was detected only sporadically in these skin specimens. In addition, the cutaneous HPV1, which is exclusively associated with benign skin lesions, was clearly less prevalent than the EV HPV types in psoriatic skin (Favre et al., 1998).

Serological data have confirmed the association between HPV5 and psoriasis. The prevalence of antibodies against the major capsid protein L1 of HPV5 was higher in patients with psoriasis than in individuals with atopic dermatitis or renal transplant recipients (Favre et al., 1998). However, the role of EV HPV types in the pathology of psoriasis is still uncertain. In fact, it is not yet clear whether the HPV is directly involved in the disease or is merely taking advantage of the keratinocyte hyperproliferation that occurs in psoriatic skin to efficiently complete its life cycle. The fact that the titre of HPV5 L1 antibodies is also elevated in non-psoriatic individuals with epidermal repair processes, *e.g.* patients with extensive burns or autoimmune bullous diseases (Favre et al., 2000), may be interpreted as supporting the second hypothesis.

Psoriasis is normally treated by photochemotherapy, which consists of administration of the photosensitising compound psolaren followed by UVA radiation (PUVA). This treatment represents a risk factor for the development of skin cancer in psoriasis patients (Bruynzeel et al., 1991; Chuang et al., 1992; Forman et al., 1989; Lindelof et al., 1991; Stern and Lange, 1988; Stern et al., 1979), but the precise mechanisms involved remain to be elucidated. It has been proposed that PUVA treatment may favour EV HPV infection, either positively influencing HPV activities (*e.g.* viral transcription and/or replication) and/or inducing immune suppression with consequent persistence of the virus (Harwood et al., 1998; Wolf et al., 2004). A recent study has shown that the prevalence of EV HPV types in plucked hairs is higher in patients with psoriasis treated with PUVA than in patients without a history of PUVA exposure (Wolf et al., 2004). These findings provide further support for the involvement of EV HPV types in the increased risk of skin cancer development in PUVA-treated patients.

5. Biology of the Cutaneous HPV Types

Although convincing evidence exists on the role of mucosal HPV types in human cancer and the molecular mechanisms underlying the transformation process (zur Hausen, 2002), little is known of the contribution of cutaneous HPV types in oncogenesis. Data from epidemiological studies have clearly demonstrated that specific cutaneous HPV types (25 EV HPV types or more) are commonly and consistently found in NMSC, but the real oncogenicity of these viruses and the molecular mechanisms involved are only starting to be addressed.

The genomic structure of the EV HPV types is very similar to that of the mucosal HPV types, the only exception being in the region encoding the early genes that lacks the E5 open reading frame. A few studies have focused on the E6 and E7 proteins, while nothing is known about the other viral proteins. It is becoming evident that, like the E6 and E7 of the oncogenic mucosal HPV types, E6 and E7 of some cutaneous HPV types have the ability to interfere with regulation of cell cycle and apoptosis and may act as major transforming proteins (Caldeira et al., 2003; Jackson et al., 2000).

It is probable that these virus-mediated events act in cooperation with UV radiation, the major risk factor for NMSC development, in promoting cellular transformation (Storey, 2002). In a normal skin, several defence mechanisms are activated following UV-induced DNA damage, in order to avoid the proliferation of abnormal cells. Cell-cycle arrest can be immediately induced in order to allow a cell to repair its damaged DNA before subsequent replication. However, when the intensity of UV exposure results in too many DNA lesions, apoptotic cascades are induced that remove irremediably damaged cells from the proliferating pool of the epidermis (Sinha and Hader, 2002). Clearly, impairments in these defence mechanisms could lead to carcinogenesis.

Recent data have shown that repairing of the UV-induced thymine dimers is compromised in cells expressing E6 protein from HPV5 and HPV18 (Giampieri and Storey, 2004). E6 from mucosal and cutaneous HPV types (EV and non-EV types) can interfere with the apoptotic pathways via association with the pro-apoptotic protein Bak, a member of the Bcl-2 family (Jackson et al., 2000). Early apoptotic signals lead Bak to form pores in the mitochondrial membrane, resulting in release of cytochrome C from the mitochondria, followed by induction of caspase apoptotic cascades (Sorenson, 2004). In normal skin, the Bak protein is highly stabilized and activated throughout the entire epidermis in response to UV irradiation. However, in the presence of HPV E6 protein, neither stabilisation of the Bak protein nor detection of the apoptotic process occurs in regenerated human epidermis following UV-B irradiation (Jackson et al., 2000). It has been demonstrated that E6 stimulates the ubiquitin-mediated degradation of Bak. This interaction between E6 and Bak, first demonstrated for the mucosal HPV18 E6 protein (Thomas and Banks, 1998), has also been shown for all E6 proteins studied (E6 of low-risk mucosal HPV11 and 6, EV types HPV5 and 8, and the cutaneous types HPV10 and 77 (Jackson and Storey, 2000; Thomas and Banks, 1999). Thus this ability of E6 proteins to bypass apoptosis via degradation of Bak appears to be a property shared between HPV types from different sub-groups. The fact that E6 from HPV10, a common cutaneous HPV type exclusively associated with benign skin lesions, is able to target Bak indicates that such activity is required for completion of the viral life cycle and may not be sufficient to confer *in vivo* oncogenic properties to the virus. For the skin HPV types, the prevention of UV-induced apoptosis might be a physiological adaptation by which they prevent death of the infected cell thus guaranteeing completion of the viral life cycle. The inhibition of Bak-induced apoptosis in the epidermis by the E6 proteins would be a means by which cutaneous HPV types could play a role in skin cancer development together with UV-induced DNA lesions, but probably this would not be a key event and some other viral oncoprotein functions would doubtless be required.

It has been clearly shown that the E6 proteins from the high-risk mucosal HPV types, such as HPV16 and 18, are able to target p53 for ubiquitin-mediated proteosomal degradation (zur Hausen, 2002). Since p53 is the major tumour-suppressor protein involved in the surveillance of genome integrity, cells expressing HPV16 E6 show accumulation of DNA damage that greatly increases the probability of HPV-infected cells evolving towards malignancy (Kessis et al., 1993). In contrast, none of the cutaneous HPV E6 proteins studied to date promote p53 degradation (Caldeira et al., 2004; Elbel et al., 1997; Jackson and Storey, 2000). However, several lines of evidence indicate that E6 proteins from EV and non-EV skin HPV types do interfere with p53 functions. For instance, those from HPV5 (EV) and HPV77 (non-EV) inhibit UV-induced apoptosis in conditions where p53 is strongly stabilized and activated (Jackson and Storey, 2000). HPV77 E6 protein attenuates the UV-induced p53-dependent transcription of the proapoptotic genes Fas, PUMAβ, APAF-1 and PIG3 (Giampieri et al., 2004). This inhibition of p53-dependent transactivation appears to be selective for the pro-apoptotic p53 target genes, since the transcription of p53-responsive genes involved in the cell cycle or regulation of p53 itself, such as $p21^{WAF1/CIP1}$ and Hdm2, is efficiently increased after UV-B irradiation. Although the mechanism for this inhibition is not known, this is a novel activity for an E6 protein whose expression effectively targets p53-dependent apoptosis without interfering with p53-mediated cell-cycle arrest (Giampieri et al., 2004).

It has also been shown that the upstream regulatory region of HPV77 that controls expression of the E6 and E7 genes contains a p53-responsive element and can be transcriptionally activated by p53 (Purdie et al., 1999). Therefore as a cutaneous type, HPV77 has developed a means to take advantage of the frequent UV activation of p53 in the skin to increase its own gene transcription, as well as a way to counteract the apoptotic function of p53. HPV77 is closely related to HPV29 and also has a high degree of DNA sequence homology with HPVs 3, 10 and 28, all restricted to benign skin lesions in the general population (de Villiers et al., 2004). HPV77 has, however, been identified in some tumours and warts from renal transplant recipients, but not in lesions from immunocompetent individuals or EV patients. Currently, HPV77 is the only HPV type whose expression has been found to be up-regulated by p53. It is also the only one found capable of discriminating between p53 target genes. It would be important to determine whether other cutaneous types share these properties, in particular the EV HPV types.

In contrast to the E7 gene of high-risk mucosal HPV16, E7 from HPV5 lacks any transforming activity in established rodent cells (Iftner et al., 1988; Yamashita et al., 1993). However, when co-expressed with an activated ras oncogene, E7 from HPV5 and HPV8 can induce transformation of primary rat embryo fibroblasts. In this assay E7 from HPV8 had lower activity than E7 from HPV5 or HPV16. E7 from HPV 5 and HPV8 associates with the tumour suppressor pRb, but with lower affinity than HPV16 E7 (Iftner et al., 1988). Thus, these *in vitro* transforming activities of HPV5 and HPV8 E7s may be explained in part by their activity in promoting the cell cycle G1/S transition via inactivation of pRb.

HPV16 E6 and E7 are able to immortalize primary human keratinocytes, which represents the first step towards neoplastic epithelial transformation (zur Hausen, 2002). No studies have been published on the ability of the EV HPV E6 and E7 proteins to induce immortalisation of keratinocytes, except for those from HPV38. HPV 38 E6 and E7 proteins are sufficient to corrupt the cell cycle and senescence programs in primary keratinocytes, leading to their immortalisation (Caldeira et al., 2003). A comparative analysis of HPV16 and HPV38 showed that both E7 proteins inactivate the tumour suppressor pRb by the same mechanism, *i.e.* binding and induction of degradation. As a result, both proteins alter cell-cycle control at the G1/S transition with similar efficiency (Caldeira et al., 2003). However, the events involved in HPV16- and HPV38-induced immortalisation appear to differ somewhat. Thus, HPV38-mediated immortalisation is always preceded by a crisis with no apparent proliferation, while no crisis is observed in the case of HPV16-infected keratinocytes (Caldeira et al., 2003). These events may be explained by differences in the biological properties of E6 and E7 from HPV38 and HPV16.

6. Conclusions

Although research on the cutaneous HPV types is progressing fast, many questions remain to be answered regarding their potential role in human carcinogenesis. The high heterogeneity of EV HPV types has hampered the development of a universal protocol for their detection. Independent investigations have used different HPV typing methods and obtained inconsistent results. Thus, on the basis of available data on the prevalence of the different EV HPV types in NMSC, it is not possible to conclude whether high-risk types exist within the EV HPV group. New techniques for detection of viral DNA as well as antibodies are needed to allow large-scale epidemiological studies of possible associations between specific EV HPV types and NMSC.

Initial data on the quantification of EV HPV DNA in tumour tissues indicate that the presence of viral genome may not be required in the later stages of carcinogenesis, underlining a difference in the molecular mechanisms mediated by cutaneous and mucosal HPV types. This difference between the two groups of HPV types may be due to subtle effects. Since cutaneous and mucosal HPV types replicate in cells at distinct anatomical sites exposed to different environmental stresses, it is not surprising that they have evolved with divergent biological properties. All HPV types rely upon the DNA replication machinery of the host cell; they have therefore developed several mechanisms to promote cellular proliferation and efficiently replicate their own DNA. Exposure of skin keratinocytes to UV radiation leads to accumulation of DNA damage that in turn induces cell cycle arrest or apoptosis to allow repair or elimination of the damaged

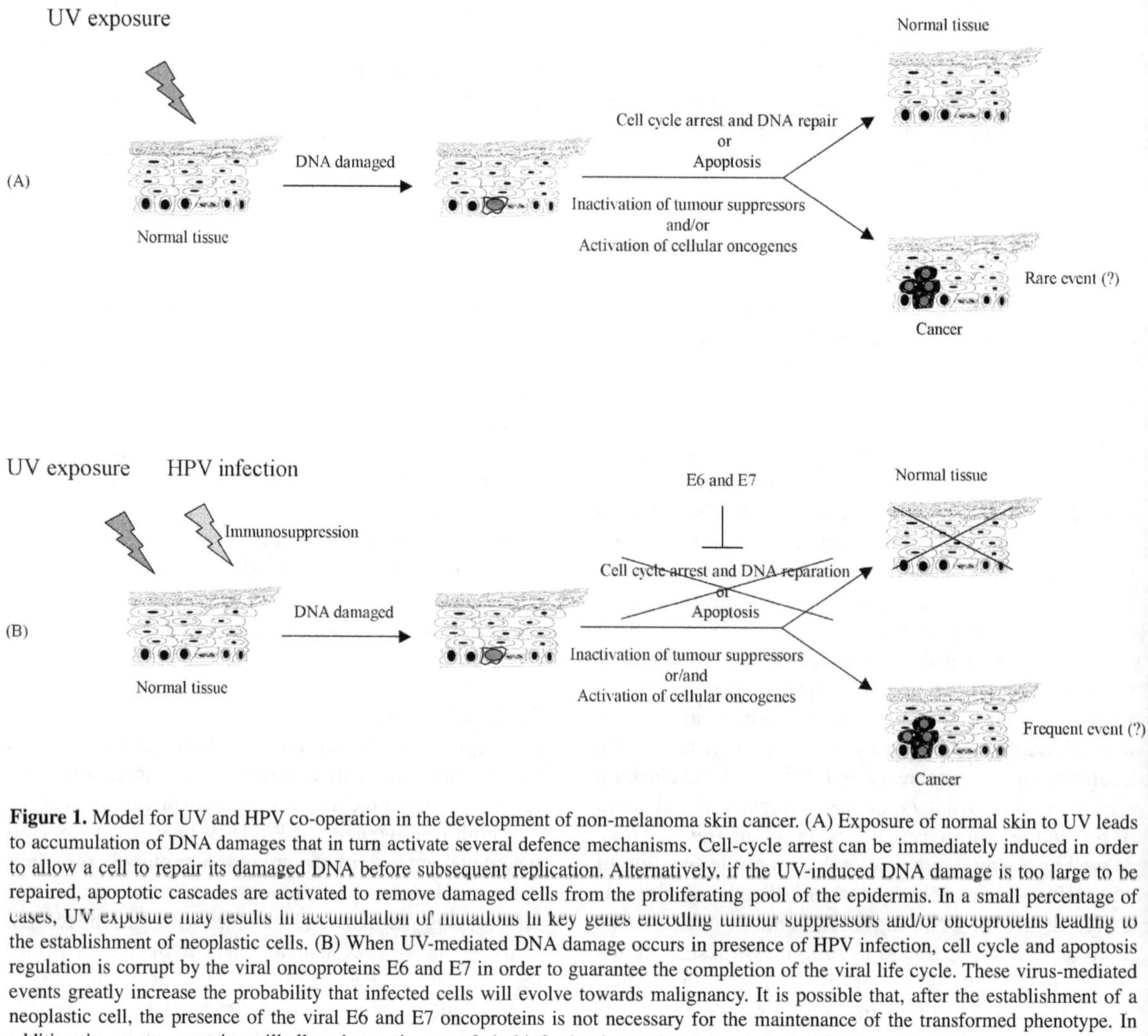

Figure 1. Model for UV and HPV co-operation in the development of non-melanoma skin cancer. (A) Exposure of normal skin to UV leads to accumulation of DNA damages that in turn activate several defence mechanisms. Cell-cycle arrest can be immediately induced in order to allow a cell to repair its damaged DNA before subsequent replication. Alternatively, if the UV-induced DNA damage is too large to be repaired, apoptotic cascades are activated to remove damaged cells from the proliferating pool of the epidermis. In a small percentage of cases, UV exposure may results in accumulation of mutations in key genes encoding tumour suppressors and/or oncoproteins leading to the establishment of neoplastic cells. (B) When UV-mediated DNA damage occurs in presence of HPV infection, cell cycle and apoptosis regulation is corrupt by the viral oncoproteins E6 and E7 in order to guarantee the completion of the viral life cycle. These virus-mediated events greatly increase the probability that infected cells will evolve towards malignancy. It is possible that, after the establishment of a neoplastic cell, the presence of the viral E6 and E7 oncoproteins is not necessary for the maintenance of the transformed phenotype. In addition, immunosuppression will allow the persistence of viral infection increasing the probability of tumour formation.

cell, respectively (Figure 1A). The cutaneous HPV types, in order to guarantee the replication of their own DNA, must be able to circumvent this adverse effect of UV radiation on keratinocyte proliferation. Thus, the development of skin cancer would result from the combined effects of the virus and of UV radiation on cellular proliferation and loss of host genome integrity. In this model, HPV activity may be required only in the initial stage of carcinogenesis, and no longer be necessary after establishment of the neoplastic phenotype (Figure 1B).

Studies on the mucosal HPV types have shown that analysis of the transforming activities of E6 and E7 is a valid tool to predict the potential oncogenicity of specific HPV types. However, E6 and E7 of only a few cutaneous HPV types have been even partially characterized to date. Further functional studies are required and must be extended to more skin HPV types (EV and non-EV). The biological findings will need to be tested in well-designed large-scale epidemiological studies, since only a close collaboration between biologists and epidemiologists will unequivocally confirm the link between the EV HPV types and NMSC, opening new prevention opportunities for this increasingly common type of cancer.

References

Ananthaswamy, H. N., Loughlin, S. M., Cox, P., Evans, R. L., Ullrich, S. E., and Kripke, M. L. (1997). Sunlight and skin cancer: inhibition of p53 mutations in UV-irradiated mouse skin by sunscreens. Nat. Med. *3*, 510-514.

Antonsson, A., Forslund, O., Ekberg, H., Sterner, G., and Hansson, B. G. (2000). The ubiquity and impressive genomic diversity of human skin papillomaviruses suggest a commensalic nature of these viruses. J. Virol. *74*, 11636-11641.

Arends, M. J., Benton, E. C., McLaren, K. M., Stark, L. A., Hunter, J. A., and Bird, C. C. (1997). Renal allograft recipients with high susceptibility to cutaneous malignancy have an increased prevalence of human papillomavirus DNA in skin tumours and a greater risk of anogenital malignancy. Br. J. Cancer *75*, 722-728.

Armstrong, B. K., and Kricker, A. (2001). The epidemiology of UV induced skin cancer. J. Photochem. Photobiol B *63*, 8-18.

Astori, G., Lavergne, D., Benton, C., Hockmayr, B., Egawa, K., Garbe, C., and de Villiers, E. M. (1998). Human papillomaviruses are commonly found in normal skin of immunocompetent hosts. J. Invest. Dermatol. *110*, 752-755.

Ateenyi-Agaba, C., Weiderpass, E., Smet, A., Dong, W., Dai, M., Kahwa, B., Wabinga, H., Katongole-Mbidde, E., Franceschi, S., and Tommasino, M. (2004). Epidermodysplasia verruciformis human papillomavirus types and carcinoma of the conjunctiva: a pilot study. Br. J. Cancer *90*, 1777-1779.

Berkhout, R. J., Bouwes Bavinck, J. N., and ter Schegget, J. (2000). Persistence of human papillomavirus DNA in benign and (pre)malignant skin lesions from renal transplant recipients. J. Clin. Microbiol. *38*, 2087-2096.

Biliris, K. A., Koumantakis, E., Dokianakis, D. N., Sourvinos, G., and Spandidos, D. A. (2000). Human papillomavirus infection of non-melanoma skin cancers in immunocompetent hosts. Cancer Lett. *161*, 83-88.

Boxman, I. L., Berkhout, R. J., Mulder, L. H., Wolkers, M. C., Bouwes Bavinck, J. N., Vermeer, B. J., and ter Schegget, J. (1997). Detection of human papillomavirus DNA in plucked hairs from renal transplant recipients and healthy volunteers. J. Invest. Dermatol. *108*, 712-715.

Boxman, I. L., Mulder, L. H., Russell, A., Bouwes Bavinck, J. N., Green, A., and Ter Schegget, J. (1999). Human papillomavirus type 5 is commonly present in immunosuppressed and immunocompetent individuals. Br. J. Dermatol. *141*, 246-249.

Boxman, I. L., Russell, A., Mulder, L. H., Bavinck, J. N., Schegget, J. T., and Green, A. (2000). Case-control study in a subtropical Australian population to assess the relation between non-melanoma skin cancer and epidermodysplasia verruciformis human papillomavirus DNA in plucked eyebrow hairs. The Nambour Skin Cancer Prevention Study Group. Int. J. Cancer *86*, 118-121.

Boxman, I. L., Russell, A., Mulder, L. H., Bavinck, J. N., ter Schegget, J., and Green, A. (2001). Association between epidermodysplasia verruciformis-associated human papillomavirus DNA in plucked eyebrow hair and solar keratoses. J. Invest. Dermatol. *117*, 1108-1112.

Boyle, J., MacKie, R. M., Briggs, J. D., Junor, B. J., and Aitchison, T. C. (1984). Cancer, warts, and sunshine in renal transplant patients. A case-control study. Lancet *1*, 702-705.

Bruynzeel, I., Bergman, W., Hartevelt, H. M., Kenter, C. C., Van de Velde, E. A., Schothorst, A. A., and Suurmond, D. (1991). 'High single-dose' European PUVA regimen also causes an excess of non-melanoma skin cancer. Br. J. Dermatol. *124*, 49-55.

Caldeira, S., Filotico, R., Accardi, R., Zehbe, I., Franceschi, S., and Tommasino, M. (2004). p53 mutations are common in human papillomavirus type 38-positive non-melanoma skin cancers. Cancer Lett. *209*, 119-124.

Caldeira, S., Zehbe, I., Accardi, R., Malanchi, I., Dong, W., Giarre, M., deVilliers, E.-M., Filotico, R., Boukamp, P., and Tommasino, M. (2003). The E6 and E7 proteins of cutaneous human papillomavirus type 38 display transforming properties. J. Virol. *77*, 2195-2206.

Chuang, T. Y., Heinrich, L. A., Schultz, M. D., Reizner, G. T., Kumm, R. C., and Cripps, D. J. (1992). PUVA and skin cancer. A historical cohort study on 492 patients. J. Amer. Acad. Dermatol. *26*, 173-177.

de Jong-Tieben, L. M., Berkhout, R. J., ter Schegget, J., Vermeer, B. J., de Fijter, J. W., Bruijn, J. A., Westendorp, R. G., and Bouwes Bavinck, J. N. (2000). The prevalence of human papillomavirus DNA in benign keratotic skin lesions of renal transplant recipients with and without a history of skin cancer is equally high: a clinical study to assess risk factors for keratotic skin lesions and skin cancer. Transplantation *69*, 44-49.

de Villiers, E. M., Fauquet, C., Broker, T. R., Bernard, H. U., and zur Hausen, H. (2004). Classification of papillomaviruses. Virol. *324*, 17-27.

de Villiers, E. M., Lavergne, D., McLaren, K., and Benton, E. C. (1997). Prevailing papillomavirus types in non-melanoma carcinomas of the skin in renal allograft recipients. Int. J. Cancer *73*, 356-361.

Elbel, M., Carl, S., Spaderna, S., and Iftner, T. (1997). A comparative analysis of the interactions of the E6 proteins from cutaneous and genital papillomaviruses with p53 and E6AP in correlation to their transforming potential. Virol. *239*, 132-149.

Favre, M., Majewski, S., Noszczyk, B., Maienfisch, F., Pura, A., Orth, G., and Jablonska, S. (2000). Antibodies to human papillomavirus type 5 are generated in epidermal repair processes. J. Invest. Dermatol. *114*, 403-407.

Favre, M., Orth, G., Majewski, S., Baloul, S., Pura, A., and Jablonska, S. (1998). Psoriasis: A possible reservoir for human papillomavirus type 5, the virus associated with skin carcinomas of epidermodysplasia verruciformis. J. Invest. Dermatol. *110*, 311-317.

Feltkamp, M. C., Broer, R., di Summa, F. M., Struijk, L., van der Meijden, E., Verlaan, B. P., Westendorp, R. G., ter Schegget, J., Spaan, W. J., and Bouwes Bavinck, J. N. (2003). Seroreactivity to epidermodysplasia verruciformis-related human papillomavirus types is associated with nonmelanoma skin cancer. Cancer Res. *63*, 2695-2700.

Ferrandiz, C., Fuente, M. J., Ariza, A., Ribera, M., and Paradelo, C. (1998). Detection and typing of human papillomavirus in skin lesions from renal transplant recipients and equivalent lesions from immunocompetent patients. Arch. Dermatol. *134*, 381-382.

Forman, A. B., Roenigk, H. H., Jr., Caro, W. A., and Magid, M. L. (1989). Long-term follow-up of skin cancer in the PUVA-48 cooperative study. Arch. Dermatol. *125*, 515-519.

Forslund, O., Antonsson, A., Nordin, P., Stenquist, B., and Hansson, B. G. (1999). A broad range of human papillomavirus types detected with a general PCR method suitable for analysis of cutaneous tumours and normal skin. J. Gen. Virol. *80*, 2437-2443.

Forslund, O., Lindelof, B., Hradil, E., Nordin, P., Stenquist, B., Kirnbauer, R., Slupetzky, K., and Dillner, J. (2004). High prevalence of cutaneous human papillomavirus DNA on the top of skin tumors but not in "Stripped" biopsies from the same tumors. J. Invest. Dermatol. *123*, 388-394.

Frisch, M., and Melbye, M. (1995). New primary cancers after squamous cell skin cancer. Amer. J. Epidemiol. *141*, 916-922.

Giampieri, S., Garcia-Escudero, R., Green, J., and Storey, A. (2004). Human papillomavirus type 77 E6 protein selectively inhibits p53-dependent transcription of proapoptotic genes following UV-B irradiation. Oncogene *23*, 5864-5870.

Giampieri, S., and Storey, A. (2004). Repair of UV-induced thymine dimers is compromised in cells expressing the E6 protein from human papillomaviruses types 5 and 18. Br. J. Cancer *90*, 2203-2209.

Greenlee, R. T., Murray, T., Bolden, S., and Wingo, P. A. (2000). Cancer statistics, 2000. CA Cancer J. Clin. *50*, 7-33.

Harwood, C. A., Spink, P. J., Surentheran, T., Leigh, I. M., Hawke, J. L., Proby, C. M., Breuer, J., and McGregor, J. M. (1998). Detection of human papillomavirus DNA in PUVA-associated non-melanoma skin cancers. J. Invest. Dermatol. *111*, 123-127.

Harwood, C. A., Surentheran, T., McGregor, J. M., Spink, P. J., Leigh, I. M., Breuer, J., and Proby, C. M. (2000). Human papillomavirus infection and non-melanoma skin cancer in immunosuppressed and immunocompetent individuals. J. Med. Virol. *61*, 289-297.

Harwood, C. A., Surentheran, T., Sasieni, P., Proby, C. M., Bordea, C., Leigh, I. M., Wojnarowska, F., Breuer, J., and McGregor, J. M. (2004). Increased risk of skin cancer associated with the presence of epidermodysplasia verruciformis human papillomavirus types in normal skin. Br. J. Dermatol. *150*, 949-957.

Hopfl, R., Bens, G., Wieland, U., Petter, A., Zelger, B., Fritsch, P., and Pfister, H. (1997). Human papillomavirus DNA in non-melanoma skin cancers of a renal transplant recipient: detection of a new sequence related to epidermodysplasia verruciformis associated types. J. Invest. Dermatol. *108*, 53-56.

Iftner, A., Klug, S. J., Garbe, C., Blum, A., Stancu, A., Wilczynski, S. P., and Iftner, T. (2003). The prevalence of human papillomavirus genotypes in nonmelanoma skin cancers of nonimmunosuppressed individuals identifies high-risk genital types as possible risk factors. Cancer Res. *63*, 7515-7519.

Iftner, T., Bierfelder, S., Csapo, Z., and Pfister, H. (1988). Involvement of human papillomavirus type 8 genes E6 and E7 in transformation and replication. J. Virol. *62*, 3655-3661.

Jablonska, S., Dabrowski, J., and Jakubowicz, K. (1972). Epidermodysplasia verruciformis as a model in studies on the role of papovaviruses in oncogenesis. Cancer Res. *32*, 583-589.

Jackson, S., Harwood, C., Thomas, M., Banks, L., and Storey, A. (2000). Role of Bak in UV-induced apoptosis in skin cancer and abrogation by HPV E6 proteins. Genes Dev. *14*, 3065-3073.

Jackson, S., and Storey, A. (2000). E6 proteins from diverse cutaneous HPV types inhibit apoptosis in response to UV damage. Oncogene *19*, 592-598.

Kahn, H. S., Tatham, L. M., Patel, A. V., Thun, M. J., and Heath, C. W., Jr. (1998). Increased cancer mortality following a history of nonmelanoma skin cancer. Jama *280*, 910-912.

Kessis, T., Slebos, R., Nelson, W., Kastan, M., Plunkett, B., Han, S., Lorincz, A., Hedrick, L., and Cho, K. (1993). Human papillomavirus 16 E6 expression disrupts the p53-mediated cellular response to DNA damage. Proc. Natl. Acad. Sci. USA. *90*, 3988-3992.

Kiviat, N. B. (1999). Papillomaviruses in non-melanoma skin cancer: epidemiological aspects. Semin Cancer Biol. *9*, 397-403.

Laffort, C., Le Deist, F., Favre, M., Caillat-Zucman, S., Radford-Weiss, I., Debre, M., Fraitag, S., Blanche, S., Cavazzana-Calvo, M., de Saint Basile, G., et al. (2004). Severe cutaneous papillomavirus disease after haemopoietic stem-cell transplantation in patients with severe combined immune deficiency caused by common gammac cytokine receptor subunit or JAK-3 deficiency. Lancet *363*, 2051-2054.

Levi, F., La Vecchia, C., Te, V. C., Randimbison, L., and Erler, G. (1998). Incidence of invasive cancers following basal cell skin cancer. Amer. J. Epidemiol. *147*, 722-726.

Levi, F., Randimbison, L., La Vecchia, C., Erler, G., and Te, V. C. (1997). Incidence of invasive cancers following squamous cell skin cancer. Amer. J. Epidemiol. *146*, 734-739.

Lindelof, B., Sigurgeirsson, B., Tegner, E., Larko, O., Johannesson, A., Berne, B., Christensen, O. B., Andersson, T., Torngren, M., Molin, L., and et al. (1991). PUVA and cancer: a large-scale epidemiological study. Lancet *338*, 91-93.

Mahe, E., Bodemer, C., Descamps, V., Mahe, I., Crickx, B., De Prost, Y., and Favre, M. (2003). High frequency of detection of human papillomaviruses associated with epidermodysplasia verruciformis in children with psoriasis. Br. J. Dermatol. *149*, 819-825.

Majewski, S., Jablonska, S., and Orth, G. (1997). Epidermodysplasia verruciformis. Immunological and nonimmunological surveillance mechanisms: role in tumor progression. Clin. Dermatol. *15*, 321-334.

Marks, R. (1995). An overview of skin cancers. Incidence and causation. Cancer *75*, 607-612.

Masini, C., Fuchs, P. G., Gabrielli, F., Stark, S., Sera, F., Ploner, M., Melchi, C. F., Primavera, G., Pirchio, G., Picconi, O., et al. (2003). Evidence for the association of human papillomavirus infection and cutaneous squamous cell carcinoma in immunocompetent individuals. Arch. Dermatol. *139*, 890-894.

Meyer, T., Arndt, R., Nindl, I., Ulrich, C., Christophers, E., and Stockfleth, E. (2003). Association of human papillomavirus infections with cutaneous tumors in immunosuppressed patients. Transpl Int. *16*, 146-153.

Newton, R. (1996). A review of the aetiology of squamous cell carcinoma of the conjunctiva. Br. J. Cancer *74*, 1511-1513.

Orth, G., Jablonska, S., Favre, M., Croissant, O., Jarzabek-Chorzelska, M., and Rzesa, G. (1978). Characterization of two types of human papillomaviruses in lesions of epidermodysplasia verruciformis. Proc. Natl. Acad. Sci. USA. *75*, 1537-1541.

Orth, G., Jablonska, S., Jarzabek-Chorzelska, M., Obalek, S., Rzesa, G., Favre, M., and Croissant, O. (1979). Characteristics of the lesions and risk of malignant conversion associated with the type of human papillomavirus involved in epidermodysplasia verruciformis. Cancer Res. *39*, 1074-1082.

Pfister, H. (2003). Chapter 8: human papillomavirus and skin cancer. J. Natl. Cancer Inst Monogr *31*, 52-56.

Pfister, H., and Ter Schegget, J. (1997). Role of HPV in cutaneous premalignant and malignant tumors. Clin. Dermatol. *15*, 335-347.

Pisani, P., Bray, F., and Parkin, D. M. (2002). Estimates of the world-wide prevalence of cancer for 25 sites in the adult population. Int. J. Cancer *97*, 72-81.

Preston, D. S., and Stern, R. S. (1992). Nonmelanoma cancers of the skin. N. Engl. J. Med. *327*, 1649-1662.

Purdie, K. J., Pennington, J., Proby, C. M., Khalaf, S., de Villiers, E. M., Leigh, I. M., and Storey, A. (1999). The promoter of a novel human papillomavirus (HPV77) associated with skin cancer displays UV responsiveness, which is mediated through a consensus p53 binding sequence. Embo J. *18*, 5359-5369.

Ramoz, N., Rueda, L. A., Bouadjar, B., Montoya, L. S., Orth, G., and Favre, M. (2002). Mutations in two adjacent novel genes are associated with epidermodysplasia verruciformis. Nat. Genet. *32*, 579-581.

Shamanin, V., zur Hausen, H., Lavergne, D., Proby, C. M., Leigh, I. M., Neumann, C., Hamm, H., Goos, M., Haustein, U. F., Jung, E. G., and et al. (1996). Human papillomavirus infections in nonmelanoma skin cancers from renal transplant recipients and nonimmunosuppressed patients [see comments]. J. Natl. Cancer Inst. *88*, 802-811.

Sinha, R. P., and Hader, D. P. (2002). UV-induced DNA damage and repair: a review. Photochem. Photobiol. Sci. *1*, 225-236.

Sorenson, C. M. (2004). Bcl-2 family members and disease. Biochim. Biophys. Acta *1644*, 169-177.

Stark, L. A., Arends, M. J., McLaren, K. M., Benton, E. C., Shahidullah, H., Hunter, J. A., and Bird, C. C. (1994). Prevalence of human papillomavirus DNA in cutaneous neoplasms from renal allograft recipients supports a possible viral role in tumour promotion. Br. J. Cancer *69*, 222-229.

Stark, S., Petridis, A. K., Ghim, S. J., Jenson, A. B., Bouwes Bavinck, J. N., Gross, G., Stockfleth, E., Fuchs, P. G., and Pfister, H. (1998). Prevalence of antibodies against virus-like particles of Epidermodysplasia verruciformis-associated HPV8 in patients at risk of skin cancer. J. Invest. Dermatol. *111*, 696-701.

Stern, R. S., and Lange, R. (1988). Non-melanoma skin cancer occurring in patients treated with PUVA five to ten years after first treatment. J. Invest. Dermatol. *91*, 120-124.

Stern, R. S., Thibodeau, L. A., Kleinerman, R. A., Parrish, J. A., and Fitzpatrick, T. B. (1979). Risk of cutaneous carcinoma in patients treated with oral methoxsalen photochemotherapy for psoriasis. N. Engl. J. Med. *300*, 809-813.

Stockfleth, E., Nindl, I., Sterry, W., Ulrich, C., Schmook, T., and Meyer, T. (2004). Human papillomaviruses in transplant-associated skin cancers. Dermatol. Surg. *30*, 604-609.

Storey, A. (2002). Papillomaviruses: death-defying acts in skin cancer. Trends Mol. Med. *8*, 417-421.

Struijk, L., Bouwes Bavinck, J. N., Wanningen, P., van der Meijden, E., Westendorp, R. G., Ter Schegget, J., and Feltkamp, M. C. (2003). Presence of human papillomavirus DNA in plucked eyebrow hairs is associated with a history of cutaneous squamous cell carcinoma. J. Invest. Dermatol. *121*, 1531-1535.

Thomas, M., and Banks, L. (1998). Inhibition of Bak-induced apoptosis by HPV-18 E6. Oncogene *17*, 2943-2954.

Thomas, M., and Banks, L. (1999). Human papillomavirus (HPV) E6 interactions with Bak are conserved amongst E6 proteins from high and low risk HPV types. J. Gen. Virol. *80 (Pt 6)*, 1513-1517.

Tieben, L. M., Berkhout, R. J., Smits, H. L., Bouwes Bavinck, J. N., Vermeer, B. J., Bruijn, J. A., Van der Woude, F. J., and Ter Schegget, J. (1994). Detection of epidermodysplasia verruciformis-like human papillomavirus types in malignant and premalignant skin lesions of renal transplant recipients. Br. J. Dermatol. *131*, 226-230.

Walder, B. K., Jeremy, D., Charlesworth, J. A., Macdonald, G. J., Pussell, B. A., and Robertson, M. R. (1976). The skin and immunosuppression. Australas J. Dermatol. *17*, 94-97.

Walder, B. K., Robertson, M. R., and Jeremy, D. (1971). Skin cancer and immunosuppression. Lancet 2, 1282-1283.

Weissenborn, S. J., Hopfl, R., Weber, F., Smola, H., Pfister, H. J., and Fuchs, P. G. (1999). High prevalence of a variety of epidermodysplasia verruciformis-associated human papillomaviruses in psoriatic skin of patients treated or not treated with PUVA. J. Invest. Dermatol. *113*, 122-126.

Wolf, P., Seidl, H., Back, B., Binder, B., Hofler, G., Quehenberger, F., Hoffmann, C., Kerl, H., Stark, S., Pfister, H. J., and Fuchs, P. G. (2004). Increased prevalence of human papillomavirus in hairs plucked from patients with psoriasis treated with psoralen-UV-A. Arch. Dermatol. *140*, 317-324.

Yamashita, T., Segawa, K., Fujinaga, Y., Nishikawa, T., and Fujinaga, K. (1993). Biological and biochemical activity of E7 genes of the cutaneous human papillomavirus type 5 and 8. Oncogene *8*, 2433-2441.

zur Hausen, H. (2002). Papillomaviruses and cancer: from basic studies to clinical application. Nat. Rev. Cancer *2*, 342-350.

From: Papillomavirus Research: From Natural History to Vaccines and Beyond. Edited by: M. Saveria Campo

Chapter 18

Papillomavirus-like Particles and Their Applications in Molecular Virology, Human Serology and Vaccines

Richard B.S. Roden*§ and Raphael P. Viscidi

Abstract

The discovery early in the 1990s that the major capsid protein of papillomaviruses, L1, assembles into virus-like particles (VLPs) has propelled prophylactic vaccine development and led to significant advances in the study of the natural history and molecular virology of this DNA tumor virus. VLPs present the conformational epitopes required for generating high titer neutralizing antibodies but are devoid of the potentially oncogenic viral genome. Therefore, VLPs are attractive candidates for a vaccine to prevent genital HPV infection. Indeed ongoing clinical vaccine trials with L1 VLPs demonstrate the induction of protective immunity. HPV VLPs have also been exploited as an antigen in enzyme-linked immunosorbant assays (ELISAs) to detect anti-virion immune responses in human sera. HPV 16 VLP-based ELISAs were the first HPV serological assays to have sufficient sensitivity and specificity to make them widely applicable to studies of the natural history of HPV infection and to demonstrate its association with neoplastic disease. Finally, studies of the molecular virology and immunology of oncogenic type papillomaviruses have been hampered by the difficulties in obtaining biochemical quantities of native virions. VLPs have been a useful surrogate for study of capsid structure, immunology and function in infection. In this chapter we summarize the dramatic impact of VLP technology in advancing our understanding of papillomavirus biology and immunology and discuss some of its limitations.

1. Application of VLP Technology to Studies of Virion Structure and Assembly

Early studies of Papillomavirus (PV) structure were dependent upon virions purified from warts because the virus failed to replicate in conventional tissue culture (Hagensee and Galloway, 1993). These studies revealed a 55-60 nm diameter non-enveloped icosahedral structures containing an approximately 8 Kb double stranded supercoiled DNA genome packaged as a nucleohistone complex (Crawford, 1969; Crawford and Crawford, 1963; Crawford and Waring, 1967; Favre et al., 1977). Biochemical studies indicated that the capsid surrounds a nucleohistone core and is composed of the major and minor capsid proteins, designated L1 and L2 respectively (Doorbar and Gallimore, 1987; Favre, 1975). The capsid has T=7 icosahedral symmetry and is composed of 72 pentameric capsomers that have a right-hand skew (Finch and Klug, 1965; Klug and Finch, 1968). Computer generated reconstructions from cryo-electron micrographs indicate that the capsomers at the 12 vertices of the virion contact five other capsomers (pentavalent), whereas the remaining sixty capsomers contact six other capsomers (hexavalent) (Baker et al., 1991; Trus et al., 1997). Thus PV virions resemble SV40 and polyoma virions (Trus et al., 1997). However, in contrast to both SV40 and polyoma (Liddington et al., 1991; Stehle et al., 1994), no high resolution X-ray crystallographic structure has been determined for an intact PV virion despite several attempts. This failure reflects the difficulties in obtaining native virus from warts or tissue culture in sufficient quantity and quality for crystallization studies (Hagensee and Galloway, 1993).

In the early 1990s several groups demonstrated that the major capsid protein L1 derived from diverse PV genotypes can assemble alone into virus-like particles (VLPs) that morphologically resemble native virions when expressed in mammalian (Hagensee et al., 1993; Zhou et al., 1993; 1991), insect (Kirnbauer et al., 1992; 1993), yeast (Sasagawa et al., 1995) or bacterial cells (Nardelli-Haefliger et al., 1997; Zhang et al., 1998) (Figure 1). Thus, these L1 VLPs provided an opportunity to better define the contribution of L1 to the capsid structure. Unlike virions, the VLPs can easily be generated in quantities sufficient for crystallization and necessary to obtain a high resolution structural model for PV virions based on X-ray diffraction analysis (Chen et al., 2001; 2000). Since VLPs are produced in systems amenable to genetic manipulation, a number of virus related processes can now be explored by mutagenesis of the capsid proteins, including virion assembly and structure-function relationships.

*For correspondence. Email roden@jhmi.edu

§Paid consultant of Knobbe, Martens, Olson and Bear

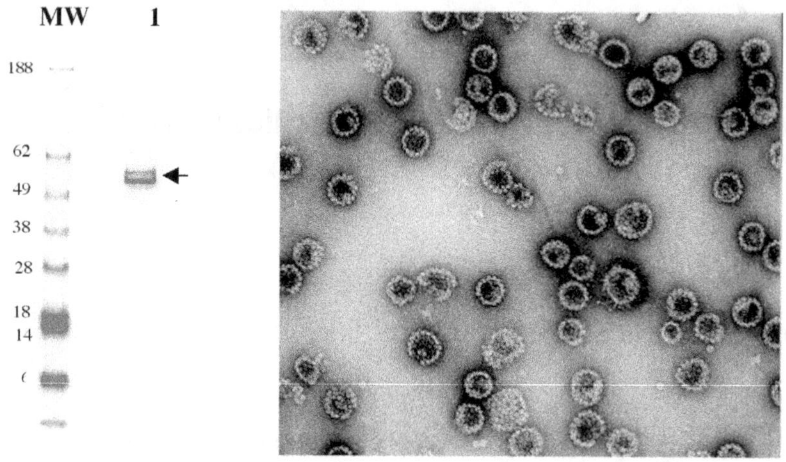

Figure 1. HPV16 L1 VLPs. A. HPV16 L1 purified from baculovirus-infected insect cells (arrow, lane 1) and molecular weight markers (MW in kDa). B. A transmission electron micrograph of HPV16 L1 VLPs (kindly provided by D. Lowy).

Cryo-electron micrograph images of virions can be reconstructed into three dimensional electron density maps. This approach has been used successfully both for the polyomaviruses and papillomavirius native virions, including HPV1, BPV1 and CRPV (Baker et al., 1991; Belnap et al., 1996; Griffith et al., 1992). While this approach verifies the conserved T=7d structure of papillomavirus, it is not clear how each capsid protein contributes to the virion structure (Belnap et al., 1997; 1996). To address the contribution of L1 and L2 to the structure of the capsid, Hagensee et al. attempted to reconstruct L1 only and L1/L2 VLPs generated by recombinant Vaccinia virus-driven overexpression. A three dimensional reconstruction from cryo-electron micrographs revealed HPV1 L1 VLP comprise 72 pentamers of L1 centered on the vertices of a T=7 icosahedral lattice. This structure appeared identical at a resolution of 35Å to the capsid structure of HPV1 virions (Baker et al., 1991), thus implying that L1 VLPs closely replicate native capsid structure. Unfortunately, the analysis failed to distinguish L1 only VLPs from those containing L2, suggesting that higher resolution analysis was necessary to identify L2 or that its occupancy was limited. However the relative heterogeneity of VLP assembly compared to native virions complicates attaining much higher resolution images. Although we achieved a resolution of 9Å for an image reconstruction of native Bovine Papillomavirus type 1 purified from cow warts, we were only able to generate reconstructions of BPV1 L1 only or L1/L2 VLPs to a greater than approximately 25 Å (unpublished data). At higher resolutions, the image reconstruction BPV1 virions revealed a blob of density at the center of the pentavalent capsomers at the capsid vertices which was absent from the hexavalent capsomers (Trus et al., 1997). Notably the T=7 structure of papillomavirus contains 60 hexavalent and only 12 pentavalent capsomers. Furthermore estimates from Coomassie-stained SDS-PAGE analyses of purified BPV1 virions suggest a ratio of 30 L1 molecules for each L2 molecule. Since there are 360 L1 molecules per virion we hypothesized that each virion contains only 12 L2 molecules and that the blob of density at the center of the 12 pentavalent capsomers represented a portion of the L2 protein. This remains unproven, and surprising since the minor capsid proteins VP2 and/or VP3 are in the center of all capsomers in polyomaviruses (Griffith et al., 1992; Liddington et al., 1991).

While full length HPV16 L1 assembles into relatively regular 72 pentamer T=7 VLPs at low pH, a deletion mutant lacking the N-terminal ten amino acids assembled in to a 12-pentamer T=1 capsid of 318Å diameter under all conditions tested (Chen et al., 2001). The X-ray crystal structure of T=1 HPV16 L1 VLPs (Chen et al., 2000) demonstrates that residues 20-282 of L1 form a classic 'jelly roll' β sandwich, similar to that described for polyoma and SV40 VP1 (Liddington et al., 1991; Stehle et al., 1994) (Figure 2). Thus despite minimal sequence similarity between polyomavirus VP1 and L1, their architecture is closely related. Elaborate L1 interstrand loops protrude from the surface of the capsomer structure. Comparison of the L1 sequences amongst different HPV genotypes reveals highly variable stretches that are interspersed with regions of sequence conservation. The regions of low sequence conservation correspond to the loops protruding from the capsid surface. The apparent lack of tissue specificity of the primary papillomavirus cell surface receptor suggests that these changes are not to determine tropism. However, several studies have identified residues that contribute to the structure of conformationally-dependent L1 neutralizing epitopes using VLPs or capsomers derived from mutant L1 proteins (Carter et al., 2003; Ludmerer et al., 1996; 1997; Roden et al., 1997a; White et al., 1999). The locations of sequence differences in L1 both map to the surface of the atomic model and are all close to the residues identified as contributing to the conformationally-dependent L1 neutralizing epitopes (Chen et al., 2000).

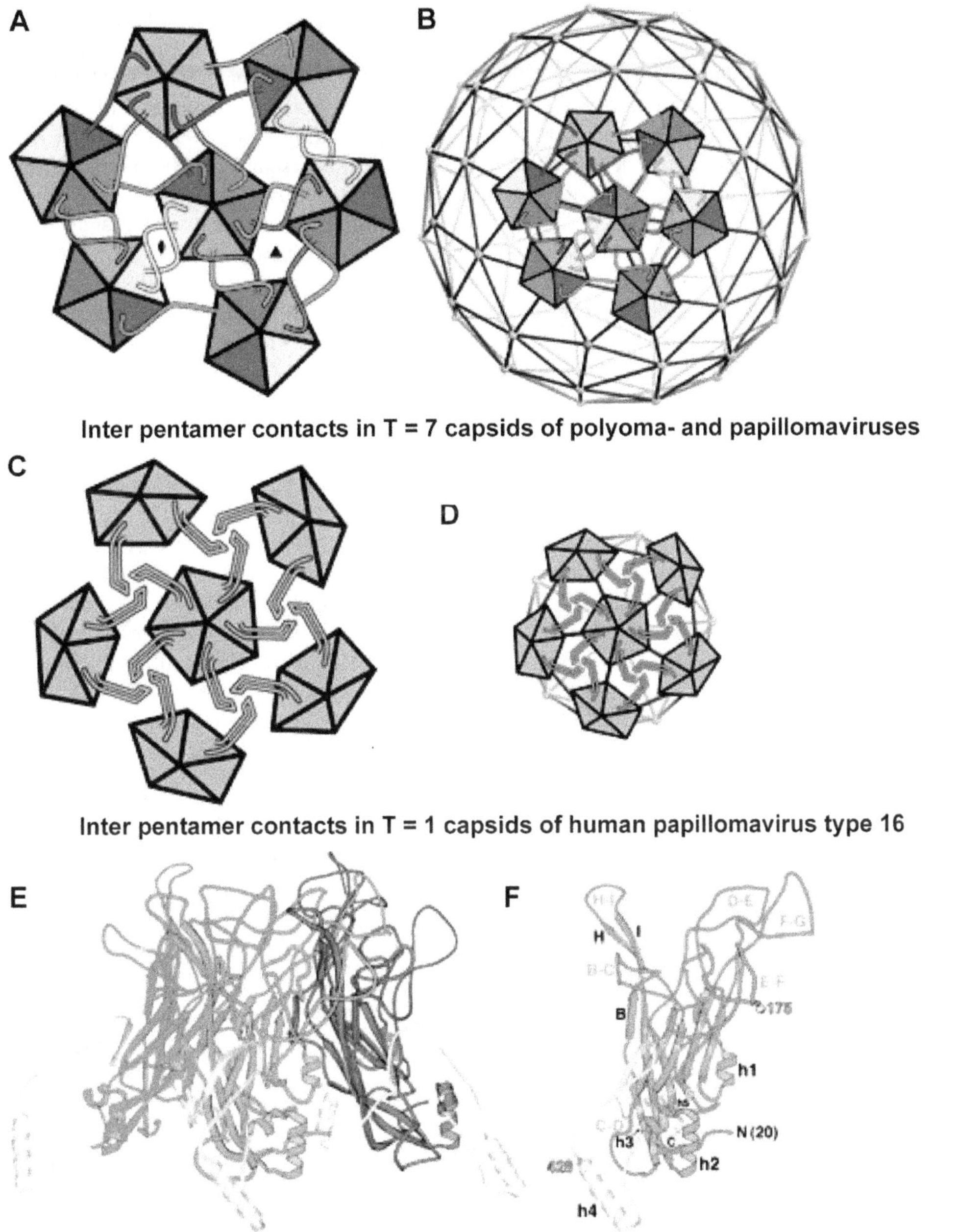

Figure 2. Modes of interaction between pentamers in various papovavirus subunit assemblies (Reprinted from (Modis et al., 2002) with permissions from EMBO Journal www.nature.com/**embo**j/, copyright (2002) Macmillan Publishers Ltd, and Dr S Harrison, Harvard University). (**A**) Pentamer contacts in SV40 (Liddington et al., 1991), polyomavirus, and in full-size (T = 7) papillomavirus particles, as shown in (Modis et al., 2002). The subunit 'body', a jellyroll -barrel (see E and F), is represented by a triangle that forms one-fifth of a pentamer. (**B**) Diagram in (A) superimposed on the full T = 7 lattice. (**C**) Pentamer contacts in small (T = 1) VLPs derived from HPV L1 (Chen et al., 2000). Note that the C-terminal arms fold back into the subunit to which they belong, rather than reaching out into a neighboring pentamer. (**D**) Diagram in (C) superimposed on a T = 1 lattice. (B) and (D) are drawn to the same scale. (**E**) Ribbon diagram of an HPV16 L1 pentamer, in the conformation found in small VLPs (Chen et al., 2001). Three subunits are shown in green, blue and red. The C-terminal arms are in gray, to indicate that these portions of the subunit rearrange when pentamers assemble into virions or into full-sized capsids. (**F**) Ribbon diagram of a single subunit, approximately in the orientation of the blue subunit in (E). Secondary structural elements are labeled, with letters B–I for β-strands and h1–h5 for the five α-helices. The N- and C-termini of the ordered structure are marked N(20) and C, respectively. Residue 20 is the first ordered residue in the small VLP structure. Approximately the last 30 residues of the subunit (the exact number varies among types) are disordered and extend into the interior of the particle. Loops between strands are labeled B–C, C–D, etc. The two cysteines that participate in interpentamer disulfide bonding in the virion or in virion-sized particles are shown in yellow, together with their residue numbers, 175 and 428.

The colour version of this figure is available on the back cover of this book and also at: http://www.caister.com/supplementary/pv

The C-terminal arm of HPV16 L1 is significantly α helical and three helices (h2-4) mediate contact with adjacent L1 monomers (Chen et al., 2000). This projection is anchored back to the central jelly roll structure via a short strand (βJ), which adds to the C edge of the CHEF sheet, and a final helix (h5) (Figure 2). This h5 helix tucks into the axial cavity of the pentamer via a hydrophobic interaction with the base of the BIDG sheet. The final 31 residues project towards the lumen but are disordered. The backbones of the five L1 monomers associate intimately with adjacent molecules to form the tightly packed donut shape of the pentamer. The tightest contacts within the pentamer occur at the higher radii and the L1 monomers splay outwards towards the floor of the coat. Thus the L1 pentamer exhibits a conical hollow that progressively opens towards the interior. This hollow is only ~14 Å in diameter at its narrowest. Contact is mediated by the G strand at the inner edge of the BIDG sheet which interacts with the CHEF sheet of the adjacent L1 monomer, before returning. The HI loop also intertwines between the FG and EF loops of the adjacent L1 monomer, thus generating the five points of the pentamer's stellate cap (Chen et al., 2000).

Clearly the atomic model of small VLPs has greatly informed our understanding of papillomavirus structure. However, there are limitations to the small VLPs as a model of virion structure including the absence of interactions with L2 and the nucleohistone core. Furthermore, the T=1 structure of the small VLP suggests that the interpentamer contacts may not be representative of those in native virions. In the small VLP, the interpentamer contact occurs as an HPV16 L1 C-terminal subdomain protrudes into the adjacent pentamer then returns to the central channel of the pentamer from which it originated (Figure 2). By contrast, the C-terminal arms of polyomavirus VP1 tether by directly invading the adjacent pentamer. Furthermore, while T=1 small VLPs contain only pentamers co-ordinated with five adjacent pentamers (i.e. pentavalent capsomers), the native T=7 virion and VLP of both polyomaviruses and papillomaviruses comprise 60 pentavalent and 12 hexavalent capsomers. Thus the 'arm invasion' model is more consistent with the true structure rather than the simple contact between L1 capsomers in the small VLPs (Figure 2). Indeed, biochemical and mutagenesis studies of T=7 VLPs also support the 'arm invasion' model.

Comparison of the apparent molecular weight of L1 on reducing and non-reducing SDS-PAGE indicates that disulphide bonding mediates the formation of L1 trimers (Sapp et al., 1998). Point mutation of each individual cysteine in HPV33 L1 revealed that two residues, cysteines 176 and 427 form a disulphide bridge and that this bond is critical to particle assembly. Likewise, point mutation of HPV11 cysetine 424 to glycine resulted in the production of capsomers but not VLPs (Li et al., 1998). Indeed, HPV VLPs can be quantitatively disassembled and re-assembled by addition and removal of reducing agents under the appropriate ionic conditions (McCarthy et al., 1998; Zhang et al., 1998). The presence of calcium ions was not required for VLP assembly (Li et al., 1998; McCarthy et al., 1998), although it is necessary for *in vitro* reassembly of BPV1 virions (Paintsil et al., 1998). Furthermore, proteolytic digestion and mutagenesis studies suggest that these cysteines tether the C terminal arm of one L1 subunit to an outer loop of an adjacent L1 subunit (Li et al., 1998; 1997; Sapp et al., 1998). Li et al. demonstrated that trypsin cleavage of HPV11 L1 at R 415 removed the C-terminal 86 residues and prevented assembly beyond capsomers (Li et al., 1997). However, the C-terminal 24 amino acids of BPV1 L1 were dispensable for the formation of VLPs (Paintsil et al., 1996). Additional evidence for this model was obtained by LC-MS/MS analysis of the tryptic fragments derived from L1 dimers separated on a non-reducing SDS-PAGE gel (Modis et al., 2002). This study revealed the corresponding peptides in HPV18 L1 linked via a disulphide bond between cysteine 175 and cysteine 429, indicating that this is conserved between three distinct evolutionary clades of papillomaviruses, represented by HPV11, HPV18 and HPV33 (Li et al., 1998; 1997; Modis et al., 2002; Sapp et al., 1998). To generate a model for the structure of T=7d papillomavirus VLP, Modis et al. fitted the X-ray crystal structure into the high resolution image reconstruction generated from native BPV1 virions (Modis et al., 2002). The two structures demonstrated an excellent fit. The only region of the HPV16 L1 small VLP atomic coordinates that failed to match the BPV1 map contained residues 402-446, which forms the intercapsomer contacts in the small VLP. Therefore, these residues were rebuilt to fit in a new conformation into unoccupied volume within the BPV1 map and constrained by the requirement of cysteine 428 in the C-terminal arm forming a disulphide bond with cysteine 175 in the EF loop of the adjacent L1, consistent with the 'invading arm' model (Modis et al., 2002). Confirmation of this model structure awaits an X-ray crystallographic analysis of native virions. Despite such a high resolution structure, the residues responsible for binding to cell surface receptor molecules are not known. Furthermore, studies using monoclonal antibodies to probe structure indicate that the capsid may undergo a significant conformational change upon interaction with the cell surface (Selinka et al., 2003). This change may relate to the closed (compact) and open (swollen) structures described by Belnap et al. for CRPV virions (Belnap et al., 1996). It is unclear how the X ray crystallographic structure of small VLPs relate to these additional structural forms of the papillomavirus capsid.

The study of VLPs has also informed our understanding of L1 assembly *in vivo*. While purified L1 can self-assemble *in vitro*, it is likely that chaperones guide this process inside the cell. Expression of HPV11

L1 in insect cells via recombinant baculovirus resulted in the accumulation of VLPs in both the cytoplasm and nucleus, although predominantly in the latter (Rose et al., 1993). Indeed L1 lacking a nuclear localization signal is able to assemble into VLPs but fails to bind DNA (Chen et al., 1998; Paintsil et al., 1996; Touze et al., 2000). Studies of the polyomavirus major and minor capsid proteins, VP1 and VP2/3, indicate that they associate in the cytoplasm and enter the nucleus as a complex. However, L2 is expressed prior to L1 in papillomas and accumulates in the nucleus prior to L1 when co-expressed in recombinant Vaccinia virus infected cells (Florin et al., 2002). Further, nuclear translocation of L2 but not L1 is blocked by treatment with proteosome inhibitor. Once in the nucleus, L2 trafficks to the subnuclear domain termed ND-10 or PML-oncogenic domain (POD) (Day et al., 1998). Although L1 alone exhibits a diffuse nuclear pattern by indirect immunofluoresence, co-expression with L2 causes their co-localization in ND-10 (Day et al., 1998). Accumulation of L1 in ND-10 occurs after L2-induced exit of Sp100 from ND-10 (Florin et al., 2002).

L2 binds to L1 capsomers but not whole VLP, suggesting that L2 co-assembles with L1 rather than being inserted into a pre-formed capsid (Okun et al., 2001). Indeed, L2 is required for efficient genome encapsidation (Roden et al., 1996a; Zhao et al., 1998; 1993), suggesting that the capsid assembles around the histone-bound genome, rather than by injection of the genome into the capsid via a portal vertex. L2 does not require disulphide bridges to interact with L1 in a VLP (Sapp et al., 1995). Discrete salt sensitive binding and salt insensitive L1–binding was observed upon disassembly of HPV33 L1/L2 VLPs (Sapp et al., 1995). Indeed two distinct L1 binding domains were described for BPV1 L2; a C-terminal L1 binding domain (residues 384-460) that interacts with L1 capsomers *in vitro*, and a central region (residues 129-246) that fails to interact with capsomers (Okun et al., 2001). The interaction between L2 residues 129-246 and BPV1 L1 were described on the basis of co-immunoprecipitation and co-localization studies. An L1 binding domain was described between residues 396-439 of HPV11 L2, consistent with the C-terminal L1 binding domain in residues 384-460 of BPV1 L2. This interaction, which occurs *in vitro* and is relatively salt insensitive, enhances *in vitro* assembly of VLPs (Finnen et al., 2003). Previous studies have commented upon enhanced or more regular *in vivo* assembly of VLPs in the presence of L2 (Kirnbauer et al., 1993). Indeed, fourfold higher hemagglutinating activity was described for BPV1 L1/L2 versus L1 only VLP preparations (Roden et al., 1995). An early study reported that HPV 16 L1 was unable to self-assemble in the absence of L2 (Zhou et al., 1991). However, the L1 of the reference strain used in these experiments was later shown to be an assembly-deficient mutant (Kirnbauer et al., 1993). Most HPV 16 L1 proteins, and all that efficiently self-assemble, contain an aspartate at position 202 rather than the histidine found in the L1 of the reference strain (Pushko et al., 1994). It is interesting to note that the presence of L2 provided for significant assembly of even this defective mutant of L1 (Zhou et al., 1991). Therefore, care must be taken in comparison between the mechanisms of *in vivo* virion assembly and the 'five-around-one' *in vitro* L1 VLP assembly mechanism (Casini et al., 2004), since L2 and chaperones are likely to profoundly influence this process (Florin et al., 2004).

L2-specific antibody neutralizes infection *in vitro* and is protective *in vivo* demonstrating that a small portion of L2 is exposed on the surface of the capsid (Campo, 1991; Campo et al., 1993; Christensen et al., 1991; Roden et al., 1994b; 2000). Several studies have attempted to identify the nature of these neutralizing epitopes using L1/L2 VLP to better define the topology of L2 (Heino et al., 1995; Kawana et al., 1998a). Neutralization data suggest that HPV16 L2 residues 108-120 and 69-81 are epitopes displayed on the surface of VLPs and virions (Kawana et al., 1999). Clearly our knowledge of L2's topology in the capsid is limited but perhaps the L1 capsomer-L2 complex (Finnen et al., 2003) or pseudovirions (Buck et al., 2004) may be suitable for X-ray crystallographic studies. Unlike structures of VLPs or capsomers, analysis of pseudovirion or true virion preparations would also clarify the interaction between the capsid and the nucleohistone core. Studies with purified capsid proteins or VLPs indicate that the C-terminal positively charged tail of L1 that includes a nuclear localization signal is also critical for binding to and packaging DNA (Touze et al., 2000). Similar sequences on both termini of L2 may also play a role in encapsidation of the viral genome (Zhou et al., 1994) as well as infection (Bousarghin et al., 2003a; 2004; Roden et al., 2001). While suggestive, the studies with VLPs and purified capsid proteins await detailed virion mutagenesis and structural studies for confirmation.

In summary, VLPs have been important reagents for studies of capsid structure and assembly. Unlike virions, the VLPs can easily be generated in quantities sufficient for crystallization and have provided a high resolution structural model for PV virions based on X-ray diffraction analysis (Chen et al., 2000). While this is a great achievement, certain aspects of the structure of papillomavirus such as inter capsomer bonding and the structures of L2 and the nucleohistone core await high resolution structural and site-directed genetic analysis of virions.

2. Utility of Papillomavirus-like Particles in Mechanistic Studies of Infection

Early studies of the mechanism of infection were dependent upon virions purified from natural lesions, since the propagation of virus in culture was very challenging (Hagensee and Galloway, 1993; Meyers et al., 1992; Meyers and Laimins, 1994). This greatly limited the experimental avenues open to study the contribution of each capsid protein and individual residues thereof to infection. Since VLPs are produced in systems amenable to genetic manipulation, a number of infection related processes could then be explored, including the interaction of the capsid with its cell surface receptor and transport of particles across the cytoplasm and nuclear import. While the use of VLPs is now being replaced with virions or pseudovirions carrying markers genes (Buck et al., 2004; Frattini et al., 1996; Roden et al., 1996a), these early studies have provided important information on the mechanism of infection.

Early studies addressed which capsid protein mediated the initial interaction between the virus and the cell surface. The presence of VLPs comprising either BPV1 L1 only or L1 plus L2 inhibited infection of mouse fibroblasts by BPV virions as assessed by focal transformation (Roden et al., 1994a). This suggests that VLPs have affinity for the same cell surface receptor(s) as virions and that L2 is not required for this interaction. Furthermore, monoclonal antibodies to L1 prevent virion binding to the cell surface, but neutralizing antisera to L2 do not (Roden et al., 1994b). HPV11 or HPV16 VLPs also inhibited BPV infection, implying that these PVs share a common cell surface receptor or a downstream processing event during infection. Supporting the former possibility is the observation that the binding of BPV virions to cell surfaces (and the uptake of HPV 6b, 11 and 16 VLPs) is inhibited in the presence of heterologous VLPs (Muller et al., 1995; Roden et al., 1994a).

BPV virions agglutinate mouse erythrocytes, but not those of other species including rat and human (Favre et al., 1974). In support of the importance of L1 in binding to the primary cell surface receptor, VLPs composed of BPV L1 alone or L1 plus L2 and VLPs derived from a number of HPV types also agglutinate erythrocytes with the same species specificity as BPV virions (Roden et al., 1995). Therefore, mouse erythrocytes but not those from other species present a PV cell surface receptor. It is unclear whether the receptor on mouse erythrocytes is the same as that on nucleated cells, but the erythrocyte receptor is also cleaved by trypsin and antibodies that block PV binding to nucleated cells also inhibit hemagglutination (Roden et al., 1995). Antibody to L1 blocked hemagglutination but antibody to L2 did not, further supporting L1 as binding to the primary surface receptor (Roden et al., 1996b).

The tropism of many viruses is determined by the species or tissue distribution of their receptors. HPV11 and HPV33 VLPs, and also labeled BPV virions, bind to a variety of cell lines of epithelial, mesenchymal and neuronal origin derived from a broad range of species, suggesting that the cell surface receptor for PVs is widely expressed and evolutionarily conserved (Muller et al., 1995; Roden et al., 1994a; Volpers et al., 1995). This is surprising in light of the strict host range of the papillomaviruses and suggests that this specificity appears to be determined by some post-binding event in the infectious cycle (Koller and Olson, 1972). However, the above analyses do not rule out the possibility that, in addition to a widely distributed primary receptor, a more specific secondary receptor may be involved in virus uptake. Indeed, evidence suggests that L2 might bind to a secondary viral receptor (Kawana et al., 2001; Yang et al., 2003a; 2003b).

It has been estimated that there are approximately 2.6 x 10^4 cell surface HPV33 VLP receptors per HeLa cell that bind with a dissociation constant of 84 pM (Volpers et al., 1995). Similar results were obtained using HPV6b VLPs and CV-1 cells (Qi et al., 1996). The binding of VLPs with the cell surface receptor is decreased at high salt concentrations and abolished above pH 7.8 consistent with a polar interaction (Volpers et al., 1995). Dissociated capsomers produced from VLPs by divalent cation chelation under reducing conditions neither bind to cell surfaces nor compete with VLP binding to cells implying that intercapsomer contacts are required for binding the receptor (Volpers et al., 1995). Trypsin treatment of cell surfaces prior to addition of particles inhibited their binding to the cells (Muller et al., 1995; Roden et al., 1995; Volpers et al., 1995) suggesting that the primary cell surface receptor has a protein component. Initially, $\alpha6\beta1$ or $\alpha6\beta4$ integrin was suggested as the primary cell surface receptor for HPV6 VLPs (Evander et al., 1997; McMillan et al., 1999; Yoon et al., 2001). However, subsequent studies in other systems have not supported a necessary role for $\alpha6$ integrin in papillomavirus infection (Giroglou et al., 2001a; Shafti-Keramat et al., 2003; Sibbet et al., 2000). Another study suggested CD16, the FcγIII as a molecule that binds to HPV16 VLPs on both immune cells and foreskin epithelium but more studies are required to validate its importance in infection (Da Silva et al., 2001). A recent study has also identified other putative receptors bound by peptides that compete HPV16 VLP binding to HeLa cells (Vera-Bravo et al., 2003).

A consensus supports the importance of cell surface gylcosaminoglycans (GAGs) in the primary interaction between papillomavirus and the cell surface. Joyce et al. demonstrated that HPV11 L1 VLPs interact with heparin and cell surface GAGs resembling heparin on the surface of keratinocytes and CHO cells (Joyce et al., 1999). Surface binding by HPV11 VLPs can be directly competed by free heparin and dextran sulphate, and this competition is influenced by both the molecular weight and charge density of the inhibitor. Sequence comparison between L1 of different genotypes suggests a conserved heparin-binding domain at the C-terminus. Enzymatic cleavage of this domain from L1 prevents its binding to both heparin and human keratinocytes. Heparinase or heparitinase pre-treatment of keratinocytes reduces

L1 binding by 80-90%. Furthermore, treatment of cells with chlorate or substituted β-D-xylosides, which result in undersecretion of GAGs, reduces VLP binding. Finally, CHO cells deficient in GAG synthesis exhibit 10% of VLP binding compared to control CHO cells. Thus the data are consistent with surface GAGs as the primary receptor for HPV11 L1 VLP (Joyce et al., 1999). Several groups find that GAGs are also critical for papillomavirus infection (Drobni et al., 2003; Giroglou et al., 2001a; Selinka et al., 2003; 2002; Shafti-Keramat et al., 2003). K562 cells, which exhibit minimal surface GAG expression, demonstrated enhanced VLP-binding upon ectopic expression of syndecan-4 or glypican-1 but most notably syndecan-1. Ectopic syndecan-1 expression also enhanced permissiveness to HPV11 infection 8-fold (Shafti-Keramat et al., 2003).

In addition to studies of cell surface binding, VLPs have also been used to explore the mechanisms of viral uptake and passage to the nucleus. Comparison of the uptake of HPV16 VLPs and BPV1 virions in the same cell reveals overlap and suggests the same mechanism of uptake (Day et al., 2003). This is supported by the inhibition of BPV1 infection in the presence of HPV16 VLPs (Roden et al., 1994a). Uptake of VLPs followed a slow time course, with a 4h t1/2 for internalization (Day et al., 2003) that is consistent with late post-binding neutralization (Christensen et al., 1995). Co-localization of BPV1 VLPs with the clathrin adaptor protein AP-2 was clearly evident at 2h post initiation of infection (Day et al., 2003). By 3h post infection, colocalization with transferrin receptor was apparent, suggestive of entry into the early endocytic pathway. Surprisingly, entry to the lysozomes, as evidenced by co-localization with Lamp-2, was observed for BPV1 VLPs stained with monoclonal antibody #9, but not 5B6 (Day et al., 2003). Another interesting finding suggests that despite the similarities in the structure of HPVs, different genotypes may enter via different pathways. HPV16 and 58 enter via clathrin-mediated endocytosis while HPV31 uptake is dependent upon caveosomes (Bousarghin et al., 2003b). Unlike the slow uptake pathway described for BPV1 (Day et al., 2003), a more rapid uptake of HPV16 and HPV33 VLPs into smooth cytoplasmic vesicles was described by ultramicroscopy studies (Volpers et al., 1995; Yang et al., 2003b). This suggests the possibility of two uptake pathways; perhaps one for infection and a second default pathway.

HPV16 L1 and L1/L2 are rapidly taken up by BPHE-1 cells with similar efficiency (Yang et al., 2003b). Interestingly, L1/L2 VLPs aligned along distinct radial tracks across the cytoplasm and within ~2 h at 37 °C arrived in the perinuclear region. However, L1 VLPs were not aligned along such radial tracks. Rather, L1 VLPs remained widely distributed throughout the whole cell and showed a less clear-cut tropism toward the nucleus during this 6-h time course. These differences in the uptake of L1/L2 and L1 VLPs suggest that L2 contributes to the transport of particles across the cytoplasm (Yang et al., 2003b). Cytochalasin B both disrupted the microfilament network by depolymerizing actin and inhibited uptake of HPV16 L1 and L1/L2 VLPs (Yang et al., 2003b), as well as BPV1 infection (Day et al., 2003; Zhou et al., 1995). Residues 25-45 of HPV16 L2 were demonstrated to bind directly to actin. Furthermore, the uptake of VLPs containing L2 deleted for residues 25-45 followed a similar course to L1 only VLPs as opposed to L1/L2 VLPs. This interaction was also critical to infection (Yang et al., 2003b). It is currently unclear if/when particles cross from the lumen of vesicles into the cytoplasm during the infectious process.

While L1/L2 VLPs rapidly arrive in the perinuclear region, they have not been described as entering the nucleus (Day et al., 2003; Zhou et al., 1995). Studies in digitonin permeabilized HeLa cells reveal that L1 capsomers are imported into the nucleus in a receptor-mediated manner (Nelson et al., 2002). Indeed, nuclear import of HPV16 L1 capsomeres was mediated by Kap alpha2beta1 heterodimers, required RanGDP and free GTP, and was independent of GTP hydrolysis. Importantly, complete HPV16 L1 VLPs were not imported into the nucleus under these conditions, suggesting that the particles disassemble within the reducing environment of the cytoplasm to release their nucleohistone core to enter the nucleus during infection (Nelson et al., 2002). This is surprising since SV40 does enter the nucleus during infection, but this difference may reflect their smaller diameter compared to papillomavirus (Clever et al., 1991; Kasamatsu and Nakanishi, 1998).

VLPs have been largely replaced by pseudotyped virions that deliver reporter DNA as a surrogate for virions in studies of infection. Early studies used a reporter DNA cross linked *in vitro* to HPV VLPs as a surrogate infectivity assay (Muller et al., 1995; Yeager et al., 2000). An alternative approach was *in vitro* re-assembly of VLPs around reporter DNAs (Kawana et al., 1998b; Touze and Coursaget, 1998). While these *in vitro* assembled pseudovirions have proved useful for neutralization assays, they have largely been abandoned in favor of *in vivo* assembled pseudovirions or true virions generated in raft culture (Buck et al., 2004; Dollard et al., 1992; Meyers et al., 1992; Roden et al., 1996a; Unckell et al., 1997; Zhou et al., 1993).

3. Use of VLPs in Seroepidemiological Studies of HPV Infection and HPV-related Disease

In addition to their utility in studying viral structure, assembly and infectious processes, VLPs have also proven useful in dissecting the host immune response to natural infection. Indeed, seroepidemiologic studies using HPV VLP ELISA assays have provided important evidence about the prevalence of natural infection in different populations, the mode and efficiency of transmission, the

Table 1. Prevalence of HPV-16 VLP antibodies in relation to cervical HPV DNA by PCR

Study Population	Age (yrs)	HPV DNA	No. Women	% Positive	Reference
Population-based cases/controls	25-59	Negative	376	26	Kjellberg et al., 1999
		Other Types	112	30	
		Type 16	52	65	
GYN and student health clinic	18-42	Negative	31	6	Kirnbauer et al., 1994
		Other Types	37	27	
		Type 16	54	59	
University students	18-40	Negative	247	19	Viscidi et al., 1997
		Other Types	101	30	
		Type 16	28	46	
University students	18-20	Negative	216	7	Carter et al., 1996
		Other Types	58	5	
		Type 16	19	53	
GYN clinic	19-83	Negative	97	12	Tachezy et al., 1999
		Other Types	24	17	
		Type 16	67	46	

relationship between HPV infection and various cancer types as well as the natural course of infection.

4. Development of VLP-based serological Assays for HPV

The development of serological assays for genital HPVs lagged considerably behind development of assays for HPV DNA detection. The principal reason for this is that HPVs cannot be propagated in tissue culture and naturally occurring genital lesions contain small amounts of viral protein. This problem was partly addressed by the propagation of some HPV types (e.g., HPV-11) in the nude mouse and SCID mouse xenograft systems, but there is no reliable source for adequate quantities of authentic viral proteins of many HPV types (Kreider et al., 1986). The first attempts to overcome this problem involved the preparation of HPV proteins in bacterial expression systems or by chemical synthesis of HPV derived peptides (Jenison et al., 1990; Strike et al., 1989; Dillner et al., 1990a; 990b; Krchnak et al., 1990). Extensive studies of the immune response to HPV were conducted with these reagents (reviewed in (Viscidi and Shah, 1992)). They have in general shown that assays using these reagents have a limited ability to distinguish between HPV infected patients and control populations. A notable exception to this generalization is the observation that sera from patients with invasive cervical cancer react strongly, and in a type specific manner, to bacterially expressed E7 protein and E7 derived synthetic peptides (Jochmus-Kudielka et al., 1989; Muller et al., 1990). In addition, in some studies, reactivity to other peptides has been found to be a marker of HPV-associated cervical cancer (Dillner et al., 1989; Hamsikova et al., 1994). The most likely explanation for the generally poor performance of these reagents is that the epitopes displayed on synthetic peptides and fusion proteins are predominately linear and the majority of antibodies elicited during natural HPV infection are directed toward conformational epitopes. Experimental proof in support of this hypothesis had to await the generation of virus like particles and the establishment, initially in 1994, of a VLP-based enzyme linked immunosorbent assay (ELISA) for detection of serum antibodies to HPV 16 (Kirnbauer et al., 1994). As compared to previous studies using L1 fusion proteins and synthetic peptides, the HPV 16 VLP ELISA showed a high concordance with detection of HPV 16 viral DNA at the cervix. Serum IgG antibodies to HPV 16 capsids were detected in 59% of women with an HPV 16 infection detected by PCR, whereas only 6% of women negative for cervical HPV DNA were seropositive. VLPs display conformation-dependent neutralizing epitopes as evidenced by their reactivity with well-characterized monoclonal antibodies (Carter et al., 2003; Ludmerer et al., 1996; 1997; Roden et al., 1997a; White et al., 1999). The importance of this class of antibodies in the human immune response to HPV is illustrated by the observation that the majority of the antibody response in human serum against HPV 16 capsids can be entirely blocked by the conformation sensitive neutralizing monoclonal antibody designated V5 (Wang et al., 1997). Studies from several laboratories have validated the sensitivity and specificity of VLP assays for HPV type 16 and other HPV types and established the intra- and interassay variability of the assays. VLP-based ELISAs have been used to describe the natural history of the HPV serum antibody response and in studies of HPV epidemiology. Key features of the performance characteristics of the assays and their application in epidemiological studies are summarized below.

4.1. Sensitivity of VLP-based ELISA

The sensitivity of VLP-based enzyme immunoassays has been measured using serum samples obtained from individuals with documented HPV infections determined by viral DNA detection methods. In general, the sensitivity of HPV-16 VLP-based enzyme immunoassays for identification of prevalent HPV infection, determined by PCR, is approximately 50% or greater (Table 1). The titers of antibody to HPV-16 VLPs are low, with the majority of seropositive samples having absorbance values close to the cut point for seropositivity.

Sensitivity appears to be strongly influenced by viral load and persistence. Women with prevalent low viral loads have ~35% seroprevalence for HPV 16, compared to ~65% seroprevalence among women with high viral load (Kirnbauer et al., 1994; Viscidi et al., 1997). Similarly, only ~25% of women with a transient HPV 16 infection are seropositive, relative to >70% seropositives among women shown to have persistently detectable HPV DNA (de Gruijl et al., 1997; Wideroff et al., 1995). The lower seroprevalence in women with weakly detectable or transiently detectable HPV DNA may result from insufficient production of viral antigen to effectively stimulate an immune response. In addition, the presence of viral DNA may not be associated with production of capsid proteins either because HPV virions are cleared before an infection is established, or infection is confined to the basal layer of the epithelium where only early genes are expressed. The use of PCR as the gold standard for HPV infection has potential shortcomings. HPV virions are resistant to desiccation and may retain viability extracellularly for up to a week (Roden et al., 1997b). Thus, detection of viral DNA by PCR may reflect the presence of viral genomes that never result in infection. Reliance on PCR for HPV diagnosis may also result in misclassification due to contamination in PCR assays, inadequate sampling, or differential sensitivity for some viral genotypes. Another explanation for the low sensitivity of serological assays is the long delay between the detection of HPV DNA by PCR and seroconversion. In a prospective study by Carter and colleagues (Carter et al., 2000), the median time to seroconversion was approximately 12 months in women with an incident HPV type 16, 18 or 6 infection. Finally, since titers of serum antibodies to HPV capsids are low following natural infection, the sensitivity of VLP-based ELISAs can be influenced by assay variability (see below).

4.2. Specificity of VLP-based ELISA

The specificity of VLP-based enzyme immunoassays has been examined in studies with animal sera of known specificity and in studies comparing the reactivity of serum from women with the same HPV type, women infected with other HPV types, and women presumably not exposed to HPV. Polyclonal antisera to different VLPs raised in rabbits have been shown by ELISA (Giroglou et al., 2001b) and hemagglutination inhibition assays (Roden et al., 1996b) to react principally with the specific VLP type used as the immunogen. Although some cross-reactivity has been demonstrated for closely related types such as, HPV types 6 and 11 and types 18 and 45, heterotypic VLP absorption of sera has a minimal effect on homotypic VLP reactivity, suggesting that shared epitopes are weakly antigenic and distinct from the immunodominant type specific epitopes (Giroglou et al., 2001b). A number of observations from studies using human serum samples support the type specificity of HPV VLP seroreactivity. The reactivity of human sera to bovine papillomavirus VLPs is negligible, suggesting that there are no genus specific epitopes displayed on intact VLPs (Heim et al., 1995). The low HPV-16 seroprevalence in children (3-5% (af Geijersstam et al., 1999; Manns et al., 1999; Marais et al., 1997)) argues against significant cross-reactivity between HPV 16 and cutaneous HPV types, which cause skin warts and occur commonly in children. The strong positive association of HPV-16 VLP seroreactivity with sexual behavior is also evidence for the lack of cross-reactivity with cutaneous HPV types. Numerous epidemiological studies have found that lifetime number of male sex partners is the major determinant of HPV-16 VLP seroreactivity in women who are HPV DNA negative (Carter et al., 1996; Castle et al., 2002a; Dillner et al., 1996; Kjaer et al., 2001; Kjellberg et al., 1999; Viscidi et al., 1997; Wang et al., 2003; Wideroff et al., 1996). For example, we observed that HPV 16 seroprevalence increased linearly with increasing number of male sex partners from 8% in women with a single sex partner to 15%, 20%, and 36% in women with 2, 3 to 5, and 6 to 9 lifetime partners, respectively (Viscidi et al., 1997). A study of young women initiating sexual activity also confirms the sexual mode of transmission of HPV-16 and suggests that women have no important extracervical sites that harbor nonsexually transmitted HPV-16 infections (Andersson-Ellstrom et al., 1996b). In this study, 94 adolescent girls were followed for 2 years with repeated serum samples, cervical specimens, and structured interviews regarding sexuality every six months. No antibodies to HPV-16 VLPs were detected among 36 women without sexual experience. In contrast, 4 of the 6 women who acquired HPV-16 DNA during the study seroconverted.

Cross-sectional studies measuring reactivity to multiple HPV VLP types have found that seroreactivity to one type is strongly associated with seroreactivity to other types (Combita et al., 2002; Wang et al., 2003; Wideroff et al., 1999). Interpretation of this data is confounded by the fact that different genital HPV types are transmitted similarly and that high risk women infected with a certain HPV type may have had infections with other HPV types in the past. However, the type specificity of VLP reactivity for individual genital HPV types is supported by the observation of a stronger association of HPV-16 seroreactivity with HPV-16 DNA than with DNA of other types (Table 1). In most of the studies cited in Table 1, women positive for DNA of other types had an HPV-16 seroprevalence similar to that of HPV DNA negative women. In addition, if there were serological cross-reactivity among HPV types, one might expect greater cross-reactivity between genetically similar HPV types, such as between HPV-16 and -31 or between HPV-18 and -45. However, in a study reported by Wideroff et al., stronger associations were not observed for genetically similar pairs (Wideroff et al., 1999). Thus, the most likely explanation for the concordant reactivity to multiple types is cumulative exposure to multiple HPV types, rather than serologic cross-reactivity between types. In fact, studies of the prevalence of cervical HPV infection using PCR have shown that coinfection or serial infection with multiple HPV types is relatively common (Franco et al., 1999; Thomas et al., 2000). However, cross sectional epidemiological studies cannot rule out the possibility of serological cross reactivity. Since type-specific antibodies are usually present in higher titer than cross reactive antibodies, the specificity of VLP-based ELISA assays can be improved by assigning a higher cut-off value for seropositivity. The use of a higher cutpoint may be justified for studies where specificity is more important than sensitivity, for example, to establish a type specific disease association.

The most convincing demonstration of the type specificity of VLP seroreactivity comes from serologic studies of women with incident HPV infection. In the study of Carter et al. (Carter et al., 1996), 67% of women with an incident HPV 16 DNA positive infection seroconverted within 15 months, while only 5% of women with incident infections positive for HPV DNA of other types and 4% of women in whom no DNA was detected developed HPV 16 antibodies over the same time period.

Finally, two other pieces of evidence support the type specificity of HPV-16 VLP seroreactivity of human sera. The HPV-16 type specific monoclonal antibody designated V5 is capable of inhibiting the reactivity of more than 75% of reactive human sera (Wang et al., 2003; Wang et al., 1997). Second, studies with a small number of human sera have shown that absorption with heterotypic VLPs has only a minor effect on the homotypic HPV-16 VLP reactivity (Combita et al., 2002).

4.3. Assay and biological variability of VLP-based ELISA

The interassay coefficient of variation for VLP-based ELISA is typically between 10-15%. A study of variability between different laboratories found good agreement, even though assay protocols were not standadardized, and thus suggests that comparisons between published studies using VLP-based ELISAs should be valid (Strickler et al., 1997). However, the sensitivity of VLP-based assays is highly dependent on having intact VLPs, as partial disruption of VLPs results in loss of reactivity of human sera (Christensen et al., 1996a; Kirnbauer et al., 1994; Wang et al., 1997). Purification methods, storage conditions, and handling during performance of ELISA assays can affect VLP integrity and may explain some of the differences among studies in HPV seroprevalence. In a study of pregnant women who provided paired serum samples approximately 2 years apart, the variability between samples was similar to that of the same sample tested twice, indicating limited biological variability over time (af Geijersstam et al., 1998a). However, pronounced gender differences in seroreactivity to HPV capsids have been noted. In a study of men and women attending sexually transmitted diseases clinics, seroreactivity to HPV 16 VLPs was significantly lower in men, despite the men reporting more lifetime sexual partners than the women (Svare et al., 1997). In a study of HPV 6 VLP reactivity, 30% of men as compared to 50% of women with genital warts were HPV 6 seropositive (Carter et al., 1995). In a population–based survey conducted in the United States, seroprevalence of HPV 16 was significantly higher in women (17.9%) than in men (7.9%). The higher seroprevalence in women is consistent with the well known greater susceptibility of women to most sexually transmitted diseases (Hooper et al., 1978; Langenberg et al., 1999; Mertz et al., 1998) and is attributable to anatomic and other biological differences between men and women. Women and men also have fundamental differences in immune response after exposure to genital tract pathogens (Grossman, 1989; Witkin et al., 2000). Biological differences between the sexes in the natural history of HPV infection are another factor likely to account for the differences in seroprevalence. Infection in men is more transient than in women (Bosch et al., 1996) and the viral load may be lower in men than women due to differences between the sexes in the extent of genital mucosal surface area or the influence of sex hormones on viral gene expression (Brown et al., 1992).

Studies of the reactivity of human sera to HPV VLPs of types other than 16 are limited, but in general have shown that the performance characteristics of VLP-based enzyme immunoassays using these types are similar to those of HPV-16 VLP assays (Dillner et al., 1996; Heim et al., 1995; Kjellberg et al., 1999). One exception to this generalization is the lower sensitivity of HPV-18 VLP-based assays; the sensitivity of HPV-18 VLP assays for HPV-18 DNA positive women is lower (20-35%), relative to the sensitivity of HPV 16 VLP assays in the same studies (50-70%) (Carter et al., 2000; Tachezy et al., 1999; Wideroff et al., 1999). It is unknown whether the lower sensitivity is due to a biological difference in the immune response to HPV-18 compared to HPV-16 or differences in the quantity or quality of the VLP protein used in the assays.

5. Natural History of Anti-capsid Antibody Response

The kinetics of serum immunoglobulin G responses to HPV type 16, 18 and 6 VLPs in relationship to HPV DNA detection has been investigated by Carter and colleagues (Carter et al., 2000). Among women with incident HPV infections, 59%, 54% and 69% seroconverted for HPV types 16, 18 and 6, respectively, within 18 months of detecting the corresponding HPV DNA. The median time to seroconversion was approximately 12 months for the three HPV types. Transient HPV DNA was associated with a failure to seroconvert following incident infection; however, some women with persistent HPV DNA never seroconverted. None of the factors other than DNA persistence examined in this study, which included site of infection, presence of other HPV DNAs, detection and treatment of squamous intraepithelial lesions, and HPV variants, were shown to contribute to failure to seroconvert. Approximately three quarters of HPV 16 and HPV 18 responses persistent throughout the follow-up period, averaging 31 months, but only about a third of HPV 6 antibody responses persisted. Another type specific difference in antibody responses was that HPV 6 seroconversions were detected coincident with the detection of HPV 6 DNA more frequently than were HPV 16 serconversions among women with HPV 16 DNA. As compared to women who were HPV DNA positive, antibodies to HPV VLPs, in women who were repeatedly HPV DNA negative, tended not to persist. Thus, among women who were HPV 16 seropositive at least once but were repeatedly HPV DNA negative, only 16% were seropositive at two or more consecutive visits.

6. Anti-capsid Antibodies of Different Class and Subclass

The serum antibody response to HPV VLPs is primarily of the IgG class, with the major isotype IgG1 and to a lesser extent IgG3 ((Wang et al., 2003) and unpublished data). The prevalence of serum IgA to HPV VLPs is much lower than that of IgG, consistent with the lower serum concentration of total IgA. In the study of Wang et al, IgA seropositivity was associated with recent number of sex partners and, among younger women, correlated with lifetime number of sexual partners. In contrast, IgG seropositivity, while strongly associated with lifetime number of sexual partners, has not been correlated with number of recent sexual partners, suggesting that an IgA response is not as biologically stable over time as an IgG response. This possibility is consistent with the shorter half life of serum IgA (~7 days) compared to IgG (~21 days). Thus, HPV IgA may be a marker of recent or active infection, as is seen for other DNA viruses that cause persistent infection (Chien et al., 2001; Karner and Bauer, 1994; Wittek et al., 1983). Serum HPV 16 IgA reactivity is also predictive of the development of high grade squamous intraepithelial lesions (HSIL) in HIV positive women (Viscidi et al., 2003). The most likely explanation for this association is that persistent infection, which is a major determinant for development of HSIL, would be expected to stimulate and maintain an IgA response. Secretory IgA as well as IgG have been detected in cervical secretions and is associated with detection of the concordant type of HPV DNA and cytological abnormalities caused by high-risk HPV types (Hagensee et al., 2000; Wang et al., 1996). IgM responses are difficult to demonstrate following natural infection, presumably because the IgM response follows shortly after a primary antigenic exposure and then disappears rapidly.

6.1. Seroepidemiological studies of the temporal relationship of HPV infection to HPV-associated cancers

A key element in the epidemiological definition of causal association is that exposure to the putative etiological agent occur prior to the disease. While cross sectional studies of HPV infection in cancer cases and controls have shown a strong association of HPV with cervical and other anogenital cancers and head and neck cancers, it has been difficult to establish the temporal relationship between exposure and cancer using DNA detection methods. The availability of a serum marker of cumulative HPV exposure, the serum IgG response to HPV VLPs, has facilitated prospective seroepidemiological studies of the role of HPV infection in human cancers. Nested case control studies employing serum banks from women living in the Nordic countries and the United States have shown that HPV 16 infection precedes the development of cervical cancer (Chua and Hjerpe, 1996; Dillner et al., 1997; Hisada et al., 2001; Lehtinen et al., 1996; Shah et al., 1997). HPV 16 seropositivity is associated with an increased risk of squamous cell carcinoma, while the risk associated with HPV 18 seropositivity tends to be higher for cervical adenocarcinoma (Dillner et al., 1997). Prospective seroepidemiological studies have also provided evidence that infection with HPV 16 confers an increased risk of developing non-cervical genital cancers, particularly vulvar and vaginal cancers, and squamous cell carcinoma of the head and neck (Bjorge et al., 1997b; Mork et al., 2001). A study combining prediagnostic serology and HPV DNA detection in subsequent cervical carcinomas found a strong type specific correlation between seropositivity and DNA type in the tumor, implying that the carcinogenic effect of HPV infection is dependent on persistence of type specific viral DNA (Sigstad et al., 2002).

6.2. Seroepidemiological studies of association of HPV with non-cervical cancers

The presence of HPV genomes has been reported from a large number of non-cervical cancers, including vulvar, vaginal, penile, anal, bladder, prostate, esophagus, oral cavity, skin, breast, and lung cancers. The consistency of studies for individual cancers has varied widely, perhaps due to technical difficulties in HPV DNA detection. In addition, for many target tissues, samples cannot be obtained from non-diseased persons, making it difficult to establish a disease specific association and to estimate the magnitude of risk. HPV VLP serology provides an easily standadardized measure of HPV exposure for studies of the association of HPV with cancer, and serum samples are frequently available or easily obtained for such studies. Case control seroepidemiological studies have demonstrated an association between HPV 16 and vulvar carcinoma and vulvar intraepithelial neoplasia (Hildesheim et al., 1997; Sun et al., 1996), and penile cancer (Strickler et al., 1998b; Van Doornum et al., 2003). A prospective study has provided evidence for an association between infection with HPV types 16 and 18 and anal and perianal skin cancers (Bjorge et al., 1997a). An etiological role for HPV in some cancers of the oral cavity and oropharynx is now widely accepted (Gillison et al., 2000; Mork et al., 2001; Schwartz et al., 1998; Shah, 1998). Seroepidemiological studies using VLP-based ELISA have contributed significantly to the evidence in support of this association. In a large multicenter cross sectional case control study of head and neck cancers, antibodies to HPV 16 VLPs were associated with risk for cancers of the oral cavity and the oropharynx (Herrero et al., 2003). An increased risk of oral cancers in HPV 16 VLP seropositive women was also observed in a smaller study (Schwartz et al., 1998). A nested case controls study of prediagnostic serum samples showed an increased risk for squamous cell carcinoma of the head and neck among HPV 16 seropositive individuals. Serologic studies have not supported a role for HPV in cancers of the prostate, bladder, lung or breast (Adami et al., 2003; Rosenblatt et al., 2003; Strickler et al., 1998a; 1998b).

Highly sensitive detection techniques, such as nested PCR, have identified HPV DNA in a substantial proportion of nonmelanoma skin cancers (Berkhout et al., 2000; de Villiers et al., 1997; Pfister and Ter Schegget, 1997). However, HPV DNA is detected frequently in healthy skin samples from individuals with and without skin cancer (Antonsson et al., 2000; Boxman et al., 1997; de Jong-Tieben et al., 2000). Because HPV DNA may be a remnant in cancers of past activities or evidence of transient infections of tumors and control subjects, measurement of type-specific humoral immune responses, which reflect viral activities over years, may be a better marker of HPV exposure to use in case control studies of HPV and skin cancer. In a case control study of skin cancer, seroreactivity in HPV types 8, 20 and 38 VLP-based ELISA assays was significantly increased in nonmelanoma skin cancer cases compared to controls (Feltkamp et al., 2003). A high seroprevalence to HPV 8 in individuals with skin cancer in comparison with the general population has also been reported in two other studies (Masini et al., 2003; Stark et al., 1998). However, seropositivity was often associated with more than one skin cancer type, suggesting that the reactivity may be associated with skin cancer in general, and perhaps induced upon skin cancer formation. Furthermore, conditions associated with epidermal repair, such as recovery from second degree burns, autoimmune bullous disease and psoriasis, are associated with generation of antibodies to HPV 5, a cutaneous HPV type also detected in nonmelanoma skin cancers (Favre et al., 2000). The role of HPV in nonmelanoma skin cancers is uncertain and defining that role will likely require longitudinal studies to define the temporal relationship between HPV infection and skin cancer.

7. Serological Studies of Humoral Protective Immunity Following HPV Infection

Mucosal-genital human papillomaviruses are very common sexually transmitted infections. Most infections are benign and clear spontaneously within 1-2 years. Following an HPV infection, approximately 50% of women develop a systemic humoral immune response directed at the major viral capsid protein, which is detectable by VLP-based ELISA. An answer to the question whether the response confers protection against re-infection with the same or related types is important for understanding the natural history of HPV infection. A large population based prospective study of HPV failed to demonstrate that serum antibodies to HPV types 16, 18 or 31 elicited by natural infection are associated with immune protection against re-infection with the homologous HPV type or heterologous types (Viscidi et al., 2004) In another study sustained high levels of antibody to HPV 16 were associated with a reduced risk for subsequent infection with HPV 16 and genetically related types (Ho et al., 2002). Comparisons between the studies are difficult because demographic characteristics of the populations, definition of immune status, duration of follow-up, and methods of analysis differed. In a study of seropositivity to VLPs of HPV types 16, 18, 31, 35 and 45 and subsequent cervical HPV infection among HIV positive and negative women, no significant difference in the risk of new infection with the homologous type was found between HPV seropositive and -seronegative women (Viscidi et al., 2005). Likewise, there was no evidence for significant group-specific protection conferred by HPV seropositivity.

These results from natural history studies conflict with the reported 100% efficacy for VLP-based vaccines administered to human volunteers (Koutsky et al., 2002). The most likely explanation is the 50-60 fold lower level of anti-VLP antibodies elicited following natural infection. In addition, for HIV-positive women some of the newly detected infections may have been reactivation of previous infection rather than an exogenously re-acquired or true incident infection. These studies should not be interpreted to mean there is no naturally induced protective immunity to HPV. Serological measurements may not identify all immune persons. Approximately 50% of HPV-infected women do not seroconvert but may nevertheless be protected against subsequent infection as a result of being primed for a rapid anamnestic antibody response. The cellular arm of the immune system may mediate immune protection but correlate poorly with measures of serum antibody. Incident infection is the most easily measured outcome for analysis of immune protection, but cervical cancer is the outcome of greatest interest. Prospective nested case control studies of cervical cancer risk have shown evidence for serological protection by HPV6 infection against HPV type 16 cervical cancer (Silins et al., 1999). The risk of cervical cancer among women seropositive for both HPV6 and HPV16 was significantly less than expected. A similar antagonism was not observed for HPV6 and HPV18 or 33 (Luostarinen et al., 1999; Luostarinen et al., 2004). One explanation for this observation is that HPV 6 and HPV 16 capsid antibodies are a marker of a cross-protective cell-mediated immune response to HPV infection. However, differential misclassification of exposure or selective (by infection) treatment of women for premalignant lesions cannot be excluded as explanation for the antagonistic interaction between HPV 6 and 16 seroreactivity.

8. Applications of VLP-based ELISA in Population Surveys

The natural history of HPV infection, with a high rate of acquisition and a high rate of spontaneous clearance, has made it difficult to ascertain the cumulative prevalence of HPV infection using DNA detection methods. HPV seropositivity, on the other hand, is a marker that is primarily linked with past exposure, as measured directly by past HPV DNA positivity or indirectly by number of lifetime sex partners. Indeed, in a multivariate model analysis, current HPV DNA was not associated with seropositivity after adjusting for past DNA positivity (Castle et al., 2002b). A number of studies have examined the prevalence of HPV in population-based cohorts. In a population–based cohort of 10,000 women in Guanacaste province, Costa Rica, overall the seroprevalence for HPV 16, 18, 31 and 45 was 15, 15, 16 and 11%, respectively (Wang et al., 2003). Although HPV 16 alone accounts for approximately half of all cervical cancers in Guanacaste (Herrero et al., 2000) and worldwide (Munoz et al., 2003) and the population-based prevalence of HPV 16 DNA is highest in this cohort, the predominance of HPV 16 is not reflected in seroprevalence and the overall and age-specific seroprevalence of HPV 16, 18 and 31 were equivalent. The reasons for the difference between type-specific DNA prevalence and seroprevalence are unclear. In Guanacaste, HPV seroprevalence reached its highest peak at 25-34 years for HPV 31 and at 35-44 years for HPV 16, 18 and 45. Seropositivity declined only slightly with age after its peak. In contrast, HPV DNA positivity peaked for the four HPV types in women less than 25 years old and declined with increasing age. The 10-20 year lag between peak DNA positivity and peak seroprevalence is probably due to low rates of seroconversion and/or the potential need for repeated exposure, and thus antigenic stimulation, to induce a detectable antibody response. The simultaneous increase in seroprevalence and decline in DNA prevalence in women over 25 years of age may indicate that there women continue to be exposed to HPV as a result of repeated or recurrent infection or reactivation of latent infection, but that levels of viral replication are below the detection limit of PCR. A national seroepidemiological survey of HPV 16 infection in the Untied States was performed on serum samples from the second phase of the third National Health and Nutrition Examination Survey (NHANES III), conducted from 1991 to 1994 (Stone et al., 2002). Overall, 13% of study participants aged 12-59 years had antibodies to HPV 16. Seroprevalence was significantly higher in women (17.9%) than in men (7.9%). Seroprevalence was 12.5% among whites, 19.1% among blacks, and 8.9% among Mexican Americans. HPV 16 seroprevalence reached a peak of 24.7% in women ages 20-29 years. In men seroprevalence peaked later at ages 30-39 (11.5%). The HPV 16 seroprevalence probably underestimates the cumulative prevalence of HPV 16 infection in the Untied States because not all persons who become infected with HPV 16 develop a serologic response and the duration of seropositivity is not well defined especially for men. Possible explanations for gender differences are discussed above in the section on biological variability of serological assays. Population-based seroprevalence studies have helped to explain population differences in the risk of cervical cancer. For example, the prevalence of HPV 16 in population-based cancer controls in Columbia and Spain was 22% and 3%, respectively (Nonnenmacher et al., 1995; 1996), which correlates with the 8-fold higher incidence of cervical cancer in Colombia than in Spain (de Sanjose et al., 1994). The etiological association between HPV infection and cervical cancer leads to the prediction of a geographic concordance between prevalence of genital HPV infection and the incidence of cervical cancer, as observed in the example of Colombia and Spain cited above. However, a lower HPV DNA prevalence has been reported for Greenlandic compared to Danish women, despite a 5 times higher incidence of cervical cancer in Greenland (Kjaer et al., 1993). This paradox was resolved by using a VLP-based ELISA to measure HPV exposure (Nonnenmacher et al., 1996). The overall seroprevalence of HPV was, as expected, higher in Greenlandic women (56%) than Danish women (41%) and the population difference was even more pronounced for women in the 20-24 year age-group (60% versus 27% in Greenlandic versus Danish women, respectively). It is therefore likely that high risk genital HPV infection at an early age is generally common in Greenland and contributes to the high rate of cervical cancer in Greenlandic women. Another area in which HPV serology has proven invaluable is studies of temporal trends in HPV infection. A population based survey of HPV 16 infection in pregnant women in Stockholm, Sweden during 1969, 1983 and 1989, showed overall seroprevalence rates of 16% in 1969, 22% in 1983 and 21% in 1989 (af Geijersstam et al., 1998b). The seroprevalence of herpes simplex virus (HSV) type 2 showed a similar trend, whereas seroprevalence rates of HSV1 were the same in these years. The results suggest that HPV 16 and HSV 2 became more generally spread in the Swedish population between 1969 and 1983 and then the spread was stable during the 1980s. The serological findings are consistent with changes in the overall rate of sexual contacts in the population. Similar trends have been reported from Finland (Laukkanen et al., 2003). For 23-31 year old women, HPV 16 incidence increased over the period 1983-97 and HPV 16 seroprevalence increased from 17% in 1983-5 to 24% in 1995-7. Changes in sexual behavior of Finnish women from 1971 to 1999 have been well documented and the consequent increase in risk-taking behavior probably exposed more and more susceptible individuals to HPV 16 and eventually resulted in the increase in HPV 16 seroprevalence. In contrast to HPV 16, HPV 6 and 11 prevalence was stable at 9-12% throughout this period, which suggests an epidemic steady state. This presumably reflects different modes of transmission and dynamics of spread for HPV 6/11 and HPV 16 infections. HPV 16 is predominately transmitted by mucosal contacts, while HPV 6/11 can also be transmitted via the skin (Wen et al., 1999). An alternative explanation for the finding is competitive advantages favoring the spread of HPV 16 over HPV 6. The major mode of transmission of genital HPVs is sexual, as evidenced by the absence of HPV DNA among virginal women (Andersson-Ellstrom et al., 1996a; Rylander et al., 1994), the fact that detection of HPV DNA is strongly correlated to a recent change of sexual partner (Hildesheim et al., 1993), and the strong association of HPV seropositivity with life–time number of sexual partners (Castle et al., 2002a; Dillner et al., 1996; Viscidi et al., 1997). However, other modes of transmission have been suggested, such as infection in utero or at birth. Studies using PCR have found HPV DNA in buccal swabs taken from children (Cason et al., 1995; Puranen et al., 1996; Tenti et al., 1999), although other studies using similar methods found no evidence of perinatal transmission of HPV (Smith et al., 1995; Watts et al., 1998)-. Because PCR is extremely sensitive, detection of HPV DNA cannot distinguish between infection and merely contamination with viral DNA. In a survey of 1031 children aged 0 to 13 years, resident in Stockholm Sweden, the HPV seroprevalence by VLP-based ELISA was 3.0% for HPV 16, 0.6% for HPV 18 and 2.76% for HPV 33 (af Geijersstam et al., 1999). The presence of HPV antibodies among some children may suggest that non-sexual transmission of these viruses can occur. However, alternative explanations also need to be considered, including maternal IgG for newborns, sexual activity in the children aged 10-13 years, cross reactivity with cutaneous HPV types, especially for the low titered serum samples, which constituted the majority of the seropositives, transfusion of blood or platelets, or sexual abuse among pre-adolescent children. With these caveats the results suggest that infection with oncogenic HPV types 16, 18 and 33 is uncommon in children.

9. Vaccines Against Papillomavirus

Prophylactic HPV vaccines aim to prevent new HPV infections by generating antibodies that neutralize the viral inoculum prior to infection, thus interrupting the transmission of HPV from one person to another. HPV may, however, escape antibody neutralization and establish breakthrough infections in some patients even after prophylactic vaccination. In addition to potential breakthrough infections, preexisting HPV infections are very common and a current cause of morbidity in many patients. The long time course from HPV infection to the development of cervical cancer provides a large window of time in which to treat these established HPV infections or HPV-associated neoplasia before they progress to invasive cancer. Thus, development of treatments for existing HPV infections and HPV-associated neoplasia is also an important endeavor. Therapeutic HPV vaccines aim to treat established HPV infections and HPV-associated malignancies by eliciting an immune response against HPV-infected cells, thus potentially preventing the development of lesions, eliminating existing lesions, and possibly even eliminating malignancies. These approaches are not mutually exclusive. Indeed, the best approach would be a combined prophylactic and therapeutic vaccine to prevent new infections and eliminate breakthrough infections and pre-existing disease.

9.1. Prophylactic vaccination with VLPs

The biology of HPV provides potential targets for prophylactic HPV vaccination. Because L1 and L2 are the structural components of the infectious viral capsids, prophylactic vaccines generally attempt to stimulate the

Table 2. Summary of published L1-based prophylactic vaccine studies in animal models

Study	Virus - animal	Dose-Adjuvant	Vaccination Schedule	ELISA Titer	Challenge	Protected
Jarrett et al., 1990a; 1990b; 1991	BPV1,2,4,5,6-Cow	virus	0, 3 wks	>1000		Yes for homologous type &1/2
Bell et al., 1994	COPV-Dog	2x inactivated virus	0, 2 wks	N/D	+ 4wks	Yes
Donnelly et al., 1996	CRPV-Rabbit	1mg L1 DNA	0, 3 wks	3,000	+4wks	Yes
Moore et al., 2002	COPV-Dog	9μg L1 DNA	0,6,12 wks	320	+3wks	Yes
Lin et al., 1993	CRPV-Rabbit	Vaccinia L1 10^6 pfu	0,4 weeks	810-1080	+5wks	Yes
Breitburd et al., 1995	CRPV-Rabbit	3x 50μg VLP-Alum	0, 2, 4 wks	5,000	+1 wk	Yes
Kirnbauer et al., 1996	BPV4-Cow	2x 150μg VLP-Alum	0, 4 wks	1,000	+2wk	Yes
Suzich et al., 1995	COPV-Dog	2x 20μg VLP-None	0,2 wks	<1,000	+2 wk	Yes
Christensen et al., 1996b	CRPV-Rabbit	3x 50μg VLP-None	0, 4, 8 wks	10,000	+2 wk	Yes
				100	+12 mo	
Jansen et al., 1995	CRPV-Rabbit	3x 1μg VLP-None	0, 4, 8 wks	10,000	+2 wk	Yes
Yuan et al., 2001	COPV-Dog	0.4μg L1 pentamers-None	0, 2 wks	>100	+2 wk	Yes

production of anti-L1 or anti-L2 antibodies that prevent infection by neutralizing HPV virions. Prophylactic vaccines have classically used live attenuated or inactivated viruses to simulate infections in patients and provoke an immune response. Seminal studies demonstrated that vaccination of cows with formalin inactivated BPV1 virions both induces high titers of neutralizing antibodies in the serum and provides protection against viral challenge (Dvoretzky et al., 1980; Jarrett et al., 1990a; 1990b) (**Table** 2). Furthermore, vaccination with inactivated canine oral papillomavirus (COPV) virions is protective for natural transmission in dogs (Bell et al., 1994). However, this approach is inappropriate in HPV vaccination because HPV cannot be efficiently propagated in cultured cells (Hagensee and Galloway, 1993) and vaccination with live attenuated or inactivated HPV would carry a risk of virally induced carcinogenesis in healthy young women, the majority of whom are not destined to develop cervical cancer. Vaccination, instead, with HPV capsid proteins, L1 and/or L2, in the absence of the viral genome, circumvents these issues.

In 1992, Kirnbauer et al. demonstrated that BPV1 L1 assembles into VLPs when it is over-expressed in insect cells and that vaccination with these particles induces titers of serum neutralizing antibodies equivalent to those generated with native virions (Kirnbauer et al., 1992). In several animal models, parenteral vaccination with animal papillomavirus L1 VLPs demonstrated the production of high titers of serum neutralizing antibodies, which were able to protect the animals from cutaneous or mucosal challenge with infectious papillomaviruses of homologous type (Breitburd et al., 1995; Kirnbauer et al., 1996; Suzich et al., 1995) (Table 2). Passive transfer of immune serum immunoglobulin into naïve animals is protective, demonstrating that serum neutralizing antibody is sufficient for immunity to experimental infection (Breitburd et al., 1995; Kirnbauer et al., 1992; Suzich et al., 1995). It is also important to note that vaccination with denatured L1 neither induces neutralizing antibodies nor protects against experimental infection (Breitburd et al., 1995; Pilacinski et al., 1986). Interestingly, several studies demonstrate that L1 capsomers are also an effective immunogen (Ohlschlager et al., 2003; Rose et al., 1998; Yuan et al., 2001). Thus VLPs and capsomers but not denatured L1 display the conformational epitopes necessary to induce high titer neutralizing antibodies and protection against experimental papillomavirus challenge.

Table 3. HPV VLP vaccine clinical trials

HPV Type	Vaccine Type	Ag	Dose (Adjuvant)	Vaccination Schedule	Sponsor	Patient Population	Study
HPV-6b	VLP	L1	1, 5, or 10 µg (none)	0, 4, 8 weeks	CSL Ltd.	Clinical trial in patients with genital warts	Zhang et al., 2000
HPV-11	VLP -- MEDI-501	L1	3, 9, 30, or 100 µg (Alum)	0, 4, 16 weeks	MedImmune	Phase I trial in healthy volunteers	Evans et al., 2001
HPV-11	VLP	L1	10, 20, 50, or 100 µg (Alum)	0, 8, 24 weeks	Merck	Phase I trial in healthy volunteers	Brown et al., 2001
HPV-16	VLP	L1	10 or 50 µg (none, Alum, or M59)	0, 4, 16 weeks	NCI	Phase I trial in healthy volunteers	Harro et al., 2001
HPV-16	VLP	L1	50 µg (none)	0, 4, 16 weeks	NCI	Phase II trial in healthy volunteers and large scale, randomized, double-blind, placebo controlled clinical trial in 21,000 women in Costa Rica	Bratti et al., 2004
HPV-16	VLP	L1	2 µg at 0, 4 weeks then 50 µg at 12 and 16 weeks (none) or 50 µg at 0 and 4 weeks (none)		NCI	Clinical trial in healthy volunteers	Nardelli-Haefliger et al., 2003
HPV-16	VLP	L1	40 µg (Alum)	0, 8, 24 weeks	Merck	Double-blind, randomized controlled clinical trial with 2392 healthy volunteers	Koutsky et al., 2002
HPV-18	VLP	L1	80 µg (Alum)	0, 8, 24 weeks	Merck	Phase I trial in healthy volunteers	Ault et al., 2004
HPV-11, 16,	VLP	L1	10, 20, 50, or 100 µg HPV-11 L1 (Alum); 10, 40, or 80 µg HPV-16 L1 (Alum)	0, 8, 24 weeks	Merck	Phase I trial in healthy volunteers	Fife et al., 2004
HPV-16, 18	VLP	L1	20 µg HPV-16 L1, 20 µg HPV-18 L1 (Alum/MPL)	0, 4, 24 weeks	GSK	Double-blind, randomized controlled clinical trial with 1113 healthy volunteers	Harper et al., 2004
HPV-16	Chimeric VLP	L1-E7			Medigene	CIN patients	unpublished

HPV, human papillomavirus; n/a, not available or not applicable; NCI, National Cancer Institute; VLP, virus like particle

In addition to parenteral vaccination, other, less invasive, routes of vaccine administration have been explored as methods of inducing a neutralizing antibody response. Oral vaccination of mice with HPV VLPs induces a systemic HPV-specific IgG and IgA antibody response (Rose et al., 1999). Nasal inoculation of mice with VLPs also generated IgG and IgA antibodies specific for HPV (Balmelli et al., 1998; Nardelli-Haefliger et al., 1997; Nardelli-Haefliger et al., 2003). These alternative vaccination methods represent less invasive vaccination methods that are capable of producing neutralizing antibodies. Other alternatives such as naked L1 DNA vaccines or live viral or bacterial vectors have also been tested (Donnelly et al., 1996; Lin et al., 1992; Nardelli-Haefliger et al., 1997) (Table 2).

Although L1 VLP vaccines can generate significant titers of specific neutralizing antibodies, several studies have demonstrated that the antibodies produced by these vaccines possess only a limited ability to neutralize papillomavirus types other than the one used for vaccination (Breitburd et al., 1995; Roden et al., 1996b; Rose et al., 1999; White et al., 1998). VLPs of different genotypes but with an L1 exhibiting >85% sequence identity show weak cross-neutralization e.g. HPV18 and HPV45 or HPV6 and HPV11 (Roden et al., 1996b; White et al., 1998). Similarly, vaccination with L1 VLPs fails to protect animals from experimental challenge from different papillomavirus types (Breitburd et al., 1995; Jarrett et al., 1990b). This lack of (or very limited) cross-neutralization and cross-protection is one major drawback of L1 VLP vaccination because there are at least 22 oncogenic HPV genotypes, which potentially represent an equivalent number of serotypes (Roden et al., 1996a; 1996b). Therefore, protection against infection by all of these types with a VLP-based vaccine would likely require a highly polymeric vaccine that could be difficult to generate.

L1 VLPs have been the major prophylactic HPV vaccine produced for use in clinical trials because of their ease of large-scale production and their ability to elicit high titers of neutralizing antibody. Importantly, in all clinical trials to date, intramuscular vaccination with HPV L1 VLPs has been shown not only to be immunogenic but also safe in women (Evans et al., 2001; Harro et al., 2001; Nardelli-Haefliger et al., 2003) (Table 3).

While L1 VLPs have provided a means to produce prophylactic HPV vaccine particles on a large scale, some doubts have been raised about the ability of parenteral vaccination to protect against sexual transmission of HPV. Because antibodies capable of neutralizing an infection must be present at the site of infection (i.e. at the cervical mucosa) in order to prevent infection, there has been some concern that parenteral vaccination, which produces a systemic antibody response, may not generate local neutralizing antibodies that reach the cervical mucosa and prevent the natural transmission of HPV. However, these concerns have been largely removed by the demonstration firstly that women vaccinated intramuscularly with HPV16 L1 VLPs had significant titers of specific antibodies in cervical secretions (Nardelli-Haefliger et al., 2003) presumably via transudation of serum IgG, and secondly a double-blind, randomized, controlled trial in 2392 women demonstrated that parenteral vaccination with HPV16 L1 VLPs protects women from HPV infection and HPV-associated CIN (Koutsky et al., 2002). In this study, the incidence of persistent HPV16 infection was 3.8 per 100 woman-years at risk in the placebo group and 0 per 100 woman-years at risk in the vaccine group, i.e. the vaccine was 100% effective (95% CI 90%-100%) in this trial for preventing persistent HPV16 infection. Additionally, all nine cases of HPV16-related CIN occurred among the placebo recipients (Koutsky et al., 2002). These studies suggest that serum antibodies can, in fact, transudate from plasma to the cervical mucosa where they protect from HPV infection. Interestingly, while this vaccine was 100% effective at preventing persistent HPV infections and CIN, some patients in the treatment group developed transient HPV infections that later resolved. These results and the observation that HPV16 L1 VLP vaccination generates L1-specific $CD4^+$ and $CD8^+$ T cells (Pinto et al., 2003) suggest the possible contribution of cellular immunity in preventing persistent infection or transient detection of deposited, but neutralized, virus. Based on early promising results, other clinical trials of VLP vaccines have been initiated, including large, randomized, double blind, placebo controlled trials of HPV16 and 18 L1 VLPs. Harper et al. recently reported a randomized trial in 1113 women in North America and Brazil using three immunizations with HPV16 and 18 L1 VLPs in AS04 adjuvant (Harper et al., 2004). This trial demonstrated 91.6% (95% CI 64.5-98.0) protection against incident infection and 100% against persistent infection (95% CI 47.0-100) with HPV16 or HPV18 after 27 months of follow up (Harper et al., 2004). Another longer term trial in Costa Rican women already characterized as part of a study of the natural history of HPV infection and cervical neoplasia is underway (Bratti et al., 2004). Assessment of the long-term protective efficacy of VLP vaccination awaits the completion of these trials.

While early data about HPV L1 VLP vaccines seem to demonstrate a significant ability to protect women from the specific HPV type used for vaccination, the problem of limited cross-neutralizing power observed in preclinical studies has also been observed in clinical studies. In the trial of 2392 women described by Koutsky et al., equal numbers of non-HPV16 related CIN were present in both the placebo and vaccine arms; however, HPV16 positive CIN only occurred in the placebo arm, demonstrating type specific protection (Koutsky et al., 2002). To overcome the limited cross-neutralizing ability of L1 VLPs and protect against multiple HPV genotypes, clinical studies have turned to the use of polyvalent L1 VLP vaccines. Both GlaxoSmithKline plc (GSK) and Merck & Co., Inc. have begun clinical trials of polyvalent HPV L1 VLP vaccines to protect against multiple HPV types (Table 3). GSK has introduced a vaccine containing both HPV16 and 18 L1 VLPs (Harper et al., 2004) and Merck has started clinical testing of a tetravalent vaccine containing HPV types 6, 11, 16, and 18 L1 VLPs, for which initial results indicate that the vaccine is well tolerated and can generate neutralizing antibody titers to each of the four HPV types in quantities similar to those produced by monovalent L1 VLP vaccines. With the continued clinical development of VLP vaccines for HPV types other than HPV16, such as HPV6b (Zhang et al., 2000) and HPV11 (Brown et al., 2001; Emeny et al., 2002; Evans et al., 2001), and of polyvalent VLP vaccines, protection may be possible against low risk HPV types. However, it is important to note that together HPV16 and HPV18 currently account for only ~70% of cervical cancer cases. This leaves ~30% of cervical cancer caused by at least 20 other HPV genotypes and the theoretical concern that additional HPV genotypes could expand to fill the niche left after effective vaccination against only HPV16 and HPV18. Therefore broad protection against all oncogenic HPV types is likely to require a highly polyvalent L1 VLP vaccine that will be complex to formulate.

Another possibility is to find a conserved protective antigen. L2 has shown promise in this regard (Campo and Jarrett, 1994; Christensen et al., 1991; Kawana et al., 1999; Lin et al., 1992; Roden et al., 2000). An HPV L2 vaccine has also been tested for its ability to elicit cross-neutralizing antibodies in women. One clinical study showed that nasal inoculation with an epitope of HPV16 L2 produces L2-specific cross-neutralizing antibodies that bind both HPV16 and HPV52 L1/L2 capsids (Kawana et al., 2003). Unfortunately, vaccination with VLPs comprising L1 and L2 results in a predominantly L1 and type-specific neutralizing antibody response (Roden et al., 2000). This suggests that it will be difficult to generate an L2-specific cross-neutralizing response in the context of the VLP unless the L1 immunodominant epitopes are substituted with L2 epitopes (Varsani et al., 2003) or the particle is chemically coated with L2 neutralizing epitopes (Chackerian et al., 2001).

9.2. Therapeutic vaccination against HPV-related disease

While the main purpose of prophylactic HPV vaccination is to generate neutralizing antibodies, therapeutic HPV vaccines attempt to elicit cellular immune responses. Increasing evidence suggests that a cell-mediated immune response is important for controlling established HPV infections and other HPV-associated diseases. For example, studies show an increased prevalence of HPV-related infection and malignancies in immunosuppressed patients with impaired cell-mediated immunity, such as transplant or AIDS patients (Halpert et al., 1986; Laga et al., 1992; Schafer et al., 1991; Sun et al., 1997). Furthermore, warts in patients treated with immunosuppressive drugs often regress when drug therapy is discontinued (Benton et al., 1989). T-cell-mediated immune responses are particularly indicated because of findings that $CD4^+$ and $CD8^+$ T cells infiltrate spontaneously regressing warts at a higher frequency than non-regressing warts (Coleman et al., 1994).

While vaccination with L1 and L1/L2 VLPs may generate cellular immune responses to L1 and/or L2, this cellular immunity is unlikely to be useful for clearing infected basal cells or HPV-associated malignancies because they do not express L1 and L2 at detectable levels. Therefore, it is important to consider the development of therapeutic vaccines that can generate specific cellular immune responses against other HPV antigens. Only some virally encoded early proteins, such as E6 and E7, are expressed at all levels of the infected epithelium and in HPV-associated malignancies. For these reasons, therapeutic HPV vaccines generally attempt to clear existing infections by generating cellular immunity against virally encoded early proteins.

When designing therapeutic vaccines, the choice of which early gene product to target as an antigen is important. For therapeutic HPV vaccines, the early viral proteins (E1, E2, E5, E6, and E7) are potential target antigens because they are expressed early in viral infection and help regulate progression of the disease. Of these proteins, E6 and E7 are most frequently targeted because they are consistently expressed in most cervical cancers and their precursor lesions, unlike some other early proteins. E6 and E7 also make good potential targets for therapeutic vaccination because they are required for the induction and maintenance of the malignant phenotype (Crook et al., 1989) and, thus, are not likely to be lost by antigen loss as tumor cells evolve to escape the immune response. Therefore, E6 and E7 represent potentially ideal antigen targets for therapeutic HPV vaccination. The optimal approach would be to develop a single vaccine with both prophylactic and therapeutic activity.

10. Chimeric VLP Vaccines

Several experimental vaccine strategies have attempted to combine the prophylactic efficacy of L1 VLP vaccines with therapeutic immunity by including early papillomaviral proteins into VLPs. These chimeric VLP strategies for combination/therapeutic vaccination may be derived by fusion of the therapeutic antigen, such as E7, E6 or E2 with either L1 or L2. Fusion with L1 has the advantage of higher copy number per particle, i.e. 360 molecules of the early papillomavirus antigen per VLP, but these fusions easily disrupt particle structure so assembly fails and neutralizing antibodies are not generated (Dale et al.,

2002; Liu et al., 2000; Muller et al., 1997; Peng et al., 1998). For example the size of the early papillomavirus antigen may be limited to approximately 60 amino acids for fusions at the C-terminus of L1 (Muller et al., 1997). This issue may be circumvented by the use of chimeric capsomers (Ohlschlager et al., 2003). Insertion of epitopes into the surface loops of L1 has also been tried, but this is limited to very small peptides and the particle structure is unpredictably disrupted by these insertions/substitutions (Chackerian et al., 1999; Slupetzky et al., 2001), although direct coupling to the surface of the VLP is an alternative approach (Chackerian et al., 2001). Fusion of the early papillomavirus antigen with L2 has the disadvantage of lower copy number per virion, i.e. approximately 12 molecules per particle. However, very large molecules can be fused to the C-terminus of L2 without disrupting its insertion within the VLP or the capsid structure. Indeed fusions of E6, E7, E2 or E7E2 can be made to the C-terminus of L2 without disrupting co-assembly with L1 (Greenstone et al., 1998). Indeed vaccination with these chimeric VLP induce equivalent neutralizing titers as seen for L1 only VLPs (Greenstone et al., 1998; Muller et al., 1997).

Although VLPs represent an exogenous antigen, it has been postulated that upon binding to a cell surface receptor, VLPs gain entry into the cytosol of infected cells and the capsid proteins L1 and L2 can be processed in the MHC class I presentation pathway (Peng et al., 1998). Vaccination with L1 VLPs induces a potent L1-specific, CD8+ T cell-mediated, perforin dependent immune response (De Bruijn et al., 1998; Peng et al., 1998; Rudolf et al., 1999). Indeed, a single vaccination with L1 VLPs without adjuvant is sufficient to protect mice from an L1-expressing tumor line (De Bruijn et al., 1998). However, expression of L1 is generally absent from cervical cancers, and high grade CIN, as well as basal epithelium harboring an HPV infection. Therefore L1 is not an appropriate target for therapeutic vaccine, although some suggestion of therapeutic effects with VLPs have been suggested (Zhang et al., 2000).

Protection from tumor challenge was tested in C57BL/6 mice by using the tumor cell line TC-1, which expresses HPV16 E6 and E7, but not the virion structural proteins, is a frequently used model to assess oncoprotein specific immunity (Wu et al., 1995). Injection of HPV16 L1/L2-HPV16 E7 chimeric VLPs, but not HPV16 L1/L2 VLPs, protected the mice from TC-1 tumor challenge, even in the absence of adjuvant (Greenstone et al., 1998). The chimeric VLPs also induced protection against tumor challenge in major histocompatibility class II-deficient mice, but not in beta2-microglobulin or perforin knockout mice implying that protection was mediated by class I-restricted cytotoxic lymphocytes (Greenstone et al., 1998). HPV16 capsids or capsomers also stimulate a MHC class I restricted CTL response with human peripheral blood lymphocytes (PBL) *in vitro* (Ohlschlager et al., 2003; Rudolf et al., 1999). The vigorous response was specific for VLP-infected target cells and was MHC class I restricted demonstrating the presence of at least one HLA-A*0201 restricted CTL epitope within the HPV-16 capsid proteins by using a VLP-'infected' HLA-A*0201 transfected human cell line as target cells (Ohlschlager et al., 2003; Rudolf et al., 1999).

The influence of vaccination route upon the cellular immune response to chimeric VLP in mice immunized intrarectally (i.r.), intravaginally (i.va.) or intramuscularly (i.m.) was assessed using P18 chimeric VLP without adjuvant (Liu et al., 2002). Cytotoxic T lymphocyte precursors specific for the P18 epitope were recovered from the spleen following i.m., i.va. or i.r. immunization with P18 VLP, and were similarly detected in Peyer's patches following i.m. or i.r. immunization (Liu et al., 2002). It is noteworthy that the effectiveness of cytotoxic T lymphocyte (CTL) induction after repetitive vaccination with the same chimeric VLP, and thus vaccine efficacy, is limited by the presence of neutralizing antibodies induced after the first application (Da Silva et al., 2001). While mucosal delivery of chimeric HPV16 L1/L2-E7 VLPs resulted in detection of systemic E7-specific CD8(+) T cells, vaccination by these routes was not able to bypass the inhibitory effect of circulating antibodies against homologous VLP types (Da Silva et al., 2003). In contrast, mice immunized and boosted with heterologous cVLPs containing HPV16 E7 showed a higher frequency of E7-specific T cells *in vitro* and displayed reduced tumor growth in a therapeutic setting compared to mice treated with homologous cVLPs (Da Silva et al., 2003).

A comparison of chimeric VLPs with the HPV16 E7 antigen fused to L1 versus chimeric VLPs with E7 fused to L2 with respect to their ability to protect mice from tumor challenge indicates that HPV16 L1-E7(1-57) is perhaps slightly superior to HPV16 L1/L2-E7(1-57) in eliciting tumor protection at equivalent doses, although both types of particles were able to protect mice (Wakabayashi et al., 2002). Both chimeric VLPs induced a specific cytotoxic T lymphocyte (CTL) response to the H2-D(b)-restricted E7 peptide (E7(49-57)) as determined by an ELISPOT assay and tetramer staining; however, immunization with the L1-E7(1-57) chimeric VLPs resulted in twofold higher CTL precursor frequencies. These results indicate that chimeric VLPs with the antigen fused to L1 maybe slightly more efficient vaccine with respect to tumor prevention than chimeric VLPs with the antigen fused to L2. However, L1 chimeric VLPs are limited by the size of the antigen that can be incorporated when compared to L1/L2 chimeric VLPs (Wakabayashi et al., 2002).

11. Immune Recognition of VLPs

HPV16 VLPs exhibit a highly ordered, close packed foreign structure (Bachmann et al., 1993) and engender a potent Th1 immune response in humans, inducing both high titer neutralizing antibody and cell-mediated immune responses (Evans et al., 2001; Harro et al., 2001). Indeed, a single L1 VLP vaccination without adjuvant affords mice CD8+ T cell dependent protection against tumor formation by the HPV16 genome-transformed cell line C3 (De Bruijn et al., 1998). The immunogenicity of other peptide or whole protein antigens is dramatically enhanced upon fusion with VLPs (Greenstone et al., 1998; Jochmus et al., 1999). Indeed, vaccination with chimeric VLP elicits a potent CTL response against a self (Nieland et al., 1999) or an otherwise weak antigen (Rudolf et al., 2001a; 2001b; 1999) and protects against tumor challenge in preclinical models (Da Silva et al., 2003; Greenstone et al., 1998; Peng et al., 1998). Notably, presentation of self antigens at high density on the surface of VLPs is sufficient to overcome B cell tolerance (Chackerian et al., 2002; 1999; 2001). To ascertain the nature of the potent immunogenicity of VLPs several investigators have studies their interaction with antigen-presenting cells and the role of innate immunity in their recognition.

Antigen-presenting cells, and especially dendritic cells (DCs), employ pattern recognition receptors including the Toll family to sense infections (Takeda et al., 2003). In response to a pathogen-associated molecular pattern, Toll family members signal via an adaptor molecule MyD88, activating NF-κB and AP-1 transcription factors and innate, pro-inflammatory responses (Akira, 2003b). Importantly, these innate responses are largely responsible for shaping adaptive immune responses to a pathogen or vaccine antigen. DCs comprise a heterogeneous cell population in naïve animals that have been classified based on surface markers, functional potential and location, as Langerhans cells in the skin, mature tissue interstitial DCs present in secondary lymph nodes, and multiple DC subsets in the spleen (Shortman and Liu, 2002).

Recent studies demonstrate that VLPs bind to DCs and stimulate their maturation including the upregulation of MHC class I, class II, CD80, CD86 and CD40 and cytokine production (Fausch et al., 2003; 2002; Lenz et al., 2001; 2003). VLPs effectively activate BMDCs to produce a range of responses, notably expression of IFN (α and γ) and IL-12 (Yang et al., 2004a). These immune mediators favor the differentiation of T helper 1 (Th1) cells, consistent with the responses of mice immunized with VLP-loaded DCs. Interestingly, production of IFN (α and γ) and IL-12 and Th1-biased immune responses to HPV16 VLPs is dependent upon MyD88, consistent with a role for the Toll/IL-1 receptor family in immune recognition of HPV16 VLP (Yang et al., 2004a).

The murine model is useful for studies of the immune response to HPV16 L1 VLPs because of the availability of gene knockout mice, but observations should be validated in the human system. Both vaccination of patients with HPV16 VLP (Harro et al., 2001) and the treatment of naïve C57BL/6 mice with HPV16 VLP-loaded BMDC induce Th1-biased responses, suggesting that mice are an appropriate model (Yang et al., 2004a). Mice represent a heterologus host for HPV16 whereas humans are homologous, yet human DC and LC respond to HPV16 VLPs in a similar manner to their murine counterparts. Notably both murine BMDCs and human myeloid DCs produce IL-12 in response to VLP and neither human nor murine Langerhans cells respond to HPV16 VLPs (Fausch et al., 2002; Lenz et al., 2003). Like IFN γ and IL-12, type I IFN plays a central role in the innate response against viral infections and is also involved in the regulation of adaptive immune responses (Biron, 2001). Type I IFNs can influence adaptive immune responses by modifying the maturation of DCs e.g. by driving DC subpopulations to promote Th1 cells. Murine BMDCs also produce type I IFN in response to HPV16 VLP. Herpes simplex virus and influenza virus infections, CpG, and poly I: C can activate DCs to produce high level of type I IFN (Cella et al., 1999a; 1999b; Siegal et al., 1999). Poly I:C induces ~10-fold higher levels of IFNα by BMDCs than HPV16 VLPs, although the latter are clearly sufficient to support a robust Th1 response (Yang et al., 2004a).

The regulation of IL-12 production is complex (reviewed by (Trinchieri, 2003)). IFN γ can enhance production of IL-12 (Hayes et al., 1998), thereby forming a positive-feedback mechanism during inflammatory and Th1 responses. IFNγ also plays a role by enhancing mRNA expression of the IL-12 receptor β1 and β2 chain (Grohmann et al., 2001; Szabo et al., 1997). Indeed, autocrine/paracrine stimulation of IL-12 production by IFNγ and upregulation of both IL-12 receptor β chains was observed in BMDC after VLP stimulation (Yang et al., 2004a). IFN production is also associated with expression of the IFN-regulated transcripts. For example BMDCs strongly upregulate IP-10, ISP15, Mx1, GARG-16, GARG-39, GARG-49, and interferon-activated genes 203 and 204 (Yang et al., 2004a).

Chemokines cause recruitment and polarization of T cells (Wong and Fish, 2003). Receptors CCR5 and CXCR3 are markers for Th1 cells. RANTES, MIP1α and MIP1β are produced by BMDCs upon HPV VLP stimulation and are ligands for CCR5 (Yang et al., 2004b). Similarly, IP-10 mRNA is upregulated by BMDC in response to HPV16 VLP and IP-10 is a ligand CXCR3 (Wong and Fish, 2003; Yang et al., 2004a). Thus BMDCs upregulate numerous chemokines that ligand the Th1 markers CCR5 and CXCR3 and Th1 associated cytokines such as IL-12 and IFN which likely drive the potent Th1-biased response of VLP immunized mice (Yang et al., 2004b). However, CCR2 is a marker for Th2 cells (Wong and Fish, 2003) and its ligands MCP-1, MCP-3, and MCP-5 are upregulated in HPV16 VLP-stimulated BMDC. The induction of these CCR2 ligands and the other Th2-associated cytokines IL-6, IL-10 and G-CSF, may contribute to the weak Th2-type IL-5 production by stimulation of T helper cells derived from mice vaccinated with VLP-loaded BMDCs and by patients vaccinated with HPV VLP (Harro et al., 2001). Further experimentation is necessary to clarify the contribution of these chemokines to VLP-induced immunity.

DCs promote innate and adaptive immunity by sensing pathogens. Recognition of HPV16 VLP initiates signaling through the MyD88 adaptor protein and innate responses, including the activation of NF-κB, NF-AT and AP-1 in RAW 264.7 cells and production of IFNα, IFNγ and IL-12 by BMDCs (Yang et al., 2004a). This dependence on MyD88 signaling is suggestive of recognition of HPV16 VLPs by the Toll family of pattern recognition receptors (Akira, 2003a). HPV L1 binds to heparan sulfate GAGs on the cell surface during infection (Joyce et al., 1999). These interactions may be related to HPV16 L1 induced signaling via Toll receptors as the oligosaccharides of hyaluronan can activate DCs via TLR4 (Termeer et al., 2002). It is also notable that RSV and MMTV have been shown to bind heparan sulphate on the cell surface (Feldman et al., 1999; Zhang et al., 2003) and are also recognized by TLR4-depedent pathways (Kurt-Jones et al., 2000; Rassa et al., 2002). Several studies indicate that MyD88 signaling is involved in the initial commitment of naïve CD4+T cells to differentiate into Th1 and Th2 lineage cells (Ahmad-Nejad et al., 2002; Jankovic et al., 2002). MyD88-deficient mice showed an inability to generate Th1 type immune responses to antigen in the presence of complete Freund's adjuvant (containing heat-killed mycobacterium tuberculosis, components of which can activate TLRs such as TLR2 and TLR4) (Jankovic et al., 2002). Indeed, MyD88 is also critical to the induction of Th1-biased immune responses including the production of IgG to HPV16 L1 VLPs (Yang et al., 2004a; 2005a; 2005b).

Rational vaccine development requires knowledge of the pathways that regulate innate and adaptive immune responses to clinically effective vaccine antigens such as HPV16 L1 VLP. Therefore future studies of immune recognition of HPV VLPs will be important to understand the clinical correlates and underlying mechanisms of protective immunity in the genital tract for the development of improved combination protective/theraputic vaccines against HPV (e.g chimeric VLPs) and other mucosal pathogens. These studies are also important for simplified (one shot), needle-free vaccination approaches that are desperately needed for vaccination in poor countries where 78% of cervical cancers occur.

12. Summary

The use of VLP technology has dramatically advanced the field of papillomavirus research. Notably, seroepidemiologic studies with HPV VLP-based ELISA have provided critical evidence to support the link between oncogenic HPV infection and cancer. However, the HPV VLP-based ELISA lacks the sensitivity and specificity necessary for a diagnostic test, a niche now filled by hybrid capture-based detection of viral DNA. Studies of virion biology and structure have advanced greatly by the use of VLPs, but most investigations now employ native virions or pseudovirions generated in culture using recombinant DNA technology. Unquestionably, the greatest contribution of VLP technology has been as a protective antigen for prophylactic vaccination. Human studies demonstrate the efficacy of VLP vaccination for protection against natural transmission of HPV16. This finding is of enormous significance, since a quarter of a million women die each year from cervical cancer. Future studies of the interaction between VLPs and the human immune system will also improve our understanding of the mechanisms of protection and immunogenicity. These studies are of particular importance for rational design of second generation HPV vaccines and vaccines against other infectious agents. Clearly issues around the prevalence of existing HPV-related disease that is not treated by VLP vaccination, the type-specificity of VLP-based protection, the need to vaccinate multiple times with a vaccine maintained in a cold chain leave room for the development a second generation pan-HPV vaccine that could be used inexpensively to vaccinate in developing nations where HPV-related disease takes such a toll.

References

Adami, H. O., Kuper, H., Andersson, S. O., Bergstrom, R., and Dillner, J. (2003). Prostate cancer risk and serologic evidence of human papilloma virus infection: a population-based case-control study. Cancer. Epidemiol. Biomarkers Prev *12*, 872-875.

af Geijersstam, V., Eklund, C., Wang, Z., Sapp, M., Schiller, J. T., Dillner, J., and Dillner, L. (1999). A survey of seroprevalence of human papillomavirus types 16, 18 and 33 among children. Int. J. Cancer *80*, 489-493.

af Geijersstam, V., Kibur, M., Wang, Z., Koskela, P., Pukkala, E., Schiller, J., Lehtinen, M., and Dillner, J. (1998a). Stability over time of serum antibody levels to human papillomavirus type 16. J. Infect. Dis. *177*, 1710-1714.

af Geijersstam, V., Wang, Z., Lewensohn-Fuchs, I., Eklund, C., Schiller, J. T., Forsgren, M., and Dillner, J. (1998b). Trends in seroprevalence of human papillomavirus type 16 among pregnant women in Stockholm, Sweden, during 1969-1989. Int. J. Cancer *76*, 341-344.

Ahmad-Nejad, P., Hacker, H., Rutz, M., Bauer, S., Vabulas, R. M., and Wagner, H. (2002). Bacterial CpG-DNA and lipopolysaccharides activate Toll-like receptors at distinct cellular compartments. Eur. J. Immunol. *32*, 1958-1968.

Akira, S. (2003a). Mammalian Toll-like receptors. Curr. Opin. Immunol. *15*, 5-11.

Akira, S. (2003b). Toll-like receptor signaling. J. Biol. Chem. *278*, 38105-38108.

Andersson-Ellstrom, A., Dillner, J., Hagmar, B., Schiller, J., Sapp, M., Forssman, L., and Milsom, I. (1996a). Comparison of development of serum antibodies to HPV16 and HPV33 and acquisition of cervical HPV DNA among sexually experienced and virginal young girls. A longitudinal cohort study. Sex. Transm. Dis. *23*, 234-238.

Andersson-Ellstrom, A., Hagmar, B. M., Johansson, B., Kalantari, M., Warleby, B., and Forssman, L. (1996b). Human papillomavirus deoxyribonucleic acid in cervix only detected in girls after coitus. Int. J. STD AIDS *7*, 333-336.

Antonsson, A., Forslund, O., Ekberg, H., Sterner, G., and Hansson, B. G. (2000). The ubiquity and impressive genomic diversity of human skin papillomaviruses suggest a commensalic nature of these viruses. J. Virol. *74*, 11636-11641.

Ault, K. A., Giuliano, A. R., Edwards, R. P., Tamms, G., Kim, L. L., Smith, J. F., Jansen, K. U., Allende, M., Taddeo, F. J., Skulsky, D., and Barr, E. (2004). A phase I study to evaluate a human papillomavirus (HPV) type 18 L1 VLP vaccine. Vaccine *22*, 3004-3007.

Bachmann, M. F., Rohrer, U. H., Kundig, T. M., Burki, K., Hengartner, H., and Zinkernagel, R. M. (1993). The influence of antigen organization on B cell responsiveness. Science *262*, 1448-1451.

Baker, T. S., Newcomb, W. W., Olson, N. H., Cowsert, L. M., Olson, C., and Brown, J. C. (1991). Structures of bovine and human papillomaviruses. Analysis by cryoelectron microscopy and three-dimensional image reconstruction. Biophys. J. *60*, 1445-1456.

Balmelli, C., Roden, R., Potts, A., Schiller, J., De Grandi, P., and Nardelli-Haefliger, D. (1998). Nasal immunization of mice with human papillomavirus type 16 virus-like particles elicits neutralizing antibodies in mucosal secretions. J. Virol. *72*, 8220-8229.

Bell, J. A., Sundberg, J. P., Ghim, S. J., Newsome, J., Jenson, A. B., and Schlegel, R. (1994). A formalin-inactivated vaccine protects against mucosal papillomavirus infection: a canine model. Pathobiology *62*, 194-198.

Belnap, D. M., Olson, N. H., and Baker, T. S. (1997). A method for establishing the handedness of biological macromolecules. J. Struct. Biol. *120*, 44-51.

Belnap, D. M., Olson, N. H., Cladel, N. M., Newcomb, W. W., Brown, J. C., Kreider, J. W., Christensen, N. D., and Baker, T. S. (1996). Conserved features in papillomavirus and polyomavirus capsids. J. Mol. Biol. *259*, 249-263.

Benton, E. C., McLaren, K., Barr, B. B., Blessing, K., Bunney, M. H., Rudlinger, R., Smith, I. W., and Hunter, J. A. (1989). Human papilloma virus infection and its relationship to skin cancer in a group of renal allograft recipients. Curr. Probl. Dermatol. *18*, 168-177.

Berkhout, R. J., Bouwes Bavinck, J. N., and ter Schegget, J. (2000). Persistence of human papillomavirus DNA in benign and (Pre)malignant skin lesions from renal transplant recipients. J. Clin. Microbiol. *38*, 2087-2096.

Biron, C. A. (2001). Interferons alpha and beta as immune regulators--a new look. Immunity *14*, 661-664.

Bjorge, T., Dillner, J., Anttila, T., Engeland, A., Hakulinen, T., Jellum, E., Lehtinen, M., Luostarinen, T., Paavonen, J., Pukkala, E., *et al.* (1997a). Prospective seroepidemiological study of role of human papillomavirus in non-cervical anogenital cancers. BMJ *315*, 646-649.

Bjorge, T., Hakulinen, T., Engeland, A., Jellum, E., Koskela, P., Lehtinen, M., Luostarinen, T., Paavonen, J., Sapp, M., Schiller, J., *et al.* (1997b). A prospective, seroepidemiological study of the role of human papillomavirus in esophageal cancer in Norway. Cancer. Res. *57*, 3989-3992.

Bosch, F. X., Castellsague, X., Munoz, N., de Sanjose, S., Ghaffari, A. M., Gonzalez, L. C., Gili, M., Izarzugaza, I., Viladiu, P., Navarro, C., *et al.* (1996). Male sexual behavior and human papillomavirus DNA: key risk factors for cervical cancer in Spain. J. Natl. Cancer Inst. *88*, 1060-1067.

Bousarghin, L., Touze, A., Combita-Rojas, A. L., and Coursaget, P. (2003a). Positively charged sequences of human papillomavirus type 16 capsid proteins are sufficient to mediate gene transfer into target cells via the heparan sulfate receptor. J. Gen. Virol. *84*, 157-164.

Bousarghin, L., Touze, A., Sizaret, P. Y., and Coursaget, P. (2003b). Human papillomavirus types 16, 31, and 58 use different endocytosis pathways to enter cells. J. Virol. *77*, 3846-3850.

Bousarghin, L., Touze, A., Yvonnet, B., and Coursaget, P. (2004). Positively charged synthetic peptides from structural proteins of papillomaviruses abrogate human papillomavirus infectivity. J. Med. Virol. *73*, 474-480.

Boxman, I. L., Berkhout, R. J., Mulder, L. H., Wolkers, M. C., Bouwes Bavinck, J. N., Vermeer, B. J., and ter Schegget, J. (1997). Detection of human papillomavirus DNA in plucked hairs from renal transplant recipients and healthy volunteers. J. Invest. Dermatol. *108*, 712-715.

Bratti, M. C., Rodriguez, A. C., Schiffman, M., Hildesheim, A., Morales, J., Alfaro, M., Guillen, D., Hutchinson, M., Sherman, M. E., Eklund, C., *et al.* (2004). Description of a seven-year prospective study of human papillomavirus infection and cervical neoplasia among 10000 women in Guanacaste, Costa Rica. Rev. Panam. Salud Publica *15*, 75-89.

Breitburd, F., Kirnbauer, R., Hubbert, N. L., Nonnenmacher, B., Trin-Dinh-Desmarquet, C., Orth, G., Schiller, J. T., and Lowy, D. R. (1995). Immunization with viruslike particles from cottontail rabbit papillomavirus (CRPV) can protect against experimental CRPV infection. J. Virol. *69*, 3959-3963.

Brown, D. R., Bryan, J. T., Rodriguez, M., and Katz, B. P. (1992). Factors associated with detection of human papillomavirus E4 and L1 proteins in condylomata acuminata. J. Infect. Dis. *166*, 512-517.

Brown, D. R., Bryan, J. T., Schroeder, J. M., Robinson, T. S., Fife, K. H., Wheeler, C. M., Barr, E., Smith, P. R., Chiacchierini, L., DiCello, A., and Jansen, K. U. (2001). Neutralization of human papillomavirus type 11 (HPV-11) by serum from women vaccinated with yeast-derived HPV-11 L1 virus-like particles: correlation with competitive radioimmunoassay titer. J. Infect. Dis. *184*, 1183-1186.

Buck, C. B., Pastrana, D. V., Lowy, D. R., and Schiller, J. T. (2004). Efficient intracellular assembly of papillomaviral vectors. J. Virol. *78*, 751-757.

Campo, M. S. (1991). Vaccination against papillomavirus. Cancer Cells *3*, 421-426.

Campo, M. S., Grindlay, G. J., O'Neil, B. W., Chandrachud, L. M., McGarvie, G. M., and Jarrett, W. F. (1993). Prophylactic and therapeutic vaccination against a mucosal papillomavirus. J Gen. Virol. *74*, 945-953.

Campo, M. S., and Jarrett, W. F. (1994). Vaccination against cutaneous and mucosal papillomavirus in cattle. Ciba Found. Symp. *187*, 61-73.

Carter, J. J., Koutsky, L. A., Hughes, J. P., Lee, S. K., Kuypers, J., Kiviat, N., and Galloway, D. A. (2000). Comparison of Human Papillomavirus Types 16, 18, and 6 Capsid Antibody Responses Following Incident Infection. J. Infect. Dis. *181*, 1911-1919.

Carter, J. J., Koutsky, L. A., Wipf, G. C., Christensen, N. D., Lee, S. K., Kuypers, J., Kiviat, N., and Galloway, D. A. (1996). The natural history of human papillomavirus type 16 capsid antibodies among a cohort of university women. J. Infect. Dis. *174*, 927-936.

Carter, J. J., Wipf, G. C., Benki, S. F., Christensen, N. D., and Galloway, D. A. (2003). Identification of a human papillomavirus type 16-specific epitope on the C-terminal arm of the major capsid protein L1. J. Virol. *77*, 11625-11632.

Carter, J. J., Wipf, G. C., Hagensee, M. E., McKnight, B., Habel, L. A., Lee, S. K., Kuypers, J., Kiviat, N., Daling, J. R., Koutsky, L. A., and et al. (1995). Use of human papillomavirus type 6 capsids to detect antibodies in people with genital warts. J. Infect. Dis. *172*, 11-18.

Casini, G. L., Graham, D., Heine, D., Garcea, R. L., and Wu, D. T. (2004). *In vitro* papillomavirus capsid assembly analyzed by light scattering. Virology *325*, 320-327.

Cason, J., Kaye, J. N., Jewers, R. J., Kambo, P. K., Bible, J. M., Kell, B., Shergill, B., Pakarian, F., Raju, K. S., and Best, J. M. (1995). Perinatal infection and persistence of human papillomavirus types 16 and 18 in infants. J. Med. Virol. *47*, 209-218.

Castle, P. E., Shields, T., Kirnbauer, R., Manos, M. M., Burk, R. D., Glass, A. G., Scott, D. R., Sherman, M. E., and Schiffman, M. (2002a). Sexual behavior, human papillomavirus type 16 (HPV 16) infection, and HPV 16 seropositivity. Sex. Transm. Dis. *29*, 182-187.

Castle, P. E., Wacholder, S., Sherman, M. E., Lorincz, A. T., Glass, A. G., Scott, D. R., Rush, B. B., Demuth, F., and Schiffman, M. (2002b). Absolute risk of a subsequent abnormal pap among oncogenic human papillomavirus DNA-positive, cytologically negative women. Cancer *95*, 2145-2151.

Cella, M., Jarrossay, D., Facchetti, F., Alebardi, O., Nakajima, H., Lanzavecchia, A., and Colonna, M. (1999a). Plasmacytoid monocytes migrate to inflamed lymph nodes and produce large amounts of type I interferon. Nat. Med. *5*, 919-923.

Cella, M., Salio, M., Sakakibara, Y., Langen, H., Julkunen, I., and Lanzavecchia, A. (1999b). Maturation, activation, and protection of dendritic cells induced by double-stranded RNA. J. Exp. Med. *189*, 821-829.

Chackerian, B., Lenz, P., Lowy, D. R., and Schiller, J. T. (2002). Determinants of autoantibody induction by conjugated papillomavirus virus-like particles. J. Immunol. *169*, 6120-6126.

Chackerian, B., Lowy, D. R., and Schiller, J. T. (1999). Induction of autoantibodies to mouse CCR5 with recombinant papillomavirus particles. Proc. Natl. Acad. Sci. USA *96*, 2373-2378.

Chackerian, B., Lowy, D. R., and Schiller, J. T. (2001). Conjugation of a self-antigen to papillomavirus-like particles allows for efficient induction of protective autoantibodies. J. Clin. Invest. *108*, 415-423.

Chen, X. S., Casini, G., Harrison, S. C., and Garcea, R. L. (2001). Papillomavirus capsid protein expression in Escherichia coli: purification and assembly of HPV11 and HPV16 L1. J. Mol. Biol. *307*, 173-182.

Chen, X. S., Garcea, R. L., Goldberg, I., Casini, G., and Harrison, S. C. (2000). Structure of small virus-like particles assembled from the L1 protein of human papillomavirus 16. Mol. Cell. *5*, 557-567.

Chen, Y., Ghim, S. J., Jenson, A. B., and Schlegel, R. (1998). Mutant canine oral papillomavirus L1 capsid proteins which form virus- like particles but lack native conformational epitopes. J. Gen. Virol. *79*, 2137-2146.

Chien, Y. C., Chen, J. Y., Liu, M. Y., Yang, H. I., Hsu, M. M., Chen, C. J., and Yang, C. S. (2001). Serologic markers of Epstein-Barr virus infection and nasopharyngeal carcinoma in Taiwanese men. N. Engl. J. Med. *345*, 1877-1882.

Christensen, N. D., Cladel, N. M., and Reed, C. A. (1995). Postattachment neutralization of papillomaviruses by monoclonal and polyclonal antibodies. Virology *207*, 136-142.

Christensen, N. D., Kreider, J. W., Kan, N. C., and DiAngelo, S. L. (1991). The open reading frame L2 of cottontail rabbit papillomavirus contains antibody-inducing neutralizing epitopes. Virology *181*, 572-579.

Christensen, N. D., Reed, C. A., Cladel, N. M., Hall, K., and Leiserowitz, G. S. (1996a). Monoclonal antibodies to HPV-6 L1 virus-like particles identify conformational and linear neutralizing epitopes on HPV-11 in addition to type-specific epitopes on HPV-6. Virology *224*, 477-486.

Christensen, N. D., Reed, C. A., Cladel, N. M., Han, R., and Kreider, J. W. (1996b). Immunization with viruslike particles induces long-term protection of rabbits against challenge with cottontail rabbit papillomavirus. J. Virol. *70*, 960-965.

Chua, K. L., and Hjerpe, A. (1996). Persistence of human papillomavirus (HPV) infections preceding cervical carcinoma. Cancer *77*, 121-127.

Clever, J., Yamada, M., and Kasamatsu, H. (1991). Import of simian virus 40 virions through nuclear pore complexes. Proc. Natl. Acad. Sci. USA *88*, 7333-7337.

Coleman, N., Birley, H. D., Renton, A. M., Hanna, N. F., Ryait, B. K., Byrne, M., Taylor-Robinson, D., and Stanley, M. A. (1994). Immunological events in regressing genital warts. Am. J. Clin. Pathol. *102*, 768-774.

Combita, A. L., Touze, A., Bousarghin, L., Christensen, N. D., and Coursaget, P. (2002). Identification of two cross-neutralizing linear epitopes within the L1 major capsid protein of human papillomaviruses. J. Virol. *76*, 6480-6486.

Crawford, L. V. (1969). Nucleic acids of tumor viruses. Adv. Virus Res. *14*, 89-152.

Crawford, L. V., and Crawford, E. M. (1963). A Comparative Study of Polyoma and Papilloma Viruses. Virology *21*, 258-263.

Crawford, L. V., and Waring, M. J. (1967). The supercoiling of papilloma virus DNA. J. Gen. Virol. *1*, 387-390.

Crook, T., Morgenstern, J. P., Crawford, L., and Banks, L. (1989). Continued expression of HPV-16 E7 protein is required for maintenance of the transformed phenotype of cells co-transformed by HPV-16 plus EJ- ras. EMBO J. *8*, 513-519.

Da Silva, D. M., Schiller, J. T., and Kast, W. M. (2003). Heterologous boosting increases immunogenicity of chimeric papillomavirus virus-like particle vaccines. Vaccine *21*, 3219-3227.

Da Silva, D. M., Velders, M. P., Nieland, J. D., Schiller, J. T., Nickoloff, B. J., and Kast, W. M. (2001). Physical interaction of human papillomavirus virus-like particles with immune cells. Int. Immunol. *13*, 633-641.

Dale, C. J., Liu, X. S., De Rose, R., Purcell, D. F., Anderson, J., Xu, Y., Leggatt, G. R., Frazer, I. H., and Kent, S. J. (2002). Chimeric human papilloma virus-simian/human immunodeficiency virus virus-like-particle vaccines: immunogenicity and protective efficacy in macaques. Virology *301*, 176-187.

Day, P. M., Lowy, D. R., and Schiller, J. T. (2003). Papillomaviruses infect cells via a clathrin-dependent pathway. Virology *307*, 1-11.

Day, P. M., Roden, R. B., Lowy, D. R., and Schiller, J. T. (1998). The papillomavirus minor capsid protein, L2, induces localization of the major capsid protein, L1, and the viral transcription/replication protein, E2, to PML oncogenic domains. J. Virol. *72*, 142-150.

De Bruijn, M. L., Greenstone, H. L., Vermeulen, H., Melief, C. J., Lowy, D. R., Schiller, J. T., and Kast, W. M. (1998). L1-specific protection from tumor challenge elicited by HPV16 virus- like particles. Virology *250*, 371-376.

de Gruijl, T. D., Bontkes, H. J., Walboomers, J. M., Schiller, J. T., Stukart, M. J., Groot, B. S., Chabaud, M. M., Remmink, A. J., Verheijen, R. H., Helmerhorst, T. J., *et al.* (1997). Immunoglobulin G responses against human papillomavirus type 16 virus- like particles in a prospective nonintervention cohort study of women with cervical intraepithelial neoplasia. J. Natl. Cancer Inst. *89*, 630-638.

de Jong-Tieben, L. M., Berkhout, R. J., ter Schegget, J., Vermeer, B. J., de Fijter, J. W., Bruijn, J. A., Westendorp, R. G., and Bouwes Bavinck, J. N. (2000). The prevalence of human papillomavirus DNA in benign keratotic skin lesions of renal transplant recipients with and without a history of skin cancer is equally high: a clinical study to assess risk factors for keratotic skin lesions and skin cancer. Transplantation *69*, 44-49.

de Sanjose, S., Munoz, N., Bosch, F. X., Reimann, K., Pedersen, N. S., Orfila, J., Ascunce, N., Gonzalez, L. C., Tafur, L., Gili, M., and et al. (1994). Sexually transmitted agents and cervical neoplasia in Colombia and Spain. Int. J. Cancer *56*, 358-363.

de Villiers, E. M., Lavergne, D., McLaren, K., and Benton, E. C. (1997). Prevailing papillomavirus types in non-melanoma carcinomas of the skin in renal allograft recipients. Int. J. Cancer *73*, 356-361.

Dillner, J., Dillner, L., Robb, J., Willems, J., Jones, I., Lancaster, W., Smith, R., and Lerner, R. (1989). A synthetic peptide defines a serologic IgA response to a human papillomavirus-encoded nuclear antigen expressed in virus-carrying cervical neoplasia. Proc. Natl. Acad. Sci. USA *86*, 3838-3841.

Dillner, J., Dillner, L., Utter, G., Eklund, C., Rotola, A., Costa, S., and DiLuca, D. (1990a). Mapping of linear epitopes of human papillomavirus type 16: the L1 and L2 open reading frames. Int. J. Cancer *45*, 529-535.

Dillner, J., Kallings, I., Brihmer, C., Sikstrom, B., Koskela, P., Lehtinen, M., Schiller, J. T., Sapp, M., and Mardh, P. A. (1996). Seropositivities to human papillomavirus types 16, 18, or 33 capsids and to Chlamydia trachomatis are markers of sexual behavior. J. Infect. Dis. *173*, 1394-1398.

Dillner, J., Lehtinen, M., Bjorge, T., Luostarinen, T., Youngman, L., Jellum, E., Koskela, P., Gislefoss, R. E., Hallmans, G., Paavonen, J., *et al.* (1997). Prospective seroepidemiologic study of human papillomavirus infection as a risk factor for invasive cervical cancer. J. Natl. Cancer Inst. *89*, 1293-1299.

Dillner, L., Moreno-Lopez, J., and Dillner, J. (1990b). Serological responses to papillomavirus group-specific antigens in women with neoplasia of the cervix uteri. J. Clin. Microbiol. *28*, 624-627.

Dollard, S. C., Wilson, J. L., Demeter, L. M., Bonnez, W., Reichman, R. C., Broker, T. R., and Chow, L. T. (1992). Production of human papillomavirus and modulation of the infectious program in epithelial raft cultures. Genes Dev. *6*, 1131-1142.

Donnelly, J. J., Martinez, D., Jansen, K. U., Ellis, R. W., Montgomery, D. L., and Liu, M. A. (1996). Protection against papillomavirus with a polynucleotide vaccine. J. Infect. Dis. *173*, 314-320.

Doorbar, J., and Gallimore, P. H. (1987). Identification of proteins encoded by the L1 and L2 open reading frames of human papillomavirus 1a. J. Virol. *61*, 2793-2799.

Drobni, P., Mistry, N., McMillan, N., and Evander, M. (2003). Carboxy-fluorescein diacetate, succinimidyl ester labeled papillomavirus virus-like particles fluoresce after internalization and interact with heparan sulfate for binding and entry. Virology *310*, 163-172.

Dvoretzky, I., Shober, R., Chattopadhyay, S. K., and Lowy, D. R. (1980). A quantitative *in vitro* focus assay for bovine papilloma virus. Virology *103*, 369-375.

Emeny, R. T., Wheeler, C. M., Jansen, K. U., Hunt, W. C., Fu, T. M., Smith, J. F., MacMullen, S., Esser, M. T., and Paliard, X. (2002). Priming of human papillomavirus type 11-specific humoral and cellular immune responses in college-aged women with a virus-like particle vaccine. J. Virol. *76*, 7832-7842.

Evander, M., Frazer, I. H., Payne, E., Qi, Y. M., Hengst, K., and McMillan, N. A. (1997). Identification of the alpha6 integrin as a candidate receptor for papillomaviruses. J. Virol. *71*, 2449-2456.

Evans, T. G., Bonnez, W., Rose, R. C., Koenig, S., Demeter, L., Suzich, J. A., O'Brien, D., Campbell, M., White, W. I., Balsley, J., and Reichman, R. C. (2001). A Phase 1 study of a recombinant viruslike particle vaccine against human papillomavirus type 11 in healthy adult volunteers. J. Infect. Dis. *183*, 1485-1493.

Fausch, S. C., Da Silva, D. M., and Kast, W. M. (2003). Differential uptake and cross-presentation of human papillomavirus virus-like particles by dendritic cells and Langerhans cells. Cancer Res. *63*, 3478-3482.

Fausch, S. C., Da Silva, D. M., Rudolf, M. P., and Kast, W. M. (2002). Human papillomavirus virus-like particles do not activate Langerhans cells: a possible immune escape mechanism used by human papillomaviruses. J. Immunol. *169*, 3242-3249.

Favre, M. (1975). Structural polypeptides of rabbit, bovine, and human papillomaviruses. J. Virol. *15*, 1239-1247.

Favre, M., Breitburd, F., Croissant, O., and Orth, G. (1974). Hemagglutinating activity of bovine papilloma virus. Virology *60*, 572-578.

Favre, M., Breitburd, F., Croissant, O., and Orth, G. (1977). Chromatin-like structures obtained after alkaline disruption of bovine and human papillomaviruses. J. Virol. *21*, 1205-1209.

Favre, M., Majewski, S., Noszczyk, B., Maienfisch, F., Pura, A., Orth, G., and Jablonska, S. (2000). Antibodies to human papillomavirus type 5 are generated in epidermal repair processes. J. Invest. Dermatol. *114*, 403-407.

Feldman, S. A., Hendry, R. M., and Beeler, J. A. (1999). Identification of a linear heparin binding domain for human respiratory syncytial virus attachment glycoprotein G. J. Virol. *73*, 6610-6617.

Feltkamp, M. C., Broer, R., di Summa, F. M., Struijk, L., van der Meijden, E., Verlaan, B. P., Westendorp, R. G., ter Schegget, J., Spaan, W. J., and Bouwes Bavinck, J. N. (2003). Seroreactivity to epidermodysplasia verruciformis-related human papillomavirus types is associated with nonmelanoma skin cancer. Cancer Res. *63*, 2695-2700.

Fife, K. H., Wheeler, C. M., Koutsky, L. A., Barr, E., Brown, D. R., Schiff, M. A., Kiviat, N. B., Jansen, K. U., Barber, H., Smith, J. F., *et al.* (2004). Dose-ranging studies of the safety and immunogenicity of human papillomavirus Type 11 and Type 16 virus-like particle candidate vaccines in young healthy women. Vaccine *22*, 2943-2952.

Finch, J. T., and Klug, A. (1965). The structure of viruses of the papilloma-polyoma type 3. Structure of rabbit papilloma virus, with an appendix on the topography of contrast in negative-staining for electron-microscopy. J. Mol. Biol. *13*, 1-12.

Finnen, R. L., Erickson, K. D., Chen, X. S., and Garcea, R. L. (2003). Interactions between papillomavirus L1 and L2 capsid proteins. J. Virol. *77*, 4818-4826.

Florin, L., Becker, K. A., Sapp, C., Lambert, C., Sirma, H., Muller, M., Streeck, R. E., and Sapp, M. (2004). Nuclear translocation of papillomavirus minor capsid protein L2 requires Hsc70. J. Virol. *78*, 5546-5553.

Florin, L., Sapp, C., Streeck, R. E., and Sapp, M. (2002). Assembly and translocation of papillomavirus capsid proteins. J. Virol. *76*, 10009-10014.

Franco, E. L., Rohan, T. E., and Villa, L. L. (1999). Epidemiologic evidence and human papillomavirus infection as a necessary cause of cervical cancer. J. Natl. Cancer Inst. *91*, 506-511.

Frattini, M. G., Lim, H. B., and Laimins, L. A. (1996). *In vitro* synthesis of oncogenic human papillomaviruses requires episomal genomes for differentiation-dependent late expression. Proc. Natl. Acad. Sci. USA *93*, 3062-3067.

Gillison, M. L., Koch, W. M., Capone, R. B., Spafford, M., Westra, W. H., Wu, L., Zahurak, M. L., Daniel, R. W., Viglione, M., Symer, D. E., *et al.* (2000). Evidence for a causal association between human papillomavirus and a subset of head and neck cancers. J. Natl. Cancer Inst. *92*, 709-720.

Giroglou, T., Florin, L., Schafer, F., Streeck, R. E., and Sapp, M. (2001a). Human Papillomavirus Infection Requires Cell Surface Heparan Sulfate. J. Virol. *75*, 1565-1570.

Giroglou, T., Sapp, M., Lane, C., Fligge, C., Christensen, N. D., Streeck, R. E., and Rose, R. C. (2001b). Immunological analyses of human papillomavirus capsids. Vaccine *19*, 1783-1793.

Greenstone, H. L., Nieland, J. D., de Visser, K. E., De Bruijn, M. L., Kirnbauer, R., Roden, R. B., Lowy, D. R., Kast, W. M., and Schiller, J. T. (1998). Chimeric papillomavirus virus-like particles elicit antitumor immunity against the E7 oncoprotein in an HPV16 tumor model. Proc. Natl. Acad. Sci. USA *95*, 1800-1805.

Griffith, J. P., Griffith, D. L., Rayment, I., Murakami, W. T., and Caspar, D. L. (1992). Inside polyomavirus at 25-A resolution. Nature *355*, 652-654.

Grohmann, U., Belladonna, M. L., Vacca, C., Bianchi, R., Fallarino, F., Orabona, C., Fioretti, M. C., and Puccetti, P. (2001). Positive regulatory role of IL-12 in macrophages and modulation by IFN-gamma. J. Immunol. *167*, 221-227.

Grossman, C. (1989). Possible underlying mechanisms of sexual dimorphism in the immune response, fact and hypothesis. J Steroid Biochem *34*, 241-251.

Hagensee, M. E., and Galloway, D. A. (1993). Growing human papillomaviruses and virus-like particles in the laboratory. Papillomavirus Report *4*, 121-124.

Hagensee, M. E., Koutsky, L. A., Lee, S. K., Grubert, T., Kuypers, J., Kiviat, N. B., and Galloway, D. A. (2000). Detection of cervical antibodies to human papillomavirus type 16 (HPV-16) capsid antigens in relation to detection of HPV-16 DNA and cervical lesions. J. Infect. Dis. *181*, 1234-1239.

Hagensee, M. E., Yaegashi, N., and Galloway, D. A. (1993). Self-assembly of human papillomavirus type 1 capsids by expression of the L1 protein alone or by coexpression of the L1 and L2 capsid proteins. J. Virol. *67*, 315-322.

Halpert, R., Fruchter, R. G., Sedlis, A., Butt, K., Boyce, J. G., and Sillman, F. H. (1986). Human papillomavirus and lower genital neoplasia in renal transplant patients. Obstet. Gynecol. *68*, 251-258.

Hamsikova, E., Novak, J., Hofmannova, V., Munoz, N., Bosch, F. X., de Sanjose, S., Shah, K., Roth, Z., and Vonka, V. (1994). Presence of antibodies to seven human papillomavirus type 16-derived peptides in cervical cancer patients and healthy controls. J. Infect. Dis. *170*, 1424-1431.

Harper, D. M., Franco, E. L., Wheeler, C., Ferris, D. G., Jenkins, D., Schuind, A., Zahaf, T., Innis, B., Naud, P., De Carvalho, N. S., *et al.* (2004). Efficacy of a bivalent L1 virus-like particle vaccine in prevention of infection with human papillomavirus types 16 and 18 in young women: a randomised controlled trial. Lancet *364*, 1757-1765.

Harro, C. D., Pang, Y. Y., Roden, R. B., Hildesheim, A., Wang, Z., Reynolds, M. J., Mast, T. C., Robinson, R., Murphy, B. R., Karron, R. A., *et al.* (2001). Safety and immunogenicity trial in adult volunteers of a human papillomavirus 16 L1 virus-like particle vaccine. J. Natl. Cancer Inst. *93*, 284-292.

Hayes, M. P., Murphy, F. J., and Burd, P. R. (1998). Interferon-gamma-dependent inducible expression of the human interleukin-12 p35 gene in monocytes initiates from a TATA-containing promoter distinct from the CpG-rich promoter active in Epstein-Barr virus-transformed lymphoblastoid cells. Blood *91*, 4645-4651.

Heim, K., Christensen, N. D., Hoepfl, R., Wartusch, B., Pinzger, G., Zeimet, A., Baumgartner, P., Kreider, J. W., and Dapunt, O. (1995). Serum IgG, IgM, and IgA reactivity to human papillomavirus types 11 and 6 virus-like particles in different gynecologic patient groups. J. Infect. Dis. *172*, 395-402.

Heino, P., Skyldberg, B., Lehtinen, M., Rantala, I., Hagmar, B., Kreider, J. W., Kirnbauer, R., and Dillner, J. (1995). Human papillomavirus type 16 capsids expose multiple type-restricted and type-common antigenic epitopes. J. Gen. Virol. *76*, 1141-1153.

Herrero, R., Castellsague, X., Pawlita, M., Lissowska, J., Kee, F., Balaram, P., Rajkumar, T., Sridhar, H., Rose, B., Pintos, J., *et al.* (2003). Human papillomavirus and oral cancer: the International Agency for Research on Cancer multicenter study. J. Natl. Cancer Inst. *95*, 1772-1783.

Herrero, R., Hildesheim, A., Bratti, C., Sherman, M. E., Hutchinson, M., Morales, J., Balmaceda, I., Greenberg, M. D., Alfaro, M., Burk, R. D., *et al.* (2000). Population-based study of human papillomavirus infection and cervical neoplasia in rural Costa Rica. J. Natl. Cancer Inst. *92*, 464-474.

Hildesheim, A., Gravitt, P., Schiffman, M. H., Kurman, R. J., Barnes, W., Jones, S., Tchabo, J. G., Brinton, L. A., Copeland, C., Epp, J., and et al. (1993). Determinants of genital human papillomavirus infection in low-income women in Washington, D.C. Sex. Transm. Dis. *20*, 279-285.

Hildesheim, A., Han, C. L., Brinton, L. A., Kurman, R. J., and Schiller, J. T. (1997). Human papillomavirus type 16 and risk of preinvasive and invasive vulvar cancer: results from a seroepidemiological case-control study. Obstet. Gynecol. *90*, 748-754.

Hisada, M., van den Berg, B. J., Strickler, H. D., Christianson, R. E., Wright, W. E., Waters, D. J., and Rabkin, C. S. (2001). Prospective study of antibody to human papilloma virus type 16 and risk of cervical, endometrial, and ovarian cancers (United States). Cancer Causes Control *12*, 335-341.

Ho, G. Y., Studentsov, Y., Hall, C. B., Bierman, R., Beardsley, L., Lempa, M., and Burk, R. D. (2002). Risk factors for subsequent cervicovaginal human papillomavirus (HPV) infection and the protective role of antibodies to HPV-16 virus-like particles. J. Infect. Dis. *186*, 737-742.

Hooper, R. R., Reynolds, G. H., Jones, O. G., Zaidi, A., Wiesner, P. J., Latimer, K. P., Lester, A., Campbell, A. F., Harrison, W. O., Karney, W. W., and Holmes, K. K. (1978). Cohort study of venereal disease. I: the risk of gonorrhea transmission from infected women to men. Am. J. Epidemiol. *108*, 136-144.

Jankovic, D., Kullberg, M. C., Hieny, S., Caspar, P., Collazo, C. M., and Sher, A. (2002). In the absence of IL-12, CD4(+) T cell responses to intracellular pathogens fail to default to a Th2 pattern and are host protective in an IL-10(-/-) setting. Immunity *16*, 429-439.

Jansen, K. U., Rosolowsky, M., Schultz, L. D., Markus, H. Z., Cook, J. C., Donnelly, J. J., Martinez, D., Ellis, R. W., and Shaw, A. R. (1995). Vaccination with yeast-expressed cottontail rabbit papillomavirus (CRPV) virus-like particles protects rabbits from CRPV-induced papilloma formation. Vaccine *13*, 1509-1514.

Jarrett, W. F., O'Neil, B. W., Gaukroger, J. M., Laird, H. M., Smith, K. T., and Campo, M. S. (1990a). Studies on vaccination against papillomaviruses: a comparison of purified virus, tumour extract and transformed cells in prophylactic vaccination. Vet. Rec. *126*, 449-452.

Jarrett, W. F., O'Neil, B. W., Gaukroger, J. M., Smith, K. T., Laird, H. M., and Campo, M. S. (1990b). Studies on vaccination against papillomaviruses: the immunity after infection and vaccination with bovine papillomaviruses of different types. Vet. Rec. *126*, 473-475.

Jarrett, W. F., Smith, K. T., O'Neil, B. W., Gaukroger, J. M., Chandrachud, L. M., Grindlay, G. J., McGarvie, G. M., and Campo, M. S. (1991). Studies on vaccination against papillomaviruses: prophylactic and therapeutic vaccination with recombinant structural proteins. Virology *184*, 33-42.

Jenison, S. A., Yu, X. P., Valentine, J. M., Koutsky, L. A., Christiansen, A. E., Beckmann, A. M., and Galloway, D. A. (1990). Evidence of prevalent genital-type human papillomavirus infections in adults and children. J. Infect. Dis. *162*, 60-69.

Jochmus, I., Schafer, K., Faath, S., Muller, M., and Gissmann, L. (1999). Chimeric virus-like particles of the human papillomavirus type 16 (HPV 16) as a prophylactic and therapeutic vaccine. Arch. Med. Res. *30*, 269-274.

Jochmus-Kudielka, I., Schneider, A., Braun, R., Kimmig, R., Koldovsky, U., Schneweis, K. E., Seedorf, K., and Gissmann, L. (1989). Antibodies against the human papillomavirus type 16 early proteins in human sera: correlation of anti-E7 reactivity with cervical cancer. J. Natl. Cancer Inst. *81*, 1698-1704.

Joyce, J. G., Tung, J. S., Przysiecki, C. T., Cook, J. C., Lehman, E. D., Sands, J. A., Jansen, K. U., and Keller, P. M. (1999). The L1 major capsid protein of human papillomavirus type 11 recombinant virus-like particles interacts with heparin and cell-surface glycosaminoglycans on human keratinocytes. J. Biol. Chem. *274*, 5810-5822.

Karner, W., and Bauer, G. (1994). Activation of a varicella-zoster virus-specific IgA response during acute Epstein-Barr virus infection. J. Med. Virol. *44*, 258-262.

Kasamatsu, H., and Nakanishi, A. (1998). How do animal DNA viruses get to the nucleus? Annu. Rev. Microbiol. *52*, 627-686.

Kawana, K., Matsumoto, K., Yoshikawa, H., Taketani, Y., Kawana, T., Yoshiike, K., and Kanda, T. (1998a). A surface immunodeterminant of human papillomavirus type 16 minor capsid protein L2. Virology *245*, 353-359.

Kawana, K., Yasugi, T., Kanda, T., Kino, N., Oda, K., Okada, S., Kawana, Y., Nei, T., Takada, T., Toyoshima, S., *et al.* (2003). Safety and immunogenicity of a peptide containing the cross-neutralization epitope of HPV16 L2 administered nasally in healthy volunteers. Vaccine *21*, 4256-4260.

Kawana, K., Yoshikawa, H., Taketani, Y., Yoshiike, K., and Kanda, T. (1998b). *In vitro* construction of pseudovirions of human papillomavirus type 16: incorporation of plasmid DNA into reassembled L1/L2 capsids. J. Virol. *72*, 10298-10300.

Kawana, K., Yoshikawa, H., Taketani, Y., Yoshiike, K., and Kanda, T. (1999). Common neutralization epitope in minor capsid protein L2 of human papillomavirus types 16 and 6. J. Virol. *73*, 6188-6190.

Kawana, Y., Kawana, K., Yoshikawa, H., Taketani, Y., Yoshiike, K., and Kanda, T. (2001). Human papillomavirus type 16 minor capsid protein L2 N-terminal region containing a common neutralization epitope binds to the cell surface and enters the cytoplasm. J. Virol. *75*, 2331-2336.

Kirnbauer, R., Booy, F., Cheng, N., Lowy, D. R., and Schiller, J. T. (1992). Papillomavirus L1 major capsid protein self-assembles into virus-like particles that are highly immunogenic. Proc. Natl. Acad. Sci. USA *89*, 12180-12184.

Kirnbauer, R., Chandrachud, L. M., O'Neil, B. W., Wagner, E. R., Grindlay, G. J., Armstrong, A., McGarvie, G. M., Schiller, J. T., Lowy, D. R., and Campo, M. S. (1996). Virus-like particles of bovine papillomavirus type 4 in prophylactic and therapeutic immunization. Virology *219*, 37-44.

Kirnbauer, R., Hubbert, N. L., Wheeler, C. M., Becker, T. M., Lowy, D. R., and Schiller, J. T. (1994). A virus-like particle enzyme-linked immunosorbent assay detects serum antibodies in a majority of women infected with human papillomavirus type 16. J. Natl. Cancer Inst. *86*, 494-499.

Kirnbauer, R., Taub, J., Greenstone, H., Roden, R., Durst, M., Gissmann, L., Lowy, D. R., and Schiller, J. T. (1993). Efficient self-assembly of human papillomavirus type 16 L1 and L1-L2 into virus-like particles. J. Virol. *67*, 6929-6936.

Kjaer, S. K., Chackerian, B., van den Brule, A. J., Svare, E. I., Paull, G., Walbomers, J. M., Schiller, J. T., Bock, J. E., Sherman, M. E., Lowy, D. R., and Meijer, C. L. (2001). High-risk human papillomavirus is sexually transmitted: evidence from a follow-up study of virgins starting sexual activity (intercourse). Cancer Epidemiol. Biomarkers Prev. *10*, 101-106.

Kjaer, S. K., Engholm, G., Dahl, C., Bock, J. E., Lynge, E., and Jensen, O. M. (1993). Case-control study of risk factors for cervical squamous-cell neoplasia in Denmark. III. Role of oral contraceptive use. Cancer Causes Control *4*, 513-519.

Kjellberg, L., Wang, Z., Wiklund, F., Edlund, K., Angstrom, T., Lenner, P., Sjoberg, I., Hallmans, G., Wallin, K. L., Sapp, M., *et al.* (1999). Sexual behaviour and papillomavirus exposure in cervical intraepithelial neoplasia: a population-based case-control study. J. Gen. Virol. *80*, 391-398.

Klug, A., and Finch, J. T. (1968). Structure of viruses of the papilloma-polyoma type. IV. Analysis of tilting experiments in the electron microscope. J. Mol. Biol. *31*, 1-12.

Koller, L. D., and Olson, C. (1972). Attempted transmission of warts from man, cattle, and horses and of deer fibroma, to selected hosts. J. Invest. Dermatol. *58*, 366-368.

Koutsky, L. A., Ault, K. A., Wheeler, C. M., Brown, D. R., Barr, E., Alvarez, F. B., Chiacchierini, L. M., and Jansen, K. U. (2002). A controlled trial of a human papillomavirus type 16 vaccine. N. Engl. J. Med. *347*, 1645-1651.

Krchnak, V., Vagner, J., Novak, J., Suchankova, A., and Roubal, J. (1990). A general procedure for evaluation of immunological relevance of synthetic peptides: peptides synthesized on paper in enzyme-linked immunosorbent assay. Anal. Biochem. *189*, 80-83.

Kreider, J. W., Howett, M. K., Lill, N. L., Bartlett, G. L., Zaino, R. J., Sedlacek, T. V., and Mortel, R. (1986). *In vivo* transformation of human skin with human papillomavirus type 11 from condylomata acuminata. J. Virol. *59*, 369-376.

Kurt-Jones, E. A., Popova, L., Kwinn, L., Haynes, L. M., Jones, L. P., Tripp, R. A., Walsh, E. E., Freeman, M. W., Golenbock, D. T., Anderson, L. J., and Finberg, R. W. (2000). Pattern recognition receptors TLR4 and CD14 mediate response to respiratory syncytial virus. Nat. Immunol. *1*, 398-401.

Laga, M., Icenogle, J. P., Marsella, R., Manoka, A. T., Nzila, N., Ryder, R. W., Vermund, S. H., Heyward, W. L., Nelson, A., and Reeves, W. C. (1992). Genital papillomavirus infection and cervical dysplasia--opportunistic complications of HIV infection. Int. J. Cancer *50*, 45-48.

Langenberg, A. G., Corey, L., Ashley, R. L., Leong, W. P., and Straus, S. E. (1999). A prospective study of new infections with herpes simplex virus type 1 and type 2. Chiron HSV Vaccine Study Group. N. Engl. J. Med. *341*, 1432-1438.

Laukkanen, P., Koskela, P., Pukkala, E., Dillner, J., Laara, E., Knekt, P., and Lehtinen, M. (2003). Time trends in incidence and prevalence of human papillomavirus type 6, 11 and 16 infections in Finland. J. Gen. Virol. *84*, 2105-2109.

Lehtinen, M., Dillner, J., Knekt, P., Luostarinen, T., Aromaa, A., Kirnbauer, R., Koskela, P., Paavonen, J., Peto, R., Schiller, J. T., and Hakama, M. (1996). Serologically diagnosed infection with human papillomavirus type 16 and risk for subsequent development of cervical carcinoma: nested case- control study. BMJ *312*, 537-539.

Lenz, P., Day, P. M., Pang, Y. Y., Frye, S. A., Jensen, P. N., Lowy, D. R., and Schiller, J. T. (2001). Papillomavirus-like particles induce acute activation of dendritic cells. J. Immunol. *166*, 5346-5355.

Lenz, P., Thompson, C. D., Day, P. M., Bacot, S. M., Lowy, D. R., and Schiller, J. T. (2003). Interaction of papillomavirus virus-like particles with human myeloid antigen-presenting cells. Clin. Immunol. *106*, 231-237.

Li, M., Beard, P., Estes, P. A., Lyon, M. K., and Garcea, R. L. (1998). Intercapsomeric disulfide bonds in papillomavirus assembly and disassembly. J. Virol. *72*, 2160-2167.

Li, M., Cripe, T. P., Estes, P. A., Lyon, M. K., Rose, R. C., and Garcea, R. L. (1997). Expression of the human papillomavirus type 11 L1 capsid protein in Escherichia coli: characterization of protein domains involved in DNA binding and capsid assembly. J. Virol. *71*, 2988-2995.

Liddington, R. C., Yan, Y., Moulai, J., Sahli, R., Benjamin, T. L., and Harrison, S. C. (1991). Structure of simian virus 40 at 3.8-A resolution. Nature *354*, 278-284.

Lin, Y. L., Borenstein, L. A., Ahmed, R., and Wettstein, F. O. (1993). Cottontail rabbit papillomavirus L1 protein-based vaccines: protection is achieved only with a full-length, nondenatured product. J. Virol. *67*, 4154-4162.

Lin, Y. L., Borenstein, L. A., Selvakumar, R., Ahmed, R., and Wettstein, F. O. (1992). Effective vaccination against papilloma development by immunization with L1 or L2 structural protein of cottontail rabbit papillomavirus. Virology *187*, 612-619.

Liu, W. J., Liu, X. S., Zhao, K. N., Leggatt, G. R., and Frazer, I. H. (2000). Papillomavirus virus-like particles for the delivery of multiple cytotoxic T cell epitopes. Virology *273*, 374-382.

Liu, X. S., Liu, W. J., Zhao, K. N., Liu, Y. H., Leggatt, G., and Frazer, I. H. (2002). Route of administration of chimeric BPV1 VLP determines the character of the induced immune responses. Immunol. Cell. Biol. *80*, 21-29.

Ludmerer, S. W., Benincasa, D., and Mark, G. E., 3rd (1996). Two amino acid residues confer type specificity to a neutralizing, conformationally dependent epitope on human papillomavirus type 11. J. Virol. *70*, 4791-4794.

Ludmerer, S. W., Benincasa, D., Mark, G. E., 3rd, and Christensen, N. D. (1997). A neutralizing epitope of human papillomavirus type 11 is principally described by a continuous set of residues which overlap a distinct linear, surface-exposed epitope. J. Virol. *71*, 3834-3839.

Luostarinen, T., af Geijersstam, V., Bjorge, T., Eklund, C., Hakama, M., Hakulinen, T., Jellum, E., Koskela, P.,

Paavonen, J., Pukkala, E., *et al.* (1999). No excess risk of cervical carcinoma among women seropositive for both HPV16 and HPV6/11. Int. J. Cancer *80*, 818-822.

Luostarinen, T., Lehtinen, M., Bjorge, T., Abeler, V., Hakama, M., Hallmans, G., Jellum, E., Koskela, P., Lenner, P., Lie, A. K., *et al.* (2004). Joint effects of different human papillomaviruses and Chlamydia trachomatis infections on risk of squamous cell carcinoma of the cervix uteri. Eur. J. Cancer *40*, 1058-1065.

Manns, A., Strickler, H. D., Wikktor, S. Z., Pate, E. J., Gray, R., and Waters, D. (1999). Low incidence of human papillomavirus type 16 antibody seroconversion in young children. Pediatr. Infect. Dis. J *18*, 833-835.

Marais, D., Rose, R. C., and Williamson, A. L. (1997). Age distribution of antibodies to human papillomavirus in children, women with cervical intraepithelial neoplasia and blood donors from South Africa. J. Med. Virol. *51*, 126-131.

Masini, C., Fuchs, P. G., Gabrielli, F., Stark, S., Sera, F., Ploner, M., Melchi, C. F., Primavera, G., Pirchio, G., Picconi, O., *et al.* (2003). Evidence for the association of human papillomavirus infection and cutaneous squamous cell carcinoma in immunocompetent individuals. Arch. Dermatol. *139*, 890-894.

McCarthy, M. P., White, W. I., Palmer-Hill, F., Koenig, S., and Suzich, J. A. (1998). Quantitative disassembly and reassembly of human papillomavirus type 11 virus like particles *in vitro*. J. Virol. *72*, 32-41.

McMillan, N. A., Payne, E., Frazer, I. H., and Evander, M. (1999). Expression of the alpha 6 integrin confers papillomavirus binding upon receptor-negative B-cells. Virology *261*, 271-279.

Mertz, K. J., McQuillan, G. M., Levine, W. C., Candal, D. H., Bullard, J. C., Johnson, R. E., St Louis, M. E., and Black, C. M. (1998). A pilot study of the prevalence of chlamydial infection in a national household survey. Sex. Transm. Dis. *25*, 225-228.

Meyers, C., Frattini, M. G., Hudson, J. B., and Laimins, L. A. (1992). Biosynthesis of human papillomavirus from a continuous cell line upon epithelial differentiation. Science *257*, 971-973.

Meyers, C., and Laimins, L. A. (1994). *In vitro* systems for the study and propagation of human papillomaviruses. Curr. Top. Microbiol. Immunol. *186*, 199-215.

Modis, Y., Trus, B. L., and Harrison, S. C. (2002). Atomic model of the papillomavirus capsid. EMBO J *21*, 4754-4762.

Moore, R. A., Nicholls, P. K., Santos, E. B., Gough, G. W., and Stanley, M. A. (2002). Absence of canine oral papillomavirus DNA following prophylactic L1 particle-mediated immunotherapeutic delivery vaccination. J. Gen. Virol. *83*, 2299-2301.

Mork, J., Lie, A. K., Glattre, E., Hallmans, G., Jellum, E., Koskela, P., Moller, B., Pukkala, E., Schiller, J. T., Youngman, L., *et al.* (2001). Human papillomavirus infection as a risk factor for squamous-cell carcinoma of the head and neck. N. Engl. J. Med. *344*, 1125-1131.

Muller, M., Gausepohl, H., de Martynoff, G., Frank, R., Brasseur, R., and Gissmann, L. (1990). Identification of seroreactive regions of the human papillomavirus type 16 protein E4, E6, E7 and L1. J. Gen. Virol. *71*, 2709-2717.

Muller, M., Gissmann, L., Cristiano, R. J., Sun, X. Y., Frazer, I. H., Jenson, A. B., Alonso, A., Zentgraf, H., and Zhou, J. (1995). Papillomavirus capsid binding and uptake by cells from different tissues and species. J. Virol. *69*, 948-954.

Muller, M., Zhou, J., Reed, T. D., Rittmuller, C., Burger, A., Gabelsberger, J., Braspenning, J., and Gissmann, L. (1997). Chimeric papillomavirus-like particles. Virology *234*, 93-111.

Munoz, N., Bosch, F. X., de Sanjose, S., Herrero, R., Castellsague, X., Shah, K. V., Snijders, P. J., and Meijer, C. J. (2003). Epidemiologic classification of human papillomavirus types associated with cervical cancer. N. Engl. J. Med. *348*, 518-527.

Nardelli-Haefliger, D., Roden, R. B., Benyacoub, J., Sahli, R., Kraehenbuhl, J. P., Schiller, J. T., Lachat, P., Potts, A., and De Grandi, P. (1997). Human papillomavirus type 16 virus-like particles expressed in attenuated Salmonella typhimurium elicit mucosal and systemic neutralizing antibodies in mice. Infect. Immun. *65*, 3328-3336.

Nardelli-Haefliger, D., Wirthner, D., Schiller, J. T., Lowy, D. R., Hildesheim, A., Ponci, F., and De Grandi, P. (2003). Specific antibody levels at the cervix during the menstrual cycle of women vaccinated with human papillomavirus 16 virus-like particles. J. Natl. Cancer Inst. *95*, 1128-1137.

Nelson, L. M., Rose, R. C., and Moroianu, J. (2002). Nuclear import strategies of high risk HPV16 L1 major capsid protein. J. Biol. Chem. *277*, 23958-23964.

Nieland, J. D., Da Silva, D. M., Velders, M. P., de Visser, K. E., Schiller, J. T., Muller, M., and Kast, W. M. (1999). Chimeric papillomavirus virus-like particles induce a murine self- antigen-specific protective and therapeutic antitumor immune response. J. Cell. Biochem. *73*, 145-152.

Nonnenmacher, B., Hubbert, N. L., Kirnbauer, R., Shah, K. V., Munoz, N., Bosch, F. X., de Sanjose, S., Viscidi, R., Lowy, D. R., and Schiller, J. T. (1995). Serologic response to human papillomavirus type 16 (HPV-16) virus-like particles in HPV-16 DNA-positive invasive cervical cancer and cervical intraepithelial neoplasia grade III patients and controls from Colombia and Spain. J. Infect. Dis. *172*, 19-24.

Nonnenmacher, B., Kruger Kjaer, S., Svare, E. I., Scott, J. D., Hubbert, N. L., van den Brule, A. J., Kirnbauer, R., Walboomers, J. M., Lowy, D. R., and Schiller, J. T. (1996). Seroreactivity to HPV16 virus-like particles as a marker for cervical cancer risk in high-risk populations. Int. J. Cancer *68*, 704-709.

Ohlschlager, P., Osen, W., Dell, K., Faath, S., Garcea, R. L., Jochmus, I., Muller, M., Pawlita, M., Schafer, K., Sehr, P., *et al.* (2003). Human papillomavirus type 16 L1 capsomeres induce L1-specific cytotoxic T lymphocytes and tumor regression in C57BL/6 mice. J. Virol. *77*, 4635-4645.

Okun, M. M., Day, P. M., Greenstone, H. L., Booy, F. P., Lowy, D. R., Schiller, J. T., and Roden, R. B. (2001). L1 interaction domains of papillomavirus L2 necessary for viral genome encapsidation. J. Virol. *75*, 4332-4342.

Paintsil, J., Muller, M., Picken, M., Gissmann, L., and Zhou, J. (1996). Carboxyl terminus of bovine papillomavirus type-1 L1 protein is not required for capsid formation. Virology *223*, 238-244.

Paintsil, J., Muller, M., Picken, M., Gissmann, L., and Zhou, J. (1998). Calcium is required in reassembly of bovine papillomavirus *in vitro*. J. Gen. Virol. *79*, 1133-1141.

Peng, S., Frazer, I. H., Fernando, G. J., and Zhou, J. (1998). Papillomavirus virus-like particles can deliver defined CTL epitopes to the MHC class I pathway. Virology *240*, 147-157.

Pfister, H., and Ter Schegget, J. (1997). Role of HPV in cutaneous premalignant and malignant tumors. Clin. Dermatol. *15*, 335-347.

Pilacinski, W. P., Glassman, D. L., Glassman, K. F., Reed, D. E., Lum, M. A., Marshall, R. F., Muscoplat, C. C., and Faras, A. J. (1986). Immunization against bovine papillomavirus infection. Ciba Found. Symp. *120*, 136-156.

Pinto, L. A., Edwards, J., Castle, P. E., Harro, C. D., Lowy, D. R., Schiller, J. T., Wallace, D., Kopp, W., Adelsberger, J. W., Baseler, M. W., *et al.* (2003). Cellular immune responses to human papillomavirus (HPV)-16 L1 in healthy volunteers immunized with recombinant HPV-16 L1 virus-like particles. J. Infect. Dis. *188*, 327-338.

Puranen, M., Yliskoski, M., Saarikoski, S., Syrjanen, K., and Syrjanen, S. (1996). Vertical transmission of human papillomavirus from infected mothers to their newborn babies and persistence of the virus in childhood. Am. J. Obstet. Gynecol. *174*, 694-699.

Pushko, P., Sasagawa, T., Cuzick, J., and Crawford, L. (1994). Sequence variation in the capsid protein genes of human papillomavirus type 16. J. Gen. Virol. *75*, 911-916.

Qi, Y. M., Peng, S. W., Hengst, K., Evander, M., Park, D. S., Zhou, J., and Frazer, I. H. (1996). Epithelial cells display separate receptors for papillomavirus VLPs and for soluble L1 capsid protein. Virology *216*, 35-45.

Rassa, J. C., Meyers, J. L., Zhang, Y., Kudaravalli, R., and Ross, S. R. (2002). Murine retroviruses activate B cells via interaction with toll-like receptor 4. Proc. Natl. Acad. Sci. USA *99*, 2281-2286.

Roden, R. B., Armstrong, A., Haderer, P., Christensen, N. D., Hubbert, N. L., Lowy, D. R., Schiller, J. T., and Kirnbauer, R. (1997a). Characterization of a human papillomavirus type 16 variant-dependent neutralizing epitope. J. Virol. *71*, 6247-6252.

Roden, R. B., Day, P. M., Bronzo, B. K., Yutzy, W. H. t., Yang, Y., Lowy, D. R., and Schiller, J. T. (2001). Positively charged termini of the L2 minor capsid protein are necessary for papillomavirus infection. J. Virol. *75*, 10493-10497.

Roden, R. B., Greenstone, H. L., Kirnbauer, R., Booy, F. P., Jessie, J., Lowy, D. R., and Schiller, J. T. (1996a). *In vitro* generation and type-specific neutralization of a human papillomavirus type 16 virion pseudotype. J. Virol. *70*, 5875-5883.

Roden, R. B., Hubbert, N. L., Kirnbauer, R., Breitburd, F., Lowy, D. R., and Schiller, J. T. (1995). Papillomavirus L1 capsids agglutinate mouse erythrocytes through a proteinaceous receptor. J. Virol. *69*, 5147-5151.

Roden, R. B., Hubbert, N. L., Kirnbauer, R., Christensen, N. D., Lowy, D. R., and Schiller, J. T. (1996b). Assessment of the serological relatedness of genital human papillomaviruses by hemagglutination inhibition. J. Virol. *70*, 3298-3301.

Roden, R. B., Kirnbauer, R., Jenson, A. B., Lowy, D. R., and Schiller, J. T. (1994a). Interaction of papillomaviruses with the cell surface. J. Virol. *68*, 7260-7266.

Roden, R. B., Lowy, D. R., and Schiller, J. T. (1997b). Papillomavirus is resistant to desiccation. J. Infect. Dis. *176*, 1076-1079.

Roden, R. B., Weissinger, E. M., Henderson, D. W., Booy, F., Kirnbauer, R., Mushinski, J. F., Lowy, D. R., and Schiller, J. T. (1994b). Neutralization of bovine papillomavirus by antibodies to L1 and L2 capsid proteins. J. Virol. *68*, 7570-7574.

Roden, R. B., Yutzy, W. I., Fallon, R., Inglis, S., Lowy, D. R., and Schiller, J. T. (2000). Minor capsid protein of human genital papillomaviruses contains subdominant, cross-neutralizing epitopes. Virology *270*, 254-257.

Rose, R. C., Bonnez, W., Reichman, R. C., and Garcea, R. L. (1993). Expression of human papillomavirus type 11 L1 protein in insect cells: *in vivo* and *in vitro* assembly of viruslike particles. J. Virol. *67*, 1936-1944.

Rose, R. C., Lane, C., Wilson, S., Suzich, J. A., Rybicki, E., and Williamson, A. L. (1999). Oral vaccination of mice with human papillomavirus virus-like particles induces systemic virus-neutralizing antibodies. Vaccine *17*, 2129-2135.

Rose, R. C., White, W. I., Li, M., Suzich, J. A., Lane, C., and Garcea, R. L. (1998). Human papillomavirus type 11 recombinant L1 capsomeres induce virus- neutralizing antibodies. J. Virol. *72*, 6151-6154.

Rosenblatt, K. A., Carter, J. J., Iwasaki, L. M., Galloway, D. A., and Stanford, J. L. (2003). Serologic evidence of human papillomavirus 16 and 18 infections and risk of prostate cancer. Cancer Epidemiol. Biomarkers Prev. *12*, 763-768.

Rudolf, M. P., Fausch, S. C., Da Silva, D. M., and Kast, W. M. (2001a). Human dendritic cells are activated by chimeric human papillomavirus type-16 virus-like particles and induce epitope-specific human T cell responses *in vitro*. J. Immunol. *166*, 5917-5924.

Rudolf, M. P., Man, S., Melief, C. J., Sette, A., and Kast, W. M. (2001b). Human T-cell responses to HLA-A-restricted high binding affinity peptides of human papillomavirus type 18 proteins E6 and E7. Clin. Cancer Res. *7*, 788s-795s.

Rudolf, M. P., Nieland, J. D., DaSilva, D. M., Velders, M. P., Muller, M., Greenstone, H. L., Schiller, J. T., and Kast, W. M. (1999). Induction of HPV16 capsid protein-specific human T cell responses by virus-like particles. Biol. Chem. *380*, 335-340.

Rylander, E., Ruusuvaara, L., Almstromer, M. W., Evander, M., and Wadell, G. (1994). The absence of vaginal human papillomavirus 16 DNA in women who have not experienced sexual intercourse. Obstet. Gynecol. *83*, 735-737.

Sapp, M., Fligge, C., Petzak, I., Harris, J. R., and Streeck, R. E. (1998). Papillomavirus assembly requires trimerization of the major capsid protein by disulfides between two highly conserved cysteines. J. Virol. *72*, 6186-6189.

Sapp, M., Volpers, C., Muller, M., and Streeck, R. E. (1995). Organization of the major and minor capsid proteins in human papillomavirus type 33 virus-like particles. J. Gen. Virol. *76*, 2407-2412.

Sasagawa, T., Pushko, P., Steers, G., Gschmeissner, S. E., Hajibagheri, M. A., Finch, J., Crawford, L., and Tommasino, M. (1995). Synthesis and assembly of virus-like particles of human papillomaviruses type 6 and type 16 in fission yeast Schizosaccharomyces pombe. Virology *206*, 126-135.

Schafer, A., Friedmann, W., Mielke, M., Schwartlander, B., and Koch, M. A. (1991). The increased frequency of cervical dysplasia-neoplasia in women infected with the human immunodeficiency virus is related to the degree of immunosuppression. Am J Obstet. Gynecol. *164*, 593-599.

Schwartz, S. M., Daling, J. R., Doody, D. R., Wipf, G. C., Carter, J. J., Madeleine, M. M., Mao, E. J., Fitzgibbons, E. D., Huang, S., Beckmann, A. M., *et al.* (1998). Oral cancer risk in relation to sexual history and evidence of human papillomavirus infection. J. Natl. Cancer Inst. *90*, 1626-1636.

Selinka, H. C., Giroglou, T., Nowak, T., Christensen, N. D., and Sapp, M. (2003). Further evidence that papillomavirus capsids exist in two distinct conformations. J. Virol. *77*, 12961-12967.

Selinka, H. C., Giroglou, T., and Sapp, M. (2002). Analysis of the infectious entry pathway of human papillomavirus type 33 pseudovirions. Virology *299*, 279-287.

Shafti-Keramat, S., Handisurya, A., Kriehuber, E., Meneguzzi, G., Slupetzky, K., and Kirnbauer, R. (2003). Different heparan sulfate proteoglycans serve as cellular receptors for human papillomaviruses. J. Virol. *77*, 13125-13135.

Shah, K. V. (1998). Do human papillomavirus infections cause oral cancer? J. Natl. Cancer Inst. *90*, 1585-1586.

Shah, K. V., Viscidi, R. P., Alberg, A. J., Helzlsouer, K. J., and Comstock, G. W. (1997). Antibodies to human papillomavirus 16 and subsequent *in situ* or invasive cancer of the cervix. Cancer Epidemiol. Biomarkers Prev. *6*, 233-237.

Shortman, K., and Liu, Y. J. (2002). Mouse and human dendritic cell subtypes. Nat. Rev. Immunol. *2*, 151-161.

Sibbet, G., Romero-Graillet, C., Meneguzzi, G., and Saveria Campo, M. (2000). alpha6 integrin is not the obligatory cell receptor for bovine papillomavirus type 4. J. Gen. Virol. *81*, 1629.

Siegal, F. P., Kadowaki, N., Shodell, M., Fitzgerald-Bocarsly, P. A., Shah, K., Ho, S., Antonenko, S., and Liu, Y. J. (1999). The nature of the principal type 1 interferon-producing cells in human blood. Science *284*, 1835-1837.

Sigstad, E., Lie, A. K., Luostarinen, T., Dillner, J., Jellum, E., Lehtinen, M., Thoresen, S., and Abeler, V. (2002). A prospective study of the relationship between prediagnostic human papillomavirus seropositivity and HPV DNA in subsequent cervical carcinomas. Br. J. Cancer *87*, 175-180.

Silins, I., Wang, Z., Avall-Lundqvist, E., Frankendal, B., Vikmanis, U., Sapp, M., Schiller, J. T., and Dillner, J. (1999). Serological evidence for protection by human papillomavirus (HPV) type 6 infection against HPV type 16 cervical carcinogenesis. J. Gen. Virol. *80 (Pt 11)*, 2931-2936.

Slupetzky, K., Shafti-Keramat, S., Lenz, P., Brandt, S., Grassauer, A., Sara, M., and Kirnbauer, R. (2001). Chimeric papillomavirus-like particles expressing a foreign epitope on capsid surface loops. J. Gen. Virol. *82*, 2799-2804.

Smith, E. M., Johnson, S. R., Cripe, T., Perlman, S., McGuinness, G., Jiang, D., Cripe, L., and Turek, L. P. (1995). Perinatal transmission and maternal risks of human papillomavirus infection. Cancer. Detect. Prev. *19*, 196-205.

Stark, S., Petridis, A. K., Ghim, S. J., Jenson, A. B., Bouwes Bavinck, J. N., Gross, G., Stockfleth, E., Fuchs, P. G., and Pfister, H. (1998). Prevalence of antibodies against virus-like particles of *Epidermodysplasia verruciformis*-associated HPV8 in patients at risk of skin cancer. J. Invest. Dermatol. *111*, 696-701.

Stehle, T., Yan, Y., Benjamin, T. L., and Harrison, S. C. (1994). Structure of murine polyomavirus complexed with an oligosaccharide receptor fragment. Nature *369*, 160-163.

Stone, K. M., Karem, K. L., Sternberg, M. R., McQuillan, G. M., Poon, A. D., Unger, E. R., and Reeves, W. C. (2002). Seroprevalence of human papillomavirus type 16 infection in the United States. J. Infect. Dis. *186*, 1396-1402.

Strickler, H. D., Burk, R., Shah, K., Viscidi, R., Jackson, A., Pizza, G., Bertoni, F., Schiller, J. T., Manns, A., Metcalf, R., *et al.* (1998a). A multifaceted study of human papillomavirus and prostate carcinoma. Cancer *82*, 1118-1125.

Strickler, H. D., Hildesheim, A., Viscidi, R. P., Shah, K. V., Goebel, B., Drummond, J., Waters, D., Sun, Y., Hubbert, N. L., Wacholder, S., *et al.* (1997). Interlaboratory agreement among results of human papillomavirus type 16 enzyme-linked immunosorbent assays. J. Clin. Microbiol. *35*, 1751-1756.

Strickler, H. D., Schiffman, M. H., Shah, K. V., Rabkin, C. S., Schiller, J. T., Wacholder, S., Clayman, B., and Viscidi, R. P. (1998b). A survey of human papillomavirus 16 antibodies in patients with epithelial cancers. Eur. J. Cancer Prev *7*, 305-313.

Strike, D. G., Bonnez, W., Rose, R. C., and Reichman, R. C. (1989). Expression in Escherichia coli of seven DNA fragments comprising the complete L1 and L2 open reading frames of human papillomavirus type 6b and localization of the 'common antigen' region. J. Gen. Virol. *70*, 543-555.

Sun, X. W., Kuhn, L., Ellerbrock, T. V., Chiasson, M. A., Bush, T. J., and Wright, T. C., Jr. (1997). Human papillomavirus infection in women infected with the human immunodeficiency virus. N Engl J Med *337*, 1343-1349.

Sun, Y., Hildesheim, A., Brinton, L. A., Nasca, P. C., Trimble, C. L., Kurman, R. J., Viscidi, R. P., and Shah, K. V. (1996). Human papillomavirus-specific serologic response in vulvar neoplasia. Gynecol. Oncol. *63*, 200-203.

Suzich, J. A., Ghim, S. J., Palmer-Hill, F. J., White, W. I., Tamura, J. K., Bell, J. A., Newsome, J. A., Jenson, A. B., and Schlegel, R. (1995). Systemic immunization with papillomavirus L1 protein completely prevents the development of viral mucosal papillomas. Proc. Natl. Acad. Sci. USA *92*, 11553-11557.

Svare, E. I., Kjaer, S. K., Nonnenmacher, B., Worm, A. M., Moi, H., Christensen, R. B., van den Brule, A. J., Walboomers, J. M., Meijer, C. J., Hubbert, N. L., *et al.* (1997). Seroreactivity to human papillomavirus type 16 virus-like particles is lower in high-risk men than in high-risk women. J. Infect. Dis. *176*, 876-883.

Szabo, S. J., Dighe, A. S., Gubler, U., and Murphy, K. M. (1997). Regulation of the interleukin (IL)-12R beta 2 subunit expression in developing T helper 1 (Th1) and Th2 cells. J. Exp. Med. *185*, 817-824.

Tachezy, R., Hamsikova, E., Hajek, T., Mikyskova, I., Smahel, M., Van Ranst, M., Kanka, J., Havrankova, A., Rob, L., Guttner, V., *et al.* (1999). Human papillomavirus genotype spectrum in Czech women: correlation of HPV DNA presence with antibodies against HPV-16, 18, and 33 virus-like particles. J. Med. Virol. *58*, 378-386.

Takeda, K., Kaisho, T., and Akira, S. (2003). Toll-like receptors. Annu. Rev. Immunol. *21*, 335-376.

Tenti, P., Zappatore, R., Migliora, P., Spinillo, A., Belloni, C., and Carnevali, L. (1999). Perinatal transmission of human papillomavirus from gravidas with latent infections. Obstet. Gynecol. *93*, 475-479.

Termeer, C., Benedix, F., Sleeman, J., Fieber, C., Voith, U., Ahrens, T., Miyake, K., Freudenberg, M., Galanos, C., and Simon, J. C. (2002). Oligosaccharides of Hyaluronan activate dendritic cells via toll-like receptor 4. J. Exp. Med. *195*, 99-111.

Thomas, K. K., Hughes, J. P., Kuypers, J. M., Kiviat, N. B., Lee, S. K., Adam, D. E., and Koutsky, L. A. (2000). Concurrent and sequential acquisition of different genital human papillomavirus types. J. Infect. Dis. *182*, 1097-1102.

Touze, A., and Coursaget, P. (1998). *In vitro* gene transfer using human papillomavirus-like particles. Nucleic Acids Res. *26*, 1317-1323.

Touze, A., Mahe, D., El Mehdaoui, S., Dupuy, C., Combita-Rojas, A. L., Bousarghin, L., Sizaret, P. Y., and Coursaget, P. (2000). The nine C-terminal amino acids of the major capsid protein of the human papillomavirus type 16 are essential for DNA binding and gene transfer capacity. FEMS Microbiol. Lett. *189*, 121-127.

Trinchieri, G. (2003). Interleukin-12 and the regulation of innate resistance and adaptive immunity. Nat. Rev. Immunol. *3*, 133-146.

Trus, B. L., Roden, R. B., Greenstone, H. L., Vrhel, M., Schiller, J. T., and Booy, F. P. (1997). Novel structural features of bovine papillomavirus capsid revealed by a three-dimensional reconstruction to 9 Å resolution. Nat. Struct. Biol. *4*, 413-420.

Unckell, F., Streeck, R. E., and Sapp, M. (1997). Generation and neutralization of pseudovirions of human papillomavirus type 33. J. Virol. *71*, 2934-2939.

Van Doornum, G. J., Korse, C. M., Buning-Kager, J. C., Bonfrer, J. M., Horenblas, S., Taal, B. G., and Dillner, J. (2003). Reactivity to human papillomavirus type 16 L1 virus-like particles in sera from patients with genital cancer and patients with carcinomas at five different extragenital sites. Br. J. Cancer *88*, 1095-1100.

Varsani, A., Williamson, A. L., de Villiers, D., Becker, I., Christensen, N. D., and Rybicki, E. P. (2003). Chimeric human papillomavirus type 16 (HPV-16) L1 particles presenting the common neutralizing epitope for the L2 minor capsid protein of HPV-6 and HPV-16. J. Virol. *77*, 8386-8393.

Vera-Bravo, R., Ocampo, M., Urquiza, M., Garcia, J. E., Rodriguez, L. E., Puentes, A., Lopez, R., Curtidor, H., Suarez, J. E., Torres, E., *et al.* (2003). Human papillomavirus type 16 and 18 L1 protein peptide binding to VERO and HeLa cells inhibits their VLPs binding. Int. J. Cancer *107*, 416-424.

Viscidi, R.P., Snyder, B., Cu-Uvin, S., Hogan, J.W., Clayman, B., Klein, R.S., Sobel, J., and Shah, K.V. (2005). Human papillomavirus capsid antibody response to natural infection and risk of subsequent HPV infection in HIV-positive and HIV-negative women. Cancer Epidemiol. Biomarkers Prev. *14*, 283-288.

Viscidi, R. P., Ahdieh-Grant, L., Schneider, M. F., Clayman, B., Massad, L. S., Anastos, K. M., Burk, R. D., Minkoff, H., Palefsky, J., Levine, A., and Strickler, H. (2003). Serum immunoglobulin A response to human papillomavirus type 16 virus-like particles in human immunodeficiency virus (HIV)-positive and high-risk HIV-negative women. J. Infect. Dis. *188*, 1834-1844.

Viscidi, R. P., Kotloff, K. L., Clayman, B., Russ, K., Shapiro, S., and Shah, K. V. (1997). Prevalence of antibodies to human papillomavirus (HPV) type 16 virus- like particles in relation to cervical HPV infection among college women. Clin. Diagn. Lab. Immunol. *4*, 122-126.

Viscidi, R. P., Schiffman, M., Hildesheim, A., Herrero, R., Castle, P. E., Bratti, M. C., Rodriguez, A. C., Sherman, M. E., Wang, S., Clayman, B., and Burk, R. D. (2004). Seroreactivity to human papillomavirus (HPV) types 16, 18, or 31 and risk of subsequent HPV infection: results from a population-based study in Costa Rica. Cancer Epidemiol. Biomarkers Prev. *13*, 324-327.

Viscidi, R. P., Snyder, B., Cu-Uvin, S., Hogan, J. W., Clayman, B., Klein, R. S., Sobel, J., and Shah, K. V. (in press). Human Papillomavirus (HPV) Capsid Antibody Response to Natural Infection and Risk of Subsequent HPV Infection in Human Immunodeficiency Virus-Positive and -Negative Women. Cancer. Epi. Biomarker Preven.

Volpers, C., Unckell, F., Schirmacher, P., Streeck, R. E., and Sapp, M. (1995). Binding and internalization of human papillomavirus type 33 virus-like particles by eukaryotic cells. J. Virol. *69*, 3258-3264.

Wakabayashi, M. T., Da Silva, D. M., Potkul, R. K., and Kast, W. M. (2002). Comparison of human papillomavirus type 16 L1 chimeric virus-like particles versus L1/L2 chimeric virus-like particles in tumor prevention. Intervirology *45*, 300-307.

Wang, S. S., Schiffman, M., Shields, T. S., Herrero, R., Hildesheim, A., Bratti, M. C., Sherman, M. E., Rodriguez, A. C., Castle, P. E., Morales, J., *et al.* (2003). Seroprevalence of human papillomavirus-16, -18, -31, and -45 in a population-based cohort of 10000 women in Costa Rica. Br. J. Cancer *89*, 1248-1254.

Wang, Z., Hansson, B. G., Forslund, O., Dillner, L., Sapp, M., Schiller, J. T., Bjerre, B., and Dillner, J. (1996). Cervical mucus antibodies against human papillomavirus type 16, 18, and 33 capsids in relation to presence of viral DNA. J. Clin. Microbiol. *34*, 3056-3062.

Wang, Z., Konya, J., Avall-Lundkvist, E., Sapp, M., Dillner, J., and Dillner, L. (1997). Human papillomavirus antibody responses among patients with incident cervical carcinoma. J. Med. Virol. *52*, 436-440.

Watts, D. H., Koutsky, L. A., Holmes, K. K., Goldman, D., Kuypers, J., Kiviat, N. B., and Galloway, D. A. (1998). Low risk of perinatal transmission of human papillomavirus: results from a prospective cohort study. Am. J. Obstet. Gynecol. *178*, 365-373.

Wen, L. M., Estcourt, C. S., Simpson, J. M., and Mindel, A. (1999). Risk factors for the acquisition of genital warts: are condoms protective? Sex. Transm. Infect. *75*, 312-316.

White, W. I., Wilson, S. D., Bonnez, W., Rose, R. C., Koenig, S., and Suzich, J. A. (1998). *In vitro* infection and type-restricted antibody-mediated neutralization of authentic human papillomavirus type 16. J. Virol. *72*, 959-964.

White, W. I., Wilson, S. D., Palmer-Hill, F. J., Woods, R. M., Ghim, S. J., Hewitt, L. A., Goldman, D. M., Burke, S. J., Jenson, A. B., Koenig, S., and Suzich, J. A. (1999). Characterization of a major neutralizing epitope on human papillomavirus type 16 L1. J. Virol. *73*, 4882-4889.

Wideroff, L., Schiffman, M., Haderer, P., Armstrong, A., Greer, C. E., Manos, M. M., Burk, R. D., Scott, D. R., Sherman, M. E., Schiller, J. T., *et al.* (1999). Seroreactivity to human papillomavirus types 16, 18, 31, and 45 virus-like particles in a case-control study of cervical squamous intraepithelial lesions. J. Infect. Dis. *180*, 1424-1428.

Wideroff, L., Schiffman, M. H., Nonnenmacher, B., Hubbert, N., Kirnbauer, R., Greer, C. E., Lowy, D., Lorincz, A. T., Manos, M. M., Glass, A. G., and et al. (1995). Evaluation of seroreactivity to human papillomavirus type 16 virus-like particles in an incident case-control study of cervical neoplasia . J. Infect. Dis. *172*, 1425-1430.

Wideroff, L., Schottenfeld, D., Carey, T. E., Beals, T., Fu, G., Sakr, W., Sarkar, F., Schork, A., Grossman, H. B., and Shaw, M. W. (1996). Human papillomavirus DNA in malignant and hyperplastic prostate tissue of black and white males. Prostate *28*, 117-123.

Witkin, S. S., Linhares, I., Giraldo, P., Jeremias, J., and Ledger, W. J. (2000). Individual immunity and susceptibility to female genital tract infection. Am J Obstet. Gynecol. *183*, 252-256.

Wittek, A. E., Arvin, A. M., and Koropchak, C. M. (1983). Serum immunoglobulin A antibody to varicella-zoster virus in subjects with primary varicella and herpes zoster infections and in immune subjects. J. Clin. Microbiol. *18*, 1146-1149.

Wong, M. M., and Fish, E. N. (2003). Chemokines: attractive mediators of the immune response. Semin. Immunol. *15*, 5-14.

Wu, T. C., Guarnieri, F. G., Staveley-O'Carroll, K. F., Viscidi, R. P., Levitsky, H. I., Hedrick, L., Cho, K. R., August, J. T., and Pardoll, D. M. (1995). Engineering an intracellular pathway for major histocompatibility complex class II presentation of antigens. Proc. Natl. Acad. Sci. USA *92*, 11671-11675.

Yang, R., Murillo, F.M., Delannoy, M.J., Blosser, R.L., Yutzy, W.H. 4th, Uematsu, S., Takeda, K., Akira, S., Viscidi, R.P., and Roden, R.B.(2005a). B Lymphocyte Activation by Human Papillomavirus-Like Particles Directly Induces Ig Class Switch Recombination via TLR4-MyD88. J. Immunol. *174*, 7912-7919.

Yang, R., Wheeler, C.M., Chen, X., Uematsu, S., Takeda, K., Akira, S., Pastrana, D.V., Viscidi, R.P,. and Roden, R.B. (2005b). Papillomavirus capsid mutation to escape dendritic cell-dependent innate immunity in cervical cancer. J. Virol. *79*, 6741-6750.

Yang, R., Murillo, F. M., Cui, H., Blosser, R., Uematsu, S., Takeda, K., Akira, S., Viscidi, R. P., and Roden, R. B. (2004a). Papillomavirus-like particles stimulate murine bone marrow-derived dendritic cells to produce alpha interferon and Th1 immune responses via MyD88. J. Virol. *78*, 11152-11160.

Yang, R., Murillo, F. M., Lin, K. Y., Yutzy, W. H. t., Uematsu, S., Takeda, K., Akira, S., Viscidi, R. P., and Roden, R. B. (2004b). Human papillomavirus type-16 virus-like particles activate complementary defense responses in key dendritic cell subpopulations. J. Immunol. *173*, 2624-2631.

Yang, R., Day, P.M., Yutzy, W.H. 4th, Lin, K.Y., Hung, C.F., Roden, R.B. (2003a). Cell surface-binding motifs of L2 that facilitate papillomavirus infection. J. Virol. *77*, 3531-3541.

Yang, R., Yutzy, W. H. t., Viscidi, R. P., and Roden, R. B. (2003b). Interaction of L2 with beta-actin directs intracellular transport of papillomavirus and infection. J. Biol. Chem. *278*, 12546-12553.

Yeager, M. D., Aste-Amezaga, M., Brown, D. R., Martin, M. M., Shah, M. J., Cook, J. C., Christensen, N. D., Ackerson, C., Lowe, R. S., Smith, J. F., *et al.* (2000). Neutralization of human papillomavirus (HPV) pseudovirions: a novel and efficient approach to detect and characterize HPV neutralizing antibodies. Virology *278*, 570-577.

Yoon, C. S., Kim, K. D., Park, S. N., and Cheong, S. W. (2001). alpha(6) Integrin is the main receptor of human papillomavirus type 16 VLP. Biochem. Biophys. Res. Commun. *283*, 668-673.

Yuan, H., Estes, P. A., Chen, Y., Newsome, J., Olcese, V. A., Garcea, R. L., and Schlegel, R. (2001). Immunization with a pentameric L1 fusion protein protects against papillomavirus infection. J. Virol. *75*, 7848-7853.

Zhang, L. F., Zhou, J., Chen, S., Cai, L. L., Bao, Q. Y., Zheng, F. Y., Lu, J. Q., Padmanabha, J., Hengst, K., Malcolm, K., and Frazer, I. H. (2000). HPV6b virus like particles are potent immunogens without adjuvant in man. Vaccine *18*, 1051-1058.

Zhang, W., Carmichael, J., Ferguson, J., Inglis, S., Ashrafian, H., and Stanley, M. (1998). Expression of human papillomavirus type 16 L1 protein in *Escherichia coli*: denaturation, renaturation, and self-assembly of virus-like particles *in vitro*. Virology *243*, 423-431.

Zhang, Y., Rassa, J. C., deObaldia, M. E., Albritton, L. M., and Ross, S. R. (2003). Identification of the receptor binding domain of the mouse mammary tumor virus envelope protein. J. Virol. *77*, 10468-10478.

Zhao, K. N., Sun, X. Y., Frazer, I. H., and Zhou, J. (1998). DNA packaging by L1 and L2 capsid proteins of bovine papillomavirus type 1. Virology *243*, 482-491

Zhou, J., Gissmann, L., Zentgraf, H., Muller, H., Picken, M., and Muller, M. (1995). Early phase in the infection of cultured cells with papillomavirus virions. Virology *214*, 167-176.

Zhou, J., Stenzel, D. J., Sun, X. Y., and Frazer, I. H. (1993). Synthesis and assembly of infectious bovine papillomavirus particles *in vitro*. J. Gen. Virol. *74*, 763-768.

Zhou, J., Sun, X. Y., Louis, K., and Frazer, I. H. (1994). Interaction of human papillomavirus (HPV) type 16 capsid proteins with HPV DNA requires an intact L2 N-terminal sequence. J. Virol. *68*, 619-625.

Zhou, J., Sun, X. Y., Stenzel, D. J., and Frazer, I. H. (1991). Expression of vaccinia recombinant HPV 16 L1 and L2 ORF proteins in epithelial cells is sufficient for assembly of HPV virion-like particles. Virology *185*, 251-257.

From: Papillomavirus Research: From Natural History to Vaccines and Beyond. Edited by: M. Saveria Campo

Chapter 19

Immunobiology of Papillomaviruses

Margaret A. Stanley*

Abstract

HPV infection is ubiquitous. In the genital tract almost all sexually active indivuals are infected at some point but most clear their infections without overt clinical disease. Lesions in which productive viral infection is occurring (ano-genital warts and CIN1) are not associated with inflammation or histological evidence of immune activity and HPV evades immune recognition by evading innate immune defences and delaying the activation of adaptive immunity. Regression of ano-genital warts is accompanied histologically by a response characteristic of cell mediated immunity, animal models support this and provide evidence that the response is regulated by CD4 T cell-dependent mechanisms. The increased prevalence of HPV infections in individuals immunosuppressed either as a consequence of organ transplantation or HIV infection demonstrates the central importance of the CD4 T cell population in the control of established HPV infections. Although it seems clear that the CD4 T cell subset is critical for the induction and regulation of the host response to HPV the nature of the effector response remains unclear. There is increasing evidence that NK cells and antigen specific CTLs recognising E2 and E6 are important effectors but CTL responses and the target antigens are still poorly understood.

1. Introduction

Papillomaviruses are ubiquitous and successful infectious agents that are characterised by absolute species specificity and tissue tropism. The infectious cycle of these viruses is tailored to the differentiation programme of the target cell – the keratinocyte – different phases of permissive viral growth accompanying the steady maturation of the keratinocyte as it progresses up the epithelium to its destiny as a terminally differentiated squame. This infectious cycle raises several important issues with respect to immune recognition.

Firstly infection and vegetative viral growth are absolutely dependent upon a complete programme of keratinocyte differentiation. Virus infects primitive basal keratinocytes, probably targeting stem cells, but high level viral expression of viral proteins and viral assembly occur only in the upper layers of the stratum spinosum and granulosum of squamous epithelia. Viral gene expression is confined to the keratinocyte and there is no evidence from clinical infections that viral genes are expressed in any cell other than keratinocytes or cells with the potential for squamous maturation. Secondly the replication cycle takes a long time, even in the best scenario the time from infection to release of virus will take about 3 weeks since this is the time taken for the basal keratinocyte to undergo complete differentiation and desquamate. In reality the period between infection and the appearance of lesions is highly variable and can vary from weeks to months (Oriel, 1971; Koutsky, 1991) indicating that the virus can effectively evade the immune system. Thirdly there is no cytolysis or cytopathic death as a consequence of viral replication and assembly. These key events for the virus occur in the differentiating keratinocyte, a cell destined for death and desquamation far from the sites of immune activity. The virus actually delays nuclear condensation in differentiating keratinocytes forming the koilocyte, the pathognomic cell of HPV infection. This may be the consequence (at least in high risk HPV infections) of the combined effects of viral E6 and E7 proteins in halting apoptosis until viral replication is completed but then the virus laden keratinocyte proceeds to its inevitable fate, death by natural causes. As a consequence of this, HPV infection is not accompanied by inflammation, there is no obvious "danger signal" to alert the immune system. This is a viral strategy that results in persistent, chronic infections as the host remains ignorant of the pathogen for long periods. The central questions therefore are, does natural infection with HPV evoke a host response, what is the nature of this response, when and how does it occur, what is the role of humoral and cell mediated immunity in the natural history of genital HPV associated disease and how do virus positive neoplasms arise in the face of host defences? The present chapter reviews some aspects of the cell mediated immune response to HPV infection.

2. Immune Responses to Virus Infections

Host defence is a partnership between innate immunity (phagocytes, soluble proteins *e.g.* cytokines, complement and epithelial barriers) together with adaptive immunity

*For correspondence. Email mas@mole.bio.cam.ac.uk.

(antibody, cytotoxic effector cells). Put simply, the innate immune system detects the pathogen and acts as first line defence, thus 90% of microbial assaults are cleared by innate responses alone. Critically, innate immunity activates the appropriate adaptive immune response that will generate antigen specific effector responses and specific memory to the insult. Thus the adaptive responses of antibody mediated humoral immunity clear free virus particles from body fluids and can prevent re-infection by virus, those of cell mediated immune (CMI) responses are essential for the clearance of virus infected cells and the generation of immune memory.

T lymphocytes are the generals of the adaptive immune system and the key players in CMI. T cells recognise antigen processed into short peptides bound to the major histocompatibility complex (MHC) proteins and presented as a membrane bound receptor complex on the cell surface. There are 2 major subsets of T cells identified by the surface markers CD4 or CD8. CD4+ T cells recognise antigen presented by class II MHC; CD8+ T cells recognise antigen presented by class I MHC. Antigen in the context of class I MHC is endogenous antigen derived, usually but not always, from intracellular synthesis of proteins broken down in the proteasome into short peptides and presented on the cell surface as an MHC/peptide complex for recognition by the T cell receptor (TCR) on the CD8+ lymphocyte. The geometry of the MHC/peptide complex is so precise that it is only recognised and bound with the correct affinity by a TCR that matches it precisely. Usually antigen plus class II is exogenous antigen taken up from the extracellular milieu and broken down in the endosome of the APC for association into the class II complex for presentation on the cell surface and recognition by the specific TCR on the CD4+ T cell. The interaction between the CD4+ T cell and the APC is very complex requiring several other receptor/ligand interactions to occur in a regulated order before the T cell is activated and starts to proliferate. In particular the T cell must receive a specific second signal from other receptors on the APC. Failure to receive the second signal can render the T cell anergic or unresponsive to any subsequent encounter with the antigen.

T cell activation results in the secretion of a repertoire of small proteins or cytokines that help and regulate other cells. The pattern of cytokine expression defines two major subsets of CD4+ T cells known as Th2 or Th1 cells

- Th2 cells secrete interleukin (IL)-4 and IL-10 (and other cytokines) and help antigen primed B cells to differentiate into plasma cells and secrete antibody for humoral responses.
- Th1 cells secrete interferon (IFN)-γ and create a milieu in which key cytotoxic effectors macrophages, natural killer (NK) cells and cytotoxic T lymphocytes (CTL) are activated generating CMI.

A third category of T cells generally described as regulatory T cells (Tregs), with the phenotype CD4+CD25+ that express the signature transcription factor Foxp3+ and usually secrete IL-10 and TGF-β, are of intense contemporary interest. Cells with this phenotype are considered to recognise self antigens and function to prevent auto-immunity. However they also regulate responses to exogenous antigens and have been implicated in chronic viral infections favouring the persistence of virus and immunopathology (for review see Rouse et al., 2004).

Whether the T cell takes the Th2 or Th1 or regulatory path is very strongly influenced by the APC as a consequence of the receptors the APC expresses and the cytokines secreted in the local milieu. These functions of the APC are in turn activated by signals received from the receptor ligand interactions between the APC and the pathogen and also by the cytokines released by the APC and other cells in the immediate vicinity. The APC and the cytokines released are the bridge between innate and adaptive immunity. The APC, in effect, "tells" the T cell what sort of defence is needed and is central to the generation of an effective and appropriate immune response and to the regulation of this response.

It seems that only dendritic cells (DC) can initiate primary immune responses and activate virgin T cells but HPV non-structural proteins (such as E6 and E7) are only expressed in keratinocytes so how can CMI responses be induced against them? This is achieved almost certainly by cross presentation of antigen. The most likely (but not the only) scenario is that antigen released from dying HPV infected cells is taken up by the DC and enters both the class I and class II processing pathways. The DC presents to the CD4+ T cell that then signals back to the DC via CD40/CD40 ligand or other receptor ligand interactions activating or licensing that DC to directly present to naïve CD8+ T cells that can then differentiate into potent CTLs and kill HPV infected cells that they detect. Since cell death is a pre-requisite for inflammation, the cardinal sign of activated innate immunity, the absence of inflammation and death in permissive papillomavirus infections is a major component of the immune evasion strategies of these viruses.

3. Cell Mediated Immunity in HPV Infections

One might ask in view of this, is there an immune response to papillomaviruses? The evidence from natural history studies and immunocompromised individuals is clearly that there is. Large cohort studies (Herrero et al., 2000) have delineated the natural history of genital HPV infections in women; these show that these infections are common in young sexually active women and that the majority of these resolve indicating that there are effective host defence mechanisms against these viruses. The increased incidence and progression of HPV infections in immuno-suppressed individuals

(Benton et al., 1996) illustrates the critical role of cell-mediated immune responses in the resolution and control of HPV infections. HIV infected patients show multiple recurrences of cervical HPV infections (Fruchter et al., 1996) and an increased incidence of genital warts (Fennema et al., 1995) which appears to reflect an increased risk of progression from sub clinical to clinical infection (Chirgwin et al., 1995). Overall the evidence from allograft recipients, inherited immunodeficiencies and HIV infected individuals suggests that it is the absolute deficit in CD4 cells which is important in HPV induced disease and associated neoplastic progression implying a central role for CD4 mediated mechanisms in the control of HPV infection.

4. Histological Studies

Clues to the nature of the cellular immune response to HPV infection have come from immunohistological studies of spontaneously regressing genital warts (Coleman et al., 1994b). Non-regressing genital warts are characterised by a lack of immune cells, the few intra-epithelial lymphocytes are CD8+ cells and mononuclear cells are present mainly in the stroma. Wart regression is characterised by a massive mononuclear cell infiltrate in both stroma and epithelium. The infiltrating lymphocytes are predominantly CD4+ cells but many CD8+ T cells are present and are concentrated in the epithelium. The lymphocytes are activated and express the IL-2 receptor: most are "antigen experienced" expressing the CD45 RO marker. The wart keratinocytes express HLA-DR and the adhesion molecule ICAM-1 implying that there has been local release of cytokines such as IFN-γ. There is upregulation of the adhesion molecules required for lymphocyte trafficking on the endothelium of the wart capillaries reinforcing the notion of local releases of proinflammatory cytokines and chemokines. These appearances are characteristic of a Th1 biased lymphocyte response and this is reinforced by the analysis of cytokine expression in the regressing lesions. There is expression of mRNA for the proinflammatory cytokines IFNγ, tumour necrosis factor (TNF)-α and the Il-12 p40 subunit. Interestingly bioactive IL-12 is expressed not only by dendritic cells and macrophages in the regressing wart but also by the infected keratinocytes (Stanley, Colemand and Scarpini, unpublished observations). No statistical differences in Langerhans cell number were seen between regressors and non-regressors (Coleman et al., 1994a) although loss of dendritic arborisations was seen in Langerhans cells (LC) in non-regressing lesions, a phenomenon reported also in HPV associated cervical lesions (Morelli et al., 1992). Cross sectional studies however, only provide a snapshot of a dynamic process and in a longitudinal study examining the action of Imiquimod™, a topical preparation with immunomodulatory activities, a decrease in mRNA for the LC marker CD1a was observed during wart regression post treatment (Tyring et al., 1998).

The response to genital warts represents host defence to HPV infection uncomplicated by the genetic instability that characterises neoplasia. The situation in HPV infected cervical epithelium is more complicated since there are two biological processes – HPV infection and neoplasia. Immunohistochemical studies of cervical low grade squamous intraepithelial neoplasia (LGSIL) suggest that, overall these lesions are immunologically quiescent with few infiltrating T cells and a decreased number of LC compared to normal ectocervical epithelium (for review see Stanley et al., 2003). It is possible that this decrease simply reflects the normal egress of antigen carrying LC from the epidermis to the draining node for antigen presentation. However virtually all cervical SIL arise in the transformation zone (TZ), often associated with squamous metaplasia, and there is evidence that LC numbers are reduced in squamous metaplasia (Roncalli et al., 1988) and in the TZ where these LC have an immature phenotype (Giannini et al., 2002). The reduced numbers of LC in SIL lesions could simply reflect the histogenesis.

High grade SIL (HGSIL) also show a decreased number of LC, a phenomenon that may be related to neoplasia since there is *in vitro* evidence that immortalised HPV infected cervical cells inhibit LC recruitment (Hubert et al., 1999). In addition expression of HPV 16 E6 in basal keratinocytes inhibits E cadherin expression and since LC/keratinocyte adhesion is mediated via cadherin:cadherin interactions this could reduce LC retention in squamous epithelium (Matthews et al., 2003). Immature CD1a+ stromal DC expressing IL-10 and TGF-β have been shown to be resident throughout the normal cervical stroma but increased in HGSIL (Kobayashi et al., 2004). These DC have been implicated in inducing immunotolerance (Wakkach et al., 2003) and their increased density in HGSIL would support the notion that the immune milieu in high grade lesions is immunosuppressive. However the local immune environment in LGSIL and HGSIL is influenced by the events of neoplastic progression. HGSIL are aneuploid, genetically unstable lesions exhibiting heterogeneity in the expression of immunologically relevant molecules (Coleman et al., 1993; 1994a; 1994b). In a comprehensive analysis of the cervical microenvironment Kobayashi and colleagues show clearly that strong pro and anti inflammatory responses are both present in HGSIL (Kobayashi et al., 2004). In particular a robust Th1 response with abundant expression of IFN-γ by CD4, CD8 and NK cells could be demonstrated but despite this HPV infected HGSIL persists, a phenomenon that may be related to the immunosuppressive environment with the recruitment of regulatory T cells together with the acquisition of resistance to apoptosis induced by cytokines such as Type I interferons and TNF-α.

5. Cytokines

Both the initiation and regulation of cell-mediated immune responses are dependent, to a very large degree, on the local cytokine milieu. The production of these molecules is not restricted to immunocytes such as lymphocytes, dendritic cells and macrophages but keratinocytes themselves are important sources. Keratinocytes constitutively secrete low levels of several cytokines (Wang et al., 1999) and can be induced by various stimuli – injury, UV irradiation, infection – to produce significant amounts (Nozaki et al., 1992). The expression of the Type 1 Interferons (IFNα, IFNβ), TNF-α and IL-10 is of particular interest in cervical HPV associated disease.

The Type 1 interferons have antiviral, antiproliferative, anti-angiogenic and immunostimulatory properties but crucially they act as a bridge between innate and adaptive immunity activating immature DC (Le Bon et al., 2002). Most DNA viruses have mechanisms for inhibiting IFN synthesis and signalling and the papillomaviruses are no exception. Microarray studies on cell lines in which HPV 31 is present as an episome show that interferon inducible genes and the transcriptional regulator Stat-1 are down regulated (Chang et al., 2000) and this seems to be related to E2 expression (Chang et al., 2002). In other work using keratinocytes transfected with HPV 16 E6/E7 expression of IFN-α inducible genes was inhibited (Nees et al., 2001). HPV 16 E7 interacts with components of the IFN signalling pathway binding to the p48 component of the interferon induced transcription complex (Barnard et al., 1999) and E6 physically interacts with the JAK kinase tyk-2 (Li et al., 1999). Paradoxically interferon inducible gene expression is upregulated in cells in which HPV 16 is integrated compared to episome only containing cells (Alazawi et al., 2002), a phenomenon that may be related to the interferon resistance that appears to accompany neoplastic progression (Bachmann et al., 2002).

Secreted TNF-α is a potent activator of immune responses and the protein in its intracellular form is constitutively expressed by keratinocytes *in vivo* and *in vitro* (Kock et al., 1990). Normal cervical keratinocytes and most LGSIL express TNF and TNF receptors but this expression is lost in HGSIL (Stanley et al., 2003; Mota et al., 1999). Il-10 is a cytokine that inhibits Th1 responses, a property mediated via its effects on APC (Enk et al., 1993). Keratinocytes can be induced to secrete IL-10 (Enk et al., 1992) and it is likely that this plays a role in regulating the LC phenotype and function in squamous epithelium. IL-10 expression at both mRNA and protein level in normal cervix and SIL was examined in a careful study by Mota and colleagues (Mota et al., 1999). No expression could be detected in normal cervix but a proportion of both low grade and high grade lesions showed epithelial expression. Expression in LSIL was usually in basal cells but many HSIL exhibited full thickness expression. In our own studies (Stanley et al., 2003) epithelial IL-10 expression could not be detected in either normal cervix or SIL but could be detected in infiltrating lymphocytes in high grade lesions.

6. NK cells

NK cells are key components of the innate immune response to viral infections. They are a sub-set of lymphocytes that kill virally infected or tumour cells lacking surface expression of MHC Class I molecules. There is evidence for defective NK cell function in HPV 16 associated disease. PBMC from patients with active HPV 16 neoplastic disease display a reduced NK cell activity against HPV 16 infected keratinocytes (Malejczyk et al., 1993). *In vitro* expression of the E7 protein of HPV 16/18 precludes the lysis of HPV transformed cells by interferon stimulated NK cells and the E6/E7 proteins of HPV 16 inhibit IL-18 expression by NK cells by binding to both IL-18 and its receptor (Cho et al., 2001; Lee et al., 2001). Persuasive evidence for the importance of NK cell activity for the control of HPV infection comes from a recent study in which patients with severe combined immunodeficiency were followed up over the long term post haemopoietic stem cell transplantation (Laffort et al., 2004). A subset of these individuals exhibited severe cutaneous papillomatosis associated with either common γc receptor cytokine subunit or Janus kinase-3 (JAK-3) deficiency. A common consequence of such signalling defects is NK cell deficiency and indeed NK cell counts were lower in these patients.

7. CD4 T cell responses

Immunohistological studies indicate clearly that regression of HPV infected lesions is associated with a Th1 response but the viral antigens that provoke this response still remain to be identified unequivocally. The evidence from both experimental models and human studies show that viral proteins are immune targets. In a murine model in which viral antigen is expressed in keratinocytes and mimics the natural route of infection, DTH responses to E6 and E7 can be shown. The ability to prime the immune system and elicit a DTH response in this system depends upon antigen dose (Chambers et al., 1994). Low levels of antigen induce immune non-responsiveness, a phenomenon associated with a decrease in the percentage of IL-2 secreting γδ T cells and a switch in Th1-Th2 cytokine expression in the CD8 subset in the draining lymph node, suggestive of a suppressor effect (Lopez et al., 2000).

Most clinical studies have concentrated on responses to high risk HPV E6 and E7 proteins or the L1 capsid protein in patients with cervical HPV associated disease, and specific T cell responses to these proteins have been identified in patients with cervical intraepithelial neoplasia (CIN), of all grades, in cross sectional studies. To date no definitive association has been established between HPV specific T cell responses to the early antigens E6 and

E7 and progression of HPV associated disease. In cross sectional studies lymphoproliferative T cell responses to E7 in CIN have been described as more common in high grade than low grade disease (Luxton et al., 1996) but usually decline in patients with carcinoma (de Gruijl et al., 1996). Using an IFNγ ELISPOT assay (van der Burg et al., 2001) frequent E7 specific responses were found in patients with HPV 16+ high grade CIN or cancer but were rarely detected in normal healthy controls. T cell proliferative responses to HPV 16 E7 were described in a longitudinal study in which patients fell into 3 groups, those who cleared disease, those with persistent disease and those with progressive disease. Paradoxically HPV 16 E7-specific responses were associated both with clearance and persistence (de Gruijl et al., 1996; 1998) a phenomenon interpreted as reflecting antigen availability resulting from either clearance or persistence. A different outcome was reported in a study by Kadish and colleagues (Kadish et al., 2002) who determined T cell proliferative responses to HPV 16 E6 and E7 peptides at 3 month intervals over a 12 month period in a cohort of women with CIN I or II. CMI responses in this group, whose lesions were not exclusively HPV 16 positive, to a specific HPV 16 E7 peptide correlated significantly with regression of disease and resolution of infection. Intriguingly these responses were not HPV type specific.

Few studies have determined the antigen specificity of lymphocytes infiltrating HPV infected lesions. Wart infiltrating lymphocytes (WIL) isolated from HPV 6 infected ano-genital warts were characterised for their response to HPV 6 E7 and L1 proteins in a lymphoproliferation assay (Hong et al., 1997). HPV 6 E7 or L1 specific WIL could be isolated from more than 75% of the patients studied (n=24). Interestingly despite the huge variation between patients in surface markers of the WIL (CD4, CD8, TCRαβ or TCRγδ) the HPV 6 E7 and L1 peptides recognised by WIL differed from those recognised by peripheral T cells.

There is increasing evidence both in animal and human papillomavirus infections that the E2 and E6 proteins are important immune targets for effective viral clearance. In the dog, during the infectious cycle of the canine oral papillomavirus (COPV), systemic T cell responses directed to E2 and E6 peptides can be detected at distinct time points (Stanley unpublished observations). These systemic responses occur at low frequency that is remarkable when the intensity of the local lymphocytic infiltrate in the regressing lesions is considered, implying that the local immune responses must be of a greater magnitude. These responses also occur in narrow time windows that coincide with periods of viral DNA amplification and are maximal at wart regression *i.e.* weeks 8-10 post infection: thereafter declining quite rapidly. In the dog immunisation with a DNA vaccine encoding COPV E2, modified to enhance protein expression, results in protection both prophylactically and therapeutically (Moore et al., 2003). In cottontail rabbit papillomavirus (CRPV) infection in the rabbit, which provides a good model for papilloma/carcinoma progression, the induction of T cell proliferative responses to the E2 protein was the best predictor of lesion regression (Selvakumar et al., 1995). Recently in this model it has been shown that immunisation with a recombinant adenovirus E2 in a therapeutic protocol induces lesion regression and prevents progression to carcinoma (Brandsma et al., 2004).

There is increasing evidence that T cell responses to E2 are important in genital HPV infections. A non intervention follow up study of women with cytological evidence of low grade CIN, showed that HPV16 E2-specific T cell responses, as measured by specific IL-2 release *in vitro*, occurred frequently at the time of lesion clearance (Bontkes et al., 1999). Good Th-1 type immunity against the E2 and E6 proteins has been detected in healthy individuals with no clinical signs of HPV16 infection (de Jong et al., 2002; Welters et al., 2003). Importantly, these Th1 type responses were found only occasionally in high-grade CIN patients and were impaired in cervical cancer patients (de Jong et al., 2004). These data suggest that a hallmark of effective immune control of HPV 16 infection in the cervix is the generation of CD4+ T cells specific for E2 and E6.

8. Cytotoxic T Cell Responses

Cell mediated cytotoxicity is the most important effector mechanism for the control and clearance of viral infections and is implemented by a range of cells including CTL and NK cells. The role of these effectors in HPV infection is the subject of intense reserach. There is now good evidence that HPV specific CTL can be detected in patients with previous (Nakagawa et al., 1997) or ongoing HPV infection (Evans et al., 1996; Alexander et al., 1996; Evans et al., 1997). Both CD4+ and CD8+ cytotoxic effectors have been shown to be involved in these responses (Nakagawa et al., 1999). CTL responses have been described in 6/10 CIN 3 patients in an assay in which PBMC restimulated with adenovirus recombinants expressing an HPV16 and 18 E6/E7 fusion protein, lysed autologous targets infected with a recombinant vaccinia virus expressing the same fusion protein (Nimako et al., 1997). No CTL responses were identified in control subjects. However the HPV status of controls and patients was not defined in this study. High affinity HPV E7 specific CTL are rare in patients with carcinoma or HGSIL but can be detected with increased sensitivity using the technology whereby fluorescently labelled HLA peptide complexes (tetramers) is employed (Youde et al., 2000).

Increased numbers of T cells are seen locally in squamous cell carcinoma of the cervix (Ghosh et al., 1994) with a dominance of CD8+ cells. In a study in which HPV 16 and 18 E6/E7 specific CTL were detected in PBMC, draining lymph nodes and tumour of cervical cancer patients, limiting dilution analysis was used to determine the frequency of the HPV specific CTL: these were present in significantly higher numbers in the tumours and lymph nodes than peripheral blood (Evans et al., 2001).

The association between HPV 16 E6 and E7 specific CTL and HPV 16 persistence was examined in a longitudinal study of women with PCR determined cervical HPV 16 infection. Lack of CTL response to E6 but not E7 correlated with persistent HPV infection suggesting that a CTL response to HPV 16 E6 is important for viral clearance and, by implication, neoplastic progression (Nakagawa et al., 2000) and there is some support for this notion. A sequence variant of HPV 16 is over-represented in HLA-B7 cervical cancer patients. This variant exhibits a single amino acid change in the HPV 16 E6 protein in a putative HLA-B7 restricted CTL epitope (Ellis et al., 1995), a change that alters the affinity of binding of the TCR/HLA-B7 peptide complex and could therefore compromise the CTL response. A study by Evans and colleagues (Evans et al., 2001) that addressed the question, of whether cervical carcinoma cells can process and present HPV E6/E7 antigen for CTL recognition, is highly relevant to this issue. HLA-A*0201 restricted CTL clones were generated against HPV 16 E6 CTL epitopes. The clones recognised E6 transfected lymphoblastoid cells but not HLA A*0201 HPV 16 E6+ carcinoma cells even when the endogenous level of E6 in the latter was enhanced by transfection. This defect in recognition was due to Class I down regulation on the tumour cells but even when Class I was upregulated by IFNγ, unless E6 expression was concomitantly increased by transfection, CTL recognition failed suggesting that both antigen processing and presentation of E6 are compromised in invasive carcinomas.

9. Immunogenetics and HPV Infection

The HLA haplotype of the individual could influence the natural history of HPV infection in that individual since, for any one protein antigen, different alleles of the MHC will present different peptides to the immune system. Thus the HLA haplotype of the individual could be a major determinant of whether infection is cleared or persists which, in the case of the oncogenic viruses, could influence the risk of neoplastic progression. If this thesis is correct then it could be reflected in different HLA frequencies in patients with persistent or chronic HPV infections compared to the appropriate control populations.

Data from the rabbit strongly supports this supposition since the regression or progression of papillomas induced by CRPV is strongly linked to the MHC DR or DQ phenotypes respectively of the animals (Han et al., 1992). Recurrent respiratory papillomatosis represents an extreme form of persistent HPV infection and an over-representation of HLA-DQ3 and DQ-11 alleles have been reported in these patients (Bonagura et al., 1994). A number of studies have examined associations between HLA haplotype and cervical carcinoma but the evidence is that the associations are between HPV type and HLA. This was first raised in a case control study of Hispanic women with cervical cancer compared to those with normal Pap smears from the South West USA. Thus certain HLA class II haplotypes including DRB1*150 - DQB1*0602 were significantly associated with HPV 16 containing cancer whereas DR13 haplotypes were negatively associated (Apple et al., 1994). The association between susceptibility to HPV 16 infection and these haplotypes has been shown in other well-controlled studies and importantly such individuals tend to have more long-term infections. Host genetic factors such as variation at the Class II locus will influence host responses to the virus and the outcome of infection.

NK cells are important in HPV infections and host genetic factors may be central to this. Target recognition by NK cells does not involve classical antigen receptors because these cells do not express T or B cell antigen receptors nor do they rearrange their T cell receptor genes. This apparent lack of antigen-specific receptors and MHC recognition originally led to the view that NK mediated killing was broad and non-MHC restricted but, it turns out, NK cells express clonally distributed receptors specific for MHC class I type molecules. Killer immunoglobulin like receptors (KIRs) are surface inhibitory receptors of this type. Upon binding to class I MHC, KIRs block NK cell activation and function but cells lacking the relevant MHC molecules are killed. Genes encoding KIRs are polymorphic and KIR haplotypes vary considerably in the number of genes that they contain (briefly reviewed in (Moretta et al., 2004). MHC haplotype as has been discussed briefly above may influence resistance or susceptibility to HPV infections and KIR haplotype may be of equal importance.

10. Five Year View

Effective anti-viral therapies for HPV infection are very much needed and immunotherapies are an obvious route but antigen-specific immunotherapies have been disappointing to date. The crucial importance of innate defences to viral infections particularly at mucosal surfaces is being revealed and offers opportunities for novel therapies. Most DNA viruses have mechanisms that down regulate Type I interferon responses and HPV is no exception, many viruses modulate interactions with toll like receptors during viral entry influencing the intensity of the adaptive immune response and theta defensins have broad spectrum antiviral efficacy. These areas offer new opportunities for antiviral therapy, are relatively unexplored with respect to HPV infections and represent a rich field for future investigation of HPV immunobiology.

References

Alazawi, W., Pett, M., Arch, B., Scott, L., Freeman, T., Stanley, M.A. and Coleman, N. (2002). Changes in cervical keratinocyte gene expression associated with integration of human papillomavirus 16. Cancer Res. *62,* 6959-6965.

Alexander, M., Salgaller, M.L., Celis, E., Sette, A., Barnes, W.A., Rosenberg, S.A. and Steller, M.A. (1996). Generation of tumor-specific cytolytic T lymphocytes from peripheral blood of cervical cancer patients by *in vitro* stimulation with a synthetic human papillomavirus type 16 E7 epitope. Amer. J. Obstet. Gynecol. *175,* 1586-1593.

Apple, R.J., Erlich, H.A., Klitz, W., Manos, M.M., Becker, T.M. and Wheeler, C.M. (1994). HLA DR DQ associations with cervical carcinoma show papillomavirus type specificity. Nat. Genet. *6,* 157-162.

Bachmann A., Hanke Brigitte, Zawatzky R., Soto U., van Riggelen, J., zur Hausen H. and Rosl F. (2002). Disturbance of tumor necrosis factor alpha-mediated beta interferon signaling in cervical carcinoma cells. J. Virol. *76,* 280-291.

Barnard, P. and McMillan, N.A. (1999). The human papillomavirus E7 oncoprotein abrogates signaling mediated by interferon-alpha. Virol. *259,* 305-313.

Benton, E.C. and Arends, M.J. (1996). Human papillomavirus in the immunosuppressed. In Papillomavirus Reviews: Current Research on Papillomaviruses. Lacey, C. (ed.), Leeds, Leeds University Press. 271-279.

Bonagura, V.R., Siegal, F.P., Abramson, A.L., Santiago Schwarz, F., O'Reilly, M.E., Shah, K., Drake, D. and Steinberg, B.M. (1994). Enriched HLA DQ3 phenotype and decreased class I major histocompatibility complex antigen expression in recurrent respiratory papillomatosis. Clin. Diagn. Lab. Immunol. *1,* 357-360.

Bontkes, H.J., de Gruijl, T.D., Bijl, A., Verheijen, R.H., Meijer, C.J., Scheper, R.J., Stern, P.L., Burns, J.E., Maitland, N.J. and Walboomers, J.M. (1999). Human papillomavirus type 16 E2-specific T-helper lymphocyte responses in patients with cervical intraepithelial neoplasia. J. Gen. Virol. *80,* 2453-2459.

Brandsma, J.L., Shlyankevich, M., Zhang, L., Slade, M.D., Goodwin, E.C., Peh, W. and Deisseroth, A.B. (2004). Vaccination of rabbits with an adenovirus vector expressing the papillomavirus E2 protein leads to clearance of papillomas and infection. J. Virol. *78,* 116-123.

Chambers, M.A., Wei, Z., Coleman, N., Nash, A.A. and Stanley, M.A. (1994). "Natural" presentation of human papillomavirus type 16 E7 protein to immunocompetent mice results in antigen specific sensitization or sustained unresponsiveness. Eur. J. Immunol. *24,* 738-745.

Chang, Y.E. and Laimins, L.A. (2000). Microarray analysis identifies interferon-inducible genes and Stat-1 as major transcriptional targets of human papillomavirus type 31. J. Virol. *74,* 4174-4182.

Chang, Y.E., Pena, L., Sen, G.C., Park, J.K. and Laimins, L.A. (2002). Long-term effect of interferon on keratinocytes that maintain human papillomavirus type 31. J. Virol. *76,* 8864-8874.

Chirgwin, K.D., Feldman, J., Augenbraun, M., Landesman, S. and Minkoff, H. (1995). Incidence of venereal warts in human immunodeficiency virus infected and uninfected women. J. Infect. Dis. *172,* 235-238.

Cho, Y. S, Kang, J. W, Cho, M., Cho, C. W, Lee, S., Choe, Y. K, Kim, Y., Choi, I., Park, S. N, Kim, S., Dinarello, C. A and Yoon, D. Y (2001). Down modulation of IL-18 expression by human papillomavirus type 16 E6 oncogene via binding to IL-18. FEBS Lett. *501,* 139-45.

Coleman, N., Birley, H.D., Renton, A.M., Hanna, N.F., Ryait, B.K., Byrne, M., Taylor Robinson, D. and Stanley, M.A. (1994a). Immunological events in regressing genital warts. Amer. J. Clin. Pathol. *102,* 768-774.

Coleman, N., Birley, H.D., Renton, A.M., Hanna, N.F., Ryait, B.K., Byrne, M., Taylor Robinson, D. and Stanley, M.A. (1994b). Immunological events in regressing genital warts. Amer. J. Clin. Pathol. *102,* 768-74.

Coleman, N., Greenfield, I.M., Hare, J., Kruger Gray, H., Chain, B.M. and Stanley, M.A. (1993). Characterization and functional analysis of the expression of intercellular adhesion molecule 1 in human papillomavirus related disease of cervical keratinocytes. Amer. J. Pathol. *143,* 355-367.

Coleman, N. and Stanley, M.A. (1994a). Analysis of HLA DR expression on keratinocytes in cervical neoplasia. Int. J. Cancer *56,* 314-319.

Coleman, N. and Stanley, M.A. (1994b). Characterization and functional analysis of the expression of vascular adhesion molecules in human papillomavirus-related disease of the cervix. Cancer *74,* 884-892.

de Gruijl, T.D., Bontkes, H.J., Stukart, M.J., Walboomers, J.M.M., Remmink, A., Verheijen, R.H.M., Helmerhorst, T.J.M., Meijer, C.J.L.M. and Scheper, R. (1996). T cell proliferative responses against human papillomavirus type 16 E7 oncoprotein are most prominent in cervical intraepithelial neoplasia patients with persistent viral infection. J. Gen. Virol. *77,* 2183-2191.

de Gruijl, T.D., Bontkes, H.J., Walboomers, J.M., Stukart, M.J., Doekhie, F.S., Remmink, A.J., Helmerhorst, T.J., Verheijen, R.H., Duggan Keen, M.F., Stern, P.L., Meijer, C.J. and Scheper, R.J. (1998). Differential T helper cell responses to human papillomavirus type 16 E7 related to viral clearance or persistence in patients with cervical neoplasia: a longitudinal study. Cancer Res. *58,* 1700-6.

de Jong, A., van Poelgeest, M.I., van der Hulst, J.M., Drijfhout, J.W., Fleuren, G.J., Melief, C.J., Kenter, G., Offringa, R. and van der Burg, S.H. (2004). Human papillomavirus type 16-positive cervical cancer is associated with impaired CD4+ T-cell immunity against early antigens E2 and E6. Cancer Res. *64,* 5449-55.

de Jong, A., van der Burg, S.H., Kwappenberg, K.M. C, van der Hulst, J.M., Franken, K.L. M C, Geluk A., van Meijgaarden, K.E., Drijfhout, J. Wouter, Kenter G., Vermeij P., Melief, C.J. M and Offringa R. (2002). Frequent detection of human papillomavirus 16 E2-specific T-helper immunity in healthy subjects. Cancer Res. *62,* 472-9.

Ellis, J.R., Keating, P.J., Baird, J., Hounsell, E.F., Renouf, D.V., Rowe, M., Hopkins, D., Duggan Keen, M.F., Bartholomew, J.S., Young, L.S. and Et, A. (1995). The association of an HPV16 oncogene variant with HLA B7 has implications for vaccine design in cervical cancer. Nat. Med. *1,* 464-70.

Enk, A.H., Angeloni, V.L., Udey, M.C. and Katz, S.I. (1993). Inhibition of LC-APC function by IL-10. A role for IL-10 in induction of tolerance. J. Immunol. *151,* 2390-7.

Enk, A.H. and Katz, S.I. (1992). Identification and induction of keratinocyte-derived IL-10. J. Immunol. *149,* 92-5.

Evans, C., Bauer, S., Grubert, T., Brucker, C., Baur, S., Heeg, K., Wagner, H. and Lipford, G.B. (1996). HLA-A2-restricted peripheral blood cytolytic T lymphocyte response to HPV type 16 proteins E6 and E7 from patients with neoplastic cervical lesions. Cancer Immunol. Immunother. *42,* 151-60.

Evans, E.M., Man, S., Evans, A.S. and Borysiewicz, L.K. (1997). Infiltration of cervical cancer tissue with human papillomavirus- specific cytotoxic T-lymphocytes. Cancer Res. *57,* 2943-50.

Evans, M., Borysiewicz, L.K., Evans, A.S., Rowe, M., Jones, M., Gileadi, U., Cerundolo, V. and Man, S. (2001). Antigen processing defects in cervical carcinomas limit the presentation of a CTL epitope from human papillomavirus 16 E6. J. Immunol. *167,* 5420-8.

Fennema, J.S., Van Ameijden, E.J., Coutinho, R.A. and Van Den Hoek, A.A. (1995). HIV, sexually transmitted diseases and gynaecologic disorders in women: increased risk for genital herpes and warts among HIV infected prostitutes in Amsterdam. AIDS. *9,* 1071-8.

Fruchter, R.G., Maiman, M., Sedlis, A., Bartley, L. and Arrastia, C.D. (1996). Multiple recurrence of cervical intraepithelial neoplasia in women with the human immunodeficiency virus. Obstet. Gynecol. *87,* 338-344.

Ghosh, A.K., Glenville, S., Bartholomew, J. and Stern, P.L. (1994). Analysis of tumour-infiltrating lymphocytes in cervical carcinoma. In: Stanley, M.A., ed, Immunology of Human Papillomaviruses, New York, Plenum. 249-253.

Giannini, S.L., Hubert, P., Doyen, J., Boniver, J. and Delvenne, P. (2002). Influence of the mucosal epithelium microenvironment on Langerhans cells: implications for the development of squamous intraepithelial lesions of the cervix. Int. J. Cancer. *97,* 654-9.

Han, R., Breitburd, F., Marche, P.N. and Orth, G. (1992). Linkage of regression and malignant conversion of rabbit viral papillomas to MHC class II genes [see comments]. Nature *356,* 66-8.

Herrero, R., Hildesheim, A., Bratti, C., Sherman, M.E., Hutchinson, M., Morales, J., Balmaceda, I., Greenberg, M.D., Alfaro, M., Burk, R.D., Wacholder, S., Plummer, M. and Schiffman, M. (2000). Population-based study of human papillomavirus infection and cervical neoplasia in rural Costa Rica. J. Natl. Cancer Inst. *92,* 464-74.

Hong, K., Greer, C.E., Ketter, N., Van Nest, G. and Paliard, X. (1997). Isolation and characterization of human papillomavirus type 6-specific T cells infiltrating genital warts. J. Virol. *71,* 6427-6432.

Hubert, P., van den Brule, F., Giannini, S.L., Franzen Detrooz, E., Boniver, J. and Delvenne, P. (1999). Colonization of *in vitro*-formed cervical human papillomavirus- associated (pre)neoplastic lesions with dendritic cells: role of granulocyte/macrophage colony-stimulating factor. Am-J-Pathol. *154,* 775-784.

Kadish, A.S., Timmins Patrick, Wang Yuexian, Ho, G.Y. F, Burk, R.D., Ketz John, He, W.u., Romney, S.L., Johnson Anne, Angeletti Ruth and Abadi Maria (2002). Regression of cervical intraepithelial neoplasia and loss of human papillomavirus (HPV) infection is associated with cell-mediated immune responses to an HPV type 16 E7 peptide. Cancer Epidemiol. Biomarkers Prev. *11,* 483-488.

Kobayashi, A., Greenblatt, R.M., Anastos, K., Minkoff, H., Massad, L.S., Young, M., Levine, A.M., Darragh, T.M., Weinberg, V. and Smith-McCune, K.K. (2004). Functional attributes of mucosal immunity in cervical intraepithelial neoplasia and effects of HIV infection. Cancer Res. *64,* 6766-6774.

Kock, A., Schwarz, T., Kirnbauer, R., Urbanski, A., Perry, P., Ansel, J.C. and Luger, T.A. (1990). Human keratinocytes are a source of tumour necrosis factor: evidence for synthesis and release upon stimulation with endotoxin or ultraviolet light. J. Exp. Med. *172,* 1609-1614

Koutsky, L. (1991). Role of epidemiology in defining events that influence transmission and natural history of anogenital papillomavirus infections [editorial;comment]. J. Natl. Cancer Inst. *83,* 978-979.

Laffort, C., Le Deist, F., Favre, M., Caillat Zucman, S., Radford Weiss, I., Debre, M., Fraitag, S., Blanche, S., Cavazzana Calvo, M., de Saint Basile, G., de Villartay, J.P., Giliani, S., Orth, G., Casanova, J.L., Bodemer, C. and Fischer, A. (2004). Severe cutaneous papillomavirus disease after haemopoietic stem-cell transplantation in patients with severe combined immune deficiency caused by common gammac cytokine receptor subunit or JAK-3 deficiency. Lancet *363,* 2051-2054.

Le Bon, A. and Tough, D.F. (2002). Links between innate and adaptive immunity via type I interferon. Curr. Opin. Immunol. *14,* 432-436.

Lee, S.J., Cho, Y.S., Cho, M.C., Shim, J.H., Lee, K.A., Ko, K.K., Choe, Y.K., Park, S.N., Hoshino, T., Kim, S., Dinarello, C.A. and Yoon, D.Y. (2001). Both E6 and E7 oncoproteins of human papillomavirus 16 inhibit IL-18-induced IFN-gamma production in human peripheral blood mononuclear and NK cells. J. Immunol. *167,* 497-504.

Li, S., Labrecque, S., Gauzzi, M.C., Cuddihy, A.R., Wong, A.H., Pellegrini, S., Matlashewski, G.J. and Koromilas, A.E. (1999). The human papilloma virus (HPV)-18 E6 oncoprotein physically associates with Tyk2 and impairs Jak-STAT activation by interferon-alpha. Oncogene *18,* 5727-5737.

Lopez, M.C. and Stanley, M.A. (2000). Cytokine profile of draining lymph node lymphocytes in mice grafted with syngeneic keratinocytes expressing human papillomavirus type 16 E7 protein. J. Gen. Virol. *81,* 1175-1182.

Luxton, J.C., Rowe, A.J., Cridland, J.C., Coletart, T., Wilson, P. and Shepherd, P.S. (1996). Proliferative T cell responses to the human papillomavirus type 16 E7 protein in women with cervical dysplasia and cervical carcinoma and in healthy individuals. J. Gen. Virol. *77,* 1585-1593.

Malejczyk J, Malejczyk M, Majewski S, Orth G, and Jablonska S. (1993) NK-cell activity in patients with HPV16-associated anogenital tumors: defective recognition of HPV16-harboring keratinocytes and restricted unresponsiveness to immunostimulatory cytokines. Int. J. Cancer, 54, 917-921.

Matthews, K., Leong, C.M., Baxter, L., Inglis, E., Yun, K., Backstrom, B.T., Doorbar, J. and Hibma, M. (2003). Depletion of Langerhans cells in human papillomavirus type 16-infected skin is associated with E6-mediated down regulation of E-cadherin. J. Virol. *77,* 8378-8385.

Moore, R.A., Walcott, S., White, K.L., Anderson, D.M., Jain, S., Lloyd, A., Topley, P., Thomsen, L., Gough, G.W. and Stanley, M.A. (2003). Therapeutic immunisation with COPV early genes by epithelial DNA delivery. Virology. *314,* 630-635.

Morelli, A.E., Ronchetti, R.D., Secchi, AD., Cufre, M.A., Paredes, A. and Fainboim, L. (1992). Assessment by planimetry of Langerhans' cell density in penile epithelium with human papillomavirus infection: changes observed after topical treatment. J. Urol. *147,* 1268-1273.

Moretta, L. and Moretta, A. (2004). Killer immunoglobulin receptors. Curr. Opin. Immunol. *16,* 626-633.

Mota, F., Rayment, N., Chong, S., Singer, A. and Chain, B. (1999). The antigen-presenting environment in normal and human papillomavirus (HPV)-related premalignant cervical epithelium. Clin. Exp. Immunol. *116,* 33-40.

Nakagawa, M., Stites, D.P., Farhat, S., Sisler, J.R., Moss, B., Kong, F., Moscicki, A.B. and Palefsky, J.M. (1997). Cytotoxic T lymphocyte responses to E6 and E7 proteins of human papillomavirus type 16: relationship to cervical intraepithelial neoplasia. J. Infect. Dis. *175,* 927-9231.

Nakagawa, M., Stites, D.P., Palefsky, J.M., Kneass, Z. and Moscicki, A.B. (1999). CD4-positive and CD8-positive cytotoxic T lymphocytes contribute to human papillomavirus type 16 E6 and E7 responses. Clin. Diagn. Lab. Immunol. *6,* 494-498.

Nakagawa, M., Stites, D.P., Patel, S., Farhat, S., Scott, M., Hills, N.K., Palefsky, J.M. and Moscicki, A.B. (2000). Persistence of human papillomavirus type 16 infection is associated with lack of cytotoxic T lymphocyte response to the E6 antigens. J. Infect. Dis. *182,* 595-598.

Nees, M., Geoghegan, J.M., Hyman, T., Frank, S., Miller, L. and Woodworth, C.D. (2001). Papillomavirus type 16 oncogenes downregulate expression of interferon-responsive genes and upregulate proliferation-associated and NF-kappaB-responsive genes in cervical keratinocytes. J-Virol. *75,* 4283-4296.

Nimako, M., Fiander, A.N., Wilkinson, G.W.G., Borysiewicz, L.K. and Man, S. (1997). Human papillomavirus-specific cytotoxic T lymphocytes in patients with cervical intraepithelial neoplasia grade III. J. Virol. 57,4855-4861.

Nozaki, S., Feliciani, C. and Sauder, D.N. (1992). Keratinocyte cytokines. Adv. Dermatol. *7,* 83-100.

Oriel, J.D. (1971). Natural history of genital warts. British Journal of Venereal Diseases. *47,* 1-13.

Roncalli, M., Sideri, M., Gie, P. and Servida, E. (1988). Immunophenotypic analysis of the transformation zone of human cervix. Lab. Invest. *58,* 141-149.

Rouse, B.T. and Suvas, S. (2004). Regulatory cells and infectious agents: detentes cordiale and contraire. J. Immunol. *173,* 2211-5.

Selvakumar, R., Ahmed, R. and Wettstein, F. O (1995). Tumor regression is associated with a specific immune response to the E2 protein of cottontail rabbit papillomavirus. Virol. *208,* 298-302.

Stanley, M.A., Scarpini, C. and Coleman, N. (2003). Cell mediated immunity and lower genital tract neoplasia . In Lower Genital Tract Neoplasia, A.B. McLean, A.Singer, H Critchley, eds. (London, UK: RCOG press), pp.27-44.

Tyring, S.K., Arany, I., Stanley, M.A., Tomai, M.A., Miller, R.L., Smith, M.H., McDermott, D.J. and Slade, H.B. (1998). A randomized, controlled, molecular study of condylomata acuminata clearance during treatment with imiquimod. J. Infect. Dis. *178,* 551-555.

van der Burg, S.H., Ressing, M.E., Kwappenberg, K.M., de Jong, A., Straathof, K., de Jong, J., Geluk, A., van Meijgaarden, K.E., Franken, K.L., Ottenhoff, T.H., Fleuren, G.J., Kenter, G., Melief, C.J. and Offringa, R. (2001). Natural T-helper immunity against human papillomavirus type 16 (HPV16) E7-derived peptide epitopes in patients with HPV16-positive cervical lesions: identification of 3 human leukocyte antigen class II-restricted epitopes. Int. J. Cancer. *91,* 612-618.

Wakkach, A., Fournier, N., Brun, V., Breittmayer, J.P., Cottrez, F. and Groux, H. (2003). Characterization of dendritic cells that induce tolerance and T regulatory 1 cell differentiation *in vivo.* Immunity *18,* 605-617.

Wang, B., Amerio, P., and Sauder, D.N. (1999). Role of cytokines in epidermal Langerhans cell migration. J. Leukocy. Biol. *66,* 33-39.

Welters, M.J., de Jong, A., van den Eeden, S.J., van der Hulst, J.M., Kwappenberg, K.M., Hassane, S., Franken, K.L., Drijfhout, J.W., Fleuren, G.J., Kenter, G., Melief, C.J., Offringa, R. and van der Burg, S.H. (2003). Frequent display of human papillomavirus type 16 E6-specific memory t-Helper cells in the healthy population as witness of previous viral encounter. Cancer Res. *63,* 636-641.

Youde, S.J., Dunbar, P.R., Evans, E.M., Fiander, A.N., Borysiewicz, L.K., Cerundolo, V. and Man, S. (2000). Use of fluorogenic histocompatibility leukocyte antigen-A*0201/ HPV 16 E7 peptide complexes to isolate rare human cytotoxic T-lymphocyte-recognizing endogenous human papillomavirus antigens. Cancer Res. *60,* 365-371.

From: Papillomavirus Research: From Natural History to Vaccines and Beyond. Edited by: M. Saveria Campo

Chapter 20

Immune Evasion in Genital Papillomavirus Infection and Cervical Cancer: Role of Cytokines and Chemokines

Sigrun Smola-Hess* and Herbert Pfister

Abstract

Epidemiological and molecular studies of the past 20 years have provided evidence that persistent infection with human papillomaviruses (HPVs) is the main aetiological factor for development of cervical cancer. Most genital papillomavirus infections are cleared within one year. Yet, infections persist in up to one percent of immunocompetent individuals and progress to cancer. Cancer development usually takes a long period of time, eventually requiring decades. The increased incidences of squamous intraepithelial lesions (SIL, variously named cervical intraepithelial neoplasia, CIN) and cervical cancer in immunocompromised individuals strongly support a central role of the immune system in controlling HPV infection and progression. This chapter focuses on cytokines and chemokines in genital papillomavirus infection and cervical cancer as major regulators of the immune response, and discusses their role in immune evasion.

1. Introduction

Cancer of the uterine cervix is of major importance to public health still being the second most common cancer in women worldwide and responsible for approximately 200,000 deaths per year. Its association with HPV infection of the lower genital tract is well established. In different studies HPV DNA has been detected in 90.7 up to 99.7 % of cervical carcinomas (Munoz et al., 2003; Walboomers et al., 1999). Among more than forty anogenital HPV types, fifteen HPV types including HPV16, 18, 31 and 45 were found to be associated with cancer progression and are classified as "high risk types" (Munoz et al., 2003). Molecular studies showing that HPV early proteins are capable of transforming human keratinocytes support the causal association between HPV infection and cancer development.

Though genital HPV infection is a major prerequisite of cervical cancer, cancer development from genital HPV infections is a rather rare event. In young women the majority (70-90 %) of HPV infections are cleared within one or two years (Evander et al., 1995; Ho et al., 1998; Moscicki et al., 1998), probably via natural immune surveillance. Only 1 % of CIN I progress to invasive cancer (Ostor, 1993). Whereas mice transgenic for HPV16 early genes develop malignant lesions within months, this process takes 10 to 20 years or even longer in humans. The natural course of HPV infection is thus influenced by additional factors. Individuals with impaired cellular immune responses, like transplant recipients undergoing immunosuppressive therapy and patients with other acquired immunodeficiencies show increased prevalences of HPV-infection or HPV-induced pre-malignant or malignant lesions (Ellerbrock et al., 2000; Halpert et al., 1986; Kiviat et al., 1993; Petry et al., 1994). Spontaneous regression of HPV-induced genital lesions was paralleled by the infiltration of lymphocytes (Coleman et al., 1994). All these observations point to a major impact of cell-mediated immunity on the fate of HPV-infections.

In most individuals the immune response against HPV is capable of controlling genital HPV infection and leads to the elimination of the virus. Under certain conditions, however, the virus or the virally infected cell may escape these immune control mechanisms even in immunocompetent individuals and persist in the host. The underlying molecular mechanisms are only beginning to be understood and the interaction between viral and host factors appears to play an important role.

Innate as well as adaptive immune responses of the host are regulated by the expression and signalling events of cytokines, chemokines and their receptors. Cytokines are either produced by mucosal keratinocytes, the host cells of HPV or the local environment including immunocompetent cells. Viral infection does not only alter the cytokine expression patterns of keratinocytes, but also interferes with cytokine signalling and thus perturbs local cell-cell communication via cytokine networks. Moreover, changes of cytokine expression and receptor-mediated signalling during malignant progression may further contribute to carcinogenesis.

This chapter discusses the role of cytokines and chemokines for the control of genital HPV infection, as

*For correspondence. Email s.smola@uni-koeln.de

well as their contribution to immune evasion mechanisms during persistence and progression to cervical cancer. The main topics of this article are

- the local cytokine milieu in the transformation zone of the uterine cervix, where most of the squamous intraepithelial lesions arise;
- the regulation of cytokine expression in HPV-infected host cells and interference of HPV with cytokine or chemokine signals;
- consequences of cytokine and chemokine expression for the regulation of viral transcription, the host cell or the immune system;
- and changes in cytokine expression and function during malignant progression.

Understanding the molecular immune mechanisms, which either control HPV infection or favour malignant progression, is necessary for a rational design of vaccines and efficient immunotherapy against HPV infection and cervical carcinoma.

2. The Mucosal Microenvironment

Human papillomaviruses infect keratinocytes within stratified epithelia. Their replicative cycle is closely adapted to the differentiation program of keratinocytes. In most cases, viral infection is eliminated by so far not very well characterised mechanisms potentially involving cell autonomous as well as innate and adaptive immune defence signals. If the attempts to eliminate the virus fail, HPV may persist in the epithelium for years or decades.

Most cervical precancerous lesions and cancers arise at the squamo-columnar junction of the uterine cervix (Abdul-Karim et al., 1982). During reproductive years, as a dynamic "metaplastic" process, there is continuous replacement of columnar epithelium by squamous epithelium at the so-called transformation zone. Proliferating reserve cells within this squamous epithelium are thought to be especially susceptible to HPV infection and to their oncogenic potential. In addition, recent studies suggest that the local cytokine and chemokine milieu of the metaplastic mucosa may create a site of reduced immune reactivity favouring HPV infection and progression to cancer (for review see (Delvenne et al., 2004)).

While keratinocytes support the HPV life cycle, these cells do not have the equipment to effectively present antigens to the immune system. However, epithelia also harbour professional antigen presenting cells (APC) like Langerhans' cells or inflammatory dendritic cells, which are not permissive for HPV. Professional APCs have the task to take up antigen, deliver the information of viral infection to the immune system and instruct effector cells to eliminate the infection. APCs have a central role in shaping the immune response of naïve T helper lymphocytes (T_H cells) into so-called "T_H1 or T_H2" direction. These helper cells can be distinguished on the basis of their cytokine expression patterns. T_H1 cells, characterised by interleukin-2 (IL-2), tumor necrosis factor-α (TNF-α) and interferon-γ (IFN-γ) production, are typical effectors of inflammatory responses. They are involved in the activation of cytotoxic T cells attacking the virally infected cells and are thought to be crucial for mounting a cell-mediated anti-papillomaviral response. In fact, in several patients a T_H1 cytokine pattern in the mucosal epithelium preceded the clearance of genital HPV infection (Scott et al., 1999). In contrast, T_H2 cells producing IL-4, IL-5, IL-6, IL-10 and IL-13 govern anti-parasitic, allergic and humoral responses. T_H2 cells may even inhibit the T_H1 response and rather impair the cell-mediated cytotoxic T cell response. It is therefore assumed that they are less protective against virally infected keratinocytes or tumor cells. It is well accepted that the production of the cytokine IL-12 by APCs plays a major role for mounting a T_H1 response (reviewed in (Trinchieri, 2003)), although recent data indicate that stimulation of Langerhans' cells via their CD40 receptor can prime a T_H1 response independently of IL-12 under certain conditions (Gorbachev and Fairchild, 2004). IFNs and IL-12 are also important to activate a natural killer cell-mediated anti-tumoral defence (for review see Wu and Lanier, 2003)).

In the following section cytokine and chemokine expression patterns in the transformation zone of the uterine cervix and their potential contribution to HPV susceptibility will be discussed.

2.1. Macrophage inflammatory protein-3α

Within the transformation zone, the density of Langerhans' cells was shown to be reduced compared with normal epithelium, however this is controversially discussed by other groups (Edwards and Morris, 1985; Giannini et al., 2002; Morris et al., 1983). Homeostasis of Langerhans' cells is regulated by the influx of precursor cells into and efflux of differentiated cells from the epithelium. Besides disturbances of in- or efflux rates, differences in Langerhans' cell numbers may reflect transdifferentiation of the precursor cells into macrophage-like cells or cell death within the epithelium. Keratinocytes critically contribute to the homeostasis of APC numbers in the epithelium. One important regulatory mechanism involves chemokine production levels. Chemokines are chemotactic cytokines that direct the migration of immune cells along a gradient (Yoshie et al., 2001). Keratinocytes of skin and mucosal epithelia constitutively release the chemokine macrophage inflammatory protein-3α (MIP-3α), variously named CCL20, LARC or Exodus-1. In inflammatory conditions MIP-3α production is strongly enhanced. This chemokine specifically attracts cells expressing the chemokine receptor CCR6. CCR6 expression is confined to immature Langerhans' cells, dermal-resident immature DCs, which potentially differentiate into Langerhans' cells and certain lymphocyte subpopulations (Larregina et al., 2001) (for review see (Schutyser et al., 2003)). In fact the observation, that the transformation zone contains fewer Langerhans' cells was paralleled by lower MIP-3α expression (Giannini et al., 2002). Thus, suppressed MIP-3α production may attract fewer Langerhans' cells and may thus favour immune evasion of HPVs. Keratinocytes can also secrete so called "inflammatory chemokines", like MCP-1 (monocyte chemoattractant protein-1) attracting "inflammatory dendritic cells", which are involved in the amplification and modulation of the inflammatory tissue response (Charbonnier et al., 1999). However, the expression of this chemokine in the transformation zone of the uterine cervix still needs further analysis.

2.2. Interleukin-10

The local epithelial cytokine milieu may not only influence numbers of APCs but can also determine their function. In the microenvironment of the transformation zone, IL-10 was shown to be expressed at higher levels than in normal cervical epithelium (Giannini et al., 2002). Lymphocyte cultures were established from the same sites, which also showed significant IL-10 expression (Jacobs et al., 2003). This class 2 cytokine displays a variety of inhibitory effects on the immune response. IL-10 profoundly impairs APC and T cell functions.

IL-10 can prevent differentiation of monocytes to (inflammatory) dendritic cells. It rather skews these cells to acquire a macrophage-like phenotype (Allavena et al., 1997; Buelens et al., 1997b). This may be one mechanism how IL-10 contributes to lower numbers of professional APCs cells in the transformation zone. Antigen loaded dendritic or Langerhans' cells normally mature in the context of local inflammatory stimuli and are bound to migrate to the lymphoid tissue. This is accomplished by concomitant up-regulation of the chemokine receptor CCR7 during maturation. In the presence of IL-10, however, the up-regulation of CCR7 is strongly impaired (D'Amico et al., 2000). IL-10 negatively affects their antigen presentation capacity, since it inhibits the up-regulation of MHC class II and the CD86 molecule, which is important for CD28 co-stimulation of T cells (Buelens et al., 1997a). IL-10 inhibits IL-12 production in dendritic cells leading to suppression of the protective T_H1 response (De Smedt et al., 1997) and may even contribute to a state of tolerance or anergy in T cells (Steinbrink et al., 1997). Molecular targets of IL-10 signalling in APCs are pro-inflammatory signalling molecules, which include the transcription factor NF-κB (nuclear factor κB) (Schottelius et al., 1999). NF-κB is central to normal APC functions, as are the mitogen-activated protein kinases (MAPK) ERK2 (extracellular signal-regulated kinase 2), SAPK/JNK (stress activated protein kinase/c-Jun N-terminal kinase) and the p38 MAPK (Sato et al., 1999).

IL-10 can promote the differentiation of so-called Type 1 regulatory T cells (Tr1) by acting on dendritic cells or directly on naïve or resting T cells (Wakkach et al., 2003), for review see (Jonuleit and Schmitt, 2003). Tr1 cells produce large amounts of IL-10, as well, and can regulate the activation of naïve and memory T cells. Of note, they negatively act back on dendritic cells and suppress T_H1- and T_H2-mediated immune responses raised against pathogens or tumors. In T cells IL-10 has also been reported to selectively inhibit the CD28 co-stimulator pathway, thereby inducing tolerance (Akdis et al., 2001). The presence of regulatory T cells in the transformation zone, however, still has to be investigated.

Thus IL-10 is a multifaceted molecule with immunosuppressive properties at various levels of the immune response. The enhanced expression of IL-10 may therefore provide a basis for immune evasion and persistence of HPV within the transformation zone of the uterine cervix.

2.3. Transforming growth factor-β

In some cell types the expression of IL-10 can be induced by transforming growth factor-β (TGF-β) (Maeda et al., 1995). Similar to IL-10, TGF-β mRNA has also been detected in the transformation zone of the uterine cervix, although the cellular source is still unclear (Giannini et al., 1998). It may have an important role within this part of the epithelium. TGF-β itself is a multifunctional cytokine with partially opposing properties. It is essential for tissue repair and integrity, which could be of major importance within the highly vulnerable squamo-columnar junction of the uterine cervix. In the immune system TGF-β has mainly suppressive and anti-inflammatory functions. Although TGF-β can activate resting immune cells, the opposite is observed for activated inflammatory cells (for review see Sporn (1999)). Based on its high evolutionary conservation, it has been suggested that TGF-β may protect the host from overwhelming and destructive inflammation and may thus serve to buffer inflammatory events at the squamo-columnar junction of the cervical epithelium. Besides different functions in macrophages, TGF-β can suppress T_H1 as well as T_H2 effector T cells (Bogdan and Nathan, 1993). This seems to be the case especially when low IL-2 levels are present. TGF-β itself may inhibit IL-2 production and effects of the T_H1 cytokine IFN-γ (reviewed in Wahl et al. (2004)). In analogy to IL-10, TGF-β can be produced by a subset of regulatory T cells (T_H3 cells), which may convey peripheral tolerance by suppressing T_H1 and T_H2 cells (for review see Jonuleit and Schmitt (2003)).

Thus, the cytokine milieu within the transformation zone appears to be rather immunosuppressive than immunostimulatory. HPV may take advantage of this naturally "immunocompromised niche" within the mucosal epithelium of the uterine cervix and may thereby ensure its own protection from host defence.

3. Cytokine Expression and Function at Early Stages of Viral Infection (in HPV-infected Keratinocytes and SIL)

3.1. Cytokines that primarily influence the infected keratinocyte

3.1.1. Type I interferons as first line of antiviral defence

Once HPV has infected the mucosal keratinocytes the virus has to cope with innate immunity. Keratinocytes have acquired cell autonomous mechanisms that handle and cope with invading microorganisms and viruses. Type I IFNs, which have been defined on the basis of their anti-viral activity, are likely to convey the first line of cellular anti-viral defence also against HPV infection. Today IFNs are recognised not only to have a role in host defence against infection but also to protect against tumorigenesis.

IFNs are structurally and functionally related cytokines. Based on their structure and biological properties, they are classified into type I and II IFNs. Type II IFN, which is solely represented by IFN-γ, is mainly produced by activated T lymphocytes and natural killer cells. In contrast, the human type I IFN family comprises various members. It is subdivided into IFN-α, IFN-β, IFN-ω classes and the more recently discovered IFN-ε, IFN-κ and IFN-λ classes. Whereas only one human IFN-β encoding gene has been described, at least 13 different IFN-α encoding genes are transcribed (van Pesch et al., 2004; for review see Kontsek et al. (2003)). The reason for these multiple type I IFN subtypes is not fully understood at present. Upon stimulation most human nucleated cell types are able to synthesize type I IFNs. Also in human keratinocytes IFN-α and IFN-β mRNAs are inducible (Fujisawa et al., 1997). Transcriptional regulation of type I IFNs involves activation of the interferon regulatory factor-3 (IRF-3) in various cell types (Schafer et al., 1998) (Figure 1). An exception seems to be the IFN-κ gene, which is also expressed in resting epidermal keratinocytes (LaFleur et al., 2001). In differentiating keratinocytes (stimulated by a Ca^{2+} shift to higher concentrations), higher levels of IFN-α and IFN-β mRNAs were detected by RT-PCR. While secreted IFN protein could not be measured in keratinocyte supernatants, intracellular IFN-α (but not IFN-β) was detected at low levels.

To exert their functions, IFNs engage transmembrane surface receptors and an intracellular signalling cascade transducing their signals to the nucleus. Like the IL-10 receptor, IFN receptors belong to the class II cytokine receptor family. Three different IFN receptors are used. The type I IFNs α, β, ω, ε and κ bind and activate IFNAR, (reviewed in (Mogensen et al., 1999)), whereas IFN-λ recognises IFNLR (Kotenko et al., 2003). IFN-γ binds to a third receptor, IFNGR (for review see Bach et al. (1997)). All IFN receptors are ubiquitously expressed and present on keratinocytes. They are composed of two different chains (R-1 and R-2), which specifically activate Jak (Janus kinase)-STAT (signal transducer and activator of transcription) pathways. The IFNAR-1 chain is pre-associated with the cytoplasmic Tyrosine kinase 2 (Tyk2). IFNAR-2 further recruits Jak1, whereas IFNγR associates with Jak1 and Jak2. Upon binding of type I IFNs to the heteromeric receptor the receptor-associated kinases are activated, followed by tyrosine phosphorylation of STAT1 and STAT2 proteins. These heterodimerise, and together with the IRF-9 (p48) they form a complex (ISGF3, IFN stimulated gene factor 3), which translocates into the nucleus. There the complex binds to specific regulatory sequences (ISRE, IFN-stimulated response elements) regulating type I IFN-mediated gene transcription (Figure 1, right panel). IFN-γ homodimers lead to dimerisation of both IFNGR-1 and IFNGR-2 chains. Subsequently recruited STAT1 proteins are phosphorylated by the associated Jak kinases, form homodimers also termed IFN-γ activated factor (GAF), ready to translocate into the nucleus and regulate transcription via the GAS element (gamma-interferon activated sites). Small amounts of STAT1 homodimers may also be formed following type I IFN stimulation (for review see Kontsek et al. (2003)). Cellular responses to IFNs are not only mediated by STAT factors and ISGF3 but also by the interferon regulatory factor IRF-1, which is itself IFN-inducible (Fujita et al., 1989). In contrast to this the IFN-γ-inducible factor SOCS1 (suppressor of cytokine signalling 1) gives a negative feedback by inhibiting JAK tyrosine kinase activity (reviewed in Kubo et al. (2003)).

In keratinocytes retrovirally infected with HPV16 E6/E7 or E6 expression constructs the intracellular IFN-α was decreased by 5-fold (Nees et al., 2001). HPV16 E6 was subsequently shown to bind to IRF-3 and thereby inhibit its transcriptional activation (Ronco et al., 1998). In contrast to non-tumorigenic hybrids established from the HPV18-positive cervical carcinoma cell line HeLa and fibroblasts, tumorigenic HPV-positive cells lacked the ability to synthesise IFN-β mRNA, also in response to TNF-α (Bachmann et al., 2002). These data suggest that HPV early proteins may keep type I IFNs in check already at the level of cytokine expression.

IFNs mediate a large panel of cellular responses such as anti-viral and anti-proliferative activities. Both IFN types inhibit the proliferative capacity and promote the differentiation of normal keratinocytes (Nickoloff et al., 1984; Saunders et al., 1996; Yaar et al., 1985). This fact is important for HPV-infected keratinocytes. HPVs lack an own DNA polymerase. For DNA replication they completely depend on the host cell's DNA replication machinery, and therefore on host cell proliferation. The anti-proliferative effect of IFNs on keratinocytes may thus be regarded as an anti-papillomaviral mechanism. Moreover, IFNs modulate innate and adaptive immune

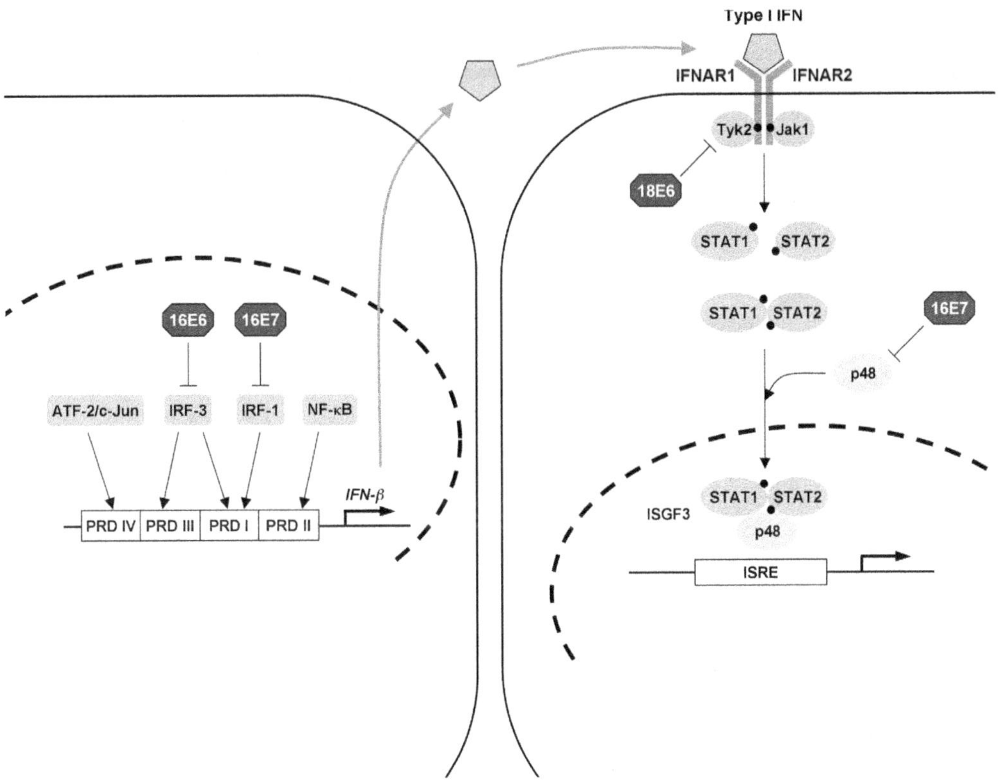

Figure 1. Interference of HPV oncoproteins with the activation of the IFN-β promoter (left panel) and with type I IFN signalling (right panel). IRFs are important regulators of IFN-β transcription. They bind to positive regulatory domains (PRD) within the IFN-β promoter. HPV16 E6 interacts with IRF-3 and HPV16 E7 with IRF-1. As a consequence the transactivation function of both IRFs is inhibited and less IFN-β transcribed. Inhibition of type 1 IFN signalling by HPV oncoproteins takes place at two levels. HPV-18 E6 functions as a negative regulator of the STAT signalling pathway activated by type I IFN. It physically interacts with Tyk2 at sites important for binding to the cytoplasmic portion of IFNAR1 and inhibits Tyk2 activation. Further downstream in the signalling pathway, HPV16 E7 binds to the p48 component of the ISGF3 complex. It inhibits nuclear translocation of this complex and thereby gene activation through IFN-stimulated response elements (ISRE). Tyrosine phosphorylation is represented by black filled circles.

responses. They enhance the antigen presentation capacity of an infected cell by increasing the expression levels of MHC class I molecules. Type II IFN additionally induces the formation of an immunoproteasome (Rivett and Hearn, 2004). Both IFN types modulate the differentiation and maturation of dendritic cells (Tough, 2004).

Besides their effect on the immune system and normal keratinocytes both IFN types directly inhibit the growth of HPV immortalised cervical epithelial cells (Agarwal et al., 1994a). In these cells the antiproliferative effect of IFNs may be potentiated by transcriptional suppression of the viral oncogenes E6 and E7 as observed in different HPV16 or 18 immortalised or cervical carcinoma cells (Nawa et al., 1990; Perea et al., 1995; Woodworth et al., 1992). However, suppression of oncogene transcription was not observed in all tested cervical carcinoma cell lines corresponding to a highly variable clinical response to IFN therapy (Kim et al., 2000). Recently, a HPV-Interferon Responsive Element-1 (HIRE-1) was identified that even stimulates transcription in the context of the complete long control region of HPV16 in a cell type specific manner (Arany et al., 2003). Another suggestion for the anti-papillomaviral activity came from a study showing that IFN-γ induces proteasome-dependent degradation of the E7 protein mediated by physical interaction with SOCS1 (Kamio et al., 2004).

As demonstrated recently, IFNs and HPV seem to influence each other in a bi-directional manner: Responses to IFNs were found to be disturbed in keratinocytes expressing genital high risk HPV E6/E7. Several studies on global gene expression in such cells unravelled a down-regulation of IFN-responsive genes (Chang and Laimins, 2000; Nees et al., 2001). In the case of HPV31 the transcription factor STAT-1, which is central to IFN signalling, was repressed. HPV proteins were shown to directly interact with components of the IFN signalling pathway. HPV18 E6 associates with Tyk2 and thereby inhibits IFN-α signalling (Li et al., 1999). Also the E7 protein was shown to interfere with IFN-α signalling. E7 targets the ISGF3 component p48 rendering this complex unable to translocate to the nucleus after IFN stimulation (Barnard and McMillan, 1999; Barnard et al., 2000). E7 also impairs IRF-1 activation thus interfering with the activation of the IFN-β promoter (Park et al., 2000;

Perea et al., 2000). It could be shown that the E7 protein binds to IRF-1 and inhibits its transactivation function by recruiting histone deacetylase (HDAC) to the promoter. From these data the net effect of IFNs on HPV-positive cells clearly depends on the expression levels of the viral oncoproteins at the time of IFN exposure. In summary, HPV oncoproteins interfere with type I IFN induction and signalling at various levels, as depicted in Figure 1.

3.1.2. Tumor necrosis factor-α

TNF-α is one of the most important mediators of inflammation. It is strongly inducible upon viral infection and, together with IFNs, which can be induced by TNF-α, it may serve as first line defence against viruses. In contrast to most other viral infections, however, HPV does not primarily cause an inflammatory response. This may be part of a general papillomaviral strategy: to hide within the epithelium without alerting the local immune system. In fact, inducibility of the pro-inflammatory cytokine TNF-α was lower in HPV-positive than in normal keratinocytes (Havard et al., 2002). Nevertheless, TNF-α may play an important role during immune-mediated regression of HPV-induced cervical lesions. TNF-α belongs to the TNF superfamily consisting of 19 members that engage 29 receptors involved in the regulation of cellular proliferation, survival, differentiation and apoptosis (for review see (Aggarwal, 2003)). Two TNF receptors (p55TNFR and p75TNFR) bind to and transmit the signals of their ligand TNF-α. In epithelial cells TNF-α signalling is mainly mediated through p55TNFR. Adaptor molecules like TRADD (TNFR associated death domain) and TRAFs (TNFR associated factors) link the receptors to signalling pathways regulating proliferation and gene activation. TNF-α signalling results in the activation of the p38 MAPK, and transcription factors belonging to AP-1 (activator protein 1) and NF-κB families. Furthermore, TNF-α signalling can act via the transcription factor C/EBPβ (NF-IL6), which physically and functionally interacts with AP-1 and NF-κB factors (Klampfer et al., 1994; Stein and Baldwin, 1993; Stein et al., 1993). All these factors are important regulators of pro-inflammatory responses to the cytokine TNF-α. In addition, the adapter molecule FADD (Fas associated death domain) links pro-apoptotic receptors of the TNFR family to the caspase activation cascade mediating apoptosis.

TNF-α signalling in HPV-positive cells appears to have a direct anti-papillomaviral effect in HPV-immortalised keratinocytes. In non-malignant HPV16-immortalised cells TNF-α is able to suppress endogenous E6/7 oncogene mRNA expression (Kyo et al., 1994). Suppression correlated with the induction C/EBPβ. In the carcinoma cell line CaSki transfected with a C/EBPβ expression vector, C/EBPβ expression levels appeared to be reciprocally correlated with those of endogenous E6/E7 mRNA (Kyo et al., 1993). Transient transfection of a C/EBPβ encoding plasmid also inhibited the activity of the HPV18 early promoter in the carcinoma cell line SW756 (Smola-Hess et al., 2001a). It was shown that the suppressive effect of C/EBPβ is mediated by a C/EBPβ isoform encoding only the C-terminal DNA-binding domain of this transcription factor, whereas isoforms encoding also the transactivation domain may even be stimulatory (Struyk et al., 2000). Thus it can be speculated that the suppressive activity of TNF-α further depends on the relative nuclear expression levels of the different C/EBPβ isoforms in a respective cell. Besides C/EBPβ, the transcription factor NF-κB, which is another important target of TNF-α signalling, was shown to bind to and repress the activity of the HPV16 long control region (Fontaine et al., 2000). In addition to this, activation and composition of the transcription factor AP-1 were suggested to play a role in TNF-α-mediated suppression of HPV early gene transcription (Soto et al., 1999).

In normal keratinocytes TNF-α rather induces reversible growth inhibition than apoptosis, similar to IFNs (Detmar and Orfanos, 1990; Symington, 1989). The effect of TNF-α on the cell cycle of HPV-immortalised cells has been controversially discussed in different studies and may depend on the respective HPV type. In one study HPV16 immortalised human foreskin keratinocytes were growth inhibited by TNF-α, whereas HPV-18-immortalised keratinocytes were resistant to this cytokine (Vieira et al., 1996). However, in other studies it was shown that both the HPV16 and HPV18 E7 proteins were sufficient to confer resistance to TNF-α-mediated growth inhibition (Basile et al., 2001; Boccardo et al., 2004). Moreover, it could be demonstrated that HPV16- and HPV18-immortalised keratinocytes or cervical carcinoma-derived cell lines, which have become insensitive to TNF-α-mediated growth inhibition, may even be stimulated to proliferate in an amphiregulin, epidermal growth factor receptor and ras dependent autocrine manner (Gaiotti et al., 2000; Woodworth et al., 1995).

Obviously, HPV has evolved strategies to protect itself against TNF-α signals, similar to its defence against IFNs. HPV regulatory proteins may also interfere with TNF-α signal transduction at various levels. Recently, it has been shown that HPV16 E6 binds to p55TNFR, decreases its interaction with TRADD and in addition to this also interacts with FADD (Filippova et al., 2004; 2002). As a consequence, expression of HPV16 E6 rendered various non-keratinocyte cell lines more resistant to TNF-α- or Fas-induced apoptosis. On the other hand it has been described that HPV16 E6 enhances susceptibility of murine fibroblasts L929 and an ovary cancer cell line to TNF-α-induced apoptosis (Liu et al., 1999; Vikhanskaya et al., 2002). The significance of these findings for HPV infected keratinocytes still has to be determined. Conflicting data also exist on the question whether and to which direction HPV alters the activity of the transcription factor NF-κB. It was shown that the HPV16 oncoproteins up-regulate NF-κB responsive genes in cervical keratinocytes (Nees et al., 2001). On the contrary other studies revealed that HPV16 E6 binds to the co-activator proteins CBP/p300 and decreases their ability to activate NF-κB responsive promoters, including the IL-8 promoter (Huang and McCance, 2002; Patel et al., 1999). Similarly, it was reported that the HPV E7 protein associates with the IκB-kinase complex and attenuates its activity (Spitkovsky et al., 2002).

3.1.3. Transforming growth factor-β

In HPV-infected tissue interesting changes have been observed regarding the expression patterns of the cytokine TGF-β1. Although high in the transformation zone, TGF-β1 expression was shown to decrease from CIN I to microinvasive carcinoma. TGF-β receptor levels stayed the same, as investigated in paraffin embedded tissue (Torng et al., 2003). A more refined analysis revealed that epithelial TGF-β1, 2 and 3 expression decreases in the basal third in CIN I, in the basal and middle thirds in CIN II and in all layers in CIN III as demonstrated by quantitative RT-PCR, also after microdissection, and by immunohistochemistry (El-Sherif et al., 2000). There are several possibilities, why epithelial TGF-β expression is reduced *in vivo* concomitant to CIN aggravation, although it is a potent negative regulator of the immune response.

As mentioned above, HPV depends on a proliferating cell for its own replication. Like interferons, TGF-β delivers potent anti-proliferative signals in epithelial cells thereby causing cell cycle arrest in the G1 phase (reviewed in (Derynck et al., 2001; Massague et al., 2000)). It leads to up-regulation of cell-cycle-inhibitory proteins, such as the cyclin dependent kinase inhibitory proteins (CKI) $p15^{INK4B}$ and $p21^{CIP1}$. TGF-β also inhibits the induction of proteins that stimulate proliferation, such as c-myc, a suppressor of CKI gene activation.

Concerning the virus itself, it could be demonstrated that TGF-β reversibly inhibits the expression of HPV16 E6 and E7 genes at the transcriptional level in various immortalised cell lines (Braun et al., 1992; Woodworth et al., 1990). Repression of viral oncogene expression correlated with inhibition of proliferation (Hasskarl et al., 2000; Rorke et al., 2000).

TGF-β activates a heteromeric complex of type I and type II serine/threonine kinase receptors belonging to the Activin-receptor-like kinase (Alk) family (for review see (ten Dijke and Hill, 2004)). In response to TGF-β, intracellular signal transducing molecules, receptor regulated R-Smad proteins (Smad 2 and 3) are recruited to the receptor complex with the help of an auxiliary protein SARA (Smad anchor for receptor activation) and are phosphorylated by the type I kinase. R-Smads again form heteromeric complexes with Smad 4. These complexes are released from the receptor and translocate to the nucleus. Although Smad 3 and 4 both can specifically bind to DNA with low affinity, efficient binding to promoters and regulation of target genes mostly requires cooperation with sequence-specific transcription factors (reviewed in (Derynck and Zhang, 2003)). Thus, TFG-β signalling diversification and specificity are determined by the levels of nuclear Smad complexes as well as by the expression and activation patterns of transcription factors, co-activators and co-repressors that cooperate with Smad proteins in transcriptional activation or repression. Recent studies shed some light into TGF-β-mediated regulation of HPV transcription. The oncoprotein Ski interacts with Smads and represses TGF-β signalling (Luo et al., 1999). However, TGF-β itself can down-regulate Ski expression levels in HPV16 immortalised keratinocytes (Baldwin et al., 2004). The same Ski oncoprotein also associates with and increases transcriptional activity of nuclear factor 1 (NF1) known to bind to the HPV16 long control region at various sites. Thus a model was proposed in which TGF-β may influence NF1 activity by down-regulating Ski expression (Baldwin et al., 2004).

The effect of TGF-β on HPV-positive keratinocytes was investigated in various studies. Some studies suggested that normal and HPV-immortalised keratinocytes express similar levels of TGF-β1 *in vitro* and exhibit similar responsiveness to the three TGF-β isoforms in monolayer as well as organotypic cultures (Braun et al., 1992). Only under conditions promoting squamous differentiation TGF-β was suggested to support autonomous growth in an autocrine amphiregulin and EGFR-dependent way (Woodworth et al., 1996). High doses of TGF-β1 even enhanced apoptosis in HPV-immortalised human ectocervical epithelial cells (Rorke and Jacobberger, 1995). However, HPV-positive cell lines derived from CIN lesions were less sensitive to the anti-proliferative effect of TGF-β than normal cervical cells, and responsiveness was further lost with increasing passages (De Geest et al., 1994). This lower sensitivity may be partially explained by the fact that the HPV16 E7 oncoprotein abrogates TGF-β-mediated inhibition of keratinocyte growth and expression of the c-myc gene, similar to other viral transforming proteins with retinoblastoma binding domains (Pietenpol et al., 1990a; 1990b). Moreover, it could be demonstrated that HPV16 E7 directly interacts with Smad 2, 3 and 4 and decreases Smad mediated transcriptional activity (Lee et al., 2002). Non-tumorigenic epithelial cells immortalised with HPV16 were found to be more sensitive to TGF-β *in vitro* than tumorigenic cells (Braun et al., 1990). Cells also became partially resistant to TGF-β-mediated HPV oncogene suppression after prolonged cultivation *in vitro* or after malignant transformation (Woodworth et al., 1990). It was therefore hypothesised that additional changes in the TGF-β signalling pathway may occur during *in vitro* progression or in cervical carcinoma cells. Taken together, these data suggested that the HPV16

E7 protein has some capacity to interfere with TGF-β signalling pathways. However, complete resistance to TGF-β-mediated growth inhibition is likely to be achieved only at later stages during malignant progression as a consequence of further changes in TGF-β signalling.

3.2. Cytokines that primarily influence the immune milieu

3.2.1. Chemokines recruiting immunocompetent cells

Depletion of Langerhans' cells was found in high grade SILs similar to the observation made in the transformation zone (al-Saleh et al., 1995; Chardonnet et al., 1989; McArdle and Muller, 1986). Decreased numbers of APCs might be due to reduced recruitment of their precursor cells into the infected epithelium. Therefore, expression of the chemokines MIP-3α and MCP-1 attracting such precursor cells has been investigated also in HPV-positive lesions. Whereas most biopsies from the ectocervix showed a uniform expression pattern for MIP-3α, this was observed in a much lower percentage of SILs (Giannini et al., 2002). MCP-1 also attracts precursors of dendritic and Langerhans' cells to the epithelium (Nakamura et al., 1995) and ectopic expression of MCP-1 can also lead to monocyte accumulation *in vivo*. *In vitro*, monocyte-enriched fractions from peripheral blood were able to suppress HPV oncogene transcription in non-tumorigenic hybrids established from the HPV18-positive cervical carcinoma cell line HeLa and fibroblasts (Rosl et al., 1994). MCP-1 was also found to be differentially expressed in HPV-infected lesions. In high grade SIL, where the HPV oncogenes E6 and E7 were strongly expressed, *in situ* hybridisation signals for MCP-1 mRNA and expression of MCP-1 protein were low (Riethdorf et al., 1996; 1998). In cases where MCP-1 was detectable, this corresponded to an inflammatory response including macrophages. A second study suggested that there is little MCP-1 expression in CIN I and CIN II lesions (Kleine-Lowinski et al., 1999) and that CIN III lesions completely lacked MCP-1 mRNA expression and macrophage infiltration. It was therefore hypothesised that HPV oncoproteins are negative regulators of MCP-1 expression. In fact HPV16 E6 and E7 were shown to suppress MCP-1 expression in primary human keratinocytes (Kleine-Lowinski et al., 2003). Interestingly, in a third study a rather high density of stromal macrophages was found in high grade SIL depleted of Langerhans' cells (al-Saleh et al., 1995). This suggests that in addition to low expression of chemokines, transdifferentiation of monocytic precursors to macrophage-like cells may take place in the stroma of SILs. Moreover, in these tissues, HLA class II molecules were up-regulated on keratinocytes. Such keratinocytes may also be able to present antigens, however, they lack the ability to co-stimulate T cells in an appropriate manner. It can therefore be concluded that professional APC seem to be replaced by cell types, which present antigens with less efficiency in HPV-infected tissues.

The regulation of other chemokines by keratinocytes expressing genital high risk HPV has been less investigated. We could show that MCP-3, RANTES (regulated on activation, normally T cell-expressed and -secreted) and IP-10 (IFN-γ-inducible protein-10), which preferentially attract monocytes, lymphocytes or NK cells, potentially playing an important role in the immune control of HPV, are only expressed at low levels in non-malignant HPV-positive keratinocytes as well as cervical carcinoma cells. The chemokine IL-8 was found to be expressed constitutively at higher levels (Altenburg et al., 1999). IL-8 has originally been described as a potent chemoattractant for neutrophils. The role of neutrophils in immunity against HPV has not been determined. Investigations in mice transgenic for HPV16 early genes rather suggest a tumor promoting than a protective role (Coussens et al., 2000). In addition to its chemotactic properties, IL-8 has recently been shown to function as a mitogenic, angiogenic and motogenic factor contributing to progression in various cancer types (for review see (Xie, 2001)). Data on the HPV-mediated regulation of the IL-8 gene are controversial. Under conditions promoting differentiation keratinocytes retrovirally infected with HPV16 E6/7 or E6 produced higher amounts of the chemokine IL-8, corresponding to higher binding activities of the transcription factors NF-κB and AP-1 (Nees et al., 2001). In contrast, it had been reported that the HPV16 E6 protein and to a lesser extent the E7 protein repressed TNF-α-induced IL-8 promoter activation under normal growth conditions; HPV16 E6 by competing with NF-κB and a second transcription factor SRC-1 (steroid receptor co-activator) for binding to the co-activator CBP (CREB binding protein); HPV E7 by binding to P/CAF, a p300/CBP-associated factor (Huang and McCance, 2002; Patel et al., 1999).

3.2.2. Cytokines indicating a T_H1-T_H2 shift or immunosuppression

Following progression from low grade to high grade SIL a pronounced shift from T_H1 to T_H2 cytokines was observed within the tissue. T_H2 cytokines are thought to be less protective against virally infected cells. Currently it is unclear, whether altered cytokine expression patterns predispose to HPV persistence, or whether they are a result of HPV infection.

A T_H2 immune deviation was suggested by *in situ* detection of cytokines using immunohistochemistry. Also an eosinophilic infiltrate in tumors indicated a T_H2 cytokine milieu, which is associated with a less effective immune response (Bostrom and Hart, 1981; van Driel et al., 1996). In the connective tissue of high grade lesions a lower density of IL-2-positive and a higher density of IL-4-positive cells were detectable beneath severely dysplastic epithelium (al-Saleh et al., 1998). Also more IL-6-positive cells were present in HSIL than in normal epithelium or the transformation zone. In the same study even a slightly higher IFN-γ expression was found in low and high grade lesions compared with normal ectocervical tissues, although IFN-γ is an indicator of T_H1 immune reactivity. IL-10 protein expression was shown to increase with severity of the lesion. In different studies IL-10 was shown to be localised in the stroma (Giannini et al., 1998; Kobayashi et al., 2004) and in few cases in the cytoplasm of epithelial cells (Mota et al., 1999). Interestingly, IL-10 protein expression in the stroma co-localised with CD1a expression, a marker for myeloid dendritic cells. Subepithelial IL-10 expression was also found to be significantly increased in CIN II and III on mRNA level (El-Sherif et al., 2001). However, in one study IL-10 mRNA was detected at lower frequencies in invasive carcinoma as compared with pre-malignant biopsies (de Gruijl et al., 1999). Using an *in vitro* mixed autologous lymphocyte and tumor cell culture system it could be demonstrated that activated T cells in the human cancer milieu predominantly express a T_H2 phenotype mediated through IL-10 and TGF-β (Sheu et al., 2001). Moreover, CD4+ and CD25+ T cells producing TGF-β were detected in the stroma of CIN II and CIN III lesions (Kobayashi et al., 2004). This led to the hypothesis that IL-10 producing dendritic cells might direct the differentiation of naïve CD4+ T cells to regulatory T cells as seen in other infectious diseases (McGuirk et al., 2002).

Although IL-12 levels seemed to increase in LSIL and HSIL, the percentage of high grade SILs expressing the IL-12p40 subunit was lower than that of low grade SILs (Giannini et al., 1998). Additionally, in another study IL-12 and IFN-γ mRNA transcripts were found at lower frequencies in invasive carcinoma biopsies than in pre-malignant lesions, which would be consistent with a defective T_H1 response in carcinomas. The T_H1 cytokine TNF-α was detected in normal basal keratinocytes but not in a number of CIN specimens (Mota et al., 1999). Furthermore, TNF-α could not be detected in cervical carcinoma cell lines (Hazelbag et al., 2001; and own unpublished results).

Apparently, not only the local immune response is biased towards a T_H2 response in (pre)neoplastic lesions. A lower systemical IL-12/IL-10 ratio was also detected in peripheral blood mononuclear cell (PBMC) culture supernatants and in plasma samples of patients with SIL (Jacobs et al., 1998). PBMC from patients with extended disease stimulated *in vitro* showed increased IL-4 and IL-10 production, whereas IL-2 production was reduced (Clerici et al., 1997). IL-2 production was also diminished after HPV-16 E6 and E7 peptide stimulation of peripheral lymphocytes from patients with high grade SIL and cervical cancer (Tsukui et al., 1996). In contrast to these observations within tumors, cervical washing fluid does not present this cytokine pattern, since higher levels of IL-12p40, IL-10, TNF-α and IL-1β in cervical cancer patients but no difference for IFN-γ were described (Tjiong et al., 1999; 2001).

In summary, with few exceptions, a switch from T_H1 to T_H2 cytokines takes place in patients with extensive HPV infection and progression to cancer indicating a defective cytotoxic T cell response. It was therefore suggested that the T_H2 phenotype is associated with the persistence of viral infection and dysplasia (Wu and Kurman, 1997).

4. Changes in Cytokine Expression and Signalling at Later Stages of Malignant Progression

Whereas squamous intraepithelial lesions develop within months after HPV infection, progression to malignancy takes 10 to 20 years or even longer. Viral integration and deregulated expression of the HPV oncoproteins E6 and E7 contribute to the genetic instability of infected host cells. During the long latency period a multistep process of carcinogenesis takes place. Cellular properties are selected that provide the basis for unlimited replicative potential and growth, evasion of apoptosis, angiogenesis, tissue invasion and metastasis (Hanahan and Weinberg, 2000), and immune evasion. These include also altered expression patterns of host cell proteins, like cytokines, their receptors or signalling factors.

4.1. Interleukin-6 and interleukin-6 receptor signalling

At late stages of malignant progression the cytokine IL-6 has been detected not only in stromal macrophages (Tartour et al., 1994), but also in high percentages of epithelial cells in invasive cancer cells *in situ*, in contrast to low percentages in CIN II and CIN III lesions (Hess et al., 2000; Srivani and Nagarajan, 2003). In one study the influence of the microenvironment on tumor cells was dissected from autochthonous cytokine production. IL-6 mRNA was detected not only in tumor cell suspensions containing varying percentages of other cell types but also in cervical carcinoma cell lines. In contrast, the cytokine IL-10 was only detectable in the mixed cell suspensions but not in tumor cell lines (Hazelbag et al., 2001). This is in accordance with our own observations that unstimulated cervical carcinomas produce IL-6 (Hess et al., 2000) but not IL-10 protein *in vitro* (unpublished data).

The presence of IL-6 positive tumor cells correlated with a poor prognosis of disease progression. These patients were found to have residual disease after treatment or recurrence of disease within 28 months (Srivani and Nagarajan, 2003). It can therefore be speculated that IL-6 is not just a marker of disease progression but also has a profound molecular impact on the fate of cervical dysplasia and carcinomas. Whereas only IL-6 mRNA but no IL-6 protein was detected in keratinocytes retrovirally infected with HPV16 E6/E7 expression constructs (Nees et al., 2001), a variety of (newly) established HPV-positive cervical carcinoma cell lines were shown to constitutively release IL-6 *in vitro* (Eustace et al., 1993; Hess et al., 2000).

IL-6 is a cytokine with pleiotropic functions. It belongs to the family of IL-6-type cytokines comprising also IL-11, leukaemia inhibitory factor, oncostatin M, and other cytokines involved in the regulation of cellular differentiation, survival, apoptosis and proliferation (for review see (Akira et al., 1993)). IL-6 is also an important regulator of the immune response. Here it displays pro- as well as anti-inflammatory properties (Tilg et al., 1997). IL-6 is an important stimulator of the acute phase response. Being a potent inducer of MCP-1 production, it orchestrates the transition from neutrophil to monocyte recruitment during inflammation (Kaplanski et al., 2003). IL-6 is regarded as a T_H2 cytokine, since it directly promotes $CD4^+$ T_H2 and inhibits T_H1 differentiation (Diehl et al., 2000). Furthermore, it stimulates B cell differentiation and antibody production.

IL-6 delivers its signals via a common receptor chain (gp130), which is also engaged by other IL-6-type cytokines (for reviews see Heinrich et al. (2003) and Kamimura et al. (2003)). It binds to the α-chain of the IL-6R (IL-6Rα, gp80) in its transmembrane or naturally occurring soluble form (sgp80). This complex then associates with gp130 and induces its dimerisation. Whereas gp130 is ubiquitously expressed, IL-6 signaling is limited by the availability of gp80 or sgp80. Dimerisation of gp130 is followed by the rapid activation of tyrosine kinases of the Jak family and subsequent activation of transcription factors of the STAT family, STAT3 and to a lesser extent STAT1. Recent data indicate that STAT3 may be involved in the IL-6-mediated transcriptional activation of the transcription factor C/EBPβ (CCAAT/enhancer binding protein β) (Niehof et al., 2001). Activated STAT3 is considered as an oncogene, since it may promote cell-cycle progression via activation of cyclin D1 and c-myc, support cellular transformation, and prevent apoptosis in various cell types (Calo et al., 2003). In other tumor cell types STAT3 has been shown to block expression of pro-inflammatory mediators, such as IFN-β, TNF-α, IL-6, RANTES and IP-10. Conversely, inhibition of STAT3 increased the expression of cytokines and chemokines (Wang et al., 2004). Besides this Jak-STAT pathway, IL-6 can activate MAPK cascades. Activation of ERK (extracellular signal regulated kinase) is initiated by the SHP-2 protein tyrosine phosphatase, mediated in a ras-dependent manner, eventually leading to activation of the transcription factor C/EBPβ by Thr-phosphorylation (Ramji and Foka, 2002). IL-6 has also been described to activate p38 MAPK as well as the c-jun N-terminal kinase pathways. The PI3K (phosphatidylinositide 3-kinase) pathway leading to the activation of protein kinase B/Akt is stimulated by IL-6 in a cell type specific manner. This pathway, which is involved in the inhibition of apoptosis and up-regulation of the anti-apoptotic molecule mcl-1, was also suggested for SiHa cells, a HPV16-positive cervical carcinoma cell line (Wei et al., 2001b). Moreover, IL-6 was described to serve as a growth factor for HPV-positive cervical carcinoma cells *in vitro*, acting via an autocrine epidermal growth factor dependent pathway that involves TGF-α and amphiregulin (Eustace et al., 1993; Iglesias et al., 1995). Studies in SiHa cells and the HPV-negative cervical carcinoma cell line C33a indicated that IL-6 can also lead to the production of VEGF (vascular endothelial growth factor) involving the STAT3 pathway (Wei et al., 2001a; 2003). Various HPV-positive cervical carcinoma cell lines producing IL-6 show low constitutive activation of the transcription factor STAT3, which may be sufficient for pro-tumorigenic effects (Hess et al., 2000). Addition of neutralising anti-IL-6 antibodies partially abolished STAT3 activation, indicating that weak autocrine IL-6 signalling contributed to this weak STAT3 activation (unpublished observation). However, from the high amounts of IL-6 produced by the carcinoma cells, much stronger STAT3 signals should have been expected. In contrast to keratinocytes transformed with HPV *in vitro*, all cervical carcinoma cells completely failed to respond to exogenously added IL-6 despite similar STAT3 protein levels in the cells. Whereas the signal transducing receptor chain gp130 was well expressed on all cells, all HPV-positive carcinoma cells showed almost undetectable mRNA and protein levels of the ligand binding chain gp80. This might not only explain the low STAT3 response (Hess et al., 2000), but also lack of HPV18 early promoter regulation in cervical carcinoma cells (Bauknecht et al., 1999). Interestingly, the IL-6 pathway could be restored by addition of sgp80 leading to a strong STAT3 and C/EBPβ response and to strong induction of the mononuclear cell attracting chemokine MCP-1 (Hess et al., 2000; Smola-Hess et al., 2001a). Unexpectedly, sgp80 led to the up-regulation of HPV18 early promoter activity in a STAT3-dependent manner in a carcinoma cell line (Smola-Hess et al., 2001a). This was surprising, since C/EBPβ, which is also activated by IL-6 signalling, may suppress this promoter (Bauknecht et al., 1999; Smola-Hess et al., 2001a). Similar to IL-6, IL-10 was also shown to activate the HPV16 upstream regulatory region involving STAT3 (Arany et al., 2002). Thus, silencing of an otherwise massive autocrine IL-6 response in cervical carcinoma cells might rather represent an important escape mechanism that prevents MCP-1 expression within the tumor.

It was so far unclear, why the carcinoma cells produced much higher quantities of IL-6 than used for autocrine activation. Of note, cervical cancer is not the only malignant tumor type expressing high amounts of IL-6. This points to a more general role of IL-6 in cancer. Recent studies have provided evidence that IL-6 may profoundly alter the functional properties of APCs. It has been observed that IL-6 can switch the differentiation of dendritic cells to a macrophage-like phenotype (Chomarat et al., 2000; Menetrier-Caux et al., 1998; Mitani et al., 2000). Recently, evidence was provided that tumor-derived factors that induce STAT3 in dendritic cells may affect their differentiation (Nefedova et al., 2004). Pro-inflammatory stimuli, including TNF-α, IL-1 and CD40 ligand (CD40L) interfere with the IL-6 effect (Hegde et al., 2004; Mitani et al., 2000). We could show that dendritic cells differentiated under the influence of IL-6 and a pro-inflammatory stimulus, appeared phenotypically mature and displayed normal or even higher expression of the co-stimulatory molecule CD86 and MHC class II on their surface. However, IL-6 caused broad functional impairment of dendritic cells (Hegde et al., 2004). While p38 MAPK activation was rather stimulated by IL-6 in dendritic cells, IL-6 led to strong suppression of their NF-κB binding activity. This transcription factor controls important functions of dendritic cells. Under the influence of IL-6, dendritic cells did not only display a reduced capacity to stimulate T helper cells. IL-6 also inhibited the production of the pro-inflammatory cytokine TNF-α and the chemokine IP-10, and suppressed the up-regulation of the chemokine CCR7. Differentiation of dendritic cells under the influence of IL-6 induced the production of autocrine IL-10. These low IL-10 levels could, however, only partially explain the immunosuppressive properties of IL-6 (Hegde et al., 2004).

Thus IL-6, a cytokine mainly regarded as pro-inflammatory, can mediate also anti-inflammatory and immunosuppressive effects. This may contribute to the fact that IL-6 expression in cervical cancer is associated with a negative disease prognosis.

4.2. Vascular endothelial growth factor production

In addition to its own effect on APCs, IL-6 may also act via secondary factors such as VEGF. VEGF is not only an important mediator of angiogenesis, a hallmark of carcinogenesis; it also exerts immunosuppressive activities on dendritic cells (reviewed in Ohm and Carbone (2001)). Similar to IL-6 and IL-10, VEGF was shown to alter dendritic cell functions via NF-κB inhibition (Oyama et al., 1998). *In vivo*, VEGF expression increased with disease progression. It was found to be expressed strongest in HSIL and invasive SCC (Dobbs et al., 1997; Guidi et al., 1995) and was therefore suggested as a prognostic factor in early cervical SCC (Cheng et al., 2000; Terlikowski et al., 2001). Interestingly, VEGF was also detected in pre-malignant lesions of transgenic mice expressing the HPV16 early genes from the keratin-14 promoter, and was progressively up-regulated during carcinogenesis comparable to the human situation (Smith-McCune et al., 1997). This indicated that HPV itself might contribute to VEGF expression. Recent studies could demonstrate that HPV16 oncoproteins can induce VEGF expression in primary human foreskin keratinocytes (Toussaint-Smith et al., 2004). The HPV16 E6 protein was sufficient for induction of the VEGF promoter in a p53-independent manner, as tested in cell lines (Lopez-Ocejo et al., 2000).

4.3. Tumor necrosis factor-α-mediated gene induction

HPV-positive cells obviously engage a variety of strategies to circumvent production of the chemokine MCP-1. In malignant cells additional mechanisms may occur. Cervical cancer cells do not only ameliorate autocrine IL-6 signalling; the capability of TNF-α to induce MCP-1 production was also found to differ profoundly between HeLa cells, tumorigenic segregants of HeLa-fibroblast hybrids and non-tumorigenic hybrid cells (Rosl et al., 1994). AP-1, a dimeric transcription factor composed of jun, fos or ATF (activating transcription factor) subunits that bind to a common DNA site, is known to be relevant for MCP-1 promoter activation. Analysis of the AP-1 composition in such hybrid cells revealed that MCP-1 is expressed only if members of the jun-family are heterodimerised with the fos-related protein fra-1 (Finzer et al., 2000), which was not the case in the malignant cells. Thus changes in the composition of the AP-1 complex may alter the gene induction profile of TNF-α during malignant progression of genital HPV lesions.

The effect of TNF-α on endogenous viral transcription also changes profoundly during malignant progression. In the HPV18-positive carcinoma cell lines HeLa and SW756 it could be demonstrated that TNF-α stimulation did not affect endogenous E6/7 mRNA expression (Rosl et al., 1994; and own unpublished results). It was suggested that alterations in TNF-α signalling, like changes in the composition of the transcription factor AP-1, may determine the outcome. This hypothesis was based on the observation that differential AP-1 composition in non-tumorigenic and tumorigenic HeLa-fibroblast hybrids corresponded to the ability of TNF-α to suppress endogenous early gene transcription or not (Soto et al., 1999). However, it could also be demonstrated that TNF-α is able to suppress HPV16 and HPV18 early promoter activity in transient reporter gene analysis performed with HeLa and SW756 cells (Kyo et al., 1994; and own unpublished results). This indicates that the anti-papillomaviral signal transduction pathway of TNF-α remains intact in the carcinoma cells, whereas regulation of the endogenous HPV early promoter might be blocked by other yet unknown mechanisms. The insensitivity of endogenous HPV early gene regulation to TNF-α may also partially account for the observed changes of cellular responses to TNF-α during progression.

4.4. CD40 up-regulation and signalling

Interestingly, MCP-1 was found to be re-expressed in most cervical carcinomas *in vivo* at the epithelial-mesenchymal interface, where tumor cells are in close contact with stromal cells (Riethdorf et al., 1996; 1998). In search of a mechanism explaining this phenomenon, we found that CD40, also a member of the TNFR-family, is strongly up-regulated in cervical carcinoma cells *in situ*. Its ligand was detected in the stroma, most likely on the surface of T helper cells (Altenburg et al., 1999). Similarly, cervical carcinoma cell lines strongly expressed CD40 *in vitro*, whereas this receptor was only weakly expressed on HPV-positive keratinocytes transformed *in vitro*. Despite much higher CD40 expression in the carcinoma cells, MCP-1 induction was significantly lower as compared with non-tumorigenic cells. However, after sensitisation with IFN-γ, another T-cell derived cytokine, MCP-1 and other chemokines (MCP-3, IP-10 and RANTES) were produced at high levels also in the carcinoma cells (Altenburg et al., 1999). Thus, two T cell derived mediators strongly synergised in chemokine induction in cervical carcinoma cells and may explain the MCP-1 expression pattern at the epithelial-mesenchymal interface. These data also corresponded to the fact that low IFN-γ mRNA expression in tumor biopsy specimens was found to be associated with poor prognosis (Pao et al., 1995; Tartour et al., 1998). In addition to chemokine induction, CD40 activation exerted tumor-promoting activities on the carcinoma cells. It could be demonstrated that activation of CD40 in carcinoma cells induced the matrix-metalloproteinase MMP-9 important for invasion and metastasis of malignant tumors (Smola-Hess et al., 2001b). These data may give at least one explanation why the receptor CD40 is up-regulated during malignant progression.

4.5. Transforming growth factor-β expression and transforming growth factor-β receptor signalling

Similar to MCP-1, the TGF-β expression patterns changed during progression from pre-malignant to the malignant stage. Although intracellular TGF-β decreased in CIN and cervical carcinoma, extracellular TGF-β was enhanced in stroma surrounding invasive cancer eventually expressed by regulatory T cells as seen in CIN (Comerci et al., 1996; Kobayashi et al., 2004). Correspondingly, *in vitro* studies showed that malignant HPV-positive cells loose their sensitivity to TGF-β-mediated growth inhibition. TGF-β is known to reduce the proliferation of normal keratinocytes and non-tumorigenic HPV 16-immortalised cell lines. Cervical carcinoma cell lines were shown to be refractory to TGF-β-induced growth inhibition (Borger et al., 2000; Braun et al., 1990; Hasskarl et al., 2000) thus potentially able to tolerate TGF-β levels found *in vivo*. It was therefore hypothesised that additional changes in the TGF-β signalling pathway may occur during progression or in cervical carcinoma cells. On the one hand, resistance to TGF-β at late stages of *in vitro* progression was attributed to the loss of type I TGF-β receptor (Mi et al., 2000). On the other hand expression of signal transducing molecules appeared to change during progression. Compared with non-malignant samples, Smad 2 expression was reduced in some cervical tumor samples possibly due to genetic alterations of the Smad 2 gene. Loss of Smad 4 expression found in some cervical tumor samples was assumed to take place at the transcriptional level (Baldus et al., 2004; Maliekal et al., 2003). These changes during malignant progression may have profound consequences for the outcome of TGF-β signalling and may favour resistance to TGF-β-mediated growth inhibition. However, it could also be shown that TGF-β signalling was not completely blocked. It has been hypothesised that TGF-β can transmit signals in cervical carcinoma cells also independently of Smads (Maliekal et al., 2004). TGF-β1 treatment increased the activity of MMP-2 and MMP-9 in normal and cervical carcinoma cells (Agarwal et al., 1994b) and c-jun mRNA could still be induced in otherwise refractory cells. Although loss of TGF-βR type I was described at late stages of *in vitro* progression (Mi et al., 2000), the former observations indicated that the anti-proliferative pathway of TGF-β was selectively affected.

5. Concluding remarks

Persistent infection of cervical epithelium with genital "high risk" HPV is the decisive risk factor and molecular prerequisite for cervical cancer. However, only a small percentage of HPV-infected women develops cancer. As summarised in this article, a variety of host factors regulating cellular communication and innate as well as adaptive immune responses, like cytokines and chemokines, have a major impact on the outcome of primary infection. Thus, it is the complex interplay between virally infected cells and host immune cells that determines the fate of the infected cells, their elimination or progression to cancer.

References

Abdul-Karim, F. W., Fu, Y. S., Reagan, J. W., and Wentz, W. B. (1982). Morphometric study of intraepithelial neoplasia of the uterine cervix. Obstet. Gynecol. *60*, 210-214.

Agarwal, C., Hembree, J. R., Rorke, E. A., and Eckert, R. L. (1994a). Interferon and retinoic acid suppress the growth of human papillomavirus type 16 immortalized cervical epithelial cells, but only interferon suppresses the level of the human papillomavirus transforming oncogenes. Cancer Res. *54*, 2108-2112.

Agarwal, C., Hembree, J. R., Rorke, E. A., and Eckert, R. L. (1994b). Transforming growth factor beta 1 regulation of metalloproteinase production in cultured human cervical epithelial cells. Cancer Res. *54*, 943-949.

Aggarwal, B. B. (2003). Signalling pathways of the TNF superfamily: a double-edged sword. Nat. Rev. Immunol. *3*, 745-756.

Akdis, C. A., Joss, A., Akdis, M., and Blaser, K. (2001). Mechanism of IL-10-induced T cell inactivation in allergic inflammation and normal response to allergens. Int. Arch. Allergy Immunol. *124*, 180-182.

Akira, S., Taga, T., and Kishimoto, T. (1993). Interleukin-6 in biology and medicine. Adv. Immunol. *54*, 1-78.

al-Saleh, W., Delvenne, P., Arrese, J. E., Nikkels, A. F., Pierard, G. E., and Boniver, J. (1995). Inverse modulation of intraepithelial Langerhans' cells and stromal macrophage/ dendrocyte populations in human papillomavirus-associated squamous intraepithelial lesions of the cervix. Virchows Arch. *427*, 41-48.

al-Saleh, W., Giannini, S. L., Jacobs, N., Moutschen, M., Doyen, J., Boniver, J., and Delvenne, P. (1998). Correlation of T-helper secretory differentiation and types of antigen-presenting cells in squamous intraepithelial lesions of the uterine cervix. J. Pathol. *184*, 283-290.

Allavena, P., Piemonti, L., Longoni, D., Bernasconi, S., Stoppacciaro, A., Ruco, L., and Mantovani, A. (1997). IL-10 prevents the generation of dendritic cells from CD14+ blood monocytes, promotes the differentiation to mature macrophages and stimulates endocytosis of FITC-dextran. Adv. Exp. Med. Biol. *417*, 323-327.

Altenburg, A., Baldus, S. E., Smola, H., Pfister, H., and Hess, S. (1999). CD40 ligand-CD40 interaction induces chemokines in cervical carcinoma cells in synergism with IFN-gamma. J. Immunol. *162*, 4140-4147.

Arany, I., Grattendick, K. G., and Tyring, S. K. (2002). Interleukin-10 induces transcription of the early promoter of human papillomavirus type 16 (HPV16) through the 5'-segment of the upstream regulatory region (URR). Antiviral Res. *55*, 331-339.

Arany, I., Grattendick, K. J., Whitehead, W. E., Ember, I. A., and Tyring, S. K. (2003). A functional interferon regulatory factor-1 (IRF-1)-binding site in the upstream regulatory region (URR) of human papillomavirus type 16. Virology *310*, 280-286.

Bach, E. A., Aguet, M., and Schreiber, R. D. (1997). The IFN gamma receptor: a paradigm for cytokine receptor signaling. Annu. Rev. Immunol. *15*, 563-591.

Bachmann, A., Hanke, B., Zawatzky, R., Soto, U., van Riggelen, J., zur Hausen, H., and Rosl, F. (2002). Disturbance of tumor necrosis factor alpha-mediated beta interferon signaling in cervical carcinoma cells. J. Virol. *76*, 280-291.

Baldus, S. E., Schwarz, E., Lohrey, C., Zapatka, M., Landsberg, S., Hahn, S. A., Schmidt, D., Dienes, H. P., Schmiegel, W. H., and Schwarte-Waldhoff, I. (2004). Smad4 deficiency in cervical carcinoma cells. Oncogene. *24*, 810-819.

Baldwin, A., Pirisi, L., and Creek, K. E. (2004). NFI-Ski interactions mediate transforming growth factor beta modulation of human papillomavirus type 16 early gene expression. J. Virol. *78*, 3953-3964.

Barnard, P., and McMillan, N. A. (1999). The human papillomavirus E7 oncoprotein abrogates signaling mediated by interferon-alpha. Virology *259*, 305-313.

Barnard, P., Payne, E., and McMillan, N. A. (2000). The human papillomavirus E7 protein is able to inhibit the antiviral and anti-growth functions of interferon-alpha. Virology *277*, 411-419.

Basile, J. R., Zacny, V., and Munger, K. (2001). The cytokines tumor necrosis factor-alpha (TNF-alpha) and TNF-related apoptosis-inducing ligand differentially modulate proliferation and apoptotic pathways in human keratinocytes expressing the human papillomavirus-16 E7 oncoprotein. J. Biol. Chem. *276*, 22522-22528.

Bauknecht, T., Randelzhofer, B., Schmitt, B., Ban, Z., and Hernando, J. J. (1999). Response to IL-6 of HPV-18 cervical carcinoma cell lines. Virology *258*, 344-354.

Boccardo, E., Noya, F., Broker, T. R., Chow, L. T., and Villa, L. L. (2004). HPV-18 confers resistance to TNF-alpha in organotypic cultures of human keratinocytes. Virology *328*, 233-243.

Bogdan, C., and Nathan, C. (1993). Modulation of macrophage function by transforming growth factor beta, interleukin-4, and interleukin-10. Ann. N. Y Acad. Sci. *685*, 713-739.

Borger, D. R., Mi, Y., Geslani, G., Zyzak, L. L., Batova, A., Engin, T. S., Pirisi, L., and Creek, K. E. (2000). Retinoic acid resistance at late stages of human papillomavirus type 16-mediated transformation of human keratinocytes arises despite intact retinoid signaling and is due to a loss of sensitivity to transforming growth factor-beta. Virology *270*, 397-407.

Bostrom, S. G., and Hart, W. R. (1981). Carcinomas of the cervix with intense stromal eosinophilia. Cancer *47*, 2887-2893.

Braun, L., Durst, M., Mikumo, R., Crowley, A., and Robinson, M. (1992). Regulation of growth and gene expression in human papillomavirus-transformed keratinocytes by transforming growth factor-beta: implications for the control of papillomavirus infection. Mol. Carcinog. *6*, 100-111.

Braun, L., Durst, M., Mikumo, R., and Gruppuso, P. (1990). Differential response of nontumorigenic and tumorigenic human papillomavirus type 16-positive epithelial cells to transforming growth factor beta 1. Cancer Res. *50*, 7324-7332.

Buelens, C., Verhasselt, V., De Groote, D., Thielemans, K., Goldman, M., and Willems, F. (1997a). Human dendritic cell responses to lipopolysaccharide and CD40 ligation are differentially regulated by interleukin-10. Eur. J. Immunol. *27*, 1848-1852.

Buelens, C., Verhasselt, V., De Groote, D., Thielemans, K., Goldman, M., and Willems, F. (1997b). Interleukin-10 prevents the generation of dendritic cells from human peripheral blood mononuclear cells cultured with interleukin-4 and granulocyte/macrophage-colony-stimulating factor. Eur. J. Immunol. *27*, 756-762.

Calo, V., Migliavacca, M., Bazan, V., Macaluso, M., Buscemi, M., Gebbia, N., and Russo, A. (2003). STAT proteins: from normal control of cellular events to tumorigenesis. J. Cell Physiol. *197*, 157-168.

Chang, Y. E., and Laimins, L. A. (2000). Microarray analysis identifies interferon-inducible genes and Stat-1 as major transcriptional targets of human papillomavirus type 31. J. Virol. *74*, 4174-4182.

Charbonnier, A. S., Kohrgruber, N., Kriehuber, E., Stingl, G., Rot, A., and Maurer, D. (1999). Macrophage inflammatory protein 3alpha is involved in the constitutive trafficking of epidermal langerhans cells. J. Exp. Med. *190*, 1755-1768.

Chardonnet, Y., Viac, J., and Schmitt, D. (1989). [Malpighian epithelia infected by DNA viruses and Langerhans cells]. Pathol. Biol. (Paris) *37*, 927-936.

Cheng, W. F., Chen, C. A., Lee, C. N., Wei, L. H., Hsieh, F. J., and Hsieh, C. Y. (2000). Vascular endothelial growth factor and prognosis of cervical carcinoma. Obstet. Gynecol. *96*, 721-726.

Chomarat, P., Banchereau, J., Davoust, J., and Palucka, A. K. (2000). IL-6 switches the differentiation of monocytes from dendritic cells to macrophages. Nat. Immunol. *1*, 510-514.

Clerici, M., Merola, M., Ferrario, E., Trabattoni, D., Villa, M. L., Stefanon, B., Venzon, D. J., Shearer, G. M., De Palo, G., and Clerici, E. (1997). Cytokine production patterns in cervical intraepithelial neoplasia: association with human papillomavirus infection. J. Natl. Cancer Inst *89*, 245-250.

Coleman, N., Birley, H. D., Renton, A. M., Hanna, N. F., Ryait, B. K., Byrne, M., Taylor-Robinson, D., and Stanley, M. A. (1994). Immunological events in regressing genital warts. Amer. J. Clin. Pathol. *102*, 768-774.

Comerci, J. T., Jr., Runowicz, C. D., Flanders, K. C., De Victoria, C., Fields, A. L., Kadish, A. S., and Goldberg, G. L. (1996). Altered expression of transforming growth factor-beta 1 in cervical neoplasia as an early biomarker in carcinogenesis of the uterine cervix. Cancer *77*, 1107-1114.

Coussens, L. M., Tinkle, C. L., Hanahan, D., and Werb, Z. (2000). MMP-9 supplied by bone marrow-derived cells contributes to skin carcinogenesis. Cell *103*, 481-490.

D'Amico, G., Frascaroli, G., Bianchi, G., Transidico, P., Doni, A., Vecchi, A., Sozzani, S., Allavena, P., and Mantovani, A. (2000). Uncoupling of inflammatory chemokine receptors by IL-10: generation of functional decoys. Nat. Immunol. *1*, 387-391.

De Geest, K., Bergman, C. A., Turyk, M. E., Frank, B. S., and Wilbanks, G. D. (1994). Differential response of cervical intraepithelial and cervical carcinoma cell lines to transforming growth factor-beta 1. Gynecol. Oncol. *55*, 376-385.

de Gruijl, T. D., Bontkes, H. J., van den Muysenberg, A. J., van Oostveen, J. W., Stukart, M. J., Verheijen, R. H., van der Vange, N., Snijders, P. J., Meijer, C. J., Walboomers, J. M., and Scheper, R. J. (1999). Differences in cytokine mRNA profiles between premalignant and malignant lesions of the uterine cervix. Eur. J. Cancer *35*, 490-497.

De Smedt, T., Van Mechelen, M., De Becker, G., Urbain, J., Leo, O., and Moser, M. (1997). Effect of interleukin-10 on dendritic cell maturation and function. Eur. J. Immunol. *27*, 1229-1235.

Delvenne, P., Hubert, P., and Jacobs, N. (2004). Epithelial metaplasia: an inadequate environment for antitumour immunity? Trends Immunol. *25*, 169-173.

Derynck, R., Akhurst, R. J., and Balmain, A. (2001). TGF-beta signaling in tumor suppression and cancer progression. Nat. Genet. *29*, 117-129.

Derynck, R., and Zhang, Y. E. (2003). Smad-dependent and Smad-independent pathways in TGF-beta family signalling. Nature *425*, 577-584.

Detmar, M., and Orfanos, C. E. (1990). Tumor necrosis factor-alpha inhibits cell proliferation and induces class II antigens and cell adhesion molecules in cultured normal human keratinocytes *in vitro*. Arch. Dermatol. Res. *282*, 238-245.

Diehl, S., Anguita, J., Hoffmeyer, A., Zapton, T., Ihle, J. N., Fikrig, E., and Rincon, M. (2000). Inhibition of Th1 differentiation by IL-6 is mediated by SOCS1. Immunity *13*, 805-815.

Dobbs, S. P., Hewett, P. W., Johnson, I. R., Carmichael, J., and Murray, J. C. (1997). Angiogenesis is associated with vascular endothelial growth factor expression in cervical intraepithelial neoplasia. Br. J. Cancer *76*, 1410-1415.

Edwards, J. N., and Morris, H. B. (1985). Langerhans' cells and lymphocyte subsets in the female genital tract. Br. J. Obstet. Gynaecol *92*, 974-982.

El-Sherif, A. M., Seth, R., Tighe, P. J., and Jenkins, D. (2000). Decreased synthesis and expression of TGF-beta1, beta2, and beta3 in epithelium of HPV 16-positive cervical precancer: a study by microdissection, quantitative RT-PCR, and immunocytochemistry. J. Pathol. *192*, 494-501.

El-Sherif, A. M., Seth, R., Tighe, P. J., and Jenkins, D. (2001). Quantitative analysis of IL-10 and IFN-gamma mRNA levels in normal cervix and human papillomavirus type 16 associated cervical precancer. J. Pathol. *195*, 179-185.

Ellerbrock, T. V., Chiasson, M. A., Bush, T. J., Sun, X. W., Sawo, D., Brudney, K., and Wright, T. C., Jr. (2000). Incidence of cervical squamous intraepithelial lesions in HIV-infected women. J. Amer. Med. Assoc. *283*, 1031-1037.

Eustace, D., Han, X., Gooding, R., Rowbottom, A., Riches, P., and Heyderman, E. (1993). Interleukin-6 (IL-6) functions as an autocrine growth factor in cervical carcinomas *in vitro*. Gynecol. Oncol. *50*, 15-19.

Evander, M., Edlund, K., Gustafsson, A., Jonsson, M., Karlsson, R., Rylander, E., and Wadell, G. (1995). Human papillomavirus infection is transient in young women: a population-based cohort study. J. Infect. Dis. *171*, 1026-1030.

Filippova, M., Parkhurst, L., and Duerksen-Hughes, P. J. (2004). The human papillomavirus 16 E6 protein binds to Fas-associated death domain and protects cells from Fas-triggered apoptosis. J. Biol. Chem. *279*, 25729-25744.

Filippova, M., Song, H., Connolly, J. L., Dermody, T. S., and Duerksen-Hughes, P. J. (2002). The human papillomavirus 16 E6 protein binds to tumor necrosis factor (TNF) R1 and protects cells from TNF-induced apoptosis. J. Biol. Chem. *277*, 21730-21739.

Finzer, P., Soto, U., Delius, H., Patzelt, A., Coy, J. F., Poustka, A., zur Hausen, H., and Rosl, F. (2000). Differential transcriptional regulation of the monocyte-chemoattractant protein-1 (MCP-1) gene in tumorigenic and non-tumorigenic HPV 18 positive cells: the role of the chromatin structure and AP-1 composition. Oncogene *19*, 3235-3244.

Fontaine, V., van der Meijden, E., de Graaf, J., ter Schegget, J., and Struyk, L. (2000). A functional NF-kappaB binding site in the human papillomavirus type 16 long control region. Virology *272*, 40-49.

Fujisawa, H., Kondo, S., Wang, B., Shivji, G. M., and Sauder, D. N. (1997). The expression and modulation of IFN-alpha and IFN-beta in human keratinocytes. J. Interferon Cytokine Res. *17*, 721-725.

Fujita, T., Reis, L. F., Watanabe, N., Kimura, Y., Taniguchi, T., and Vilcek, J. (1989). Induction of the transcription factor IRF-1 and interferon-beta mRNAs by cytokines and activators of second-messenger pathways. Proc. Natl. Acad. Sci. USA. *86*, 9936-9940.

Gaiotti, D., Chung, J., Iglesias, M., Nees, M., Baker, P. D., Evans, C. H., and Woodworth, C. D. (2000). Tumor necrosis factor-alpha promotes human papillomavirus (HPV) E6/E7 RNA expression and cyclin-dependent kinase activity in HPV-immortalized keratinocytes by a ras-dependent pathway. Mol. Carcinog. *27*, 97-109.

Giannini, S. L., Al-Saleh, W., Piron, H., Jacobs, N., Doyen, J., Boniver, J., and Delvenne, P. (1998). Cytokine expression in squamous intraepithelial lesions of the uterine cervix: implications for the generation of local immunosuppression. Clin. Exp. Immunol. *113*, 183-189.

Giannini, S. L., Hubert, P., Doyen, J., Boniver, J., and Delvenne, P. (2002). Influence of the mucosal epithelium microenvironment on Langerhans cells: implications for the development of squamous intraepithelial lesions of the cervix. Int. J. Cancer *97*, 654-659.

Gorbachev, A. V., and Fairchild, R. L. (2004). CD40 Engagement Enhances Antigen-Presenting Langerhans Cell Priming of IFN-{gamma}-Producing CD4+ and CD8+ T Cells Independently of IL-12. J. Immunol. *173*, 2443-2452.

Guidi, A. J., Abu-Jawdeh, G., Berse, B., Jackman, R. W., Tognazzi, K., Dvorak, H. F., and Brown, L. F. (1995). Vascular permeability factor (vascular endothelial growth factor) expression and angiogenesis in cervical neoplasia. J. Natl. Cancer Inst *87*, 1237-1245.

Halpert, R., Fruchter, R. G., Sedlis, A., Butt, K., Boyce, J. G., and Sillman, F. H. (1986). Human papillomavirus and lower genital neoplasia in renal Transplant. patients. Obstet. Gynecol. *68*, 251-258.

Hanahan, D., and Weinberg, R. A. (2000). The hallmarks of cancer. Cell *100*, 57-70.

Hasskarl, J., Butz, K., Whitaker, N., Ullmann, A., Durst, M., and Hoppe-Seyler, F. (2000). Differential cell cycle response of nontumorigenic and tumorigenic human papillomavirus-positive keratinocytes towards transforming growth factor-beta1. J. Mol. Med. *78*, 94-101.

Havard, L., Delvenne, P., Frare, P., Boniver, J., and Giannini, S. L. (2002). Differential production of cytokines and activation of NF-kappaB in HPV-transformed keratinocytes. Virology *298*, 271-285.

Hazelbag, S., Fleuren, G. J., Baelde, J. J., Schuuring, E., Kenter, G. G., and Gorter, A. (2001). Cytokine profile of cervical cancer cells. Gynecol. Oncol. *83*, 235-243.

Hegde, S., Pahne, J., and Smola-Hess, S. (2004). Novel immunosuppressive properties of interleukin-6 in dendritic cells: inhibition of NF-kappaB binding activity and CCR7 expression. Faseb J. *online publication Jul 9, doi:10.1096/fj.03-0969fje*.

Heinrich, P. C., Behrmann, I., Haan, S., Hermanns, H. M., Muller-Newen, G., and Schaper, F. (2003). Principles of interleukin (IL)-6-type cytokine signalling and its regulation. Biochem. J. *374*, 1-20.

Hess, S., Smola, H., Sandaradura De Silva, U., Hadaschik, D., Kube, D., Baldus, S. E., Flucke, U., and Pfister, H. (2000). Loss of IL-6 receptor expression in cervical carcinoma cells inhibits autocrine IL-6 stimulation: abrogation of constitutive monocyte chemoattractant protein-1 production. J. Immunol. *165*, 1939-1948.

Ho, G. Y., Bierman, R., Beardsley, L., Chang, C. J., and Burk, R. D. (1998). Natural history of cervicovaginal papillomavirus infection in young women. N. Engl. J. Med. *338*, 423-428.

Huang, S. M., and McCance, D. J. (2002). Down regulation of the interleukin-8 promoter by human papillomavirus type 16 E6 and E7 through effects on CREB binding protein/p300 and P/CAF. J. Virol. *76*, 8710-8721.

Iglesias, M., Plowman, G. D., and Woodworth, C. D. (1995). Interleukin-6 and interleukin-6 soluble receptor regulate proliferation of normal, human papillomavirus-immortalized, and carcinoma-derived cervical cells *in vitro*. Amer. J. Pathol. *146*, 944-952.

Jacobs, N., Giannini, S. L., Doyen, J., Baptista, A., Moutschen, M., Boniver, J., and Delvenne, P. (1998). Inverse modulation of IL-10 and IL-12 in the blood of women with preneoplastic lesions of the uterine cervix. Clin. Exp. Immunol. *111*, 219-224.

Jacobs, N., Renard, I., Al-Saleh, W., Hubert, P., Doyen, J., Kedzia, W., Boniver, J., and Delvenne, P. (2003). Distinct T cell subsets and cytokine production in cultures derived from transformation zone and squamous intraepithelial lesion biopsies of the uterine cervix. Amer. J. Reprod. Immunol. *49*, 6-13.

Jonuleit, H., and Schmitt, E. (2003). The regulatory T cell family: distinct subsets and their interrelations. J. Immunol. *171*, 6323-6327.

Kamimura, D., Ishihara, K., and Hirano, T. (2003). IL-6 signal transduction and its physiological roles: the signal orchestration model. Rev. Physiol. Biochem. Pharmacol *149*, 1-38.

Kamio, M., Yoshida, T., Ogata, H., Douchi, T., Nagata, Y., Inoue, M., Hasegawa, M., Yonemitsu, Y., and Yoshimura, A. (2004). SOC1 inhibits HPV-E7-mediated transformation by inducing degradation of E7 protein. Oncogene *23*, 3107-3115.

Kaplanski, G., Marin, V., Montero-Julian, F., Mantovani, A., and Farnarier, C. (2003). IL-6: a regulator of the transition from neutrophil to monocyte recruitment during inflammation. Trends Immunol. *24*, 25-29.

Kim, K. Y., Blatt, L., and Taylor, M. W. (2000). The effects of interferon on the expression of human papillomavirus oncogenes. J. Gen. Virol. *81*, 695-700.

Kiviat, N. B., Critchlow, C. W., Holmes, K. K., Kuypers, J., Sayer, J., Dunphy, C., Surawicz, C., Kirby, P., Wood, R., and Daling, J. R. (1993). Association of Anal. dysplasia and human papillomavirus with immunosuppression and HIV infection among homosexual men. AIDS *7*, 43-49.

Klampfer, L., Lee, T. H., Hsu, W., Vilcek, J., and Chen-Kiang, S. (1994). NF-IL6 and AP-1 cooperatively modulate the activation of the TSG-6 gene by tumor necrosis factor alpha and interleukin-1. Mol. Cell Biol. *14*, 6561-6569.

Kleine-Lowinski, K., Gillitzer, R., Kuhne-Heid, R., and Rosl, F. (1999). Monocyte-chemo-attractant-protein-1 (MCP-1)-gene expression in cervical intra-epithelial neoplasias and cervical carcinomas. Int. J. Cancer *82*, 6-11.

Kleine-Lowinski, K., Rheinwald, J. G., Fichorova, R. N., Anderson, D. J., Basile, J., Munger, K., Daly, C. M., Rosl, F., and Rollins, B. J. (2003). Selective suppression of monocyte chemoattractant protein-1 expression by human papillomavirus E6 and E7 oncoproteins in human cervical epithelial and epidermal cells. Int. J. Cancer *107*, 407-415.

Kobayashi, A., Greenblatt, R. M., Anastos, K., Minkoff, H., Massad, L. S., Young, M., Levine, A. M., Darragh, T. M., Weinberg, V., and Smith-McCune, K. K. (2004). Functional attributes of mucosal immunity in cervical intraepithelial neoplasia and effects of HIV infection. Cancer Res. *64*, 6766-6774.

Kontsek, P., Karayianni-Vasconcelos, G., and Kontsekova, E. (2003). The human interferon system: characterization and classification after discovery of novel members. Acta Virol. *47*, 201-215.

Kotenko, S. V., Gallagher, G., Baurin, V. V., Lewis-Antes, A., Shen, M., Shah, N. K., Langer, J. A., Sheikh, F.,

Dickensheets, H., and Donnelly, R. P. (2003). IFN-lambdas mediate antiviral protection through a distinct class II cytokine receptor complex. Nat. Immunol. *4*, 69-77.

Kubo, M., Hanada, T., and Yoshimura, A. (2003). Suppressors of cytokine signaling and immunity. Nat. Immunol. *4*, 1169-1176.

Kyo, S., Inoue, M., Hayasaka, N., Inoue, T., Yutsudo, M., Tanizawa, O., and Hakura, A. (1994). Regulation of early gene expression of human papillomavirus type 16 by inflammatory cytokines. Virology *200*, 130-139.

Kyo, S., Inoue, M., Nishio, Y., Nakanishi, K., Akira, S., Inoue, H., Yutsudo, M., Tanizawa, O., and Hakura, A. (1993). NF-IL6 represses early gene expression of human papillomavirus type 16 through binding to the noncoding region. J. Virol. *67*, 1058-1066.

LaFleur, D. W., Nardelli, B., Tsareva, T., Mather, D., Feng, P., Semenuk, M., Taylor, K., Buergin, M., Chinchilla, D., Roshke, V., *et al.* (2001). Interferon-kappa, a novel type I interferon expressed in human keratinocytes. J. Biol. Chem. *276*, 39765-39771.

Larregina, A. T., Morelli, A. E., Spencer, L. A., Logar, A. J., Watkins, S. C., Thomson, A. W., and Falo, L. D., Jr. (2001). Dermal-resident CD14+ cells differentiate into Langerhans cells. Nat. Immunol. *2*, 1151-1158.

Lee, D. K., Kim, B. C., Kim, I. Y., Cho, E. A., Satterwhite, D. J., and Kim, S. J. (2002). The human papilloma virus E7 oncoprotein inhibits transforming growth factor-beta signaling by blocking binding of the Smad complex to its target sequence. J. Biol. Chem. *277*, 38557-38564.

Li, S., Labrecque, S., Gauzzi, M. C., Cuddihy, A. R., Wong, A. H., Pellegrini, S., Matlashewski, G. J., and Koromilas, A. E. (1999). The human papilloma virus (HPV)-18 E6 oncoprotein physically associates with Tyk2 and impairs Jak STAT activation by interferon-alpha. Oncogene *18*, 5727-5737.

Liu, Y., Tergaonkar, V., Krishna, S., and Androphy, E. J. (1999). Human papillomavirus type 16 E6-enhanced susceptibility of L929 cells to tumor necrosis factor alpha correlates with increased accumulation of reactive oxygen species. J. Biol. Chem. *274*, 24819-24827.

Lopez-Ocejo, O., Viloria-Petit, A., Bequet-Romero, M., Mukhopadhyay, D., Rak, J., and Kerbel, R. S. (2000). Oncogenes and tumor angiogenesis: the HPV-16 E6 oncoprotein activates the vascular endothelial growth factor (VEGF) gene promoter in a p53 independent manner. Oncogene *19*, 4611-4620.

Luo, K., Stroschein, S. L., Wang, W., Chen, D., Martens, E., Zhou, S., and Zhou, Q. (1999). The Ski oncoprotein interacts with the Smad proteins to repress TGFbeta signaling. Genes Dev. *13*, 2196-2206.

Maeda, H., Kuwahara, H., Ichimura, Y., Ohtsuki, M., Kurakata, S., and Shiraishi, A. (1995). TGF-beta enhances macrophage ability to produce IL-10 in normal and tumor-bearing mice. J. Immunol. *155*, 4926-4932.

Maliekal, T. T., Anto, R. J., and Karunagaran, D. (2004). Differential activation of Smads in HeLa and SiHa cells that differ in their response to transforming growth factor-beta. J. Biol. Chem. *279*, 36287-36292.

Maliekal, T. T., Antony, M. L., Nair, A., Paulmurugan, R., and Karunagaran, D. (2003). Loss of expression, and mutations of Smad 2 and Smad 4 in human cervical cancer. Oncogene *22*, 4889-4897.

Massague, J., Blain, S. W., and Lo, R. S. (2000). TGFbeta signaling in growth control, cancer, and heritable disorders. Cell *103*, 295-309.

McArdle, J. P., and Muller, H. K. (1986). Quantitative assessment of Langerhans' cells in human cervical intraepithelial neoplasia and wart virus infection. Amer. J. Obstet. Gynecol. *154*, 509-515.

McGuirk, P., McCann, C., and Mills, K. H. (2002). Pathogen-specific T regulatory 1 cells induced in the respiratory tract by a bacterial molecule that stimulates interleukin 10 production by dendritic cells: a novel strategy for evasion of protective T helper type 1 responses by Bordetella pertussis. J. Exp. Med. *195*, 221-231.

Menetrier-Caux, C., Montmain, G., Dieu, M. C., Bain, C., Favrot, M. C., Caux, C., and Blay, J. Y. (1998). Inhibition of the differentiation of dendritic cells from CD34(+) progenitors by tumor cells: role of interleukin-6 and macrophage colony-stimulating factor. Blood *92*, 4778-4791.

Mi, Y., Borger, D. R., Fernandes, P. R., Pirisi, L., and Creek, K. E. (2000). Loss of transforming growth factor-beta (TGF-beta) receptor type I mediates TGF-beta resistance in human papillomavirus type 16-transformed human keratinocytes at late stages of *in vitro* progression. Virology *270*, 408-416.

Mitani, H., Katayama, N., Araki, H., Ohishi, K., Kobayashi, K., Suzuki, H., Nishii, K., Masuya, M., Yasukawa, K., Minami, N., and Shiku, H. (2000). Activity of interleukin 6 in the differentiation of monocytes to macrophages and dendritic cells. Br. J. Haematol. *109*, 288-295.

Mogensen, K. E., Lewerenz, M., Reboul, J., Lutfalla, G., and Uze, G. (1999). The type I interferon receptor: structure, function, and evolution of a family business. J. Interferon Cytokine Res. *19*, 1069-1098.

Morris, H. H., Gatter, K. C., Stein, H., and Mason, D. Y. (1983). Langerhans' cells in human cervical epithelium: an immunohistological study. Br. J. Obstet. Gynaecol *90*, 400-411.

Moscicki, A. B., Shiboski, S., Broering, J., Powell, K., Clayton, L., Jay, N., Darragh, T. M., Brescia, R., Kanowitz, S., Miller, S. B., *et al.* (1998). The natural history of human papillomavirus infection as measured by repeated DNA testing in adolescent and young women. J. Pediatr. *132*, 277-284.

Mota, F., Rayment, N., Chong, S., Singer, A., and Chain, B. (1999). The antigen-presenting environment in normal and human papillomavirus (HPV)-related premalignant cervical epithelium. Clin. Exp. Immunol. *116*, 33-40.

Munoz, N., Bosch, F. X., de Sanjose, S., Herrero, R., Castellsague, X., Shah, K. V., Snijders, P. J., and Meijer, C. J. (2003). Epidemiologic classification of human papillomavirus types associated with cervical cancer. N. Engl. J. Med. *348*, 518-527.

Nakamura, K., Williams, I. R., and Kupper, T. S. (1995). Keratinocyte-derived monocyte chemoattractant protein 1 (MCP-1): analysis in a transgenic model demonstrates MCP-1 can recruit dendritic and Langerhans cells to skin. J. Invest Dermatol. *105*, 635-643.

Nawa, A., Nishiyama, Y., Yamamoto, N., Maeno, K., Goto, S., and Tomoda, Y. (1990). Selective suppression of human papilloma virus type 18 mRNA level in HeLa cells by interferon. Biochem. Biophys Res. Commun. *170*, 793-799.

Nees, M., Geoghegan, J. M., Hyman, T., Frank, S., Miller, L., and Woodworth, C. D. (2001). Papillomavirus type 16 oncogenes downregulate expression of interferon-responsive genes and upregulate proliferation-associated and NF-kappaB-responsive genes in cervical keratinocytes. J. Virol. *75*, 4283-4296.

Nefedova, Y., Huang, M., Kusmartsev, S., Bhattacharya, R., Cheng, P., Salup, R., Jove, R., and Gabrilovich, D. (2004). Hyperactivation of STAT3 is involved in abnormal differentiation of dendritic cells in cancer. J. Immunol. *172*, 464-474.

Nickoloff, B. J., Basham, T. Y., Merigan, T. C., and Morhenn, V. B. (1984). Antiproliferative effects of recombinant alpha- and gamma-interferons on cultured human keratinocytes. Lab. Invest *51*, 697-701.

Niehof, M., Streetz, K., Rakemann, T., Bischoff, S. C., Manns, M. P., Horn, F., and Trautwein, C. (2001). Interleukin-6-induced tethering of STAT3 to the LAP/C/EBPbeta promoter suggests a new mechanism of transcriptional regulation by STAT3. J. Biol. Chem. *276*, 9016-9027.

Ohm, J. E., and Carbone, D. P. (2001). VEGF as a mediator of tumor-associated immunodeficiency. Immunol. Res. *23*, 263-272.

Ostor, A. G. (1993). Natural history of cervical intraepithelial neoplasia: a critical review. Int. J. Gynecol. Pathol. *12*, 186-192.

Oyama, T., Ran, S., Ishida, T., Nadaf, S., Kerr, L., Carbone, D. P., and Gabrilovich, D. I. (1998). Vascular endothelial growth factor affects dendritic cell maturation through the inhibition of nuclear factor-kappa B activation in hemopoietic progenitor cells. J. Immunol. *160*, 1224-1232.

Pao, C. C., Lin, C. Y., Yao, D. S., and Tseng, C. J. (1995). Differential expression of cytokine genes in cervical cancer tissues. Biochem. Biophys Res. Commun. *214*, 1146-1151.

Park, J. S., Kim, E. J., Kwon, H. J., Hwang, E. S., Namkoong, S. E., and Um, S. J. (2000). Inactivation of interferon regulatory factor-1 tumor suppressor protein by HPV E7 oncoprotein. Implication for the E7-mediated immune evasion mechanism in cervical carcinogenesis. J. Biol. Chem. *275*, 6764-6769.

Patel, D., Huang, S. M., Baglia, L. A., and McCance, D. J. (1999). The E6 protein of human papillomavirus type 16 binds to and inhibits co-activation by CBP and p300. Embo J. *18*, 5061-5072.

Perea, S. E., Lopez-Ocejo, O., Garcia-Milian, R., and Arana, M. J. (1995). Interferon-alpha elicits downregulation of human papillomavirus 18 mRNA in HeLa cells by selective repression of endogenous viral transcription. J. Interferon Cytokine Res. *15*, 495-501.

Perea, S. E., Massimi, P., and Banks, L. (2000). Human papillomavirus type 16 E7 impairs the activation of the interferon regulatory factor-1. Int. J. Mol. Med. *5*, 661-666.

Petry, K. U., Scheffel, D., Bode, U., Gabrysiak, T., Kochel, H., Kupsch, E., Glaubitz, M., Niesert, S., Kuhnle, H., and Schedel, I. (1994). Cellular immunodeficiency enhances the progression of human papillomavirus-associated cervical lesions. Int. J. Cancer *57*, 836-840.

Pietenpol, J. A., Holt, J. T., Stein, R. W., and Moses, H. L. (1990a). Transforming growth factor beta 1 suppression of c-myc gene transcription: role in inhibition of keratinocyte proliferation. Proc. Natl. Acad. Sci. USA. *87*, 3758-3762.

Pietenpol, J. A., Stein, R. W., Moran, E., Yaciuk, P., Schlegel, R., Lyons, R. M., Pittelkow, M. R., Munger, K., Howley, P. M., and Moses, H. L. (1990b). TGF-beta 1 inhibition of c-myc transcription and growth in keratinocytes is abrogated by viral transforming proteins with pRB binding domains. Cell *61*, 777-785.

Ramji, D. P., and Foka, P. (2002). CCAAT/enhancer-binding proteins: structure, function and regulation. Biochem. J. *365*, 561-575.

Riethdorf, L., Riethdorf, S., Gutzlaff, K., Prall, F., and Loning, T. (1996). Differential expression of the monocyte chemoattractant protein-1 gene in human papillomavirus-16-infected squamous intraepithelial lesions and squamous cell carcinomas of the cervix uteri. Amer. J. Pathol. *149*, 1469-1476.

Riethdorf, S., Riethdorf, L., Richter, N., and Loning, T. (1998). Expression of the MCP-1 gene and the HPV 16 E6/E7 oncogenes in squamous cell carcinomas of the cervix uteri and metastases. Pathobiology *66*, 260-267.

Rivett, A. J., and Hearn, A. R. (2004). Proteasome function in antigen presentation: immunoproteasome complexes, Peptide production, and interactions with viral proteins. Curr. Protein Pept Sci. *5*, 153-161.

Ronco, L. V., Karpova, A. Y., Vidal, M., and Howley, P. M. (1998). Human papillomavirus 16 E6 oncoprotein binds to interferon regulatory factor-3 and inhibits its transcriptional activity. Genes Dev. *12*, 2061-2072.

Rorke, E. A., and Jacobberger, J. W. (1995). Transforming growth factor-beta 1 (TGF beta 1) enhances apoptosis in human papillomavirus type 16-immortalized human ectocervical epithelial cells. Exp. Cell Res. *216*, 65-72.

Rorke, E. A., Zhang, D., Choo, C. K., Eckert, R. L., and Jacobberger, J. W. (2000). TGF-beta-mediated cell cycle arrest of HPV16-immortalized human ectocervical cells correlates with decreased E6/E7 mRNA and increased p53 and p21(WAF-1) expression. Exp. Cell Res. *259*, 149-157.

Rosl, F., Lengert, M., Albrecht, J., Kleine, K., Zawatzky, R., Schraven, B., and zur Hausen, H. (1994). Differential regulation of the JE gene encoding the monocyte chemoattractant protein (MCP-1) in cervical carcinoma cells and derived hybrids. J. Virol. *68*, 2142-2150.

Sato, K., Nagayama, H., Tadokoro, K., Juji, T., and Takahashi, T. A. (1999). Extracellular signal-regulated kinase, stress-activated protein kinase/c-Jun N-terminal kinase, and p38mapk are involved in IL-10-mediated selective repression of TNF-alpha-induced activation and maturation of human peripheral blood monocyte-derived dendritic cells. J. Immunol. *162*, 3865-3872.

Saunders, N., Dahler, A., Jones, S., Smith, R., and Jetten, A. (1996). Interferon-gamma as a regulator of squamous differentiation. J. Dermatol. Sci. *13*, 98-106.

Schafer, S. L., Lin, R., Moore, P. A., Hiscott, J., and Pitha, P. M. (1998). Regulation of type I interferon gene expression by interferon regulatory factor-3. J. Biol. Chem. *273*, 2714-2720.

Schottelius, A. J., Mayo, M. W., Sartor, R. B., and Baldwin, A. S., Jr. (1999). Interleukin-10 signaling blocks inhibitor of kappaB kinase activity and nuclear factor kappaB DNA binding. J. Biol. Chem. *274*, 31868-31874.

Schutyser, E., Struyf, S., and Van Damme, J. (2003). The CC chemokine CCL20 and its receptor CCR6. Cytokine Growth Factor Rev. *14*, 409-426.

Scott, M., Stites, D. P., and Moscicki, A. B. (1999). Th1 cytokine patterns in cervical human papillomavirus infection. Clin. Diagn. Lab. Immunol. *6*, 751-755.

Sheu, B. C., Lin, R. H., Lien, H. C., Ho, H. N., Hsu, S. M., and Huang, S. C. (2001). Predominant Th2/Tc2 polarity of tumor-infiltrating lymphocytes in human cervical cancer. J. Immunol. *167*, 2972-2978.

Smith-McCune, K., Zhu, Y. H., Hanahan, D., and Arbeit, J. (1997). Cross-species comparison of angiogenesis during the premalignant stages of squamous carcinogenesis in the human cervix and K14-HPV16 transgenic mice. Cancer Res. *57*, 1294-1300.

Smola-Hess, S., de Silva, U. S., Hadaschik, D., and Pfister, H. J. (2001a). Soluble interleukin-6 receptor activates the human papillomavirus type 18 long control region in SW756 cervical carcinoma cells in a STAT3-dependent manner. J. Gen. Virol. *82*, 2335-2339.

Smola-Hess, S., Schnitzler, R., Hadaschik, D., Smola, H., Mauch, C., Krieg, T., and Pfister, H. (2001b). CD40L induces matrix-metalloproteinase-9 but not tissue inhibitor of metalloproteinases-1 in cervical carcinoma cells: imbalance between NF-kappaB and STAT3 activation. Exp. Cell Res. *267*, 205-215.

Soto, U., Das, B. C., Lengert, M., Finzer, P., zur Hausen, H., and Rosl, F. (1999). Conversion of HPV 18 positive non-tumorigenic HeLa-fibroblast hybrids to invasive growth involves loss of TNF-alpha mediated repression of viral transcription and modification of the AP-1 transcription complex. Oncogene *18*, 3187-3198.

Spitkovsky, D., Hehner, S. P., Hofmann, T. G., Moller, A., and Schmitz, M. L. (2002). The human papillomavirus oncoprotein E7 attenuates NF-kappa B activation by targeting the Ikappa B kinase complex. J. Biol. Chem. *277*, 25576-25582.

Sporn, M. B. (1999). TGF-beta: 20 years and counting. Microbes Infect. *1*, 1251-1253.

Srivani, R., and Nagarajan, B. (2003). A prognostic insight on *in vivo* expression of interleukin-6 in uterine cervical cancer. Int. J. Gynecol. Cancer *13*, 331-339.

Stein, B., and Baldwin, A. S., Jr. (1993). Distinct mechanisms for regulation of the interleukin-8 gene involve synergism and cooperativity between C/EBP and NF-kappa B. Mol. Cell Biol. *13*, 7191-7198.

Stein, B., Cogswell, P. C., and Baldwin, A. S., Jr. (1993). Functional and physical associations between NF-kappa B and C/EBP family members: a Rel domain-bZIP interaction. Mol. Cell Biol. *13*, 3964-3974.

Steinbrink, K., Wolfl, M., Jonuleit, H., Knop, J., and Enk, A. H. (1997). Induction of tolerance by IL-10-treated dendritic cells. J. Immunol. *159*, 4772-4780.

Struyk, L., van der Meijden, E., Minnaar, R., Fontaine, V., Meijer, I., and ter Schegget, J. (2000). Transcriptional regulation of human papillomavirus type 16 LCR by different C/EBPbeta isoforms. Mol. Carcinog. *28*, 42-50.

Symington, F. W. (1989). Lymphotoxin, tumor necrosis factor, and gamma interferon are cytostatic for normal human keratinocytes. J. Invest Dermatol. *92*, 798-805.

Tartour, E., Gey, A., Sastre-Garau, X., Lombard Surin, I., Mosseri, V., and Fridman, W. H. (1998). Prognostic value of intratumoral interferon gamma messenger RNA expression in invasive cervical carcinomas. J. Natl. Cancer Inst *90*, 287-294.

Tartour, E., Gey, A., Sastre-Garau, X., Pannetier, C., Mosseri, V., Kourilsky, P., and Fridman, W. H. (1994). Analysis of interleukin 6 gene expression in cervical neoplasia using a quantitative polymerase chain reaction assay: evidence for enhanced interleukin 6 gene expression in invasive carcinoma. Cancer Res. *54*, 6243-6248.

ten Dijke, P., and Hill, C. S. (2004). New insights into TGF-beta-Smad signalling. Trends Biochem. Sci. *29*, 265-273.

Terlikowski, S., Lenczewski, A., Sulkowska, M., Famulski, W., Sulkowski, S., and Kulikowski, M. (2001). Tissue expression of VEGF as a prognostic factor in early cervical squamous cell carcinoma. Folia Histochem Cytobiol *39 Suppl 2*, 112-113.

Tilg, H., Dinarello, C. A., and Mier, J. W. (1997). IL-6 and APPs: anti-inflammatory and immunosuppressive mediators. Immunol. Today *18*, 428-432.

Tjiong, M. Y., van der Vange, N., ten Kate, F. J., Tjong, A. H. S. P., ter Schegget, J., Burger, M. P., and Out, T. A. (1999). Increased IL-6 and IL-8 levels in cervicovaginal secretions of patients with cervical cancer. Gynecol. Oncol. *73*, 285-291.

Tjiong, M. Y., van der Vange, N., ter Schegget, J. S., Burger, M. P., ten Kate, F. W., and Out, T. A. (2001). Cytokines in cervicovaginal washing fluid from patients with cervical neoplasia. Cytokine *14*, 357-360.

Torng, P. L., Chan, W. Y., Lin, C. T., and Huang, S. C. (2003). Decreased expression of human papillomavirus E2 protein and transforming growth factor-beta1 in human cervical neoplasia as an early marker in carcinogenesis. J. Surg Oncol. *84*, 17-23.

Tough, D. F. (2004). Type I interferon as a link between innate and adaptive immunity through dendritic cell stimulation. Leuk. Lymphoma *45*, 257-264.

Toussaint-Smith, E., Donner, D. B., and Roman, A. (2004). Expression of human papillomavirus type 16 E6 and E7 oncoproteins in primary foreskin keratinocytes is sufficient to alter the expression of angiogenic factors. Oncogene *23*, 2988-2995.

Trinchieri, G. (2003). Interleukin-12 and the regulation of innate resistance and adaptive immunity. Nat. Rev. Immunol. *3*, 133-146.

Tsukui, T., Hildesheim, A., Schiffman, M. H., Lucci, J., 3rd, Contois, D., Lawler, P., Rush, B. B., Lorincz, A. T., Corrigan, A., Burk, R. D., *et al.* (1996). Interleukin 2 production *in vitro* by peripheral lymphocytes in response to human papillomavirus-derived peptides: correlation with cervical pathology. Cancer Res. *56*, 3967-3974.

van Driel, W. J., Hogendoorn, P. C., Jansen, F. W., Zwinderman, A. H., Trimbos, J. B., and Fleuren, G. J. (1996). Tumor-associated eosinophilic infiltrate of cervical cancer is indicative for a less effective immune response. Hum. Pathol. *27*, 904-911.

van Pesch, V., Lanaya, H., Renauld, J. C., and Michiels, T. (2004). Characterization of the murine alpha interferon gene family. J. Virol. *78*, 8219-8228.

Vieira, K. B., Goldstein, D. J., and Villa, L. L. (1996). Tumor necrosis factor alpha interferes with the cell cycle of normal and papillomavirus-immortalized human keratinocytes. Cancer Res. *56*, 2452-2457.

Vikhanskaya, F., Falugi, C., Valente, P., and Russo, P. (2002).

Human papillomavirus type 16 E6-enhanced susceptibility to apoptosis induced by TNF in A2780 human ovarian cancer cell line. Int. J. Cancer *97*, 732-739.

Wahl, S. M., Swisher, J., McCartney-Francis, N., and Chen, W. (2004). TGF-beta: the perpetrator of immune suppression by regulatory T cells and suicidal T cells. J. Leukoc Biol. *76*, 15-24.

Wakkach, A., Fournier, N., Brun, V., Breittmayer, J. P., Cottrez, F., and Groux, H. (2003). Characterization of dendritic cells that induce tolerance and T regulatory 1 cell differentiation *in vivo*. Immunity *18*, 605-617.

Walboomers, J. M., Jacobs, M. V., Manos, M. M., Bosch, F. X., Kummer, J. A., Shah, K. V., Snijders, P. J., Peto, J., Meijer, C. J., and Munoz, N. (1999). Human papillomavirus is a necessary cause of invasive cervical cancer worldwide. J. Pathol. *189*, 12-19.

Wang, T., Niu, G., Kortylewski, M., Burdelya, L., Shain, K., Zhang, S., Bhattacharya, R., Gabrilovich, D., Heller, R., Coppola, D., *et al.* (2004). Regulation of the innate and adaptive immune responses by Stat-3 signaling in tumor cells. Nat. Med. *10*, 48-54.

Wei, L. H., Kuo, M. L., Chen, C. A., Cheng, W. F., Cheng, S. P., Hsieh, F. J., and Hsieh, C. Y. (2001a). Interleukin-6 in cervical cancer: the relationship with vascular endothelial growth factor. Gynecol. Oncol. *82*, 49-56.

Wei, L. H., Kuo, M. L., Chen, C. A., Chou, C. H., Cheng, W. F., Chang, M. C., Su, J. L., and Hsieh, C. Y. (2001b). The anti-apoptotic role of interleukin-6 in human cervical cancer is mediated by up-regulation of Mcl-1 through a PI 3-K/Akt pathway. Oncogene *20*, 5799-5809.

Wei, L. H., Kuo, M. L., Chen, C. A., Chou, C. H., Lai, K. B., Lee, C. N., and Hsieh, C. Y. (2003). Interleukin-6 promotes cervical tumor growth by VEGF-dependent angiogenesis via a STAT3 pathway. Oncogene *22*, 1517-1527.

Woodworth, C. D., Chung, J., McMullin, E., Plowman, G. D., Simpson, S., and Iglesias, M. (1996). Transforming growth factor beta 1 supports autonomous growth of human papillomavirus-immortalized cervical keratinocytes under conditions promoting squamous differentiation. Cell Growth Differ. *7*, 811-820.

Woodworth, C. D., Lichti, U., Simpson, S., Evans, C. H., and DiPaolo, J. A. (1992). Leukoregulin and gamma-interferon inhibit human papillomavirus type 16 gene transcription in human papillomavirus-immortalized human cervical cells. Cancer Res. *52*, 456-463.

Woodworth, C. D., McMullin, E., Iglesias, M., and Plowman, G. D. (1995). Interleukin 1 alpha and tumor necrosis factor alpha stimulate autocrine amphiregulin expression and proliferation of human papillomavirus-immortalized and carcinoma-derived cervical epithelial cells. Proc. Natl. Acad. Sci. USA. *92*, 2840-2844.

Woodworth, C. D., Notario, V., and DiPaolo, J. A. (1990). Transforming growth factors beta 1 and 2 transcriptionally regulate human papillomavirus (HPV) type 16 early gene expression in HPV-immortalized human genital epithelial cells. J. Virol. *64*, 4767-4775.

Wu, J., and Lanier, L. L. (2003). Natural killer cells and cancer. Adv. Cancer Res. *90*, 127-156.

Wu, T. C., and Kurman, R. J. (1997). Analysis of cytokine profiles in patients with human papillomavirus-associated neoplasms. J. Natl. Cancer Inst *89*, 185-187.

Xie, K. (2001). Interleukin-8 and human cancer biology. Cytokine Growth Factor Rev. *12*, 375-391.

Yaar, M., Karassik, R. L., Schnipper, L. E., and Gilchrest, B. A. (1985). Effects of alpha and beta interferons on cultured human keratinocytes. J. Invest Dermatol. *85*, 70-74.

Yoshie, O., Imai, T., and Nomiyama, H. (2001). Chemokines in immunity. Adv. Immunol. *78*, 57-110.

From: Papillomavirus Research: From Natural History to Vaccines and Beyond. Edited by: M. Saveria Campo

Chapter 21

HPV Vaccines

Michelle Giles and Suzanne M. Garland*

Abstract

Vaccines against human papillomavirus (HPV) may be prophylactic, therapeutic or a combination. Currently, prophylactic multivalent virus like particle (VLP) vaccines comprising the major capsid protein L1 are in Phase III clinical trials with preliminary results from Phase II trials indicating safety, plus 100% efficacy from persistent infection, as well as from infectious complications such as cervical dysplasia. The success of prophylactic vaccines is underpinned by their induction of neutralizing antibodies to the viral capsid, although we await long-term follow-up, to evaluate their prevention of neoplasia. Numerous therapeutic vaccines targeting nonstructural early viral antigens and including approaches such as peptide, viral-vector, DNA and dendritic cell vaccines are under investigation. Therapeutic vaccines are very much in their infancy compared with prophylactic vaccines, as although many show immunogenicity, this does not generally translate into reliable clinical responses. A number of these vaccines have been studied in phase I/II trials in humans, namely in women with cervical intraepithelial neoplasia (CIN), vulval intraepithelial neoplasia (VIN) and advanced cervical cancer. Although it is likely that a prophylactic vaccine will be licensed in the very near future, there are still many challenges to implementation of an effective HPV vaccine program, not the least of which will be cost effectiveness in settings where there are preventative screening programmes. Furthermore such challenges will differ from developed to developing countries. Challenges include unanswered questions such as the best age group to target, whether both genders should be vaccinated, the duration of efficacy, how vaccination will alter current recommendations for conventional cervical screening, whether there will be HPV genotype replacement, or genotype resistance, whether any cross protection will exist between genotypes, and how many genotypes can be placed in one vaccine. Of primary importance is whether the vaccine will be accessible and affordable to women in developing countries, where there is the greatest burden of HPV related disease. In addition, for successful vaccine implementation and uptake, there will need to be education of physicians, the general public, as well as those at government levels. The success of such vaccine programs will probably take many years to realize. Nevertheless, the prospect of a vaccine, which protects against the most common viral sexually transmitted infection and thereby protects against the complications of HPV infection such as cervical cancer, is extremely exciting. With current practices, the world projected rates of the burden of invasive cancer is over one million cases per year by 2050. Therefore, primary prevention by the introduction of a successful HPV prophylactic vaccine should be one of the most significant public health initiatives of this decade.

1. Introduction

Papillomaviruses infect not just humans, but other vertebrates such as horses, cattle and rabbits, although they have a restricted host range and show exquisite tissue tropism. They are relatively difficult to culture, because their complete complex lifecycle only occurs in terminally differentiated squamous epithelial cells. Consequently obtaining large volumes of antigen for serological assays or vaccine development has been difficult. Development of virus-like particles (VLPs) has however revolutionized current vaccine development (see Chapter 18).

The clinical spectrum of HPV disease ranges from asymptomatic infection, to benign warts (also known as condylomata acuminata and primarily caused by low risk HPV 6 and 11) to metastasising malignancy. Oncogenic HPVs have been associated with neoplasia in many sites including oropharyngeal (20%) and oesophageal (10-20%) cancers, non-melanotic skin cancer (80-90%), cervical (>99%), vaginal (50%), anal (85%), vulval (35%) and penile (25%) cancers (see Chapters 3, and 15-17).

The immune system normally controls viral infections by neutralizing a virus with specific antibody or by killing virally infected cells. Antibodies are made by B cells, which recognize antigenic determinants on macromolecules. After encountering an antigen, B cells (other than with a few defined exceptions) require help from T helper cells to enter the differentiation program.

*For correspondence. Email suzanne.garland@rch.org.au

The T helper cell recognizes an antigen and interacts with the B cell to enable specific antibody production. T cells, unlike B cells, require the antigen to be presented to them in association with major histocompatibility complex (MHC) molecules. There are two types of MHC molecules; class I molecules expressed to varying extents on all somatic cells except red blood cells and class II molecules expressed on professional antigen presenting cells (B cells, dendritic cells and macrophages). Cytotoxic T cells are the cells that can selectively kill cells infected with the virus. Endogenous antigen derived from viral protein can be presented on the cell surface in association with MHC class I molecules. This MHC class I/antigen complex is recognized by T lymphocytes with the CD8 surface marker. Upon recognition of the epitope in association with the MHC class I molecule, the CD8 cells are activated and usually with cytokine help from CD4 cells, become cytotoxic effector cells which can kill any cell expressing the epitope such as a virally infected cell. Skin keratinocytes express only MHC class I molecules, but can have the expression of class II molecules switched on by IFN-γ. Therefore T helper cells will not be activated by HPV infected keratinocytes. The epithelial cells are not killed by the infection and no HPV proteins or particles are released to be taken up by the antigen presenting cells. If the HPV infected skin or mucosa is damaged by ablative therapy (common therapy for HPV related warts or dysplasia) then the dead tissue is phagocytosed by macrophages and the HPV protein can then be presented to the immune system. Whilst those adaptive immune responses have been widely studied, the importance of initiation of the innate immune system is only now being recognized (Chapter 19). There is likely a genetic contribution with chemokine receptor polymorphisms as determinants of disease progression (Chapter 20). The importance of cells of the innate immune system such as dendritic cells and macrophages has been highlighted by the effectiveness of pharmacological agents that modulate the functions of these cells, such as the imidazoquinolone compounds used in the treatment of HPV associated genital warts (Stanley, 2002).

HPV infection does not cause a systemic infection; it does not kill the keratinocyte and it induces no or a poor, slow, local inflammatory response. However, regression of warts is accompanied by a CD4 T cell dependent delayed type hypersensitivity response. In animal models spontaneous regression of papillomas is accompanied by massive lymphocyte infiltration of the infected epithelium (see Chapter 19). Neutralizing antibodies are found in both regressing and progressing lesions suggesting these do not influence the regression process. Humoral immunity does not appear to be as important as cell-mediated immunity in clearing HPV infection, but rather is important in protecting an individual from becoming infected. Of note, in natural infection, only around 50% of those becoming HPV DNA positive show a neutralizing antibody response. The remainder of individuals clear infection without developing a measurable antibody response (Carter et al., 2000). Humoral immunodeficiency does not increase susceptibility to development of HPV lesions (Benton et al., 1992). In contrast, those with altered CD4 T cell function, such as solid organ transplant recipients and patients infected with human immunodeficiency virus (HIV) have an increased prevalence of HPV infection and related disease such as cervical intraepithelial neoplasia. Similarly *in vitro* treatment of animals with agents that suppress their cellular immunity and reduce lymphocyte invasion have diminished regression. If cell mediated immunity fails to induce lesion regression and viral clearance then persistent infection results. It is persistent (or chronic) rather than transient HPV infection, which has been shown to correlate with the development of dysplasia.

2. Vaccines

There are potentially three stages of oncogenic anogenital HPV infection in which vaccine mediated immune intervention strategies could be implemented. The first is a prophylactic vaccine which aims to elicit a neutralizing antibody mediated inhibition of HPV binding to the cell surface receptor or a subsequent step in the cell invasion pathway. The second potential role of HPV vaccination is after infection when the virus is replicating in the basal cells of the epithelium. A therapeutic vaccine directed toward early viral proteins would aim to eliminate infection. Finally, if the virus has integrated into the host and transformed the cell, another therapeutic vaccine would be aimed at controlling growth of invasive tumors.

Currently, there are two different types of vaccines: prophylactic vaccine that would elicit an antibody response and prevent infection, and therapeutic vaccine which would induce a specific T cell mediated response leading to regression of already existing lesions. Antibodies to the viral capsid proteins L1 and L2 can neutralize extracellular virus and, in animal models, not only protect against primary infection, but also provide resistance to challenge with high dose of homologous virus. Neutralizing antibodies are also long lasting with passive transfer offering protection against experimental challenge. HPV E6 and E7 gene products are attractive candidates for a therapeutic vaccine as these are expressed in all HPV induced lesions and are required to maintain the proliferative state. Because HPV is difficult to obtain in sufficient quantities from *in vitro* culture systems, an attenuated virus vaccine is unlikely. Development of an HPV vaccine relies on expression of recombinant viral antigen in cell systems or the use of overlapping synthetic peptides corresponding to individual gene products.

Virus-like particles (VLPs) are DNA free and can be produced in yeasts, for example, *Saccharomyces cerevisiae* or higher eukaryotic cells via vaccinia virus or Semliki Forest virus vectors. The VLPs are morphologically indistinguishable from the complete native virion, yet provide a source of epitopes in a conformational correct format, typically genotype specific, and elicit an antibody response (see Chapter 18 for a full description of the properties of VLPs). VLPs are attractive candidates for a prophylactic vaccine.

A therapeutic vaccine needs to be able to prime cytotoxic T lymphocytes directed to epitopes derived from intracellularly processed viral proteins presented by MHC class I complexes. The target proteins should be ones expressed in the suprabasal epithelium rather than in the differentiated layer of HPV infected epithelium, as this would not eliminate the underlying self-renewing stem cells. E1 and E2 proteins are the major antigens expressed in the suprabasal epithelium; however these are expressed in very small quantities and may escape specific CTL attack. Ideally a therapeutic vaccine would be combined with a means of up-regulating E1 and E2 expression.

HPV16 E6 and E7 proteins are potential targets for cytotoxic T cell responses and for stimulating immunity with a therapeutic vaccine as they are constitutively expressed in high-grade squamous intraepithelial lesions (HSIL) and cancer of the cervix and are never deleted in HPV transformed cells. Continued expression of the oncogenes E6 and E7 is essential for progression to and maintenance of the malignant phenotype, so these are potential targets for a vaccine aimed at controlling HPV induced tumors. HPV induced tumor cells have devised ways of evading the hosts immune system; for example by point mutations in the CTL epitopes, loss of expression of MHC class I alleles or by down regulating endogenous peptide processing for MHC presentation. Strategies to overcome these problems include development of a vaccine which is engineered *ex vivo* to secrete cytokines or combining the target protein for example, E7 to the lysosomal and endosomal components of the MHC class II pathway using a lysosomal-associated membrane protein (LAMP) sorting signal. This last method has been used in mice with vaccinia virus containing a chimeric E7/LAMP-1 gene, which generated greater antibody response than E7 alone.

3. Animal Models

Since papillomaviruses demonstrate extreme species specificity, there are difficulties in developing animal models for HPV. Animal models are important to assess if a vaccine elicits an immune response, the nature of this response, whether it protects against primary infection or from challenge, or whether it induces regression or inhibits progression of established lesions. It is possible to overcome the problem of species specific infection by transferring HPV to animals indirectly, for example by expressing parts of the viral genome in transplantable syngeneic cells.

Animal papillomavirus models useful in the development of an HPV vaccine include bovine papillomavirus (BPV), cottontail rabbit papillomavirus (CRPV), canine oral papillomavirus (COPV) and rhesus monkey papillomavirus (RhPV). Papillomavirus specific CTL-mediated tumor regression has been demonstrated in such animal models, but their role in mediating regression in human disease has not been proven.

3.1. Bovine models

Bovine papillomavirus (BPV) infects cutaneous or mucosal sites of cattle. Like HPV there are multiple related types of BPV, but unlike HPV-associated lesions, the lesions caused by some BPVs have dermal and epidermal components (see Chapter 23). The lesions induced by BPV-4, which is a mucosatrophic virus infecting the oropharynx and oesophagus of cattle, are wholly epithelial and can progress to carcinoma. However BPV4 disappears during the malignant conversion and cocarcinogens are required for progression (Gaukroger et al., 1993). Vaccination with purified BPV or extracts of BPV related warts is able to induce prophylactic immunity. For example, animals vaccinated with BPV2 were protected from BPV2 or from the closely related BPV1, but not against BPV4 and vice versa. These findings of serotype specificity suggest that for an HPV vaccine to be widely protective it must incorporate antigens from several of the commonest infecting oncogenic viral types (Jarrett et al., 1990). Protection from infection was achieved with the L1 protein of BPV-2 (Jarrett et al., 1991), the L1 protein or the L2 protein of BPV-4 (Campo et al., 1993) and with BPV-4 VLP (Kirnbauer et al., 1996). Notably, vaccination with a single peptide from the N-terminus of L2 confers total immunity (Campo et al., 1997), with development of neutralizing antibodies (Gaukroger et al., 1996). Studies with BPV in cattle have also shown promise in the development of a therapeutic vaccine. Cattle administered BPV2 L2 vaccine developed lesions but underwent regression of these even before similar lesions had developed in the unvaccinated control group. Biopsy of the lesions confirmed massive infiltration with lymphocytes, which may have been activated by the vaccine (Jarrett et al., 1991). Similarly, vaccination with BPV-4 E7 induced early regression of mucosal papillomas (Campo et al., 1993).

3.2. Rabbit models

As in other species rabbits have a specific papillomavirus, which can cause disease. Cottontail rabbit papillomavirus (CRPV) can cause skin papillomas, which may regress spontaneously or persist and progress to squamous cell carcinoma. Animal studies in cottontail rabbits have demonstrated the protective effects of DNA L1 capsid immunized cottontail rabbits against rabbit papillomavirus skin infection. A recombinant baculovirus system expressing the L1 papillomavirus protein or L1 plus L2 was used in insect cells as a source of VLPs. Inoculated rabbits developed fewer papillomas and none developed cancer within one year of vaccination (Breitburd et al., 1995). In the rabbit model, relative long lasting protection against skin papillomas post inoculation with CRPV VLP has also been observed (Christensen et al., 1996). CRPV with mutations leading to the loss of the E1, E2, E6, E7 and L1 genes lose their papilloma inducing potential (Nasseri et al., 1989; Wu et al., 1994). This means that these are all potential vaccine targets. In addition, rabbit oral papillomavirus (ROPV) which is a mucosatrophic virus causing lesions on the oral mucosa of the domestic rabbit as well as the genital tract holds promise for future studies in animal vaccine trials (Embers et al., 2002).

3.3. Canine models

In dogs, canine oral papillomavirus can cause buccal, lingual and gingival papillomas that do not progress to malignancy. Baculovirus expressed recombinant canine oral papillomavirus (COPV) L1 spontaneously assembles into VLP and when injected intradermally into the footpad of dogs renders them completely resistant to challenge with COPV (Suzich et al., 1995). Protection is antibody mediated and can be transferred to naïve recipients. This study demonstrates that intradermal vaccination can produce mucosal surface protection. Successful immunization was strictly dependent upon native L1 protein conformation and L1 type. Other studies in canine models using intra-epithelial vaccination with COPV L1 DNA by particle-mediated delivery have confirmed cell-mediated lymphoproliferative and humoral antibody responses and protection against disease (Stanley et al., 2001). In addition, a canine model of respiratory papillomatosis using COPV has been established (Jahan-Parwar et al., 2003).

3.4. Rhesus monkey papillomavirus

A papillomavirus (RhPV1) has been isolated from penile carcinoma and other genital tract lesions such as squamous cell carcinoma of the cervix in sexually interactive rhesus monkeys (Kloster et al., 1988; Ostrow et al., 1990). This genome has been sequenced and is closely related to HPV 16 (Ostrow et al., 1991). Although this primate is closely related to humans, and would therefore seem an ideal model for HPV-associated oncogenesis and vaccine studies, it has been difficult to proceed with research due to a lack of a source of infectious virus, as well as the costs and ethical issues relating to the use of primates in research.

4. Prophylactic Vaccines

Purely prophylactic vaccines against HPV must be administered to individuals prior to exposure to the virus and must be able to stimulate a sufficiently high neutralizing antibody response, preferably at the mucosal surface to be effective. VLPs are candidates for prophylactic vaccines and studies in animals have confirmed protection against live virus challenge after immunization (Kirnbauer et al., 1996; Christensen et al., 1996).

When expressed from eukaryotic expression vectors such as recombinant vaccinia, baculovirus, Semliki Forest virus or yeast, the major capsid protein, L1 of HPV self assembles into virus like particles (VLPs). These VLPs lack the minor capsid protein L2 and viral DNA but resembles natural virions sufficiently to approximate the antigenic characteristics of wild type virus. Conformational epitopes on intact VLPs are critical to induce neutralizing (IgG1 subclass) antibody and for successful vaccination since denatured L1 protein fails to generate neutralizing antibody or protect against viral challenge. These prophylactic vaccines seem to function at the prevention of entry level of viral infection. Although L1 VLPs also induce cell mediated immunity (as measured by lymphoproliferation and cytokine assays) animal models suggest that these vaccines will not be effective therapeutically (Moore, 2004).

Phase I double blind placebo controlled studies of recombinant HPV L1 VLP vaccines in healthy adults have demonstrated almost universal seroconversion in vaccinated subjects (Brown et al., 2001; Evans et al., 2001; Harro et al., 2001). Moreover the anti VLP antibody responses have been substantially higher than that resulting from natural infection. The study by Harro demonstrated the dominant antibody response to be IgG1 subclass and was shown to be neutralizing by an HPV-16 pseudovirion neutralizing assay (Roden et al., 1996). Antibody responses when vaccine was given without adjuvant or with MF59 were dose dependent, however vaccine given with alum as the adjuvant was not dose dependent. The study by Evans demonstrated anti-VLP antibodies to be neutralizing in an *in vitro* RT-PCR-based neutralization assay (Smith et al., 1995). Interestingly, in the study by Evans, some cross-reactivity between HPV-11 and HPV16 VLP was observed, but whether this will translate into cross-protection clinically remains to be seen. Similarly, peripheral blood mononuclear cells from HPV 16 L1 VLP recipients has been shown to respond *in vitro* to L1 VLP from HPV-18, HPV 31 and HPV-53, suggesting the presence of conserved T-cell epitopes in L1 across different HPV types (Pinto et al., 2004). The clinical significance of these findings remains to be seen. In summary, the data from these studies showed that HPV VLP vaccines are safe, well tolerated and highly immunogenic generating both binding and neutralizing antibodies.

Importantly however is the question as to whether neutralizing antibodies alone are sufficient for protection or whether a strong cell mediated response is also required. Chimeric VLPs where L1 is fused to a peptide sequence of the HPV oncogenes has been developed and been shown to induce strong cytotoxic T lymphocyte responses in mice and humans (Chapter 18; Greenstone et al., 1998; Kaufmann et al., 2001; Nigland et al., 1999; Pong et al., 1998). However, in primates only weak humoral and T cell responses were induced to the chimeric component (Dale et al., 2002). This type of vaccine may have a future role in preventing lesion development post viral exposure, but improvement in responses to the chimeric component will need to occur if this type of vaccine is to be useful clinically. VLPs alone without adjuvants have also been shown to generate cell mediated responses (Marals et al., 1999).

In 2002, Koutsky et al. (2002) reported the results of a randomized double blind placebo controlled trial of a HPV-16 VLP proof of principle vaccine trial in young women. The HPV-16 L1 VLP vaccine was expressed in yeast (*Saccharomyces cerevisiae*) and adsorbed onto amorphous aluminium hydroxyphosphate sulfate adjuvant without preservative. The primary end point was persistent HPV-16 infection (HPV DNA detected in two consecutive samples at least four months apart) and the primary analysis was limited to women who were negative for HPV-16 DNA and HPV-16 antibodies at enrolment and month 7. A three dose intramuscular regimen of the HPV vaccine reduced the incidence of persistent HPV infection; all 41 cases of new HPV-16 infection, including 9 cases of HPV-16 related cervical intraepithelial neoplasia, occurred in placebo recipients (vaccine efficacy 100%). Of women who received the vaccine 99.7% seroconverted and the mean titre of antibodies was 58.7 times as high as the mean titre of antibodies in women at enrolment with serologic evidence of natural infection. This was a monovalent vaccine and what remains to be demonstrated is the sustainability of response, impact on cervical cancer and effects of a multivalent vaccine that would protect against the other common types of HPV implicated in cervical cancer. Unpublished data from a phase two bivalent VLP HPV 16,18 vaccine efficacy trial reported similar excellent results, with antibody titres developing in 99.8% of those receiving vaccine and at levels many fold greater (80-100 fold) than natural infection. Moreover vaccine recipients were not only protected from incident and persistent HPV 16,18 infection, but HPV related cervical disease. For HPV 16,18 related CIN there was complete protection. (Harper et al., 2004).

Many participants in these vaccination trials are taking the oral contraceptive pill, thereby suppressing ovulation. Of note, it has been demonstrated that specific IgG and total IgG and IgA titres in the cervical secretions decrease around the time of ovulation in women not taking the oral contraceptive pill, yet remain relatively constant in women taking such contraception (Nardelli-Haefliger et al., 2003). The significance of these changes during the menstrual cycle is unknown given the very high levels of antibody titres induced by vaccination compared with natural infection alone.

Several phase III clinical vaccine efficacy trials with multivalent VLP based vaccines are currently underway, with results due in several years. It is expected that vaccine manufacturers will file for license late in 2005, with launch of commercial sales of a multivalent vaccine in 2006.

Apart from VLP based vaccine approaches, synthetic peptides are also being studied. A peptide consisting of 13 amino acids 108-120 of HPV 16 L2, a highly conserved region among genital HPVs, contains a cross-neutralization epitope (Kawana et al., 1999). Nasal immunization of mice with a synthetic peptide corresponding to the HPV 16 L2-108/120 is capable of inducing systemic and mucosal cross-reactive antibodies to HPV 6, 11, 16 and 18.(Kawana et al., 2001) A phase I trial in healthy volunteers of this peptide administered nasally found no serious local or systemic complications and at doses of 0.5mg 80% of recipients generated anti-L2 antibodies binding to both HPV 16 and 52.(Kawana et al., 2003) Although still in its infancy compared with other prophylactic vaccines, this peptide vaccine strategy may have potential as a broad-spectrum prophylactic vaccine against genital HPV.

5. Therapeutic Vaccines

It has been estimated globally that there are around 100 million women already infected with high risk HPVs, with 5 million having persistent infections, which with time, if left untreated, may cause anogenital cancers (Frazer, 2002). Therefore, there is an urgent need to develop an effective therapeutic vaccine.

The spectrum of endpoints for a therapeutic HPV vaccine may include eradication of genital warts, prevention of progression of low or high-grade cervical squamous intraepithelial lesions or elimination of residual malignant cells in patients with cervical cancer. Therapeutic vaccines are aimed at eradicating or reducing already infected cells. Once infection is established it is unlikely that an antibody based immune response alone will be sufficient to eradicate infected cells. Therefore a basic principle of therapeutic HPV vaccines is to achieve upregulation of MHC Class I alleles and stimulate cytotoxic T lymphocytes as effector cells that will recognize and target infected cells. Immunization with early proteins of HPV in animal models may be effective as a therapeutic strategy. It is now well recognized that HPV associated tumors express E6 and E7 oncoproteins, thus providing an ideal target to direct cytotoxic T lymphocytes. Vaccination directed at the high-risk E6 and E7 oncoproteins is an attractive alternative. One vaccine strategy is to induce specific CD8 cytotoxic T lymphocytes to the specific peptide epitopes of oncogenic HPV E6 and E7 proteins.

Numerous approaches have been used to generate HPV-specific cytotoxic T lymphocyte responses and trialled in animal models. Whilst it is recognized that CD8 T cells are potent mediators of viral clearance and tumour eradication, it is now clear that CD4 T-helper cells are required to sustain effective CD8 T cell activity. Therefore it is suggested that therapeutic vaccines should induce both strong CD4 and CD8 T cell reactivity, preferably over a long period of time. For example, the expression of HPV 16 E7 fused to a nonhaemolytic portion of the *Listeria monocytogenes* virulence factor, listeriolysin O (LLO), using vaccinia virus as a live vector, has induced an immune response capable of causing regression of established immortalized tumors in mice (Lamikanra et al., 2001). The basis of this is that *Listeria monocytogenes* naturally infects antigen-presenting cells and delivers antigen directly to both MHC Class I and MHC Class II pathways of antigen presentation. Fifty percent of the mice given this vaccine remained tumor free two months after tumor inoculation. Another *Listeria monocytogenes* vaccine, Lm-ActA-E7, has demonstrated the ability to generate antitumor T-cell responses and cause regression of HPV positive tumors in a murine model (Sewell et al., 2004). Similarly, vaccination of mice with a Venezuelan equine encephalitis replicon particles expressing E7 induced CD8 T cell response and prevented tumor development in all of the mice and effectively eliminated 7-day established tumors in 67% of tumor bearing mice (Velders et al., 2001). Recombinant Semliki Forest virus expressing HPV-16 oncoproteins E6 and E7 has also been described as inducing specific cytotoxic T lymphocyte and anti tumor responses (Daemen et al., 2003).

In the rabbit model, infection of the skin with cottontail rabbit papillomavirus induces benign papillomas or warts, which may progress to carcinoma over time. Immunisation of these rabbits with the nonstructural early proteins E1 and E2 induces a T helper response, leading to development of fewer papillomas and more rapid regression than non-vaccinated rabbits (Selvakumar et al., 1995). The rabbit model has also been used to assess gene gun-mediated intracutaneous vaccination with papillomavirus E6 and E7 genes. Papillomas on the rabbits were allowed to grow for three months without any treatment intervention. An immunization protocol administering DNA plasmids encoding either the E6 or E7 gene or vector only was instituted at three months and the rabbits followed for two years. There was no delay in cancer development in the E6-vaccinated rabbits but there was some delay in cancer development in the E7-vaccinated group. Also, the E7-vaccinated group with primary skin carcinomas had fewer metastases compared with the control group (Han et al., 2002). This same group of researchers has also vaccinated rabbits with a combination of gene gun-mediated intracutaneous DNA plasmids encoding cottontail rabbit papillomavirus E1, E2, E6 and E7. The rabbits vaccinated with a combination of the genes were protected from high dose viral challenge (Han et al., 1999). In rabbits with papillomavirus-induced skin lesions prior to the administration of this vaccine, DNA vaccination prevented or delayed the development of carcinoma (Han et al., 2000).

There is a paucity of information on the clinical efficacy of a therapeutic HPV vaccine on benign HPV lesions such as anogenital warts. Anogenital warts are a common clinical problem and associated with low risk HPV types, most commonly HPV 6 and 11. In an uncontrolled trial of HPV6b VLP L1 protein vaccine, subjects developed VLP specific antibodies, a delayed type hypersensitivity response to L1 and 76% of subjects demonstrated complete regression of genital warts over a 20-week observation period (Zhang et al., 2000). A retrospective review of responses in 14 patients enrolled in an open-label trial examining the efficacy and immunogenicity of HPV16E7 protein/heat shock protein (hsp) 65 vaccine reported at week 24 post vaccination that 3 of the 14 had complete resolution of their warts, 10 reported a reduction in size and 1 reported an increase in size of warts (Goldstone et al., 2002). As the immune response elicited to HPV is type-specific, the therapeutic effect noted here might well be a result of innate immunity induced by the heat shock protein. In a different study a fusion protein vaccine consisting of HPV 6 L2E7 was administered to 27 subjects with genital warts in an open label study. Five subjects completely cleared their warts within eight weeks. The remainder of the group were offered conventional therapy. Thirteen of the 27 subjects who had their warts cleared either by vaccination alone or by conventional therapy did not have any recurrence (Lacey et al., 1999).

6. Clinical Studies of Human Papillomavirus Vaccines in Pre-Invasive and Invasive Cancer

Cervical cancer is the second commonest cause of cancer death in women worldwide. HPV DNA has been identified in more than 99% of tumour biopsy specimens and oncogenic HPVs are the aetiological agent of cervical cancer in a multistage process together with various cofactors. The high-risk types 16 and 18 are associated with over 70% of cervical cancers. Detection and treatment of the precursor lesions (high-grade dysplasias or cervical intraepithelial neoplasia grade three) form the basis of cervical screening programs. However, 80% of cancer deaths worldwide occur in countries with inadequate resources to establish effective cervical cytology Papanicolaou screening (Pap test) programs or to treat invasive disease. Because of this there is a need for alternative strategies to reduce the global burden of disease and mortality.

When considering therapeutic HPV vaccines for cervical dysplasia or carcinoma, it is important to distinguish the state of the virus in low-grade squamous intraepithelial lesions (LSIL) as compared to high-grade intraepithelial lesions (HSIL). In low-grade lesions all early proteins are expressed, the lesions are genetically stable and in most, with time, regress with an appropriate host immune response. An effective therapeutic vaccine should result in lesion clearance without recurrence in an immunocompetent person. In many high-grade lesions and in cervical cancers however, viral gene expression is dysregulated, with constitutive expression of the oncogenes E6 and E7 essential for progression to, and maintenance of, malignancy. Therefore, there are really only two antigenic targets for a therapeutic vaccine against high-grade lesions or malignancy, although there may be more for low-grade lesions.

HPV 16 is most commonly associated with HSIL and found in 50% of all cervical cancers worldwide (Luxton and Shepherd, 2001). HPV 16 and 18 E6 and E7 proteins have been the focus of recent clinical vaccine trials.

One of the problems inherent to the development of a therapeutic vaccine for cervical cancer is the genetic instability of the cancer cells and the potential for immune escape mechanisms to evolve. The extent to which this occurs may lead to progression despite a response to the vaccine or even partial regression of the lesion.

6.1. Peptide and protein based vaccines

Peptide and full-length viral E6/E7 protein-based vaccine candidates are attractive as they are relatively inexpensive to synthesize and are relatively non-toxic. To be effective though, these vaccines require identification of the HLA-restricted peptide epitope of HPV-16 and HPV-18 E6 and E7. Clinical trials have reported mixed results to date. One trial was in 15 patients with HPV-16 positive cervical carcinoma and who received vaccination with synthetic peptides representing two HPV 16 E7 encoded, HLA-A* 0201 restricted cytotoxic T lymphocyte epitopes and a pan-HLA-DR-binding T-helper epitope in adjuvant (Ressing et al., 2000). No cytotoxic T lymphocyte response against the HPV 16 E7 peptide was detectable and there was no evidence of clinical benefit. A similar trial in 19 cervical cancer patients with advanced disease used two HPV16 E7 peptides and one helper peptide emulsified in adjuvant. Again, there was little evidence of clinical benefit (van Driel et al., 1999). However, in a further trial, women with high-grade cervical and vulvar intraepithelial neoplasia who were HPV-16 positive were treated with escalating doses of a vaccine containing a 9-amino acid peptide encoded by the E7 gene emulsified with incomplete Freund's adjuvant (Muderspach et al., 2000). Only three of the 18 patients cleared their dysplasia after vaccine, but an increase in S100+ dendritic cells was noted in 6 out of 6 patients tested. Virological assays showed two thirds of patients cleared the viral DNA from cervical scrapings by the fourth vaccine injection, but all biopsy samples were still positive by *in situ* hybridization after vaccination. Three patients cleared their dysplasia after vaccine and six patients had partial regression as measured by colposcopy. Overall, the peptide vaccines seem well tolerated by patients, but further work is required to improve the immune response generated.

To effectively utilize peptides, the HLA genotype of the patient must be known. To overcome this barrier, investigators have used full-length E6 or E7 proteins or fused them with other proteins. A small study of five women with severe cervical dysplasia, HPV 16 positive, were injected three times with HPV16 E7 full length protein linked to the first 1087 amino acids of *Haemophilus influenzae* protein D, adjuvant ASO2B. The injections were well tolerated, and despite limited regression of the lesions, massive CD4 and CD8 T cell lymphocytic infiltration occurred (Simon et al., 2003). A further Phase I/II trial of HPV16 E7 fusion protein with the same adjuvant demonstrated increased interferon secretion in the vaccinated group compared with placebo and generation of long-term T cell immunity and IgG production (Hallez et al., 2004).

Heat shock protein of *Mycobacterium bovis* bacilli Calmette-Guerin (BCG) is a potent activator of dendritic cells and inducer of T cell responses. Linkage of this heat shock protein to E7 as a fusion protein is a potent immunotherapeutic in murine models and is under investigation currently in patients with HPV-16 positive high grade squamous intraepithelial lesions (Bosch et al., 2001). BCG heat shock protein has been fused with E7 protein from HPV-16 and studied in an open-label trial in patients with persistent HSIL. Also, HPV 16 E6 E7 fusion protein joined with a saponin-based adjuvant ISCOMATRIX has been trialled in healthy volunteers in a phase I observer-blind randomized, placebo-controlled dose escalation study. This therapeutic was well tolerated and induced strong cellular and humoral responses in a high proportion of vaccinees, and viral load overall decreased (Barnden et al., 2004). In addition, results from an open-label trial of recombinant heat shock protein from BCG and HPV E7 protein have demonstrated benefits in pediatric patients with recurrent respiratory papillomatosis insofar as it reduces the frequency of required surgery (Derkay et al., 2004). Phase III trials are required.

6.2. Viral-vector vaccines

There is growing experience with viral-vector HPV vaccines in animal models. The use of viral vectors is an effective way of introducing targeted genes to the host; the antigens can be spliced into recombinant expression vectors, allowing for their increased cellular production, processing and presentation by an antigen presenting cell for generation of MHC Class I restricted cytotoxic T lymphocyte responses. Vaccinia vaccine has a long track record of safety having been employed in large-scale smallpox vaccination programs. It also has a lytic life cycle minimizing any potential risk associated with oncogenic insertion.

Phase I/II trials of a recombinant vaccinia virus expressing HPV 16, 18 E6 and E7 proteins (TA-HPV) have been published (Borysiewicz et al., 1996; Kaufmann et al., 2002). These studies assessed safety and immunogenicity in eight patients with late stage cervical cancer and twenty-nine patients with stage Ib or IIa cervical cancer. Vaccination was well tolerated. Eleven patients developed HPV specific antibody response and HPV specific CTLs were found in five patients. The HPV CTL response was seen at only one point in time and was not sustained. Antibody alone is unlikely to be sufficient to control disease so further development is needed to utilize adjuvants or delivery systems to optimize Th1-type CTL responses, rather than Th2-type responses. Because the vaccine was largely used in an adjuvant setting and numbers were small, clinical effectiveness could not be determined.

Another phase II study of vaccinia-expressed HPV 16 and 18 E6 and E7 was trialled in 12 women aged 42 to 54 years with high-grade HPV positive vulval or vaginal intraepithelial neoplasia of up to 15 years duration. Five of the 12 (42%) patients had 50% reduction in lesion diameter over 24 weeks, with one patient having complete regression. Overall, 83% of women showed some improvement with an average decrease in lesion size of 40% (Baldwin et al., 2003). A similar vaccine has also been studied in 18 women with HPV 16-positive high-grade vulval intraepithelial neoplasia. Follow up was for 24 weeks with lesion measurement and histological analysis. Eight patients demonstrated a reduction in lesion size of at least 50% and a further four patients reported improvement in clinical symptoms. Interestingly, viral load was reduced or eliminated in 75% of responders, but also in 60% of non-responders (Davidson et al., 2003). Non-responders demonstrated an increase in CD4 and CD8 cells post vaccination, but at lower levels overall compared with responders.

The modified virus Ankara (MVA)-E2 recombinant vaccinia virus has been trialled in 36 women with cervical intraepithelial neoplasia (CIN 1, 2 and 3 lesions) associated with HPV infection (Corona Gutierrez et al., 2004). The vaccine was injected directly into the uterus once every week over 6 weeks. Reduction of lesions was monitored by colposcopy and cytologic analysis. Immune responses were assessed by measuring antibody titres against MVA E2 virus and the E2 protein and by cytotoxic activity against cancer cells bearing HPV DNA. Thirty-four of the 36 patients showed complete elimination of the precancerous lesions after treatment with the MVA E2 vaccine. All patients developed antibodies against the MVA E2 vaccine and vaccination generated a specific cytotoxic response against HPV transformed cells. In addition, half of the vaccinated patients showed no evidence of HPV infection after vaccination. The other half demonstrated persistent HPV DNA, but at a lower viral load than in the unvaccinated group. These results form an exciting prospect for a locally delivered potentially therapeutic vaccine.

Other viral delivery systems such as recombinant adenovirus or alphavirus have been described in animal models, but are yet to be reported in human clinical trials.

6.3. DNA vaccines

Immunization with naked DNA is another strategy for vaccine development. Plasmid DNA encoding HPV non-structural proteins E6 and E7 can be given to a host to stimulate an antigen-specific humoral and cellular immune response. By circumventing vectors, DNA vaccines may be a cost effective and safe alternative for a therapeutic vaccine. To date, the majority of work with HPV DNA vaccines has been performed in animal models, the focus being to find techniques that enhance immunogenicity and promote antigen processing by intracellular targeting. For example, HPV 16 E7 has been linked to the *Mycobacterium tuberculosis* heat shock protein 70 (HSP70) and administered to mice as a DNA vaccine (Trimble et al., 2003). This vaccine generated an increase in E7-specific CD8+ T cell precursors and a strong anti-tumor effect against E7-expressing tumor in vaccinated mice. It has subsequently been found that administration of this DNA vaccine via gene gun, compared with syringe or biojector administration, generated the highest number of E7-specific CD8+ T cell response and required the least comparative dose. This has led to proposals for Phase I/II clinical trials in patients with HPV-16 associated HSIL of the cervix and in patients with advanced HPV-associated head and neck squamous cell carcinoma (Trimble et al., 2003).

Plasmid DNA encapsulated in biodegradable polymer microparticles has been studied in a phase I trial involving 12 patients with HPV 16 anal infection and dysplasia and a HLA-A2 haplotype. The investigational agent was well tolerated in all subjects and 25% experienced partial histological responses (Klencke et al., 2002). This is encouraging as it suggests that although most work with therapeutic vaccines has been done to date in the area of cervical dysplasia and carcinoma, there is potential for application to other HPV associated malignancies at other sites.

In a double blind randomized placebo controlled trial in women with cervical intraepithelial neoplasia (CIN) 2/3, of plasmid-DNA-encoding fragments derived from E6 and E7 proteins of HPV 16 and 18, resolution of CIN 2/3 was promoted in women less than 25 years of age. When all ages were considered, the proportion of subjects who resolved was higher in the treatment arm but this was not statistically significant (Garcia et al., 2004).

Other techniques have been employed in an attempt to direct the target antigen (such as HPV 16 E7) to the desired intracellular compartment in order to enhance antigen processing and presentation to T cells. Some of these strategies have included using lysosome-associated membrane protein (LAMP-1), calreticulin (CRT) and the translocation domain (dII) of *Pseudomonas aeruginosa* exotoxin A (ETA).(Kim et al., 2004) Vaccination with DNA encoding HPV E7 antigen linked to any of these molecules leads to a significant enhancement of E7-specific CD8+ T cell immune response, the most potent being with CRT, even in the setting of down regulated MHC class I expression (Kim et al., 2004). No human clinical trials have been reported using this DNA vaccine to date.

6.4. DC-based vaccines

Dendritic cells are important for induction and regulation of immune response and are an alternative way of generating large numbers of autologous antigen presenting cells. Dendritic cells are unique as they are capable of priming naïve T cells. However they have a limited life span, hindering their long-term ability to prime antigen-specific T cells. Autologous full length protein antigen-loaded dendritic cells have been trialled (Ferrara et al., 2003; Nonn et al., 2003). Autologous dendritic cells loaded with HPV E7 protein can induce T cell responses in healthy individuals by *in vitro* stimulation and evoke responses in tumor infiltrating lymphocyte populations of cervical cancer patients (Nonn et al., 2003). When administered to 15 patients with advanced cervical cancer, the vaccine was well tolerated with no local or systemic side effects recorded, but no objective clinical response was observed (Ferrara et al., 2003). An interesting case report in the literature describes a woman with metastases secondary to recurrent HPV-18 associated cervical cancer given subcutaneously administered autologous mature, monocyte-derived dendritic cells pulsed with HPV-18 E7 oncoprotein (Santin et al., 2002). Serial CT scans of the lung metastases showed no progression over the 13 months of therapy. Compared with biopsy prior to immunotherapy, after 10 months of injections, fine needle biopsy revealed fibrosis and macrophage infiltration, but no viable tumor cells.

Due to the limited life span of dendritic cells, attempts to prolong the dendritic cell life have included coadministering DNA encoding antigen (HPV 16 E7) with DNA encoding inhibitors of apoptosis (Kim et al., 2003). Vaccines combining these with intracellular targeting strategies has been trialled in mice, but not yet reported in humans. They have demonstrated increased E7 specific CD8+ T cells (up to 13 fold increase), increased CD8+ T cell functional avidity, increased E7-specific CD4+ Th1 cell response, enhanced tumor treatment ability and strong long-term tumor protection when compared with mice vaccinated without the DNA encoding inhibitors of apoptosis (Kim et al., 2003).

Activation and maturation of dendritic cells is regulated by Toll-like receptors. These receptors recognize molecular patterns and can be mimicked by adjuvants such as CpG oligonucleotide sequences (mimics bacterial DNA), Montanide (LPS), polyI:C (viral dsRNA) or imidazochinolines (immune response modifiers) (Schreckenberger and Kaufmann, 2004). Dendritic cells respond to CpG DNA by producing a broad range of cytokines and chemokines. The overall effect is activation of cells involved in the innate immune response. Animal models suggest that this may be beneficial in cancer immunotherapy and early clinical trials suggest that CpG DNA can be safely administered to humans (Wooldridge et al., 2003), but further research is required.

6.5. Immune escape

One of the challenges of therapeutic vaccination is the potential for cervical cancer to develop immune escape mechanisms. Although humoral and cell mediated immune responses are produced by vaccines, they may not work because the tumour escapes the immune response. The progression of cancer may well reflect not only the transforming potential of the particular HPV type but the enhanced ability to avoid the immune response in allowing the virus to persist longer and have a high probability of neoplastic progression (O'Brien and Campo, 2003). Down regulation or loss of MHC class I occurs in almost all cervical cancers (Garrido et al., 1997). This means that even if a vaccine is able to produce CTL specific response it may not be effective if the peptide targets are not displayed on the cells. MHC class I can be up regulated by IFNγ and so combination therapies may need to be designed. Similarly, another proposed mechanism of immune escape is the modulation of the cytokine milieu, such that the dominant cytokine response is a Th2-type rather than Th1. Therefore vaccine design must take into consideration an assessment of the class I status and the antigen presenting capability of the patient's cancer cells.

Novel, new vectors are being developed for HPV vaccines. Freeze-dried HPV-16 L1 transgenic tomato has been trialled in mice and offered a stable, economical and easily administered vaccine (Rose et al., 2004). Plasmids carrying modified E7 gene fused to genes from plant viruses has also been trialled in mice proving to elicit an immune response and regression of neoplasia (Franconi et al., 2004). In addition, vectors expressing chimeric hepatitis B surface antigen fused to a HPV-16 E7 protein, have been developed and administered to mice leading to both humoral and cellular immune response and protecting mice against isogenic tumor cells loaded with E7 protein (Baez-Asua et al., 2004).

7. Challenges to a HPV Vaccine

There are many challenges to implementation of an HPV vaccination program. The most likely vaccines to be licensed in the near future are multivalent prophylactic vaccines directed against the high-risk HPV types (16 and 18) and the low risk HPV types (6 and 11) or multiple high-risk types. The main groups these vaccines have been trialled in to date are young females aged 16 to 24 years. There is a need to evaluate the immune response in younger people of both genders. However, as this is a prophylactic vaccine it would be preferable to vaccinate, and therefore produce a protective immune response, prior to HPV exposure, which is most likely to occur after sexual debut. Therefore, the most appropriate age group for vaccination may be between 10-12 years, prior to becoming sexually active. Depending on the long term durability of the vaccine, it may even be considered as part of the childhood immunisation programme. It is likely that some members of the public may be uncomfortable vaccinating young girls against a sexually transmitted infection and education regarding this and the important relationship to cervical dysplasia and cancer will need to be carefully addressed.

Another challenge to a vaccination program if this age group were targeted is whether a school-based or clinic-based program should be recommended. A school-based immunization program would be able to provide an infrastructure in which to vaccinate adolescents and incorporating a three dose HPV vaccination schedule into the academic year would likely increase compliance compared with the inherent problems of adolescents booking appointments at a health care facility and attending for three scheduled visits. It is anticipated there may be some social and political opposition to a school-based immunization program against a sexually transmitted infection.

To date the benefits of an HPV vaccine have targeted women. We await data from trials regarding efficacy in older women. Another important question to consider with licensing of a vaccine is whether males should be included in such a program. Although HPV is most commonly associated with cervical cancer, it also plays a role in other anogenital cancers and dysplasias such as of the vulva, anus and penis, as well as a subset of head and neck cancers (Herrero et al. 2003). The precedent of vaccinating males as well as females, even though the greatest benefit may be demonstrable in women, has been established in other vaccine preventable infectious diseases such as rubella virus. In order to eliminate congenital rubella syndrome many countries have an established policy of also vaccinating males (Slater et al., 1996). Therefore reducing HPV infection and carriage in males may actually help reduce the burden of disease in women, particularly if unvaccinated. Whether in fact there will be gender specific differences in efficacy is yet to be defined, as the recent results of the herpes simplex glycoprotein D vaccination demonstrated (Stanberry et al., 2002). Mucosal antibodies in genital secretions are likely to be important in prevention of HPV transmission. Consequently with the keratinized nature of the genital skin in males, such protection may not be shown.

As already mentioned, the greatest burden of disease from cervical cancer is in the developing world where incidence and mortality rates are high. With any vaccine development it is important to consider its application to this population. In the developing world access to cervical screening programs is poor (often around 5% or nonexistent) so the prospect of an effective vaccine holds great promise. The challenges however would be to manufacture the vaccine at a cost which is affordable to resource-poor countries and to provide infrastructure, which enables women in these countries to access the vaccine. In the developing world, vaccines need to be inexpensive and easily delivered. The potential need for three injections may limit its use too, so future research should analyze whether delivery of the vaccine can be improved, with the number of injections reduced or development of a non-injectable form, preferably without the need for a tight cold chain. Alternatively, other modes of administration such as oral administration should be further explored. To date, a study in mice has shown that HPV 16 and HPV 18 VLPs are immunogenic when administered orally and that coadministering these antigens with *Escherichia coli* heat-labile enterotoxin mutant R192G or CpG DNA significantly improves anti-VLP humoral responses in peripheral blood and in genital mucosal secretions (Gerber et al., 2001).

If one could still induce a neutralizing antibody response without VLP assembly then cheaper and perhaps more stable vaccines could be developed. Capsomeres formed from L1 protein of COPV in *E. coli* have reacted with conformation-specific anti-COPV antibodies and protected immunised animals from high-dose viral challenge (Yuan et al., 2001).

The duration of efficacy of HPV vaccination is not currently known. This information will be important in making future decisions regarding the need for boosters and/or recommended frequency of Pap tests over a woman's lifetime. In general, if an effective vaccine reduces the incidence of CIN and cancer, any screening test will have improved negative predictive value and worse positive predictive value. This, along with consideration of cost, may lead to changes in recommendations for age to start and stop screening, as well as the frequency.

Induction of simultaneous protection against many HPV types, with multivalent vaccines, is currently in phase III clinical trials. The limit to number of antigens, which can be included in the one vaccine, is currently unknown. It is also unclear whether an individual with past infection to one type may have limited response to other HPV types contained in the vaccine.

More than 20 types of papillomavirus have been identified and associated with cervical cancer. Data to date suggests that vaccination against one genotype will not protect against infection by another. In addition, the clinical effect of vaccine in women already infected or in women already with cervical dysplasia is not well understood. Therefore, even with the licensing of a HPV vaccine other forms of screening which currently exist such as the Pap test need to continue. Over time however it may be that the frequency of Pap tests in vaccinated women can be reduced and with the anticipated reduction in frequency of abnormal Pap smears, a substantial burden from the health care system may be reduced. Although it seems worldwide that HPV 16 and 18 consistently account for around 70% of cervical cancers, the prevalence of genotypes thereafter may well have geographical variation creating a further challenge for vaccine manufacturers. In a recent meta-analysis of HPV prevalence, it was found that HPV type 16 and 18 were most prevalent in all countries studied, but subsequent genotypes varied in prevalence depending on region. For example in Asia, after HPV type 16 and 18, HPV type 58 was the next most common, although in Africa HPV type 45 was the next most common and HPV type 33 in Europe (Clifford et al., 2003).

Other questions yet to be answered include whether vaccination of women with prevalent high risk HPVs would have their neutralising antibodies boosted resulting in reduction in auto- inoculation and consequent viral persistence, and reduction in transmission to males by neutralising shed virus. Central to all the challenges facing a HPV vaccine is education of the community. To achieve widespread societal acceptance there must be an improvement in recognition of the frequency of HPV infection, the ubiquity of infection and its associated diseases including cancer and an effort to de-stigmatise and demystify the area. Another interesting challenge will be whether to market the vaccine as an "anti-cancer" or an "anti-sexually transmitted infection" vaccine.

8. Cost Benefit of a HPV Vaccine

In terms of clinical application, a prophylactic HPV vaccine to prevent cervical cancer is the most advanced (both monovalent against type 16 and polyvalent against type 16/18 with or without types 6 and 11 currently undergoing phase III clinical trials). Cervical cancer is the second most common cause of cancer in women worldwide. Approximately 470,000 cases of cervical cancer and 235,000 deaths occurred worldwide in 2000.

In countries with high quality Pap screening programs with broad coverage of the at risk population, the incidence and mortality rates from cervical cancer have fallen significantly. Unfortunately in developing countries where much of the burden of disease lies, access to screening programs is poor, screening programs are inadequate or nonexistent. Prophylactic vaccination programs therefore may hold the potential for dramatic improvements in incidence of cervical cancer.

When modelling cost effectiveness for vaccination programs it is critical to have accurate information such as incidence of disease, natural history, cost of vaccination, cost of current treatment, efficacy of vaccine, likely uptake rate of vaccine and durability of protection from vaccination.

Before licensing a vaccine it is important to assess the cost effectiveness. Modelling of cost benefit has been performed and applied to populations in the developed world (Sanders and Taira, 2003). One such analysis considered a hypothetical cohort of adolescent girls in the United States aged 12 years. This age was selected based on previous mathematical modelling and was thought to cover 97% of females prior to their sexual debut. This analysis assumed a universal vaccination strategy of a vaccine specific to high-risk types of HPV (as are in current development in clinical vaccine trials) rather than targeting vaccination to "high risk" groups because of the broad range of risk factors and the difficulty identifying these behaviours. The model followed the girls over their lifetime and included probabilities of occurrence and progression of HPV, of squamous intraepithelial lesions and of cervical cancer as well as the probability of death, costs and quality of life associated with various health states. Wherever possible, probability estimates were based on previously reported high quality studies in the literature. Other assumptions made in this model included 70% uptake rate, that unvaccinated girls would receive the current standard of care and vaccinated girls would continue to have Pap tests at the same interval. In addition, as the longevity of protection from HPV vaccination is unknown, this model assumed that successful vaccination (three injections) would confer immunity for 10 years and that repeated boosters every 10 years would be required to maintain the vaccine's efficacy. Adjustments for quality of life associated with current health, HPV, squamous intraepithelial lesions and with cervical cancer and its treatment were incorporated.

This analysis found that a prophylactic vaccine against high-risk HPV types is more expensive than current practice, but results in greater quality-adjusted life expectancy. HPV vaccination of 12-year-old girls based on this model would improve their life expectancy by 2.8 days or 4.0 quality-adjusted life days at a cost of $246 relative to current practice (incremental cost effectiveness $22,755 per quality adjusted life year or QALY). This compares favourably to vaccinations against measles, mumps, rubella and pertussis each of which save 2.7, 3.0, 0.3 and 3.3 life days respectively (Hinman and Koplan, 1984; White et al., 1985; Wright and Weinstein, 1998). Vaccinating this cohort would avert greater than 220,000 cases of HPV, greater than 112,000 cases of SIL, more than 3,000 cases of cervical cancer and more than 1,300 cervical-cancer deaths over the cohorts lifetime. The number needed to vaccinate to prevent one case of cervical cancer would be 600. The number needed to vaccinate to prevent one case of cervical cancer would fall to 250 if the current vaccine could provide lifelong immunity negating the need for a booster. If the age of vaccination were changed to 15 years vaccination would cost an additional $40,440 per additional QALY gained. In addition, a strategy in which vaccinated women were able to have less frequent Pap tests would also reduce costs of such a program. The results were also sensitive to changes in other variables, including varying the underlying incidence of HPV. Sensitivity analyses found that the HPV vaccine would be cost effective even assuming vaccine efficacy as low as 40%, or that boosters were required every 3 years (Sanders and Taira, 2003).

Other mathematical models have been developed to examine the potential health and economic effects of an HPV vaccine in the setting of existing screening in the United States(Kulasingam and Myers, 2003). In the study by Kulasingam and Myers, a Markov model was used to estimate the lifetime (age 12-85) costs and life expectancy of a hypothetical cohort of women screened for cervical cancer. Three strategies were compared: (1) vaccination only; (2) conventional cytological screening only; and (3) vaccination followed by screening. Assumptions in this model include a vaccine targeting a defined proportion of high-risk HPV types (targeted against 70% of oncogenic HPV types including HPV 16 and 18), 90% vaccine efficacy, 10 year duration of protective effect, 100% compliance with primary screening, follow up and treatment, a sensitivity of 55.6% and specificity of 95.7% of conventional cytology for detection of CIN 2-3, a 100% sensitivity and specificity of colposcopy for CIN 2-3 and women with treated CIN received annual screening for the remainder of their lifetime or until hysterectomy for other indications. Screening intervals of 1,2,3 and 5 years and the starting ages for screening of 18, 22, 24, 26 and 30 years were chosen for two of the strategies (conventional cytological screening only and vaccination followed by screening).

With this mathematical modelling, vaccination only or adding vaccination to screening conducted every 3 and 5 years was not cost effective. The most cost effective strategy was vaccination plus biennial screening delayed until age 24 years (US$44,889 incremental cost per life year saved) compared with 3 yearly screening only beginning age 18 years. A key to overall vaccine effectiveness in this analysis is adequate protection during the years of peak oncogenic HPV incidence. The cost effectiveness of vaccination plus delayed screening was highly sensitive to age of vaccination, duration of vaccine efficacy and cost of vaccination.

Other authors have performed similar studies (Goldie et al., 2004). Goldie's model showed that the most effective strategy with an incremental cost-effectiveness ratio of less than US$60,000 per QALY is one combining vaccination at age twelve years with three yearly Pap tests beginning at age 25. This strategy would reduce the absolute lifetime risk of cervical cancer by 94% compared with no intervention. The next best strategy was vaccination aged 12 with Pap tests every 5 years commencing age 21. These results were sensitive to a change in assumptions such as underlying patterns of cervical cancer screening, duration of vaccine efficacy and natural history of HPV infection in older women. In developed countries it would appear that a program of vaccination that allows a later age of conventional cytological screening initiation and a less frequent screening interval is likely to be cost effective.

Projected clinical benefits, cost effectiveness and implementation in developing countries require analysis. Even with successful prophylactic vaccination it will take at least 10 years before vaccinated young girls reach the age of joining screening programmes and therefore begin to have an impact on screening activity. Consequently, with vaccination a diminishing number of women would remain at risk. It would be envisaged that for vaccinated women there would be less intense screening (increased interval and perhaps starting as a later age) and result in reduction of cervical abnormalities of all grades. This should result in budgetary savings, which should offset the costs of vaccination.

9. Conclusion

There has been much work in the past decade in the area of HPV vaccines. The development of VLP based HPV prophylactic vaccines offers exciting possibilities in the very near future in primary prevention of cervical cancer. With high uptake these vaccines have the potential to prevent up to 70% of cervical cancers. Prior to introduction however, consideration needs to be given to issues such as public education and debate to demystify HPV infection and increase the likelihood of successful implementation. Other important issues to address will be how to achieve a successful vaccination program in the developing world where the potential for greatest public health impact remains. Numerous therapeutic HPV vaccines are also in development, although lag behind in their demonstrable clinical benefit. It is anticipated in the future we will see a vaccine available to the community, which protects against the most common viral sexually transmitted infection and the complications of HPV infection such as anogenital dysplasia and cancer.

References

Baez-Asua, A., Juarez, V., Pasolli, A., zur Hausen, H., and Cid-Arregui, A. (2004). Genetic immunization to HPV-16 E7 with herpes simplex and adenovirus vectors expressing hepatitis B surface antigen E7 fusion proteins. In 21st International Papillomavirus Conference (Mexico City).

Baldwin, P. J., van der Burg, S. H., Boswell, C. M., Offringa, R., Hickling, J. K., Dobson, J., Roberts, J. S., Latimer, J. A., Moseley, R. P., Coleman, N., Stanley, M. A., and Sterling, J. C. (2003). Vaccinia-expressed human papillomavirus 16 and 18 e6 and e7 as a therapeutic vaccination for vulval and vaginal intraepithelial neoplasia. Clin. Cancer Res. *9*, 5205-5213.

Barnden, M., Papalia, L., Davis, R., Millar, C., Nikolic, N., O'Brien, K., Wang, L., McKenzie, A., White, O., Frazer, I., and Maraskovsky, E. (2004). Antigen-specific CD4+ and CD8+ T cell responses to HPV 16 E6E7 ISCOMATRIX vaccine in healthy volunteers. In 21st International Papillomavirus Conference (Mexico City).

Benton, C., Shahidullah, H., and Hunter, J. A. A. (1992). Human papilloma virus in the immunosuppressed. Papillomavirus *3*, 23-26.

Borysiewicz, L. K., Fiander, A., Nimako, M., Man, S., Westmoreland, D., Evans, A. S., M., A., and et, a. l. (1996). A recombinant vaccinia virus encoding human papilloma virus types 16 and 18 E6 and E7 proteins as immunotherapy for cervical cancer. Lancet *347*, 1523-1527.

Bosch, F. X., Rohan, T., Schneider, A., Frazer, I., Pfister, H., Castelisague, X., de Sanjose, S., Moreno, V., and et, a. l. (2001). Papilloma research update: Highlights of the Barcelona HPV 2000 International papilloma conference. J. Clin. Pathol. *64*, 163-175.

Breitburd, F., Kirnbauer, R., Hubbert, N. L., Nonnenmacher, B., Trin-Dinh-Desmarquet, C., Orth, G., Schiller, J. T., and Lowy, D. R. (1995). Immunization with viruslike particles from cottontail rabbit papillomavirus (CRPV) can protect against experimental CRPV infection. J. Virol. *69*, 3959-3963.

Brown, D. R., Bryan, J. T., Schroeder, J. M., Robinson, T. S., Fife, K. H., Wheeler, C. M., Barr, E., Smith, P. R., and et, a. l. (2001). Neutralisation of human papillomavirus type 11(HPV 11) by serum from women vaccinated with yeast derived HPV-11 L1 virus like particles: correlation with competitive radioimmunoassay titer. J. Infect. Dis. *184*, 1183-1186.

Campo, M.S., Jarrett, W.F.H., Grindlay, G.J., Chandrachud, L.M., McGarvie, G.M. and O'Neil, B.W. (1993). Prophylactic and therapeutic vaccination against a mucosal papillomavirus. J. Gen. Virol. 74, 945-953.

Campo, M.S., O'Neil, B.W., Grindlay, G.J., Curtis, F., Knowles, G. and Chandrachud, L.M. (1997) A peptide encoding a B-cell epitope from the N-terminus of the capsid protein L2 of bovine papillomavirus-4 prevents disease. Virology, 234, 261-266.

Carter, J. J., Koutsky, L. A., Hughes, J. P., Lee, S. K., Kuypers, J., Kiviat, N., and Galloway, D. A. (2000). Comparison of human papillomavirus types 16, 18 and 6 capsid antibody responses following incident infection. J. Infect. Dis. *181*, 1911-1919.

Christensen, N. D., C.A., R., Cladel, N. M., Han, R., and Krelder, J. W. (1996). Immunisation with virus-like particles induces long term protection of rabbits against challenge with cottontail rabbit papillomavirus. J. Virol. *70*, 960-965.

Clifford, G. M., Smith, J. S., Plummer, M., Munoz, N., and Franceschi, S. (2003). Human papillomavirus types in invasive cervical cancer worldwide: a meta-analysis. Br. J. Cancer *88*, 63-73.

Corona Gutierrez, C. M., Tinoco, A., Navarro, T., Contreras, M. L., Cortes, R. R., Calzado, P., Reyes, L., and et, a. l. (2004). Therapeutic vaccination with MVA E2 can eliminate precancerous lesions (CIN1, CIN 2 and CIN 3) associated with infection by oncogenic human papillomavirus. Hum. Gene Ther *15*, 421-431.

Daemen, T., Riezebos-Brilman, A., Bungener, L., Regts, J., Dontje, B., and Wilschut, J. (2003). Eradication of established HPV 16-transformed tumors after immunisation with recombinant Semliki Forest virus expressing a fusion protein E6 and E7. Vaccine *21*, 1082-1088.

Dale, C. J., Liu, X. S., De Rose, R., Purcell, D. F., Anderson, J., Xu, Y., Leggatt, G. R., Frazer, I. H., and et, e. l. (2002). Chimeric human papilloma virus-simian/human immunodeficiency virus virus-like-particle vaccines: immunogenicity and protective efficacy in macaques. Virology *301*, 176-187.

Davidson, E. J., Boswell, C. M., Sehr, P., Pawlita, M., Tomlinson, A. E., McVey, R. J., Dobson, J., Roberts, J. S., Hickling, J., Kitchener, H. C., and Stern, P. L. (2003). Immunological and clinical responses in women with vulval intraepithelial neoplasia vaccinated with a vaccinia virus encoding human papillomavirus 16/18 oncoproteins. Cancer Res. *63*, 6032-6041.

Derkay, C. S., Arnold, J., Bower, C., McClay, J., Smith, R., Van Burik, J., Wiatrak, B., Wohl, D., Berger, B., and Neefe, J. (2004). HspE7 treatment of pediatric recurrent respiratory papillomatosis (RRP): final results of an open-label trial. In 21st International Papillomavirus Conference (Mexico City).

Embers, M. E., Budgeon, L. R., Pickel, M., and Christensen, N. D. (2002). Protective immunity to rabbit oral and cutaneous papillomaviruses by immunisation with short peptides of L2, the minor capsid protein. J. Virol. *76*, 9798-9805.

Evans, T. G., Bonnez, W., Rose, R. C., Koenig, S., Demeter, L., Suzich, J. A., O'Brien, D., Campbell, M., White, W. I., and et, a. l. (2001). A phase I study of recombinant viruslike particle vaccine against human papillomavirus type II in healthy adult volunteers. J. Infect. Dis. *183*, 1485-1493.

Ferrara, A., Nonn, M., Sehr, P., Schreckenberger, C., Pawlita, M., Durst, M., Schneider, A., and Kaufmann, A. M. (2003). Dendritic cell-based tumor vaccine for cervical cancer II: results of a clinical pilot study in 15 individual patients. J. Cancer Res. Clin. Oncol *129*, 521-530.

Franconi, R., Massa, S., Simeone, P., Illiano, E., Muller, A., Benvenuto, E., and Venuti, A. (2004). Plant viruses can offer new strategies to increase the efficacy of therapeutic DNA vaccines against HPV. In 21st International Papillomavirus Conference (Mexico City).

Frazer, I. (2002). Vaccines for papillomavirus infection. Virus Res. *89*, 271-274.

Garcia, F., Petry, K. U., Muderspach, L., Gold, M. A., Braly, P., Crum, C. P., Magill, M., Silverman, M., Urban, R. G., Hedley, M. L., and Beach, K. J. (2004). ZYC101a for treatment of high-grade cervical intraepithelial neoplasia: a randomized controlled trial. Obstet. Gynecol. *103*, 317-326.

Garrido, F., Ruiz-Cabello, F., Cabrera, T., Perez-Villar, J. J., Lopez-Botet, M., Duggan-Keen, M., and Stern, P. L. (1997). Implications for immunosurveillance of altered HLA class I phenotypes in human tumours. Immunol. Today *2*, 89-95.

Gaukroger, J. M., Bradley, A., Chandrachud, L., Jarrett, W. F., and Campo, M. S. (1993). Interaction between bovine papillomavirus type 4 and cocarcinogens in the production of malignant tumors. J. Gen. Virol. *74*, 2275-2280.

Gaukroger, J.M., Chandrachud, L.M., O'Neil, B.W., Grindlay, G.J., Knowles, G. and Campo, M.S. (1996) Vaccination of cattle with bovine papillomavirus type 4 L2 elicits the production of virus neutralizing antibodies. J. Gen. Virol., 77, 1577-1583.

Gerber, S., Lane, C., Brown, D. M., Lord, E., DiLorenzo, M., Clements, J. D., Rybicki, E., Williamson, A. L., and Rose, R. C. (2001). Human papillomavirus virus-like particles are efficient oral immunogens when coadministered with Escherichia coli heat-labile enterotoxin mutant R192G or CpG DNA. J. Virol. *75*, 4752-4760.

Goldie, S. J., Kohli, M., Grima, D., Weinstein, M. C., Wright, T. C., Bosch, F. X., and Franco, E. (2004). Projected Clinical Benefits and Cost-effectiveness of a Human Papillomavirus 16/18 Vaccine. J. Natl. Cancer Inst. *96*, 604-615.

Goldstone, S. E., Palefsky, J. M., Winnett, M. T., and Neefe, J. R. (2002). Activity of HspE7, a novel immunotherapy in patients with anogenital warts. Dis Colon Rectum *45*, 502-507.

Greenstone, H. L., Nisland, J. D., de Visser, K. E., De Brujin, M. L., Kimbauer, R., Rodan, R. B., Lowy, D. R., Kast, W. M., and Schiller, J. T. (1998). Chimeric papillomavirus-like particles elicit antitumour immunity against the E7 oncoprotein in an HPV16 tumor model. Proc. Natl. Acad. Sci. USA. *95*, 1800-1805.

Hallez, S., Simon, P., Mandoux, F., Doyen, J., Noel, J. C., Beliard, A., Capelle, X., Buxant, F., Fayt, I., Lagrost, A. C., and et, a. l. (2004). Phase I/II trial of immunogenicity of a human papillomavirus (HPV) type 16 E7 protein-based vaccine in women with oncogenic HPV-positive cervical intraepithelial neoplasia. Cancer Immunol. Immunother *53*, 642-650.

Han, R., Cladel, N. M., Reed, C. A., Peng, X., Budgeon, L. R., Pickel, M., and Christensen, N. D. (2000). DNA vaccination prevents and/or delays carcinoma development of papillomavirus-induced skin papilloma on rabbits. J. Virol. *74*, 9712-9716.

Han, R., Cladel, N. M., Reed, C. A., Peng, X., and Christensen, N. D. (1999). Protection of rabbits from viral challenge by gene gun-based intracutaneous vaccination with a combination of cottontail rabbit papillomavirus E1, E2, E6 and E7 genes. J. Virol. *73*, 7039-7043.

Han, R., Peng, X., Reed, C. A., Cladel, N. M., Budgron, L. R., Pickel, M. D., and Christensen, N. D. (2002). Gene gun-mediated intracutaneous vaccination with papillomavirus E7 delays cancer development of papillomavirus-induced skin papillomas on rabbits. Cancer Detect Prev *26*, 458-467.

Harper, D.M., Franco, E.L., Wheeler, C., Ferris, D.G., Jenkins, D., Schuind, A., Zahaf, T. et al. (2004). Efficacy of a bivalent L1 virus-like particle vaccine in prevention of infection with human papillomavirus types 16 and 18 in young women: a randomised controlled trial. Lancet *364*, 1757-1765.

Harro, C. D., Pang, Y. Y., Roden, R. B., Hildesholm, A., Wang, Z., Reynolds, M. J., Mast, T. C., Robinson, R., and et, a. l. (2001). Safety and immunogenicity trial in adult volunteers of a human paipllomavirus 16 L1 virus-like particle vaccine. J. Natl. Cancer Inst. *93*, 284-292.

Herrero, R., Castellsague, X., Pawlita M, et al. (2003). Human papillomavirus and oral cancer: the International Agency for Research on Cancer multicenter study. J. Natl. Cancer Inst 95: 1772-1783.

Hinman, A. R., and Koplan, J. P. (1984). Pertussis and pertussis vaccine: reanalysis of benefits, risks, and costs. JAMA *251*, 3109-3113.

Jahan-Parwar, B., Chhetri, D. K., Hart, S., Bhuta, S., and Berke, G. S. (2003). Development of a canine model for recurrent respiratory papillomatosis. Ann. Otol Rhinol Laryngol *112*, 1011-1013.

Jarrett, W. F., O'Neil, B. W., Gaukroger, J. M., Smith, K. T., Laird, H. M., and Campo, M. S. (1990). Studies on vaccination against papillomaviruses: the immunity after infection and vaccination with bovine papillomaviruses of different types. Vet. Rec *126*, 473-475.

Jarrett, W. F., Smith, K. T., O'Neil, B. W., Gaukroger, J. M., Chandrachud, L. M., Grindlay, G. J., McGarvie, G. M., and Campo, M. S. (1991). Studies on vaccination against papillomaviruses: prophylactic and therapeutic vaccination with recombinant structural proteins. Virology *184*, 33-42.

Kaufmann, A. M., Nieland, J., Schinz, M., Nonn, M., Gabelsberger, J., Meissner, H., Muller, R. T., Jochmus, I., Gissmann, L., Schneider, A., and Durst, M. (2001). HPV16 L1E7 chimeric virus-like particles induce specific HLA-restricted T cells in humans after *in vitro* vaccination. Int. J. Cancer *92*, 285-293.

Kaufmann, A. M., Stern, P. L., Rankin, B., Sommer, H., Nuessler, V., Schneider, A., Adams, M., and a.l., e. (2002). Safety and immunogenicity of TA-HPV, a recombinant vaccinia virus expressing modified human papillomavirus HPV-16 and HPV-18 E6 and E7 genes, in women with progressive cervical cancer. Clin. Cancer Res. *8*, 3676-85.

Kawana, K., Kawana, Y., Yoshkawa, H., Taketani, Y., Yoshiike, K., and Kanda, T. (2001). Nasal immunization of mice with peptide having a cross-neutralization epitope on minor capsid protein L2 of human papillomavirus type 16 elicit systemic and mucosal antibodies. Vaccine *19*, 1496-1502.

Kawana, K., Yasugi, T., Kanda, T., Kino, N., Oda, K., Okada, S., Kawana, Y., Nei, T., Takada, T., Toyoshima, S., Tsuchiya, A., Kondo, K., Yoshikawa, H., Tsutsumi, O., and Taketani, Y. (2003). Safety and immunogenicity of a peptide containing the cross-neutralization epitope of HPV 16 L2 administered nasally in healthy volunteers. Vaccine *21*, 4256-4260.

Kawana, K., Yoshikawa, H., Taketani, Y., Yoshiike, K., and Kanda, T. (1999). Common neutralizing epitope in minor capsid protein L2 of human papillomavirus type 16 and 6. J. Virol. *73*, 6188-6190.

Kim, J. W., Hung, C. F., Juang, J., He, L., Kim, T. W., Armstrong, D. K., Pai, S. I., Chen, P. J., Lin, C.T., Boyd, D. A., and Wu, T. C. (2004). Comparison of HPV DNA vaccines employing intracellular targeting strategies. Gene Ther *11*, 1011-1018.

Kim, T. W., Hung, C. F., Boyd, D., Juang, J., He, L., Kim, J. W., Hardwick, J. M., and Wu, T. U. (2003). Enhancing DNA vaccine potency by combining a strategy to prolong dendritic cell life with intracellular targeting strategies. J. Immunol. *171*, 2970-2976.

Kirnbauer R., Chandrachud, L.M., O'Neil, B.W., Wagner, E.R., Grindlay, G.J., Armstrong, A., McGarvie, G.M., Schiller. J.T., Lowy, D.R. and Campo, M.S. (1996) Virus-like particles of bovine papillomavirus type 4 in prophylactic and therapeutic immunization. Virology 219, 37-44.

Klencke, B., Matijevic, M., Urban, R. G., Lathey, J. L., Hedley, M. L., Berry, M., Thatcher, J., Weinberg, V., Wilson, J., Darragh, T., Jay, N., Da Costa, M., and Palefsky, J. M. (2002). Encapsulated plasmid DNA treatment for human papillomavirus 16-associated anal dysplasia: a Phase I study of ZYC101. Clin. Cancer Res. *8*, 1028-1037.

Kloster, B. E., Manias, D. A., Ostrow, R. S., Shaver, M. K., McPherson, S. W., Rangen, S. R., Uno, H., and Faras, A. J. (1988). Molecular cloning and characterisation of the DNA of two papillomaviruses from monkeys. Virology *166*, 30-40.

Koutsky, .LA., Ault, K.A., Wheeler, C.M., Brown, D.R., Barr, E., Alvarez, F.B., Chiacchierini, L.M., and Jansen, K.U. (2002). Proof of principle study investigators. A controlled trial of a human papillomavirus type 16 vaccine. N. Engl. J. Med. *34*, 1645-1651.

Kulasingam, S. L., and Myers, E. R. (2003). Potential Health and Economic Impact of Adding a Human Papillomavirus Vaccine to Screening Programs. JAMA *290*, 781-9.

Lacey, C. J., Thompson, H. S., Monteiro, E. F., O'Neill, T., Davies, M. L., Holding, F. P., Fallon, R. E., and Roberts, J. S. (1999). Phase IIa safety and immunogenicity of a therapeutic vaccine, TA-GW in persons with genital warts. J. Infect. Dis. *179*, 612-618.

Lamikanra, A., Pan, Z. K., Isaacs, S. N., Wu, T. C., and Paterson, Y. (2001). Regression of established human papillomavirus type 16 (HPV-16) immortalized tumors *in vivo* by vaccinia viruses expressing different forms of HPV-16 E7 correlates with enhanced CD8(+) T-cell responses that home to the tumor site. J. Virol. *75*, 9654-9664.

Luxton, J., and Shepherd, P. (2001). Human papillomavirus antigens and T-cell recognition. Curr. Opin. Infect. Dis. *14*, 139-143.

Marals, D., Passmore, J. A., Maclean, J., Rose, R., and Willamson, A. L. (1999). A recombinant human papillomavirus (HPV) type 16 L1 vaccinia virus murine challenge model demonstrates cell mediated immunity against HPV virus-like particles. J. Gen. Virol. *80*, 2471-2475.

Moore, R. A. (2004). Canine papillomavirus- a mucosal model of human disease. Papillomavirus Report *15*, 55-62.

Muderspach, L., Wilczynski, S., Roman, L., Bade, L., Felix, J., Small, L. A., Kast, W. M., Fascio, G., Marty, V., and Weber, J. (2000). A phase I trial of a human papillomavirus (HPV) peptide vaccine for women with high-grade cervical and vulvar intraepithelial neoplasia who are HPV 16 positive. Clin. Cancer Res. *6*, 3406-3416.

Nardelli-Haefliger, D., Wirthner, D., Schiller, J. T., Lowy, D. R., Hildesheim, A., Ponci, F., and De Grandi, P. (2003). Specific antibody levels at the cervix during the menstrual cycle of women vaccinated with human papillomavirus 16 virus-like particles. J. Natl. Cancer Inst *95*, 1128-1137.

Nasseri, M., Meyers, C., and Wettstein, F. O. (1989). Genetic analysis of CRPV pathogenesis: the L1 open reading frame is dispensable for cellular transformation but is required for papilloma formation. Virology *170*, 321-325.

Nigland, J. D., Da Silva, D. M., Valders, M. P., de Visser, K. E., Schiller, J. T., Mullr, M., and Kast, W. M. (1999). Chimeric papillomavirus-like particles induce a murine self-antigen specific protective and therapeutic antitumor immune response. J. Cell Biochem. *73*, 145-152.

Nonn, M., Schinz, M., Zumbach, K., Pawlita, M., Schneider, A., Durst, M., and Kaufmann, A. M. (2003). Dendritic cell-based tumor vaccine for cervical cancer I: *in vitro* stimulation with recombinant protein-pulsed dendritic cells induces specific T cells to HPV 16 E7 or HPV 18 E7. J. Cancer Res. Clin. Oncol *129*, 511-520.

O'Brien, P. M., and Campo, M. S. (2003). Papillomaviruses: a correlation between immune evasion and oncogenicity? Trends Microbiol. *11*, 300-305.

Ostrow, R. S., LaBresh, K. V., and Faras, A. J. (1991). Characterisation of the complete RhPV 1 genomic sequence and an integration locus from a metastatic tumor. Virology *181*, 424-429.

Ostrow, R. S., McGlennen, R. C., Shaver, M. K., Kloster, B. E., Houser, D., and Faras, A. J. (1990). A rhesus monkey model for sexual transmission of a papillomavirus isolated from a squamous cell carcinoma. Proc. Natl. Acad. Sci. USA. *87*, 8170-8174.

Pinto, L. A., Viscidi, R., Harro, C. D., Wallace, D., Trivett, M., Lowy, D. R., Schiller, J. T., Berzofsky, J. A., and Hildesheim, A. (2004). Cellular immune responses to L1 from heterologous HPV types following vaccination with HPV-16 L1 virus-like particles. In 21st International Papillomavirus Conference (Mexico City).

Pong, S., Frazer, I. H., Fernando, G. J., and Zhou, J. (1998). Papillomavirus-like particles can deliver defined CTL epitopes to the MHC class I pathway. Virology *240*, 147-157.

Ressing, M. E., van Driel, W. J., Brandt, R. M., Kenter, G. G., Bauknecht, T., Fleuren, G. J., Hoogerhout, P., and et, a. l. (2000). Detection of T helper responses, but not of human papillomavirus-specific cytotoxic T lymphocyte responses, after peptide vaccination of patients with cervical carcinoma. J. Immunother *23*, 255-266.

Roden, R. B. S., Greensione, H. L., Kimbauer, R., Booy, F. P., Jessie, J., Lowy, D. R., and Schiller, J. T. (1996). In vitro generation and type-specific neutralisation of a human papillomavirus type 16 virion pseudotype. J. Virol. *70*.

Rose, R. C., Lane, C., Clements, J., and Mason, H. (2004). Oral immunogenicity of freeze-dried HPV16 L1 transgenic tomato. In 21st International Papillomavirus Conference (Mexico City).

Sanders, G. D., and Taira, A. V. (2003). Cost effectiveness of a potential vaccine for *Human papillomavirus*. Emerg. Infect. Diseases 9, 37-48.

Santin, A. D., Bellone, S., Gokden, M., Cannon, M. J., and Parham, G. P. (2002). Vaccination with HPV-18 E7-pulsed dendritic cells in a patient with metastatic cervical cancer. N. Engl. J. Med. *346*, 1752-1753.

Schreckenberger, C., and Kaufmann, A. M. (2004). Vaccination strategies for the treatment and prevention of cervical cancer. Curr. Opin. Oncol *16*, 485-491.

Selvakumar, R., Borenstein, L. A., Lin, Y.-L., Ahmed, R., and Wettstein, f. O. (1995). Immunization with nonstructural proteins E1 and E2 of cotton tail rabbit papillomavirus stimulates regression of virus-induced papillomas. J. Virol. *69*, 602-605.

Sewell, D. A., Douven, D., Pan, Z. K., Rodriguez, A., and Paterson, Y. (2004). Regression of HPV-positive tumors treated with a new Listeria monocytogenes vaccine. Arch. Otolaryngol Head Neck Surg *130*, 92-97.

Simon, P., Buxant, F., Hallez, S., Burny, A., Fayt, I., Anaf, V., and Noel, J. C. (2003). Cervical response to vaccination against HPV16 E7 in case of severe dysplasia. Eur. J. Obstet. Gynecol. Reprod. Biol. *109*, 219-223.

Slater, P. E., Roitman, M., Leventhal, A., and Anis, E. (1996). Control of rubella in Israel: progress and challenge. Public Health Rev. *24*, 183-192.

Smith, L. H., Foster, C., Hitchcock, M. E., Leiserowitz, G. S., Hall, K., Isseroff, R., Christenson, N. D., and Kreider, J. W. (1995). Titration of HPV-11 infectivity and antibody neutralisation can be measured *in vitro*. J. Invest Dermatol *105*, 438-444.

Stanberry, L. R., Spruance, S. L., Cunningham, A. L., Bernstein, D. I., Mindel, A., Sacks, S., Tyring, S., Aoki, F. Y., Slaoui, M., Denis, M., and et, a. l. (2002). Glycoprotein-D-adjuvant vaccine to prevent genital herpes. N. Engl. J. Med. *347*, 1652-1661.

Stanley, M. A. (2002). Imiquimod and the imidazoquinolones: mechanism of action and therapeutic potential. Clin. Exp. Dermatol *27*, 571-577.

Stanley, M. A., Moore, R. A., Nicholls, P. K., Santos, E. B., Thomsen, L., Parry, N., Walcott, S., and Gough, G. (2001). Intra-epithelial vaccination with COPV L1 DNA by particle-mediated delivery protects against mucosal challenge with infectious COPV in beagle dogs. Vaccine *19*, 2783-2792.

Suzich, J. A., Ghim, S. J., Palmer Hill, F. J., White, W. I., Tamura, J. K., Bell, J. A., Newsome, J. A., Jenson, A. B., and Schlegel, R. (1995). Systemic immunization with papillomavirus L1 protein completely prevents the development of viral mucosal papillomas. Proc. Natl. Acad. Sci. USA. *92*, 11553-11557.

Trimble, C., Lin, C. T., Hung, C. F., Pai, S., J., J., He, L., Gillison, M., Pardoll, D., Wu, L., and Wu, T. C. (2003). Comparison of the CD8+ T cell responses and antitumor effects generated by DNA vaccine administered through gene gun, biojector and syringe. Vaccine *21*, 4036-4042.

van Driel, W. J., Ressing, M. E., Kenter, G. G., Brandt, R. M., Krul, E. J., van Rossum, A. B., and et, a. l. (1999). Vaccination with HPV-16 peptides of patients with advanced cervical carcinoma: clinical evaluation of a phase I-II trial. Eur. J. Cancer *35*, 946-952.

Velders, M. P., McElhiney, S., Cassetti, M. C., Eiben, G. L., Higgins, T., Kovacs, G. R., Elmishad, A. G., Kast, W. M., and Smith, L. R. (2001). Eradication of established tumors by vaccination with Venezuelan equine encephalitis virus replicon particles delivering human papillomavirus 16 E7 RNA. Cancer Res. *61*, 7861-7867.

White, C. C., Koplan, J. P., and Orenstein, W. A. (1985). Benefits, risks and costs of immunisation for measles, mumps and rubella. Amer. J. Public Health *75*, 739-744.

Wooldridge, J., Link, B. K., Weisdorf, D. J., et al. (2003). Phase I study of oligodeoxynucleotide CpG 7909 in patients with previously treated non-Hodgkin's Lymphoma. In Proceedings of the American Society of Clinical Oncology (ASCO).

Wright, J. C., and Weinstein, M. C. (1998). Gains in life expectancy from medical interventions-standardizing data on outcomes. N. Engl. J. Med. *339*, 380-386.

Wu, X., Xiao, W., and Brandsma, J. L. (1994). Papilloma formation by cottontail rabbit papillomavirus required E1 and E2 regulatory genes in addition to E6 and E7 transforming genes. J. Virol. *68*, 6097-6102.

Yuan, H., Estes, P. A., Chen, Y., Newsome, J., Olcese, V. A., Garcea, R. L., and Schlegel, R. (2001). Immunization with a pentameric L1 fusion protein protects against papillomavirus infection. J. Virol. *75*, 7848-7853.

Zhang, L. F., Zhou, J., Chen, S., Cai, L. L., Bao, Q. Y., Zheng, F. Y., Lu, J. Q., Padmanabha, J., Hengst, K., Malcolm, K., and Frazer, I. H. (2000). HPV6b virus like particles are potent immunogens without adjuvant in man. Vaccine *18*, 1051-1058.

From: Papillomavirus Research: From Natural History to Vaccines and Beyond. Edited by: M. Saveria Campo

Chapter 22

HPV Vaccines in Plants: An Appetising Solution to Control Infection and Associated Cancers

Rosella Franconi and Aldo Venuti*

Abstract

The cultivation of plants with specific properties has been the foundation of medicine for millennia. Modern biotechnology may one day extend their medicinal uses to include the delivery of vaccines. Because of their low production costs and facility of use, plant-derived vaccines provide an interesting perspective. The simple demands for solar light, water and minerals make plants an economic system for the production of heterologous proteins, supplying eukaryotic post-translational modifications (*e.g.* glycosylation, phosphorylation), eliminating the risks of contaminations with animal pathogens and allowing oral administration. Other advantages stem from the possibility to express the protein of interest in reservoir organs, of production on wide scale and to omit purification passages, particularly for oral administration. In this case, crushed seeds, endosperma, or fruits could be used for direct administration (edible vaccine). The list of plant-derived antigens, potentially usable as vaccines, continues to lengthen and includes pathogenic viral and bacterial proteins. In this chapter recent advances in the production of prophylactic and therapeutic plant vaccines against papillomaviruses are reported. Some aspects of the public debate on the genetically modified organism will be discussed too.

1. Preface

A night in the first century A.C.

> "......*Allata est tamen gustatio valde lauta; nam iam omnes discubuerant praeter ipsum Trimalchionem, cui locus novo more primus servabatur. Ceterum in promulsidari asellus erat Corinthius cum bisaccio positus, qui habebat olivas in altera parte albas, in altera nigras. Tegebant asellum duae lances, in quarum marginibus nomen Trimalchionis inscriptum erat et argenti pondus* ...."

(The Satyricon of Petronius Chapter XXXI)

The same night at the beginning of 21st century D.C:

> "........Then the preliminary course was served in very elegant style. For all were now at table except Trimalchio, for whom the first place was reserved, by a reversal of ordinary usage. Among the other hors d'oeuvres stood a little ass of Corinthian bronze with a packsaddle holding transgenic tomatoes, green tomatoes on one side for antiviral vaccination, red mini-tomatoes on the other for anti-malaria vaccination. The animal was flanked right and left by silver dishes, on the rim of which Trimalchio's name was engraved and the weight........"

(modified from the translation of Alfred R. Allinson, 1930)

Apart from the language, what is the main difference between the two nights?

Edible vaccines.

How far are we from the second night?

In this chapter we will try to answer to this question by presenting the advances made in this field and their application for HPV vaccines.

2. Introduction

Beside their invaluable role in the eradication of human calamitous diseases (*i.e.* smallpox) vaccines have resulted in a cost–benefit ratio of 1:10 (Center for Disease Control, 1998) showing their great economic value in healthcare. The development and implementation of new vaccines require usually a high input budget, which is an insurmountable obstacle in economically depressed areas where medical infrastructure and coordinated healthcare policies are lacking and preventative measures are even more crucial. Therefore, new biotechnology strategies are faced by a major challenge: to make low cost innovations for the developing world.

Plants have been used for therapeutic purposes since ancient times, mostly by providing medicinal compounds that have been extracted and used to treat illness. Nowadays, plant molecular farming provides new therapeutic possibilities combining the innovations

*For correspondence. Email venuti@ifo.it

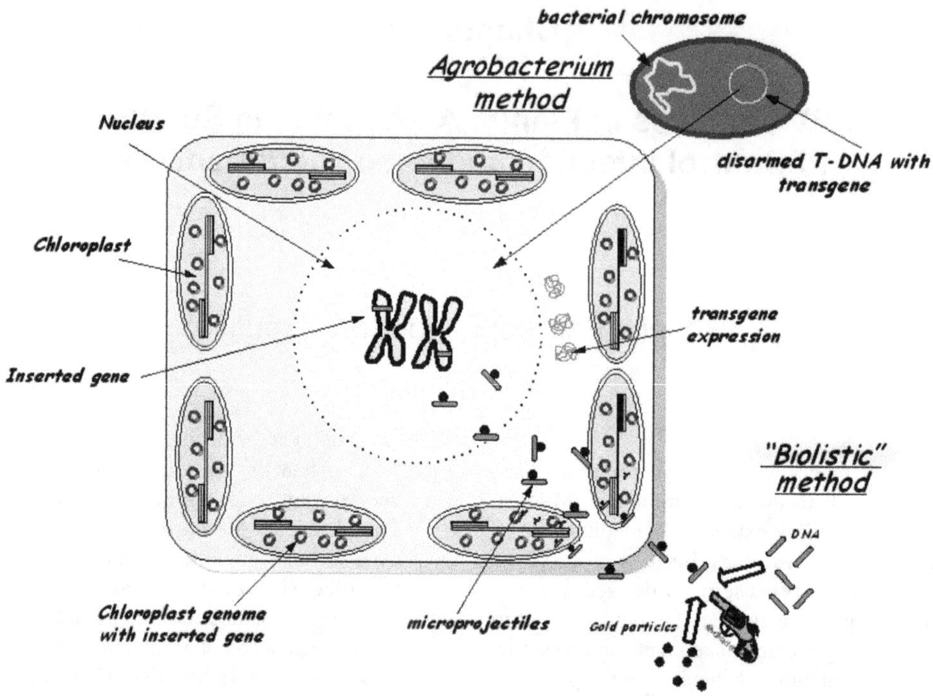

Figure 1. Schematic representation of nuclear or chloroplast transformation. Nuclear transformation is performed by using the plant pathogen *Agrobacterium tumefacien* or by the biolistic method. With the first method, the T-DNA, in which the ectopic gene has been inserted, is transferred into the nuclear chromosomal DNA causing chromosomal integration at random sites (red bars) and, usually, resulting in low transgene expression. With the second method, it is possible to obtain nuclear transformation (generally resulting in multiple gene insertions) as well as plastid transformation. Chloroplast transformation is performed by microprojectile bombardment with chloroplast vectors that allow site-specific integration of transgenes via homologous recombination (not shown). Due to high ploidy of transgenes (5000-10000 copies per cell, maternally transmitted without any segregation) high expression level of the foreign genes can be achieved by this way.

in medical science and plant biology to create affordable pharmaceutical products.

Research over the past 15 years has significantly increased our knowledge of gene regulation and protein synthesis in plant. More than 100 recombinant proteins have been produced in a range of different species and the first plant-derived biopharmaceutical proteins (antibodies, vaccines, etc.) are now approaching commercial approval (Giddings, 2001; Daniell et al., 2001a; Ma et al., 2003; Twyman et al., 2003).

There are several reasons to produce therapeutic molecules in plants.

- The simple demands for solar light, water and minerals make plants an easier and more economical system for the production of heterologous proteins than industrial facilities using fermentation technology. It is estimated that recombinant proteins can be produced in plants at 2–10% of the cost of microbial fermentation systems and at 0.1% of the cost of mammalian cell cultures. Yields of 0.1–1.0% total soluble protein (TSP) are sufficiently competitive with other expression systems to make recombinant plants economically viable (Hood et al., 2002). Moreover, scale-up technology is available for harvesting and processing plants or plant products on a large (potentially agricultural) scale.
- Plants are considered to be much safer than both microbes and animals because they generally lack human pathogens (such as prions or viruses), oncogenic DNA sequences and endotoxins. This minimizes health risks and contributes to reduce production costs.
- The protein synthesis pathway is conserved between plants and animals. Possessing an endomembrane system, they supply eukaryotic post-translational modifications (*e.g.* glycosylation and phosphorylation) to recombinant proteins that can fold and assemble efficiently. This is a great advantage over bacterial expression systems, in which many proteins fail to fold properly and are either degraded (resulting in low yields or absence of protein) or accumulate as insoluble inclusion bodies.
- Subcellular targeting can be exploited to increase the yield of recombinant proteins.
- Several types of recombinant protein can be used in unprocessed or partially processed plant material, therefore removing many of the downstream costs (> 85% of costs). Many topical preparations are already derived from plants and recombinant subunit vaccines produced in plants can be administered by the consumption of raw or part processed fruits and

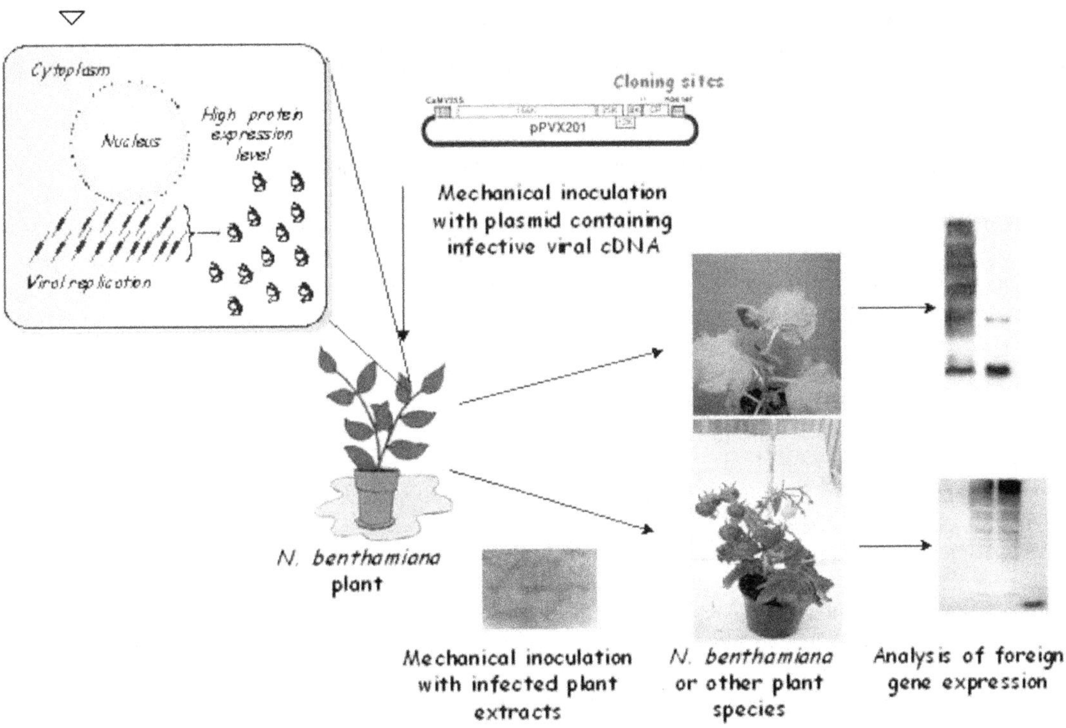

Figure 2. Transient expression mediated by recombinant plant virus. Primary infection of established plants can be achieved with *in vitro* transcribed viral RNA or with a plasmid containing the recombinant viral cDNA cloned into a plant expression cassette. The figure shows a potato virus X (PVX)-derived vector (pPVX201) where the expression cassette is under the control of the cauliflower mosaic virus 35S promoter (CaMV 35S). In this vector the 166 K gene encodes a protein required for PVX replication; 25K, 12K, and 8K represent three partially overlapping ORFs, known as the 'triple block'; CP is the PVX coat protein gene and NOS-ter represents the transcription terminator from the nopaline synthase gene of *Agrobacterium tumefaciens*. Virus replication occurs inside the plant cells (schematically represented in the left drawing, where high expression level is indicated) followed by systemic spread in most plant tissues. Purified recombinant virus or infected plant extracts can be often used to re-infect plants of the same or of different species, in order to obtain higher amounts of the protein of interest. Expression of the heterologous protein is detected by different methods (usually immunoblotting and ELISA). In particular, the figure shows the immunoblotting of HPV16 E7 protein in *Nicotiana benthamiana* plants.

vegetables. For pharmaceutical proteins that need to be purified before use, simple strategies can be developed to reduce downstream processing costs.

- Prolonged stable storage can be easily achieved by exploiting the ability to target the recombinant proteins to natural storage organs (e. g. seeds or tubers).

Early experiments were carried out in model plants species but more recently important crop plants have been demonstrated to be effective production platforms (Fischer et al., 2004).

Two are the main systems to produce vaccine antigens in plant, stable nuclear or chloroplast transformation and transient expression. Stable transformation produces a genetic line that can be propagated vegetatively or by seeds and represents the method used for the published edible vaccine clinical trials to date (Tacket et al., 1998; 2000). Transient expression can be achieved either by vacuum infiltration or by using a recombinant plant virus that causes the plant to express the antigen by systemic infection (Koprowski and Yusibov, 2001). The main advantage using a plant–virus system is that virus replication amplifies gene copy number, typically resulting in a much higher level of expression than with stable transformation. However, the plant–virus systems might suffer from instability and loss of cloned gene if larger than 1 kb. In addition, the need to inoculate individually each plant makes large-scale production rather labour intensive. The principal aspects of these technologies are discussed below and summarized in Figures 1 and 2.

2.1. Stable genetic transformation (transgenic plants)

As a result of stable transformation, the recombinant DNA integrates into the genome of either the nucleus or the chloroplast of the plant cell.

2.1.1. Nuclear transformation.

Nuclear transformation is most often achieved by using the Ti (tumour-inducing) plasmid of the plant pathogen *Agrobacterium tumefaciens*. *A. tumefaciens*, during the natural process of infection, transfers a segment delimited by the left and right border sequences of the Ti plasmid (T-DNA) into the genome of the host (with a process similar to bacterial conjugation) (Zambryski, 1988). When plant tissues are in contact with transgenic Agrobacteria, the T-DNA in which the ectopic gene has been inserted, is transferred into the plant genome causing chromosomal integration at random sites (Figure 1). Strains of *Agrobacterium* have been modified to eliminate the virulence genes retaining those that mediate efficient transfer of the DNA that is cloned between the specific border sequences. *Agrobacterium* transformation cannot be achieved in some agronomically important plant species (*e.g.* soybean and most cereal grains); for these plants, a 'biolistic' method (microprojectile bombardment) is frequently used (Sanford et al., 1993). The microprojectiles are constituted of DNA coated on micron-sized gold particles and are propelled into plant cells with compressed helium. By 'biolistic' or *Agrobacterium* methods foreign DNA integrates into the nuclear genome producing stable transformants in which the transgene is expressed and transmitted to the progeny in typical Mendelian fashion. In general the technology is limited by low expression levels for nuclear integrated transgenes. Nevertheless the biolistic method usually results in higher copy-number integration of the transgene than *Agrobacterium,* and, therefore, in enhanced transgene expression (Hobbs et al., 1993). It is important to point out that excessive copies or very high level expression of nuclear genes may cause gene silencing (Finnegan and McElroy, 1994; Vaucheret, et al. 1998), thus there is a need to select transgenic plants carrying only one or few copies of the transgene.

2.1.2. Chloroplast transformation

The biolistic method is also used to introduce transgenes into the genome of the chloroplast (Svab and Maliga, 1993). Plants have multiple copies of chloroplasts in each cell (Figure 1). The chloroplast offers a number of attractive advantages, including high-level transgene expression (Daniell et al., 2002a), multi-gene engineering in a single transformation event (DeCosa et al., 2001; Ruiz et al., 2003; Daniell et al., 2002b), transgene containment via maternal inheritance (Daniell, 2002c), lack of gene silencing (DeCosa et al., 2001), lack of positional effect due to site-specific transgene integration (Daniell et al., 2002a). In nuclear transformation random integration of transgenes via non-homologous recombination results in nuclear transgenic lines with extremely high variability of foreign gene expression and this requires screening of hundreds of independent transformation events. Such a screening is not necessary in chloroplast transgenic lines because of uniform foreign gene expression in chloroplast transgenic lines due to site specific integration of transgenes via homologous recombination of flanking sequences present in chloroplast transformation vectors. High chloroplast genome copy number in plant cells contributes to enhanced recombinant protein expression, as shown for cholera toxin (CT) B subunit (Daniell et al., 2001b) and for an insecticidal protein that was expressed at up to 45% of total protein in chloroplasts of mature tobacco leaves (De Cosa et al., 2001).

Most examples of chloroplast-based molecular farming have been reported in tobacco (where the technology for chloroplast gene transfer is most advanced) and in other few species. However the potential for chloroplast transformation in other crop plants is strong. Stable transgene expression was also described in the chloroplast-derived chromoplasts of tomato fruit (Ruf et al., 2001). Recently a highly efficient process for carrot *(Daucus carota L.)* has been successfully achieved by using non-green tissues as explants, regenerated via somatic embryogenesis (Kumar et al., 2004). Through this way it is expected to engineer the plastid genome of several major crops in which regeneration is mediated through somatic embryogenesis such as cereal crops (wheat, rice, corn, sugarcane), legumes (soybean, alfalfa), oil crops (sunflower, olive), cash crops (cotton, coffee, tea, rubber, flex, cork oak, pines), vegetables (eggplant, cucumber, cassava, chili pepper, asparagus etc.), fruits (apple, cherry, banana, plantain, melons, grape, guava), nuts (cashew, walnuts, peanuts), and trees (date palm etc.). Although chloroplast transformation is not yet a routine procedure, at least 25 different foreign genes have been expressed in transplastomic plants (Daniell et al., 2002a). One limitation to this technology is that some eukaryotic proteins that require specialized cellular processing machinery like glycosylation might not be suitable for chloroplasts. It has been recently shown that the enteropathogen, *Campylobacter jejuni,* has a N-linked general protein glycosylation pathway (Pgl) that modifies many of the organism's proteins. The *pgl* cluster was found to be fully functional when transferred in *Escherichia coli* as monitored by glycosylation of a co-expressed protein (Wacker et al., 2002). Efforts are also being made to introduce the *pgl* cluster into the plant chloroplast to overcome the above limitation (Daniell, unpublished).

2.2. Transient expression

As an alternative to stable transformation, several plant-based transient expression systems are available. Transient expression by the vacuum infiltration of tobacco leaves with recombinant *A. tumefaciens* can be used to verify expression and also to produce large quantities of protein (Kapila et al., 1997). Virus-mediated expression of heterologous proteins in tobacco has emerged as a powerful tool for the production and delivery of pharmaceutical proteins including antibodies and subunit vaccines (Koprowski and Yusibov, 2001; Awram et al., 2002), in alternative to the time-consuming regeneration of transgenic plants. The gene coding for the protein of interest is incorporated in the virus genome, which then infects the host plant (Figure 2). Host plants are cultivated to the proper stage before being inoculated with the engineered virus. As the virus spreads and replicates, numerous copies of the target gene are produced by the host plant and high levels of protein expression can be achieved in a relatively short time. Viral vectors show many advantages such as speed of expression, high yield, reduced cost and high throughput. Viral vector are usually not transmitted by pollen, preventing gene contamination of other plants. Moreover, when the foreign gene encodes a product that is toxic or interferes with the plant metabolism (and hence cannot be expressed as transgenic) virus-mediated expression represents the only possibility to express that gene. The main approach consists of designing viral vectors that behave as wild type virus but are modified to carry and express a heterologous sequence that encodes a protein of interest under the control of a duplicated subgenomic coat protein promoter. In addition foreign proteins or peptides encoding for the interesting epitopes, can be genetically fused to the coat protein to obtain chimeric viral particles, expressing the foreign epitopes on their surface. Several efficient vectors consist essentially in fully functional viruses that, usually, retain infectivity, are stable, have the ability to move systemically within the host (generally 2-3 weeks), and produce infectious viral particles. Very often it is possible to increase the amount of the foreign protein produced simply by performing several passage of re-infection (in the same or on different plant host) by utilizing virus infected plant extracts (Figure 2). Virus-derived vectors are based primarily on RNA viruses, such as the tobacco mosaic virus (TMV) and the potato virus X (PVX). Sometimes, as in the case of PVX, infectious DNA plasmids carrying the cDNA of the plant virus are available and this simplifies the mechanical infection procedure (Fig.2). Some characteristics of PVX (absence of animal vectors and the ability to infect several plant species) make it an effective and safe instrument for transient expression in plants (Chapman et al., 1992). Optimal use of RNA virus vectors is limited by incomplete understanding of replication and 'payload' genes might add constrains to regain replicative fitness. Many researchers have attempted to improve the efficiency of foreign protein production by improving the viral vectors or the plant host (Gleba et al., 2004). Several biopharmaceuticals that have been obtained using plant viral vectors are now in clinical trials (Awram et al., 2002). Some cancer biomedicals have already been obtained using plant viral vectors. A tumour idiotype-specific scFv epitope from a mouse B cell lymphoma, was produced at high levels in *N. benthamiana* by a modified TMV vector. It was secreted in a folded form (through a plant-derived signal peptide) in the extracellular compartments (McCormick et al., 1999). The purified scFv, utilized as therapeutic lymphoma vaccine in subcutaneous immunization, induced an anti-idiotype immune response and protected mice from challenge by a lethal dose of syngeneic tumor cells. Interestingly, mice that received the scFv alone, without adjuvant, showed a high degree of protection, equivalent to that showed by mice immunized with the scFv plus the QS-21 adjuvant (McCormick et al., 1999) indicating that either the proper conformation or some other unknown factor provided by the plant-expression system, improved the efficacy of the immunogen. The same adjuvant-like effect was noticed when other plant-produced human scFvs (cloned from tumor biopsy cells), purified from the interstitial fraction (IF) were tested in mice for appropriate anti-idiotype response (McCormick et al., 2003). These plant-produced scFvs are currently undergoing phase I clinical trials. A colorectal cancer antibody (Verch et al., 1998) and a colorectal cancer antigen (Verch et al., 2004) have been also produced in *N. benthamiana* by a TMV–based vector. The purified plant-derived tumour antigen was able to stimulate T cells and indicated the presence of some adjuvant-like effect. Our recent data indicate that the same adjuvant-like effect is obtained in immunizations with crude plant extract containing the E7 protein. We showed that, beside the ability to evoke an antibody response, recombinant plant-derived vaccines as '*in planta* formulation' without adjuvants are able to elicit also a protective Th1 cell response in mice (Franconi et al., 2002). This activity might be further improved by increasing the amount of the E7 protein in the extracts.

2.3. Optimization of foreign protein expression

To optimize yields of both transient and stable expression, both transcription and translation can be maximized. Moreover, signal sequences can be used to target proteins for accumulation in specific areas of plant cells or organs.

The secretory pathway is a more suitable compartment for folding and assembly than the cytosol. The endoplasmic reticulum (ER) provides an oxidizing environment and an abundance of molecular chaperones (*i.e.* BiP), but few proteases. In the absence of further targeting information, the expressed protein is secreted to the apoplast. Depending on its size, the protein can be retained therein or might leach from the cell, with important implications for production systems based on cell-suspension cultures. We recently achieved PVX-mediated E7 protein expression in the secretory compartment and in the apoplast by using a plant-derived signal sequence. Targeting to the plant secretory pathway enhanced about five-fold the amount of the E7 protein produced and mice immunised with this plant extract showed a fast and increased immune response and a better tumor protection (Franconi et al., unpublished).

Targeting of proteins to the secretory pathway might also result in their exudation in roots and leaves. Rhizosecretion offers easier extraction of proteins than from leaves. Expression levels can be increased even further if the protein is retrieved to the ER lumen using an H/KDEL C-terminal tetrapeptide tag (Twyman et al., 2003). Molecules retrieved in this way are not modified in the Golgi apparatus. This is an important issue since there are differences in post-translational modifications between mammals and plants. In fact, while early glycosylation steps are similar in plants and in mammals, later steps are different. Plant-derived recombinant human proteins tend to have the carbohydrate groups $\beta(1 \rightarrow 2)$-xylose and $\alpha(1 \rightarrow 3)$-fucose, which are absent in mammals, but lack the terminal galactose and sialic acid, residues that are found on many native human glycoproteins. It has been demonstrated that plant N-glycans are immunogenic in humans (Bardor et al., 2003). Recently it has been shown that it is possible to manipulate specific glycosyl transferases in plant cells to yield human-like glycosylation pattern on recombinant proteins (Gomord and Faye, 2004). However, for the formulation of more potent vaccines, plant–derived glycan-chain molecules might represent an advantage rather than a disadvantage.

Molecules retained in the ER will have mannose glycans but not plant-specific xylose and fucose residues. Other approaches consist in introducing chloroplast or vacuolar targeting sequences. In this way, proteins, which are potentially toxic for the plant, could be more easily expressed.

3. Edible Vaccines

3.1. Reasons to use edible plant vaccines

Vaccines consisting of only one or a few 'subunit' proteins of a pathogen are considered much safer than live attenuated vaccines. Recombinant DNA technology provides the means to clone antigen genes from the pathogen and transfer them into less dangerous organisms, where they are expressed and purified for injection. The plant biotechnology expansion during the past 15 years has made possible the use of transgenic plants as viable alternatives to cell-culture systems for the production of recombinant subunit vaccines. In theory, scale-up of production using agricultural practice with a stable transgenic line would entail simply increasing the cultivated acreage. Although the vaccine antigens could be as readily purified for injection from the plant tissue (as from other expression system such as yeast or insect cells), edible plants offer a palatable oral delivery system that would avoid the costly purification process required of injectable vaccines. Moreover, as mentioned before, plant recombinant expression systems are much less likely to contain prions or other human pathogens. One of the major disadvantages of oral delivery is that the digestive system could degrade the antigens, and thus might limit the use of oral subunit vaccines to gut-stable or encapsulated antigens. In this regard, the use of a plant expression system allows the accumulation of vaccine antigen within the intracellular compartment of the plant cell, which will itself provide a natural protective shield. Edible vaccines are now considered to be more stable and practical in the harsh environment of the gastrointestinal tract than purified vaccine antigens (Daniell, 2001a). Furthermore, the vaccine antigen expression can be directed into the plant intracellular compartment where protein is stored in a more stable condition (Walmsley and Arntzen, 2000). A plant-based multicomponent vaccine that expressed a cholera toxin fusion with two different disease antigens in transformed potatoes has been developed and protects mice against infectious virus and bacterial diseases (Yu and Langridge, 2001). Recently, the subunit B of labile toxin (LT-B) of enterotoxigenic *E. coli.* was expressed at a high level in corn and the LT-B delivered in the seed elicited protective immune responses (Streatfield et al., 2001). The LT-B in transgenic corn results in an increased mucosal immune response compared with that achieved with bacterial LT-B when fed to mice. This interesting finding is due to the fact that the antigen is protected from degradation in the gut (Streatfield et al., 2001). These promising results open the way to generate a vaccine with a built-in adjuvant in plants, in which the vaccine can be targeted to the mucosal immune system and ameliorates the quality of antigen-specific immune responses for greater protection efficacy.

3.2. How the edible vaccine induces immunity

Plant-derived oral vaccines would not be delivered in fresh form, but processed to yield uniform, stable batches with well-defined antigen content. Protection of antigens from degradation in the gut is needed for optimal delivery, and this should be provided by the plant tissue matrix (although treatments might affect the encapsulation provided by the plant cell wall). Important questions regarding mucosal immunity remain to be answered such as the possibility to induce tolerance, that probably could be avoided by enhancing expression levels, use proper adjuvant and reducing the number of boosts. Another aspect to be considered, closely related to the development of an oral anti-cancer vaccine, is the ability to evoke a cell-mediated immune response. In this case, using appropriate adjuvants or better methods to target subunit vaccines to appropriate antigen presenting cells (APC) it might be possible to overcome the apparent limitation that mucosal immunization gives rise only to antibody responses.

Once the oral vaccine has overcome the gut barrier, it has to induce an immune response from the common mucosal immune system (CMIS), which serves to sample antigens and to protect the epithelia that line the gastrointestinal, respiratory and genitourinary tracts (Ogra et al., 2001). Generally, foreign antigens and pathogens are encountered through normal physiological functions such as ingestion and inhalation, and the host thus has evolved specialised lymphoid tissue in order to

Table I. Edible plant hosts used for biopharmaceutical production

Plants	Advantages	Disadvantages
Simple plants		
Chlamydomonas reinhardtii, Lemna	Edible, clonal propagation, secretion into medium, genetic stability, containment	Scalability
Leafy crops		
Tobacco	Species with low alkaloid content are edible; high yield, efficient transformation system rapid scale-up	Low protein stability in harvested material; need of alkaloid removal
Alfalfa, clover	Edible leaves (uncooked); useful for edible animal vaccines, high yield and high protein content, clonal propagation,	Low protein stability in harvested material, presence of oxalic acid; potential for outcrossing in the field
Lettuce	Edible, useful for human vaccines	Low protein stability in harvested material
Cereals		
Maize, rice	Protein stability during storage, high yield, easy to transform and manipulate	
Wheat, barley	Protein stability during storage	Low yields, difficult to transform and manipulate
Legumes		
Soybean	Economical, high biomass, Well-suited for animal vaccines	Low expression levels, inefficient transformation system
Pea	Edible, high protein content in seeds	Low expression levels
Fruits and vegetables		
Banana	Eaten raw; clonally propagated	Little data on protein expression; expensive to grow in the greenhouse
Potato	Edible tubers; proteins stable in storage tissues; clonally propagated (low potential for outcrossing in the field)	Unpalatable; cooking might hamper immunogenicity of the vaccine
Carrot	Edible, proteins stable in storage tissues	
Tomato	Fruit is edible, containment in greenhouses. Industrial greenhouse culture and fruit processing well established	More expensive to grow, temperature sensitive, must be chilled after harvest

fight mucosal antigens and pathogens. These organised lymphoid tissues, including Peyer's patch (PP), appendix and solitary lymphoid nodules, collectively termed gut-associated lymphoid tissues (GALT), and nasopharyngeal-associated lymphoreticular tissue (NALT) contain all of the immunocompetent cells, including B cells, Th cells and CTLs, necessary for the development of effector and memory B and T cells upon antigen presentation by DCs and macrophages. Thus, the CMIS is integral to the development of a mucosal/oral vaccine.

The surface of PP is covered by a unique epithelial layer known as follicle-associated epithelium (FAE). The FAE is enriched with specialised antigen-sampling cells known as microfold (M) cells for the presence of irregular and shortened microvilli (Owen and Jones, 1974). M cells form an apparent pocket at the basal membrane site, a pocket that contains T cells, B cells, macrophages and DCs. It is believed that antigen uptake by M cells does not result in the degradation of the antigen, but rather in the delivery of the intact antigen to the underlying APC (Owen and Bhalla, 1983). In addition to the transport of lumenal antigens, M cells play a central role in the induction of immune response since they serve as a port of entry for pathogens. M cells preferentially transport particulate antigens such as viruses and bacteria from the intestinal lumen to the underlying follicle. Invasive strains of *Salmonella* and reovirus initiate the infection by invading the M cells of the PP (Wolf et al., 1981; Weinstein et al., 1998). Therefore, M cells are considered a gateway to the mucosal immune system. The PP are anatomically characterised as possessing a dome configuration, with the area just below the FAE known to be enriched with immunoglobulin A (IgA)-committed B cells, Th1 and Th2 lymphocytes, macrophages and DCs, all necessary for the induction of antigen- specific immune responses. After the uptake and delivery of antigen via M cells, the antigens are immediately processed and presented by DCs as APCs (Kelsall and Strober, 1996).

At least three DC subpopulations have been described in PP with specialized functions. In particular DC1 are capable of inducing Th1 differentiation for the subsequent development of cell-mediated immunity (CMI), while DC2 induce Th2 cells for the generation of IgA immune responses at mucosal effector sites (Banchereau, 2000; Iwasaki and Kelsall, 2001). After presentation of processed antigen by the distinctive subset of DCs the activation of T and B cells result in maturation of the latter into antibody secreting plasma cells. Mucosal immune responses are characterized by the production of secretory immunoglobulin A (S-IgA) and its specific transport across the epithelium to the mucosal surface. S-IgA acts as a first line of defense against pathogen colonization and infection by binding bacterial cells or virus particles and thus preventing their specific interactions with receptors on the mucosal cell surface. The PPs are also populated by B cells of the IgG type that are stimulated to produce serum IgG (Ogra et al., 2001). Thus, in principle, ingestion of plants expressing appropriate vaccine antigens can result in specific protective mucosal S-IgA and serum IgG as well as CMI responses against pathogens that display these antigens. Furthermore the activity of edible vaccines could be tailored (*i.e.* conformational antigens, intrinsic adjuvants) to the different component of the CMSI in order to tune the immune response on mucosal/serum or cell mediated responses.

3.3. Choice of host plant for edible vaccines

Many aspects have to be considered when choosing a host plant for the production of recombinant proteins (Menkhaus et al., 2004). The most appropriate expression system must be determined on a case-by-case basis: some plant hosts naturally accumulate a greater quantity of protein than others, or may yield the appropriate product while others fail. The vastly different presence of proteases in various crops and their location within the plant or cell, should also be considered. The plant host systems suitable for edible vaccines are listed in Table 1.

Tobacco (*Nicotiana tabacum*) is a dicotyledonous species and represents a well-established expression system for molecular farming (Fischer et al., 2004). The major advantages of tobacco include the well-established technology for gene transfer (both nuclear and chloroplast) and expression, high biomass yield, prolific seed production and the existence of a large-scale processing infrastructure. Because tobacco is neither a food nor a feed crop, there is little risk that tobacco material will contaminate either the food or feed chains. The chemical composition of tobacco leaf vary quite significantly among cultivars: many tobacco cultivars produce high levels of toxic alkaloids, but low-alkaloid varieties can be used for the production of pharmaceutical proteins. Due to its abundant production, tobacco leaf has been the main site for harvesting the expressed protein. Fresh leaves are generally blended and homogenized for protein extraction. As an alternative, recombinant proteins, including antibody derivatives, have been produced in tobacco cell cultures. Tobacco has been adopted as a platform system by several biotec companies and at least two companies (Planet Biotechnology Inc and the Meristem Therapeutics) have plant-derived pharmaceuticals undergoing phase II clinical trials.

Lettuce is also being investigated as a production host for edible recombinant vaccines, and has been used in one series of clinical trials for a vaccine against hepatitis B subunit vaccine (Kapusta et al., 1999). One of the main disadvantages of all leafy crops (tobacco, alfalfa, soybean and lettuce) is that in these species the recombinant proteins are synthesized in an aqueous environment and are often unstable, resulting in low yields. The leaves must be frozen or dried for transport, or processed soon after harvest to extract useful amounts of the product. On the contrary, the accumulation of recombinant proteins in seeds (cereals, rice, wheat and maize, and legumes, pea and soybean) allows long-term storage at ambient temperature. Seeds are able to accumulate proteins by virtue of specialized storage compartments (protein bodies and storage vacuole), which are derived from the secretory pathway and are protected from proteolytic degradation. Maize is now the main commercial production crop for recombinant proteins and vaccines for its high biomass yield, and for the easiness of transformation and scale-up. Thus, seed crops can be produced in huge quantities and preserved and shipped over long distances at room temperature, allowing a wider scope for global vaccination.

Protein expression in fruit and vegetable crops is advantageous since the edible organs can be consumed uncooked, unprocessed or partially-processed, making them ideal for the production of recombinant biopharmaceuticals. Potatoes have been frequently used as a convenient model system because transgenic lines can be efficiently produced using *Agrobacterium*-mediated DNA delivery. Potatoes have been used for the production of several proteins and represent, up to now, the major system for vaccine production. Transgenic potato tubers have been administered to humans in at least three clinical trials carried out to date. However, a more palatable system is preferred, as the vaccine must be consumed uncooked to prevent heat denaturation of the antigens. Tomatoes are more palatable than potatoes and have other advantages, including high biomass yield and an increased containment that is offered by growth in greenhouses. Tomato was used to express orally immunogenic respiratory syncytial virus (RSV) fusion (F) protein in fruit (Sandhu et al., 2000).

Carrot is emerging as a promising edible species for molecular farming. It is one of the most important vegetables used worldwide for human and animal consumption. The carrot plant is biennial and in the first year the plant produces the fleshy taproot, which is edible and only if left in the ground, plants flower in the second year. Carrot chloroplast genome is transmitted strictly through maternal inheritance. Thus carrot is environmentally safe and is doubly protected against transgene flow via pollen and seeds to achieve zero-tolerance on transgene flow advocated for food crops. There is no need to cook carrots and this would preserve the structural integrity of therapeutic proteins.

Banana is an attractive host for the production of vaccines as it is widely cultivated in developing countries and transformation systems for this plant has been reported (May et al., 1995).

Oilcrops are useful hosts for protein production because the oil bodies can be exploited to simplify protein isolation (Seon et al., 2002). Finally, there are more diverse plant species that have been used to produce recombinant proteins, which can be easily contained and transformed. They include a unicellular green alga, *Chlamydomonas reinhardtii,* in which chloroplast-targeted transgenes were used to express antibodies (Mayfield et al., 2003), the aquatic plant Lemna (duckweed) (Gasdaka et al., 2003) and the moss *Physcomitrella patens* (Decker and Reski, 2004).

3.4. Clinical trials with edible plant vaccines

Studies in animals and humans have shown that ingestion of transgenic plants containing vaccine proteins causes production of antigen-specific antibodies in serum and mucosal secretion. To date, three published studies of edible plant vaccines in human subjects (Tacket et al., 1998, 2000; Kapusta et al., 1999) have shown promising results using three different recombinant antigens, each stimulating serum and in some cases mucosal antibody responses. Two vaccines were directed against agents of gastroenteritis and the third one against hepatitis B virus.

The first ever clinical study to examine an edible plant vaccine (Tacket et al., 1998) was performed at the Center for Vaccine Development (Baltimore, MD). The vaccine was designed against the labile toxin of enterotoxigenic *E. coli.* The B subunit of this toxin, as well as that of Cholera toxin, is non-toxic and mediates the binding to GM1 gangliosides present on epithelial cell surfaces (Sixma et al., 1991). Both LT-B (Mason et al., 1998) and CT-B (Arakawa et al., 1998) expressed in transgenic potatoes produced toxin-protective gut antibody responses after ingestion by mice, showing that the plant-produced proteins folded correctly and assembled native GM1-binding pentameric complexes. LT-B potatoes were used in a double-blind study, the volunteers ate either 50 or 100 g of raw LT-B potatoes (containing ~375 or ~750 μg LT-B) or non-transgenic placebo potatoes, administered in three weekly doses. Ten out of 11 subjects in the LT-B groups developed at least four-fold increases in levels of serum LT-neutralizing anti-LT-B antibodies. Five of ten volunteers showed increases in anti-LT-B IgA in stool samples. This pivotal study proved that ingestion of transgenic plant material could be used effectively to stimulate both IgG and IgA potentially protective antibody responses in humans.

The second clinical study was conducted against a viral enteritis agent, the Norwalk virus. Recombinant viral capsid proteins of this virus as well as of other viruses including papillomavirus, can assemble virus-like particles (VLPs) that can be efficiently sampled by M cells in the gut. These VLPs mimic the authentic virus particles (Jiang et al., 1992), reacting with sera from infected humans, and resisting gut degradation (Ball et al., 1999). When VLPs expressed in transgenic tobacco and potato were fed to mice they stimulated serum IgG and gut IgA specific for Norwalk virus (Mason et al., 1996). Transgenic potatoes for the Norwalk virus capsid protein (NVCP) were used in clinical trials (Tacket et al., 2000). Only ~50% of the NVCP subunits were assembled as VLPs in the potato cells, probably a result of low concentration. Twenty adult volunteers ingested either two or three doses each of 150 g raw transgenic potato containing 215–750 μg NVCP (expression was variable). 19 of the 20 subjects showed significant increases in the numbers of specific anti-NVCP antibody-secreting cells of the IgA subtype, and six out of 20 developed increases in IgG antibody-secreting cells. These results demonstrated that orally delivered plant-expressed VLPs are able to stimulate immune responses.

The third clinical study was conducted against the HbsAg, the main envelope protein of hepatitis B virus (HBV). When expressed in transgenic tobacco, HBsAg makes VLPs with an average size of 22 nm (Mason et al., 1992) and structurally similar to VLPs produced in transgenic yeast, which is the material used in the currently used injectable hepatitis B vaccine (McAleer et al., 1984). The tobacco HBsAg was immunogenic when injected into mice, and stimulated all IgG subclasses and IgM (Thanavala et al., 1995). Electron microscopy of sectioned HBsAg transgenic potato leaf showed that HBsAg VLPs accumulate in ER-derived vesicles (Kong et al., 2001), thus providing encapsulation and possibly protection from stomach digestion. These transgenic potatoes produced anti-IgG responses on oral delivery in mice (Kong et al., 2001). The published clinical study was performed with HBsAg expressed in transgenic lettuce and delivered by ingestion. Oral delivery stimulated serum anti-HBsAg IgG in naive individuals (Kapusta et al., 1999) with a titre above the 10 international units per litre (IU L-1) that is accepted to correlate with protection (Kapusta et al., 1999). It is likely that HbsAg was intracellularly encapsulated as it was found in potato leaf, thus creating a particulate structure that may enhance sampling by M cells of the GALT (Ogra et al., 2001). These data confirmed the importance of the correct antigen expression in the plant tissue in order to elicit a valid immune response on oral delivery in humans.

4. Edible Vaccines Against HPV

HPV infection is usually cleared by the host without consequence but in a number of people the infection becomes latent. In these individuals other factors may lead to the progression to chronic disease with clinical/histological appearance of low or high grade Squamous Intraepithelial Lesions (SIL). These clinical conditions may regress but may progress to cancer (Tindle, 2002). In these stages of the disease a therapeutic intervention has the efficacy to cure the disease (Low-High SIL) and to prevent tumour development. In the US, Europe and Japan about 5-6 million women a year are diagnosed with a pre-neoplastic lesion caused by HPV, and the number of affected women in developing countries is much larger. There are major activities driven by large pharmaceutical companies to develop prophylactic vaccines that induce HPV-neutralizing antibodies aiming to prevent the infection and, consequently, to reduce the incidence of cancer. Due to the long latency period between infection and cancer, the benefits of a prophylactic vaccination would be visible after decades. Thus it is necessary to work also on therapeutic vaccines targeting those individuals who are already infected or that have precancerous lesions caused by oncogenic strains of HPV. A successful immunotherapeutic would rapidly clear these lesions, decreasing the incidence of surgical interventions and may provide protection against future exposure. Several therapeutic and prophylactic HPV-specific vaccines have been generated and their production is discussed in more details elsewhere in the book. Some of them are currently in phase II and III clinical trials and the first promising results have been reported (Koutsky et al., 2002; Billich, 2003). Nevertheless it is important to focus on second-generation vaccines, through technologies that may render these vaccines more suitable for developing countries. Edible vaccines match all the requisites for both prophylactic and therapeutic successful vaccination in developing country.

4.1. Prophylactic vaccine

As already mentioned, a number of different antigens has already been produced in plants and their oral delivery led to the induction of a protective immune response. Thus it can be argued that production of an edible plant-derived papillomavirus vaccine is feasible. Moreover it has been demonstrated that papillomavirus capsid antigens are immunogenic when delivered orally (Gerber et al., 2001; Rose et al., 1999), most likely because they are recognized by the immune system due to their virus-like structure. As already reported in this chapter, while the majority of antigens in the intestine are ignored by the immune system, M cells located in the Peyer's patches selectively take up particulate antigens such as virus particles and transport them from the intestine to underlying follicles. The activation of a mucosal immune response leads to the production of secretory IgA antibodies but also of serum IgG antibodies. A prerequisite for the induction of a mucosal immune response against viral capsid antigens by edible vaccines, however, is that the proteins are presented to the M cells in a particulate state. This requires that the capsid proteins are able to assemble into VLPs or capsomeres within the transgenic plants and that upon ingestion and passage through the low pH of the stomach, these particles remain intact. In general, it can be expected that the sturdy plant cell walls provide a protective environment for the expressed antigens. In addition, it has been also demonstrated that the papillomavirus capsid antigens remain immunogenic after exposure to low pH through oral delivery (Gerber et al., 2001; Rose et al., 1999).

Since the first demonstration that Hepatitis B virus surface antigen and Norwalk virus capsid protein can assemble into VLPs when expressed in transgenic plants (Mason et al., 1992, 1996), a number of recent studies demonstrate that VLPs from HPV can be also produced in a variety of plant species including tobacco, potato and tomato (Warzeka et al., 2003; Biemelt et al., 2003; Varsani et al., 2003; Franconi et al., unpublished).

The L1 major capsid protein of HPV 11 was utilized to produce transgenic plants. To enhance L1 expression in plant, internal polyadenylation signals, intron splice recognition sequences and mRNA destabilizing elements were identified and removed (Warzeka et al., 2003). A plant codon-optimized version of the HPV11 L1 gene, with or without its nuclear localization signal (NLS), was first transiently expressed in tobacco. Full-length L1 was localized in the plant cell nuclei. Constructs, either with or without NLS, were then used to transform potato plants, and a few transgenic regenerants were obtained. The expression of the full-length L1 protein seemed to affect plant viability, while removal of the carboxy-terminal NLS sequence allowed L1 expression and self-assembling into VLPs. The plant-expressed L1 retained immunological properties comparable to those of native HPV virions and ingestion of transgenic L1 potato was associated with activation of an anti-VLP humoral immune response in mice that was qualitatively similar to that induced by VLP parenteral administration. This response was enhanced significantly by subsequent oral boosting with purified insect cell-derived VLPs, showing

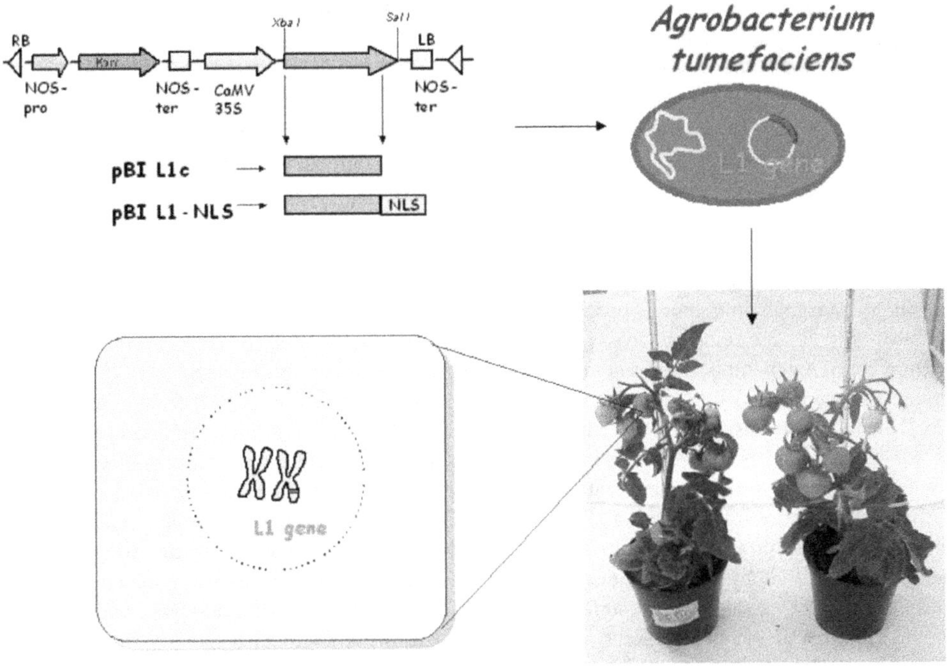

Figure 3. HPV16 L1 transgenic *Lycopersicon esculentum* (Micro-Tom): schematic representation of the experimental procedures. The pBI L1c and pBI L1-NLS are the recombinant plasmids for *Agrobacterium*-mediated transformation containing the major capsid protein gene (L1) of HPV16 without or with the original L1 Nuclear Localization Signal (NLS), respectively. The viral sequences were cloned as *Xba I - Sal I* fragments under the control of the CaMV 35S promoter. The plasmids were transferred in *Agrobacterium tumefaciens* that was thereafter utilised to transform the miniature tomato (Micro-Tom). Transgenic plants with L1 gene insertion were obtained after regeneration on selective medium and stabilized after repeated cycles of self-pollination. RB: right border; LB: left border; NOS-pro and NOS-ter: transcription promoter and terminator from the nopaline synthase gene of *Agrobacterium tumefaciens;* Kan^r: kanamycine resistance gene; CaMV 35S: cauliflower mosaic virus 35S promoter.

that plant-derived VLPs generated antigen-specific memory cells. The requirement for coadministered adjuvant for induction of immune response was probably related to VLP low concentration (20 ng/g fresh tuber) and low oral dose, ~ 100 ng, corresponding to 1/10 the amount previously determined to represent the minimal oral dose of purified insect cell-derived VLPs without adjuvant (Warzeka et al., 2003).

The L1 gene of the 'high risk' HPV 16 was used to construct transgenic tobacco and potato plants (Biemelt et al., 2003). In this case, all attempts to express either the original, unmodified L1 gene or an L1 gene with a codon usage optimized for expression in plants failed. Surprisingly, small amounts of the protein were detected using an L1 gene optimized for expression in human cells. Introduction of the translational enhancer Ω derived from the tobacco mosaic virus strongly increased transcript stability and resulted in accumulation of L1 protein to approximately 0.5 to 0.2% of total soluble protein in transgenic tobacco and potato plants, respectively. The plant-derived L1 protein displayed conformation-specific epitopes and assembled into VLPs. Plant-derived VLPs (40 ng/dose per mouse) were able to induce a humoral immune response similar to that of VLPs derived from baculovirus-infected insect cells. Feeding of tubers from transgenic potatoes to mice induced an anti-L1 antibody response in 3 out of 24 mice, although this response was only transient in two of them. However, half of the animals fed with transgenic tubers developed L1 specific-antibodies after boosting with an injected sub-immunogenic dose, indicating that oral ingestion of L1-positive tubers had primed a humoral response against the L1 protein.

Another work showed that it is not necessary to perform significative modification of the HPV16 L1 gene to get L1/VLPs expression in plants (Varsani et al., 2003). The unmodified HPV 16 L1 gene, with and without nuclear localisation signals, was used to transform *Nicotiana tabacum* cv. Xanthi plants. Both constructs were capable of expressing low amounts of L1 protein that self-assembled into capsomers and VLPs. Unmodified L1 forms VLPs that are structurally and antigenically identical to those produced by recombinant baculovirus in insect cells. Moreover, concentrated extracts from transgenic plant could be used to elicit a specific anti-HPV-16 L1 response in rabbits (either subcutaneous or intramuscular immunizations).

In our laboratories we recently succeeded in obtaining transgenic tomato expressing the HPV 16 L1 gene. The unmodified L1 gene, with or without NLS, was cloned into a pBI-derived vector under the control of the CaMV 35S promoter (Figure 3). The plasmids were transferred into the A. tumefaciens strain EHA105 by electroporation and used for leaf disc transformation of the miniature Lycopersicon esculentum cultivar Micro-Tom (microtomato) (Meissner et al., 1997). Tomato is more palatable for human consumption than potato and tobacco. Micro-Tom represents a powerful system since it is possible to analyze simultaneously many plants grown in the greenhouse (due to their small size) and to introduce easily a transgene into more appropriate commercial tomato cultivars by standard crosses. We obtained a large number of regenerants and did not notice any adverse effect on regeneration due to the presence of NLS. Several plants were positive for L1 mRNA expression. As for tobacco and potato, the quantity of L1 was low but, by utilizing appropriate modification to stabilize transcripts and/or subcellular targeting signals, we are working towards the production of an edible product be able to induce oral immunity (Franconi et al., unpublished). In conclusion, no general rules for L1 expression have emerged. Major problems in L1 production seem to be related to toxicity and to low yield of capsomers/VLP per gram in the plant system. Modifications in viral sequences such as optimization of codon usage, removal of nuclear leader sequence as well as fusion with sequences known to stabilize the transcripts might be effective in obtaining higher L1 expression. The studies have been conducted in different plant/virus models and the best conditions to obtain VLPs must be checked for each VLP and plant species. Nevertheless two proof-of-concept studies have clearly demonstrated that HPV recombinant edible vaccines are able to induce the production of specific IgG, indicating the feasibility of this approach (Biemelt et al., 2003; Warzeka et al., 2003). However, the immunogenicity of edible VLP vaccines, although detectable, needs further improvement that can be achieved by enhancement of transgene expression.

4.2. Therapeutic vaccine

Experimental therapeutic vaccines are being developed to treat persisting HPV infection, based on viral non-structural proteins (*i.e.* E6 and E7), and designed to evoke antigen-specific T-cell-directed immune effector mechanisms. Many different potential HPV therapeutic vaccines eliminate tumours in animal models and some evoke specific cell-mediated immune responses in early phase human trials (see other chapters in the book). Definitive evidence for clinical efficacy and definition of surrogate markers of efficacy await further study. The major biological aspects of HPV lesions that hamper therapeutic vaccination are the poor presentation of viral antigens that are expressed at low levels, and the poor trafficking of effector T-cell populations to non-inflamed mucosal/skin sites. Effective methods to overcome these challenges are at present lacking. However, papillomavirus-specific immune responses can be induced in humans, enhancement of innate immunity seems to assist the natural resolution of persistent infection with papillomavirus, and vaccines that promote trafficking of effectors preferentially to the epithelia are under development, allowing cautious optimism that effective immunotherapy for papillomavirus infection can be achieved. Early phase clinical trials cannot clearly define clinical benefits (Frazer, 2004). Where these are indicated, they might, in part, reflect the induction of innate immunity, as the responses seen are not specific to the disease caused by the papillomavirus genotype to which the vaccine is targeted.

Most studies in the literature show that edible vaccines can readily stimulate humoral immune responses but there are little few studies on T-cell responses. As stated before mucosal immunization produces antigen-specific Th1- and/or Th2-type responses by distinct subtypes of DCs that induce the differentiation of Th1 or Th2 type cells from naïve precursors. Depending on the nature of antigen, adjuvant and antigen delivery vehicle used, different responses can be activated. For example, administration of intracellular pathogens such as *Salmonella* led to the formation of Th1 cells producing IFN-γ, IL-2 and TNFα (Hess et al., 1996). In this regard, bacterial infection activates DCs via initial stimulation of the toll-like receptors (TLRs: *e.g.* TLR2 and TLR4) and induces production of mainly Th1-type inducing cytokines such as IL-12 (Reis et al., 1999). Therefore, TLR-stimulated DCs tend to direct T cell differentiation to the Th1 cell type. Some plants (*Nicotiana benthamiana*) possess adjuvating activity to a Th1 stimulation perhaps via TLR (Franconi et al., 2002) when delivered by parenteral route. Freeze-dried *N. benthamiana* leaves, with increasing amounts of the HPV antigens E7, mixed to feedstuff, has been employed in the mouse model demonstrating no toxicity of these dried crude extract (Franconi et al. unpublished) and the feasibility of such approach. Besides the intrinsic adjuvanting activity, other transgenic plants could be employed together with the appropriate adjuvant. There are indications that ISCOMs in oral administration induce strong CMI (Mowat et al., 1999). ISCOMs are cage-like particles, approximately 40nm in size, that are formed spontaneously when cholesterol is mixed with Quil A saponin, an immunostimulatory adjuvant derived from the bark of *Quillaja saponaria*. In addition it has been recently reported that orally administered QS-21, a pure fraction of Quil A saponin with low toxicity, showed adjuvant activity for the induction of both systemic and mucosal immunity to co-administered protein antigen (Boyaka et al., 2001). Finally the possibility to utilize CpG DNA motif as CMI adjuvating molecule has to be taken into account, as it was reported that oral administration of hepatitis B surface antigen or tetanus toxoid (TT) together with CpG in mice induced predominantly

Th1-mediated IgG2a type responses, and cell-mediated systemic immunity (McCluskie et al., 2001).

Nevertheless the research has to be more oriented toward edible vaccine with intrinsic CMI adjuvanting acitivity rather than in adjuvating compounds to be added to the edible products. Only an edible vaccine with intrinsic adjuvating activity will fully match the requirement for a cheap and safe vaccine for the developing countries.

At the present the production of edible vaccine that induce CMI response to HPV infected cells are in progress but it remains uncertain whether they will be able to eradicate pre-existing HPV infection/transformation. Further studies on immunotherapy are requested and the development of surrogate markers is needed to ascertain the efficacy of a HPV therapeutic vaccine. Nevertheless, the stimulation of innate immunity seems to play a central role in the therapeutic activity of the vaccine and plant products seem to be promising in stimulating innate and adaptive immunity. This indication justifies the pursuit of further studies on edible therapeutic vaccines.

5. GMO Debate and Safety Concerns

Plants that produce edible vaccines are genetically modified (GM) organisms and, in the public perception, put the environment at risk through either unintended release or genetic crossing with related wild type plants. Many public bodies stress the importance of risk assessment and environmental impact of GM plants, although there is a paucity of scientific or evidence-based data to actually substantiate risk.

The potential environmental impact of transgenic plants producing vaccines has to be determined for each single preparation (although the expressed genes are not expected to confer a selective advantage on transgenic plants) since the variety of genes that can be inserted into plants means that no two GMOs are the same. To make large-scale production of plant-derived vaccines the consequence on the environmental release has to be considered. In general an edible vaccine would contain molecules that do not represent a health risk either to animals or to humans. Nevertheless the recent occurrence of contamination of the food supply by plant-derived pharmaceuticals proteins highlights the need to produce pharmaceuticals in non-food or feed crops. Concerns about transfer of genes to non-target organisms remain to be addressed. However, it is likely that advances in plant biotechnology over the next decade will confront many of these issues. One such advance is the development of chloroplast transformation. As previous described the chloroplast genome is maternally inherited. This means that the transgene and any protein expressed following chloroplast transformation are not present in pollen, thereby reducing the risk of transmission of the transgene to neighbouring crops or weed species by cross-pollination. Alternatively, transient expression by plant virus vector would keep transmission to a minimum, however risk of contamination by recombinant plant virus limits the utilisation of this technology to confined areas. Finally another important aspect is the choice of plant species like tobacco and the tomato cultivar Micro-Tom that are non-food and non-feed crops, easy to manipulate through genetic engineering (both nucleus and chloroplast), and especially distinguishable from normal food crops. It is not superfluous to underline that plants producing vaccines must be considered as pharmaceuticals and never food for humans or animals. Expression of vaccines in commonly consumed foods does not mean that they will become a component of regular diets, otherwise repeated exposures to an oral antigen could produce immunological tolerance. The plant (or food) is not the final pharmaceutical product, just as microbial or yeast pharmaceutical production systems do not represent the final product. Rather, the plant represents one step in a complex, multi-step pharmaceutical production process. As medicines, edible vaccines have to be produced in controlled conditions, following traditional regulatory requirements and should be administered appropriately.

6. Conclusion

Many different subunit vaccine antigens in a number of plant systems have been reported to be immunogenic when delivered orally in humans and other animals inducing, in some cases, protection from virulent challenge. To fully understand the potential of edible vaccines much more work is needed. The low expression of antigens in transgenic plants, the protection of antigens from gut degradation, the potential for developing oral tolerance, and the induction of cell mediated immunity are the main aspects on which further researches are focused. The involvement of a growing number of academic and industrial researchers in the field of plant vaccine technologies will probably lead to eventual success in addressing these questions.

Nevertheless edible vaccines are a reality and the promising results from clinical trials warrant their utilization in the near future for a number of infectious and non-infectious diseases. HPV edible vaccines can be considered a valid second generation of prophylactic vaccines particularly suitable for developing countries, whereas therapeutic HPV vaccines will require more basic studies before their utilization. From an optimistic point of view, a combined prophylactic and therapeutic vaccination is expected in the short term to prevent and accelerate the elimination of early HPV infection and in the long run to eradicate the disease.

7. Postface

The authors invite the readers to a dinner where (HPV) edible vaccines will be served in ten years time (formal dress not required).

References

Allison, A.R. (1930). The Satyricon/from the Latin of Petronius; translated and introduced by Alfred R. Allison. (New York: The Panurge Press).

Arakawa, T., Yu, J., Chong, D.K., and Langridge, W.H. (1998). Efficacy of a food plant based oral cholera toxin B subunit vaccine. Nat. Biotechnol. *16*, 292–297.

Awram, P., Gardner, R.C., Forster, R.L., and Bellamy, A.R. (2002). The potential of plant viral vectors and transgenic plants for subunit vaccine production. Adv. Virus Res. *58*, 81-124.

Ball, J.M., Graham, D.Y., Opekun, A.R., Gilger, M.A., Guerrero, R.A., and Estes, M.K. (1999) Recombinant Norwalk virus-like particles given orally to volunteers: phase I study. Gastroenterol. *117*, 40–48.

Banchereau, J., Briere, F., Caux, C., Davoust, J., Lebecque, S., Liu, Y.J., Pulendran, B., and Palucka, K. (2000). Immunobiology of dendritic cells. Annu. Rev. Immunol. *18*, 767–811.

Bardor, M., Faveeuw, C., Fitchette, A.C., Gilbert, D., Galas, L., Trottein, F., Faye, L., and Lerouge, P. (2003). Immunoreactivity in mammals of two typical plant glyco-epitopes, core α (*1*,3)-fucose and core xylose. Glycobiology *13*, 427-434.

Biemelt, S., Sonnewald, U., Galmbacher, P., Willmitzer, L., and Muller, M. (2003). Production of human papillomavirus type 16 virus-like particles in transgenic plants. J. Virol. *77*, 9211–9220.

Billich, A. (2003). HPV vaccine MedImmune/GlaxoSmithKline. Curr. Opin. Investig. Drugs *4*, 210–213.

Boyaka, P.N., Marinaro, M., Jackson, R.J., van Ginkel, F.W., Cormet-Boyaka, E., Kirk, K.L., Kensil, C.R. and, McGhee, J.R. (2001). Oral QS-21 requires early IL-4 help for induction of mucosal and systemic immunity. J. Immunol. *166*, 2283-2290.

Chapman, S., Kavanagh, T., and Baulcombe, D. (1992). Potato virus X as a vector for gene expression in plants. Plant J. *2*, 549-557.

Center for Disease Control. (1998). Preventing emerging infectious diseases: a strategy for the 21st century. Morb. Mortal. Wkly. Rep. *47*, 1-4.

Daniell, H., Streatfield, S.J., and Wycoff, K. (2001a). Medical molecular farming: production of antibodies, biopharmaceuticals and edible vaccines in plants. Trends Plant Sci. *6*, 219-226.

Daniell, H., Lee, S.B., Panchal, T, and Wiebe, P.O. (2001b). Expression of cholera toxin B subunit gene and assembly as functional oligomers in transgenic tobacco chloroplasts. J. Mol. Biol. *311*, 1001–1009.

Daniell, H., Khan, M.S., and Alison, L. (2002a). Milestones in chloroplast genetic engineering: an environmentally friendly era in biotechnology. Trends Plant Sci. *7*, 84–91.

Daniell, H., and Dhingra, A. (2002b). Multigene engineering: dawn of an exciting new era in biotechnology. Curr. Opin. Biotechnol. *13*, 136–41.

Daniell, H. (2002c). Molecular strategies for gene containment in transgenic crops. Nat Biotechnol. *20*, 581–6.

DeCosa, B., Moar, W., Lee, S.B., Miller, M., and Daniell, H. (2001). Overexpression of the Bt Cry2Aa2 operon in chloroplasts leads to formation of insecticidal crystals. Nat. Biotechnol. *19*, 71–74.

Decker, E.L., and Reski, R. (2004). The moss bioreactor. Curr Opin. Plant Biol. *7*, 166-170.

Finnegan, J. and McElroy, D. (1994). Transgene inactivation: plants fight back! Biotechnology *12*, 883–888.

Fischer, R., Stoger, E., Schillberg, S., Christou, P., and Twyman, R. M. (2004). Plant-based production of biopharmaceuticals. Curr. Opin. Plant Biol. *7*, 152-158.

Franconi, R., Di Bonito, P., Dibello, F., Accardi, L., Muller, A., Cirilli, A., Simeone, P., Donà, G., Venuti, A., and Giorgi, C. (2002). Plant-derived human papillomavirus 16 E7 oncoprotein induces immune response and specific tumor protection. Cancer Res. *62*, 3654-58.

Frazer, I.H. (2004). Prevention of cervical cancer through papillomavirus vaccination. Nat. Rev. Immunol. *4*, 46-54.

Gasdaska, J.R., Spencer, D., and Dickey, L. (2003). Advantages of therapeutic protein production in the aquatic plant Lemna. Bioprocess J. *2*, 49-56.

Gerber, S., Lane, C., Brown, D.M., Lord, E., DiLorenzo, M., Clements, J.D., Rybicki, E., Williamson, A.L, and Rose, R.C. (2001). Human papillomavirus virus-like particles are efficient oral immunogens when coadministered with Escherichia coli heat-labile enterotoxin mutant R192G or CpG DNA. J. Virol. *75*, 4752–4760.

Giddings, G. (2001). Transgenic plants as protein factories. Curr. Opin. Biotechnol. *12*, 450–454.

Gleba, Y., Marillonnet, S., and Klimyuk, V. (2004). Engineering viral expression vectors for plants: the 'full virus' and the deconstructed virus' strategies. Curr. Opin. Plant Biol. *7*, 182-188.

Gomord, V. and Faye, F. (2004). Posttranslational modification of therapeutic proteins in plants. Curr. Opin. Plant Biol. *7*, 171–181.

Hess, J., Ladel, C., Miko, D., and Kaufmann, S.H. (1996). Salmonella typhimurium-aroA infection in gene-targeted immunodeficient mice: major role of CD4þ TCR-alpha beta cells and IFN-gamma in bacterial clearance independent of intracellular location. J. Immunol. *156*, 3321–3326.

Hobbs, S., Warkentin, T.D., and DeLong, C.M. (1993). Transgene copy number can be positively or negatively associated with transgene expression. Plant Mol. Biol. *21*, 17–26.

Hood, E.E., Woodard, S.L., and Horn, M.E. (2002). Monoclonal antibody manufacturing in transgenic plants – myths and realities. Curr. Opin. Biotechnol. *13*, 630–635.

Iwasaki ,A. and Kelsall, BL. (2001). Unique function of CD11bþ, CD8þ, and double-negative Peyer's patch dendritic cells. J Immunol. *166*, 4884–4890.

Jiang, X., Wang, M., Graham, D.Y., and Estes, M.K. (1992). Expression, self-assembly, and antigenicity of the Norwalk virus capsid protein. J. Virol. *66*, 6527–6532

Kapila, J., DeRycke, R., VanMontagu, M., and Angenon, G. (1997). An Agrobacterium-mediated transient gene expression system for intact leaves. Plant Sci.*122*, 101-108.

Kapusta, J. Modelska, A., Figlerowicz, M., Pniewski, T., Letellier, M., Lisowa, O., Yusibov, V., Koprowski, H., Plucienniczak, A., and Legocki, A.B. (1999). A plant-derived edible vaccine against hepatitis B virus. FASEB J. *13*, 1796–1799.

Kelsall, B.L. and Strober, W. (1996). Distinct populations of dendritic cells are present in the subepithelial dome and T cell regions of the murine Peyer's patch. J. Exp. Med. *183*, 237–247.

Kong, Q., Richter L., Yang, Y.F., Arntzen, C.J., Mason, H.S., and Thanavala, Y. (2001). Oral immunization with hepatitis B surface antigen expressed in transgenic plants. Proc. Natl. Acad. Sci. USA *98*, 11539–11544.

Koprowski, H. and Yusibov, V. (2001). The green revolution: plants as heterologous expression vectors. Vaccine *19*, 2734-2741.

Koutsky, L.A., Ault, K.A., Wheeler, C.M., Brown, D.R., Barr, E., Alvarez, F.B., Chiacchierini, L.M., and Janse,n K.U. (2002). A controlled trial of a human papillomavirus type 16 vaccine. N. Engl. J. Med. *347*, 1645–1651.

Kumar, S., Dhingra, A., and Daniell, H. (2004). Plastid-expressed beatine aldehyde dehydrogenase gene in carrot cultured cells, roots and leaves confers enhanced salt tolerance. Plant Physiology *136*, 1-12.

Ma, J.K., Drake, P.M.W., and Christou, P. (2003). The production of recombinant pharmaceuticals proteins in plants. Nat. Rev. Genet. *4*, 794-805.

Mason, H.S., Lam, D.M., and Arntzen, C.J. (1992). Expression of hepatitis B surface antigen in transgenic plants. Proc. Natl. Acad. Sci. USA *89*, 11745–11749.

Mason, H.S., Ball, J.M., Shi, J.J., Jiang, X., Estes, M.K., and Arntzen, C.J. (1996). Expression of Norwalk virus capsid protein in transgenic tobacco and protein and its oral immunogenicity in mice. Proc. Natl. Acad. Sci. U. S. A *93*, 5335–5340.

Mason, H.S., Haq, T.A., Clements, J.D., and Arntzen, C.J. (1998). Edible vaccine protects mice against *E. coli* heat-labile enterotoxin (LT): potatoes expressing a synthetic LT-B gene. Vaccine *16*, 1336–1343.

May, G.D. et al. (1995). Generation of transgenic banana (*Musa acuminata*) plants via agrobacterium-mediated transformation. Biotechnology *13*, 487–492

Mayfield, S.P., Franklin, S.E., and Lerner, R.A. (2003). Expression and assembly of a fully active antibody in algae. Proc. Natl. Acad. Sci. USA *100*, 438-442.

McAleer, W.J., Buynak, E.B., Maigetter, R.Z., Wampler, D.E., Miller, W.J., and Hilleman, M.R. (1984). Human hepatitis B vaccine from recombinant yeast. Nature *307*, 178–180.

McCluskie, M.J., Weeratna, R.D., and Davis, H.L. (2001). The potential of oligodeoxynucleotides as mucosal and parenteral adjuvants. Vaccine *19*, 2657–2660.

McCormick, A.A., Kumagai, M.H., Hanley, K., Turpen, T.H., Hakim, I., Grill, L.K., Tusè, D., Levy, S., and Levy, R. (1999). Rapid production of specific vaccines for lymphoma by expression of the tumor-derived single-chain Fv epitopes in tobacco plants. Proc. Natl. Acad. Sci. USA *96*, 703-708.

McCormick, A.A., Reinl, S.J., Cameron, T.I., Vojdani, F., Fronefield, M., Levy, R., and Tusè, D. (2003). Individualized human scFv vaccines produced in plants: humoral anti-idiotype responses in vaccinated mice confirm relevance to the tumor Ig. J. Immunol. Methods *278*, 95-104.

Meissner, R., Jacobson, Y., Melamed, S., Levyatuv, S., Shalev, G., Ashri, A, Elkind, Y., and Levy, A. (1997). A new model system for tomato genetics. Plant J. *12*, 1465-1472.

Menkhaus, T.J., Bai, Y., Zhang, C., Nikolov, Z.L., and Glatz, C.E. (2004). Considerations for the recovery of recombinant proteins from plants. Biotechnol. Prog. *20*, 1001-1014.

Mowat, A.M., Smith, R.E., Donachie, A.M., Furrie, E., Grdic, D., and Lycke, N. (1999). Oral vaccination with immune stimulating complexes. Immunol. Lett. *65*, 133–140.

Ogra, P.L., Faden, H., and Welliver, R.C. (2001). Vaccination strategies for mucosal immune responses. Clin. Microbiol. Rev. *14*, 430–445.

Owen, R.L.and Jones, A.L. (1974). Epithelial cell specialization within human Peyer's patches; and ultrastructural study of internal lymphoid follicles. Gastroenterol. *66*, 189–203.

Owen, R.L.and Bhalla, D.K. (1983). Cytochemical analysis of alkaline phosphatase and esterase activities and of lectin-binding and anionic sites in rat and mouse Peyer's patch M cells. Am. J. Anat. *168*, 199–212.

Reis, E., Sousa, C., Sher, A., and Kaye, P. (1999). The role of dendritic cells in the induction and regulation of immunity to microbial infection. Curr. Opin. Immunol. *11*, 392–399.

Rose, R.C., Lane, C., Wilson, S., Suzich, J.A., Rybicki, E., and Williamson, A.L. (1999). Oral vaccination of mice with human papillomavirus virus-like particles induces systemic virus-neutralizing antibodies. Vaccine *17*, 2129–2135.

Ruf, S., Hermann, M., Berger, I.J., Carrer, H., and Bock, R. (2001). Stable genetic transformation of tomato plastids and expression of a foreign protein in fruit. Nat. Biotechnol. *19*, 870–875.

Ruiz, O.N., Hussein, H., Terry, N., and Daniell, H. (2003). Phytoremediation of organomercurial compounds via chloroplast genetic engineering. Plant Physiol. *132*, 1–9.

Sandhu, J.S. Krasnyanski, S.F., Domier, L.L., Korban, S.S., Osadjan, M.D., and Buetow, D.E. (2000). Oral immunization of mice with transgenic tomato fruit expressing respiratory syncytial virus-F protein induces a systemic immune response. Transgenic Res. *9*, 127–135.

Sanford, J.C., Smith, F.D., and Russell, J.A. (1993). Optimizing the biolistic process for different biological applications. Methods Enzymol. *217*, 483–509.

Seon, J.-H., Szarka, J.S., and Moloney, M.M. (2002). A unique strategy for recovering recombinant proteins from molecular farming:affinity capture on engineered oilbodies. J. Plant Biotechnol. *4*, 95-101.

Sixma, T.K., Pronk, S.E, Kalk, K.H., Wartna, E.S., van Zanten, B.A., Witholt, B., and Hol, W.G. (1991). Crystal structure of a cholera toxin-related heat-labile enterotoxin from *E. coli*. Nature *351*, 371–377.

Streatfield, S.J., Jilka, J.M., Hood, E.E., Turner, D.D., Bailey, M.R., Mayor, J.M., Woodard, S.L., Beifuss, K.K., Horn, M.E., Delaney, D.E., Tizard, I.R., and Howard, J.A. (2001). Plant-based vaccines: unique advantages. Vaccine *19*, 2742–2748.

Svab, Z. and Maliga,, P. (1993). High-frequency plastid transformation in tobacco by selection for a chimeric aadA gene. Proc. Natl. Acad. Sci. USA *90*, 913–917.

Tacket, C.O., Mason, H.S., Losonsky, G., Clements, J.D., Levine, M.M., and Arntzen, C.J. (1998). Immunogenicity in humans of a recombinant bacterial-antigen delivered in transgenic potato. Nat. Med. *4*, 607-609.

Tacket, C.O., Mason, H.S., Losonsky, G., Estes, M.K., Levine, M.M., and Arntzen, C.J. (2000). Human immune responses to a novel Norwalk virus vaccine delivered in transgenic potato. J. Infect. Dis. *182*, 302-305.

Thanavala, Y., Yang, Y.F., Lyons, P, Mason, H.S., and Arntzen, C. (1995). Immunogenicity of transgenic plant-derived hepatitis B surface antigen. Proc. Natl. Acad. Sci. USA *92*, 3358–3361.

Tindle, W.R. (2002). Immune evasion in human papillomavirus-associated cervical cancer. Nature Rev. Cancer 2, 1-7.

Twyman, R.M., Stoger, E., Schillberg, S., Christou, P., and Fischer, R. (2003). Molecular farming in plants: host systems and expression technology. Trends Biotechnol. *21*, 570-578.

Varsani, A., Williamson, A.L., Rose, R.C., Jaffer, M., and Rybicki, E.P. (2003). Expression of Human papillomavirus type 16 major capsid protein in transgenic Nicotiana tabacum cv. Xanthi. Arch Virol. *148*, 1771–1786.

Vaucheret, H., Beclin, C., Elmayan, T., Feuerbach, F., Godon, C., Morel, J.B., Mourrain, P., Palauqui, J.C., and Vernhettes, S. (1998). Transgene-induced gene silencing in plants. Plant J. *16*, 651–659.

Verch, T., Yusibov, V., and Koprowski, H. (1998). Expression and assembly of a full-length monoclonal antibody in plants using a plant-virus vector. J. Immunol. Methods *220*, 69-75.

Verch, T., Hooper, D.C., Kiyatkin, A., Steplewski, Z., and Koprowski, H. (2004). Immunization with a plant-produced colorectal cancer antigen. Cancer Immunol. Immunother. *53*, 92–99.

Wacker, M., Linton, D., Hitchen, P.G., Nita-Lazar, M., Haslam, S.M., North, S.J., Panico, M., Morris, H.R., Dell, A., Wren, B.W., and Aebi, M. (2002). N-linked glycosylation in Campylobacter jejuni and its functional transfer into *E. coli*. Science *298*, 1790–1793.

Walmsley, A.M., and Arntzen, C.J. (2000). Plants for delivery of edible vaccines. Curr. Opin. Biotechnol. *11*, 126–129.

Warzecha, H., Mason, H.S., Lane, C., Tryggvesson, C., Rybicki, E., Williamson, AL., Clements, J.D., and Rose, R.C. (2003). Oral Immunogenicity of Human papillomavirus-like particles expressed in potato. J. Virol. *77*, 8702–8711.

Weinstein, D.L., O'Neill, B.L., Hone, D.M., and Metcalf, E.S. (1998). Differential early interactions between *Salmonella enterica* serovar Typhi and two other pathogenic *Salmonella* serovars with intestinal epithelial cells. Infect Immun. *66*, 2310–2318.

Wolf, J.L., Rubin, D.H., Finberg, R., Kauffman, R.S., Sharpe, A.H., Trier, J.S., and Fields, B.N. (1981). Intestinal M cells: a pathway for entry of reovirus into the host. Science *212*, 471–472.

Yu, J. and Langridge, W.H. (2001). A plant-based multicomponent vaccine protects mice from enteric diseases. Nat. Biotechnol. *19*, 548–552.

Zambryski, P. (1988). Basic process underlying Agrobacterium-mediated DNA transfer to plant cells. Annu. Rev. Genet. *22*, 1–30.

From: Papillomavirus Research: From Natural History to Vaccines and Beyond. Edited by: M. Saveria Campo

Chapter 23

Bovine Papillomavirus: Old system, New Lessons?

M. Saveria Campo*

Abstract

Bovine papillomavirus (BPV) is perhaps the most extensively studied animal papillomavirus. BPVs induce exophytic papillomas of cutaneous or mucosal epithelia in cattle. The papillomas are benign tumours and generally regress without eliciting any serious clinical problems in the host, but occasionally persist and provide the focus for malignant transformation to squamous cell carcinoma, particularly in the presence of environmental cofactors. This has been experimentally demonstrated for cancer of the urinary bladder and cancer of the upper alimentary canal in cattle feeding on bracken fern.

Six BPV types (BPV-1-6) have been characterised associated with specific lesions with different histopathological characteristics. Recently the biology of BPV-5 has been reassessed, and novel types have been found, the biological characterisation of which is awaited with great interest.

BPV has been studied both as an infectious agent in its own right and as a model in which to investigate the interaction of papillomavirus with its natural host and with environmental cofactors. It has also provided a powerful model for vaccination against human papillomavirus (HPV). It continues to provide information applicable to HPV: new functions recently discovered for BPV proteins have been confirmed for HPV proteins.

1. Introduction - BPV, An Old System

Papillomaviruses (PVs) are strictly species-specific and, even in experimental conditions, do not infect any other host than their natural one. The only known case of cross-species infection is the infection of horses and other equids by BPV-1 and BPV-2 (see Chapter 24). Given the apparently insurmountable problem of species-specificity, no animal model of HPV infection exists and papillomavirologists have had to rely for direct experimentation on animal papillomavirus systems.

BPV has been invaluable in the investigation of virus biology, the direct link between virus infection and neoplasia, the relationship between virus and host and environment, the host immune response to the virus, and in the development of the first anti-papillomavirus vaccines. Additionally, and equally importantly, BPV is an agent of disease in farm animals and therefore of considerable veterinary and agricultural importance (Campo, 2002).

2. Heterogeneity of BPV

The six well characterised BPVs were originally classified into two subgroups, based on their genome structure and recognised pathology: subgroup A and subgroup B. Subgroup A comprised BPV-1, BPV-2 and BPV-5. These BPVs are commonly defined as fibropapillomaviruses, that is, viruses that infect both the epithelium and the underlying derma, giving rise to fibropapillomas (Figure 1, upper panel A). Subgroup B comprised BPV-3, BPV-4 and BPV-6, purely epitheliotropic BPVs which infect only the epithelium and induce true papillomas (Figure 1, upper panel B). Papillomaviruses have recently been re-classified (de Villiers et al., 2004) (Chapter 2) following the Greek letter nomenclature used for other virus families. According to the new nomenclature, which will be followed from now on, the epitheliotropic BPV-3, BPV-4 and BPV-6 are defined as Xi-papillomaviruses and BPV-1 and BPV-2 as Delta-papillomaviruses. The genome of BPV-5 appears to share homology with both Xi-PVs and Delta-PVs (Bloch and Breen, 1997) and interestingly BPV-5 appears to have a dual pathology causing both fibropapillomas and epithelial papillomas (Bloch et al., 1994). These two observations have lead to the re-classification of BPV-5 as the single member of the Epsilon-PV genus (de Villiers et al., 2004). It is clear that this virus deserves further investigation.

Thirteen additional novel BPVs have been reported recently (Ogawa et al., 2004), more than trebling the heterogeneity of BPVs. It is likely that further investigations will reveal the existence of even more BPVs and it will be very interesting to establish whether their pathology follows Delta-, Xi- or Epsilon-PVs.

*For correspondence. Email s.campo@vet.gla.ac.uk

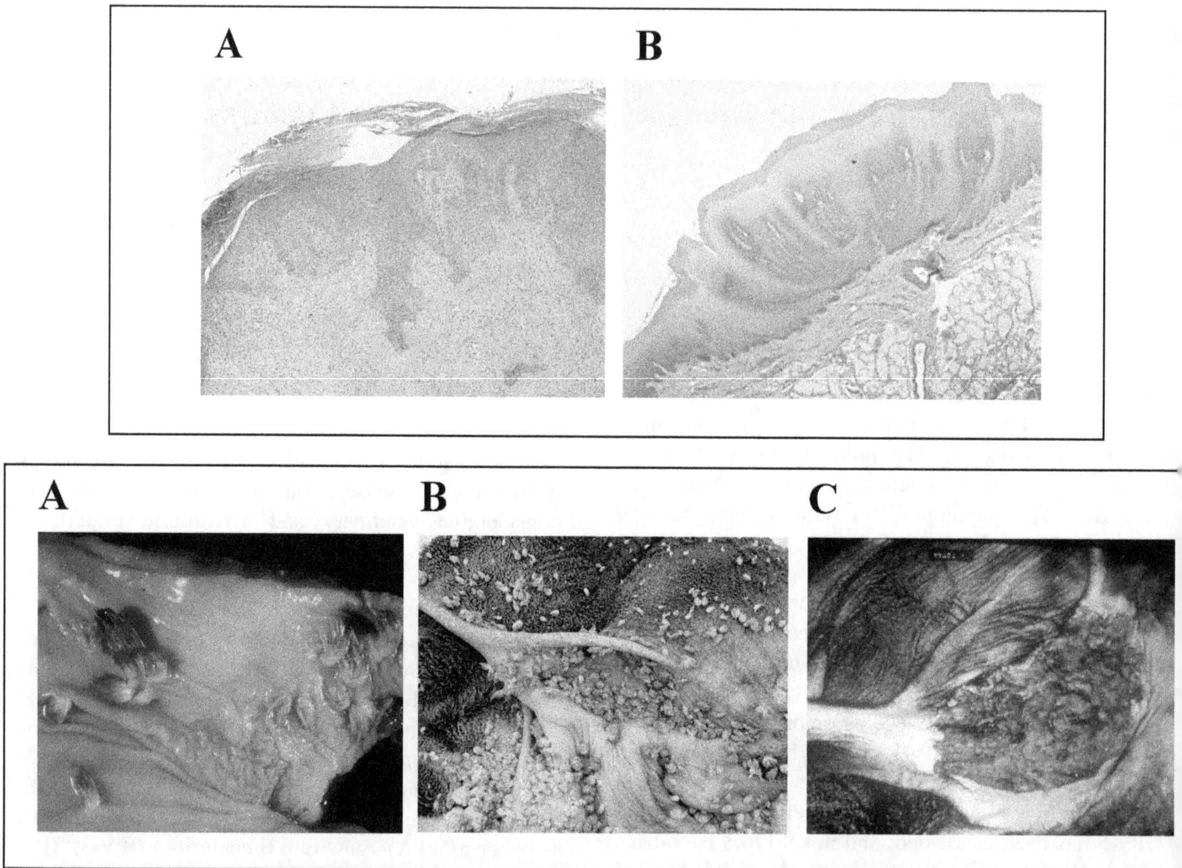

Figure 1. Upper panel: Histology of BPV-induced tumours. A, fibropapilloma induced by BPV-1; B, epithelial papilloma induced by BPV-6. Lower panel. Neoplastic progression of BPV-4 papillomas. A, single papillomas of the oesophagus in a healthy immunocompetent animal; B, extensive florid papillomatosis of the rumen in an animal immunosuppressed by bracken fern; C, squamous cell carcinoma of the rumen in an animal immunosuppressed by bracken fern.

3. The Virion, the Viral Genome and Viral Transcripts

3.1. The virion

Independently from the site or type of lesion, the BPV virion has a constant morphology and structure. Like other PV virions, the BPV virion is a non-enveloped icosahedral structure of 55-60 nm diameter, which forms paracrystalline arrays in the nucleus of infected cells; it contains the double stranded covalently closed circular DNA complexed with cellular histones. The virion is composed of the major L1 and minor L2 capsid proteins. The detailed structure of the BPV virion has been determined recently and an atomic model has been generated which shows that the C-terminus of L1 is exposed on the surface of the virion and is thus likely to have a role in infection and immunogenicity (Modis et al., 2002).

3.2. The viral genome

The genome of BPV-1, BPV-2 and BPV-5 is a double stranded closed circle of approximately 8000 nucleotides (nt), but only approximately 7300 nt in BPV-3, BPV-4 and BPV-6. The genome is divided into 3 canonical regions: a long control region (LCR, containing the necessary cis-regulatory elements for the replication and transcription of the viral DNA), the region containing the early genes (encoding non-structural proteins) and the region containing the late genes (encoding the capsid proteins) (**Figure** 2). The genomic plan of the BPV-1, BPV-2 and BPV-5 is similar to that of most other PVs, whereas BPV-3, BPV-4 and BPV-6 lack the E6 gene, which has been replaced by the E5 gene (Figure 2). The E5 gene and its encoded protein were originally defined as E8. The realisation that most (although not all) of the functions of this protein are shared with BPV-1 E5, prompted its redefinition as E5 (Morgan and Campo, 2000).

3.3. Transcriptional control

Another difference between BPV-1, -2 and -5 and BPV-3, -4 and -6 is the organisation of the LCR. The LCR of the former group of viruses contains 12 E2 DNA binding sites (BS), whereas the LCR of the latter group contains only 4 E2BS, with an arrangement very similar to that of the LCR of genital HPVs (see Chapter 4 for a diagrammatic representation of the LCR of genital HPVs).

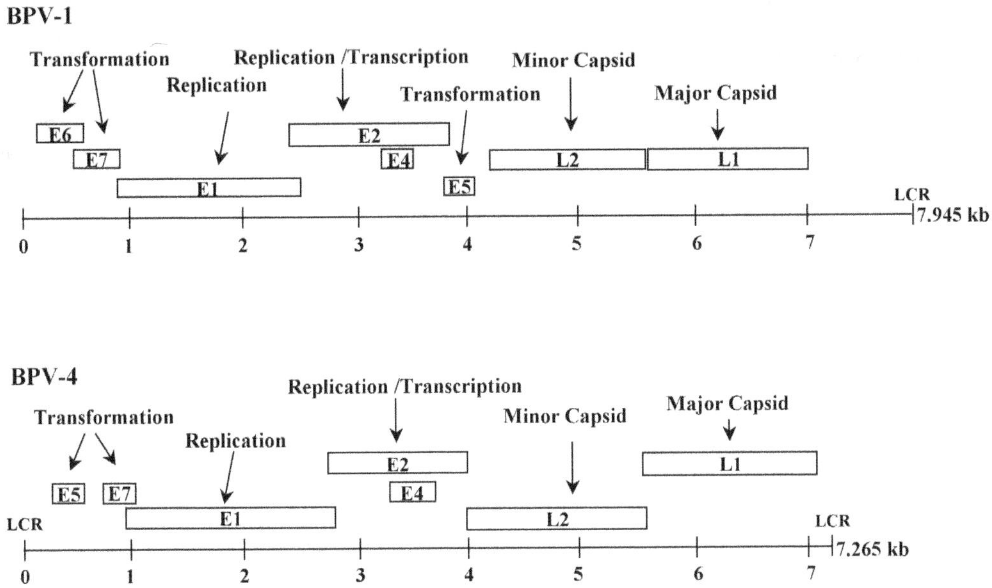

Figure 2. Diagrammatic genomic organisation of BPV-1 and BPV-4. The viral genomes are represented as linear, with the open reading frames (ORF) as rectangles. The function of the proteins encoded by the individual ORF is indicated.

E2 is the viral transcription regulator, and its binding to the LCR activates or represses transcription of the viral genes (see Chapter 6 for a detailed description of E2 functions). In the BPV-4 system, E2 has different affinities for its cognate binding sites and binding to the different sites results in different outcomes. E2 binding to the TATA box-distal sites 3 and 4 results in transcriptional activation, whereas E2 binding to the TATA-proximal sites 1 and 2 leads to transcriptional repression in an epithelial specific manner (Jackson and Campo, 1995; Morgan et al., 1998).

The LCR of BPV-4 contains also positive and negative E2-independent regulatory elements both in the promoter and in the enhancer regions (Jackson and Campo, 1991; Morgan et al, 1999; Vance et al., 1999; Vance et al., 2001), which interact with a number of cellular transcription factors (McCaffery and Jackson, 1994; Jackson and Campo, 1995). As is the case in HPVs, regulation of BPV-4 gene transcription is a very finely tuned circuit where positive and negative, viral and cellular controlling elements all contribute to the spatially and temporally regulated expression of the viral genes. If this delicate balance is subverted, the outcome is often cell transformation to the neoplastic state.

3.4. Viral transcripts

Expression of the viral genes takes place through a complex pattern of RNA splicing events which process mature mRNA from polycistronic precursor RNA both in Delta- and in Xi-BPVs (Yang et al., 1985; Campo et al., 1994a). As a general rule, the late transcripts encoding the structural proteins are found only in the more differentiated layers of warts and other productive lesions (see Chapter 15), while the RNAs encoding the early proteins are found both in warts and in transformed cells.

4. BPV Proteins in Papillomas and Transformed Cells

4.1. Early proteins

Delta-BPVs encode three oncoproteins, E5, E6 and E7, whereas the Xi-BPVs lack the E6 gene and only encode E5 and E7. The relative importance of each protein in cell transformation varies between different BPVs. E5 is the major oncoprotein of BPV-1 (see Chapter 8), followed by E6, with a more modest role played by E7. The transforming proteins of BPV-4 are E5 and E7, with E5 promoting the acquisition of anchorage independent growth and E7 inducing growth advantage and extended life span (Pennie et al., 1993).

4.2. E6

BPV-1 E6 is itself a transcriptional activator; it interacts with and inhibit the transcription co-activator CBP/p300 and in so doing inhibits also the function of p53 (Zimmerman et al., 2000), thus achieving the same result as the E6 proteins of Alpha-HPVs, namely the inactivation of the apoptotic and cell cycle arrest functions of p53 (see Chapter 9). E6 binds the focal adhesion protein paxillin and blocks the interaction between paxillin and several other focal adhesion proteins, including vinculin and focal adhesion kinase (Tong et al., 1997). Disruption of focal adhesions thus likely underlies the observed anchorage-independent growth and disruption of the actin cytoskeleton induced by E6. Furthermore, E6 interacts with AP-1, the trans-Golgi network-specific clathrin adaptor complex and this interaction is likely to affect cellular processes involving clathrin-mediated trafficking pathway (Tong et al., 1998).

4.3. E7

BPV-1 E7 cooperates with E5 as well as E6 in cell transformation, as judged by anchorage-independent growth of mouse cells (Bohl et al., 2001). It is not yet clear how BPV-1 E7 contributes to cell transformation as it lacks the canonical p105Rb-binding domain LXCXE (see below; Narechania et al., 2004). On the contrary, like most other E7 proteins, BPV-4 E7 contains two CXXC motifs and the p105Rb-binding domain LHCEE. Mutations that disrupt the CXXC motifs, or the p105Rb-binding domain, drastically reduce or abolish its transforming activity. However, in common with CRPV, HPV-5 and HPV-8 E7 proteins, BPV-4 E7 does not possess a CKII phosphorylation site (Campo, 1997a).

4.4. E5

The E5 proteins of BPV-1 and BPV-4 are probably the best studied oncoproteins of these viruses. They will be discussed only briefly here as they are described in great depth and detail in Chapter 8. Both E5 proteins are very hydrophobic with a high leucine content and similar hydrophobicity profiles; they are both localised in the cell endomembrane compartments, particularly the Golgi apparatus, and both bind the 16k ductin/subunit c of the vacuolar H^+ ATPase. This interaction results in the inhibition of gap junction intercellular communication and in the lack acidification of endosomes and Golgi apparatus.

Gap junctions are channels for small molecular weight secondary messengers, important in the homeostatic control in a tissue; if a transformed cell is released from the control of the surrounding normal cells, it can proliferate freely and give rise to an expanding transformed clone. Accordingly, the lack of gap junction communication in papillomavirus transformed cells is probably an early event in transformation and, by isolating the newly infected cell from the surrounding normal sisters, allows other transforming events to become established.

The inhibition of acidification of endosomes and Golgi apparatus leads to disruption of cellular protein processing and sorting (see below and Chapter 8), resulting in the retention and re-cycling of undegraded activated growth factor receptors from endosomal compartments. Additionally BPV-1 E5 interacts directly with and activates the PDGF receptor. Therefore, the mitogenic response is potentiated, directly, by E5 interacting with the PDGF receptor, and, indirectly, by E5 interacting with the 16k protein, and in so doing inhibiting receptor down regulation. Moreover, E5 activates several protein kinases involved in the control of cell cycle, thus causing a general dysregulation of the normally tightly controlled programme of cell proliferation (Chapter 8). For instance, the activation of the G1-S cyclin A-cdk2 complex underlies the ability of BPV-4 E5 to induce anchorage-independent growth both in established murine cells and in primary bovine PalF cells (O'Brien and Campo, 1998; O'Brien et al., 1999; Zago et al., 2004; see Chapter 8 for additional relevant references).

4.5. Consequences of the lack of E6 in Xi-BPVs

While BPV-1 can transform primary foetal bovine cells (PalF) by itself and thus encodes all the required genetic information for cell transformation (our unpublished observations), BPV-4 alone is insufficient to induce transformation of PalF cells. However, the complete E5 and E7 ORFs of BPV-4 co-operate with an activated *ras* oncogene in the morphological transformation of PalF cells, although the cells are neither immortal nor tumorigenic (Jaggar et al., 1990). Xi-BPVs are unique among the papillomaviruses in not possessing an E6 ORF (Jackson et al., 1991). Thus for these viruses an E6 protein is not required for a successful infection cycle, or for papilloma progression to carcinoma in the case of BPV-4 (see below). This raises the question of whether E6 functions are not required by BPV-4, or whether these functions are supplied by another viral or host protein. The demonstration that HPV-16 E6 confers immortality to PalF cells transformed by BPV-4 E7 and *ras*, whether E5 is present or not, shows that HPV-16 E6 does indeed provide functions which are not supplied by the BPV-4 proteins (Pennie et al., 1993). Even after immortalisation by HPV-16 E6, transformed PalF cells are not tumorigenic. Tumorigenicity is achieved by the further introduction into PalF cells of a mutant p53 gene (Scobie et al., 1997), thus indicating that dysfunction of p53 is required for the ultimate step in cell transformation. The impact of the lack of an E6 gene on the abrogation of p53 functions necessary for cell transformation is highlighted by the ability of BPV-4 DNA to transform p53-null fibroblasts by itself. Further addition of HPV-16 E6 increases the number of transformed colonies (Scobie et al., 1997). Therefore, in the absence of E6-like functions, the controlling activity of p53 must be either bypassed or abrogated by cofactors during papilloma development and neoplastic progression to carcinoma. This is confirmed by the observation that the p53 gene is mutated in at least some papillomas and carcinomas (Scobie et al., 1997).

4.6. *In vivo* protein expression

The distribution of BPV-1 and BPV-4 E5 and E7 proteins in bovine warts is similar but not identical (Anderson et al., 1997; Bohl et al., 2001; Araibi et al., 2004). E5 of both viruses is expressed in the cytoplasm of both basal and superficial differentiating keratinocytes. BPV-4 E5 is present in papillomas at early stages of development but its expression is extinguished as the papilloma ages (Anderson et al., 1997); this temporally regulated expression, together with the established functions of E5 (see above), supports the hypothesis that BPV-4 E5 is needed during the early stages of viral infection, while its presence in the more differentiated layers suggests an involvement of E5 in the late stages of the virus life cycle (see Chapter 13 for a discussion of E5 as a "late" protein).

BPV-1 E7 is only observed in the cytoplasm and nucleoli of cells in the basal and lower spinous layers (Bohl et al., 2001), whereas BPV-4 E7 is expressed in all layers of papillomas at all stages (consistent with its necessary role in cell transformation) and its cellular localisation changes from nuclear in the basal and suprabasal layers, to cytoplasmic in the more differentiated upper layers (Anderson et al., 1997). The significance of the cytoplasmic localization of BPV-4 E7 in the differentiated layers is unclear, but it has been suggested that E7's frequent co-expression with E5 in the lower layers serves to modulate the cellular response of basal epithelial cells to E5 (Bohl et al., 2001).

The E4 protein of BPV-4 is expressed only in the cytoplasm of keratinocytes that support the productive phase of viral DNA replication (Anderson et al., 1997), implicating it in the late stages of the virus life cycle, in agreement with the E4 proteins of other papillomavirus (see Chapter 7 for a full description of the function of E4).

The structural proteins L1 and L2 are expressed in the nucleus of the differentiated keratinocytes (Anderson et al., 1997). During virion morphogenesis L2 binds to viral DNA, favouring its encapsidation (Fay et al., 2004 and references herein). L1 and possibly also L2 mediate virus attachment to the cell receptor (see Chapters 12 and 18 for further details). Both proteins encode virus neutralising epitopes and, in the case of BPV-2 L2, tumour rejection epitopes; both proteins have been successfully used as vaccines (see below; Campo, 1997b).

5. BPV and Benign Disease

The natural history of BPV infection has been extensively studied both in the field and in experimental conditions (Jarrett, 1985). The stages of experimental disease are indistinguishable from those encountered in the field.

Infection by Delta-PVs (fibropapillomaviruses) leads to an initial transformation of the subepithelial fibroblasts followed by epithelial plexiform acanthosis and then papillomatosis, while infection by (epitheliotropic) Xi-PVs induces epithelial papillomas without fibroblast involvement (Figure 2, upper panel B; Jarrett, 1985). A large survey of bovine papillomatosis in Scotland established that BPV-1 causes teat and penile fibropapillomas; BPV-2, common cutaneous warts and oesophageal fibropapillomas; BPV-4, papillomas of the upper gastrointestinal (GI) tract; BPV-6, frond papillomas of the teats, and BPV-5, rice grain fibropapillomas of the udder. Workers in Australia have found that BPV-5 has a dual pathology (Bloch et al., 1994) and thus epithelial papillomas should be added to the spectrum of lesions induced by this virus. More work should be done on BPV-5 to establish if genetic differences between the Scottish and Australian isolates may be responsible for differences in the observed pathogenesis. BPV-3 was isolated in Australia from an epithelial papilloma of the skin (Pfister et al., 1979), and very little is known about its natural history. The tumours induced by BPV-3 are characterised by the absence of papillomatous fronds which are typical of tumours induced by BPV-4 and BPV-6. It should be noted that infection of the oesophagus by BPV-2 is abortive and a dead-end process in which cell transformation is achieved but no infectious virus progeny is produced, despite the presence and replication of multiple episomal copies of the viral genome.

Normally, the papillomas regress, but some animals are unable to reject the infection and succumb to widespread cutaneous or mucosal involvement. These forms of papillomatosis are problematic and of economical significance.

5.1. Papilloma of the teats and udder

The teats and udders of cows are subjected to infection by three different types of BPVs: BPV-1, BPV-5 and BPV-6 (Campo, 2002). This disease, especially if caused by BPV-6, is not only a health problem but has also economic consequences. One characteristic of BPV-6 papillomas is their secondary and even tertiary spread around the primary tumours. This spread is clinically significant, as cows with teat papillomas cannot be milked, young calves cannot suckle, and often the peduncolated papillomas snap off, the sites become infected and mastitis may ensue with distortion of the milk canals. Occasionally, entire herds have to be culled if the papillomatosis does not regress. The reasons why BPV infection of the teats is more prolonged and less prone to resolution than BPV infection of the skin are not known and more work is needed to clarify them. Regression of BPV-6 papillomas has not been observed in experimental situations.

5.2. Fibropapilloma of the penis

These lesions, caused by BPV-1, occur both in the penile and preputial epithelia and can spread along the perineum and up toward the back of the bull; they can become necrotic and cause loss of reproductive functions. Although the penile fibropapilloma goes through the stages delineated for the fibropapillomas of the teats and udder (Jarrett, 1985), its life span is longer, especially if the tumour is large. The reasons for this persistence are not known and the tumour causes relevant economic damage.

5.3. Florid papillomatosis of the upper GI tract

Papillomas of the upper GI tract are induced by BPV-4. All sites from the tongue to the stomach can be affected. In healthy cattle the papillomas are few and normally regress after approximately twelve months. In chronically immunosuppressed animals regression does not take place and the papillomas spread and persist (Figure 2, lower panel). Chronic immunosuppression in cattle commonly results from exposure to immunosuppressive chemicals present in bracken fern (*Pteridium aquilinum*), on which the animals graze in the marginal lands of North-West Scotland and in several other countries (Borzacchiello et al., 2003a, Campo, 2002), but can be due to other causes, including infection with bovine viral diarrhoea virus (Tsirimonaki et al., 2003). Animals that develop such extensive papillomatosis have difficulty in eating and breathing and have to be culled. If the papillomatosis is not too severe, the animals survive but are at great risk of developing squamous cell carcinomas (see below).

6. BPV and Cancer

Persistent PV-induced lesions are at risk of progression to cancer. Cancer, including PV-associated cancer, is a multi-factorial disease and several steps are required before full neoplastic transformation is achieved. Bracken fern has been identified as a major environmental co-factor in BPV-induced carcinogenesis in cattle. Bracken fern contains immunosuppressants and a number of mutagens.

Bracken-induced immunosuppression is associated with two marked haematological changes. The first of these is a dramatic fall in polymorphonuclear leucocytes. If unchecked, this leads to severe acute immunosuppression with invasion of the bloodstream by alimentary bacteria and death from septicaemia. This is the well described veterinary syndrome of Acute Bracken Poisoning (ABP). The second effect of bracken feeding is a chronic drop in circulating lymphocytes. Even during periods of bracken withdrawal the lymphocyte count remains very low.

Field cases of upper and lower GI tract cancer are found to occur at a high frequency in areas such as the Nasampolai Valley of Kenya and the Western Highlands of Scotland where cattle graze on bracken-infested land (Plowright et al., 1971; Jarrett et al., 1978). In addition, bracken-eating cattle develop chronic enzootic haematuria, urinary bladder cancers and chromosomal abnormalities (Borzacchiello et al., 2001a, Borzacchiello et al., 2003b, Jarrett et al., 1978, Lioi et al., 2004, Moura et al., 1988, Stocco dos Santos et al., 1998).

6.1. Cancer of the upper GI tract

In cattle feeding on bracken fern, the BPV-4-induced papillomas of the upper GI tract are at a high risk of progressing to cancer. The contribution of viral, immunological and chemical factors in progression of papillomas to carcinomas was first established in the field (Jarrett et al., 1978) and then in experimental conditions (Campo et al, 1994b). In both circumstances, healthy cattle infected with BPV-4, grazing on grass or kept on a diet of hay, reject their papillomas after approximately one year from infection. In contrast, in bracken-eating immunosuppressed cattle the papillomas spread throughout the oro- and nasopharynx, often coalescing in numerous large clumps (Figure 2, lower panel B), do not undergo regression and progress to cancer (Figure 2, lower panel C). The animals maintain their tumours for the rest of their life: in experimental conditions, one animal died 13 years after infection and still had papillomas which had spread from the mouth to the lower oesophagus and the rumen. In the same experiment, two animals infected with BPV-4 and fed with bracken developed foci of carcinoma in the oesophagus, which infiltrated the subjacent tissue, and multiple polyps, adenomas and adenocarcinomas in the duodenum, jejunum and colon (Campo et al., 1994c).

These epidemiological and molecular studies have allowed the determination of the events that take place during carcinogenesis. In addition to the action of the viral oncoproteins the number of the cellular receptors for epidermal growth factor increases, the ras gene is activated and the p53 gene is mutated (Campo, 2002). These transforming events, involving cellular genes, are ascribed to the bracken fern mutagens, but, although this has been proved *in vitro* (Beniston et al., 2001; Cairney and Campo, 1995; Connolly et al., 1998; Pennie and Campo, 1992), it remains to be established *in vivo*. In support of a role for bracken mutagens in carcinogenesis, two of the most potent mutagens of the plant (quercetin and ptaquiloside) are found in urine, serum and milk of bracken-eating cattle (Alonso-Amelot et al., 1996; Lioi et al., 2004).

6.2. Cancer of the urinary bladder

Cattle feeding on bracken fern are also affected by cancer of the urinary bladder (Campo, 1997a; 2002). Urinary bladder cancer comprises two main types, carcinoma of the urothelium, as seen in humans, and haemangioendotheliomas of the subjacent capillaries. Often the two types of tumour occur together in the same bladder. The involvement of bracken and BPV-1 or BPV-2 in bladder carcinogenesis has been recognised for a long time (Olson et al., 1959; 1965), recently the virus-bracken interaction has been reproduced experimentally (Campo et al., 1992) and a carcinogenic process, similar in outline to the one established for BPV-4 has been recognised. BPV-1/2 infects the epithelium of the urinary bladder (perhaps as a secondary infection deriving from infection of the paragenital area) and establishes an abortive infection, with no production of virus, as is the case for the fibropapillomas of the alimentary canal (see above). The viral DNA in the bladder lesions is still infectious and capable of initiating a replicative cycle in the permissive environment of the skin, as extracts from urinary bladder cancers could induce skin warts (Olson et al., 1965). The viral oncoprotein E5 is expressed (Borzacchiello et al., 2003b), and in complex with activated PDGF receptor (G. Borzacchiello, University of Naples, personal communication) the ras gene is activated (Campo, 2002, Shahin et al., 1998), and expression of the tumour suppressor *fragile histidine tetrads* (*FHIT*) locus is down-regulated (Borzacchiello et al., 2001). The immunosuppression induced by bracken fern prevents tumour rejection and the fern mutagens contribute to genome destabilisation. Interestingly fragile sites are often disrupted by integration of HPV DNA in cervical cancers (Butler et al., 2000) and alterations of *FHIT* expression has been observed in many cervical carcinomas (Takizawa et al., 2003).

7. Are BPV-induced, Bracken-triggered Cancers a Model for Human Carcinogenesis?

7.1. Cancer of the oesophagus in humans

Carcinoma of the upper GI tract is one of the most frequent forms of human cancer: oesophageal cancer ranks sixth in developing countries and eighteenth in developed countries (Globocan, 2001). Mortality is high in both developed and developing countries. In Latin America there is little or no difference between the incidence and death rates of oesophageal cancer (Day and Varghese, 1994). Although in the western world alcohol and tobacco are deemed to be the main causative agents of this type of cancer (Graham et al., 1990), these factors seem to play a lesser role in other countries (Demirer et al., 1990).

Several HPV types have also been found with variable prevalence in oral cancers (Syrjanen, 2003) and oesophageal cancers and precancers (Syrjanen, 2002; see Chapter 15 of this book). HPV is more prevalent in oesophageal cancer from the developing world than the developed world. The highly variable geographical incidence of oesophageal cancer and of HPV detection in the cancer suggests a multi-factorial aetiology and the involvement of environmental dietary co-factors (Syrjanen, 2002; Chapter 15).

In humans too, exposure to bracken fern has been linked to oesophageal cancer in several parts of the world (Alonso-Amelot and Avendano, 2001; 2002; Alonso-Amelot et al., 1996; Hirayama, 1979; Marliere et al., 1999; Villalobos-Salazar et al., 1995). Bracken fern is eaten and used as a herbal remedy in South America, Japan, China, Korea and other eastern countries. Exposure to bracken fern directly through consumption or possibly inhalation, or indirectly through milk from bracken eating cattle has been epidemiologically linked to human oesophageal and gastric cancer in Western Europe, South America and Japan. The relative risk reported for bracken exposure and oesophageal cancer ranges from 2.6 in Wales (Galpin et al., 1990), 3.7 in Japan (Hirayama, 1979) to 8.12 in Brazil (Marliere et al., 2002) independent of other factors such as age, gender, smoking and tobacco consumption. Chromosomal abnormalities have been found both in bracken-grazing cattle (Lioi et al., 2004; Moura et al., 1988) and in bracken-eating people (Recouso et al., 2003). There is a geographical overlap between consumption of bracken fern, infection of the alimentary canal mucosa by HPV and oesophageal cancer in the Minas Gerais region of Brazil (our unpublished observations). Bracken, milk from bracken fed cows and bracken spores have all been shown to be carcinogenic in experimental animals (Santos et al., 1987; 1992) and DNA adducts have been found in the upper GI tissue of mice fed bracken fern (Povey et al., 1996).

These findings suggest that some cases of cancer of the upper GI tract in humans may have the same aetiology as in cattle, *i.e.* papillomavirus and bracken, and open up the possibility that the molecular mechanisms elucidated for cell transformation and cancer in cattle operate similarly in humans. If this is the case, the bovine system acquires even greater relevance to human papillomavirus infection and cancer.

7.2. Cancer of the urinary bladder in humans

The involvement of HPV in cancer of the urinary bladder in humans is still an unresolved issue. HPV DNA has been found in a number of cases of urinary tract cancer, sometime ascribed to secondary spread from penile or vulval condylomas. However, the involvement of HPV in this type of human cancer has been disputed (Oliver et al., 1998). Recent evidence supports the idea that the virus is instrumental only in a proportion, perhaps a minority, of urinary tract cancer in selected populations, in combination with other infections such as Bilharzia (Khaled et al., 2001; 2003). This would explain the contradictory results and leave the door open for a role of HPV in human bladder cancer, perhaps as a co-factor.

8. BPV-4 E7 and Quercetin

Given the important role played by bracken fern in carcinogenesis in cattle, it is of great interest to identify the factors involved and their cellular targets. Several mutagenic and/or carcinogenic substances have been isolated from bracken, including ptaquiloside and quercetin (Alonso-Amelot and Avendano, 2002). Ptaquiloside, a norsesquiterpene glucoside, induces tumours and haematuria in rats and produces chromosomal aberrations in Chinese hamster cells; breakdown products of ptaquiloside, detected in milk of bracken-eating cows (Alonso-Amelot et al., 1996; Lioi et al., 2004), exhibit mutagenic activity. The flavonoid quercetin (3,3',4',5,7-pentahydroxyflavone) is mutagenic in both prokaryotic and eukaryotic cells, can act as an initiator in a two-stage transfomation assay *in vitro* and induces chromosomal aberrations in mammalian cells. In addition, quercetin activates or inhibits protein kinases and phosphatases thus interfering with critical steps in signal transduction

The contribution of quercetin to neoplastic cell transformation has been established in primary foetal bovine cells (PalF). In addition to the oncogenes of BPV-4, these cells need an activated ras gene and, as BPV-4 does not possess an E6 gene, the E6 oncogene of HPV-16 to be established as immortal transformed clones. Addition of a mutated p53 gene, leads to the formation of tumorigenic clones (see above). In this cell system, quercetin can substitute for more than one oncogene (Pennie and Campo, 1992). In the presence of quercetin, E7 is the only BPV-4 oncogene required for full transformation, as quercetin substitutes for E5 in conferring anchorage-independence, for E6 in inducing immortalisation and for mutant p53 in inducing tumorigenicity (Cairney and Campo, 1995). The potent effects of quercetin upon primary cells in tissue culture strongly suggest that quercetin is, if not *the* active, one of the most active components of bracken in carcinogenesis of the upper GI tract.

The effect of quercetin on cell transformation is underpinned by transcriptional activation of the BPV-4 LCR; a *cis*-acting element has been mapped in the LCR which responds to quercetin (Connolly et al., 1998). LCR activation, in turn, stimulates expression of the E7 oncoprotein. E7 inhibits one of the negative regulators of the cell cycle, $p27^{kip1}$, and in so doing allows quercetin-damaged cells to proliferate, rather than arresting in G1, and thus spread and maintain any genetic damage caused by quercetin (Beniston et al., 2001).

Ptaquiloside, another potent BF mutagen, has been implicated in neoplasia but its role in cell transformation and its relationship with papillomavirus are unknown .

9. BPV Latency

Like many viruses, BPVs can establish a latent infection. The viral genome can be often found in normal epithelia with no clinical sign of disease (Ogawa et al., 2004) both in tumour-bearing and in clinically normal hosts. Normal epithelia are the accepted site of latent infection, and indeed the reactivation of BPV at sites of trauma suggests that viral DNA is present in these sites in latent form, and that damage of the epithelium, possibly through production of inflammatory cytokines and stimulation of cell proliferation, induces expression of viral genes leading to papilloma formation (Campo et al., 1994b). However, epithelia may not be the only site of latent papillomavirus. BPV DNA is present in episomal form in circulating lymphocytes of cattle (Campo et al., 1994b; F. Roperto, University of Naples, personal communication) and latent BPV infection of lymphocytes has been established in experimental cattle (Stocco dos Santos et al., 1998). The detection of BPV DNA in cattle lymphocytes by three independent groups in three different, geographically separated countries is a strong indication that the presence of viral DNA in these cells is not accidental but likely to have biological implications. HPV DNA has been found in blood cells and lymph nodes of women with urogenital HPV infection and cervical cancer (Kedzia et al., 1992; Pao et al., 1997; 1991; Payne et al., 1993; Tseng et al., 1999) but the significance of these observations is not clear.

Intriguingly, papillomavirus virus-like particles have been shown *in vitro* to bind strongly to a variety of immune cells, including dendritic cells, B cells, monocytes, and macrophages, and it has been hypothesised that this interaction is likely to be important in the immune response to PV capsids (Da Silva et al., 2001; Chapter 12).

These findings present the very attractive, but still purely speculative, possibilities that blood cells represent another site of latent virus and that, additionally, the relationship between PV and immune cells can modulate the host immune response to virus. Clearly more work needs to be done in this respect.

10. Immune Response to BPV

Viruses and their hosts are in a delicately balanced relationship. Viruses must be able to overcome the host immune response to replicate and produce infectious progeny. Nevertheless, despite the virus ability to evade immune surveillance, eventually the host mounts an effective immune response and virus and virus-infected cells are eliminated.

The immune response of cattle to BPV is surprisingly poor (Campo, 1998). Animals may carry massive tumours, actively producing virus in large quantities, but cattle do not respond easily to BPV antigens during the course of infection and anti-BPV antibodies are seldom detected. This is the case for all the BPV types investigated, ruling out either the virus type or the site of infection for the poor humoral response. The failure of the immune system to recognise either incoming virus or progeny virus is due to fact that all the virus life cycle is restricted to the epithelium and therefore is not in contact with the immune system (O'Brien and Campo, 2002; Tindle, 2002). This interpretation is supported by the fact that field animals with ulcerated and bleeding tumours do have high titres of natural anti-BPV antibodies and good antibody responses can be obtained after intramuscular inoculation with purified virus or viral proteins, confirming that only when the papilloma is damaged, or a threshold of unknown nature is reached via immunisation, viral antigens come into contact with immune cells (Campo, 1998). Weak T- and B-cell responses to capsid proteins or to the transforming protein E7 can be observed in some animals at later stages of infection and appear to be associated with papilloma rejection (Campo, 1998).

During rejection of BPV-4 papillomas, large masses of activated lymphocytes accumulate in the derma underlying the papilloma. In these clusters $CD4^+$ lymphocytes are the predominant subtype, followed by γδ T-cells and $CD8^+$ lymphocytes. In contrast, γδ and $CD8^+$ lymphocytes predominate in the basal layer and among the keratinocytes (Knowles et al., 1996; Campo, 1998). The contribution of the individual lymphocyte subtypes to papilloma regression remains to be established but a similar differential distribution of $CD4^+$ and $CD8^+$ lymphocytes has been reported for regressing genital warts in humans and regressing papillomas in rabbits (Chapters 19 and 25 respectively). The overall temporal and topological similarity between regressing lesions in different papillomavirus systems confirms the generality of the observations.

The poor immune response to BPV is likely to be the main reasons for the persistence of infection: even in immunocompetent hosts, the papillomas persist for many months before regression takes place. It has recently become clear that, in addition to the passive immunoescape due to the viral life cycle being confined to the epithelium, papillomaviruses have evolved active ways of hiding from the host immune system (O'Brien and Campo, 2002). Among these there is down-regulation of the Major Histocompatibility Complex class I (MHC I) (see below).

11. BPV Vaccines

One of the major goals of papillomavirus research is the development of vaccines which will either prevent viral infection (prophylactic vaccines) or accelerate rejection of papillomas (therapeutic vaccines) thus leading to a decreased incidence of human anogenital cancer, in which HPV infection plays a pivotal role (Chapter 21).

The observation that, although the immune response to BPV infection is poor, a prompt and prolonged immune response is elicited when cattle are immunised with BPV proteins has lead to the successful development of anti-BPV vaccines.

Prophylactic vaccination was achieved with BPV-1 virions and virus neutralising antibodies were elicited by vaccination with BPV-1 L1 protein (Campo, 1997b; and references therein). Following these earlier studies, BPV-2 and BPV-4 were chosen as emblematic for skin and mucosal papillomavirus, respectively, and because, as described above, both are implicated in cancer. The first series of vaccines consisted of intramuscular vaccination with virus. These "classical" vaccines result in the production of neutralizing antibodies and complete protection from subsequent challenge. They are however type-specific conferring protection only against the homologous virus (Campo, 1997b). This serotype specificity suggests that for an HPV vaccine to be widely protective it must incorporate antigens from several of the commonest infecting high risk HPV types.

11.1. BPV-2

As for BPV-1, a subunit vaccine based on the L1 protein of BPV-2 elicits neutralising antibodies and confers protection from infection. This is in accordance with the structure of the BPV virion, in which the C-terminus of L1 is displayed on the virion surface (Modis et al., 2002). On the contrary, a BPV-2 L2 vaccine does not afford protection from viral challenge but induces early regression of warts; antibodies against the antigen are generated in the vaccinated animals, but they are not neutralizing. Large infiltrates of lymphocytes and macrophages are observed in the regressing warts, indicative of a cell-mediated immune response (Campo, 1997b). L2-induced wart regression challenges the old, and perhaps naïve, belief that viral structural protein vaccines would only provide protection from infection.

11.2. BPV-4

A more extensive analysis of vaccination and immune response has been conducted in the BPV-4 system (Campo, 1997b).

11.2.1. Prophylactic vaccination

Vaccination with BPV-4 L2 induces protection from challenge by BPV-4 (Campo et al., 1993). Similar results have been obtained following vaccination of rabbits with CRPV L2 (see Chapter 25) indicating that L2-induced protection may be a general phenomenon. Immunity induced by BPV-4 L2 is long lasting and the vaccinated animals are refractory to a second virus challenge more than a year after vaccination. Immunity is conferred by the N-terminus of L2 (Chandrachud et al., 1995) and the vaccinated calves develop virus neutralising antibodies (Gaukroger et al., 1996). Three immunodominant B-cell epitopes have been identified in the N-terminus of L2 (Knowles et al., 1997), and a single 20 amino acid long peptide (residue 131-151) corresponding to one of the epitopes is sufficient for full protection from infection (Campo et al., 1997). These results are consistent with the observation that the N-terminus of L2 is exposed on the surface of the BPV-1 virion (Liu et al., 1997) and therefore accessible by the immune system. Interestingly, the BPV-4 L2 epitopes are homologous to and cross-react with epitopes identified in the L2 protein of several genital HPVs (Campo et al., 1997) leading to the hypothesis that vaccines based on HPV L2 proteins may elicit cross-type protective immunity in humans (Roden et al., 2000).

Papillomavirus capsid proteins can self-assemble in empty virus-like particles (VLPs) when expressed in eukaryotic cells. VLPs are both structurally and antigenically similar to virus, and present conformational neutralising epitopes (Chapter 18). BPV-4 VLPs containing either the L1 and L2 proteins or only the L1 protein are very potent prophylactic vaccines (Kirnbauer et al., 1996), while their status as therapeutic vaccines is less certain. Protective immunity against BPV-4 can therefore be achieved either with L2 alone or with L1 alone.

However, although vaccination with either L1 or L2 induces the production of neutralising antibodies, not all virus neutralisation events occur at the cell surface. BPV-

4 is neutralised by anti-L2 sera after cell entry (Gaukroger et al., 1996; Sibbet et al., 2000), but the mechanisms of this intracellular neutralisation are not known. Therefore, although BPV-4 L2 provides a powerful prophylactic vaccine, it protects not against infection but against disease.

11.2.2. Therapeutic vaccination

Vaccination of calves with the BPV-4 E7 protein does not prevent infection. However in the vaccinated animals papilloma development is slower and the duration of papillomatosis shorter than in control animals. In vaccinated calves papillomas do not reach the final stage of development and start regressing at an earlier stage (Campo et al., 1993). The vaccinated calves produce high titre antibodies directed to three immunodominant B-cell epitopes (Chandrachud et al., 1994) and develop a strong cellular immune response to two T-cell epitopes (McGarvie et al., 1995).

Naturally regressing papillomas are infiltrated by large numbers of activated lymphocytes (see above) but it is not known against which viral antigens the lymphocytes are directed, or whether the cellular immune response in vaccinated calves is similar to the one that mediates natural papilloma rejection.

12. BPV, New Lessons

Many of the functions of papillomavirus proteins have been elucidated in BPV first and then confirmed for HPV. This continues to be the case, as exemplified below.

12.1. E5 and MHC class I down-regulation

MHC I plays a critical role in immune surveillance as it is responsible for the presentation of antigenic peptides to effector T-cells. Once the heavy chain (HC) of MHC I associates with β2-microglobulin and peptides, the complex is transported from the endoplasmic reticulum through the Golgi apparatus to the plasma membrane for recognition by T-cells (Cresswell, 2000).

E5 is the major transforming protein of BPV (see Chapter 8 for a full description of E5's cell transforming activities). A recently discovered function of BPV E5 is the down-regulation of MHC I (Ashrafi et al., 2002, Marchetti et al., 2002). Down-regulation of MHC I by BPV E5 takes place at multiple levels: transcription of the MHC I HC gene is reduced, the MHC I HC peptide is degraded (Ashrafi et al., 2002) and the MHC I complex is sequestered in the Golgi cisternae and is prevented from reaching the cell surface (Marchetti et al., 2002). Retention of MHC I in the Golgi cisternae is due, at least in part, to the E5-induced alkalinisation of the Golgi apparatus as a similar reduction of surface MHC I is observed in cell treated with ionophores that prevent the acidification of endomembranes (Marchetti et al., 2002). Importantly, BPV E5 down-regulation of MHC I is not observed only in cultured cells but also in bovine papillomas (Araibi et al., 2004). This strongly supports the notion that BPV E5 helps the establishment of a successful infection not only through cell transformation but also by down-regulating MHC I, and thus allowing the infected cells to evade host immunosurveillance.

These new facets of BPV E5 have prompted similar investigations on HPV E5. Indeed E5 proteins of both mucosal (HPV-16 and HPV-6) (Ashrafi et al., 2005) and cutaneous viruses (HPV-2) (Cartin and Alonso, 2003) prevent the transport of MHC I to the cell surface. However, down-regulation of MHC I by HPV E5 is less severe that that brought about by BPV E5, as neither transcription of HC gene nor stability of HC peptide are affected (Ashrafi et al., 2005). Thus, although down-regulation of surface MHC I is a common function of E5 proteins, it appears that there is a correlation between their transforming "strength" and the extent to which they interfere with MHC I (O'Brien and Campo, 2003).

12.2. E2 and mitotic chromosomes

BPV-1 E2 associates with mitotic chromosomes (Abroi et al., 2004; Bastien and McBride, 2000; Ilves et al., 1999) (see Chapter 6) via its amino terminal portion allowing its DNA binding domain to remain associated with viral genomes. In so doing, BPV E2 mediates the association of the viral genomes with the mitotic chromosomes resulting in efficient distribution of the BPV genomes into the daughter cells. The E2 proteins of HPV-11, -16, and -18 associate with the mitotic spindle during mitosis and HPV-11 E2 also associates with centrosomes (Van Tine et al., 2004). Although the mechanism underlying partition of the viral genomes between daughter cells may be different between BPV E2 and HPV E2 (Van Tine et al., 2004), it is clear that the E2 protein is pivotal in ensuring that the viral genomes are properly segregated during mitosis.

12.3. E2, L2 and ND10

A further role for E2 in the regulation of the viral life cycle has come from the observation that BPV-1 E2 can enhance the packaging of plasmid DNA into pseudovirions (Zhao et al., 2000) and appears to be directed to ND10 domains by the minor capsid protein L2 (Day et al., 1998; Heino et al., 2000). ND10 are nuclear domains associated with virus replication, and localisation of E2 to these structures is suggestive of its role in promotion of virion assembly.

The localisation of L2 to ND10 has been confirmed for HPV-33 L2 (Florin et al., 2002). The translocation of L2 to ND10 causes their reorganisation, with dispersal of some, and the recruitment of other, of the ND10-associated proteins (Florin et al., 2002) suggesting that the ND10-associated proteins play an important part in the life cycle of the virus and in its relationship with the host cell.

The interaction between E2 and L2, established recently for HPV-16 too (Okoye et al., 2005) has functional consequences. L2 selectively inhibits the transcriptional activation function, but not the DNA replication function, of HPV16 E2 (Heino et al., 2000; Okoye et al., 2005). The effect of L2 on E2 also has the potential to be crucial for regulation of the viral life cycle: L2-induced repression of E2 transcriptional functions would allow E2 to focus on DNA amplification and thus ultimately control virion production.

12.3. E7, pRb and fibropapillomas

The E7 protein interacts with, and inhibits, the pRb tumour suppressor protein through the canonical domain LXCXE (see Chapter 10). This domain is shared by all E7 proteins with the exception of the E7 proteins of artiodactyl fibropapillomaviruses (Narechania et al., 2004). It has been suggested that the absence of the pRb-binding domain contributes to the characteristic pathology of these viruses, that is, induction of fibropapillomas (Figure 1, upper panel) (Narechania et al., 2004). This is an interesting hypothesis that deserves exploration. It will be particularly interesting to understand how fibropapillomavirus E7 contributes to cell transformation and which cellular proteins it binds to. Although no HPV has been found so far with an E7 lacking the pRb-binding domain, the possibility remains that also among HPVs there are members which do not interact with pRb.

12.4. HPV-18 and quercetin

Quercetin is a potent co-factor of BPV-4 E7 in cell transformation: it substitutes for E5 in inducing anchorage independence, for E6 in inducing immortalization and for mutant p53 in inducing tumorigenicity. This synergy is molecularly underpinned by the transcriptional transactivation of the BPV-4 LCR with increased expression of E7, which drives the cells through the cell cycle, thus fixing and expanding quercetin-induced mutations and chromosomal abnormalities (see above). A similar situation is found with keratinocytes transformed by HPV-18 E6 and E7.* In these cells, quercetin activates the HPV-18 LCR via an initiator-like quercetin responsive element, homologous to the one found in the LCR of BPV-4, and increases transcription of the E7 gene. E7 inhibits the negative cell cycle regulator $p27^{kip1}$, preventing the normal quercetin-induced cell cycle arrest in G1 and driving the cells through the cell cycle (Beniston and Campo, 2005). Therefore the effect of quercetin on the LCR and the E7 protein of HPV-18 parallels that on the LCR and E7 protein of BPV-4, and supports the involvement of HPV and bracken fern in oesophageal carcinogenesis in humans (see above).

13. Conclusions

The BPV system has been ground breaking in the recognition of the oncogenic nature of the virus, the elucidation of the relationship between virus and co-factors and the development of anti-viral vaccines. The HPV vaccines currently in clinical trials are based on vaccines developed against BPV (and CRPV), including VLP-based vaccines (see Chapters 18, 19, 21) and E6/E7/L2-based prime-boost vaccines (Davidson et al., 2004). Although not all the facets of the bovine model system can be applied to the human situation, comparative pathology studies between BPV and HPV-induced diseases have contributed in no small measure to the understanding of HPV-induced cancer in humans. The latest insights in BPV E5, E2 and E7 functions show that BPV can still lead the way and contribute significantly to our understanding of virus biology. There are still many things to be learned from BPV. It is not an exaggeration to say that the study of BPV has blurred the dividing line between veterinary and human medicine.

Acknowledgements

The author would like to acknowledge support from the Medical Research Council, the Association for International Cancer Research, the World Cancer Research Fund International and Cancer Research UK. The author is a Life Fellow of CR UK.

References

Abroi, A., Ilves, I., Kivi, S. and Ustav, M. (2004). Analysis of Chromatin Attachment and Partitioning Functions of Bovine Papillomavirus Type 1 E2 Protein. J. Virol. *78*, 2100-2113.

Alonso-Amelot, M.E. and Avendano, M. (2001). Possible association between gastric cancer and bracken fern in Venezuela: an epidemiologic study. Int. J. Cancer. *91*, 252-259.

Alonso-Amelot, M.E. and Avendano, M. (2002). Human carcinogenesis and bracken fern: a review of the evidence. Curr. Med. Chem. *9*, 675-686.

Alonso-Amelot, M.E., Castillo, U., Smith, B.L. and Lauren, D.R. (1996). Bracken ptaquilosode in milk. Nature *382*, 587.

Anderson, R.A., Scobie, L., O'Neil, B.W., Grindlay, G.J. and Campo, M.S. (1997).Viral proteins of bovine papillomavirus type 4 during the development of alimentary canal tumours. Veter. J. *154*, 69-78.

Araibi, E., Marchetti, B., Ashrafi, G.H. and Campo, M.S. (2004). Down-regulation of MHC class I in bovine papillomas. J. Gen. Virol. *85*, 2809-2814.

Ashrafi, G., Haghshenas, M.R., Marchetti, B., O'Brien, P.M. and Campo, M.S. (2005). The E5 protein of human papillomavirus type 16 selectively down-regulates surface HLA class I. Int. J. Can. *113*, 276-283. Published online: 25 Aug 2004.

Ashrafi, G.H., Tsirimonaki, E., Marchetti, B., O'Brien, P.M., Sibbet, G.J., Andrew, L. and Campo, M. S. (2002). Down-regulation of MHC class I by bovine papillomavirus E5 oncoproteins. Oncogene *21*, 248-259.

Bastien, N. and McBride AA. (2000). Interaction of the papillomavirus E2 protein with mitotic chromosomes. Virology *270*, 124-134.

Beniston, R. and Campo, M.S. (2005) HPV-18 transformed cells fail to arrest in G1 in response to quercetin treatment. Virus Research *109*, 203-209. Published on line 20 January 2005.

Beniston, R., Morgan, I.M., O'Brien, V. and Campo, M.S. (2001). Quercetin, E7 and p53 in papillomavirus oncogenic cell transformation. Carcinogenesis *22*, 1069-1076.

Bloch, N. and Breen, M. (1997). Bovine papillomavirus type 5: partial sequence and comparison with other bovine papillomaviruses. Virus Genes *14*, 171-174.

Bloch, N., Sutton, R.H. and Spradbrow, P.B. (1994). Bovine cutaneous papillomas associated with bovine papillomavirus type 5. Archives Virology *138*, 373-377.

Bohl, J., Hull, B. and Vande Pol, S.B. (2001) Cooperative transformation and coexpression of bovine papillomavirus type 1 E5 and E7 proteins. J. Virol. *75*, 513-521.

Borzacchiello, G., Ambrosio, V., Galati, P., Poggiali, F., Venuti, A. and Roperto, F. (2001). The pagetoid variant of urothelial carcinoma *in situ* of urinary bladder in a cow. Vet. Pathol. *38*, 113-116.

Borzacchiello, G., Ambrosio, V., Roperto, S., Poggiali, F., Tsirimonakis, E., Venuti, A., Campo, M.S. and Roperto, F. (2003a). Bovine papillomavirus type 4 in oesophageal papillomas of cattle from the south of Italy. J. Comp. Path. *128*, 203-206.

Borzacchiello, G., Iovane, G., Marcante, M.L., Poggiali, F., Roperto, F., Roperto, S. and Venuti, A. (2003b). Presence of bovine papillomavirus type 2 DNA and expression of the viral oncoprotein E5 in naturally occurring urinary bladder tumours in cows. J. Gen. Virol. *84*, 2921-2926.

Butler, D., Collins, C., Mabruk, M., Barry Walsh, C., Leader, M.B. and Kay, E.W. (2000). Deletion of the FHIT gene in neoplastic and invasive cervical lesions is related to high-risk HPV infection but is independent of histopathological features. J. Pathol. *192*, 502-510.

Cairney, M. and Campo, M.S. (1995). The synergism between bovine papillomavirus type 4 and quercetin is dependent on the timing of exposure. Carcinogenesis *16*, 1997-2001.

Campo, M.S. (1997a). Bovine papillomavirus and cancer. Veterin. J. 154, 175-188.

Campo, M.S. (1997b). Vaccination against papillomavirus in cattle. Clinics in Dermatology, *15*, 275-283.

Campo, M.S. (1998). Persistent infection by bovine papillomavirus. In "Persistent Viral Infections". Eds R. Ahmed and I.S.Y. Chen, (John Wiley and Sons, ed), pp503-516.

Campo, M.S. (2002). Animal models of papillomavirus pathogenesis. Virus Research *89*, 249-261.

Campo, M.S., Jarrett, W.F.H., Barron, R., O'Neil, B.W. and Smith, K.T. (1992). Association of bovine papillomavirus type 2 and bracken fern with bladder cancer in cattle. Cancer Res. *52*, 6898-6904.

Campo, M.S., Jarrett, W.F.H., Grindlay, G.J., Chandrachud, L.M., McGarvie, G.M. and O'Neil, B.W. (1993). Prophylactic and therapeutic vaccination against a mucosal papillomavirus. J. Gen. Virol. *74*, 945-953.

Campo, M.S., Anderson, R.A., Cairney, M. and Jackson, M.E. (1994a). Bovine papillomavirus type 4: from transcriptional control to control of disease. in "128th SGM Symposium - Viruses and cancer". A.C. Minson Editor, Cambridge University Press, 47-70.

Campo, M.S., Jarrett, W.F.H., O'Neil, B.W. and Barron, R.J. (1994b). Latent papillomavirus infection in cattle. Res. Vet. Sci. *56*, 151-157.

Campo, M.S., O'Neil, B.W., Barron, R.J. and Jarrett, W.F. (1994c). Experimental reproduction of the papilloma-carcinoma complex of the alimentary canal in cattle. Carcinogenesis *15*, 1597-1601.

Campo, M.S., O'Neil, B.W., Grindlay, G.J., Curtis, F., Knowles, G. and Chandrachud, L.M. (1997) A peptide encoding a B-cell epitope from the N-terminus of the capsid protein L2 of bovine papillomavirus-4 prevents disease. Virology *234*, 261-266.

Cartin, W. and Alonso, A. (2003). The human papillomavirus HPV2a E5 protein localizes to the Golgi apparatus and modulates signal transduction. Virology *314*, 572-579.

Chandrachud, L.M., Grindlay, G.J., McGarvie, G.M., O'Neil, B.W., Wagner, E.R., Jarrett, W.F.H. and Campo, M.S. (1995) Vaccination of cattle with the N-terminus of L2 is necessary and sufficient for preventing infection by bovine papillomavirus type 4. Virology *211*, 204-208.

Chandrachud, L.M., O'Neil, B.W., Jarrett, W.H.F., Grindlay, G.J., McGarvie, G.M. and Campo, M.S. (1994) Humoral immune response to the E7 protein of bovine papillomavirus type 4 and identification of B-cell epitopes. Virology *200*, 98-104.

Connolly, J., Morgan, I.M., Jackson, M.E. and Campo, M.S. (1998). The BPV-4 co-carcinogen quercetin induces cell cycle arrest and upregulates the LCR of BPV-4. Oncogene *16*, 2739-2746.

Cresswell, P. (2000). Intracellular surveillance: Controlling the assembly of MHC class I-peptide complexes. Traffic *1*, 301-305.

Da Silva, D.M., Velders, M.P., Nieland, J.D., Schiller, J.T., Nickoloff, B.J., and Kast, W. M. (2001). Physical interaction of human papillomavirus virus-like particles with immune cells. Int. Immunol. *13*, 633-641.

Davidson, E.J., Faulkner, R.L., Sehr, P., Pawlita, M., Smyth, L.J., Burt, D.J., Tomlinson, A.E., Hickling, J., Kitchener, H.C. and Stern, P.L. (2004) Effect of TA-CIN (HPV 16 L2E6E7) booster immunisation in vulval intraepithelial neoplasia patients previously vaccinated with TA-HPV (vaccinia virus encoding HPV 16/18 E6E7). Vaccine *22*, 2722-2729.

Day, N. and Varghese, C. (1994). Oesophegeal cancer. Cancer Surveys vol *19/20*, pp43-54.

Day, P.M., Roden, R.B., Lowy, D.R. and Schiller, J.T. (1998). The papillomavirus minor capsid protein, L2, induces localization of the major capsid protein, L1, and the viral transcription/replication protein, E2, to PML oncogenic domains. J. Virol. *72*, 142-50.

de Villiers, E-M., Fauquet, C., Broker, T.R., Bernard, H-U. and zur Hausen, H. (2004). Classification of papillomaviruses. Virology *324*, 17-27.

Demirer, T., Icli, F., Uzunalimoglu, O. and Kucuk, O. (1990). Diet and stomach cancer incidence: a case control study in Turkey. Cancer Surveys *65*, 2344-2348.

Fay, A., Yutzy, W.H. 4th, Roden, R.B. and Moroianu, J. (2004) The positively charged termini of L2 minor capsid protein required for bovine papillomavirus infection function separately in nuclear import and DNA binding. J. Virol. *78*, 13447-13454.

Florin, L., Schafer, F., Sotlar, K., Streeck, R.E. and Sapp, M. (2002). Reorganization of Nuclear Domain 10 Induced by Papillomavirus Capsid Protein L2. Virology *295*, 97-107.

Galpin, O.P., Whitaker, C.J., Whitaker, R. and Kassab, J.Y. (1990). Gastric cancer in Gwynedd. Possible links with bracken. British J. Cancer *61*, 737-740.

Gaukroger, J.M., Chandrachud, L.M., O'Neil, B.W., Grindlay, G.J., Knowles, G. and Campo, M.S. (1996) Vaccination of cattle with bovine papillomavirus type 4 L2 elicits the production of virus neutralizing antibodies. J. Gen. Virol. *77*, 1577-1583.

Globocan (2001). Cancer Incidence, Mortality and Prevalence Worldwide, Version 1.0. IARC CancerBase No. 5, Lyon, IARC Press.

Graham, S., Haughey, B. and Marshall, J. (1990). Diet in the epidemiology of gastric cancer. Nutrition and Cancer *13*, 19-34.

Heino, P., Zhou, J. and Lambert, P.F. (2000). Interaction of the papillomavirus transcription/replication factor, E2, and the viral capsid protein, L2. Virology *276*, 304-314.

Hirayama, T. (1979). Diet and Cancer. Nutr. Cancer *1*, 67-81.

Ilves, I., Kivi, S. and Ustav, M. (1999). Long-term episomal maintenance of bovine papillomavirus type 1 plasmids is determined by attachment to host chromosomes, which is mediated by the viral E2 protein and its binding sites. J. Virol. *73*, 4404-4412.

Jackson, M.E. and Campo, M.S. (1991) Positive and negative E2-independent regulatory elements in the long control region of bovine papillomavirus type 4. J. Gen. Virol. *72*, 877-883.

Jackson, M.E. and Campo, M.S. (1995) Both viral E2 protein and cellular factor PEBP2 regulate transcription via E2 consensus binding sites within the bovine papillomavirus long control region. J. Virol., *69*, 6038-6046.

Jackson, M.E., Pennie, W.D., McCaffery, R.E., Smith, K.T., Grindlay G.J. and Campo, M.S. (1991) The B subgroup bovine papillomaviruses lack an identifiable E6 open reading frame. Molecular Carcinogenesis, *4*, 382-387.

Jaggar, R.T., Pennie, W.D., Smith, K.T., Jackson, M.E. and Campo, M.S. (1990) Cooperation between bovine papillomavirus type 4 and ras in the morphological transformation of primary bovine fibroblasts. J. Gen. Virol. *71*, 3041-3046.

Jarrett, W.F.H., McNeal, P.E., Grimshaw, T.R., Selman, I.E. and McIntyre, W.I.M. (1978). High incidence area of cattle cancer with a possible interaction between an environmental carcinogen and a papillomavirus. Nature *274*, 215-217.

Jarrett, W.F.H. (1985). The natural history of bovine papillomavirus infections. Advances in Viral Oncology *5*, 83-102.

Kedzia, H., Gozdzicka-Jozefiak, A., Wolna, M. and Tomczak, E. (1992). Distribution of human papillomavirus 16 in the blood of women with uterine cervix carcinoma. Eur. J. Gynaecol. Oncol. *13*, 522-526.

Khaled, H., Raafat, A., Mokhtar, N., Zekri, A.R. and Gaballah, H. (2001). Human papilloma virus infection and overexpression of p53 protein in bilharzial bladder cancer. Tumori *87*, 256-261.

Khaled, H.M., Bahnassi, A.A., Zekri, A-R.N., Kassem, H.A. and Mokhtar, N. (2003). Correlation between p53 mutations and HPV in bilharzial bladder cancer. Urologic Oncology: Seminars and Original Investigations *21*, 334-341.

Kirnbauer, R., Chandrachud, L.M., O'Neil, B.W., Wagner, E.R., Grindlay, G.J., Armstrong, A., McGarvie, G.M., Schiller, J.T., Lowy, D.R. and Campo, M.S. (1996) Virus-like particles of bovine papillomavirus type 4 in prophylactic and therapeutic immunization. Virology *219*, 37-44.

Knowles, G., Chandrachud, L.M., O'Neil, B.W., Grindlay, G.J. and Campo, M.S. (1997). Linear B-cell epitopes in the N-terminus of L2 of bovine papillomavirus type 4. Res. Vet. Sci. *62*, 289-291.

Knowles, G., O'Neil, B.W. and Campo, M.S. (1996) Phenotypic characterization of lymphocytes infiltrating regressing papillomas. J. Virol. *70*, 8451-8458.

Lioi, M.B., Barbieri, R., Borzacchiello, G., Dezzi, S., Roperto, S., Santoro, A., Russo, V. and Roperto, F. (2004). Chromosome Aberrations in Cattle with Chronic Enzootic Haematuria. J. Comp. Path. *131*, 233-236.

Liu, W. J., Gissmann, L., Sun, X. Y., Kanjanahaluethai, A., Muller, M., Doorbar, J., and Zhou, J. (1997). Sequence close to the N-terminus of L2 protein is displayed on the surface of bovine papillomavirus type 1 virions. Virology *227*, 474-483.

Marchetti, B., Ashrafi G.H., Tsirimonaki E., O'Brien P.M. and Campo, M.S. (2002). The papillomavirus oncoprotein E5 retains the major histocompatibility class I in the Golgi apparatus and prevents its transport to the cell surface. Oncogene *21*, 7808-7816.

Marliere, C., Galvao, M.A.M., Wathern, P. and Freitas, R.N. (1999). Bracken fern consumption and oesophageal and gastric cancer in the Ouro Preto region, MG, Brazil. Bracken fern: toxicity, biology and control. Eds. JA Taylors and RT Smith., pp144-149

Marliere, C., Wathern, P., Castro, M.C., O'Connor, P. and Galvao, M.A. (2002). Bracken fern (Pteridium aquilinum) ingestion and oesophageal and stomach cancer. IARC Science Publications *156*, 379-380.

McCaffery, R.E. and Jackson, M.E. (1994) An element binding a C/EBP-related transcription factor contributes to the negative regulation of the bovine papillomavirus type 4 long control region. J. Gen. Virol. *75*, 3047-3056.

McGarvie, G.M., Grindlay, G.J., Chandrachud, L.M., O'Neil, B.W., Jarrett, W.F.H. and Campo, M.S. (1995) T-cell responses to BPV-4 E7 in vaccinated and non vaccinated cattle during the course of BPV-4 infection and identification of T-cell epitopes. Virology, *206*, 504-510.

Modis, Y., Trus, B.L. and Harrison, S.C. (2002). Atomic model of the papillomavirus capsid. EMBO J. *21*, 4754-4762.

Morgan, I.M. and Campo, M.S. (2000) Recent developments in bovine papillomaviruses. In Papillomavirus Report *11*, 127-132.

Morgan, I.M., Grindlay, G.J. and Campo, M.S. (1998) Epithelial specific transcriptional regulation of the bovine papillomavirus 4 promoter by E2. J. Gen. Virol. *79*, 501-508.

Morgan, I.M., Grindlay, G.J. and Campo, M.S. (1999) The bovine papillomavirus type 4 long control region contains an epithelium specific transcriptional enhancer. J.Gen. Virol. *80*, 23-27.

Moura, J.W., Stocco dos Santos, R.C., Dagli, M.L., D'Angelino, J.L., Birgel, E.H. and Becak, W. (1988). Chromosome aberrations in cattle raised on bracken fern pasture. Experientia *44*, 785-788.

Narechania, A., Terai, M., Chen, Z., DeSalle, R. and Burk, R. D. (2004). Lack of the canonical pRB-binding domain in the E7 ORF of artiodactyl papillomaviruses is associated with the development of fibropapillomas. J. Gen. Virol. *85*, 1243-1250.

O'Brien, P.M. and Campo, M.S. (2003). Papillomaviruses: a correlation between immune evasion and oncogenicity? Trends in Microbiology *11*, 300-305.

O'Brien, P.M., and Campo, M.S. (2002). Evasion of host immunity directed by papillomavirus-encoded proteins. Virus Research *1-2*, 103-118.

O'Brien, V. and Campo, M.S. (1998) BPV-4 E5 transforms NIH 3T3 cells, up-regulates cyclin A and cyclin A-associated kinase activity and de-regulates expression of the cdk inhibitor $p27^{KIP1}$. Oncogene *17*, 293-301.

O'Brien, V., Ashrafi, G.H., Grindlay, G.J., Anderson, R.A. and Campo, M.S. (1999) A mutational analysis of the transforming functions of the E5 protein of bovine papillomavirus type 4. Virology *254*, 385-394.

Ogawa, T., Tomita, Y., Okada, M., Shinozaki, K., Kubonoya, H., Kaiho, I. and Shirasawa, H. (2004). Broad-spectrum detection of papillomaviruses in bovine teat papillomas and healthy teat skin. J. Gen. Virol. *85*, 2191-2197.

Okoye, A., Cordano, P., Taylor, E.R., Morgan, I.M., Everett, R. and Campo, M.S. (2005). Human papillomavirus 16 L2 inhibits the transcriptional activation function, but not the DNA replication function, of HPV16 E2. Virus Research, *108*, 1-14. Published online 27 August 2004.

Oliver, R., Breuer, J., Nouri, A.M.E. and Campo, M.S. (1998). Malignancy-associated papillomaviruses: could they be factors affecting morphology of human bladder cancer? Cancer Surveys *31*, 29-47.

Olson, C., Pamukcu, A.M. and Brobst, D.F. (1965). Papilloma-like virus from bovine urinary bladder tumors. Cancer Research *25*, 840-849.

Olson, C., Pamukcu, A.M., Brobst, D.F., Kowalczyk, T., Satter, E.J. and Price, J.M. (1959). A urinary bladder tumor induced by a bovine cutaneous papilloma agent. Cancer Research *19*, 779-782.

Pao, C., Hor, J.J., Yang, F.P., Lin, C.Y. and Tseng, C.J. (1997). Detection of human papillomavirus mRNA and cervical cancer cells in peripheral blood of cervical cancer patients with metastasis. J. Clin. Oncol. *15*, 1008-1012.

Pao, C., Lin, S-S., Lin, C-Y., Maa, J-S., Lai, C-H. and Hsieh, T-T. (1991). Identification of human papillomavirus DNA sequences in peripheral blood mononuclear cells. Am. J. Clin. Path. *95*, 540-546.

Payne, D., Tyring, S.K., Doherty, M.G., Daya, D. and Chan, T-S. (1993). Detection of Human Papillomavirus Type 16 in Plasma Cells. Gynecologic Oncology *48*, 406-412.

Pennie, W.D. and Campo, M.S. (1992). Synergism between bovine papillomavirus type 4 and the flavonoid quercetin in cell transformation *in vitro*. Virology *190*, 861-5.

Pennie WD, Grindlay GJ, Cairney M and Campo MS. (1993) Analysis of the transforming functions of bovine papillomavirus type 4. Virology *193*, 614-620.

Pfister, H., Linz, U., Gissmann, L., Huchthausen, B., Hoffmann, D. and zur Hausen, H. (1979). Partial characterization of a new type of bovine papilloma virus. Virology *96*, 1-8.

Plowright, W., Linsell, C.A. and Peers, F.G. (1971). A focus of rumenal cancer in Kenyan cattle. British J. Cancer 25, 72-80.

Povey, A., Potter, D. and O'Connor, P.J. (1996). 32P-post-labelling analysis of DNA adducts formed in the upper gastrointestinal tissue of mice fed bracken extract or bracken spores. Brit. J. Cancer *74*, 1342-1348.

Recouso, R., Stocco dos Santos, R.C., Freitas, R., Santos, R.C., de Freitas, A.C., Brunner, O., Becak, W. and Lindsey, C.J. (2003). Clastogenic effetct of bracken fern (Pteridium aquilinum v. arichnoideum) diet in peripheral lymphocytes of human consumers: preliminary data. Veterinary and Comparative Oncology *1*, 22-29.

Roden, R. B., Yutzy, W. H. t., Fallon, R., Inglis, S., Lowy, D. R., and Schiller, J. T. (2000). Minor capsid protein of human genital papillomaviruses contains subdominant, cross-neutralizing epitopes. Virology *270*, 254-257.

Santos, R., Brasiliero-Filho, G. and Hojo, E.S. (1987). Induction of tumours in rats by bracken fern from Ouro Preto (Minas Gerais, Brazil). Brazilian J. Med. Biol. Res. *20*, 73-77.

Santos, R., Brasiliero-Filho, G. and Silva, M.E. (1992). Tumorigenicity of boiling water extracts of bracken fern. Cienc. Tecnol. Aliment. *12*, 72-76.

Scobie, L., Jackson, M.E. and Campo, M.S. (1997) The role of exogenous p53 and E6 oncoproteins in *in vitro* transformation by bovine papillomavirus type 4: significance of the absence of an E6 ORF in the BPV-4 genome., J. Gen. Virol. *78*, 3001-3008.

Shahin, M., Moore, M.R., Worrall, S., Smith, B.L., Seawright, A.A. and Prakash, A.S. (1998). H-ras activation is an early event in the ptaquiloside-induced carcinogenesis: Comparison of Acute and Chronic Toxicity in Rats. Biochem. Biophys, Res. Commun. *250*, 491-497.

Sibbet, G., Romero-Graillet, C., Meneguzzi, G., and Campo, M. S. (2000). alpha6 integrin is not the obligatory cell receptor for bovine papillomavirus type 4. J. Gen. Virol. *81*, 327-334.

Stocco dos Santos, R., Lindsey, C., Ferraz, O., Pinto, J., Mirandola, R., Benesi, F., Birgel, E., Pereira, C. and Becak, W. (1998). Bovine papillomavirus transmission and chromosomal aberrations: an experimental model. J. Gen. Virol. *79*, 2127-2135.

Syrjanen, K. (2002). HPV infections and oesophageal cancer. J. Clin. Patho. *55*, 721-728.

Syrjanen, S. (2003). Human papillomavirus infections and oral tumors. Medical Microbiology and Immunology *192*, 123-128.

Takizawa, S., Nakagawa, S., Nakagawa, K., Yasugi, T., Fujii, T., Kugu, K., Yano, T., Yoshikawa, H. and Taketani, Y. (2003). Abnormal FHIT expression is an independent poor prognostic factor for cervical cancer. British J. Cancer *88*, 1213-1216.

Tindle, R. (2002). Immune evasion in human papillomavirus-associated cervical cancer. Nature Reviews Cancer *2*, *59*-65.

Tong, X., Boll, W., Kirchhausen, T. and Howley, P.M. (1998) Interaction of the bovine papillomavirus E6 protein with the clathrin adaptor complex AP-1. J. Virol. *72*, 476-482.

Tong, X., Salgia, R., Li, J.L., Griffin, J.D. and Howley, P.M. (1997) The bovine papillomavirus E6 protein binds to the LD motif repeats of paxillin and blocks its interaction with vinculin and the focal adhesion kinase. J. Biol. Chem. *272*, 33373-33376.

Tseng, C-J., Pao, C.C., Lin, J-D., Soong Y-K., Hong, J-H. and Hsueh, S. (1999). Detection of Human Papillomavirus

Types 16 and 18 mRNA in Peripheral Blood of Advanced Cervical Cancer Patients and Its Association With Prognosis. J. Clin. Oncol. *17*, 1391-1396.

Tsirimonaki, E., O'Neil, B.W., Williams, R. and Campo, M.S. (2003). Extensive Papillomatosis of the Bovine Upper Gastrointestinal Tract. J. Comp. Path. *129*, 93-99.

Vance KW, Campo MS and Morgan IM (1999) An enhanced epithelial response of a papillomavirus promoter to transcriptional activators. J. Biol. Chem. *274*, 27839-27844.

Vance KW, Campo MS and Morgan IM (2001) A novel silencer element in the BPV-4 promoter represses the transcriptional response to the papillomavirus E2 protein. J. Virol. *75*, 2829-2838.

Van Tine, B.A., Dao, L.D., Wu, S-Y., Sonbuchner, T.M., Lin, B.Y., Zou, N., Chiang, C-M., Broker, T.R. and Chow, L.T. (2004). Human papillomavirus (HPV) origin-binding protein associates with mitotic spindles to enable viral DNA partitioning. PNAS *101*, 4030-4035.

Villalobos-Salazar, J., Mora, J., Meneses, A. and Pashov, B. (1995). The carcinogenic effect of bracken spores. Bracken, an environmental issue. Eds. J.A. Taylor and R.T. Smith, pp102-103.

Yang, W.C., Okayama, H. and Howley, P.M. (1985) Bovine papilloavirus contains multiple transforming genes. Proc. Natl. Acad. Sci. *82*, 1030-1034.

Zago, M., Campo, M.S. and O'Brien, V. (2004) Cyclin A expression and growth in suspension can be uncoupled from p27 de-regulation and ERK activity in cells transformed by BPV-4 E5. J. Gen. Virol. *85*, 3585-3595.

Zhao, K., Hengst, K., Liu, W.J., Liu, Y.H., Liu, X.S., McMillan, N.A. and Frazer, I.H. (2000). BPV-1 E2 protein enhances packaging of full-length plasmid DNA in BPV-1 pseudovirions. Virology *272*, 382-393.

Zimmermann, H., Koh, C.H., Degenkolbe, R., O'Connor, M.J., Muller, A., Steger, G., Chen, J.J., Lui, Y., Androphy, E. and Bernard, H.U. (2000) Interaction with CBP/p300 enables the bovine papillomavirus type 1 E6 oncoprotein to downregulate CBP/p300-mediated transactivation by p53. J. Gen. Virol. *81*, 2617-23.

From: Papillomavirus Research: From Natural History to Vaccines and Beyond. Edited by: M. Saveria Campo

Chapter 24

Bovine Papillomaviruses and Equine Sarcoids

Lubna Nasir* and Stuart W.J. Reid

Abstract

Equine sarcoids are locally aggressive, non regressing, fibroblastic skin tumours representing the most common dermatological neoplasm reported in equidae. It is widely accepted that the Bovine Papillomavirus (BPV) types 1/2 are associated with the pathogenesis of sarcoid disease. The majority of lesions contain detectable viral nucleic acids and express the BPV 1/2 major transforming protein, E5, but appear to be non-productive for infectious virions. Here we review the evidence supporting association of sarcoids with bovine papillomavirus (BPV) and the nature of the infecting viral genomes and consider the practical implications of BPV infection for diagnostic and therapeutic purposes.

1. Equine Sarcoids

The term 'sarcoid' was first used by Jackson in 1936 to describe a locally invasive, fibroblastic skin tumour affecting horses, donkey and mules. Although this definition has been applied for over 60 years, various terms have been used to describe sarcoids including warts, papillomas, fibromas, fibrosarcomas, schwannomas (neurofibromas) and recurring granulation tissue. However, the equine sarcoid is recognised as a distinct pathological disease which can be distinguished clinically and histopathogically from the above mentioned lesions.

Sarcoid are benign non-metastatic lesions, however lesions may infiltrate locally into the surrounding tissues (Lane, 1977; Tarwid et al., 1985), often occurring at sites of previous injury or scarring (Ragland et al., 1970; Torrontentegui and Reid, 1994). Tumours can exist as single or multiple lesions in different forms ranging from small wart-like lesions to large ulcerated fibrous growths (Figure 1). Based on the clinical appearance, five different types of sarcoids can be recognised: Occult sarcoids present as hairless circular areas of skins; Verrucous lesions exhibit a wart-like appearance; Fibroblastic sarcoids present as ulcerated fleshy masses; Nodular sarcoids consist of firm masses lying underneath the skin and Mixed sarcoids show a combination of features of verrucous, fibroblastic and nodular sarcoids (Pascoe and Knottenbelt, 1999). Histopathologically however, the different types of lesions are often difficult to distinguish as they exhibit similar histological changes including dermal proliferation of spindle shaped fibroblasts forming whorls, epidermal hyperplasia and hyperkeratosis and rete peg formation (Pascoe and Knottenbelt, 1999).

Sarcoids represent the most common neoplasm in equids with reported incidences ranging from 12.9 to 67% of all equine tumours (Sullins et al., 1989). In a population of donkeys Reid et al. (1994) reported an estimated incidence of 0.6 cases per hundred animal-years. Sites of skin that are commonly affected include the head, ventral abdomen, legs and the paragenital region (Jackson, 1936; Ragland et al., 1970; Lane, 1977; Fretz and Barber, 1980; Sullins et al., 1986; Torrontentegui and Reid, 1994). To date, no sex predilection has been

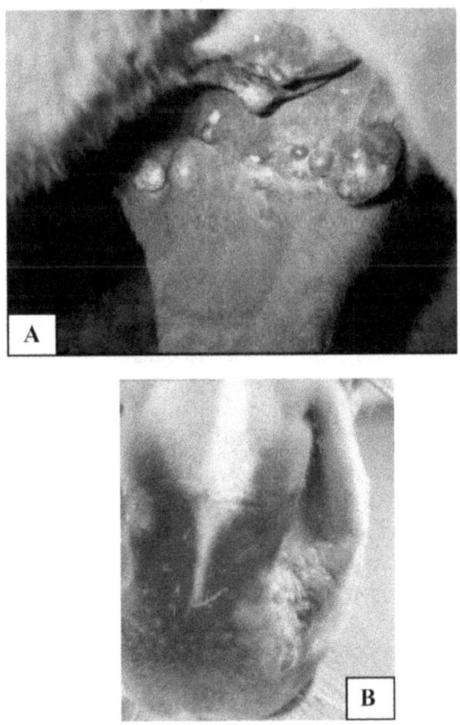

Figure 1. **A.** Fibroblastic sarcoids affecting the medial aspect of the hind legs of a horse. **B.** Verrucose sarcoid located on the external nares of a horse.

*For correspondence. Email l.nasir@vet.gla.ac.uk

reported, however an increased incidence has been reported in younger animals (Mohammed et al., 1992; Reid et al., 1994; Torrontentegui and Reid, 1994; Reid and Mohammed, 1997).

Currently there is no permanent effective therapy for the treatment of sarcoids, although some clinicians have reported pragmatic success with topical unlicensed applications (Knottenbelt and Walker, 1994). Commonly employed treatments include cryotherapy, surgical excision and local immune modulation often combined with chemotherapy (Goodrich et al., 1998). Lesions of the limb are more difficult to treat that those on the surface of the body and a complete cure is less commonly achieved in animals presenting with multiple sarcoids. Furthermore, the anatomical location of sarcoids can exclude the use of surgery, for example, sarcoids of the eyelid. Sarcoids frequently display hyperproliferation or recurrence following surgical intervention in up to 50% of cases (Tarwid, 1985), often necessitating further treatment. Conversely many private practitioners treat sarcoids successfully by a policy of non-intervention, which may represent a specific population of sarcoids that remain quiescent or are rare spontaneous regressors and there is some anecdotal evidence for this (Goodrich et al., 1998).

2. Evidence for a Viral Aetiology

Several lines of evidence support the involvement of an infectious agent in the pathogenesis of equine sarcoid tumours. The first report by Jackson (Jackson, 1936) describing the equine sarcoid suggested a viral origin for the tumour, based on its appearance and pattern of spread by contact with riding equipment. This was further supported by epidemiological evidence of outbreaks of sarcoids in populations of horses (Ragland et al., 1966). Subsequently, more substantial evidence was provided by transmission studies in which inoculation with either sarcoid tissue or cell-free supernatant from minced tumours onto the scarified skin of sarcoid-free horses resulted in the appearance of tumours at the inoculation site which were morphologically indistinguishable from naturally occurring sarcoids (Voss, 1969). Similarly, inoculation with bovine papillomavirus (BPV) in non-affected horses, as well as in horses with existing sarcoids, resulted in the growth of sarcoid-like tumours (Olson and Cook, 1951; Ragland and Spencer, 1969; 1969; Lancaster et al., 1977). The experimentally induced sarcoids showed spontaneous regression and BPV-neutralising antibodies could be detected in horses. However, spontaneous regression is rarely encountered in naturally occurring sarcoid tumours with no evidence of BPV-neutralising antibodies (Ragland and Spencer 1969; Lancaster et al., 1977; Brostrom 1979).

Although the early transmission experiments suggested that bovine papillomavirus (BPV) may be the causative agent (Olson and Cook 1951; Ragland and Spencer, 1969), no papillomavirus has been isolated from clinical cases. Studies on a cell line derived from an equine sarcoid (the Mc-1 cell line) and on a cell line derived from a tumour induced by inoculation of Mc-1 cells (the T-77-4 cell line) into a foal affected by combined immunodeficiency revealed the presence of virus particles containing high molecular weight RNA genomes and reverse transcriptase activity (England et al., 1973; Fatemi-Nainie et al., 1982; 1984; Cheevers et al., 1982). However, the virus associated with Mc-1 cells and their derivatives was a non-oncogenic, replication defective virus presumed to be an endogenous equine retrovirus and a causative relationship between this virus and equine sarcoids was not established (Cheevers et al., 1986).

3. Viral Genes and Viral Gene Expression

Despite the failure to isolate papillomavirus from sarcoids, there is now a large body of evidence strongly supporting the hypothesis that BPV is the etiological agent of equine sarcoids. Since the initial suggestion of such a relationship between BPV and equine sarcoids by Olson and Cook (1951), groups in the USA, Australia, continental Europe and the UK have reported the presence of the closely related viruses BPV-1 and BPV-2 in equine sarcoids. Early studies detected BPV DNA in sarcoids from horses and donkeys using DNA hybridisation techniques (Lancaster et al., 1979; Amtmann et al., 1980). In more recent years, PCR-based detection methods, offering greater sensitivity, have been used to demonstrate the presence of BPV DNA in sarcoids (Trenfield et al., 1985; Angelos et al., 1991; Lory et al., 1993; Bloch et al., 1994; Reid et al., 1994). The reported detection rate varies between study groups, from 73% (Bloch et al., 1994) to 88-91% (Martens et al., 2001a and b) to 96%-100% (Carr et al., 2001a; 2001b; Otten et al., 1993). Both BPV-1 and BPV-2 have been detected in sarcoid tumours with BPV-1 as the predominant type. Contrary to the finding of BPV DNA in bovine peripheral blood cells (see Chapter 23), BPV DNA has not been detected in blood samples obtained from horses (Nasir et al., 1997) or in non-sarcoid equine tumours or equine papillomas (Otten et al., 1993; Carr et al., 2001a; 2001b). The absence of BPV DNA in equine blood cells may be related to the abortive nature of the infection. However, viral DNA has been found in some cases of dermatitis and the significance of this is as yet unknown (Angelos et al., 1991; Chambers et al., 2003). In the study by Chambers et al. (2003), BPV DNA was detected in 4 of 5 granulomatous dermatitis lesions. Although it is possible that this was due to a misclassification of the lesions, each of the samples was independently evaluated by two pathologists to confirm the initial diagnosis. It has recently been shown that human papillomavirus (HPV) type 5 may contribute to the pathology of the inflammatory skin condition psoriasis, and has led to

Table 1. BPV-1 sequence variants identified in a panel of Swiss and British equine sarcoids. [1]BPV sequence isolated from a Scottish bovine penile papilloma. [2]Black Underlined Nucleotide changes do not cause changes in the E5 amino acid sequence but those in underlined grey do. [3]Genbank nucleotide and amino acid sequences are from STD database.

Sequence type	No. of Samples	BPV type	Genbank Acession No.	DNA Sequence Variation[2] (genbank vs sample [3])	Amino Acid Sequence Changes (genbank vs sample [3]) The entire E5 aminoacid sequence is shown
Swiss I	19	BPV1	AY232257	3936 CTA 3938 CT**G**	None
Swiss II	2	BPV1	AY232258	3936 CTATTA 3941 CT**GT**TC	1 … 20 … 44 genbank: MPNLWFLLFLGLVAAMQLLLLLFLLLFFLVYWDHFECSCTGLPF sarcoid: MPNLWFLLFLGLVAAMQLLLFLFLLLFFLVYWDHFECSCTGLPF
Swiss III	2	BPV1	AY232259	3921 GCA 3924 / 3936 CTA 3938 TCA / CT**G**	1 … 20 … 44 genbank: MPNLWFLLFLGLVAAMQLLLLLFLLLFFLVYWDHFECSCTGLPF sarcoid: MPNLWFLLFLGLVASMQLLLLLFLLLFFLVYWDHFECSCTGLPF
Swiss IV	1	BPV1	AY232260	3933 CTGCTA 3938 / 3948 TTACTCTTG 3956 CT**C**CT**G** / ATGCT**GC**TG 3999 GGT 4001 / 4011 TAA 4013 AAT / T**G**A	1 … 20 … 44 genbank: MPNLWFLLFLGLVAAMQLLLLLFLLLFFLVYWDHFECSCTGLPF sarcoid: MPNLWFLLFLGLVAAMQLLLLLFMLLFFLVYWDHFECSCTNLPF
Swiss V	3	BPV2	AY232264	No variation	None
British I	7	BPV1	AY232261	None	
British II	1	BPV1	AY232263	3936 CTA 3938 CT**G**	None

the suggestion that these lesions may be a reservoir of HPV (Majewski et al., 1999; Favre et al., 1998). Thus the presence of BPV in granulomatous dermatitis support the hypothesis that certain inflammatory skin lesions may harbour papillomavirus. However, it is currently unclear what role the granulomatous inflammation might play, if any, in sarcoid pathogenesis. Despite the consistent finding of papillomavirus DNA in the sarcoid lesions, papillomavirus particles have not been demonstrated, and the disease is therefore considered to be a non-productive infection in which the viral DNA exists episomally (Amtmann et al., 1980; Lancaster, 1981).

Several studies have evaluated BPV gene expression in equine sarcoids using RT-PCR and Western blotting. Nasir and Reid (1999) examined 20 equine sarcoids containing BPV-1 DNA and demonstrated BPV specific RNA in all samples. Most of the studies concerning gene expression have largely focussed on expression of the major BPV transforming gene, E5. Carr and co-workers (2001b) analysed 23 sarcoids by Western blot and demonstrated the presence of the BPV E5 protein in all tumours (including one in which the amount of viral DNA was too low for detection) whereas E5 was absent in all of the non-sarcoid samples examined. Similarly we have reported E5 expression in 85% of sarcoids (Chambers et al., 2003). In addition to E5, the study by Nixon et al. demonstrated that the BPV-1 E6 transforming gene is also expressed in approx. 72% of sarcoids (Nixon et al., 2005). Collectively, these studies clearly provide evidence that the viral genes are expressed and that therefore the presence of the viral DNA is not accidental.

4. Viral Sequence Variants

The early studies using hybridisation techniques for the detection of BPV DNA in sarcoids (Trenfield et al., 1985; Angelos et al., 1991) provided evidence for sequence variation within the BPV genome in sarcoids, suggesting that the virus may not necessarily be BPV but a virus closely related to BPV. Detailed sequence analysis of BPV DNA isolated from sarcoids has revealed the presence of distinct equine sarcoid specific variants (Otten et al., 1993; Reid et al., 1994; Chambers et al., 2003). Reid et al. found two minor differences in the sequence of the BPV E5 open reading frame in donkey sarcoids compare with the published bovine sequences. In contrast, another report suggests absolute homology between the BPV E5 sequences in sarcoids and the published bovine BPV sequences (Carr et al., 2001a). However, the same nucleotide substitution, A3939G, previously detected in BPV-1 sequences isolated from donkey sarcoids (Reid et al., 1994) has recently been reported in E5 ORFs isolated from sarcoids (Chambers et al., 2003). The study by Chambers et al. is the most detailed analysis to date of sequence variants associated with sarcoids. In this study 6 different BPV-1 E5 variant sequences were identified in 32 sarcoids. Seven sarcoids contained E5 sequences identical to the published BPV-1 sequence, whereas the remaining 25 sarcoids contained E5 variants (Table 1). Interestingly, all five variants of BPV type 1 contain the same nucleotide change A3938G. Whilst this nucleotide substitution has no affect on the amino acid residue, the variation is associated with an increase in codon usage. The codon 3936CTA3938 of wild type BPV-1 has a usage

Table 2. Nucleotide variation and frequency of the codon usage associated with E5 in BPV-1 and sarcoid variant E5 in equine cells detected in E5 variants isolated from equine sarcoids.

BPV-1 E5	Codon freq.	Variant E5	Codon freq.
^{3936}CTA	5.9	CTG	44.8
^{3939}TTA	5.6	TTC	24.0
^{3921}GCA	13.2	TCA	10.4
^{3933}CTG	44.8	CTC	22.0
^{3948}TTA	5.6	ATG	23.0
^{3951}CTC	22.4	CTG	44.8
^{3954}TTG	12.5	CTG	44.8
^{3999}GGT	10.6	AAT	16.3
^{4011}TAA	1.1	TGA	1.2

of 8.3%, however the change A3938G increases codon usage in equine cells to 46.8%, a 5- fold increase. This suggests that the variants containing A3938G are more efficiently translated in equine cells. Since this nucleotide substitution has been detected in the majority of sarcoids, it may be that A3938G variants preferentially infect equine cells. Indeed, much of the variation seen in the E5 ORF sequence variants is associated with an increase in codon usage as shown in Table 2. Furthermore, variation has also been detected in additional regions of the BPV-1 genome in sarcoids, namely within the regulatory Long Control Region and within the ORF encoding the E2 transcriptional regulator (L. Nasir, unpublished data). Whilst these data do not clearly point to the existence of an equine-associated BPV-1 variant, it is interesting that many of the studies have detected nucleotide substitutions in the viral DNA isolated from sarcoids. It is possible that the variant sequences are associated with particular clinical presentations or that altered expression of the coding protein may contribute to sarcoid pathogenesis. However, to date the functional significance of BPV variants have not been elucidated.

In contrast, intratypic variation has been extensively studied in HPVs, particularly HPV-16. Such studies have shown that some variants are associated with functional differences. Sequence variation within the transcriptional regulatory sites of papillomaviruses have been shown to result in enhanced activity of the promoter (Dong et al., 1994; Chen et al., 1997; Veress et al., 1999), thereby potentially facilitating tumour progression. Similarly, some LCR variants have been shown to be associated with a more aggressive clinical behaviour (Kurvinen et al., 2000). The generation of invasive tumors from high-grade precursor lesions has also been associated with an HPV E6 variant comprising a T-to-G transition at nucleotide 350 corresponding to amino acid L83V (Zehbe et al., 1988; Andersson et al., 2000). Thus, more detailed investigations of BPV sequence in equine sarcoids variants and elucidation of the functional consequences of genome variation are warranted.

5. BPV Infection in Horses and Cattle

Collectively, the early transmission experiments and the more recent molecular techniques employed to detect viral DNA and RNA, provide substantial evidence to support BPV as the aetiological agent in sarcoid pathogenesis. Normally, PVs are strictly species-specific, and hence the equine sarcoid represents a biologically interesting and challenging tumour as it is the only documented case of natural cross-species PV infection (Gorman 1985; Lancaster et al., 1977). Interestingly, the infection of BPV in cattle differs markedly from the outcome of infection in horses. Infection by BPV-1/2 in cattle results in the formation of cutaneous fibropapillomas, benign tumours with a fibroblastic and an epithelial component productive for infectious progeny virus. As in most PV infections, BPV induced papillomas regress in response to an appropriate effector immune response. Although the precise mechanisms that contribute to the clearance of PV infected cells are not clear, a cell-mediated immune response is the major contributor to the regression of PV associated disease. Both CD4 and CD8 T lymphocytes are present in the immune infiltrate of regressing papillomas (Coleman et al., 1994; Knowles, 1996; Okabayashi et al., 1991), and major histocompatibility antigen (MHC) class I-restricted CD8+ cytotoxic T cells (CTL) against the major transforming proteins of PV are important in clearance of virus infected cells (Frazer et al., 1996). In contrast to the normal course of BPV disease in cattle, infection with BPV-1/2 in horses results in sarcoids, lesions which are non permissive for virus production (Gorman, 1985; Amtmann et al.,1980), locally aggressive and non regressing (Ragland et al., 1970). However, sarcoid regression can be stimulated in animals with a moderate tumour burden by intra-lesional inoculation of BCG (Vanselow et al., 1988) or by the application of immunomodulatory compounds (Rush and Flaminio, 2000), both of which stimulate local cellular immunity by providing a favourable cytokine environment. Therapeutic vaccination with autologous tumour cells can induce clearance of recurrent sarcoid lesions (Kinnunen et al., 1999), correlating with early results showing that experimentally induced sarcoids also regress (Ragland and Spencer, 1969; Lancaster et al., 1977). Therefore, under appropriate conditions, sarcoid lesions can be induced to regress, suggesting that the persistence of naturally occurring equine sarcoids may be due to an impaired immune response to the infecting PV.

6. Genetic Predisposition to Sarcoids

Whilst substantial evidence supports a role for BPV infection in sarcoid aetiology, other factors have also been implicated in the disease process. A long- term study in the USA examining the association of sarcoid development with breed, showed that the frequency of sarcoids in Quarter horses was nearly twice that of Thoroughbreds. In contrast, the frequency of sarcoids in Standardbreds was less than half that of Thoroughbreds horses (Angelos et al., 1988) supporting a genetic predisposition to the disease. A familial predisposition to equine sarcoid has also been described (James 1968; Ragland et al., 1966).

Other studies have shown a strong association between risk of sarcoid development and specific equine MHC class II alleles. When the frequency of equine MHC class II haplotypes were examined in Thoroughbred and Standardbred horses in the USA, a highly significant association between the MHC class II haplotype W3 and MHC class I haplotype B1 was detected in the Thoroughbred population. These findings were the first to suggest an association between predisposition to sarcoids and particular MHC haplotypes (Meredith et al., 1986). Subsequently, the class II W13 haplotype was shown to be strongly associated with sarcoids in Swedish halfbreds (Brostrom et al., 1988) and Swiss Warmbloods horses (Gerber et al., 1988). A further Swedish study also showed an association between increased recurrence of sarcoids following surgery in horses with the class II W13 haplotype, and association between early onset of sarcoids and the class I A5 haplotype (Brostrom, 1995).

7. The Role of the Immune Response and Sarcoids

To date a systematic investigation of the immune response to the equine sarcoid has not been carried out. The recognised association between sarcoid prevalence and the serologically defined MHC class I haplotype A3 and, particularly, class II (DW13) confirm that presentation of antigenic peptides to T cells is important in mediating protection. Specific MHC class II alleles may be associated with an impaired immune response to BPV. Associations between certain MHC class II genes and the development of tumours induced in rabbits by CRPV (Han et al., 1992) and in human cervical carcinoma associated with HPV types 16 or 18 are recognised (Wank and Thomssen, 1991; Breitburd et al., 1996).

The role of the immune response in determining the outcome of PV infections is well established. In most cases, regression of PV lesions occurs following activation of the host immune response. However, several immune evasion mechanisms that may contribute to persistence and progression of PV-associated disease have also been described (O'Brien and Campo 2002). Sarcoids are non-regressing, in contrast to many other lesions caused by PV infection suggesting that expression of the BPV proteins in equine cells may evoke similar immune evasion mechanisms. The observation that, under appropriate conditions, sarcoids can be induced to regress (Ragland and Spencer, 1969; Lancaster et al., 1977), further supports that the persistence of naturally occurring equine sarcoids may be due to an impaired immune response.

Recently, E5, the major oncoprotein of BPV, has been shown to cause retention of MHC class I molecules within the Golgi apparatus and thus prevent transport of MHC class I complex to the cell surface both *in vitro* and *in vivo* (Ashrafi et al., 2002; Marchetti et al., 2002; Araibi et al., 2004; 2005). MHC I is required for the presentation of viral antigens to $CD8^+$ T-lymphocytes for the elimination of infected cells. Therefore the E5 protein may provide a direct way in which the virus avoids the immune system during papillomaviral infection. E5 is expressed in over 75% of sarcoid lesions suggesting that expression is maintained throughout the disease (Nasir and Reid 1999; Carr et al., 2001; Chambers et al., 2003). Furthermore, sarcoids are associated with E5 sequence variants. However, very little is known about the immune response to equine sarcoids and hence the significance of these evasion mechanisms to prolonged BPV persistence is not known. Whether E5 or the E5 variants can down-regulate MHC class I in equine cells thereby contributing to immune evasion remains to be established.

8. Other Causative Factors

To date, most studies on sarcoids have focussed on the viral aetiology of the tumour, however several investigators have examined the role of cell cycle regulatory proteins. Bucher et al., (1996) failed to detect tumour suppressor p53 gene mutations in equine sarcoid tumours, suggesting that p53 does not play a significant role in the pathogenesis of sarcoids. This was further corroborated by an investigation of sarcoids in donkeys (Nasir et al., 1998). However, more recently, aberrant perinuclear localisation of p53 protein has been demonstrated in 44% of equine sarcoid lesions (Martens et al., 2001b), implying that mutational independent inactivation of p53 commonly occurs in sarcoids. A recent study by Nixon et al., (2005) has shown that in sarcoid cells p53 is stabilised and expressed not only in the nucleus but also in a perinuclear location. Although present in high levels, the transactivation function of p53 is abrogated providing evidence for loss of p53 function.

Over-expression of CDK-2 has been correlated with tumour size, development, and progression in certain human carcinomas (Ito et al., 1998) and increased expression of cyclin A has been detected in a range of tumours including hepatocellular carcinoma and squamous cell carcinoma (Murakami et al., 2000; Oliver and MacDonald, 2000; Nozoe et al., 2002). In BPV-4, the E5 protein can activate cyclin A and CDK-2, promoting entry into the cell cycle, but, paradoxically increases also the level of the CDK inhibitor $p27^{kip1}$ (Venuti and Campo

2002; O'Brien and Campo 1998; O'Brien et al., 2001). However, Nixon et al. (2005) showed that, whereas all the sarcoids examined expressed E5, no over-expression of cyclin A, $p27^{kip1}$ or CDK-2 could be detected, and the levels of cell proliferation, assessed by Ki-67 expression, were similar to those of normal skin. The low levels of cell proliferation and the lack of expression of cell cycle regulators implies that sarcoid molecular pathology is generally not associated with abnormal cell cycle control mechanisms. The loss of p53 function suggests that sarcoid lesions may in fact be associated with abnormalities in apoptotic pathways, where the balance of cell growth and cell death is perturbed.

In HPV infections, malignant transformation of the cervical mucosa can be accomplished by activation of the enzyme telomerase (Klingelhutz et al., 1996). Telomerase is a ribonucleoprotein DNA polymerase containing an RNA component that directs the synthesis of telomeric DNA repeats (Blasco and Hahn, 2003), allowing the attainment of cellular immortality by overcoming telomere erosion. Telomerase is expressed at low or undetectable levels in most somatic tissues, but is present in normal germ line cells and a wide range of cancers (Kim et al., 1994) including HPV associated cervical cancers (Reddy et al., 2001). The role of telomerase in equine sarcoids has recently been investigated by Argyle et al. (2003). In this study, twenty one equine sarcoids and five equine papilloma tissue samples were evaluated for telomerase activity. Whilst telomerase activity was detected in equine immortalised cells no activity could be demonstrated in sarcoids or papillomas. HPV E6 can induce telomerase activity in primary human keratinocytes and mammary epithelial cells (Stoppler et al., 1997; Anderson et al., 1997; Veldman et al., 2001) and when co expressed with E7, even higher level of telomerase expression are achieved (Klingelhutz et al., 1996). Conversely, repression of either E6 or E7 in HeLa cells has been shown to inhibit telomerase activity (DeFillippis et al., 2003). However, the data presented by Argyle et al. (2003) would suggest that BPV-1 E6 is not capable of inducing telomerase activity at least in equine cells. Whether BPV-1 E6 can activate telomerase in bovine cells remains to establish.

9. Implications of BPV Infection in Diagnosis

The diverse clinical manifestation of equine sarcoids and similarity with other skin neoplasms warrants the development of a definitive diagnostic marker. The consistence finding of BPV DNA in sarcoid biopsies implicates BPV DNA as such a marker. The potential value of BPV as a minimally invasive diagnostic test for sarcoids has recently been investigated in a panel of superficial swabs and scraping from 92 sarcoids. Whilst no viral DNA could be detected in non-sarcoid lesions, BPV DNA was detected in 88-91% of samples with a sensitivity significantly lower than compared to the clinical diagnosis of sarcoids (Martens et al., 2001). Based on this study, PCR based identification of BPV DNA is not suitable as a diagnostic indicator. The detection of viral protein expression in samples apparently negative for viral DNA (Carr et al., 2001) further corroborates that the sensitivity of PCR-based tests are less than optimal. The finding of BPV DNA in other skin lesions including once case of fibropapilloma, fibrosarcoma and several cases of granulatomous dermatitis (Angelos et al., 1991; Chambers et al., 2003) also undermine BPV DNA testing for diagnostic purposes. It may be that sarcoid tissue extracts contain inhibitors of PCR or that the high connective tissue content of sarcoids may hamper PCR amplification or alternatively the regions of sarcoid lesions used for preparation of DNA do not harbour viral DNA. Interestingly, BPV DNA has been detected in apparently normal skin margins of sarcoids and the frequency of detection decreases as the resection margin increases. Furthermore, animals with a surgical margin containing BPV have a greater probability of local recurrence (Martens et al., 2001). The development of specific BPV-1 antibodies and in-situ hybridisation techniques and their evaluation in sarcoids and non sarcoid lesions may provide improved sensitivity and specificity, although such techniques would require tumour excision and fixation and hence are not minimally invasive. Given the above, currently the most appropriate diagnosis for sarcoids remains a histological and clinical assessment.

10. BPV as a Potential Target for Therapy

The association of a causative viral agent does raise the possibility of employing anti-viral therapies in the treatment of sarcoids, including vaccination against BPV. This may be particularly beneficial in populations with a high incidence of sarcoids, for example where large numbers of animals are stabled together for long periods of time. It has previously been shown in cattle that prophylactic vaccination against the virus capsid proteins of BPV can prevent infection and disease and therapeutic vaccination against the E7 protein can stimulate regression of established papillomas (Campo et al., 1993; Jarrett et al., 1991), hence supporting the feasibility of vaccination against BPV to reduce or eliminate disease (for review see Campo 1997b).

It has been demonstrated recently that vaccination against HPV-16 E5 can eliminate tumour growth in mice supporting E5 as a target antigen for therapeutic vaccination (Lui et al., 2000; Chen et al., 2004). Given that E5 expression is frequently detected in sarcoid, vaccination against E5 is worthy of consideration. However, considerable research would be needed in order to determine the validity of such an approach.

References

Amtmann, E., Muller, H. and Sauer, G. (1980). Equine connective tissue tumors contain unintegrated bovine papillomavirus DNA. J. Virol. *35*, 962-964.

Anderson, S., Shera, K., Ihle., J, Billman.. L., Goff, B., Greer, B., Tamimi, H., McDougall, J., and Klingelhutz, A. (1997). Telomerase activation in cervical cancer. Am. J. Pathol. *151*, 25-31.

Andersson, S., Alemi, M., Rylander, E., Strand, A., Larsson, B., Sallstrom, J., and Wilander, E. (2000). Uneven distribution of HPV 16 E6 prototype and variant (L83V) oncoprotein in cervical neoplastic lesions. Br. J. Cancer *83*, 307-310.

Angelos, J.A., Marti, E., Lazary, S. and Carmichael, L.E. (1991). Characterisation of BPV-like DNA in equine sarcoids. Archiv. Virol. *119*, 95-109.

Angelos, J.A., Oppenheim, Y., Rebhun, W., Mohammed, H. and Antczak, D.F. (1988). Evaluation of breed as a risk factor for sarcoid and uveitis in horses. Anim. Genet. *19*, 417-425.

Araibi EH , Marchetti B, Ashrafi GH and Campo MS. (2004) Down-regulation of MHC class I in bovine papillomas. J. Gen. Virol. 85, 2809-2814.

Argyle D, Ellsmore V, Gault EA, Munro AF and Nasir L. (2003). Equine telomeres and telomerase in cellular immortalisation and ageing. Mech. Ageing Dev. 124, 759-763.

Argyle, D., Ellsmore, V., Gault, E.A., Munro, A.F. and Nasir, L. (2003). Equine telomeres and telomerase in cellular immortalisation and ageing. Mech. Ageing Dev. *124,* 759-763.

Ashrafi, G.H., Hagnashenas, M.R., Marchetti, B., O'Brien, P.M. and Campo, M.S. (2005). The E5 protein of human papillomavirus type 16 selectively down-regulates surface HLA class I. Int. J. Cancer, 113, 276-283. Published online: 25 Aug 2004.

Ashrafi, G.H., Tsirimonaki, E., Marchetti, B., O'Brien, P.M., Sibbet, G.J., Andrew, L. and Campo, M.S. (2002). Down-regulation of MHC class I by bovine papillomavirus E5 oncoproteins. Oncogene *21*, 248-259.

Blasco, M.A., and W.C. Hahn. (2003). Evolving views of telomerase and cancer. Trends Cell Biol. *13*, 289–294.

Bloch, N., Breen, M. and Spradbrow, P.B. (1994). Genomic sequences of bovine papillomaviruses in formalin-fixed sarcoids from Australian horses revealed by polymerase chain reaction. Vet. Microbiol. *41*, 163-172.

Breitburd, F., Ramoz, N., Salmon, J. and Orth, G. (1996). HLA control in the progression of human papillomavirus infections. Sem. Can. Biol. *7*, 359-371.

Brostrom, H. (1989). Surface antigens on equine sarcoid cells and normal dermal fibroblasts as assessed by xenogenic antisera. Res. Vet. Sci. *46*, 172-179.

Brostrom, H. (1995). Equine sarcoids. A clinical and epidemiological study in relation to equine leucocyte antigens (ELA). Acta. Vet. Scand. *36*, 223-236.

Brostrom, H., Bredberg-Raden, U., England, J., Obel, N. and Perlmann, P. (1979). Cell mediated immunity in horses with sarcoid tumours against sarcoid cells *in vitro*. Am. J. Vet. Res. *40*, 1701-1706.

Brostrom, H., Fahlbrink, E., Dubath, M-L. and Lazary, S. (1988). Association between equine leucocyte antigens (ELA) and equine sarcoid tumours in the population of Swedish halfbreeds and some of their families. Vet. Immunol. Immunopathol. *19*, 215-223.

Bucher, K., Szalai, G., Marti, E., Griot-Wenk, M.E., Lazary, S. and Pauli, U. (1996). Tumour suppressor gene p53 in the horse: identification, cloning, sequencing and a possible role in the pathogenesis of equine sarcoid. Res. Vet. Sci. *61*, 114-119.

Campo, M.S. (1997a). Bovine papillomavirus and cancer. Vet. J. *154*, 175-188.

Campo, M.S. (1997b). Vaccination against papillomavirus in cattle. Clin. Dermatol. *15*, 275-283.

Campo, M.S., Jarrett, W.F.H., Grindlay, G.J., Chandrachud, L.M., McGarvie, G.M. and O'Neil, B.W. (1993). Prophylactic and therapeutic vaccination against a mucosal papillomavirus. J. Gen. Virol. *74*, 945-953.

Carr, E.A., Theon, A.P., Madewell, B.R., Griffey, S.M. and Hitchcock M.E. (2001b). Bovine papillomavirus DNA in neoplastic and non-neoplastic tissues obtained from horses with and without sarcoids in the western United States. Am. J. Vet. Res. *62*, 741-744.

Carr, E.A., Theon, A.P., Madewell, B.R., Hitchcock, M.E., Schlegel, R. and Schiller J.T. (2001a). Expression of a transforming gene (E5) of bovine papillomavirus in sarcoids obtained from horses. Am. J. Vet. Res. *62*, 1212-1217.

Chambers, G., Ellsmore, V.A., O'Brien, P.M., Reid, S.W.J., Love, S., Campo, M.S. and Nasir, L.(2003). Sequence variants of bovine papillomavirus E5 detected in equine sarcoids. Virus Res. *96*, 141-145.

Cheevers, W.P., Fatemi-Nainie, S. and Anderson, L.W. (1986). Spontaneous expression of an endogenous retrovirus by the equine sarcoid-derived MC-1 cell line. Am. J. Vet. Res. *47*, 50-52.

Cheevers, W.P., Roberson, S.M. Brassfield, A.L., Davis W.C. and Crawford T.B. (1982). Isolation of a retrovirus from cultured equine sarcoid tumour cells. Am. J. Vet. Res. *43*, 804-806.

Chen, Y.F., Lin, C.W., Tsao, Y.P. and Chen, S.L. (2004). Cytotoxic-T-lymphocyte human papillomavirus type 16 E5 peptide with CpG-oligodeoxynucleotide can eliminate tumor growth in C57BL/6 mice. J. Virol. *78*, 1333-1343.

Chen, Z., Storthz, K.A. and Shillitoe, E.J. (1997). Mutations in the long control region of human papillomavirus DNA in oral cancer cells, and their functional consequences. Cancer Res. *57*, 1614-1619.

Coleman, N., Birley, H.D., Renton, A.M., Hanna, N.F., Ryait, B.K., Byrne, M., Taylor-Robinson, D. and Stanley, M.A. (1994). Immunological events in regressing genital warts. Am. J. Clin. Path. *102*, 768-774.

DeFilippis, R.A., Goodwin, E.C., Wu, L. and DiMaio, D. (2003). Endogenous human papillomavirus E6 and E7 proteins differentially regulate proliferation, senescence, and apoptosis in HeLa cervical carcinoma cells. J. Virol. *77*, 1551–1563.

DiMaio, D. and Mattoon, D. (2001). Mechanisms of cell transformation by papillomavirus E5 proteins. Oncogene *20*, 7866-7873.

Dong, X.P., Stubenrauch, F., Beyer-Finkler, E. and Pfister, H. (1994).Prevalence of deletions of YY1-binding sites in episomal HPV 16 DNA from cervical cancers. Int. J. Cancer *58*, 803-808.

England, J.J., Watson, R.E. Jr. and Larson, K.A. (1973). Virus-like particles in an equine sarcoid cell line. Am. J. Vet. *Res.* 34, 1601-1603.

Faccini, A.M., Cairney, M., Ashrafi, G.H., Finbow, M.E., Campo, M.S. and Pitts J.D. (1996). The bovine papillomavirus type 4 E8 (E5) protein binds to ductin and causes loss of gap junctional intercellular communication in primary fibroblasts. J. Virol. *70*, 9041-9045.

Fatemi-Nainie, S., Anderson, L.W. and Cheevers W.P. (1984). Culture characteristics and tumourigenicity of the equine sarcoid derived MC-1 cell line. Am. J. Vet. Res. *45*, 1105-1108.

Fatemi-Nainie, S., Anderson, L.W. and Cheevers, W.P. (1982). Identification of a transforming retrovirus from cultured equine dermal fibrosarcoma. Virol. *120*, 490-494.

Favre, M., Orth, G., Majewski, S., Baloul, S., Pura, A, and Jablonska, S. (1998). Psoriasis: A possible resevoir for human papillomavirus type 5, the virus associated with skin carcinomas of epidermodysplasia veruciformis. J. Invest. Dermatol. *110*, 31-317.

Frazer, I.H. (1996). Immunology of papillomavirus infection. Curr. Opinion Immunol. *8*, 484-491.

Fretz, P.B. and Barber, S.M. (1980).Prospective analysis cryosurgery as the sole treatment for equine sarcoids. Vet. Clin. North. Am. Small. Anim. Pract. *10*, 847-859.

Gerber, H., Dubath, M-L. and Lazary L. (1988). Association between predisposition to equine sarcoid and MHC in multiple-case families. In Equine Infectious diseases V – Proceedings of the Sixth International Conference. D.K. Powel , ed. (University Press of Kentucky), pp. 272-277.

Giannoudis, A. and Herrington, C.S. (2001). Human papillomavirus variants and squamous neoplasia of the cervix. J. Pathol. *193*, 295-302.

Goldschmidt, M.H. and Hendrick, M.J. (2002). Equine sarcoid. In Tumours in Domestic Animals, D.J. Meuten, ed. (Iowa State Press), pp. 88-89.

Goodrich, L., Gerber, H., Marti, E. and Antczack, D.F. (1998). Equine Sarcoids. Vet. Clin. North. Am. Equine Pract. *14*, 607-623.

Han, R., Breitburd, F., Marche, P.N. and Orth, G. (1992). Linkage of regression and malignant conversion of rabbit viral papillomas to MHC class II genes. Nature *356*, 66-68.

Ito, K., Sasano, H., Yoshida, Y., Sato, S. and Yasiam, A. (1998). Immunohistochemical study of cyclins D+E and cyclin dependent kinases in human endometrial cancer, Anticancer Res. *18*, 1661-1664.

Jackson, C. (1936). The incidence and pathology of tumours of domesticated animals in South Africa. Onderstepoort J. Vet. Sci. Animal Indust. *6*, 378-385.

James, V.S. (1968). A family tendency to equine sarcoid. S.W. Veterinarian *21*, 235-236.

Jarrett, W.F.H., Smith, K.T., O'Neil, B.W., Gaukrodger, J.M., Chandrachud, L.M., Grindlay, G.J., McGarvie, G.M. and Campo, M.S. (1991). Studies on vaccination against papillomaviruses: Prophylactic and therapeutic vaccination with recombinant structural proteins. Virol. *184*, 33-42.

Kim, N.W., Piatyszek, M.A., Prowse, K.R., Harley, C.B., West, M.D., Ho, P.L., Coviello, G.M., Wright, W.E., Weinrich, S.L., Shay, J.W. (1994). Specific association of human telomerase activity with immortal cells and cancer. Science *266*, 2011–2015.

Kinnunen, RE, Tallberg, T, Stenback, H and Sarna, S (1999). Equine sarcoid tumour treated by autogenous tumour vaccine. Anticancer Res. *19*, 3367-3374.

Klingelhutz, A.J, Foster, A.J. and McDougall, J.K (1996). Telomerase activation by the E6 gene product of human papillomavirus type 16. Nature *380*, 79-82.

Knottenbelt, D.C. and Walker, J.A. (1994). Treatment of the Equine Sarcoid. Equine Vet. Educ. *6*, 72-75.

Knowles, G., O'Neil, B.W. and Campo, M.S. (1996). Phenotypic characterization of lymphocytes infiltrating regressing papillomas. J. Virol. *70*, 8451-8458.

Kurvinen, K., Yliskoski, M., Saarikoski, S., Syrjanen, K., Syrjanen, S. (2000) Variants of the long control region of human papillomavirus type 16. Eur. J. Cancer 11, 1402-1410.

Lancaster, W.D. (1981). Apparent lack of integration of bovine papillomavirus DNA in virus induced equine and bovine tumour cells and virus transformed mouse cells. Virol. *108*, 251-255.

Lancaster, W.D., Olson, C. and Meinke, W. (1997). Bovine papilloma virus: presence of virus-specific DNA sequences in naturally occurring equine tumors. Proc. Natl. Acad. Sci. (USA), *74*, 524-528.

Lancaster, W.D., Theillen, G.H. and Olson, C. (1979). Hybridisation of bovine papillomavirus type 1 and type 2 DNA to DNA from virus-induced hamster tumours and naturally occurring equine tumours. Intervirol. 11, 227-233.

Lane, J.G. (1977). The treatment of equine sarcoids by cryosurgery. Equine Vet. J. *9*, 127-133.

Lazary, S., Gerber, H., Glatt, G. and Straub, R. (1985). Equine leucocyte antigens in sarcoid-affected horses. Eq. Vet. J. *17*, 283-286.

Liu, D. W., Y. P. Tsao, C. H. Hsieh, J. T. Hsieh, J. T. Kung, C. L. Chiang, S. J. Huang, and Chen. S L. (2000). Induction of CD8 T cells by vaccination with recombinant adenovirus expressing human papillomavirus type 16 E5 gene reduces tumor growth. J. Virol. *74*, 9083-9089.

Lory, S., von Tscharner, C., Marti, E., Bestetti, G., Grimm, S. and Waldvogel, A. (1993). In situ hybridisation of equine sarcoids with bovine papillomavirus. Vet. Rec. *132*, 132-133.

Majewski, S., Jablonska, S, Favre, M., Ramoz, N. and Orth, G. (1999). Papillomavirus and autoimmunity in psoriasis. Immunol. Today *10*, 475-476.

Marchetti, B., Ashrafi, G.H., Tsirimonaki, E., O'Brien, P.M. and Campo, M.S. (2002) The papillomavirus oncoprotein E5 retains the major histocompatibility class I in the Golgi apparatus and prevents its transport to the cell surface. Oncogene *21*, 7808-7816.

Martens, A., De Moor, A. and Ducatelle, R. (2001b). PCR detection of bovine papilloma virus DNA in superficial swabs and scrapings from equine sarcoids. Vet. J. *161*, 280-286.

Martens, A., De Moor, A., Demeulemeester, J. and Peelman, L. (2001a). Polymerase chain reaction analysis of the surgical margins of equine sarcoids for bovine papillomavirus DNA. Vet. Surg. *30*, 460-467.

Marti, E., Lazary, S., Antczak, D.F. and Gerber, H. (1993). Report of the first international workshop on equine sarcoid. Eq. Vet. J. *25*, 397-407.

Meredith, D., Elser, A.H., Wolf, B., Soma, L.R., Donawick, W.J. and Lazary S. (1986). Equine leukocyte antigens: relationships with sarcoid tumours and laminitis in two pure breeds. Immunogenetics *23*, 221-225.

Mohammed, H.O., Rebhun, W.C. and Antczack, D.F. (1992). Factors associated with the risk of developing sarcoid tumours in horses. Eq. Vet. J. *24*, 165.

Murakami, S. Tateyama, A. Rungsipipat, K. Uchida, R. And Yamaguchi, R. (2000) Immunohistochemical analysis of cyclin A, cyclin D1, and p53 in mammary tumours, squamous cell carcinomas, and basal cell tumours of dogs and cats. J. Vet. Med. Sci. *62,* 743-750.

Nasir, L. and Reid S.W.J. (1999). Bovine papillomaviral gene expression in equine sarcoid tumours. Virus Res. *61*, 171-175.

Nasir, L., McFarlane, S.T. and Reid S.W.J (1998). Mutational status of the tumour suppressor gene (p53) in donkey sarcoid tumours. Vet. J. *157*, 99-101.

Nasir, L., McFarlane, S.T., Torrontegui, B.O. and Reid, S.W.J. (1997). Screening for bovine papillomavirus in peripheral blood cells of donkeys with and without sarcoids. Res. Vet. Sci. *63*, 289-290.

Nixon, C., Chambers, G., Ellsmore, V., Campo, M.S., Burr, P., Argyle, D.J., Reid, S.W.J. and Nasir, L. (2005). Expression of cell cycle associated proteins cyclin A, CDK-2, $p27^{kip1}$ and p53 in equine sarcoids. Cancer Letters. *221*, 237-245.

Nozoe, T,. Korenaga, D, Futatsugi, M,. Saeki, H., Ohga, T. and Sugimachi, K. (2002). Cyclin A expression in superficial squamous cell carcinoma of the oesophagus and co-existing infiltrated lymphocyte follicle, Cancer Lett. *188,* 221-229.

O'Brien, J. Grindlay, S. and Campo, M.S. (2001). Cell transformation by the E5/E8 protein of bovine papillomavirus type 4, J. Biol. Chem. *276*, 33861-33868.

O'Brien, P.M. and Campo, M.S. (2002) Evasion of host immunity directed by papillomavirus-encoded proteins. Virus Research, 88, 103-117.

O'Brien, V. and Campo, M.S. (1998). BPV 4 E8 transforms NIH3T3 cells, up-regulates cyclin A, and cyclin A associated kinase activity, and deregulates expression of the CDK inhibitor $p27^{KIP-1}$, Oncogene, *17*, 293-301.

Okabayashi, M., Angell, M.G., Christensen, N.D. and Kreider, J.W. (1991). Morphometric analysis and identification of infiltrating leucocytes in regressing and progressing Shope rabbit papillomas. Int. J. Cancer *49*, 919-923.

Oliver, R. and MacDonald, D. (2000). Comparison of BrdU and Cyclin A as markers of the S-Phase in oral pre-cancerous lesions, J. Oral. Path. Med. *29,* 426-431.

Olson, C. Jr. and Cook, R.H. (1951). Cutaneous sarcoma-like lesions of the horse caused by the agent of bovine papilloma. Proc. Soc. Exp. Biol. Med. *77*, 281-284.

Otten, N., Von Tscharner, C., Lazary, S., Antczak, D.F. and Gerber, H. (1993). DNA of bovine papillomavirus type 1 and type 2 in equine sarcoids – PCR detection and direct sequencing. Archiv. Virol. *132*, 121-131.

P. Pascoe R.R. and Summers P.M. (1981). Clinical survey of tumours and tumour-like lesions in horses in south east Queensland. Eq. Vet. J. *13*, 235-239.

Pascoe, R.R. and Knottenbelt, D.C. (1999). Manual of equine dermatology. pp.244-250. London WB Saunders.

Ragland, W.H., Keown, G.H. and Spencer,G.R. (1970). Equine sarcoid. Eq. Vet. J. *2*, 2-11.

Ragland, W.L. and Spencer, G.R. (1969). Attempts to relate bovine papillomavirus to the cause of equine sarcoid: equidae inoculated intradermally with bovine papillomavirus. Am. J. Vet. Res. *30*, 743-752.

Ragland, W.L., Keown G.H. and Gorham J.R. (1966). An epizootic of equine sarcoid. Nature *210*, 1399.

Rapp, L., Liu, Y., Hong, Y., Androphy, E.J. and Chen, J.J. (1999). The bovine papillomavirus type 1 E6 oncoprotein sensitises cells to tumour necrosis factor alpha-induced apoptosis. Oncogene *18*, 607-615.

Reddy, V.G., Khanna, N., Jain, S.K., Das, B.C., Singh, N., (2001). Telomerase-A molecular marker for cervical cancer screening. Int. J. Gynecol. Cancer *11*, 100-106.

Reid, S.W.J. and Mohammed, H.O. (1997). Longitudinal and cross-sectional studies to evaluate the risk of sarcoid associated with castration. Canadian J. Vet. Res. *61*, 89-93.

Reid, S.W.J., Smith, K.T. and Jarrett, W.H.F. (1994). Detection, cloning, and characterisation of papillomaviral DNA present in sarcoid tumours of Equus Asinus. Vet. Record *135*, 430-432.

Robl, M.G. and Olson, C. (1968). Oncogenic action of bovine papilloma virus in hamsters. Can. Res. *28*, 1596-1604.

Rush, BR and Flaminio, M.J. (2000). Immunomodulation in horses. Vet. Clin. North Am. Eq. Pract. *16*, 183-197.

Stoppler, H., Hartman, D.P., Sherman, L., Schlegel, R. (1997), The human papillomavirus type 16 E6 and E7 oncoproteins dissociate cellular telomerase activity from the maintenance of telomere length. J. Biol. Chem. *272*, 13332-13337.

Sullins, K.E., Roberts, S.M., Lavach, J.D. and Severin, G.A. (1986). Equine Sarcoid. Eq. Pract., *8*, 21-27.

Tarwid, J.N., Fretz, P.B., and Clark, E.G. (1985). Equine sarcoids: a study with emphasis on pathological diagnosis. The Compendium of Continuing Education. *7*, 293-300.

Torrontegui, B.O. and Reid, S.J. (1994). Clinical and pathological epidemiology of the equine sarcoid in a referral population. Eq. Vet. Educ. *6*, 85-88.

Trenfield, K., Spradbrow, P.B. and Vaneslow, B. (1985). Sequences of papillomavirus DNA in equine sarcoids. Eq. Vet. J. *17*, 449-452.

Vanselow, B.A., Abetz, I. and Jackson, A.R. (1988). BCG emulsion immunotherapy of equine sarcoid. Eq. Vet. J. *20*, 444-447.

Veldman, T., Horikawa, I., Barrett, J.C. and Schlegel, R. (2001). Transcriptional activation of the telomerase hTERT gene by human papillomavirus type 16 E6 oncoprotein. J. Virol. *75*, 4467-4472.

Venuti, A. and Campo, M.S. (2002). The E5 protein of papillomaviruses. In Progress in Medical Virology: Papillomaviruses. D.J. McCance, ed. (Elsevier) pp.141-162.

Veress, G., Szarka, K., Dong, X.P., Gergely, L. and Pfister, H. (1999). Functional significance of sequence variation in the E2 gene and the long control region of human papillomavirs type 16. J. Gen. Virol. *80*, 1035-1043.

Voss, J.L. (1969). Transmission of equine sarcoid. Am. J. Vet. Res. *30*, 183-191.

Wank, R. and Thomssen, C. (1991). High risk of squamous cell carcinoma of the cervix for women with HLA-DQw3. Nature *352*, 723-725.

Watson, R.E. and Larson, K.A. (1974). Detection of tumour specific antigens in an equine cell line. Infec. Immun. *9*, 714-718.

Zehbe, Y. Wilander I., E,. Delius H, and. Tommasino M. (1998). Human papillomavirus 16 E6 variants are more prevalent in invasive cervical carcinoma than the prototype. Cancer Res. *58*, 829-833.

From: Papillomavirus Research: From Natural History to Vaccines and Beyond. Edited by: M. Saveria Campo

Chapter 25

Rabbit Viral Papillomas and Carcinomas: Model Diseases for Human Papillomavirus Infections and Associated Carcinogenesis

Françoise Breitburd*, Mathieu Nonnenmacher, Jérôme Salmon and Gérard Orth

Abstract

Rabbit papillomaviruses are models of the natural history of diseases caused by oncogenic and non-oncogenic human papillomaviruses (tissue specificity, latency, regression, persistence, malignant progression). The Shope cottontail rabbit papillomavirus (CRPV) induces skin warts in domestic rabbits that may regress or persist depending on the host. Persistent warts may progress to invasive carcinoma. The domestic rabbit oral papillomavirus causes oral and, in experimental conditions, genital papillomas, that usually regress. CRPVa and b subtypes, with strikingly divergent regulatory region and E6 and E7 oncogenes, induce warts with distinct immunogenic properties and significant differences in levels of vegetative viral DNA replication and in the risk for malignant progression. CRPV genetics, molecular biology of CRPV-induced warts and carcinomas, rabbit immune responses to infection and tumors, evaluation of vaccines, antivirals and immunomodulators, contributed to our understanding of human papillomavirus diseases. The rabbit model disclosed that the polymorphism of the major histocompatibility complex class II genes and CD4+ cells as well as variations in biologic and immunogenic properties of viral oncoproteins played key roles in determining wart evolution. This model was instrumental in proving the efficacy of preventive and therapeutic vaccines. The rabbit viruses share an E8 gene with some human cutaneous PVs and should provide a model to unravel the role of this gene in the virus life cycle.

1. Introduction

The (Shope) cottontail rabbit papillomavirus (CRPV) was discovered by R. E. Shope, in 1933, as the etiologic agent of the naturally occurring warts or « horns » of Western cottontail rabbits (Shope, 1933). The demonstration by Rous and Beard, in 1935, that the warts induced by the Shope papillomavirus could convert into carcinomas (Rous and Beard, 1935) was the first demonstration of the role of a papillomavirus in the development of an invasive squamous cell carcinoma (SCC). The Shope virus provided the first model to study viral carcinogenesis in mammals which led to important concepts such as virus masking in tumors and synergy between a virus and chemical carcinogens in cancer development (for reviews, see Orth et al., 1977; Kreider and Bartlett, 1981; Wettstein, 1987; Wettstein, 1990; Breitburd et al., 1997). The Shope virus has also been the viral vector of an early, though unsuccessful, human gene therapy model (Friedmann, 2001)

CRPV was the first oncogenic virus and one of the first animal viruses purified to homogeneity, allowing the early establishment of physical, chemical and biological properties of the viral particles (Beard et al., 1955). Several important properties of papillomaviruses have been first discovered in the CRPV model. Among these are the infectivity of viral DNA (Ito, 1960), the linkage between viral replication and terminal differentiation of keratinocytes (Noyes and Mellors, 1957; Orth et al., 1971; Croissant et al., 1972), the induction of cellular DNA synthesis in differentiating papilloma cells (Croissant and Orth, 1967; 1970; Orth and Croissant, 1970), the existence of a papillomavirus group-specific antigen (Orth et al., 1978), and the role of the E6 and E7 viral genes in the development of papillomas and carcinomas (Georges et al., 1984; Nasseri and Wettstein, 1984a; Danos et al., 1985; Wettstein, 1987; Wettstein, 1990). The similarity between the malignant transformation of rabbit papillomas and the rare invasive and highly destructive behavior of some human condylomata acuminata had been stressed very early (Rous and Beard, 1935). CRPV-induced warts may regress or evolve towards carcinoma and their natural history resembles that of genital intraepithelial neoplasia caused by « high risk » genital human papillomavirus (HPV) genotypes (Lowy and Howley, 2001). The CRPV-induced disease may also constitute an animal model for epidermodysplasia verruciformis (EV), a rare human genetic condition characterized by the development of

*For correspondence. Email fbreit@pasteur.fr.

cutaneous SCCs associated with a subset of EV-specific HPV genotypes (HPV5, HPV8) (Orth, 1987; Orth et al., 2001).

CRPV-induced tumors are still the most amenable animal model to analyze the immunologic mechanisms involved in regression of PV-induced lesions, to define strategies for immunoprevention and immunotherapy of PV-associated tumors, to understand the genetic factors, including immunogenetics, that condition the expression of the oncogenic potential of papillomaviruses, and to unravel the role, still debated, of HPVs in skin carcinogenesis (Orth, 2004).

2. Biologic Properties of Rabbit Papillomaviruses

2.1. Host and tissue specificity

Highly keratinized and pigmented skin warts are often observed in cottontail rabbits (*Sylvilagus floridanus*) trapped in the Midwestern states bordering the Mississippi River (Shope, 1933; Syverton, 1952). Experimental inoculation of a wart extract onto scarified skin results in the transmission of the disease to cottontail rabbits, snowshoe rabbits (*Lepus americanus*), jack rabbits (*Lepus californicus*) and domestic rabbits (*Oryctolagus cuniculus*) (Beard and Rous, 1935). The epithelium of the lip, tongue and gum or the conjunctival epithelium keratinizing under vitamin A deprivation are not susceptible to CRPV infection (Kreider and Bartlett, 1981). Worth stressing, in transgenic rabbits harboring multiple copies of integrated CRPV DNA, viral transcripts are only detected in skin and skin tumors (Peng et al., 1993). Whether this tissue specificity involves different levels of viral DNA methylation remains to be further studied (Wettstein and Stevens, 1983; Peng et al., 1995).

Oral papillomatosis in domestic rabbits is caused by a papillomavirus unrelated to CRPV (Parsons and Kidd, 1936), the rabbit oral PV (ROPV), which has been recently completely sequenced (Christensen et al., 2000a). It was found that rabbit genital tissues were susceptible to ROPV infection, although less so than the undersurfaces of the tongue by 1 to 2 orders of magnitude. Viral DNA replication and virus production were detected in subrenal or subcutaneous grafts to athymic mice of infected tongue or genital tissue (penile sheath, vulva, vagina) from newborn or adult rabbits (Christensen et al., 1996a; Harvey et al., 1998), or in penile proliferations experimentally induced in adult immunocompetent rabbits (Harvey et al., 1998). ROPV-induced proliferations quickly regress by 50 to 60 days postinfection in 100% of infected rabbits. *In vitro* studies showed that ROPV E6 and E7 proteins were weak oncogenes (Hu et al., 2004b).

Comparison of the life cycles of different animal and human papillomaviruses, on the basis of initiation of early events, onset of viral genome amplification and expression of capsid proteins, indicated that the life cycle of ROPV most resembles those of human mucosal PV types (Peh et al., 2002). ROPV and CRPV may thus provide a unique animal model system for the study of both skin and genital HPV associated diseases, including tissue specificity, tumor regression or progression, host immune responses and preventive and therapeutic vaccine strategies.

2.2. Target cells of CRPV infection

Skin treatment resulting in epidermal hyperplasia, such as applications of a mixture of turpentine and acetone, increases susceptibility to CRPV infection. In these conditions, rapid regeneration of the epidermis destroyed by scarification takes place, originating from hair follicle shafts (Friedewald, 1944). It has been recently shown that the target cells of CRPV are most probably stem cells of the hair follicles (Schmitt et al., 1996). Localized in the bulge nestled below the sebaceous gland in the intermediate part of the hair follicles, stem cells repopulate hair bulbs at the beginning of the anagen phase of the cycling hair follicle and colonize also damaged interfollicular epidermis upon wound healing (Alonso and Fuchs, 2003).

The amount of virus required to produce a papilloma in half of the animals inoculated by scarification onto normal skin and the lowest amount of virus inducing a papilloma have been evaluated to about 8×10^7 particles and 5.6×10^6 particles, respectively (Beard, 1956). The inoculation of virus on scarified hyperplastic skin and protection of the inoculated area with a layer of paraffined gauze was reported to increase at least 100-fold the sensitivity of the bioassay (Friedewald, 1944).

Inoculation of highly diluted CRPV suspension induces latent infection that can be activated to cause papilloma formation by stimulating cell proliferation through subtherapeutic photodynamic therapy (Amella et al., 1994), mechanic irritation or UV-C irradiation (Zhang et al., 1999).

2.3. Evolution of CRPV infection

CRPV-induced papillomas may regress (Beard and Rous, 1934) or may persist, either remaining benign or converting into invasive carcinomas (Rous and Beard, 1935). In cottontail rabbits infected in natural conditions, wart regression is observed in about 20% of the animals within the first 6 months after trapping and in about one third of the animals after 6 months. Malignant transformation of certain warts occurs after 6 months in about 25% of the animals with persistent warts, and metastatic tumors are observed in the majority of them (Syverton, 1952).

In cottontail and domestic rabbits infected under experimental conditions, the first lesions are observed between 1 and 5 weeks after infection. A proliferative phase of 6 to 8 weeks is followed by a stationary phase. In some animals, wart regression takes place 2 to 8 weeks after wart outgrowth. The frequency of regression is low in cottontail rabbits (less than 10%) (Syverton, 1952), and varies between 10 and 70% in domestic rabbits (Kidd, 1938; Syverton, 1952; Evans, 1963; McMichael, 1967; Kreider and Bartlett, 1981; Han et al., 1992). As a rule, all papillomas of the same animal regress at the same time. In some cases, however, regression may be partial or transient (Kidd, 1938). This phenotype was recently shown to be linked to the genetic constitution of the host and to the genetic heterogeneity of the viral isolates prepared from pooled warts collected from naturally infected cottontail rabbits (Salmon et al., 1997; 2000). In persistor animals, malignant transformation of one or several warts is observed between 6 and 12 months in about 25% of cottontail rabbits and in up to 75% of domestic rabbits (Syverton, 1952). Histologic features of the primary carcinomas and of the tumors metastasizing to lymph nodes and lungs vary from those of a malignant papilloma to those of an anaplastic carcinoma (Rous and Beard, 1935).

The CRPV papilloma and carcinoma complex provides an invaluable model system to investigate changes in viral and cellular gene expression associated with malignant progression since pairs of benign papillomas and carcinomas can be obtained from the same host.

Malignant transformation of papillomas is accelerated and its frequency increased when scarified skin is exposed to a chemical carcinogen (dimethylbenzanthracene) in a few days following CRPV infection (Rogers and Rous, 1951), or when tar or methylcholanthrene are applied onto papillomas (Rous and Friedewald, 1944). Conversely, CRPV has dramatic effects on the benign papillomas induced by repeated applications of tar on the skin of domestic rabbits. Growth and persistence of such papillomas depend on continued tar treatment, and their malignant transformation is rarely observed. When administered intravenously to such rabbits, CRPV localizes on tarred skin, eliciting carcinomas forthwith as well as a great variety of papillomas (Rous and Kidd, 1938). The virus behaves then as a transforming agent, suggesting that CRPV is a carcinogen in the course of the natural transformation of the warts it induces.

Allografts of primary or metastatic carcinomas to adult domestic rabbits are usually unsuccessful (Kidd and Rous, 1940), whereas intramuscular graft of carcinomas to newborn rabbits is easy (Smith et al., 1952). A CRPV-induced tumor, the VX2 carcinoma, was however successfully established and serially transplanted in adult rabbits (Kidd and Rous, 1940), and another tumor isolated in newborn rabbits, the VX7 carcinoma, was subsequently serially transplanted in adult rabbits (Rogers et al., 1960).

2.4. CRPV recoverability from tumors

Warts are easily transmitted from papillomas of naturally infected cottontail rabbits (Shope, 1933). However, the amount of viral particles in warts varies greatly (a thousand-fold) among animals, the maximum yield being 1mg of purified viral particles per gram of warts (Beard, 1956). CRPV may be recovered from papillomatous cysts developed from grafts of cottontail rabbit skin chips infected with CRPV, placed under the renal capsule of athymic mice (Christensen and Kreider, 1990).

Papillomas are not usually transmitted from domestic rabbit warts (Shope, 1933). However, the production of low amounts of virus in a majority of domestic rabbit warts was demonstrated by increasing the sensitivity of the bioassay (Friedewald, 1944). Moreover, CRPV strains allowing serial transmission of papillomas in domestic rabbits have been isolated (Shope, 1935; Selbie and Robinson, 1948; Salmon et al., 2000). The early recognized recoverable strains showed delayed wart outgrowth and similar frequency of wart malignant conversion (Shope, 1935) whereas the recently identified CRPVb subtype displays a less pronounced oncogenic potential (Salmon et al., 2000; Salmon, 2003; J. Salmon, O. Croissant, F. Breitburd and G. Orth, unpublished results).

Infectious virus has never been recovered from primary or metastatic carcinomas of cottontail or domestic rabbits (Syverton, 1952). The production of small amounts of virus in some cancers was demonstrated, however, by the detection of viral antibodies in rabbits bearing transplantable carcinomas (Kidd and Rous, 1940; Smith et al., 1952; Rogers et al., 1960; Orth et al., 1978), and by the isolation of infectious CRPV from transplanted domestic rabbit carcinomas to newborn rabbits (Smith et al., 1952; Rogers et al., 1960).

3. Organization of the CRPV Genome

The CRPV particle, capsid and genome share the common structure of all PVs (Howley and Lowy, 2001), and, in contrast with previous reports, CRPV capsomeres are arranged on a T=7*dextro* icosahedral lattice with the same handedness (right hand) as other PVs studied (Belnap et al., 1996).

3.1. Specific features of the CRPV genome organization

Whereas the overall organization of the CRPVa and b genomes is similar to that of other PVs (Giri et al., 1985; Howley and Lowy, 2001) (Figure 1), it is unique, however, by the greater size of the E6 ORF which encodes two E6 proteins, the long E6 or LE6 (273/276 amino acids) and the short E6 or SE6 (176/179 amino acids) (Danos et al., 1985; Wettstein, 1990; Salmon et al., 1997). This is in contrast to the single E6 protein of about 150 amino acids found for other PVs (Howley and Lowy, 2001). A second unusual feature is the localization of the major

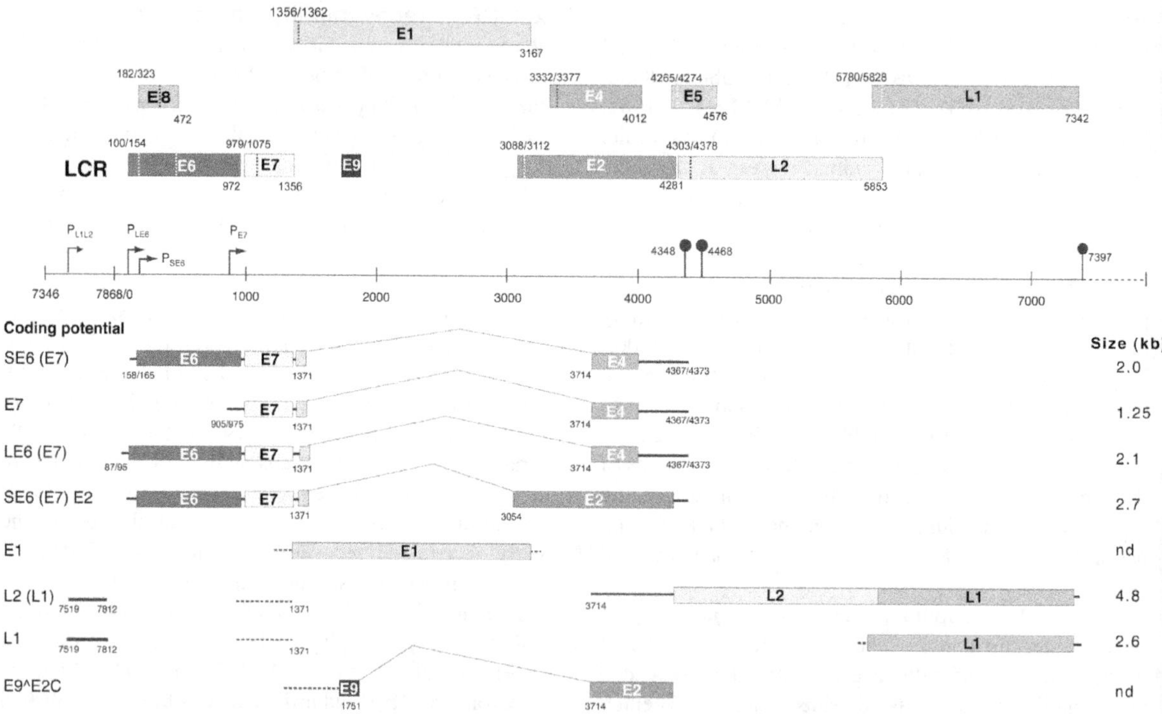

Figure 1. CRPV genetic and transcription maps.The map of CRPV ORFs is indicated on the top. The position of the first ATG in each ORF and of the second ATG of E6 ORF is indicated by a vertical dashed line. The positions of the early (PE) and late (PL) promoters and of the polyadenylation sites (closed circles) are indicated beneath the ORFs. The long control region (LCR) is located between nucleotide positions 7346 and 99. The 5′ part of the LCR is represented twice, on the left and on the right (dashed line) of the map. Transcripts are depicted in the lower part of the figure. Transcripts encompassing the E1 ORFs are represented as a single box. Undetermined sequences are indicated by dashed lines. The coding potential of RNAs is given on the left side of the figure. Proteins probably not translated are indicated in parentheses. The size of RNAs is indicated in kb on the right of the figure (nd, not determined). Data were taken from Danos et al., 1985; Giri et al., 1985; Wettstein, 1990; Breitburd et al., 1997; Jeckel et al., 2003; Salmon, 2003; Nonnenmacher, 2004.

polyadenylation site of the early transcripts within the E5 ORF (Wettstein, 1990; Meyers et al., 1992) (Figure 1), most probably rendering unlikely the expression of a functional E5 protein. A third feature is the presence of an E8 ORF colinear to the 5′ part of the E6 ORF (Giri et al., 1985). The E8 ORF potentially encodes a 50-amino acid hydrophobic protein with features resembling the E5 protein of the bovine PV type 1 and of the HPV genotypes belonging to the alpha PV genus (de Villiers et al., 2004). The ROPV also presents a large E6 ORF and an E8 ORF overlapping the E6 ORF, pointing to a specific trait of rabbit PVs (Christensen et al., 2000a), and suggesting that species diversification of PVs preceded tissue adaptation. A small ORF designated E9, most likely corresponding to the E8 ORF of other PVs (Howley and Lowy, 2001), has been identified as part of a spliced E9^E2C transcript cloned recently from domestic and cottontail rabbit skin papillomas (Jeckel et al., 2003) (Figure 1). This messenger RNA encodes ten aminoacids of the E9 gene fused to the hinge region and the C terminus of the E2 protein (Figure 1) (Jeckel et al., 2003), a structure described for the E8^E2C proteins of other PVs (Howley and Lowy, 2001).

3.2. Genetic studies

DNA extracted from CRPV virions, from domestic rabbit papillomas and carcinomas, as well as molecularly cloned CRPV DNA, excised from plasmid sequences and recircularized, induce papillomas upon inoculation to rabbit epidermis (Ito, 1960; 1962; Nasseri et al., 1989; Brandsma et al., 1991; Georges et al., 1992; Salmon et al., 2000). This has allowed mutagenesis analysis of the genes required for the formation of a papilloma. The E6, E7 and E1 ORFs, the dimerization and DNA binding domain or the transactivating domain of E2, as well as the L1 ORF, were found to be required for the induction of warts while the L2, E5, E4 and E9 ORFs were found dispensable (Nasseri et al., 1989; Brandsma et al., 1991; 1992; Wu et al., 1994; Jeckel et al., 2002; 2003). Worth stressing, mutations in either LE6 or SE6 abolished papilloma formation (Meyers et al., 1992). Discrepant data were obtained for the E8 ORF which was found to be required for papilloma formation and development (Nonnenmacher, 2004; M.Nonnenmacher, J. Salmon, Y. Jacob, F. Breitburd and G. Orth, unpublished results), or dispensable for papilloma formation and required for papilloma development (Hu et al., 2002a), most likely because of different experimental conditions

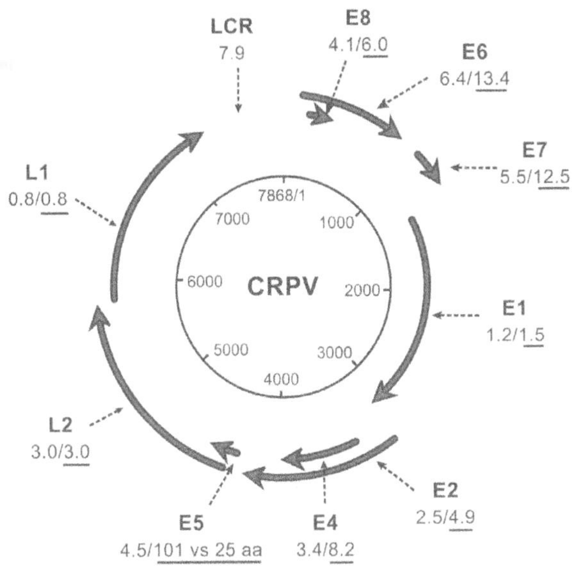

Figure 2. Percentage of divergent nucleotides and amino acids (underlined) between CRPVa and CRPVb subtypes (Salmon et al., 1997; 2000).

(Nonnenmacher, 2004; Nonnenmacher et al., unpublished results). The E4 ORF is dispensable for papilloma formation and development but the onset of viral DNA amplification and the L1 gene expression are abolished in papillomas induced by a mutant CRPV genome incapable of expressing the E4 protein (Peh et al., 2004). CRPV genomes with mutations affecting the E9 start codon or the correct splicing of the E9^E2C transcript induce papillomas with the same efficacy as wild type genomes, and with similar viral genome copy number and viral transcription levels. This was not expected from the repressor properties of E9^E2C for *in vitro* viral DNA transcription and replication (Jeckel et al., 2003).

4. Genetic Variability of CRPV

The first evidence for a variation of CRPV has been the isolation of strains recoverable from domestic rabbits (Shope, 1935; Friedewald and Kidd, 1944; Selbie and Robinson, 1948). The existence of cancer-producing variants was postulated (Rous, 1936) but could not be substantiated (Rous, 1943; Smith et al., 1952; Rogers et al., 1960).

4.1. The CRPV variants

Available sequence data had suggested a limited (less than 0.5%) intratypic variability among different CRPV isolates, leading to the identification of five variants. CRPV a1 (CRPV-KS) is the sequenced CRPV prototype originating from Kansas cottontail rabbit warts (Favre et al., 1982; Giri et al., 1985; Salmon et al., 2000). CRPVa2 (CRPV-WA) corresponds to the Washington B strain isolated from naturally infected Whidbey Island cottontail rabbits (Evans and Ito, 1966; Brandsma et al., 1991; Wu et al., 1994; Harry and Wettstein, 1996) and CRPVa3 (VX7-8kb) was isolated from the VX7 carcinoma derived from a papilloma induced by the recoverable strain of CRPV (Rogers et al., 1960; Georges et al., 1992). CRPVa4 (Salmon et al., 1997; 2000) and CRPVa5 (CRPV-HE) (Han et al., 1998) were obtained from Kansas cottontail rabbit warts. The complete sequence of CRPVa1 (Giri et al., 1985; Salmon et al., 1997; 2000), CRPVa3 (P. Cassonnet and G. Orth, unpublished results) and CRPVa4 (Salmon et al., 1997; Salmon et al., 2000) and the partial sequence of CRPVa2 (Brandsma et al., 1992; Han et al., 1998) and CRPVa5 (Han et al., 1998) revealed a striking variability of the E5 ORF among the variants. Nucleotide insertions and deletions resulted in 10 contiguous amino acid changes for CRPVa2 and in a shorter E5 protein of 62 amino acids for CRPVa3 and CRPVa5 instead of the 101 amino acids for CRPVa1 and CRPVa4. Mutations introduced within the E5 ORF showed that although the E5 ORF is not required for papilloma formation, it may affect the efficiency of the process (Brandsma et al., 1992). The CRPVa3 encodes a truncated E1 protein (325 instead of 601 amino acids) and was found to be no longer infectious (Georges et al., 1992). If papillomas exceptionally arose after gene gun inoculation, CRPVa3 sequences were found integrated into the cell DNA (M. Nonnenmacher, C. Pons, F. Breitburd and G. Orth, unpublished data).

4.2. The CRPVb subtype

The existence of a second CRPV subtype (CRPVb) with a significant nucleotide divergence in the long control region (LCR) and the E6/E7 region was recently disclosed (Salmon et al., 1997; 2000) (Figure 2). A high intratypic variability has been reported also for the HPV5 genotype associated with epidermodysplasia verruciformis, contrasting with the lower extent of variation among HPV16 isolates (8.8% vs. 1.2 and 1.3% in the E6 and L1 ORFs) (Deau et al., 1993; Kawase et al., 1996).

4.2.1. Main features of the CRPVb genome

The CRPVb genome (7867 bp) is 1 base pair shorter than the genome of the prototypical CRPVa1. The CRPVa4 variant, found as a minor component in the same isolate, differed only by 5 nucleotide changes (0.06%) from the CRPVa1, further substantiating the genetic stability of the CRPVa subtype (Salmon et al., 1997; 2000). The CRPVb differed by 241 (3.1%) positions, warranting the recognition of a new subtype (Salmon et al., 1997; de Villiers et al., 2004). The divergence of the E6 and E7 oncoproteins of CRPVa1 and CRPVb involves a great number of mutations (4.4%) and a high proportion (83.5%) of non synonymous mutations. Mutations that changed stop codons in both E6 and E7 ORFs, and in-frame sequence insertions that involve the duplication of adjacent sequences (18 and 6 nucleotides for E6 and E7 ORFs, respectively), result in an increase in size of 3

```
       LE6
CRPVa  MENCLPRSLEKLQQILQISLEDLPFGCIFCGKLLGAAEKQLFKCTGLCIVWHKGWPYGTCRDCTVLSCAL
CRPVb  ----------------H-------L--------------E---------------------N--------
                                   SE6
CRPVa  DLYCHLALTAPALEAEALVGQEISSWFMRCTVCGRRLTIPEKIELRARNCTLCCIDKGQYFQWRGHCSSC
CRPVb  -F-Y------T-------------C---------------------------------------------

CRPVa  KLSDQGDLGGYPPSPG      SRCGECDECCVPDLTHLTPVDLEELGLYPGPEGTYPDLVDLGPGVFGE
CRPVb  -------F--------LAPSPG-C--D--V-R-----N-------------S---A---------D----

CRPVa  EDEEGGGLFDSFEEEDPGPNQCGCFFCTSYPSGTGDTDINQGPAGAAGIALQSDPVCFCENCINFTEFR*
CRPVb  ----------------LR--L-------R-----DG-----------E-----GS-----D---C-*
       E7
CRPVa  MIGRTPKLSELVLGETAEALSLHCDEALENLSDDDEEDHQDRQVFIERPYAVSVPCKRCRQTISFVCVCA
CRPVb  ---------------------------IDD--------------YRD----------S---V---I---G

CRPVa  PEAIRTLNRLLSASLSLVCPECCN*
CRPVb  ------------------------RY*
```

Figure 3. Alignment of CRPVa and CRPVb LE6, SE6 and E7 amino acid sequences, as deduced from nucleotidic sequences (Salmon et al., 1997; 2000). Dashes indicate conserved amino acids. Sequences resulting from a duplication or from the mutations of stop codons are boxed.

amino-acids for E6 and of 2 amino acids for E7 proteins (Salmon et al., 1997) (Figure 3). In the CRPVb LCR (7.9% nucleotide divergence), a nucleotide substitution in the putative promoter sequence for the late transcripts (Wettstein et al., 1987), an additional binding site for the viral E2 protein, and the deletion of a 32-bp sequence containing the cap site of CRPVa LE6 transcripts (Danos et al., 1985; Wettstein, 1990) could affect the transcription of the viral genome. A mutation in the stop codon results in an 8-amino-acid–longer CRPVb E4 protein, and a nucleotide deletion reduces the coding capacity of the E5 gene from 101 to 25 amino acids (Salmon et al., 2000) (Figure 2).

Probably due to some constraint required for its function, the E1 protein shows little variation (1.5% variable amino acids) (Salmon et al., 2000) and no amino acid changes are found within the 50-amino-acid sequence containing a zinc finger common to LE6 and SE6 proteins and within the region of the E7 protein containing the binding site for the pRB protein (Salmon et al., 1997; Howley and Lowy, 2001). Similarly, the amino acid changes are all conservative in the E8 protein which is involved in the control of wart development (Harry and Wettstein, 1996; Hu et al., 2002a; Nonnenmacher, 2004; Nonnenmacher et al., unpublished results).

The contrast between the number of variable amino acids in the E6 and E7 proteins (around 12%) and in the L1 major capsid protein (0.8%) (Figure 2) (Salmon et al., 1997) is an unusual feature among PVs. The variability of the E6 and L1 ORFs is generally within the same order of magnitude and the basis for phylogenetic calculations (de Villiers et al., 2004). The L1 sequence usually taken for the construction of phylogenetic trees is identical for CRPVa and CRPVb. This would mean that CRPVa and CRPVb diverged much later than inferred from the variation of E6 and E7. Sequence insertions or deletions are thus likely to have played a major role in the diversification of PV types, in addition to the accumulation of point mutations (Salmon et al., 1997; de Villiers et al., 2004).

4.2.2. Biological properties of the CRPVb

Like CRPVa, the genome of CRPVb induces papillomas in rabbits that may evolve towards carcinomas. In rabbits homozygous for a defined major histocompatibility complex (MHC) class II DRA and DQA haplotype, warts induced by CRPVb regress whereas those induced by CRPVa persist, indicating that CRPVa and CRPVb induced warts display different immunogenic properties (Salmon et al., 1997; 2000; Salmon, 2003). In rabbits homozygous for other specific haplotypes, the warts induced by the two subtypes either persist or regress, pointing to the role of the capacity of the host to mount an efficient immunologic response (Salmon et al., 2000; Salmon, 2003). By swapping different subgenomic fragments between the two subtypes, it was found that the LCR/E6/E7 region contains the immunodominant epitopes targeted in regression (Salmon et al., 2000) and that such determinants are localized within the E6 region (Nonnenmacher, 2004). It was also found that the C-terminal end of the E6 protein, in particular a single amino acid change (G259E), specifically controls wart persistence or regression in EIII/J inbred rabbits (Hu et al., 2002b).

In contrast to cottontail rabbit papillomas where virus replication is readily detected *in situ*, cells with detectable amplified viral DNA or accumulated L1 capsid antigens are seldom observed in domestic rabbit warts induced by CRPVa. Their frequency is found to be greater by one order of magnitude for warts induced by CRPVb (Salmon et al., 2000). Chimeric genomes with swapped LCR or genomes mutated in the late promoter sequence had no effect on the expression of late viral functions (Nonnenmacher, 2004).

Chimeric CRPVb genomes with a CRPVa LCR/E6/E7 region induce warts with a higher risk of malignant conversion than chimeric CRPVa genomes with a CRPVb LCR/E6/E7 region (Salmon et al., 2000), which confirms the difference in the oncogenic potential of the two viruses and involves the variable region in this difference. The approach, however, could not discriminate further which of the LCR, the E6 or the E7 gene modulates the oncogenic potential of the CRPV, mainly because of the variations observed in the infectivity of the different chimeric constructs (Nonnenmacher, 2004).

5. Molecular Biology of CRPV-induced Tumors

5.1. Physical state and transcription of the CRPV genome

In apparently normal epidermis upon latent infection, CRPV DNA may be detected by PCR (Amella et al., 1994), and also transcripts of the E1 ORF, which is expected to be involved in the persistence of viral DNA episomes (Zhang et al., 1999). Upon activation of latent infection and subsequent wart outgrowth following UV irradiation, E6 and E7 transcripts are detected within 1 week of irradiation preceding papilloma formation observed after 3 to 5 weeks, whereas sites remaining free of papillomas are found DNA positive and E6/E7 RNA negative (Zhang et al., 1999). After complete wart regression, viral transcripts are no longer detected either, but viral DNA persists in a latent stage for prolonged periods of time (Selvakumar et al., 1997). Such long lasting inapparent infections most likely account for the frequent development of HPV-associated lesions in immunosuppressed patients and for the high recurrence of laryngeal papillomas in affected individuals (Lowy and Howley, 2001).

In papillomas and carcinomas, the CRPV genome is predominantly or exclusively maintained as extrachromosomal (10 to 100 copies per cell in non-virus-producing tumors), but was found to be integrated into the cell DNA in VX2 and VX7 carcinoma (Stevens and Wettstein, 1979; Georges et al., 1984; Wettstein, 1990; Georges et al., 1992). Two major messenger RNA species are detected in all tumors (Nasseri et al., 1982; Georges et al., 1984; Nasseri and Wettstein, 1984b) and encode the SE6 protein and the E7 protein (Nasseri and Wettstein, 1984b; Danos et al., 1985; Wettstein, 1990) (Figure 1). Minor RNA species encode the LE6 protein and the full length transactivating E2 protein, and two unspliced rare species were found colinear with E1 (Danos et al., 1985; Wettstein, 1987; 1990) (Figure 1). An E9^E2C cDNA has been cloned from domestic and cottontail rabbit warts, in which a small ORF, designated E9, is spliced to sequences encoding the hinge and C-terminal region of E2 (Figure 1) (Jeckel et al., 2003), a structure described for the E8^E2C repressor protein of bovine PV type 1 (Howley and Lowy, 2001). The messenger RNAs encoding LE6 and SE6 proteins are transcribed from two distinct promoters (Danos et al., 1985) and the proteins are initiated at the first and second ATG of the E6 ORF, respectively (Barbosa and Wettstein, 1987) (Figure 1). The E7 mRNA is transcribed from a promoter localized in the 3′ end of the E6 ORF (Danos et al., 1985; Wettstein, 1990) (Figure 1). Late messenger RNAs encoding the L1 and L2 capsid proteins are readily found in virus-producing cottontail rabbit papillomas and are transcribed from a promoter localized in the 5′ region of the LCR (Nasseri and Wettstein, 1984a; Wettstein, 1987) (Figure 1). L1 and L2 transcripts have been localized by *in situ* hybridization in the upper, terminally differentiating layers of cottontail rabbit papillomas (Zeltner et al., 1994) where vegetative viral DNA replication takes place (Orth et al., 1971).

Analysis of the transcription of the E6 and E7 genes in benign or malignant tumors induced by CRPVa or CRPVb disclosed no difference between the expression and the structure of the CRPVa and CRPVb SE6 and E7 transcripts. In contrast, as expected from the deletion in CRPVb LCR, the LE6 transcripts showed untranslated 5′ ends of different sizes, that is, about 60 nucleotides and 10 nucleotides for CRPVa and CRPVb, respectively, which could affect the efficacy of translation of the protein (Nonnenmacher, 2004).

5.2. Expression and properties of CRPV proteins

Except for capsid antigens in papillomas, the expression of CRPV-induced proteins is too low to be detected in infected tissue or carcinoma. CRPVa and CRPVb genomes were not found to either transform primary rabbit epidermal keratinocytes or extend their lifespan *in vitro* (Nonnenmacher, 2004). If expressed individually from a retroviral vector, CRPVa LE6, SE6 and E7 proteins, but not E5 protein, confer anchorage independent growth to NIH 3T3 cells and transform Sf1 Ep cells, a cottontail rabbit skin epithelial cell line (Meyers et al., 1992). E8 was not found to be a transforming protein (Harry and Wettstein, 1996). However, results demonstrating that CRPV E5 and E8 are oncogenes were obtained with C127 cells when using a CRPVa5 isolate encoding a shorter (62 amino acids) E5 protein with extensive amino acid changes, and when transfecting the E8 gene in the presence of platelet-derived growth factor (Han et al., 1998).

If expressed each from a pCiNeo vector, the CRPVa and CRPVb LE6, SE6 or E7 proteins, but not the E8 proteins, confer anchorage independent growth to C127 cells and transform the non tumorigenic HaCaT human keratinocytes into cell lines forming growing cysts in athymic mice (Nonnenmacher, 2004). The CRPVa SE6 protein induces a significantly greater stimulation of *in vitro* and *in vivo* cell proliferation and malignant histological features are only observed for the cysts induced by CRPVa LE6 transformed HaCaT cells (Nonnenmacher, 2004).

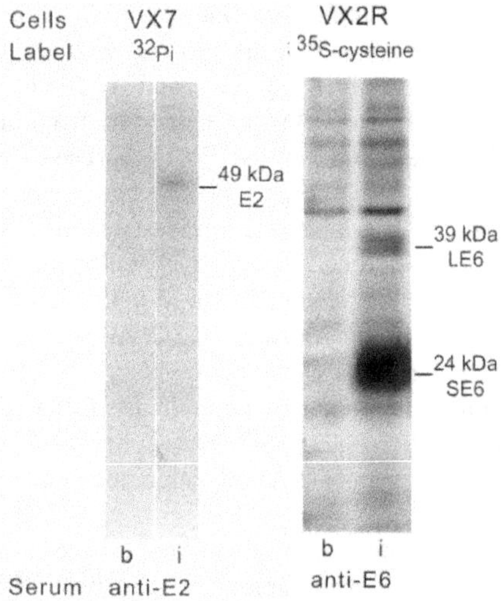

Figure 4. Detection of CRPVa LE6 and SE6 proteins in VX2R cells and of CRPVa E2 in VX7 cells. Proteins were immunoprecipitated from ^{35}S cysteine-labeled extracts of VX2R cells, derived from the VX2 carcinoma (Georges et al., 1985) or from ^{32}P–labeled extracts of VX7 cells, derived from the VX7 carcinoma (F. Breitburd and G. Orth, unpublished) by using antibodies raised against CRPVa E6 and E2 proteins (i) or antisera blocked with the corresponding antigen (b). The immunoprecipitated LE6 (39 kDa), SE6 (24 kDa) and E2 (49kDa) proteins are indicated. Immune and blocked antisera were made available to us by Dr. M. Barbosa and Prof. F.O. Wettstein (Barbosa and Wettstein, 1988a; 1988b).

These data point to a higher oncogenic potential of the SE6 and LE6 proteins of CRPVa that may account for the higher risk of malignant conversion of CRPVa-induced papillomas in rabbits (Salmon et al., 2000). Moreover, a significantly higher growth stimulation is observed for CRPVb E7 as compared to CRPVa E7 (Nonnenmacher, 2004), which may suggest that a stronger deregulation of the negative control of the cell cycle could support the higher level of viral DNA replication observed in CRPVb induced papillomas (Salmon et al., 2000).

The LE6, SE6, E8 and E7 proteins are translated from polycistronic messenger RNAs, and their relative expression remains unknown. None of the start codon has a favourable environment (Kozak, 1999). To test the efficacy of translation from the start codons of LE6, SE6, E8 or E7 in their original context, under the transcriptional control of the LCR, the subgenomic regions from the 5′ end of the LCR to each start codon were cloned upstream of a luciferase reporter gene into a pGL3 expression vector (Nonnenmacher, 2004). Vectors were transfected in VX2R cells that express levels of E2 transcripts comparable to those detected in rabbit papillomas (Georges et al., 1985; Nonnenmacher, 2004). Of all proteins, the LE6 protein is the most efficiently translated in spite of its lower level of transcription. In addition, the CRPVa LE6 expression is about 4-fold higher than that of CRPVb LE6, indicating that the structure of the LE6 transcript is likely to affect the efficacy of the translation of the protein (Nonnenmacher, 2004). The expression of CRPVa and CRPVb SE6 and E8 proteins shows no significant difference but CRPVa E7 is expressed at a significant higher level than CRPVb E7 (about two-fold). The higher oncogenic potential of CRPVa may thus result from both the higher levels of expression and the differences in the biological properties of the E6 and E7 oncoproteins due to the polymorphism of both the LCR and the proteins (Nonnenmacher, 2004).

CRPV LE6 is a non-phosphorylated protein associated with the nuclear matrix and SE6 is a cytoplasmic phosphoprotein (Barbosa and Wettstein, 1988b). Both contain more Cys-X-X-Cys zinc-binding motifs, 8 and 6 respectively, as compared to the 4 motifs found for E6 proteins of other PVs (Danos and Yaniv, 1987; Howley and Lowy, 2001). None of the CRPV E6 proteins binds to the p53 protein (Harry and Wettstein, 1996) but they were found to associate with the hDlg/SAP97 tumor suppressor (Du et al., 2005). The CRPV E7 protein shares some of the properties of the E7 protein of HPV16: both proteins bind the retinoblastoma suppressor protein (pRB), disrupt the complex between pRB and the transcription factor E2F, and transactivate the adenovirus E2 promoter (Haskell et al., 1993). LE6 and SE6 proteins have been detected in VX2 carcinoma-derived cell lines (Georges et al., 1985; Breitburd et al., 1997) and the E2 protein, identified as a phosphorylated species (Barbosa and Wettstein, 1988a), was detected in a VX7 carcinoma-derived cell line (F. Breitburd, M. Barbosa, F. Wettstein, G. Orth, unpublished results) (Figure 4).

Like the E8^E2C proteins of other PVs (Howley and Lowy, 2001), the CRPV E9^E2C protein was found to be a repressor of viral minimal promoters, a function requiring E9 sequences, and a repressor of the E1/E2 transactivation of the replication of a CRPV origin construct in a tissue culture system but these activities had no detectable effect *in vivo* (Jeckel et al., 2003).

E8 knock-out CRPVa and CRPVb genomes induce very few, delayed and slow-growing papillomas, in spite of E6 and E7 gene transcription levels similar to those observed in wild type CRPV induced papillomas. This points to a major role of E8 protein in stimulating the growth of stem cells required for the formation of a papilloma (Nonnenmacher, 2004; Nonnenmacher et al., unpublished results). The E8 proteins of CRPVa and CRPVb, like the E5 protein of HPV16, were found to interfere with the MAP kinase pathway, by performing a yeast two-hybrid screen of a HaCaT human keratinocyte cDNA library. This could suggest that CRPV E8 and HPV E5 proteins may have similar functions in virus-cell interactions, in addition to their structural similarity (Nonnenmacher, 2004; Nonnenmacher et al., unpublished results).

Little difference, if any, is found in the expression of viral oncogenes between papillomas and carcinomas (Schneider-Maunoury et al., 1987; Wettstein, 1987), pointing to the involvement of changes in cellular gene expression in malignant progression. Twenty-three differentially expressed genes were identified by suppression subtractive cDNA hybridization of pairs of papillomas and carcinomas from individual rabbits. Quantitative RT-PCR, in combination with RNA-RNA *in situ* hybridization and laser capture microdissection, confirmed a constant upregulation of transcripts for the multifunctional adaptor protein 14-3-3 ζ and the Y-box binding transcription factor YB-1, whereas transcripts for proteins involved in cell motility and tissue invasion, m-type calpain 2 and NB thymosin β, together with casein kinase 1 α, chaperonin and annexin I, were upregulated in most cases. Small interfering RNA-mediated silencing of genes encoding YB-1, m-type calpain or NB thymosin β, decreased the capacity of a CRPV harboring rabbit keratinocyte line to invade a matrigel (Huber et al., 2004).

As observed for other PVs, the CRPV L1 capsid protein alone or the CRPV L1 and L2 proteins are capable of auto-assembling into virus-like particles (VLPs), when expressed in prokaryotic or eukaryotic expression systems under the control of strong promoters. Such particles, devoid of infectious genetic material, display morphologic features close to those of virions and present antigenic determinants of native viral particles. Capable of stimulating immune defenses against papillomavirus infections, VLPs constitute good prophylactic vaccines (Breitburd and Coursaget, 1999; Schiller, 1999; Lowy and Frazer, 2003; Chapter 18 of this volume). By their capacity to incorporate foreign DNA molecules, VLPs form pseudovirions or pseudotypes that may be used for gene transfer experiments and for the identification of cross-reactive epitopes for vaccine design (Bousarghin et al., 2002; Combita et al., 2002; Lowy and Frazer, 2003).

6. Immune Responses to CRPV Infection

6.1. Humoral responses

Cottontail and domestic rabbits carrying naturally occurring or experimentally induced papillomas are partially or completely resistant to reinfection. Sera of rabbits with papillomas contain antibodies capable of neutralizing the virus (Shope, 1933; Kidd et al., 1936; Shope, 1937). The neutralizing epitopes have been found associated with intact viral capsids and constitute non cross-reactive epitopes among PVs (Christensen and Kreider, 1990; 1991; Breitburd et al., 1995). They have been localized on both L1 and L2 capsid proteins (Christensen and Kreider, 1991; Christensen et al., 1991; Lin et al., 1992; 1993a). In addition to type-specific antibodies, sera of rabbits bearing the VX7 carcinoma contain antibodies reacting with antigenic determinants masked in intact particles and shared by CRPV and HPV type 1 (HPV1) (Orth et al., 1978). These CRPV antigenic determinants are recognized by antisera prepared against dissociated HPV1 particles but not by antibodies raised against intact HPV1 particles (Orth et al., 1978). This observation provided the first demonstration of the existence of a group-specific antigen common to all papillomaviruses

The humoral immune responses to each of the CRPV structural and non structural proteins have been characterized at different stages of wart evolution (Lin et al., 1993b). Antibodies to conformational epitopes of the L1 major capsid protein are detected in essentially all rabbits with papillomas 4 to 6 months after infection, whereas the response to linear epitopes, as detected by Western blotting and Elisa tests using bacterial fusion proteins, is only found in less than one third of the rabbits. Antibodies to the L2 minor capsid protein are only detected in some cottontail rabbits. No major difference is observed in the responses to viral structural proteins between rabbits with regressing or persisting papillomas. In contrast, a drastic increase in the response to L1 and L2 fusion proteins is observed as papillomas progress to carcinomas (Lin et al., 1993b), although transcription of the L1 and L2 genes is not found increased upon progression (Selvakumar et al., 1994). It may be assumed that invasion and necrosis of tumor cells favor the liberation of viral and tumor antigens in the bloodstream and their recognition by the immune system.

A response to the non structural proteins E1, E2, E6, and E7, as tested by Western blotting using bacterial fusion proteins, is observed in only 10 to 30% of the rabbits with papillomas, initially raising and, at the time of malignant conversion, declining for E6 and E7 and remaining level for E1 and E2. All animals are negative for antibodies to E4 and E5 proteins. A much higher response frequency to the E2 protein is observed in regressor rabbits (90%) as compared to rabbits with papillomas (29%) (Selvakumar et al., 1995a).

6.2. Cell-mediated responses

Cellular immune reactions against viral structural antigens were tested *in vivo* by intracutaneous tests (Höpfl et al., 1993). Almost all regressor rabbits and about half of the progressor rabbits showed a positive response against intact and/or disrupted viral particles. Time course of swelling and cellular infiltrates resembled a delayed-type hypersensitivity reaction. Recall reactions remained detectable from several months to more than 2 years after regression (Höpfl et al., 1993).

In vitro lymphoproliferation tests were performed using whole blood or peripheral blood mononuclear cells and CRPV particles or bacterial L1 or L2 fusion proteins as antigens (Selvakumar et al., 1994; Höpfl et al., 1995). A trend of greater reactivity to intact CRPV particles was observed in regressor rabbits when compared to progressors (Höpfl et al., 1995). Low proliferative responses to L1 bacterial fusion protein were detected in about 50% of rabbits with papillomas and were not found for L2 fusion protein. Rabbits with carcinomas responded more frequently (about 80%) and more strongly to L1 and about half of them responded to L2, pointing to the same trend in T cell and humoral responses (Selvakumar et al., 1994).

Lymphoproliferative assays performed using CRPV non-structural proteins showed infrequent responses to E6 and E7 proteins and none to E1 protein, in either persistor or regressor rabbits. In contrast, as found for antibody responses, a proliferative response to E2 protein was detected in a significant proportion of regressor rabbits (about 45%) (Selvakumar et al., 1995a). E2 protein may thus be one of the targets involved in wart regression.

6.3. A role for cell-mediated immunity in wart regression

Early studies had shown that regression of CRPV-induced warts was not influenced by antiviral antibodies (Kidd et al., 1936) and had stressed that histological findings for papilloma regression were essentially identical with those observed when transplanted tissues retrogress (Kidd, 1938). Several lines of evidence pointed to a role of cell-mediated immune responses. Spontaneously regressing rabbit papillomas were found infiltrated by macrophages and lymphocytes (Kidd, 1938), as also observed in humans with regressing flat warts (Tagami et al., 1985). Rabbit papilloma regression was stimulated by vaccination with freshly excised autologous or heterologous papilloma cells (Evans et al., 1962). When skin fragments infected with CRPV *in vitro* were regrafted to donors (Kreider, 1963) or when rabbits were inoculated with CRPV DNA (Evans and Ito, 1966), papillomas developed in persistor but not in regressor rabbits. Treatment with prednisolone (McMichael, 1967) or cyclosporin A (Hu et al., 2005) was found to inhibit papilloma regression .

The mechanisms of wart regression have been difficult to study because of the lack of easily available inbred strains of rabbits (Cohen, 1987) and the paucity of reagents identifying lymphocyte markers (Gordon and Kindt, 1992). Thus, lymphocyte passive transfer experiments could not be achieved (Evans, 1963) and the existence of specific cytotoxic T cell responses has not been demonstrated yet. So far, two inbred rabbit strains have been reported in CRPV studies, the persistor B/J rabbits (Seto et al., 1987) and the EIII/JC rabbits prone to respond most efficiently to vaccination against CRPV early proteins (Han et al., 2000b; Hu et al., 2002a).

Regressing rabbit papillomas were reported to contain four times as many infiltrating leukocytes as progressing tumors and that these infiltrates were concentrated immediately beneath the basal membrane and often obliterated the lowest epithelial cell layers (Kreider and Bartlett, 1981). Most (about 70%) of these infiltrating mononuclear cells were shown to be T cells that expressed class II antigens (Okabayashi et al., 1991). Intraepidermal T cells in regressing warts were found to represent a cycling T cell population whereas most dermal T cells formed a non cycling cell population in both progressing and regressing warts (Okabayashi et al., 1991). Keratinocytes were found to express frequently class II antigens in regressing papillomas (Okabayashi et al., 1991), in contrast to normal epidermal keratinocytes. Moreover, proliferating papilloma cells were found to be evenly distributed in all layers of progressing papillomas whereas they were concentrated in the basal and lower layers of regressing papillomas. These data suggested that both intraepidermal T cell activation and suppression of tumor cell proliferation might play a role in papilloma regression (Okabayashi et al., 1993). CD8+ T cells infiltrating papillomas regressing late after infection (17 to 26 weeks) have been reported but, due to the lack of a relevant antiserum, CD4+ T cells could not be detected (Selvakumar et al., 1997).

Both T lymphocytes and macrophages infiltrating regressing warts were found to express tumor necrosis factor (TNF) transcripts and apoptotic changes were detected in keratinocytes scattered in all epithelial layers, instead of a few layers just beneath the horny layer in progressing papillomas (Hagari et al., 1995). Thus, TNF produced by infiltrating mononuclear cells might play a role in wart regression.

So far, the data support the involvement of delayed-type hypersensitivity responses in papilloma regression, as also suggested for genital warts (Coleman et al., 1994). The role of cytotoxic T cell responses in papilloma regression requires further study.

7. Evaluation of Preventive and Therapeutic Vaccine Strategies

The rabbit model is one of the best and most amenable to test vaccine strategies for the prevention of PV infections and the eradication of established PV-induced tumors. Early immunization experiments performed with wart extracts showed effective protection against infection and neutralizing antibodies were demonstrated in the sera of vaccinated animals (Shope, 1937). Vaccines made of freshly excised autologous or homologous rabbit papillomas were shown to substantially increase the rate of wart regression (Evans et al., 1962). This warranted the use of the rabbit model to evaluate the efficacy of different vaccine approaches.

7.1. Structural protein vaccines

Prophylactic CRPV vaccines were designed to elicit neutralizing antibodies directed against the L1 and L2 capsid proteins. L1 and L2 proteins expressed as bacterial fusion proteins were used as subunit vaccines (Christensen et al., 1991; Lin et al., 1992; 1993a) and recombinant vaccinia viruses expressing L1 protein as live vaccines (Lin et al., 1992). Virus-like particles (VLPs) presenting conformational epitopes recognized by neutralizing antibodies were obtained by auto-assembly of L1 alone or L1 and L2 proteins expressed in high amounts from recombinant baculovirus or yeast expression vectors (Breitburd et al., 1995; Jansen et al., 1995; Christensen et al., 1996b). Alternatively, the CRPV L1 DNA sequence inserted in an expression vector downstream of the cytomegalovirus enhancer and promoter was inoculated as a polynucleotide vaccine in a DNA-based immunization protocol (Donnelly et al., 1996) or by gene gun intracutaneous injection (Sundaram et al., 1997).

Under these many forms, the CRPV L1 protein was fully protective (Christensen and Kreider, 1991; Lin et al., 1992; 1993a; Breitburd et al., 1995; Jansen et al., 1995; Christensen et al., 1996b; Donnelly et al., 1996; Sundaram et al., 1997). A long-term protection was demonstrated for CRPV VLP vaccines (Christensen et al., 1996b). The L2 protein used as a fusion protein proved also protective and induced neutralizing antibodies but to a lesser extent than the L1 protein (Christensen et al., 1991; Lin et al., 1992).

Incorporation of the L2 protein into L1 VLPs resulted in a better protection towards massive infection (Breitburd et al., 1995). No cross-protection was observed when using BPV1 (Breitburd et al., 1995) or HPV11 VLPs (Christensen et al., 1996b). Protection required a full-length non denatured L1 protein (Christensen and Kreider, 1991; Lin et al., 1993a) or intact VLPs (Breitburd et al., 1995), indicating that PV-group-specific antigenic determinants unmasked in disrupted particles were not protective. High titers of anti capsid antibodies were elicited by VLP vaccination, as demonstrated by ELISA and hemagglutination inhibition tests (Roden et al., 1995). The presence of neutralizing antibodies in VLP-vaccinated rabbits was demonstrated by passive transfer experiments (Breitburd et al., 1995; Jansen et al., 1995).

An attenuated live recombinant vesicular stomatitis virus (VSV) expressing high levels of CRPV L1 protein from an upstream site in the VSV genome completely protects against CRPV-induced papilloma formation after a single dose of vaccine. This, together with an inexpensive production, the efficacy of live viruses to induce long term immunity, including cell-mediated and antibody responses, and the ability to provide intranasal boost of mucosal immunity, should represent a most affordable approach for vaccination in developing countries (Roberts et al., 2004).

A potential for cross neutralization was reported for subdominant epitopes of genital HPV L1 (Combita et al., 2002) or L2 proteins (Kawana et al., 1999; Roden et al., 2000), but whether it is sufficient to afford cross-protection *in vivo* remained an open question. Two such overlapping short peptides spanning the amino acids 94-112 and 107-122 of CRPV and ROPV L2 proteins protect from challenge with the cognate virus but not with viral DNA, suggesting the involvement of an antibody-mediated neutralization. Despite serum reactivity to the heterologous peptides, little cross-protection was observed from challenge with the heterologous virus (Embers et al., 2002). Whether a cross protection could be obtained by increasing the size of the peptides or by incorporating these epitopes into a multiple antigen peptide-vaccine remains to be studied.

7.2. Non structural protein vaccines

It was thought that vaccines aimed at stimulating cell-mediated immune responses against the viral E1, E2, E6 or E7 non structural proteins could be also protective by eliminating infected cells expressing these proteins, as early as they emerged. When given prior to challenge, none of these vaccines prevented wart growth in rabbits (Lathe et al., 1989; Selvakumar et al., 1995b; Orth and Breitburd, unpublished results). However, LE6 protein expressed from a recombinant vaccinia virus and E1 and E2 fusion proteins significantly accelerated regression and increased the number of regressor rabbits (Lathe et al., 1989; Selvakumar et al., 1995b; Orth and Breitburd, unpublished results). An inconstant effect was observed with an E7 recombinant vaccinia virus vaccine (Lathe et al., 1989; Orth and Breitburd, unpublished results). Highly variable *in vitro* lymphoproliferative responses have been observed in rabbits vaccinated by intramuscular or gene gun inoculation of DNA plasmids expressing CRPV non structural proteins (Sundaram et al., 1998; Han et al., 1999a; 1999b). Despite the more efficient gene gun intradermal inoculation in promoting protection against CRPV challenge (Han et al., 1999b; 2000b), E1, E2, E6 or E7 DNA plasmids alone provide no (E2, E7) or only partial (E1, E2, E6) protection. In contrast, gene gun inoculation of the four plasmids in combination completely protects 7 of 9 outbred and all EIII/JC inbred rabbits from viral challenge (Han et al., 1999b; 2000b) and prevents or delays carcinoma development from established papillomas (Han et al., 2000a). These data confirm the multiplicity of epitopes involved in regression.

Antiviral compounds or immunomodulators are partially effective in the control of wart growth and recurrence (Christensen et al., 2000b). The cure of papillomas and the prevention of their recurrences can be obtained by the association of an antiviral compound to a specific immune stimulation like cidofovir given intralesionally between the third and boost inoculation of the E1, E2, E6 and E7 DNA vaccine (Christensen et al., 2001). The partial efficacy of an E6 DNA vaccine (Sundaram et al., 1998) is significantly enhanced by priming the cutaneous vaccination sites with a plasmid expressing mouse granulocyte–macrophage colony-stimulating factor (GM-CSF) inoculated 3 days before the first and second vaccination and at the time of the boost (Leachman et al., 2000). In these conditions, the cytokine induces differentiation and local recruitment of MHC class II-positive professional antigen-presenting cells, including macrophages, dermal dendritic cells and Langherans cells (Leachman et al., 2000). Major improvements concern papilloma formation (disease-free sites rising from 50% to 67%), papilloma volume (90% reduction), delay in outgrowth (from 25 to 75 days) and rate of papilloma regression (from 13 to 71%) (Leachman et al., 2000). The efficacy of E6 vaccination is also further increased by fusing an ubiquitin monomer to the E6 gene to enhance antigen processing and presentation via the MHC class I pathway. Complete protection is obtained by combining the four ubiquitin-fused CRPV E1, E2, E6 and E7 genes (Leachman et al., 2002). Gene gun vaccination with a CRPV E8 plasmid partially protects outbred and EIII/JC inbred rabbits from CRPV challenge, promotes papilloma regression in EIII/JC rabbits (Hu et al., 2002a) which is further stimulated by priming with GM-CSF DNA (Hu et al., 2004a).

Most effective live vaccines were only tested for E1 and E2 proteins which do not display oncogenic properties. Vaccination with a recombinant *Listeria monocytogenes* secreting CRPV E1 protein stimulates regression of CRPV-induced papillomas, controls papilloma growth and protects against CRPV DNA challenge in more than 75% of infected rabbits (Jensen et al., 1997). Primary immunization with a recombinant adenovirus expressing CRPV E2 protein, given intradermally at about 50 sites adjacent to CRPV inoculation sites, reduces the number of papilloma–forming sites (from 17 of 45 to 4 of 45). Booster immunization prevents formation of new papillomas and stimulates regression of all papillomas. No CRPV DNA is detected at inoculated sites free of papillomas or at papilloma regression sites (Brandsma et al., 2004).

The CRPV-associated VX2 carcinoma was used to evaluate strategies for the eradication of malignant tumors. Perilesional interleukin-2 treatment of a VX2 head-and-neck carcinoma model in rabbits resulted in complete regression of locally treated and untreated tumors in 4 out of 12 rabbits, and of metastases in 3 of those. Tumor cells injected in cured animals were rejected, demonstrating local and systemic therapeutic effects of the treatment (Van Es et al., 2000). Interestingly, probably because E2 proteins are rather conserved among papillomaviruses and their regulatory functions involve DNA binding sites present in all PV genomes, a recombinant vaccinia virus expressing the bovine PV E2 protein (MVA E2) induced regression of large VX2 tumors in 80% of rabbits by stimulating macrophage antibody-dependent cytotoxicity whereas normal vaccinia virus had no effect (Rosales et al., 2000). It must be stressed that MVA E2 inhibited also the growth of HeLa cells grafted to athymic mice, involving the repressor activity of E2 for HPV18 promoter (Valdez Graham et al., 2000). Furthermore, in a clinical trial, MVA E2 vaccine given locally eliminated precancerous cervical lesions (CIN1 to CIN3) associated with oncogenic HPVs in 34 out of 36 patients, 50% of whom cleared HPV DNA while the remaining 50% had a greatly reduced viral load. Patients developed antibody to the MVA 2 vaccine and vaccination generated a specific cytotoxicity against HPV-transformed cells (Corona Gutierrez et al., 2004).

Vaccine studies in the CRPV rabbit model point to many approaches and their combination in multistep protocols that might be efficient as therapeutic strategies to eradicate established tumors. It is clear, however, that regression is not observed in all vaccinated animals. This may reflect genetically controlled differences in the capacity to mount a cell-mediated immune response to non-structural PV antigens.

8. Immunogenetic Control of Wart Evolution

Spontaneous regression of warts precludes the development of cancer and is a major event in the control of the oncogenic potential of CRPV. It is likely to be due to cell-mediated immune responses directed against some of the non structural viral proteins expressed in tumor cells, as shown by vaccine studies. Available data point to the possible involvement of a delayed-type hypersensitivity reaction, in which helper T cells and macrophages are likely to play a major part (Murphy and Reiner, 2002). However, cytotoxic T cell responses may have to be considered also since cytotoxic lymphocytes recognizing the HPV16 E6 and E7 oncoproteins have been demonstrated (Ressing et al., 1999). The polymorphism of the genes involved in immune responses may be responsible for the variable capacity of rabbits to reject warts. Many of these genes are clustered within the major histocompatibility complex.

8.1. The major histocompatibility complex in humans and rabbits

The human major histocompatibility complex (MHC), termed also the human leukocyte antigen (HLA) complex, comprises many genes of known or unknown functions, among which those encoding three distinct classes of products involved in immune responses (Beck and Trowsdale, 2000). The class I gene family encodes cell surface glycoproteins that associate non covalently with β2-microglobulin and are involved in the presentation of processed peptides to CD8+ T cells. Class I molecules are found on the surface of all cell types. Class II genes encode polymorphic α and β chains which associate into heterodimeric glycoproteins. The class II molecules present antigenic peptides to CD4+ T lymphocytes. They are expressed primarily on professional antigen presenting cells (APC) (dendritic cells, B lymphocytes, macrophages). The extensive polymorphism of HLA class I and class II molecules is concentrated within the region of the peptide binding groove providing a repertoire of peptide binding capacity. The class II region contains also genes encoding proteins involved in the processing (LMP2, LMP7) and the transport (TAP1, TAP2) of peptides bound and presented by class I molecules. The class III region includes genes encoding components of the complement (C2, Bf, C4), heat shock proteins (Hsp70), tumor necrosis factor and lymphotoxin (TNF and LT). In humans, MHC class I and II genes form two discrete regions separated by class III genes (Beck and Trowsdale, 2000).

The rabbit leukocyte antigen (RLA) complex was first divided into two loci, A and D. Five alleles were defined at locus A by allograft rejection and four alleles at locus D by mixed lymphocyte reactions (Cohen, 1987). Molecular studies have identified RLA class I, class II and class III genes, due to their great extent of DNA sequence homology with human HLA genes (Rebière et al., 1987; Marche et al., 1989; Sittisombut et al., 1989; Chouchane and Kindt, 1992; Chouchane et al., 1993; Sittisombut, 1996; Hermel et al., 1999). The RLA class I genes comprise 8 to 14 genes, depending on the haplotype. So far, four transcripts and only one protein have been identified (Wilkinson et al., 1982; Rebière et al., 1987; Gordon and Kindt, 1992). The class II genes are likewise very similar to those of humans. Homologues of HLA-DR, -DQ, -DP genes, HLA -DO B and -DN A genes and HLA DM genes have been recognized (LeGuern et al., 1985; Sittisombut and Knight, 1986; LeGuern et al., 1987; Sittisombut et al., 1988; Laverriere et al., 1989; Sittisombut et al., 1989; Chouchane et al., 1993; Sittisombut, 1996; Hermel et al., 1999). A number of class III genes have been identified, including the TNF and LT genes (Shakhov et al., 1990). Contrasting with the organization of human MHC, RLA class I and class II genes are contiguous in rabbits (Chouchane and Kindt, 1992) and class III genes have been only recently colocalized with class I and class II genes on chromosome 12q1.1 (Rogel-Gaillard et al., 2001). Molecular DNA typing of rabbit MHC genes based on a restriction fragment length polymorphism (RFLP) of class I and DRA and DQA genes has allowed to define 13 haplotypes (Marche et al., 1989; Gordon and Kindt, 1992). More recently, an additional polymorphism, characterized by insertion/deletion of large DNA fragments, has been described in the region of class II genes (Chouchane and Kindt, 1992).

Our aim has been to find out whether the variability of the clinical evolution of warts in rabbits - regression, persistence or malignant transformation - could be related to the polymorphism of rabbit MHC genes.

8.2. Polymorphism of rabbit MHC genes and wart evolution

8.2.1. Linkage of class II DRA and DQA genes to wart evolution in non selected rabbits

Seventy-seven outbred New Zealand White rabbits were infected with two doses of CRPV, 5-and 50-fold the minimal dose of CRPV inducing warts in 100% of rabbits (Han et al., 1992). Skin warts developed after 3-4 weeks and regression was observed 1 to 6 weeks later in 42 animals. A late regression was observed in 12 rabbits after 8 to 14 weeks. Eleven of the 23 persistor rabbits developed invasive carcinomas. Significant associations were observed between wart evolution and the RFLP of RLA class II DRA and DQA genes. Of the four *Eco*RI fragments defining DRA RFLP alleles A to D, and of the seven *Pvu*II fragments defining DQA RFLP alleles A to G, two alleles in linkage disequilibrium, DRA-B and DQA-E, were found strongly linked to wart regression. A DQA allele, DQA-G, was found associated with a 4.5-times higher risk of wart malignant conversion in the year following infection (Han et al., 1992; Breitburd et al., 1996). This indicates a genetic control of wart evolution, involving genes localized in and/or linked to the class II region of the MHC.

To find out whether this immunogenetic control could be connected with the antigen binding and presentation function of the $\alpha 1$ domain of class II molecules (Germain, 1994), the nucleotide sequence of the exon 2 of the different DRA and DQA RFLP alleles was determined and the encoded amino acid sequences deduced (Han et al., 1994). No amino acid polymorphism was found in the $\alpha 1$ domain of the DRα chain. In contrast, 27 of the 82 amino acids of the DQ$\alpha 1$ domain were found variable, defining five amino acid sequence alleles. The most divergent DQ$\alpha 1$ allele was that associated with a higher risk of cancer. Alignment of rabbit and human DQA exon 2 alleles disclosed that amino acid variations occur at positions assumed to be important for peptide binding in humans. However, the DQA RFLP allele (DQA-E) linked to regression and another allele (DQ-B) not linked to regression shared the same $\alpha 1$ domain. These data do not support that the variable capacity to reject warts is directly linked to the polymorphism of the exon 2 of RLA-DRA and -DQA genes but rather point to the involvement of some closely linked polymorphic genes, such as DRB or DQB, encoding the β chain of class II molecules (Han et al., 1994).

8.2.2. Restriction fragment length polymorphism of MHC genes in rabbits homozygous for RLA- DRA and –DQA genes

In order to further characterize the linkage between wart regression and RLA genes, the response to CRPV infection was studied in families of 18-20 rabbits homozygous for four DRA-DQA RFLP haplotypes, the two found linked to regression (DRA-B/DQA-E) or with a higher rate of malignant conversion (DRA-C/DQA-G), and two not known to be associated with a specific mode of wart evolution (DRA-D/DQA-B and DRA-A/DQA-E) (Salmon et al., 2001; Salmon, 2003). RFLP markers of RLA class I genes (19-1-3', pR27), class II (DRB, DQB, DPB) genes and class III (TNF, LT) genes were analyzed for each rabbit. Rabbits within the same family were found homozygous for these extended RLA haplotypes comprising nine markers of MHC genes. Seventy to 75% of the rabbits showed the mode of wart evolution expected for the DRA-DQA haplotype linked to systemic wart regression (R1 haplotype) or wart persistence (P haplotype). Partial regression, a mode of wart evolution seldom observed in heterozygous rabbits, was observed in 80% of rabbits with the DRA-D/DQA-B haplotype (PR haplotype) (Salmon et al., 2000; 2001; Salmon, 2003) and 70% of the rabbits with the fourth haplotype showed systemic wart regression (R2 haplotype). These data further support an immunogenetic control of wart evolution but do not exclude that a sequence polymorphism of the antigen-binding domain of RFLP alleles could account for the discrepant wart evolution observed in about 25% of the animals of each family.

8.2.3. The partial regression phenotype

Partial regression preferentially associated with the PR haplotype is characterized by the rejection of a great majority of the warts and the persistence of a few of them that may evolve towards carcinoma (Salmon et al., 1997; 2000; 2001; Salmon, 2003). Nucleotide sequencing of the CRPV genome used in these studies disclosed the genetic heterogeneity of the virus isolate obtained from Kansas cottontail rabbit warts. A new CRPV subtype, CRPVb, with a significant E6 and E7 nucleotide and amino acid variation, as compared to the prototypical CRPV, was recognized as the major component of the inoculum (Salmon et al., 1997; 2000). Sequencing of the CRPV E6 and E7 ORFs in warts prior and after partial regression showed that warts associated with CRPVb were eliminated and that persistent warts were associated with CRPVa4, a variant close to the prototype CRPVa1 (Giri et al., 1985), present as a minor component in the inoculum (Salmon et al., 1997). Except for about 20% of them, rabbits with a P haplotype were unable to reject warts associated with either CRPV subtype and rabbits with an R haplotype eliminated both subtypes (Salmon et al., 2001; Salmon, 2003; Salmon et al., unpublished results). These data support that an allelic restriction of the binding and presentation of some variable CRPV antigenic peptides by MHC molecules plays a key role in determining wart evolution.

8.2.4. MHC class II DQ molecules as a major determinant in wart regression

To further define the molecular bases of the immunogenetic control of wart evolution, rabbits homozygous for each of the four R1, R2, P or PR haplotypes and showing the expected or a discrepant mode of wart evolution, were analyzed for the sequence polymorphism of the antigen binding domains of class II DRB, DQA, DQB molecules and the DRB and class I molecules expressed in warts and leukocytes. Sequencing of exons encoding peptide binding domains of expressed class I (exons 2 and 3) and class II (exon 2) molecules disclosed a high percentage of variable amino acid positions in DQ α1(31%), DQ β1 (12%) and DR β1 (26%) domains and allowed to identify three alleles of DQA and DQB genes and two alleles of DRB1 gene (J. Salmon, N. Chanteloup, F. Viard, P. Coudert, F. Breitburd and G. Orth, unpublished results). Specific DQ alleles were found associated with R, P, and PR haplotypes whereas the same DRB1 allele was shared by R2, P and PR haplotypes pointing to a linkage of wart evolution with the DQ locus (J. Salmon, O. Croissant, F. Breitburd and G. Orth, unpublished results). For each of R1, P and PR haplotypes, one specific class I sequence was found predominantly expressed in warts and 13% to 17.5% variable amino acid positions were observed among the three class I sequences. A minor transcript expressed in PR rabbits most likely corresponds to a non-classical class I gene (Salmon, Chanteloup et al., unpublished results). No specific class I or class II alleles were found associated with discrepant wart evolution, pointing to the involvement of additional mechanisms in the control of wart evolution (Salmon et al., 2001; 2002; Salmon, 2003; Salmon, Croissant et al., unpublished results).

9. CRPV-induced Tumors, a Model for Immunogenetics of HPV-associated Neoplasia

Specific human PV genotypes, mostly HPV16 and 18, play a central role in the development of cervical intraepithelial neoplasia (CIN) and invasive carcinoma of the uterine cervix (Lowy and Howley, 2001). Since 1991, several groups have reported positive or negative associations of particular HLA class II alleles with cervical neoplasia, some of them showing papillomavirus-type specificity (Wank and Thomssen, 1991; Apple et al., 1994; Breitburd et al., 1996; Sastre-Garau et al., 1996; Stern, 1996; Hildesheim and Wang, 2002; Chapter 19 of this volume). This suggested that susceptibility or resistance to cervical neoplasia was related to the polymorphism of genes in the MHC class II region.

The conflicting reports on HLA associations with CIN or invasive cervical carcinoma probably arise from the ethnic origin and selection of the patient and control groups, differences in peptide binding efficacy of different HLA molecules, behavioral and environmental factors, and HPV intratype variability (Wang and Hildesheim, 2003). The available data already suggest, however, that certain HLA class II haplotypes may influence the immune responses to specific HPV-encoded epitopes and affect the risk of development of high-grade CIN and invasive cervical cancer (Ressing et al., 1999). Evidence for the immunogenetic control of CRPV-induced tumors (Salmon et al., 2001; 2002; Salmon, 2003) further supports the importance of immunogenetics in the control of HPV-associated carcinogenesis. Interestingly, positive and negative associations of specific HLA DR-DQ haplotypes have been also identified in patients suffering from epidermodysplasia verruciformis (Ramoz et al., unpublished results). Rabbits homozygous for extended MHC class I and class II haplotypes and their specific responses to infection by two significantly different strains of CRPV constitute an invaluable tool to progress in the characterization of the mechanisms involved in the evolution of HPV-induced disease.

10. Insight into the Mechanisms of Wart Regression or Persistence

Wart regression in rabbits has been reported to involve CD8+ T lymphocytes but CD4+ T cells were not detected (Selvakumar et al., 1997). Immunohistochemical studies show, however, that early wart regression in R rabbits is associated with early infiltration of CD4+ cells in the basal and parabasal wart layers, later followed by the infiltration of CD8+ cells and macrophages in all wart layers. In contrast, persistent warts in P rabbits are characterized by the absence of CD4+ cells and the inconstant presence of CD8+ cells and macrophages (Salmon et al., 2001; 2002; Salmon, 2003; Salmon, Croissant et al., unpublished results). Class II antigens are not expressed by normal epidermal keratinocytes but can be detected upon PV infection (Breitburd et al., 1996). In R rabbits, regression is associated with a generalized expression of DQ antigens by wart keratinocytes whereas in P rabbits, persistent warts show no or moderate DQ antigen expression (Salmon et al., 2001; 2002; Salmon, 2003; Salmon, Croissant et al., unpublished results). Class I antigens are detected in the deeper layers of normal epidermis and may be lost upon malignant progression of PV induced lesions (Breitburd et al., 1996). A generalized expression of class I antigens is observed in all layers of warts from R and P rabbits, upon regression or persistence. In PR rabbits, however, class I antigen expression is lost in most persistent warts (Salmon et al., 2001; 2002; Salmon, 2003; Salmon, Croissant et al., unpublished results). The data indicate, so far, that DQ antigens and CD4+ cells play a major part in wart evolution, together with CD8+ cells and class I antigens, and that this depends both on the CRPV strain and the haplotype of the host (Salmon et al., 2002; Salmon, 2003; Salmon, Croissant et al., unpublished results). As indicated by the discrepant modes of wart evolution observed in 25% of the rabbits, additional mechanisms, unlinked to MHC, are likely to be involved in wart regression or persistence and remain to be investigated.

11. Future Prospects: Genetic Susceptibility to HPV Infection and Carcinogenesis

Inoculation of wild type, chimeric or mutated CRPV genomes to the same animal has revealed clearly that rabbits differ in their susceptibility to infection, even at high doses (our unpublished observations). Little is known of the host genes involved in the control of latent infection, the transition from latent to productive infection and from persistence to malignant progression. Two genes (*EVER* 1 and *EVER* 2) have been recently identified as involved in the susceptibility to epidermodysplasia verruciformis (EV) (Ramoz et al., 1999; Ramoz et al., 2002). EV is a rare disease characterized by an abnormal susceptibility to infection by EV HPVs and by a high risk to develop non melanoma skin cancers associated with HPV5 (Orth, 1987; Orth et al., 2001). EV HPVs have a worldwide distribution and cause widespread unapparent infections only detectable by highly sensitive nested PCR approaches (Orth, 2004), which contrasts with the high copy number of episomal viral genomes and abundant E6 and E7 transcripts found in EV benign lesions and carcinomas. Homozygous mutations in either of the two *EVER* genes have been identified in EV patients (Ramoz et al., 2002) that result in the development and lifelong persistence of disseminated macular or flat wart-like EV lesions with a high risk for malignant conversion when associated with EV HPV oncogenic genotypes (Orth, 1987; Orth et al., 2001; Orth, 2004).

Sequences related to the *EVER* genes have been detected in rabbits and short sequences of 400 bp (*EVER*1) and 900 bp (*EVER*2) have been cloned (Zygadlo, 2002). This should enable to probe a rabbit cDNA library to isolate rabbit *EVER* genes. Further studies should determine whether a polymorphism or mutations of rabbit *EVER* genes may be related to a higher susceptibility to CRPV infection, to wart evolution towards regression or persistence, independent from MHC gene polymorphism, and whether further alterations affect these genes in CRPV associated carcinomas. In addition to study the function still unknown of *EVER* genes, the rabbit model should also provide an excellent experimental approach to further identify cellular genes involved in the control of papillomavirus infection and carcinogenesis of the skin.

Acknowledgements

The authors wish to thank Dr Odile Croissant for the many critical and fruitful discussions, Christian Pons for his skilful experimental assistance, Dr Michel Favre and Patricia Cassonnet for their support and Danielle Senlecques for her great expertise in performing the artwork and the organization of the manuscript.

References

Alonso, L., and Fuchs, E. (2003). Stem cells of the skin epithelium. Proc. Natl. Acad. Sci. USA *100*, 11830-11835.

Amella, C. A., Lofgren, L. A., Ronn, A. M., Nouri, M., Shikowitz, M. J., and Steinberg, B. M. (1994). Latent infection induced with cottontail rabbit papillomavirus. A model for human papillomavirus latency. Am. J. Pathol. *144*, 1167-1171.

Apple, R. J., Erlich, H. A., Klitz, W., Manos, M. M., Becker, T. M., and Wheeler, C. M. (1994). HLA DR-DQ associations with cervical carcinoma show papillomavirus-type specificity. Nat. Genet. *6*, 157-162.

Barbosa, M. S., and Wettstein, F. O. (1987). Transcription of the cottontail rabbit papillomavirus early region and identification of two E6 polypeptides in COS-7 cells. J. Virol. *61*, 2938-2942.

Barbosa, M. S., and Wettstein, F. O. (1988a). E2 of cottontail rabbit papillomavirus is a nuclear phosphoprotein translated from an mRNA encoding multiple open reading frames. J. Virol. *62*, 3242-3249.

Barbosa, M. S., and Wettstein, F. O. (1988b). The two proteins encoded by the cottontail rabbit papillomavirus E6 open reading frame differ with respect to localization and phosphorylation. J. Virol. *62*, 1088-1092.

Belnap, D. M., Olson, N. H., Cladel, N. M., Newcomb, W. W., Brown, J. C., Kreider, J. W., Christensen, N. D., and Baker, T. S. (1996). Conserved features in papillomavirus and polyomavirus capsids. J. Mol. Biol. *259*, 249-263.

Beard, J., Sharp, D., and Eckert, E. (1955). Tumor viruses. Adv. Virus Res. *3*, 149-197.

Beard, J. W., and Rous, P. (1934). A virus-induced mammalian growth with the characters of a tumor (the Shope rabbit papilloma). II. Experimental alterations of the growth of the skin: morphological considerations: the phenomena of retrogression. J. Exp. Med. *60*, 723-740.

Beard, J. W., and Rous, P. (1935). Effectiveness of Shope papilloma virus in various American rabbits. Proc. Soc. Exp. Biol. Med. *33*, 191-193.

Beard, J. W. (1956). The fallacy of the concept of virus "masking": A review. Cancer Res. *16*, 279-291.

Beck, S., and Trowsdale, J. (2000). The human major histocompatibility complex: Lessons from the DNA sequence. Annu. Rev. Genomics Hum. Genet. *1*, 117-137.

Bousarghin, L., Combita-Rojas, A. L., Touze, A., Mehdaoui, S. E., Sizaret, P. Y., Bravo, M. M., and Coursaget, P. (2002). Detection of neutralizing antibodies against human papillomaviruses (HPV) by inhibition of gene transfer mediated by HPV pseudovirions. J. Clin. Microbiol. *40*, 926-932.

Brandsma, J. L., Yang, Z.-H., Barthold, S. W., and Johnson, E. A. (1991). Use of a rapid, efficient inoculation method to induce papillomas by cottontail rabbit papillomavirus DNA shows that the *E7* gene is required. Proc. Natl. Acad. Sci. USA *88*, 4816-4820.

Brandsma, J. L., Yang, Z.-H., DiMaio, D., Barthold, S. W., Johnson, E., and Xiao, W. (1992). The putative E5 open reading frame of cottontail rabbit papillomavirus is dispensable for papilloma formation in domestic rabbits. J. Virol. *66*, 6204-6207.

Brandsma, J. L., Shlyankevich, M., Zhang, L., Slade, M. D., Goodwin, E. C., Peh, W., and Deisseroth, A. B. (2004). Vaccination of rabbits with an adenovirus vector expressing the papillomavirus E2 protein leads to clearance of papillomas and infection. J. Virol. *78*, 116-123.

Breitburd, F., Kirnbauer, R., Hubbert, N. L., Nonnenmacher, B., Trin-Dinh-Desmarquet, C., Orth, G., Schiller, J. T., and Lowy, D. R. (1995). Immunization with viruslike particles from cottontail rabbit papillomavirus (CRPV) can protect against experimental CRPV infection. J. Virol. *69*, 3959-3963.

Breitburd, F., Ramoz, N., Salmon, J., and Orth, G. (1996). HLA control in the progression of human papillomavirus infections. Semin. Cancer Biol. *7*, 359-371.

Breitburd, F., Salmon, J., and Orth, G. (1997). The rabbit viral skin papillomas and carcinomas: a model for the immunogenetics of HPV-associated carcinogenesis. Clin. Dermatol. *15*, 237-247.

Breitburd, F., and Coursaget, P. (1999). Human papillomavirus vaccines. Semin. Cancer Biol. *9*, 431-444.

Chouchane, L., and Kindt, T. J. (1992). Mapping of the rabbit MHC reveals that class I genes are adjacent to the DR subregion and defines an insertion/deletion-related polymorphism in the class II region. J. Immunol. *149*, 1216-1222.

Chouchane, L., Brown, T. J., and Kindt, T. J. (1993). Structure and expression of a nonpolymorphic rabbit class II gene with homology to *HLA-DOB*. Immunogenetics *38*, 64-66.

Christensen, N., and Kreider, J. (1990). Antibody-mediated neutralization *in vivo* of infectious papillomaviruses. J. Virol. *64*, 3151-3156.

Christensen, N., Cladel, N., Reed, C., Budgeon, L., Welsh, P., Patrick, S., and Kreider, J. (1996a). Laboratory production of infectious stocks of rabbit oral papillomavirus. J. Gen. Virol. *77*, 1793-1798.

Christensen, N. D., and Kreider, J. W. (1991). Neutralization of CRPV infectivity by monoclonal antibodies that identify conformational epitopes on intact virions. Virus Res. *21*, 169-179.

Christensen, N. D., Kreider, J. W., Kan, N. C., and DiAngelo, S. L. (1991). The open reading frame L2 of cottontail rabbit papillomavirus contains antibody-inducing neutralizing epitopes. Virology *181*, 572-579.

Christensen, N. D., Reed, C. A., Cladel, N. M., Han, R., and Kreider, J. W. (1996b). Immunization with viruslike particles induces long-term protection of rabbits against challenge with cottontail rabbit papillomavirus. J. Virol. *70*, 960-965.

Christensen, N. D., Cladel, N. M., Reed, C., and Han, R. (2000a). Rabbit oral papillomavirus complete genome sequence and immunity following genital infection. Virology *269*, 451-461.

Christensen, N. D., Pickel, M. D., Budgeon, L. R., and Kreider, J. W. (2000b). *In vivo* anti-papillomavirus activity of nucleoside analogues including cidofovir on CRPV-induced rabbit papillomas. Antiviral Research *48*, 131-142.

Christensen, N. D., Han, R., Cladel, N. M., and Pickel, M. D. (2001). Combination treatment with intralesional cidofovir and viral DNA vaccination cures large cottontail rabbit papillomavirus-induced papillomas and reduces recurrences. Antimicrob. Agents Chemother. *45*, 1201-1209.

Cohen, C. (1987). Histocompatibility and transplantation. In The rabbit in contemporary immunological research, S. Dubisky, ed. (Harlow, UK: Longman Scientific and Technical), pp. 13-26.

Coleman, N., Birley, H. D. L., Renton, A. M., Hanna, N. F., Ryait, B. K., Byrne, M., Taylor-Robinson, D., and Stanley, M. A. (1994). Immunological events in regressing genital warts. Am. J. Clin. Pathol. *102*, 768-774.

Combita, A. L., Touzé, A., Bousarghin, L., Christensen, N. D., and Coursaget, P. (2002). Identification of two cross-neutralizing linear epitopes within the L1 major capsid protein of human papillomaviruses. J. Virol. *76*, 6480-6486.

Corona Gutierrez, C. M., Tinoco, A., Navarro, T., Contreras, M. L., Cortes, R. R., Calzado, P., Reyes, L., Posternak, R., Morosoli, G., Verde, M. L., and Rosales, R. (2004). Therapeutic vaccination with MVA E2 can eliminate precancerous lesions (CIN1, CIN2 and CIN3) associated with infection by oncogenic human papillomavirus. Hum. Gene Ther. *15*, 421-431.

Croissant, O. and Orth, G. (1967). Demonstration by autoradiography of DNA synthesis in cells in the process of keratinization of the papilloma induced by Shope virus in the domestic rabbit. C.R. Acad. Sci. Hebd. Seances Acad. Sci. D. *265*, 1341-1344

Croissant, O. and Orth, G. (1970). Control of mitotic activity in cells in the process of keratinization of growing papillomas induced by Shope virus in rabbits. C.R. Acad. Sci. Hebd. Seances Acad. Sci. D. *270*, 2609-2612.

Croissant, O., Dauguet, C., Jeanteur, P., and Orth, G. (1972). Application of the molecular hybridization technic *in situ* for the demonstration with the electron microscope, of vegetative replication of viral DNA in the papillomas induced with Shope virus in cottontail rabbits. C.R. Acad. Sci. Hebd. Seances Acad. Sci D. *274*, 614-617.

Danos, O., Georges, E., Orth, G., and Yaniv, M. (1985). Fine structure of the cottontail rabbit papillomavirus mRNAs expressed in the transplantable VX2 carcinoma. J. Virol. *53*, 735-741.

Danos, O., and Yaniv, M. (1987). E6 and E7 gene products evolved by amplification of a 33-amino-acid peptide with a potential nucleic-acid-binding structure. Cancer Cells *5*, 145-149.

de Villiers, E., Fauquet, C., Broker, T., Bernard, H., and zur Hausen, H. (2004). Classification of papillomaviruses. Virology *324*, 17-27.

Deau, M. C., Favre, M., Jablonska, S., Rueda, L. A., and Orth, G. (1993). Genetic heterogeneity of oncogenic human papillomavirus type 5 (HPV5) and phylogeny of HPV5 variants associated with epidermodysplasia verruciformis. J. Clin. Microbiol. *31*, 2918-2926.

Donnelly, J. J., Martinez, D., Jansen, K. U., Ellis, R. W., Montgomery, D. L., and Liu, M. A. (1996). Protection against papillomavirus with a polynucleotide vaccine. J. Infect. Dis. *173*, 314-320.

Du, M., Fan, X., Hanada, T., Gao, H., Lutchman, M., Brandsma, J.L., Chishti, A.H., and Chen, J.J. (2005). Association of cottontail rabbit papillomavirus E6 oncoproteins with the hDlg/SAP97 tumor suppressor. J. Cell. Biochem. *94*, 1038-1045.

Embers, M. E., Budgeon, L. R., Pickel, M., and Christensen, N. D. (2002). Protective immunity to rabbit oral and cutaneous papillomaviruses by immunization with short peptides of L2, the minor capsid protein. J. Virol. *76*, 9798-9805.

Evans, C. A., Gorman, L. R., Ito, Y., and Weiser, R. S. (1962). Antitumor immunity in the Shope papilloma-carcinoma complex of rabbits. I. Papilloma regression induced by homologous and autologous tissue vaccines. J. Natl. Cancer Inst. *29*, 277-285.

Evans, C. A. (1963). Immunological studies of the Shope papilloma-carcinoma complex of rabbits. Acta Unio Int. Contra Cancrum *19*, 110-112.

Evans, C. A., and Ito, Y. (1966). Antitumor immunity in the Shope papilloma-carcinoma complex of rabbits. III. Response to reinfection with viral nucleic acid. J. Natl. Cancer Inst. *36*, 1161-1166.

Favre, M., Jibard, N., and Orth, G. (1982). Restriction mapping and physical characterization of the cottontail rabbit papillomavirus genome in transplantable VX2 and VX7 domestic rabbit carcinomas. Virology *119*, 298-309.

Friedewald, W., and Kidd, J. (1944). The recoverability of virus from papillomas produced therewith in domestic rabbits. J. Exp. Med. *79*, 591-605.

Friedewald, W. F. (1944). Certain conditions determining enhanced infection with the rabbit papilloma virus. J. Exp. Med. *80*, 65-76.

Friedmann, T. (2001). Stanfield Rogers: insights into virus vectors and failure of an early gene therapy model. Molecular Therapy *4*, 285-288.

Georges, E., Croissant, O., Bonneaud, N., and Orth, G. (1984). Physical state and transcription of the cottontail rabbit papillomavirus genome in warts and transplantable VX2 and VX7 carcinomas of domestic rabbits. J. Virol. *51*, 530-538.

Georges, E., Breitburd, F., Jibard, N., and Orth, G. (1985). Two Shope papillomavirus-associated VX2 carcinoma cell lines with different levels of keratinocyte differentiation and transplantability. J. Virol. *55*, 246-250.

Georges, E., Pehau-Arnaudet, G., and Orth, G. (1992). Molecular and biological characterization of cottontail rabbit papillomavirus variant DNA sequences integrated in the VX7 carcinoma. Virology *186*, 750-759.

Germain, R. N. (1994). MHC-dependent antigen processing and peptide presentation providing ligands for T lymphocyte activation. Cell *76*, 287-299.

Giri, I., Danos, O., and Yaniv, M. (1985). Genomic structure of the cottontail rabbit (Shope) papillomavirus. Proc. Natl. Acad. Sci. USA *82*, 1580-1584.

Gordon, M. R., and Kindt, T. J. (1992). Rabbit immune system. In Encyclopedia of immunology, I. M. Roitt, ed. (London, San Diego: Academic Press), pp. 1296-1298.

Hagari, Y., Budgeon, L. R., Pickel, M. D., and Kreider, J. W. (1995). Association of tumor necrosis factor-α gene expression and apoptotic cell death with regression of Shope papillomas. J. Invest. Dermatol. *104*, 526-529.

Han, R., Breitburd, F., Marche, P. N., and Orth, G. (1992). Linkage of regression and malignant conversion of rabbit viral papillomas to MHC class II genes. Nature *356*, 66-68.

Han, R., Breitburd, F., Marche, P. N., and Orth, G. (1994). Analysis of the nucleotide sequence variation of the antigen-binding domain of DRα and DQα molecules as related to the evolution of papillomavirus-induced warts in rabbits. J. Invest. Dermatol. *103*, 376-380.

Han, R., Cladel, N. M., Reed, C. A., and Christensen, N. D. (1998). Characterization of transformation function of cottontail rabbit papillomavirus E5 and E8 genes. Virology *251*, 253-263.

Han, R., Reed, C. A., Cladel, N. M., and Christensen, N. D. (1999a). Intramuscular injection of plasmid DNA encoding cottontail rabbit papillomavirus E1, E2, E6 and E7 induces T cell-mediated but not humoral immune responses in rabbits. Vaccine *17*, 1558-1566.

Han, R. C., Cladel, N. M., Reed, C. A., Peng, X. W., and Christensen, N. D. (1999b). Protection of rabbits from viral challenge by gene gun-based intracutaneous vaccination with a combination of cottontail rabbit papillomavirus E1, E2, E6, and E7 genes. J. Virol. *73*, 7039-7043.

Han, R., Cladel, N. M., Reed, C. A., Peng, X. W., Budgeon, L. R., Pickel, M., and Christensen, N. D. (2000a). DNA vaccination prevents and/or delays carcinoma development of papillomavirus-induced skin papillomas on rabbits. J. Virol. *74*, 9712-9716.

Han, R., Reed, C. A., Cladel, N. M., and Christensen, N. D. (2000b). Immunization of rabbits with cottontail rabbit papillomavirus E1 and E2 gene: protective immunity induced by gene-gun mediated intracutaneous delivery but not by intramuscular injection. Vaccine *18*, 2937-2944.

Harry, J. B., and Wettstein, F. O. (1996). Transforming properties of the cottontail rabbit papillomavirus oncoproteins LE6 and SE6 and of the E8 protein. J. Virol. *70*, 3355-3362.

Harvey, S., Cladel, N., Budgeon, L., Welsh, P., Griffith, J., Lang, C., and Christensen, N. (1998). Rabbit genital tissue is susceptible to infection by rabbit oral papillomavirus : an animal model for a genital tissue-targeting papillomavirus. J. Virol. *72*, 5239-5244.

Haskell, K. M., Vuocolo, G. A., Defeo-Jones, D., Jones, R. E., and Ivey-Hoyle, M. (1993). Comparison of the binding of the human papillomavirus type-16 and cottontail rabbit papillomavirus E7-proteins to the retinoblastoma gene product. J. Gen. Virol. *74*, 115-119.

Hermel, E., Han, M., Hague, B., Kindt, T., and Monaco, J. J. (1999). Isolation and mapping of the rabbit *DM* genes. Immunogenetics *49*, 295-302.

Hildesheim, A., and Wang, S. S. (2002). Host and viral genetics and risk of cervical cancer: a review. Virus Res. *89*, 229-240.

Höpfl, R. M., Christensen, N. D., Angell, M. G., and Kreider, J. W. (1993). Skin test to assess immunity against cottontail rabbit papillomavirus antigens in rabbits with progressing papillomas or after papilloma regression. J. Invest. Dermatol. *101*, 227-231.

Höpfl, R. M., Christensen, N. D., Angell, M. G., and Kreider, J. W. (1995). Leukocyte proliferation *in vitro* against cottontail rabbit papillomavirus in rabbits with persisting papillomas/cancer or after regression. Arch. Dermatol. Res. *287*, 652-658.

Howley, P. M., and Lowy, D. R. (2001). Papillomaviruses and their replication. In Fields Virology, D. M. Knipe, P. M. Howley, D. E. Griffin, R. A. Lamb, M. A. Martin, B. Roizman and S. E. Straus, eds. (Philadelphia, USA: Lippincott Williams and Wilkins), pp. 2197-2229.

Hu, J., Han, R., Cladel, N., Pickel, M., and Christensen, N. (2002a). Intracutaneous DNA vaccination with the E8 gene of cottontail rabbit papillomavirus induces protective immunity against virus challenge in rabbits. J. Virol. *76*, 6453-6459.

Hu, J., Cladel, N. M., Wang, Z., Han, R., Pickel, M. D., and Christensen, N. D. (2004a). GM-CSF enhances protective immunity to cottontail rabbit papillomavirus E8 genetic vaccination in rabbits. Vaccine *22*, 1124-1130.

Hu, J. F., Cladel, N. M., Pickel, M. D., and Christensen, N. D. (2002b). Amino acid residues in the carboxy-terminal region of cottontail rabbit papillomavirus E6 influence spontaneous regression of cutaneous papillomas. J. Virol. *76*, 11801-11808.

Hu, J. F., Cladel, N. M., Budgeon, L. R., and Christensen, N. D. (2004b). Characterization of three rabbit oral papillomavirus oncogenes. Virology *325*, 48-55.

Hu, J. Peng, X., Cladel, N.M., Pickel, M.D. and Christensen, N.D. (2005). Large cutaneous rabbit papillomas that persist during cyclosporin A treatment can regress spontaneously after cessation of immunosuppression. J. Gen. Virol. *86*, 55-63.

Huber, E., Vlasny, D., Jeckel, S., Stubenrauch, F., and Iftner, T. (2004). Gene profiling of cottontail rabbit papillomavirus-induced carcinomas identifies upregulated genes directly involved in stroma invasion as shown by small interfering RNA-mediated gene silencing. J. Virol. *78*, 7478-7489.

Ito, Y. (1960). A tumor-producing factor extracted by phenol from papillomatous tissue (Shope) of cottontail rabbits. Virology *12*, 596-601.

Ito, Y. (1962). Relationship of components of papilloma virus to papilloma and carcinoma cells. Cold Spring Harbor Symposia on Quantitative Biology *27*, 387-394.

Jansen, K. U., Rosolowsky, M., Schultz, L. D., Markus, H. Z., Cook, J. C., Donnelly, J. J., Martinez, D., Ellis, R. W., and Shaw, A. R. (1995). Vaccination with yeast-expressed cottontail rabbit papillomavirus (CRPV) virus-like particles protects rabbits from CRPV-induced papilloma formation. Vaccine *13*, 1509-1514.

Jeckel, S., Huber, E., Stubenrauch, F., and Iftner, T. (2002). A transactivator function of cottontail rabbit papillomavirus E2 is essential for tumor induction in rabbits. J. Virol. *76*, 11209-11215.

Jeckel, S., Loetzsch, E., Huber, E., Stubenrauch, F., and Iftner, T. (2003). Identification of the E9/E2C cDNA and functional characterization of the gene product reveal a new repressor of transcription and replication in cottontail rabbit papillomavirus. J. Virol. *77*, 8736-8744.

Jensen, E. R., Selvakumar, R., Shen, H., Ahmed, R., Wettstein, F. O., and Miller, J. F. (1997). Recombinant *Listeria monocytogenes* vaccination eliminates papillomavirus-induced tumors and prevents papilloma formation from viral DNA. J. Virol. *71*, 8467-8474.

Kawana, K., Yoshikawa, H., Taketani, Y., Yoshiike, K., and Kanda, T. (1999). Common neutralization epitope in minor capsid protein L2 of human papillomavirus types 16 and 6. J. Virol. *73*, 6188-6190.

Kawase, M., Orth, G., Jablonska, S., Blanchet-Bardon, C., Rueda, L., and Favre, M. (1996). Variability and phylogeny of the L1 capsid protein gene of human papillomavirus type 5: contribution of clusters of nonsynonymous mutations and of a 30-nucleotide duplication. Virology *221*, 189-198.

Kidd, J. G., Beard, J. W., and Rous, P. (1936). Serological reactions with a virus causing rabbit papillomas which become cancerous. J. Exp. Med. *64*, 63-77.

Kidd, J. G. (1938). The course of virus-induced rabbit papillomas as determined by virus, cells, and host. J. Exp. Med. *67*, 551-573.

Kidd, J. G., and Rous, P. (1940). A transplantable rabbit carcinoma originating in a virus-induced papilloma and containing the virus in masked or altered form. J. Exp. Med. *71*, 813-838.

Kozak, M. (1999). Initiation of translation in prokaryotes and eukaryotes. Gene *234*, 187-208.

Kreider, J. W. (1963). Studies on the mechanism responsible for the spontaneous regression of the Shope rabbit papilloma. Cancer Res. *23*, 1593-1599.

Kreider, J. W., and Bartlett, G. L. (1981). The Shope papilloma-carcinoma complex of rabbits: a model system of neoplastic progression and spontaneous regression. Adv. Cancer Res. *35*, 81-110.

Lathe, R., Kieny, M.-P., Dott, K., Gauthier, C., Clertant, P., Cuzin, F., Breitburd, F., Orth, G., and Meneguzzi, G.

(1989). Vaccination against polyoma- and papillomavirus-induce tumors using vaccinia recombinants expressing non structural proteins. In Vaccines for Sexually Transmitted Diseases, A. Meheus and R. E. Spier, eds. (London: Butterworths), pp. 166-176.

Laverriere, A., Kulaga, H., Kindt, T. J., LeGuern, C., and Marche, P. N. (1989). A rabbit class II MHC gene with strong similarities to *HLA-DRA*. Immunogenetics *30*, 137-140.

Leachman, S. A., Tigelaar, R. E., Shlyankevich, M., Slade, M. D., Irwin, M., Chang, E., Wu, T. C., Xiao, W., Pazhani, S., Sundaram, P., et al. (2000). Granulocyte-macrophage colony-stimulating factor priming plus papillomavirus E6 DNA vaccination: Effects on papilloma formation and regression in the cottontail rabbit papillomavirus-rabbit model. J. Virol. *74*, 8700-8708.

Leachman, S. A., Shylankevich, M., Slade, M. D., Levine, D. K., Sundaram, R., Xiao, W., Bryan, M., Zelterman, D., and Tiegelaar, R. E. (2002). Ubiquitin-fused and/or multiple early genes from cottontail rabbit papillomavirus as DNA vaccines. J. Virol. *76*, 7616-7624.

LeGuern, C., Marche, P. N., and Kindt, T. J. (1985). Molecular evidence for five distinct MHC class II α genes in the rabbit. Immunogenetics *22*, 141-148.

LeGuern, C., Weissman, J. D., Marche, P. N., Jouvin-Marche, E., Laverriere, A., Bagnato, M. R., and Kindt, T. J. (1987). Sequence determination of a transcribed rabbit class II gene with homology to *HLA-DQα*. Immunogenetics *25*, 104-109.

Lin, Y.-L., Borenstein, L. A., Selvakumar, R., Ahmed, R., and Wettstein, F. O. (1992). Effective vaccination against papilloma development by immunization with L1 or L2 structural protein of cottontail rabbit papillomavirus. Virology *187*, 612-619.

Lin, Y.-L., Borenstein, L. A., Ahmed, R., and Wettstein, F. O. (1993a). Cottontail rabbit papillomavirus L1 protein-based vaccines: protection is achieved only with a full-length, nondenatured product. J. Virol. *67*, 4154-4162.

Lin, Y.-L., Borenstein, L. A., Selvakumar, R., Ahmed, R., and Wettstein, F. O. (1993b). Progression from papilloma to carcinoma is accompanied by changes in antibody response to papillomavirus proteins. J. Virol. *67*, 382-389.

Lowy, D. R., and Howley, P. M. (2001). Papillomaviruses. In Fields Virology, D. M. Knipe, P. M. Howley, D. E. Griffin, R. A. Lamb, M. A. Martin, B. Roizman and S. E. Straus, eds. (Philadelphia, USA: Lippincott Williams & Wilkins), pp. 2231-2264.

Lowy, D. R., and Frazer, I. H. (2003). Prophylactic human papillomavirus vaccines. J. Natl. Cancer Inst. *31*, 111-116.

Marche, P. N., Rebiere, M. C., Laverriere, A., English, D. W., LeGuern, C., and Kindt, T. J. (1989). Definition of rabbit class I and class II MHC gene haplotypes using molecular typing procedures. Immunogenetics *29*, 273-276.

McMichael, H. (1967). Inhibition by methylprednisolone of regression of the Shope rabbit papilloma. J. Natl. Cancer Inst. *39*, 55-65.

Meyers, C., Harry, J., Lin, Y.-L., and Wettstein, F. O. (1992). Identification of three transforming proteins encoded by cottontail rabbit papillomavirus. J. Virol. *66*, 1655-1664.

Murphy, K. M., and Reiner, S. L. (2002). The lineage decisions of helper T cells. Nat. Rev. Immunol. *2*, 933-944.

Nasseri, M., Wettstein, F. O., and Stevens, J. G. (1982). Two colinear and spliced viral transcripts are present in non-virus producing benign and malignant neoplasms induced by the Shope (rabbit) papilloma virus. J. Virol. *44*, 263-268.

Nasseri, M., and Wettstein, F. O. (1984a). Differences exist between viral transcripts in cottontail rabbit papillomavirus-induced benign and malignant tumors as well as non-virus-producing and virus-producing tumors. J. Virol. *51*, 706-712.

Nasseri, M., and Wettstein, F. O. (1984b). Cottontail rabbit papillomavirus-specific transcripts in transplantable tumors with integrated DNA. Virology *138*, 362-367.

Nasseri, M., Meyers, C., and Wettstein, F. O. (1989). Genetic analysis of CRPV pathogenesis: The L1 open reading frame is dispensable for cellular transformation but is required for papilloma formation. Virology *170*, 321-325.

Nonnenmacher, M. (2004) Molecular bases of the oncogenic potential of papillomaviruses: role of viral variation and functional analysis of the E8 oncoprotein of the cottontail rabbit papillomavirus, PhD, University Paris 7, Paris, France.

Noyes, W. F., and Mellors, R. C. (1957). Fluorescent antibody detection of the antigens of the Shope papilloma virus in papillomas of the wild and domestic rabbit. J. Exp. Med. *106*, 555-562.

Okabayashi, M., Angell, M. G., Christensen, N. D., and Kreider, J. W. (1991). Morphometric analysis and identification of infiltrating leucocytes in regressing and progressing Shope rabbit papillomas. Int. J. Cancer *49*, 919-923.

Okabayashi, M., Angell, M. G., Budgeon, L. R., and Kreider, J. W. (1993). Shope papilloma cell and leukocyte proliferation in regressing and progressing lesions. Am. J. Pathol. *142*, 489-496.

Orth, G. and Croissant, O. (1970). Analyse par autoradiographie de la synthèse d'ADN induite par le virus de Shope dans les cellules en voie de kératinisation des papillomes du lapin domestique. International CNRS Colloquia *183*, 173-177.

Orth, G., Jeanteur, P., and Croissant, O. (1971). Evidence for and localization of vegetative viral DNA replication by autoradiographic detection of RNA-DNA hybrids in sections of tumors induced by Shope papilloma virus. Proc. Natl. Acad. Sci. USA *68*, 1876-1880.

Orth, G., Breitburd, F., Favre, M., and Croissant, O. (1977). Papillomaviruses: possible role in human cancer. Cold Spring Harb. Conf. Cell Prolif. *4*, 1043-1068.

Orth, G., Breitburd, F., and Favre, M. (1978). Evidence for antigenic determinants shared by the structural polypeptides of (Shope) rabbit papillomavirus and human papillomavirus type 1. Virology *91*, 243-255.

Orth, G. (1987). Epidermodysplasia verruciformis. In The papillomaviruses, P. M. Howley and N. P. Salzman, eds. (New York: Plenum Press), pp. 199-243.

Orth, G., Favre, M., Majewski, S., and Jablonska, S. (2001). Epidermodysplasia verruciformis defines a subset of cutaneous human papillomaviruses. J. Virol. *75*, 4952-4953.

Orth, G. (2004). Human papillomaviruses and the skin: More to be learned. J. Invest. Dermatol. *123*, XI-XIII.

Parsons, R. J., and Kidd, J. G. (1936). A virus causing oral papillomatosis in rabbits. Proc. Soc. Exp. Biol. Med. *35*, 441-443.

Peh, W. L., Middleton, K., Christensen, N., Nicholls, P., Egawa, K., Sotlar, K., Brandsma, J., Percival, A., Lewis, J., Liu, W. J., and Doorbar, J. (2002). Life cycle heterogeneity in animal models of human papillomavirus-associated disease. J. Virol. *76*, 10401-10416.

Peh, W. L., Brandsma, J. L., Christensen, N. D., Cladel, N. M., Wu, X., and Doorbar, J. (2004). The viral E4 protein is required for the completion of the cottontail rabbit papillomavirus productive cycle *in vivo*. J. Virol. *78*, 2142-2151.

Peng, X., Olson, R., Christian, C., Lang, C., and Kreider, J. (1993). Papillomas and carcinomas in transgenic rabbits carrying EJ-ras DNA and cottontail rabbit papillomavirus DNA. J. Virol. *67*, 1698-1701.

Peng, X., Lang, C., and Kreider, J. W. (1995). Methylation of cottontail rabbit papillomavirus DNA and tissue-specific expression in transgenic rabbits. Virus Res. *35*, 101-108.

Ramoz, N., Rueda, L.-A., Bouadjar, B., Favre, M., and Orth, G. (1999). A susceptibility locus for epidermodysplasia verruciformis, an abnormal predisposition to infection with the oncogenic human papillomavirus type 5, maps to chromosome 17qter in a region containing a psoriasis locus. J. Invest. Dermatol. *112*, 259-263.

Ramoz, N., Rueda, L. A., Bouadjar, B., Montoya, L. S., Orth, G., and Favre, M. (2002). Mutations in two adjacent novel genes are associated with epidermodysplasia verruciformis. Nat. Genet. *32*, 579-581.

Ramoz, N., Buée, B., Jablonska, S., Favre, M., and Orth, G. (unpublished results). Specific HLA DR-DQ haplotypes linked to epidermodysplasia verruciformis.

Rebière, M. C., Marche, P. N., LeGuern, C., and Kindt, T. J. (1987). Molecular studies of the rabbit major histocompatibility complex gene. In The Rabbit in Contemporary Immunological Research, S. Dubiski, ed. (Essex: Longman Scientific Research), pp. 42-64.

Ressing, M. E., Toes, R. E. M., Shoenberger, S. E., Brandt, R. M. P., de Jong, J. H., Kast, W. M., Offringa, R., and Melief, C. J. M. (1999). Anti-tumor immunotherapy with synthetic peptides representing tumor-associated T-cell epitopes: implications for peptide-based vaccination of cervical cancer. In Vaccines for human papillomavirus infecttion and anogenital disease, R. W. Tindle, ed. (Austin: R.G. Lands Company), pp. 53-68.

Roberts, A., Reuter, J. D., Wilson, J. H., Baldwin, S., and Rose, J. K. (2004). Complete protection from papillomavirus challenge after a single vaccination with a vesicular stomatitis virus vector expressing high levels of L1 protein. J. Virol. *78*, 3196-3199.

Roden, R. B. S., Hubbert, N. L., Kirnbauer, R., Breitburd, F., Lowy, D. R., and Schiller, J. T. (1995). Papillomavirus L1 capsids agglutinate mouse erythrocytes through a proteinaceous receptor. *69*, 5147-5151.

Roden, R. B., Yutzy, W. H. t., Fallon, R., Inglis, S., Lowy, D. R., and Schiller, J. T. (2000). Minor capsid protein of human genital papillomaviruses contains subdominant, cross-neutralizing epitopes. Virology *270*, 254-257.

Rogel-Gaillard, C., Piumi, F., Billault, A., Bourgeaux, N., Save, J.-C., Urien, C., Salmon, J., and Chardon, P. (2001). Construction of a rabbit bacterial artificial chromosome (BAC) library: application to the mapping of the major histocompatibility complex position 12q1.1. Mamm. Genome *12*, 253-255.

Rogers, S., and Rous, P. (1951). Joint action of a chemical carcinogen and a neoplastic virus to induce cancer in rabbits. Results of exposing epidermal cells to carcinogenic hydrocarbon at time of infection with the Shope papilloma virus. J. Exp. Med. *93*, 459-488.

Rogers, S., Kidd, J. G., and Rous, P. (1960). Relationships of the Shope papilloma virus to the cancers it determines in domestic rabbits. Acta Unio Int. Contra Cancrum *16*, 129-130.

Rosales, C., Graham, V. V., Rosas, G. A., Merchant, H., and Rosales, R. (2000). A recombinant vaccinia virus containing the papilloma E2 protein promotes tumor regression by stimulating macrophage antibody-dependent cytotoxicity. Cancer Immunol. Immunother. *49*, 347-360.

Rous, P., and Beard, J. W. (1935). The progression to carcinoma of virus-induced rabbit papillomas (Shope). J. Exp. Med. *62*, 523-548.

Rous, P. (1936). The virus tumors and the tumor problem. Am. J. Cancer *28*, 233-272.

Rous, P., and Kidd, J. G. (1938). The carcinogenic effect of a papilloma virus on the tarred skin of rabbits. I. Description of the phenomenom. J. Exp. Med. *67*, 399-422.

Rous, P. (1943). The nearer causes of cancer. JAMA *122*, 573-581.

Rous, P., and Friedewald, W. F. (1944). The effect of chemical carcinogens on virus-induced rabbit papillomas. J. Exp. Med. *79*, 511-538.

Salmon, J., Ramoz, N., Cassonnet, P., Orth, G., and Breitburd, F. (1997). A cottontail rabbit papillomavirus strain (CRPVb) with strikingly divergent E6 and E7 oncoproteins: an insight in the evolution of papillomaviruses. Virology *235*, 228-234.

Salmon, J., Nonnenmacher, M., Cazé, S., Flamant, P., Croissant, O., Orth, G., and Breitburd, F. (2000). Variation in the nucleotide sequence of cottontail rabbit papillomavirus a and b subtypes affects wart regression and malignant transformation and the level of viral replication in domestic rabbits. J. Virol. *74*, 10766-10777.

Salmon, J., Chanteloup, N., Viard, F., Coudert, P., Flamant, P., Croissant, O., Orth, G., and Breitburd, F. (2001). Linkage of domestic rabbit wart evolution to MHC DQ haplotype and cottontail rabbit papillomavirus strain and involvement of both CD4+ and CD8+ T lymphocytes in regression. Paper presented at: 19th International Papillomavirus Conference (Florianopolis, Brazil).

Salmon, J., Flamant, P., Croissant, O., Orth, G., and Breitburd, F. (2002). Immunogenetic control of domestic rabbit wart evolution. Insight into the mechanisms of regression or persistence. Paper presented at: 20th International Papillomavirus Conference (Paris, France).

Salmon, J. (2003) Factors and effectors controlling the evolution of tumors caused by an oncogenic papillomavirus: the rabbit model, PhD, University Paris 7, Paris, France.

Sastre-Garau, X., Loste, M. N., Salomon, A. V., Favre, M., Mouret, E., de la Rochefordière, A., Durand, J. C., Tartour, E., Lepage, V., and Charron, D. (1996). Decreased frequency of HLA-DRB1*13 alleles in French women with HPV-positive carcinoma of the cervix. Int. J. Cancer *69*, 159-164.

Schiller, J. T. (1999). Papillomavirus-like particle vaccines for cervical cancer. Mol. Med. Today *5*, 209-215.

Schmitt, A., Rochat, A., Zeltner, R., Borenstein, L., Barrandon, Y., Wettstein, F. O., and Iftner, T. (1996). The primary target cells of the high-risk cottontail rabbit papillomavirus colocalize with hair follicle stem cells. J. Virol. *70*, 1912-1922.

Schneider-Maunoury, S., Georges, E., and Orth, G. (1987). Structure and expression of the cottontail rabbit papillomavirus (CRPV) genome and of *myc* and *ras* genes during the progression of domestic rabbit tumors. Cancer Cells *5*, 229-236.

Selbie, F. R., and Robinson, R. H. M. (1948). Serial transmission of infectious papillomatosis in the domestic rabbit. Br. J. Cancer *1*, 371-379.

Selvakumar, R., Borenstein, L. A., Lin, Y. L., Ahmed, R., and Wettstein, F. O. (1994). T-cell response to cottontail rabbit papillomavirus structural proteins in infected rabbits. J. Virol. *68*, 4043-4048.

Selvakumar, R., Ahmed, R., and Wettstein, F. O. (1995a). Tumor regression is associated with a specific immune response to the E2 protein of cottontail rabbit papillomavirus. Virology *208*, 298-302.

Selvakumar, R., Borenstein, L. A., Lin, Y.-L., Ahmed, R., and Wettstein, F. O. (1995b). Immunization with nonstructural proteins E1 and E2 of cottontail rabbit papillomavirus stimulates regression of virus-induced papillomas. J. Virol. *69*, 602-605.

Selvakumar, R., Schmitt, A., Iftner, T., Ahmed, R., and Wettstein, F. O. (1997). Regression of papillomas induced by cottontail rabbit papillomavirus is associated with infiltration off CD8+ cells and persistence of viral DNA after regression. J. Virol. *71*, 5540-5548.

Seto, A., Fu, N., Matsuda, S., Eguchi, T., Miyoshi, I., and Ito, Y. (1987). Introduction of inbred rabbits for rescach on viral tumors. Cancer Cells *5*, 367-371.

Shakhov, A. N., Kuprash, D. V., Azizov, M. M., Jongeneel, C. V., and Nedospasov, S. A. (1990). Structural analysis of the rabbit TNF locus, containing the genes encoding TNF-β (lymphotoxin) and TNF-α (tumor necrosis factor). Gene *95*, 215-221.

Shope, R. E. (1933). Infectious papillomatosis of rabbits (with a note on the histopathology by E.W. Hurst). J. Exp. Med. *58*, 607-624.

Shope, R. E. (1935). Serial transmission of virus of infectious papillomatosis in domestic rabbits. Proc. Soc. Exp. Biol. Med. *32*, 830-832.

Shope, R. E. (1937). Immunization of rabbits to infectious papillomatosis. J. Exp. Med. *65*, 219-231.

Sittisombut, N., and Knight, K. L. (1986). Rabbit major histocompatibility complex. I. Isolation and characterization of three subregions of class II genes. J. Immunol. *136*, 1871-1875.

Sittisombut, N., Mordacq, J., and Knight, K. L. (1988). Rabbit MHC. II. Sequence analysis of the R-DP α- and β-genes. J. Immunol. *140*, 3237-3243.

Sittisombut, N., Tissot, R. G., and Knight, K. L. (1989). Rabbit major histocompatibility complex. III. Multiple class II DRβ genes and restriction fragment length polymorphism of the class II α and β genes. J. Immunogenetics *16*, 63-75.

Sittisombut, N. (1996). The major histocompatibility complex of the rabbit. In The Major Histocompatibility Complex of Domestic Animal Species, L. B. Schook and S. J. Lamont, eds. (Boca Raton: CRC Press, Inc.), pp. 269-292.

Smith, W. E., Kidd, J. G., and Rous, P. (1952). Experiments on the cause of the rabbit carcinomas derived from virus-induced papillomas. I. Propagation of several of the cancers in sucklings with etiological tests. J. Exp. Med. *95*, 299-317.

Stern, P. (1996). Immunity to human papillomavirus-associated cervical neoplasia. Adv. Cancer Res. *69*, 175-211.

Stevens, J. G., and Wettstein, F. O. (1979). Multiple copies of Shope virus DNA are present in cells of benign and malignant non-virus-producing neoplasms. J. Virol. *30*, 891-898.

Sundaram, P., Tigelaar, R. E., and Brandsma, J. L. (1997). Intracutaneous vaccination of rabbits with the cottontail rabbit papillomavirus (CRPV) L1 gene protects against virus challenge. Vaccine *15*, 664-671 (Erratum in Vaccine 16: 655, 1998).

Sundaram, P., Tigelaar, R. E., Xiao, W., and Brandsma, J. L. (1998). Intracutaneous vaccination of rabbits with the E6 gene of cottontail rabbit papillomavirus provides partial protection against virus challenge. Vaccine *16*, 613-623.

Syverton, J. T. (1952). The pathogenesis of the rabbit papilloma-to-carcinoma sequence. Ann. N.Y. Acad. Sci. *54*, 1126-1140.

Tagami, H., Aiba, S., and Rokugo, M. (1985). Regression of flat warts and common warts. Clin Dermatol *3(4)*, 170-178.

Valdez Graham, V., Sutter, G., Jose, M. V., Garcia-Carranca, A., Erfle, V., Moreno Mendoza, N., Merchant, H., and Rosales, R. (2000). Human tumor growth is inhibited by a vaccinia virus carrying the E2 gene of bovine papillomavirus. Cancer *88*, 1650-1662.

Van Es, R. J. J., Baselman, A. H. C., Koten, J. W., Van Dijk, J. E., Koole, R., and den Otter, W. (2000). Perilesional IL-2 treatment of a VX2 head-and-neck cancer model can induce a systemic anti-tumour activity. Anticancer Res. *20*, 4163-4170.

Wang, S. S., and Hildesheim, A. (2003). Viral and host factors in human papillomavirus persistence and progression. J. Natl. Cancer Inst. *31*, 35-40.

Wank, R., and Thomssen, C. (1991). High risk of squamous cell carcinoma of the cervix for women with HLA-DQw3. Nature *352*, 723-725.

Wettstein, F. O., and Stevens, J. G. (1983). Shope papilloma virus DNA is extensively methylated in non-virus-producing neoplasms. Virology *126*, 493-504.

Wettstein, F. O. (1987). Papillomaviruses and carcinogenic progression: cottontail rabbit (Shope) papillomavirus. In The *Papovaviridae*: The Papillomaviruses, N. P. Salzman and P. M. Howley, eds. (New York: Plenum Press), pp. 167-186.

Wettstein, F. O., Barbosa, M. S., and Nasseri, M. (1987). Identification of the major cottontail rabbit papillomavirus late RNA cap site and mapping and quantitation of an E2 and minor E6 coding mRNA in papillomas and carcinomas. Virology *159*, 321-328.

Wettstein, F. O. (1990). State of viral DNA and gene expression in benign vs. malignant tumors. In Papillomaviruses and human cancer, H. Pfister, ed. (Boca Raton, FL: CRC Press), pp. 155-179.

Wilkinson, J. M., Tykocinski, M. L., Coligan, J. E., Kimball, E. S., and Kindt, T. J. (1982). Rabbit MHC antigens: occurrence of non-beta-2-microglobulin-associated class I molecules. Mol. Immunol. *19*, 1441-1451.

Index*

A

Animal models 237, 261-262, 373-383, 389-394, 399-413

Apoptosis 75-80, 88-90, 104-106, 115-119, 193-199, 218-223, 269-273, 326-330

B

Bovine papillomavirus (BPV) 373-383, 390-394
- heterogeneity 373
- immune response 380
- latency 380
- fibropapillomas 374, 377, 383
- papillomas 377
- upper GI tract
  - cancer 378
  - papillomas 378
- urinary bladder cancer 378-379
- vaccines 381-382
- viral proteins 375-376

C

CD4 T cells 260-262, 295-298, 311-315, 329-330, 342-349, 408-413

CD8 T cells 260-262, 295-298, 312-315, 342-349, 408-413

CD40 331-332

Cell
- cycle 75, 86-92, 133-141,154-163
- differentiation 41-48, 136-141, 145-165, 193-203, 216-222, 322-331
- growth 88-92, 139
- immortalization 139-140, 154-160
- mediated immunity (CMI) 311-313, 380-381; 407-408
- transformation 97-108, 115-124, 133-141, 193-200, 269-273

Cervical
- cancer 3-6, 19-33
- intraepithelial neoplasia (CIN) 19-33, 146-162, 314-316, 321-332, 341-352,

c-fos 106, 205

Chemokines 297-299, 321-333

CIN-612 cells 145-165, 206

c-jun 106, 134, 215

Cottontail rabbit papillomavirus (CRPV) 399-413
- evolution of infection 400-401
- immune response 407-408
- MHC 410-412
- subtypes/variants 402-405
- vaccines 408-410
- viral genome organisation 401-402
- viral proteins 405-407

Cutaneous HPV 269-274

Cyclin A and cyclin A/cdk2 62-63, 91, 100-103, 118, 140-141, 376, 394

Cyclin D and cyclin D/cdk4 62, 135-140

Cyclin E and cyclin E/cdk2 54, 62-63, 91, 135-141, 154-155

Cytokines 260, 313-314, 321-332

Cytoskeleton 85-90

Cytotoxic T cells 322, 342

E

E1 and E1-interactors 53-64, 73, 196-197
- CRM-1 62
- cyclin E and cyclin E/cdk2 62-63
- DNA polymerase α 61
- E2 60
- histone H1 61
- p53 61
- RPA 61

E2 and E2-interactors 43, 55-58, 60, 73-80, 148, 196-197
- AMF1 78
- BRCA 78
- Brd4 78
- PARP 78
- pCAF 78
- p53 78
- SMN 78
- TBP 78
- TFIIB 78
- TopBP1 78

E4 and E4-interactors 83-92, 148-149, 195, 198-199
- actin 87
- keratins 88-89
- mitochondria 89-90
- ND10 90

E5 and E5-interactors 98-108, 199-200, 213-216
- 16k subunit c/V-ATPase 101-103
- CBF 102
- cyclin A and cyclin A/cdk2 100-103
- c-src 99-101
- E2F 102
- EGF/EGF-R 104-106
- ERK 102, 106
- MHC/HLA 101-103, 106-107
- $p21^{Cip1}$ 106
- $p27^{KIP1}$ 102-103
- PDGF-R 98-99
- PI3K 100-101, 106

* The index is not exhaustive. Items and their corresponding pages have been listed only when the item is addressed extensively.

E6 and E6-interactors 115-124, 195, 197-198, 216-219
- Bak 117
- Dlg 117, 121
- E6-AP 119
- MAGI 117,120
- MUPP-1 117, 120
- p300/CBP 117
- p53 117-118
- PDZ domains 120, 123
- Scrib 117, 121

E7 and E7-interactors 133-141, 154-155, 195, 197-198, 216-217, 219-221
- AP-1 137
- c-jun 134
- E2F 134, 138
- Jun-B 134
- p105Rb 134-135
- TAFs 138
- TBP 138

EGF/EGF-R 104-106
Epidemiology
- cervical cancer 19-33
- oesophageal cancer 229-244

Epidermodysplasia verruciformis 270-273
Equine sarcoids 390-394
Erk 102, 181, 330
Estrogen 222-223

G

Glycosaminoglycans (GAG) 177-178
Golgi apparatus 90, 98-106, 362, 376-382, 393

H

Heparan sulphate protoglycan (HSPG) 177-178
Heterogeneity 11-16
High risk HPV 3-6, 11-16, 20-33

I

IFN 162, 224, 259-262, 298-299, 312-316, 322-332, 324-326
Immune
- evasion 258-259, 321-332, 349
- response 258-261, 380-382, 393, 407-408

Immunobiology/immunogenetics 311-316
Infection 175-187, 283-285
Interleukins 311-316, 321-332

J

Jun-B 134, 215

L

L1/L2 175-187, 279-299, 381-382, 408-409
Late events 193-205
Late regulatory elements 203-204
Latency 257
Low risk HPV 3-6, 11-16, 20-33
LXCXE 135

M

Macrophage inflammatory protein 322-323
Major histocompatibility complex (MHC) class I/II 97, 100-108, 258-262, 297-298, 312-316, 323-331, 342-349, 381-382, 404-413
Microtubule organising center (MTOC) 185
Microtubules 182-186

N

Natural history
- cervical cancer 21-23
- oesophageal
  - cancer 229-244
  - papilloma 230-234

ND10/PML 54, 77, 85-92, 186-187, 199-201, 382
NK cells 106-107, 312-316
Non-melanoma skin cancer 269-274

O

Oesophageal cancer 229-244
Organotypic cultures 145-165

P

$p21^{cip1}$ 61, 63, 140, 154-155, 197-198, 258, 327
$p27^{KIP1}$ 102-103,140, 155, 198, 380, 383, 393-394
PDGF/PDGF-R 98-99
Phylogeny/taxonomy 11-16
Plasmid maintenance 53-64, 73-80
Progression 156-157, 257, 378-379, 400-401
Psoriasis 271-272

Q

Quercetin 379-380, 383

R

Raft cultures 145-165
Recurrent respiratory papillomatosis (RRP) 255-262
Risk factors/co-factors/genetic predisposition 27-30, 160-161, 235-236, 378-380, 393, 410-413
- cervical cancer 27-30
- oesophageal cancer 235-236

cattle cancer 378-380
equine sarcoids 393
rabbit cancer 410-413
RNA splicing/processing 41-48, 193-206
HnRNP 205
SF2/ASF 203-204
SnRNP 203-204
U2AF 203-204

S

Screening studies
cervical cancer 31-32
oesophageal cancer 232-241
Serology 242, 285-292
Signal transduction 180-181
Squamous intraepithelial lesion (SIL) 19-33, 311-316, 321-332,
Subtypes/variants 14-15, 391-392, 403-404

T

TCR 312-316
TGFβ/TGFβ-R 323, 327-328
Th1/Th2 260-262, 297-299, 312-315, 322-330, 348-349, 361-368
TNFα 326-327, 331
Transgenic mice 213-224
Transgenic plants 358-369
chloroplast transformation 360
nuclear transformation 359-360

U

UV rays 269, 272-274

V

Vaccines 32, 261, 292-297, 341-352, 358-369
animal models 343-344
cattle 293, 343, 381-382
dog 293, 315, 344
rabbit 293, 343-344, 408-410
rhesus monkey 344
cost benefits 351-352
dendritic cell vaccines 349
DNA vaccines 348-349
edible vaccines 362-369
peptide/protein vaccines 347
prophylactic vaccines 292-297, 344-345, 366-368, 381-382, 408-409
therapeutic vaccines 297, 345-346, 368-369, 381-382, 409-410
viral vectors 347-348
virus-like particles (VLP) 344-345
VEGF 330-331
Viral gene expression/transcription 41-48
LCR/URR 41-48, 150-151
transcript stability 48
transcriptional
co-activators 77
control 374-375
enhancer 44
promoter 43-44, 149
regulation 74
silencer 45
termination 47
Viral genome
methylation 46
recombination 47
replication 53-64
origin 54, 59
segregation 75-79
Virion structure and assembly 279-283
Virus binding 176-180
Virus life cycle 83, 147-148, 175-187, 193-205
Virus-like particles (VLP) 175-187, 279-299

W

W12 cells 145-165, 206

Current Books of Interest

Title	Year
Acanthamoeba	2009
Bacterial Secreted Proteins: Secretory Mechanisms and Role in Pathogenesis	2009
Lactobacillus Molecular Biology: From Genomics to Probiotics	2009
Mycobacterium: Genomics and Molecular Biology	2009
Real-Time PCR: Current Technology and Applications	2009
Clostridia: Molecular Biology in the Post-genomic Era	2009
Plant Pathogenic Bacteria: Genomics and Molecular Biology	2009
Microbial Production of Biopolymers and Polymer Precursors	2009
Plasmids: Current Research and Future Trends	2008
Pasteurellaceae: Biology, Genomics and Molecular Aspects	2008
Vibrio cholerae: Genomics and Molecular Biology	2008
Pathogenic Fungi: Insights in Molecular Biology	2008
Helicobacter pylori: Molecular Genetics and Cellular Biology	2008
Corynebacteria: Genomics and Molecular Biology	2008
Staphylococcus: Molecular Genetics	2008
Leishmania: After The Genome	2008
Archaea: New Models for Prokaryotic Biology	2008
RNA and the Regulation of Gene Expression	2008
Legionella Molecular Microbiology	2008
Molecular Oral Microbiology	2008
Epigenetics	2008
Animal Viruses: Molecular Biology	2008
Segmented Double-Stranded RNA Viruses	2008
Acinetobacter Molecular Biology	2008
Pseudomonas: Genomics and Molecular Biology	2008
Microbial Biodegradation: Genomics and Molecular Biology	2008
The Cyanobacteria: Molecular Biology, Genomics and Evolution	2008
Coronaviruses: Molecular and Cellular Biology	2007
Real-Time PCR in Microbiology: From Diagnosis to Characterisation	2007
Bacteriophage: Genetics and Molecular Biology	2007
Candida: Comparative and Functional Genomics	2007
Bacillus: Cellular and Molecular Biology	2007
AIDS Vaccine Development: Challenges and Opportunities	2007
Alpha Herpesviruses: Molecular and Cellular Biology	2007
Pathogenic *Treponema*: Molecular and Cellular Biology	2007
PCR Troubleshooting: The Essential Guide	2006
Influenza Virology: Current Topics	2006
Microbial Subversion of Immunity: Current Topics	2006
Cytomegaloviruses: Molecular Biology and Immunology	2006
Papillomavirus Research: From Natural History To Vaccines and Beyond	2006

Caister Academic Press www.caister.com

www.ingramcontent.com/pod-product-compliance
Lightning Source LLC
LaVergne TN
LVHW062109200926
848484LV00007B/1624